INTRODUCTION TO
PROTEINS

STRUCTURE, FUNCTION, AND MOTION

SECOND EDITION

CHAPMAN & HALL/CRC
Mathematical and Computational Biology Series

Aims and scope:

This series aims to capture new developments and summarize what is known over the entire spectrum of mathematical and computational biology and medicine. It seeks to encourage the integration of mathematical, statistical, and computational methods into biology by publishing a broad range of textbooks, reference works, and handbooks. The titles included in the series are meant to appeal to students, researchers, and professionals in the mathematical, statistical and computational sciences, fundamental biology and bioengineering, as well as interdisciplinary researchers involved in the field. The inclusion of concrete examples and applications, and programming techniques and examples, is highly encouraged.

Series Editors

N. F. Britton
Department of Mathematical Sciences
University of Bath

Xihong Lin
Department of Biostatistics
Harvard University

Nicola Mulder
University of Cape Town
South Africa

Maria Victoria Schneider
European Bioinformatics Institute

Mona Singh
Department of Computer Science
Princeton University

Proposals for the series should be submitted to one of the series editors above or directly to:
CRC Press, Taylor & Francis Group
3 Park Square, Milton Park
Abingdon, Oxfordshire OX14 4RN
UK

Chapman & Hall/CRC Mathematical and Computational Biology Series

INTRODUCTION TO PROTEINS

STRUCTURE, FUNCTION, AND MOTION

SECOND EDITION

AMIT KESSEL • NIR BEN-TAL

CRC Press
Taylor & Francis Group
Boca Raton London New York

CRC Press is an imprint of the
Taylor & Francis Group, an **informa** business

CRC Press
Taylor & Francis Group
6000 Broken Sound Parkway NW, Suite 300
Boca Raton, FL 33487-2742

Printed on acid-free paper
Version Date: 20180322

International Standard Book Number-13: 978-1-4987-4717-2 (Hardback)

Library of Congress Cataloging-in-Publication Data

Names: Kessel, Amit, author. | Ben-Tal, Nir, author.
Title: Introduction to proteins : structure, function, and motion
Amit Kessel and Nir Ben-Tal.
Description: Second edition. | Boca Raton, Florida : CRC Press, [2018] |
Series: Chapman & Hall/CRC mathematical and computational biology series
Identifiers: LCCN 2017052036| ISBN 9781498747172 (hardback) |
ISBN 9781315113876 (ebook)
Subjects: LCSH: Proteins. | Proteins--Structure-activity relationships.|
Physical biochemistry.
Classification: LCC QP551 .K465 2018 | DDC 572/.6--dc23
LC record available at https://lccn.loc.gov/2017052036

Visit the Taylor & Francis Web site at
http://www.taylorandfrancis.com

and the CRC Press Web site at
http://www.crcpress.com

Printed and bound by CPI Group (UK) Ltd, Croydon, CR0 4YY

Contents

List of Figures

List of Tables

List of Boxes

Preface

Proteins are highly complex molecules that are actively involved in the most basic and important aspects of life. These include metabolism, movement, defense, cellular communication, and molecular recognition. Accordingly, protein science is at the very center of biological research, and is applied to disciplines such as medicine, agriculture, biotechnology, and even unconventional warfare.

In the last few decades, with the development of accurate and sophisticated means of molecular structure determination, it has become clear that the functions of macromolecules in general and of proteins in particular are direct results of structure and structural dynamics. It has similarly become evident that, to obtain a true understanding of protein function, structure and dynamics, it is necessary to both qualitatively and quantitatively characterize the dominant physical forces that act on proteins at the atomic level. These insights have prompted the emergence of a new field in biological sciences, termed 'structural biophysics'.

This book aims to provide the reader with a detailed description of protein structure and dynamics, combined with an in-depth discussion of the relationship between these aspects and protein function. We approach these topics through the lens of structural biophysics, focusing on the molecular interactions and thermodynamic changes that transpire in highly complex biological systems. There are several types of textbooks describing protein structure and function. Biochemistry textbooks emphasize the functional aspects of proteins and provide rather general descriptions of structures and the structure-function relationships. Structural biology textbooks provide extensive descriptions of protein structures, and also refer to the structure-function relationships with varying degrees of detail. However, energy-related aspects are often avoided. Molecular biophysics textbooks focus on molecular interactions and thermodynamic aspects of protein structures, but tend not to delve deeply into structural and dynamic aspects or into the structure-function relationships. Our book refers to all of the aforementioned aspects, and attempts to provide a unified view. Our energy-oriented approach is manifested throughout the book, whether we discuss structure, dynamics, or specific functions of proteins, such as catalysis or signal transduction. An extensive discussion of the energetics of protein structure is also given in a chapter dedicated to this topic.

Most textbooks that describe the structure-function relationships in proteins do so by using specific examples or protein types. This approach provides a broad view of protein activity, but may be insufficient to enable the reader to draw general conclusions. Here, when possible, we attempt to provide clear outlines of the principles of protein action (as we understand them). This is done throughout the book, but particularly in the last three chapters describing membrane-bound proteins, protein-ligand interactions, and enzyme-mediated catalysis. These topics, as well as any topic involving protein structure and function, are difficult to explain by using text and simple graphics. We therefore use the following means to convey to the reader the full experience of protein science:

1. Numerous high-resolution figures that portray real three-dimensional (3D) protein structures. Readers with a professional background in structural biology can also use the (free) PyMOL session files that we provide for each structural figure. These files, which are available online, can be viewed and manipulated using free PyMOL software.

2. Animations of biochemical processes. The animations are fully accessible by QR codes, which can be scanned directly and easily from the book by any smartphone (see more in 'Changes in second edition' section). The animations can also be found at the following URL: http://ibis.tau.ac.il/wiki/nir_bental/index.php/Proteins_Book

The central dogma of structural biology is the dependence of function on structure. Yet, some proteins, termed 'intrinsically unstructured proteins' (IUPs), are inherently devoid of a regular three-dimensional structure, and still have numerous functions. IUPs have been studied extensively in the last few years, yet are not mentioned in most textbooks. We dedicate a chapter of the book to these proteins, in order to provide a more complete and realistic view of the proteome and explore the full repertoire of protein functions.

Much of the knowledge on the relationship between protein structure and function became available only with the advent of technologies for the determination or prediction of proteins' three-dimensional structures. Accordingly, we provide a concise description of the main experimental and computational methods used today for studying protein structure and dynamics. In this respect, we mention various Internet-based resources, such as databases, algorithms, software and webservers, which are widely used and fully accessible to the reader. Moreover, as mentioned above, we emphasize that protein science is not only of academic interest. Indeed, it has been applied in various industrial, medical, and agricultural fields. In our book we discuss three of these applications: the industrial use of enzymes, protein engineering, and the rational design of pharmaceutical drugs that target specific proteins. We believe that our broad coverage of the different facets of protein science makes our book relevant to both students and scientists of protein-related fields.

Introduction to Proteins: Structure, Function, and Motion is intended for various audiences. First, the book can be used by undergraduate or graduate students of biochemistry, structural biology, computational biophysics, bioinformatics, and biotechnology, as an introduction to protein structure. In that sense, it may serve as a stand-alone textbook for basic- to intermediate-level courses in structural biology. For such purposes, we provide exercises related to theory and practice. Sample answers, as well as a set of PowerPoint slide shows that incorporate the figures presented in this book, are also available for qualifying instructors. Second, we expect that the parts of the book that provide detailed discussions of energetic, dynamic and evolutionary aspects of proteins will be of special interest to postgraduate scientists and industry professionals. To make it easier for these two groups of readers to find their texts of interest, we have, in some cases, separated the basic material from more advanced discussions by putting advanced material in numbered boxes. Finally, the book refers to many everyday issues related to proteins and enzymes, such as medical disorders, drugs, toxins, chemical warfare, and animal behavior. We hope that our coverage of these topics will create interest among some non-professional science enthusiasts as well.

The following is a general outline of the book:

Chapter 1, 'Introduction', includes three parts. The first provides an overview of proteins' main functions and their importance to various fields, e.g., medicine and the drug industry. The second explains the central 'structure-dynamics-function' paradigm in proteins, thus providing the general rationale of the book. The third part describes the non-covalent forces acting on macromolecules, an overview that provides the reader with the necessary background to understand the notions presented later on in the book. Finally, the general layout of the book is presented.

Chapter 2, 'Protein Structure', describes in detail the different levels of protein structure. The physico-chemical properties of amino acids are described at length. The descriptions of secondary, tertiary, and quaternary structures that follow emphasize the structural principles achieved by the observed architectures. Other factors affecting both protein structure and function — i.e., non-natural amino acids, enzymatic cofactors, prosthetic groups, and post-translational modifications — are also described, with emphasis on the structure-function relationship. All of these topics are exemplified using specific proteins. For instance, *protein kinase A (PKA)*, a central enzyme in cellular communication, is used to demonstrate some of the main advantages of quaternary structure. Pyruvate dehydrogenase, a large enzyme complex involved in carbohydrate metabolism, is used to demonstrate the roles of cofactors and prosthetic groups in protein function. The end of the chapter discusses a group of proteins that play relatively simple roles inside and outside cells, forming large fibrous structures. We discuss some well-studied examples such as *collagen*, the principal protein of connective tissues, and *keratin*, a protein that provides toughness to horns, nails and claws.

Chapter 3, 'Methods of Structure Determination and Prediction', describes the main methods used today for structure determinations, and their applications. First, methods based on particle and wave diffraction or scattering are described. These include *X-ray crystallography*, *neutron scattering*, *electron scattering*, and *electron microscopy*. We then discuss spectroscopic methods, including *nuclear magnetic resonance (NMR) spectroscopy*, *electron paramagnetic resonance (EPR) spectroscopy*, and *circular dichroism (CD)*. This discussion is followed by a description of computational methods for predicting protein structure, which can be classified into two main groups, or approaches. The first, 'physical' approach relies on mathematical descriptions of the physical forces acting on a protein's atoms. We elaborate on several well-known methods corresponding to this approach, including *molecular dynamics* and *simulated annealing*. The second, 'comparative' approach, the most prominent of which is *homology modeling*, relies on sequence comparisons and statistical data. In covering this topic, we dedicate a great deal of the discussion to analyzing the advantages and disadvantages of each method and the cases in which a given method is most applicable. Finally, we present the current tools for comparing the different methods and evaluating their efficiency.

Chapter 4, 'Energetics and Protein Stability', discusses the thermodynamic aspects of protein structure. It begins with an overview of the basic thermodynamic variables, the means by which they can be measured or calculated, and their interpretation in molecular systems. In discussing the latter, we refer to biological processes that can be characterized using thermodynamic variables. These include metabolic processes, protein folding, and protein-ligand interactions. The second section of the chapter discusses the main physical forces in a system with respect to their influence on protein structure. In the third and fourth sections we examine two cases in which the theoretical principles discussed are applied. The first is the adaptation of unicellular organisms to extreme environments, and the second is the use of protein engineering to enhance the industrial uses of enzymes.

Chapter 5, 'Protein Dynamics', expands the structure-function paradigm by incorporating structural dynamics. Two aspects of protein dynamics are discussed: protein folding, and folded (native) state dynamics. In addressing protein folding, we present the current views on how proteins acquire their three-dimensional structures. This field has been studied extensively, and we present the main conclusions. In addition, we discuss some well-known pathologies involving protein misfolding, such as *cystic fibrosis*, *Parkinson's disease*, and *mad cow disease*. Next, we discuss changes that can occur in a protein's native structure over time, and illustrate their functional importance on different levels. In this context, we elaborate on allostery as a key cellular approach for regulating protein function through manipulation of a protein's dynamic properties. We discuss different models and mechanisms of allostery and use specific proteins to demonstrate them. For example, we refer to the medically important enzyme *dihydrofolate reductase (DFHR)*, which has been shown to be subject to long-distance allosteric effects. The oxygen-carrying protein hemoglobin is used to provide a detailed example of multi-level changes in protein dynamics induced by allosteric regulators.

Chapter 6, 'Intrinsically Unstructured Proteins', focuses on a group of proteins that seem to deviate from the 'globular' behavior presented in the previous chapters. These proteins, called intrinsically unstructured proteins (IUPs), are characterized by the absence of a regular tertiary structure. IUPs have evolved to fulfill many different functions that do not require a permanent structure, and even benefit from the lack thereof. As in Chapter 2, we discuss the principal properties of IUPs, with emphasis on the structure-function relationship.

Chapter 7, 'Membrane-Bound Proteins', focuses on a subtype of globular proteins that are located near and inside cellular membranes. These proteins constitute 20% to 30% of the genome and play numerous roles in cellular physiology. Unlike water-soluble globular proteins, membrane-bound proteins are surrounded by a lipid environment, and are therefore subjected to different forces, and consequently behave differently. The first part of this chapter overviews the structure, organization, and function of biological membranes. In particular, it discusses membrane asymmetry and the variability of membrane composition (and hence, the variability of membrane properties) among different organisms. The second part analyzes membrane proteins, emphasizing common sequence- and structure-related themes, as well as folding energetics. The third part discusses the important issue of protein-membrane interactions, which has implications for both structures and functions of membrane proteins. Finally, to illustrate the structure-function relationship in membrane proteins, we focus on *G-protein coupled receptors (GPCRs)*, a group of receptors that serve as targets of most pharmacological drugs. We discuss in detail the *β-adrenergic receptor*, the structure of which has recently been determined in its active state. Membrane proteins are notoriously difficult to crystallize, and are therefore desirable targets for structure prediction. Throughout this chapter, we mention key computational approaches for locating membrane proteins within genomes, for predicting their topology, and for predicting their full three-dimensional structures.

Chapter 8, 'Protein-Ligand Interactions', demonstrates the structure-function relationship in proteins by addressing proteins' most important ability, i.e., binding to other molecules. After a short overview of the functional aspects of this ability, we discuss past and present theories on binding and their thermodynamic implications. We then analyze protein binding on a molecular level, by focusing on the properties of protein binding sites. One such property is electrostatic potential, which we discuss using the example of *acetylcholinesterase (AChE)*. AChE is a major enzyme responsible for the correct functioning of the nervous system and is, therefore, also a major target of various nerve agents and toxins. Its action is extremely fast, in part because of the mechanism of 'electrostatic steering', which the enzyme uses to draw its natural substrate into a catalytic site. The chapter subsequently illustrates the principles discussed above of protein-ligand binding by addressing the example of protein-protein interactions. Finally, we discuss the rational design of pharmaceutical drugs, which is a key practical application of protein-ligand interactions.

Chapter 9, 'Enzymatic Catalysis', discusses enzymes, which are probably the most sophisticated proteins in terms of function and molecular mechanism. In contrast to most biochemistry and protein structure and function books, this chapter provides a wide-angle, yet detailed description of various topics related to enzymes, including types of enzymes and reactions, molecular mechanisms, thermodynamics and kinetics, specificity, regulation, and 'real-world' applications, both medical and industrial. In accordance with the central theme of the book, Chapter 9 emphasizes structure-function relationships in enzyme catalysis, including some aspects that are usually ignored in other books, e.g., quantum tunneling and vibrational effects. Finally, the reader is provided with animations of the numerous chemical reactions and enzymatic mechanisms described in this chapter. These animations are easily accessible via in-text QR codes that can be scanned using a smartphone. The animations can also be found at the following URL: http://ibis.tau.ac.il/wiki/nir_bental/index.php/ Proteins_Book

In this book, we use numerous proteins as examples, demonstrating the various topics and principles discussed. Some proteins are mentioned in several different contexts, to reflect the multiple ways in which proteins can be studied and analyzed. For example, *hemoglobin* is used to demonstrate quaternary structure, pathologies stemming from structure-altering mutations, and the role of dynamics in allosteric regulation. Another example is the cancer-related *ras* protein, used to demonstrate different types of post-translational modifications.

Changes in Second Edition

The field of proteins is vast, and in the seven years that have passed since the publication of the first edition of this book, many exciting discoveries have emerged, covering virtually every topic discussed herein. We have updated all of our discussions accordingly, with emphasis on the following:

Membrane proteins (Chapter 7) — Three-dimensional structures of membrane proteins have always been more difficult to determine compared with the structures of water-soluble proteins. Dramatic progress has been made in structure determination methods, resulting in numerous new structures of highly important membrane proteins, including receptors, channels, transporters, and enzymes. Furthermore, in certain types of membrane proteins, such as GPCRs, the new structures have provided new functional insights. For example, many of the new GPCR structures were determined in a partially or fully activated mode. Such structures were almost completely absent when the first edition of this book came out. Accordingly, in the current edition, we are able to elaborate on the activation processes of GPCRs. In addition, the availability of new structures of GPCRs belonging to classes B, C and F enables us to describe the features that these proteins share with the more common class A GPCRs, as well as their differences.

Methods for studying proteins (Chapter 3) — Recent technological breakthroughs in cryo-electron microscopy (cryo-EM) and small-angle X-ray scattering (SAXS) are reshaping structural biology. First, thanks to dramatic improvement in the resolution of cryo-EM, scientists have been able to determine the structures of many proteins that are difficult to crystallize and therefore inaccessible to X-ray diffraction (e.g., membrane proteins). Second, cryo-EM and SAXS have facilitated the structural determinations of large protein complexes, which are often the proteins' functional forms. Third, the data extracted from these methods can be used as constraints that guide computational structure predictions of proteins. Thus, the availability of such data has revolutionized the field of computational structure prediction. To reflect these developments, we have expanded Chapter 3 significantly by elaborating on the uses of cryo-EM and SAXS. We also provide an extensive discussion of new hybrid computational methods, which integrate different approaches for the purpose of computational structure prediction.

Enzyme catalysis (Chapter 9) — The relationship between structure and function in proteins reaches its highest level of sophistication in enzymatic catalysis. Enzymes are also key components in all life-forms, and are involved in virtually all life processes. For this reason, we addressed enzymes and enzymatic catalysis in the first edition, emphasizing their metabolic roles. However, since enzymes are routinely described in biochemistry textbooks, we refrained from elaborating on this topic.

Yet, the recent proliferation of structural and biophysical analysis methods has led to a better understanding of the structures, energetics, molecular dynamics, and chemical mechanisms of enzymes. This made it possible in the current edition to apply our physicochemical approach to enzymatic catalysis, as we had done previously for the other aspects of protein structure and function. Thus, we have added a completely new chapter to the book, which provides an extensive description of various aspects of enzymes, including types and classifications, metabolic roles, molecular mechanisms, kinetics, energetics, dynamics, the use of cofactors, inhibition, related diseases, engineering, and practical uses in medicine and in other industries. Our discussion of catalytic mechanisms integrates structural, dynamic, and thermodynamic aspects, including quantum phenomena, which are ignored in most textbooks. This comprehensive coverage provides the reader with what we believe is an unprecedented view of enzymes. We decided to make this chapter the last one in the book, because all the important principles of enzymatic catalysis result from phenomena described in the previous chapters: preorganized structure (Chapter 2), dynamic qualities (Chapter 5), and protein-ligand interactions (Chapter 8). Similarly, applications that involve enzymes are also touched on in earlier chapters; for example, enzyme engineering is largely based on general protein engineering and drug design principles, which are described in Chapter 8.

In addition to updating the book and expanding its scope, we also aimed to make it more reader-friendly. Thus, the second edition includes a larger number of figures, and also incorporates two new technical features that make the learning experience more enjoyable and efficient:

Animations — Certain processes, such as multistep chemical reactions and conformational changes in macromolecules, are difficult to describe using static images alone. We have therefore created animations of these processes, each of which is linked to the corresponding book page via a QR code, which the reader can scan using a smartphone. Scanning the QR code immediately links the smartphone to the Internet location of the animation, enabling the reader to watch the animation while reading the book. The animations can also be found at the following URL: http://ibis.tau.ac.il/wiki/nir_bental/index.php/Proteins_Book

PyMOL session files — The book contains numerous images of 3D protein structures, which are used to explain structural, dynamic, and functional phenomena. While these images are very informative, readers and instructors often wish they could look at the displayed proteins from different angles or use different molecular representations to get a better understanding of the ideas that an image aims to illustrate. PyMOL is free software that can be used to perform such manipulations, provided that the user has a file containing the protein's 3D coordinates. In the current edition, we provide PyMOL session files (.pse) for many of the structural images in the book. Each session file allows the reader to use PyMOL to open a molecular representation of the protein, exactly as it is shown in the corresponding book image. The reader can then use the representation as a starting point for further changes and manipulations. The provided PyMOL session files include, in most cases, pre-defined elements of the respective proteins (secondary structures, ligands, electrostatic potential maps, annotations of polar interactions, etc.), which make it easier for the readers to manipulate the proteins.

Acknowledgments

The authors would like to thank the following people:

- For review, consultation, and helpful discussions (in alphabetical order): Abdussalam Azem, Amnon Horovitz, Avner Schlessinger, Aya Narunsky, Dan Tawfik, Eric Marz, Leslie Kuhn, Matan Kalman, Rachel Kolodny, Sarel J. Fleishman, Sharon Rozovsky, Steven Bottomley, and Turkan Haliloglu.

- For graphics assistance: Varda Vexler, Iddo Better, Maya Schushan, and Gal Masrati.

- For cover design: Dan Latovicz and Gal Masrati.

- For animations design: Gad and Elon Yariv.

- For editorial assistance: Karen Marron.

- For LaTeX 2_ε formatting: Elio Arturo Farina.

A. K. would also like to thank Rachel and Natan Stempler, as well as Yifat, Eyal, Yoav, and Na'ama Kaufman for their ongoing personal support.

Authors

Dr. Amit Kessel obtained his master's degree in experimental biochemistry at Tel-Aviv University, studying the innate response of human blood cells to pathogenic bacteria. During his Ph.D. studies he trained as a computational biologist, investigating the molecular basis of peptide-membrane interactions and the mechanisms of antibacterial peptides. In his postdoctoral research at Columbia University, Dr. Kessel continued studying proteins at the molecular level, focusing on various physico-chemical aspects of protein-protein interactions. In 2010 he co-founded *ES-IS Technologies*, a company that designed novel enzymatic solutions for the pharmaceutical industry, and he headed the company's R&D department. He is currently involved in several academic and industrial biotech initiatives, focusing on experimentally guided protein structure prediction and on the in vitro construction of protein-based nanoparticles.

Prof. Nir Ben-Tal obtained his bachelor's degree in biology, chemistry, and physics at the Hebrew University and his D.Sc. in chemistry at the Technion Israel Institute of Technology. He carried out his postdoctoral training as a computational biophysicist at Columbia University, and later joined the Department of Biochemistry and Molecular Biology at Tel Aviv University. His research covers various aspects of computational biology with a focus on structural bioinformatics and the protein universe. In particular, his laboratory has predicted the 3-dimensional structures of a number of transmembrane proteins, thereby providing molecular insight into their mechanisms. His lab also develops the ConSurf web-server for the detection of functional regions by mapping evolutionary data onto protein structures (jointly with the Mayrose and Pupko labs).

Department of Biochemistry and Molecular Biology
The George S. Wise Faculty of Life Sciences
Tel-Aviv University

Laboratory homepage: http://bental.tau.ac.il
Email: bental@tauex.tau.ac.il

Physical Quantities and Constants

QUANTITIES

Quantity	Name	Symbol	SI units	Related units
Base units				
length	meter	m	m	
mass	kilogram	kg	kg	gram $= 10^{-3}\,\text{kg}$
				$\text{Da}^{*a} = 1.66 \times 10^{-27}\,\text{kg}$
time	second	s	s	
temperature	kelvin	K	K	
amount of substance	mole	mol	mol	
electric current	ampere	A	A	
Derived units				
force	newton	N	m·kg·s^{-2}	
energy	joule (N·m)	J	$\text{m}^2\text{·kg·s}^{-2}$	calorie $= 4.18\,\text{J}$
pressure	pascal (N·m^{-2})	Pa	$\text{m}^{-1}\text{·kg·s}^{-2}$	
electric charge	coulomb	C	s·A	
electric potential	volt	V	$\text{m}^2\text{·kg·s}^{-3}\text{·A}^{-1}$	
Celsius temperature	degree Celsius	°C	$K - 273.15$	
capacitance	farad (C·V^{-1})	F	$\text{m}^{-2}\text{·kg}^{-1}\text{·s}^4\text{·A}^2$	
permittivity^{*b}	farad per meter	ε; F·m^{-1}	$\text{m}^{-3}\text{·kg}^{-1}\text{·s}^4\text{·A}^2$	
electric dipole moment	coulomb–meter	C·m	s·A·m	Debye $= 3.34 \times 10^{-30}\,\text{C} \cdot \text{m}$

[*a] Dalton, a.k.a. molecular weight or atomic mass unit (a.m.u), is defined as 1/12 of the mass of an unbound, neutral ^{12}C atom.

[*b] A measure of how much a medium resists the propagation of an electric field.

CONSTANTS

Constant	Symbol	Value	Definition
Avogadro's constant	N_A	6.022×10^{23} mol^{-1}	Number of particles in 1 mol of substance
Electronic charge	e	1.602×10^{-19} C	Electric charge of an electron
Faraday's constant	$F = N_A e$	96,485 C·mol^{-1}	Electric charge per mole of electrons
Universal gas constant[*a]	R	8.314 J·K^{-1}·mol^{-1} (1.989 cal·mol^{-1}·K^{-1})	Relation of energy to temperature at mol scale
Boltzmann's constant	$k_B = R/N_A$	1.381×10^{-23} J·K^{-1} (3.3×10^{-24} cal·K^{-1})	Relation of energy to temperature at single particle scale
Planck's constant	h	6.626×10^{-34} J·s	Relation between photon energy and frequency
Permittivity of free space	ε_0	8.854×10^{-12} F·m^{-1}	Permittivity of vacuum
Dielectric constant[*b]	$\varepsilon_r = \varepsilon/\varepsilon_0$	1, 2, 3, ...	Permittivity of a substance compared to that of vacuum
Coulomb's constant	k_e	8.987×10^9 N·m^2·C^{-2}	Proportionality in equations relating electric variables ($1/4\pi\varepsilon_0$)

[*a] Also known as 'molar gas constant'.
[*b] Also known as 'relative permittivity'.

List of Acronyms

2,3 BPG	2,3 Bis-phosphoglycerate
5-FdUMP	5-Fluorodeoxyuracil monophosphate
α-KG	Alpha-ketoglutarate
AA-DH	Amino acid dehydrogenases
ABC	ATP-binding cassette transporter
ACBP	Acyl-coenzyme A binding protein
ACE	Angiotensin converting enzyme
ACh	Acetylcholine
AChE	Acetylcholinesterase
ACoA	Acetyl-CoA
ACTH	Adrenocorticotropic hormone
AC	Adenylyl cyclase
ADA	Adenosine deaminase
ADD	Attention deficit disorders
ADH	Anti-diuretic hormone
AdoCbl	Adenosylcobalamin
ADP	Adenosine diphosphate
AD	Alzheimer's disease
AFP	Antifreeze protein
AKR	Aldo-keto reductase
ALS	Amyotrophic lateral sclerosis
ALT	Alanine aminotransferase
AMBER	Assisted model building with energy refinement
ANM	Anisotropic network model
ANS	Autonomic nervous system
APBS	Adaptive Poisson-Boltzmann solver
APCs	Antigen-presenting cells
API	Active pharmaceutical ingredient
AP	Alkaline phosphatase
AP	Antibody precipitation
AQP0	Assembly of lens aquaporin-0
ASEdb	Alanine scanning energetics database
ASICs	Application-specific integrated circuits
AST	Aspartate aminotransferase
ATP	Adenosine triphosphate
BBB	Blood-brain barrier
bHLH	Basic helix-loop-helix
BLIP	Beta-lactamase inhibitor protein

BPTI	Pancreatic trypsin inhibitor
BRENDA	Braunschweig Enzyme Database
BSE	Mad cow disease
BTX	Botulinum toxin
CAB	Carbonic anhydrase B
CADD	Computer-aided drug design
CAL-B	*Candida antarctica* lipase B
cAMP	Cyclic adenosine monophosphate
CaM	Calmodulin
CaSRs	Ca^{2+}-sensing receptors
CBP	CREB-binding protein
CCD	Charge-coupled device
CCO	Cytochrome C oxidase
CD4	Complementarity determinant 4
CD	Circular dichroism
CF	Cystic fibrosis
CFF	Consistent force field
CFTR	Cystic fibrosis transmembrane conductance regulator
cGMP	Cyclic guanosine monophosphate
CHARMM	Chemistry at Harvard Macromolecular Mechanics
CI2	Chymotrypsin inhibitor 2
CJD	Creutzfeldt-Jakob disease
CK, CPK	Creatine phosphokinase
CK-MB	Creatine kinase-myocardium-bound
CL	Cardiolipin
CNS	Central nervous system
CoA	Coenzyme A
COSY	Correlation spectroscopy
COX	Cyclooxygenase
CPA	Carboxypeptidase A
CPEB	Cytoplasmic polyadenylation element binding protein
CRF	Corticotropin-releasing factor
CRFR1	CRF receptor subtype 1
CRPs	Collagen-related proteins
cryo-EM	Electron cryomicroscopy
CR	Cysteine-rich
CS	Continuum solvent
CTX	Cholera
DAG	Diacylglycerol
DDD	Direct detection device
DHAP	Dihydroxyacetone phosphate
DHF	Dihydrofolate
DHFR	Dihydrofolate reductase
DMF	Dimethylformamide
DMSO	Dimethylsulfoxide
DNA	Deoxyribonucleic acid

DSC	Differential scanning calorimetry
DT	Diphtheria
Ea	Activation energy
ECD	Extracellular domain
ECM	Extracellular matrix
EC	Enzyme classification or commission
EF-2	Protein synthesis elongation factor 2
EGFR	Epidermal growth factor receptor
EICs	Evolutionary inferred couplings
EINS	Elastic incoherent scattering
EM	Electron microscopy
ENM	Elastic network model
EPR	Electron paramagnetic resonance
ERT	Enzyme replacement therapy
ER	Endoplasmic reticulum
ESI	Electrospray ionization
EVH1	Enabled VASP homology 1
F1,6BP	Fructose 1,6-bisphosphate
F2,6BP	Fructose 2,6-bisphosphate
F6P	Fructose 6-phosphate
Fab	Antigen-binding region of antibody
FAD	Flavin adenine dinucleotide
FAS	Fatty acid synthase
FEP	Free energy perturbation
FIH	Inhibiting factor
FMN	Flavin adenine mononucleotide
FM	Fluid mosaic
FRET	Fluorescence resonance energy transfer
FTIR	Fourier transform infrared
FT	Fourier transform
G3P	Glyceraldehyde 3-phosphate
G6PD	Glucose-6-phosphate dehydrogenase
GABA	Gamma-amino butyric acid
GAFF	General AMBER force field
GB	Generalized Born [model]
GCGR	Glucagon receptor
GDT	Global distance test score
GEF	GDP/GTP exchange factors
GES	Goldman-Engelman-Steitz [method]
GF	Growth factor
GGT, γGT	Gamma-glutamyl transpeptidase
GIP	Glucose-dependent insulinotropic polypeptide
GI	Gastrointestinal
GLP-1	Glucagon-like peptide 1
GNM	Gaussian network model
GPCR	G protein-coupled receptor

GRF	Growth hormone-releasing factor
GRK	G protein-coupled receptor kinase
GROMACS	Groningen machine for chemical simulations
GSH	Glutathione
GSLs	Glycosphingolipids
GTP	Guanosine-5′-triphosphate
HbA	Hemoglobin A
HbI	Shellfish hemoglobin
HbS	Hemoglobin S
HDX-MS	Hydrogen/deuterium exchange mass spectrometry
HD	Huntington's disease
HeLa	Henrietta Lacks cells
Hh	Hedgehog
HLH	Helix-loop-helix
HMM	Hidden Markov Models
HOESY	Heteronuclear Overhauser effect spectroscopy
HPA	Hypothalamic-pituitary-adrenal
HPLC	High-performance liquid chromatography
HSP60	Heat shock protein 60
Hyp	Hydroxyproline
IAA	Iodoacetic acid
IC	Intracellular
ICL	Intracellular loop
IDPs	Intrinsically disordered proteins
IDRs	Intrinsically disordered regions
Ig	Immunoglobulin
IgD	Immunoglobulin domain
IgF	Immunoglobulin fold
IgSF	Immunoglobulin superfamily
IHP	Inositol hexaphosphate
IINS	Inelastic incoherent scattering
IMP	Integrative modeling platform
IMS	Intermembrane space
ITC	Isothermal calorimetry
Itk	Interleukin-2 tyrosine kinase
IUBMB	International Union of Biochemistry and Molecular Biology
IUPAC	International Union of Pure and Applied Chemistry
IUPs	Intrinsically unstructured proteins
kAP	Ketomethylene ACE Inhibitors
KcsA	pH-dependent bacterial K^+ channel
KD	Kyte-Doolittle [scale]
KEGG	Kyoto Encyclopedia of Genes and Genomes
KNF	Koshland-Némethy-Filmer [model]
LBHBs	Low-barrier hydrogen bonds
LC/MS	Liquid chromatography/mass spectrometry
LDH	Lactate dehydrogenase

LHRH	Luteinizing hormone-releasing hormone
LIE	Linear interaction energy
LPS	Lipopolysaccharides
Ltn	Human lymphotactin
MACiE	Mechanism, annotation and classification in enzymes
MALDI	Matrix-assisted laser desorption/ionization
MAO	Monoamine oxidase
MAP	Mitogen-activated protein kinase
MARCKS	Myristoylated alanine-rich C-kinase substrate
MCSS	Multiple copy simultaneous search
MC	Monte Carlo
MDMA	3,4-Methylenedioxymethamphetamine
MD	Molecular dynamics
MES	Minimal ensemble search
mGluR	Metabotropic glutamate receptor
MHC	Major histocompatibility complex
MIP	Major intrinsic protein
M–M	Michaelis-Menten
MM	Molecular mechanics
MM/MD	Molecular mechanical/molecular dynamic
MM-GBSA	Molecular mechanics generalized Born surface area
MoREs	Molecular recognition elements
MoRFs	Molecular recognition features
mRNA	Messenger RNA
MS/MS	Tandem mass specrometry
MSA	Multiple sequence alignment
MS	Mass spectrometry
MWC	Monod-Wyman-Changeux model
NAD(H)	Nicotinamide dinucleotide
NADP(H)	Nicotine adenine dinucleotide phosphate
NES	Nuclear export signals
NE	Norepinephrine
NIH	National Institutes of Health
NK	Natural killer
NLS	Nuclear localization signals
NMA	Normal mode analysis
NMR	Nuclear magnetic resonance
NM	Normal mode
NOESY	Nuclear Overhauser effect spectrometry
NOE	Nuclear Overhauser effect
NPC	Nuclear pore complex
NSAIDs	Non-steroidal anti-inflammatory drugs
NTPs	Nucleoside triphosphates
NTS1	Neurotensin 1
NUPs	Natively unfolded proteins
OPLS-AA	All-atom OPLS

OPLS-UA	United-atom OPLS
OPLS	Optimized potentials for liquid simulations
OP	Organophosphate
PABA	Para-aminobenzoic acid
PACAP	Pituitary adenylyl cyclase activating polypeptide
PAPS	Phosphoadenosine phosphosulfate
PBE	Poisson-Boltzmann equation
PCDDB	Protein Circular Dichroism Data Bank
PCE	Protein continuum electrostatics
PC	Phosphatidylcholine
PDB	Protein Data Bank
PDGF-R	Platelet-derived growth factor receptor
PDH	Pyruvate dehydrogenase
PDI	Protein disulfide isomerase
PD	Parkinson's disease
PEA	Phenylethylamine
PEG	Polyethylene glycol
PEPCK	Phosphoenolpyruvate carboxykinase
PE	Phosphatidylethanolamine
PFK-1	Phosphofructokinase-1
PFK-2	Phosphofructokinase-2
PGA	2-Phosphoglycerate
PG	Phosphatidylglycerol
PG	Prostaglandin
PheDH	Phenylalanine dehydrogenase
PH	Pleckstrin homology
PIP2	Phosphatidylinositol bisphosphate
P_i	Inorganic phosphate
PI	Phosphatidylinositol
PI	Phosphoinositol
PKA	Protein kinase A
PKC	Protein kinase C
PKU	Phenylketonuria
PK	Pyruvate kinase
PLC	Phospholipase C
PLP	Pyridoxal phosphate
PMF	Potential of mean force
pmf	Proton motive force
PMP	Protein model portal
PMP	Pyridoxamine phosphate
PNP	Purine nucleoside phosphorylase
POPE	Palmitoyl-oleoyl phosphatidylethanolamine
PPII	Polyproline II helix
PPII	Type II polyproline helix
PPI	Peptidyl prolyl *cis-trans* isomerase
PreSMos	Pre-structured motifs

PRE	Paramagnetic relaxation enhancement
PrP	Proteinaceous infectious particle
PSMs	Phenol-soluble modulins
PSSM	Position-specific scoring matrix
PS	Phosphatidylserine
PTH	Parathyroid hormone
PTM	Post-translational modification
PTP	Protein tyrosine phosphatase
PTX	Pertussis
pY	Phosphorylated Tyr residue
QINS	Quasi-elastic incoherent scattering
QM/MM	Quantum mechanics/molecular mechanics
QM	Quantum-mechanical
QSAR	Quantitative structure-activity relationship
RAA	Renin-angiotensin-aldosterone [system]
RCSB	Research collaboratory for structural bioinformatics
RDA	Recommended dietary allowance
RDC	Residual dipolar coupling
rER	Rough endoplasmic reticulum
RGSs	Regulators of G protein signaling
RLKs	Receptor-like kinases
RNA	Ribonucleic acid
SAM	S-Adenosyl methionine
SANS	Small-angle neutron scattering
SARA	SMAD anchor for receptor activation
SAR	Structure-activity relationship
SAXS	Small angle X-ray scattering
SA	Surface area
SBD	SMAD-binding domain
SCID	Severe combined immunodeficiency
SCOP	Structural classification of proteins
SDR	Short-chain dehydrogenase/reductase
SD	Succinate dehydrogenase
sER	Smooth endoplasmic reticulum
sGOT	Serum glutamic oxaloacetic transaminase
sGPT	Serum glutamic pyruvic transaminase
SH2	Src-homology 2
SH3	Src-homology 3
sHSPs	Small heat-shock proteins
SH	Reduced thiol groups
SKEMPI	Structural kinetic and energetic Database of Mutant Protein Interactions
SLiMs	Short linear motifs
SMO	Smoothened protein
SM	Sphingomyelin
SNRIs	Serotonin and norepinephrine reuptake inhibitors
SOD	Superoxide dismutase

SRCD	Synchrotron radiation circular dichroism
SR	Sarcoplasmic reticulum
SSRIs	Selective serotonin reuptake inhibitors
ST-EPR	Saturation transfer EPR
SUMO	Small ubiquitin-like modifier
TCR	T-cell receptor
TEM	Transmission electron microscope
TF	Human transcription factor
THC	Tetrahydrocannabinol
THF	Tetrahydrofolate
TIM	Triosephosphate isomerase
TLP	Thermolysin-like protease
TM	Transmembrane
TMAO	Trimethylamine N-oxide
TMD	Transmembrane domain
TOCSY	Total correlation spectroscopy
TPCK	Tosyl-L-phenylalanine chloromethyl ketone
TPP	Thiamine pyrophosphate
TR-SAXS	Time-resolved small angle X-ray scattering
TRH	Thyroxin-releasing hormone
tRNA	Transfer RNA
Trx	Thioredoxin
TSEs	Transmissible spongiform encephalopathies
TTS	Tertiary two-state model
ULF	Unit length filament
VFT	Venus fly trap
VIP	Vasoactive intestinal peptide
VMD	Virtual molecular dynamics/visual molecular dynamics
WASP	Wiskott-Aldrich syndrome protein
WAXS	Wide-angle X-ray scattering
WH2	WASP-homology domain 2
wwPDB	Worldwide Protein Data Bank

CHAPTER 1

Introduction

1.1 IMPORTANCE OF PROTEINS IN LIVING ORGANISMS

1.1.1 Life, proteins and mysterious forces

What is life? If you could ask an early 18th century scientist this question, he or she would probably mention *'vital forces'* — mysterious, metaphysical energies that inhabit 'organic' matter, and that keep organisms alive and functional. This concept is very old, preceding the philosophers of ancient Greece, perhaps even Egypt. Yet, despite the numerous scientific revelations of the last millennia in physics, chemistry and physiology, such ideas advocating the metaphysical uniqueness of living matter continued to be widely accepted until as recently as 150 years ago [1]. The change came in the beginning of the 19th century, with the gradual spreading of mechanistic theories regarding nature and physiology [2]. These philosophies posited that all life-related phenomena can be explained by the same physical and chemical principles that rule the inanimate world [3]. An important breakthrough of this approach was achieved by Louis Pasteur (1822–1895), who demonstrated that the chemical process of converting sugar into alcohol (i.e., fermentation) was a result of the growth of microorganisms. In doing so, Pasteur established a link between life processes and chemical reactions. Pasteur's work was followed by studies of scientists such as Marcellin Berthelot and Eduard Buchner [4] (Figure 1.1), who demonstrated that it was possible to achieve fermentation, in addition to other life-related processes, in the absence of microorganisms, by using substances extracted from those microorganisms[*1]. These substances were termed *'enzymes'* [5], which means 'in yeast', and although their chemical nature was at first unclear, they were later found (in all cases) to be proteins. These proteins acted as catalysts, i.e., they accelerated chemical reactions within cells and tissues without changing their nature. The discovery of enzyme activity led to a major turning point in scientific thinking: **Life was no longer considered to be a result of mysterious and vague phenomena acting on organisms, but instead the consequence of numerous chemical processes made possible by proteins**[*2]. Indeed, this notion became the cornerstone of modern biochemistry and molecular biology. Since the work of Pasteur and his successors, researchers have identified many other functionally important proteins, in addition to enzymes. Perhaps the best-known example of an important non-catalytic protein is hemoglobin, an animal protein functioning

---

[*1]For his work, Buchner received a Nobel Prize in chemistry, considered to be the first awarded to a biochemist.

[*2]The word 'protein' (*primarius* in Latin, πρώτειος in Greek) means 'of the first rank' [6,7]. This term was coined by Berzelius and used for the first time in a publication by Mulder [8], in 1838.

in carrying oxygen from the lungs to body organs and tissues, as well as carrying CO_2, a metabolic waste product, back to the lungs. The genetic revolution that started in the second half of the 20th century — which has led to the deciphering of DNA structure, as well as the genetic code — has even further elucidated the nature of proteins. Specifically, we now know that proteins are more than just 'molecular machines' active within cells and tissues; they are also the primary products of genes, responsible (among other things) for the expression of genetic information.

In this book we aim to convey the principles of protein action to the reader, using a mechanistic approach. The book is primarily intended for readers who have some basic background in biochemistry and cell biology. For those readers who are new to biological sciences, we use the following subsections to explain the general architecture of living organisms, and the major roles of their proteins.

FIGURE 1.1   **Eduard Buchner, winner of the 1907 Nobel Prize in chemistry.** The image is taken from [9] (originally from [10]).

1.1.2   Molecular organization of living organisms

Earth is populated by a huge diversity of organisms, the number of which is estimated to be in the millions [11]. Despite this diversity, which is manifested in morphology, behavior, diet, and modes of reproduction, there is one universal trait shared by all organisms; they are all made of cells [2,12,13]. Indeed, the basic cellular structure can be found in bacteria and yeasts, which are made up of single cells (i.e., unicellular); in simple invertebrates that contain several tens of nearly identical cells that all share the same function; and in mammals (including humans), whose bodies contain trillions of morphologically and biochemically distinct cells. These cells are organized in a hierarchical manner as tissues and organs, and carry out distinct functions. Humans were completely unaware of the existence of cells until the invention of the light microscope around the 17th century. That is because even the largest cells are at least five times smaller than the resolution capacity of the human eye.

It is customary to separate the population of biological cells (and the organisms they form) into two principal types. The first type, termed *prokaryotic*, is small (~1 μm = 10^{-6} m), and lacks any visible internal organization. A prokaryotic cell consists of a lipid membrane (the *plasma membrane*) engulfing an inner aqueous environment (the *cyto-*

plasm). The cytoplasm is where all life processes take place, and it is separated from the external environment of the cell by the plasma membrane. This separation, however, is not absolute; the membrane selectively allows the uptake of required molecules from the environment into the cell and the excretion of waste products. In addition, the membrane 'senses' the outside environment and relays important information into the cell. Prokaryotic organisms include all bacteria, which are the most abundant form of life on Earth. Bacterial cells also have cell walls, which physically protect them from the external environment.

The second and more advanced type of cell is termed '*eukaryotic*'. Eukaryotic cells are much larger than prokaryotic cells, with a diameter ranging between 10 μm and ~100 μm. **In addition to containing aqueous fluid (cytosol), the eukaryotic cytoplasm also includes inner compartments (organelles) that specialize in carrying out distinct cellular processes** (Figure 1.2). The *nucleus* is the organelle containing the cell's genetic material. A region in the nucleus called the *nucleolus* specializes in constructing some of the cell's biosynthetic machinery. *Mitochondria* are the cell's power stations, extracting chemical energy from food and storing it as accessible energy currency (ATP, see below). The *endoplasmic reticulum* (*ER*), a closed membranous structure extending from the nucleus towards the periphery, is responsible for the synthesis and modification of membrane proteins as well as of proteins destined for secretion or for other organelles. Protein synthesis and modification take place in a region of the ER termed the '*rough endoplasmic reticulum*' (*rER*). As the major biosynthetic center of the cell, the ER is also responsible for building most of the cell's lipids. This process takes place in a different region of the ER, termed the '*smooth endoplasmic reticulum*' (*sER*). The *Golgi apparatus* is a collection of membranous sacs near the periphery of the cell. It receives lipids and proteins from the ER and sorts them for distribution to different cellular locations, as well as for secretion. Eukaryotic cells may be further separated into plant and animal cells. Plant cells have several additional features, which are absent in animal cells: The *cell wall* of a plant cell resides peripherally to the plasma membrane, and provides mechanical support to the cell. The second feature, the *vacuole*, plays several roles. First, it stores nutrients, waste products, and pigments. Second, it participates in degradation of cellular components. Finally, it regulates cell size, pH, and turgor pressure. *Chloroplasts*, which are also present in algae, perform *photosynthesis*, a highly complex process in which solar energy is harnessed for the synthesis of carbohydrates from atmospheric CO_2. In other words, chloroplasts convert inorganic carbon into organic form. In doing so, plants and algae supply fuel and building blocks for higher organisms, such as fish, insects, reptiles, birds, and mammals.

In addition to the larger and more complex organelles, eukaryotic cells also contain *vesicles* of different kinds, which perform various functions. For example, *lysosomes* function as waste disposal units; they contain hydrolytic (i.e., degrading) enzymes that decompose outdated molecules, organelles, and chemicals that have penetrated the cell. Another type of vesicle, the *peroxisome*, contains oxidizing enzymes. The oxidation acts on different molecules for different reasons, e.g., the neutralization of drugs and toxins. Finally, *transport vesicles* allow the cell to transfer proteins and lipids between the different organelles, to integrate certain proteins within the plasma membrane, to externalize other proteins (exocytosis), and to internalize extracellular proteins (endocytosis and phagocytosis).

The separation of organisms into prokaryotes and eukaryotes is not just structural; it also represents the evolution of life on the planet. Although we are not sure when exactly life began on Earth, there is evidence that unicellular life forms with prokaryotic cell structures existed as early as 3.2 to 3.8 billion years ago [15–18]. These simple organisms went

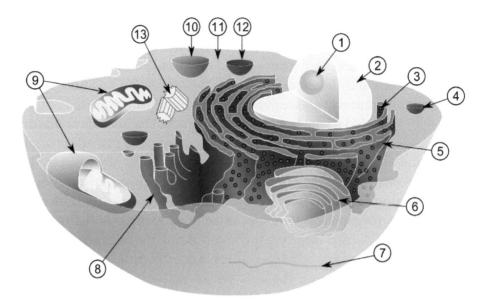

FIGURE 1.2   **An illustration of the inner organization of a eukaryotic animal cell.** (1) Nucleolus. (2) Nucleus. (3) Ribosome. (4) Vesicle. (5) Rough endoplasmic reticulum (rER). (6) Golgi apparatus. (7) Cytoskeleton. (8) Smooth endoplasmic reticulum (sER). (9) Mitochondria. (10) Vacuole. (11) Cytosol. (12) Lysosome. (13) Centriole. The image is taken from [14].

through numerous evolutionary cycles of mutations and selection, leading to a morphologically and metabolically diverse collection of bacteria. Despite these changes, it seems that for 2 billion years or so there was no change in the basic prokaryotic structure of the bacterial organisms populating Earth. Indeed, the first known eukaryotic organisms appeared only 1.5 billion years ago. These cells contained very simple inner compartments, which, according to the accepted theory were formed by infolding of intracellular membranes into hollow bodies. These compartments developed into the nucleus, ER and Golgi apparatus. Mitochondria and chloroplasts emerged later. Judging by the characteristics of these organelles, Lynn Margulis has suggested that both mitochondria and chloroplasts originated from ancient bacteria capable of oxidative metabolism and photosynthesis (respectively), and internalized by primitive eukaryotic cells (the *endosymbiotic theory* [19]). According to this theory, these bacteria somehow escaped digestion by the host cell, and over the eons gradually lost their independent characteristics. Thus, they turned into cellular organelles that are dependent on the host, but still capable of carrying out their ancestors' principal metabolic functions [19,20].

Unicellular organisms populated the Earth exclusively until about 1 billion years ago, when the first simple multicellular organisms started to appear. About 600 million years ago, a sharp rise in the number of different types of complex multicellular species occurred. The exact reason for this *'Cambrian explosion'* is still a matter of speculation. However, at roughly the same time, the atmosphere oxygen levels reached their maximum [21], and it is assumed that the two events are connected. That is, because complex multicellular organisms consume large quantities of oxygen, they were not able to form until oxygen levels in the atmosphere reached a certain threshold value. In any case, this occurrence started a chain of events, which finally led to the formation of tissues and organs in higher organisms.

Why is the cellular structure so important for maintaining life? There may be several answers to this question, but at the most basic level, the advantage of cells is that they enable the organisms they build to distinguish themselves from the environment. That is to say, **the cellular structure creates an inner environment that differs in its physical and chemical properties from the outer environment.** The manifestation of this distinction is what we call 'life processes', i.e., the ability of the cell to extract energy from its environment, build complex materials and degrade waste, grow, divide, move, etc. (see Table 1.1). Given that living organisms are made of the same atoms as inanimate matter, it may seem strange that cells are chemically unique. Indeed, carbon, hydrogen, oxygen, nitrogen, and sulfur, which are the common atoms of living tissues, all come from either the crust or atmosphere of our planet. However, there is a difference between chemical composition and molecular composition. That is, **the uniqueness of biological cells is not expressed in their atomic composition, but rather in the way these atoms are organized in the form of molecules. Whereas the cells' inanimate environment is made of simple molecules such as water (H_2O), gases (O_2, N_2, CO_2), metals and minerals, cells include, in addition to the above, complex molecules.** In particular, cells are rich in highly complex molecules termed *macromolecules*, which may contain thousands to millions of atoms. **As we will see later, macromolecules are built from basic organic building blocks, all of which have unique properties that make the existence of macromolecules (and life) possible — the tendency to self-assemble. That is, these small organic molecules tend to chemically react and physically interact with each other to form larger and more complex molecules.**

There are three types of macromolecules: proteins, nucleic acids and carbohydrates (Figure 1.3). These are responsible for the most basic aspects of life processes. Nucleic acids, i.e., DNA and RNA, function in the encoding and expression of the cell's own genetic information. Complex carbohydrates function as energy stores in animals (glycogen) and plants (starch); as constituents of the cell wall in plants (cellulose) and of the exoskeleton of insects (chitin); and as a sophisticated means of molecular recognition. In addition, the above macromolecular building blocks (nucleotides and monosaccharides) and their chemical derivates fulfill many other roles at the cellular, tissue, and organ levels.

Proteins also play a variety of important roles in cells and tissues, and their unique properties distinguish them from nucleic acids, lipids and carbohydrates. Proteins are the most extensively studied, and perhaps the most interesting molecules in life sciences. Why is this the case? First, they are the most abundant macromolecules in cells, making up as much as ~50% of the cell's total dry mass [22,23]. Moreover, the number of different functional proteins in cells and tissues is much higher than the number of other macromolecules. Although RNA molecules have been discovered in recent years to be much more diverse than previously believed and to carry out certain cellular functions such as regulation [24] and catalysis [25,26], they have yet to match the diversity of proteins.

Protein diversity, the importance of which is explained below, is particularly pronounced in eukaryotic cells, due to gene splicing and post-translational processing [27]. For example, the human body is estimated to contain only 20,500 [28]*1 genes but ~100,000 different proteins [30], and biochemical methods of protein detection suggest that each cell may express up to 15,000 distinct proteins [31–33]. The total number of protein types in nature has yet to be determined, although estimates do exist. For example, in specific organisms whose genomes have been sequenced, it is possible to determine the total number of different proteins according to the number of *open reading frames* in the organisms' DNA. The number of proteins produced thus far by this method is on the order of millions [34,35]. Another estimate, carried out according to the (estimated) number of species on Earth, suggests a much greater number of proteins: 10^{10}–10^{12} [30]. In any case, these estimates indicate that large

*1 According to a new study the number is even lower, 19,000 genes. [29]

numbers of proteins exist in nature. **This conclusion is highly significant, as it suggests that proteins are functionally diverse.**

TABLE 1.1   **Shared characteristics of all life forms.** Steven Bottomley (Curtin University), personal communication.

The tenets of life on Earth
• A highly organized, dynamic, and complex cellular system of enzyme-catalyzed chemical transformations.
• Ability to extract, transform, and use energy from the environment.
• Self-assembly of simple building blocks into complex molecules and structures.
• Ability to self-replicate.
• Ability to sense and respond to the environment.
• Ability to evolve.

1.1.3   Proteins have numerous biological roles

Indeed, **proteins carry out numerous roles, and are involved in virtually all life processes in biological organisms.** These roles can be grouped into a few types, which will be briefly discussed in the following subsections. Some of the functions overlap, as can be expected when dealing with such a complex system. More elaborate descriptions can be found in biochemistry [37] and cell biology [38] textbooks.

1.1.3.1   Catalysis of metabolic processes

Living organisms maintain a wide range of metabolic processes that allow them to grow or divide, extract energy from foodstuff, build complex materials, decompose waste products, detoxify harmful substances, etc. **These metabolic processes, which are responsible for sustaining life in all organisms, involve thousands of chemical reactions that cells and tissues execute both simultaneously and consecutively.** Many of these reactions occur readily, as their products are more stable (i.e., have less free energy[*1]) than their reactants. However, the molecular needs of the organism dictate that these reactions must be completed within a time scale of 10^{-5} to 10^2 seconds [39]. In stark contrast, many chemical reactions have much longer half-lives, which may span from minutes to millions of years [40]. Obviously, during their long evolution processes, living organisms have developed means of accelerating the chemical reactions occurring within them.

According to the currently accepted model, a chemical reaction transforms reactants into products via a short-lived and high-energy *transition state* (Figure 1.4). The energy required for a reactant to be converted into its transition state is called the *activation energy* (Ea), and the rate of the reaction depends directly on the magnitude of this energy, as well as on the temperature. This relationship is captured by the well-known *Arrhenius equation* [41]:

$$k \propto e^{(-Ea/RT)} \tag{1.1}$$

(where k is the reaction rate, R is the universal gas constant (1.989 cal/(molK)), and T is the absolute temperature (in K)).

[*1]There are different types of energy (thermal, kinetic, electric, etc.), but they all add up to what is called *free energy*, which, in other words, is the total energy in the system. This concept is described in detail in Chapter 4.

(a)   (b)

(c)

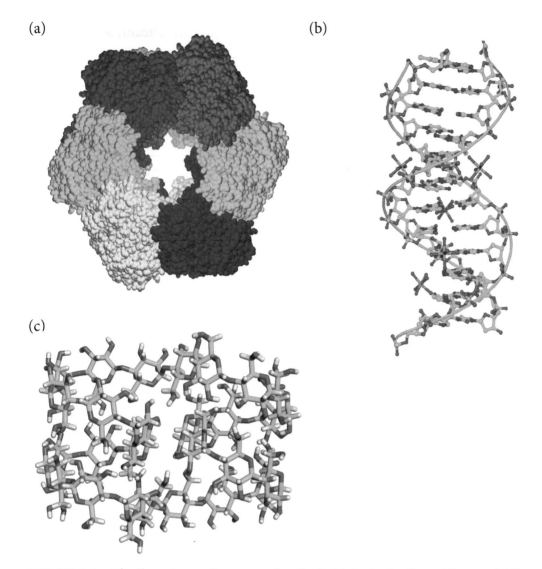

FIGURE 1.3 **The three types of macromolecules in biological cells and tissues.** (a) Proteins. The enzyme glutamine synthase is shown, with atoms presented as spheres. Each of the six subunits is colored differently. (b) Nucleic acids. The B-DNA double helix is shown, with covalent bonds shown as sticks. The color code follows the convention: carbon atoms are green, oxygen atoms are red, hydrogen atoms are white, nitrogen atoms are blue, and phosphorus atoms are orange. (c) Complex carbohydrates. The image shows cycloamylose (a.k.a. cyclodextrin), an enzymatic product of starch. Cycloamylose is a cyclic oligosaccharide, in which the α-D-glucopyranoside units are linked via $1 \longrightarrow 4$ glycosidic bonds [36]. The atoms are presented as in (b).

In other words, the reactant(s) must gain energy that is equal to or higher than the activation energy, in order to be able to reach the transition state and turn into product(s).

One way to enable a reactant to gain sufficient activation energy is to increase the temperature to a very high value. This approach is infeasible in living organisms, however, as they exist and function within a very limited temperature range. Thus, the only way for an organism to accelerate its metabolic reactions is to lower the activation energy required, i.e., to stabilize the transition states of its reactions (Equation (1.1)). This can be done by *catalysts*, i.e., chemical species that accelerate reactions without changing themselves. Different elements, such as transition metals, may serve as catalysts. Many of these chemical

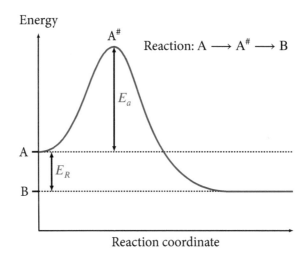

FIGURE 1.4 **Energetics of a chemical reaction.** The plot depicts a spontaneous chemical reaction, in which a hypothetical reactant (A) is transformed into a more energetically stable product (B). The corresponding drop in energy is represented by E_R. The reaction involves a transition state ($A^{\#}$), which is higher in energy than the reactant by Ea (the activation energy).

species are available in the various environments on our planet, and could therefore be easily harvested by living organisms. Surprisingly, however, these chemical catalysts were not adopted by biological organisms as catalysts. Rather, organisms have developed a wide range of proteins that can accelerate (catalyze) chemical reactions. These catalytic 'nanomachines' are termed *'enzymes'*.

Why were enzymes selected over simple catalysts during evolution? There are several possible reasons, but the most likely is the *high specificity of enzymes* towards their intended reactants (a.k.a. *substrates*). In contrast to simple catalysts such as metals, which can accelerate many different chemical reactions, each enzyme accelerates only a specific type of reaction, involving a specific substrate or substrates (Figure 1.5). The specificity results from the complex three-dimensional structure of the enzyme; this structure includes an *active site* that is specifically designed to bind only the enzyme's intended substrate (via noncovalent interactions, see Section 1.3 below), as well as to execute only the intended reaction. The amazing specificity of enzymes constitutes a huge advantage in the highly diverse chemical environment within living organisms; it allows them to *control the rates of each of their metabolic reactions* by controlling the enzymes executing them.

1.1.3.2 Energy transfer

Cells and tissues carry out a diverse set of processes, which are needed for sustaining life in all organisms. These processes include molecular biosynthesis, transport of chemicals across biological membranes, and movement. Most of these processes are not spontaneous and require an input of energy in order to take place. **This means that cells must be able to harness environmentally available energy in order to stay alive.** Indeed, a large portion of each cell's metabolic activity is dedicated to extracting and processing energy from available sources. **Despite being metabolically complex, all living organisms use at least one of two available sources of energy: chemical (i.e., food), and electromagnetic (i.e., solar radiation).** In both cases, a complex, multi-component machinery within the cell is responsible for utilizing the available energy source. Some of the components of this machinery

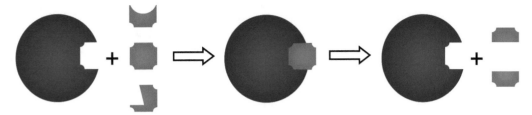

FIGURE 1.5 **The specificity of enzymes.** A highly schematic representation of an enzyme's specificity to its intended substrate. In the depicted reaction, the enzyme (blue circle) facilitates the cleavage of a substrate (in orange) into two identical products. As the scheme shows, the enzyme will act only on its intended substrate, even in the presence of other chemical entities. This results from the three-dimensional structure of the enzyme, which creates a geometric match between its active site and the structure of the intended substrate (but not the other possible substrates).

are enzymes performing catabolic reactions, in which foodstuff molecules are simultaneously degraded and oxidized in order to extract the chemical energy stored in them (see Chapter 9 for details). However, there are also other protein components that function in the efficient transfer of the extracted energy between different cellular compartments, or along a certain distance within a single compartment. These proteins are not enzymes, as they transfer energy in its pure form (i.e., as electrons or electromagnetic radiation), without causing any chemical change. The energy is finally stored as ATP, which is the most accessible form of energy in biological systems, and is therefore often referred to as the *'universal energy currency'*[*1]. However, the utilization of raw energy from the source to form ATP is done gradually (for regulation purposes) using different energy-converting processes. In the following subsection we will go over the two principal forms of energy utilization in living organisms.

1.1.3.2.1 Respiration

Organisms referred to as *'chemotrophs'* extract the chemical energy stored in foodstuff by oxidizing it. This task is carried out by a series of enzyme-mediated catabolic reactions, organized as pathways (e.g., glycolysis and the Krebs cycle). The oxidation releases the energy in the form of electrons, but these high-energy electrons are not used directly to form ATP. Instead, they are used to create an *electrochemical gradient of protons* (H^+) across the plasma membrane (in prokaryotes), or the inner mitochondrial membrane (in eukaryotes) (Figure 1.6). The proton gradient can be viewed as another temporary means of storing energy. Maintaining the gradient is possible thanks to the lipid component of the membrane, which is, in essence, impermeable to the electrically charged protons. The energy of the proton gradient can be released on demand through the formation of a transient opening within the membrane, which dissipates the gradient. The opening is in fact an ion channel, which is part of a large protein complex called *'ATP synthase'*. This interesting and highly important enzyme specializes in harnessing the energy of the proton gradient (via the flow of protons through its channel) to produce ATP. Thus, the initial process of foodstuff oxidation is coupled to ATP formation.

[*1]The central role of ATP as an energy currency is reflected, among other things, by the fact that enzymatic reactions utilizing this molecule were among the first to evolve on Earth. [42]

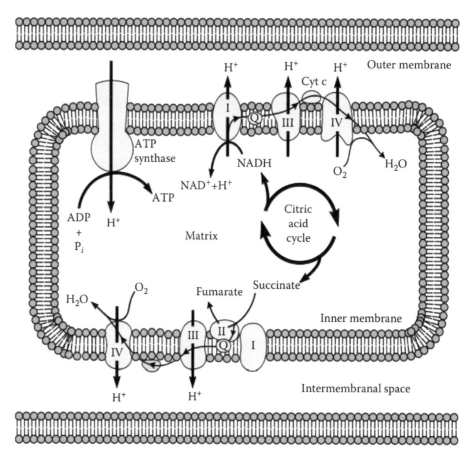

FIGURE 1.6 **Cellular respiration within mitochondria.** The schematic illustration shows the main players in this multi-component system. The citric acid (Krebs) cycle completes the aerobic oxidation of foodstuff inside the mitochondrial matrix. The high-energy electrons released by this process are first stored as NADH and FADH$_2$, and are then passed on to the first component of the electron transport chain embedded within the inner mitochondrial membrane. The electrons pass through several cytochromes and finally pass to molecular oxygen, which is reduced to water. The electron chain contains three protein complexes (I, III, IV), which use the energy released upon electron transport to pump protons from the matrix into the intramembrane space. The energy stored in the gradient is released when the latter dissipates via ATP synthase, a protein complex that couples proton movement to ATP synthesis from ADP and inorganic phosphate. The image is taken from [43].

As explained, the electrochemical proton gradient forms across the plasma membranes of bacteria, or the inner mitochondrial membranes. In any case, this means that the system carrying out this task must exist almost exclusively within those membranes. Indeed, the system in charge of this feat is a highly sophisticated complex of proteins (*cytochromes*) and small molecules. The system is referred to as the '*electron-transport chain*', a name that originates from its principal role: the transfer of foodstuff-borne, high-energy electrons to molecular oxygen, and the use of the energy released from this process to build the electrochemical proton gradient. Whereas all proteins of the chain are capable of passing on electrons, only three components of the chain have the ability to harness the energy released from this process to create the proton gradient. These three components are multi-protein complexes, functioning as proton pumps. Upon the transfer of electrons, these pumps actively move protons from one side of the membrane to the other, against their electrochemical gradient. The electrons passing through the chain are finally transferred to molecular oxygen, thus reducing it to water (H$_2$O). For this reason, the entire process is often referred to as '*cellular respiration*'.

1.1.3.2.2 Photosynthesis

Organisms named *'phototrophs'* use the electromagnetic radiation of the sun directly as an energy source. They, too, store the absorbed energy as ATP, but in addition, they use it to convert the biologically inaccessible form of carbon, CO_2 into its organic and fully accessible form, glucose ($C_6H_{12}O_6$). This process, called 'photosynthesis', was discovered in 1772 by Joseph Priestley, but is still studied intensively even today. Scientists believe this process first appeared ~2.5 billion years ago in blue-green bacteria [44], and that it bears full responsibility for the emergence of molecular oxygen in Earth's ancient atmosphere. Thus, **photosynthesis has also been indirectly responsible for the appearance of aerobic (oxygen-consuming) organisms**. The emergence of these organisms was a major step forward in the history of Earth's biochemistry. First, aerobic respiration is ~16 times more efficient than anaerobic respiration in generating energy (i.e., ATP) [45]. Second, the appearance of molecular oxygen allowed thousands of new biochemical reactions to occur [46], creating new physiologically important molecules such as steroids, alkaloids, and isoflavonoids [42]. Today, photosynthesis is carried out in cyanobacteria, algae, and plants. In the latter, the process occurs within *chloroplasts*, and is carried out by a highly complex system containing ~100 different proteins [47]. The first part of the process, called the *'light reaction'*, uses solar energy to form ATP and NADPH, a reducing agent. The second part of the process (the *dark reaction*) uses those two molecules to reduce CO_2 to glucose.

The light reaction begins with the collection and focusing of solar energy. This reaction is carried out by a network of proteins and pigments, which is referred to as an *'antenna'*, and resides within an internal chloroplast membrane. The proteins of this network contain small organic molecules (*pigments*) specializing in the absorbance of light energy and transferring it along via electron excitation. The main pigment in green plants is *chlorophyll II*, which gives these plants their color. In addition, there are pigments called *carotenoids*, which absorb light energy of different wavelengths than that absorbed by chlorophyll. Energy collected by a few hundred chlorophylls is channeled to a couple of central chlorophyll molecules called the *'special pair'* [48]. There, the focused power of light creates an electrochemical potential of 1.1 V, which is large enough to oxidize the special pair, i.e., to delocalize four electrons. The rest of the process is similar to cellular respiration. That is, the high-energy electrons are transferred through a chain of cytochromes, and the released energy is used to create a proton gradient. The latter is then used to generate ATP and NADPH.

1.1.3.3 Gene expression

Genetic information resides within every cell in our bodies (except for red blood cells). This information is stored inside the nucleus in the form of DNA, a chemical code. DNA is a polymer; it is made of building blocks termed *'nucleotides'*. Nucleotides join together to form a strand of DNA. Two such strands join together to form a double-helical shape, which constitutes the 'mature' form of DNA. There are numerous types of nucleotides, but only four of these build DNA. The genetic information unique to each organism resides in the exact sequence of the nucleotides composing its DNA. At first glance, the DNA strand may look like a continuous thread of nucleotides, devoid of any meaning. However, this sequence is organized as separate functional units called *'genes'*. Each gene contains instructions for a different element, important for the existence of the cell. Although each cell contains the full set of genes (i.e. the *genome*), only some of the genes are actually expressed at any given moment[*1], and in multicellular organisms, only a subset of genes is expressed in each cell. These genes make the cells of our body different from one another. Indeed, **the selective expression of genes enables each cell to acquire its unique properties, and specialize in carrying out certain tasks, depending on the tissue and organ to which it belongs.** The expression of a gene is induced by an inherent plan in the cell, or by an external signal. In single-cell organisms this signal is environmental, e.g., a change in the temperature, pH, salinity, pressure, food availability, etc. In multicellular organisms, where cells experience virtually unchanged environmental conditions, the 'external' signals affecting gene expression are usually sent by other cells (via hormones or other chemical messengers), in accordance with the needs of the body.

The process of gene expression executes the instructions embedded in each gene, which are written in DNA language. The execution of the genetic instructions is carried out by proteins, some of which are enzymes that catalyze reactions; others have different roles. The overall process of gene expression includes two main steps:

1. **Transcription** – In this step, a copy of the gene to be expressed is formed. This step takes place as a result of external or internal signals. The signals activate proteins called *'transcription factors'*, which recognize and bind to the gene. Then, a key enzyme called *'RNA polymerase'* joins the complex of proteins building up on the gene. This enzyme moves along the DNA strand and creates a copy of the nucleotide sequence in the form of an RNA strand, a process called transcription. RNA is a type of nucleic acid similar to DNA, but includes only one strand, and is used for different purposes in the cellular context. In prokaryotes, the entire process of gene expression is carried out in the cytoplasm, and the RNA copy formed during transcription moves on to the next step. In eukaryotes, transcription happens in the nucleus, and the RNA copy created in this step is then processed to remove unnecessary information (*'RNA splicing'*); only when RNA splicing is complete is the RNA copy transported to the cytoplasm for the next step.

2. **Translation** – For most of the genome, the information within the RNA copy of a single gene is used for building a corresponding sequence of amino acids, the building blocks of proteins. The amino acid chain created by this process folds into a functional protein. **Thus, the information within the 'average' gene constitutes instructions for making a specific protein and the organism's genome codes for the complete set of proteins functioning in that organism (i.e., the proteome).** Some genes are the exception; they code for RNA species that are not translated into proteins, but rather fulfill different, specific, cellular roles [24–26,49]. One of these roles involves translation process itself. This process is carried out by a huge protein-RNA complex, the *ribosome*. Thus, the construction of proteins requires existing proteins, which sounds like a causality paradox. Scientists posit that the first proteins to be made on Earth were synthesized using RNA molecules alone. These *catalytic RNA* molecules supposedly performed other tasks as well, which are carried out today by proteins, including the RNA molecules' own replication. This so-called *'RNA world'* theory [50] is yet to be proven, but if true, it may explain the ribosome-involving paradox mentioned above. Other roles of genes that do not code for proteins are regulatory. That is, these genes are transcribed to RNA molecules that affect the expression of other genes, thus regulating cellular function at the genetic level [24,49]. Such regulation affects many aspects of cellular physiology, including differentiation and organism development.

[*1]See description of the gene expression process below.

In addition to the proteins that participate directly in gene expression, there are other proteins that regulate these processes, to make sure the right genes are expressed in accordance with the type of cell, its status, and its environment.

1.1.3.4 Transport of solutes across biological membranes

As described above, the core of the plasma membrane is made of lipids, which render the membrane an excellent physical barrier for biomolecules and ions, most of which are hydrophilic, i.e. lipid-insoluble. Nevertheless, biological cells must maintain a routine exchange of chemicals with their environment in order to survive. Energy sources and building blocks must enter the cell, whereas waste products and other specific molecules must be able to leave it. These exchanges are made possible by proteins specializing in molecular transport across membranes. **Since the activity of proteins can be controlled (see Subsection 1.1.2), using them as a sole means of transport allows the cell to regulate the entire process, and actively determine its own chemical composition.** This issue is discussed in detail in Chapter 7, which focuses on membrane-bound proteins.

1.1.3.5 Cellular communication

Cells are capable of communicating with one another, whether they are single-cell organisms occupying the same ecosystem, or tissue cells residing inside the same multicellular body. Cell-cell communication is usually carried out via chemical messengers, i.e., molecules secreted from the signaling cell and acting on the target cell (Figure 1.7). When the communicating cells are close, the chemical messenger diffuses from the signaling cell to the target cell. However, when they are part of a multicellular body, and are far away from each other, the communication is carried out via the body's circulatory system (i.e., blood). This signaling system is called the *endocrine system*, and the chemical messengers involved are called *'hormones'*.

In contrast to ion transport, communication does not usually involve the physical entry of an external molecule (i.e., the chemical messenger) into the target cell [52]. **Rather, the target cell internalizes the message using complex signaling systems, which begin with membrane-bound receptors that specifically bind the messenger.** Steroid hormones, which are lipid-soluble, are the exception to this rule. They enter the cell, bind to a cytoplasmic receptor, and migrate into the nucleus, where they take part in the execution of the information they carry. In all other cases, the binding of a chemical messenger to a membrane-bound receptor on the target cell induces a chain of molecular events inside the cytoplasm, and sometimes in the nucleus as well (Figure 1.7a). This chain of events is highly complex; it involves different components that act as transducers, amplifiers, messengers and sensors, and the incoming signal leads to the activation of some and the inactivation of others. These events ultimately lead to one or more changes in the cell's behavior. The change may be simple, e.g., an increase in the extent of production or degradation of certain chemicals; secretion or absorption of molecules, etc. However, the change may also be dramatic, e.g., cellular division, growth, and even suicide.[*1]

Proteins constitute a hefty part of signaling processes. First, the extracellular messen-

---

[*1]This extreme event, called *'programmed cell death'*, includes two main types: *apoptosis* and *autophagy* (types I and II, respectively) [53]. Whereas the former always leads to cell death, the latter, triggered mainly by starvation, may under certain conditions lead to the opposite, i.e., cell rehabilitation.

(a)

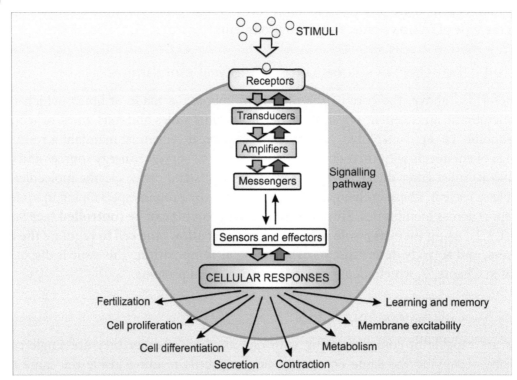

(b)

ger itself (i.e., the stimulus in Figure 1.7a) may be a protein or peptide (short protein segment). For example, the hormone *vasopressin* is a peptide secreted from cells in the pituitary glands of animals following dehydration or blood loss, and affects different cells in the body. Its effect on kidney cells decreases water loss during urine formation, whereas its effect on smooth muscle cells around arteries leads to their contraction and, as a result, elevation of blood pressure [54]. Second, membrane-bound receptors that bind the chemical messenger are also proteins. Messenger-receptor binding induces a structural change in the receptor, which is transmitted to intracellular components of the signaling system (transducers, amplifiers, intracellular messengers, sensors, and effectors in Figure 1.7a). There are numerous such components; some are proteins, whereas others are small organic molecules. For example, the *heterotrimeric G-proteins*, which constitute a central part of many signaling systems, are large proteins made of three units (Figure 1.7b). These proteins act as transducers; they relay the initial message from the membrane-bound receptor to the secondary components of the system inside the cytoplasm. G-proteins have a unique mechanism involving GTP hydrolysis (see Chapter 7, Subsection 7.5.2.2). That is, their active form binds the nucleotide GTP and hydrolyzes it after a fixed period of time to GDP and orthophosphate (P_i). Since only the GTP-bound form of the protein is active, these proteins are used in signal transduction pathways as molecular 'clocks' that limit the duration of the signal. Other important protein components of signaling are cytosolic enzymes, which become either activated or inhibited as a result of the relayed signal. A prominent group of such enzymes are small monomeric GTPases, which include *ras*, *rho*, *ran*, and *rab*. These act similarly to G-proteins in the sense that they bind GTP and remain active until the latter is hydrolyzed, but they are much smaller than G-proteins and have only one unit. Nevertheless, the participation of these enzymes in signal transduction leads to some key cellular processes, such as vesicle trafficking, cytoskeletal dynamics, cell polarity, membrane fusion, chromosome segregation, and nuclear transport [55].

FIGURE 1.7 **The main players in a typical signal transduction cascade.** (Opposite) (a) A general scheme showing the principal components in a signal transduction cascade. Stimuli (e.g., hormones, neurotransmitters, or growth factors) act on cell-surface receptors, which activate transducers to relay the signal into the cell. The transducers use amplifiers to generate internal messengers, which either act locally or diffuse throughout the cell. These messengers then engage sensors that are coupled to the effectors responsible for activating cellular responses. Note that the order of participation of pathway components may vary across different signaling pathways. For example, messengers may be used to activate amplifiers instead of being produced by them (see panel (b)). The green arrows indicate ON mechanisms, during which information flows down the pathway, and the red arrows indicate opposing OFF mechanisms that switch off the different steps of the signaling pathway. Virtually all of the components mentioned above may be proteins. The image is taken from [51]. (b) The cAMP-PKA cascade. Binding of an external chemical messenger (hormone, neurotransmitter, etc.) to a membrane-bound protein receptor induces the activation of an enzyme called a G-protein, which acts as a transducer. This activation makes one of the protein's subunits detach from the other, bind to the enzyme adenylyl cyclase (AC), and activate it. Activated AC catalyzes the conversion of ATP into cyclic AMP, which acts as an intracellular messenger. It binds to and activates the enzyme amplifier PKA, which in turn phosphorylates a large set of cytoplasmic proteins. The phosphorylated proteins may activate other cellular components, or perform a certain function (that is, they may act as sensors and/or effectors). In any case, this signal transduction eventually leads to changes in the cell's behavior, i.e., to a biological response.

The entire signaling process can be viewed as a molecular version of a relay race, in which a protein at the top of the signaling chain activates the following one, and so on, until the last protein in the chain induces the required change in the cell. The difference between a relay race and signal transduction is that the latter is not linear, but resembles a cascade or network. That is, many components of the signaling chain act as amplifiers; they activate (and/or inhibit) more than one target, resulting in the amplification of the initial signal. For example, the enzyme *protein kinase A* (*PKA*), which takes part in the signaling of many hormones, can activate tens of other different cellular proteins, which act as sensors and/or effectors [56] (Figure 1.7b). **Thus, one extracellular messenger binding to the target cell is capable of activating or inhibiting numerous cellular components, and such action is needed to create a significant change in the cell's behavior.**

1.1.3.6 Molecular recognition

Cellular communication, as described above, is based on interaction between the chemical messenger and its cognate receptor. In most cases, such interactions are highly specific, and this specificity prevents cellular receptors from being activated by the wrong molecules. The capacity of one molecule to bind another in a specific manner is at the basis of many biological processes, and is generally referred to as '*molecular recognition*' (see a detailed discussion in Chapter 8). Molecular recognition may take place between two individual molecules, between a chemical messenger and its receptor, and even between two molecules that are bound to the surfaces of different cells. Cell-cell recognition may lead to a variety of consequences. For example, adjacent cells tend to stop dividing when they come into contact with each other. This phenomenon, known as '*contact inhibition*', is responsible for the normal structure and organization of tissues in multicellular cells. Indeed, the formation of a cancerous tumor is usually accompanied by a loss of contact inhibition, which allows the growing tumor to invade neighboring tissues. Another well-known example of cell-cell recognition is the transient contact formed among cells of the immune system. In this case, the purpose of the contact is transmitting information regarding an invading pathogen (disease-causing organism), which requires the initiation of an immune reaction involving cell division and production of antibodies (Figure 1.8).

Proteins are involved in most molecular recognition processes. This is probably because they are complex enough to create the specific contacts required by such processes. Many recognition proteins include carbohydrate groups, which confer further complexity, and therefore enhance specificity potential. Such protein conjugates are called '*glycoproteins*'. Lipid molecules carrying carbohydrate groups (*glycolipids*) may also be involved in recognition. For example, the *ABO antigens* on the surface of our blood cells are glycoproteins and glycolipids differing in carbohydrate composition. Another well-known example of protein-dependent molecular recognition is the infection of epithelial cells (i.e., cells that line organs) by the influenza virus [57]. The surface of this virus contains a protein called '*hemagglutinin*', which recognizes sialic acid (an amino sugar) on the membranes of target cells. Binding of the hemagglutinin to the sialic acid creates physical contact between the virus and the target cell, which facilitates fusion between their membranes and internalization of the virus by the target (now host) cell. This example is particularly interesting, as it demonstrates the coevolution of pathogens and their hosts, during which the former 'learn' how to exploit natural mechanisms of the latter (in this case, the built-in recognition and internalization mechanisms of mammalian cells). Other pathogens, such as bacteria, have developed different means of tricking the host's molecular recognition machinery. These primarily include bacterial toxins that are secreted into the host's blood stream and recognize specific host elements, rendering them inactive. The inactivated element is in most cases a protein that is involved in a central recognition-dependent process. Famous examples include the inactivation of a protein synthesis elongation factor (EF-2) by the *diphtheria toxin*, and the inactivation of the aforementioned enzyme adenylyl cyclase by the *cholera toxin* [58]. Such inactivation is usually beneficial for the infection and proliferation processes of the pathogen, but due to its physiological effects on the host, may cause disease and death, long before the pathogen takes over the body.

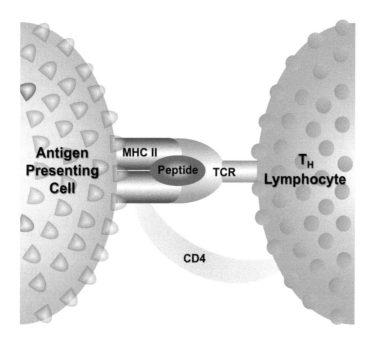

FIGURE 1.8 **Proteins involved in the interactions between an antigen-presenting cell and a T helper (T$_H$) lymphocyte.** Antigen-presenting cells (APCs), such as macrophages, can internalize bacteria, cleave their proteins to peptides, and present these peptides on their surfaces while bound to some of their own proteins, called "major histocompatibility complex type II" (MHC-II). The T$_H$ lymphocyte, a type of immune cell, can recognize the APC-foreign antigen complex using its own cell-surface proteins. Specifically, the T-cell receptor (TCR) of the lymphocyte recognizes the particular foreign antigen presented by the APC, whereas another protein, called CD4 (complementarity determinant 4), recognizes the MHC-II protein. There are other recognition proteins that are not presented here. The recognition process described above allows the T$_H$ lymphocyte to alert the rest of the immune system about the presence of a pathogen inside the body, which leads to the mounting of a full-scale immune response against the threat.

1.1.3.7 Defense

Cells contain different mechanisms for fighting invading pathogens. Some of these mechanisms are simple and based on enzymes, which recognize and destroy the foreign molecules (e.g., double-stranded RNA, which is present only in viruses). In animals, long evolutionary processes have led to the development of a highly-efficient defense system against pathogens [59]. This is called the *'immune system'* because it is capable of remembering pathogens it has already encountered, and reacts so rapidly that the infected animal does not even know it has been compromised. The immune system consists of two branches. The first, the *humoral branch*, is mainly based on antibodies, molecules that are formed by white blood cells called *B-lymphocytes*, and which patrol the body via the circulatory system, locating foreign elements by specifically interacting with them. The second branch, referred to as the 'cellular branch', includes *T-lymphocytes* (some of which have been mentioned above). These are white blood cells, formed inside bone marrow, and they can be found in the circulatory system, lymph system, tissues, and some bodily fluids (e.g., milk). The efficiency of the immune system results from the specialization of each of its components, and from the tight cooperation between them.

Antibodies specialize in highly specific binding to foreign elements; a particular antibody binds an element whose molecular structure is geometrically and physicochemically compatible with its binding site. Such elements may be individual molecules, but also molecular configurations on the surfaces of viruses or bacteria. The molecule or molecular configuration recognized by the antibody is called an 'antigen'. Each antibody can recognize one specific antigen, and the immense number of antibodies in our body allows us to fight a corresponding number of foreign elements. When an antibody binds to a small antigen (toxin, virus), the binding may, in some cases, neutralize the foreign element. However, when an antibody binds an antigen on the surface of a large pathogen (bacterium, parasite), the binding itself does not harm the pathogen, but rather 'tags' it for the cellular response. This response includes T-lymphocytes, and *natural killer (NK) cells*. It also includes *phagocytes* — cells, such as the aforementioned macrophage, that are capable of 'swallowing' the pathogen, degrading it, and presenting its antigens to other parts of the immune system. In fact, phagocytes are not really a part of the immune system, but rather belong to the *innate system*, a defensive response we are born with, and which is active at all times. Nevertheless, phagocytes cooperate tightly with different components of the immune system, as described below.

Activation of lymphocytes during the immune response leads to several different events: activation of B-lymphocytes stimulates them to massively produce antibodies specific for the pathogen already recognized, thus enhancing the efficiency of the response. Activation of T_H-lymphocytes leads to enhancement and regulation of the entire immune response, and to the creation of 'immune memory'. Activation of T_c-lymphocytes leads to the destruction of host cells already infected by a virus, and which may compromise the entire body. As described in Subsection 1.1.3.6 above, a T-lymphocyte recognizes its cognate antigen via a membrane-bound receptor it possesses (TCR, see Figure 1.8). The TCR of each T-lymphocyte can recognize only one specific antigen, as in the case of antibodies. However, whereas antibodies are capable of recognizing any type of antigen, the TCR can only recognize peptide antigens. The latter are produced by the degradation of bacterial proteins by phagocytes or the degradation of viral proteins by infected host cells. In addition, the TCR recognizes its cognate antigen only when the latter is presented on the surface of a cell by an MHC molecule (see Subsection 1.1.3.6 above).

Many of the components of the immune system are proteins, including the TCR, antibodies and MHC molecules. Interestingly, these functionally distinct molecules belong to the same protein group, referred to as 'immunoglobulins'. Proteins belonging to this group possess the same principal structure, which we will discuss in Chapter 2. In addition to the TCR and MHC, the immunoglobulin group also includes other lymphocytic membrane-bound molecules, which help regulate the immune response. The humoral branch of the immune system includes, in addition to antibodies, a set of proteins referred to as the 'com-

plement system'. Like antibodies, proteins of the complement system reside in the blood, and bind to pathogens. However, the rest of the process is different from the case of antibodies; after binding to the pathogen, the complement proteins bind to each other, forming a killing complex. This complex acts as a molecular drill, which perforates the membrane of the pathogen or leads to its death by other, indirect means. Finally, proteins are also involved in the long-range communication between immune cells, and between them and other body cells. Such proteins, called '*cytokines*', are chemical messengers that are secreted by activated lymphocytes or by other white blood cells. These not only relay messages between lymphocytes, but also enhance the anti-viral capabilities of other cells (e.g., in the case of the cytokine called '*interferon*').

1.1.3.8   Forming intracellular and extracellular structures

Some proteins may serve as building blocks of large intracellular and extracellular structures. The most prominent structures of this type are the *cytoskeleton* and the *extracellular matrix*. Both may seem relatively passive at first sight, but as we will see, the truth is far from that. The cytoskeleton is a network of fibers within the eukaryotic cytoplasm. It is made up of the proteins *actin* and *tubulin*, in addition to other proteins that are present in certain cell types [38]. This complex structure has many roles. First, it determines the shape of the cell and provides it with mechanical support. Second, the protein building blocks of the cytoskeleton are in a constant process of joining and detaching the fibers, in accordance with the needs of the cell. Thus, the cytoskeleton also gives the cell the ability to change its shape and even move [60–62]. This capacity is specifically important for cells specializing in functions that require shape shifting, such as phagocytes. For example, macrophages change their shape dramatically in order to engulf pathogens or foreign bodies, a process leading to their internalization. Changes in cytoskeletal fibers are responsible for both macrophage movement and formation of endocytic vesicles [63].

Third, the cytoskeleton affects the inner organization of the cell; cellular organelles are in physical contact with cytoskeletal fibers. Consequently, the location of each organelle is affected by the way the cytoskeleton is organized. Moreover, cytoskeletal fibers can function as railway tracks, on which organelles move throughout the cells using *motor proteins* (e.g., *myosin*) [64,65]. Finally, certain elements of the cytoskeleton are involved in the formation of the *nuclear lamina*. It should be mentioned that most of the information we have on the cytoskeleton comes from eukaryotic cell research. New studies show that bacterial cells, which have always been thought to lack internal organization, also include polymeric fibers that are organized similarly to the eukaryotic cytoskeleton [66].

The extracellular matrix (*ECM*) is a tangled network of fibrous proteins such as *collagen* and *elastin*, which are embedded in a gelatinous material composed primarily of complex sugar molecules [67,68]. These components are built by *fibroblasts* (cells that form *connective tissues*) and secreted to form the network. In some tissues such as bone, minerals are added to the protein-sugar network, making it physically tougher. In vertebrates, the ECM builds connective tissues, which function as a mechanical support for the entire body. Although this type of tissue is common, the extent of its importance varies across different organs. For example, in cartilage, bone, skin, and tendon, the connective tissue is the prime component (Figure 1.9a), whereas in the brain it is of secondary importance. The ECM constitutes most of the volume of connective tissues, with the cells sparsely checkered into it. This is one of the differences between connective and epithelial tissues. In the latter, the cells are tightly arranged against each other (Figure 1.9b). The composition of the ECM is similar across the different types of tissues, although there are certain components unique to certain tissues. Bone tissue contains a calcium-phosphorus compound called 'hydroxyapatite', which is the main ingredient conferring toughness to the bone. As mentioned above, collagen and elastin are the principal components of the ECM. Collagen, the most widespread protein in mammals, constitutes 25% of their total protein mass. It is also the main component of the ECM, with the role of keeping tissue organization and conferring it with mechanical strength. Elastin, as the name implies, confers elasticity to the ECM (see Chapter 2 for more details on collagen and elastin). Other proteins also present in the ECM mainly function in connecting different network elements to each other or to nearby cells. For example, *fibronectin* plays both roles, interacting with different elements, one of which is the common membrane-bound protein *integrin*.

Despite being separate entities, the cytoskeleton and the ECM are physically connected to each other. Integrin, which is directly bound to fibronectin, is connected on its other side to actin fibers of the cytoskeleton via cytoplasmic proteins such as *talin* and *vinculin*. **Thus, extracellular events may be transmitted into the cell via direct physical contact between different cytoplasmic, membrane-bound, and extracellular proteins** [69-71]. This property illustrates the active and complex roles of the ECM beyond providing the cell with mechanical protection and strength, two roles once considered to be the only functions of the ECM. Indeed, **the physical contact with nearby cells allows the ECM to affect central processes inside them, which may be critical for their survival and overall development.** The importance of mechanical force transduction into cells via the ECM-cytoskeletal system has also been demonstrated on the physiological level, e.g., on the development and maintenance of bone, blood vessels, and muscles; regulation of blood pressure, and motility of cells [69,72-74]. Finally, ECM proteins are also known to bind soluble growth factors (GF) and regulate their distribution, activation, and presentation to cells [75]. As a result, ECM proteins can take part in processes mediated by GF, such as cellular growth and tissue development.

1.1.3.9 Cell- and tissue-specific functions

Most of the roles described above for proteins are universal[*1], but the prominence of particular protein functions may vary across different types of cells. For example, biosynthetic enzymes are present in all cells, but in liver cells they are significantly more active than, e.g., in bone cells [76]. This is because the liver is a major biosynthetic organ in animals. Another role played primarily by the liver is detoxification, that is, the chemical neutralization and/or clearance of potentially harmful substances ingested by the body, especially fat-soluble drugs and toxins. The detoxification process is carried out in the sER. It involves different proteins and occurs in two phases. In *phase I*, the foreign compound is enzymatically oxidized, hydrolyzed, or cyclicized. The most famous protein in this group is an oxidizing enzyme called 'cytochrome P450'. In *phase II*, different hydrophilic groups (e.g., sulfate, uridine diphosphate, and glucuronate) are attached to the already modified compound, to further increase its water solubility, and thus enhance its efficient removal from the body via urine or bile.

[*1]In humans, the proteins carrying out these 'universal' roles constitute 44% of the protein-coding genome [76].

(a)

(b)

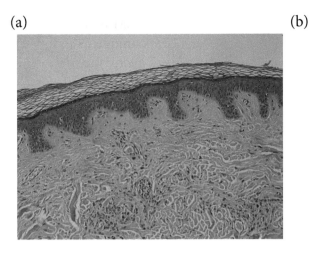

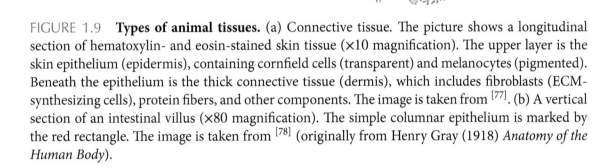

Central lacteal

Smooth muscle fibers

Reticular tissue

Columnar epithelium

FIGURE 1.9 **Types of animal tissues.** (a) Connective tissue. The picture shows a longitudinal section of hematoxylin- and eosin-stained skin tissue (×10 magnification). The upper layer is the skin epithelium (epidermis), containing cornfield cells (transparent) and melanocytes (pigmented). Beneath the epithelium is the thick connective tissue (dermis), which includes fibroblasts (ECM-synthesizing cells), protein fibers, and other components. The image is taken from [77]. (b) A vertical section of an intestinal villus (×80 magnification). The simple columnar epithelium is marked by the red rectangle. The image is taken from [78] (originally from Henry Gray (1918) *Anatomy of the Human Body*).

Other examples of organs in which certain protein roles are more dominant include the following [76]:

- Skeletal and cardiac muscles – energy utilization.

- Adipose tissues – storage and degradation of complex lipids.

- The gastrointestinal tract – nutrient breakdown, transport, and metabolism, host protection, and tissue morphology maintenance.

Some proteins perform roles that are virtually unique to specific organs. Actin and myosin are common proteins in cells, but in muscle cells they are organized in a unique way, which facilitates the mechanical contraction of the entire organ. Furthermore, some protein functions are observed only in certain organisms. For example, the membranes of multicellular organisms contain proteins that physically connect adjacent cells, thus forming body tissues. Needless to say, unicellular organisms do not contain such proteins, although they do have surface receptors that enable them to communicate with other nearby bacteria. Some of the proteins that connect tissue cells also function as junctions that pass chemical compounds or information from one cell to another, thereby facilitating the function of the tissue. Finally, the oxygen carrier hemoglobin, which is present only in multicellular organisms, allows cells that are far away from the lungs to oxidize foodstuff for energy (i.e., to respire).

1.1.4   Physiological and evolutionary importance of proteins

The implications of having a diverse set of proteins in any organism become clear when we inspect the different levels of protein function. The first level relates to the type of process carried out by each protein in terms of reaction chemistry and substrate type (in enzymes). However, when multiple proteins are active within the same environment, they mutually affect each other, and may form functional networks. This is the second level of function, the cellular level, and it is also affected by the expression levels of the proteins and their localization in the cell. Finally, the combined work of cellular and tissue proteins affects the organism's overall function, its unique physiological properties, and its ability to react to different environmental conditions. Thus, the huge diversity of proteins in nature is directly responsible for the large diversity of organisms. This relationship implies that the evolution of organisms on Earth has been tightly connected to the molecular evolution of proteins.

This hypothesis has been tested by a comparative study carried out on 38 different eukaryotic organisms, differing in their level of complexity (evaluated by their number of cells) [79]. The specific goal of the study was to see if there is any correlation between organism complexity and the development of protein groups identified with a specific function (each such group is referred to as a 'family'). The extent of development of each protein family was represented by the number of new 'family members'. The results of the study demonstrated two different developmental patterns: most protein families underwent 'conservative' development, i.e., development that was accompanied by genome enlargement, but no change in the organism's level of complexity. In some cases, this pattern of development led to enhancement of the organism's compatibility with its environment. Families that underwent conservative development mainly included proteins involved in basic cellular processes: metabolism, DNA replication, and gene expression. In contrast, a small number of protein families underwent 'progressive' development, i.e., development that was statistically correlated with the level of complexity of the entire organism. The prominent proteins in these families were involved in cell- and tissue-specific functions, in intercellular communication (e.g., molecular recognition and signal transduction proteins), or in functions specific to vertebrates, such as those related to the immune system. **To conclude, protein evolution has indeed contributed to the large diversity of organisms on Earth.** However, not all protein families have contributed equally to this diversity. In particular, and as might be expected, the proteins that have contributed most are those whose functions impart functional uniqueness to the organism, as compared to its ancestors.

1.1.5   Medical, industrial, and social importance of proteins

1.1.5.1   Proteins as drug targets

Given the numerous roles of proteins, one can conclude that the overall health of an organism depends on the normal function of its proteins, and that any significant loss of this function may lead to the development of a pathological process. Indeed, changes in the activity of proteins due to hereditary factors or exposure to toxins or radiation lie at the bases of many pathologies, such as metabolic disease and cancer. It is therefore not surprising that **proteins constitute ~80% of current pharmaceutical targets** [80,81]. Most pharmaceutical drugs act by binding to an enzyme, receptor, ion channel, transport protein, or nuclear protein, and reversibly changing its activity. Good examples are drugs of the selective serotonin reuptake inhibitors (*SSRI*) family (e.g., Prozac®), which are used primarily for treating depression and anxiety [82,83]. SSRIs are considered to be efficient drugs with relatively mild side effects. Both of these advantages are attributed to their ability to selectively inhibit a brain transport protein responsible for the reuptake of the neurotransmitter serotonin, a protein that pumps serotonin from the nerve synapse back into the secreting neuron; thus, SSRIs stop serotonin from acting on target receptors. This may seem to be an odd therapeutic approach, considering how important the reuptake of neurotransmitters is for brain function. Indeed, the reuptake of neurotransmitters prevents overstimulation of their target receptors, and this function should remain active in healthy people. However, in people suffering from depression, the synaptic steady-state levels of serotonin are low, so inhibiting its reuptake can actually help these people fight depression.

Functionally, most proteins targeted by drugs are classified as G protein-coupled receptors (GPCRs; common cell surface receptors that relay many physiologically-important signals) or enzymes [80]. In fact, enzyme-targeting drugs that act as inhibitors constitute 25% of the drug market. The structures and functions of GPCRs are described in Chapter 7. Other aspects of protein-drug interactions are described in Chapters 8 and 9, with the latter focusing on drugs acting as enzyme inhibitors.

1.1.5.2 Proteins as toxin targets

Proteins are also major targets for toxins and drugs of abuse. This is far from being surprising, as both often act similarly to pharmaceutical drugs, especially the latter. Drugs of abuse are in most cases small organic molecules whose shapes and chemical properties resemble those of endogenous neurotransmitters. The similarity allows the drug to compete with a neurotransmitter on binding to its target protein, whether a receptor, a transport protein responsible for the neurotransmitter's reuptake, or an enzyme responsible for the neurotransmitter's neutralization. By doing this, the drug disturbs the natural function of the nervous system, and leads to symptoms such as hallucinations, stimulation, and euphoria. For example, narcotic drugs such as *morphine* and *heroin* are chemically similar to a group of neurotransmitters called *'opioids'*, which include *endorphins* and *enkephalins*. The main physiological function of opioids is to lower the sensation of pain during physical trauma, by elevating the brain's pain threshold. They do so by binding to their cognate brain receptor, which regulates pain perception. When a narcotic drug binds to the opioid receptor it overstimulates it, causing a strong sensation of physical pleasure and euphoria. Moreover, consistent use of narcotics (and many other drug types) leads to receptor desensitization and other biochemical phenomena that cause addiction and long-term damages to bodily functions. Since both pharmaceutical drugs and drugs of abuse act on the same biological targets, the latter are sometimes used as the former, under medical supervision. For example, morphine, a known illegal drug, is used by doctors as an extremely potent painkiller in cases where conventional drugs lose their edge.

Toxins are chemicals that harm biological organisms upon exposure. The toxic effect is always harmful, although it may inflict different degrees of damage, and have acute or chronic effects. Toxins are diverse in terms of their chemistry and origin. They may result from natural sources, such as snake venom, or artificial ones, such as industrial pollution. As drugs, many toxins target proteins to exert their effects. Naturally produced toxins are known in many cases to change the activity of receptors, enzymes, or transport proteins in the peripheral nervous system. Such *neurotoxins* have evolved to serve as very efficient tools that help organisms to defend themselves or to prey on other organisms. For example, *tetrodotoxin*, produced by the puffer fish, blocks voltage-dependent Na⁺ channels in voluntary muscle cells [84]. The inhibition of these channels results in muscle weakness, which may reach the level of flaccid paralysis of the muscles. The latter is potentially life-endangering, because the breathing muscles are voluntary, and thus affected by the toxin.

Some of the naturally-produced toxins are proteins, as in the case of *ricin*, the *botulinum toxin* (Botox®), and the infamous *anthrax toxin* [85]. Ricin, which is present in castor beans, is one of the most lethal (natural) toxins known to man. It contains two components; the first enables the toxin molecule to enter a cell by 'hitching' onto its natural internalization mechanism (i.e., *endocytosis*), whereas the second component functions as an enzyme, whose action neutralizes cellular ribosomes and therefore shuts down protein synthesis [86]. This leads to cell death and severe damage to the surrounding tissue. The symptoms, as well as their onset and severity, depend on the way the toxin enters the body. For example, inhaled ricin is toxic in particularly small quantities and induces respiratory distress, fever, and nausea, whereas ingested ricin requires a significantly higher dosage and leads to extensive impairment of gastrointestinal functions [87]. In any case, ricin poisoning induces death in four to five days, if untreated. Unintentional exposure to ricin is highly unlikely, and cases of poisoning are usually deliberate. For example, ricin is thought to have been used in 1978 to assassinate the Bulgarian dissident Georgi Markov [88]. Markov, who lived in London, was pricked by an umbrella while walking in the street. The umbrella was equipped with a small platinum capsule carrying ricin, which was injected into Markov's body. Markov soon became ill and died a few days later [88].

Similarly to drugs of abuse, some toxins may also serve as therapeutic agents, as in the case of botulinum toxin (Botox®). This toxin, produced by bacteria in spoiled meat, may cause lethal paralysis by inhibiting the release of the neurotransmitter *acetylcholine* at the neuromuscular junction (see more on acetylcholine physiology in Box 8.1). Botox is an extremely potent toxin, but when injected in very small amounts, it may actually alleviate muscle spasms that accompany illnesses such as Parkinson's disease. In addition, Botox is very popular in the cosmetics industry, where it is used to smooth wrinkles in the skin. Recently, ricin too has been integrated into medical practice, where its cytotoxic abilities are used to fight and kill cancer cells [86].

The dual nature of many physiologically active substances such as toxins and drugs of abuse has been recognized for millennia. For example, this duality is reflected in the ancient Greek word for 'drug' (*'pharmakon'*), which refers both to 'remedy' and 'poison' [89,90]. Additional examples of harmful substances that are used in medicine as therapeutic agents are given in Chapter 8, Box 8.1.

1.1.5.3 Industrial applications of proteins

Proteins in general, particularly enzymes, are currently used for commercial and industrial purposes [91,92]. Enzymes are used as laundry detergents, food additives, and biological sensors, as well as catalysts in fuel-alcohol production, food, fat, and oil processing, synthesis of pharmaceutical drugs or agricultural products, and textile applications [92]. The growing demand for enzymes is driven by the need for catalysts that (1) act specifically on a desired substrate (thus avoiding side reactions and formation of toxic intermediates), (2) create a product with a desired stereochemical configuration[*1], and (3) achieve high efficiency under mild temperature, pH, salinity, and other conditions. The industrial uses of enzymes are discussed in detail in the last part of Chapter 9. Non-enzyme proteins are also used in industry for various applications. For example, antibody-like proteins are a clean and efficient means of isolating specific chemicals from a reaction mixture [93].

1.2 STRUCTURAL COMPLEXITY AND ITS EFFECT ON PROTEIN FUNCTION

The previous section demonstrated the functional diversity of proteins, which leads to the question of how proteins achieve this scope of diversity. The answer, in short, is: *their structural complexity*. To understand the meaning of this assertion, let us use an old example, Lego. Anyone who used to play this game surely remembers its basic concept; a small number of simple but distinct building blocks (bricks) are used to build a larger number of 'functionally' distinct structures (e.g., house, tractor, windmill, etc.). All Lego bricks are made of the same material (plastic), but each has a distinct shape. The shape of an individual brick affects the shape of the structures it builds, both directly and by limiting the number of other brick types to which it can connect. If we have an unlimited supply of bricks we can build many different structures, whether we use only 2 types of bricks, or alternatively, 10 types. However, it will much easier to find a 'functional' Lego structure (a house, tractor, windmill, etc.) among the structures built using 10 brick types than among the structures built using only 2 brick types. This is because a large number of brick types can form more shape combinations than a small number of brick types, which makes the former more *structurally complex*. Similarly, a group of structurally complex structures is likely to include more types of functional structures compared with a group of structurally simple structure. Thus, **an object with inherent structural complexity has a high potential for functional diversity.** One may argue that many shapes can function as a residence or for mobility, and therefore functional diversity does not necessarily stem from structural diversity. This claim can be counter-argued by stating that while different shapes may serve the same *general* function, they do not necessarily execute it in the same *exact* manner. For example, both car and tractor shapes can be used for mobility. However, whereas the former is used for transportation alone, the latter can also be used for digging or towing. Thus, the structure- and shape-function correlation abides.

Does the above example hold for proteins? Like Lego structures, proteins are also polymers, i.e., built from a limited number of basic building blocks. In the case of proteins, these building blocks are organic molecules called amino acids. These connect to each other to form long linear chains [94] (Figure 1.10a). Like the different types of Lego bricks, the different types of amino acids are also similar but not identical; there are 20 different amino acids, each characterized by a unique set of physicochemical properties, such as shape and electric charge. Interestingly, these properties lead to attraction between certain amino acids, and repulsion between others. For example, a positively charged amino acid tends to be attracted to a negatively charged one, whereas amino acids that are devoid of any electric

---

[*1]For each asymmetric carbon in a molecule there are two possible *configurations*, S and R. These are two completely opposite ways of organizing the same chemical groups around the asymmetric carbon. The molecule corresponding to each of the two configurations is called an '*enantiomer*', and the two enantiomers are mirror images of each other.

charge attract each other. When a sufficient number of amino acids attach to one another to form a protein chain, the collective set of attractive and repulsive forces between them drives the chain to fold in three-dimensional space (Figure 1.10b). This folding process results in a compact three-dimensional structure, which balances all the forces between the amino acids building it (Figure 1.10c). In other words, **the specific three-dimensional structure of each protein is that which achieves the best physical proximity between mutually attracted amino acids, and the most distance between mutually repelled amino acids.**

What makes the structure of each protein specific is not only the types of amino acids in the chain, but the unique order of these amino acids along the chain (i.e., the *amino acid sequence*). This is because a specific sequence creates a unique pattern of mutual attractive and repulsive forces between the amino acids, and this leads to a folding process resulting in one specific three-dimensional structure. Since different proteins have different sequences, they are expected to fold into different three-dimensional structures [95]. So, as in the case of Lego structures, the structural complexity of a given protein also depends on the number of different building blocks joined together, and on the way they are joined. However, the complexity potential of proteins far exceeds that of Lego structures, or any other man-made apparatus in existence. To get an idea of the extent of this potential, let us remember that each position along the protein chain may contain any of the 20 possible types of amino acids. This makes the number of different possible protein sequences N^{20} (where N is the number of amino acid positions in the sequence). As proteins typically include hundreds, and even thousands of amino acids, this number is very large, which makes the number of different possible protein structures very large as well. As we will see in Chapter 2, different sequences may yield similar structures, so the number of structures in reality is smaller than the number of potential structures. Still, this number is very large. **To conclude, proteins are inherently complex due to the large number of ways in which their building blocks can be combined.**

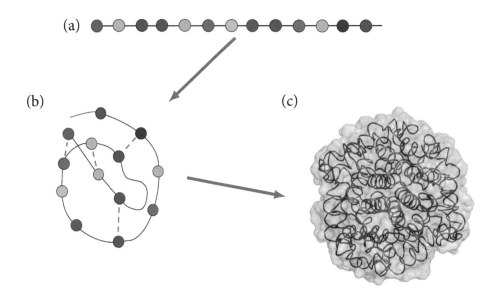

FIGURE 1.10 **The principal structure and folding of proteins.** (a) The protein's amino acid sequence, shown schematically as beads on a string. The different colors represent the physical and chemical diversity of amino acids. (b) A folding intermediate, in which some of the total possible amino acid interactions are satisfied (represented by dotted lines). (c) The final (native) structure of the protein.

The other assertion we made in our Lego example is that each of the structures built serves a different function; that is, the structural complexity of Lego translates into functional diversity. Does that conclusion hold for proteins as well? Most scientists believe it does; **the exact three-dimensional structure of a protein directly determines its function. Thus, similar yet non-identical protein structures usually fulfill different functions.** This assertion is based on numerous studies carried out over a century, but is most heavily influenced by the studies of three key biochemists. The first, Emil Fischer, suggested as early as 1894 that the activity of an enzyme results from its spatial-structural compatibility with its natural substrate [96]. The second biochemist, Linus Pauling, investigated protein denaturation, i.e., the loss of enzyme activity as a result of extreme environmental changes (temperature, pH, salinity, etc.). On the basis of lab observation, Pauling and his partner Mirsky concluded that the loss of activity of these proteins results from changes in their structure [97]. The third biochemist, Christian Anfinsen, proved this conclusion directly, by experimenting on the enzyme ribonuclease [98]. Today, the structure-function link is a central paradigm of molecular biology, and is at the basis of a distinct scientific field, *structural biology*. This link is discussed throughout the book, but to get the general idea, let us use the example of enzymes, in which the structure-function relationship has been studied most extensively. A more thorough discussion is given in Chapter 9, which is dedicated to enzymes.

We have seen that, unlike simple non-protein catalysts, which do not exhibit specificity, each enzyme acts on a particular substrate, thanks to a built-in, substrate-compatible binding site. Enzyme-substrate compatibility is made possible by the specific three-dimensional fold of the protein chain. The specific fold positions chemical groups of the protein, i.e., amino acid groups, at a specific distance and angle from the substrate's chemical groups [99] (Figure 1.11). The attractive forces between enzyme and substrate groups hold the latter in place within the binding site (see Chapter 9 for details). Other molecules in the enzyme's cellular environment are chemically and geometrically different from the substrate. Therefore, they are incompatible with the binding site, and would not be able to bind to it even if they could somehow gain access. As mentioned in Subsection 1.1.3.1, in some cases enzyme-substrate compatibility is high enough to enable the enzyme to discriminate between two stereoisomers of the same molecule (i.e., enantiomers). Such enzymes are said to be *'stereospecific'*. The compatibility of the enzyme with its substrate is also important to the step following binding, i.e., catalysis. Whereas some enzyme groups hold the substrate in place within the binding site, others act on it chemically in a way that accelerates its conversion into the reaction product [100] (see Chapter 9 for details). These chemical groups are unique in their ability to destabilize the substrate, or alternatively, to stabilize the transition state of the reaction, far beyond the capability of the solvent to do so. As explained in Chapter 9, the catalytic groups of the enzyme do their work using different strategies, such as formation of transient covalent bonds with the substrate, transfer of protons or electrons to or from it, electrostatic polarization of its chemical groups, etc.

In conclusion, the activity of each enzyme is made possible by its specific three-dimensional fold. All enzymes are proteins, and are thus built in principally the same way. However, since each enzyme has a different amino acid sequence, its exact three-dimensional architecture is different from those of other enzymes. As a result, its binding site's geometry and chemistry are also unique, which means the enzyme will bind a different substrate, and most likely perform a different type of catalysis than the others.

The above discussion demonstrates the importance of noncovalent bonds in proteins. In fact, noncovalent bonds have an important role in all macromolecules. These molecules, being polymers, are built with covalent bonds that keep their building blocks attached to each other in the form of a chain. However, **it is the noncovalent interactions that drive a chain to fold into a stable, specific three-dimensional structure**, as described here for proteins[*1]. The use of noncovalent interactions provides proteins with an important advantage; **these interactions are weak enough to allow the folded structure to constantly change conformation[*2] within a limited range, i.e., to be dynamic.** This phenomenon has been neglected for many years due to the lack of instrumentation fast enough to capture the different steady-state conformations of proteins. However, more recent studies using state-of-the-art instruments show not only that proteins exhibit dynamics, but also that **such dynamics is important for protein function.** A well-known example of this phenomenon is the working of the protein hemoglobin, discussed in detail below.

The example of enzymes given above demonstrates another important function of noncovalent interactions in proteins, namely, facilitating the protein's binding to its substrate or any other ligand. Aside from rendering the binding specific, noncovalent interactions render it reversible, which prevents the enzyme from getting 'stuck' with the first ligand or substrate it binds. This book will discuss in detail all of the three phenomena mentioned above, i.e., fold stabilization, dynamics, and ligand binding, emphasizing their dependency on various physical forces and effects. In order to make it easier for the reader to understand these issues, we first discuss noncovalent interactions and forces.

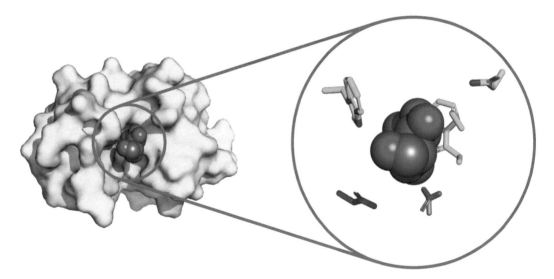

FIGURE 1.11 **Structure-function relationship in the binding site of an enzyme (lysozyme).** *Left*: The overall structure of an enzyme (PDB entry 9lyz). The contours of the enzyme are shown as a yellow surface. The substrate (blue spheres) is shown within the binding site (red circle). *Right*: A magnification of the binding site, showing the substrate and some of the enzyme's amino acids that participate in the binding (green sticks) and catalysis (orange sticks) of the substrate.

[*1]Noncovalent interactions in proteins are often cooperative. That is, the formation of one interaction makes the following interactions stronger, thus encouraging the formation of complex structures (see also Chapter 4). This phenomenon is considered to be one of the hallmarks of protein structure and function, as it contributes not only to the stabilization of protein structure but also to ligand binding.

[*2]The *conformation* of a molecule denotes the spatial arrangement of its atoms. Different molecular conformations can have the same connectivity (i.e., the pattern of covalent bonds between the atoms), but differ in the direction that some of the atoms face. Thus, molecules can switch between different conformations without breaking covalent bonds, just by moving their atoms. This is made possible by the thermal energy in the surroundings of the molecules, which is converted into kinetic energy (see Box 1.2 below for details).

1.3 NONCOVALENT INTERACTIONS BETWEEN ATOMS IN BIOMOLECULES

Noncovalent interactions are very common in macromolecules (including proteins), due to the relative weakness of these interactions. Indeed, whereas covalent bonds require an energy input of 65 to 175 kcal/mol to be broken [101], the energy content of a noncovalent bond is only up to a few kcal/mol. As explained above, the weakness of these interactions allows proteins to change conformation and bind ligands, two properties that are highly important for their function. Although they are weak, the high number of noncovalent interactions still allows them to stabilize three-dimensional protein structures. The nature and strength of noncovalent interactions in a protein are affected by the chemical nature of the protein's environment. It is therefore customary when discussing protein structure to take into consideration not only the protein and any ligand it may bind (both are referred to as the 'solute'), but also the molecules constituting their environment (referred to as the 'solvent').

There are two main types of noncovalent interactions (see also Figure 1.12):

1. Electrostatic interactions – These are interactions that occur between electrically charged atoms, and include both attractive and repulsive forces. The charges may be full or partial[*1], and partial charges may be fixed or induced (see details below). One class of electrostatic interactions, *van der Waals* interactions, occur between induced partial charges. This type of interaction has several unique features (e.g., it occurs between any pair of atoms that are close enough to each other) and is therefore described separately from other electrostatic interactions.

2. Nonpolar interactions – These attractive interactions, resulting from the hydrophobic effect, are most noticeable between atoms or chemical groups that are devoid of charge.

In the following subsections we discuss in detail the physicochemical features of the two interaction types. Due to the unique features of van der Waals interactions and their mathematical treatment, we will describe them separately from other electrostatic interactions. As mentioned above, noncovalent interactions are central to virtually any aspect of protein structure and function. Therefore, these interactions will also be discussed in detail in other chapters, in the following contexts:

Chapter 2: the groups and substructures in proteins that interact via noncovalent forces.

Chapter 4: the quantitative contribution of noncovalent interactions to protein structure.

Chapter 8: protein-ligand binding via noncovalent interactions.

Chapter 9: the importance of noncovalent interactions to enzymatic catalysis.

[*1]As shown in Figure 1.12 and explained in Box 1.1 below, partial charges exist as dipoles. A dipole can form, for example, when covalently bonded atoms possess significantly different electronegativities [102]. In such cases, negative charge is built up on the more electronegative atom, making it more negative than its bonded counterpart.

(a)

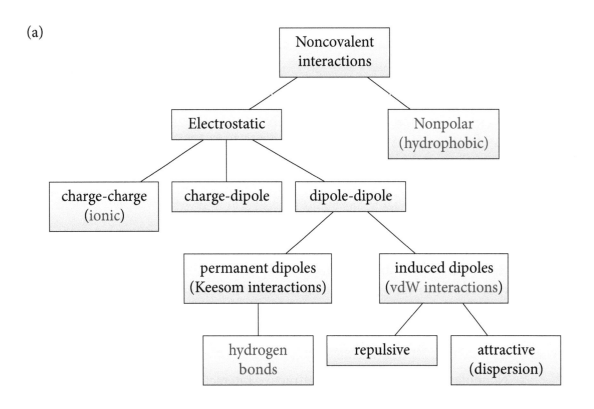

(b)

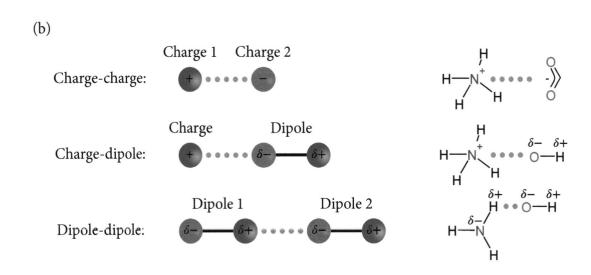

1.3.1 Electrostatic interactions

1.3.1.1 Introduction

Two atoms that are electrically charged interact electrostatically with each other (Figure 1.13). The nature of these interactions depends on the signs of the charges; charges of the same sign repel each other, and charges of opposite sign attract each other. Each of the interacting charges generates an electric field, which surrounds it. The interaction between the two charges results from the way the electric field emanating from one charge affects the other charge [103]. Since both full and partial charges may be involved, there are various types of electrostatic interactions between atoms, including charge-charge, charge-dipole, and dipole-dipole (Figure 1.12), where the latter two may involve fixed or induced dipoles. In proteins, the most common interactions (except for van der Waals interactions, which will be discussed separately) are:

1. **Ionic interaction** occurs between fully charged atoms. When the atom-atom distance is less than 4 Å, the ionic interaction is referred to as an *'ion pair'* or *'salt bridge'* [104].

2. **Hydrogen 'bond'** occurs between two electric dipoles, one of which includes a hydrogen atom. Less standard hydrogen bonds may involve a full charge and a dipole.

Electrostatic interactions, and hydrogen bonds in particular, are very common in proteins, and play different roles, such as imparting specificity to the protein structure or to its association with other molecules [105]; contributing to protein folding; and aiding in enzymatic catalysis [106]. In proteins, these interactions may occur as follows:

1. Between charged atoms belonging to different chemical groups within the protein.

2. Between charged atoms on the protein surface and charges in the environment (e.g., the cytoplasm, membrane, interstitial fluid, or other bodily fluids).

3. Between charged atoms of the protein and those present on its ligand.

In the following two sections we focus on the physical principles underlying electrostatic interactions. Ionic interactions can be described by the basic concepts of electrostatics, whereas hydrogen bonds require the consideration of other characteristics. Accordingly, we will begin with a description of the principles underlying both interaction types, and then discuss separately the additional qualities pertaining to hydrogen bonds alone. Table 1.2, which is given at the end of this subsection, summarizes key quantitative features of the main electrostatic interactions in Figure 1.12.

FIGURE 1.12 **Main noncovalent interactions found in proteins.** (Opposite) (a) The relationship between the interaction types (see more detailed description in the main text). The dominant interactions in proteins are colored in blue. vdW: van der Waals. (b) Types of charges involved in attractive electrostatic interactions. Plus and minus symbols indicate full positive and negative charges (respectively) on the atoms; $\delta+$ and $\delta-$ indicate (permanent) partial positive and negative charges (respectively). *Left*: A schematic representation of the interactions. The covalent bonds are marked by bars, and the electrostatic interaction is marked by the orange dotted line. *Right*: Examples of the interactions, between the following groups: a positively charged ammonium and a negatively charged carboxylate (top), a positively charged ammonium and a hydroxyl dipole (middle), and an amino dipole and a hydroxy dipole (bottom). Induced dipoles are not shown and will be discussed in Subsection 1.3.2.

(a)

(b)

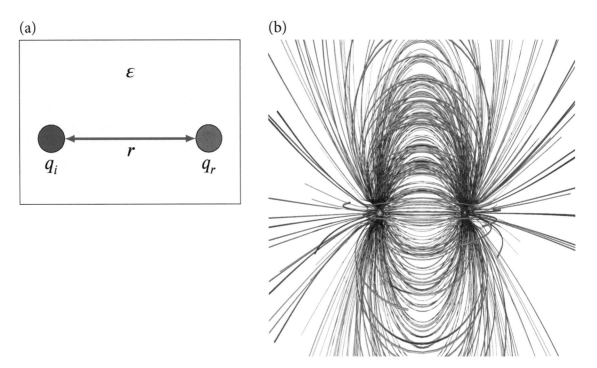

FIGURE 1.13 **Electrostatic interaction between two point charges of opposite sign.** (a) The charges are depicted as two points, with r denoting the distance separating them. One of the charges, q_i, is positive (blue), whereas the other, q_j, is negative (red). The gray box represents the medium, of dielectric constant ε, in which the charges are situated. Note that in cases in which the charges are also chemical entities (atoms, molecules), they too have dielectric constants. (b) The electrostatic field between two interacting charges of +1 and −1 units (right–positive, left–negative). The charges are separated by 10 Å in a medium of dielectric 80. The blue and red lines denote positive and negative electrostatic field lines. The electrostatic field was calculated and presented using BAL-LView [107].

1.3.1.2 Basic principles

1.3.1.2.1 Coulomb interactions

The potential energy[*1] of interaction between two point charges in a uniform medium is described by *Coulomb's law*:

$$U_{\text{Coul}} = \frac{q_i q_j}{\varepsilon_r r_{ij}} \tag{1.2}$$

(where U_{Coul} is the potential interaction energy (multiply by *332 (kcal/mol)Å/q_e^2* to obtain the energy in kcal/mol), q_i and q_j are the interacting charges (in electron charges), r_{ij} is the distance between the charges (in Å), and ε_r is the relative dielectric constant of the medium).

Equation (1.2) demonstrates the three quantities on which the interaction energy depends:

1. The magnitude of the interacting charges.

2. The distance between the charges.

3. The dielectrics of the medium (see Box 1.1 for detail).

[*1] The *potential energy* of a noncovalent interaction describes the energy resulting from interaction between two species (atoms, groups, molecules) in a *static* system. It is therefore only one component of the total (i.e., free) energy of the system; the latter accounts also for changes in the interaction resulting from dynamics of both the interacting species and their environment (i.e., *entropic* changes). A more extensive description of the potential energy of a system and its relation to the free energy is given in Chapter 4.

The dependency of the interaction on the second and third quantities provides important information on its physical nature. First, the energy depends inversely on the distance between the charges (r_{ij}). That is, the charges must be separated by a relatively large distance in order for the interaction to become insignificant. Thus, **electrostatic interactions are long-ranged**. For comparison, attractive van der Waals interactions, which are discussed below, depend on the inverse of the sixth power of the distance, making them short-ranged. Indeed, electrostatic interactions in proteins may be significant even at distances of up to a few tens of Ångströms [106]. By 'significant' we mean an energy value equal to or greater than RT (≈ 0.6 kcal/mol) (see Box 1.2 for details).

The dependency of electrostatic interactions on the dielectric constant provides a great deal of information about the environmental aspect of the interactions. The *dielectric constant* represents the ability of the medium to mask (i.e., reduce) the energy of the interaction between charges (see Box 1.1). In a vacuum, there is no masking at all, and the dielectric constant is 1. The resulting potential energy of the Coulomb interaction is in this case maximal. For example, two opposite charges with a magnitude of 1 electron charge, placed in a vacuum 4 Å away from each other, will have a Coulomb interaction energy of −83 kcal/mol (Table 1.1). Water represents the other extreme; there is ample masking of electrostatic interactions, as reflected by the high dielectric constant (~80). The electrostatic interactions in water will therefore be 80 times weaker than in a vacuum. The cytoplasm of cells is generally considered to be an aqueous, high-dielectric environment. However, as explained in Box 1.1, measurements indicate that the dielectric constant inside cells is ~60 at the most [108]. Lipid media, such as biological membranes, have dielectric constants of 2 to 30 [109,110], higher than that of a vacuum but still considered low. Thus, the Coulomb interaction in these media will be considerably stronger than in water. Even inside the protein, the dielectrics may change, which makes things much more complicated, as described in the following section.

Finally, note that although r_{ij} and ε may change the strength of the Coulomb interaction, they cannot change its nature, i.e., attractive or repulsive. That depends only on the signs of the interacting charges.

BOX 1.1 THE DIELECTRIC CONSTANT AND POLARIZATION

The dielectric constant is a general property of a medium, representing its ability to polarize in response to an externally applied electric field. On the molecular level, polarization of a medium involves reorientation of its molecules, to optimize their interaction with the electric field. This is similar to the needle of a compass reorienting when positioned in a magnetic field. Not all media are capable of becoming polarized, only those containing molecular dipoles. A dipole is formed when covalently bonded atoms possess significantly different electronegativities [102]. In such cases, a negative charge is built up on the more electronegative atom, making it more negative than its bonded counterpart. Molecules containing a bond dipole are considered 'polar'. Molecules containing more than one dipole can be considered polar as well, provided that the dipoles are oriented such that they do not cancel each other out. This is the case in the water molecule, the most common component in living organisms (Figure 1.1.1). The oxygen atom in water is more electronegative (3.5) than the hydrogen atom (2.1). Therefore, each of the two O−H bonds in water is a dipole, with the oxygen atom being partially negative compared to the hydrogen atom.

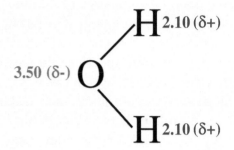

FIGURE 1.1.1 **The polarity of a water molecule.** The electronegativity value of each atom is noted. The most electronegative atom and its bonded partner are noted by red and blue numbers, respectively; $\delta+$ and $\delta-$ denote the positive and negative poles of each bond.

The magnitude of a dipole is expressed as its *moment (μ)*. In the case of two charges of opposite sign but equal magnitude, the moment dipole is:

$$\mu_{\text{bond}} = qr \tag{1.1.1}$$

(where q is the magnitude of the charges (in Coulombs), r is the separation between them (in meters), and μ is expressed in Coulomb $\times$ meter. Here we use a more convenient unit for μ, the Debye (D), where $1\,\text{D} = 3.336 \times 10^{-30}$ Coulomb $\times$ meter).

For example, each of the O−H bonds of a water molecule has a dipole moment of 1.51 D. In the water molecule, the two O−H bonds are separated by an angle of 104.45° (Figure 1.1.2).

Since the two hydrogen atoms face the same general direction, i.e., away from the oxygen atom, the entire molecule is a dipole, which should be stronger than the dipoles of the two separate O−H bonds. The *molecular dipole (d)* can be calculated by weighting the two bond dipoles according to the angle between them (θ):

$$d = 2r \cos\left(\frac{\theta}{2}\right) \tag{1.1.2}$$

The moment of the molecular dipole is as follows:

$$\mu_{\text{mol}} = qd \tag{1.1.3}$$

For water, $\theta = 104.45°$, $r = 0.96$ Å, and $\mu_{\text{mol}} = 1.85$ D, expressing the high polarity of this molecule.

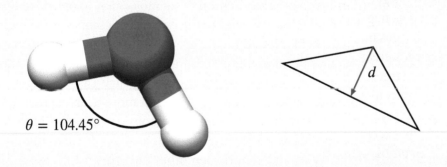

FIGURE 1.1.2 **The geometry of a water molecule.** θ is the angle between the two O−H bonds. The direction of the molecular dipole (d) is shown on the right.

Methane represents the opposite end of the polarity scale. This molecule includes four C–H bonds (Figure 1.1.3), and is completely nonpolar, for two reasons. First, unlike oxygen, carbon is close to hydrogen in terms of electronegativity. Moreover, each of the C–H bonds faces a different direction, so the weak C–H dipoles that do exist essentially cancel each other out. As a result, methane has a polarity of 0 D.

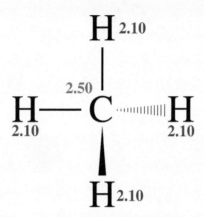

FIGURE 1.1.3 **The nonpolar nature of methane.** The electronegativity values and colors are presented as in Figure 1.1.1. The hydrogen atom bonded by the solid wedge points out of the plane of the paper or screen, towards the observer, whereas the hydrogen atom bonded by the hashed wedge points into the plane of the paper or screen, away from the observer.

As explained above, the polarity of any medium can also be described by its dielectric (ε). So, which represents polarity better, μ or ε? In fact, they both describe the same phenomenon, but on different scales; μ represents the polarity of a single bond or molecule, whereas ε is an *average* quantity representing the polarity of a bulk medium. Thus, the two values offer two different ways of relating to the medium's polarity, depending on the requirements of the model chosen to represent it. We have seen that a single water molecule possesses a significant dipole moment. Indeed, the dielectric of bulk water is the largest observed in biological systems ($\varepsilon \approx 80$).

Electrostatic interactions depend on the dielectric of the medium in which they occur (see Box 1.3 and the main text). Pairwise (Coulomb) interactions between solutes depend inversely on ε, meaning that they are weakened by polar media. In such media, the solvent dipoles reorient in order to optimize their interactions with the electric field emanating from each of the solutes. In water, for example, the reorientation allows the (partially) charged oxygen atoms to face positively charged solutes, whereas the hydrogen atoms face negatively charged solutes (Figure 1.1.4). When the system includes two adjacent charges interacting with each other, the reoriented water dipoles around each of the charges create a screening effect, which weakens the Coulomb interactions between the charges. This process does not happen to a large extent in nonpolar media, since the molecules of these media contain weak dipoles, and therefore do not provide the interacting charges with sufficient screening. It should be mentioned, though, that with the exception of a vacuum ($\varepsilon = 1$), no medium is completely nonpolar. For example, octanol, which contains one polar OH group attached to a long nonpolar hydrocarbon chain, has a dielectric of 10.3 [111]. In fact, even liquid alkanes, which are completely devoid of electronegative atoms, have some polarity ($\varepsilon = 2$). Still, compared with water, alkanes and long-chain alcohols are considered nonpolar.

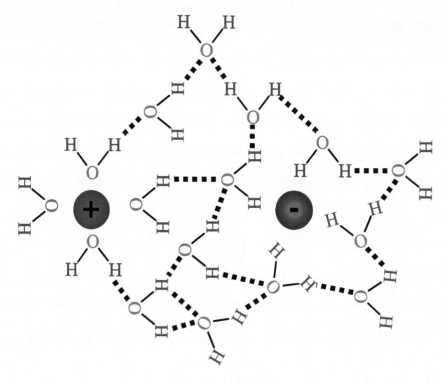

FIGURE 1.1.4 **Charge screening by surrounding water molecules.** The dotted lines represent hydrogen bonds. The angle separating the two O−H bonds in water molecules is 104.45°. Here we used a smaller value for graphic purposes.

Biological systems include media with different degrees of polarity. The cytoplasm is based on water, which allows intracellular molecules to be efficiently screened and thus function independently. However, as noted above, the dielectric inside cells is still lower than that of water, and according to measurements is estimated to be up to ~60 [108]. A less polar environment would allow these molecules to interact nonspecifically with each other, aggregate, and lose their biological function. It is therefore not surprising that water is often termed 'the universal solvent'. Indeed, despite the large morphological, genetic, and biochemical variety of organisms on Earth, they are all made mostly of water, and cannot exist without it[*a]. Water-based (aqueous) environments can also be found in the extracellular matrix, and in bodily fluids. In contrast, the plasma membrane and the inner membranes of eukaryotic cells are largely nonpolar ($\varepsilon \approx 2$), increasing the electrostatic interactions ~30 fold compared with the cytoplasm. As demonstrated in Chapter 7, membrane-bound proteins have evolved to take advantage of this phenomenon, in order to remain stable and active.

[*a]The general importance of water to the physical world has been recognized since antiquity. The most famous reference in this regard is probably the statement of the Greek philosopher *Thales of Miletus* (6th century BCE), that water is the 'nature of matter' [112].

BOX 1.2 THE SIGNIFICANCE OF *RT*

RT is the heat energy present in a system, e.g., an aqueous solution. At room temperature ($T = 25\,°C = 298\,K$), with $R = 1.989\,cal/(mol \times K)$, the product yields a value of 0.6 kcal/mol. The heat energy is absorbed by molecules within the solution and converted into kinetic energy (K):

$$K \propto k_B T \tag{1.2.1}$$

(where k_B is the Boltzmann universal constant).

The above implies a similarity between R and k_B. Indeed, they are essentially the same, but whereas k_B refers to 1 molecule, R refers to 1 mol (Avogadro's number of molecules/atoms (N_A)):

$$R = N_A k_B \tag{1.2.2}$$

The kinetic energy provided by the adsorbed heat to the solvent molecules manifests as internal motions, such as atomic vibrations and movements of regions of the molecule, as well as external motions, i.e., those of the entire molecule in solution (Brownian motion). The higher the kinetic energy of a molecule, the larger the motions associated with it:

$$K = \frac{1}{2}mv^2 \tag{1.2.3}$$

(where m is the mass and v is velocity).

The capacity of the kinetic energy to induce both internal and external motions explains the general dependency of K on T in Equation (1.2.1). For example, when the internal motions of each molecule are neglected, and only its 'rigid-body' movements are considered, the dependency factor is 1.5. Then, we can say that each water molecule has a kinetic energy of about $k_B T$, which translates to RT if the energy is considered per mole.

RT is significant not only in its physical sense and its relationship to the kinetic energy, but also as a general reference for energy changes in molecular systems. Indeed, it is considered as the 'noise' always present in solution, and therefore, **energy changes that are lower than *RT* are considered negligible.**

1.3.1.2.2 *Electrostatic interactions also include a polarization component*

The dielectric dependence of electrostatic interactions draws attention to the fact that electric charges within the solute (i.e. fixed charges) interact not only with each other (Figure 1.14a, left) but also with their environment (Figure 1.14a, right). This is easy to understand when considering a simple ion in water. As described in Box 1.1, the electric field of the ion polarizes the surrounding water molecules, which reorganize to optimize their interactions with the field. Put simply, they mask the charge electrostatically. Such an effect of a charge on its environment is called 'polarization', and the energy it creates is referred to as 'polarization energy' [113,114]. The same phenomenon also happens inside proteins (or other macromolecules); a charge that is part of the protein polarizes the charges around it, which generates favorable polarization energy. However, if the charge's environment is devoid of other charges, i.e., includes nonpolar chemical groups, there is nothing to mask the charge, and the resulting polarization energy is highly unfavorable. Thus, the interaction between a charge and its environment, i.e., the polarization energy, is not always favorable. This is in contrast to the pairwise (Coulomb) interaction between adjacent charges, which is *always* favorable between charges of opposite sign.

To summarize, **electrostatic interactions between charges of opposite sign include a favorable Coulomb component but also a polarization component, which may or may not be favorable, depending on the immediate environment of the interacting charges**. The potential energy of the total electrostatic interaction (U_{elec}) is therefore a sum of the Coulomb (U_{Coul}) and polarization (U_{pol}) components:

$$U_{elec} = U_{Coul} + U_{pol} \tag{1.3}$$

This can be illustrated quantitatively using a simple case of two monovalent ions of opposite charge (e.g., Na^+ and Cl^-) in water. When the ions are far from each other, each is completely surrounded by water molecules, which electrostatically mask its charge. However, if we bring the two ions to a distance of only 3 Å from each other, which is similar to their distance in a salt crystal, the water molecules between them are pushed aside, and as a result, the dielectric constant in this region drops considerably. Both the proximity of the two charges and the drop in dielectrics stabilize the Coulomb interaction between the two charges by 55 kcal/mol[*1]. However, the drop in dielectrics also creates an unfavorable change of 57 kcal/mol in the polarization energy. Accordingly, the total potential energy change in the electrostatic interactions in the system, as a result of bringing the two charges together, is unfavorable by 2 kcal/mol.

As explained above, the potential energy (U) is only one component of the total (free) energy of the system (G). Thus, when describing noncovalent interactions in a system, we are interested in the free energy of the interactions. Since the free energy incorporates additional factors beyond those described by the potential energy, calculating it is more complicated. This issue is described in Box 1.3.

What about electrostatic interactions between fixed charges (of opposite sign) within proteins? Are they favorable? This issue is a bit more complicated than the case of two simple ions in solution, and is discussed in detail in Chapter 4. For now, it will suffice to say that measuring the energy of such interactions is technically difficult, and calculating it yields significantly different values, depending on the exact method used to calculate the polarization energy. The complexity of proteins' geometrical shapes further factors into this difficulty. The different approaches to calculating the polarization free energy, and the reasons why they yield different results, are described in Box 1.3.

[*1]Calculated using DelPhi [115]. For simplicity, the radius of each of the two ions was set to 1 Å.

(a)

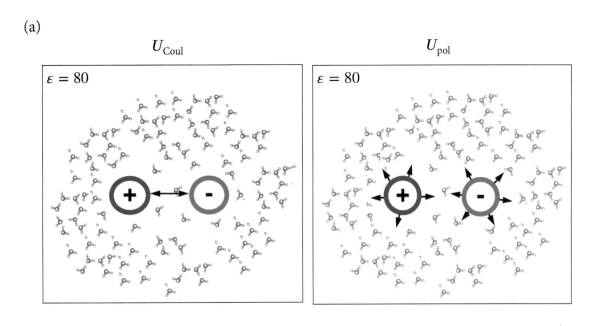

(b)

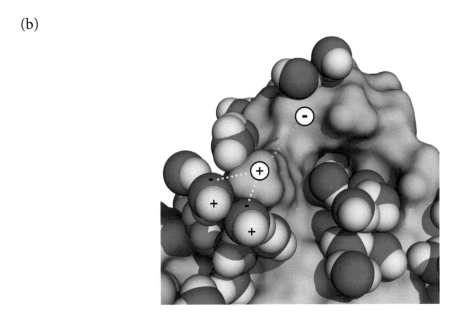

FIGURE 1.14 **Components of the total electrostatic interaction.** (a) Schematic illustration of the two types of interactions involving two simple ionic charges (blue and red circles) inside water: pairwise (Coulomb) interactions between the two charges (left), and the (polarization) interactions between each of the charges and the aqueous environment (right). U_{Coul} and U_{pol} denote the potential energies of the two types of interactions, respectively. (b) The same types of interactions, when the interacting charges (white circles with plus or minus signs) are in a protein. The surface area of the protein is shown in blue, along with the first layer of solvent water molecules around it (red and white spheres). Two types of electrostatic interactions are presented: between charges within the protein (green dotted line), and between protein and solvent charges (yellow dotted lines).

1.3.1.2.3 Solute-solute vs. solute-solvent interactions

The need to take polarization effects into account applies not only to interactions between fixed charges inside the protein, but also to interactions between charges in the protein and the solvent (Figure 1.14b). Most cellular proteins reside in an aqueous (water-like) environment. Such environments include the cytoplasm and the fluids filling the various organelles. Electrostatically, an aqueous environment consists of the partial charges of water molecules (see Box 1.1), and the full charges of solvated ions. As explained in the previous subsection, protein charges tend to interact favorably with polar media due to electrostatic masking. Some cellular proteins, however, reside inside the lipid core of the plasma membrane, which is nonpolar. In such cases, the electrostatic interaction between protein charges and their environment is unfavorable, which is why these membrane proteins tend to include chemical groups devoid of any significant charges (i.e., nonpolar).

Accounting for the solvent in computational studies of proteins is a difficult task, not only because of the need to address polarization effects, but also because water is a dynamic solvent, which makes protein-solvent interactions change constantly. All of these effects must be accounted for in order to describe the behavior of the protein reliably. These issues are addressed in Box 1.3.

BOX 1.3 CALCULATING THE ELECTROSTATIC FREE ENERGY

In a system containing charged solutes immersed in a solvent, the potential electrostatic free energy (U_{elec}) results from two types of interactions: the (sum of) pairwise interactions between the solutes, and the solvent polarization (see main text for details). The potential energy of pairwise interactions (U_{Coul}) can be described using the Coulomb equation (Equation (1.2)). Solvent polarization results from the interactions of solvent charges with the charges of the solute [113,114,116]. In polar media such as water, these interactions involve the reorientation of water dipoles around the solute. As explained in Box 1.1 and in the main text, the reorientation of the water dipoles optimizes their interactions with the electrostatic field emanating from the charges of the solute. In principle, any of the solute-water interactions can be described using the Coulomb equation, and the total polarization (potential) energy can then be obtained by integrating over all interactions. In reality, using this approach is unrealistic, because Coulomb's law refers to point charges, whereas biological solutes (e.g., proteins) are bulky, and usually contain many charges. As a result, the system contains at least two different dielectric regions: the low dielectric solute ($2 < \varepsilon < 20$), and the high dielectric aqueous solvent ($\varepsilon \approx 80$).

In addition, the potential energy provides only partial information on the electrostatic interactions, as it describes them in a static system in which all atoms are 'frozen'. In reality, both the solute and the aqueous solvent surrounding it are dynamic, which means that the Coulomb interactions between them change constantly. To describe the electrostatic interactions fully, their *free energy* (G_{elec}) must be calculated, as it accounts also for the changes in interactions resulting from system dynamics[*a].

[*a]In a broader sense, the free energy of a system includes both its potential energy and its *entropy*, a thermodynamic quantity representing the number of possible configurations of the system. This topic is described in detail in Chapter 4.

Thus, calculating G_{elec} requires the consideration of solute-water interactions in all possible configurations of the system — that is, in all solute conformations and with all possible water arrangements. In biological systems containing macromolecules, this task is in many cases computationally infeasible.

To solve this problem, an alternative approach can be used. Instead of calculating each protein-solvent interaction separately, the total effect of solute charges on the solvent is calculated. This is done by replacing the all-atom (explicit) description of the solvent with an implicit one; the solvent is described using a single parameter, its dielectric constant. The approach is called the 'continuum-solvent (CS) model' [117-120]. It is based on the assumption that the dielectric, being an average property, embodies all the features of the solvent relevant to its electrostatic interaction with the solute charges (see Box 1.1). A key advantage of this model is that it avoids the computational burden of the explicit description.

The interaction energy between the solute charges and the surrounding dielectric body is calculated in two steps:

First step: The *electrostatic potential* (Φ) is calculated using the *Poisson equation*. This equation describes the dependence of Φ on the solute's charge density (ρ) and the dielectric constant (ε) [117]:

$$\nabla \left[\varepsilon(r) \nabla \Phi(r) \right] = -4\pi \rho_{(r)} \tag{1.3.1}$$

Note that ρ replaces the detailed description of the solute's charges, used in Coulomb's equation.

What is the meaning of Φ? The charged solute (q_1) emits an electric field in all directions (Figure 1.3.1). A probe charge (q_2) positioned within the field senses the field and has electrostatic energy (G_{elec}). The electrostatic potential, Φ, is defined as the energy of that charge:

$$\Phi = \frac{G_{elec}}{q_2} \tag{1.3.2}$$

Φ depends on the proximity to the source charge, but also on the local dielectric. In the solute-solvent system, each point in space has a value of Φ, which depends on the charge and dielectric values at that point. Indeed, in the Poisson equation, Φ, ρ, and ε depend on the position vector, r. This complex dependency cannot be solved analytically, except in cases where the solute has a simple geometry, such as a charged sphere or plane. In all other cases, numerical solutions are used. The approach here is to make an initial estimate of Φ, and use it as a basis for further calculations that provide more accurate estimations. After numerous iterations, the calculation converges to yield the final value of Φ (Figure 1.3.2). One of the popular forms of solving the Poisson equation is the *finite difference method* [121], in which the protein-solvent system is mapped onto a three-dimensional grid. Each grid point is assigned charge and dielectric values, which are used to calculate Φ.

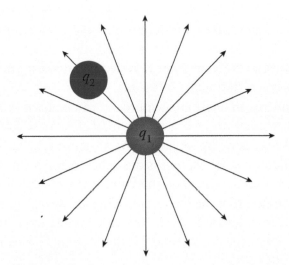

FIGURE 1.3.1 **The electrostatic potential.** The potential (Φ) is the energy acquired by a probe charge (q_2) positioned within an electrostatic field (black arrows) that is emitted by a source charge (q_1). This is also a way to measure the potential.

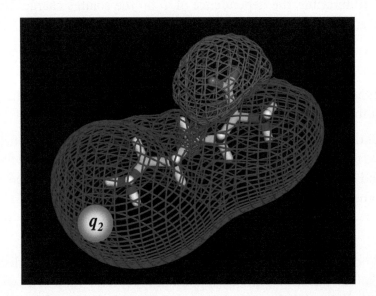

FIGURE 1.3.2 **An equipotential surface representation of Φ around the amino acid Lys.** The blue and red nets represent equipotential surfaces, i.e., all points in space in which Φ is $+2$ or $-2k_BT/e$, respectively.

Second step: The system's *electrostatic free energy* (G_{elec}) is calculated by integrating the values of ($\rho \cdot \Phi$) over all grid points:

$$G_{elec} = \frac{1}{2} \int \rho(r)\Phi(r)\,dr \qquad (1.3.3)$$

Since the calculation is based on an average property of the system, the calculated quantity is the free energy (G_{elec}), not just the potential energy (U_{elec}) (see Chapter 4 for details). Thus, **the CS model allows one to calculate free energies**

without the need to sample the entire configurational space of the system, as explained above.

The CS model may also account for the electrostatic interactions between the solute and physiological salt ions in the solvent. In this case, the Poisson equation is supplemented with an additional expression that refers to the salt charge density, using the *ionic strength* (I) and the *Boltzmann constant* (k_B). The underlying assumption is that the salt ions equilibrate rapidly and adjust their locations in space to the distribution of the source charges of the macromolecule. Thus, their local concentration follows the Boltzmann distribution. The full expression is termed the *'Poisson-Boltzmann (PB) equation'*:

$$\nabla \left[\varepsilon(r)\nabla\Phi(r) \right] - \varepsilon(r) \left(\frac{8\pi q^2 I}{\varepsilon k_B T} \right) \sinh \left[\Phi(r) \right] = -4\pi\rho(r) \qquad (1.3.4)$$

(where q is the charge of the ions).

The ionic strength, I, is related to the charges and concentrations of all ions present in the solution:

$$I = \frac{1}{2} \sum C_{(i)} Z_{(i)}^2 \qquad (1.3.5)$$

(where $C_{(i)}$ is the molar concentration of ion i, Z_i is its charge number, and the sum is taken over all ions in the solution). The ionic strength of biological cells is estimated to range between 50 and 250 mM [122].

There are a number of computational tools capable of solving the PB equation; some are accessible free-of-charge on the Internet. Two popular tools in this category are DelPhi [123] and the adaptive Poisson-Boltzmann solver (APBS), which can be used in a stand-alone mode, or as a plug-in in program packages, such as GRASP [124], visual molecular dynamics (VMD) [125] and PyMOL [126]. In addition, pre-calculated electrostatic potential maps can be viewed by a variety of molecular viewers such as those mentioned above, as well as by protein modeling packages such as UCSF Chimera [127]. For inexperienced scientists and students we recommend automatic web-based servers, such as the protein continuum electrostatics (PCE) server [128]*a. These and other molecular graphic tools are able to represent the electrostatic potential on the surfaces of proteins and other biomolecules, as depicted in Figure 1.3.1b. As explained in Chapter 2, this form of representation is very helpful in relating structure to function in many proteins. For example, some proteins use positively charged residues to associate electrostatically with the negatively charged DNA or negative side of the plasma membrane. It is difficult to find such regions using a simple representation of the protein. However, in an electrostatic potential map of the protein, such regions are easily spotted as distinct 'clouds' of positive potential. A powerful demonstration of the use of such representation to investigate structure-function relationships has been given by Murray and Honig [129].

---

*ahttp://mobyle.rpbs.univ-paris-diderot.fr/cgi-bin/portal.py#forms::PCE-pot

1.3.1.3 Hydrogen bonds

Electrostatic interactions may involve partial charges constituting an electric dipole. In biological macromolecules we find many interactions between permanent dipoles (Keesom interactions [130]) or between full charges and dipoles, with the most common type being the hydrogen bond. The standard hydrogen bond involves two dipoles [131] (Figure 1.15a):

1. D−H, where D is a *hydrogen donor*, an atom that is significantly more electronegative than carbon and hydrogen (e.g., oxygen or nitrogen). As we will see in the next subsection, less standard (and weaker) hydrogen bonds may involve heavy atoms such as sulfur, which are not as electronegative as oxygen or nitrogen, but have substantial electron density.

2. A−C, where C is carbon and A is a *hydrogen acceptor*, an electronegative atom having non-bonding polarized orbitals (e.g., the same atom types as D). Again, in non-standard, weaker hydrogen bonds, D may be other electron-dense species, e.g., an aromatic group containing a cloud of π electrons (see next subsection).

Since the hydrogen of the D−H dipole is partially positive, and the acceptor atom A is partially negative, the hydrogen and the acceptor atom are attracted to each other. As a result, the distance separating them is shorter than the sum of their van der Waals radii. This has led to the traditional view of the interaction as a 'bond', in which there is a type of sharing of the hydrogen atom between D and A. In reality, however, the hydrogen atom stays covalently attached to D, and there is no real sharing.

As in ion interactions, the strength of a hydrogen bond depends on the types of interacting atoms (specifically, the heavy, electronegative atoms participating in the hydrogen bond), and on the interatomic distance. While hydrogen bonds in molecules display a variety of lengths, most tend to be within a limited range [132,133]. For example, the typical D···A and H···A distances are 2.8 to 3.0 Å [133] and ~2 Å [134], respectively (Table 1.2). However, in contrast to ionic interactions, the strength of hydrogen bonds depends also on the orientation of the dipoles of the donor and acceptor, as reflected by the angles between them (Figure 1.15b). In proteins and small molecules the angles characterizing hydrogen bonds vary, albeit within a certain range due to geometric constraints imposed by the other parts of the molecule [132,133].

Given the bond length and angle, the potential electrostatic energy of a hydrogen bond can be approximated by the following equation, which corresponds to Figure 1.15c):

$$U = \frac{\mu_i \mu_j (2 \cos \theta_i \cos \theta_j - \sin \theta_i \sin \theta_j)}{\varepsilon_r r_{ij}^3} \tag{1.4}$$

(where U is the potential energy (multiply by 14.4 to obtain the energy in kcal/mol), μ_i and μ_j represent the strengths of the two interacting dipoles (in Debye), r_{ij} is the distance between their centers, ε_r is the relative dielectric constant, and θ_i, θ_j are the angles between each of the interacting atoms and the imaginary axis connecting the two centers; see Figure 1.15c).

Equation (1.4) shows that the electrostatic energy depends inversely on the third power of the distance (r_{ij}^3). The optimal interaction energy is produced by a head-to-tail arrangement of the two dipoles (Figure 1.15d). In such a geometry, the energy is equal to $-2\mu_1\mu_2/\varepsilon_r r_{ij}^3$. It should be mentioned that Equation (1.4) is only an approximation. First, it is based on classical electrostatics and excludes any quantum chemical effects. Second, it does not take into account the temperature dependence of the dipole. Third, most of the hydrogen bonds in proteins are not collinear, due to factors we will not go into at this point.

The importance of hydrogen bond geometry is supported by the fact that algorithms using geometric considerations are much more efficient in predicting energies of hydrogen bonds, as compared with algorithms that are based on classical electrostatic calculations alone [134]. In the former type of algorithm, the interaction energy is usually represented by a geometry-dependent expression, produced statistically from known protein structures. In addition to interatomic distance and angle, the electron densities of the two electronegative atoms are important as well: the lower the electron density of the donor atom, and higher the electron density of the acceptor atom, the greater the magnitude of the interaction energy. In extreme cases, the donor is positively charged (e.g., the $C=N^+H_2$ group of the amino acid arginine), or the acceptor is negatively charged (e.g., the COO^- group of the amino acids glutamate and aspartate). Such bonds are called 'ionic hydrogen bonds'. Finally, as in any electrostatic interaction, the interaction energy depends on the polarity of the medium. In water, hydrogen bonds have an energy of ~-1 kcal/mol [135], whereas inside macromolecules the energy depends on the immediate chemical environment of the charge, but does not exceed a few kcal/mol [136] (see also Chapter 4).

Hydrogen bonds are extremely common in proteins [132,137,138], and they are among the factors determining the unique architecture of these molecules. Most hydrogen bonds are formed between a hydroxyl ($-OH$) or amino ($-NH_2$) group of the donor, and a nitrogen or oxygen atom of the acceptor. In addition, water molecules caged inside the protein tend to take part in hydrogen bonds as well, interacting with chemical groups of the protein chain. Sometimes, hydrogen bonds involve $C-H$ as the donor group. Since the electronegativity of the carbon atom is lower than that of oxygen or nitrogen atoms, the strength of such bonds is about half of the strength of 'regular' hydrogen bonds [139], and they tend to appear only in nonpolar environments (i.e., the protein core or the membrane). Finally there are low-barrier hydrogen bonds (LBHBs), which occur between donor and acceptor atoms that are separated by less than ~2.5 Å and have similar pKa values [140] (see Box 2.2 in Chapter 2 for an explanation of pKa). In this state, the energy barrier for proton transfer between the two atoms is very low, and the hydrogen bond itself is very strong, estimated to be about half the strength of a single covalent bond. LBHBs are often formed within the catalytic sites of enzymes, where they play an important role in the stabilization of the reaction's transition state.

1.3.1.4 Other types of electrostatic interactions

Salt bridges and standard hydrogen bonds constitute the bulk of electrostatic interactions within proteins. In addition, there are chemical species (atoms, groups) capable of participating in other electrostatic interactions, as described below and summarized in Table 1.3. While most of these interactions are generally considered to be weaker than those discussed above, including standard hydrogen bonds, their microenvironment inside the protein (dielectric, polarization by adjacent groups) may raise their energy. **Like hydrogen bonds, most of these interactions are strongly affected by stereoelectronic considerations in addition to pure electrostatics [105], and are therefore directional (i.e., geometry-dependent). This directionality in turn contributes to the specificity of the interactions, which is highly important in biological systems (see Chapters 4 and 8).** The chemical species and the interactions in which they participate are as follows:

(a)

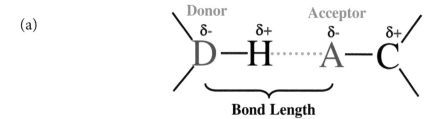

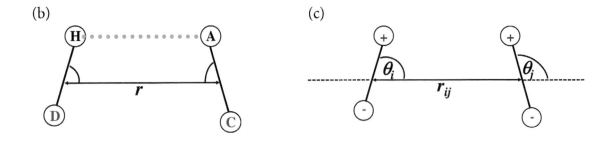

(b)

(c)

(d)

Aromatic rings and π interactions: Aromatic groups are common in proteins and appear in four of the twenty natural amino acid types that constitute proteins (see Chapter 2). Furthermore, they are highly common in many of the small molecules that proteins bind (*ligands*), such as hormones, neurotransmitters and pharmaceutical drugs. Aromatic rings are planar and have a very characteristic electronic configuration; they contain resonant double bonds, which include σ-orbital electrons within the ring plane, and π-orbital electrons above and below the plane (Figure 1.16aI). The latter provide the ring with a partially negative electric charge, which allows it to interact with other aromatic rings (see Chapter 2, Subsection 2.2.1), polar atoms and groups (e.g., amides [145], which are dipolar and also contain π electrons), metals (*cation-π interactions* [146], Figure 1.16aII) and even participate (as acceptors) in weak hydrogen bonds (Figure 1.16aIII) [147,148]. The latter, called *π-hydrogen bonds*, are geometrically more flexible than standard hydrogen bonds [149].

Sulfur: Sulfur appears in two of the twenty natural amino acids that build proteins, as well as in many organic molecules that interact with proteins. Examples include the following: (*i*) vitamins that form enzyme cofactors (e.g., coenzyme A, created from vitamin B_5, see Chapter 9), (*ii*) protein-bound organic complexes (e.g., the Fe–S clusters in respiratory proteins), and (*iii*) pharmaceutical drugs (e.g., penicillin-group antibiotics). In contrast to oxygen and nitrogen, the sulfur atom has low electronegativity (2.5), which should prevent it from participating in hydrogen bonds. However, the two lone electron pairs on sulfur allow it to participate in electrostatic interactions, including hydrogen bonds. Hydrogen bonds involving sulfur are of intermediary strength compared with standard hydrogen bonds, and occur between thiol (S–H) groups as donors and O and N-containing groups as acceptors [150]. Another type of sulfur-mediated electrostatic interaction involves the π electrons of aromatic groups [141,151]. Quantum-mechanical calculations, as well as surveys of thiol-π couples in proteins, show that the geometric positioning of the two interacting groups is limited to certain configurations (Figure 1.16b). This limitation is probably a means of avoiding electrostatic repulsion between the two lone pairs of the thiol group and the π electron cloud of the aromatic group. However, while one of the three most

FIGURE 1.15 **The hydrogen bond.** (Opposite) (a) *Top*: A general depiction of the hydrogen bond between a donor D and an acceptor A. $\delta+$ and $\delta-$ indicate the signs of the partial positive and negative charges (respectively) on the atoms. The covalent bonds are marked by bars, and the hydrogen bond is marked by the orange dotted line. *Bottom*: examples of hydrogen bonds found in proteins. The R groups signify the moieties bound to the reactive groups; in all cases the heavy atom in the reactive group is bound to a carbon atom in the R group. (b) The three geometric parameters used to characterize hydrogen bonds (see main text for details). The bond is shown schematically, with the four atoms involved represented as spheres, and each of the covalent bonds between them represented as a bar. (c) The parameters used for calculating the potential energy of two interacting electric dipoles. The scheme is similar to (b), except that the two dipoles are positioned with their heads in parallel, and the two θ angles are equivalent instead of complementary. (d) The electrostatically-preferable dipole-dipole interaction geometry. As explained in the main text, Quantum-Mechanical effects might favor a non-linear configuration. In addition, entropy considerations would favor non-linear configurations, as there are more such configurations for each fixed angle, which is non-zero. Atoms and bonds are represented as in (b).

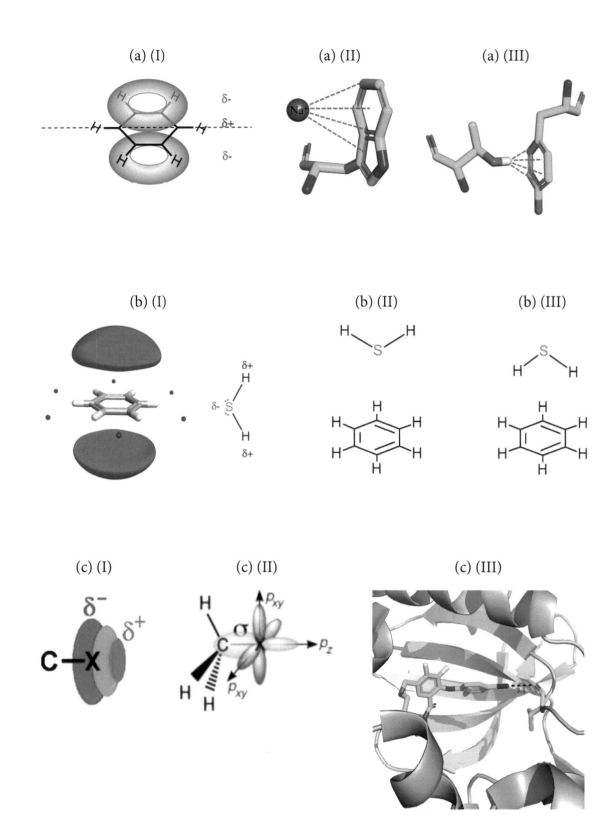

common configurations can easily be explained by simple dipole-dipole interactions (Figure 1.16bI), the other two (Figure 1.16bII and III) suggest different types of favorable interactions. Indeed, sulfur is also able to interact with electron-dense species via its low-lying $\sigma*$ orbitals [152–154].

Lone electron pairs and $n \rightarrow \pi*$ interactions: Lone electron pairs on atoms such as oxygen and nitrogen can interact favorably with the antibonding orbital ($\pi*$) of adjacent species. Such interactions, involving protein carbonyl and amide groups, have been demonstrated [155,156], and their energy has been estimated at a value of ~5% to 25% of that of a standard hydrogen bond.

Halogens and X-bonds: The halogens that are most prevalent in biological systems (F, Cl, Br, I) do not normally appear in proteins, but can sometimes be introduced enzymatically during an inflammatory response (e.g., asthma [157]) [158]. In contrast, halogens are quite common in protein ligands, such as the iodine-containing thyroid hormones [159], as well as in pharmaceutical drugs (e.g., the antibiotic vancomycin and anti-cancer drugs that inhibit protein kinases [160]) [143]. In drugs, the halogen is usually F or Cl (the heavier Br and I are less commonly used). The general view is that halogens are very poor hydrogen bond acceptors, even F, which is the most electronegative halogen [161]. However, halogens are capable of participating in another favorable electrostatic interaction, called an 'X-bond' [143,144,162,163]. This interaction occurs between a carbon-bound Cl, Br or I atom (C−X) and an electron-dense species, typically, the lone pairs of O, N or S atoms, whole charges on groups containing these atoms, and the π electrons of aromatic rings. The interaction results from

FIGURE 1.16 **Weak electrostatic interactions in proteins.** (Opposite) (a) (I) Electrostatic properties of aromatic rings. Delocalization of π electrons in aromatic rings (in this case benzene) creates a partially negative charge above and below the ring plane (marked by the dashed line), and a partially positive charge at the ring plane. (II) Cation-π interactions between an Na$^+$ ion (purple sphere) and an indole ring (shown as sticks with carbon atoms colored green and nitrogen atoms colored blue) in the protein lysozyme (PDB entry 1lpi). (III) π-hydrogen bond between an OH group and a phenol ring (both shown as sticks) in the protein glutathione transferase (PDB entry 6gst). The atoms are colored as in II, with the hydrogen atom colored white. (b) Interactions between thiol groups (represented here by SH$_2$) and aromatic rings. The three geometric configurations shown in the figure are known to be common in proteins [141]. Configuration I represents simple electrostatic interactions between (*i*) the lone electron pair of S (shown as red dots) and the positive potential around the plane of the aromatic ring (the potential's extrema are shown as blue spheres (created using TorchLite[*1] [142])), and (*ii*) the partial positive charges of the thiol's hydrogen atoms ($\delta+$ signs) and the π electron clouds above and below the ring's plane (red shapes, created using TorchLite). In configurations II and III the sulfur atom is right above the plane of the aromatic ring. These configurations are stabilized primarily by interactions between the sulfur's low-lying $\sigma*$ orbitals and the aromatic ring's π electron cloud. (c) Interactions between halogens and electronegative species. (I) Schematic representation of electron distribution on the halogen. Reprinted with permission from [143]. Copyright (2013) American Chemical Society. (II) The σ-hole resulting from redistribution of the valence electron in the p$_z$-atomic orbital (blue) to form the covalent C−X σ-bond (yellow) of a halomethane (X−Me) molecule (taken from [144]). (III) Interactions between a carbonyl oxygen (red) in the enzyme MEK kinase and an iodine atom (purple) in its ligand (PDB entry 3dv3). Adapted with permission from [143]. Copyright (2013) American Chemical Society.

the uneven distribution of electrons along the C−X σ bond, creating a positive electrostatic potential on the side of the halogen that is opposite the bond (Figure 1.16cI). This potential can then interact favorably with the negative potential of the electron-dense species. The polarity of the C−X bond results from the orbital configuration of the halogen (Figure 1.16cII). Specifically, the p_z orbital of the halogen's valence shell is collinear with the C−X σ bond and facing away from it. The electron of this orbital participates in the formation of the σ bond, leaving the orbital depopulated and partially exposing the high effective charge of the halogen's nucleus [144]. This partially positive region of the halogen atom is called the 'σ-hole' [164,165]. Since p_z faces away from the C−X bond, a positive potential is created opposite the bond, i.e., facing outwards (Figure 1.16cI). This makes X-bonds directional with preference to linearity [143], similarly to hydrogen bonds. Indeed, the σ-hole in an X-bond can be viewed as analogous to the hydrogen-bond donor. The magnitude of the C−X positive potential increases with the size of the halogen (F < Cl < Br < I). The polarizability of F is so small that this halogen does not contain a significant σ-hole. Another factor that increases the potential and resulting X-bond energy is the group containing the C−X carbon, with electron-withdrawing groups (e.g., aromatic rings) having a positive effect. X-bonds in proteins have been documented in numerous cases [143,144,158] (e.g., Figure 1.16cIII) and have been shown to contribute to ligand binding and biological function (see notable examples in [143]). An interesting question is why F, which cannot accept hydrogen bonds or even form significant X-bonds, appears on various drug molecules. The answer has to do with the large electronegativity of F, which draws electrons from the carbon atom to which it is bound. As a result, the bond between the two becomes stronger and the part of the molecule near the carbon has a lower electron density. These two effects decrease the probability of the drug molecule being chemically or metabolically changed. Thus, replacing the hydrogen in a C−H bond with an F is an elegant way to protect the drug from chemical or metabolic inactivation. Since H and F have similar size, the replacement does not interfere with binding of the drug molecule to its target protein.

1.3.2   Van der Waals interactions

The distribution of electrons around atom nuclei is subjected to transient quantum fluctuations. The result of the fluctuations is an electronic dipole at the atom level, which constantly changes its direction. This in turn leads to constant changes in the electric field around the atom, and when two neighboring atoms are ~7 Å apart, the changes in the electric field of one induce opposite changes in electric field of the other. In other words, a randomly produced electronic dipole in one atom polarizes an adjacent atom and induces it to form an opposite dipole. The extent of the induction depends directly on the radius of the polarized atom, and inversely on its electronegativity [103]. The two opposite dipoles interact electrostatically, and as a result attract each other (Figure 1.17). This attractive force is called the 'London force' (after Fritz London), and since it results from the induced dipoles, it is also referred to as a 'dispersion force' [171,172]. As the atoms approach each other, their electronic shells begin to electrically repel each other, in accordance with *Pauli's exclusion principle*, stating that electrons cannot spatially overlap. (Repulsion between the nuclei of the two atoms also exists, but it is much less significant than the electronic repulsion.) The collec-

TABLE 1.2 **A summary of the the main types of electrostatic interactions found in proteins.** Van der Waals interactions are also described here because of their electrostatic nature. q is the full charge (in electron charges), μ is the dipole moment (in Debye), r_{ij} is the distance between the charges (in Å), and ε_r is the relative dielectric constant of the medium. The θ angles in the second and third equations are defined in Figure 1.15b. To obtain the energy in kcal/mol, the following prefactors should be used: **332** for the first equation, **69.1** for the second equation, and **14.4** for the third equation.

Interaction	Example	Potential energy	Distance dependence	Typical distance (Å)	Typical strength in vacuum [*a] (kcal/mol)
Charge-charge (ionic)	Salt bridge	$U = \dfrac{q_i q_j}{\varepsilon_r r_{ij}}$	$1/r$	< 4 (salt bridge)	~ 80
Charge-fixed dipole	Hydrogen bond	$U = \dfrac{q_i \mu_j (\cos \theta)}{\varepsilon_r r_{ij}^2}$	$1/r^2$	2.8–3.0 [*b]	0–7 [*c]
Fixed dipole – fixed dipole	Hydrogen bond	$U = \dfrac{\mu_i \mu_j (2 \cos \theta_i \cos \theta_j - \sin \theta_i \sin \theta_j)}{\varepsilon_r r_{ij}^3}$	$1/r^3$		
Induced dipole – induced dipole	Van-der Waals interactions	$U = \dfrac{A_{ij}}{r_{ij}^{12}} - \dfrac{B_{ij}}{r_{ij}^6}$	$1/r^6$ (attractive) $1/r^{12}$ (repulsive)	3.5 [*d]	0.1–0.5 [*e]

[*a] In proteins, the strength of these interactions is difficult to determine, and different values have been suggested by different studies (e.g., [136]; see Chapter 4 for details).

[*b] [133]. In the case of hydrogen bonds, if the bond is described as donor-hydrogen-acceptor (D−H···A), the reported values correspond to the D···A distance. The corresponding H···A distance is typically 2 Å [134].

[*c] [166–168].

[*d] [169].

[*e] [167,170].

tive attractive and repulsive forces described above are referred to as '*van der Waals interactions*', although this name is often used to describe the attractive force alone. **Since van der Waals interactions result from basic atomic characteristics, they occur between any two adjacent atoms.** In macromolecules such as proteins, atoms are often packed against each other such that the distances among them produce optimal van der Waals energy. That is, the atom-atom distances are short enough to produce attractive interactions, but not significant repulsive interactions (see the minimum in Figure 1.17, which usually corresponds to ~3.5 Å). Accordingly, each atom has been assigned a *van der Waals radius*, which is equal to the optimal distance from the atom nucleus to the outer shell of a neighboring atom. This value is used in graphical representations of atoms to outline their contours. When two atoms form a covalent bond, the process involves overcoming the repulsive van der Waals forces. As a result, the distance between covalently bound atoms is shorter than the sum of their van der Waals radii. This is also true for two atoms that are engaged in a hydrogen bond.

There is no single expression describing van der Waals interactions accurately. Instead, it is customary to use the empirical Lennard-Jones expression to describe the potential energy

TABLE 1.3   **Less common yet important electrostatic interactions found in proteins.**

Chemical species	Interaction	Interaction features
Aromatic rings	π–π	• Involves two aromatic rings • Occurs between partially negative and partially positive regions in the rings • Highly geometry-dependent
	cation-π	• Often involve cationic metals
	π hydrogen bonds	• The π electrons serve as the hydrogen bond acceptors • Weaker, yet geometrically more flexible than standard hydrogen bonds
Sulfur and thiol	Hydrogen bonds	• Involve the thiol (SH) group as a weak donor • May involve the two sulfur lone pairs as very weak acceptors
	Thiol-π	• Involve dipole-dipole interactions • May also involve interactions between the π electrons and the sulfur's low-lying $\sigma*$ orbitals • Highly geometry-dependent
Lone electrons	$n \longrightarrow \pi*$	• Involves the lone electrons of electronegative atoms and the antibonding orbitals of other atoms • Estimated energy: ~5%–25% of a standard hydrogen bond
Halogens	X-bonds	• Electrostatic, occurring between a C−Cl/Br/I group and an electron-dense species (fully or completely charged) • Involve the internal dipole in the carbon-bound halogen • Strength depends on halogen: Cl < Br < I

of the van der Waals interaction (U_{vdW}) between atoms i and j (see also Table 1.2):

$$U_{\text{vdW}} \approx \frac{A_{ij}}{r_{ij}^{12}} - \frac{B_{ij}}{r_{ij}^{6}} \qquad (1.5)$$

(where r_{ij} is the distance between the atom nuclei, and A_{ij} and B_{ij} are constants of the repulsive and attractive interactions, respectively; the constants represent the specific physical chemistry of each atom pair).

As Equation (1.5) shows, the attractive van der Waals forces depend on the 6th power of the interatomic distance, which means that **van der Waals interactions are short-ranged.** Again, we should remember that the Lennard-Jones expression is empirical, and that the constants A_{ij} and B_{ij} are produced by fitting. Most of the values currently used in protein research have been produced by a statistical study in which the Lennard-Jones expression was calibrated according to interatomic interactions within 15 proteins [173]. Van der Waals interactions are weak even compared to other noncovalent interactions (0.1–0.5 kcal/mol [167,170], see Table 1.2), which is to be expected considering that van der Waals interactions involve induced dipoles. However, since they occur between any pair of adjacent atoms, **the cumulative contribution of van der Waals interactions is significant in macromolecules.**

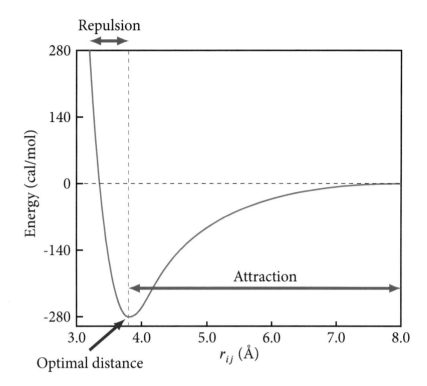

FIGURE 1.17 **Van der Waals interactions.** The van der Waals interaction energy as a function of the distance between the nuclei within an argon dimer. The energy was calculated using an empirical potential [175]. The long-ranged (electrostatic) attraction, resulting from the dispersion force between the atoms, and the repulsion, resulting from Pauling's exclusion principle at short distance, are noticeable. The image is taken from [176].

Moreover, since their energy is highly dependent on the interatomic distance, it is stronger in proteins, which are tightly packed (see Chapter 2), than in looser media such as water. Indeed, estimations indicate that van der Waals interactions involving a methylene ($-CH_2-$) group have energy levels of -1.8 and -3.1 kcal/mol in water and in the protein interior, respectively [174].

1.3.3 Nonpolar interactions and hydrophobic effect

Water, as explained in Box 1.1, has unique properties that enable it to serve as the universal solvent [177]. Specifically, water molecules possess an electric dipole, in which the oxygen atom carries a partial negative charge, whereas both hydrogen atoms carry a partial positive charge. As a result, individual water molecules tend to hydrogen-bond with each other in a way that connects their dipoles into one large network. This property is manifested in the high surface tension [178] and boiling temperature of bulk water. Although the water molecules are interconnected, bulk water is dynamic; the individual molecules tend to detach from and re-attract to the network rapidly. This increases the inherent disorder of bulk water, or in other words, its entropy. As mentioned above, the second law of thermodynamics states that the entropy of the universe tends to increase with time, which means that high-entropy states in nature are stable (see Chapter 4 for detail). Thus, **the structure of bulk water is doubly stabilized: the individual dipoles in the structure interact favor-**

ably with each other, but the entire structure remains dynamic enough to maintain high entropy.

The insertion of a solute into water disturbs the water's stable yet dynamic structure. In the case of a polar solute, the surrounding water molecules hydrogen-bond to it, thereby partially compensating for the loss of local water-water bonds. However, in the case of a nonpolar solute, such solute-solvent interaction cannot happen. Instead, the surrounding water molecules reorganize around the solute, forming a cage-like structure, in which they hydrogen-bond to each other [179]. Although some stabilization is gained from these bonds, the ordered structure formed around the solute significantly lowers the entropy of the system, which creates a net destabilizing effect. This process repeats itself with the insertion of additional nonpolar solute molecules, further lowering the system's entropy. However, if the individual nonpolar molecules, while diffusing freely in solution, associate with each other, the surrounding water molecules reorganize and form a large cage-like structure around the solute aggregate (Figure 1.18a). The surface area of the large water cage is smaller than the sum of surface areas of the individual cages around each of the nonpolar molecules (Figure 1.18b). Since the drop in entropy due to the formation of a cage-like water structure directly depends on the dimensions of this structure, **the system favors a single aggregate of nonpolar solute over the unattached molecules**.

Indeed, **nonpolar solutes tend to aggregate in water**, as anyone who has watched oil droplets in a pot of water knows. This phenomenon is referred to as the '*hydrophobic effect*' [180], and the interactions between the nonpolar solute molecules are called '*nonpolar interactions*'. It should be noted that these are not classical atom-atom interactions, but instead an indirect effect resulting from properties of the solvent [181]. For this reason, it is difficult to know the exact magnitude of the nonpolar 'interaction', although it is known to correlate with the dimensions of the 'interacting' molecules [182]. Such correlations and the underlying nonpolar energy can be observed in experiments that measure the partitioning of model molecules between water and a nonpolar medium (e.g., cyclohexane)[*1][180]. When small molecules are involved (up to 20 carbon atoms), the (nonpolar) interaction free energy correlates with the molecular volume or number of carbon atoms [182]. However, when a large molecule, such as a protein, is involved, the energy correlates best with another dimension-related parameter, the surface area. In the latter case, the correlation constant has been found to be ~25 cal/mol for every $Å^2$ of the molecule involved in the interaction [184–186] (see Chapter 4 for details):

$$\Delta G_{np} \approx -0.025\Delta\text{SA} \tag{1.6}$$

(where ΔG_{np} is the nonpolar interaction free energy (in kcal/mol) and ΔSA is the surface area of the molecule involved in the interaction).

The dependency demonstrates that **any molecule with a surface area is capable of participating in nonpolar interactions**. This is an important conclusion, as many molecules include polar groups and also have a large surface area. Such molecules are therefore capable of both polar and nonpolar interactions, and it is the balance between the two that determines the overall tendency of the molecule to be either hydrophilic or hydrophobic (i.e., 'water-loving' or 'water-hating'). Since nonpolar interactions result indirectly from solvent effects, the distance dependence of the interaction is unknown. However, studies in small molecules suggest an exponential dependence within a distance range of 0 to 100 Å [103].

[*1]Note, however, that in the case of highly nonpolar molecules, such an approach is problematic due to low water solubility [183].

(a)

(b)

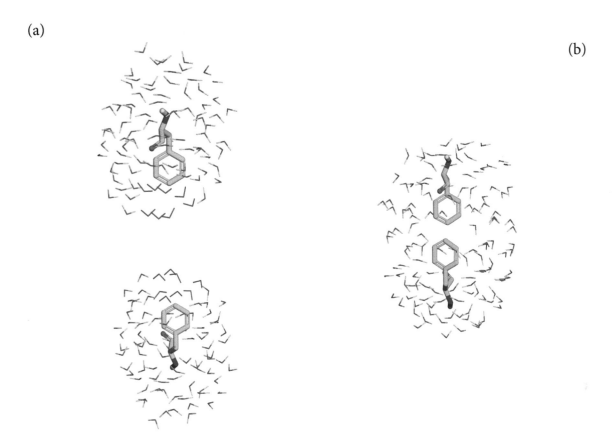

FIGURE 1.18 **The hydrophobic effect.** The effect is demonstrated by the amino acid phenylala-nine (Phe), which possesses a polar part containing amide and carbonyl groups, and a large nonpolar benzene ring. (a) Two Phe molecules in solvent. Each of the molecules is surrounded by a hydration shell, in which the water molecules are more ordered than in the bulk water. The Phe are shown as sticks and the surrounding water molecules as lines. Water molecules that separate the two Phe molecules are in blue. (b) Association of the two Phe molecules pushes the water molecules sepa-rating them (in blue) into the bulk, and a single hydration shell is formed around the complex. The order of the released water molecules decreases, which means that the overall entropy of the system increases.

1.3.4   Conclusions

Biological macromolecules are stabilized by different types of bonds and interactions. Strong covalent bonds construct the backbones of macromolecules. However, the function of a large molecule depends on the exact way in which its backbone folds, which depends on weak noncovalent interactions. These interactions can be separated into different types, but ultimately, they all have an electrostatic basis; van der Waals interactions are based on induced (electrostatic) atomic dipoles, whereas nonpolar interactions are based on solvent molecular dipoles. A complete and reliable representation of these interactions and forces can be achieved in principle by describing the electronic distribution of all atoms in the investigated system. Such a description is currently available only for small systems, us-ing quantum-mechanical (QM) calculations. Biological systems usually include hundreds of thousands to millions of atoms, and therefore cannot be described by QM calculations. Instead, scientists must rely on simple, approximate descriptions of each interaction. Alter-natively, it is possible to use QM calculations on small confined regions of biological macro-molecules, such as the active sites of enzymes or binding sites of receptors [187,188]. With the constant growth in available computer power, such calculations are likely to extend much farther, and one day may even encompass whole proteins and other macromolecules.

1.4   SUMMARY

- All life forms on Earth are cell-based.

- Prokaryotes are unicellular organisms whose inner environments are un-compartmentalized, yet display limited functional organization. Eukaryotes, which may be either unicellular or multicellular, contain functionally distinct inner compartments.

- A cell creates an internal environment that differs in its physical and chemical properties from the external environment.

- The distinction between the cell's internal environment and its external environment is expressed at the molecular level; cells contain complex molecules, whereas inanimate matter is made only of elements and simple molecules.

- Of the four basic macromolecules of cells, proteins are the most functionally diverse. This diversity is a direct result of proteins' high structural complexity, which in turn results from their polymeric nature and the chemical diversity of their building blocks, amino acids. These two features create numerous ways in which proteins may fold into three-dimensional structures. The exact structure of a protein provides its unique function.

- The three-dimensional fold of a protein is maintained by a set of three basic types of noncovalent interactions: electrostatic, van der Waals, and nonpolar. These interactions are only marginally higher in magnitude than the thermal energy RT, allowing proteins to be dynamic, which is crucial for their function.

1.5   ORGANIZATION OF BOOK

This book focuses on three central aspects of protein function: structure, energetics, and motion. Most of the discussion relates to the common types of cellular proteins, i.e., those characterized by a globular shape and high water solubility. These proteins reside in the cytoplasm, organelles, and extracellular space. They are discussed in Chapters 1 through 5. Some proteins, which tend to form fibers, are characterized by simpler structures with repetitive features. They are discussed in Chapter 2. Other proteins, which exist in a structure-less, yet active form, are discussed in Chapter 6. Finally, proteins that reside within the plasma membrane (or inner membranes of the cell) are globular, yet lipid-soluble. These proteins constitute about 20% of all cellular proteins (~7,000 in humans [189]), and are discussed in Chapter 7. Chapters 8 and 9 are dedicated to two central aspects of protein function, ligand binding (Chapter 8) and catalysis of chemical reactions (Chapter 9). Ligand binding is one of the most basic functions of proteins, whereas catalysis is the most sophisticated. Both aspects are excellent examples of the protein structure-function relationship.

EXERCISES

1.1 All living organisms are made of cells. Suggest reasons for this phenomenon.

1.2 Explain the advantages of molecular complexity, and how living organisms achieve it.

1.3 How many unique polypeptide chains with 60 amino acids are there?

1.4 Suggest a reason why evolution has led to the stabilization of proteins via noncovalent interactions, rather the much stronger covalent bonds.

1.5 Provide a short, qualitative description of all intermolecular forces that could associate between amino (NH_3^+) and carbonyl ($C=O$) groups.

1.6 Use your own words to describe the van der Waals interaction plot (Figure 1.17).

1.7 Does the hydrophobic effect involve direct inter-atomic force? Explain.

1.8 Two proteins bind each other noncovalently in an aqueous solution, with a total interaction surface (i.e., the interface of both binding partners) of 500 Å^2. A pH change leads to conformational changes in both proteins, which results in a decrease of their interface to 300 Å^2. Estimate the resulting change in the nonpolar interaction energy, using the empirical method described in the text.

1.9 Write mathematical expressions describing the corresponding energies of electrostatic (Coulomb) and van der Waals interactions. What can you deduce from the expressions on the range of each interaction?

1.10 Briefly explain the significance of the dielectric constant to electrostatic interactions.

1.11 Describe the two main differences between ionic interactions and hydrogen bonds.

REFERENCES

1. U. Lagerkvist. *The Enigma of Ferment: From the Philosopher's Stone to the First Biochemical Nobel Prize.* World Scientific Publishing Co., London, 2006.
2. P. Nurse. The great ideas of biology. *Clin. Med.*, 3(6):560–8, 2003.
3. L. N. Magner. History of Physiology, 2001.
4. E. Buchner. Alkoholische Gährung ohne Hefezellen. *Ber. Dtsch. Chem. Ges.*, 30:117–124, 1897.
5. W. Kühne. Über das Verhalten verschiedener organisirter und sog. ungeformter Fermente. *Verhandlungen des naturhistorisch-medicinischen Vereins zu Heidelberg. (Neue Folge)*, 1:190–193, 1877.
6. H. Hartley. Origin of the word 'protein'. *Nature*, 168(4267):244, 1951.
7. H. B. Vickery. The origin of the word protein. *Yale J. Biol. Med.*, 22(5):387–93, 1950.
8. G. J. Mulder. Over Proteine en hare Verbindingen en Ontledingsproducten. *Natuur- en scheikundig Archief*, 6:87 162, 1838.
9. Nobel Foundation. Eduard Buchner. Wikipedia, the free encyclopedia. http://en.wikipedia.org/wiki/File: Eduardbuchner.jpg, 1907.
10. Nobel Foundation. Eduard Buchner. Biographical. http://www.nobelprize.org/nobel_prizes/chemistry/ laureates/1907/buchner-bio.html. The Nobel Prize in Chemistry 1907.
11. R. M. May. How Many Species Are There on Earth? *Science*, 241(4872):1441–1449, 1988.
12. H. Lodish, A. Berk, C. A. Kaiser, M. Krieger, M. P. Scott, A. Bretscher, H. Ploegh, and P. Matsudaira. *Molecular Cell Biology*. W. H. Freeman and Company, New York, 6th edition, 2007.
13. D. E. Koshland Jr. Special essay. The seven pillars of life. *Science*, 295(5563):2215–2216, 2002.

14. Messer-Woland and Szczepan. Diagram of a typical animal cell. Wikipedia, the free encyclopedia, 2006.

15. J. W. Schopf. Microfossils of the Early Archean Apex chert: new evidence of the antiquity of life. *Science*, 260:640–6, 1993.

16. J. W. Schopf and B. M. Packer. Early Archean (3.3-billion to 3.5-billion-year-old) microfossils from War- rawoona Group, Australia. *Science*, 237:70–3, 1987.

17. S. J. Mojzsis, G. Arrhenius, K. D. McKeegan, T. M. Harrison, A. P. Nutman, and C. R. Friend. Evidence for life on Earth before 3,800 million years ago. *Nature*, 384(6604):55–9, 1996.

18. L. J. Ducka, M. Gliksona, S. D. Goldinga, and R. D. Webb. Microbial remains and other carbonaceous forms from the 3.24 Ga Sulphur Springs black smoker deposit, Western Australia. *Precambrian Res.*, 154(3– 4):205–220, 2007.

19. L. Margulis. *Origin of Eukaryotic Cells*. Yale University Press, New Haven, CT, 1970.

20. W. Martin and M. Muller. The hydrogen hypothesis for the first eukaryote. *Nature*, 392(6671):37–41, 1998.

21. T. W. Dahl, E. U. Hammarlund, A. D. Anbar, D. P. Bond, B. C. Gill, G. W. Gordon, A. H. Knoll, A. T. Nielsen, N. H. Schovsbo, and D. E. Canfield. Devonian rise in atmospheric oxygen correlated to the radiations of terrestrial plants and large predatory fish. *Proc. Natl. Acad. Sci. USA*, 107(42):17911–5, 2010.

22. K. A. Dill, K. Ghosh, and J. D. Schmit. Physical limits of cells and proteomes. *Proc. Natl. Acad. Sci. USA*, 108(44):17876–82, 2011.

23. D. S. Goodsell. Inside a living cell. *Trends Biochem. Sci.*, 16(6):203–6, 1991.

24. E. G. H. Wagner, S. Altuvia, and W. Nellen. Regulatory RNA, 2001.

25. S. A. Strobel and J. C. Cochrane. RNA catalysis: ribozymes, ribosomes, and riboswitches. *Curr. Opin. Chem. Biol.*, 11(6):636–43, 2007.

26. F. Walter and E. Westhof. Catalytic RNA. In *Encyclopedia of Life Sciences*. John Wiley & Sons, Ltd., 2001.

27. C. A. Orengo and J. M. Thornton. Protein families and their evolution-a structural perspective. *Annu. Rev. Biochem.*, 74:867–900, 2005.

28. M. Clamp, B. Fry, M. Kamal, X. Xie, J. Cuff, M. F. Lin, M. Kellis, K. Lindblad-Toh, and E. S. Lander. Distinguishing protein-coding and noncoding genes in the human genome. *Proc. Natl. Acad. Sci. USA*, 104(49):19428–33, 2007.

29. L. Ezkurdia, D. Juan, J. M. Rodriguez, A. Frankish, M. Diekhans, J. Harrow, J. Vazquez, A. Valencia, and M.L. Tress. The shrinking human protein coding complement: are there now fewer than 20,000 genes? *arXiv:1312.7111*, 2013.

30. I. G. Choi and S. H. Kim. Evolution of protein structural classes and protein sequence families. *Proc. Natl. Acad. Sci. USA*, 103(38):14056–61, 2006.

31. S. D. Patterson. How Much of the Proteome Do We See with Discovery-Based Proteomics Methods and How Much Do We Need to See? *Curr. Proteomics*, 1(1):3–12, 2004.

32. J. E. Celis, M. Ostergaard, N. A. Jensen, I. Gromova, H. H. Rasmussen, and P. Gromov. Human and mouse proteomic databases: novel resources in the protein universe. *FEBS Lett.*, 430(1–2):64–72, 1998.

33. R. Bravo and J. E. Celis. Up-dated catalogue of HeLa cell proteins: percentages and characteristics of the major cell polypeptides labeled with a mixture of 16 14C-labeled amino acids. *Clin. Chem.*, 28(4 Pt 2):766– 81, 1982.

34. S. Yooseph, G. Sutton, D. B. Rusch, A. L. Halpern, S. J. Williamson, et al. The Sorcerer II Global Ocean Sampling expedition: Expanding the universe of protein families. *PLoS Biol.*, 5:e16, 2007.

35. J. C. Venter, K. Remington, J. F. Heidelberg, A. L. Halpern, D. Rusch, J. A. Eisen, D. Wu, I. Paulsen, K. E. Nelson, W. Nelson, D. E. Fouts, S. Levy, A. H. Knap, M. W. Lomas, K. Nealson, O. White, J. Peterson, J. Hoffman, R. Parsons, H. Baden-Tillson, C. Pfannkoch, Y. H. Rogers, and H. O. Smith. Environmental genome shotgun sequencing of the Sargasso Sea. *Science*, 304(5667):66–74, 2004.

36. E. M. Martin Del Valle. Cyclodextrins and their uses: a review. *Process Biochem.*, 39(9):1033–1046, 2004.

37. A. Lehninger, D. L. Nelson, and M. Cox. *Lehninger Principles of Biochemistry*. W. H. Freeman and Company, New York, 5th edition, 2008.

38. B. Alberts, A. Johnson, J. Lewis, M. Raff, K. Roberts, and P. Walter. *Molecular Biology of the Cell*. Garland Science Texbooks, New York & London, 5th edition, 2007.

39. A. Bar-Even, E. Noor, Y. Savir, W. Liebermeister, D. Davidi, D. S. Tawfik, and R. Milo. The moder- ately efficient enzyme: evolutionary and physicochemical trends shaping enzyme parameters. *Biochemistry*, 50(21):4402–10, 2011.

40. R. Wolfenden and M. J. Snider. The depth of chemical time and the power of enzymes as catalysts. *Acc. Chem. Res.*, 34(12):938–945, 2001.

41. S. A. Arrhenius. Über die Dissociationswärme und den Einfluß der Temperatur auf den Dissociationsgrad der Elektrolyte. *Z. Phys. Chem.*, 4:96–116, 1889.

42. K. M. Kim, T. Qin, Y. Y. Jiang, L. L. Chen, M. Xiong, D. Caetano-Anolles, H. Y. Zhang, and G. Caetano-Anolles. Protein domain structure uncovers the origin of aerobic metabolism and the rise of planetary oxygen. *Structure*, 20(1):67–76, 2012.

43. Fvasconcellos. Schematic diagram of the mitochondrial electron transport chain. Wikipedia, the free encyclopedia. http://en.wikipedia.org/wiki/File:Mitochondrial_electron_transport_chain%E2%80%94Etc4.svg, 2007.

44. J. Farquhar, M. Peters, D. T. Johnston, H. Strauss, A. Masterson, U. Wiechert, and A. J. Kaufman. Isotopic evidence for Mesoarchaean anoxia and changing atmospheric sulphur chemistry. *Nature*, 449(7163):706–9, 2007.

45. D. C. Catling, C. R. Glein, K. J. Zahnle, and C. P. McKay. Why O_2 is required by complex life on habitable planets and the concept of planetary "oxygenation time". *Astrobiology*, 5(3):415–38, 2005.

46. J. Raymond and D. Segre. The effect of oxygen on biochemical networks and the evolution of complex life. *Science*, 311(5768):1764–7, 2006.

47. J. F. Allen and W. Martin. Evolutionary biology: out of thin air. *Nature*, 445(7128):610–2, 2007.

48. S. Iwata and J. Barber. Structure of photosystem II and molecular architecture of the oxygen-evolving centre. *Curr. Opin. Struct. Biol.*, 14(4):447–53, 2004.

49. K. V. Morris and J. S. Mattick. The rise of regulatory RNA. *Nat. Rev. Genet.*, 15(6):423, 2014.

50. W. Gilbert. Origin of life: The RNA world. *Nature*, 319:618, 1986.

51. M. J. Berridge. Introduction. In *Cell Signalling Biology*, chapter 1, pages 1–69. Portland Press, 2014.

52. M. A. Hanson and R. C. Stevens. Discovery of New GPCR Biology: One Receptor Structure at a Time. *Structure*, 17:8–14, 2009.

53. W. Bursch, A. Ellinger, C. Gerner, U. Frohwein, and R. Schulte-Hermann. Programmed cell death (PCD). Apoptosis, autophagic PCD, or others? *Ann. N. Y. Acad. Sci.*, 926:1–12, 2000.

54. S. Nussey and S. Whitehead. *Endocrinology: An Integrated Approach.* BIOS Scientific Publishers Ltd, Oxford, England, 2001.

55. A. F. Neuwald. The glycine brace: a component of Rab, Rho, and Ran GTPases associated with hinge regions of guanine- and phosphate-binding loops. *BMC Struct. Biol.*, 9:11–28, 2009.

56. J. B. Shabb. Physiological substrates of cAMP-dependent protein kinase. *Chem. Rev.*, 101(8):2381–411, 2001.

57. R. Wagner, M. Matrosovich, and H. D. Klenk. Functional balance between haemagglutinin and neuraminidase in influenza virus infections. *Rev. Med. Virol.*, 12(3):159–66, 2002.

58. Q. Deng and J. T. Barbieri. Molecular mechanisms of the cytotoxicity of ADP-ribosylating toxins. *Annu. Rev. Microbiol.*, 62:271–88, 2008.

59. A. E. Pedersen. Immunity to Infection, 2001.

60. M. P. Sheetz. Cell control by membrane-cytoskeleton adhesion. *Nat. Rev. Mol. Cell Biol.*, 2(5):392–6, 2001.

61. M. D. Ledesma and C. G. Dotti. Membrane and cytoskeleton dynamics during axonal elongation and stabilization. *Int. Rev. Cytol.*, 227:183–219, 2003.

62. M. S. Bretscher. Getting membrane flow and the cytoskeleton to cooperate in moving cells. *Cell*, 87(4):601–6, 1996.

63. D. Yarar, C. M. Waterman-Storer, and S. L. Schmid. A dynamic actin cytoskeleton functions at multiple stages of clathrin-mediated endocytosis. *Mol. Biol. Cell*, 16(2):964–75, 2005.

64. E. Rodriguez-Boulan, G. Kreitzer, and A. Musch. Organization of vesicular trafficking in epithelia. *Nat. Rev. Mol. Cell Biol.*, 6(3):233–47, 2005.

65. V. Allan and R. Vale. Movement of membrane tubules along microtubules in vitro: evidence for specialised sites of motor attachment. *J. Cell Sci.*, 107 (Pt 7):1885–97, 1994.

66. J. Pogliano. The bacterial cytoskeleton. *Curr. Opin. Cell Biol.*, 20(1):19–27, 2008.

67. R. Berisio, L. Vitagliano, L. Mazzarella, and A. Zagari. Recent progress on collagen triple helix structure, stability and assembly. *Protein Pept. Lett.*, 9(2):107–16, 2002.

68. A. Bhattacharjee and M. Bansal. Collagen structure: the Madras triple helix and the current scenario. *IUBMB Life*, 57(3):161–72, 2005.

69. M. A. Schwartz. The Force Is with Us. *Science*, 323:588–589, 2009.

70. R. O. Hynes. Integrins: bidirectional, allosteric signaling machines. *Cell*, 110(6):673–87, 2002.

71. K. R. Legate, S. A. Wickstrom, and R. Fassler. Genetic and cell biological analysis of integrin outside-in signaling. *Genes Dev.*, 23(4):397–418, 2009.

72. A. W. Orr, B. P. Helmke, B. R. Blackman, and M. A. Schwartz. Mechanisms of mechanotransduction. *Dev. Cell*, 10(1):11–20, 2006.

73. D. E. Discher, D. J. Mooney, and P. W. Zandstra. Growth factors, matrices, and forces combine and control stem cells. *Science*, 324(5935):1673–7, 2009.

74. B. Geiger, J. P. Spatz, and A. D. Bershadsky. Environmental sensing through focal adhesions. *Nat. Rev. Mol. Cell Biol.*, 10(1):21–33, 2009.

75. R. O. Hynes. The extracellular matrix: not just pretty fibrils. *Science*, 326(5957):1216–9, 2009.

76. M. Uhlen, L. Fagerberg, B. M. Hallstrom, C. Lindskog, P. Oksvold, A. Mardinoglu, A. Sivertsson, C. Kampf, E. Sjostedt, A. Asplund, I. Olsson, K. Edlund, E. Lundberg, S. Navani, C. A. Szigyarto, J. Odeberg, D. Djureinovic, J. O. Takanen, S. Hober, T. Alm, P. H. Edqvist, H. Berling, H. Tegel, J. Mulder, J. Rockberg, P. Nilsson, J. M. Schwenk, M. Hamsten, K. von Feilitzen, M. Forsberg, L. Persson, F. Johansson, M. Zwahlen, G. von Heijne, J. Nielsen, and F. Ponten. Proteomics. Tissue-based map of the human proteome. *Science*, 347(6220):1260419, 2015.

77. Kilbad. Normal Epidermis and Dermis with Intradermal Nevus 10x. Wikipedia, the free encyclopedia. http://en.wikipedia.org/wiki/File:Normal_Epidermis_and_Dermis_with_Intradermal_Nevus_10x.JPG, 2008.

78. H. Gray. Vertical section of a villus from the dog's small intestine X 80. Wikipedia, the free encyclopedia. http://en.wikipedia.org/wiki/File:Gray1059.png, 1918.

79. C. Vogel and C. Chothia. Protein family expansions and biological complexity. *PLoS Comput. Biol.*, 2(5):e48, 2006.

80. J. G. Robertson. Enzymes as a special class of therapeutic target: clinical drugs and modes of action. *Curr. Opin. Struct. Biol.*, 17(6):674–9, 2007.

81. J. Weigelt, L. D. McBroom-Cerajewski, M. Schapira, Y. Zhao, and C. H. Arrowmsmith. Structural genomics and drug discovery: all in the family. *Curr. Opin. Chem. Biol.*, 12:32–39, 2008.

82. G. J. Siegel. *Basic Neurochemistry: Molecular, Cellular and Medical Aspects*. Lippincott Williams & Wilkins, Philadelphia, 6th edition, 1999.

83. R. B. Russell and D. S. Eggleston. New roles for structure in biology and drug discovery. *Nat. Struct. Biol.*, 7 Suppl:928–30, 2000.

84. L. Karalliedde. Animal toxins. *Br. J. Anaesth.*, 74(3):319–27, 1995.

85. R. P. Hicks, M. G. Hartell, D. A. Nichols, A. K. Bhattacharjee, J. E. van Hamont, and D. R. Skillman. The medicinal chemistry of botulinum, ricin and anthrax toxins. *Curr. Med. Chem.*, 12(6):667–90, 2005.

86. M. J. Lord, N. A. Jolliffe, C. J. Marsden, C. S. Pateman, D. C. Smith, R. A. Spooner, P. D. Watson, and L. M. Roberts. Ricin. Mechanisms of cytotoxicity. *Toxicol. Rev.*, 22(1):53–64, 2003.

87. The Centers for Disease Control and Prevention (CDC). Facts About Ricin. http://www.bt.cdc.gov/agent/ricin/facts.asp, 2015. In *Emergency preparedness and response*.

88. S. Olsnes. The history of ricin, abrin and related toxins. *Toxicon*, 44(4):361–70, 2004.

89. M. A. Rinella. *Pharmakon: Plato, Drug Culture, and Identity in Ancient Athens*. Lexington Books, Lanham, MD, 2010.

90. J. Derrida. Plato's Pharmacy. In *Dissemination (trans. Barbara Johnson)*, pages 61–172. The Athlone Press, London, 1981.

91. A. Schmid, F. Hollmann, J. B. Park, and B. Bühler. The use of enzymes in the chemical industry in Europe. *Curr. Opin. Biotechnol.*, 13(4):359–366, 2002.

92. O. Kirk, T. V. Borchert, and C. C. Fuglsang. Industrial enzyme applications. *Curr. Opin. Biotechnol.*, 13(4):345–51, 2002.

93. T. Hey, E. Fiedler, R. Rudolph, and M. Fiedler. Artificial, non-antibody binding proteins for pharmaceutical and industrial applications. *Trends Biotechnol.*, 23(10):514–22, 2005.

94. F. Sanger, E. O. Thompson, and R. Kitai. The amide groups of insulin. *Biochem. J.*, 59(3):509–18, 1955.

95. R. Sasidharan and C. Chothia. The selection of acceptable protein mutations. *Proc. Natl. Acad. Sci. USA*, 104(24):10080–5, 2007.

96. E. Fischer. Einfluss der Configuration auf die Wirkung der Enzyme. *Ber. Dtsch. Chem. Ges.*, 27(3):2985–2993, 1894.

97. A. E. Mirsky and L. Pauling. On the Structure of Native, Denatured, and Coagulated Proteins. *Proc. Natl. Acad. Sci. USA*, 22(7):439–447, 1936.

98. C. B. Anfinsen. Principles that Govern the Folding of Protein Chains. *Science*, 181(96):223–230, 1973.

99. J. R. Knowles. Enzyme catalysis: not different, just better. *Nature*, 350(6314):121–4, 1991.

100. A. Gutteridge and J. M. Thornton. Understanding nature's catalytic toolkit. *Trends Biochem. Sci.*, 30(11):622–629, 2005.

101. K. E. van Holde, W. C. Johnson, and P. S. Ho. *Principles of Physical Biochemistry*, 2006.

102. L. Pauling. The Nature of the Chemical Bond. IV. The Energy of Single Bonds and the Relative Electronegativity of Atoms. *J. Am. Chem. Soc.*, 54(9):3570–3582, 1932.

103. M. Laberge. Intrinsic protein electric fields: basic noncovalent interactions and relationship to protein-induced Stark effects. *Biochim. Biophys. Acta*, 1386(2):305–30, 1998.

104. S. Kumar and R. Nussinov. Close-range electrostatic interactions in proteins. *ChemBioChem*, 3(7):604–617, 2002.

105. E. Persch, O. Dumele, and F. Diederich. Molecular recognition in chemical and biological systems. *Angew. Chem. Int. Ed.*, 54(11):3290–327, 2015.

106. F. Fogolari, A. Brigo, and H. Molinari. The Poisson-Boltzmann equation for biomolecular electrostatics: a tool for structural biology. *J. Mol. Recognit.*, 15(6):377–92, 2002.

107. A. Moll, A. Hildebrandt, H. P. Lenhof, and O. Kohlbacher. BALLView: an object-oriented molecular visualization and modeling framework. *J. Comput. Aided Mol. Des.*, 19(11):791–800, 2005.

108. F.-X. Theillet, A. Binolfi, T. Frembgen-Kesner, K. Hingorani, M. Sarkar, C. Kyne, C. Li, P. B. Crowley, L. Gierasch, G. J. Pielak, A. H. Elcock, A. Gershenson, and P. Selenko. Physicochemical properties of cells and their effects on intrinsically disordered proteins (IDPs). *Chem. Rev.*, 114(13):6661–6714, 2014.

109. G. Valincius, F. Heinrich, R. Budvytyte, D. J. Vanderah, D. J. McGillivray, Y. Sokolov, J. E. Hall, and M. Losche. Soluble amyloid beta-oligomers affect dielectric membrane properties by bilayer insertion and domain formation: implications for cell toxicity. *Biophys. J.*, 95(10):4845–61, 2008.

110. Y. Kimura and A. Ikegami. Local dielectric properties around polar region of lipid bilayer membranes. *J. Membr. Biol.*, 85(3):225–31, 1985.

111. M. Kawagoe and T. Ishimi. On the properties of organic liquids affecting the crazing behaviour in glassy polymers. *J. Mater. Sci.*, 37(23):5115–5121, 2002.

112. Aristotle. *Metaphysics*. Volume 1, page 983b, 350 BCE.

113. K. A. Sharp and B. Honig. Electrostatic interactions in macromolecules: theory and applications. *Annu. Rev. Biophys. Biophys. Chem.*, 19(1):301–332, 1990.

114. M. Born. Volumes and heats of hydration of ions. *Z. Phys.*, 1:45–48, 1920.

115. A. Nicholls and B. Honig. A Rapid Finite Difference Algorithm, Utilizing Successive Over-Relaxation to Solve the Poisson-Boltzmann Equation. *J. Comput. Chem.*, 12:435–445, 1991.

116. M. K. Gilson. Introduction to continuum electrostatics, with molecular applications. *Biophysics Textbooks online*, 2006.

117. B. Honig and A. Nicholls. Classical electrostatics in biology and chemistry. *Science*, 268(5214):1144–1149, 1995.

118. B. Roux and T. Simonson. Implicit solvent models. *Biophys. Chem.*, 78(1–2):1–20, 1999.

119. N. A. Baker. Improving implicit solvent simulations: a Poisson-centric view. *Curr. Opin. Struct. Biol.*, 15(2):137–143, 2005.

120. M. Feig and C. L. Brooks. Recent advances in the development and application of implicit solvent models in biomolecule simulations. *Curr. Opin. Struct. Biol.*, 14(2):217–224, 2004.

121. B. Honig, K. Sharp, and A. Suei-Yang. Macroscopic models of aqueous solutions: biological and chemical applications. *J. Phys. Chem.*, 97:1101–1109, 1993.

122. H. S. Haraldsdottir, I. Thiele, and R. M. Fleming. Quantitative assignment of reaction directionality in a multicompartmental human metabolic reconstruction. *Biophys. J.*, 102(8):1703–11, 2012.

123. N. A. Baker, D. Sept, S. Joseph, M. J. Holst, and J. A. McCammon. Electrostatics of nanosystems: Application to microtubules and the ribosome. *Proc. Natl. Acad. Sci. USA*, 98(18):10037–10041, 2001.

124. A. Nicholls, K. A. Sharp, and B. Honig. Protein folding and association: insights from the interfacial and thermodynamic properties of hydrocarbons. *Proteins*, 11(4):281–96, 1991.

125. W. Humphrey, A. Dalke, and K. Schulten. VMD: visual molecular dynamics. *J. Mol. Graph.*, 14(1):33–8, 1996.

126. W. L. DeLano. The PyMOL Molecular Graphics System. http://www.pymol.org, 2002.

127. E. F. Pettersen, T. D. Goddard, C. C. Huang, G. S. Couch, D. M. Greenblatt, E. C. Meng, and T. E. Ferrin. UCSF Chimera: a visualization system for exploratory research and analysis. *J. Comput. Chem.*, 25(13):1605–12, 2004.

128. M. A. Miteva, P. Tuffery, and B. O. Villoutreix. PCE: web tools to compute protein continuum electrostatics. *Nucleic Acids Res.*, 33(suppl 2):W372–5, 2005.

129. D. Murray and B. Honig. Electrostatic control of the membrane targeting of C2 domains. *Mol. Cell*, 9(1):145–54, 2002.

130. W. H. Keesom. The second viral coefficient for rigid spherical molecules, whose mutual attraction is equivalent to that of a quadruplet placed at their centre. *Proc. K. Ned. Akad. Wet.*, 18(1):636–646, 1915.

131. G. C. Pimentel and A. L. McClellan. *The Hydrogen Bond*. W. H. Freeman and Company, San Francisco, CA, 1960.

132. I. K. McDonald and J. M. Thornton. Satisfying hydrogen bonding potential in proteins. *J. Mol. Biol.*, 238(5):777–793, 1994.

133. C. Bissantz, B. Kuhn, and M. Stahl. A medicinal chemist's guide to molecular interactions. *J. Med. Chem.*, 53(14):5061–84, 2010.

134. T. Kortemme, A. V. Morozov, and D. Baker. An orientation-dependent hydrogen bonding potential improves prediction of specificity and structure for proteins and protein-protein complexes. *J. Mol. Biol.*, 326(4):1239–59, 2003.

135. J. M. Scholtz, S. Marqusee, R. L. Baldwin, E. J. York, J. M. Stewart, M. Santoro, and D. W. Bolen. Calorimetric determination of the enthalpy change for the alpha-helix to coil transition of an alanine peptide in water. *Proc. Natl. Acad. Sci. USA*, 88(7):2854–8, 1991.

136. C. N. Pace, H. Fu, K. Lee Fryar, J. Landua, S. R. Trevino, D. Schell, R. L. Thurlkill, S. Imura, J. M. Scholtz, K. Gajiwala, J. Sevcik, L. Urbanikova, J. K. Myers, K. Takano, E. J. Hebert, B. A. Shirley, and G. R. Grimsley. Contribution of hydrogen bonds to protein stability. *Protein Sci.*, 23(5):652–661, 2014.

137. D. F. Sticke, L. G. Presta, K. A. Dill, and G. D. Rose. Hydrogen bonding in globular proteins. *J. Mol. Biol.*, 226(4):1143–1159, 1992.

138. P. J. Fleming and G. D. Rose. Do all backbone polar groups in proteins form hydrogen bonds? *Protein Sci.*, 14(7):1911–1917, 2005.

139. S. Scheiner, T. Kar, and Y. Gu. Strength of the $C\alpha H\cdots O$ hydrogen bond of amino acid residues. *J. Biol. Chem.*, 276(13):9832–7, 2001.

140. W. W. Cleland. Low-barrier hydrogen bonds and enzymatic catalysis. *Arch. Biochem. Biophys.*, 382(1):1–5, 2000.

141. A. L. Ringer, A. Senenko, and C. D. Sherrill. Models of S/π interactions in protein structures: Comparison of the H_2S–benzene complex with PDB data. *Protein Sci.*, 16(10):2216–2223, 2007.

142. T. Cheeseright, M. Mackey, S. Rose, and A. Vinter. Molecular Field Extrema as Descriptors of Biological Activity: Definition and Validation. *J. Chem. Inf. Model.*, 46(2):665–676, 2006.

143. S. Sirimulla, J. B. Bailey, R. Vegesna, and M. Narayan. Halogen interactions in protein-ligand complexes: implications of halogen bonding for rational drug design. *J. Chem. Inf. Model.*, 53(11):2781–91, 2013.

144. M. R. Scholfield, C. M. Zanden, M. Carter, and P. S. Ho. Halogen bonding (X-bonding): a biological perspective. *Protein Sci.*, 22(2):139–52, 2013.

145. M. Harder, B. Kuhn, and F. Diederich. Efficient stacking on protein amide fragments. *ChemMedChem*, 8(3):397–404, 2013.

146. D. A. Dougherty. The cation-π interaction. *Acc. Chem. Res.*, 46(4):885–93, 2013.

147. C. A. Hunter, J. Singh, and J. M. Thornton. π-π interactions: The geometry and energetics of phenylalanine-phenylalanine interactions in proteins. *J. Mol. Biol.*, 218(4):837–846, 1991.

148. M. Levitt and M. F. Perutz. Aromatic rings act as hydrogen bond acceptors. *J. Mol. Biol.*, 201(4):751–754, 1988.

149. K. P. Gierszal, J. G. Davis, M. D. Hands, D. S. Wilcox, L. V. Slipchenko, and D. Ben-Amotz. π-Hydrogen Bonding in Liquid Water. *J. Phys. Chem. Lett.*, 2(22):2930–2933, 2011.

150. P. Zhou, F. Tian, F. Lv, and Z. Shang. Geometric characteristics of hydrogen bonds involving sulfur atoms in proteins. *Proteins*, 76(1):151–63, 2009.

151. G. Duan, V. H. Smith, and D. F. Weaver. Characterization of aromatic-thiol π-type hydrogen bonding and phenylalanine-cysteine side chain interactions through *ab initio* calculations and protein database analyses. *Mol. Phys.*, 99(19):1689–1699, 2001.

152. M. Iwaoka, S. Takemoto, M. Okada, and S. Tomoda. Weak Nonbonded $S\cdots X$ (X=O, N, and S) Interactions in Proteins. Statistical and Theoretical Studies. *Bull. Chem. Soc. Jpn.*, 75(7):1611–1625, 2002.

153. M. Iwaoka and N. Isozumi. Possible roles of $S\cdots O$ and $S\cdots N$ interactions in the functions and evolution of phospholipase A_2. *Biophysics*, 2:23–34, 2006.

154. B. R. Beno, K. S. Yeung, M. D. Bartberger, L. D. Pennington, and N. A. Meanwell. A Survey of the Role of Noncovalent Sulfur Interactions in Drug Design. *J. Med. Chem.*, 58(11):4383–438, 2015.

155. G. J. Bartlett, A. Choudhary, R. T. Raines, and D. N. Woolfson. $n \rightarrow \pi*$ interactions in proteins. *Nat. Chem. Biol.*, 6(8):615–620, 2010.

156. G. J. Bartlett, R. W. Newberry, B. VanVeller, R. T. Raines, and D. N. Woolfson. Interplay of hydrogen bonds and $n \rightarrow \pi*$ interactions in proteins. *J. Am. Chem. Soc.*, 135(49):18682–8, 2013.

157. W. Wu, M. K. Samoszuk, S. A. Comhair, M. J. Thomassen, C. F. Farver, R. A. Dweik, M. S. Kavuru, S. C. Erzurum, and S. L. Hazen. Eosinophils generate brominating oxidants in allergen-induced asthma. *J. Clin. Invest.*, 105(10):1455–63, 2000.

158. P. Auffinger, F. A. Hays, E. Westhof, and P. S. Ho. Halogen bonds in biological molecules. *Proc. Natl. Acad. Sci. USA*, 101(48):16789–94, 2004.

159. A. Wojtczak, V. Cody, J. R. Luft, and W. Pangborn. Structure of rat transthyretin (rTTR) complex with thyroxine at 2.5 Å resolution: first non-biased insight into thyroxine binding reveals different hormone orientation in two binding sites. *Acta Crystallogr. Sect. D*, 57(8):1061–1070, 2001.

160. A. R. Voth and P. S. Ho. The role of halogen bonding in inhibitor recognition and binding by protein kinases. *Curr. Top. Med. Chem.*, 7(14):1336–48, 2007.

161. J. D. Dunitz and R. Taylor. Organic fluorine hardly ever accepts hydrogen bonds. *Chem. Eur. J.*, 3:89–98, 1997.

162. P. Politzer, J. S. Murray, and T. Clark. Halogen bonding and other sigma-hole interactions: a perspective. *Phys. Chem. Chem. Phys.*, 15(27):11178–89, 2013.

163. O. Hassel. Structural aspects of interatomic charge-transfer bonding. *Science*, 170(3957):497–502, 1970.

164. T. Clark, M. Hennemann, J. S. Murray, and P. Politzer. Halogen bonding: the sigma-hole. Proceedings of "Modeling interactions in biomolecules II", Prague, September 5th–9th, 2005. *J. Mol. Model.*, 13(2):291–6, 2007.

165. P. Politzer, J. S. Murray, and P. Lane. σ-Hole bonding and hydrogen bonding: Competitive interactions. *Int. J. Quantum Chem.*, 107(15):3046–3052, 2007.

166. E. V. Anslyn and D. A. Dougherty. Binding Forces. In *Modern Physical Organic Chemistry*, chapter 3. University Science Books, 2005.

167. P. Atkins and J. de Paula. Molecular Interactions. In *Physical Chemistry*, chapter 18. W. H. Freeman and Company, 9th edition, 2009.

168. N. Ben-Tal, D. Sitkoff, I. A. Topol, A.-S. Yang, S. K. Burt, and B. Honig. Free energy of amide hydrogen bond formation in vacuum, in water, and in liquid alkane solution. *J. Phys. Chem. B*, 101(3):450–457, 1997.

169. G. A. Petsko and R. Dagmar. Bonds that stabilize folded proteins. In *Protein Structure and Function*, Primers in Biology, chapter 1–4. Sinauer Associates, Inc., 2003.

170. G. Zubay. *Biochemistry*. William C. Brown Publishers, Dubuque, IA, 3rd edition, 1993.

171. F. London. The General Theory of Molecular Forces. *Trans. Faraday Soc.*, 33:8–26, 1937.

172. F. Z. London. Zur Theorie und Systematik der Molekularkrafte. *Physik*, 63:245, 1930.

173. R. L. Baldwin. Energetics of protein folding. *J. Mol. Biol.*, 371(2):283–301, 2007.

174. T. Lazaridis, G. Archontis, and M. Karplus. Enthalpic contribution to protein stability: insights from atom-based calculations and statistical mechanics. *Adv. Protein Chem.*, 47:231–306, 1995.

175. R. A. Aziz. A highly accurate interatomic potential for argon. *J. Chem. Phys.*, 99:4518–25, 1993.

176. Poszwa. Interaction energy of argon dimer. Wikipedia, the free encyclopedia, 2005.

177. K. A. Sharp. Water: Structure and Properties, 2001.

178. N. K. Adam. *The Physics and Chemistry of Surfaces*. Oxford University Press, 1941.

179. H. S. Frank and M. W. Evans. Free Volume and Entropy in Condensed Systems III. Entropy in Binary Liquid Mixtures; Partial Molal Entropy in Dilute Solutions; Structure and Thermodynamics in Aqueous Electrolytes. *J. Chem. Phys.*, 13(11):507–532, 1945.

180. W. Kauzmann. Some factors in the interpretation of protein denaturation. *Adv. Protein Chem.*, 14:1–63, 1959.

181. P. L. Privalov and S. J. Gill. Stability of protein structure and hydrophobic interaction. *Adv. Protein Chem.*, 39:191–234, 1988.

182. D. Chandler. Interfaces and the driving force of hydrophobic assembly. *Nature*, 437(7059):640–7, 2005.

183. A. L. Ferguson, P. G. Debenedetti, and A. Z. Panagiotopoulos. Solubility and molecular conformations of n-alkane chains in water. *J. Phys. Chem. B*, 113(18):6405–14, 2009.

184. W. C. Wimley, T. P. Creamer, and S. H. White. Solvation energies of amino acid side chains and backbone in a family of host-guest pentapeptides. *Biochemistry*, 35(16):5109–5124, 1996.

185. C. Chothia. Hydrophobic bonding and accessible surface area in proteins. *Nature*, 248(446):338–339, 1974.

186. J. A. Reynolds, D. B. Gilbert, and C. Tanford. Empirical Correlation Between Hydrophobic Free Energy and Aqueous Cavity Surface Area. *Proc. Natl. Acad. Sci. USA*, 71(8):2925–2927, 1974.

187. H. M. Senn and W. Thiel. QM/MM studies of enzymes. *Curr. Opin. Chem. Biol.*, 11(2):182–7, 2007.

188. A. Warshel. Computer simulations of enzyme catalysis: methods, progress, and insights. *Annu. Rev. Biophys. Biomol. Struct.*, 32:425–43, 2003.

189. M. Baker. Structural biology: The gatekeepers revealed. *Nature*, 465(7299):823–826, 2010.

CHAPTER 2

Protein Structure

2.1 INTRODUCTION

In 1960 the British biochemist John Kendrew used a method called '*X-ray diffraction*' to 'photograph' myoglobin at a 2-Å resolution, and became the first person to determine the three-dimensional structure of a protein [1]. A short while later, Max Perutz, Kendrew's colleague at Cambridge University, determined the structure of a similar, yet more complex protein, hemoglobin [2]. For these feats the two scientists were awarded the 1962 Nobel Prize in chemistry. Since then, the structures of tens of thousands of different proteins have been determined at high resolution. Today, these structures are freely accessible over the Internet for anyone interested, and their investigation enables scientists to better understand the architectural, functional and energetic principles of proteins.

2.1.1 Hierarchy in protein structure

At first glance, most proteins look like chaotic crowdings of atoms. A closer look, however, reveals complex structures organized in a hierarchical manner [3] (Figure 2.1). The first level of this hierarchy, referred to as the '*primary structure*', is the ordered sequence of amino acids composing the protein chain. Certain segments within this chain tend to fold into simple shapes, such as helices, loops, etc. These structures are referred to as '*secondary elements*', and collectively constitute the second level of the protein hierarchy, the *secondary structure*. Secondary elements are local, and (except for loops) proceed along one axis of the protein chain. The overall chain tends to fold further into a compact, three-dimensional *tertiary structure*, which constitutes the third level of the hierarchy. As explained in Chapter 1, the tertiary structure is the most stable form of the protein, since it optimizes the various attraction forces among the different amino acids that compose the chain. Moreover, the tertiary structure is also the biologically active form of the protein, and its disruption renders the protein partially or completely inactive. Therefore, the tertiary structure is often referred to as the '*native structure*' of the protein.

These three levels of structural hierarchy exist in all proteins, although deviations from the classical 'rules' of the tertiary structure can be observed in some proteins. For example, *fibrous* proteins tend to acquire an elongated form, which includes an arrangement of secondary elements, yet is devoid of the characteristic complex three-dimensional fold (see Section 2.7 below). Other proteins take this idea to the extreme and avoid ordered structure altogether, at least part of the time (see Chapter 6). The third hierarchical level of structure may be common to most proteins, but it is not necessarily the final level. Some proteins in-

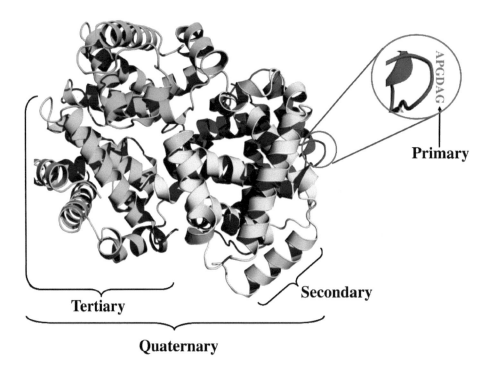

FIGURE 2.1 **The four levels of protein structure.** The levels are depicted on the structure of the hemoglobin protein, which includes four different chains (shown in different colors). The first level is the amino acid sequence of the chains, depicted here as an olive-colored one-letter code (see main text below; see blow-up on the top right side of the figure). The second level includes the helical (spring-like) segments of the protein, as well as the connecting loops. Other proteins may include other secondary elements, in which the chain is more extended (i.e., less compact) than in helices. The third level includes the complete three-dimensional organization of each of the chains. Finally, the fourth level includes the arrangement of the different chains. The structure was taken from the Protein Data Bank (PDB, entry 1hho).

clude more than one chain. In such cases, each chain folds separately into a tertiary structure, and then joins the others to form a biologically active complex. This type of organization constitutes the fourth level of structural hierarchy, and is referred to as *'quaternary structure'*. It is important to distinguish between the quaternary structure and complexes that form when cellular proteins interact physically with other members of their biochemical pathways. Such temporary complexes are not considered to be quaternary structures, since the individual proteins composing them are also active when separated.

2.1.2 Coenzymes and prosthetic groups

The amino acid chain is the primary and central component of the protein, but not necessarily the only component. Some proteins may include other atoms or small molecules, which are required for the proteins' function and/or stability. These *cofactors* are chemically diverse and can be organic molecules or elements, either metallic (e.g., zinc) or non-metallic (e.g., selenium). Some of them bind to the protein chain temporarily, whereas others, referred to as *'prosthetic groups'*, are integral parts of the protein [4]. The latter are tightly attached to the protein, sometimes even covalently. Enzymes, for example, have long been known to require cofactors in order to carry out their functions. As will be explained in detail in Chapter 9, such cofactors, termed *'coenzymes'*, are in most cases small organic molecules derived from vitamins. The common coenzymes NADH and $FADH_2$ are well-known examples. These two molecules, derived, respectively, from the B-complex vitamins niacin (B_3) and riboflavin (B_2), serve as carriers and/or donors of high-energy electrons in reduction-oxidation (*redox*) reactions.

Some enzymes use several different cofactors. As an example, let us look at *pyruvate dehydrogenase (PDH)*, which is a key enzyme in carbohydrate catabolism [5]. The catabolic metabolism of carbohydrates begins with *glycolysis*, a 10-step biochemical pathway that turns one glucose molecule into two molecules of pyruvate [6], and continues with the *citric acid (Krebs) cycle* [7], which further degrades and oxidizes the remnants of pyruvate to CO_2. PDH works between those two major pathways, and it is responsible for the activation of pyruvate, so as to allow it to enter the Krebs cycle. Specifically, PDH catalyzes the oxidative decarboxylation of pyruvate to *acetyl-CoA (ACoA)*, its activated form (Figure 2.2a). PDH is not a single enzyme, but rather a three-component complex, with each component participating in a different step in the activation of pyruvate, and each using a different coenzyme (Figure 2.2b). The first component uses *thiamine pyrophosphate (TPP)*, a derivate of *thiamine* (a.k.a. *vitamin B_1*) (Figure 2.2c). TPP enables the first component of PDH to decarboxylate pyruvate, a three-carbon molecule, into acetaldehyde, a two-carbon group (which is in the form of a hydroxyl-ethyl group when attached to TPP). The third carbon is released as CO_2. The biological importance of TPP is reflected in the outcome of its deficiency: people who do not obtain enough thiamine in their diet often suffer from a disease called *beriberi*, which harms several body systems [8].

The second component of PDH uses *lipoic acid*. This prosthetic group is covalently attached to the enzyme (see below). The active part of lipoic acid is a ring structure that contains two covalently bonded sulfur atoms (i.e., an S—S bond, or disulfide) (Figure 2.2c). This part catalyzes the oxidation of the substrate's hydroxyl-ethyl group and its transfer to the third cofactor, *coenzyme A (CoA)*, thus yielding the principal product of the reaction, acetyl-CoA. CoA is a derivate of *pantothenic acid*, also known as *vitamin B_5* (Figure 2.2c). The binding of the two-carbon substrate to lipoic acid involves the reduction-opening of the S—S bond into two thiol groups (—SH), a coupled oxidation of the substrate's hydroxyl-ethyl group, and its attachment to one of the thiols. Thiol groups are chemically reactive, as reflected in their susceptibility to toxic derivates of the element *arsenic* [9] (see Box 2.1). In order to allow PDH to remain active, the two thiol groups must be re-oxidized and re-form the S—S bond. This role is played by the third (and last) component of PDH, which uses the coenzyme flavin adenine dinucleotide (FAD). Finally, FAD itself is reconstituted by NAD^+, which is consequently reduced to NADH, the third product of the reaction (CO_2 is the first one). It is interesting to note that the principal PDH reaction, as complex as it may be, is not unique; it is also used by *α-ketoglutarate dehydrogenase*, a Krebs cycle enzyme. The catalyzed reaction is identical to that of PDH, but the substrate is a different keto-acid (α-ketoglutarate instead of pyruvate). As we shall see later, the reuse of 'successful' protein functions is a key evolutionary strategy.

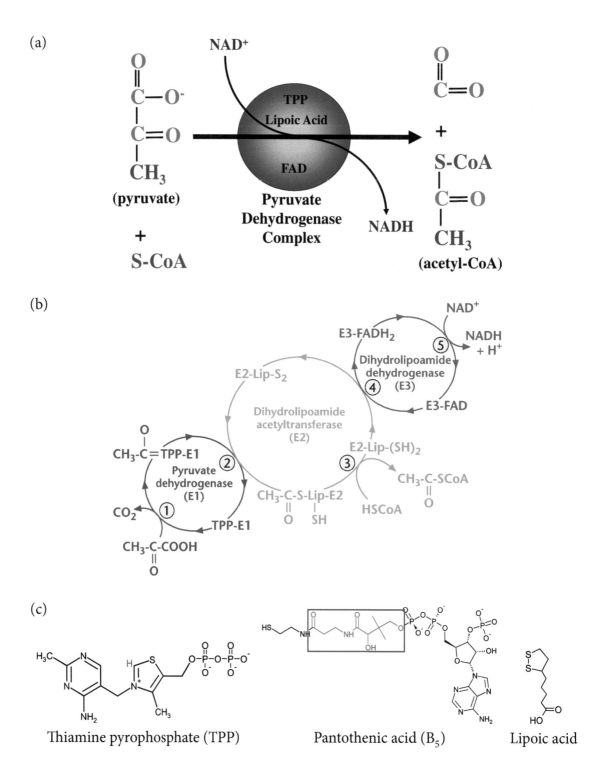

FIGURE 2.2 **The pyruvate dehydrogenase (PDH) complex.** (a) The reaction catalyzed by PDH. S-CoA stands for coenzyme A, TPP for thiamine pyrophosphate, NAD for nicotine-adenine dinucleotide, and FAD for flavin adenine dinucleotide. (b) The mechanism of PDH. The image was taken from [5]. (c) The chemical structures of TPP, lipoic acid, and coenzyme A.

BOX 2.1 ARSENIC POISONING, POLITICAL STRUGGLES, AND NAPOLEON BONAPARTE

Arsenic (As) is an element found in water, soil, and air [9]. It appears in both organic and inorganic forms and in different oxidation states, of which the tri- and penta-valent states are considered most dangerous to human health (Figure 2.1.1). Arsenic poisoning may occur in different ways, but the most common is through drinking contaminated water. Indeed, water-related poisoning cases have been documented worldwide, especially in Taiwan, Chile, Mexico, India and Pakistan. The toxic effects of arsenic may occur through one of two different mechanisms, depending on the form and chemistry properties of the arsenic species. Inorganic (penta-valent) arsenic is believed to act by competing with phosphate on the binding to various proteins [9]. This is made possible by the great similarity between the two species (Figure 2.1.2): they are of similar size (arsenate is only 4% larger than phosphate), are both pentavalent, and have the same geometry, pKa[*a] and charge [10]. In contrast, organic arsenicals act in a completely different way, by blocking protein thiol groups, thus neutralizing several enzymes in which these groups are functionally important. Pyruvate dehydrogenase (PDH) is one of the primary metabolic enzymes implicated in arsenic poisoning. The arsenic-containing compound blocks the thiol groups of lipoic acid, one of PDH's coenzymes (Figure 2.1.3). Physiologically, acute arsenic poisoning manifests as violent abdominal cramping, diarrhea and vomiting, often followed by death from shock [11]. In contrast, chronic poisoning leads to multi-organ pathologies, which may include skin thickening and pigmentation changes, gastrointestinal problems (including cancer), esophageal bleeding, enlargement of the spleen, anemia, bone marrow depression, liver disease, and more [12,13]. The symptoms usually include weakness, confusion and paralysis.

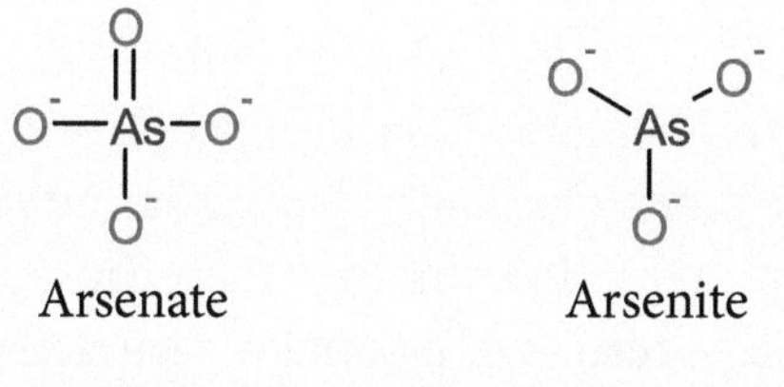

Arsenate   Arsenite

FIGURE 2.1.1 **The most common forms of arsenic.** The two molecules shown are the pentavalent arsenate (left) and trivalent arsenite.

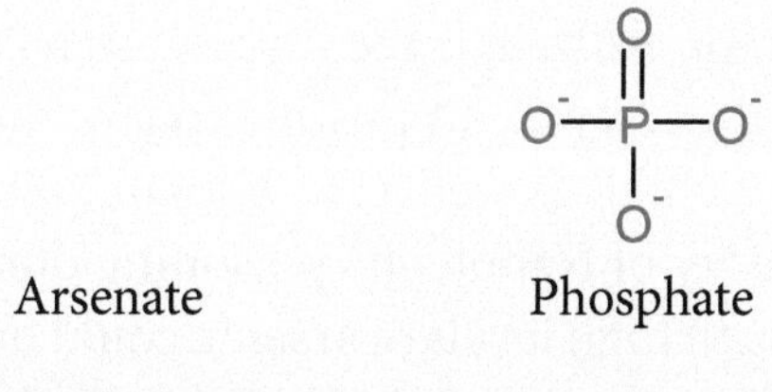

Arsenate   Phosphate

FIGURE 2.1.2 **The similarity between inorganic arsenate and phosphate.**

[*a]See Box 2.2 for an explanation on the nature of pKa.

Arsenic-based compounds have been used since ancient times, for both beneficial and sinister purposes. For example, low doses of arsenic have been used as medication against the sexually transmitted disease *syphilis*. During the Middle Ages and the Renaissance period, arsenic trioxide, nicknamed 'the white powder', was used quite frequently for assassination of political opponents [11]. Among the accused of using arsenic for such purposes were members of the infamous *Borgia family* of Italy. The popularity of the poison was a result of its availability, and the difficulty to recognize its effects (at least up to the 19th century), among other qualities. Indeed, acute arsenic poisoning most often passed as food poisoning, due to the harsh gastrointestinal distress, whereas the effects of chronic poisoning could be attributed to other diseases [11].

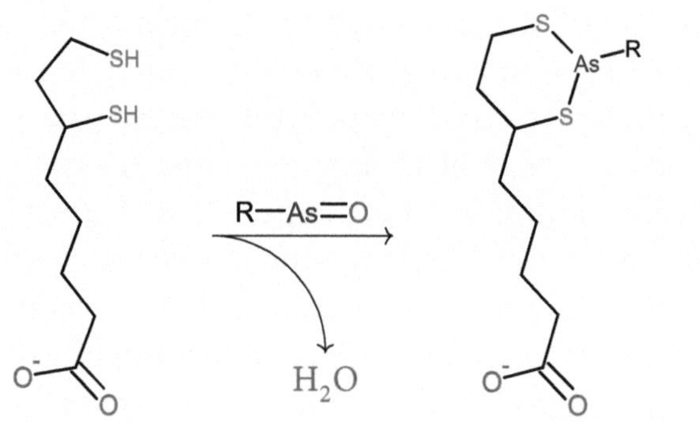

Free lipoic acid   Blocked lipoic acid

FIGURE 2.1.3 **Pyruvate dehydrogenase neutralization by arsenicals.** The proposed mechanism of action is blockage of the two thiol groups of lipoic acid by covalent bonding to the arsenical (R–As=O).

The common use of arsenic-based poisons, especially among royalty and politicians, led to quite a few 'conspiracy theories' implicating arsenic poisoning in the deaths of some famous historic figures. Such was the case with *Napoleon Bonaparte I* of France (Figure 2.1.4), who died in 1821 while in exile on the Island of Saint Helena [14]. The official cause of death of Napoleon was stomach cancer. In 1960, a forensic scientist examined a hair sample taken from Napoleon a day after he died. The examination discovered levels of arsenic many times higher than those in the hair of normal people [14]. These results, as well as results of an analysis carried out years later [15,16], raised the suspicion that Napoleon's cause of death was not stomach cancer, but in fact arsenic poisoning [17]. Although it could not be ascertained whether the poisoning was accidental or the result of homicide, the theory spread quickly and became popular worldwide. Recently, a team of scientists from Italy reexamined the hair more accurately, using a nuclear reactor. Although the examination confirmed the high levels of arsenic in the hair taken from Napoleon posthumously, these levels were present also in hair taken from him as a child, as well as in his wife's and son's hair [18]. These findings seem to refute the theory of Napoleon's poisoning during exile. Nevertheless, the question remains of how such high levels of arsenic could have accumulated in the hair of the Bonaparte family. The answer might be very simple; low-dose arsenic potions were widely used in the 19th century as a tonic or aphrodisiac. Napoleon, like other people of his time, might have ingested the toxic element quite willingly, as a drug or just for the 'kick' of it.

FIGURE 2.1.4 **Napoleon on his imperial throne (by Jean Auguste Dominique Ingres).** The picture was taken from [19].

Chemical groups that are bound tightly to the protein chain are referred to as 'prosthetic groups', and they are often used as a basis for the grouping and naming of proteins. For example, membrane-bound or secreted[*1] proteins tend to bind a large number of sugar moieties (*glycans*). They are therefore referred to as '*glycoproteins*'. Similarly, proteins that bind lipid groups are called '*lipoproteins*'. These proteins are bound to large spherical fatty bodies, which are used for the transport of fatty acids and cholesterol between different organs. Myoglobin and hemoglobin, mentioned at the beginning of the chapter, each contain an organic group called '*heme*', which includes an iron atom. They are therefore referred to as '*hemoproteins*'. Hemoglobin transports oxygen and carbon dioxide between the lungs and peripheral tissues, whereas myoglobin serves as a temporary reservoir of oxygen in muscles. In both proteins, the heme group functions in binding oxygen, and is essential for protein function.

Finally, some proteins bind small ions, e.g., metals or halogens, which are important for function despite their small dimensions. Metals in particular are important for different proteins, many of which are enzymes. We will discuss the role of metals later, in relation to post-translational modifications.

2.2 PRIMARY STRUCTURE

The primary structure of a protein is the exact ordering (i.e., sequence) of the amino acids that form its chain. **The exact sequence of the protein is very important, as it determines the final fold, and therefore the function, of the protein** [21]. We will discuss primary struc-

---

[*1]Proteins that constitute ~10% of the human genome [20], and which include certain hormones and local mediators, antibodies, digestive enzymes, coagulation factors and growth factors.

ture in two steps. In the first we will get to know amino acids and their physicochemical properties, and in the second we will see how the amino acids connect to each other, and how this changes their properties.

2.2.1 Amino acids and their properties

2.2.1.1 Amino acid structure

All amino acids possess a common structure that includes a central carbon atom called the 'α-carbon', or simply 'C_α', surrounded by four substituents: a hydrogen atom, an amino group ('α-amino'), a carboxyl group ('α-carboxyl'), and a fourth group referred to as a '*side chain*' (Figure 2.3). The α-carboxyl group has a low pKa (~2), and therefore tends to be deprotonated and negatively charged at physiological pH (~7)[*1]. The α-amino group, on the other hand, has a high pKa (9–10), and therefore tends to be protonated and positively charged at physiological pH. As we will see later, these charges are nullified when each amino acid is integrated within the protein chain, except for the charges located on the two ends of the protein sequence. It is customary to divide each amino acid into two parts. The first includes all non-side chain atoms, that is, C_α, its hydrogen atom, the α-carboxyl group, and the α-amino groups. This part is called the '*backbone*', and is identical across all amino acids. The second part consists of the side chain, which is different in each amino acid. In other words, the side chain group is what differentiates between amino acids. Due to the uniqueness of the side chain residue, amino acids are often referred to as '*residues*' when incorporated within the protein chain. In the following sections we will review in detail the specific properties of amino acid side chains. Before doing so however, it is necessary to discuss one additional property of these molecules, which, though unrelated to the specific nature of the amino acid side chain, still affects protein structure and function. This property is the *configuration* of amino acids.

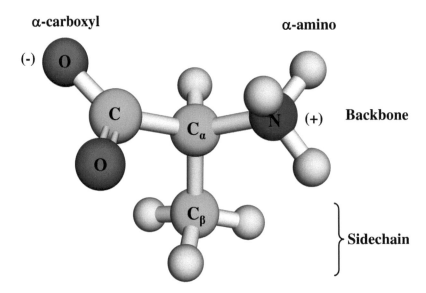

FIGURE 2.3 **The principal structure of amino acids.** Atoms and bonds are depicted as balls and sticks, respectively. Carbon atoms are shown in green, oxygen atoms in red, hydrogen atoms in white, and the nitrogen atom in blue.

[*1] See Box 2.2 for an explanation about the nature of pKa.

BOX 2.2 THE MEANING OF pKa

Some chemical compounds behave as *acids* or as *bases*. One of the simplest definitions of such compounds is Brønsted and Lowry's, which, in essence, refers to acids as molecules that release protons into the solution in which they reside, and to bases as molecules that absorb protons from solution. The *'deprotonation'* of an acid (RH) proceeds until equilibrium is reached between the acid and its *conjugated base* (R⁻):

$$RH \longleftrightarrow R^- + H^+ \tag{2.2.1}$$

Like any other equilibrium process, deprotonation is also characterized by an *equilibrium constant (Ka)*, which depends on the equilibrium concentrations of the reactants and products:

$$Ka = \frac{[R^-][H^+]}{[RH]} \tag{2.2.2}$$

(where the square parentheses represent molar concentration).

Ka is an important parameter, serving as a measure of an acid's strength, i.e., its tendency to release protons; the higher the Ka, the stronger the acid (Figure 2.2.1). However, the values of Ka may be very large or very small, which is a bit inconvenient. Thus, scientists prefer to use a related parameter, pKa, defined as minus the base 10 logarithm of Ka:

$$pKa = -\log Ka \tag{2.2.3}$$

The $-\log$ operation makes the numeric difference between strong and weak acids conveniently small. Thus, the pKa can be defined as a logarithmic measure of the proton affinity of an acid; the stronger the acid, the lower its pKa (Figure 2.2.1). The degree of deprotonation of an acid depends on the pH (i.e., $-\log[H^+]$) of the solution: the larger the pH compared to the pKa, the higher the degree of deprotonation. This is because high pH values correspond to abundance of hydroxide ions (OH^-) in solution, which scavenge the acid's protons and thus shift the equilibrium of the acid's deprotonation towards the products. The degree of deprotonation of any acid can be calculated using the Henderson–Hasselbalch equation [22], provided that the pH and the pKa values are known:

$$pH = pKa + \log \frac{[R^-]}{[RH]} \tag{2.2.4}$$

The equation, which follows from Equation (2.2.2), shows that when the pH of the solution equals the pKa of the acid, exactly 50% of the acid is deprotonated.

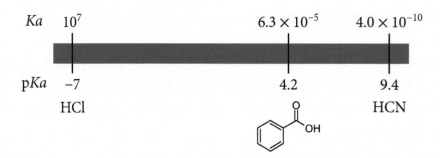

FIGURE 2.2.1 **Ka and corresponding pKa values of different acids.** The Ka and pKa scales proceed from the strongest acids on the left to the weakest on the right. The acids shown are, from left to right, hydrochloric acid (HCl), benzoic acid, and hydrocyanic acid.

The pKa of an acid is a measurable quantity, and can be determined using titration of the acid with a strong base, e.g., sodium hydroxide (NaOH). In this process, the protons produced by the dissociated acid are scavenged by the hydroxide ions from the dissociated base:

$$\begin{array}{rcl} RH & \longleftrightarrow & R^- + H^+ \\ NaOH & \longrightarrow & Na^+ + OH^- \\ \hline RH + NaOH & \longrightarrow & NaR + H_2O \end{array} \qquad (2.2.5)$$

This leads to a gradual rise in the solution's pH, which can be tracked as the base is being added. In the simplest case, i.e., when both the acid and the base are monoprotic, one equivalent of added base leads to the deprotonation of exactly one equivalent of acid, as depicted by the *titration curve* in Figure 2.2.2a. The pKa is the pH value of the solution measured when exactly half of the acid has been titrated, i.e., when half of the base equivalents have been added. As shown in the figure, the curve has a flat region in the middle, which corresponds to a pH range in which the pH changes very little, although the base is constantly being added. This phenomenon, known as '*buffering*', happens when a large enough quantity of the acid has been deprotonated. Typically, the buffering capability of an acid appears when approaching the halfway point of the deprotonation process. In pH terms, the buffering starts when the pH is ~1 unit lower than that of the halfway point, and stops when the pH is ~1 unit higher than this point. Thus, the halfway point of deprotonation, which is also the value of the pKa, should be roughly at the middle of this range. This phenomenon comes in handy when dealing with polyprotic acids (e.g., amino acids), especially when the number of protonated groups is unknown. In such cases, each of the different groups has its own pKa, and becomes fully deprotonated at a different pH value. Assuming that the pKa values of the groups are not highly similar, they are titrated sequentially, each requiring one equivalent of base to become fully deprotonated. This creates a multi-step titration curve, such as the one presented in Figure 2.2.2b. As mentioned above, a scientist can identify the pKa values of the titrated groups[*a] by finding the middle points of all the flat regions of the curve.

[*a] And thus, also their number, and sometimes even their chemical nature.

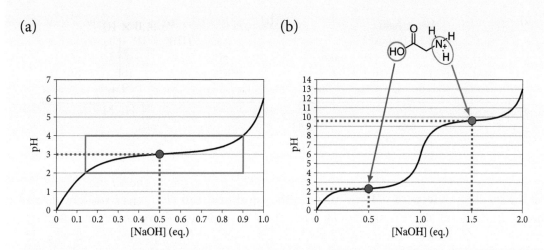

FIGURE 2.2.2 **Titration curves.** (a) The idealized titration curve of a monoprotic acid. The curve represents the change in pH value as a result of the added base. The red rectangle marks the range in which the acid behaves as a buffer, and the green sphere shows the pKa point. (b) The titration curve of glycine, a diprotic amino acid. The pKa values corresponding to the carboxyl and amino groups of the molecules, 2.3 and 9.6 (respectively), are marked. The chemical structure of glycine is shown at the upper part of the figure, with protonated groups marked by circles.

Finally, the deprotonation of an acid affects its electric charge. In Equation (2.2.1), the acid is electrically neutral when protonated, and becomes negatively charged when it loses the proton. This pattern characterizes many organic acids, e.g., carboxylic acid:

$$COOH \longleftrightarrow COO^- + H^+ \tag{2.2.6}$$

Other acids may behave in the opposite way. That is, they are positively charged when protonated, and become neutral when undergoing deprotonation. This happens, e.g., in the case of ammonia:

$$NH_2-H^+ \longleftrightarrow NH_2 + H^+ \tag{2.2.7}$$

Such acids usually behave like bases, i.e., their pKa is higher than physiological pH, and they therefore tend to be protonated at this pH.

Thus, it is necessary to know both the pKa and the chemical structure of an acid (or base) in order to be able to infer its ionization state.

The effects of the pH and the pKa of an acid or base on its equilibrium and electric charge are summarized in Figure 2.2.3.

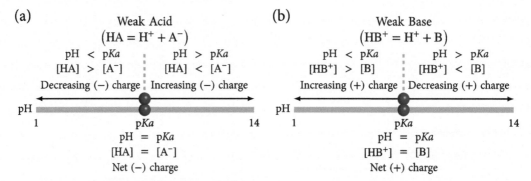

FIGURE 2.2.3 **Effects of pH and p*Ka* on the equilibria and charges of weak acids and bases.** In acids (a), decreasing the pH below the p*Ka* leads to a decrease in the concentration of A^-, which means the molecule becomes less negative. The opposite happens when increasing the pH beyond the p*Ka*: the A^- form becomes dominant, and the molecule's negative charge increases. In weak bases (b), decreasing the pH below the p*Ka* leads to an increase in the concentration of HB^+, which means the molecule becomes more positive. When the pH is increased beyond the p*Ka* the opposite happens. Courtesy of Steven Bottomley.

Molecules that contain multiple protic groups (i.e., groups that can undergo protonation or deprotonation) can exist in different ionization states, depending on the pH and the p*Ka* of each group. At a certain pH, called the '*isoelectric point*', or *pI*, the negative and positive charges on the molecule are equal in number, and the net charge is zero. When the pH is lower than the pI, the molecule is positively charged, whereas at pH values higher than the pI values the molecule is negatively charged (Figure 2.2.4). Thus, knowing the pI of biomolecules such as proteins is of interest to scientists, as it allows them to separate such molecules from one another on the basis of their net charge at a certain pH. Such separation is usually done using methods such as *ion exchange chromatography*, in which a mixture of molecules is passed through a column containing beads with either a positive or negative charge. When the mixture passes through the column, molecules whose charge is opposite to that of the beads bind to the latter and are retained inside the column. The rest go through the column. After the column is washed, the pH can be changed to neutralize or reverse the charges of the bound molecules, which allows them to detach from the beads and exit the column. By using buffers of different pH values, scientists can gradually isolate different molecules from the same mixture by adjusting their electric charges.

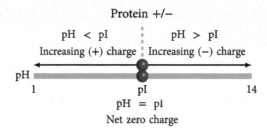

FIGURE 2.2.4 **pI and the effect of pH on the net charges of proteins.** Courtesy of Steven Bottomley.

2.2.1.2 Configurations of amino acids

Many organic molecules contain one or more carbon atoms that are bonded to four different substituents. Such atoms are referred to as 'chiral', or 'asymmetry' centers. This term reflects the fact that the mirror image of such a chemical group cannot be superimposed on the original one. In other words, each chiral center in a molecule has two non-superimposable forms, each reflecting a different arrangement of the same substituents around the chiral center. This unique arrangement is referred to as the *absolute configuration* of the chiral center, and the two mirror images as the center's *optical isomers* or *enantiomers* [23,24]. To distinguish between the two alternative isomers, one is termed 'R', whereas the other is termed 'S'. Biochemists often prefer to address the configuration of a molecule according to its relation to the configuration of the simplest chiral carbohydrate, glyceraldehyde. According to this labeling system, one isomer is referred to as 'D', and the other is referred to as 'L'. For convenience, we will use the D/L system. Another term reflecting spatial arrangement of chemical groups and atoms is the *molecular conformation*, a concept we encountered in Chapter 1. As explained, whereas the transition from one configuration of a molecule to another requires the breaking of covalent bonds, the transition between different conformations only requires rotation of the atoms around the (single) bonds.

The configuration of amino acids is determined by the arrangement of substituents around C_α (Figure 2.4), although some amino acids contain other chiral carbons. The configuration of individual amino acids is of great importance, as it determines the spatial properties of the entire protein comprising them. Specifically, this property is important for the ability of proteins to participate in molecular recognition events. As described in Chapter 1, many proteins are stereospecific; that is, their function is based on their capability to accurately recognize a single configuration of their respective ligands. This means that the three-dimensional structure of the ligand must match that of the protein's binding site, i.e., the two structures must complement each other stereochemically. Since the complementarity is spatial, it can only be achieved when the binding site has certain three-dimensional properties; these properties are determined by the configuration of the amino acids building the site.

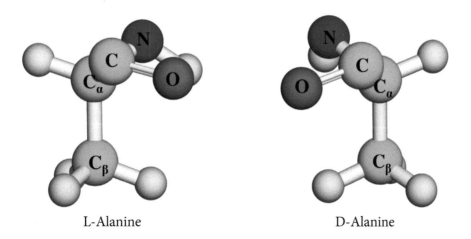

L-Alanine   D-Alanine

FIGURE 2.4 **The L- and D-configurations of the amino acid alanine.**

Indeed, whereas individual amino acids in nature may possess either the D- or L-configurations, **amino acids within proteins almost exclusively possess the L.** This *homochirality* helps proteins to ensure spatial complementarity between their binding sites and their ligands, and is a result of the way cells build their proteins. The process begins with the loading of individual amino acids on their respective *transfer RNA (tRNA)* molecules, and proceeds when each of the tRNA-amino acid complexes binds to the *ribosome*, i.e., the cellular protein-synthesis machine. As it turns out, the enzymes that carry out the first stage of the protein-building process recognize only those amino acids that are in the L-configuration. Thus, only L-amino acids are incorporated into proteins. The origin of this homochirality is unknown, but studies show that it may result from the properties of tRNA, especially the D-configuration of its ribose moieties [25,26].

Although homochirality is highly prevalent in proteins, it is not absolute; certain proteins and peptides do contain D-amino acids. This may occur in two ways, depending on the organism:

1. In microorganisms: Certain unicellular organisms possess a non-ribosomal peptide-synthesizing system, which can use D-amino acids as substrates [27,28]. Such systems include a number of enzymes that work together, each catalyzing a different stage in the synthesis. The enzymes also act as templates for the selection of amino acids (in ribosomal synthesis, the mRNA serves as such a template). Non-ribosomal systems are responsible in microorganisms for the building of biologically active peptides, such as those building the cell wall, as well as peptides acting as antibiotics. The latter can be exemplified by valinomycin, a cyclic peptide that contains a four-residue repeating sequence. Two of the residues in the sequence have the D-configuration, whereas the other two are L-amino acids [29].

2. In multicellular organisms: Peptides (and sometimes also proteins) containing D-amino acids can also be found in more complex organisms, from snails to humans [30]. These peptides and proteins play important physiological roles that are associated with development, aging, defense, and neurotransmission. Multicellular organisms do not possess non-ribosomal systems for protein synthesis, and therefore cannot incorporate existing D-amino acids into protein chains while they are being built [30,31]. Instead, such organisms may use specific enzymes to convert certain L-amino acids in an existing protein into the D-configuration [32–34]. Many peptides containing amino acids with D-configurations are produced in the skin of amphibians; these peptides act as a means of defense against predators. The peptide *dermorphin*, for example, which is produced by the South American frog *Phyllomedusa sauvagii*, is built from seven residues, one of which is D-alanine (*Tyr-**D-Ala**-Phe-Gly-Tyr-Pro-Ser-NH$_2$*) [35,36]. The peptide has a morphine-like activity (with an effect many times stronger than morphine's [37]), and the frogs use it to repel potential predators. Although enzymes typically catalyze the process in which L-residues are converted into their D-isomers in multicellular organisms, in one case, that of *L-aspartate*, this conversion occurs spontaneously, during ageing. The process is thought to involve an interaction between the negatively charged side chain of aspartate and the α-amino group of the following residue [30].

What is the evolutionary incentive for inclusion of D-amino acids in proteins and peptides? Many peptides containing D-amino acids participate in defense roles. This group includes secreted peptides (e.g., valinomycin and gramicidin) that have antibiotic activity against neighboring bacteria, as well as psychoactive peptides (e.g., dermorphin) that produce toxic effects on larger predators. In any case, the mode of action of these peptides involves their insertion into another organism, which also possesses defense mechanisms. Such mechanisms are often based on stereospecific recognition, e.g., in the case of proteolytic enzymes and antibodies. In that sense, constructing a peptide with at least one D-amino acid is highly beneficial, as it prevents the host's defense system from recognizing and degrading the peptide. D-amino acids may also promote a unique structure that allows the peptide to fulfill a specific function. This is nicely demonstrated by the antibacterial peptide *gramicidin* [38]. The peptide acts against a host bacterium by creating a wide ion channel within its plasma membrane. This causes massive ion loss, dissipation of the plasma membrane's electric potential, and loss of key metabolic functions [39,40]. Together, these effects eventually lead to the death of the attacked bacterium. The formation of a wide channel requires a special peptide structure called a *β-helix* [41], a structure that is made possible by the unique sequence of gramicidin, which includes alternating L- and D-amino acids. As we will discuss later, it is not unusual for protein chains to organize as helical structures; however, the geometry of the typical 'α-helix' structure is different from that of the β-helix. In particular, a single α-helix is too compact to create an ion channel on its own, let alone one that is wide.

2.2.1.3 Side chain properties

In contrast to the backbone, which is chemically identical in all amino acids, the chemical nature of each amino acid's side chain is unique. Given this property, 20 distinct amino acids can be identified in natural proteins. Table 2.1 and Figure 2.5 present the chemical characteristics of each of the 20 different side chains and their full molecular structures, respectively. At first glance it seems that each amino acid has a completely different side chain chemistry. However, a closer look reveals certain chemical groups, such as amide, carboxyl and hydroxyl that are present in more than one side chain. Thus, the 20 amino acids can be classified into a small number of groups, which differ in certain properties, the most important being *polarity*. As explained in Chapter 1, a polar molecule contains at least one hydrogen atom that is bonded to a heavier, considerably more electronegative atom (oxygen, nitrogen, etc.). The uneven distribution of electrons in such bonds creates an electric dipole. In nonpolar molecules, in contrast, hydrogen atoms may only be bonded to carbon atoms, since the electronegativity difference between the two is too small to create a dipole. As we will see below, the sulfur-hydrogen bond (S−H), which appears in the amino acid cysteine, constitutes an exception to the rule stated above, because of the electronic structure of sulfur. Polarity is highly important in biomolecules, since it determines their ability to interact with other molecules. This is also true for amino acid side chains; those that are polar are able to interact electrostatically with polar entities, such as other protein residues, organic molecules in the vicinity of the protein, and even the aqueous solvent itself (Figure 2.6).

Such interactions are highly important in the biological context, for the following reasons:

1. Polar interactions take part in determining the three-dimensional fold of the protein.

2. The tendency of polar residues to interact with water molecules surrounding the protein enhances the solubility of the protein. As we will see later, this tendency also affects the likelihood of protein residues to appear at the core or periphery of the protein [42,43].

3. Some polar residues within proteins participate in ligand binding, which is the basis for the functions of nearly all proteins (e.g., enzyme-substrate, receptor-hormone, and antibody-antigen binding).

4. Some polar interactions between enzymes and their substrates participate in catalysis [44]. As explained in Chapter 9, the high electron density on polar atoms allows some of them to act as nucleophiles, that is, to interact with partially positive carbon atoms of the substrate, so as to facilitate, e.g., the breaking of its labile bonds. In addition, polar interactions may indirectly contribute to catalysis by polarizing catalytic residues.

As Figure 2.5 shows, it is customary to group amino acids into five types:

I. Nonpolar (Figure 2.5a): The side chains of these amino acids contain only carbon and hydrogen atoms, except for one special case, which will be explained later. Also, note that the side chain of proline is fused to the backbone (see more below).

II. Polar-uncharged (Figure 2.5b): The side chains of these amino acids are electrically neutral, but still carry partial charges, which create electric dipoles.

III. Polar-charged (Figure 2.5c): The side chains of these residues are charged at physiological pH. The charge is either −1 or +1 units.

IV. Aromatic (Figure 2.5d): The side chains of these residues contain one or two aromatic rings.

V. Glycine (Gly) (Figure 2.5e): Glycine is the one amino acid that does not have a side chain.

The following subsections describe in detail the properties of the side chains of the different groups of amino acids, with emphasis on their locations within proteins, their interactions with other residues, and their functional importance. A summary of these features is given in Table 2.2. A more detailed description of interactions between protein residues can be found in [45]. This description was produced by extensive statistical analysis of the behaviors of amino acids in proteins of known structure. This information also appears online at: http://www.biochem.ucl.ac.uk/bsm/sidechains/index.html.

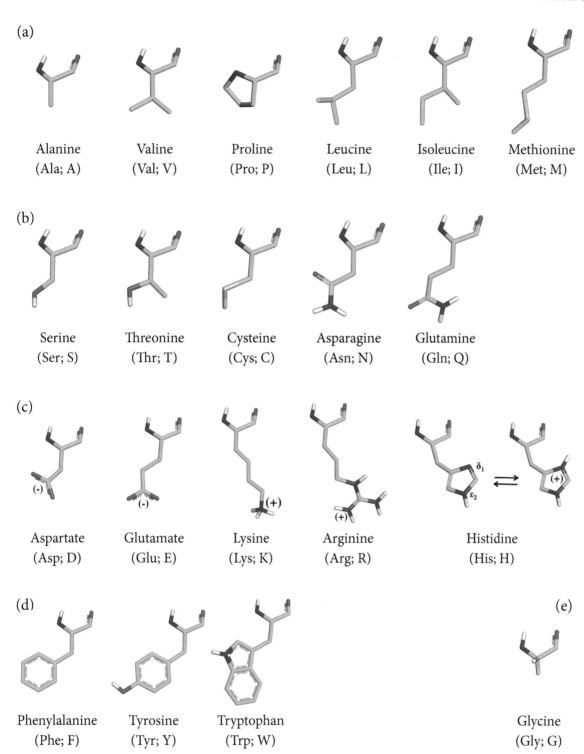

FIGURE 2.5 **The molecular structures of the 20 amino acid types found in proteins.** The amino acids are grouped into five basic types according to side chain properties. (a) Nonpolar. (b) Polar-uncharged. (c) Polar-charged. (d) Aromatic. (e) Glycine. The amino acids are shown as sticks and colored by atoms type (carbon – green, hydrogen – white, oxygen – red, nitrogen – blue, sulfur – yellow). The α-carboxyl and α-amide groups are presented as they are in proteins, i.e., as electrically neutral carbonyl and amide, respectively. The ionization states of the side chain groups correspond to physiological pH (~7). In the case of His, the charged and neutral forms have similar probabilities at pH 7, with a weak preference to the neutral form. Therefore, both forms are shown. Note that whereas Cys is traditionally considered polar, as shown here, some categorize it as mildly nonpolar (see main text for details).

TABLE 2.1 **The 20 natural amino acids and chemical descriptions of their side chains.**

Group	Full Name	Three-Letter Name	One-Letter Name	Side Chain Group
No side chain	Glycine	Gly	G	—
Nonpolar	Alanine	Ala	A	Methane
	Valine	Val	V	Propane
	Leucine	Leu	L	2-Methyl-propane
	Isoleucine	Ile	I	Butane
	Proline	Pro	P	Pyrrolidine
	Methionine	Met	M	(Methyl-sulfanyl)ethane
Polar-uncharged	Serine	Ser	S	Methanol
	Threonine	Thr	T	Ethanol
	Cysteine	Cys	C	Methanethiol
	Asparagine	Asn	N	Acetamide
	Glutamine	Gln	Q	Propanamide
Electrically charged	Glutamate	Glu	E	Propanoate
	Aspartate	Asp	D	Acetate
	Lysine	Lys	K	Butan-1-amine
	Arginine	Arg	R	1-Propyl-guanidine
	Histidine	His	H	4-Methyl-1H-imidazole
Aromatic	Phenylalanine	Phe	F	Methyl-benzene
	Tyrosine	Tyr	Y	4-Methyl-phenol
	Tryptophan	Trp	W	3-Methyl-1H-indole

TABLE 2.2 **Main structural and functional features of specific amino acids in proteins.**

Amino Acid	Structural and Functional Features in Proteins
Glycine	• Confers flexibility to proteins
Proline	• Creates rigid kinks in proteins
Methionine	• Participates in weak polar interactions via sulfur's nonbonding electron pair • Possible antioxidant • Participates in enzymatic metal catalysis
Serine and threonine	• Important to regulation of protein and cellular functions via phosphorylation • Important to solubility, protection and recognition of membrane and secreted proteins via glycosylation • Act as nucleophiles in covalent catalysis in enzymes

Amino Acid	Structural and Functional Features in Proteins
Cysteine	• Appeared late in evolution • Participates in weak hydrogen bonds and σ-π interactions • Binds metals covalently • Contributes to protein stabilization, protection, folding, and signaling via disulfide bond formation • Participates in signal transduction (farnesylation, palmitoylation) • Participates in enzymatic covalent, redox and metal catalysis • Antioxidant • Target of alkylating agents (toxins, lab reagents)
Asparagine	• Important to solubility, protection and recognition of membrane and secreted proteins via glycosylation
Glutamate and aspartate	• Interact electrostatically with cationic amino acids, ligands, and metals • Bind metals covalently • Participate in enzymatic acid, base, and metal catalysis • Important to regulation of blood clotting when γ-carboxylated (glutamate)
Lysine	• Interacts electrostatically with anionic amino acids and ligands • Binds cofactors via Schiff base • Stabilizes and protects proteins by forming isopeptide bonds • Stabilizes protein complexes by forming covalent crosslinks • Participates in enzymatic acid and base catalysis • Important to regulation of cellular processes by acetylation, ubiquitinylation, and SUMOylation
Arginine	• Interacts electrostatically with anionic amino acids and ligands • Participates in enzymatic catalysis by pKa modulation and stabilization of anionic transition states
Histidine	• Appeared late in evolution • Binds metals covalently • Participates in enzymatic acid, base, and metal catalysis
Aromatic amino acids	• Appeared late in evolution • Form gates in ion channels and transporters • Important to ligand binding via van der Waals, nonpolar, hydrogen bonding, π-π and π-cations • UV light absorption (protein characterization in lab)
Tyrosine	• Important to regulation of protein and cellular functions via phosphorylation • Facilitates protein secretion, viral entry into cells and metal binding via sulfation • Participates in enzymatic acid-base, redox, and radical-based catalysis
Tryptophan	• Has the largest side chain of all amino acids • Has low frequency in proteins • Participates in electron transport • Fluorescent (lab characterization of conformational changes and ligand binding)

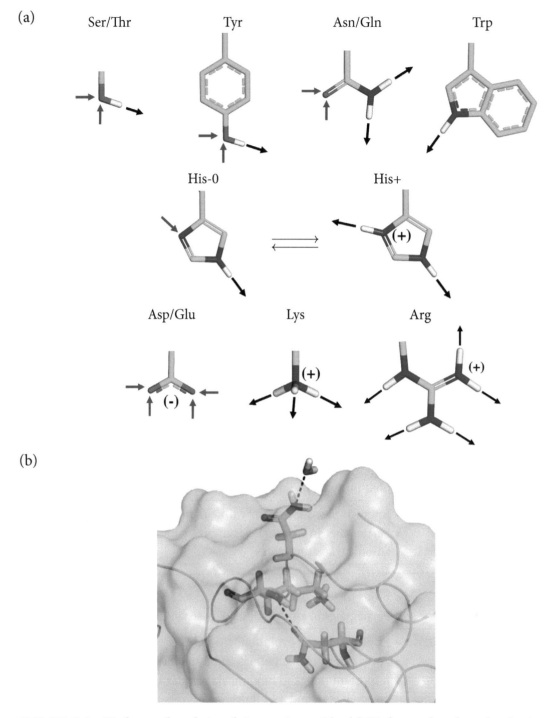

FIGURE 2.6 **Hydrogen bonds involving amino acids.** (a) Hydrogen bond tendencies in polar amino acid side chains. The hydroxyl (Ser/Thr), phenol (Tyr), carboxamide (Asn/Gln), imidazole (His), carboxylate (Asp/Glu), ammonium (Lys), guanidinium (Arg), and indole (Trp) are presented as in Figure 2.5. Hydrogen bond accepting and donating interactions are marked as pink and black arrows, respectively. Note that the two pairs of lone electrons on oxygen atoms enable these atoms to accept two hydrogen bonds, whereas the single electron pair on basic nitrogen atoms (as in uncharged His) enables them to accept only one hydrogen bond. As in Figure 2.5, the ionization states of the side chain groups correspond to physiological pH (~7), and His is shown in its two probable states. (b) The surface of the enzyme acetylcholinesterase (PDB entry 1acj) is shown in cyan. Two hydrogen bonds are shown (black dashed lines). The first is between an Asn residue of the protein and a solvent water molecule. The other hydrogen bond is between two Lys residues inside the protein. All interacting residues are presented as sticks and colored by atom type.

2.2.1.3.1 Glycine

Glycine does not have a side chain, a property that has two implications for the amino acid's location in the protein. First, glycine's small dimensions enable it to be located in crowded regions of the protein that cannot accommodate other residues. Second, because of glycine's lack of a side chain, the region around this amino acid in the protein chain is flexible. Accordingly, glycine has a tendency to be located in regions of the protein where increased flexibility is desirable or can be tolerated.

2.2.1.3.2 Nonpolar amino acids

The nonpolar group includes the highly common [46] amino acids *alanine* (Ala; A), *valine* (Val; V), *leucine* (Leu; L), *isoleucine* (Ile; I), *methionine* (Met; M), and *proline* (Pro; P). The side chains of these amino acids, with the exception of methionine, contain only carbon and hydrogen atoms. The methionine side chain contains a divalent sulfur atom, which, like oxygen, appears in the 16th column of the periodic table (*chalcogens*), and thus possesses an electronic structure similar to that of oxygen. We would, therefore, expect the presence of the sulfur atom to render the side chain of methionine polar. Nevertheless, methionine is considered to be nonpolar for two reasons. First, the electronegativity of sulfur (2.5) is lower than that of oxygen (3.5), due to differences of dimensions [47], and is identical to the electronegativity of the carbon atom [48]. Second, in the methionine side chain, the sulfur is flanked on both sides by two carbon atoms, which prevent the formation of an electric dipole.

Although methionine is overall nonpolar, it is still capable of interacting weakly with polar molecules (including aromatic rings; see [49] and references therein), via its non-bonding electron pair [50,51]. In addition, there seems to be an interesting duality in the sulfur atom; it interacts with both nucleophilic (e.g., OH) and electrophilic (e.g., the hydrogen of NH) groups [52]. These unique properties of methionine may be the reason why it was retained during protein evolution. In other words, its attributes seem to have been worth the burden of incorporating the sulfur atom into cellular biosynthetic systems. The uniqueness of methionine also manifests in other ways. For example, methionine side chains that are exposed on the protein surface may undergo oxidation and turn into sulfoxides. Since this process does not seem to harm protein function, and can be reversed by the enzyme sulfoxide reductase, it has been suggested that methionine acts as a built-in antioxidant in proteins, which prevents the oxidation of other residues in the protein by scavenging free radicals [53]. Another beneficial property of methionine is its capacity to interact with cationic metals. Again, this capacity is due to the sulfur's non-bonding electrons. The size of the sulfur atom allows its outer-shell electrons to be held only weakly by its nucleus, which makes them accessible for interactions with nearby cations. Indeed, methionine is known to interact with iron (Fe), zinc (Zn), copper (Cu), and molybdenum (Mo) cations in metalloenzymes. This interesting group of enzymes uses metals to carry out oxidation-reduction reactions, which are important to cellular metabolism. Finally, methionine is also important as an individual amino acid; it is the biosynthetic precursor of Cys (another amino acid), as well as certain phospholipids and cellular metabolites such as *S-adenosyl methionine*, a 'universal' donor of methyl groups in anabolic reactions.

Another unique amino acid in the nonpolar group is proline. Its side chain is covalently attached to the α-amino group, forming a *pyrrolidine* ring. The implications of this structure are discussed in later sections. Certain proteins include proline-rich sequences. Such sequences possess unique geometric and chemical properties that enable them to participate in protein-protein recognition [54]. This issue will also be discussed below.

Experimentally determined structures of proteins show that nonpolar residues reside mainly inside the core of the protein, where they can interact with each other [42,43] (Figure 2.7). In this sense, the amino acids in the nonpolar group differ from one another in the extent to which they are 'nonpolar' (i.e., hydrophobic): after glycine, alanine is the least hydrophobic; valine has an intermediary hydrophobicity, whereas leucine and isoleucine are highly hydrophobic. Moreover, the branched side chains of valine, leucine, and isoleucine allow these residues to be tightly packed within the protein core, while optimizing the van der Waals and nonpolar interactions between them. Yet, the preference of nonpolar residues for the protein core is not absolute, and they can also be found on the surface. Comparative analysis shows that nonpolar residues are highly common among proteins of all kingdoms of life (i.e., Archaea, Eubacteria, and Eukaryotes) [55].

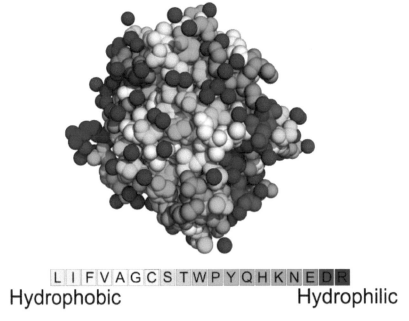

L I F V A G C S T W P Y Q H K N E D R
Hydrophobic   Hydrophilic

FIGURE 2.7 **Organization of polar and nonpolar amino acids in proteins.** Protein atoms are shown as spheres, colored according to polarity. The color code (at the bottom of the figure) is according to the Kessel-Ben-Tal Scale [56], obtained by calculations of amino acid transfer between aqueous and lipid media. Solvent (water) molecules are also shown as spheres, colored magenta. The figure shows that most of the polar residues are in the protein periphery, where they can interact electrostatically with the solvent.

2.2.1.3.3 *Polar-uncharged amino acids*

The polar-uncharged group includes serine (Ser; S), threonine (Thr; T), cysteine (Cys; C), glutamine (Gln; Q), and asparagine (Asn; N). Polar residues have a clear preference for the surface of the protein, where they can hydrogen-bond to each other or to the surrounding water molecules of the solvent (Figure 2.7). Polar residues may also appear inside the protein core, in which case they usually fulfill a certain function, e.g., enzymatic catalysis. This issue is discussed in Chapter 4.

2.2.1.3.3.1. Serine and threonine

Serine and threonine each include a hydroxyl group (OH) in their side chains. This group can donate one hydrogen bond and accept two bonds (via the oxygen's two lone electron pairs, see Figure 2.6a). The hydroxyl serves as a target for two types of post-translational modifications. The first is *phosphorylation*, which is one of the major mechanisms used for signal transduction, as well as a means for regulating the activity of enzymes [57]. The other modification is *glycosylation* (i.e., attachment of carbohydrate moieties), which many membrane-bound and secreted proteins undergo. The carbohydrate moieties are attached to the hydroxyl's oxygen atom (*O-linked glycosylation*).

The hydroxyl group serves in many enzymes as a nucleophile during catalysis (see Chapter 9 for further details on covalent catalysis). Ser, being a primary alcohol, is more reactive in this capacity than Thr, which is a secondary alcohol. Acting as a nucleophile requires the hydroxyl group to deprotonate to its anionic form (O^-). Such a process does not tend to happen spontaneously at physiological pH, due to the high pKa of the hydroxyl (~13, see Table 2.3). However, inside the protein's active site the deprotonation of serine is facilitated by certain residues in its vicinity. The residues interact with serine's hydroxyl group in a way that lowers its pKa. This happens, e.g., in enzymes of the *serine protease* group [58,59], such as trypsin and chymotrypsin, which function in food digestion. It also happens in enzymes of the *serine esterases* group, such as acetylcholinesterase and butyrylcholinesterase. The principal residue acting to deprotonate serine is histidine, which acts as a general base (see Box 8.1 and Chapter 9). Thr, despite having a hydroxyl-containing side chain similar to that of Ser, is less likely to act in such a mechanism, since its additional methyl group makes the hydroxyl less accessible, and therefore less reactive.

TABLE 2.3   **pKa values of amino acid side chains.**

Residue	Deprotonation Process [*a]	pKa_{int} [*b]	pKa_{prot} [*c]
Serine	$R-OH \leftrightarrow R-O^- + H^+$	~13	
Threonine	$R-OH \leftrightarrow R-O^- + H^+$	~13	
Arginine	$R_1=NH_2^+ \leftrightarrow R_1=NH + H^+$	12.3[*d]	
Lysine	$R-NH_3^+ \leftrightarrow R-NH_2 + H^+$	10.4	10.5 ± 1.1
Tyrosine	$R-OH \leftrightarrow R-O^- + H^+$	9.8	10.3 ± 1.2
Cysteine	$R-SH \leftrightarrow R-S^- + H^+$	8.6	6.8 ± 2.7
Histidine	$R_1=NH^+-R_2 \leftrightarrow R_1=N-R_2 + H^+$	6.5	6.6 ± 1.0
Glutamate	$R-COOH \leftrightarrow R-COO^- + H^+$	4.3	4.2 ± 0.9
Aspartate	$R-COOH \leftrightarrow R-COO^- + H^+$	3.9	3.5 ± 1.2

[*a]R denotes the rest of the residue.

[*b]Intrinsic side chain pKa, i.e., the pKa of the amino acid side chain when it is fully exposed to the aqueous solvent, not bonded to any chemical species, and unaffected by any formal charge(s). The values were measured using a host-guest pentapeptide [60], that is, a peptide with the sequence Ala-Ala-X-Ala-Ala, where X is any of the 20 natural amino acids. Data taken from Pace et al. [61], except for serine and threonine.

[*c]Side chain pKa values of amino acids that are part of the folded protein. The average values and their standard deviations were derived from measurements of 541 ionizable groups from 78 different proteins [62]. Data taken from Pace et al. [61], except for serine and threonine.

[*d]Note, however, that recent potentiometry and NMR spectroscopy measurements suggest that the pKa of arginine's side chain is actually higher, i.e., 13.8 [63].

2.2.1.3.3.2. Cysteine

The side chain of cysteine contains a thiol group (a.k.a. sulfhydryl). The polarity of this side chain is controversial; on the one hand, the sulfur atom has an electronegativity of 2.5 [48], which should make the S−H bond roughly as polar as the C−H bond. On the other hand, sulfur has four unpaired electrons on its outer shell, providing it with a partial negative charge. Thus, cysteine may be considered (slightly) polar, not because it draws S−H bond electrons towards the sulfur atom, but rather because it has a certain electronic structure. The low polarity of cysteine compared to that of the other amino acids in this group makes it a poor hydrogen bond donor and acceptor, although some studies suggest that cysteine is involved in weak- to intermediate-strength hydrogen bonds in proteins, which are structurally and functionally important (see [64] and references therein). In such hydrogen bonds the thiol group usually acts as a donor, where the acceptor is either an O or N-containing group [64] or a π electron cloud in aromatic groups [65,66]. Moreover, it is thought that sulfur can interact with π electrons of aromatic groups also via its low-lying $\sigma*$ orbitals [50,51,67] (see Chapter 1, Subsection 1.3.1.4). Cysteine's low polarity also accounts (in part) for its lower tendency to appear on the surface of the protein [68,69]. Indeed, analysis of 61 protein structures suggests that 90% of cysteine residues are buried inside the protein core [69]. Counterintuitively, the other reason for this tendency has to do with cysteine's high reactivity. Under oxidizing conditions, the thiol groups of two adjacent cysteine residues tend to lose their hydrogen atoms[*1], and form a covalent bond between their two sulfur atoms [71] (Figure 2.8). In biological systems, this reaction is in most cases enzyme-catalyzed (see below). The bond that is created is referred to as a 'disulfide bond', 'disulfide bridges', or simply 'S−S bond'. The dimer formed by the two bonded cysteine residues is called 'cystine'. The S−S group is less polar than each of the free thiol groups [72], and that accounts for the tendency of cysteine groups to appear inside the protein's hydrophobic core rather than on its hydrophilic surface. The ease with which the side chains of cysteine become oxidized also has to do with the dimensions of the sulfur atom. Being large (e.g., in comparison to nitrogen and oxygen), the volume of the sulfur atom allows the negative charge of the reaction's transition state to be efficiently distributed, and that stabilizes it. As discussed in Chapters 1 and 9, stabilization of the transition state increases the reaction rate. The stabilization of the deprotonated form of cysteine also accounts for the fact that this amino acid's intrinsic pKa (value of 8.6) is lower than that of its oxygen-containing 'twin', serine (pKa value of ~13; see Table 2.3).

Disulfide bonds have important implications for protein structure [73]. They contribute to the stabilization of the protein's folded chain, drive its correct folding, and reduce the chance of aggregation by limiting partial unfolding of the chain. However, their distribution in nature is heterogeneous; they are more common in eukaryotes [74] and appear almost exclusively in cell surface, secreted, and mitochondrial proteins [75,76][*2]. This distribution has to do with the conditions required for the formation of disulfides.

In a typical cell, the cytosol and nucleus contain large concentrations of reduced NADPH and *glutathione (GSH)*. These molecules, along with certain enzymes, participate

---

[*1]In the overall reaction one sulfur atom loses a hydride ion (H$^-$) while the other loses a proton (H$^+$). However, the enzymatic mechanism of this transformation is complex and involves different intermediates and cofactors. For example, in sulfhydryl oxidases the reaction involves a flavin cofactor and an internal redox active disulfide group [70]. Also, different oxidants (e.g., O$_2$ and H$_2$O$_2$) may serve as the electron acceptors.

[*2]In some rare cases disulfide bonds can also be found in cytosolic proteins. These include certain enzymes with a thiol oxidation step in their catalytic cycles (e.g., [77]), as well as some redox-regulated proteins (e.g., [78]). Also, disulfide bonds might be formed transiently within the cytosol as part of certain biochemical processes.

in reduction-oxidation (redox) reactions that are important for preventing oxidative damage in cells and tissues [78]. NADPH is also required for some key biosynthetic processes in cells. Redox reactions may proceed either in the direction of reduction or in the direction of oxidation, but as a result of the high NADPH concentrations in the cytosol (or prokaryotic cytoplasm), the dominant direction is reduction. The overall reductive nature of the cytosol is unfavorable for the formation of disulfide bonds. As a result, **stable S−S bonds do not form in cytosolic proteins**, even if the two thiol groups involved are close enough. In contrast, secreted proteins and outward-facing parts of cell surface proteins experience oxidative conditions, which allow disulfide bonds to form. Such bond formation already happens when the protein is processed in the ER lumen, a compartment that is topologically equivalent to the external environment of the cell [79]. There, formation of S−S bonds is catalyzed by two enzymes that act consecutively: *protein-disulfide isomerase (PDI)* [80] and *sulfhydryl oxidase (Ero1)* [81]*1. The electrons released by the formation of the disulfide bond are passed to PDI, then to Ero1, and then to two O_2 molecules, turning them into $2 H_2O_2$. PDI is an isomerase (see Chapter 9); as such, it also allows disulfide bonds initially created between non-native cysteine pairs to exchange until the correct connectivity is achieved [85]. Some protein disulfide bonds may also form in the Golgi system [86]. Reducing agents such as glutathione can be found in the lumen of the ER and Golgi, but their concentrations there are many times lower than those in the cytosol [87]. The presence of disulfide bonds in secreted and cell surface proteins makes sense; it is the extracellular environment that is most dangerous to protein stability, due to the large temperature and pH ranges, as well as the presence of free radicals and other oxidizing agents.

As mentioned above, disulfide bonds in eukaryotic proteins are also created in the mitochondria, specifically, in the intermembrane space (IMS). This is done using a chain of IMS proteins that relays the electrons released by disulfide formation to O_2, turning it into $2 H_2O$ [88]*2. This process allows IMS proteins to fold by creating structural disulfide bonds within them, and it also makes the transport of these proteins into the mitochondria more efficient, as the folded proteins can no longer return to the cytosol through the outer membrane. In Gram-negative bacteria, protein disulfide bonds are formed predominantly within the periplasmic space lying between the inner and outer membranes [74]. In Gram-positive bacteria and in Archaea, however, the exact process by which disulfide bonds are formed is less understood.

Overall, the function of disulfides stresses the evolutionary need for a thiol-containing amino acid. Serine, which is identical to cysteine except for the sulfur atom, cannot form equivalent bonds (i.e., dioxides (O−O)), since the hydroxyl group is a weaker acid than the thiol group. The versatility of cysteine and the sophistication it confers to proteins may explain its relatively late appearance in proteins during their molecular evolution [89,90]*3. Serine, considered to be one of the earliest amino acids to appear in proteins, has similar chemistry to cysteine, but lacks some of its added sophistication and beneficial properties. This may also explain why proteins have evolved to become more cysteine-rich, a trend that still endures [91].

Biochemistry books tend to mention disulfide bonds almost exclusively with respect

---

*1Other ER enzymes implicated in disulfide bond formation include peroxiredoxin [82], vitamin K epoxide reductase [83], and glutathione peroxidases [84].

*2The last two proteins in this system are part of the mitochondrial respiratory chain.

*3This has also been suggested for the aromatic amino acids; they too display physicochemical and functional sophistication, and appeared late in protein evolution (see Subsection 2.2.1.3.4 below).

(a)   (b)

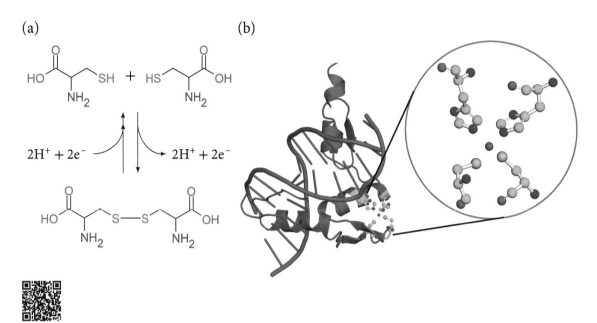

FIGURE 2.8   **Roles of cysteine residues in proteins.** (a) Formation of a disulfide bond between two cysteine amino acids. The thiol (sulfhydryl) groups involved are colored in gold. The overall reaction involves the transfer of two protons and two electrons from the cysteines' thiol groups to an acceptor. When the reaction is catalyzed by sulfhydryl oxidase the acceptor is O_2, which is reduced to H_2O_2 [92] (not shown). As explained in the main text, the mechanism of this reaction is complex and involves different intermediates [70]. (b) Cys as a metal-binding residue. The figure on the left shows a complex comprising a protein (red) and DNA (light blue). The protein contains a 'zinc finger' motif as a way of attaching to the negatively charged DNA molecule. The figure on the right is a magnification of the motif, revealing four residues coordinating the zinc cation, two of which are Cys residues. The zinc atom is presented as a gray sphere, and the residues as balls and sticks, colored according to atom type.

to the stabilization of protein structure. Recently-accumulated data suggest that in some cases disulfides may also be functionally important. This is the case with insulin, an animal protein-based hormone, which plays a central role in regulating metabolism. Insulin is secreted from the pancreas following food consumption, and functions in facilitating the transport of food-derived carbohydrates from circulation into the body's cells. In addition, insulin activates metabolic pathways responsible for carbohydrate utilization inside the cells, for energy or biosynthetic purposes. To the general public, insulin is primarily known as a treatment given worldwide for diabetes. All of the functions of insulin depend on its ability to bind to its cognate cell-surface receptor and activate it. The activation sends a biochemical message into the cell, which either activates or inactivates enzymes involved in the processes mentioned above. The structure of insulin contains three disulfide bonds. Though each bond contributes differently to the stability of the insulin structure, all three are crucial for receptor binding [93].

In the above example, disulfide bonds act indirectly, by allowing a signaling protein to form. However, disulfide bonds have also been implicated in cellular function and regulation more directly, as sensors and signaling agents of oxidative stress [65,94,95]. These functions are facilitated by the capacity of disulfide bonds to form and break reversibly, in accordance with the environmental concentrations of oxidants and reductants, such that the disulfide bond becomes a type of 'oxidative signaling switch'. This role of the disulfide bond may be important for monitoring (and perhaps counteracting) the levels of oxidants both in the external environment of the cell and inside its mitochondria.

Like serine, cysteine is also capable of functioning as a nucleophile in enzyme-mediated catalysis (see Chapter 9 for further details on covalent catalysis). This function is observed, e.g., in enzymes of the *cysteine protease* group, such as *papain* [96], which possess a catalytic array of residues similar to that of serine proteases, but with cysteine as the principal nucleophile [97]. In fact, the thiol group is a stronger nucleophile than the hydroxyl group of serine. This, again, is due to the large size of the sulfur atom, resulting in weaker attraction of outer-shell electrons by the nucleus. Indeed, a survey of the MACiE enzyme database [98], conducted by Thornton and coworkers [99], shows that 36% of the Cys residues in the database act in covalent catalysis (see Chapter 9). Cysteine also functions in enzymatic redox reactions. Most of these enzymes perform electron transfer between substrates using cofactors, which may be organic (e.g., FAD and NAD$^+$) or inorganic (e.g., heme, transition metals). Other oxidoreductases, however, may use a reactive residue instead, usually cysteine. In this respect, the advantage of using cysteine stems from both the high reactivity of the thiol group, and the ability of the sulfur atom to acquire oxidation states ranging between -2 and $+6$ [100]. These properties enable cysteine residues to participate in versatile redox reactions, associated with metabolism, defense against oxidative damage (see Box 2.3), and even cellular signal transduction [101,102].

The reactivity of the thiol group also renders cysteine residues a target for the binding of different atoms and molecules. Such molecules include the *farnesyl* and *palmitoyl* groups, which are lipid chains used by cells as prosthetic groups for the anchoring of certain proteins to the plasma membrane [103]. As we will see later, the anchoring is often a step in a larger signal transduction process. Another example of binding that involves cysteine is the attachment of metals to proteins by coordinate bonds^{*1}. The thiol group of cysteine residues is very efficient in coordinating d-block metals (e.g., zinc, iron, copper) in many proteins and metal-containing enzymes (e.g., [104]). For instance, many DNA-binding proteins use a local structure called a *'zinc finger'* to attach onto the DNA molecule (Figure 2.8b). The zinc finger structure is based on a cationic zinc atom coordinated to several protein residues, one or two of which may be cysteines. The positive charge of the zinc ion, combined with the three-dimensional structure of the 'finger', enables these proteins, which often function as transcription factors, to bind functional regions within the DNA molecule [105]. Interestingly, in other proteins, zinc fingers may participate in protein-protein interactions [106], again illustrating the evolutionary tendency for the same successful 'tricks' to be selected for different purposes. It should be mentioned, though, that the chemical reactivity of cysteine might sometimes harm the protein, e.g., when it serves as the target of toxic compounds. This is the case with *iodoacetic acid (IAA)*, which physically blocks thiol groups, and may harm the activity of enzymes whose catalytic mechanisms include cysteine residues. Indeed, IAA is known to inhibit glycolysis by blocking a catalytic cysteine residue in the enzyme glyceraldehyde-3-phosphate dehydrogenase [107].

---

^{*1}Coordinate bonds are covalent bonds in which the shared electrons come from one of the two bonded atoms.

BOX 2.3 CYSTEINE AS AN ANTIOXIDANT

Cells and tissues are constantly exposed to oxidizing agents that inflict damage on cellular components and may endanger the health of the entire organism. These *oxidants* may come from two sources. The first is endogenous, and includes free radicals that are produced either as byproducts of metabolic processes or as a means of defense against pathogens. For example, the *superoxide radical* (O_2^-) is constantly formed as a byproduct of mitochondrial respiration. This molecule, as well as another oxidant, *hydrogen peroxide* (H_2O_2), is also produced by phagocytes as part of a mechanism of killing internalized bacteria. Interestingly, hydrogen peroxide has recently been identified as an intercellular chemical messenger [108]. The other source of harmful oxidants is exogenous, and usually comes from industrial products and byproducts in the air, food, and water.

A rise in the levels of oxidants in our body creates what scientists refer to as 'oxidative stress'. Oxidative stress may lead to the damaging of biomolecules, as follows:

1. Proteins: Oxidation of body proteins may harm their function directly, by chemically changing catalytic residues, or indirectly, by inducing structural changes incompatible with the proteins' biological function. Protein cofactors may also be oxidized, leading to functional impairment. Such impairment occurs, for example, in diabetes, when hemoglobin is oxidized. *Diabetes mellitus* involves the massive production of *ketone bodies*, which create oxidative stress both directly and indirectly. The oxidation of the *heme* cofactor of hemoglobin by these oxidants renders the protein inactive, which leads to secondary damage in tissues due to hypoxia. From there, the way to tissue necrosis, gangrene, and even organ failure is not long.

2. DNA: Oxidation of the cell's genetic material may result in a mutation, which in turn can be lethal or lead to cancerous transformation of the cell.

3. Plasma membrane: Oxidation of either the lipid or protein components of the plasma membrane may compromise its function as a selective barrier between the cytoplasm and the extracellular environment. In addition, by reducing the flexibility of the membrane, such oxidation may lead to membrane disintegration and to premature death of the cell. This phenomenon is observed in people deficient in the enzyme *glucose-6-phosphate dehydrogenase* (G6PD), which normally functions in the production of NADPH, an antioxidant (see the following paragraph for details). When these people are exposed to strong oxidants (e.g., in food or drugs), they may develop *oxidative stress* due to the low levels of NADPH. The stress is particularly harsh in red blood cells, which rely solely on G6PD for producing NADPH. The resulting damage to the plasma membranes of the red blood cells leads to their death, which manifests physiologically as massive bleeding. This situation is clinically referred to as 'hemolytic anemia'.

As mentioned above, the human body is constantly exposed to oxidants. However, the body is innately equipped with effective ways of dealing with the dangers of oxidative stress. These are collectively referred to as *'antioxidants'*. Some of the antioxidant mechanisms involve the amino acid Cys. The two main systems in this respect are the thioredoxin and the glutathione systems [101]. Both systems use reduced thiol groups (SH) of two cysteine residues, in order to reduce biomolecules damaged by oxidative stress. The first stage of this process includes the reduction of the oxidized biomolecule, coupled to the oxidation of the two thiol groups into a disulfide (S—S) bond. In the second stage, the thiol groups are reconstructed back to their reduced form by an enzymatic process that uses NADPH as a coenzyme. Thus, the constant activity of the antioxidation systems relies on the constant production of NADPH.

Thioredoxin (Trx) is a common enzyme appearing in virtually all types of life forms. In animals, it is present in the cytosol, nucleus, and mitochondria [109]. Trx represents an enzyme family of the same name, the members of which are either antioxidants or regulators of cellular and metabolic processes [110]. The enzyme includes two adjacent cysteine residues that act on oxidatively damaged proteins. The antioxidation process can be summed up as follows:

$$\text{Thioredoxin-2·}\mathbf{SH} + \text{Protein}_{(ox)} \longrightarrow \text{Thioredoxin-}\mathbf{S-S} + \text{Protein}_{(red)} \qquad (2.3.1)$$

$$\text{Thioredoxin-}\mathbf{S-S} + \text{NADPH} \longrightarrow \text{Thioredoxin-2·}\mathbf{SH} + \text{NADP}^+ \qquad (2.3.2)$$

(where 'Protein$_{(ox)}$' and 'Protein$_{(red)}$' represent the oxidized and reduced forms of the target protein, respectively, and '2·$\mathbf{SH}$' and '$\mathbf{S-S}$' represent the reduced and bonded forms of the cysteines).

The other antioxidant system includes the short peptide glutathione. This molecule includes only one cysteine. Therefore, two molecules are required for the antioxidation process, which is overall very similar to the one described for Trx:

$$2\mathbf{GSH} + \text{Protein}_{(ox)} \longrightarrow \mathbf{GS-SG} + \text{Protein}_{(red)} \qquad (2.3.3)$$

$$\mathbf{GS-SG} + \text{NADPH} \longrightarrow 2\mathbf{GSH} + \text{NADP}^+ \qquad (2.3.4)$$

(where GSH is the reduced form of glutathione, and GS—SG is the oxidized form, including two peptides connected by a disulfide bond).

Interestingly, glutathione's cysteine residue tends to be in its reduced form under cellular conditions. However, due to the large concentration of the peptide in the cytosol, the entire reaction is shifted towards the products, which requires the cysteine groups to undergo deprotonation.

2.2.1.3.3.3. Glutamine and asparagine

Glutamine and asparagine each contain a *carboxamide* ($CONH_2$) group within their side chains. In fact, the only difference between their side chains is that glutamine contains one extra methylene group ($-CH_{2-}$), making it more flexible than asparagine. The carboxamide group can serve as both hydrogen bond donor (two bonds donated by the amino subgroup[*1], see Figure 2.6a) and acceptor (two bonds accepted by the carbonyl subgroup), and indeed, glutamine and asparagine are often involved in hydrogen bond networks within proteins and with the solvent. Like serine and threonine, asparagine also serves as a target for post-translational glycosylation. However, in the case of asparagine, the glycosyl group attaches to the carboxamide's nitrogen (*N*-linked glycosylation) rather than to a hydroxyl oxygen (*O*-linked glycosylation).

2.2.1.3.4 Polar-charged (ionizable) amino acids

Polar-charged amino acids, which make up (on average) ~30% of the amino acids in proteins[91], include glutamate (Glu; E), aspartate (Asp; D), histidine (His; H), lysine (Lys; K), and arginine (Arg; R). Each of these contains an electrically charged side chain at physiological pH. Like other polar residues, these also appear mostly on the protein surface, where they interact electrostatically with other residues and/or with the surrounding water. Their net charge increases their reactivity, which explains why they are also very common as catalytic residues in enzymes[44,112]. In this capacity, charged residues may act as nucleophiles, electrostatically stabilize the transition state, polarize the substrate to make it more labile to catalysis, and even activate other catalytic residues (see Chapter 9).

The side chain of a polar-charged amino acid contains a chemical group, which can undergo protonation and deprotonation. The dominant ionization state is determined according to the difference between the pH of the solution and the p*Ka* of the group, as explained in Box 2.2. In biological systems, the pH is almost always around 7. Due to the chemical differences between the five polar-charged residues, they have different p*Ka* values, which means that some of them are protonated, whereas the others are deprotonated. While each of the five residues carries an electric charge under these conditions, the charge is negative in some and positive in the others. On the basis of these differences, the charged residues can be divided into two groups: acidic and basic.

2.2.1.3.4.1. Acidic amino acids

The acidic group of polar-charged amino acids includes glutamate and aspartate. Each of these amino acids is referred to as 'acidic' because its side chain contains a carboxylic group (COOH), the p*Ka* of which is lower than physiological pH (4.3 in glutamate and 3.9 in aspartate). As a result, glutamate and aspartate tend to function as acids, i.e., to deprotonate at pH 7. As in the case of asparagine and glutamine, aspartate and glutamate differ only in one side chain methylene group. In proteins, these residues tend to be involved in ionic interactions with basic residues, with positively charged groups that are present in the ligand or substrate, and with metal cations, Ca^{2+} in particular. Despite their low p*Ka*, glutamate and

---

[*1]The nitrogen's lone electron pair is a poor hydrogen bond acceptor because it interacts with neighboring π electrons to form the group's electronic resonance[111].

aspartate are sometimes capable of taking part in general acid-base catalysis[*1], as in the case of the enzyme triosephosphate isomerase (TIM) [113]. Specifically, a glutamate residue in the active site of this enzyme has been shown to function as a base, abstracting a proton from the substrate in the first step of catalysis.

Glutamate and aspartate may serve as targets for post-translational modifications; aspartate may undergo hydroxylation on C_β (*β-hydroxylation*), whereas glutamate may be *γ*-carboxylated. The latter is particularly important in blood clotting proteins (e.g., prothrombin), in which the added carboxylate group enhances the Ca^{2+}-binding capability of glutamate. The binding of Ca^{2+} ions is required for the regulation of the clotting process.

2.2.1.3.4.2. Basic amino acids

This group includes lysine, arginine, and histidine. The members of this group are referred to as 'basic' because their respective side chains each include a nitrogen-containing group whose p*Ka* is higher than physiological pH. As a result, these residues tend to act as bases, i.e., to hold on to their protons at pH 7. The protonated state of these residues is positively charged. Lysine's side chain contains an amino group on C_ε[*2], with a p*Ka* of 10.4 (Table 2.3). Arginine's side chain contains a *guanidine* group[*3] with a p*Ka* of 12.3 and possibly higher[*4]. Histidine is a more complex case (see below). As the main carriers of positive charge, lysine and arginine tend to interact with glutamate and aspartate, as well as with negatively charged groups in the protein's ligand or substrate (e.g., phosphoryl groups [114]). Indeed, in enzymes the two residues are often involved in the electrostatic stabilization of reaction intermediates [99].

The side chains of both lysine and arginine are chemically reactive and tend to participate in the formation of covalent bonds, which are functionally important in proteins. For example, the *ε*-amino group of lysine can undergo the following reactions (see Section 2.6 below for details):

- Schiff base ($C=N^+$) formation: This bond is formed between lysine side chains and aldehydes. The reaction is used for attaching aldehyde-containing prosthetic groups to proteins. Such groups include the enzyme cofactors *biotin* and *pyridoxine*, as well as the pigment *retinal*, which is part of the visual photoreceptor *rhodopsin*.

- Isopeptide bond formation [115][*5]: These bonds form between the lysine side chain and the carbonyl group of a nearby asparagine or aspartate side chain. Such bonds have been found in cell surface proteins, elements of bacterial pili, cellular assembly proteins, and even viral capsid proteins. The role of isopeptide bonds seems to be similar to that of disulfide bonds: to confer resistance to mechanical and thermal stress, and in some cases even protect against proteolysis (enzymatic degradation of proteins).

[*1]Catalytic mechanisms in enzymes that involve the transfer of protons to/from active site amino acids (see Chapter 9 for details).

[*2]Referred to as an 'ε-amino' group.

[*3]The positively charged group, which is dominant under physiological pH, is called '*guanidinium*'.

[*4]Recent potentiometric and NMR spectroscopic measurements suggest that the p*Ka* of arginine's side chain (free or in a tripeptide) is actually closer to 14 [63]. In any case, the high p*Ka* value of this side chain results from the stabilization of the positive charge by electronic resonance (see Chapter 1, Subsection 1.3.1.4).

[*5]The name refers to the similarity of these bonds to the amide bonds that attach amino acids to form the protein chain, and which are called 'peptide bonds' (see Subsection 2.2.2 below).

- Interprotein crosslink formation: These bonds form between lysine side chains in one protein molecule, and aldehyde groups in another protein molecule. Such bonds stabilize structural proteins such as collagen and elastin, which form large complexes (see Section 2.7 below). In this case, however, the lysine side chain must undergo oxidation or hydroxylation to form the bonds.

- Acetylation, ubiquitination, and SUMOylation: These group-attachment reactions are used by cells to regulate key processes. The reactions are discussed in detail in Section 2.6 below.

The side chain of lysine may also function as a general acid or base in enzymatic acid-base catalysis [44]. For example, in the enzyme L-asparaginase, the group's function as a general base leads to the activation of an adjacent threonine residue, making the latter an efficient nucleophile. It is not entirely clear how lysine's high pKa allows it to function as a general acid, but it probably has to do with its immediate environment within the enzyme's active site.

The guanidine group of arginine has a particularly complex nature. This group can serve as a donor of five hydrogen bonds (Figure 2.6a), and its electronic resonance effectively distributes the positive charge all over the group. The resonance also makes the guanidinium group planar and rigid. In these qualities, arginine differs from lysine, whose charge is much more localized, and which is more flexible. Accordingly, the binding capabilities of the two amino acids are different from each other; whereas both often bind and stabilize negatively charged species in binding sites, arginine is used where the stabilizing charge needs to be spread over a large rigid plane (e.g., to stabilize large charged species such as a phosphate group), whereas lysine is used where a flexible, point-like stabilizing charge is needed. Arginine has many biological roles, such as stabilization of negatively charged ligands and enzyme substrates and transition states (see Chapter 9), pKa modulation of adjacent residues, post-translational modifications, and, to a lesser extent, acid-base catalysis (see Chapter 9).

Histidine's side chain contains an *imidazole* group, which is heterocyclic and includes two nitrogen atoms ($N_{\delta 1}$ and $N_{\varepsilon 2}$). The imidazole is planar and possesses electronic resonance; together, these two qualities render the histidine side chain aromatic. Hence, the electrons of the two nitrogen atoms in the ring may exist in two alternative states. In the 'frozen' state shown in Figure 2.5c, $N_{\varepsilon 2}$ is bound to the ring's carbon atoms via two single bonds. This allows $N_{\varepsilon 2}$ to form a third bond with a hydrogen atom, and still remain electrically neutral. $N_{\delta 1}$ is also bound to two carbon atoms, but since one of the bonds is double, its protonation adds a positive charge to the group. The tendency of the imidazole group to deprotonate is stronger than the deprotonation tendencies of lysine's ε-amino group and arginine's guanidine group; accordingly, the pKa of the histidine side chain (6.5) is lower than the pKa values of the side chains of lysine and arginine (Table 2.3). This means that at physiological pH, ~50% of the histidine residues are protonated[*1], making them both potential donors and acceptors of protons (and hydrogen bonds[*2]). **Indeed, histidine is used extensively by enzymes as both a general acid and a base during catalysis** [44,99] (see Chapter 9, Subsection 9.3.3.5 for further details). The best known example is probably the cat-

---

[*1]Note, however, that the immediate environment of histidine inside proteins residues affects their protonation state to a large extent [116] (see also below).

[*2]In the electrically neutral form of histidine the protonated nitrogen ($N_{\varepsilon 2}$ in Figure 2.5c) serves as a hydrogen bond donor and the deprotonated nitrogen ($N_{\delta 1}$ in Figure 2.5c) serves as a hydrogen bond acceptor. In the positively charged form of histidine both nitrogen atoms serve as donors (Figure 2.6a).

alytic mechanism of serine proteases [58], mentioned above. Due to the electronic resonance in the imidazole group, any of the nitrogen atoms may be the one that undergoes protonation or deprotonation, although with different probabilities. The nitrogen that does not bind a proton is a nucleophile, and that allows histidine (like cysteine) to bind metal cations, such as Zn^{2+}, Cu^{2+}, and Fe^{2+}, via coordinate bonds. Because of all these properties, histidine is the residue most frequently involved in enzyme-mediated catalysis, and the one that has the highest tendency to participate directly in catalysis [44,112]. Interestingly, the enzymatic biosynthesis pathway of His appeared relatively late in evolution (~3.2 billion years ago), which may explain the late appearance of proteins that depend on it (e.g., metalloproteases, whose binding sites contain His-coordinated metals) [117].

2.2.1.3.4.3. Environmental modulation of residual pKa

The discussion above demonstrates the importance of residues' ionization states in protein structure and activity [118]. Most noticeably, changes in the ionization states of enzymes' catalytic residues can lead to changes in these enzymes' activity. As Table 2.3 explains, the ionization state of any chemical species capable of undergoing deprotonation is determined by the difference between its p*Ka* and the pH of the solution. Despite the general homeostasis present in living organisms, pH levels within cells, tissues, and organs may differ within a limited range. Heterogeneity of pH values is observed in different body systems, and also at the subcellular level. Indeed, the cytosol, ER, and nucleus all have a pH of 7.2, whereas the pH values for mitochondria, peroxisomes and lysosomes are 8.0, 7.0, and 4.7, respectively [119] (Figure 2.9). In fact, the pH value may change even within different compartments of the same organelle. For example, switching from the *cis* part of the Golgi apparatus to the *trans* is accompanied by a drop in pH, from 6.7 to 6. Similarly, a drop of 1.6 pH units has been measured during the formation of lysosomes from early endosomes [119].

The p*Ka* values of polar-charged residues are also amenable to change. That is, the intrinsic p*Ka* values of individual amino acids might change when they become incorporated as residues within a protein structure, due to the change in the chemical nature of their im-

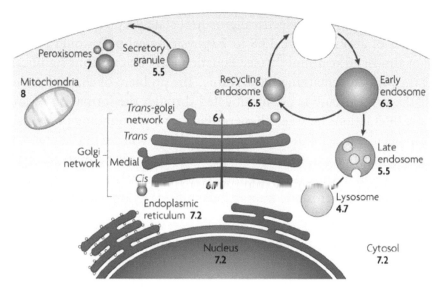

FIGURE 2.9   **pH values of the different compartments in a mammalian cell.** Taken from [119]

mediate environment in comparison to bulk water [120–123]. This change, which may reach 8 pH units [120], can be determined, e.g., by measuring the pH dependence of chemical shifts using NMR. Such shifts in pH tend to result primarily from the following environmental effects [61]:

1. *Desolvation (the Born effect)*

    Ionizable amino acids are usually located on the surface of proteins. However, statistical surveys of numerous protein structures show that some of these amino acids do become buried inside the protein core. This occurs because of structural and functional requirements of the protein (see Chapter 4 for further details). For example, such amino acids may appear in the active sites of enzymes, the binding sites of receptors or antibodies, etc. Moreover, mutational studies show that the burial of a charged amino acid inside a protein core does not necessarily destabilize the protein's structure or even abolish its activity [121]. As explained in Chapter 1, the exposure of electric charges to a hydrophobic environment is highly unfavorable [124]. This is because the nonpolar environment lacks the electric dipoles needed to electrostatically mask the buried charges [125]. Since ionizable amino acids do become buried in a nonpolar core during protein folding, the tolerance of proteins to this process is likely to result from the failure of the ionizable amino acids to become ionized. Indeed, the large energy cost of burying the charged amino acids creates an opposing effect to the ionization of their side chains. As a result, an ionizable amino acid that becomes buried tends to stay in an electrically neutral state, which is expressed as a shift in pKa as compared with the amino acid's intrinsic pKa value. **In the case of an acidic amino acid, the shift in pKa will be upwards, whereas in the case of a basic amino acid, the shift will be downwards.** Such a shift has been demonstrated, e.g., in the staphylococcal nuclease, where a Val→Asp mutation at position 66 inside the hydrophobic core of the enzyme leads to a five-unit increase in the pKa_{int} of the latter, keeping it electrically neutral [126]. pKa shifts of similar magnitudes were obtained in another study, which introduced 22 lysine residues into internal positions in the same protein [123].

    Arginine behaves somewhat differently from the other ionizable residues. The pKa of this amino acid is greater than 12 (possibly closer to 14 [63]), which makes downwards pKa shifts that lead to an electrically-neutral state very rare [127]. Indeed, a decrease in the pKa of arginine from 13.8 to 7 (physiological pH) would require an energy input of almost 10 kcal/mol [63] (see Equation (2.1) below), thus significantly destabilizing the protein containing the arginine. The problem of charged arginine residues inside the protein core is often mitigated through the formation of favorable electrostatic interactions between these residues and surrounding polar residues and/or water molecules (see following subsection). As we will see in Chapter 4, this is a general strategy that proteins use to stabilize buried charged amino acids. However, it is especially useful in the case of arginine residues, which are able to participate in numerous types of electrostatic interactions, including ion pairing, hydrogen bonding, and π-interactions [63].

    The tendencies mentioned above are also observed in amino acids that reside on the surfaces of integral membrane proteins (see Chapter 7), as these amino acids are normally exposed to the highly nonpolar core of the lipid membrane. Again, such exposure of charged amino acids would result in a large energy penalty, which would typically drive them towards an electrically-neutral state.

2. *Coulomb interactions*

When the ionizable amino acid is in proximity to another charged species (amino acid, metal, etc.), the pairwise electrostatic (Coulomb) interaction between the two species may affect the ionization state of the ionizable amino acid. This may happen in two ways:

(a) The ionization of the amino acid creates a charge opposite to that of the adjacent species. In this case, the favorable Coulomb interaction between them will encourage the ionization process. **Accordingly, the pKa of the ionizable amino acid will shift. Specifically, it will decrease if the residue is acidic, and will increase if the residue is basic**.

(b) The ionization of the amino acid creates a charge of same sign as that of the adjacent species. In this case, the unfavorable Coulomb interaction will oppose the ionization process. Here, the effect will be similar to that of charge desolvation: **an increase in pKa will be observed in acidic residues, and a decrease in pKa will be observed in basic residues**.

The effect of Coulomb interactions on pKa_{int} is particularly important in the case of ionizable amino acids that serve catalytic roles [128]. This is nicely demonstrated in the case of serine and cysteine proteases. In these two enzyme families, the ability of the catalytic serine or cysteine residues to act as nucleophiles requires them to undergo deprotonation and become negatively charged. Since both serine and cysteine have high pKa values (~13 and 8.6, respectively; see Table 2.3), deprotonation is facilitated only with the help of a nearby histidine, which acts as a general base. Indeed, the transfer of a proton from the Ser or Cys to the adjacent His charges the two residues oppositely, and the resulting (favorable) Coulomb interaction stabilizes their ionizable forms. Histidine is not the only amino acid that can serve in this capacity; cases are known in which a shift in the pKa_{int} of acidic residues is facilitated by nearby lysine or arginine side chains, which stabilize the negative charge. This happens, for example, in some redox enzymes that rely on the deprotonation of a catalytic cysteine, forming a negatively charged thiolate group [101]. Another example of the effect of a Coulomb interaction on pKa_{int} is given in studies of RNase Sa [129].

Such effects can be demonstrated quantitatively (although inaccurately) using simple electrostatics [130]; assuming that two amino acids with a single opposite charge are positioned 5 Å apart in an environment of dielectric 4, their interaction energy can be calculated using Coulomb's law (see Equation (1.2)). The calculation yields an energy value of −33.2 kcal/mol. The interaction requires that the two residues be ionized, although they may be uncharged, depending on their pKa and the pH of the environment. Keeping the two amino acids charged may require a shift in their pKa, which involves a free energy penalty. If the penalty is overcompensated for by the attractive interaction, the two amino acids will remain charged and form a salt bridge. The relationship between free energy and pKa is given by:

$$\Delta G^0 = 2.3RT \times pKa \qquad (2.1)$$

Thus, in the above example, an attractive energy of 33.2 kcal/mol can facilitate a pKa shift of up to 12 units.

In sum, if the interaction between the two amino acids in the salt bridge is strong enough, it can force the residues to remain in their ionized states, even if these states are incompatible with the residues' inherent protonation or deprotonation tendencies in a particular environment. It is noteworthy that because of the large desolvation free energy penalty associated with the transfer of charges into hydrophobic environments, it is sometime beneficial for the amino acids to remain in their neutral forms, replacing the salt bridge with hydrogen bond(s). This is especially true for titratable residues that are buried in the hydrophobic core of the protein.

3. *Hydrogen bonds*

The effect of hydrogen bonding on the pKa of an ionizable amino acid is similar to the Coulomb effects described above. That is, when the interaction involving the charged amino acid is favorable, it will encourage the latter to ionize, and *vice versa*. However, unlike the simple Coulomb interaction, optimal hydrogen bonds do not necessarily involve a charged amino acid; the degree to which a certain hydrogen bond is favorable depends on the unique chemistry of the interacting species. For example, if the interaction is to occur between a serine and an adjacent carbonyl group, the former must be protonated, i.e., uncharged. Thus, **if the interaction is more favorable with the protonated form of the amino acid, its pKa_{int} will likely increase, and if the interaction is more favorable with the deprotonated form, its pKa_{int} will decrease.** Examples of such effects can be found in studies of RNase T1 [131]. Other examples are given by Li and coworkers [132]. These studies suggest that the pKa shifts induced by a single hydrogen bond can attain a value of up to 1.6 for carboxyl groups, and can reach even higher values for sulfhydryl groups.

The three factors described above are not mutually exclusive. That is, an ionizable amino acid inside a protein may experience two or all three of these effects. Since desolvation, Coulomb interactions and hydrogen bonds are all electrostatic in nature, the combined effect experienced by the amino acid can be calculated by solving the Poisson-Boltzmann equation (PBE) [133,134], which accounts for both pairwise and polarization effects (see Box 1.3 in Chapter 1). The resulting shift in the amino acid's pKa_{int} can then be derived from the calculated data [135]. Indeed, most PBE solvers today have a built-in pKa calculation mode. For the inexperienced scientist or student we recommend two user-friendly web servers that are accessible online: PDB2PQR [136]*1 and H++ [137,138]*2.

2.2.1.3.4.4. Effects of the residues' ionization state on protein properties

We have seen how the ionization states of polar-charged residues affect local, often catalytic properties of proteins. In addition, a residue's ionization state may have more global effects. For example, in an environment with low pH, most acidic residues are electrically neutral, whereas most basic residues are positively charged. The overall protein charge will therefore be positive. In high-pH environments, the exact opposite occurs: Most acidic residues are negatively charged, most basic residues are electrically neutral, and the overall protein charge is negative. In both these extreme cases, the *water solubility* of the protein is expected to be high, due to electrostatic repulsion between adjacent protein molecules of the same

---

*1 PDB2PQR: http://nbcr-222.ucsd.edu/pdb2pqr_2.0.0
*2 H++: http://biophysics.cs.vt.edu/H++

type of charge (positive-positive at low pH and negative-negative at high pH). Thus, protein solubility is maximal at extreme pH values, and moderate around the middle of the scale [139]. Physiological pH is roughly around the middle (~7 [119]), which may seem odd at first, since it means proteins evolved to be less soluble. In fact, this evolutionary pattern makes a lot of sense: At physiological pH the solubility of proteins is low enough to make their interactions reversible (in accordance with biological needs), but not low enough to promote nonspecific interactions, which would lead to sedimentation.

In addition, residues' ionization states at extreme pH values are expected to be different from their ionization states under physiological conditions. Thus, exposure to such extreme environments can produce two unwanted results. First, it may disrupt polar interactions that normally stabilize the structure of the protein. Second, it may abolish functional capabilities that depend on the residues' ionization states.

To summarize, the evolutionary process has led to the emergence of amino acids with pKa values that provide optimal stability, solubility, and functionality, under physiological conditions. As mentioned above, some proteins reside and function in subcellular compartments whose pH values are different from physiological pH (e.g., pH = ~5 or lower in lysosomes [119]). The amino acid compositions of such proteins allow them to remain functional and stable at the pH ranges to which they are exposed. Correspondingly, at physiological pH these proteins are less active. However, not only does this phenomenon not harm the function of the cell, it is crucial for its existence. Lysosomal enzymes are hydrolytic, i.e., they specialize in degrading different molecules [140]. While this function is important for the cell's ability to dispose of obsolete components and internalized pathogens, the hydrolytic activity is potentially hazardous to the cell. Leakage of such enzymes into the cytosol has the potential to initiate degradation of cellular metabolites and cause damage to other organelles. Fortunately, lysosomal enzymes are optimally active only at a pH range of 5.2 to 5.5 [140], which means that even if they leak into the cytosol, the amount of damage they might inflict is minimal.

2.2.1.3.5 Aromatic amino acids

This group includes phenylalanine (Phe; F), tyrosine (Tyr; Y) and tryptophan (Trp; W).

2.2.1.3.5.1. Common properties

The aromatic amino acids are named so because their respective side chains each include an aromatic group: *benzene* in phenylalanine, *phenol* in tyrosine, and *indole* in tryptophan. This means that polarity, which is the principal classification parameter of amino acids, is not considered in the aromatic group. Indeed, phenylalanine is overall nonpolar, whereas tyrosine and tryptophan include polar groups (OH and NH, respectively). Since both also have large side chains, they are also capable of nonpolar interactions. This duality allows the aromatic residues to appear in different locations within proteins and to interact with different chemical groups. Another common trait derived from the aromatic nature of these residues is their ability to absorb ultra-violet radiation [141].

Aromatic residues tend to appear in pairs within proteins, where they interact with each other [142]. Structural analysis has shown that aromatic interactions have a strong tendency to appear in two forms: offset stacking, and perpendicular [143] (Figure 2.10). As explained in

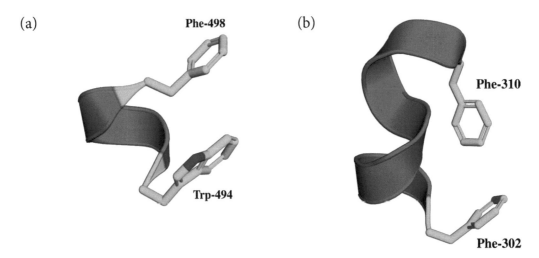

FIGURE 2.10 **Common π-π (aromatic) interactions in proteins.** (a) Offset-stacking (PDB entry 1myr); (b) Perpendicular (PDB entry 1hpm). The two examples are based on Figure 2 in [147]. The atoms are colored according to type (see Figure 2.5). Atoms and bonds of the interacting amino acids are shown as sticks, whereas the main chain connecting them is presented as a ribbon.

Chapter 1, an aromatic ring has electron 'clouds' above and below the ring plane, which render it partially negative. When the electron distribution in the various orbitals is represented as point charges, the interaction between aromatic residues appears to be electrostatic in nature [143], yet short-ranged, and proportional to r^{-5}. Such interactions, referred to as 'π-π interactions', appear to be responsible for the presence of aromatic clusters within proteins, as well as for the stacking of aromatic rings with amide groups that also contain delocalized π electrons. It should be mentioned that histidine is also capable of π-π interactions. Indeed, histidine is in many ways an aromatic amino acid, but is classified as a charged one because of its high polarity. The high electron density of aromatic rings enables them to function as hydrogen bond acceptors as well, although such interactions are only half as strong as regular hydrogen bonds [144]. In addition to π-π interactions, the aromatic rings of phenylalanine, tyrosine, and tryptophan may also participate in *cation-π interactions*, in which their partially negative electron clouds interact electrostatically with the cationic side chains of lysine or arginine. Such interactions are also important for protein-protein binding [145]. Finally, aromatic amino acids are enriched in regions of proteins that function as gates, e.g., in ion channels, transporters, and enzymes [146]. Their prevalence in such structures probably has to do with the planarity and rigidity of their side chains, which provide the specific geometry required for these structures' functionality. **Aromatic amino acids appeared relatively late during evolution [89,90], and the variety of roles they carry out is probably a direct result of their unique structural and electronic properties (see above and in Subsection 2.2.1.3.4.3 below), which make the interactions in which they are involved highly geometry-dependent.** As explained in Chapters 4 and 8, these geometry-dependent interactions are important for determining proteins' unique structures and functions. Indeed, the sophistication of aromatic amino acids may explain why they are believed to be among the last amino acids to appear in proteins during evolution.

2.2.1.3.5.2. Tyrosine

The side chain of tyrosine includes a phenol ring with a hydroxyl group, which can participate in hydrogen bonds, both as a donor or an acceptor (Figure 2.6a). As we have seen in the cases of serine and threonine, hydroxyl groups do not tend to deprotonate and ionize under physiological pH. However, in the case of tyrosine, the aromatic ring stabilizes the *phenolate* ion, which lowers the side chain's pKa to 9.8 (Table 2.3). Like serine and threonine, tyrosine can also become phosphorylated on its hydroxyl group; such phosphorylation usually serves regulatory purposes. Animal proteins can be phosphorylated by two different kinases. The first acts on serine and threonine side chains, whereas the other acts on tyrosine side chains alone. In addition to phosphorylation, tyrosine residues may also undergo *sulfation*, i.e., the attachment of a sulfate group. This modification is thought to be important in protein secretion, viral entry into cells, and metal binding [148].

Like the other hydroxyl-containing amino acids, Tyr may participate in covalent catalysis as well, though it has a stronger preference for other catalytic mechanisms, such as stabilization of intermediates, acid-base catalysis, and mechanisms involving hydride ions [99]. Unexpectedly, in some enzymes Tyr has also been found to undergo a process in which it is converted into a stable, catalytically active radical (see [4] and references therein). This finding is very interesting, as evolution has assigned radical-based catalytic mechanisms to a specific, non-protein, organic cofactor, namely, cobalamine (a vitamin B_{12} derivative). In the enzyme ribonucleotide reductase, the tyrosyl radical is formed as a result of a reaction between iron ions and oxygen [149].

2.2.1.3.5.3. Tryptophan

Among all amino acids, tryptophan has the largest side chain. It contains an indole group, which itself consists of a benzene ring fused to a *pyrrole* ring. The non-bonding pair of electrons on the indole's nitrogen ($N_{\varepsilon 1}$) contributes to the aromaticity of tryptophan, and therefore, $N_{\varepsilon 1}$ is not basic. That is, it cannot bind an additional proton as do regular amino groups, and it cannot serve as a hydrogen bond acceptor. Still, the single proton bound to this nitrogen allows it to serve as a hydrogen bond donor (Figure 2.6a). In addition, the large surface area of the indole group allows the tryptophan to participate in van der Waals and nonpolar interactions. Because of these properties, tryptophan residues tend to appear in binding sites of enzymes and antibodies, where they take part in the design of the sites' geometry, as well as in protein-ligand interactions [150–153]. In addition, tryptophan residues may participate in more specific functions, such as catalysis and electron transport. As an aromatic residue, the tryptophan can absorb ultra-violet radiation. However, in contrast to phenylalanine and tyrosine residues, it can also emit fluorescent radiation following excitation with the right wavelength. The fluorescent emission of tryptophan depends on the immediate environment of the residue, and biochemists use this phenomenon to track structural changes or ligand binding events in proteins. Since tryptophan (along with cysteine) is the least common amino acid in proteins [46], it may be used to focus on specific regions in the protein when studying the above processes.

2.2.1.4 Amino acid derivates in proteins

Individual (non-protein) amino acids and their chemical derivates can be found in nature in numerous forms. Most of these molecules are chemically derived from the 20 basic types of protein-associated residues. In some cases, a derivate is produced by removing the amino group of a molecule, leaving a carbohydrate skeleton. Such metabolites — including pyruvate, which is formed from alanine; oxaloacetate, which is formed from aspartate; and α-ketoglutarate, which is formed from glutamate —usually serve as metabolic intermediates. As such, they can either be fully oxidized to produce energy[*1] or be converted to other molecules. Yet, other amino acid derivates play important biological roles, as summarized in Table 2.4. For example, the hormone *epinephrine*, the neurotransmitter *serotonin*, and the allergy mediator *histamine* are derived from tyrosine, tryptophan, and histidine, respectively. Tyrosine is also the source for the *thyroid* hormones T_3 and T_4, which are responsible for regulating metabolic rate. Another interesting amino acid derivate is *allicin*, which is produced from cysteine by the enzyme *alliinase* in garlic and onion[*2] [154]. Allicin has many biological effects, including antimicrobial and anti-cancer activity. Therefore, it has been explored as a possible pharmaceutical agent, e.g., against malignant tumors [155]. Allicin is a *reactive sulfur species* whose antimicrobial and anti-cancer activity is manifested in the promotion of oxidative stress. In normal eukaryotic cells, the oxidative effects of allicin are counteracted by the glutathione-NADPH system (see above), which explains why these cells are unharmed by allicin (although they are biologically affected by it in other ways [154]).

There are also some proteins that contain amino acid derivates (Figure 2.11). These derivates can be grouped into two types:

1. *Derivates formed after translation:* This group includes amino acids that are incorporated by the ribosomal system into proteins but are subsequently subjected, after the translation process, to covalent modifications by various enzymes. One of the well-known examples is *hydroxyproline*, produced by hydroxylation of proline on C_4, in structural proteins of the extracellular matrix. The latter proteins include *collagen* of animal tissues [87], as well as some proteins that reside inside the cell walls of plants (especially extensin) [156]. The importance of proline hydroxylation for collagen will be explained later when we discuss structural proteins. Another example, which was mentioned above, is *γ-carboxyglutamate*, produced by carboxylation of C_γ of glutamate residues [87]. As explained, this derivate, which appears in Ca^{2+}-binding proteins, contains two side chain carboxylate groups, which enhance the residue's cation-binding capability. The binding is biologically important; certain proteolytic enzymes that participate in mammalian blood clotting (*prothrombin, profactor IX, profactor X* [157]), and that include this modification on 10 to 12 glutamate residues, use the bound Ca^{2+} to adhere electrostatically to the negatively charged membranes of blood platelets. There, these proteases form complexes with other proteins, and the ensuing chain of enzymatic reactions results in the formation of a blood clot. The carboxylation of glutamate residues is carried out by the enzyme *γ-glutamyl carboxylase*, which uses *vitamin K* as a cofactor and O_2 as a co-substrate. Indeed, vitamin K

[*1]While amino acids are regularly oxidized in cells, ATP production from these molecules is less efficient compared with ATP production from carbohydrates and lipids; therefore, amino acids are seldom used as fuel for energy production.

[*2]This is the overall process. The alliinase reaction creates allicin specifically from the cysteine derivates alliin and isoalliin (in garlic and onion, respectively), which are formed separately.

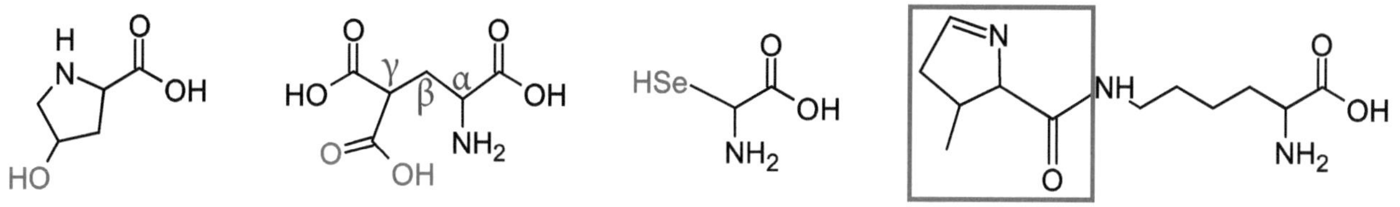

Hydroxyproline   γ-Carboxy-glutamate   Seleno-cysteine   Pyrrolysine

FIGURE 2.11 **Common amino acid derivates found in proteins.** The groups that were added to the original amino acids are marked in red.

deficiency is known to result in bleeding, due to failure in forming blood clots. A more detailed discussion of post-translational modifications of protein residues will be given at the end of the chapter.

2. *Derivates formed before translation:* This group includes only two amino acids: *selenocysteine*, which appears in all three kingdoms of life, and *pyrrolysine*, which appears in Archaea. In contrast to the derivates of the first group, these are formed pre-translationally, and incorporated in proteins according to genetic instructions [158]. For this reason, they are sometimes referred to as the 21st and 22nd amino acids. Selenocysteine is chemically identical to cysteine accept for the *selenium* atom (Se)[*1], which replaces the sulfur in cysteine [159]. It has been identified primarily in enzymes catalyzing redox reactions, such as those that reverse oxidative damage [101]. However, studies suggest that selenoproteins[*2] also participate in other important cellular and physiological functions (Figure 2.12), and can contribute to the prevention of various pathologies [160–162]. In many of the redox reactions involving selenocysteine the *selenol* group (SeH) functions as a nucleophile, similarly to the thiol group (SH) of cysteine, yet more effectively [163]. The superior effectiveness of selenol as a nucleophile in these reactions is traditionally thought to be the result of the lower pKa of selenocysteine (~5.2 [164]) as compared to cysteine (8.6). The pKa difference means that the SeH group of selenocysteine deprotonates into its nucleophilic form (Se$^-$) more easily than the SH group of cysteine under physiological (or more acidic) pH. During the catalytic cycle of selenocysteine-dependent enzymes, the residue may function also as the attacked group, and it has been shown that in this capacity, too, selenocysteine is more effective than cysteine (see review in [163]). Finally, it has recently been shown that selenocysteine has a lower tendency than cysteine to become over-oxidized during catalysis to an acid form (SeO$_2$$^-$) [165]. Because such oxidation leads the enzyme to become irreversibly inactivated, it seems that the presence of selenocysteine instead of cysteine not only allows the enzyme to produce higher reaction rates but also protects it from inactivation.

[*1]Selenium is a non-metal *trace element*, that is, an element required in only small amounts for health (< milligrams per day). This group also includes the metals iron, cobalt, copper, zinc, manganese, nickel, chromium and molybdenum (see Subsection 2.6.9 and Chapter 9), the metalloid boron, and the halogens iodine and bromine. Trace elements are distinguished from the major dietary elements (sodium, potassium, sulfur, chlorine, calcium, magnesium and phosphorus), which are typically required at amounts of grams per day.

[*2]Selenium-containing proteins, belonging to over 100 families [160].

TABLE 2.4 **Biological roles of non-protein aminoacid derivates and analogues.** Abbreviations: GABA – γ-aminobutyrate; SAM – S-adenosylmethionine. Adapted from [166].

Amino Acid Derivate	Structure	Biological Role
Alanine		
β-alanine		• A component of vitamin B_5 and coenzyme A • A component of antioxidant dipeptides carnosine, carcinine, anserine, and balenine
Arginine		
Ornithine		• Amino acid degradation and ammonia detoxification (urea cycle intermediate) • Proline, glutamate, and polyamine biosynthesis • Mitochondrial integrity • Wound healing
Agmatine		• Modulator of various biological processes, like neurotransmission, ion transport, NO synthesis, and polyamine metabolism
Nitric oxide (NO)		• Regulation of hemodynamics and blood pressure through vasodilation (widening of blood vessels) • Innate immunity (as antibacterial agent produced by macrophages) • Neurotransmission (learning and memory through long-term potentiation)
Citrulline		• Anti-oxidation • Arginine synthesis • Osmoregulation • Amino acid degradation and ammonia detoxification (urea cycle intermediate) • Nitrogen reservoir
Creatine		• Skeletal muscle action (ATP replenishment after strenuous activity) • Anti-oxidation
Aspartate		
NMDA		• Neurotransmission (excitation, withdrawal)

Amino Acid

Derivate	Structure	Biological Role
Cysteine		
Taurine		• Digestion (conjugation of bile acids to form bile salts) • Anti-oxidation and regulation of cellular redox state • Osmoregulation • Modulation of calcium homeostasis
Glutathione	Glu-Cys-Gly	• Major endogenous antioxidant • Inflammatory response (participation in leukotriene synthesis) • Detoxification (conjugation to hydrophobic toxins in liver) • Metabolism (processes involving DNA, proteins, and hormones
Allicin		• Antimicrobial activity • Anti-cancer activity
Glutamate		
GABA		• Neurotransmission (inhibiting neuronal excitability) • Regulation of muscle tone • Inhibition of T-cell response and inflammation
Glutathione	Glu-Cys-Gly	• See cysteine above
Glutamine		
Theanine		• Anti-oxidation • Neurotransmission (increasing brain levels of GABA, dopamine, serotonin, and glycine) • Neuroprotective effect
Glycine		
Trimethylglycine (betaine)		• Homocysteine methylation (methionine regeneration; homocysteine detoxification) • One-carbon unit metabolism (synthesis of bioactive molecules and SAM) • Osmoregulation
Heme		• Oxygen binding and transport • Electron binding and transport • CO production
Sarcosine		• Intermediate in glycine metabolism • Possible treatment of certain mental disorders
Glutathione	Glu-Cys-Gly	• See cysteine above

Amino Acid

Derivate	Structure	Biological Role
Histidine		
Histamine		• Mediation of allergic reaction • Vasodilation • Brain acetylcholine secretion • Regulation of gut function
Lysine		
Carnitine		• Energy production by lipid oxidation
Methionine		
Homocysteine		• Cysteine biosynthesis • Methionine regeneration • Risk factor for cardiovascular diseases
SAM		• One-carbon unit metabolism (synthesis of different bioactive molecules)
Serine		
Choline		• Neurotransmission and muscle control (acetylcholine) • Membrane structure (phosphatidylcholine, sphingomyelin) • Trimethylglycine (betaine), methionine, sarcosine and SAM synthesis
D-serine		• Brain neurotransmission (activation of NMDA receptors)
Tryptophan		
Serotonin		• Neurotransmission (mood, appetite, sleep, memory, learning) • Hemostasis and blood clotting (as vasoconstrictor) • Inhibiting production of inflammatory cytokines and superoxide radical
Niacin		• Redox metabolism (as component of NADH and NADPH)
Melatonin		• Hormonal mediation of day-night cycles • Anti-oxidation • Possible effect on immune system and inflammatory response

Amino Acid Derivate	Structure	Biological Role
Tyrosine		
Dopamine		• Neurotransmission (reward-motivation, motor control, vasodilation, regulation of prolactin secretion)
Epinephrine and norepinephrine		• Neurotransmission (exercise, 'fight-or-flight', emotional response, memory) • Muscle activity • Sugar metabolism
Melanin		• Pigmentation • Anti-oxidation
T3 and T4		• Regulation of metabolic rate (thyroid hormones)

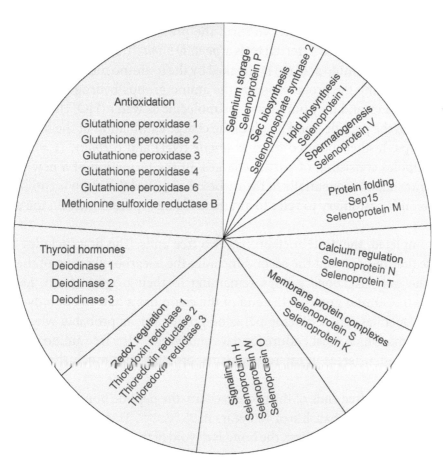

FIGURE 2.12 **Putative functions of the human selenoproteome** [160–162]. Courtesy of Sharon Rozovsky.

Interestingly, despite its resemblance to cysteine, selenocysteine is in fact produced from serine. The entire process includes three steps. First, serine is loaded by the right enzyme on a specific transfer RNA called '$tRNA_{sec}$'. Then, the serine is modified into selenocysteine while on $tRNA_{sec}$. Finally, the tRNA-selenocysteine conjugate binds to the right messenger RNA (mRNA) codon within the ribosome. As is well known, the genetic code includes 61 codons for the 20 amino acids, as well as 3 'stop' codons. How, then, can selenocysteine find its place in the protein sequence? As it turns out, UGA, which is normally a stop codon, also serves as the selenocysteine codon. The difference between this codon and the 'normal' stop codon is that the former has a unique 'stem-loop' structure, which is recognized by the tRNA-selenocysteine conjugate, but not by the elements responsible for stopping translation.

Pyrrolysine is a derivate consisting of a lysine residue with its $N\varepsilon$ atom covalently attached to a *pyrroline* group via an amide bond. It appears in the Archaean enzyme methyltransferase, which produces methane from methylamine compounds. According to the suggested mechanism, the $N\varepsilon$ atom (now in the form of imine) participates in the activation of the enzyme's substrate and its positioning in the optimal way for catalysis [167]. The incorporation of pyrrolysine into proteins probably also relies on a stop codon, but this process is much less understood than that of selenocysteine.

2.2.2 Peptide bond

The diverse chemical types of side chains provide amino acids with unique properties. The backbone, in contrast, is identical across all amino acids. It is this part of the molecule that connects amino acids to each other to form the protein chain. The connection is via an amide bond, more commonly referred to as a 'peptide bond'. The bond is formed when the α-carboxyl carbon of an amino acid (*n*) is attacked by the α-amino nitrogen of amino acid $n+1$ (Figure 2.13a). During the process, one of the amino group's hydrogen atoms binds to the carboxyl's OH group, and they both leave the molecule as water (HOH). More importantly, the formation of the peptide bond involves the charge elimination of the former amino and carboxyl groups. In their bonded form as *amide (NH)* and *carbonyl (C=O)* groups, they are merely dipoles consisting of partial charges. In the formation of a new protein chain, amino acids are added sequentially to the carboxyl end until the chain is complete. As mentioned above, it is customary to refer to amino acids incorporated within the protein chain as '*residues*'.

Once completed, the protein chain has two free ends. The first includes the α-amino group of the first residue, and the second includes the α-carboxyl group of the last residue. These are the only backbone groups remaining in their original form, i.e., electrically charged. The free amino group at the end of the chain has a half-life of only several weeks before it disintegrates into a diketopiperazine group. This is probably why in proteins of higher organisms, which must endure for days and weeks[*1], the free amino group is blocked by an acetyl group, whereas in microorganisms, proteins with much shorter life spans are not acetylated [170].

In contrast to the free ends of the protein chain, the peptide bonds connecting the other residues are stable, with a half-life of 400 years in 25 °C [171]. This stability is a direct result of the peptide bond's properties. First, the bond is devoid of the full charges that were originally

---

[*1]In human cells the half-life of most proteins is ~1 to 20 hours [168], although some proteins, *termed long-lived proteins*, may remain intact and functional for weeks, months, and even years [169]. Many of these proteins are fibrous, and therefore tend to have a fortified structure (see Subsection 2.7 below).

present in the amino and carboxyl groups. Moreover, it is stabilized by electronic resonance. The resonance is due to the fact that the electrons of the C=O double bond can shift towards the C−N group (Figure 2.13b). As a result, this group may appear in the form of C=N⁺ at least part of the time. Since the electrons move constantly, the double bond cannot be located to any of the above forms, but instead it is 'spread' on the three atoms (O−C−N), or in other words, delocalized. Alternatively, the C−N bond is said to be 'partially double'. This has a few important implications. First, it makes the bond planar and more stable than a single bond[*1]. Second, in contrast to canonical double bonds, the peptide bond is not completely rigid, but can acquire two configurations (see next section for details). Third, the moment of the N−H and C=O dipoles on both sides of the bond is particularly strong (~3.5 D), twice as strong as the dipole of a water molecule. As a result, the N−H and C=O groups are capable of interacting electrostatically with charged side chains, with polar groups of the protein's natural ligand, and with each other. The interaction of these groups with each other via hydrogen bonds has a considerable effect on the folding of the protein chain, as will be explained in the next section.

Inside cells, the peptide bond is not created spontaneously by separate amino acids, but via the complex ribosomal system, which includes mRNA, tRNA, the ribosome, and other assisting proteins. The use of this system not only increases the efficiency of protein biosynthesis, but also makes sure each protein is built according to strict genetic instructions. This process forms a chain of amino acids connected via their backbones (Figure 2.13c). According to convention, the protein chain is always presented with the α-amino group of the first residue on the left, and the α-carboxyl group of the last residue on the right. These two groups are referred to as the 'N-terminus' and 'C-terminus', respectively, or for short, N' and C' (respectively). Another convention relates to the length of the protein chain; a protein that consists of less than 50 residues, is referred to as a 'peptide', whereas longer proteins are referred to as 'polypeptides' or simply 'proteins'.

Although most biological functions are carried out by polypeptide chains, there are many biological processes that also involve peptides. These peptides are sometimes called 'bioactive' or 'physiologically active' peptides. Here are some examples:

1. Neuropeptides: This group includes peptides that act as neurotransmitters. Examples of neuropeptides include *endorphins* and *enkephalins*, which act as painkillers in animals following injuries.

2. Hormones: This group includes a diverse set of peptides. For example, *insulin* and *glucagon* are pancreatic peptides that regulate energy-related (metabolic) processes in the animal body. Another hormonally-active peptide, *vasopressin*, acts during dehydration to reduce water loss and increase blood pressure [173]. These effects are highly important in preventing the body from undergoing hypovolemic shock.

3. Antibacterial peptides: This is a diverse group of peptides, produced by a large range of organisms. These peptides act as a simple and fast means of defense against potentially harmful bacteria. The mechanisms used by antibacterial peptides are also diverse. For example, the fungal peptide alamethicin kills bacterial cells by forming low-selectivity ion channels in their plasma membranes, leading to the membranes' depolarization. Some antibacterial peptides are produced by non-ribosomal systems, and include D-amino acids.

[*1]It should be noted that inside the folded protein there are local strains and atom-atom interactions that may compel certain peptide bonds to deviate from planarity by over 20° [172].

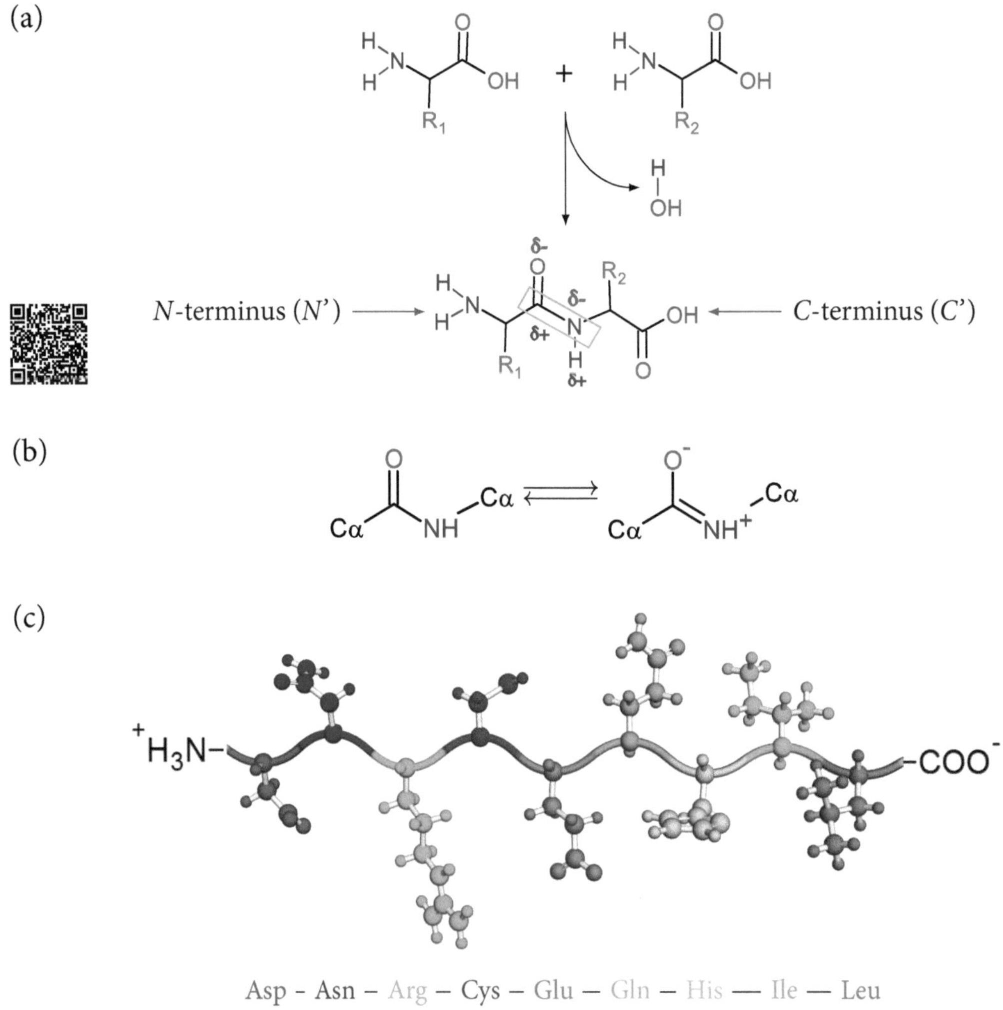

(a)

(b)

(c)

Asp - Asn – Arg – Cys – Glu – Gln – His — Ile — Leu

FIGURE 2.13 **The peptide bond.** (a) The chemical formation of the peptide bond (marked by the golden rectangle). (b) The electronic resonance of the peptide bond results from the fact that the bond alternates between three states. (c) The general organization of the backbone (presented schematically as a ribbon) and amino acid side chains (presented explicitly as sticks and balls) of the polypeptide chain (top). Each of the amino acids is colored differently. The amino acid sequence of the protein is traditionally presented from left (*N*-terminus) to right (*C*-terminus). The sequence can also be presented simply by stating the abbreviated names of the amino acids (bottom) or using their corresponding one-letter codes: DNRCENHIL.

Polypeptide chains differ significantly in their primary structures; they range from ~100 to tens of thousands of residues in length. The largest protein that can be found in the *UniProt* database of proteins [174]*1 is *titin*, which consists of 35,213 residues. The number of polypeptide chains may also vary across proteins. However, one thing remains the same in all proteins: **the amino acid sequence of the polypeptide chain contains all the information it needs to fold into its unique three-dimensional (tertiary) structure**. This was elegantly demonstrated by Christian Anfinsen in one of biology's most famous experiments [21]. Anfinsen put the enzyme *ribonuclease* in a solution of *urea* and *mercaptoethanol* at high temperature. This was done to induce loss of activity in the protein (i.e., *denatu-*

---

*1http://www.uniprot.org

ration) by disrupting the noncovalent forces and the disulfide bonds stabilizing the protein's three-dimensional fold [175]. Then, he cooled the solution and filtered out the denaturing agents. Surprisingly, the activity of the enzyme was restored along with other physical properties, showing that it can refold and regain function without any help[*1]. As explained in Chapter 1, the information stored in the amino acid sequence of the polypeptide chain manifests as the noncovalent forces existing between the different residues. These forces, in turn, drive the protein chain to fold accurately. Formation of the native fold of the protein requires satisfaction of many residue-residue interactions. This is because each individual amino acid has only a weak preference for interacting with other specific residues. Thus, folding specificity is achieved by satisfying the interaction preference of many amino acids. Inside cells, small proteins fold independently, whereas large proteins are assisted mainly by large complexes called *'molecular chaperones'*. Nevertheless, the assistance is kinetic in nature, and does not add new information needed for the protein to fold. All chaperones do is create an environment in which the protein can fold without 'distractions' due to interactions with other entities, and without getting 'stuck' in intermediary conformations along the way.

In the following sections we will discuss the folding of the protein chain, which entails folding first into simple local structures, and then into the overall, highly complex structure, which also possesses biological activity. Aspects related to the thermodynamics (i.e., energetics) of folding are discussed in Chapter 4, whereas kinetic aspects are detailed in Chapter 5.

2.3 SECONDARY STRUCTURE

Proteins possess complex structures in three-dimensional space. However, segments of the protein chain tend to fold into simple local structures, which extend along the axis of the segment. These simple folds constitute the secondary structure of the protein, which is the basis for the global three-dimensional fold of the entire chain. The folding process of the protein chain involves all of the protein's atoms. However, whereas the folding of the backbone determines the overall shape of the protein, the locations of side chain atoms fine-tune this shape into its accurate form. We will begin our discussion with the folding of the backbone. In order to understand this process, we need to ask two questions:

1. *What drives backbone folding?* We have already answered this question; the driving force is the noncovalent interaction between the residues constituting the chain. A detailed description of the individual interactions and their relative importance for folding will be given in Chapter 4.

2. *What limits backbone folding?* That is to say, what prevents the polypeptide chain from acquiring any conformation in three-dimensional space? To answer this, we first need to take a look at the bonds that constitute the hinges of the chain, since these bonds determine the freedom of movement of the chain at their respective locations. Two types of covalent bonds are present in the chain: single bonds, the movement around which is free *per se*[*2], and double bonds, which are rigid. The backbone of the polypeptide chain includes two types of bonds that can be considered as hinges. The first is the

---

[*1] For this work, Anfinsen received the 1970 Nobel Prize in Chemistry [176].

[*2] Barring conformations that involve clashes of neighboring atoms.

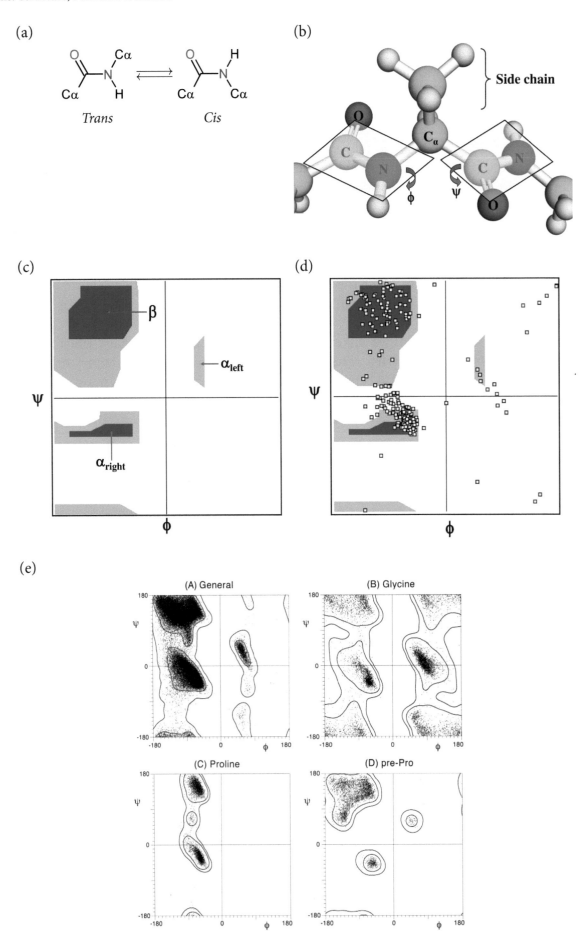

peptide bond. As explained above, this bond is only partially double due to its electronic resonance (Figure 2.13b). As a result, the bond is not completely rigid, and its dihedral angle (ω) may acquire values corresponding to two configurations: *cis* and *trans* (Figure 2.14a)[*1]. Although both configurations are feasible, the *trans* configuration is 1,000 times more stable than the *cis*, which means the latter is only seldom observed. An exception to this rule is observed in proline; due to the special structure of this residue, its *trans* configuration is only a few times more stable than the *cis*, which means that the *cis* configuration can be observed in some of the proline residues in proteins. Because of this characteristic, proline allows the protein to acquire conformations that would otherwise be impossible, and opens new functional possibilities for the protein [179]. Moreover, a shift between two proline configurations may take place when the protein is in the folded state, and although this shift is slow, it still has a significant effect on the protein's dynamics. The isomerization between the *cis* and *trans* configurations in the folded protein is catalyzed by the enzyme *peptidyl prolyl cis-trans isomerase*.

The two other hinges in the protein backbone are on both sides of the peptide bond, these are the C_α–N and C_α–C bonds. The dihedral (or torsion) angles of these bonds are termed ϕ and ψ, respectively (Figure 2.14b). Both of these bonds are single, and their rotations are therefore limited only by clashes of side chain and backbone atoms. This means that the versatility of protein folds is determined mainly by these angles, and finding their 'allowed' values may tell us something about the conformational space accessible to the protein backbone. Such thoughts probably crossed G. N. Ramachandran's mind in 1963, when he was

FIGURE 2.14 **Freedom of movement of the protein backbone.** (Opposite) (a) The *cis* and *trans* configurations of the peptide bond. (b) The two rotatable single bonds flanking the peptide bond (C_α–N and C_α–C). The angles of rotation around the bonds are marked as ϕ and ψ, respectively. A ϕ angle of 0° corresponds to a conformation in which the N–H and C–R bonds point in the same direction, whereas an angle of 180° corresponds to a conformation in which they point in opposite directions. The values of the ψ angle correspond (in the same way) to the C–R and C=O bonds. The two shaded rhombi specify the planes of the peptide bonds. (c) The Ramachandran plot. Blue areas represent 'allowed' ϕ and ψ combinations. Green areas represent partially allowed ϕ and ψ combinations, that is, conformations that involve tolerable steric clashes between atoms. Such conformations occur if smaller van der Waals radii are used in the calculation. White areas represent ϕ and ψ combinations that are disallowed (for all residues except glycine). The location of the two dominant conformations, the right-handed α-helix (α_{right}) and the β-sheet (β), and the less common, left-handed α-helix (α_{left}), are marked. (d) Experimentally obtained Ramachandran plot of the enzyme triosephosphate isomerase (PDB entry 1amk). Each point on the map is a combination of $\phi + \psi$ values, corresponding to a single amino acid. The points outside the allowed areas are of Gly residues. The maps in (c) and (d) were produced using the Virtual Molecular Dynamics (VMD) program [177], (e) Comparison between the 'general' Ramachandran plots and those of glycine and proline (taken from [178]). The 'general' plot includes 81,234 non-glycine and non-proline residues. The glycine and proline plots include 7,705 glycine and 4,415 proline residues, respectively. Each marked region in the plots includes an inner shape denoting the corresponding residue's favored area and an outer shape denoting its allowed area.

[*1]As explained above, deviation from purely planar *cis* and *trans* configurations can be observed in the folded protein; such deviation results from local strains and interactions [172].

attempting to calculate all the allowed ϕ and ψ values using models of dipeptides [180,181] (see also a perspective by Rose [182]). In these models, each atom was represented by a hard sphere with a van der Waals radius. Ramachandran's results are represented by a two-dimensional map that bears his name to this day, *Ramachandran's plot* (Figure 2.14b). The plot shows areas in ϕ and ψ space, which according to Ramachandran's calculations, do not involve any atomic clashes within the polypeptide chain, and should therefore be allowed. The map can be calculated for each of the 20 amino acid types, but since 18 of 20 types have similar maps — the exceptions are proline and glycine (see below) — the Ramachandran plot can also be presented in a general form. **As the Ramachandran plot shows, atomic clashes alone eliminate ~75% of the theoretically available ϕ, ψ space.** Ramachandran's predictions preceded the emergence of experimental methods for determination of protein structure. When these finally arrived, and various protein structures were determined experimentally, Ramachandran's predictions turned out to be true; the ϕ and ψ values derived statistically from those structures clustered within the 'allowed' areas depicted by his map[*1] (Figure 2.14d). Note that glycine and proline behave differently from the other amino acids. Glycine, having no side chain at all, can acquire ϕ and ψ values well beyond the allowed areas of the rest of the amino acids, whereas proline is geometrically limited because of the fusion of its backbone and side chain (Figure 2.14e).

When the allowed ϕ and ψ values are applied to a model of the protein chain, simple local folds emerge, such as coils, loops, and extended shapes. These are the elements of the protein's secondary structure, which are in fact recurrent patterns of allowed ϕ and ψ angles. The secondary elements in proteins are the foundation of the tertiary structure, but as we will see later, they also have other roles: enabling atoms to be packed efficiently, solving protein stability issues, and even forming biologically active structures. Not surprisingly, studies show that a protein's secondary structure correlates with both the tertiary structure and the activity of the protein [184]. For this reason, secondary structures, although local, are used today for the purposes of protein classification, tertiary structure prediction, and understanding of protein activity. The two main secondary elements in Ramachandran's plot, the α-helix and β-sheet, were already proposed in 1950 and 1951 by Linus Pauling and Robert Corey, on the basis of hydrogen bonds models [185–187] (see also review by Eisenberg [188]).

In the following subsections we review the main secondary elements in proteins, and their importance to proteins' overall structures. A summary of the geometric features of the main elements is given in Table 2.5. Before delving into this information, it is recommended for the reader to go over the main forms of graphic representation of molecular structures and their properties (Box 2.4).

BOX 2.4 GRAPHIC REPRESENTATIONS OF PROTEINS

Proteins are studied by using a variety of experimental and computational tools. These often produce three-dimensional structures, which can be further used for analysis. The efficiency of structural analysis depends to a large extent on the methods used and the experience of the scientist. However, these elements are aided by the right

---

[*1] Although they were not evenly distributed in those areas [183].

representation of the protein structure. There are numerous ways of representing a macromolecule, such as a protein, graphically; each approach highlights different information [189,190]. The most basic representation is the *space-fill* representation, showing atoms as spheres, usually with a van der Waals radius (Figure 2.4.1). This type of representation is clear and colorful, and is therefore often used to portray molecules in movies and other non-scientific media. However, aside from depicting the general shape of the protein, it does not provide any additional information about the macromolecule, and therefore is seldom used alone in scientific research and literature. In order to be able to observe more details, one can present only the covalent bonds between atoms, as *wires* or *sticks* (Figure 2.4.2). This type of representation reveals the connectivity of atoms in the protein. However, the numerous bonds, all presented explicitly, burden the scientist with details, only a few of which are actually needed for addressing a specific question. Therefore, it is customary to apply this type of representation only when focusing on certain parts of the protein (e.g., selected residues) that are of interest.

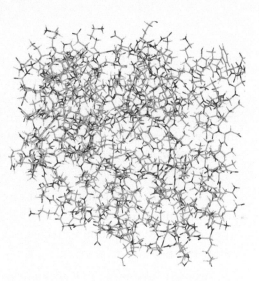

FIGURE 2.4.1 **Space-fill representation of a protein.** The atoms of triosephosphate isomerase (PDB entry 8tim) are represented as spheres with van der Waals radii, and colored according to the CPK convention. The figure was produced using PyMOL [191].

FIGURE 2.4.2 **Wire-frame representation.** The covalent bonds in triosephosphate isomerase (PDB entry 8tim) are shown as wires, colored by the type of atoms they connect. The atoms are not shown. The figure was produced using PyMOL [191].

The shortcomings of the wire representation highlight the importance of simplicity for analyzing the inner architecture of macromolecules in general, and proteins in particular. Such simplicity is achieved to a large extent by the *ribbon* representation, in which only the protein backbone is shown. The backbone is represented as a ribbon connecting its atoms or, alternatively, only the C_α atoms (Figure 2.4.3). This form of representation illustrates not only the general fold of the protein, but also local secondary elements of the chain. For this reason it is very commonly used in the scientific literature. When constructing a ribbon representation, molecular representation software often identifies secondary elements on the basis of the specifications already present in the atomic coordinates file (see Chapter 3). However, many programs also have a built-in algorithm that can deduce secondary elements from the atomic coordinates alone. The various secondary elements are usually depicted in different shapes and colors.

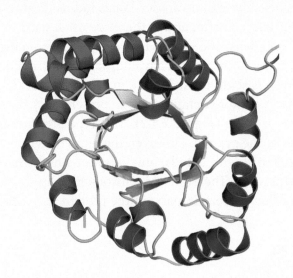

FIGURE 2.4.3 **Ribbon representation.** The atoms and bonds of triosephosphate isomerase (PDB entry 8tim) are not shown. Instead, the backbone is represented as a ribbon with different colors, corresponding to the secondary structure of the chain in that region: α-helices are colored in red, β-strands in yellow, and loops in green, as well as the disordered parts of the chain. The shape of the ribbon was calculated as the line going through all the C_α atoms of the protein. The figure was produced using PyMOL [191].

The activity of a protein almost always depends on the protein's ability to recognize and bind molecular elements in its vicinity. Therefore, scientists are often interested in properties of the protein surface, which is in contact with the surrounding environment. It is possible to calculate and represent the surface of a protein (Figure 2.4.4) by rolling a virtual water-size probe around the atomic contours of the protein [192,193] (Figure 2.4.5). Different types of surfaces can be calculated this way, such as the molecular surface and the water-accessible surface. The latter is particularly useful in enabling the scientist to identify parts of the protein that constitute potential binding or active sites. To identify these sites, the scientist would first look for depressions and crevices on the surface that look suitable for such a purpose. Identifying functional sites only by their geometry is difficult, and scientists therefore use data relating to additional properties of the protein regions in question. For example, many binding sites have electrostatic potential[*a] that is complementary to that of their ligands or substrates [125] (see Chapter 8 for details).

By calculating the potential and coloring the protein surface according to the calculated values (Figure 2.4.6), it may be possible to narrow the search for binding sites to specific regions in the protein. Some molecular graphics programs lack the capacity to calculate or present electrostatic potentials. In such cases, it is possible to present the electric charges of protein atoms instead. It should be noted, however, that the potential provides a far superior representation compared with the individual charges, since

---

[*a]See Box 1.3 for explanation about the meaning of the electrostatic potential.

electrostatic potential already embodies the mutual effect of the charges on each other and the local dielectric of the region.

The uses of electrostatic coloring are not limited to surface representation. First, the electric field emanating from the protein charges can be shown (Figure 2.4.7). Unlike the potential, the field is a vector, and can therefore provide additional information about the direction of the electrostatic forces in the system. Second, the potential around specific residues or atoms can also be shown (see Figure 1.3.2 in Box 1.3). This may be helpful when, for example, the effect of a point mutation on the electric properties of its environment needs to be analyzed.

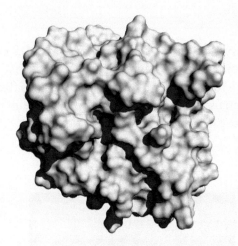

FIGURE 2.4.4 **Surface representation.** The water-accessible surface of triosephosphate isomerase (PDB entry 8tim) is shown, illustrating the indentations and crevices that may function as binding sites. The figure was produced using PyMOL [191].

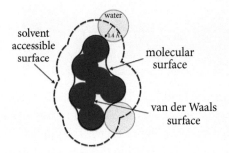

FIGURE 2.4.5 **Surface area types in molecules.** The black spheres represent the atoms making up the molecule. The gray sphere is the water probe used to calculate the solvent-accessible and molecular surfaces. (Taken from [194])

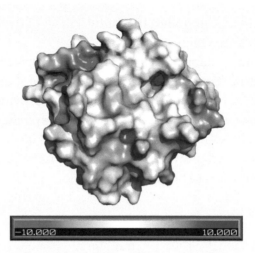

FIGURE 2.4.6 **Surface representation of γ-chymotrypsin (PDB entry 1afq), colored according to electrostatic potential.** The figure was produced using PyMOL [191]. Negative potentials ($0k_{\mathrm{B}}T/e > \phi > -10k_{\mathrm{B}}T/e$) are red, positive potentials ($0k_{\mathrm{B}}T/e < \phi < 10k_{\mathrm{B}}T/e$) are blue, and neutral potentials are white (see color code at the bottom). The electrostatic potential was calculated using APBS [195].

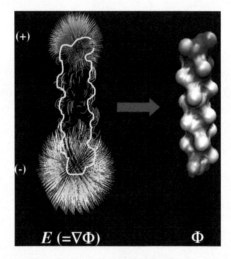

FIGURE 2.4.7 **The electrostatic field (E).** The electrostatic field of a spiral-shaped peptide (α-helix, see main text below) is shown on the left, and represented as lines emanating from the peptide (field lines are colored as in Figure 2.4.6). The peptide is represented implicitly by its contour, with the + and − signs representing the charges at its N- and C-termini, respectively. The charges result from the electric dipole, which characterizes peptides having this shape (see Figure 2.15 below). The orange sphere position on the contours of the peptide is a charged probe used to calculate the electrostatic potential at that point (shown on the right). As the bottom of the figure specifies, the electrostatic field is the derivative of the potential in three-dimensional space. The figure was produced using VMD [177].

Surfaces can be colored according to other parameters besides the electrostatic potential, such as evolutionary conservation (Figure 2.4.8), or degree of chain flexibility. When combined with other data, such information is invaluable.

To conclude, each of the graphic representation types holds certain advantages, which is why the best representation often combines some of these types into one picture (Figure 2.4.9). For example, the entire protein can be represented as a semi-transparent surface, allowing the viewer to see the inside organization of the chain, represented as a ribbon. The residues of interest in that protein would be presented explicitly as spheres or sticks and balls. Finally, interactions involving the specific residues can be depicted directly or via a mesh representation of the electrostatic potential.

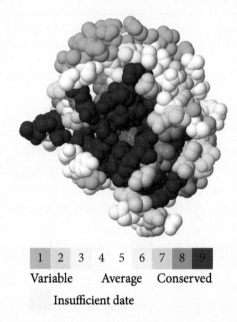

| 1 | 2 | 3 | 4 | 5 | 6 | 7 | 8 | 9 |

Variable · · · · · Average · · · · · Conserved

Insufficient date

FIGURE 2.4.8 **Coloring according to evolutionary conservation.** The protein in the figure (triosephosphate isomerase, PDB entry 1amk) is shown in space-fill representation, with each residue colored according to its level of evolutionary conservation (cyan – lowest, maroon – highest; see the full color code in the figure). The most conserved region is in the middle, where the natural substrate of the enzyme is bound. Conservation levels were calculated and presented using the ConSurf server (http://consurf.tau.ac.il) [196,197].

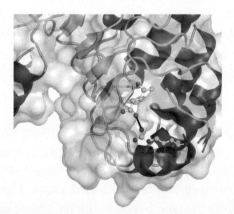

FIGURE 2.4.9 **Combined representation types.** Ribonuclease inhibitor (PDB entry 2bnh) is shown as described in the main text above. The residues presented explicitly form a hydrogen bond, shown as a black dashed line. The figure was produced using PyMOL [191].

TABLE 2.5   **Geometric features of helices and strands in proteins.**

Element Type	Element	Handedness	ϕ, ψ	Residues per turn	Rise per residue	Radius	H-Bonds
Helix	α	Right	$-57°, -47°$	3.6	1.5 Å	2.3 Å	$i \to i+4$
	3.10	Right	$-49°, -26°$	3.0	2.0 Å	1.9 Å	$i \to i+3$
	π	Right	$-57°, -70°$	4.4	1.1 Å	2.8 Å	$i \to i+5$
	PPII	Left	$-75°, +145°$	3.0	3.1 Å	1.4 Å	—
β-strand	Parallel		$-119°, +113°$		3.4 Å		Inter-strand
	Antiparallel		$-139°, +135°$		3.2 Å		Inter-strand

2.3.1 α-helix

2.3.1.1   Geometry

The α-helix is the secondary element that scientists first hypothesized to exist in proteins. Helical (i.e., spiral) structures are common in biomolecules (e.g., the double helix of DNA), and indeed, early work done by Pauling in 1953 already suggested such a shape for protein segments, although complete structures of globular proteins were yet to be determined [185,187]. The α-helix is very common in globular proteins; on average, about one-third of the residues in such proteins acquire the α-helical conformation [198]. It is a local structure, which may comprise between 5 and 40 residues. This right-handed helix corresponds to backbone ϕ and ψ values of $-57°$ and $-47°$, respectively [199]. One of the important characteristics of the α-helix is its compactness; each helical turn contains three or four residues, making the distance between the C_α atoms of adjacent residues only 1.5 Å. Because of this compactness, the side chains of the residues cannot be accommodated within the helix, and instead face the periphery, with a lateral spacing of 100° between them (Figure 2.15a). As we will see later, the compactness of α-helices is a major factor in allowing proteins to fit within the highly-crowded cytoplasm[*1].

2.3.1.2   Intramolecular interactions

Another key characteristic of α-helices is the repetitive pattern of backbone hydrogen bonds along the helical axis (Figure 2.15b). Specifically, each bond is formed between a backbone carbonyl group (C=O) and a backbone amide group (N−H) located four positions downstream in the sequence (i.e., in the C' direction). The sequence spacing is a result of each turn of the helix accommodating between three and four residues. These backbone hydrogen bonds are of great importance to protein stability, as will be explained in Subsection 2.3.4 below.

Ramachandran's plot shows that the α-helical conformation is geometrically possible. However, it does not tell us what drives protein segments to acquire this conformation, or in other words, what stabilizes the helix. Biochemistry textbooks traditionally attribute this stability to the backbone hydrogen bonds mentioned above. However, such bonds also exist in the unfolded form of the protein, between the backbone amide and carbonyl groups and the surrounding water molecules. It therefore seems unlikely that these bonds are the major stabilizing force [202,203]. One could argue that the local dielectric within the protein

---

[*1]About 40% of the cytoplasm is dry material, and the concentration of macromolecules is 300 to 400 g/L [200,201].

is lower than the dielectric of water, which should strengthen the Coulombic component of the hydrogen bond (see Chapter 1, Subsection 1.3.1.2). While this is true, studies show that the low local dielectric creates polarization effects that destabilize the hydrogen bond (see Chapter 4, Subsection 4.2.3 for details). This issue has been studied extensively and it seems that α-helices in proteins are in fact stabilized by nonpolar and van der Waals interactions involving the side chains of residues, particularly their C_β atoms [204,205] (Figure 2.15c). These interactions are likely to involve the surrounding environment as well, since isolated α-helices in water are usually unstable [206].

The carbonyl-amide hydrogen bonds occur mostly because both groups are electric dipoles, in which the carbonyl oxygen is partially negative and the amide hydrogen is partially positive (Figure 2.15d). In each of the single peptide bonds of the helix, the carbonyl amide groups of the same peptide unit always face opposite directions. This turns the peptide bond into a *microdipole*, with a strength of 3.5 D [207]. In a canonical α-helix, all the carbonyl dipoles face the C' of the helix, whereas all amide dipoles face the N'. In other words, all the microdipoles of the helical segment join together to form a *macrodipole* [208], with a strength of 5.0 D [207], and in which the N' is partially positive and the C' is partially negative (Figure 2.15e). Since all amide and carbonyl groups of the helix are hydrogen-bonded to each other, with the exception of the four groups at each terminus, the macrodipole tends to concentrate at the termini, instead of being 'smeared' along the entire length of the helix (see Figure 2.4.7 in Box 2.4 above). The charge concentration at each terminus is of a magnitude of 0.5 to 0.7 electron charges. The α-helix dipole is not just an artifact of the charge distribution over the helical shape; it plays an important role in protein activity: solved protein structures often include α-helices in binding sites, where the positively charged N' faces negatively charged groups of the ligand. The macrodipole of an α-helix also affects the helix's amino acid sequence; the N' of an α-helix tends to be populated with acidic glutamate and aspartate residues, which can interact favorably with the partially positive charge and further stabilize the helix [209,210]. Similarly, basic Lys and Arg residues tend to occupy the C'. Inside the helix, residues of opposite charge tend to appear adjacently either within the sequence or in three-dimensional space (i.e., one above the other), so they can form favorable salt bridges.

2.3.1.3 Amphipathic α-helices

Some helices have a unique pattern of distribution, in which polar and nonpolar residues face opposite directions (Figure 2.16a). Such helices are termed 'amphipathic', because of their physicochemical duality. Because of the α-helical periodicity, residues that are in adjacent turns on the same helix face are about four positions apart in sequence. As a consequence, amphipathic helices include either a polar or nonpolar residue every three or four positions (Figure 2.16b), which enables scientists to identify such helices on the basis of the protein sequence[*1]. Amphipathic helices are important to the packing of globular proteins. A globular protein has a nonpolar core and a polar surface. Helices located at the core are therefore mostly composed of nonpolar residues. However, helices at the surface face the core on one side and the solvent on the other, which means they have to be amphipathic to match both environments (Figure 2.16c).

[*1]Without prior knowledge of their three-dimensional structure.

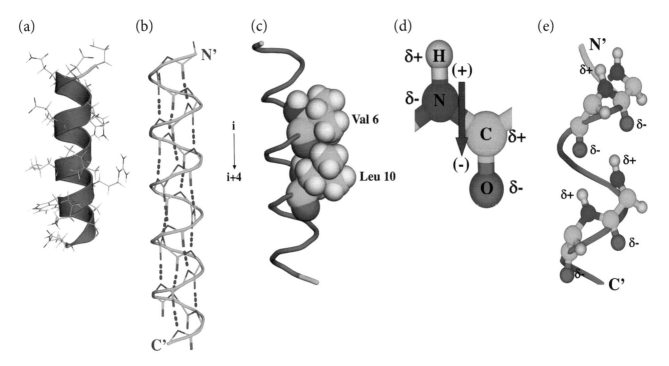

(a) (b) (c) (d) (e)

FIGURE 2.15 **The right-handed α-helix.** (a) The general organization of the helix. The backbone is depicted as a red ribbon (demonstrating the spiral-like conformation), whereas side chains are shown explicitly, in bond-stick representation. (b) The backbone hydrogen bonds pattern of the α-helix. The backbone is presented as sticks (side chains are not shown), with the hydrogen bonds between the carbonyl and amide groups represented by blue dashed lines. The pattern of the hydrogen bonds is specified on the right, whereas i denotes any hydrogen-bonded carbonyl group. (c) Helix-stabilizing interactions. The helix is depicted as a thin ribbon, except for the residues occupying positions 6 and 10, which are presented explicitly, as spheres. The two residues face the same direction on the helix, and one is positioned a turn above the other. (d) The electronic microdipole of the peptide backbone. The partial charges of the atoms are symbolized by the Greek letter δ. The direction of the dipole is marked by the purple arrow. (e) The electronic macrodipole of the α-helix. For convenience, the figure presents only the residues constituting one helix face.

2.3.2 Non-α-helices

The overall fold of the polypeptide chain may sometimes cause distortions in α-helices, and create less favorable geometries. Still, some of these geometries can be found in proteins, and occasionally serve specific purposes. These so-called '*non-α-helices*' (Figure 2.17) are described in the following subsections.

2.3.2.1 3_{10}-helix

The name of this helix originates from the fact that it includes 3 residues and 10 backbone atoms per turn. The 3_{10}-helix is narrower and longer than the α-helix (Figure 2.17b). Its conformation in proteins corresponds to ϕ and ψ angles of $-49°$, and $-26°$, respectively [199]. The narrow shape is problematic for atomic packing and may lead to clashes. Another problem of the 3_{10}-helix is the $i \rightarrow i + 3$ hydrogen bond pattern, which is energetically unfavorable, since the backbone dipoles are not optimally aligned. Accordingly, the 3_{10}-helix is seldom observed as a secondary element (about ~4% of protein residues occupy this configuration [198]), although it sometimes appears at the edges of α-helices, where it occupies a single

(a)

(c)

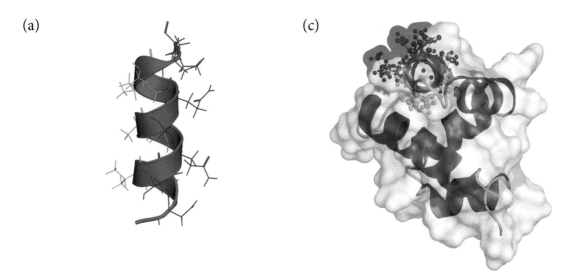

(b) $Asp_1-Thr_2-Val_3-Thr_4-Gln_5-Ala_6-Ala_7-Ser_8-Gln_9-Val_{10}-Leu_{11}-Asp_{12}-Lys_{13}$

FIGURE 2.16 **The amphipathic α-helix.** (a) The spatial distribution of polar (blue) and nonpolar (green) residues in an amphipathic α-helix. (b) The sequence periodicity of amphipathic α-helices. The sequence shown corresponds to the helix in (a). (c) The locations of amphipathic helices in proteins. The surface and secondary structure of NK-lysine is shown (PDB entry 1nkl), with the amphipathic helix (residues 24 to 37) presented as sticks and balls. The helix is located at the periphery of the protein, with polar residues (blue) facing the aqueous solvent, whereas nonpolar residues (green) face the hydrophobic core.

turn. There, it often serves as a connector between adjacent α-helices or between an α-helix and a β-strand [198]. Another role suggested for the 3_{10}-helix is serving as an intermediate in the folding/unfolding of α-helices [211]. Finally, in some interesting cases, the 3_{10}-helix seems to rectify other deviations formed within α-helices. For example, the antimicrobial peptide *alamethicin* has an overall helical structure, which allows it to bind to bacterial membranes. A proline side chain located within the segment disrupts α-helical hydrogen bonds, which may interfere with its membrane-active functions (the reason will be explained in Subsection 2.3.4 below). However, thanks to the 3_{10}-helix in that region, the backbone carbonyl and amide groups are still able to hydrogen-bond, though in a less favorable way.

2.3.2.2 π-helix

The π helix has ϕ and ψ backbone angles of $-57°$ and $-70°$ (respectively) [199], rendering it shorter and wider than the canonical α-helix (Figure 2.17c), with 4.4 residues per turn on average. These characteristics make the π-helix less favorable than the α-helix. First, the van der Waals contacts in the π-helix are not optimal. Second, its characteristic hydrogen bond pattern ($i \rightarrow i + 5$) is less energetically stable than that of the α-helix. In fact, the π-helix is so unfavorable, that for many years it was considered to be a purely hypothetical shape. As structure determination methods became more accurate, this type of helix was found in some proteins, although it is still rare [212] (about 0.02% of protein residues occupy this configuration [198]). Like the 3_{10}-helix, the π-helix also tends to appear at the edges of α-helices, where it incorporates no more than a few residues. The role of the π-helix is not clear, although it has been suggested that, like the 3_{10}-helix but to a lesser extent, it serves as an intermediate in the folding and unfolding of α-helices [211]. In addition, a study of different protein structures has implicated the unique conformation of π-helices in the formation or stabilization of protein binding sites [213].

2.3.2.3 Type II polyproline helix (PPII)

Proline-rich segments of the polypeptide chain tend to fold into a helical shape, referred to as a *'polyproline helix'*. The proline side chain is covalently bonded to the α-amino group, which has a few implications for the conformation of the chain in that region. First, the covalent bond limits the ϕ angle to values of $-65° \pm 15°$ [54]. Second, the location of C_δ in the pyrrolidine ring of the proline side chain limits the ψ values of the residue preceding the proline to the β area of the Ramachandran map. The product of these combined factors is a helical conformation. Although this conformation was originally found in proline-rich sequences, it can appear in others as well [214] (e.g., short poly-glutamine stretches [215]). There are two polyproline helical conformations. The first, termed *'PPI'*, is a compact right-handed helix ($\phi, \psi, \omega = -75°, +160°, 0°$ [216]), in which all peptide bonds are in the *cis* configuration. This conformation is of little importance to proteins, as it tends to appear only when they are put in organic solvents. In contrast, the second conformation, termed *'PPII'*, can be observed in proteins in their natural aqueous environment. The PPII conformation was first observed in fibrous proteins such as collagen, which form coiled coils [217] (see Section 2.7 below), and later in globular proteins. PPII is a left-handed helix, in which all peptide bonds are in their *trans* configuration ($\phi, \psi, \omega = -75°, +145°, +180°$ [183,218]) (Figure 2.17d). The helix contains three residues per turn, which makes it longer than the α-helix and devoid of backbone hydrogen bonds. For this reason, the PPII helix has always been hard to find in proteins, and in many cases segments with this conformation were considered to be unstructured. However, advances in computational analytic methods have made it possible to identify PPII helices in many proteins, including proteins that do not normally possess an ordered structure [219]. The possible reason for the presence of PPII helices in such proteins is discussed in Chapter 6, which focuses on unstructured proteins. In folded proteins PPII helices tend to be amphipathic, and therefore reside at the periphery of the globular structure [54].

PPII helices seem to be important for several functions of globular proteins, including signal transduction, transcription, movement, and the immune response (see [216] and references therein). For example, proline-rich sequences in a PPII conformation are recognized and bound by the *SH3 domain* [220–222] (Figure 2.17e), a functional module present in numerous proteins that are involved in cellular communication and other complex processes [223]. These proteins use their SH3 domains to recognize PPII helices in their target proteins, to which they pass on the signal. The details of this interaction are discussed in Chapter 8. Finally, there is evidence that PPII helices are common in denatured proteins, i.e., proteins that have partially or completely lost their ordered structures, due to harsh changes in their environmental conditions (temperature, pH, salinity, etc.) [224]. This issue is discussed in Chapter 4. These helices have also been implicated in the formation of pathological amyloid fibrils (e.g., [225]).

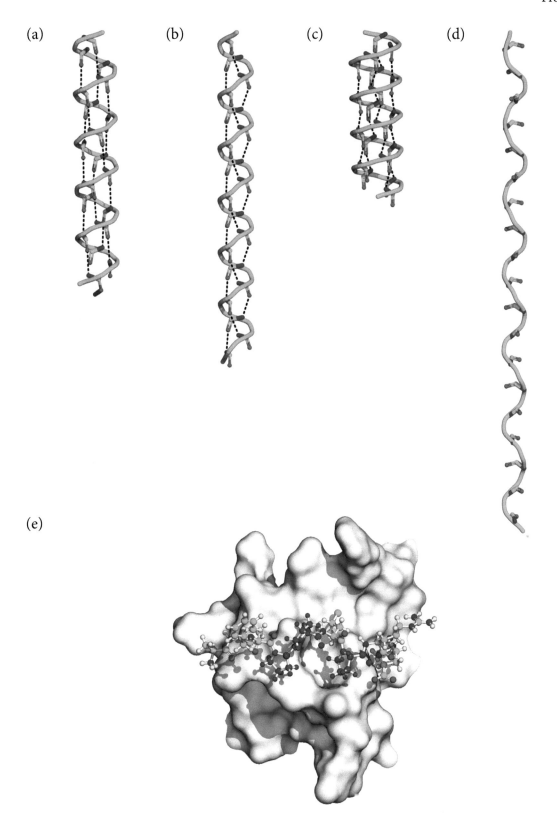

FIGURE 2.17 **Types of helices found in proteins.** (a) through (d) ribbon representations of α-helix (a), 3_{10}-helix (b), π-helix (c) and polyproline helix (d). The helices are composed of 20 alanine residues. The backbone carbonyl and amide groups are shown, as well as the hydrogen bonds they form (black dashed lines). (e) The interaction between the SH3 domain of the protein C-Src (as a white surface) and a proline-rich PPII helix (as sticks and balls, with proline residues in magenta and all other atoms in conventional atom colors). The structure was taken from the Protein Data Bank (1prl).

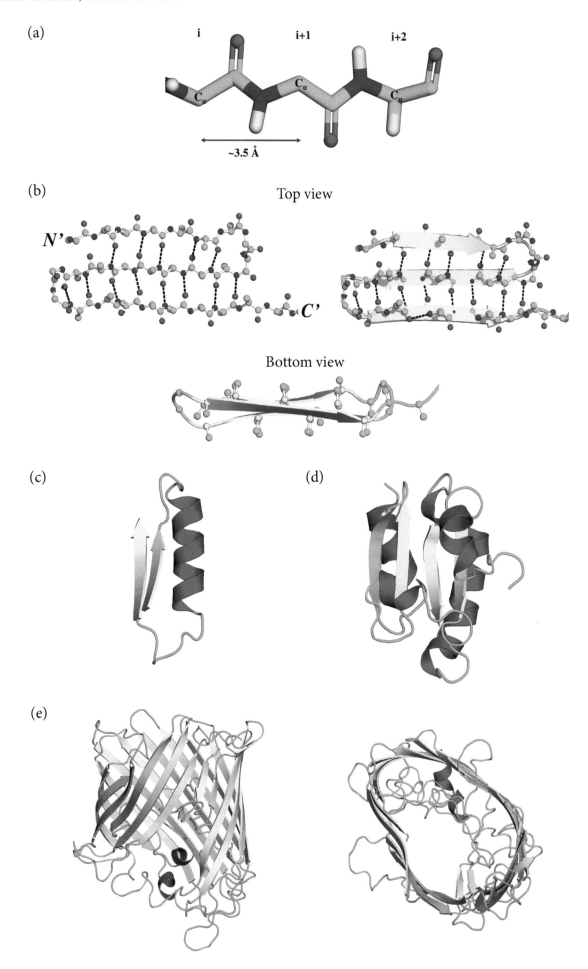

(a)

i   i+1   i+2

~3.5 Å

(b)   Top view

N'   C'

Bottom view

(c)   (d)

(e)

2.3.3 β conformation

Some segments of the polypeptide chain may acquire a less compact conformation, termed '*β*' (as opposed to the *α*-helix), and each segment possessing this conformation is referred to as a '*β-strand*' (Figure 2.18a). Such strands are typically between 5 and 10 residues long. In the *β* conformation, the distance between the C_α atoms of adjacent residues is 3.2 to 3.4 Å[*1], a distance that is over two times greater than the distance between residues in the *α* conformation. For this reason, the *β* conformation is sometimes referred to as '*extended*'. However, the polypeptide chain is not fully extended, and instead tends to zigzag up and down. This is the source of another term that is often used to describe the *β* conformation: '*pleated*'. *β*-strands are somewhat less common than *α*-helices in proteins, with about 20% of residues occupying this conformation [198]. In contrast to helices, the extended conformation does not allow intramolecular hydrogen bonds or significant stabilizing van der Waals contacts within the same strand. As a result, *β*-**strands rarely appear in proteins in their isolated form, and instead tend to arrange alongside each other to form a flat structure, referred to as a '*β*-sheet'** (Figure 2.18b). *β*-sheets may appear in two forms. The first is formed by *β*-strands ($\phi, \psi = -139°, +135°$) that are separated from each other in the polypeptide sequence by only a few residues. These residues arrange as short loops called '*β-turns*', thus allowing the strands to rearrange alongside each other. Since the direction of adjacent strands is opposite (due to the turn), this type of *β*-sheet is referred to as '*anti-parallel*'. The second form of *β*-sheet is formed by *β*-strands ($\phi, \psi = -119°, +113°$) that are separated within the sequence by more than a few residues. In fact, the segment containing these residues is long enough to allow the second strand to align alongside the first such that both face the same direction. This form of *β*-sheet is referred to as '*parallel*'. The most common type of parallel *β*-sheet is called a '*β-α-β loop*', since the segment separating the two strands acquires an *α*-helix conformation, connected from both sides to the *β*-strands by short loops (Figure 2.18c). In our later discussion of the tertiary structure we shall see that there are other structural motifs connecting secondary elements, and that they may serve a defined function in the protein.

Like *α*-helices, *β*-sheets can be amphipathic too. However, in amphipathic *β*-sheets, polar and nonpolar residues occupy every second position, instead of every third or fourth position. Another difference between the two secondary structures relates to backbone hydrogen bonds. Like *α*-helices, *β*-sheets include intramolecular hydrogen bonds between amide and carbonyl groups of the backbone. However, whereas in *α*-helices such bonds involve

FIGURE 2.18　**The *β* conformation.** (Opposite) (a) A three-residue protein segment in *β* conformation. The typical zigzag (extended) shape of the backbone is shown, with the carbonyl and amide groups pointing upwards and downwards, and with the side chains (not shown) pointing to and away from the viewer. The distance between sequential C_α atoms is denoted. (b) Structure and hydrogen bond pattern (dashed lines) of anti-parallel *β*-sheets. The strands are often depicted in the literature as wide arrows (top right). The side chains of residues in a *β*-sheet face away from (orthogonal to) the sheet's plane (bottom; for clarity, only the C_α and C_β atoms are shown). (c) A parallel two-stranded *β*-sheet in a form of a structural motif called *β-α-β* (PDB entry 1tph). (d) A twisted *β*-sheet, taken from the structure of thioredoxin (PDB entry 2trx). (e) The *β*-barrel structure of porins, presented from the side (left) and top (right) (PDB entry 1a0s).

[*1] The exact value depends on the structural context of the strand (parallel or anti-parallel).

backbone groups that are three or four positions apart, in β-sheets the interacting groups are more distant, because they originate from different strands (Figure 2.18b, top view). Also, since the interacting groups are perpendicular to the chain axis, their microdipoles do not form a macrodipole along the sheet, as they do in helices. The hydrogen bond geometry differs somewhat between the parallel and anti-parallel β-sheets. Some argue that the latter is more stable, although this is still a controversial matter. Finally, in most β-sheets, the strands are not entirely 'flat', but rather are twisted (up to 30° per residue), which slightly deforms the sheet structure (Figure 2.18d). As explained later in Chapter 5, this deformation is thought to have been selected during evolution to prevent non-specific aggregation of β-sheets [226]. This is important, as such aggregates tend to be toxic to cells and lead to neurodegenerative diseases and other pathologies.

Hydrogen bonds between backbone groups are of great importance to protein stability (see following section). Therefore, it is crucial for the protein to fold in a way that maximizes the number of such bonds. The β-sheet structure pairs most of its backbone groups in hydrogen bonds. However, half the backbone groups of the strands at the edges of the sheet are unpaired, due to a lack of bonding partners. Evolution has solved this problem by driving β-sheets to fold into larger structures. For example, a hollow cylindrical structure called a 'β-barrel' is formed when the β-sheet folds on itself [227] (Figure 2.18e). Indeed, this continuous structure leaves no strand unpaired. The hollow shape of β-barrels is perfectly suitable for their biological function; for example, Gram-negative bacteria use β-barrels extensively as low-selectivity channels, to transport various compounds through the bacterial outer membrane. It should be mentioned, though, that not all β-barrels are hollow. For example, in the TIM-barrel, described in detail below (Figure 2.27), the side chains of the β-strands pack inside the inner space. In most cases, β-barrels are composed of either anti-parallel or mixed sheets. There are also larger structures based on β-barrels, as we shall later see.

2.3.4   Why helices and sheets?

α-helices and β-sheets are very common in proteins and play a dominant role in determining protein architecture. One might wonder why these shapes evolved and not others. The first reason that comes to mind is globular proteins' need for compactness, without which these proteins would be unable to exist in their thousands within the highly-crowded cytoplasm [200], let alone remain water-soluble. As explained above, helices and sheets are very efficient in packing protein atoms tightly, so as to keep the proteins highly compact [228]. Still, there must be other shapes capable of efficiently packing atoms, which means there must be another factor favoring α-helices and β-sheets. Many studies, both experimental and theoretical, suggest this factor to be the hydrogen-bonding capability of these particular shapes. Protein folding buries most nonpolar residues within the core, while placing most polar residues on the surface. This structural organization enables polar side chains to interact with the polar solvent (water), while keeping nonpolar side chains away from it. However, the backbone of each protein residue still includes polar amide (N−H) and carbonyl (C=O) groups, regardless of the residue's side chain polarity. Thus, when nonpolar residues get buried inside the hydrophobic core of the protein during folding, the polar backbone groups get buried there as well. This means they change their immediate environment from hydrophilic to hydrophobic, or in other words, undergo *desolvation*. As explained in Chapter 1 and Subsection 2.2.1.3.3.3 above, the exposure of full and partial electric charges to a hydrophobic environment destabilizes the entire system, making this process unfavorable [124].

The energy cost of transferring a charged sphere from water (w) to a nonpolar (np) environment (Figure 2.19) can be calculated using a derivate of Coulomb's law called the *Born model of solvation*:

$$\Delta E = E_{np} - E_w = 166 \left(\frac{q^2}{r} \right) \left(\frac{1}{\varepsilon_{np}} - \frac{1}{\varepsilon_w} \right) \tag{2.2}$$

(where E is the electrostatic energy in kcal/mol, q is the magnitude of the charge in units of electron charge, r is the sphere radius, and ε is the dielectric constant of the media).

For example, if we use the model to approximate the transfer energy of a monovalent cation ($q = 1$ unit, $r = 1$ Å) from water ($\varepsilon_w = 80$) to a nonpolar environment of $\varepsilon_{np} = 4$, Equation (2.2) gives a value of 26.3 kcal/mol, which is higher than the total energy stabilizing the protein! It follows from the calculation that placing a charged amino acid, such as arginine or lysine, in the protein core is highly unfavorable. Indeed, charged residues get buried in the core only when they are needed to perform biological functions, in which case they are stabilized by salt bridges and/or hydrogen bonds with other amino acids. In this respect, the detection of a titratable amino acid in the protein core is indicative of a functional region.

Going back to our main theme, one might argue that the above calculation does not really represent the burial of backbone amide and carbonyl groups within the protein core, for three reasons. First, the two backbone groups are dipoles, made of partial charges rather than a single full charge. Second, the dielectric value used here for the hydrophobic environment ($\varepsilon_{np} = 4$) does not necessarily represent that of the protein core. Indeed, higher values have been suggested for certain regions of the core [229], although experiments in proteins indicate that the estimated average value of ε_p ranges between 2 to 4, and these values are well accepted. Third, the calculation above considers only the Coulombic component of the electrostatic free energy. As explained in Chapter 1, the total electrostatic free energy of a system also includes a polarization component.

In order to deal with these arguments and provide a more accurate description of the electrostatic free energy cost of burying backbone groups within the protein core, the Honig group used the continuum-solvent model (see Box 1.3) [230]. Their calculation suggested that the transfer of a single peptide bond with non-bonded backbone groups (N−H, O=C) from the aqueous phase into the nonpolar core destabilizes the system by 6.4 kcal/mol. This means that the transfer of a 20-residue polypeptide segment is expected to increase the energy by 128 kcal/mol, a value ~10 times higher than the energy needed for the stabilization of the entire protein. However, when the backbone groups are hydrogen-bonded (N−H--O=C), e.g., within an α-helix, the energy cost of transferring them to the protein core drops to 2.1 kcal/mol per single peptide unit. Although this transfer involves some destabilization, the energy value is small enough to be overcompensated by other interactions, which stabilize the helix. Indeed, **analyses of known three-dimensional structures of proteins show that ~90% of their polar groups are hydrogen-bonded** [231]**, and that these groups are more evolutionarily conserved than non-hydrogen-bonding polar residues** [232]. As we will see in Chapter 4, **the specific structural context of many hydrogen bonds inside proteins further stabilizes them, making them energetically favorable**.

Backbone hydrogen bonds may be formed by different types of local folds, but α-helices and β-sheets seem to form these bonds most efficiently. In other words, contrary to what

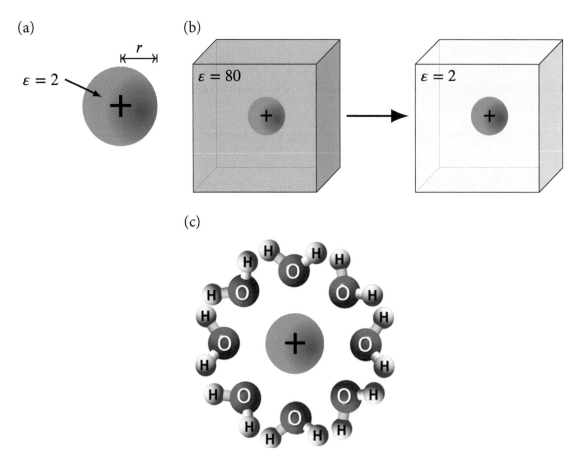

FIGURE 2.19 **The Born model for the transfer of a charged sphere between two media of different polarities.** (a) The cation is represented as a sphere of charge q, radius r and internal dielectric 2. (b) The cation is transferred from an aqueous solution of dielectric constant (ε) of 80 (blue box) to a nonpolar medium of dielectric 2 (yellow box). (c) A schematic description of the arrangement of water molecules around the cation, illustrating the concept of electrostatic masking.

is traditionally argued in many textbooks, intramolecular hydrogen bonds do not occur in order to stabilize α-helices and β-sheets. Rather, **these secondary structures are used to allow hydrogen bonds to form, thus reducing the high desolvation cost associated with the transfer of the protein's backbone groups from the aqueous phase into the low-dielectric core of the protein.** Indeed, ~50% of protein residues, and virtually all residues within the core, exist as part of an α-helix or β-sheet.

Why has evolution selected a polymer (i.e., protein) that contains a polar backbone, given that a hypothetical protein-like polymer with a completely nonpolar backbone would be much more stable? First, such a polymer might be too stable to be unfolded and degraded when no longer needed, and it might also lack the internal dynamics required for function (see Chapter 5). Moreover, a completely nonpolar polymer would be insoluble in the aqueous environment that surrounds most proteins (cytosol, intercellular fluid, blood, etc.). Such a protein would interact non-specifically with adjacent nonpolar proteins and create non-functional (and potentially toxic) sediments. Finally, polar backbone groups are functionally important; they directly participate in specific binding of substrates and ligands (see Chapter 8), and may even take part in catalysis (see Chapter 9). This functional specificity may also be promoted indirectly, through the ability of the polar backbone groups to limit the conformational space available to the polypeptide chain. Thus, the energetic stability of proteins is sacrificed for function and specificity.

In conclusion, it seems that proteins fold in a way that satisfies three requirements:

1. Tight packing of atoms is needed to maintain compactness and optimize stabilizing interactions.

2. Efficient pairing of backbone amide and carbonyl groups in hydrogen bonds is needed to reduce the destabilizing effect of these groups' desolvation.

3. Creating functional substructures (binding or active sites).

Different types of local folds may be capable of fulfilling each of these requirements, but only α-helices and β-sheets seem to be able to efficiently fulfill all three. This is probably the reason for these folds becoming the most common secondary elements of proteins.

2.3.5   Reverse turns

Reverse turns, as the name implies, are segments of the polypeptide chain, responsible for changing its direction [233,234]. Their structural importance is two fold: they connect between secondary elements, i.e., α-helices and β-strands, and by changing direction they facilitate formation of the globular form of the protein. Moreover, reverse turns tend to reside at the periphery of the protein, which exposes them to the solvent. As a result, they are capable of interacting with different elements in the vicinity of the protein, such as the protein's natural ligand or substrate. This capability is further enhanced by the numerous chemical modifications that turn residues often undergo; these modifications increase the residues' chemical diversity and allow for fine-tuning of protein-ligand interactions.

Reverse turns are a versatile group, including α-turns, β-turns, γ-turns, and loops. We will focus on the second and last, which are common in proteins.

2.3.5.1   β-turn

This is the most common type of reverse turn in proteins. The β-turn is structured as a short loop, usually connecting anti-parallel β-strands [234], hence the name (Figure 2.20a)[*1]. Despite being a loop (loops are typically flexible and have low evolutionary conservation, as elaborated below), the β-turn has an ordered structure consisting of four residues. Among those, the residue in the second position is usually *cis*-proline (Pro), whereas the fourth is glycine (Gly) [236] (Figure 2.20b). Evolutionary conservation of residues in a protein structure often implies a function of some kind, best fulfilled by these residues. In this case it is easy to see how the two residues play their role. Pro, with its side chain covalently attached to the backbone, creates the kink required for the turn to change direction. Gly, which is completely devoid of a side chain, confers flexibility to the backbone, thus allowing it to accommodate the kink created by the proline. The other two positions of the turn are less conserved, though they are commonly occupied by asparagine and aspartate [237–239]. Another characteristic of the β-turn is the backbone hydrogen bond often created between the first (i) and fourth ($i + 3$) residues. This bond is feasible because the side chains of the residues face away from the turn [233]. The entire structure is stabilized by side chain interactions between the adjacent strands. These include all types of interactions: aromatic, polar,

---

[*1]A β-turn connecting two anti-parallel strands is called a '*β-hairpin*' [235].

nonpolar, and salt bridges [233]. There are nine types of β-turns, differing in the orientation of the peptide bond connecting the second and third residues [237,240] (Table 2.6). The two most common types are I and II.

TABLE 2.6 **Classification of β-turns based on the geometry of their second and third positions.** Angle values are in degrees. [235]

Turn Type	Second Position		Third Position	
	ϕ	ψ	ϕ	ψ
I	−60	−30	−90	0
I′	+60	+30	+90	0
II	−60	+120	+80	0
II′	+60	−120	−80	0
IV	−61	+10	−53	+17
VIa1	−60	+120	−90	0
VIa2	−120	+120	−60	0
VIb	−135	+135	−75	+160
VII	−60	−30	−120	+120

2.3.5.2 Loops

Loops are segments of the polypeptide chain that are longer than turns and do not have a regular secondary structure (Figure 2.20c). The lack of structure allows loops to be flexible, but also means that their backbone groups are mostly unpaired in hydrogen bonds [241]. As a result, loops are excluded from the hydrophobic core of the protein, and tend to appear on the surface. This quality, in turn, is responsible for the natural selection of polar residues as the most common building blocks of loops. Loops can be disordered, and yet seem to play important roles in the function of many proteins, often in building ligands' binding sites. One of the best characterized examples is the antigen-binding site of antibodies, which is discussed in Subsection 2.4.2 below.

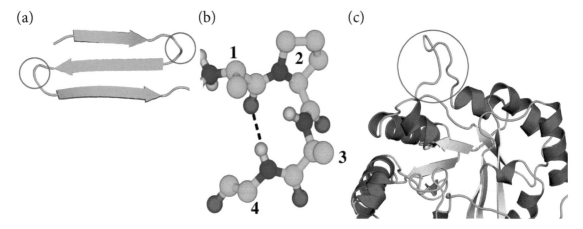

FIGURE 2.20 **Reverse turns and loops.** (a) The location of β-turns in β-sheets. The turns are marked by red circles. (b) The structure of the β-turn. The numbers denote the four principal amino acid positions within the β-turn structure. (c) A loop, marked by the red circle.

2.3.6 Secondary structure preferences of amino acids

In the previous sections we have witnessed the unique properties of helices, sheets, and turns, in terms of their three-dimensional arrangements and intramolecular interactions. Since the 20 amino acid types have different physical and chemical characteristics (shape, dimensions, polarity, etc.), the unique properties of the different secondary elements impose some constraints on which residues they can include. These constraints already became apparent in Ramachandran's predictions of the allowed ϕ and ψ angles, which were based solely on considerations of van der Waals clashes [180]. Other studies supporting this conclusion can be grouped into three types:

1. Statistical analyses of amino acid frequencies in experimentally determined protein structures. These include the earliest studies in this field, such as the well-known study of Chou and Fasman [242].

2. Mutational analyses of residues within secondary elements of proteins (e.g., [243–245]).

3. Structural-thermodynamic studies of short peptides. These model peptides, also referred to as 'host-guest peptides', have the same sequence except for one position, which contains a different amino acid in each peptide (e.g., [246]).

These studies show that the different secondary elements have preferences for certain amino acids. Whereas some of the preferences are fairly obvious, others are difficult to explain, and the rationalizations proposed by different scientists are highly controversial. One of the main difficulties is that the preferences, although detectable, are relatively weak. As a result, they may be influenced by various factors, primarily the immediate environment of the residue in question [247].

In the following subsection we will discuss the amino acid preferences of α-helices and β-sheets, focusing on those that are less controversial. The preference of β-turns has already been explained in the previous section.

2.3.6.1 α-helix

The preferred amino acids inside α-helices are Ala, Glu-0 (uncharged glutamate), Leu, Arg+ (charged arginine), Met, and Lys+ (charged lysine) [248] (Figure 2.21). These preferences can be explained by the fact that each of these residues possesses one or more of the following favorable side chain properties:

1. Small to intermediate size. As discussed earlier, the helical conformation is the most compact of all secondary elements. As such, it is expected to include primarily residues whose side chains are of small to intermediate size, as these have the least chance of clashing with each other. Indeed, this requirement is consistent with the prevalence of Ala, Glu, and Met in α-helices. Yet, other small amino acids such as Ser and Asp have considerably different helical propensities.

2. Low loss of entropy upon α-helix formation. Folding of a polypeptide segment into an α-helical conformation results in a decrease of entropy (i.e., degrees of freedom) of amino acid side chains, a process that is thermodynamically unfavorable (see Chapter 4 for details). Thus, amino acids that do not lose much entropy upon α-helix formation are more likely to appear in helices [249]. These amino acids include the following:

(a) Amino acids with small side chains. Such amino acids have very little entropy to begin with, and therefore very little to lose upon helix formation.

(b) Amino acids with linear side chains. A long linear side chain, e.g., as in Lys and Met, can protrude outwardly from the helix and therefore retain much of its original entropy even upon helix formation. In contrast, β-branched side chains like those of Val and Ile lose much entropy upon helix formation, which explains the low preference of α-helices for these amino acids. This idea is supported by the study of Makhatadze and coworkers [250], who found that having a branched side chain reduces the α-helical preferences by a factor of ~1.6. Leu and Arg also have branched side chains, but the branching is more distant from the backbone than it is in Val and Ile. Therefore, Leu and Arg can retain much of their original entropy even in helices, a quality that is reflected in their high prevalence in helices.

3. Hydrophobicity. Amino acids with nonpolar side chains stabilize α-helices via nonpolar and van der Waals interactions, provided that they are adjacent on the same face of the helix. (Figure 2.15c) This can explain the high preference of α-helices for Ala, Leu, and Met. However, the increase in the helix's overall hydrophobicity due to the presence of amino acids with nonpolar side chains also leads to exposure of polar backbone and side chain groups to a nonpolar environment, resulting in overall destabilization. This could explain the preference of helices for Glu-0 and Asp-0 over Glu and Asp, respectively.

In contrast to the above, the low prevalence of proline in α-helices is easy to explain. The side chain of proline is fused to the backbone, and this fusion has the following structural implications:

1. The fusion creates a kink of the backbone in that region [251], thereby preventing a clash between the pyrrolidine ring and the carbonyl oxygen of the residue located four positions upstream [252]. The kink prevents the backbone from maintaining the helical geometry, or in other words, 'breaks' the α-helix.

2. The kink induces some changes in local hydrogen bonds, and these changes destabilize the α-helix. For example, the C_δ–H group of the pyrrolidine ring hydrogen-bonds to the carbonyl oxygen of the residue located three, four, or five positions upstream [252].

3. The bonding of the proline side chain to its backbone amide group turns the latter into a tertiary amine. This group has no hydrogen, and therefore cannot participate (as a donor) in a helix-stabilizing hydrogen bond.

Because of the above, proline residues tend not to be present in the cores of α-helices, although they may appear in their termini. Membrane proteins are the exception; there, proline residues can be found quite often inside membrane-crossing α-helices, for reasons discussed in Chapter 7.

After Pro, Gly is the least frequently-observed residue inside α-helices. This is probably due to its lack of a side chain, although the exact cause for the absence of Gly is not entirely clear. The traditional explanation is that the lack of a side chain confers too much flexibility to the backbone, which is incompatible with the fixed ϕ and ψ values of the helical conformation [253]. It is likely that the flexibility of glycine indeed contributes to its low frequency inside helices. However, simulations show that this flexibility is only ~20% higher than that associated with an alanine residue in the same place, which makes it unlikely to be the only reason [254]. Indeed, these simulations implicate the inability of glycine to participate in side chain interactions, which, as we have seen earlier, are the primary stabilizing factor of α-helices [204]. The flexibility of Gly does, however, make it perfect for the helical termini. Indeed, Gly is over-represented at the C-termini of helices, and is considered a 'helix terminator' [255].

Amino acids other than glycine also seem to be preferred by α-helices in a sequence-dependent manner. These preferences can be distinguished according to the location of the preferred residues:

1. Helix termini. Certain amino acids tend to occupy the termini of helices (*helix caps*) [210,255,256]. For example, Asn, and to a lesser extent Ser and Asp, are over-represented at the N-termini of helices [255]. This over-representation has been rationalized mainly by the ability of these residues to form side chain–main chain hydrogen bonds [257]. Indeed, the side chains of these residues act as hydrogen-bonding 'surrogates' of the missing main chain carbonyl groups at the N-terminus, thereby stabilizing the helix. The negatively charged Asp also provides a more general electrostatic stabilization of the N-terminus, by masking the positive helix dipole. This rationale is also applicable to the C-terminus, albeit to a lesser extent; the residues that tend to cap the C-terminus act as hydrogen-bonding surrogates of the missing main chain amide groups, and the negative dipole in this area is stabilized electrostatically by positively charged amino acids.

2. Inside the helix. Charged residues of the opposite sign are preferable on adjacent helix turns (three or four positions apart), where they can interact favorably with each other. Likewise, charged residues of the same sign are unlikely to appear three or four positions apart in helices, as such positioning would put one side chain right on the top of the other, leading to strong electrostatic repulsion that could destabilize the entire helix.

2.3.6.2 β conformation

Unlike α-helices, β-strands show strong amino acid preferences only at their termini. At the C-termini of both parallel and anti-parallel β-strands, the preferred amino acids are Asp, Asn, Ser, and Pro [258]. Interestingly, the first three are also the preferred amino acids at the N-termini of α-helices [255]. As mentioned above, this preference has been rationalized by the ability of these residues to form side chain–main chain hydrogen bonds [257]. As the lack of main chain hydrogen bonds at the termini is also a problem in β-strands, the same rationale may also apply here. The preferred amino acids at the N-termini of β-sheets are Lys and Arg. Since both carry a positive charge at physiological pH, it has been proposed that their preferability at the N-terminus, which itself is partially charged due to a macrodipole, serves the purpose of fortifying the dipole, thereby promoting electrostatic interactions between the latter and other chemical species [258].

As mentioned earlier, the core of a β-sheet has much lower amino acid preference compared with the termini [198,258,259]. The (slightly) preferred residues in this region are Val, Ile, Tyr, Phe, Trp, and Thr [199]. These residues differ from one another in their properties,

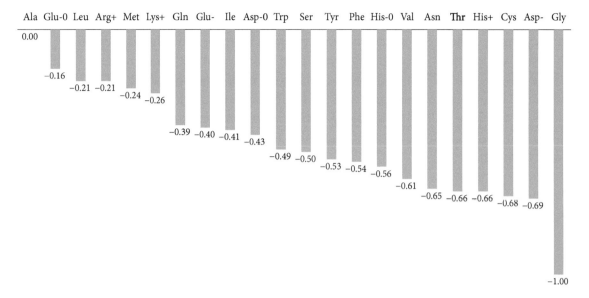

FIGURE 2.21 **α-helix preferences of the 20 types of amino acids.** The scale has been determined by Pace and Scholtz [248] on the basis of 11 sets of experimental data. The scale is presented as a histogram, where the amino acid preferences are shown relative to Ala (the most preferred amino acid), which has been set to zero. The more negative the value, the lower the propensity of the corresponding amino acid to appear in an α-helix. The preference for proline, which is much lower than the preferences of the other amino acids, is noted separately. +, − and 0 denote the electric charge on the amino acid.

which makes it difficult to rationalize the advantage they may confer to β-sheets. In fact, rationalizing the preference for any residue in β-sheets is difficult, as the latter do not seem to fold in isolation, and therefore cannot be subjected to structural-thermodynamic studies of the type carried out for α-helices [258]. One property that does seem to be common to most of the residues mentioned above is the large size of their side chains. The tolerance of β-sheets to large residues is understandable, as the β conformation is less compact than the α conformation, and is therefore likely to be more accommodating to larger residues. In addition, the side chains protrude up and down with respect to the backbone plane, and the chances of atomic clashes are therefore much smaller than they are in the α-helix. The side chain of Thr is not as large as the side chains of the other preferred residues, but it is branched. This property makes Thr unsuitable for α-helices, mainly because of the lack of space, but the residue can still be accommodated by β-sheets.

These properties may explain why large and/or branched residues are tolerated by β-sheets, but they do not explain why these residues are preferred. There have been several attempts to explain this issue. For example, it has been suggested that the large and/or branched side chains of residues such as Trp, Tyr, Val, and Ile sterically prevent the formation of hydrogen bonds between the solvent and adjacent peptide groups, and by doing so stabilize the entire structure [260]. Other factors have been suggested as well, yet none seems to fully explain β-sheet propensities of amino acids. An interesting possibility may be that these residues are preferred not because they stabilize the β structure *per se*, but rather because they prevent the formation of α-helices in that region. Assuming that a given region is populated by nonpolar residues, and must therefore assume an ordered conformation, preventing the formation of an α-helix in that region would promote the only alternative, i.e., the formation of a β structure.

2.4 TERTIARY STRUCTURE

In the previous section we reviewed the most basic folds of the polypeptide chain. These appear locally, and their repetitive patterns enable them to extend along the axis of the polypeptide segment. In addition, we have seen how the properties of these structures allow them to participate in the design of the overall protein fold, and in some cases, in its function as well. However, the secondary elements are in most cases too simple to execute the complex functions of proteins. This requires a higher degree of organization, which is achieved by the folding of the entire polypeptide chain, i.e., its *tertiary structure* (sometime also referred to as *ternary structure*) (Figure 2.1). Whereas secondary structures are similar in all proteins, there are two very different types of tertiary structures (Figure 2.22):

1. **Globular:** In these proteins the polypeptide chain changes direction often, thus creating an overall spherical (globular) structure. Most of the cellular proteins are globular, and they carry out a diverse set of functions, including most of the functions described in Chapter 1. Most of the globular proteins are also water-soluble, residing in the cytosol or in fluids of the multicellular body. Some globular proteins are lipid-soluble and reside inside the plasma membrane or (in eukaryotes) in the intracellular membranes.

2. **Fibrous:** The polypeptide chain of a fibrous protein creates an elongated, water-insoluble form. Fibrous proteins usually carry out simpler functions compared to their globular counterparts. These functions include the construction of large intracellular or extracellular structures, which provide mechanical support to cells and tissues, physical protection, or other tissue-specific functions (e.g., elasticity). Fibrous proteins are described in detail in Section 2.7 below.

Although most proteins possess some form of tertiary structure, some seem to be devoid of it altogether. Surprisingly, these *'intrinsically unstructured proteins'* (*IUPs*), which are discussed at length in Chapter 6, also play important and diverse biological roles. Nevertheless, the most diverse and sophisticated bio-processes are carried out by globular proteins. Since this book emphasizes structure-function relationships, we focus our discussion of protein structure in this chapter, as well as our discussion of protein energetics and dynamics in the two following chapters, on globular proteins, specifically those that are water-soluble. Membrane-bound proteins are globular, yet many of their characteristics are different from those of their water-soluble counterparts. We therefore discuss these proteins separately, in Chapter 7.

Regardless of protein type, the function a protein fulfills has always been considered to be a direct consequence of its three-dimensional structure. This is because lower-level structures are typically not considered to be complex enough for the required tasks (see Chapter 1 for details). Although this assumption is widely accepted, and even constitutes the central paradigm of modern structural biology, there are some observations that may call it into question. First, despite the structural complexity of globular proteins, the functions of these proteins are often executed by a small set of amino acids within an active site. This may imply that the rest of the structure is unnecessary. Indeed, Stanford scientists have succeeded in recreating the catalytic function of the enzyme *cytochrome c oxidase* by using a synthetic active site [261]. The active site was constructed from organic groups with chemical properties very similar to those of the enzyme's catalytic residues. In a different study, researchers constructed a functional enzyme from an 'alphabet' of only nine amino acids [262].

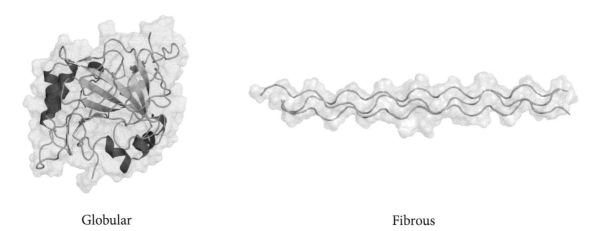

Globular   Fibrous

FIGURE 2.22 **Globular versus fibrous proteins.** The globular protein (left) is represented by the enzyme carbonic anhydrase (PDB entry 1ray), and the fibrous type (right) is represented by the structural protein collagen (PDB entry 1bkv). The secondary structures and the general shape of the protein are shown by a ribbon representation and by a semi-transparent surface representation, respectively.

These studies seem to cast some doubt over the necessity of a complex structure. However, as we will see later in this chapter, the overall structure of a protein does not merely function as a scaffold for the right placing of catalytic (or otherwise active) residues in space; it also includes structural units that couple the principal function of the protein to the biological context. Thus, regulatory units couple the activity of the protein to intracellular and/or extracellular signals, binding units couple it to the activity of other cellular elements, and transport or trafficking units are responsible for directing the protein to a certain cellular compartment, in which it can act most efficiently, sometimes in conjunction with other proteins. In many cases, regulation of protein activity is carried out by changes in the protein's conformation, which are transmitted over large distances from one place in the protein to the location of the active sites (see Chapter 5 for details). Again, this requires a complex structure with many alternative conformations.

The necessity of a complex tertiary structure is also questioned in light of the abundance of non-structured regions in proteins, particularly in eukaryotes [263]. Although the functions of these regions are not always known, it seems that in most cases they are responsible for providing the overall structure with flexibility that is required for its function, such as in the induced fit of enzymes to their substrates. A tougher case to explain is that of intrinsically unstructured proteins (IUPs), mentioned above. As we will see in Chapter 6, most of these proteins are unstructured only for part of the time and often assume ordered conformations when binding to their target molecules.

To summarize, the tertiary structure of a protein is of crucial importance to its ability to function properly within its biological environment. However, this does not mean that all parts of all proteins must have an ordered structure; indeed, in some cases it is the lack of structure that allows the protein to fulfill its function in the best way. In the following subsection we will review the tertiary structures of proteins, focusing on three main aspects:

1. Principal properties

2. Architecture

3. Evolution

2.4.1 Basic properties of tertiary structure

2.4.1.1 Structural properties required for complex function

Globular proteins are responsible for sophisticated functions such as enzyme-mediated catalysis, transport of molecules, signal transduction, defense, and regulation. These require the following properties:

1. **Compactness:** Proteins of the cytosol exist in a highly crowded environment [200]. Therefore, the structure of such a protein must be dense enough to allow the protein to coexist alongside the other components of the cytosol, and still retain its ability to diffuse freely.

2. **Solubility:** Cytosolic proteins are surrounded by an aqueous environment, which requires them to be water-soluble in order to avoid aggregation and sedimentation.

3. **Ability to form binding and active sites:** Virtually all protein functions involve molecular recognition and binding. In the case of enzymes, the binding is followed by catalysis. Both functions require certain chemical groups in the protein to be located at specific distances and angles from groups of the ligand or substrate.

All the properties mentioned above are a direct result of the globular shape of the polypeptide chain, or in other words, its ability to change direction. This is because all three properties require protein residues that are separated in sequence to be brought together into a confined space. Considering that proteins are linear polymers, having a globular shape is the only way this can be done.

2.4.1.2 Core versus surface

The evolution of biological macromolecules, including proteins, has taken place in an aqueous environment. Inside living organisms, this environment includes the cytosol, interstitial fluids, and the various fluids of the multicellular body, such as blood, saliva, lymph, and fluids of the gastrointestinal and urogenital systems. As a result, macromolecules have evolved to be highly water-soluble, which allows them to diffuse freely in the aqueous environment, while avoiding non-specific binding to other macromolecules (aggregation). Proteins contain both polar and nonpolar residues. To remain water-soluble, they must fold so as to allow polar residues to be on their surface, while burying nonpolar residues inside their core (Figure 2.23). This is indeed the case, although the partitioning of polar and nonpolar residues is not absolute. To begin with, polar residues, particularly the uncharged ones, can be found in the cores of virtually all proteins [69,264]. The burial of these residues may be energetically unfavorable, but as discussed in detail in Chapter 4, it is nevertheless beneficial, both structurally [265,266] and functionally [128]. In addition, the nonpolar environment of the protein core often encourages ionizable residues such as histidine to assume their uncharged state, thus lowering the energetic cost of burial [61].

Secondly, although the protein surface is overall polar[*1], it contains nonpolar 'patches' (Figure 2.23). A possible advantage of these patches is that they provide the protein with

---

[*1]Although all polar residue types have been identified on the surfaces of proteins, mutational studies have found that glutamate, aspartate and serine make the largest contributions to proteins' water solubility [267]. Lysine and arginine make much smaller contributions, despite their high polarity. This disparity is probably due to (*i*) the large nonpolar parts in the side chains of lysine and arginine, and (*ii*) the poor hydration of these

a way of binding to other elements in the cytosol, while still retaining its water solubility and avoiding aggregation. The different positioning of polar and nonpolar residues in the protein is to a large extent thanks to the secondary elements mentioned above; β-sheets are very efficient in the burial of nonpolar residues inside the hydrophobic core of the protein, whereas amphipathic α-helices allow the simultaneous externalization of polar residues and internalization of nonpolar ones [269] (see Figure 2.16c and detailed explanation in Subsection 2.3.1 above).

In addition to differing in their water solubility, the surfaces and cores of proteins differ in their density. The core is tightly packed, almost to the level of crystals [270,271], whereas the surface is more spacious. Still, the core may include empty spaces of 30 to 100 Å^3 in size [272]. These spaces have been often described as 'packing defects', which compromise protein stability [273]. However, both experimental [272,274] and computational [275] studies carried out in recent years suggest that these empty spaces might actually serve some purpose. For example, in myoglobin, intraprotein spaces seem to create a path for the substrate (oxygen) from the surface to its binding site in the core. In addition, it has been suggested that the spaces are important in mediating global conformational changes within proteins.

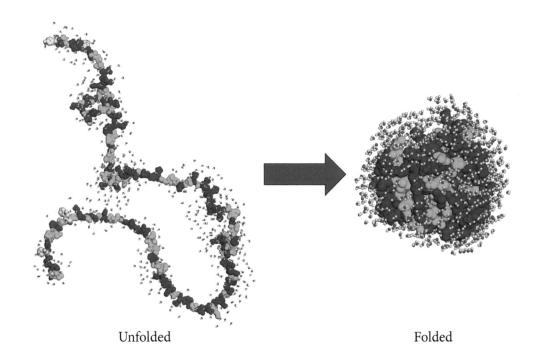

Unfolded   Folded

FIGURE 2.23 **Solvent exposure of various protein regions in the folded and unfolded states.** Polar and nonpolar residues are colored in magenta and green, respectively. The solvent (water) molecules surrounding the protein are represented as spheres. In the folded state, nonpolar residues reside mainly in the protein core away from the solvent, whereas in the unfolded state both polar and nonpolar residues are exposed to the solvent. Still, the folded state contains patches of hydrophobic residues, as explained in the main text. Water molecules were added to the protein computationally by the PDB_hydro server [276] (Delarue group, Institut Pasteur [277]).

---

residues' amino and guanidino moieties, compared to the high level of hydration of aspartate's and glutamate's carboxylate groups [268].

2.4.1.3   Stabilizing forces

The process of protein folding brings together amino acid residues that are very distant from one another in the protein sequence, and thereby enables these residues to interact with one another. In globular proteins there are thousands of noncovalent interactions (van der Waals, electrostatic, nonpolar) between residues [278]. The large number of interactions is of great importance; it serves to stabilize the entire structure, but since each of the interactions is weak (compared with a covalent bond), the structure is still dynamic, which is crucial for its function (see Chapter 5). Nonpolar interactions are obviously strongest in the protein core, which is composed mainly of nonpolar residues. The tight atomic packing in the core also optimizes van der Waals interactions. Electrostatic interactions are most common on the surface of the protein, where polar residues can interact with each other or with the surrounding water molecules (and ions) of the solvent. As mentioned above, polar residues are not limited to the surface, and can also be found, although in much lower numbers, at the core. This issue is highly controversial, as it is unclear whether such residues stabilize or destabilize the core, and to what extent. A more detailed discussion of this issue is given in Chapter 4. Finally, some membrane-bound and secreted proteins are also stabilized by covalent bonds, most of which are disulfide (S−S) bonds between adjacent cysteine residues [76]. As explained at the beginning of the chapter, disulfide bonds can form only under oxidizing conditions, and therefore (as a rule) do not appear in cytosolic proteins. The other types of covalent bonds are rare, and usually appear in structural proteins, such as collagen (see Section 2.7 below).

2.4.2   Architecture of proteins

All globular proteins possess the basic structural properties explained in the previous section. However, these properties do not necessarily manifest in the same way. In other words, each protein has its own unique fold, which satisfies the basic structural requirements mentioned above, and at the same time allows the protein to execute its own specific function. Therefore, in order to understand the structure-function relationship in proteins, we must address the different folds found in proteins, i.e., their architecture. For clarity of presentation, we divide the architecture of proteins into three basic levels:

1. Simple motifs

2. Complex folds

3. Domains

These levels are reviewed in the following subsections.

2.4.2.1   Simple folding motifs

Folding motifs are simple common combinations of secondary elements. For this reason, they are sometimes referred to as 'supersecondary structures'. Their simplicity allows identical motifs to appear in proteins of completely different structure and function [280]. As a result, it is often difficult to surmise which type of evolutionary process led to the appearance of the same motif in different proteins. More specifically, a given set of proteins with a shared motif may be descendants of a single ancestral protein that included the motif and

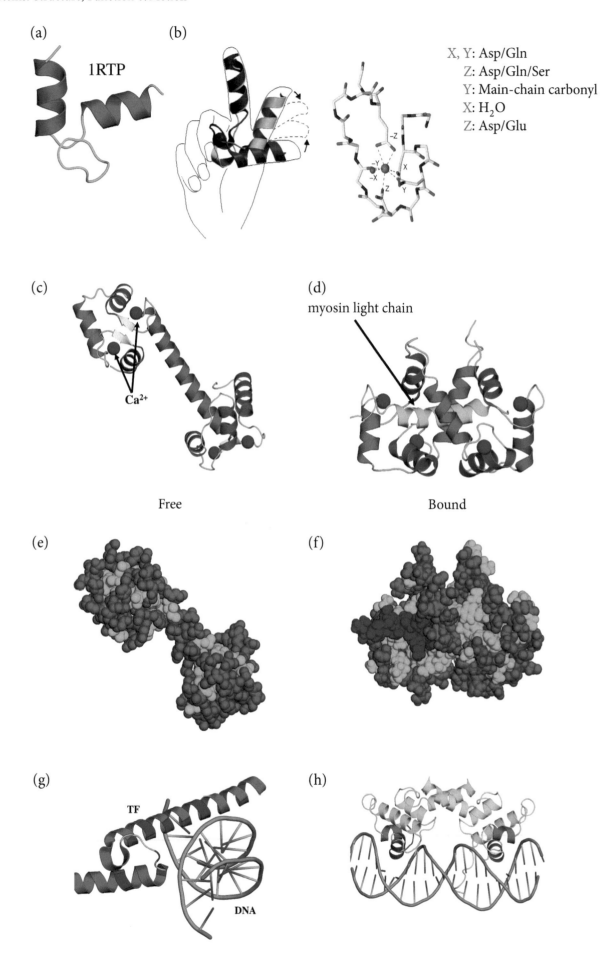

(a) 1RTP

(b)

X, Y: Asp/Gln
Z: Asp/Gln/Ser
Y: Main-chain carbonyl
X: H₂O
Z: Asp/Glu

(c)

Ca²⁺

Free

(d) myosin light chain

Bound

(e)

(f)

(g) TF DNA

(h)

diverged over the ages, leaving the motif unchanged (*divergent evolution* [281]). Alternatively, the proteins may have different ancestors, which acquired the motif independently due to its structural stability or functional advantages (*convergent evolution* [282]).

Since folding motifs are combinations of the same secondary elements, it is easy to group them into three types: *α*, *β*, and mixed [283]. In the following subsections we will review some well-known examples of these motifs. Note that some of the simple motifs are over-represented in certain common complex motifs, which will be described in this subsection as well.

2.4.2.1.1 α motifs

α motifs include only combinations of *α*-helices. One of the simplest motifs in this group is the *helix-loop-helix* (*HLH*), also referred to as '*α-loop-α*', or '*α-α corner*'. It includes two *α*-helices connected by a loop (Figure 2.24a). The HLH motif is very common, although it may serve different roles in different proteins. This phenomenon, termed '*parsimony*', is a central evolutionary strategy. That is, once a 'successful' structure is formed during evolution, it tends to be used in different ways. To illustrate this phenomenon in the HLH motif, we focus on two of its roles: Ca^{2+} sensing and DNA binding.

1. **Sensing intracellular levels of calcium ions (Ca^{2+})**

   Calcium ions (Ca^{2+}) play different roles in biology, including participation in intracellular signal transduction [279]. In this specific signal transduction pathway, a transient surge of Ca^{2+} ions in the cytosol leads to the activation of different downstream processes, depending on the type of cell. For example, in muscle cells this event induces the contraction of muscle fibers during exercise. Other examples of Ca^{2+}-dependent cellular processes include the fertilization of ova by sperm, and the release of neurotransmitters by nerve cells.

   Ca^{2+}-dependent pathways usually begin with the binding of a chemical messenger, such as a hormone or neurotransmitter, to a receptor on the surface of the target cell. The pathway continues with the activation of intracellular proteins, such as G-proteins and enzymes, while creating secondary messengers that mediate signal transfer. Some of the proteins activated during this process are Ca^{2+} channels in the plasma

FIGURE 2.24 **The helix-loop-helix (HLH) motif.** (Opposite) (a) The isolated motif. (b) The HLH motif as part of the EF-hand structure (adapted from [279]). The figure on the left demonstrates the 'hand' analogy of the motif, whereas the figure on the right shows the calcium binding residues. (c) Calmodulin (CaM) in its free state (PDB entry 1osa). The molecule contains four HLH motifs, each binding one Ca^{2+} ion (blue sphere). (d) CaM bound to a peptide derived from myosin light chain kinase (PDB entry 2bbm). (e) and (f) Free and bound states of CaM, with polar and nonpolar residues colored in red and green, respectively. The myosin chain in (f) is colored in blue. The figure shows how CaM folds so as to surround the bound peptide with nonpolar residues. (g) Human transcription factor (TF) Max in complex with DNA (PDB entry 1hlo). Max contains an HLH motif, which positions the second helix such that it can come into close contact with the DNA. Helices are colored red, except for the basic residues, which are colored blue. (h) The structurally similar helix-turn-helix (HTH) motif of the *λ* phage repressor (PDB entry 1lmb). The dimeric protein and the DNA to which the protein is bound are shown as ribbons. The chains of the protein are in green and cyan, where the two HTH motifs (one in each chain) are colored red.

membrane and in the membranes of Ca^{2+}-storing compartments within the cell. These primarily include the ER and sarcoplasmic reticulum (SR) of muscle cells. Although the concentration of Ca^{2+} within these compartments is high, the Ca^{2+} ions are confined within them, and are therefore inaccessible to cytosolic components. The activation of Ca^{2+} channels in the plasma membrane and in the membranes of these compartments leads to a massive influx of Ca^{2+} into the cytosol. In resting cells, Ca^{2+} levels are very low (10 to 100 nM), and therefore cannot invoke any biochemical processes. However, during the activation of the channels, Ca^{2+} levels can rise by up to four orders of magnitude [284–286]. This event is very short-lived, as the excess of Ca^{2+} is immediately pumped out of the cell or into the ER and SR. The elimination of excess Ca^{2+} is of major importance, since prolonged exposure of cytosolic components to Ca^{2+} triggers mechanisms that lead to programmed cell death (apoptosis). However, short-term exposure to the ion during normal signal transduction leads to the activation of numerous 'healthy' proteins.

Most of the proteins activated by the Ca^{2+} surge are enzymes, and their activation either allows the signal to proceed or creates the required end effect. The binding of Ca^{2+} to many of these proteins is mediated via a specific HLH motif called the 'EF-hand' [279]. The motif is referred to as a 'hand' because the two helices and the connecting loop resemble human fingers and the spaces between them, respectively (Figure 2.24b). The term 'EF' refers to the fifth (E) and sixth (F) helices of the protein in which the motif was originally identified, which were included in the motif. The EF-hand serves as a molecular 'switch' in some enzymes, allowing them to react to the signal-induced rise in cytosolic Ca^{2+} levels. The binding of Ca^{2+} occurs in the loop region, and can be mediated by the following chemical groups:

(a) The carboxylate groups of glutamate and aspartate

(b) The amide group of glutamine

(c) The hydroxyl group of serine

(d) Backbone carbonyl groups

(e) Water molecules inside the protein

Interestingly, the binding causes a relatively small movement of the helices. However, at the level of the entire protein, this movement is translated into a global conformational change.

Some enzymes functioning within cells as part of Ca^{2+}-mediated signal transduction do not contain the 'EF-hand motif'. These enzymes rely on another protein, *calmodulin (CaM)*, to bind the calcium ions and activate them; thus, CaM serves as these enzymes' cytosolic Ca^{2+} sensor. Upon binding Ca^{2+}, calmodulin undergoes a conformational change and can subsequently bind its target proteins. Many of these targets are enzymes that act on other proteins when activated by calmodulin. For example, enzymes called *Ca^{2+}/calmodulin (CaM)-dependent kinases* phosphorylate target proteins on serine or threonine following activation by calmodulin. In CaM-dependent kinases, the catalytic activity is normally inhibited due to blockage or distortion of the catalytic site by an autoinhibitory domain [287]. Binding of Ca^{2+}-activated calmodulin to the enzyme relieves the inhibition by disrupting the interactions between the autoinhibitory and the catalytic domains. In smooth muscle, one of these enzymes, called 'myosin kinase', induces muscle contraction by phosphorylating the muscle protein *myosin*. At least 180 different calmodulin-activated proteins have been identified so far, a number that attests to the importance of calmodulin in cellular signal transduction [288].

The structure of calmodulin includes two identical 'lobes', which are connected by a long α-helix (Figure 2.24c). Each lobe consists of two EF-hand motifs, which means that each calmodulin is capable of binding four Ca^{2+} ions. The conformational change following binding of these ions has two results. First, buried nonpolar residues are exposed to the solvent. Second, the shape of calmodulin changes from 'open' to 'closed' [289] (Figure 2.24d). Both of these changes allow calmodulin to bind and activate its target protein. Many of the residues exposed during the conformational change are methionines, whose flexible, hydrophobic side chains allow CaM to accommodate the bulky aromatic side chains of different target proteins [290] (Figures 2.24e and f). When the Ca^{2+} ions are pumped back to their cellular reservoirs, the calmodulin molecules go back to their original conformations, which makes their target proteins resume an inactive state and, in effect, terminates the signal. This phenomenon of a conformational change induced by binding to a small molecule or ion is a central paradigm of protein action, and is discussed in detail in Chapters 5 and 8. Interestingly, the protein *troponin C*, which resides in striated muscle, is structurally very similar to calmodulin, and plays the same role in striated muscle that calmodulin plays in smooth muscle [285].

2. **DNA binding**
A completely different role played by the HLH motif is DNA binding. Proteins that bind DNA fulfill functions in DNA replication, as well as in expression of cellular genes. Gene expression involves transcription factors, which are required to recognize specific nucleotide sequences within the gene. The structure of DNA includes two intertwined helices that create major and minor grooves (Figure 1.3b). The ability of certain proteins to interact specifically with the DNA molecule and 'read' its sequence requires tight binding between the two molecules, which involves penetration of the protein into the grooves. Over the course of evolution, a variety of structures with this capability have emerged; one such structure is the HLH motif. As Figure 2.24g shows, one of the helices in the motif is inserted into the major DNA groove, which in turn facilitates direct interaction between the amino acid residues of the helix and the nucleotides that make up the groove. Since many of the DNA-interacting amino acids are basic, this motif is often referred to as 'basic HLH' ('bHLH'). Why is the entire motif needed instead of the interacting helix alone? As it turns out, the other helix and the interconnecting loop are important for the positioning of the first helix in a way that enables it to penetrate the groove. Thus, the specificity of the interaction results from both the sequence of the helix and its orientation with respect to the DNA molecule.

Additional examples of DNA-binding structures, other than the HLH motif, include the following:

- **Zinc finger** (see Subsection 2.2.1 and Figure 2.8 above). The name of this structure comes from its finger-like shape, containing a zinc cation in its middle. The structure of a zinc finger includes a β-hairpin packed against an α-helix.

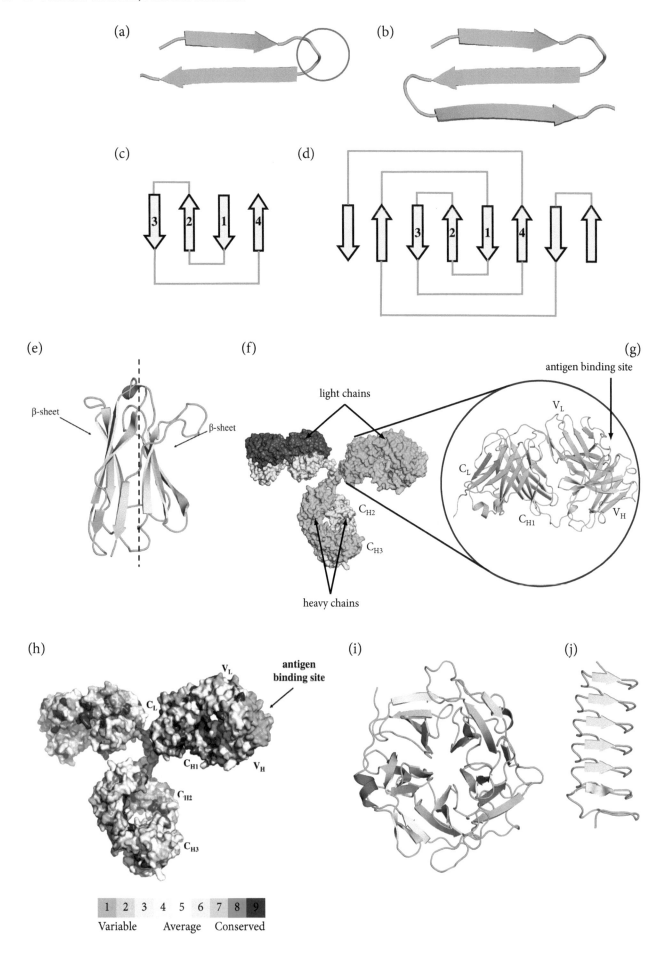

(a)

(b)

(c)

(d)

(e)

β-sheet

β-sheet

(f)

light chains

C_{H2}

C_{H3}

heavy chains

(g)

antigen binding site

V_L

C_L

C_{H1}

V_H

(h)

V_L

C_L

antigen binding site

C_{H1}

V_H

C_{H2}

C_{H3}

(i)

(j)

| 1 | 2 | 3 | 4 | 5 | 6 | 7 | 8 | 9 |

Variable   Average   Conserved

The zinc ion is held in place by coordinate bonds with histidine and cysteine residues. This motif is very common in proteins that regulate gene expression, and while it usually serves for DNA and RNA binding, it can also bind amino acids in protein-binding domains [106]. This dual functionality demonstrates the parsimonious nature of evolution, as shown above for the HLH motif.

- **Helix-turn-helix (HTH).** This motif is usually associated with phage repressors (e.g., the λ phage repressor, Figure 2.24h), but has been found in numerous other proteins that regulate gene expression, in organisms that range from viruses to humans. It is similar to the HLH motif, although the linker connecting its helices tends to be shorter (a 'turn'). Despite their similarity, the HTH and HLH motifs are distinct from each other.

2.4.2.1.2 β-motifs

The simplest β-motif is the *β-hairpin*, which consists of two anti-parallel β-strands connected by a turn (Figure 2.25a). Extension of this motif with additional β-strands creates a *β-meander*, whose name refers to the direction changes made by the polypeptide every few residues (Figure 2.25b). These two motifs allow the polypeptide chain to fold into a compact structure, but do not seem to play a specific functional role *per se*. A slightly more complex motif is the *Greek key*, which is a β-sheet composed of four anti-parallel strands (Figure 2.25c). The pattern formed by the strands is reminiscent of paintings found on relics from ancient Greece, hence the name. In this pattern, strands 1 and 2 are connected by short loops, whereas strands 3 and 4 are connected by a long loop, which allows strand 3 to complete the sheet, and strand 4 to hydrogen-bond with strand 1. Like the two previously discussed motifs, the Greek key has not been implicated in a specific function so far, although it may appear in more complex folds, such as the *jelly-roll* (Figure 2.25d) and the *β-sandwich* (Figure 2.25e). The latter constitutes the basis for a common complex fold, the *immunoglobulin (Ig) fold* (Figures 2.25f–h), which will be described in Subsection 2.4.2.2 below. Figures 2.25i and j depict other β-motifs: the *β-propeller* and the *β-helix*, respectively.

FIGURE 2.25 **β motifs.** (Opposite) (a) and (b) Three-dimensional structure of the β-hairpin and β-meander motifs (respectively), represented implicitly by ribbons. (c) and (d) Schematic representation of the Greek key and jelly-roll motifs (respectively). (e) Three-dimensional structure of the β-sandwich motif (PDB entry 1igt). (f) and (g) The immunoglobulin fold (IgF). The full atomic structure of IgG and an enlargement of one of its antigen-binding 'arms' (secondary elements only) are shown, respectively (PDB entry 1igt). The C_H (constant-heavy), C_L (constant-light), V_H (variable-heavy), and V_L (variable-light) domains, as well as the antigen-binding site, are denoted. (h) The evolutionary conservation of IgG, as calculated by the ConSurf server (http://consurf.tau.ac.il) [196,197] (cyan – lowest, maroon – highest; see color code in figure). The hypervariable antigen-binding site is marked, in addition to other regions. (i) The β-propeller motif (PDB entry 1gyd). (j) The β-helix motif (PDB entry 1ezg).

2.4.2.1.3 Mixed motifs

Mixed motifs include both α and β elements. The most common of these motifs (and probably also of all supersecondary motifs) is probably the β-α-β motif, which can be found in parallel β-sheets (Figure 2.26a). The motif has two main advantages, which are most likely the reason for its prevalence. First, the positioning of the α-helix against the two β-strands facilitates efficient packing of nonpolar residues within the protein core. Second, the loop connecting the first strand and the α-helix is capable of ligand binding. In these cases, the β-α-β motif is part of a larger fold building the binding site, and we will therefore discuss it in the next subsection, which describes complex folds.

2.4.2.2 Complex folds

The term 'fold' was coined in 1973 by Rao and Rossmann in relation to nucleotide-binding proteins [291], but was never really defined. In most cases, the term is used to describe a specific arrangement of the polypeptide chain, created by a certain combination of simple motifs. Protein *domains*, discussed in the next section, are often characterized by a certain fold. For this reason, the two terms are sometimes confused. Here, we discuss protein 'folds' in terms of specific structures, and 'domains' in terms of general organization, function, and evolution. The number of complex folds is vast, which makes it impossible to discuss them all. However, some folds, referred to as 'superfolds' [292–294], are common (e.g., [295]) (see below for details). In the following subsections we will discuss four superfolds; the first, the *immunoglobulin (Ig) fold*, is based on a β structure (β-sandwich), whereas the other three are based on the same mixed motif, β-α-β [283]. Of the latter, two, the *Rossmann fold* and the *P-loop fold*, function in nucleotide binding. The third, the *TIM barrel fold*, constitutes the active site of numerous enzymes.

2.4.2.2.1 The immunoglobulin (Ig) fold

In the previous subsection we mentioned the Greek key, an all-β motif that appears in many proteins. One of the structural forms often taken by this motif is the β-sandwich. It is formed by two different β-sheets stacked one on the top of the other (Figure 2.25e), hence the name. This arrangement involves interactions between the side chains of the stacked sheets. These interactions stabilize the protein, and provide an efficient way of packing nonpolar residues at the protein core. The stability of this structure makes it a good scaffold for functional sites. The β-sandwich motif builds one of the most common complex folds in proteins, the *immunoglobulin fold*, called that because it was first identified in antibodies, also known as 'immunoglobulins' (Igs) [296]. In fact, this fold appears in numerous proteins of higher organisms, which are involved in molecular recognition. Some of these proteins are associated with the immune system, e.g., the T-cell receptor and the various MHC molecules (see Chapter 1 for functional descriptions), whereas others are membrane-bound proteins mediating cell-cell binding. Despite the structural similarity, many proteins containing the immunoglobulin fold have quite different sequences, and are therefore collectively referred to as the 'Ig superfamily' (IgSF) [297]*1. This superfamily is one of the most common protein groups in vertebrates. In recent years, Ig-like folds have been identified in proteins having no sequence similarity to those of the IgSF. These include receptors to *cytokines*2, as well

---

*1 The difference between protein families and superfamilies is explained in Subsection 2.4.2.4 below.
*2 Chemical messengers of the immune system.

as the cell-cell binding proteins *cadherin* and *fibronectin* [296]. The distribution of IgSF-like proteins is even larger than that of IgSF proteins, and at least one IgSF-like protein has been found in bacteria.

To illustrate the functional potential of the β-sandwich, let us look at IgG, a type of antibody. This protein contains four polypeptide chains, two large ones and two short. These are referred to as 'heavy' and 'light', respectively. The two heavy chains are bonded to each other via disulfide bonds, forming the shape of the letter Y (Figure 2.25f). Each of the light chains is attached via disulfide bonds to the upper portion of a heavy chain. Both heavy and light chains are composed of segments of the polypeptide chains organized as β-sandwiches, and connected sequentially by linkers. As mentioned in Chapter 1, the animal body contains numerous antibodies that have the same general structure, yet each is unique in its ability to recognize a different foreign element. In the Y-shaped structure of the antibody, the 'stem' is the part that is nearly identical among all antibodies, whereas the 'arms', forming the antigen-binding sites (Figure 2.25g), differ. As might be expected, this difference results from the amino acid variations in this region (Figure 2.25h). This is why it is referred to as the *'hyper-variable region'*. The hyper-variable region includes both heavy and light chains, which together form the antigen-binding site. Specifically, the binding site is built from two Ig folds, one contributed by the heavy chain, and the other by the light chain (Figure 2.25g). Again, each Ig fold is basically a β-sandwich, including nine strands and loops. Interestingly, the specificity of antigen binding is determined by the six loops in the structure (three from each chain), rather than the strands. This is one of the most specific interactions in nature, primarily due to its electrostatic component [298].

The use of loops for creating a highly specific binding site may seem strange at first, as loops are highly flexible and their sequence has a low degree of evolutionary conservation, a quality often identified with functionality. In fact, these properties make loops best suited for this role. Loops have greater tolerance for the sequence variability of the antigen-binding site than do ordered structures. This is because the flexible nature of loops can accommodate the resulting structural changes, whereas ordered structures are limited to their principal shapes. The process creating this variability is one of the most fascinating processes in biology. The process, called *'gene rearrangement'*, generates multiple combinations of the genes coding for the variable parts of antibodies. This process is particularly prominent in loops, for the reasons explained above. The inherent structural flexibility of the loops increases the variability of the antigen-binding site even beyond that which is conferred by sequence variations [299]. That is, the loops are capable of undergoing spontaneous conformational changes, which change the shape of the binding site, and as a result, allow the same antibody to bind different antigens [300].

As mentioned above, the Ig fold is very common, appearing also in proteins unrelated to the immune system. The prevalence of this fold is probably due to additional advantages that it possesses, such as a stable nonpolar core, high resistance to proteolysis, and ability to interact with other folds [296].

2.4.2.2.2 *Rossmann fold*

The so-called 'Rossmann fold' was described in 1974 by Michael Rossmann and coworkers, who identified it as being responsible for the binding of dinucleotide coenzymes in different proteins [301,302]. In fact, this conclusion was based on the analysis of four NADH-binding enzymes. In addition to finding a common fold, Rossmann and colleagues' analysis also revealed that, in all enzymes possessing the fold, the nucleotide coenzyme had the same conformation and orientation with respect to the polypeptide chain. Since then, the Rossmann fold has been detected in numerous enzymes that use the dinucleotide coenzymes NADH, NADPH, or $FADH_2$ (Figure 2.26b). As explained in Chapter 1, these enzymes are involved in central cellular processes, such as energy production, biosynthesis, photosynthesis, and chemical detoxification. The prevalence of the fold is not surprising considering that the dinucleotides mentioned above are themselves among the most common cofactors in enzymes.

The Rossmann fold is an extension of the β-α-β motif. Specifically, it is built from two β-α-β-α-β units, with the strands ordered in a 654-123 pattern (Figure 2.26c). This yields two β-sheets, each having three strands connected by two α-helices. The helices are packed against the strands, and the sheets are connected by a long linker. The first helix ($\alpha_{1\rightarrow2}$) contains the sequence motif Gly-X-X-X-Gly/Ala (X denotes any residue, but is usually nonpolar), which strengthens the interaction between the helix and the first strand (β_1) [303]. Interestingly, the sequence motif Gly-X-X-X-Gly can also be found in membrane-bound proteins, where it can mediate interactions between adjacent helices within the protein (see Chapter 7). The two Gly residues are four positions apart in the sequence, and are therefore located on the same helix face. Having no side chain, they allow the two interacting helices to get very close to each other, only 6 Å apart. This proximity optimizes van der Waals interactions and allows the C_α−H group of one helix to hydrogen-bond to a backbone carbonyl group (C=O) in the other [304]. In the Rossmann fold, the effect of the Gly-X-X-X-Gly/Ala motif is very similar; the Gly residues allow the $\alpha_{1\rightarrow2}$ helix and the β_1 sheet to get very close to each other, which in turn optimizes their interactions. In this case, however, the interactions also include nonpolar interactions between Val-6 and Val-8 of β_1, and Leu-22 of $\alpha_{1\rightarrow2}$ (Figure 2.26d). Leu-31 of the β_2 strand also participates in these interactions, further stabilizing the fold. In some NADPH-binding proteins, the first glycine residue may be replaced by alanine, serine, or even proline, in order to increase the distance between the interacting helix and strand. This substitution is required in order to accommodate the phosphate group of the coenzyme, which is absent in NADH and $FADH_2$ (ribose-2′-PO_4^{2-}) [305].

The stability of the polypeptide region that includes the $\alpha_{1\rightarrow2}$ helix and the β_1-sheet is of great importance, as the loop connecting the two components is where the pyrophosphate group ($-PO_3^-$−O−PO_3^-−) of the coenzyme binds (Figure 2.26e). The binding is mediated by hydrogen bonds and ionic interactions between the oxygen atoms of the pyrophosphate group and the protein's side chain and backbone groups. The loop contains a *consensus*[*1] sequence (Gly-10-X-Gly-12-X-X-Gly-15), where Gly-12 allows the loop to come into close contact with the coenzyme's pyrophosphate group [303]. The amide group of Gly-12 also carries the partial positive charge of the ($\alpha_{1\rightarrow2}$) helix dipole, which allows the amide to interact electrostatically with the pyrophosphate [306]. Gly-15, the last residue of the consensus, is also the first residue of the Gly-X-X-X-Gly sequence motif of the $\alpha_{1\text{-}2}$ helix. As explained above, the motif facilitates close contact between the $\alpha_{1\text{-}2}$ helix and the β_1 strand. The other parts of the coenzyme interact with other loops, helices and strands in the fold [305]. Usually, the adenine unit of the coenzyme is involved in van der Waals and nonpolar interactions, whereas the nicotinamide (in NADH/NADPH) or flavin (in $FADH_2$) groups participate in multiple specific hydrogen bonds.

Some Rossmann fold proteins include only one copy of the β-α-β-α-β motif, with ad-

---

[*1]Conserved sequence motif.

ditional β-strands that hydrogen-bond with the first one. The existence of these structures implies that only the first β-sheet of the two mentioned above is actually important for binding of the nucleotide coenzyme [305]. This is supported by the binding interactions described above. Another source of variability among Rossmann folds in proteins has to do with specificity towards the coenzyme. The fold that binds NAD(P)H belongs to a different domain family than the one binding FADH$_2$. It stands to reason that the structural differences between these two variations of the Rossmann fold should be located at the regions binding the nicotinamide and flavin groups of NAD(P)H and FADH$_2$, respectively. Indeed, this is confirmed by the enzyme glutathione reductase, which binds both NADH and FADH$_2$ using two different Rossmann fold domains. Structural analysis shows that the cores of the two respective domains are structurally similar (though they share only 22% of their sequences), whereas the more peripheral parts, responsible for nucleotide specificity, are less similar.

2.4.2.2.3 P-loop fold

The overall high similarity between the NAD(P)H and FADH$_2$ folds suggests a common origin. Yet, as explained above, common structural features can be observed even in folds that are unrelated, i.e., in structures that have very low sequence similarity. This seems to be the case in the mononucleotide-binding fold, that is, the fold responsible for binding ATP and GTP in many enzymes. The fold, referred to as a 'P-loop', is one of the oldest [117] and most common [307] folds in nature. Its prevalence is probably due to the central role of ATP as an energy source for metabolic reactions, and of GTP as a molecular switch in signal transduction pathways[*1]. Although the P-loop fold (Figure 2.26f) is similar to the Rossmann fold, there are some differences. First, the nucleotide-binding loop of the P-loop is longer, and includes all three Gly residues mentioned above. As a result, the loop-nucleotide interaction is tighter than in the Rossmann fold (Figure 2.26g). Second, the presence of three negatively charged phosphate groups in the mononucleotide (Figure 2.26h) requires stronger electrostatic masking than the masking provided by the backbone hydrogen bonds in the Rossmann fold [308]. Electrostatic masking in the P-loop is provided by two cations: the ε-amino group of a lysine residue located immediately after the third Gly in the consensus sequence, and an Mg^{2+} ion (Figure 2.26g). Finally, in ATPases and GTPases, the mononucleotide is the substrate, which means that the P-loop fold is part of the active site, not the coenzyme site, as in the case of Rossmann enzymes.

The evolutionary relationship between the Rossmann fold and Rossmann-like folds (e.g., 'P-loop') is in most cases unclear. The two fold types are often included in one group referred to as 'Rossmannoids'. Some scientists believe that all Rossmannoid folds originate from a single common ancestral fold, which had nucleotide binding capabilities, in addition to a non-specific catalytic activity. This hypothetical fold is thought to have diverged during evolution into different Rossmannoid folds specializing in binding different nucleotides, whereas the proteins carrying it diverged into enzymes with different biochemical activities (oxidoreductases, dehydrogenases, ATPases, and others) (Figure 2.26i).

[*1] As explained in Chapter 1, GTP is used by GTP-binding proteins, act in signal transduction pathways as molecular 'clocks' that limit the time of the signal.

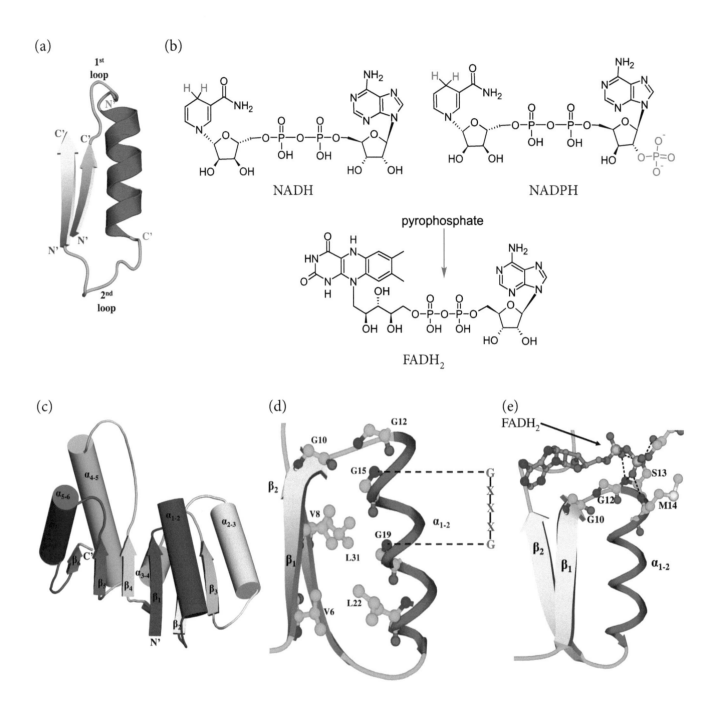

FIGURE 2.26 **The β-α-β motif and Rossmannoid folds.** (a) The structure and topology of the β-α-β motif. (b) The dinucleotide coenzymes NADH, NADPH, and FADH$_2$. (c) The Rossmann fold in the enzyme malate dehydrogenase (PDB entry 1cme, residues 1 through 146). For clarity, helices are depicted as solid cylinders, loops are smoothed, and the entire chain is colored by the sequential β-α units. Strands and helices are numbered. (d) The β_1-loop-α_{1-2} structure in sarcosine oxidase (PDB entry 1el5). The interactions stabilizing the structure are detailed in the main text. The G-X-X-X-G glycine residues are marked. (e) Interactions between the β_1-α_{1-2} loop of sarcosine oxidase and the pyrophosphate unit of its FADH$_2$ coenzyme.

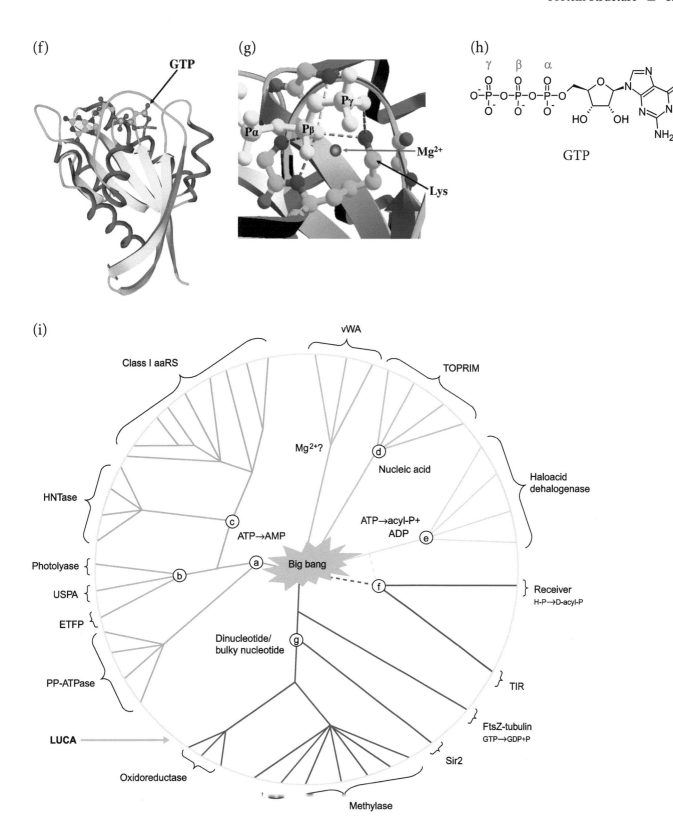

FIGURE 2.26 **The β-α-β motif and Rossmannoid folds.** (Continued) (f) The Rossmannoid fold in p21$_{ras}$ (PDB entry 1jah). The location of the mononucleotide coenzyme is marked. (g) Interactions between GTP (h), and the corresponding loop of p21$_{ras}$. (i) Hypothetical diagram describing the divergence of all known Rossmannoid proteins from their ancestor (taken from [309]).

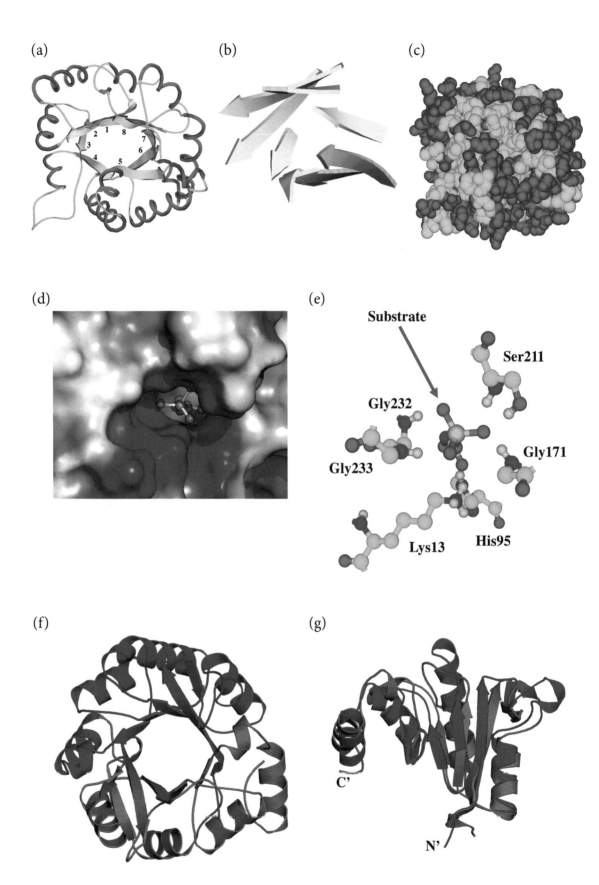

2.4.2.2.4   TIM barrel fold

The TIM barrel is also one of the oldest and most common folds [113,310], observed in ~10% of known protein structures [311]. As in the case of Rossmannoid proteins, proteins that possess the TIM barrel fold may share very little sequence similarity, which makes this fold also interesting in the evolutionary sense. The name of the fold comes from the first protein in which it was identified, the glycolytic enzyme *triosephosphate isomerase*. The fold, comprising 250 residues on average, is doughnut-shaped and has 8 β-strands in the center, surrounded by 8 α-helices in the periphery (Figure 2.27a). The β-strands form a barrel (Figure 2.27b), hence the name. However, in contrast to the β-barrel fold mentioned above, the TIM barrel is not really hollow, and its center is densely packed with nonpolar side chains, particularly the β-branched side chains of valine and isoleucine (Figure 2.27c). The interactions between the β-strands and α-helices are nonpolar as well, which means the fold provides good shielding from the external aqueous environment. Another difference between the TIM barrel and the β-barrel is that in the latter the strands are anti-parallel or mixed [227], whereas in the former the loops between the β-strands are long enough to allow a completely parallel formation. At the sequence level, the fold can be viewed as a concatenation of eight β-α units, which is why it is also called an '8-β/α barrel'. Alternatively, it can be described as a concatenation of five β-α-β units. Thus, the TIM barrel and the Rossmann fold are built from the same basic motif but in a different organization, which makes their three-dimensional structures different.

As mentioned above, the TIM barrel fold appears in numerous proteins having low sequence similarity. Moreover, in many proteins the fold is fused to other folds, located upstream or downstream, and sometimes even in the middle of the TIM barrel. For this reason, scientists suspect the fold to be a result of so-called evolutionary *'gene shuffling'*. In cases where the additional fold appears in the middle of the TIM barrel, the additional fold is located within one of the loops emanating from the C' of the TIM barrel's strands (these loops are called '$\beta \rightarrow \alpha$'). The loops may be originally very long, and therefore the presence of an additional fold within them has a minimal effect on the rest of the TIM barrel fold. The inclusion of additional folds within the TIM barrel significantly increases its functional repertoire, in terms of catalysis, binding other proteins, and oligomerization.

FIGURE 2.27   **The TIM barrel fold.** (Opposite) (a) The topology of the fold (PDB entry 8tim). The strands are numbered by their order. (b) A side view of the β-strands constituting the barrel shape. (c) A sphere representation of the fold, showing the tight packing of nonpolar (cyan) residues within the core. (d) The phosphate-binding site. The surface of the protein is colored according to electrostatic potential, calculated using APBS [195] in the absence of the phosphate group, represented by balls and sticks. Red represents negative potential ($0k_BT/e > \phi > -50k_BT/e$); blue represents positive potential ($0k_BT/e < \phi < 50k_BT/e$); and white represents neutral potential. The positive electrostatic potential of the binding site is complementary to the negatively charged phosphate group. (e) The specific interactions of the substrate in its binding site. The figure reveals Lys-13 and His-95 to be the primary sources of the positive electrostatic potential at the phosphate-binding site, as well as the backbone NH groups of Gly-171, Ser-2111, Gly-232 and Gly-233. These residues come from different segments of the chain, but the three-dimensional fold brings them close together, near strand β_8. (f) and (g) Formation of a TIM barrel fold in the protein HisF by duplication of the gene coding for half of the barrel. The complete structure is shown in (g) with the two half-barrels colored differently. In (f), the two half-barrels are superimposed and shown from the side.

A statistical survey carried out on a database of 900 proteins containing the TIM barrel fold has demonstrated that most of them function as enzymes [310]. Interestingly, these enzymes were found to belong to 5 of the 6 enzyme groups (see Chapter 9), and catalyze 64 different biochemical reactions. In some of these cases, the TIM barrel fold itself participates in the catalysis, with its catalytic residues acting as general acids or bases. Since TIM barrel enzymes catalyze different reactions, the conservation of the catalytic residues is low. The catalytic residues can be located in any of the eight β-strands, but are always on the C' of the strands, i.e., on the $\beta \rightarrow \alpha$ loops [312]. This is yet another example of the importance of loops in forming active and binding sites, as we have already seen in the case of the Ig fold. In both cases, the loops are responsible for creating sequence, conformational, and therefore functional diversity within a fold that has the same shape in all proteins in which it appears [313]. Interestingly, the loops emanating from the α-helices of the TIM barrel towards the N' of the strands (called '$\alpha \rightarrow \beta$') seem to be important for the stability of the fold, rather than its function.

Although the catalytic residues of the TIM barrel fold are not conserved, proteins possessing the fold seem to share several features. One such feature is binding of a phosphate group within the protein's substrate or coenzyme (NADPH, flavin monophosphate (FMN), or pyridoxal phosphate (PLP)). The residues involved in the phosphate binding tend to concentrate around the β_8 strand [310]. Most of them extend over a segment of the polypeptide chain that includes loops 7 and 8, and the N' of helix α_8. Moreover, the phosphate group tends to have the same orientation, regardless of its 'parent' molecule. The binding site is characterized by a positive electrostatic potential, attracting the negatively charged phosphate [314] (Figure 2.27d). The binding mostly involves backbone groups (Figure 2.27e), which explains why the locations of the binding residues are conserved, though the residues themselves are not (aside from the tendency to have glycine residues, which enable the protein and ligand to get very close to each other). Another common feature of TIM barrel proteins is binding of divalent metal cations (Mg^{2+}, Mn^{2+}, and Zn^{2+}), although the location of the binding residues is not conserved across proteins. Finally, in some TIM barrel protein families, certain conserved electrostatic interactions regulate the pKa of catalytic residues, or residues that stabilize negatively charged transition states [313].

The evolutionary process that led to the formation of the TIM barrel fold has yet to be clarified. The presence of this fold in so many enzymes with different sequences suggests a process in which independent proteins converged into a structural fold that provides stability and function [282]. However, the conservation of the location of the catalytic residues, as well as the evidence for gene shuffling, implies the opposite, i.e., divergence of an ancient ancestral fold into the many different functional forms of the TIM barrel fold we observe today. One way to increase our understanding of the evolutionary process of a certain fold is to focus on a single family, i.e., a group of proteins that share the fold and that also have very similar sequences. Studies carried out on TIM barrel families suggest that the basic structural unit of the fold is not the $(\beta\alpha)_8$ barrel, but rather the $(\beta\alpha)_4$ 'half barrel' (Figure 2.27f,g) (see details in [312]).

Two other common mixed folds are shown in Figure 2.28.

(a) (b)

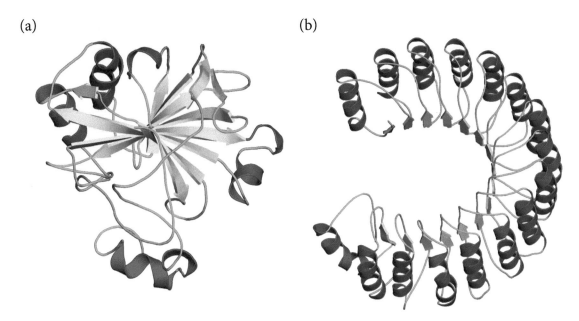

FIGURE 2.28 **Mixed folds.** (a) The β-saddle with surrounding helices, demonstrated in the carbonic anhydrase (PDB entry 1ray). (b) The horseshoe fold, demonstrated in ribonuclease inhibitor (PDB entry 2bnh).

2.4.2.2.5 How many folds are there?

Statistical surveys [282] have demonstrated a 'power law' regarding the distribution of different folds in the known protein universe. That is, most folds appear in a small set of proteins, whereas a few folds (*superfolds* [292–294]), seem to appear in numerous proteins [295,315]*1. Known examples of the latter include some of the folds we have encountered earlier, such as the jelly-roll, immunoglobulin (Ig), Rossmann, P-loop, TIM barrel, and globin-like fold (as in hemoglobin and myoglobin, Figure 2.1), as well as others, such as the β-trefoil and ferredoxin-like folds (Figure 2.29). The prevalence of superfolds suggests that they confer some kind of evolutionary advantage over other folds [282,318–320]. These advantages may include the following:

- Structural stability – Folds that are particularly stable extend the life, and therefore the activity, of the proteins they build. A fold's stability may result from its capacity to achieve compact packing of amino acids or to create secondary structures efficiently.

- Functional efficiency – Certain folds may be able to create binding and active sites more easily than others. Such folds would therefore have an advantage in binding ligands and/or catalyzing chemical reactions, which are the hallmarks of protein function. For example, we have witnessed the ability of the Ig and TIM barrel folds to accommodate binding and active sites (respectively) of similar geometry in numerous proteins, and still confer specificity to those sites.

- 'Foldability' – Certain folds may be able to fold more accurately and faster than others [321].

---

*1 It has been shown that the ten most common superfolds account for more than third of the genes in a typical genome [316,317].

In any case, these advantages would make the folds that possess them 'compatible' with numerous amino acid sequences [322]. The presence of the same folds in proteins of different sequences suggests that there are far more sequences than folds. This demonstrates a central paradigm in the protein universe: **Structure is more conserved than sequence** [323,324]. Since every sequence codes to a protein with a specific function, it can be argued that **structure is also more conserved than function.** We have already encountered an example of a fold that corresponds to multiple different functions: the TIM barrel, which characterizes numerous enzymes, which catalyze completely different reactions[*1]. The high conservation of protein folds suggests that the fold population in nature is not very large. Although the exact number of folds is unknown, studies provide rough estimates of 10^3 to 10^4 [292,294,325,326]. In any case, these numbers are much lower than the total number of distinct proteins in nature [327].

Our discussion of folds naturally leads us to the highest level of tertiary structure, the *domain*, since the core of each domain is characterized by a certain fold.

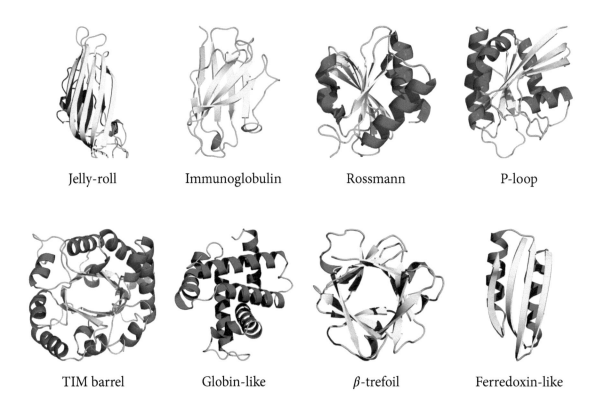

Jelly-roll	Immunoglobulin	Rossmann	P-loop
TIM barrel	Globin-like	β-trefoil	Ferredoxin-like

FIGURE 2.29 **Examples of superfolds.** The PDB entries corresponding to the structures are: 2stv (jelly-roll), 1igt (immunoglobulin), 1jg2 (Rossmann), 1jah (P-loop), 8tim (TIM barrel), 1hho (globin-like), 1bfg (β-trefoil), and 1sc6 (ferredoxin-like).

[*1] Although the different biochemical reactions catalyzed by enzymes of the same superfamily tend to have a common catalytic mechanism (see Subsection 2.4.3.3 below).

2.4.2.3 Domains

2.4.2.3.1 Definition and biological importance

An 'ideal' protein domain is defined structurally as follows:

- A repetitive, compact and stable region of the protein.

- Has its own arrangement of secondary elements[*1], and functions as a semi-independent folding unit [328,329] (Figure 2.30a).

- On average contains 100 to 250 residues.

In reality, however, the definition of domains used by structural biologists varies considerably [330]. Indeed, the various domain prediction algorithms used today emphasize different qualities and employ different structure classification schemes. In addition, there is considerable difficulty in defining domain boundaries. Thus, analysis of the same protein by different tools often leads to different predictions regarding the number and identity of the protein's domains. According to estimates, ~67% of prokaryotic proteins and ~80% of eukaryotic proteins include more than one domain [331,332]. Sequence analysis of about two million proteins shows that although a protein may include up to 12 domains [333], 95% of multi-domain proteins include only 2 to 5 domains [332].

Domains can be defined either by structure or by sequence [329,334,335]. Moreover, many domains possessing a characteristic fold have been found to have a specific function as well (e.g., Figure 2.30b). Accordingly, current protein classification databases tend to take into account the domain's function, not only its structure and sequence. The functions carried out by domains are diverse, but (as expected) the common domains are involved in the most basic functions of cells, such as protein biosynthesis [336], metabolism, and regulation [337]. Consequently, **domains are considered today to be the basic functional units of proteins**.

(a)   (b)

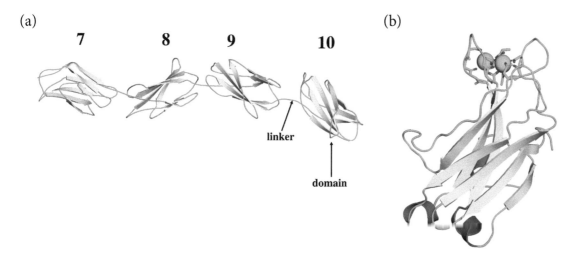

FIGURE 2.30 **Domains.** (a) Domains 7 through 10 in fibronectin typIII (PDB entry 1fnf). (b) The C_2 (Ca^{2+}-binding) domain of the signal transduction enzyme PKC^β (PDB entry 1a25). The Ca^{2+} ions are represented by blue spheres, and the residues binding them by sticks.

[*1]In other words, has a unique fold.

2.4.2.3.2 Domains as evolutionary building blocks

The accepted model describing the formation of new domains from existing ones is that of Ohno [338]. According to the model, a gene coding for a functional protein unit (i.e., domain) is duplicated, and the new product diverges into a domain of different structure and (eventually) function [333]. This process is responsible for the formation of at least 70% of prokaryotic domains, and up to 90% of eukaryotic domains [331]. The process can occur in two ways. In the first, the products of the duplicated gene diverge within the same species to fulfill new functions. These are referred to as 'paralogues', and they can differ significantly in their sequence, structure, and function. In the second case, the divergence of the duplication products accompanies the formation of a new species. These are referred to as 'orthologues', and they are usually similar, since they perform the same functions, just in different organisms. Thus, **in addition to being the basic functional units in proteins, domains are also the basic evolutionary units**.

The frequency of the duplication and divergence processes is demonstrated by the appearance of two-fold pseudosymmetry in many domains (real symmetry involves different polypeptide chains). The symmetry in these cases results from fusion of the two duplication products and their development as half-domains [280]. Such pseudosymmetry can be seen in the HisF enzyme; it has a $(\beta\alpha)_8$ (TIM barrel) fold, with both half-domains remaining similar in shape (r.m.s.d. $= 1.6$ Å[*1]) despite the divergence of their respective sequences (Figure 2.27g). Although domain duplication has been a frequent event in protein evolution, only a few domains have been duplicated extensively[*2] [282]. These are built from superfolds such as the Rossmann and TIM barrel folds, and tend to be involved in the most central functions of cells [339]. Curiously, domains that are involved in protein biosynthesis have not undergone extensive growth in numbers during evolution, whereas those involved in metabolism and regulation have. This makes sense, as protein biosynthesis is a very basic function of cells, and does not require extensive divergence. Conversely, metabolism and regulation contribute to the complexity of the organism, and are therefore expected to change considerably during evolution [282].

With the recent sequencing of the genome of many organisms and application of various protein classification methods, it has become possible to separate all known domains into families, according to their evolutionary kinship. Studies that have done so indicate that domain distribution follows a power law. That is, a few families feature many domains, and many families include only a few domains (e.g., [340]). Also, it seems that large genomes undergo more extensive domain duplication than small genomes do, an observation that is compatible with the demonstrated prominence of gene duplication in evolutionary processes [333].

BOX 2.5 MEASURES OF PROTEIN STRUCTURAL SIMILARITY

Knowing how to measure the similarity between different aligned structures of proteins is very important for structural biologists. The most popular method for carrying out such measurements is calculation of the *root mean square deviation (r.m.s.d.)* [341,342].

[*1] The meaning of r.m.s.d. is explained in Box 2.5.
[*2] They produce many daughter domains.

In this method, the two structures to be compared are superimposed so the equivalent atoms in the two structures are aligned (Figure 2.5.1), and the r.m.s.d. is calculated as follows:

$$\text{r.m.s.d.} = \left(\frac{1}{N} \sum d_i^2\right)^{\frac{1}{2}} \tag{2.5.1}$$

where N is the number of equivalent atoms compared between the two structures, and d_i is the distance (in Å) between the atoms in pair i (one atom from each protein). Since the position of each atom can be described by a set of three Cartesian coordinates (x, y, z), d^2 for the pair of compared atoms in proteins a and b can be described in terms of the atoms' Cartesian coordinates:

$$d^2 = \left(x_a - x_b\right)^2 + \left(y_a - y_b\right)^2 + \left(z_a - z_b\right)^2 \tag{2.5.2}$$

The r.m.s.d. is a measure of the *difference* between the compared structures: identical structures have r.m.s.d. of zero, and the r.m.s.d. value increases with the degree of structural dissimilarity. However, the r.m.s.d. value depends not only on the similarity of the two structures, but also on the choice of the atoms being compared. Choosing only backbone (or just C_α) atoms leads to a much smaller r.m.s.d. value than when all atoms of the structures are compared. That is because many proteins share a general outline (depicted by the backbone), but differ in their side chain conformations, which have many more degrees of freedom. Thus, by focusing on the backbone, structural biologists can classify proteins much more easily, and track their evolutionary path. For that reason, it is customary to calculate r.m.s.d. values on the basis of C_α atoms alone.

Theoretically, identical proteins should have an r.m.s.d. value of zero. However, differences in the conditions under which structures are determined may lead to C_α differences of up to 0.5 Å even when the proteins are in the same functional state (for example, an active state of the receptor; see Chapter 3). For this reason, r.m.s.d. values that are equal to or smaller than 0.5 Å are usually considered negligible. Finally, the r.m.s.d. tends to depend on secondary structure; ordered secondary elements tend to be structurally conserved to a greater extent than loops (Figure 2.5.1).

In some cases, r.m.s.d. is not the most suitable method for comparing between protein structures. One of these cases is when the compared proteins are of different size (i.e., sequence lengths). Such cases are encountered, e.g., in template-based methods for protein structure prediction, that is, methods that predict the structure of a protein according to its sequence similarity to other proteins whose structure is already known (i.e., *templates*; see Chapter 3 for details). The target protein and its templates often have different sizes, and their regions of similarity span only parts of their full sequences. In such cases, the r.m.s.d. is calculated only for those parts of the protein that are being compared, while ignoring the *coverage*, i.e., the degree of sequence length according to which the proteins *can* be compared. To illustrate this problem, consider protein A, which is being compared to two other proteins, B and C. Their size and similarity are as follows:

- Protein B – same size as A; r.m.s.d. = 2 Å.

- Protein C – twice the size of A; r.m.s.d. = 1 Å (in their overlapping parts).

According to the r.m.s.d. method, protein *C* will be considered more structurally similar to protein *A* than protein *B* is. Obviously, this is not necessarily true, as in the second case the overall alignment between the two proteins is poor. To solve this problem, several alternative measures have been designed that incorporate the degree of coverage between the compared proteins in the overall scoring. For example, one can calculate the number of substructures in the two compared proteins in which the distance between the compared atoms is below a certain threshold. This is the case with the *global distance test (GDT_TS) score* [343,344]. This method uses several thresholds (e.g., 1, 2, 4, and 8 Å), and calculates the score as the average coverage with these thresholds. A high-accuracy version of this method, (GDT_HA) uses smaller thresholds.

One problem with the GDT approach is that the absolute magnitude of the score becomes less meaningful as the compared proteins become smaller. This issue is addressed by another popular method for calculating protein structure similarity, termed the *'template modeling' (TM) score* [345]. This method normalizes the score so there is no bias to the length of the target protein. Moreover, the calculation considers all residues, not just those whose r.m.s.d. falls below certain distance thresholds. The values of the TM-score range between 0 and 1, where:

- TM-score = 1 for perfectly matched structures.

- TM-score > 0.5 for structures having the same fold.

- TM-score < 0.17 for random, unrelated structures.

The above scores are especially useful in ranking methods for protein structure prediction and refinement, as is done in the biannual CASP contest (see Chapter 3 for details). However, in day-to-day work, especially when the two compared proteins are roughly the same size, r.m.s.d. is an adequate and easy method for quickly assessing the degree of structural similarity. We will therefore use it for the rest of the book.

FIGURE 2.5.1 **The dependence of r.m.s.d. on secondary structure.** Twelve NMR structures of myoglobin (PDB entry 1myf). The figure clearly shows that the secondary structures are much more ordered than the loops.

2.4.2.3.3 Domains and modularity

Most proteins in nature contain multiple domains [333]. If we assume that the basic function of each domain is fixed[*1], proteins can be described as modular entities, whose exact combination of domains creates a complex function. Indeed, this seems to be the case for most proteins. For example, the DNA-binding domain *WHD* can be found in two different proteins. In the first, which regulates the biosynthesis of fatty acids, it borders a regulative domain that binds acyl-CoA[*2]. When a fatty acid is being built, the binding between the regulatory domain and the acyl-CoA intermediate activates the WHD domain, which in turn binds to the gene responsible for the biosynthesis and shuts it down. Thus, the role of the WHD domain in this protein is to enable the negative feedback regulating the fatty acid biosynthetic pathway. In the second protein, an endonuclease, the WHD domain is combined with a DNA-cleaving domain. The role of the WHD domain in this case is to bring the other domain close enough to its intended site of action on the DNA molecule. In conclusion, **the context-dependent combination of domains gives rise to proteins that use a small set of functional units to carry out a much larger set of different complex activities** [331,346]. This is yet another example of the parsimonious nature of the evolutionary process.

The functional diversification conferred by protein domains can be further extended in two ways [333]:

1. By connecting the same domains in different topologies. This is done at the gene level and requires the position of the domain-coding exons to be shifted (i.e., *exon shuffling*). Such a process is not accessible to all domains; some domains tend to appear together in different proteins, and in these cases the domains' order of appearance is usually conserved [332]. In other words, a hypothetical domain couple A and B would usually appear in proteins as either A–B or B–A, but not both.

2. By connecting the same domain in the same topology, but different geometry.

The use of domains as evolutionary 'building blocks' is particularly prominent in proteins that form complex networks, such as those involved in signal transduction. These proteins include a variety of binding or catalytic domains; such variety is needed in order to achieve the sophistication and complexity required for the ramified cellular networks. As would be expected, the number of domains within an organism's *proteome*[*3] correlates with the organism's relative position on the evolutionary ladder [339,347,348].

Consider, for example, two of the most common domains in signal transduction proteins, *SH2* and *SH3* [333]. These domains were originally found in the protein *src*, and were named after it (*SH = src homology*). Both domains recognize and bind certain elements in other proteins. SH2 recognizes phosphorylated tyrosine residues. Its presence enables the protein to recognize membrane-bound receptors, which are activated by phosphorylation of certain tyrosine residues in their cytosolic domains. SH3, mentioned at the beginning of this chapter, recognizes proline residues within certain amino acid sequences [220–222]. Both SH2 and SH3 are frequently found in combination with other domains, thus enabling them to act on specific target proteins (Figure 2.31). For example, in Lck, an enzyme of the src

---

[*1] That is to say, that the characteristic function of the domain does not depend on the protein containing it.
[*2] A long-chain intermediate in the fatty acid biosynthetic pathway.
[*3] All the proteins expressed in an organism.

family that participates in the activation of T-lymphocytes following antigen binding, both SH2 and SH3 appear along with another domain that has kinase activity (i.e., phosphorylation). This combination of domains enables Lck to activate many other proteins as part of the transduction of the original signal from the T-cell receptor to other cellular elements [349]. Another protein, Tec, belongs to a different enzyme family but also participates in the phosphorylation of proteins as part of signal transduction in lymphocytes [350]. Like Lck, Tec also contains the SH2 and SH3 domains. However, it also contains a *PH (pleckstrin homology)* domain, which specializes in the binding of certain phospholipids. This sends Tec to certain locations in the plasma membrane, where it can act on protein elements there.

An interesting trend in the evolution of domains belonging to network proteins has been the increase in binding valence. For example, in transcription factors the evolutionary process has enabled domains that bind same proteins (i.e., *homodimerization*) to bind different proteins as well (*heterodimerization*) [351]. Such a process turned many proteins involved in signal transduction into *hub proteins*, i.e., proteins capable of binding multiple (other) proteins, which considerably increased the complexity of cellular signaling networks.

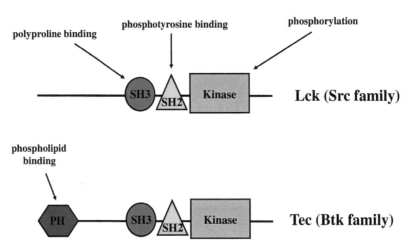

FIGURE 2.31 **Domain composition of the proteins Lck and Tec.** The function of each domain is specified.

Domains that appear in numerous proteins are called 'superdomains'. In addition, over a third of known proteins contain domains that tend to appear in conjunction with certain other domains [333]. These are referred to as 'supradomains'. It is thought that such co-occurrence of domains is a result of coupled duplication of the domains' respective genes. The evolutionary coupling between two (or more) domains is often observed in cases where the coupled domains participate together in the construction of a binding or active site. One example of this type of domain is the P-loop NTP hydrolase, found in 26 different proteins (most of which are translation factors) in conjunction with a domain participating in protein translation. The prevalence of supradomains seems to be quite high [351].

The central role of domains in protein evolution is reflected in their inclusion as a major classification parameter in current protein classification methods. For example, the Pfam database [352], which is fully accessible via the Internet[*1], groups all known proteins into domain families, according to their sequences. For each protein, the domain composition

---

[*1]Pfam: http://pfam.sanger.ac.uk/ and PDBfam (assignments of Pfam domains to PDB sequences): http://dunbrack2.fccc.edu/protcid/pdbfam/

is presented, along with structural and functional information corresponding to each domain. Pfam is currently the largest domain database using sequence information, with over 16,300 domain families. As we will see in the following section, other protein classification tools rely on both sequence and structural information, and are therefore able to identify proteins that belong to a certain domain family, even if they have low sequence similarity to the rest of the proteins in that family. As a result, these databases identify fewer distinct families than Pfam does. A more recent offshoot of Pfam, called iPfam [353]*1, catalogues high-resolution domain-domain and domain-ligand interactions by using data from both Pfam and the PDB.

2.4.2.4   Protein classification

2.4.2.4.1   Importance of classification

Protein scientists' main goal has always been to fully understand the relationships between the sequence, structure, and function of proteins. One way of achieving that goal is to track the evolution of the functional units in proteins, since this evolutionary path goes from the simple to the complex. This can be done by comparing the amino acid sequences of proteins and locating *homologous* segments, that is, segments whose sequences are very similar. As is well known, the various proteins existing today are the product of numerous divergence events starting from a relatively small group of ancestral proteins, and continuing via mutations within their amino acid sequences. Thus, **by comparing the sequences of numerous proteins belonging to the same evolutionary path, scientists are able to construct a model depicting this path, and to use it to assign specific functions to certain amino acids and/or substructures.** However, the longer the divergence period from protein *A* to protein *B*, the less similar their sequences are expected to be. As a result, sequence comparison only identifies relationships between proteins that have recently diverged. **To identify distant relatives, it is necessary to compare proteins also at the structure level, as structure tends to be more conserved than sequence.** Thus, by combining sequence-related information with structural classification of proteins, it is possible to obtain a more extensive and organized framework that can be used to understand sequence-structure-function relationship.

Protein classification relies on a few basic definitions and category types. The first category type, a *family*, denotes a group of proteins that share $\geq 40\%$ of their sequence, reflecting their common evolutionary origin. Again, such close relatives, referred to as 'homologues', can be traced using computational algorithms that follow sequence similarity, and in many cases, their evolutionary path can be reconstructed. Since sequence determines structure, which in turn determines function, proteins with high sequence similarity often also share the same activity. For example, enzymes belonging to the same family catalyze the same chemical reaction type.

When one is dealing with proteins that have only 20% to 30% sequence identity, a common origin can be suspected, but not determined with confidence. The sequence similarity is too low for these proteins to be included in the same family, but since they usually have a similar three-dimensional structure, they might be related in some way. Thus, such proteins are referred to as members of the same 'superfamily' [297]. Again, since structure determines function, at least some of these proteins have similar activities. This is particularly prominent

---

*1iPfam: http://ipfam.org/

in enzymes, where members of a superfamily catalyze chemical reactions that are different, yet share the same catalytic mechanism (see details in Subsection 2.4.3.3 below).

Some proteins are formed by divergence so large that the only feature they have in common is the general fold, expressed in the types of secondary elements, their number, their relative orientation, and the way they are connected. In such a case it is not possible to establish the evolutionary relationship between these proteins (called *analogues*). Instead, it is customary to group the proteins under a general category called *'fold'*. A final protein category type is called a *'class'* [354], defined as a group of proteins whose members have the same overall organization of secondary elements, but not necessarily with the same orientation and/or connectivity. It goes without saying that the evolutionary relationships between these proteins, if they exist at all, cannot be determined, and the grouping together of class members mainly serves a practical purpose in their classification process.

In the following subsection we will see how the categories defined above are used in protein classification. It is important to remember that this process is usually applied at the domain level, which, as we have seen earlier, is often the true evolutionary unit of proteins. In other words, we relate to domain families, superfamilies, etc. It should be noted that while classifications dominate our view of protein space, it is also possible to represent relationships among proteins using maps and networks (reviewed in [355])

2.4.2.4.2 Classification tools

There are currently several protein databases that classify protein domains according to the categories mentioned above. The most popular of these databases are *SCOP* [334] and *CATH* [356], both of which are freely accessible on the Internet[*1]. SCOP (Structural Classification Of Proteins) [334] was developed by Alexei Murzin and coworkers. SCOP uses a five-category hierarchy (Figure 2.32a), from the general to the specific. The first category, *Class*, relates to the general structure of the protein, and includes the following types (Figure 2.32b):

1. Proteins comprising only α-helices (*All α*).

2. Proteins comprising only β-sheets (*All β*).

3. Proteins comprising parallel β-sheets with α-helices connecting the strands (α/β).

4. Proteins comprising anti-parallel β-sheets and α-helices located at different regions of the protein ($\alpha + \beta$).

5. Multi-domain proteins.

6. Membrane and cell-surface proteins and peptides.

7. Small proteins.

8. Proteins consisting of 2 or 3 helices wound around each other (*coiled coils*).

9. Low-resolution structures.

10. Peptides and protein segments.

11. Engineered proteins and artificial sequences.

[*1]SCOP: http://scop.mrc-lmb.cam.ac.uk/scop/; CATH: http://www.cathdb.info/

Evidently, the classification within this category is inconsistent. Furthermore, the *Class* category refers very generally to the structures of its members while ignoring their sequences. As a result, this category gives us no real information about the evolutionary relationships among its protein members.

The second SCOP category is *Fold*. This category, like the *Class* category, ignores sequences, and therefore provides no information about evolutionary kinship. The subsequent categories in the SCOP hierarchy consider both structure and sequence, and therefore are more accurate, and also provide evolutionary information. Specifically, the third category is *Superfamily*, the members of which are likely to be related, although it is difficult to verify their relationships. The fourth is *Family*, which includes evolutionarily-related proteins. This is the only SCOP category in which an automatic classification is carried out using a computer algorithm. The other categories rely on manual classification, based on the intuition and experience of Murzin and coworkers. The fifth and last SCOP category is *Domain*, which includes the individual proteins (each with its unique three-dimensional structure), divided according to the species in which they were found. To illustrate the SCOP classification method, let us use the example of TIM (triosephosphate isomerase), which we have already encountered. The SCOP classification of TIM is as follows (Figure 2.32c):

1. Class: α/β

2. Fold: *TIM α/β barrel* (out of 134 different folds)

3. Superfamily: *triosephosphate isomerase* (out of 32 superfamilies)

4. Family: *triosephosphate isomerase*

5. Domain: *TIM* (appears in 17 different species)

Note in Figure 2.32c how the structural differences between different groups are larger as the category is more general. In the case of TIM, the different *Folds* within the α/β *Class* are significantly different, whereas the different *Superfamilies* are similar. In other cases, such as the *globins*, even families within the same *Superfamily* are significantly different (Figure 2.32d).

The other popular protein classification database is CATH [356], which was developed by Christine Orengo and Janet Thornton. CATH is named after its first four categories (Figure 2.32e):

1. Class: Like the *Class* category in SCOP, this category relates to the differences in secondary elements between proteins, while ignoring their orientation and connectivity. However, the types included within this category are more general than in SCOP (mainly α, mainly β, mixed, and those with a few secondary elements).

2. Architecture: Relates to the overall shape of the domain structure as determined by the orientations of the secondary elements, while ignoring their connectivity.

3. Topology: Similar to the *Fold* category in SCOP, i.e., relates to the overall organization of the secondary structure, while also considering the connectivity of the secondary elements.

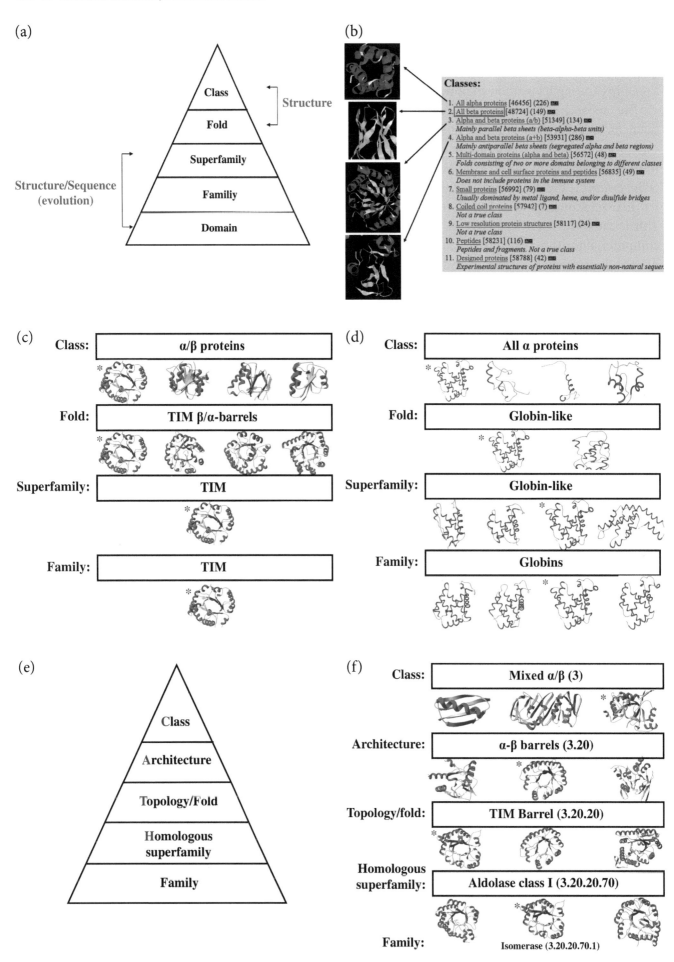

4. Homologous superfamily: like the *Superfamily* category in SCOP. The term 'homologous' here does not express high sequence identity, but rather that the members of this group are considered to have a common origin.

5. Sequence family: Like the *Family* category in SCOP, with the homology threshold being 35%. In addition, CATH provides lists of proteins having sequence identity of > 60% (very similar structures), > 95% (nearly identical structures), and 100% (identical structures).

6. Domain: Like the *Domain* category in SCOP.

The CATH classification is carried out as follows [357]: First, high-resolution protein structures are taken from the Protein Data Bank (PDB, see Chapter 3) and classified into *Sequence families* according to sequence similarity. A representative structure is taken from each family and automatically parsed into its domains. The domains are then classified into the different *Classes*. This is also done automatically, except for complicated cases, which require human discretion. The computer algorithm carrying out the classification relies on the percentage composition of secondary elements, percentage of secondary-structure contacts, secondary-element alternation, and the percentage of parallel β-strands. Within each *Class*, the proteins are further classified into folds and superfamilies based on their structures. Only then, the classification into the different *Architecture* groups is carried out, manually. To illustrate the process, we will use the same example of TIM used above (Figure 2.32f):

1. Class: mixed (α-β).
2. Architecture: α-β *barrel* (out of 14 different types).
3. Topology: *TIM barrel* (out of 13 different topologies/folds).
4. Homologous superfamily: *Aldolase class I* (out of 39 superfamilies).
5. Sequence family: *Isomerase* (out of 39 families).

A third interesting classification method is implemented by ECOD (evolutionary classification of protein domains) [358], a freely available database[*1] developed by Nick Grishin's group. Like SCOP and CATH, ECOD employs hierarchical classification of domains. However, unlike SCOP and CATH, it groups domains primarily by evolutionary relationships (i.e., homology), rather than structural topology. Thus, it can detect relationships between domains even when they are structurally different. Such relationships are often missed by SCOP and CATH. The evolutionary emphasis also allows ECOD to detect functionally-important regions in the classified domains. Another positive feature of ECOD is that it is updated weekly, to reflect new additions to the Protein Data Bank.

FIGURE 2.32 **Protein classification.** (Opposite) (a) The hierarchical classification of SCOP. (b) Detailed view of SCOP's classes. The first four classes are demonstrated by graphic examples. (c) Illustration of SCOP classification with the example of triosephosphate isomerase (TIM). Representative structures are shown in each category, and the one corresponding to TIM is marked with an asterisk. (d) Illustration of SCOP classification with the example of myoglobin. For clarity, structures within the same *Superfamily* and *Family* are shown in the same orientation, determined by superimposition. (e) The CATH classification hierarchy. (f) Illustration of CATH classification with the example of TIM, as in (c). The numbers in brackets are CATH codes for each group.

[*1]http://prodata.swmed.edu/ecod/

2.4.2.4.3   Which tool is best?

SCOP and CATH share some common features, yet differ in others (see detailed comparison in [357]). Both are hierarchical, and both group proteins in two steps: the first is based on sequence similarity, and the second is based on structure similarity. One of the differences between these tools is the distinction between $\alpha + \beta$ and α/β proteins, which is made by SCOP but not CATH. This distinction is important, as explained below. Another important difference between the two databases is the degree of automation in the classification process. SCOP's classification process relies mainly on human discretion, and uses automated procedures only for the grouping of proteins into families, according to sequence similarity. In contrast, CATH uses automation extensively and applies human discretion only at certain points. There are fully automated classification tools, such as FSSP [359], which we will not address here. The degree of automation is most important in the stage of separating each of the proteins into domains. This is because the definition of domains is subjective, and both tools use different definitions. Other classification tools use algorithms that rely on parameters such as surface area, degree of interaction between residues, and hydrophobicity. As a result, it is not uncommon for one tool to miss a domain identified by another tool. In this sense, it seems that CATH tends to assign a larger number of domains than SCOP (justifiably or otherwise)[*1][330], because the definition it uses for domains is strictly structural, whereas SCOP also uses a comparative approach. That is, SCOP also inquires as to whether the domain in question is observed in other superfamilies as well. If not, chances are it is not a real domain. Since SCOP achieves better scores than CATH in some cases but worse scores in others, it is very difficult to draw any conclusions regarding the preferability of either approach. A similar problem arises in the case of the *Fold* (*Topology*) assignment, as the definition of folds is also arbitrary. However, in this case the tendencies of the two classification tools are the opposite to those in the *Domain* assignment. That is, most of the highly populated *Topology* groups in CATH are further classified into *Fold* groups by SCOP. The reason for this tendency probably has to do with CATH's structural approach, which may group two folds into one as long as they share a certain geometric feature (for example, a parallel β-sheet with two α-helices on both sides is present in all proteins that CATH considers as possessing the Rossmann fold). Another difference between CATH and SCOP is the *Architecture* category, which is included only in the former.

In conclusion, despite the clear differences between CATH and SCOP, it is virtually impossible to determine unequivocally which is the best tool, since there are no fixed rules regarding the definition of the structural units in proteins. SCOP's tendency to include considerations that are beyond the mere structure is helpful for the integration of evolutionary information into the classification process. However, the emphasis of SCOP on human discretion makes it somewhat cumbersome. CATH provides important structural information on the protein, and even contains an additional classification category [357]. However, its reliance on automated classification procedures may lead to errors in borderline cases.

We should also consider the possibility that finding a foolproof method for classifying proteins into discrete structures is impossible to begin with. That is, the structural space of proteins may not include discrete folds (as in SCOP) or topologies (as in CATH), but rather a continuous range of structures [360–365]. This is suggested by different studies, in which numerous protein structures have been compared at the sub-domain level. The studies indicate that even seemingly-different structures may share common features such as sec-

---

[*1]Only 70%–80% of the domains classified in SCOP and CATH have similar domain boundaries [330].

ondary structure arrangement. For example, **Kolodny and coworkers have demonstrated that protein structural space contains both discrete and continuous parts** [366]. By constructing networks of SCOP domains based on sequence and structural similarity, the authors found that all-α, all-β, and $\alpha + \beta$ class domains mostly populate the discrete parts of protein space, whereas α/β domains mostly populate the continuous parts. **Folds in the continuous parts of protein space are related to each other by sub-domain sequence-structural themes**[*1]. Many of these themes are recurring and appear in different combinations in different domains. This pattern suggests that such **sub-domain themes may be the real evolutionary/functional building blocks of proteins.** To address this possibility and to allow for broader detection of domain relationships, the SCOP team recently developed *SCOP2* [367][*2], which replaced SCOP's simple, tree-like classification with a more complex, network-based one.

As mentioned above, α/β domains are highly connected to one another within the protein's structural space, and tend to mix-and-join their sequence-structural themes [366]. α/β domains are also known to be older [327], more stable, more frequently involved in domain fusion events, and more functionally diverse than the other domain classes. The functional diversity of α/β domains may result from their higher tendency (as compared with other domain classes) to mix-and-join their sequence-structural themes, though it is yet to be determined why they have this tendency.

2.4.2.5   Knotted proteins

The folds described above, whether complex or simple, display elegant shapes and forms, some of which contain internal symmetry. The elegance of these forms, along with the common grouping of secondary elements in distinct regions of the protein, allows us to imagine how such a structure could fold from an open chain into a compact structure. However, a study of carbonic anhydrase B (CAB) in 1977 [368] and a later survey of known protein structures [369] revealed a phenomenon that did not seem to be in line with the notion of the compact elegance of folds – CAB's chain was folded onto itself, forming an internal knot. Since then, over 1,300 proteins have been found with different types of knots [370] (see Figure 2.33 for examples), showing that CAB was not an isolated case. Furthermore, the knots have been shown to be preserved within and between protein families despite their large sequence divergence [371].

The rarity of protein knots[*3] can be rationalized in different ways, from local geometric aspects of the protein chain [372] to the probable difficulties in creating a knot during folding [373][*4]. Why, then, were knots formed in the first place and preserved in certain proteins? One suggestion is that internal knots allow the cell to extend the lifetime of certain proteins by making it harder for them to unfold and enter the proteasome [375] (see Subsection 2.6.4.3 below for details on the ubiquitin-proteasome degradation system in cells). Other suggested roles include enhancing thermal, mechanical, and/or kinetic stability of proteins, stabilization of protein transporters, help in shaping enzymatic binding sites, and altering enzymatic activity (see [374] and references therein).

[*1]The size of such motifs is typically a few dozen residues.

[*2]http://scop2.mrc-lmb.cam.ac.uk/

[*3]Knotted proteins constitute ~0.8% of all proteins with known structure [370].

[*4]Indeed, the folding of knotted proteins is slower compared to that unknotted ones [374] and most likely requires the assistance of chaperonins *in vivo* [373].

There are several freely available online tools that can assist in predicting and characterizing protein knots. These include the following:

KnotProt [370]: A database of proteins with knots (URL: http://knotprot.cent.uw.edu.pl/introduction).

pKnot [376]: A web server for detecting protein knots in given structures or sequences (URL: http://pknot.life.nctu.edu.tw/).

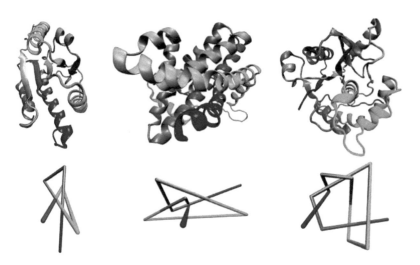

FIGURE 2.33   **Examples of knots in proteins.** The following proteins are shown, from left to right: YBEA methyltransferase (PDB entry 1ns5), Class II ketol-acid reductoisomerase (PDB entry 1yve), and ubiquitin hydrolase UCH-L3 (PDB entry 1xd3). For clarity, the structure is colored using the rainbow scheme with the *N*-terminus in red and the *C*-terminus in blue. The upper panel shows a ribbon representation of the proteins' main chains. The lower panel shows a more reduced representation to make the knots clearer. The image was taken from [375].

2.4.3   Evolutionary conservation of structure and function in proteins

2.4.3.1   Interests of individual versus those of species

In the previous subsections we have seen the tight relationship between structure and function. This relationship implies that once a functional structure is formed, it is most likely to be preserved for the viability of the organism. Evolution seems to contradict this interpretation, as it proceeds through constant changes of the organism's proteins via random mutations in their respective genes[*1]. Most single-point mutations lead only to small local changes in the structure of the protein, around the mutated position [377–379], reflecting the robustness of proteins. However, given sufficient time, the numerous mutations accompanying the evolutionary process are bound to change the structure of the protein, and along with it its activity. As mutational events are random, changes in the structure and activity of a given protein have equal likelihood of being to the benefit of the organism or to its

---

[*1]Note that 'mutation' means any change of amino acid, including changes that lead to the loss of the protein/organism and are therefore missing from the evolutionary record. Mutations that survive, either because they do not affect protein function or because their effect can be tolerated or lead to new function, are called *fixations*. Here, we usually use the term 'mutation' to describe fixation.

detriment. In the latter case, the changes may lead to serious pathological processes. Evolution, which works at the *species* level, solves this problem by using the famous *natural selection* mechanism, which does not allow organisms with too severely damaged genes to survive and to propagate the mutation. At the level of the individual organism, however, the occurrence of random mutations in cellular or tissue proteins is highly problematic. For example, deletion mutations in the regulatory domains of cell-division-promoting proteins might lead to uncontrollable divisions, and as a result to cancer [380]. In fact, even a simple point mutation leading to a single amino acid replacement might manifest as a disease. As explained above, single-point mutations usually do not lead to major changes in the conformation of single protein chains. However, if the mutation replaces a residue that plays a key role in enzymatic catalysis or in ligand binding, then the mutation can easily lead to protein malfunction and a resulting disease. Alternatively, the mutation can introduce a residue that allows the protein to interact favorably with others and form a complex structure with toxic properties. Indeed, this is the case with the well-known disease called 'sickle-cell anemia', which is caused by a single mutation in the protein *hemoglobin*. The extensive research carried out on hemoglobin during the last decades has greatly boosted our understanding of key biochemical and biophysical mechanisms. For this reason, hemoglobin is frequently used in textbooks as a general 'case study', as will be done here as well. In the following subsection we will use the accumulated information to explain sickle-cell anemia down to the molecular level and demonstrate the detrimental potential of single-point mutations. In Chapter 5 we will go back to hemoglobin in order to understand one of the most important protein-related mechanisms in biology, allostery.

Hemoglobin and sickle-cell anemia

Hemoglobin is a globular protein that almost entirely fills the cytoplasm of red blood cells. It functions in carrying molecular oxygen (O_2) from the lungs of animals to their peripheral tissues, as well as carrying the metabolic byproduct carbon dioxide (CO_2) back to the lungs, where it can be exhaled. The protein consists of four polypeptide chains (Figure 2.34a) termed α_1, β_1, α_2, and β_2; each is made of seven (or eight) α-helices connected by loops (Figure 2.34b). The four chains are structurally very similar, but differ in their sequence. Each chain binds a prosthetic group called 'heme'[*1], the function of which is to bind oxygen at the lungs and release it at the peripheral tissues[*2]. Some people have a single mutation on chromosome 11, which causes hemoglobin to organize as long curvy fibers. When these are created, the entire red blood cell is stretched into the shape of a sickle, hence the name of the disease. The stretching of the plasma membrane makes it fragile, and, as a consequence, the lifetime of the cells is reduced from 3 months to only 10–20 days. At the physiological level, the consequence of this reduction in lifetime is a constant state of *anemia*, i.e., shortage of 'blood' (in this case, blood cells). Accordingly, the inflicted person suffers from fatigue and shortness of breath; both are symptoms of anemia. The sickle cells often become physically stuck inside small blood vessels, which might cut off the blood supply to different organs. This leads to complications far worse than the original anemic state, depending on the afflicted organ. Aside from causing severe muscle pain, the insufficient blood supply to major organs such as the brain, lungs, kidneys, and even spleen may lead to their failure, and in some cases even to death.

[*1]May also be spelled 'haem'.
[*2]The molecular physiology of this process is discussed in Chapter 5.

When the amino acid sequence of normal and sickle hemoglobin (HbA and HbS, respectively) was studied, they were found to differ in only one amino acid residue, at position 6 of the β_2 chain (Figure 2.34c). Normal hemoglobin contained glutamate at this position, whereas the sickle form contained valine. To understand the meaning of this replacement, it was necessary to study the three-dimensional structure of the protein. As it turned out, the position affected by the mutation is at the periphery of the proteins, and the mutation leads to a change in the way adjacent hemoglobin molecules interact with each other (Figure 2.34d, top left). Specifically, the branched side chain of valine at this position in HbS creates a protrusion, which geometrically fits right into a hydrophobic depression in the β_1 chain of a nearby hemoglobin (Figure 2.34d, bottom right). This enables the two hemoglobin molecules to adhere to each other, and when the process repeats itself with many other hemoglobin molecules, a long sickle-shaped fiber is formed [381]. This process, which happens mostly when the red blood cell is in a low-oxygen environment[*1], i.e., in peripheral tissues, stabilizes the non-binding form of hemoglobin (see detailed in Chapter 5 below). Normal hemoglobin (HbA) does not form fibers, because it cannot interact with adjacent hemoglobin molecules. As mentioned above, normal hemoglobin contains glutamate at position 6 of the β_2 chain. The carboxylate side chain of the glutamate is negatively charged, and is therefore repelled by the hydrophobic depression in the other molecule. An interesting question is, why was the mutant form retained throughout evolution? The answer is probably that heterozygotes for the mutation are more resistant to *malaria* [382]. Indeed, the mutation is very common in malaria-endemic regions, such as sub-Saharan Africa, India, the Middle East and the Caribbean islands. Resistance to malaria is also conferred by other diseases that compromise red blood cells, such as thalassemia [383] and glucose 6-phosphate deficiency (G6PD, see Box 2.3). Moreover, it is believed that these diseases share at least some of the molecular mechanisms through which resistance to malaria is achieved, e.g., enhanced formation of harmful oxygen radicals [384].

2.4.3.2 Structure conservation: evolutionary mechanisms

The natural selection mechanism should filter out from the evolutionary process proteins that are functionally damaged by mutations, provided that the damage is severe enough to compromise the organism. Since all proteins today are the result of a long evolutionary process involving numerous mutations, we can assume that the mutations leading to the present form of each protein were (at least) not harmful. The case of sickle hemoglobin described above demonstrates the devastating potential of a single mutation. Is this a representative case? It seems not; global structural and functional changes as a result of point mutations are uncommon. One reason is that only a small number of protein residues are of crucial importance to stability and function [385]. An impressive example can be found in the case of *protein G*, which resides on the surface of pathogenic *streptococcus* bacteria. This protein includes two domains, G_A and G_B, which bind to different proteins in the host's serum (i.e., blood fluid). The importance of the binding is probably in camouflaging the bacteria from the host's immune system. The two domains differ in their structure: the structure of G_A includes a three-α-helical bundle, whereas that of G_B includes a combination of an α-helix with a four-strand β-sheet. In a recent study, mutations were introduced into the domains in order to increase their sequence similarity [386]. Surprisingly, the structures of

---

[*1]Under low oxygen conditions the hydrophobic depression, to which Val-6 binds, becomes exposed, thus enabling the binding.

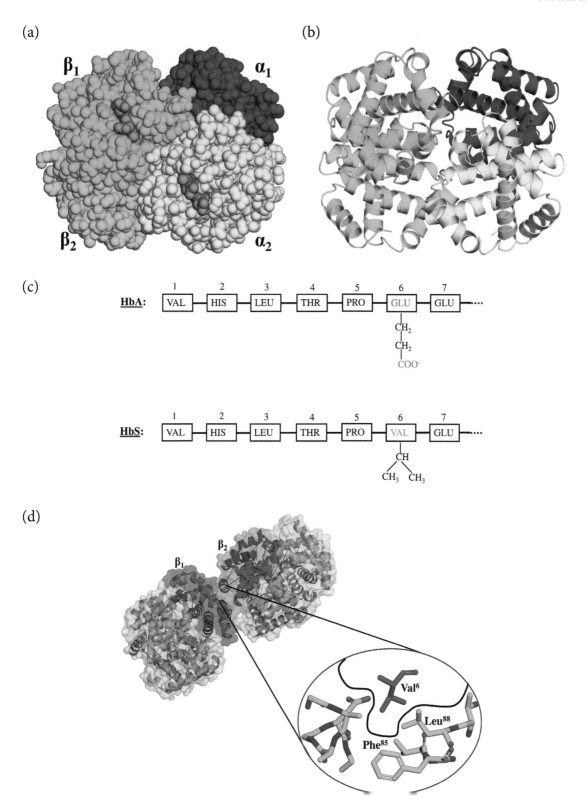

FIGURE 2.34 **Hemoglobin and sickle-cell anemia.** (a) The quaternary structure of normal hemoglobin (HbA). The four subunits are denoted. (b) The secondary structure of HbA. (c) The difference in the sequence of the β_2 subunit between normal (HbA) and mutated (HbS) hemoglobin. The side chains of amino acids at position 6 of HbA and HbS are shown in detail to demonstrate physicochemical differences. (d) The interaction between subunits β_1 and β_2 (in red and blue, respectively) of different HbS molecules. The overall binding topology between the two molecules is shown on the left, whereas the exact nonpolar interaction between Val-6 of β_2 and the hydrophobic pocket of β_1 is magnified on the right.

the two domains were preserved even when the mutations created an 88% sequence identity between them. It therefore seems that the remaining 12% of each sequence, which in this case consists of merely seven residues, is responsible for the structure of the corresponding domain. Even more dramatic support for this conclusion was given by a later study of the same group, in which one fold was transformed into another by a single mutation [387].

How relevant are the results of this experiment to the global evolutionary trends of residue conservation? One way to find out is by comparing sequences of orthologous proteins belonging to different branches of the evolutionary 'tree', and inspecting the locations of the mutated residues. In other words, we can see what types of changes proteins can endure without being structurally and functionally compromised, and which of the protein parts do not tolerate any large changes. Indeed, such studies have characterized many of the primary evolutionary trends and mapped the different regions of globular proteins according to their relative degrees of evolutionary conservation. Their conclusions are as follows:

1. **Functionally important regions of proteins are often more conserved than other regions**. Functionally important regions include residues involved in catalysis, ligand binding, transfer of electrons/protons, etc. It should be noted that in some cases, namely, when functional versatility is required (e.g., in the antigen-binding sites of antibodies), this rule of thumb regarding conservation of functional regions does not hold, but these constitute the minority of cases.

2. **The protein core is more conserved than its surface** [388,389]. This probably results from the core being more densely packed than the surface, and therefore less capable of adapting structurally to a mutation. Indeed, replacing core residues systematically in a way that changes their packing density reveals a high correlation between the stability of the protein and the capacity of its residues to fill voids in the core [273]. The high conservation of the core enables the protein to preserve its overall structure, while its surface is free to change some functional features, such as the specificity of the ligand binding site, and thus to produce different versions of the same protein in the organism, capable of working a little differently [390,391]. In fact, proteins that have only 50% sequence similarity may still share 90% of their core structure!

   In cases where mutations do occur in the core, they are usually compensated for by other effects, so the overall structure of the core in unperturbed. For example, an additional mutation (or mutations) can counteract the effect of the first. Such mutations are referred to as 'coupled' or 'correlated' (see Box 3.3 in Chapter 3). Another way to compensate for the spatial disturbance created by a mutation is conformational changes of nearby residues. This mechanism is less efficient, as core residues are themselves tightly packed, and are therefore limited in their capacity to move. It has been suggested that conformational changes can compensate for changes amounting to 60% of the sequence, but not more [389]. When the original mutation creates a very large disturbance, compensatory insertion/deletion mutations may by required, assuming that the disturbance can be rectified at all.

3. **Secondary elements are more conserved than loops.** The reason for this phenomenon is also a matter of speculation. It could be a result of the tendency of the more densely-packed helices and sheets to appear inside the protein core. In addition, loops are more flexible by nature, and can therefore quite easily compensate for the effect of a mutation by undergoing the required conformational changes.

4. **Conservative mutations are frequently allowed.** These are mutations that do not significantly change the physical-chemical features of the amino acid side chain, and therefore create only a minor structural or functional disturbance. Such mutations may include the following: Glu→Asp, Gln→Asn, Val→Leu, etc. This phenomenon, which is particularly prominent in superfolds [312], allows proteins with different sequences to create the same folds, in accordance with the *structure is more conserved than sequence* paradigm that was introduced above.

In sum, proteins can tolerate mutational changes up to a certain point, beyond which further changes disturb the overall structure and cannot be counteracted by compensatory mechanisms. On the evolutionary scale, the mutations build up until they induce structural/functional changes, which are the basis for creating functional specificity towards ligands and/or environmental conditions.

2.4.3.3 Evolution of function

The evolutionary conservation mechanisms described above are in accordance with the central *sequence → structure → function* paradigm. Nevertheless, the evolutionary development of function may sometimes seem to deviate from this paradigm, as follows:

- Proteins with no common origin and different sequence and/or structure may still perform the same function, provided that they have the same catalytic residues [280]. A well-known example is the *catalytic triad* Ser-His-Asp/Glu, which is shared by the proteolytic enzymes *chymotrypsin* and *subtilisin*.

- Proteins with the same fold may serve different functions following the replacement of their catalytic residues [392,393], or when their cellular location or environmental conditions change. For example, the protein phosphoglucose isomerase acts as an enzyme in the cytosol, but as a nerve growth factor and cytokine when outside the cell [394]. Similarly, the heat-shock protein HtrA has been found to act as a proteolytic enzyme under high temperatures, but as a molecular chaperone under low temperatures [395].

What, then, are the principles guiding the evolutionary development of function in proteins? Again, investigation of enzyme superfamilies may be very helpful in that sense, as these tend to include members of common origin, which have quite different sequences but still maintain similar fold and function. Studies of such superfamilies demonstrate the following principles:

1. **Enzymes belonging to the same superfamily usually share the same principal catalytic mechanism.** This phenomenon results from the evolutionary conservation of the catalytic residues. The individual enzymes differ primarily in the secondary catalytic stages, or in their specificity towards their substrates, coenzymes, or regulating ligands [390,391]. The evolutionary tendency to preserve the general catalytic mechanism while changing the specificity towards a substrate/ligand explains how enzymes of the same origin in different organs/tissues (paralogues) can carry out the same chemical reactions, but on different substrates and with different regulators, in accordance with the metabolic profile of the specific organ [396].

2. Differences in specificity of enzymes towards their substrates may result from simple mutations of the binding residues (e.g., malate/lactate dehydrogenase [397]), but may

also be a consequence of more complex changes. For example, mutations of peripheral residues building the binding site may induce conformational changes in this region, changing its shape or the distribution of charges. As a result, the binding site might lose its geometric and electrostatic compatibility with the substrate, and in some cases even change its specificity to another substrate.

3. Changes of residues on the protein surface may also affect the protein's activity, for example, by changing the connections between functional domains.

As discussed earlier, it is believed that functional evolution in proteins proceeds via gene duplication and divergence of the products (Ohno's model [338]). This may seem to contradict the accepted view of the evolutionary process, in which each gene keeps changing until it achieves a beneficial and required function. In fact, the duplication process solves this problem, as the duplication product (i.e., the gene copy) is free to keep developing without the cell being compelled to pay the price for the loss of its original gene's function. Still, the process of shifting from one function to another by the duplication product seems too drastic from the evolutionary point of view. Some enzymes are *promiscuous*, i.e., capable of catalyzing multiple different reactions or acting on multiple different substrates [398] (see Chapters 8 and 9). It has recently been suggested that such 'promiscuity' could have facilitated the sharp change in function of the protein products of duplicated genes. According to this suggestion, enzymes originally capable of catalyzing two different reactions served as 'intermediates' in the functional shift, and after their duplication the chosen function was made permanent by further mutations [398,399].

2.4.4   Water molecules inside proteins

Protein folding is often described as a process of removal of water from nonpolar residues destined for the hydrophobic core. While this description is essentially correct, experimentally-solved protein structures often include individual water molecules inside the core [400,401]*1. As mentioned earlier, the packing of residues within the core is not perfect, and it is not uncommon to find voids, which are referred to as 'packing defects'. Voids less than 20 Å^3 in volume are completely empty, whereas those of 40 Å^3 or more tend to include water molecules [273,404,405]. This is not a random phenomenon; structural analysis shows that such water molecules interact electrostatically with polar residues in their vicinity and also mediate interactions between residues [406,407] (Figure 2.35). **These interactions lower the energetic penalty 'paid' by the protein as a result of the burial of polar groups inside its hydrophobic core** [408] (see Equation (2.2) in Subsection 2.3.4). Interaction networks involving clusters of core polar residues exist exactly for this purpose. However, the residues are not always positioned closely enough to achieve good electrostatic masking. In such cases, water molecules serve as 'bridges' that allow the separated residues to maintain their interactions. Indeed, the appearance of water molecules in protein voids correlates with the polarity of the residues lining the voids [409]. Further stabilization is achieved by the van der Waals interactions of water molecules with these residues, whether they are polar or not.

---

*1 Since the exchange of these water molecules with the bulk solvent is slower (ns–μs [402]) than that of water molecules on the surface of the protein (ps–ns [403]), core water molecules often appear in crystal structures. For this reason, they are usually referred to as *structural water*.

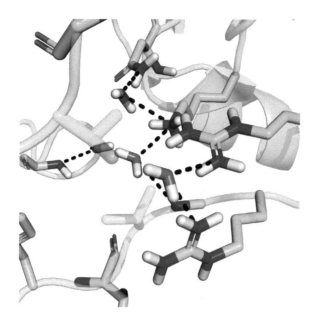

FIGURE 2.35 **Polar interactions mediated by intraprotein water molecules.** The interactions are demonstrated by horseradish peroxidase C1a (PDB entry 1w4y).

In addition to conferring structural and energetic benefits, core water molecules may also contribute directly to the function of the protein. The contributions include the following:

1. **Promotion of catalysis:** The capacity of water to serve as both a hydrogen-bond donor and acceptor enables core water molecules to facilitate biochemical reactions catalyzed by the enzymes [410] (see Chapter 9 for details).

2. **Assisting transport:** Some water molecules present within (protein) ion transporters have been shown to participate in the formation of a transport path [411]. This happens, e.g., in the bacterial protein bacteriorhodopsin, which uses solar energy to pump protons across the plasma membrane [411]. In contrast to ion channels, such transporters do not include a continuous body of water, and the transport of ions within them requires the latter to 'jump' from one polar residue to another within the protein core (see Chapter 7). Again, water molecules may serve as bridges between polar residues whose distance is too great for the ion to jump across.

3. **Assisting in folding:** Protein folding involves the creation of new interactions between residues. Electrostatic interactions are long-ranged, but also depend on the spatial organization of the interacting groups. Thus, during folding, some of the residues that will soon interact with each other are still too far from each other to do so, and it is thought that water molecules aid the folding by temporarily bridging these interactions [412]. Similarly, water molecules within the folded protein may aid conformational changes by bridging interactions during the change [110]. This idea is supported by studies showing that water increases the folded state dynamics of proteins [414,415], which is highly important for their function (see Chapter 5).

4. **Assisting in interactions with ligands:** Water molecules may bridge electrostatic interactions between polar groups of the protein and those of the ligand, as described above for folding [416,417]. Indeed, protein-protein interfaces tend to include fixed water molecules (i.e., not part of the bulk) in a concentration of

1 per 1,000 Å^2 [417], although their distribution is inhomogeneous and much higher around polar residues [418]. A more detailed discussion of protein-protein interfaces will be given in Chapter 8.

The discussion above demonstrates the structural and functional importance of water molecules bound to strategic locations inside proteins. Indeed, such structural water molecules are often found to be evolutionarily conserved, similarly to key binding and/or catalytic residues (e.g., [419,420]). Such water molecules are found in specific locations inside binding and catalytic sites of evolutionarily related (homologous) proteins and serve the same function.

2.5 QUATERNARY STRUCTURE

2.5.1 Introduction

Many proteins, perhaps even most [421], contain multiple polypeptide chains interacting with one another, usually noncovalently. The overall spatial arrangement of these chains is referred to as the protein's 'quaternary structure', and the interactions between them are called 'quaternary interactions' [422]. The chains are often called 'subunits'. The terms used to refer to proteins that have quaternary structures generally end with the suffix 'mer' and reflect the following two qualities:

1. The identity of the polypeptide chains. When the chains are identical the protein is referred to as a 'homomer', and when they are different it is called a 'heteromer'. Most proteins having quaternary structure are homomers, and contain an internal symmetry [423]. In fact, **it has been estimated that over 50% of all proteins form homomers** [424].

2. The number of polypeptide chains: Proteins with 2, 3, 4, 5, and 6 chains are called 'dimer', 'trimer', 'tetramer', 'pentamer', and 'hexamer', respectively. When the number of chains exceeds 2, the general name 'oligomer' may be used. **Most oligomers are small and include an even number of chains** [423,424].

Thus, the protein hemoglobin, which includes four non-identical subunits (Figure 2.34a), may be referred to as a 'hetero-tetramer', or more generally as a 'hetero-oligomer'.

In some cases, although not often, the degree of 'obligation' of the chains to the oligomer is included in the naming. When the oligomeric state alone is the biologically active form of the complex, the oligomer is referred to as 'obligate' or 'obligatory' [425]. 'Non-obligate' or 'non-obligatory' oligomers, in contrast, are biologically active in both monomeric and oligomeric states. Another distinction often made with respect to oligomeric proteins is between the transience and permanence of binding. Non-obligatory complexes that are in a constant association-dissociation equilibrium are said to be 'transient'[*1], whereas oligomers that require a molecular trigger to change their oligomeric state (e.g., signals coming from outside the cell) are referred to as 'permanent' [425]. Note, however, that the permanence of the oligomer may also be affected by environmental conditions such as pH, temperature,

---

[*1]Such transient interactions often form the basis for functional subcellular networks. The complexes involved in these networks are sometimes described as the 'quinary structure' of the proteins forming them [426–428]. As the name implies, this is supposed to be the next structural level achieved by proteins, after quaternary structure.

etc. Also note that the distinction between obligation and permanence of binding can some-times become blurred, and the two concepts are often confused in the literature.

When an oligomer consists of different chains, it is sometimes difficult to determine whether it is a hetero-oligomer of the same protein, or several different proteins working as a complex. This is usually decided according to the degree of obligation presented by the oligomer. That is, if only the oligomer state is active *in vivo* (i.e., an obligate oligomer), the oligomer is considered as one, multi-chain protein. Conversely, if one or more of the separate chains are also active, it is considered a complex of different proteins. It should be noted, however, that some proteins that act as oligomers have subunits that function in isolation, although to a lesser extent.

One interesting question regarding oligomeric proteins is how they developed in the first place. Transient interactions between protein segments may occur inside or outside the cell unaided, but for a protein to become a permanent oligomer, it must undergo a change at the genetic level. There are several genetic mechanisms that could have facilitated the transition of monomers into oligomers. The process probably started with mutations that changed the geometric and electrostatic nature of the proteins' surfaces, to make them more likely to interact with other proteins [429], that is, to turn them into *interfaces*. Such mutations could have been of different types: insertion, deletion, substitution, or recombination, and their number could have been quite low. This is because protein-protein interactions usually rely on a small number of key interfacial residues [430–432]. Although protein-protein interfaces may have emerged only in certain proteins, it is likely that at some point simple structures that had turned into interfaces became independent binding units. Once such 'oligomer-ization units' were formed, they might have attached to existing proteins by genetic fusion, conferring oligomerization capabilities. In other words, such units became *oligomerization domains*, attached in a combinatorial manner to other domains in different proteins. The en-tire process, as described here, has been suggested, e.g., for transcription factors using the helix-loop-helix (HLH) motif for DNA recognition (see Subsection 2.4.2.1.1 above) [433]. Specifically, it was suggested that the oligomerization domain appearing in these proteins was at some point duplicated and inserted into other, unrelated genes. If this is true, this mechanism could explain the prevalence of the HLH motif in nature.

2.5.2 Characteristics

2.5.2.1 Dimensions and complexity

Different proteins comprise different numbers of subunits. Most oligomeric proteins have only a few, yet some have large numbers of subunits. For example, the F_0 component of the famous ATP-forming enzyme ATP-synthase includes 13 different subunits (Figure 2.36). Very large numbers of subunits can usually be found in viral envelope proteins, which may include up to 60 polypeptide chains.

2.5.2.2 Symmetry

Internal symmetry is one of the prominent tendencies of oligomeric proteins [421]. Globular proteins, such as enzymes, receptors and antibodies, usually have a simple point group sym-metry (i.e., two to six folds) [434] (Figure 2.37a), whereas proteins of the plasma membrane or viral envelope, as well as proteins forming soluble fibers in the cytosol (e.g., microtubules), tend to have a high-fold rotational symmetry. Symmetry is probably another consequence of evolution's general tendency towards parsimony, as it allows the cell to achieve a higher function using copies of the same structural unit. It is therefore not surprising that this phenomenon appears in proteins of all three kingdoms of nature. Analysis of protein structures suggests that asymmetrical oligomers appeared only later in evolution [435]. As mentioned earlier, symmetry is often the result of duplication and divergence processes [280].

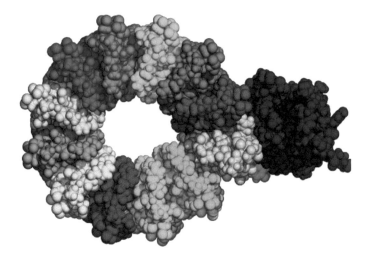

FIGURE 2.36   **The multi-subunit character of the enzyme F_0-ATP synthase (PDB entry 1c17).** The subunits are marked with different colors.

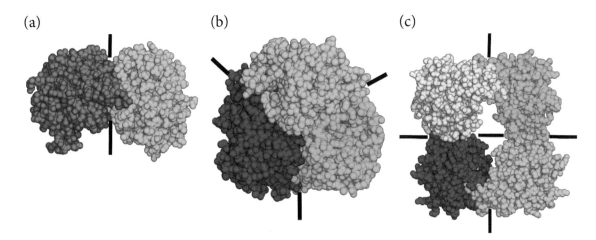

FIGURE 2.37   **Symmetry in homo-oligomers.** (a) Two-fold symmetry in dimeric triosephosphate isomerase (PDB entry 1amk). The binding surfaces of the two subunits are identical and rotated 180° with respect to each other, making the interface 'isologous'. (b) Three-fold symmetry in trimeric chloramphenicol acetyltransferase (PDB entry 3cla). The interface includes two different binding surfaces, i.e., it is 'heterologous'. (c) Dihedral symmetry in tetrameric β-tryptase (PDB entry 1a01), including two isologous interfaces. Dihedral symmetry appears in most tetrameric and hexameric proteins. The axes of symmetry are denoted in each structure.

2.5.2.3 Subunit interactions

Quaternary structures are stabilized primarily by noncovalent interactions, allowing dynamic equilibrium to exist between different oligomeric states of the same protein [421]. As in the case of protein folding (see Chapter 4), the formation of the quaternary structure is driven by the hydrophobic effect [436], and rendered specific by electrostatic interactions. The latter tend to occur between cooperative interaction networks in the interfaces between subunits, with arginine often serving as a bridging element [209]. Still, the interface is mostly nonpolar. We have seen that protein subunits are often formed by duplication of genes coding for monomeric proteins. Since monomeric protein surfaces are more polar than subunit interfaces, there must have been an evolutionary process making the surface less polar. One of the genetic mechanisms involved in this process is thought to be *domain swapping* [437]. In this proposed mechanism, segments in each of two identical proteins interact with their complementary regions in the other protein, which allows the two proteins to interact with each other while maintaining their tertiary structures (Figure 2.38). Mutations in the regions of the chain linking the segments could make the complex more stable than its individual components, thus perpetuating the complex. A small number of mutations suffice for this purpose, which makes the mechanism 'cheap' and efficient [280]. Although this mechanism is speculative, it has been proposed for at least 40 proteins, including barnase and the diphtheria toxin [438]. It is noteworthy that the mechanism also works at the sub-domain level, where two domains interact by swapping an amino acid segment (e.g., a helix or loop) between them [439].

In some proteins the quaternary structure is also stabilized by covalent bonds. In membrane-bound or secreted globular proteins these are mostly disulfide bonds, whereas in certain extracellular fibrous proteins the bonds tend to involve the side chain of lysine. These proteins include collagen (Lys-Lys), elastin (Lys-Gln), and fibrillin (Lys-Glu).

As discussed in Chapter 8, which deals with protein-ligand interactions, the interfaces between protein subunits differ from protein-protein interfaces. Nevertheless, as both cases involve interfacing of polypeptide chains, the physical-chemical principles of the interactions occurring in the two types of interfaces are very similar.

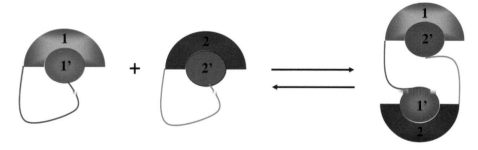

FIGURE 2.38 **Domain-swapping mechanism.** The figure schematically presents two identical proteins, each having two domains: 1 + 1′, and 2 + 2′. Since the proteins are identical, their domains may undergo an evolutionary process of swapping, as described in the main text. The figure is adapted from [437].

2.5.3 Advantages of quaternary structure

Like any evolutionary process, the processes leading to the appearance of quaternary structures must have involved survival advantages to the organisms in which they appeared. Here are some of the advantages conferred by this level of structure:

1. **Active site diversity**

    It is estimated that one out of six of all protein oligomers functioning as enzymes include active sites built from multiple different subunits [429]. A known example is aspartyl protease from the AIDS-causing virus HIV-1 (Figure 2.39). The requirement for more than one subunit provides another level of regulation on the activity of the enzyme. Furthermore, active sites that are built from multiple different chains are more versatile than those built from a single chain within the frame of tertiary structure.

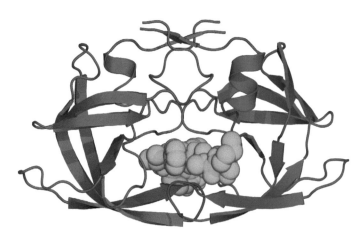

FIGURE 2.39 **The active site of HIV-1 aspartyl protease (PDB entry 1a8k).** The enzyme is shown in ribbon representation, and the peptide ligand as an atom-sphere model. The two polypeptide chains building the entire enzyme (including the binding site) are colored differently.

2. **Time and space coupling of metabolically related processes**

    The main biochemical pathways in cells, such as glycolysis and the Krebs cycle, are carried out in several steps. Different steps are usually catalyzed by different enzymes. However, in some cases, a single oligomeric enzyme executes the steps by using different subunits. This phenomenon is exemplified by the enzyme *fatty acid synthase* (*FAS*), which synthesizes fatty acids from acetyl-CoA (see Chapter 9, Subsection 9.1.5.2.3 for details). In bacteria and plants, each of the biosynthetic steps is carried out by a different enzyme (together referred to as FAS-II), whereas in fungi and higher eukaryotes one enzyme (FAS-I) does it all [440,441]. FAS is a seven-subunit protein; each of the first six subunits is responsible for a different step in the pathway, and the seventh subunit transfers the substrate among the subunits until the process is completed. Such a mechanism is much more efficient than using separate enzymes, as it makes the substrate immediately accessible to each of the reactions, saving the diffusion time from one enzyme to the next in line. A multi-functional enzyme may also take up less space than multiple enzymes, which is particularly important in the dense cytoplasm.

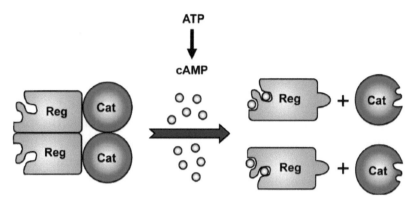

FIGURE 2.40   **A schematic depiction of the enzyme protein kinase A (PKA) and its activation by the secondary messenger cAMP.** The catalytic and regulatory subunits of PKA are marked as 'Cat' and 'Reg', respectively. cAMP is represented by cyan spheres. Binding of cAMP induces the detachment of catalytic and regulatory subunits, allowing the former to bind to other proteins and phosphorylate them.

3. **Regulation of enzyme activity**

   Metabolic enzymes are tightly regulated, for obvious reasons. One of the most common forms of regulation is *allostery*, in which binding of a small organic molecule to the enzyme changes its activity (see Chapter 5 for details). Many enzymes contain catalytic and regulatory domains, with the latter regulating the activity of the former. The regulatory domain is also where the small organic regulators bind. Specifically, the binding induces conformational changes in the regulatory subunit, which affect the activity of the catalytic subunit for better or for worse. A well-known example is the enzyme *protein kinase A (PKA)*, which serves as a key component in one of the hormone-activated signal transduction pathways that regulate numerous cellular processes [442]. Indeed, over 100 different proteins are known to be affected by the activity of this enzyme [443]. PKA acts by phosphorylating certain serine and threonine residues in its target proteins. The phosphorylation is carried out by two catalytic subunits, and is regulated by two regulatory subunits (Figure 2.40). In the resting state of the enzyme the catalytic subunits are inactive. This is because the catalytic subunits are 'blocked' by segments of the regulatory subunits, which are similar (sequence-wise) to the natural substrate of the enzyme[*1]. When the right hormonal signal arrives, a small molecule called *cyclic AMP (cAMP)* binds to the regulatory subunits of PKA, and by inducing conformational changes makes them detach from the catalytic subunits. This lifts the blockage from the catalytic subunits and allows them to bind to their natural substrates and phosphorylate them.

4. **Stability**

   As most biochemists know, aggregated proteins are more stable than those isolated in solution. Protein complexes are particularly stable configurations of bound proteins; they are more organized, and more likely to be biologically active, than the amorphic, inactive aggregate. Protein complexes are more stable than isolated proteins for two reasons. First, the contacts between polypeptide chains restrain their inherent motions, thereby decreasing the likelihood of a loss of structure. Second, in the bound

---

[*1] Except for the serine and threonine residues.

state, the chains are less accessible to proteolytic enzymes. This is nicely illustrated by the zinc-binding form of the hormone *insulin* (Zn-insulin) [434]. Insulin is synthesized in the *pancreas*, and may be stored there until it is needed. In this particular state, the insulin exists as a hexamer, i.e., as six units bound to one another, as well as to two Zn^{2+} ions. The hexamer is particularly stable, which allows it to remain intact inside the pancreas for a few days. When we eat carbohydrates, the insulin is secreted into the blood stream, where it acts to shuttle glucose into cells and maximize the utilization of excess sugar. In the blood, the insulin hexamer separates into its individual chains, each of which has a lifetime of only a few minutes. The short lifetime limits the duration of the hormonal activity of insulin, thus preventing unwanted physiological responses of the body to the excess concentration of the hormone (for example, desensitization). Although this example relates to a homo-oligomer, the same principle also holds for heteromeric protein complexes. In fact, increasing protein oligomerization is one of the strategies that microorganisms living under high temperatures use to increase the stability of their proteins [444].

5. **Formation of large structures**
   Many of the proteins that form large structures in cells (e.g., the cytoskeleton) are oligomers. This is not coincidental; the oligomerization enables cells to construct very complex and elongated structures using a relatively small repertoire of repetitive units [421].

6. **Enhancing protein translation efficiency**
   The translation of RNA into polypeptides is a highly efficient process. Nevertheless, errors may appear during translation, and the chances for such errors increase significantly as the translated chain gets longer. The use of oligomeric proteins enables cells to construct functional units using short polypeptide chains, instead of one very long chain [429]. This strategy significantly decreases the number of errors during translation, and thus the likelihood of producing a defective protein.

2.6   POST-TRANSLATIONAL MODIFICATIONS

2.6.1   Introduction

In Chapter 1 we discussed the reasons for the huge diversity of proteins, focusing on the very large number of arrangements of the 20 natural amino acids that are possible within the protein sequence. Nevertheless, it is still difficult to explain the immense number of proteins existing in nature, which is estimated to be in the millions [445,446], particularly when the number of genes in various organisms is at least an order of magnitude lower. Therefore, there must be another mechanism contributing to the structural and functional diversity of proteins. We already encountered amino acid derivates that are created post-translationally by enzymes. For example, γ-carboxyglutamate, which has an additional COO^- group, functions in the binding of Ca^{2+} ions by blood-clotting proteins. In fact, numerous proteins undergo chemical modifications that render them unique [447]. Such covalent changes are referred to as '*post-translational modifications*' (*PTMs*). They are especially common in eukaryotes; indeed, 5% of the eukaryotic genome codes for enzymes involved in PTM [87]. The modifications are diverse; over 200 different types of PTMs are currently known, involving

15 of the 20 natural amino acids [448]. It has even been argued that PTMs increase the number of unique amino acids in proteins from 20 to about 140 [449]. Indeed, assuming that the ~20,000 protein-coding genes in the human genome are translated into ~10^5 different sequences (due to alternative splicing, mRNA editing, etc.), PTMs are expected to change the number of different proteins to ~10^6 [450,451]! But more importantly, PTMs affect the functions of the modified proteins, including substrate and ligand binding, responsiveness to allosteric regulation, etc. This phenomenon, which works beyond the DNA level[*1], explains the relatively small difference in genome size between higher organisms such as humans, and much simpler ones, such as the nematode worm.

PTMs can be divided into two main types: addition or replacement of a chemical group, and proteolytic cleavage. Most PTMs belong to the former category, i.e., they involve the transfer of a chemical group from one molecule to the protein. Accordingly, during evolution different molecules have become 'professional' donors of such groups. Such molecules include *S-adenosyl-methionine* (*SAM*), which donates methyl groups, acetyl-CoA, which donates an acetate group, etc. [4]. Despite the fact that PTMs are covalent, in many cases they are reversible, as they are readily undone by specific cellular enzymes upon the arrival of the right signal. Out of the ~200 known PTMs, the following five are most common:

1. **Phosphorylation** – attachment of a phosphate group (Figure 2.41a).

2. **Glycosylation** – attachment of carbohydrate groups (glycans) (Figure 2.41b).

3. **Acylation** – attachment of carboxylic acids with fatty chains (Figure 2.41c).

4. **Alkylation** – attachment of fatty chains (Figure 2.41d).

5. **Oxidation**

In addition, membrane-bound and secreted proteins tend to contain disulfide bonds, as discussed earlier in Subsection 2.2.1.3.3.2. Table 2.7 shows some of the general mechanisms used by PTMs in affecting the behavior of proteins. In the following sections we will elaborate on these mechanisms and their importance in the cellular context. Further discussion can be found in Walsh's review paper [87] and book [448].

TABLE 2.7 **General mechanisms of PTM effects on proteins.**

Mechanism	Example
Change of physical-chemical properties	Glycosylation increases stability and water solubility
Regulation of activity	Hormone-induced phosphorylation turns activity of many enzymes on and off
Cellular trafficking	Acylation serves as membrane anchor
Regulation of half-life	Ubiquitinylation tags proteins for proteolysis

[*1] And is therefore called *'epigenetic'*.

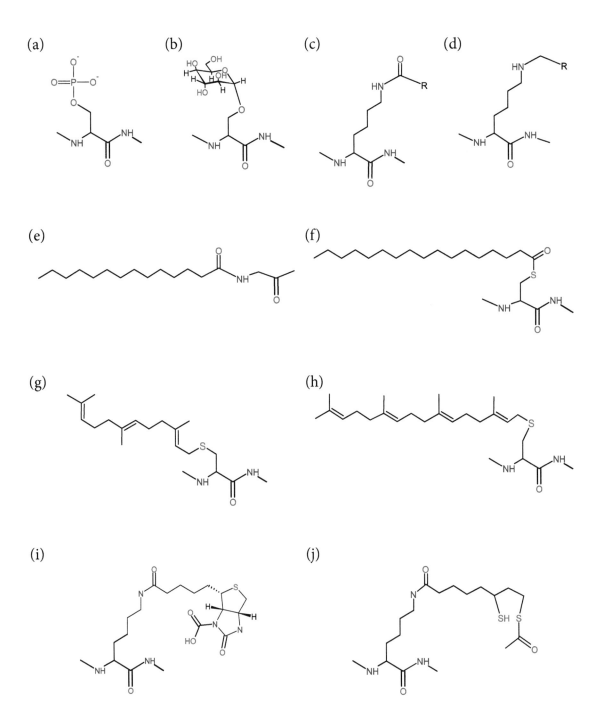

FIGURE 2.41 **Post-translational modifications in proteins.** (a) Phosphorylation of serine's side chain. (b) *O*-glycosylation of serine's side chain. (c) *N*-acylation of lysine's side chain, with R denoting any chemical group. (d) *N*-alkylation of lysine's side chain. (e) *N*-myristoylation of glycine's amino terminus. (f) *S*-palmitoylation of cysteine's side chain. (g) *S*-farnesylation of cysteine's side chain. (h) *S*-geranyl-geranylation of cysteine's side chain. (i) *N*-carboxybiotinylation of lysine's side chain. (j) *N*-lipoylation of lysine's side chain.

2.6.2 Phosphorylation

Protein phosphorylation is a highly common process, in both prokaryotes and eukaryotes [452]. In higher eukaryotes, phosphorylation may involve serine (Figure 2.41a), threonine, or tyrosine residues; in mammals, the frequency of phosphorylation in the first two is nine times higher than in the third [453]. The scope of phosphorylation in higher eukaryotes is thought to reach ~30% of the cellular proteins. Phosphorylation in bacteria and unicellular fungi may also involve histidine and aspartate residues. Protein phosphorylation is a major means of regulating enzyme activity in cells, with the phosphate group serving as an on/off switch. Proteins regulated in this way are involved in metabolism, cross-membrane transport, gene expression, and movement [452]. The regulation may be direct, involving phosphorylation of the key protein or enzyme of the process, or indirect, involving phosphorylation of signal transduction elements that control the activity of the former. The resulting change in the activity of proteins may reach 10^5 to 10^6 fold!

Phosphorylation is carried out by enzymes called *'phosphoryl transferases'*, or, more commonly, *'kinases'*[*1] (see Chapter 9, Subsection 9.1.5 for details). Kinases make up the largest and most diverse group of PTM enzymes, comprising 500 members in humans alone [454]. Kinases can be separated into two main groups. The first and most extensive includes kinases that phosphorylate serine and threonine residues, whereas the second includes kinases that phosphorylate tyrosine residues, usually as part of signal transduction processes involving growth factors. The latter group includes the membrane-bound receptors for these factors, as well as downstream cytosolic proteins, such as *src*. While kinases belong to either the Ser and Thr or Tyr groups, their protein substrates may undergo phosphorylation on both types of residues. For example, the protein *Abl*, which itself functions as a kinase, is phosphorylated on nine tyrosine residues, one serine residue, and one threonine residue [455]. Such amenability to phosphorylation increases the modularity and regulation potential of the protein, as the different phosphorylation states may correspond to different functional states of the protein. As mentioned above, many kinases are part of signal transduction pathways, and therefore are also subjected to phosphorylation. Moreover, some kinases, such as those functioning as membrane-bound growth factor receptors, phosphorylate themselves (i.e., *autophosphorylate*) upon binding of their ligands, which stimulates the intrinsic catalytic activity [452]. Specifically, ligand binding to the extracellular side of the membrane-bound receptor induces formation of a dimer[*2], which in turn promotes the phosphorylation of one monomer by the other (*trans-autophosphorylation*) on its intracellular side. Despite the fact that phosphorylation involves covalent bonding, in the cellular context it is reversible, and can be readily undone by enzymes called *phosphatases*, upon the arrival of the right signal. Again, this property allows the phosphate group to serve as a molecular switch.

Many studies have focused on the mechanism of action of phosphorylation (see review in [452]). The different mechanisms revealed (reviewed below) are mostly based on the phosphate group having a bulky tetrahedral shape with a double negative charge. These properties allow the group to participate in electrostatic interactions with other protein charges; these interactions may be attractive or repulsive, short-range or long-ranged, and in hydrogen bond networks. The main attractive interactions involving phosphate groups are with

---

[*1] The historic name *'phosphorylases'* is also used in the literature.

[*2] Receptor kinases may form both active and inactive dimers. Ligand binding is thought to induce conformational changes within the dimer, which allow it to undergo autophosphorylation [456,457].

the bulky, planar, and positively charged guanidine side chain of arginine. The electrostatic interaction between the two serves an important functional role in molecular recognition. Protein phosphate groups also interact with lysine, tyrosine, serine, threonine, asparagine, and histidine residues, and in prokaryotes, also with metal ions. The specific mechanisms through which phosphorylation affects protein function are as follows:

1. **Inducing conformational changes:** The addition of the bulky, electrically charged phosphate group changes the pattern of noncovalent forces within the protein, and may therefore induce changes in its conformation (e.g., [458]). The interactions of the phosphate group are not limited to neighboring residues, but may include residues located 15 Å away from it. An alternative way of viewing this effect is to envision the protein as a physical entity surrounded by an electrostatic field that emanates from the full and partial charges of the protein. The introduction of the negatively charged phosphate group into the protein creates a perturbation in the field, manifested as changes in both the magnitude and spatial distribution of the energy of the protein. As we will see in Chapter 5, proteins are dynamic, and constantly change conformation. **The dominant conformation is always the one of lowest energy. Thus, when phosphorylation of the protein perturbs its electrostatic field, the protein responds by acquiring a different conformation, one that is able to maintain low energy despite the perturbation** [459,460]. This view may explain why some of the conformational changes documented in proteins following phosphorylation are extensive, and involve residues far away from the phosphorylation site. For example, in the enzyme *glycogen phosphorylase* the addition of a phosphate group to Ser-14 induces conformational changes that translate this residue 50 Å away from its original location! Similarly, a double phosphorylation in the inactivation domain of the *potassium channel* leads to loss of tertiary structure due to proximity of the phosphate groups to two negatively charged glutamate residues. Since protein structure and function are related, structural changes in a phosphorylated protein may lead to corresponding changes in its activity. For example, phosphorylation-induced structural changes in protein kinases may increase their activity either by forming the substrate binding site or by leading to better alignment of the catalytic residues with other catalytic species in the active site or with substrate residues [461].

2. **Affecting ligand binding:** When phosphorylation occurs in a binding site, the added group and its charge may change the affinity between the protein and its ligand, as well as the specificity of binding. Whether the bound ligand is the protein's substrate or a regulator, the change in binding is expected to change the activity of the protein. The change in binding may be positive or negative. For example, phosphorylation of Ser-113 in the enzyme *isocitrate dehydrogenase* lowers the enzyme's activity by creating a steric and electrostatic barrier to the binding of the negatively charged substrate.

3. **Directly affecting catalytic residues:** Binding of phosphate groups to catalytic residues in an active site obviously interferes with these residues' ability to participate in catalysis, but it may also affect other catalytic residues in their vicinity, by changing the spatial and electrostatic properties of the active site.

2.6.3 Glycosylation

The attachment of carbohydrate groups (glycans) to protein residues is also highly common. Until recently, glycosylation was assumed to take place only in eukaryotic membrane-bound and secreted proteins. Today, however, glycosylation is also known to occur in many eukaryotic nuclear and cytoplasmic proteins [462,463], as well as in bacteria [464] and Archaea [465]. Moreover, it seems that the products of this process in bacteria are more diverse than those in eukaryotes. However, since the glycosylation process has been studied mainly in eukaryotic membrane and secreted proteins, we will limit our discussion to these types of proteins. Glycosylation takes place on serine (Figure 2.41b), threonine, or asparagine residues. The process starts in the ER, concomitantly to the insertion of the soon-to-become membrane or secreted protein into the lumen of this compartment. Carbohydrate chains (oligosaccharides) that are already mounted on membrane-bound lipids are transferred to the carboxamide nitrogen of the asparagine side chain. This process, referred to as *'N-linked glycosylation'*, is particularly common in eukaryotes, and is context-dependent [466]. That is, it occurs on asparagine residues that are part of the sequence Asn-X-Ser/Thr (X is any residue except Pro). The chain in its original form is composed of 14 carbohydrate units, including *glucose*, *mannose*, and *N-acetylglucosamine*, with the latter serving as the site of attachment of the chain to the protein. This chain is subjected to enzyme processing, including cleavage of carbohydrate units, their rearrangement, and attachment of *galactose* units [87]. When the process is completed, the protein is transported to the Golgi apparatus via vesicles, and the short carbohydrate chains undergo further processing. In addition, new carbohydrate chains are added to the protein. However, each of the latter is attached to the side chain hydroxyl oxygen of Ser and Thr residues (*O-linked glycosylation*). In contrast to the glycosylation of asparagine residues, that of Ser and Thr residues seems to be context-independent, and the carbohydrate chains produced in the process are shorter. The multiple glycosylation sites in proteins, combined with the numerous ways of organizing the carbohydrate units in the chains, allow glycosylation to create a unique 'personal signature' for each protein, often used for the molecular recognition of the protein by its receptor or ligand. Glycosylation of nuclear and cytoplasmic proteins involves an *O-linked β-N-acetylglucosamine* (*O-GlcNAc*) group, and is carried out by certain cytoplasmic enzymes.

Glycosylation processes within cells create two forms of protein-carbohydrate conjugates. In the first, called *'glycoproteins'*, the protein is the dominant component, in terms of size. Glycoproteins function as membrane-bound proteins (receptors, transporters, etc.) or as secreted elements performing different functions. Examples of these proteins include the following:

1. Extracellular enzymes

2. Hormones

3. Antibodies

4. Nutritional proteins (e.g., the milk protein casein)

5. Transport proteins (e.g., serum albumin in the blood)

6. Mucins: proteins that provide mechanical strength to the mucus enveloping epithelial cells (e.g., those that line the gastrointestinal, respiratory, and reproductive tracts).

The second type of protein-carbohydrate conjugate is called a *'proteoglycan'*; in this conjugate, the carbohydrates are the dominant component. The carbohydrate component in a proteoglycan is made of a long sugar fiber called *'hyaluronate'* (that includes a repetitive sequence of up to 50,000 disaccharides), from which protein chains protrude in opposite directions. These protein chains are also decorated with carbohydrate chains of different types, such as *keratan sulfate* and *chondroitin sulfate*, which include up to 60 carbohydrate units. Proteoglycans can be found in the extracellular space of connective tissues, such as bone and cartilage, where they function mainly in conferring strength and resilience to the tissue. Like glycoproteins, proteoglycans may also appear in mucus that protects exposed epithelial tissues.

The scope of protein glycosylation attests to its importance, and although not all of its roles have been discovered, there are some known advantages it provides [467]:

1. Protection. The sugar coating provides physical protection to proteins against deleterious elements in their environment. Such protection is especially important for secreted proteins, which are completely exposed to proteolytic enzymes and chemically active substances in the extracellular environment. Membrane-bound proteins, which are partly exposed to the external environment, also enjoy this protection. In bacteria, which lack glycosylation capabilities, membrane-bound proteins are protected by the cell wall.

2. Structural stabilization. Studies show that the extent to which a protein undergoes glycosylation correlates with its stability, and, inversely, with its internal dynamics [468–470]. These correlations result from noncovalent interactions (hydrogen bonds, hydrophobic contacts, and CH-π interactions) between the carbohydrate chains, or between the chains and protein polar groups [471]. In addition, it has been suggested that glycan-mediated stabilization may also result from solvent effects [472]; the frequent dipole moment fluctuations of water molecules amplify the inherent fluctuations in the solvated proteins [473–475], leading to a decrease in their stability [476]. Glycosylation physically masks proteins from the surrounding water, and should therefore increase their stability by neutralizing the water's fluctuations before they reach the proteins [472]. However, this protection is a double-edged sword, as it also decreases the natural dynamics of the protein. This means that the enzyme's activation energy increases, leading to a reduction of its catalytic efficiency [477].

   A computational study has suggested a completely different mechanism of glycosylation-induced structural stabilization [478]. According to this mechanism, the carbohydrate groups interfere with interactions between protein residues in the unfolded state. In other words, by destabilizing the unfolded state, glycosylation may increase the tendency of the protein to fold.

   A related and important question is whether glycosylation changes the native conformation of the (non-glycosylated) protein. This issue has been studied both by comparing protein structures in the PDB and by carrying out molecular dynamics simulations. However, while some studies indicate significant (yet not extreme) conformational effects of glycosylation [479], others suggest the opposite [468].

3. Increasing solubility. The multiple hydroxyl groups on the carbohydrate moieties significantly increase the solubility of the glycosylated protein. This, as well as the large

size of the glycosyl groups and their significant hydration, also reduces the risk of non-specific aggregation [480].

4. Enhancing folding. Certain carbohydrate moieties on glycosylated proteins are recognized by molecular chaperones in the ER, resulting in more efficient folding of the protein chain [471]. Protein that undergo faulty glycosylation are translocated back to the cytosol and degraded [481].

5. Enhancing molecular recognition processes. As already explained above, the inherent complexity of carbohydrate chains allows the proteins carrying them to have their own molecular signatures [472], which makes their interactions with other biomolecules highly specific. These interactions occur between enzymes and their substrates, receptors and their ligands, antibodies and their antigens, etc. In the case of glycosylated membrane proteins, the binding may also mediate cell-cell interactions, which are highly important to biological processes such as cellular division and differentiation, the immune response, neurotransmission, tissue construction, and inflammation. Studies confirm that in such processes the specific glycosylation pattern — i.e., the types of carbohydrate units, their number and their organization — affects the process significantly [467].

Protein glycosylation has been also implicated in other aspects of protein action, including oligomerization and aggregation [482,483], as well as ER quality control and trafficking [484].

2.6.4 Acylation

Acylation is the process by which protein residues become attached to acyl chains, i.e., carbon chains (C_n) that are mostly hydrophobic but include a carboxylic head group. The acyl chains that are commonly involved in this process include *acetyl* (C_2), *myristoyl* (C_{14}), and *palmitoyl* (C_{16}). In addition, the small protein ubiquitin may be attached to proteins as an acyl moiety. Like phosphorylation, acylation is also reversible within cells, thanks to *deacylases*, enzymes that readily detach acyl chains from proteins, pending the right signal. The common protein acylation processes are as follows.

2.6.4.1 ε-N-acetylation

In some proteins, such as *p53* and *histone*, multiple lysine residues undergo reversible acetylation [485] of their side chains' ε nitrogen (Figure 2.41c), with acetyl-CoA being the acetyl group donor. This modification functions in regulating key cellular processes such as DNA replication and repair, gene expression, the cell cycle, chromatin modeling, nuclear transport, cell motility, and chaperone-assisted protein folding [486]. Therefore, enzymes catalyzing acetylation (acetyltransferases) are important targets for pharmaceutical drugs, including drugs that are used to treat cancer and neurodegenerative diseases, as well as to reprogram stem cells.

On the molecular level, the prominent effect of acetylation is neutralizing the positive charge on the lysine side chain. In addition, studies suggest that the acetyl group may also be involved in molecular recognition. Indeed, the bromodomain of transcription factors is known to recognize this group. This way, the acetylation of histone[*1] for example, may be a way of recruiting transcription factors to that region, and initiating gene transcription [487].

The cell-cycle-regulating protein p53 is one of the central factors affecting cellular fate. When cellular DNA is damaged by potentially hazardous mutations, p53 is activated. It prevents the cell from dividing, a process that might lead to cancerous transformation, and also induces the action of DNA repair proteins. If the damage is beyond repair, p53 induces apoptosis to prevent the onset of cancer in the multicellular body. Accordingly, p53 is defined as a *'tumor suppressor protein'*, and has even been referred to as 'the guardian of the genome' [488]. p53 works by regulating the transcription of genes involved in the above-mentioned processes (DNA repair, apoptosis, and cell cycle arrest) [489]. Being a multi-domain protein, p53 can execute its effects by interacting with different signaling proteins, as well as with DNA [490]. Due to its immense influence on cell fate, the activity of p53 must be tightly regulated. As explained in Subsection 2.6.4.3 below, one way to prevent excess activity of p53 is to tag it for degradation using the small protein ubiquitin. An opposite process, making sure that p53 is not degraded when needed, is the acetylation of lysine residues in p53 by the transcriptional co-activator p300. Indeed, acetylation prevents the tagging of p53 with ubiquitin, thus prolonging its lifetime.

2.6.4.2 *N'*-myristoylation and *S*-palmitoylation

These two modifications, which involve the attachment of a long fatty acid chain to the protein, allow the protein to bind to the plasma membrane, usually as part of a signal transduction pathway that involves membrane-bound proteins. Promotion of membrane binding also allows these modifications to participate in regulating subcellular localization of different proteins. Myristoylation involves the transfer of myristic acid (C_{14}) from a donor, myristoyl-CoA, to the α-amino nitrogen of glycine at the *N'* of the protein (Figure 2.41e) [491]. Since proteins are constructed with methionine at their *N'*, myristoylation requires the removal of this residue first. The anchoring of a protein to the plasma membrane using a myristoyl chain may be contingent on a signal that induces conformational changes in the protein and leads to the externalization of the fatty acid moiety [492]. The process of palmitoylation is similar to myristoylation, except that the chain involved is palmitic acid (C_{16}), which is transferred to a sulfur atom in the side chain of cysteine (Figure 2.41f). Thus, in addition to promoting membrane binding, this modification is also involved in regulating active cysteine residues in enzymes [493]. The palmitoylation of many signal transduction proteins has been studied, including that of cytosolic proteins such as the GTP-hydrolase *ras* [494] and the tyrosine kinase *lck*, as well as permanently membrane-bound proteins, such as the chemokine receptor *CCR*. Given these roles, and the fact that palmitoylation is reversible inside the cell and can therefore act as an on/off switch, this modification has been implicated in regulating key cellular processes, as well as in the onset of related diseases [495]. The latter include cancer, diabetes, and different diseases of the nervous system, e.g., schizophrenia, Alzheimer's disease and Huntington's disease [195].

[*1] The protein on which DNA is wound inside the nucleus of eukaryotic cells. This winding achieves compact packing ($\sim 3 \times 10^5$ fold) of the cell's genetic material.

2.6.4.3 Ubiquitination and SUMOylation

Ubiquitin is a small (76 residues, 8 kDa) and highly conserved protein. Ubiquitin molecules constantly become attached to cellular proteins, a process that has a few dramatic consequences. The first and most dramatic is the dispatching of the ubiquitylated protein for proteolytic degradation by a protein complex called a *'proteasome'*, as part of the quality control of cells on their biochemical processes [496–498]. Compared with lysosomal degradation, which is mostly intended for internalized elements (e.g., peptide hormones, clotting factors, and antibodies), and *autophagy*, which degrades proteins mainly to supply fuel for energy production [499], proteasomal degradation is more specific, and therefore requires tagging of the proteins intended for degradation. Thus, the cell can determine the half-life of any of its proteins separately, according to its role and the environmental conditions. For example, cell-cycle regulation proteins such as p53 exist for only minutes, whereas more constantly active proteins, such as the muscle proteins *actin* and *myosin*, remain intact for days. Finally, the eye lens protein *crystallin*, which acts as a constant light refractor, has a half-life of years. Protein degradation via the ubiquitin pathway may also occur as a result of an external signal that leads to activation or inhibition of certain biochemical pathways, or may be due to genetic or environmental damage to the folding ability of proteins.

Numerous studies conducted in recent decades have demonstrated the central involvement of the ubiquitin-proteasome system in the most basic cellular processes, such as division, differentiation and development, response to stress or extracellular factors, formation of neural networks, regulation of membrane-bound receptors and transport proteins, DNA repair, transcriptional regulation, and organelle formation [496]. In addition, the system has been found to be important in physiological processes, such as memory formation and the immune or inflammatory response. Indeed, the discovery of this system led Aaron Ciechanover, Avram Hershko, and Irwin Rose to be awarded the 2004 Nobel Prize in chemistry. Another testament to the importance of the ubiquitin-proteasomal system relates to its extensive involvement in pathologies. For example, in *Liddle syndrome*, a disease manifested as high blood pressure from childhood, a genetic defect in sodium channels of the kidneys prevents these channels from being identified by a central element of the ubiquitin system, which leads to their excessive activity [500]. Another pathology related to the ubiquitin system is *cystic fibrosis (CF)*, expressed in chronic blockage of the respiratory tracts, and in digestive problems. CF results from different genetic problems in the CFTR protein, which functions as a chloride channel. Most cases of the disease result from a mutation that prevents the CFTR channel from folding properly. As a consequence, most copies of the channel protein are kept inside the ER lumen, destined for degradation [501]. Finally, at least one of the elements of the ubiquitin system seems to be involved in cancer (see Subsection 2.6.11 below).

The process of tagging a single protein for degradation is carried out by a multimeric chain of ubiquitin. The first unit is usually transferred to the ε-amino group of a lysine's side chain, and the following units bind to each other. The tagging process is complex, and involves three different enzymes (E1, E2, and E3). In the first step, ubiquitin is attached to E1 in a thioester bond, using ATP as an energy source. In the second step, E2 transfers the ubiquitin from E1 to the protein substrate, while it is attached to E3. Finally, the ubiquitin chain is recognized by the proteasome complex, and the protein carrying it is degraded. This process implicates E3 as the element conferring specificity towards the target protein. The recognition process is anything but simple, and in most cases requires activation of

E3, done either by modification of the target protein or of E3 itself, or by attachment of auxiliary proteins to the target protein. Some of these activation processes are induced by changes in the target protein, whereas others are induced by cellular signals. In the case of transcription factors, detachment from the DNA molecule is sometimes sufficient for them to be recognized by E3, so they are degraded as soon as their work is done, not before.

As mentioned above, attachment of ubiquitin to proteins may lead to different consequences. We have seen how attachment of a long chain of ubiquitin units leads to degradation. However, in several different proteins, attachment of a single ubiquitin unit has been found to act similarly to phosphorylation, i.e., to change the proteins' activity, meaning that ubiquitylation functions as a type of molecular switch, allowing modulation of activity [502].

Another post-translational process similar to ubiquitination is SUMOylation, i.e., the covalent (yet reversible) attachment of *small ubiquitin-like modifier (SUMO) proteins* to lysine side chains in target proteins [503,504]. The bond is formed enzymatically between the ε-amino group of the lysine side chain and the terminal carboxyl group of the SUMO protein (an *isopeptide bond*). SUMOylation is essential for the viability of most eukaryotes and plays important roles in different processes, such as nuclear transport, chromatin remodeling, apoptosis, transcriptional regulation, defense against DNA damage and proteotoxic stress (e.g., heat shock [505]), cell migration, and cell-cycle regulation [504]. Thus, despite the similarity between SUMO and ubiquitin, as well as between the enzymatic machineries of SUMOylation and of ubiquitination , the former has more diverse consequences than the latter. The molecular mechanism of SUMOylation seems to be diverse as well; the added SUMO protein seems to function by (*i*) masking binding sites, (*ii*) inducing conformational changes in the target protein, which alter its activity, and/or (*iii*) serving as a hub for the recruitment of other proteins. Of the three mechanisms, the latter is the most common in SUMOylated pathways, and is mediated by noncovalent interactions between the SUMO protein and SUMO-binding domains in the recruited proteins.

2.6.5 Alkylation

The process of alkylation involves the attachment of carbon groups, most of which are hydrophobic chains, to different locations in proteins. The three types of alkylation are as follows.

2.6.5.1 Methylation

Methylation is a relatively rare event in proteins, and involves the attachment of methyl groups ($-CH_3$), usually to the side chain nitrogen atoms of lysine (Figure 2.41d) or arginine. The number of groups attached to each side chain is in most cases up to three. Methylated carbon, oxygen, and sulfur atoms can also be found in proteins, although these events are extremely rare. Methylations are carried out by enzymes called '*methyl-transferases*', some of which are residue-specific. Methyl groups are simple compared to carbohydrate or acyl groups. However, since several methyl units can be added to one site, they can still confer some degree of geometric complexity to the protein, which can be used for molecular recognition. In histone, methylations, as well as acetylations and other modifications, seem to act as an *epigenetic code*, used to activate or silence gene transcription.

Unlike acylations, lysine methylations are irreversible. This is because there are no enzymes to undo them, and the bond is too strong to be broken spontaneously. Nevertheless, in some cases the methylamine group is oxidized by the enzyme *methyl de-aminase* to an imine group ($-N=CH_2$), which is unstable and spontaneously hydrolyzes back to lysine's original amino group [506]. Another difference between methylation and acylation is that the former does not eliminate the electric charge of lysine, and therefore is not expected to change the conformation of the protein significantly, if at all. In fact, the accepted view is that the effect of methylation is indirect, involving spatial interference with the occurrence of other modifications.

2.6.5.2 *S*-prenylation

Prenylation is a process of transferring an *isoprene*-based chain to the sulfur atom of cysteine. The transferred chain is either the 15-carbon *farnesyl* (Figure 2.41g), or the 20-carbon *geranyl-geranyl* (Figure 2.41h). The role of prenylation, like that of acylation, is to attach cytosolic proteins to the plasma membrane, usually as part of a signaling pathway. In fact, there are several common characteristics to these cysteine modifications (*S*-palmitoylation and *S*-prenylation), although they differ in the reversibility of the bond created; the *thioether* created by prenylation is more stable than the *thioester* formed by palmitoylation, and as a result, the thioether bond remains intact [87]. In the cellular context, prenylation is particularly important in the membrane attachment of GTP hydrolases such as *ras* and *rho*, as well as of *G-proteins*.

2.6.5.3 Adenylation

The process of adenyl attachment targets tyrosine residues in proteins, where the phenol group of the residue's side chain binds to the 5′ of ATP, to form Tyr-AMP. Adenylation is used to regulate the activity of some proteins, as in the case of glutamine synthetase, an enzyme participating in nitrogen metabolism [507].

2.6.6 Hydroxylation and oxidation

Hydroxylation of proteins is carried out on proline, lysine, asparagine and glutamate residues. Hydroxylation may happen under a variety of circumstances, and for various reasons. For example, the structural protein collagen contains 5-OH-Lys, 4-OH-Pro, and 3-OH-Pro derivates, which are required for the stabilization of the protein [87] (see Section 2.7 below). Hydroxylation of a residue is sometime a preparatory step for additional modifications such as methylation or *O*-glycosylation, as happens in 5-OH-Lys residues in collagen. Hydroxylation of asparagine residues (at the third position) is rare, yet observed in some proteins, such as HIF, a transcription factor that is activated during *hypoxia* [508]. After being activated, HIF induces the expression of certain genes that help the tissue to deal with the hypoxic state. For example, expression of the gene coding for the hormone *erythropoietin* boosts the production of red blood cells, thereby increasing the transport of precious oxygen to the tissue. Interestingly, under normal conditions, HIF has a very short lifetime, as it is constantly degraded by ubiquitin-induced proteolysis. This is because the abundance of oxygen allows HIF to undergo proline hydroxylation readily, which increases its chances of being recognized by the ubiquitin system 1000-fold. However, when the conditions are hypoxic, the oxygen shortage prevents efficient hydroxylation of HIF, and the rate of its degradation is lowered considerably. The prolonged lifetime of HIF in such cases allows it to help the tissue to fight hypoxia.

Oxidative modifications are not limited to hydroxylation. For example, a sulfate group (SO_3^-) may be added to tyrosine residues [509]. This PTM is carried out by the Golgi apparatus, with *phosphoadenosine phosphosulfate (PAPS)* being the sulfate donor. Sulfation has been studied in the chemokine receptor CCR5 during its cellular transport to the plasma membrane. CCR5 is attached to four sulfate groups, and it seems that both the shape and the charge of the groups are important for substrate recognition by the mature receptor.

The scope of protein oxidation depends on the environment, and inversely correlates with the concentration of antioxidants within the cell. In multicellular organisms the oxidation level is constant most of the time, and increases with aging or certain pathological processes, such as cancerous transformation of the cell.

2.6.7   Proteolysis

This type of PTM includes the following:

- The removal of the *N*-terminal methionine (or formyl-methionine) in newly synthesized proteins.

- The removal of *N*-terminal signal sequences in proteins that are sent to certain subcellular organelles.

- Activation of secreted proteins, e.g., the hormone insulin or hydrolytic enzymes (*zymogens*). The latter are formed inside the cell as inactive precursors, and are cleaved at certain sites upon their secretion, thus rendering them active. The purpose of such a mechanism of activation is probably to avoid having an active hydrolase within the cell, for obvious reasons.

2.6.8   Amidation

This process includes the attachment of an amino group to the C-terminal carboxyl group of the protein. The purpose of this modification is probably to neutralize the negative charge of the carboxyl group. Many peptide hormones, such as the hypothalamic neuropeptide *thyroxin-releasing hormone (TRH)*, are subjected to amidation, as are some animal toxins.

2.6.9   Addition of metal ions

Metal cations, such as iron (Fe^{2+} and Fe^{3+}), zinc (Zn^{2+}), copper (Cu^{2+} and Cu^+), magnesium (Mg^{2+}), manganese (Mn^{2+} and Mn^{3+}), molybdenum (Mo^{3+}, Mo^{4+} and Mo^{6+}), cobalt (Co^{2+}), and nickel (Ni^+) constitute integral elements of some proteins, collectively referred to as '*metalloproteins*'. Unlike other biological ions such as sodium, potassium and chlorine (*electrolytes*), which are very common in biological organisms, the above metals, except calcium and magnesium, are *trace elements*. That is, they are required in minute quantities (milligrams or less). The metal cation can be bound to protein residues either directly or indirectly, as part of a larger non-protein group (e.g., the heme group of hemoglobin). In addition, it can appear individually within the protein, or as part of a cluster of ions (e.g., iron-sulfur clusters). Chen and Kurgan have carried out a comprehensive survey of protein-ligand interactions in ~7,700 proteins whose structures have been determined [510]. Their results demonstrate a few recurring trends in protein-bound metals. For example, in

95% of the protein-metal complexes, the binding between the two partners involves electrostatic interactions and coordinate bonds, with 31% of complexes involving only the latter. As might be expected, the electrostatic interactions between proteins and metals usually involve the acidic residues Asp and Glu.

Protein-metal covalent coordination usually involves Zn^{2+}, Ca^{2+}, and Mg^{2+} (15%). The binding between the protein and a single metal ion may involve up to six residues inside the protein binding pocket. The dominant residue in such interactions is by far His (via its δ_1/ε_2 atoms), and to a lesser extent Cys (via its S_γ atom) [510]. Oxygen atoms from various residues (including the polypeptide backbone) are also heavily involved in metal coordination. This finding is particularly interesting; it implies that **the interaction between metals and Asp/Glu, which has traditionally been considered to be purely electrostatic, may, at least partially, be coordinative in nature** [510]. The above residues do not appear randomly in protein binding pockets, but rather in the following combinations of two to four residues:

1. Four-residue combinations: $(Cys)_4$, $(Cys)_3(His)$, $(Cys)_2(His)_2$, $(Asp)_2(His)_2$, and $(Asp)(His)_3$, with the largest number of metal ions coordinated by the $(Cys)_4$ group.

2. Three-residue combinations: $(Cys)_3$, $(Cys)_1(His)_2$, $(Asp)_3$, $(Asp)_2(Glu)$, $(Asp)_2(His)$, $(Asp)(Glu)_2$, $(Asp)(Glu)(His)$, $(Asp)(His)_2$, $(Glu)_2(His)$, $(Glu)(His)_2$ and $(His)_3$, with the largest number of metal ions coordinated by the $(Asp)(His)_2$ and $(His)_3$ groups.

3. Two-residue combinations: $(Asp)_2$, $(Asp)(Glu)$, $(Asp)(His)$, $(Glu)_2$, $(Glu)(His)$ and $(His)_2$.

Protein-bound metal ions have several roles, all of which rely on the positive charge of the metal, its ability to bind and release electrons, or its tendency to participate in coordinate bonds. The main roles of protein-bound metals are described below. A summary of the cellular and physiological roles of specific metal cations is given in Table 2.8.

2.6.9.1 Stabilization of protein structure

Local structures within proteins may be stabilized by the presence of a metal ion coordinated to protein residues. For example, the '*zinc finger*' motif, discussed above, is stabilized by the coordination of Zn^{2+} to histidine and/or cysteine residues (see Subsection 2.2.1.3.3.2, Figure 2.8b).

2.6.9.2 Ligand binding

Protein-bound metals may bind to ligand atoms and groups via ionic or coordinate bonds. Examples include the binding of molecular oxygen to the Fe^{2+} ion of hemoglobin (see Box 5.2), and the binding of negatively charged phosphate groups of ATP to Mg^{2+} in many metalloenzymes.

2.6.9.3 Electron transport

Redox enzymes often use metal ions, usually Fe^{3+}, Fe^{2+}, Cu^{2+} and Cu^+ [511], as transient binding sites for electrons that are passed along a route within the protein (see more details in Chapter 9, Subsection 9.3.3.4). The same is true for some of the proteins in the electron transport chains involved in cellular respiration and photosynthesis.

2.6.9.4 Enzymatic catalysis

Metal-binding enzymes (*metalloenzymes*) use metal cations such as Mg^{2+}, Cu^{2+}, Zn^{2+}, Ca^{2+}, Mn^{2+}, Cr^{3+}, Co^{2+}, Fe^{2+} and Fe^{3+} [511] as part of their catalytic machinery. This use results from the capacity of these metals to bind other atoms in coordination or to interact with them electrostatically. The roles of the metals in catalysis include electrostatic stabilization of the reaction's transition state (or other anionic species in the active site), as well as activation of the substrate, coenzyme, or water molecules in the active site by electronic polarization. These roles, collectively referred to as 'metal ion catalysis', are discussed in detail in Chapter 9, Subsection 9.3.3.4. Most metals in metalloenzymes are trace elements (see Subsection 2.2.1 above).

TABLE 2.8 **Cellular and physiological roles of key cationic metals in living organisms (**[512,513] **and references therein).** Examples of proteins (in italics), processes, or structural motifs associated with some of the roles are in parentheses. Abbreviations: CCO – cytochrome c oxidase; CP450 – cytochrome P450; GS – glutamine synthetase; GT – glycosyltransferase; ND – NADH dehydrogenase; PC – plastocyanin; PEPCK – phosphoenolpyruvate carboxykinase; PK – pyruvate kinase; SD – succinate dehydrogenase; SOD – superoxide dismutase.

Metal	Cellular and Physiological Roles	Nutritional Deficiency
$Fe^{2+/3+}$ (iron)	• Enzymatic catalysis (electron transport, substrate or cofactor binding and stabilization) • Energy metabolism (*ND*, *SD*) and production (electron transport chain) • DNA synthesis (*ribonucleotide reductase*) • Photosynthesis (electron transport chain) • Oxygen transport (*hemoglobin*, *myoglobin*) • Nitrogen fixation • Anti-oxidation (*catalase*, *peroxidases*) • Detoxification of drugs and toxins (*CP450*) • Brain development in infants • Nitrogen fixation by certain bacteria (*nitrogenase*)	Anemia
$Cu^{+/2+}$ (copper)	• Enzymatic catalysis (electron transport, substrate or cofactor binding and stabilization) • Energy production (*CCO* in electron transport chain) • Synthesis of connective tissue proteins, red blood cells, melanin, certain hormones and neurotransmitters • Iron metabolism (*ceruloplasmin*) • Bone mineralization • Anti-oxidation (*SOD*) • Oxygen transport in invertebrates (*hemocyanin*) • Photosynthesis in plants (*PC* in electron transport chain)	Hematological and neurological disorders (very rare)

Metal	Cellular and Physiological Roles	Nutritional Deficiency
Mg^{2+} (magnesium)	• Enzymatic catalysis (substrate or cofactor binding, stabilization and activation) • DNA replication, repair, and stabilization • Energy production and biosynthesis • Nerve and muscle function • Formation of bones and teeth • Photosynthesis in plants (*chlorophyll*)	Neuromuscular, cardiovascular and metabolic dysfunction (rare)
Zn^{2+} (zinc)	• Enzymatic catalysis (substrate or cofactor binding, stabilization and activation in over 300 enzymes [514]) • DNA replication and transcription (zinc fingers in *DNA/RNA polymerases* and *transcription factors*) • Stabilization of cell membranes • Development of skeletal and reproductive systems • Wound healing (*matrix enzymes and proteins*) • Immune response • Brain function and learning • Blood pH buffering and CO_2 transport (*carbonic anhydrase*) • Cellular signaling and neurotransmission • Programmed cell death	Multiple, from hair loss, impotence and diarrhea to impaired growth and development, and susceptibility to infections
Ca^{2+} (calcium)	• Enzymatic catalysis (substrate or cofactor activation) • Cellular signaling (e.g., *calmodulin*) • Muscle contraction • Neurotransmission • Blood clotting (*coagulation factors*) • Structural element in bones and teeth	• Rickets • Clotting problems • Osteoporosis
Mn^{2+} (manganese)	• Enzymatic catalysis (substrate or cofactor binding and stabilization) • Energy production and biosynthesis (*PK, PEPCK, GS*) • Nitrogen metabolism (*arginase* in urea cycle) • Anti-oxidation (*SOD*) • Wound healing (*prolidase* in collagen formation) • Bone development (*GT* in proteoglycan synthesis) • Photosynthesis in plants (water splitting center)	• Bone demineralization • Impaired growth
Co^{2+} (cobalt)	• Enzymatic catalysis (substrate or cofactor binding and stabilization) • Vitamin B_{12}-dependent processes, e.g., DNA synthesis and amino acid metabolism	—

Metal	Cellular and Physiological Roles	Nutritional Deficiency
Mo$^{3+/4+/6+}$ (molybdenum)	• Enzymatic catalysis (electron transport, substrate binding, stabilization and activation) • Purine nucleotides breakdown (*xanthine oxidase*) • Amino acid metabolism (*sulfite oxidase*) • Metabolism and clearance of drugs and toxins (*aldehyde oxidase*) • Nitrogen fixation by certain bacteria (*nitrogenase*)	—
Cr^{3+} (chromium)	• Glucose transport into cells (by potentiation of insulin action)	• Impaired glucose tolerance • Increased insulin requirement • Diabetes

2.6.10   Mixed modifications

The modifications described above are not protein-specific. Many proteins may undergo modifications of different types, carried out to achieve specific functions [87]. To illustrate this, we will go back to the signal transduction protein *ras*, which we have already encountered a few times. *Ras*, a cytosolic protein, must migrate to the plasma membrane in order to be activated by growth factor receptors, which also function as tyrosine kinases. The attachment of *ras* to the plasma membrane requires no less than four different modifications [515]. First, *ras* is farnesylated on a cysteine side chain, which creates a hydrophobic membrane anchor in the protein. Then, two adjacent cysteine residues are palmitoylated, thus providing two additional hydrophobic anchors. The next step is the proteolytic removal of the last three residues of the protein, turning the farnesylated cysteine into the new C' of the protein. Finally, the α-carboxyl group of this cysteine is methylated to eliminate its negative charge. This prevents the carboxyl group from destabilizing the protein's interaction with the negatively charged lipids of the plasma membrane.

Some microorganisms, such as certain pathogenic bacteria, use PTMs for attacking other organisms. Such microorganisms have developed during their long evolution an arsenal of molecular 'weapons' for attacking their hosts and, at the same time, defending themselves against the host's immune system. Many of those weapons are protein toxins that are secreted within the host and attack different systems. Protein toxins may act in different ways, although many of them are enzymes that covalently change some of the host's proteins, and as a result harm their normal activity. Here are some of the modifications and their effects on the host [87]:

1. ADP-ribosylation: The attachment of adenine diphosphate (ADP) to protein residues is carried out by different toxins, such as the *cholera*, *diphtheria*, *pertussis*, and *botulinum* toxins [516–518]. The ADP donor is the common coenzyme nicotine adenine dinucleotide (NAD$^+$). The ADP group may be attached to several different nucleophiles in the proteins, and act through different mechanisms (Table 2.9).

2. Glycosylation: This modification may act in different ways. For example, glycosylation of Thr-35 in *ras* neutralizes the protein's GTPase activity [519].

3. Glutamine deamidation: This modification turns glutamine into glutamate. In the case of *rho*, deamidation of Gln-61 in the active site abolishes catalytic activity [520].

The biological consequences of a given modification depend on the role of the target protein. Thus, neutralization of eEF-2 leads to the cessation of protein synthesis in the affected cell, whereas that of *rho* leads to disassembly of the actin network. Nevertheless, all modifications serve a similar general purpose, namely, to facilitate the entry and colonization of the bacterium within the host. Interestingly, many bacterial toxins seem to target a GTPase of some type. This is understandable considering the central role of these enzymes in the regulation of key cellular processes.

Finally, many metabolic enzymes are attached to non-protein (prosthetic) groups, essential for the enzymatic activity [521]. In most cases, the role of the prosthetic group is to bind molecules or groups of a certain chemical nature, and/or to transfer them to other parts of the enzyme. This issue is discussed in detail in Chapter 9. Here are two examples related to key metabolic enzymes:

1. Biotin: This prosthetic group specializes in the binding of a carboxyl group, and therefore appears in enzymes that catalyze carboxylation reactions. For example, in *pyruvate carboxylase*, biotin transfers a carboxyl group from bicarbonate (HCO_3^-) to *pyruvate*, turning it into *oxaloacetate*. This reaction is the first step in the process of building glucose in the animal body during fasting. The biotin group is attached within the protein to the ε-nitrogen atom of a lysine residue, and can transfer the carboxyl group over ~20 Å. The same principal reaction also takes place in *acetyl-CoA carboxylase*, which turns acetyl-CoA into malonyl CoA, in the first stage of fatty acid biosynthesis.

2. Lipoic acid: This group appears in *α-ketoacid dehydrogenases*, such as the glycolytic enzyme pyruvate dehydrogenase and the Krebs cycle enzyme α-ketoglutarate dehydrogenase. Like biotin, this group is also attached to lysine residues, but its function is to carry substrate-borne acyl groups between different domains of the enzyme. The reduction of the disulfide bond in lipoic acid is coupled to the oxidation of the substrate's hydroxyethyl group. The resulting opening of the disulfide bond and attachment of the free SH group to the substrate enables lipoic acid to transfer the substrate from one region of the enzyme to another, for the continuation of the biochemical reaction.

2.6.11 Pathological aspects of post-translational modifications

2.6.11.1 Cancer

The above discussion demonstrates that PTM influence processes such as signal transduction, gene expression, and molecular recognition of external elements by cells. Changes in these processes happen, e.g., in the cancerous transformation of cells and tissues, which implies that PTMs and cancer may be related. This is indeed the case. In fact, the accurate characterization of PTMs in cells is expected in the near future to assist in the identification of molecular markers of different cancer types, and possibly in the development of approaches to fight the disease [522]. The two types of PTMs most strongly identified with cancerous

TABLE 2.9 **Some of the bacterial toxins using ADP-ribosylation on host proteins** [516,517].

Toxin	Source	Affected Protein(s)	Modified Residue	Molecular Effect	Physiological Effect
Pertussis (PTX)	*B. pertussis*	G_i [*a]	Cys	Lifting off inhibition of adenylate cyclase in different cell types, resulting in excessive production of cAMP	• Decreased phagocyte action • Lymphocytosis [*e] • Low blood sugar levels • Hypotension
Cholera (CTX)	*V. cholerae*	G_s [*a]	Arg	Over-secretion of water and ions into intestinal lumen	• Diarrhea • Dehydration
Botulinum type C3 (BTX)	*C. botulinum*	rho [*b]	Asn	Inhibition of acetylcholine release at peripheral cholinergic synapses	• Flaccid muscular paralysis
Diphtheria (DT)	*C. diphtheriae*	eEF-2 [*c]	Diphthamide [*d]	Inhibition of protein synthesis	• Sore throat • Fever • Swelling of head and neck

[*a] A type of G-protein, i.e., large heterotrimeric GTPases involved in signal transduction.
[*b] A small monomeric GTPase.
[*c] An elongation factor in eukaryotic protein translation.
[*d] A chemical derivate of histidine.
[*e] An increased number of circulating lymphocytes.

transformation are phosphorylation and glycosylation. As mentioned above, phosphorylation is a common molecular switch in signal transduction pathways. This role is manifested in membrane receptors undergoing autophosphorylation, as well as cytosolic effectors such as *ras, phospholipase C$_\gamma$ (PLC$_\gamma$)* and *phosphoinositol (PI) 3-kinase*. These bind to the cytosolic end of the membrane receptor, and as a result become phosphorylated as well. Signal transduction proteins located further downstream in the pathway may also use phosphorylation to relay the signal (e.g., *mitogen-activated protein (MAP) kinase*). These complex processes often lead to dramatic consequences (cellular division, growth or suicide), so it is not difficult to imagine how a defect in their regulation may eventually lead to cancer [523]. Indeed, several different proteins involved in signal transduction have been implicated in the cancerous transformation of cells, and many of these proteins undergo or carry out phosphorylation [524]. For example, growth factor receptors that are missing their extracellular (ligand-binding) domains become constitutively active, and consequently transform the cell. Knowledge about such processes has been used for developing different strategies for fighting cancer. For example, antibodies designed to target the extracellular part of the growth factor receptor have been successful in suppressing the growth of cancer cells [525]. Another successful treatment is the anti-cancer medication imatinib (Gleevec®), which targets the PDGF receptor [526].

Glycosylation is also identified with cancerous transformation. Cells undergoing transformation are known to change the carbohydrate composition of their surfaces so it becomes more similar to the composition in embryonic cells [527]. This phenomenon is so common that cancer-specific antigens (i.e., molecular markers) often turn out to be carbohydrate moieties attached to proteins or lipids. The transformation also involves a decrease in the number of carbohydrate moieties of the cell. Since the carbohydrates covering the cell are involved in anchoring it to the tissue, this decrease allows the transformed cell to leave the tissue and migrate to other parts of the body [528]. This process is referred to as 'metastasis'.

The change in carbohydrate composition may confer another advantage to the cancerous cell. As mentioned earlier, proper glycosylation of cell-surface proteins contributes to these proteins' ability to interact with other cells and molecules. In the case of cancer cells this is a major disadvantage, as it makes them more 'visible' to the cells and proteins of the immune system. Evidence suggests that the change in cancer cells' surface carbohydrate composition allows them to avoid such detection, thereby increasing their chances of survival. [529]. Ironically, this same unique carbohydrate composition has made it possible for scientists to develop a new strategy of fighting cancer, i.e., constructing a vaccine. In this approach, the cancer is treated like an infection: the patient is injected with the carbohydrate structures unique to the cancer carried, boosting the immune system's capacity to identify the aberrant cells [530]. This approach has high potential, which scientists and physicians are now trying to translate into medical results.

Finally, studies show that the ubiquitin system may also be involved in cancerous transformation, as in the case of the cell-cycle-regulation protein p53. As already explained in Subsection 2.6.4.1 above, p53 is a major player in keeping cells from dividing once their DNA has become damaged. Under normal conditions, p53 has a very short half-life, of several minutes, due to its frequent ubiquitination by MDM2 (a type of E_3 ubiquitin ligase; see Subsection 2.6.4.3 above), and its subsequent degradation. However, when the cell is stressed, the tagging of p53 for destruction is suppressed, thus prolonging the protein's life. Such suppression occurs, e.g., through binding of the transcriptional co-activator p300, which is essential for the transcriptional function of p53. The binding site of p300 in p53 overlaps with that of MDM2, thus preventing the ubiquitination of p53 [490]. In addition, p300 acetylates p53, which also protects the latter (see Subsection 2.6.4.1 above). In any case, the reduction in p53 degradation at the cellular level leads to heightened regulation of the transcription of genes involved in DNA repair, apoptosis, and cell cycle arrest [489]. It has been found that MDM2 is overexpressed in cancer cells, which keeps p53 at low concentrations, and allows the cell to keep accumulating mutations.

2.6.11.2 Age-related illnesses

PTMs are involved in additional pathologies other than cancer. Some of these pathologies, such as type II diabetes, vascular diseases, rheumatism, and neurodegenerative diseases, are considered to be age-related, at least in part. Although the molecular mechanisms underlying the various pathologies are different, they have all been found to be affected by *sirtuins* [531,532]. These proteins mediate the (positive) effect of calorie restriction on life expectancy in animal models, a phenomenon that is thought to work through lowering the incidence of age-related diseases. The effect of sirtuins depends on the type of disease, as well as on the type of sirtuin. For example, SIRT1 lowers the frequency of some age-related pathologies, whereas SIRT2 increases the frequency of some degenerative pathologies, such as *Parkinson's* disease. In addition, the protein targets of sirtuins and the biochemical pathways in which they participate differ among organisms. For example, mammalian SIRT1 acts on transcription factors or regulators (e.g., p53 and *NF-κB*), thus affecting metabolic and stress-related pathways [531]. In particular, the effect of SIRT1 on the protein PGC-1α induces the formation of new mitochondria in the cell, which is of great medical importance; first, the increase in the number of mitochondria speeds up glucose oxidation, which in turn leads to increased insulin sensitivity. This makes SIRT1 a potential target for diabetes drugs. Second, ageing is accompanied by a drop in the number of mitochondria within cells. It is therefore possible that SIRT1 acts to slow ageing by inducing mitochondria biosynthesis.

Recent studies show that sirtuins are post-translational modifying enzymes. Most sirtuins deacetylate lysine residues in their target proteins, thus activating or deactivating these proteins. Other sirtuins ADP-ribosylate their target proteins, thus inactivating them. Interestingly, despite the fact that these two modifications are different, both involve the cleavage of NAD^+ in the first stage of the reaction [531].

2.6.12 Identifying post-translational modifications

Many of the PTMs known today have been discovered accidentally, using standard molecular methods, such as replacement of a modified residue by point mutation [533]. Intentional search for covalently modified residues in a protein is more difficult, as it requires the specific protein form carrying the modification to be isolated in large enough amounts for biochemical analysis. Since many PTMs are carried out in a modular manner (e.g., phosphorylation of different residues), it is not uncommon for the same cell or even the same organelle to have different forms of the same protein, differing in the chemical nature of a single residue. In such cases, each of the different forms will exist in solution in very small amounts, requiring extremely sensitive methods for detection. One such method is *mass spectrometry* (*MS*). In this method, the sample of proteins in solution is ionized; it is then exposed to an electric field, which facilitates the separation of the different proteins according to their mass-to-charge ratio [534,535] (see more in Chapter 3). As PTMs change at least one of those properties, MS can be used for isolating chemically unique forms of a single protein. Some applications of the method make it possible to quantify the different forms of the protein in the sample, and therefore to characterize the degree to which a certain PTM is observed under different physiological conditions. To make the process more efficient, the different proteins may first be separated by *high-performance liquid chromatography* (*HPLC*), and in some cases, cleaved enzymatically or chemically to short peptides, which are then analyzed by MS.

Another common method of separating proteins by their covalent modifications is *two-dimensional (2D) gel electrophoresis* [536,537]. The principles of protein and peptide separation by gels are detailed in basic biochemical literature (e.g., [24]). In 2D electrophoresis, the proteins are separated first by mass and then by charge, where the two separations are carried out along perpendicular axes on the same gel. After the separation, each of the protein forms can be collected from the gel and investigated individually by using other methods. In addition to MS and gel electrophoresis, methods such as column chromatography and antibody precipitation (AP) may be used. Finally, proteolytic cleavage of proteins can be identified by Edman degradation, which determines the *N*-terminal residue of the polypeptide chain.

2.7 FIBROUS PROTEINS

In the subsections above we witnessed the highly sophisticated structure of globular proteins. This inherent complexity is the reason why most cellular functions, which are extremely diverse, are carried out by this type of protein. Nevertheless, cells and tissues also contain proteins that do not possess the characteristic properties of globular proteins, primarily the globular structure. Rather, they have an elongated shape (see Figure 2.22 above). These tend to have structural functions. That is, they serve as building blocks for large fiber-based structures inside or outside the cell. Such proteins are referred to as *'fibrous'*. As might be expected, fibrous proteins differ from globular ones not only in terms of their shape and biological roles, but also in other properties, such as water solubility, atomic packing and density, hydrophobicity, dynamics, and folding energetics. In the following subsections we will review what is currently known about fibrous proteins, emphasizing the above aspects, as well as the structure-function relationship exhibited by these proteins.

2.7.1 Fiber-based structures inside and outside cells

The cellular and physiological roles of fibrous proteins mainly involve the formation of large structures, organized as fibers inside or outside the cell. Some of these fiber-shaped structures are not built exclusively from fibrous proteins, and may also include globular proteins as building blocks. It is difficult to describe the organization and function of such structures by addressing only their fibrous components, and we therefore include all of their components in our discussion. The distinction between fiber-forming and fibrous proteins is further discussed in Subsection 2.7.2 below.

2.7.1.1 Mechanical support

Eukaryotic cells contain fiber-like structures that provide them with physical support and allow them to withstand naturally applied mechanical forces. These forces include pulling, compression, and shearing forces. The two major ultra-structures that fulfill these roles are the *cytoskeleton*, which resides inside the cell, and the *extracellular matrix*. A brief description of these two structures has been provided in Chapter 1. In the following subsections we focus on the various components of each structure. In both cases, we refer to structures associated with eukaryotic cells. It should be mentioned, though, that protein-based polymers composing a cytoskeletal system have recently been identified in bacteria as well [538].

2.7.1.1.1 Cytoskeleton

The cytoskeleton is composed of three types of fibers (Figure 2.42), each fulfilling a different function.

1 **Microfilaments**

Microfilaments are thin (8 nm) fibers made of the protein *actin* (Figures 2.42a and 2.43). They are capable of modulating their length by adding or subtracting actin monomers to one of their termini. Microfilaments may appear in different forms, depending on the proteins they interact with. The following are examples of actin fiber types:

(a) Cortex fibers – As their name implies, these fibers reside at the periphery of the cell, and are composed of crossed fibers forming a gel-like structure. The role of cortex microfilaments is to give the cell its general shape.

(b) Stress fibers – A stress fiber is a bundle of anti-parallel actin fibers, capable of contracting thanks to their interaction with another set of fibers. The latter are thick and made of the motor protein myosin II. The cytoplasmic stress fibers bind to certain structures inside the plasma membrane, which are bound on the membrane's extracellular side to matrix components. Thus, stress fibers enhance the adhesion capability of the cell to its substrate (see following subsection).

(c) Filopodia (pseudo-limbs) – A bundle of parallel fibers that push local regions of the plasma membrane towards the extracellular environment, in the form of elongated leg-like extensions. Since actin fibers can modulate their length, filopodia may appear in different sizes. Certain epithelial cells use filopodia to increase their surface area. A well-known example is the intestinal epithelium, which uses such filopodia (called *microvilli*) to increase its surface area in order to enhance nutrient absorbance.

Actin has other biological roles, one of which is to form the contractile units of skeletal muscles (see Subsection 2.7.1.3 below). In addition, it is involved in nuclear-associated functions, such as transcription, RNA processing, and transport through the nuclear lamina [539,540].

2. **Intermediate filaments**
Intermediate filaments are tough fibers that protect cells and tissues against mechanical environmental insults [543]. In vitro studies show that intermediate filaments are particularly well adjusted for this type of function; they can withstand pressures of ~1,000 dyn/cm^2, ~5 times greater than what microfilaments can withstand [544], without losing flexibility; moreover, as the pressure increases, so does their resistance to it [545]. The term 'intermediate' comes from their size, which is intermediary compared to the other two types of cytoskeletal fibers (Figure 2.42b). The mechanical protection provided by intermediate filaments is especially important in cells that are exposed to mechanical stress, including the intestinal and skin epithelia, as well as the oral mucosa, cornea, and muscle.

Intermediate filaments may be formed by different types of proteins, depending on the tissue type. In epithelia, intermediate filaments are formed by *keratin* [546,547] (Figure 2.44a). The organization of epithelia is particularly interesting in the epidermis layer of skin. In fact, skin epidermis includes several layers of cells (Figure 2.44b), with the innermost layer composed of *keratinocytes*. These metabolically-active cells are in essence capable of dividing indefinitely. Some of these cells differentiate while physically moving towards the external side of the epidermis. This process creates the other four layers of epithelial cells (from inside to outside), differing in their degree of differentiation, metabolic activity, and amount of keratin. The differentiation process involves the disappearance of the nucleus and organelles, increase in keratin content, and the abolishment of metabolic activity. It ends after 26 to 42 days, with the activation of programmed cell death (apoptosis). At this stage, the dead cells are full of keratin ('keratinized') and reside at the outermost layer of the skin. This layer is highly efficient in protecting the inner layers against physical attrition, and serves as a barrier preventing the penetration of water and chemical or biological agents into the tissue.

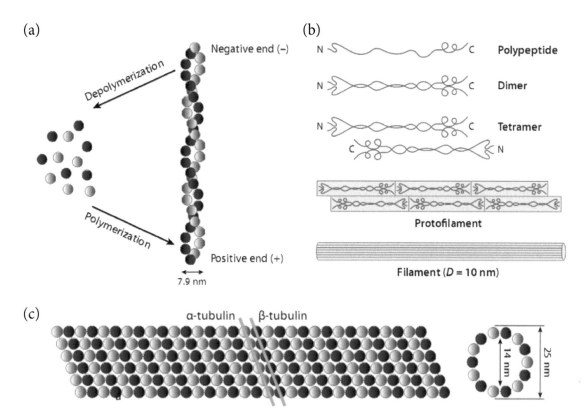

FIGURE 2.42 **Constituents of the cytoskeleton.** The figure depicts the three types of cytoskeletal fibers, and their assembly from the basic units (taken from [541]). (a) Microfilaments made of F-actin. Although F-actin forms structural fibers, it is a globular protein (see Subsection 2.7.2 in the main text for details). The fiber is dynamic; each of the F-actin monomers, depicted as spheres, can join the fiber from its (+) side or depart from it at the (−) side. (b) Intermediate filaments made of α-keratin. In contrast to F-actin, α-keratin (shown as an elongated blue line) is a fibrous protein. As a result, it oligomerizes sideways to create a supersecondary (dimeric) structure called a 'coiled coil'. Two such dimers can align head-to-tail to form a tetrameric structure, and such structures can assemble along a single axis to form a 'protofilament'. Finally, a number of protofilaments may join in a parallel formation to create a filament. (c) Microtubules made of α/β-tubulin. Like F-actin, tubulin is also a globular protein. The orange lines delineate the α and β tubulin arrangements transversely to the fiber's axis. The image on the right shows the microtubule fiber from an angle perpendicular to the left image.

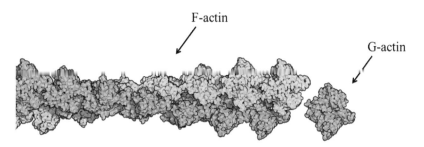

FIGURE 2.43 **The structure of actin microfilaments.** The microfilament fibers are made of globular (G) actin. Within the fiber structure, actin is referred to as 'filamentous' (F-actin). The figure was taken from [542] courtesy of David Goodsell (Scripps Institute).

The overall process is also responsible for creating animal hair, nails, horns, hooves, and feathers.

As explained, keratin constitutes a tough element in epithelial cells, and its organization as a mesh allows it to resist pressures optimally. When this ultrastructure falls apart, which happens in some diseases, the cytoplasm mollifies, leading to the appearance of boils [548]. In addition to being present in the cytoskeleton, intermediate filaments also appear in the nucleus. Nuclear intermediate filaments are composed of the protein *lamin*, and seem to be responsible for the mechanical stability of the nuclear lamina. Microscopic staining of intermediate filaments shows that these fibers extend from the cell's periphery to its nucleus, which implies that the two extreme regions may communicate in some way [549]. In fact, intermediate filaments are connected to the Golgi apparatus, mitochondria, transport vesicles, and other elements of the cytoskeleton. Therefore, it is hypothesized that intermediate filaments may participate in the inner organization of the cell [549]. The mechanical protection provided by intermediate filaments transcends the boundaries of the single cell; the fibers are connected to cell-cell junctions and to cell-extracellular matrix (ECM) junctions (*desmosomes* and *hemi-desmosomes*, respectively). Thus, intermediate filaments can be viewed as a network of fibers extending throughout the entire tissue. This makes their role as mechanical supporters much easier, as they can distribute the forces applied to them over the entire surface of the tissue, instead of over each cell separately.

3. **Microtubules**

The third and innermost category of cytoskeletal fibers comprises *microtubules*, which are hollow and are built from the protein *tubulin* (α/β). These rigid 'pipes' are about 2.5 times thicker than microfilaments and intermediate filaments (Figure 2.42c), and provide support for the latter two. Like microfilaments, microtubules are also constructed by the assembly of protein monomers along their principal axis. In animal cells, this process begins from a defined body, the *centrosome*, which is located near the nucleus [79].

Tissue protection or stabilization by intracellular fibers can sometimes be carried out *ad hoc* as a response to a change in the tissue's condition. This is the case with clotting, a process that involves the water-soluble protein *fibrinogen* [552]. During the clotting process, proteolytic enzymes activate each other in a cascade, with the last member, *thrombin*, turning fibrinogen into the water-insoluble protein fibrin. The latter organizes as a dense mesh of fibers (i.e., the blood clot, Figure 2.45), which seals the wound and prevents further blood loss. Later, the clot also participates in tissue healing.

2.7.1.1.2 Extracellular matrix (ECM)

Proteins of the extracellular matrix also participate in tissue protection and support. As mentioned in Chapter 1, the ECM is a large network of proteins and polysaccharides, which interact with each other and with membrane-bound proteins (Figure 2.46). Each of the proteins composing the ECM has different properties, and therefore plays a different role (Table 2.10). Collagen is probably the most extensively studied protein in this respect. Highly common in vertebrates[*1], it constitutes the major component of connective tissues, such as

---

[*1]In humans, collagen constitutes 33% of the total protein and 67% of the skin's dry weight [554].

(a)

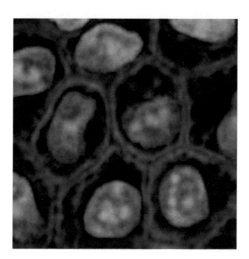

(b)

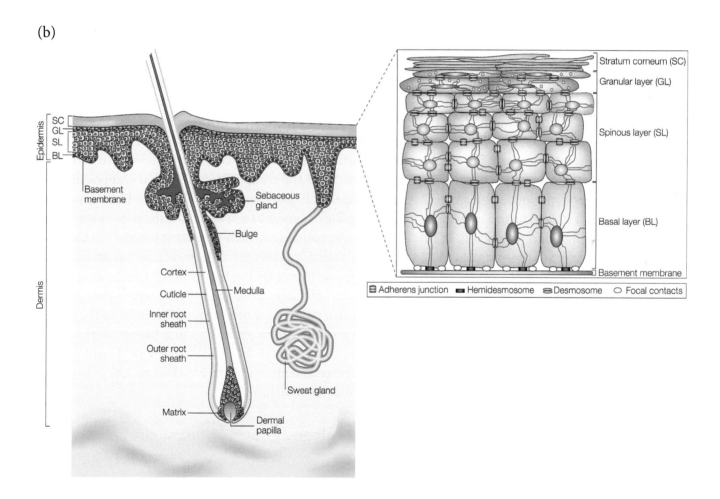

FIGURE 2.44 **Keratin intermediate filaments.** (a) Fluorescence scanning laser confocal microscopy of keratin intermediate filaments in epithelial cells. The keratin intermediate filament constitutes most of the red stain in the image, whereas nuclear DNA is colored in green. The image is taken from [550]. (b) The structure of skin. The entire structure is shown on the left. A blow-up of the epidermal layer is shown on the right, along with the major protein-based structures. The image was taken from [551].

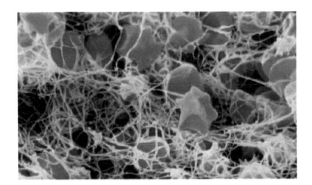

FIGURE 2.45 **A blood clot.** The fibrin mesh composing the clot is shown in white. The trapped red blood cells and platelets can also be seen. The image was taken from [553].

skin, cartilage, bone, tendons, ligaments, teeth, and the walls of blood vessels [555,556]. It is also a long-lived protein, with a half-life of ~100 to 200 years, depending on the tissue and type of collagen [557,558]. There are currently 28 known types of vertebrate collagen [554], with types I through IV being the most common. Some of the types are tissue-specific, whereas others are not. For example, collagen type I, which is the most common, is present in virtually all connective tissues except cartilage, whereas type II is present in cartilage and intervertebral disks. Type III is present in skin, blood vessels and the intestine, whereas type IV is present in the basal lamina[*1]. In tissues, collagen joins the protein *elastin* as well as polysaccharides and minerals to form a strong extracellular mesh responsible for the structural integrity of body organs [559,560]. The specific properties of each type of connective tissue result from the structure of its collagen (fiber, sheet, etc.) and the composition of related compounds. For example, in flexible tissues (skin, cartilage), collagen joins elastin and considerable numbers of polysaccharides, whereas in bone, collagen serves as a scaffold for the deposition of calcium and phosphate, in the form of *hydroxyapatite*, which renders it tough [561]. Collagen is formed within the cell as a precursor called *'procollagen'*. This molecule has unorganized N' and C', the former containing a signal sequence telling the cell to divert the molecule to the secretory pathway. Only after secretion of the procollagen to the extracellular space are the termini enzymatically cleaved, at which point the molecule is converted into the mature form of collagen (*tropocollagen*). Tropocollagen then assembles in a hierarchical manner into a complex structure. Types I through III, V, VII, XI, XXIV, and XXVII tend to form rope-like structures (fibrils), which further assemble into fibers (Figure 2.47). The other types of collagen may form networks of other structures.

2.7.1.2 Tissue organization and cell-environment communication

As explained, the two major cellular structures based on structural proteins are the cytoskeleton and the ECM. Although they are separated by the plasma membrane, these two ultra-structures are in contact with each other via membrane-crossing proteins, such as *integrins*[*2], which are bound to cytoskeletal and ECM elements on their opposite termini (Figure 2.48). This contact achieves two important goals. First, it helps in anchoring the cell

---

[*1]Base membrane; a thin sheet of extracellular matrix that separates epithelia from the underlying connective tissues.

[*2]Integrins constitute a diverse family of dimeric membrane-bound receptors, which play a central role in tissue organization and cellular survival via cell-cell and cell-ECM interactions [564].

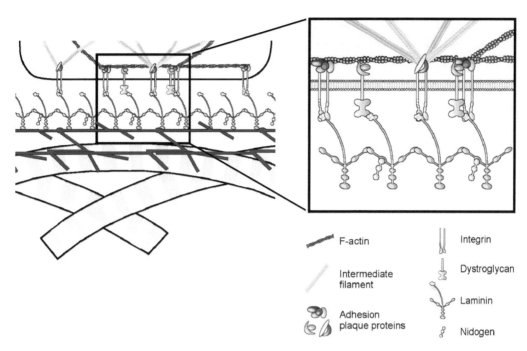

FIGURE 2.46 **Organization of the extracellular matrix.** The image on the upper-left side of the figure shows a cell (light-pink-filled body) and the extracellular matrix in which it lies. The thick, striped, elongated shapes are collagen fibers. The rest of the shapes in the figures are molecules composing intra- and extracellular fibers, as detailed in the index at the lower-right side of the figure. The image on the upper-right side of the figure is a blow-up of the extracellular matrix region that lies near the plasma membrane. It also shows the membrane-crossing molecules mediating the intra- and extracellular fiber networks. The figure was adapted from [562].

to the tissue. For example, *focal contacts* are formed between intracellular actin-myosin-based stress fibers and the ECM. The contraction ability of the stress fibers contributes significantly to the attachment of the cells to the tissue [79]. Actin fibers are also involved in cell-cell interactions in epithelia, via their indirect interaction with the membrane-crossing protein *cadherin*. The cellular contacts mediated by cadherin proteins are called *adherence junctions*. Intermediate filaments also participate in cell-cell and cell-ECM interactions, via desmosomes and hemi-desmosomes, as explained earlier.

The cytoskeletal-ECM contact also facilitates communication, i.e., the relaying of biochemical and mechanical signals from the external environment to the cell. For example, integrin-mediated contact between microfilaments and the ECM leads to the clustering of integrin molecules in a confined area of the membrane, which in turn leads to the activation of a tyrosine kinase called *FAK*. The latter is involved in signal transduction that depends on the type of substrate on which the cells proliferate. Indeed, studies show that interactions between cells and the ECM significantly affect the shape, proliferation, differentiation, and even death of the former [563–567]. This type of communication, thought to be responsible for tissue organization, operates in parallel to chemical-based communication, that is, communication based on chemical messengers such as hormones, growth factors, and cytokines, which bind to specific cellular receptors. However, this functional dichotomy is not absolute; some integrins are known to bind chemical messengers such as the neurotrophins, molecules participating in the regulation of the nervous, blood, and immune systems [568].

TABLE 2.10 **Types of extracellular matrix (ECM) proteins and their functions.** ECM: extracellular matrix. The table was adapted from [563].

Protein	Function
Collagen (fibrillar)	• Forming structural scaffolds • Stiffness control and tension resistance • Binding of adhesion molecules and some growth factors
Collagen (non-fibrillar)	• Aiding ECM organization and stability • Aiding fibrillar collagen formation • Modulation of cellular migration and proliferation • Creating physical barriers for solute penetration to tissue
Fibrin	• Forming blood clots • Stiffness control and tension resistance • Binding of adhesion molecules
Elastin	• Providing elastic recoil to tissues
Proteoglycans	• Compression resistance • Hindering transport of water and macromolecules • Binding of growth factors and chemokines

2.7.1.3 Motion

Fiber-forming proteins participate in several mechanisms of motion. These mechanisms operate on different levels of organization, including the physiological, the cellular, and the subcellular levels. The following sections review some of the main mechanisms.

2.7.1.3.1 Physiological motion

Motion at the physiological level is best exemplified by animal muscle contraction. The basic contraction unit, the *sarcomer*, is based on two fiber-forming proteins. The first is actin, which we have already encountered as a cytoskeletal protein capable of forming fibers of different lengths by adding or subtracting monomers (Figure 2.43). In the sarcomer, actin forms thin fibers of fixed length. The second protein, myosin II, forms thick fibers. The mechanical contraction of muscle is the result of these two fibrous structures sliding against each other using the chemical energy of ATP, as in stress fibers.

2.7.1.3.2 Cellular motion

Whole-cell motions of single-cell organisms can appear in two very different forms. The first results from the movement of elongated organelles that extend from the cell's surface outwardly, and function like motors. The structure of these organelles is based on the protein tubulin, and they may be short and numerous (*cilia*), or long and few (*flagella*; Figure 2.49). Such structures also appear in multicellular organisms; in these cases, instead of moving entire cells, the structures help in pushing fluids or solids away from tissue-anchored cells. The second type of whole-cell motion involves changes in the shape of the cell (*amoeboid motion*), and is based on actin microfilaments. The dynamic nature of microfilaments, resulting from the constant addition and/or subtraction of actin monomers, allows the cell to

(a)

(b)

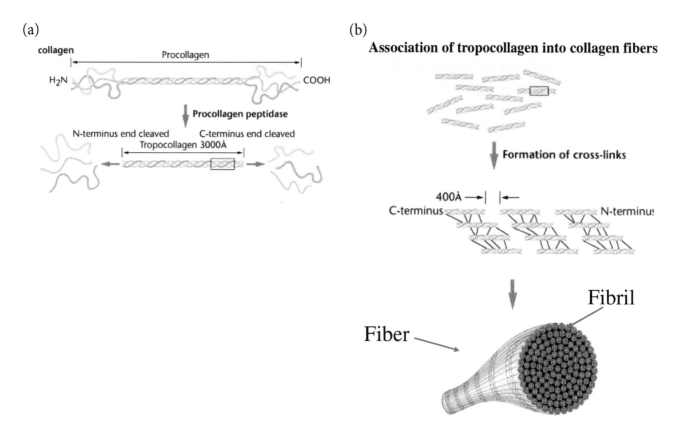

FIGURE 2.47 **Collagen maturation.** The figure shows the proteolytic processing of procollagen into tropocollagen (a) and the assembly of tropocollagen units into collagen fibers (b). The figure was adapted from [560].

create elongated extensions of the plasma membrane [79]. These extensions change the shape of the cell, but can also produce motion. In this case, the extensions are called 'pseudopodia', and the net effect of motion is created when the other parts of the cell follow them. This type of motion is particularly important for the function of *macrophages*, phagocytic cells that identify pathogens, internalize them, and then kill them using various toxic chemicals they produce. The internalization process involves the formation of pseudopodia, which engulf the pathogen.

2.7.1.3.3   Subcellular motion

Actin and tubulin are also involved in motion of different elements within cells. For example, actin fibers participate in the formation of the membrane separating the two daughter cells at the end of mitosis (*cytokinesis*). These fibers are a target for several natural toxins. For example, *cytochalasin D* is a fungal alkaloid that induces the disintegration of actin fibers, thus preventing the cell from moving or completing its mitotic cycle. Such toxins are used by scientists as important tools for investigating the cytoskeleton.

Another classic example of motion inside cells is the transport of intracellular 'cargo', i.e., organelles, vesicles, and protein complexes [571]. These are anchored to 'motor' proteins, which facilitate their movement along microfilaments and microtubule fibers, similar to the motion of a train along its rails. Motion along actin fibers is carried out using the motor proteins *myosin Va and VI* (in opposite directions), whereas motion along microtubules uses *dynein* and *kinesin*. The two types share some of their 'cargo', and include many crossing points. For example, in *exocytosis*, secretory vesicles start their way to the plasma membrane

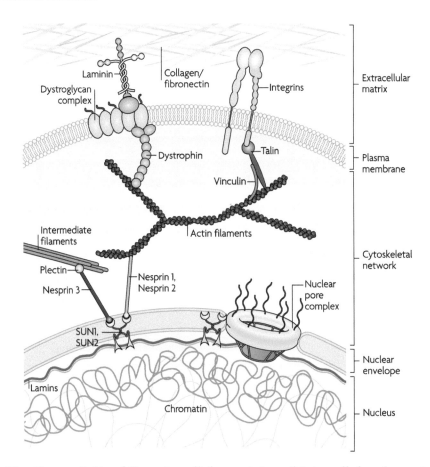

FIGURE 2.48 **Connectivity of the extracellular matrix and intracellular elements.** The figure shows the continuous protein network starting from collagen and fibronectin fibers outside the cell, and proceeding to membrane-crossing elements such as integrins, to intracellular elements such as actin and intermediate filaments of the cytoskeleton, and finally to elements inside the cell nucleus, such as lamins. The figure was taken from [569].

along microtubules, using kinesin. When approaching the periphery of the cell, the vesicles are transferred to myosin Va, which carries them along actin fibers to their destination. The opposite process, *endocytosis*, is carried out similarly, and involves the motor proteins myosin VI and dynein [571]. As might be expected, these processes are crucial for the viability of the cell, and damaging them results in lethal pathologies, such as amyotrophic lateral sclerosis (ALS) [572] (see also Chapter 5, Box 5.1).

Other types of intracellular motion involving microtubules are the building of the mitotic spindle [573] and the plant cell wall. Like microfilaments, microtubules also undergo constant assembly and disassembly. This process is targeted by certain natural toxins, such as *colchicine*, *taxol*, and *vinblastine*. Since these toxins get in the way of mitotic spindle construction, they are used as pharmaceutical drugs against various pathologies that involve unwanted cellular division. Thus, colchicine is used against gout, whereas taxol and vinblastine are used against cancer.

2.7.1.4   External structures

Some types of protein fibers are formed within the organism, yet fulfill their roles outside its body. This is the case with *silk*, which is produced by insects [574,575]. Silk fibers are used as hunting traps (spider webs), as well as for building nests and cocoons. Silk is very light

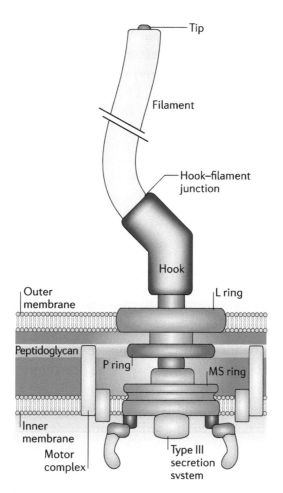

FIGURE 2.49 **A schematic depiction of the bacterial flagellum.** The different parts of the flagellum are marked, with emphasis on their connectivity and topological organization within the different regions of the bacterial cell. The figure was taken from [570].

and flexible, but at the same time very strong. In fact, gram for gram, silk fibers are stronger than any other known biological material, and are even stronger than steel. Because of these properties, silk is being considered as a potential component of future industrial materials that require both lightness and strength, e.g., bullet-proof vests and surgical sutures [576,577].

The strength of silk is particularly pronounced in spider webs, which are capable of catching large insects in flight, and holding them. There are seven different types of spider webs, divided according to their designated roles (capture, reinforcement, gluing, safety line, etc.). Some webs may include several different structural proteins (e.g., *fibroin*) [578]. These proteins exist in a liquid state within specific glands in the spider's body, where each gland specializes in a different type of web. During secretion of the liquid, changes in pH and ionic concentration decrease the water solubility of the protein significantly, which in turn leads to its assembly as non-soluble fibers.

2.7.1.5 Other roles

The traditional notion that fibrous proteins merely provide mechanical support to cells and tissues, or facilitate cellular motion, has changed in recent years, as other, more sophisticated roles have begun to emerge. For example, we have already seen how microtubules, in addition to building a supporting network inside the cell, also facilitate the trafficking of organelles and cellular division. However, one could say that in these roles the fibrous protein also essentially functions as a structural building block. It is notable, then, that recent studies have implicated some fibrous proteins in roles that do not relate to cellular shape or mechanical activity. This is nicely exemplified by keratin. This protein belongs to a family of proteins that are encoded by ~54 genes and include two major types:

- Type I – includes K9–K23 and Ha1–Ha8.

- Type II – includes K1–K8 and Hb1–Hb6.

The expression of the two types of keratin is highly regulated and depends on cell type and degree of differentiation [579,580], characteristics that in themselves reflect the biological importance of keratin. And indeed, studies focusing on the keratin family have revealed different roles played by these proteins [547]. The following subsections provide a short description of the main roles.

2.7.1.5.1 Regulation of skin pigmentation

Skin color is determined by the pigment *melanin*. The production and distribution of the pigment is a complex process that involves two types of cells: *melanocytes*, which reside in the basal layer of the epidermis (Figure 2.44b), and the adjacent *keratinocytes* [581]. Melanin is first produced (from the amino acid tyrosine) by melanocytes, inside organelles called *melanosomes*. Then, it is transported to nearby keratinocytes and internalized in a process similar to phagocytosis. Inside the keratinocytes, melanin granules localize near the nucleus, where they can protect the genetic material against the ultra-violet radiation of the sun. Keratin has been implicated in this process, on the basis of several observations. First, the 'mottled pigmentation' phenomenon, involving hyper- and hypo-pigmentation of the skin, is linked to mutations in keratin genes [582]. Second, keratin has been found to interact with dynein [583], which is known to be involved in the transport of melanin to the nucleus of keratinocytes [584]. Finally, studies have demonstrated that intermediate-filament-related proteins, including keratin, interact with motor proteins that facilitate the movement of organelles along microtubules [585]. Although the exact mechanism through which keratin affects pigmentation is yet to be found, the above observations suggest that the protein is involved in the transport of melanin granules to the nucleus [547].

2.7.1.5.2 Regulation of hair follicle growth cycle

Hair follicles reside in the thick, inner dermis layer of the skin, which also includes blood and lymph vessels, sweat and sebaceous glands, as well as nerve endings (Figure 2.44b). Despite being small, the follicle is one of the most interesting structures in terms of histological complexity. It is composed of eight epithelial layers differing in their degree of differentiation, and its life cycle includes three steps: rapid growth, regression[*1], and metabolic rest [586]. Mice that do not express the K17 keratin type suffer from a partial loss of hair [587]. Moreover, cultured keratinocytes taken from these mice are particularly susceptible to cycloheximide- and TNF-α-induced apoptosis[*2]. These studies and others suggest that keratin promotes hair growth by inhibiting apoptosis [588].

In conclusion, **keratin seems to function at the biochemical level in addition to fulfilling a structural role in providing mechanical support to cells and tissues.**

[*1]During this phase, the lower part of the tissue is eliminated by apoptosis.

[*2]Cycloheximide is an antibiotic that inhibits protein biosynthesis in eukaryotic cells. TNF-α is a biologically produced chemical involved in numerous cellular processes, such as inflammation and apoptosis.

2.7.1.5.3 Regulation of epithelial cell regeneration via protein synthesis

Tissue injury in animals automatically activates a set of healing processes in the cells surrounding the damaged area. In skin, liver, brain, and muscle tissue, these processes involve significant changes in the expression of intermediate filaments [589]. Although the exact involvement of the intermediate filaments in the healing process is not entirely clear, it has been suggested that they enhance cells' capacity to migrate to the place of injury, by changing the cells' flexibility and the viscosity of their cytoplasm. As wound healing is accompanied by massive production of proteins, it has been suggested that the K17 keratin type may also assist in the healing of epithelial tissues by promoting protein synthesis in the activated cells [590]. Indeed, cells devoid of K17 experience a 20% drop in protein production.

   In conclusion, **intermediate filaments in epithelia seem to be involved in processes that are associated with stress-related responses, signal transduction regulating protein synthesis or apoptosis, and the trafficking of organelles and/or vesicles inside cells** [547].

2.7.2   Fiber-forming versus fibrous proteins

The proteins mentioned above have different roles, yet share one common property: they all act as building blocks of larger polymeric structures. Accordingly, it is customary to refer to those proteins as '*structural*'. The structures formed by structural proteins are often organized as long fibers, creating the erroneous notion that all structural proteins are fibrous. In fact, many of these proteins are globular, as in the case of the microfilament-related protein actin[*1] and the microtubular protein tubulin. Some fiber-shaped structures are built from proteins that are themselves elongated in shape and have some properties that are significantly different from those of globular proteins. This group of *fibrous proteins* includes some of the fiber-forming proteins mentioned in the above subsections, such as the intermediate-filament-forming protein α-keratin, the extracellular matrix protein collagen, and the silk-forming protein fibroin. Our discussion in the following sections focuses on such proteins.

2.7.3   Structural differences between globular and fibrous proteins

Globular and fibrous proteins are substantially different from each other in their structural properties, beyond the obvious shape difference. First, globular proteins have a nonpolar core and a polar surface, whereas many fibrous proteins have the opposite topology. As a result, the former are water-soluble, whereas the latter are water-insoluble. The low solubility of many fibrous proteins may seem like a major disadvantage. However, in contrast to globular proteins, which need to be soluble in order to create transient contacts with other molecules without forming aggregates, fibrous proteins act mainly as building blocks for large stable structures, a function that does not require water solubility. The unique topology remains also when these proteins form large fibers, such that these fibers are nonpolar as well (see more below).

   The second marked difference between the structures of globular and fibrous proteins lies in their structural hierarchy. Globular proteins may include both α and β elements within the same molecule, and these serve as a basis for higher-level structures (supersecondary and tertiary), needed for the complex function of the protein. In contrast, fibrous

---

[*1] Actin has two forms: G (globular) and F (fibrillar), corresponding to the monomeric and polymeric forms, respectively (Figure 2.43). Still, both forms are essentially globular.

proteins include either α or β elements, usually arranged as a supersecondary structure that creates the repetitive nature of the fiber [591,592]. This property is exemplified in the following three 'classic' fibrous proteins [71]:

1. α-Keratin, composed of two α-helices wound around each other in a supersecondary structure referred to as a *'coiled coil'* [593]*1 (Figure 2.50a).

2. Collagen, composed of three non-α helices wound around each other in a structure referred to as a *'triple helix'* (Figure 2.50b).

3. Silk fibroin, composed of β-sheets bound to each other via flexible linkers. Other silk-forming proteins have a similar structure (Figure 2.50c).

The complex tertiary structure characterizing globular proteins is usually absent in fibrous proteins, since the latter proteins do not require the sophistication conferred by such complexity. Interestingly, fibrous proteins do have quaternary structures, as they tend to assemble into fibrils, and then fibers. However, in contrast to the quaternary structures of globular proteins, the structures of fibrous proteins tend to include both noncovalent and covalent bonds. Such bonds come in the form of disulfide bridges or other bonds that do not appear at all in globular proteins.

(a)   (b)   (c)

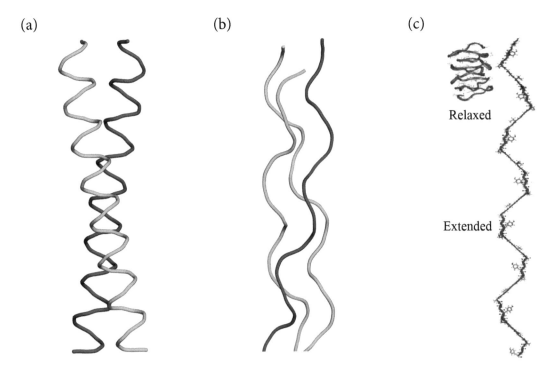

Relaxed

Extended

FIGURE 2.50 **Supersecondary structure in fiber-forming proteins.** (a) The coiled coil structure of α-keratin. (b) The triple helix structure of collagen. (c) The complex β structure of spider silk forming protein from *Nephila clavipes*. Models of both relaxed and extended forms of the protein are shown. The image was taken from [594].

---

*1 α-Keratin is present in the skin, hair, wool, nails, claws, horns, and hooves of mammals. β-Keratin is present in bird beaks and feathers, as well as in reptilian scales. In contrast to α-keratin, β-keratin has a twisted β-sheet structure.

2.7.4 Structure-function relationships in helical proteins α-keratin and collagen

α-Keratin and collagen are two fibrous proteins with some common characteristics. First, they are both helical in shape. Second, they both constitute major components in their respective tissues (epithelial and connective, respectively). Third, they both confer strength to the tissue. However, the type of strength rendered differs between the two proteins; α-keratin confers toughness, whereas collagen confers tensile strength. This difference results from the unique structure of each of the two proteins. In the following sections we will review these structures and focus on the specific determinants responsible for the differences mentioned above.

2.7.4.1 α-Keratin

α-Keratin is the principal protein of cytoskeletal intermediate filaments in mammalian epithelia. Its presence confers toughness to the tissue, and allows it to resist mechanical pressure. Although we do not have a high-resolution structure of the protein, ample data have been collected in the last 50 years or so, including X-ray diffraction data [595]. These data indicate a left-handed coiled coil, which is composed of two right-handed α-helices that are wound around each other (Figures 2.6–2.11a) [596]. The coiled coil is ~450 Å long, and has a pitch of ~150 Å. The two helices forming this structure have a characteristic seven-residue repetitive motif (*heptad repeat*) [597–599], with the first and fourth positions usually populated by nonpolar residues [600], placing them on the same face of the helix (Figure 2.51b). These residues create a tilted hydrophobic stripe along each helix in the coiled coil structure (Figure 2.51c), which mediates the important nonpolar component of the interactions between the two helices. The other positions in the heptad repeat contain mainly residues that have a preference for α-helices. Since many of these tend to be nonpolar, the coiled coil is overall hydrophobic, i.e., water-insoluble. Except for the heptad repeat, other sequence segments have been implicated in the dimerization of the helices [601,602]. The fact that 13 of the amino acids in these segments are highly conserved supports the idea that they are important for maintaining keratin's structure.

The coiled coil structure of α-keratin is highly compact, which requires close physical contact between the two interacting helices. **The proximity is facilitated by the high frequency of Gly and Ala residues in the keratin sequence**. Gly is devoid of a side chain, and Ala has the smallest side chain of all amino acids. In addition to the noncovalent interactions between the helices in the coiled coil, keratin also features disulfide bridges that fortify the structure, thus making it extremely tough. These bonds create the distinction between the different types of keratin: The 'hard' α-keratin, which is present in nails, horns, and hoofs, contains a large number of disulfide bridges, whereas the 'soft' α-keratin of skin and muscle contains a smaller number of these bonds [546]. **Thus, epithelial tissues can modulate their own mechanical properties by regulating the covalent crosslinking within resident keratin.**

α-Keratin tends to form large, higher-level fibrous structures. First, two coiled coil units organize as a four-unit *protofibril*; then, two protofibrils interact to form an eight-unit structure. These units associate one on top of the other along the helix axis, until a 32-unit bundle is formed. This bundle constitutes the principal length unit of the fiber, and is referred to as a *'unit length filament'* (ULF) (Figure 2.51d). The long α-keratin fibers, which appear in electron micrographs, are made of these units. **In 'soft' keratin, the ULFs are able to slide against each other, thus allowing the fiber to increase its length by 350%** [603].

The above illustrates how a simple coil-based motif can be used for creating mechanically rigid structures of varying toughness, which stabilize animal tissues and organs. However, this is only one example of what can be done with a coiled coil. Indeed, this general motif is one of the most common structural arrangements in proteins, and it is by no means limited to structural or fibrous proteins [595]. Its capacity to integrate two separate helices into one structure has made it perfect for promoting dimerization of proteins and domains. A well-known example is the *leucine zipper* structure (a coiled coil), formed by the dimerization domain of the yeast transcriptional activator GCN4. The stability of the coiled coil is also utilized to form structures that can span long distances. When such structures protrude from the surface of a cell or virus, they can promote recognition processes. Examples include the stalk region of non-fimbrial adhesins, and viral fusion proteins. Other roles assigned to coiled coils are described by Parry et al. [595].

2.7.4.2 Collagen

Collagen, the most abundant protein in our body, and the main protein component in our connective tissues, has several roles:

1. Helping tissues endure mechanical pressure.

2. Forming a scaffold on which other macromolecular deposits assemble. These include laminin networks, proteoglycan, and cellular receptors. Such components are involved in processes of cellular adhesion to other cells or to tissue differentiation, tissue development, and organ integrity.

3. Providing tissues with characteristic properties, such as tensile strength in skin, tendons and ligaments, or toughness in bone tissue.

These roles manifest in the structures formed by the 28 known types of collagen in different tissues [560]. For example, in cartilage, tendons, and ligaments, collagen forms a rope-like structure that confers a tensile strength of about 10^8 pascal[*1] to the tissue. In contrast, in bone and teeth, it serves as a substrate for mineral deposition (mainly calcium), which hardens those tissues. We will focus on the former type, i.e., fibrillar collagens. As mentioned earlier, fibrillar collagens tend to assemble in a hierarchical manner to form fibers. This *self-assembly* capability is encoded into their molecular structure.

Models for the structure of collagen have been proposed since 1940. The first one [604] suggested that collagen consists of a single helix. It was only later (1951) that Pauling and Corey [217] realized that **the basic collagen unit consists of not one but three helices**, and that some of the **groups within this 'triple helix' are paired in interhelical hydrogen bonds**. In addition, they proposed that some of the peptide bonds in the molecule acquire the *cis* configuration. The latter assumption was refuted three years later by Ramachandran and Kartha [605], who determined from fiber diffraction data that **the three chains forming the collagen molecule are left-handed PPII helices**, and that all peptide bonds in the molecule are in the *trans* configuration [554]. Ramachandran and Kartha's results also suggested the presence of two hydrogen bonds stabilizing the structure. The final refinement of this model, which created the currently accepted model, came in 1955 as a result of several different studies [606-608]. These studies indicated, among other things, the presence of only a single type of hydrogen bond between the strands, involving backbone carbonyl and amide

---

[*1] one pascal $\equiv$ 1 newton/m^2 $\approx 9.86 \times 10^{-6}$ atmospheres.

(a)

(c)

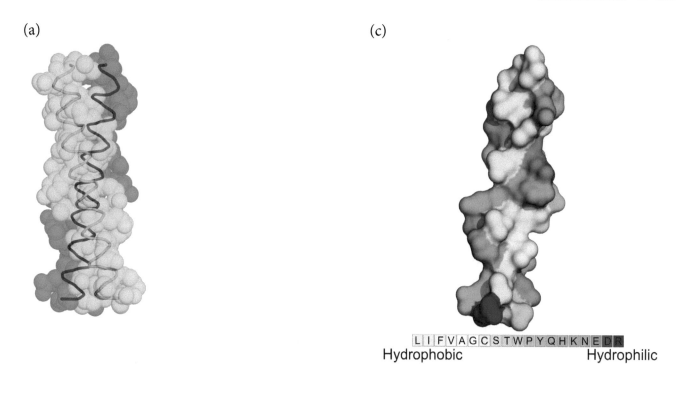

LIFVAGCSTWPYQHKNEDR
Hydrophobic   Hydrophilic

(b)

(d)

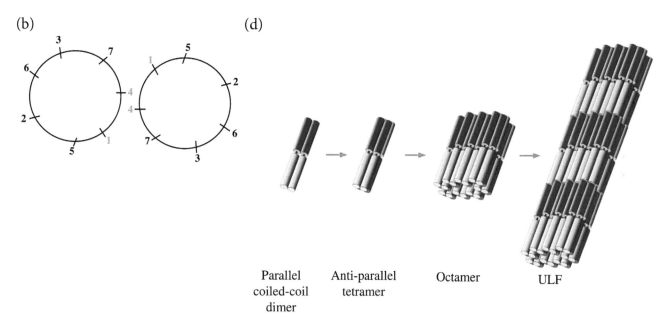

Parallel   Anti-parallel   Octamer   ULF
coiled-coil   tetramer
dimer

FIGURE 2.51   **The structure of α-keratin.** (a) The basic coiled coil structure of α-keratin. Since a high-resolution structure of keratin is unavailable, we used the structure of another coiled coil, that of the leucine zipper, formed by the dimerization domain of the yeast transcriptional activator GCN4 (PDB entry 1zik). The latter is an accepted model of coiled coils [595]. The image shows the structure as semi-transparent spheres, with each helix colored differently and the intertwined back-bones of the two helices visible. (b) A helical wheel projection of the two helices in the coiled coil structure, showing the distribution of the heptad residues (marked by their positions, from 1 to 7) along the helix axis. (c) The tilted hydrophobic stripe along the axis of each helix in the coiled coil structure. The surface of one of the chains in the structure of GCN4 is colored using the Kessel-Ben-Tal hydrophobicity scale [56] (see Figure 2.7 for details). The nonpolar stripe is colored in yellow. (d) Assembly of the unit length filament (ULF). The process is described in the main text. The figure was adapted from [549].

groups (N−H⋯O=C) (see more below). As in all other proteins, the next stage should have been the shift from low-resolution data, such as that provided by fiber diffraction, to high-resolution structures. In collagen, however, obtaining high-resolution data proved to be a major problem, due to the large size of the protein, its low solubility, and its complex hierarchical structure. As a result, most of the data that followed came from high-resolution structures of collagen-related proteins (CRPs) (e.g., [609–611]). The structure that has emerged is that of a 14-Å, right-handed triple helix with a 7/2 helical pitch (20 Å axial repeat) (Figure 2.52a,b)[*1]. The individual helices extend over 1,000 residues [560] and have a PPII helical conformation, i.e., with no intrahelical hydrogen bonds. As mentioned above, there are hydrogen bonds between the helices. There are different types of collagen; in some all three helices are identical, whereas in others they are not.

In most fibrils, the triple helices assemble into higher-level ultra-structures [612–614] (Figure 2.47). The high stability of collagen results from all interactions in these ultra-structures, i.e., the interactions between the individual helices, adjacent triple helices, and adjacent fibrils. Interestingly, the interactions within the triple helix do not seem to be sufficient; **the triple helix unit (tropocollagen) becomes stable only after the formation of the fibril** [615]. Recent studies have yielded models describing the arrangement of tropocollagen units within the fibril (e.g. [614,616]). They show that the triple helices form microfibrils, which interact with one another to form a rope-like structure, i.e., the fibril. At least some of these interactions are covalent [617].

The ability of collagen to fulfill its important roles in connective tissues results from its unique properties, the most important of which are its high stability, mechanical strength, and capacity to specifically bind other molecules [554]. The first two are a consequence of the highly efficient assembly of collagen into its fibrillar ultra-structure, which in turn results from the properties of the basic triple helical structure of tropocollagen. The formation of such a structure requires the polypeptide chain to have specific stereochemical properties. These are provided by the unique sequence of collagen, which includes the repetitive motif *Gly-X-Y*, where X and Y are any residue, but often proline and hydroxyproline (4′-OH Pro, or Hyp [87]), respectively. The following paragraphs describe the ways in which the motif facilitates the formation of the triple helix and provides it with some of the specific traits needed for self-assembly.

1. Creating the correct geometry by promoting PPII helices formation (Pro; Hyp).
   The fact that proline and hydroxyproline are highly likely to occupy the X and Y positions (occupancy frequencies of 28% and 38%, respectively) [554] provides the sequence with a PPII helical conformation [54]. This conformation offers at least two advantages for the creation of the collagen triple helix. First, **the PPII conformation is geometrically compatible with, and even required for, the formation of the triple helix**. The pyrrolidine ring, present only in the side chain of proline, creates a kink in the polypeptide chain. In globular proteins such kinks are usually a major disadvantage, as they break the helical conformation. In collagen, however, the kink is what produces the chain curvature that is needed for the three helices to fit around each other. Second, the PPII helical conformation is much more flexible than, e.g., the α-helix, which decreases the entropic cost of collagen folding [618]. It should be men-

---

[*1]Native collagen, which includes a much lower fraction of Pro compared to the CRPs used in these studies, is predicted to have a 10/3 helical pitch (28.6 Å axial repeat) [607]. Thus, the actual geometry of collagen is expected to depend on its exact sequence [554].

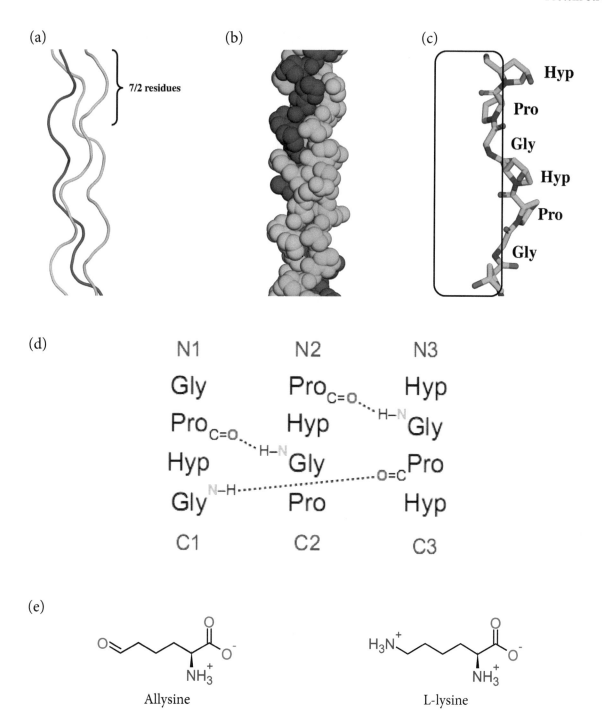

FIGURE 2.52 **The structure of collagen.** (a) Ribbon representation of the collagen triple helix (PDB entry 1qsu). Each chain is colored differently, and a segment encompassing one turn of the coil (7/2 residues) is marked. (b) A space-fill model of the triple helix. (c) The topology of the triple helix. The glycine residues of each helix face the center of the triple helix unit, whereas proline and hydroxyproline (Hyp) residues face outwards. (d) A schematic representation of the hydrogen bonds within the triple helix. A four-residue segment of each helix is depicted, from the amino terminus (top) to the carboxy terminus (bottom). Hydrogen bonds are depicted as dotted lines connecting the N–H and C=O backbone groups of the residues, which are also marked schematically. The image was taken from [554]. (e) Allysine (left), compared to L-Lys (right).

tioned that since some of the Pro and Hyp residues may appear in the *cis* configuration, the formation of the triple helix requires those peptide bonds to be isomerized to the *trans* configuration first [554].

2. Creating proximity between helices (Gly).
The glycine residue in the first position of the motif faces the center of the triple helix (Figure 2.52c). As we have already witnessed in the case of the α-keratin coiled coil, the absence of a side chain in glycine is a great advantage in structures that require close contact between polypeptide chains. However, whereas keratin can also include alanine residues in these positions, collagen's triple helix can only be formed in its native structure when glycine is present. Indeed, **most deleterious mutations in the collagen gene appear in the codon for glycine residues.** The close proximity among the three helices in collagen strengthens van der Waals and nonpolar interactions, which support the entire structure. **The absence of a side chain in glycine also makes it easier for the structure to form additional hydrogen bonds** (see the following section).

3. Triple helix stabilization (Gly, Hyp).
Early studies on collagen have demonstrated a correlation between the presence of hydroxyproline at position Y in the Gly-X-Y motif of collagen, and the thermal stability of the protein [619,620]. This observation led to the conclusion that **proline hydroxylation is a stabilizing factor in the structure of collagen.** The process of proline hydroxylation is catalyzed inside the ER lumen by the enzyme prolyl hydroxylase, which utilizes *ascorbic acid (vitamin C)* as a coenzyme [621]. Indeed, vitamin C deficiency leads to destabilization of collagen's structure, and to the subsequent disintegration of connective tissue in the animal body (Box 2.6). Most scientists at the time accepted the idea that Hyp stabilizes the triple helix; however, a controversy arose as to how it does so. One of the first ideas proposed was that the OH group of hydroxyproline participates in hydrogen bonds between the helices. However, when the three-dimensional structure of the triple helix was determined experimentally, it turned out that the residue occupying position Y turns away from the axis of the structure (Figure 2.52c). It was therefore understood that direct hydrogen bonds between this residue and the backbone groups in neighboring helices were not likely to be formed. An alternative explanation implicated water molecules around the triple helix, which were proposed to mediate the Hyp-involved hydrogen bonds [622,623]. Although some studies support this suggestion (see review by [560]), replacement studies in collagen-related proteins challenge it [554]. For example, when Hyp was replaced by 4-fluoroproline, which does not form hydrogen bonds (with or without the mediation of water molecules) [624], the triple helix became hyper-stable [625,626]. It was thus suggested that **Hyp stabilizes the triple helix not by inductive effects**[*1]**, but rather by stereoelectronic effects**[*2] **related to the conformation of the pyrrolidine ring** [627].

Aside from its geometric advantage in creating proximity among the helices of the collagen triple helix, the repetitive presence of Gly in collagen's sequence also increases the availability of backbone polar groups for hydrogen bond formation. In-

---

[*1]That is, the redistribution of charge density through σ bonds in a molecule due to the introduction of a polar substituent.

[*2]That is, effects resulting from the alignment of electronic orbitals.

deed, structures of collagen-related proteins show interhelical hydrogen bonds of the type $N-H_{(Gly)} \cdots O=C_{(X)}$, and possibly also $C\alpha-H_{(Gly/Y)} \cdots O=C_{(X/Gly)}$ bonds [606–608]. The former type was investigated by Maleev and coworkers [628], who estimated its strength to be about 1.6 kcal/mol.

4. Stabilization of collagen ultra-structure by covalent crosslinking (Lys).
   The stability of collagen depends not only on the interactions within the triple helix, but also on the interactions among the triple helices within each fibril. As a matter of fact, mutations that disrupt interactions of the latter type lead to pathologies far worse than mutations interfering with interactions of the former type [560]. This suggests that the body is far better off losing collagen altogether than allowing it to form incorrectly in deformed structures, which might affect the entire tissue. The strongest factor stabilizing the collagen ultra-structure is probably the covalent crosslinks between neighboring triple helices within or between microfibrils. The **crosslinks are formed between lysine or hydroxylysine residues, with the help of the enzyme lysyl oxidase** [617]. This process, called *'oxidative deamination'*, includes the following [629]. (1) Removal of the ε-amino group of the lysine side chain on one fibril. (2) Oxidation of the resulting terminus into an aldehyde group, thus forming *allysine* (Figure 2.52e) (or *hydroxyallysine*). (3) Creating a covalent bond between the aldehyde group and either the ε-amino group of a lysine side chain, or the aldehyde group of a modified lysine, in an adjacent fibril. Interestingly, **an excess of crosslinks, which is the result of ageing, makes collagen brittle** [630], thus decreasing its capability to fulfill its biological role. A recent study suggests that hydroxyproline-mediated hydrogen bonds contribute to fibril-fibril interactions as well [631]. Despite the covalent bonds supporting the structure of collagen, it is not as tough as α-keratin fibers in nails, hooves, claws, and horns, since the latter contain a large number of disulfide bonds, which greatly harden the entire structure (see previous subsection). This is why α-keratin and collagen have been evolutionarily selected for different roles: the former for conferring toughness, and the latter for conferring tensile strength.

5. Preventing aggregation (Pro, Hyp, Gly).
   Previously we discussed hydrophobicity as a property shared by many fibrous proteins. Although this property facilitates the principal functions of such proteins, it might impede their capacity to form biologically active fibers or networks, because it also drives them to aggregate and precipitate before the fiber is formed. This problem is particularly pronounced in proteins such as collagen, which include long sequences even before going through the first steps of assembly. **The enrichment of collagen with Pro, Hyp, and Gly helps to decrease aggregation** in two ways. First, the hydroxyl group on Hyp increases the polarity of the residue, and since it faces the outside of the triple helix, it increases the polarity of the entire structure. Second, proteins that are known to aggregate usually assume the β-sheet conformation, as in the case of amyloid proteins and peptides [632] (see Box 5.1 in Chapter 5). Gly and Pro have a very low propensity to appear in β-sheets [633,634], and it has been proposed that the general prevalence of these residues in fibrous proteins might have evolved to tackle the aggregation problem [554].

BOX 2.6 VITAMIN C DEFICIENCY

Around the 15th century, when Europeans embarked on long maritime voyages, a previously rare disease became very common aboard ships. This 'sailor's disease', or 'scurvy', manifested as fatigue, stiffness, loose teeth and bleeding gums, skin lesions, and blood blisters on the skin, back, arms, legs and buttocks [8,635]. The reason for the disease was initially a mystery, but following the 'miraculous' recovery of sick people who ate fresh fruits (particularly citrus) and/or green vegetables, it became clear that the symptoms were related to a deficiency of a certain factor normally present in these foods. Finding this factor became a pressing matter, as fruits and vegetables could not be kept fresh during long cruises. Since this factor was abundant in citrus fruits, it was suspected to be acidic in nature. However, when the use of diluted sulfuric acid did not lead to recovery, it was realized that acidity alone was insufficient. The answer to the scurvy problem came in 1747, thanks to the clinical trials held by British Royal Navy surgeon James Lind [636], following which British ships were routinely supplied with lemons and/or lime juice. However, only in the beginning of the 1930s was the cause of scurvy identified as a lack of *ascorbic acid*, i.e., *vitamin C* (Figure 2.6.1). As vitamin C is crucial for collagen synthesis and secretion into connective tissues (see below), its deficiency results in instability of gums, bone, and other connective tissues, as well as disintegration of capillaries, leading to bleeding.

FIGURE 2.6.1   **Vitamin C (ascorbic acid).** The molecule is colored according to atom type.

Vitamin C, discovered in 1927 by the Hungarian-American biochemist and Nobel Prize laureate Szent-Györgyi, is an essential micronutrient in humans and other primates, and in guinea pigs [635]. The biological importance of vitamin C is diverse. First, it is required for the biosynthesis of the following molecules:

1. Collagen: Vitamin C is needed for proline hydroxylation in procollagen. As explained in the main text, the formation of 4′-OH hydroxyproline is crucial for the

construction of stable collagen fibers or networks in animal connective tissues. The hydroxylation of collagen Lys residues is also important for stability.

2. Carnitine: This compound is essential for the production of energy in mitochondria using fatty acid β-oxidation. Specifically, carnitine is needed for transporting fatty acid from the cytosol into the mitochondrial matrix, where the oxidation process takes place. The inner mitochondrial membrane does not have a transporter for fatty acids, but it does have a transporter for carnitine. This role of vitamin C may explain the first clinical symptoms of scurvy, i.e., fatigue and lethargy, which might be the result of decreased fatty acid-based production of energy.

3. Neurotransmitters: There are some neurotransmitters whose biosynthesis requires vitamin C. Norepinephrine[*a] (NE), a chemical that also serves as a hormone, is associated with the famous 'fight-or-flight' response, as well as with various (mostly excitatory) brain functions. NE is synthesized from the amino acid tyrosine in three steps. The last step, which is carried out by *dopamine β hydroxylase*, requires vitamin C. Another key neurotransmitter that may rely on vitamin C is serotonin (5-HT), which is associated with numerous brain functions, such as mood, appetite, and sensory perception. Its involvement in mood and pain modulation has made it an important target for drugs acting against depression, anxiety, and chronic pain. Serotonin is synthesized from the amino acid tryptophan in two steps. The first one, catalyzed by *tryptophan hydroxylase*, requires tetrahydrobiopterin as a coenzyme. It has been suggested that vitamin C is required for the recycling of this coenzyme from its oxidized state.

In addition, vitamin C is a potent antioxidant that helps the body deal with the devastating effects of oxidative stress. Its role in this capacity suggests that deficiency in vitamin C may increase the risk of diseases that involve production of free radicals, such as *cardiovascular disease, cataracts* and cancer. Third, vitamin C has been implicated in the optimization of immune system function by enhancing the action of neutrophils, lymphocytes, and macrophages [637,638][*b]. Finally, it has been suggested that vitamin C enhances the intestinal absorption of non-heme iron [640].

Today, vitamin C deficiency primarily occurs in developing countries, along with other food-related problems. In developed countries it is rare[*c], although it may occur in smokers and alcohol- or drug-addicted people, whose diets tend to be poor in fruits and vegetables, as well as in the elderly, due to poor dietary habits [635].

[*a] Also called noradrenaline.

[*b] Although the common belief that vitamin C acts as an efficient prophylactic measure against the common cold has recently been refuted [639].

[*c] The recommended dietary allowance (RDA) of vitamin C is merely 10 mg per day.

2.8 SUMMARY

- Proteins have hierarchical structures, which include three, and sometimes four levels.

- The primary structure of a protein is the exact ordering of amino acids forming its chain. Amino acids are small organic molecules that each consist of a chiral carbon with four substituents. Of those, only the fourth, i.e., the side chain, differs among amino acids.

- Proteins contain 20 common amino acids. In some cases, chemical derivates of these can also be found, the bulk of which are formed post-translationally. The 20 amino acid types are traditionally separated into nonpolar, polar, charged, and aromatic groups. Polarity, the main classification characteristic, determines the location of the amino acid within the protein, as well as its ability to interact with other molecules and groups.

- The exact amino acid sequence of each protein drives it to fold into its own unique and biologically active three-dimensional fold, also known as the 'tertiary structure'. However, segments of the protein chain may acquire their own local folds, which are much simpler and usually take the shape of a spiral (α-helix), an extended shape (β conformation), or a loop. These local folds are termed 'secondary elements', and form the protein's secondary structure.

- The tertiary structure of the protein consists of different combinations of secondary elements, some of which are simple (i.e., motifs), whereas others are more complex. Repetitive parts of the protein chain, which have their own three-dimensional folds (usually evolutionarily conserved) and can be attributed some function, are called 'domains'. Domains are considered today to be the evolutionary and functional building blocks of proteins.

- The current repertoire of protein domains is thought to have evolved by duplication and divergence of ancestral folds. However, several highly common domains seem to have evolved from different proteins that converged into structurally stable or functionally efficient general structures. Both mechanisms make protein structure more conserved than protein sequence.

- There are different ways to classify proteins. Current popular tools rely on the sequence, structure, function, and even evolution of proteins as categories of classification. Some of the tools are automatic, whereas others use human intervention.

- Proteins that include more than one polypeptide chain are said to have a 'quaternary structure'. This last level of protein hierarchy confers quite a few structural and functional advantages over single-chain proteins, which is probably the reason why it emerged in evolution.

- Some proteins, such as those involved in central metabolism or in signal transduction, form transient interaction networks. The transient complexes formed by these proteins are sometimes referred to as the proteins' 'quinary structure'.

- Many proteins, primarily enzymes, contain organic or elemental components needed for their activity and/or stability. These cofactors, often derived from vitamins, are chemically diverse and may bind to their respective proteins reversibly (coenzymes) or irreversibly (prosthetic groups).

- The already large and diverse functional repertoire of proteins is, in many cases, further expanded by post-translational modifications (PTMs), which usually entail the addition of a chemical group to the protein, but may also involve removal or chemical modification of existing groups, as well as cleavage of the protein main chain. These PTMs not only make 'new' proteins, but also enable cells to tighten their regulation over their resident proteins, in accordance with environmental and/or genetic conditions.

- Fibrous proteins form large fiber-shaped structures inside and outside cells, the most pronounced of which are the cytoskeleton and the extracellular matrix, respectively. These mostly protect cells against mechanical pressure, and may also be involved in motion, at the level of the entire cell, organ, or intracellular organelles. Nevertheless, recent studies show that at least some fibrous proteins also participate in much more sophisticated roles, such as regulation of biochemical and physiological processes.

- Besides having an elongated shape, fibrous proteins tend to be more hydrophobic than globular proteins. Their overall structure includes one type of secondary element organized in a repetitive manner. Two or more such polypeptide chains assemble along one axis to form what is referred to as a 'supersecondary structure'. Common mammalian supersecondary structures are made of two or three helices wound around each other. The compact nature of these structures is facilitated by the prevalence of proline and glycine in their sequences, as well as by post-translational modifications carried out to serve specific purposes.

- Fibrous proteins tend to assemble into fibers or networks, which enhance their stability and biological activity. These higher-level (ultra) structures are often stabilized by covalent crosslinks. By modulating (1) the process by which fibrous proteins assemble into these ultra-structures, and (2) the number of covalent crosslinks, tissues regulate their own mechanical properties, such as strength, flexibility, resilience, etc. This is nicely demonstrated by the two principal fibrous proteins in animal epithelia and connective tissues, α-keratin and collagen (respectively).

EXERCISES

2.1 A. Specify the two basic types of hetero-groups in proteins.

   B. Explain the general functions of hetero-groups.

   C. Give three examples of hetero-groups and briefly describe their specific functions.

2.2 A. Specify which chemical groups in the amino acid arginine may undergo protonation and/or deprotonation.

   B. Draw the titration curve of arginine's side chain with NaOH.

2.3 What is the main criterion used to separate natural amino acids into groups? Explain why.

2.4 Cysteine is considered a polar amino acid, and yet, it is often found inside the protein core. Explain why.

2.5 Two residues, arginine and lysine, are in water, so the centers of their guanidinium and amino groups (respectively) are positioned 2 Å from each other.

    A. Estimate the residues' electrostatic interaction energy. Explain why this calculation is only an estimate.

    B. Would you expect the p*Ka* of the two residues' side chains in isolation to change as a result of their proximity? Explain why and calculate the extent of the change (in p*Ka* units).

    C. Is the p*Ka* change calculated in *B* large enough to alter the residues' charge state? If not, suggest a way to accomplish this alteration.

2.6 The enzyme *hypothetase* hydrolyzes a covalent bond in the substrate using a nucleophilic attack. Taking other known hydrolases as examples, which of the 20 natural amino acid residues is most likely to serve as the enzyme's nucleophile?

2.7 Briefly explain the two basic mechanisms used by organisms to fight oxidative damage.

2.8 Explain the mechanisms that enable amino acid derivatives to appear in proteins.

2.9 When the ϕ and ψ values of residues in experimentally determined proteins are collected, some of them reside outside the 'allowed' regions of the Ramachandran plot (Figure 2.14d). Explain why.

2.10 What are the two most prevalent secondary elements in proteins? Explain why.

2.11 What are the respective functions of backbone hydrogen bonds and nonpolar interactions in α-helices?

2.12 During the folding of a metalloenzyme, a zinc ion (Zn^{2+}), originally surrounded by water, becomes trapped inside the protein core. Assuming that the cation is a sphere with a radius of 1.4 Å, and that the dielectric constant of the protein core is 2, estimate the change in electrostatic energy accompanying the process.

2.13 Predict which secondary structure the following sequence is most likely to acquire:

    Ala-Leu-Met-Glu-Gln-Ile-Ala-Arg-Met-Gln-Leu-Glu-Ala-Ser-Met-Lys

2.14 Explain how each of the secondary elements in the immunoglobulin motif fulfills its functional role.

2.15 Explain how proteins with different sequences may still possess similar three-dimensional structures.

2.16 What are the main evolutionary advantages of quaternary structure?

2.17 Name two or three post-translational modifications of proteins that have been implicated in the development or behavior of cancerous cells.

2.18 Explain in general terms how phosphorylation may change protein activity.

2.19 Specify the main features that distinguish fibrous proteins from globular proteins.

2.20 List a few of the characteristic roles of fibrous proteins.

2.21 Explain the distinction between structural and fibrous proteins.

2.22 Compare between α-keratin and collagen in terms of source tissue, subcellular localization, structure, and function.

2.23 Explain the molecular basis for the difference between soft and hard keratins.

2.24 How would you treat a patient suffering from scurvy?

REFERENCES

1. J. C. Kendrew, R. E. Dickerson, B. E. Strandberg, R. G. Hart, D. D. Davis, D. C. Phillips, and V. C. Shore. The three dimensional structure of myoglobin. *Nature*, 185:422–427, 1960.

2. M. F. Perutz, M. G. Rossmann, A. F. Cullis, H. Muirhead, G. Will, and A. C. T. North. Structure of haemoglobin: A three-dimensional Fourier synthesis at 5.5 Å resolution, obtained by X-ray analysis. *Nature*, 185:416–422, 1960.

3. G. M. Crippen. The tree structural organization of proteins. *J. Mol. Biol.*, 126(3):315–32, 1978.

4. J. B Broderick. *Coenzymes and Cofactors*. John Wiley & Sons, Ltd., 2001.

5. M. S. Patel and L. G. Korotchkina. Pyruvate dehydrogenase complex. In *Encyclopedia of Life Sciences*. John Wiley & Sons, Ltd., 2006.

6. E. T. Harper and R. A. Harris. Glycolytic pathway. In *Encyclopedia of Life Sciences*. John Wiley & Sons, Ltd., 2005.

7. K. F. LaNoue. Citric Acid Cycle. In *Encyclopedia of Life Sciences*. John Wiley & Sons, Ltd., 2001.

8. K. J. Carpenter. History of Nutritional Science. In *Encyclopedia of Life Sciences*. John Wiley & Sons, Ltd., 2001.

9. M. F. Hughes. Arsenic toxicity and potential mechanisms of action. *Toxicol. Lett.*, 133(1):1–16, 2002.

10. D. S. Tawfik. Accuracy-rate tradeoffs: how do enzymes meet demands of selectivity and catalytic efficiency? *Curr. Opin. Chem. Biol.*, 21:73–80, 2014.

11. R. P. Smith. Arsenic: A Murderous History. *Center for Environmental Health Sciences* (http://www.dartmouth.edu/~rpsmith/index.html), 2008.

12. S. Kapaj, H. Peterson, K. Liber, and P. Bhattacharya. Human health effects from chronic arsenic poisoning – a review. *J. Environ. Sci. Health. A Tox. Hazard. Subst. Environ. Eng.*, 41(10):2399–428, 2006.

13. A. H. Hall. Chronic arsenic poisoning. *Toxicol. Lett.*, 128(1–3):69–72, 2002.

14. S. Forshufvud, H. Smith, and A. Wassen. Arsenic content of Napoleon I's hair probably taken immediately after his death. *Nature*, 192:103–5, 1961.

15. B. Weider and J. H. Fournier. Activation analyses of authenticated hairs of Napoleon Bonaparte confirm arsenic poisoning. *Am. J. Forensic Med. Pathol.*, 20(4):378–82, 1999.

16. P. F. Corso, J. T. Hindmarsh, and F. D. Stritto. The death of Napoleon. *Am. J. Forensic Med. Pathol.*, 21(3):300–5, 2000.

17. D. E. Jones and K. W. Ledingham. Arsenic in Napoleon's wallpaper. *Nature*, 299(5884):626–7, 1982.

18. W. J. Broad. Hair Analysis Deflates Napoleon Poisoning Theories. http://www.nytimes.com/2008/06/10/science/10napo.html. *The New York Times*, 2008.

19. J. A. D. Ingres. Napoleon sur son trone Impériale. Wikipedia, the free encyclopedia. http://en.wikipedia.org/wiki/File:Ingres,_Napoleon_on_his_Imperial_throne.jpg, 1806.

20. M. Uhlen, L. Fagerberg, B. M. Hallstrom, C. Lindskog, P. Oksvold, A. Mardinoglu, A. Sivertsson, C. Kampf, E. Sjostedt, A. Asplund, I. Olsson, K. Edlund, E. Lundberg, S. Navani, C. A. Szigyarto, J. Odeberg, D. Djureinovic, J. O. Takanen, S. Hober, T. Alm, P. H. Edqvist, H. Berling, H. Tegel, J. Mulder, J. Rockberg, P. Nilsson, J. M. Schwenk, M. Hamsten, K. von Feilitzen, M. Forsberg, L. Persson, F. Johansson, M. Zwahlen, G. von Heijne, J. Nielsen, and F. Ponten. Proteomics. Tissue-based map of the human proteome. *Science*, 347(6220):1260419, 2015.

21. C. B. Anfinsen. Principles that Govern the Folding of Protein Chains. *Science*, 181(96):223–230, 1973.

22. L. J. Henderson. Concerning the relationship between the strength of acids and their capacity to preserve neutrality. *Am. J. Physiol.*, 21:173–179, 1908.

23. J. E. McMurry and E. E. Simanek. *Fundamentals of Organic Chemistry*. Brooks/Cole, Florence, KY, 6th edition, 2007.

24. K. E. van Holde, W. C. Johnson, and P. S. Ho. *Principles of Physical Biochemistry*. Pearson Education, New Jersey, US, 2nd edition, 2006.

25. J. Martyn Bailey. RNA-directed amino acid homochirality. *FASEB J.*, 12(6):503–7, 1998.

26. K. Tamura and P. Schimmel. Chiral-selective aminoacylation of an RNA minihelix. *Science*, 305(5688):1253, 2004.

27. D. Konz and M. A. Marahiel. How do peptide synthetases generate structural diversity? *Chem. Biol.*, 6(2):R39–48, 1999.

28. S. L. Clugston, S. A. Sieber, M. A. Marahiel, and C. T. Walsh. Chirality of peptide bond-forming condensation domains in nonribosomal peptide synthetases: the C5 domain of tyrocidine synthetase is a (D)C(L) catalyst. *Biochemistry*, 42(41):12095–104, 2003.

29. M. Pinkerton, L. K. Steinrauf, and P. Dawkins. The molecular structure and some transport properties of valinomycin. *Biochem. Biophys. Res. Commun.*, 35(4):512–8, 1969.

30. N. Fujii. D-amino acids in living higher organisms. *Orig. Life Evol. Biosph.*, 32(2):103–27, 2002.

31. G. Kreil. D-amino acids in animal peptides. *Annu. Rev. Biochem.*, 66:337–45, 1997.

32. A. Jilek, C. Mollay, C. Tippelt, J. Grassi, G. Mignogna, J. Mullegger, V. Sander, C. Fehrer, D. Barra, and G. Kreil. Biosynthesis of a D-amino acid in peptide linkage by an enzyme from frog skin secretions. *Proc. Natl. Acad. Sci. USA*, 102(12):4235–9, 2005.

33. S. D. Heck, W. S. Faraci, P. R. Kelbaugh, N. A. Saccomano, P. F. Thadeio, and R. A. Volkmann. Posttranslational amino acid epimerization: enzyme-catalyzed isomerization of amino acid residues in peptide chains. *Proc. Natl. Acad. Sci. USA*, 93(9):4036–9, 1996.

34. S. D. Heck, P. R. Kelbaugh, P. R. Kelly, P. R. Thadeio, N. A. Saccomano, J. G. Stroh, and R. A. Volkmann. Disulfide bond assignement of omega-agatoxin-IVb and omega-agatoxin-IVc: discovery of a d-serine residue in omega-agatoxin-IVb. *J. Am. Chem. Soc.*, 116:10426–10436, 1994.

35. P. C. Montecucchi, R. de Castiglione, S. Piani, L. Gozzini, and V. Erspamer. Amino acid composition and sequence of dermorphin, a novel opiate-like peptide from the skin of *Phyllomedusa sauvagei*. *Int. J. Pept. Protein Res.*, 17(3):275–83, 1981.

36. K. Richter, R. Egger, and G. Kreil. D-alanine in the frog skin peptide dermorphin is derived from L-alanine in the precursor. *Science*, 238(4824):200–2, 1987.

37. M. Broccardo, V. Erspamer, G. Falconieri Erspamer, G. Improta, G. Linari, P. Melchiorri, and P. C. Montecucchi. Pharmacological data on dermorphins, a new class of potent opioid peptides from amphibian skin. *Br. J. Pharmacol.*, 73(3):625–31, 1981.

38. O. S. Andersen. Gramicidin channels. *Annu. Rev. Physiol.*, 46:531–48, 1984.

39. S. E. Blondelle, K. Lohner, and M. Aguilar. Lipid-induced conformation and lipid-binding properties of cytolytic and antimicrobial peptides: determination and biological specificity. *Biochim. Biophys. Acta*, 1462(1–2):89–108, 1999.

40. R. E. Hancock and D. S. Chapple. Peptide antibiotics. *Antimicrob. Agents Chemother.*, 43(6):1317–23, 1999.

41. B. A. Wallace. Common structural features in gramicidin and other ion channels. *BioEssays*, 22(3):227–34, 2000.

42. S. Miller, J. Janin, A. M. Lesk, and C. Chothia. Interior and surface of monomeric proteins. *J. Mol. Biol.*, 196(3):641–56, 1987.

43. G. D. Rose, A. R. Geselowitz, G. J. Lesser, R. H. Lee, and M. H. Zehfus. Hydrophobicity of amino acid residues in globular proteins. *Science*, 229(4716):834–8, 1985.

44. A. Gutteridge and J. M. Thornton. Understanding nature's catalytic toolkit. *Trends Biochem. Sci.*, 30(11):622–629, 2005.

45. J. Singh and J. M. Thornton. *Atlas of Protein Side-Chain Interactions*, volume I,II. IRL Press, Oxford, 1992.

46. S. Hormoz. Amino acid composition of proteins reduces deleterious impact of mutations. *Sci. Rep.*, 3:2919, 2013.

47. R. T. Sanderson. Electronegativity and bond energy. *J. Am. Chem. Soc.*, 105:2259–2261, 1983.

48. J. T. Brosnan and M. E. Brosnan. The sulfur-containing amino acids: an overview. *J. Nutr.*, 136(6 Suppl):1636S–1640S, 2006.

49. R. J. Zauhar, C. L. Colbert, R. S. Morgan, and W. J. Welsh. Evidence for a strong sulfur-aromatic interaction derived from crystallographic data. *Biopolymers*, 53(3):233–48, 2000.

50. M. Iwaoka, S. Takemoto, M. Okada, and S. Tomoda. Weak Nonbonded S⋯X (X=O, N, and S) Interactions in Proteins. Statistical and Theoretical Studies. *Bull. Chem. Soc. Jpn.*, 75(7):1611–1625, 2002.

51. M. Iwaoka and N. Isozumi. Possible roles of S⋯O and S⋯N interactions in the functions and evolution of phospholipase A$_2$. *Biophysics*, 2:23–34, 2006.

52. D. Pal and P. Chakrabarti. Non-hydrogen bond interactions involving the methionine sulfur atom. *J. Biomol. Struct. Dyn.*, 19(1):115–28, 2001.

53. R. L. Levine, L. Mosoni, B. S. Berlett, and E. R. Stadtman. Methionine residues as endogenous antioxidants in proteins. *Proc. Natl. Acad. Sci. USA*, 93(26):15036–40, 1996.

54. A. Rath, A. R. Davidson, and C. M. Deber. The structure of "unstructured" regions in peptides and proteins: role of the polyproline II helix in protein folding and recognition. *Biopolymers*, 80(2–3):179–85, 2005.

55. J. Liu and B. Rost. Comparing function and structure between entire proteomes. *Protein Sci.*, 10(10):1970–9, 2001.

56. A. Kessel and N. Ben-Tal. Free energy determinants of peptide association with lipid bilayers. *Curr. Top. Membr.*, 52:205–253, 2002.

57. L. N. Johnson, M. E. Noble, and D. J. Owen. Active and inactive protein kinases: structural basis for regulation. *Cell*, 85(2):149–58, 1996.

58. J. Kraut. Serine proteases: structure and mechanism of catalysis. *Annu. Rev. Biochem.*, 46:331–58, 1977.

59. L. Hedstrom. Serine protease mechanism and specificity. *Chem. Rev.*, 102(12):4501–24, 2002.

60. R. L. Thurlkill, G. R. Grimsley, J. M. Scholtz, and C. N. Pace. pK values of the ionizable groups of proteins. *Protein Sci.*, 15(5):1214–18, 2006.

61. C. N. Pace, G. R. Grimsley, and J. M. Scholtz. Protein ionizable groups: pK values and their contribution to protein stability and solubility. *J. Biol. Chem.*, 284(20):13285–9, 2009.

62. J. M. Grimsley, G. R. Scholtz and C. N. Pace. A summary of the measured pK values of the ionizable groups in folded proteins. *Protein Sci.*, 18(1):247–251, 2009.

63. C. A. Fitch, G. Platzer, M. Okon, B. E. Garcia-Moreno, and L. P. McIntosh. Arginine: Its p*Ka* value revisited. *Protein Sci.*, 24(5):752–61, 2015.

64. P. Zhou, F. Tian, F. Lv, and Z. Shang. Geometric characteristics of hydrogen bonds involving sulfur atoms in proteins. *Proteins*, 76(1):151–63, 2009.

65. G. Duan, V. H. Smith, and D. F. Weaver. Characterization of aromatic-thiol π-type hydrogen bonding and phenylalanine-cysteine side chain interactions through *ab initio* calculations and protein database analyses. *Mol. Phys.*, 99(19):1689–1699, 2001.

66. A. L. Ringer, A. Senenko, and C. D. Sherrill. Models of S/π interactions in protein structures: Comparison of the H$_2$S–benzene complex with PDB data. *Protein Sci.*, 16(10):2216–2223, 2007.

67. B. R. Beno, K. S. Yeung, M. D. Bartberger, L. D. Pennington, and N. A. Meanwell. A Survey of the Role of Noncovalent Sulfur Interactions in Drug Design. *J. Med. Chem.*, 58(11):4383–438, 2015.

68. J. M. Thornton. Disulphide bridges in globular proteins. *J. Mol. Biol.*, 151(2):261–87, 1981.

69. G. J. Lesser and G. D. Rose. Hydrophobicity of amino acid subgroups in proteins. *Proteins*, 8(1):6–13, 1990.

70. E. J. Heckler, P. C. Rancy, V. K. Kodali, and C. Thorpe. Generating disulfides with the quiescin sulfhydryl oxidases. *Biochim. Biophys. Acta*, 1783(4):567–577, 2008.

71. D. Voet, J. G. Voet, and C. W. Pratt. *Fundamentals of Biochemistry: Life at the Molecular Level.* John Wiley & Sons, Inc., 2nd edition, 2005.

72. A. J. Saunders, G. B. Young, and G. J. Pielak. Polarity of disulfide bonds. *Protein Sci.*, 2(7):1183–4, 1993.

73. D. Fass. Disulfide bonding in protein biophysics. *Annu. Rev. Biophys.*, 41:63–79, 2012.

74. I. Bosnjak, V. Bojovic, T. Segvic-Bubic, and A. Bielen. Occurrence of protein disulfide bonds in different domains of life: a comparison of proteins from the Protein Data Bank. *Protein Eng. Des. Sel.*, 27(3):65–72, 2014.

75. J. Riemer, N. Bulleid, and J. M. Herrmann. Disulfide formation in the ER and mitochondria: two solutions to a common process. *Science*, 324(5932):1284–7, 2009.

76. R. Bhattacharyya, D. Pal, and P. Chakrabarti. Disulfide bonds, their stereospecific environment and conservation in protein structures. *Protein Eng. Des. Sel.*, 17(11):795–808, 2004.

77. K. Sea, S. H. Sohn, A. Durazo, Y. Sheng, B. F. Shaw, X. Cao, A. B. Taylor, L. J. Whitson, S. P. Holloway, P. J. Hart, D. E. Cabelli, E. B. Gralla, and J. S. Valentine. Insights into the Role of the Unusual Disulfide Bond in Copper-Zinc Superoxide Dismutase. *J. Biol. Chem.*, 290(4):2405–2418, 2015.

78. C. M. Grant. Role of the glutathione/glutaredoxin and thioredoxin systems in yeast growth and response to stress conditions. *Mol. Microbiol.*, 39(3):533–41, 2001.

79. B. Alberts, A. Johnson, J. Lewis, M. Raff, K. Roberts, and P. Walter. *Molecular biology of the cell*. Garland Science Texbooks, New York & London, 4th edition, 2002.
80. B. Wilkinson and H. F. Gilbert. Protein disulfide isomerase. *Biochim. Biophys. Acta*, 1699(1–2):35–44, 2004.
81. B. P. Tu and J. S. Weissman. The FAD- and O_2-dependent reaction cycle of Ero1-mediated oxidative protein folding in the endoplasmic reticulum. *Mol. Cell*, 10(5):983–94, 2002.
82. E. Zito, E. P. Melo, Y. Yang, A. Wahlander, T. A. Neubert, and D. Ron. Oxidative protein folding by an endoplasmic reticulum-localized peroxiredoxin. *Mol. Cell*, 40(5):787–97, 2010.
83. N. Wajih, S. M. Hutson, and R. Wallin. Disulfide-dependent protein folding is linked to operation of the vitamin K cycle in the endoplasmic reticulum. A protein disulfide isomerase-VKORC1 redox enzyme complex appears to be responsible for vitamin K1 2,3-epoxide reduction. *J. Biol. Chem.*, 282(4):2626–35, 2007.
84. V. D. Nguyen, M. J. Saaranen, A. R. Karala, A. K. Lappi, L. Wang, I. B. Raykhel, H. I. Alanen, K. E. Salo, C. C. Wang, and L. W. Ruddock. Two endoplasmic reticulum PDI peroxidases increase the efficiency of the use of peroxide during disulfide bond formation. *J. Mol. Biol.*, 406(3):503–15, 2011.
85. M. Depuydt, J. Messens, and J. F. Collet. How proteins form disulfide bonds. *Antioxid. Redox Signal.*, 15(1):49–66, 2011.
86. V. K. Kodali and C. Thorpe. Oxidative protein folding and the quiescin-sulfhydryl oxidase family of flavoproteins. *Antioxid. Redox Signal.*, 13(8):1217–30, 2010.
87. C. T. Walsh, S. Garneau-Tsodikova, and G. J. Gatto Jr. Protein posttranslational modifications: the chemistry of proteome diversifications. *Angew. Chem. Int. Ed.*, 44(45):7342–72, 2005.
88. N. Mesecke, N. Terziyska, C. Kozany, F. Baumann, W. Neupert, K. Hell, and J. M. Herrmann. A disulfide relay system in the intermembrane space of mitochondria that mediates protein import. *Cell*, 121(7):1059–69, 2005.
89. E. N. Trifonov. The triplet code from first principles. *J. Biomol. Struct. Dyn.*, 22(1):1–11, 2004.
90. X. Liu, J. Zhang, F. Ni, X. Dong, B. Han, D. Han, Z. Ji, and Y. Zhao. Genome wide exploration of the origin and evolution of amino acids. *BMC Evol. Biol.*, 10:77, 2010.
91. I. K. Jordan, F. A. Kondrashov, I. A. Adzhubei, Y. I. Wolf, E. V. Koonin, A. S. Kondrashov, and S. Sunyaev. A universal trend of amino acid gain and loss in protein evolution. *Nature*, 433(7026):633–8, 2005.
92. G. Faccio, O. Nivala, K. Kruus, J. Buchert, and M. Saloheimo. Sulfhydryl oxidases: sources, properties, production and applications. *Appl. Microbiol. Biotechnol.*, 91(4):957–66, 2011.
93. S. G. Chang, K. D. Choi, S. H. Jang, and H. C. Shin. Role of disulfide bonds in the structure and activity of human insulin. *Mol. Cells*, 16(3):323–30, 2003.
94. M. S. Paget and M. J. Buttner. Thiol-based regulatory switches. *Annu. Rev. Genet.*, 37:91–121, 2003.
95. I. Azimi, J. W. Wong, and P. J. Hogg. Control of mature protein function by allosteric disulfide bonds. *Antioxid. Redox Signal.*, 14(1):113–26, 2011.
96. I. G. Kamphuis, J. Drenth, and E. N. Baker. Thiol proteases. Comparative studies based on the high-resolution structures of papain and actinidin, and on amino acid sequence information for cathepsins B and H, and stem bromelain. *J. Mol. Biol.*, 182(2):317–29, 1985.
97. J. N. Higaki, L. B. Evnin, and C. S. Craik. Introduction of a cysteine protease active site into trypsin. *Biochemistry*, 28(24):9256–63, 1989.
98. G. L. Holliday, D. E. Almonacid, G. J. Bartlett, N. M. O'Boyle, J. W. Torrance, P. Murray-Rust, J. B. Mitchell, and J. M. Thornton. MACiE (Mechanism, Annotation and Classification in Enzymes): novel tools for searching catalytic mechanisms. *Nucleic Acids Res.*, 35(suppl 1):D515–20, 2007.
99. G. L. Holliday, J. B. Mitchell, and J. M. Thornton. Understanding the functional roles of amino acid residues in enzyme catalysis. *J. Mol. Biol.*, 390(3):560–77, 2009.
100. C. Jacob, G. I. Giles, N. M. Giles, and H. Sies. Sulfur and selenium: the role of oxidation state in protein structure and function. *Angew. Chem. Int. Ed.*, 42(39):4742–58, 2003.
101. L. E. Netto, M. A. de Oliveira, G. Monteiro, A. P. Demasi, J. R. Cussiol, K. F. Discola, M. Demasi, G. M. Silva, S. V. Alves, V. G. Faria, and B. B. Horta. Reactive cysteine in proteins: protein folding, antioxidant defense, redox signaling and more. *Comp. Biochem. Physiol. C Toxicol. Pharmacol.*, 146(1–2):180–93, 2007.
102. L. B. Poole and K. J. Nelson. Discovering mechanisms of signaling-mediated cysteine oxidation. *Curr. Opin. Chem. Biol.*, 12:18–24, 2008.
103. B. Wittmann-Liebold and T. Choli-Papadopoulou. Proteins: Postsynthetic Modification – Function and Physical Analysis. In *Encyclopedia of Life Sciences*. John Wiley & Sons, Ltd., 2007.
104. N. J. Pace and E. Weerapana. Zinc-Binding Cysteines: Diverse Functions and Structural Motifs. *Biomolecules*, 4(2):419–434, 2014.
105. H. Xu and S. W . Morrical. Protein Motifs for DNA Binding. In *eLS*. John Wiley & Sons, Ltd., 2010.

106. R. Gamsjaeger, C. K. Liew, F. E. Loughlin, M. Crossley, and J. P. Mackay. Sticky fingers: zinc-fingers as protein-recognition motifs. *Trends Biochem. Sci.*, 32(2):63–70, 2007.

107. M. I. Sabri and S. Ochs. Inhibition of glyceraldehyde-3-phosphate dehydrogenase in mammalian nerve by iodoacetic acid. *J. Neurochem.*, 18(8):1509–14, 1971.

108. S. G. Rhee, S. W. Kang, W. Jeong, T. S. Chang, K. S. Yang, and H. A. Woo. Intracellular messenger function of hydrogen peroxide and its regulation by peroxiredoxins. *Curr. Opin. Cell Biol.*, 17(2):183–9, 2005.

109. G. Spyrou, E. Enmark, A. Miranda-Vizuete, and J. Gustafsson. Cloning and expression of a novel mammalian thioredoxin. *J. Biol. Chem.*, 272(5):2936–41, 1997.

110. G. Powis and W. R. Montfort. Properties and biological activities of thioredoxins. *Annu. Rev. Biophys. Biomol. Struct.*, 30:421–55, 2001.

111. G. L. Patrick. *An Introduction to Medicinal Chemistry*. Oxford University Press, 5th edition, 2013.

112. G. J. Bartlett, C. T. Porter, N. Borkakoti, and J. M. Thornton. Analysis of Catalytic Residues in Enzyme Active Sites. *J. Mol. Biol.*, 324(1):105–121, 2002.

113. J. A. Gerlt and F. M. Raushel. Evolution of function in (beta/alpha)8-barrel enzymes. *Curr. Opin. Chem. Biol.*, 7(2):252–64, 2003.

114. R. B. Pearson and B. E. Kemp. Protein kinase phosphorylation site sequences and consensus specificity motifs: tabulations. *Methods Enzymol.*, 200:62–81, 1991.

115. H. J. Kang and E. N. Baker. Intramolecular isopeptide bonds: protein crosslinks built for stress? *Trends Biochem. Sci.*, 36(4):229–37, 2011.

116. J. A. Vila, Y. A. Arnautova, Y. Vorobjev, and H. A. Scheraga. Assessing the fractions of tautomeric forms of the imidazole ring of histidine in proteins as a function of pH. *Proc. Natl. Acad. Sci. USA*, 108(14):5602–7, 2011.

117. K. M. Kim, T. Qin, Y. Y. Jiang, L. L. Chen, M. Xiong, D. Caetano-Anolles, H. Y. Zhang, and G. Caetano-Anolles. Protein domain structure uncovers the origin of aerobic metabolism and the rise of planetary oxygen. *Structure*, 20(1):67–76, 2012.

118. U. Heinemann, J. Ay, O. Gaiser, J. J. Muller, and M. N. Ponnuswamy. Enzymology and folding of natural and engineered bacterial beta-glucanases studied by X-ray crystallography. *Biol. Chem.*, 377(7–8):447–54, 1996.

119. J. R. Casey, S. Grinstein, and J. Orlowski. Sensors and regulators of intracellular pH. *Nat. Rev. Mol. Struct. Biol.*, 11:50–61, 2010.

120. S. Szaraz, D. Oesterhelt, and P. Ormos. pH-induced structural changes in bacteriorhodopsin studied by Fourier transform infrared spectroscopy. *Biophys. J.*, 67(4):1706–12, 1994.

121. D. G. Isom, B. R. Cannon, C. A. Castaneda, A. Robinson, and E. B. Garcia-Moreno. High tolerance for ionizable residues in the hydrophobic interior of proteins. *Proc. Natl. Acad. Sci. USA*, 105(46):17784–17788, 2008.

122. A. C. Robinson, C. A. Castaneda, J. L. Schlessman, and E. B. Garcia-Moreno. Structural and thermodynamic consequences of burial of an artificial ion pair in the hydrophobic interior of a protein. *Proc. Natl. Acad. Sci. USA*, 111(32):11685–90, 2014.

123. D. G. Isom, C. A. Castañeda, B. R. Cannon, and E. B. García-Moreno. Large shifts in pKa values of lysine residues buried inside a protein. *Proc. Natl. Acad. Sci. USA*, 108(13):5260–5265, 2011.

124. C. Tanford and J. G. Kirkwood. Theory of Protein Titration Curves. I. General Equations for Impenetrable Spheres. *J. Am. Chem. Soc.*, 79:5333, 1957.

125. K. A. Sharp and B. Honig. Electrostatic interactions in macromolecules: theory and applications. *Annu. Rev. Biophys. Biophys. Chem.*, 19(1):301–332, 1990.

126. D. A. Karp, A. G. Gittis, M. R. Stahley, C. A. Fitch, W. E. Stites, and E. B. Garcia-Moreno. High apparent dielectric constant inside a protein reflects structural reorganization coupled to the ionization of an internal Asp. *Biophys. J.*, 92(6):2041–53, 2007.

127. M. J. Harms, J. L. Schlessman, G. R. Sue, and E. B. García-Moreno. Arginine residues at internal positions in a protein are always charged. *Proc. Natl. Acad. Sci. USA*, 108(47):18954–18959, 2011.

128. T. K. Harris and G. J. Turner. Structural basis of perturbed pKa values of catalytic groups in enzyme active sites. *IUBMB Life*, 53(2):85–98, 2002.

129. D. V. Laurents, B. M. Huyghues-Despointes, M. Bruix, R. L. Thurlkill, D. Schell, S. Newsom, G. R. Grimsley, K. L. Shaw, S. Trevino, M. Rico, J. M. Briggs, J. M. Antosiewicz, J. M. Scholtz, and C. N. Pace. Charge-charge interactions are key determinants of the pK values of ionizable groups in ribonuclease Sa (pI = 3.5) and a basic variant (pI = 10.2). *J. Mol. Biol.*, 325(5):1077–92, 2003.

130. A. Cooper. Thermodynamics of Protein Folding and Stability. In *Protein: A Comprehensive Treatise*, volume 2, pages 217–270. JAI Press Inc., Stamford, CT, 1999.

131. R. L. Thurlkill, G. R. Grimsley, J. M. Scholtz, and C. N. Pace. Hydrogen bonding markedly reduces the pK of buried carboxyl groups in proteins. *J. Mol. Biol.*, 362(3):594–604, 2006.

132. H. Li, A. D. Robertson, and J. H. Jensen. Very fast empirical prediction and rationalization of protein pKa values. *Proteins*, 61(4):704–21, 2005.

133. B. Honig and A. Nicholls. Classical electrostatics in biology and chemistry. *Science*, 268(5214):1144–1149, 1995.

134. B. Honig, K. Sharp, and A. Suei-Yang. Macroscopic models of aqueous solutions: biological and chemical applications. *J. Phys. Chem.*, 97:1101–1109, 1993.

135. D. Bashford and M. Karplus. pKa's of ionizable groups in proteins: atomic detail from a continuum electrostatic model. *Biochemistry*, 29(44):10219–25, 1990.

136. T. J. Dolinsky, J. E. Nielsen, J. A. McCammon, and N. A. Baker. PDB2PQR: an automated pipeline for the setup of Poisson-Boltzmann electrostatics calculations. *Nucleic Acids Res.*, 32(suppl 2):W665–7, 2004.

137. J. C. Gordon, J. B. Myers, T. Folta, V. Shoja, L. S. Heath, and A. Onufriev. H++: a server for estimating pKas and adding missing hydrogens to macromolecules. *Nucleic Acids Res.*, 33(suppl 2):W368–71, 2005.

138. R. Anandakrishnan and A. Onufriev. Analysis of basic clustering algorithms for numerical estimation of statistical averages in biomolecules. *J. Comput. Biol.*, 15(2):165–84, 2008.

139. E. J. Cohn and J. T. Edsall. *Proteins, Amino Acids and Peptides*. Hafner Publishing Co., New York, 1943.

140. J. F. Dice. Lysosomal degradation of proteins. In *Encyclopedia of Life Sciences*. John Wiley & Sons, Ltd., 2007.

141. D. B. Wetlaufer. Ultraviolet spectra of proteins and amino acids. *Adv. Protein Chem.*, 17:303–390, 1962.

142. S. K. Burley and G. A. Petsko. Aromatic-aromatic interaction: a mechanism of protein structure stabilization. *Science*, 229(4708):23–8, 1985.

143. C. A. Hunter, J. Singh, and J. M. Thornton. π-π interactions: The geometry and energetics of phenylalanine-phenylalanine interactions in proteins. *J. Mol. Biol.*, 218(4):837–846, 1991.

144. M. Levitt and M. F. Perutz. Aromatic rings act as hydrogen bond acceptors. *J. Mol. Biol.*, 201(4):751–754, 1988.

145. P. B. Crowley and A. Golovin. Cation-pi interactions in protein-protein interfaces. *Proteins*, 59(2):231–9, 2005.

146. H. X. Zhou and J. A. McCammon. The gates of ion channels and enzymes. *Trends Biochem. Sci.*, 35:179–85, 2009.

147. R. Bhattacharyya, U. Samanta, and P. Chakrabarti. Aromatic-aromatic interactions in and around alpha-helices. *Protein Eng.*, 15(2):91–100, 2002.

148. G. S. Baldwin, M. F. Bailey, B. P. Shehan, I. Sims, and R. S. Norton. Tyrosine modification enhances metal-ion binding. *Biochem. J.*, 416(1):77–84, 2008.

149. A. Jordan and P. Reichard. Ribonucleotide reductases. *Annu. Rev. Biochem.*, 67:71–98, 1998.

150. U. Samanta, D. Pal, and P. Chakrabarti. Environment of tryptophan side chains in proteins. *Proteins*, 38(3):288–300, 2000.

151. B. Ma and R. Nussinov. Trp/Met/Phe hot spots in protein-protein interactions: potential targets in drug design. *Curr. Top. Med. Chem.*, 7(10):999–1005, 2007.

152. B. N. Bullock, A. L. Jochim, and P. S. Arora. Assessing helical protein interfaces for inhibitor design. *J. Am. Chem. Soc.*, 133(36):14220–14223, 2011.

153. A. M. Watkins and P. S. Arora. Anatomy of β-strands at protein–protein interfaces. *ACS Chem. Biol.*, 9(8):1747–1754, 2014.

154. J. Borlinghaus, F. Albrecht, C. M. Gruhlke, D. I. Nwachukwu, and J. A. Slusarenko. Allicin: Chemistry and Biological Properties. *Molecules*, 19(8), 2014.

155. E. Appel, A. Rabinkov, M. Neeman, F. Kohen, and D. Mirelman. Conjugates of daidzein-alliinase as a targeted pro-drug enzyme system against ovarian carcinoma. *J. Drug Target*, 19(5):326–35, 2011.

156. M. C. Cannon, K. Terneus, Q. Hall, L. Tan, Y. Wang, B. L. Wegenhart, L. Chen, D. T. Lamport, Y. Chen, and M. J. Kieliszewski. Self-assembly of the plant cell wall requires an extensin scaffold. *Proc. Natl. Acad. Sci. USA*, 105(6):2226–31, 2008.

157. B. Furie, B. A. Bouchard, and B. C. Furie. Vitamin K-dependent biosynthesis of gamma-carboxyglutamic acid. *Blood*, 93(6):1798–808, 1999.

158. Y. Zhang, P. V. Baranov, J. F. Atkins, and V. N. Gladyshev. Pyrrolysine and selenocysteine use dissimilar decoding strategies. *J. Biol. Chem.*, 280(21):20740–51, 2005.

159. T. C. Stadtman. Selenocysteine. *Annu. Rev. Biochem.*, 65:83–100, 1996.

160. D. L. Hatfield, P. A. Tsuji, B. A. Carlson, and V. N. Gladyshev. Selenium and selenocysteine: roles in cancer, health, and development. *Trends Biochem. Sci.*, 39(3):112–20, 2014.

161. V. M. Labunskyy, D. L. Hatfield, and V. N. Gladyshev. Selenoproteins: molecular pathways and physiological roles. *Physiol. Rev.*, 94(3):739–77, 2014.

162. J. Liu and S. Rozovsky. Membrane-Bound Selenoproteins. *Antioxid. Redox Signal.*, 23(10):795–813, 2015.

163. R. J. Hondal, S. M. Marino, and V. N. Gladyshev. Selenocysteine in thiol/disulfide-like exchange reactions. *Antioxid. Redox Signal.*, 18(13):1675–89, 2013.

164. R. E. Huber and R. S. Criddle. Comparison of the chemical properties of selenocysteine and selenocystine with their sulfur analogs. *Arch. Biochem. Biophys.*, 122(1):164–73, 1967.

165. G. W. Snider, E. Ruggles, N. Khan, and R. J. Hondal. Selenocysteine confers resistance to inactivation by oxidation in thioredoxin reductase: comparison of selenium and sulfur enzymes. *Biochemistry*, 52(32):5472–81, 2013.

166. G. Wu. Amino acids: metabolism, functions, and nutrition. *Amino Acids*, 37(1):1–17, 2009.

167. J. A. Krzycki. Function of genetically encoded pyrrolysine in corrinoid-dependent methylamine methyltransferases. *Curr. Opin. Chem. Biol.*, 8(5):484–91, 2004.

168. E. Eden, N. Geva-Zatorsky, I. Issaeva, A. Cohen, E. Dekel, T. Danon, L. Cohen, A. Mayo, and U. Alon. Proteome half-life dynamics in living human cells. *Science*, 331(6018):764–768, 2011.

169. R. J. W. Truscott, K. L. Schey, and M. G. Friedrich. Old Proteins in Man: A Field in its Infancy. *Trends Biochem. Sci.*, 41(8):654–664, 2016.

170. R. Wolfenden and M. J. Snider. The depth of chemical time and the power of enzymes as catalysts. *Acc. Chem. Res.*, 34(12):938–945, 2001.

171. A. Radzicka and R. Wolfenden. Rates of Uncatalyzed Peptide Bond Hydrolysis in Neutral Solution and the Transition State Affinities of Proteases. *J. Am. Chem. Soc.*, 118:6105–6109, 1996.

172. D. S. Berkholz, C. M. Driggers, M. V. Shapovalov, R. L. Dunbrack Jr, and P. A. Karplus. Nonplanar peptide bonds in proteins are common and conserved but not biased toward active sites. *Proc. Natl. Acad. Sci. USA*, 109(2):449–53, 2012.

173. S. Nussey and S. Whitehead. *Endocrinology: An Integrated Approach*. BIOS Scientific Publishers Ltd, Oxford, England, 2001.

174. The UniProt Consortium. The Universal Protein Resource (UniProt). *Nucleic Acids Res.*, 35(suppl 1):D193–7, 2007.

175. A. E. Mirsky and L. Pauling. On the Structure of Native, Denatured, and Coagulated Proteins. *Proc. Natl. Acad. Sci. USA*, 22(7):439–447, 1936.

176. R. F. Service. Problem solved* (*sort of). *Science*, 321(5890):784–6, 2008.

177. W. Humphrey, A. Dalke, and K. Schulten. VMD: visual molecular dynamics. *J. Mol. Graph.*, 14(1):33–8, 1996.

178. S. C. Lovell, I. W. Davis, W. B. Arendall, P. I. W. de Bakker, J. M. Word, M. G. Prisant, J. S. Richardson, and D. C. Richardson. Structure validation by Cα geometry: Φ, Ψ and Cβ deviation. *Proteins: Struct., Funct., Bioinf.*, 50(3):437–450, 2003.

179. A. H. Andreotti. Native state proline isomerization: an intrinsic molecular switch. *Biochemistry*, 42(32):9515–24, 2003.

180. G. N. Ramachandran, C. Ramakrishnan, and V. Sasisekharan. Stereochemistry of polypeptide chain configurations. *J. Mol. Biol.*, 7:95–9, 1963.

181. G. N. Ramachandran and V. Sasisekharan. Conformation of polypeptides and proteins. *Adv. Protein Chem.*, 23:283–438, 1968.

182. G. D. Rose. Perspective. *Protein Sci.*, 10(8):1691–3, 2001.

183. S. Hovmoller, T. Zhou, and T. Ohlson. Conformations of amino acids in proteins. *Acta Crystallogr. Sect. D*, 58(Pt 5):768–76, 2002.

184. T. Przytycka, R. Aurora, and G. D. Rose. A protein taxonomy based on secondary structure. *Nat. Struct. Biol.*, 6(7):672–82, 1999.

185. L. Pauling and R. B. Corey. Two hydrogen-bonded spiral configurations of the polypeptide chain. *J. Am. Chem. Soc.*, 72:5349, 1950.

186. L. Pauling and R. B. Corey. The pleated sheet, a new layer configuration of polypeptide chains. *Proc. Natl. Acad. Sci. USA*, 37(5):251–6, 1951.

187. L. Pauling, R. B. Corey, and H. R. Branson. The structure of proteins; two hydrogen-bonded helical configurations of the polypeptide chain. *Proc. Natl. Acad. Sci. USA*, 37(4):205–11, 1951.

188. D. Eisenberg. The discovery of the alpha-helix and beta-sheet, the principal structural features of proteins. *Proc. Natl. Acad. Sci. USA*, 100(20):11207–10, 2003.

189. D. S. Goodsell. Visual methods from atoms to cells. *Structure*, 13(3):347–54, 2005.

190. S. I. O'Donoghue, D. S. Goodsell, A. S. Frangakis, F. Jossinet, R. A. Laskowski, M. Nilges, H. R. Saibil, A. Schafferhans, R. C. Wade, E. Westhof, and A. J. Olson. Visualization of macromolecular structures. *Nat. Methods*, 7(3 Suppl):S42–55, 2010.

191. W. L. DeLano. The PyMOL Molecular Graphics System. http://www.pymol.org, 2002.

192. K. A. Sharp, A. Nicholls, R. F. Fine, and B. Honig. Reconciling the magnitude of the microscopic and macroscopic hydrophobic effects. *Science*, 252(5002):106–9, 1991.

193. S. A. Sridharan, A. Nicholls, and B. Honig. A new vertex algorithm to calculate solvent accessible surface area. *Biophys. J.*, 61:A174, 1992.

194. L. Mitra, N. Smolin, R. Ravindra, C. Royer, and R. Winter. Pressure perturbation calorimetric studies of the solvation properties and the thermal unfolding of proteins in solution: experiments and theoretical interpretation. *Phys. Chem. Chem. Phys.*, 8(11):1249–65, 2006.

195. N. A. Baker, D. Sept, S. Joseph, M. J. Holst, and J. A. McCammon. Electrostatics of nanosystems: Application to microtubules and the ribosome. *Proc. Natl. Acad. Sci. USA*, 98(18):10037–10041, 2001.

196. F. Glaser, T. Pupko, I. Paz, R. E. Bell, D. Bechor-Shental, E. Martz, and N. Ben-Tal. ConSurf: identification of functional regions in proteins by surface-mapping of phylogenetic information. *Bioinformatics*, 19(1):163–4, 2003.

197. M. Landau, I. Mayrose, Y. Rosenberg, F. Glaser, E. Martz, T. Pupko, and N. Ben-Tal. ConSurf 2005: the projection of evolutionary conservation scores of residues on protein structures. *Nucleic Acids Res.*, 33(suppl 2):W299–302, 2005.

198. B. Offmann, M. Tyagi, and A. G. de Brevern. Local protein structures. *Curr. Bioinform.*, 2(3):165–202, 2007.

199. T. E. Creighton. *Proteins: Structures and Molecular Properties*. W. H. Freeman and Company, New York, 2nd edition, 1992.

200. S. B. Zimmerman and S. O. Trach. Estimation of macromolecule concentrations and excluded volume effects for the cytoplasm of *Escherichia coli*. *J. Mol. Biol.*, 222(3):599–620, 1991.

201. F.-X. Theillet, A. Binolfi, T. Frembgen-Kesner, K. Hingorani, M. Sarkar, C. Kyne, C. Li, P. B. Crowley, L. Gierasch, G. J. Pielak, A. H. Elcock, A. Gershenson, and P. Selenko. Physicochemical properties of cells and their effects on intrinsically disordered proteins (IDPs). *Chem. Rev.*, 114(13):6661–6714, 2014.

202. A. Cammers-Goodwin, T. J. Allen, S. L. Oslick, K. F. McClure, J. H. Lee, and D. S. Kemp. Mechanism of Stabilization of Helical Conformations of Polypeptides by Water Containing Trifluoroethanol. *J. Am. Chem. Soc.*, 18(13):3082–3090, 1996.

203. J. A. Vila, D. R. Ripoll, and H. A. Scheraga. Physical reasons for the unusual alpha-helix stabilization afforded by charged or neutral polar residues in alanine-rich peptides. *Proc. Natl. Acad. Sci. USA*, 97(24):13075–9, 2000.

204. A. S. Yang and B. Honig. Free energy determinants of secondary structure formation: I. alpha-Helices. *J. Mol. Biol.*, 252(3):351–65, 1995.

205. D. N. Ermolenko, J. M. Richardson, and G. I. Makhatadze. Noncharged amino acid residues at the solvent-exposed positions in the middle and at the C terminus of the alpha-helix have the same helical propensity. *Protein Sci.*, 12(6):1169–76, 2003.

206. K. A. Dill, S. B. Ozkan, M. S. Shell, and T. R. Weikl. The protein folding problem. *Annu. Rev. Biophys.*, 37:289–316, 2008.

207. P. T. van Duijnen and B. T. Thole. Cooperative effects in alpha-helices: An ab initio molecular-orbital study. *Biopolymers*, 21(9):1749–1761, 1982.

208. W. G. Hol, P. T. van Duijnen, and H. J. Berendsen. The alpha-helix dipole and the properties of proteins. *Nature*, 273(5662):443–6, 1978.

209. S. Kumar and R. Nussinov. Close-range electrostatic interactions in proteins. *ChemBioChem*, 3(7):604–617, 2002.

210. R. Aurora and G. D. Rose. Helix capping. *Protein Sci.*, 7(1):21–38, 1998.

211. R. Armen, D. O. Alonso, and V. Daggett. The role of α-, 3_{10}-, and π-helix in helix $\longrightarrow$ coil transitions. *Protein Sci.*, 12(6):1145–57, 2003.

212. M. N. Fodje and S. Al-Karadaghi. Occurrence, conformational features and amino acid propensities for the pi-helix. *Protein Eng.*, 15(5):353–8, 2002.

213. T. M. Weaver. The pi-helix translates structure into function. *Protein Sci.*, 9(1):201–6, 2000.

214. A. Kentsis, M. Mezei, T. Gindin, and R. Osman. Unfolded state of polyalanine is a segmented polyproline II helix. *Proteins*, 55(3):493–501, 2004.

215. B. W. Chellgren, A. F. Miller, and T. P. Creamer. Evidence for polyproline II helical structure in short polyglutamine tracts. *J. Mol. Biol.*, 361(2):362–71, 2006.

216. J. C. Horng and R. T. Raines. Stereoelectronic effects on polyproline conformation. *Protein Sci.*, 15(1):74–83, 2006.

217. L. Pauling and R. B. Corey. The structure of fibrous proteins of the collagen-gelatin group. *Proc. Natl. Acad. Sci. USA*, 37(5):272–81, 1951.

218. A. A. Adzhubei and M. J. Sternberg. Left-handed polyproline II helices commonly occur in globular proteins. *J. Mol. Biol.*, 229(2):472–93, 1993.

219. Z. Shi, R. W. Woody, and N. R. Kallenbach. Is polyproline II a major backbone conformation in unfolded proteins? *Adv. Protein Chem.*, 62:163–240, 2002.

220. B. Fazi, M. J. Cope, A. Douangamath, S. Ferracuti, K. Schirwitz, A. Zucconi, D. G. Drubin, M. Wilmanns, G. Cesareni, and L. Castagnoli. Unusual binding properties of the SH3 domain of the yeast actin-binding protein Abp1: structural and functional analysis. *J. Biol. Chem.*, 277(7):5290–8, 2002.

221. Y. G. Gao, X. Z. Yan, A. X. Song, Y. G. Chang, X. C. Gao, N. Jiang, Q. Zhang, and H. Y. Hu. Structural insights into the specific binding of huntingtin proline-rich region with the SH3 and WW domains. *Structure*, 14(12):1755–65, 2006.

222. M. Lewitzky, M. Harkiolaki, M. C. Domart, E. Y. Jones, and S. M. Feller. Mona/Gads SH3C binding to hematopoietic progenitor kinase 1 (HPK1) combines an atypical SH3 binding motif, R/KXXK, with a classical PXXP motif embedded in a polyproline type II (PPII) helix. *J. Biol. Chem.*, 279(27):28724–32, 2004.

223. B. J. Mayer. SH3 domains: complexity in moderation. *J. Cell Sci.*, 114(Pt 7):1253–63, 2001.

224. N. C. Fitzkee, P. J. Fleming, H. Gong, N. Panasik Jr, T. O. Street, and G. D. Rose. Are proteins made from a limited parts list? *Trends Biochem. Sci.*, 30(2):73–80, 2005.

225. E. W. Blanch, L. A. Morozova-Roche, D. A. Cochran, A. J. Doig, L. Hecht, and L. D. Barron. Is polyproline II helix the killer conformation? A Raman optical activity study of the amyloidogenic prefibrillar intermediate of human lysozyme. *J. Mol. Biol.*, 301(2):553–63, 2000.

226. J. S. Richardson and D. C. Richardson. Natural beta-sheet proteins use negative design to avoid edge-to-edge aggregation. *Proc. Natl. Acad. Sci. USA*, 99(5):2754–9, 2002.

227. N. Nagano, E. G. Hutchinson, and J. M. Thornton. Barrel structures in proteins: automatic identification and classification including a sequence analysis of TIM barrels. *Protein Sci.*, 8(10):2072–84, 1999.

228. K. A. Dill. Dominant forces in protein folding. *Biochemistry*, 29(31):7133–55, 1990.

229. C. N. Schutz and A. Warshel. What are the dielectric "constants" of proteins and how to validate electrostatic models? *Proteins*, 44(4):400–17, 2001.

230. N. Ben-Tal, A. Ben-Shaul, A. Nicholls, and B. Honig. Free-energy determinants of alpha-helix insertion into lipid bilayers. *Biophys. J.*, 70(4):1803–12, 1996.

231. P. J. Fleming and G. D. Rose. Do all backbone polar groups in proteins form hydrogen bonds? *Protein Sci.*, 14(7):1911–1917, 2005.

232. C. L. Worth and T. L. Blundell. Satisfaction of hydrogen-bonding potential influences the conservation of polar sidechains. *Proteins*, 75(2):413–29, 2009.

233. A. M. Marcelino and L. M. Gierasch. Roles of beta-turns in protein folding: from peptide models to protein engineering. *Biopolymers*, 89(5):380–91, 2008.

234. G. D. Rose, L. M. Gierasch, and J. A. Smith. Turns in peptides and proteins. *Adv. Protein Chem.*, 37:1–109, 1985.

235. B. L. Sibanda, T. L. Blundell, and J. M. Thornton. Conformation of beta-hairpins in protein structures. A systematic classification with applications to modelling by homology, electron density fitting and protein engineering. *J. Mol. Biol.*, 206(4):759–77, 1989.

236. M. Monne, M. Hermansson, and G. von Heijne. A turn propensity scale for transmembrane helices. *J. Mol. Biol.*, 288(1):141–5, 1999.

237. E. G. Hutchinson and J. M. Thornton. A revised set of potentials for beta-turn formation in proteins. *Protein Sci.*, 3(12):2207–16, 1994.

238. K. Guruprasad and S. Rajkumar. Beta-and gamma-turns in proteins revisited: a new set of amino acid turn-type dependent positional preferences and potentials. *J. Biosci.*, 25(2):143–56, 2000.

239. H. J. Hsu, H. J. Chang, H. P. Peng, S. S. Huang, M. Y. Lin, and A. S. Yang. Assessing computational amino acid beta-turn propensities with a phage-displayed combinatorial library and directed evolution. *Structure*, 14(10):1499–510, 2006.

240. P. N. Lewis, F. A. Momany, and H. A. Scheraga. Chain reversals in proteins. *Biochim. Biophys. Acta (BBA) – Protein Structure*, 303(2):211–229, 1973.

241. G. D. Rose. Prediction of chain turns in globular proteins on a hydrophobic basis. *Nature*, 272(5654):586–90, 1978.

242. P. Y. Chou and G. D. Fasman. Empirical predictions of protein conformation. *Annu. Rev. Biochem.*, 47:251–276, 1978.

243. M. Blaber, X. J. Zhang, and B. W. Matthews. Structural basis of amino acid alpha helix propensity. *Science*, 260(5114):1637–40, 1993.

244. A. Horovitz, J. M. Matthews, and A. R. Fersht. Alpha-helix stability in proteins. II. Factors that influence stability at an internal position. *J. Mol. Biol.*, 227(2):560–8, 1992.

245. F. Avbelj and R. L. Baldwin. Role of backbone solvation in determining thermodynamic beta propensities of the amino acids. *Proc. Natl. Acad. Sci. USA*, 99(3):1309–13, 2002.

246. A. Chakrabartty, T. Kortemme, and R. L. Baldwin. Helix propensities of the amino acids measured in alanine-based peptides without helix-stabilizing side-chain interactions. *Protein Sci.*, 3(5):843–52, 1994.

247. D. A. Beck, D. O. Alonso, D. Inoyama, and V. Daggett. The intrinsic conformational propensities of the 20 naturally occurring amino acids and reflection of these propensities in proteins. *Proc. Natl. Acad. Sci. USA*, 105(34):12259–64, 2008.

248. C. N. Pace and J. M. Scholtz. A helix propensity scale based on experimental studies of peptides and proteins. *Biophys. J.*, 75(1):422–7, 1998.

249. T. P. Creamer and G. D. Rose. Side-chain entropy opposes alpha-helix formation but rationalizes experimentally determined helix-forming propensities. *Proc. Natl. Acad. Sci. USA*, 89(13):5937–41, 1992.

250. J. M. Richardson, M. M. Lopez, and G. I. Makhatadze. Enthalpy of helix-coil transition: missing link in rationalizing the thermodynamics of helix-forming propensities of the amino acid residues. *Proc. Natl. Acad. Sci. USA*, 102(5):1413–8, 2005.

251. F. S. Cordes, J. N. Bright, and M. S. Sansom. Proline-induced distortions of transmembrane helices. *J. Mol. Biol.*, 323(5):951–60, 2002.

252. P. Chakrabarti and S. Chakrabarti. C−H⋯O hydrogen bond involving proline residues in alpha-helices. *J. Mol. Biol.*, 284(4):867–73, 1998.

253. J. A. D'Aquino, J. Gomez, V. J. Hilser, K. H. Lee, L. M. Amzel, and E. Freire. The magnitude of the backbone conformational entropy change in protein folding. *Proteins*, 25(2):143–56, 1996.

254. K. A. Scott, D. O. V. Alonso, S. Sato, A. R. Fersht, and V. Daggett. Conformational entropy of alanine versus glycine in protein denatured states. *Proc. Natl. Acad. Sci. USA*, 104(8):2661–2666, 2007.

255. J. S. Richardson and D. C. Richardson. Amino acid preferences for specific locations at the ends of alpha helices. *Science*, 240(4859):1648–52, 1988.

256. L. Serrano and A. R. Fersht. Capping and alpha-helix stability. *Nature*, 342(6247):296–299, 1989.

257. L. G. Presta and G. D. Rose. Helix signals in proteins. *Science*, 240(4859):1632–41, 1988.

258. F. Farzadfard, N. Gharaei, H. Pezeshk, and S. A. Marashi. Beta-sheet capping: signals that initiate and terminate beta-sheet formation. *J. Struct. Biol.*, 161(1):101–10, 2008.

259. R. L. Baldwin. Energetics of protein folding. *J. Mol. Biol.*, 371(2):283–301, 2007.

260. Y. Bai and S. W. Englander. Hydrogen bond strength and beta-sheet propensities: the role of a side chain blocking effect. *Proteins*, 18(3):262–6, 1994.

261. J. P. Collman, N. K. Devaraj, R. A. Decreau, Y. Yang, Y. L. Yan, W. Ebina, T. A. Eberspacher, and C. E. Chidsey. A cytochrome C oxidase model catalyzes oxygen to water reduction under rate-limiting electron flux. *Science*, 315(5818):1565–8, 2007.

262. K. U. Walter, K. Vamvaca, and D. Hilvert. An active enzyme constructed from a 9-amino acid alphabet. *J. Biol. Chem.*, 280(45):37742–6, 2005.

263. J. Liu, H. Tan, and B. Rost. Loopy proteins appear conserved in evolution. *J. Mol. Biol.*, 322(1):53–64, 2002.

264. J. Kim, J. Mao, and M. R. Gunner. Are acidic and basic groups in buried proteins predicted to be ionized? *J. Mol. Biol.*, 348(5):1283–98, 2005.

265. I. K. McDonald and J. M. Thornton. Satisfying hydrogen bonding potential in proteins. *J. Mol. Biol.*, 238(5):777–793, 1994.

266. D. N. Bolon and S. L. Mayo. Polar residues in the protein core of *Escherichia coli* thioredoxin are important for fold specificity. *Biochemistry*, 40(34):10047–53, 2001.

267. S. R. Trevino, J. M. Scholtz, and C. N. Pace. Amino acid contribution to protein solubility: Asp, Glu, and Ser contribute more favorably than the other hydrophilic amino acids in RNase Sa. *J. Mol. Biol.*, 366(2):449–60, 2007.

268. K. D. Collins. Charge density-dependent strength of hydration and biological structure. *Biophys. J.*, 72(1):65–76, 1997.

269. L. Lins, A. Thomas, and R. Brasseur. Analysis of accessible surface of residues in proteins. *Protein Sci.*, 12(7):1406–17, 2003.

270. M. H. Klapper. On the nature of the protein interior. *Biochim. Biophys. Acta*, 229(3):557–66, 1971.

271. I. D. Kuntz and G. M. Crippen. Protein densities. *Int. J. Pept. Protein Res.*, 13(2):223–8, 1979.

272. M. Brunori and Q. H. Gibson. Cavities and packing defects in the structural dynamics of myoglobin. *EMBO Rep.*, 2(8):674–9, 2001.

273. M. Bueno, L. A. Campos, J. Estrada, and J. Sancho. Energetics of aliphatic deletions in protein cores. *Protein Sci.*, 15(8):1858–72, 2006.

274. B. Vallone and M. Brunori. Roles for holes: are cavities in proteins mere packing defects? *Ital. J. Biochem.*, 53(1):46–52, 2004.

275. J. Cohen, K. Kim, P. King, M. Seibert, and K. Schulten. Finding gas diffusion pathways in proteins: application to O_2 and H_2 transport in CpI [FeFe]-hydrogenase and the role of packing defects. *Structure*, 13(9):1321–9, 2005.

276. C. Azuara, E. Lindahl, P. Koehl, H. Orland, and M. Delarue. PDB_Hydro: incorporating dipolar solvents with variable density in the Poisson-Boltzmann treatment of macromolecule electrostatics. *Nucleic Acids Res.*, 34(suppl 2):W38–42, 2006.

277. Marc Delarue group at Institut Pasteur. http://lorentz.immstr.pasteur.fr/website/present.html, 2017.

278. A. R. Fersht, A. Matouschek, and L. Serrano. The folding of an enzyme. I. Theory of protein engineering analysis of stability and pathway of protein folding. *J. Mol. Biol.*, 224(3):771–82, 1992.

279. A. Lewit-Bentley and S. Rety. EF-hand calcium-binding proteins. *Curr. Opin. Struct. Biol.*, 10(6):637–43, 2000.

280. L. N. Kinch and N. V. Grishin. Evolution of protein structures and functions. *Curr. Opin. Struct. Biol.*, 12(3):400–8, 2002.

281. N. V. Grishin. Fold change in evolution of protein structures. *J. Struct. Biol.*, 134(2–3):167–85, 2001.

282. C. A. Orengo and J. M. Thornton. Protein families and their evolution-a structural perspective. *Annu. Rev. Biochem.*, 74:867–900, 2005.

283. C. I. Branden and J. Tooze. *Introduction to Protein Structure*. Garland Publishing, New York, 2nd edition, 1999.

284. M. J. Berridge, P. Lipp, and M. D. Bootman. The versatility and universality of calcium signalling. *Nat. Rev. Mol. Cell Biol.*, 1(1):11–21, 2000.

285. A. C. Guyton and J. E. Hall. *Textbook of medical physiology*. W. B. Saunders Company, Philadelphia, 9th edition, 1996.

286. F. J. Stevens and Y. Argon. Protein folding in the ER. *Semin. Cell Dev. Biol.*, 10(5):443–54, 1999.

287. M. T. Swulius and M. N. Waxham. Ca^{2+}/calmodulin-dependent protein kinases. *Cell. Mol. Life Sci.*, 65(17):2637, 2008.

288. K. L. Yap, J. Kim, K. Truong, M. Sherman, T. Yuan, and M. Ikura. Calmodulin target database. *J. Struct. Funct. Genomics*, 1(1):8–14, 2000.

289. J. Evenas, S. Forsen, A. Malmendal, and M. Akke. Backbone dynamics and energetics of a calmodulin domain mutant exchanging between closed and open conformations. *J. Mol. Biol.*, 289(3):603–17, 1999.

290. O. Keskin, A. Gursoy, B. Ma, and R. Nussinov. Principles of protein-protein interactions: what are the preferred ways for proteins to interact? *Chem. Rev.*, 108(4):1225–44, 2008.

291. S. T. Rao and M. G. Rossmann. Comparison of super-secondary structures in proteins. *J. Mol. Biol.*, 76(2):241–56, 1973.

292. C. A. Orengo, D. T. Jones, and J. M. Thornton. Protein superfamilies and domain superfolds. *Nature*, 372(6507):631–4, 1994.

293. C. E. Brenner, C. Chothia, and T. J. Hubbard. Population statistics of protein structures: lessons from structural classifications. *Curr. Opin. Struct. Biol.*, 7(3):369–76, 1997.

294. A. F. Coulson and J. Moult. A unifold, mesofold, and superfold model of protein fold use. *Proteins*, 46(1):61–71, 2002.

295. Y. Ishihama, T. Schmidt, J. Rappsilber, M. Mann, F. U. Hartl, M. J. Kerner, and D. Frishman. Protein abundance profiling of the *Escherichia coli* cytosol. *BMC Genomics*, 9:102, 2008.

296. A. N. Barclay. Membrane proteins with immunoglobulin-like domains: a master superfamily of interaction molecules. *Semin. Immunol.*, 15(4):215–23, 2003.

297. M. O. Dayhoff. *Atlas of Protein Sequence and Structure*. National Biomedical Research Foundation, Washington, DC, 1965.

298. N. Sinha and S. J. Smith-Gill. Electrostatics in protein binding and function. *Curr. Protein Pept. Sci.*, 3(6):601–14, 2002.

299. L. C. James, P. Roversi, and D. S. Tawfik. Antibody multispecificity mediated by conformational diversity. *Science*, 299(5611):1362–7, 2003.

300. L. C. James and D. S. Tawfik. Conformational diversity and protein evolution: a 60-year-old hypothesis revisited. *Trends Biochem. Sci.*, 28(7):361–8, 2003.

301. M. G. Rossmann, A. Liljas, C. I. Branden, and L. J. Banaszak. Evolutionary and structural relationships among dehydrogenases. In P. D. Boyer, editor, *The enzymes*, volume 11, Part A, pages 61–102. Academic Press, New York, 3rd edition, 1975.

302. M. G. Rossmann, D. Moras, and K. W. Olsen. Chemical and biological evolution of nucleotide-binding protein. *Nature*, 250(463):194–9, 1974.

303. G. Kleiger and D. Eisenberg. GXXXG and GXXXA motifs stabilize FAD and NAD(P)-binding Rossmann folds through Cα−H···O hydrogen bonds and van der waals interactions. *J. Mol. Biol.*, 323(1):69–76, 2002.

304. A. Senes, I. Ubarretxena-Belandia, and D. M. Engelman. The Cα−H···O hydrogen bond: a determinant of stability and specificity in transmembrane helix interactions. *Proc. Natl. Acad. Sci. USA*, 98(16):9056–61, 2001.

305. C. R. Bellamacina. The nicotinamide dinucleotide binding motif: a comparison of nucleotide binding proteins. *FASEB J.*, 10(11):1257–69, 1996.

306. O. Dym and D. Eisenberg. Sequence-structure analysis of FAD-containing proteins. *Protein Sci.*, 10(9):1712–28, 2001.

307. E. V. Koonin and M. Y. Galperin. *Sequence-Evolution-Function: Computational Approaches in Comparative Genomics*. Kluwer Academic Publishers, Boston, MA, 2003.

308. K. Kinoshita, K. Sadanami, A. Kidera, and N. Go. Structural motif of phosphate-binding site common to various protein superfamilies: all-against-all structural comparison of protein-mononucleotide complexes. *Protein Eng.*, 12(1):11–4, 1999.

309. L. Aravind, R. Mazumder, S. Vasudevan, and E. V. Koonin. Trends in protein evolution inferred from sequence and structure analysis. *Curr. Opin. Struct. Biol.*, 12(3):392–9, 2002.

310. N. Nagano, C. A. Orengo, and J. M. Thornton. One fold with many functions: the evolutionary relationships between TIM barrel families based on their sequences, structures and functions. *J. Mol. Biol.*, 321(5):741–65, 2002.

311. D. Nagarajan, G. Deka, and M. Rao. Design of symmetric TIM barrel proteins from first principles. *BMC Biochemistry*, 16:18, 2015.

312. R. K. Wierenga. The TIM-barrel fold: a versatile framework for efficient enzymes. *FEBS Lett.*, 492(3):193–8, 2001.

313. D. R. Livesay and D. La. The evolutionary origins and catalytic importance of conserved electrostatic networks within TIM-barrel proteins. *Protein Sci.*, 14(5):1158–70, 2005.

314. S. Raychaudhuri, F. Younas, P. A. Karplus, C. H. Faerman, and D. R. Ripoll. Backbone makes a significant contribution to the electrostatics of alpha/beta-barrel proteins. *Protein Sci.*, 6(9):1849–57, 1997.

315. F. Pearl, A. Todd, I. Sillitoe, M. Dibley, O. Redfern, T. Lewis, C. Bennett, R. Marsden, A. Grant, D. Lee, A. Akpor, M. Maibaum, A. Harrison, T. Dallman, G. Reeves, I. Diboun, S. Addou, S. Lise, C. Johnston, A. Sillero, J. Thornton, and C. Orengo. The CATH Domain Structure Database and related resources Gene3D and DHS provide comprehensive domain family information for genome analysis. *Nucleic Acids Res.*, 33(suppl 1):D247–51, 2005.

316. A. Cuff, O. C. Redfern, L. Greene, I. Sillitoe, T. Lewis, M. Dibley, A. Reid, F. Pearl, T. Dallman, A. Todd, R. Garratt, J. Thornton, and C. Orengo. The CATH hierarchy revisited-structural divergence in domain superfamilies and the continuity of fold space. *Structure*, 17(8):1051–62, 2009.

317. K. Khafizov, C. Madrid-Aliste, S. C. Almo, and A. Fiser. Trends in structural coverage of the protein universe and the impact of the Protein Structure Initiative. *Proc. Natl. Acad. Sci. USA*, 111(10):3733–8, 2014.

318. C. Chothia and M. Gerstein. Protein evolution. How far can sequences diverge? *Nature*, 385(6617):579, 581, 1997.

319. J. Söding and A. N. Lupas. More than the sum of their parts: on the evolution of proteins from peptides. *BioEssays*, 25(9):837–846, 2003.

320. R. A. Goldstein. The structure of protein evolution and the evolution of protein structure. *Curr. Opin. Struct. Biol.*, 18(2):170–177, 2008.

321. R. A. Goldstein, Z. A. Luthey-Schulten, and P. G. Wolynes. Optimal protein-folding codes from spin-glass theory. *Proc. Natl. Acad. Sci. USA*, 89(11):4918–22, 1992.

322. B. E. Shakhnovich, N. V. Dokholyan, C. DeLisi, and E. I. Shakhnovich. Functional fingerprints of folds: evidence for correlated structure-function evolution. *J. Mol. Biol.*, 326(1):1–9, 2003.

323. A. M. Lesk and C. Chothia. How different amino acid sequences determine similar protein structures: the structure and evolutionary dynamics of the globins. *J. Mol. Biol.*, 136(3):225–70, 1980.

324. C. Chothia and A. M. Lesk. The relation between the divergence of sequence and structure in proteins. *EMBO J.*, 5(4):823–6, 1986.

325. C. Chothia. Proteins. One thousand families for the molecular biologist. *Nature*, 357(6379):543–4, 1992.

326. S. Govindarajan, R. Recabarren, and R. A. Goldstein. Estimating the total number of protein folds. *Proteins: Struct., Funct., Bioinf.*, 35(4):408–414, 1999.

327. I. G. Choi and S. H. Kim. Evolution of protein structural classes and protein sequence families. *Proc. Natl. Acad. Sci. USA*, 103(38):14056–61, 2006.

328. J. S. Richardson. The anatomy and taxonomy of protein structure. *Adv. Protein Chem.*, 34:167–339, 1981.

329. E. V. Koonin, Y. I. Wolf, and G. P. Karev. The structure of the protein universe and genome evolution. *Nature*, 420(6912):218–23, 2002.

330. R. Kolodny, L. Pereyaslavets, A. O. Samson, and M. Levitt. On the universe of protein folds. *Annu. Rev. Biophys.*, 42:559–82, 2013.

331. C. Chothia, J. Gough, C. Vogel, and S. A. Teichmann. Evolution of the protein repertoire. *Science*, 300(5626):1701–3, 2003.

332. J. H. Han, S. Batey, A. A. Nickson, S. A. Teichmann, and J. Clarke. The folding and evolution of multidomain proteins. *Nat. Rev. Mol. Cell Biol.*, 8(4):319–30, 2007.

333. C. Vogel, M. Bashton, N. D. Kerrison, C. Chothia, and S. A. Teichmann. Structure, function and evolution of multidomain proteins. *Curr. Opin. Struct. Biol.*, 14(2):208–16, 2004.

334. A. G. Murzin, S. E. Brenner, T. Hubbard, and C. Chothia. SCOP: a structural classification of proteins database for the investigation of sequences and structures. *J. Mol. Biol.*, 247(4):536–40, 1995.

335. S. Dietmann and L. Holm. Identification of homology in protein structure classification. *Nat. Struct. Biol.*, 8(11):953–7, 2001.

336. E. V. Koonin. Comparative genomics, minimal gene-sets and the last universal common ancestor. *Nat. Rev. Microbiol.*, 1(2):127–36, 2003.

337. D. Lee, A. Grant, R. L. Marsden, and C. Orengo. Identification and distribution of protein families in 120 completed genomes using Gene3D. *Proteins*, 59(3):603–15, 2005.

338. S. Ohno. *Evolution by Gene Duplication*. Springer-Veralg, Berlin, Germany, 1970.

339. J. A. Ranea, D. W. Buchan, J. M. Thornton, and C. A. Orengo. Evolution of protein superfamilies and bacterial genome size. *J. Mol. Biol.*, 336(4):871–87, 2004.

340. J. Qian, N. M. Luscombe, and M. Gerstein. Protein family and fold occurrence in genomes: power-law behaviour and evolutionary model. *J. Mol. Biol.*, 313(4):673–81, 2001.

341. W. Kabsch. A discussion of the solution for the best rotation to relate two sets of vectors. *Acta Crystallogr. Sect. A*, 34(5):827–828, 1978.

342. W. Kabsch. A solution for the best rotation to relate two sets of vectors. *Acta Crystallogr. Sect. A*, 32(5):922–923, 1976.

343. A. Zemla. LGA: A method for finding 3D similarities in protein structures. *Nucleic Acids Res.*, 31(13):3370–4, 2003.

344. A. Zemla, C. Venclovas, J. Moult, and K. Fidelis. Processing and analysis of CASP3 protein structure predictions. *Proteins*, Suppl 3:22–9, 1999.

345. Y. Zhang and J. Skolnick. Scoring function for automated assessment of protein structure template quality. *Proteins*, 57(4):702–10, 2004.

346. C. Apic, J. Gough, and S. A. Teichmann. Domain combinations in archaeal, eubacterial and eukaryotic proteomes. *J. Mol. Biol.*, 310(2):311–25, 2001.

347. C. Vogel, S. A. Teichmann, and C. Chothia. The immunoglobulin superfamily in *Drosophila melanogaster* and *Caenorhabditis elegans* and the evolution of complexity. *Development*, 130(25):6317–28, 2003.

348. J. A. Ranea, A. Grant, J. M. Thornton, and C. A. Orengo. Microeconomic principles explain an optimal genome size in bacteria. *Trends Genet.*, 21(1):21–5, 2005.

349. E. H. Palacios and A. Weiss. Function of the Src-family kinases, Lck and Fyn, in T-cell development and activation. *Oncogene*, 23(48):7990–8000, 2004.

350. U. Schmidt, N. Boucheron, B. Unger, and W. Ellmeier. The role of Tec family kinases in myeloid cells. *Int. Arch. Allergy Immunol.*, 134(1):65–78, 2004.

351. E. Bornberg-Bauer, F. Beaussart, S. K. Kummerfeld, S. A. Teichmann, and J. Weiner 3rd. The evolution of domain arrangements in proteins and interaction networks. *Cell. Mol. Life Sci.*, 62(4):435–45, 2005.

352. R. D. Finn, J. Mistry, B. Schuster-Bockler, S. Griffiths-Jones, V. Hollich, T. Lassmann, S. Moxon, M. Marshall, A. Khanna, R. Durbin, S. R. Eddy, E. L. Sonnhammer, and A. Bateman. Pfam: clans, web tools and services. *Nucleic Acids Res.*, 34(suppl 1):D247–51, 2006.

353. R. D. Finn, B. L. Miller, J. Clements, and A. Bateman. iPfam: a database of protein family and domain interactions found in the Protein Data Bank. *Nucleic Acids Res.*, 42(D1):D364–73, 2014.

354. M. Levitt and C. Chothia. Structural patterns in globular proteins. *Nature*, 261(5561):552–8, 1976.

355. N. Ben-Tal and R. Kolodny. Representation of the Protein Universe using Classifications, Maps, and Networks. *Isr. J. Chem.*, 54(8–9):1286–1292, 2014.

356. C. A. Orengo, A. D. Michie, S. Jones, D. T. Jones, M. B. Swindells, and J. M. Thornton. CATH–a hierarchic classification of protein domain structures. *Structure*, 5(8):1093–108, 1997.

357. C. Hadley and D. T. Jones. A systematic comparison of protein structure classifications: SCOP, CATH and FSSP. *Structure*, 7(9):1099–112, 1999.

358. H. Cheng, R. D. Schaeffer, Y. Liao, L. N. Kinch, J. Pei, S. Shi, B. H. Kim, and N. V. Grishin. ECOD: an evolutionary classification of protein domains. *PLoS Comput. Biol.*, 10(12):e1003926, 2014.

359. L. Holm, C. Ouzounis, C. Sander, G. Tuparev, and G. Vriend. A database of protein structure families with common folding motifs. *Protein Sci.*, 1(12):1691–8, 1992.

360. I. N. Shindyalov and P. E. Bourne. An alternative view of protein fold space. *Proteins*, 38(3):247–60, 2000.

361. A. S. Yang and B. Honig. An integrated approach to the analysis and modeling of protein sequences and structures. III. A comparative study of sequence conservation in protein structural families using multiple structural alignments. *J. Mol. Biol.*, 301(3):691–711, 2000.

362. R. Kolodny, D. Petrey, and B. Honig. Protein structure comparison: implications for the nature of 'fold space', and structure and function prediction. *Curr. Opin. Struct. Biol.*, 16(3):393–8, 2006.

363. J. Skolnick, A. K. Arakaki, S. Y. Lee, and M. Brylinski. The continuity of protein structure space is an intrinsic property of proteins. *Proc. Natl. Acad. Sci. USA*, 106(37):15690–15695, 2009.

364. L. Xie and P. E. Bourne. Detecting evolutionary relationships across existing fold space, using sequence order-independent profile-profile alignments. *Proc. Natl. Acad. Sci. USA*, 105(14):5441–6, 2008.

365. A. Andreeva, A. Prlic, T. J. Hubbard, and A. G. Murzin. SISYPHUS–structural alignments for proteins with non-trivial relationships. *Nucleic Acids Res.*, 35(suppl 1):D253–9, 2007.

366. S. Nepomnyachiy, N. Ben-Tal, and R. Kolodny. Global view of the protein universe. *Proc. Natl. Acad. Sci. USA*, 111(32):11691–6, 2014.

367. A. Andreeva, D. Howorth, C. Chothia, E. Kulesha, and A. G. Murzin. SCOP2 prototype: a new approach to protein structure mining. *Nucleic Acids Res.*, 42(D1):D310–4, 2014.

368. J. S. Richardson. β-Sheet topology and the relatedness of proteins. *Nature*, 268(5620):495–500, 1977.

369. M. L. Mansfield. Are there knots in proteins? *Nat. Struct. Mol. Biol.*, 1(4):213–214, 1994.

370. M. Jamroz, W. Niemyska, E. J. Rawdon, A. Stasiak, K. C. Millett, P. Sulkowski, and J. I. Sulkowska. KnotProt: a database of proteins with knots and slipknots. *Nucleic Acids Res.*, 43(D1):D306–14, 2015.

371. J. I. Sulkowska, E. J. Rawdon, K. C. Millett, J. N. Onuchic, and A. Stasiak. Conservation of complex knotting and slipknotting patterns in proteins. *Proc. Natl. Acad. Sci. USA*, 109(26):E1715–23, 2012.

372. R. C. Lua and A. Y. Grosberg. Statistics of Knots, Geometry of Conformations, and Evolution of Proteins. *PLoS Comput. Biol.*, 2(5):e45, 2006.

373. A. L. Mallam and S. E. Jackson. Knot formation in newly translated proteins is spontaneous and accelerated by chaperonins. *Nat. Chem. Biol.*, 8(2):147–53, 2012.

374. P. F. Faisca. Knotted proteins: A tangled tale of Structural Biology. *Comput. Struct. Biotechnol. J.*, 13:459–68, 2015.

375. P. Virnau, L. A. Mirny, and M. Kardar. Intricate knots in proteins: Function and evolution. *PLoS Comput. Biol.*, 2(9):e122, 2006.

376. Y. L. Lai, C. C. Chen, and J. K. Hwang. pKNOT v.2: the protein KNOT web server. *Nucleic Acids Res.*, 40(W1):W228–31, 2012.

377. R. Liu, W. A. Baase, and B. W. Matthews. The introduction of strain and its effects on the structure and stability of T4 lysozyme. *J. Mol. Biol.*, 295(1):127–45, 2000.

378. B. W. Matthews. Genetic and structural analysis of the protein stability problem. *Biochemistry*, 26(22):6885–8, 1987.

379. F. M. Richards and W. A. Lim. An analysis of packing in the protein folding problem. *Q. Rev. Biophys.*, 26(4):423–98, 1993.

380. P. Blume-Jensen and T. Hunter. Oncogenic kinase signalling. *Nature*, 411(6835):355–65, 2001.

381. D. W. Rodgers, R. H. Crepeau, and S. J. Edelstein. Pairings and polarities of the 14 strands in sickle cell hemoglobin fibers. *Proc. Natl. Acad. Sci. USA*, 84(17):6157–61, 1987.

382. M. Aidoo, D. J. Terlouw, M. S. Kolczak, P. D. McElroy, F. O. ter Kuile, S. Kariuki, B. L. Nahlen, A. A. Lal, and V. Udhayakumar. Protective effects of the sickle cell gene against malaria morbidity and mortality. *Lancet*, 359(9314):1311–2, 2002.

383. D. J. Weatherall. The Thalassemias: Disorders of Globin Synthesis. In M. A. Lichtman, T. J. Kipps, U. Seligsohn, K. Kaushansky, and J. T. Prchal, editors, *Williams Hematology*, chapter 47. McGraw-Hill Education / Medical, 8th edition, 2010.

384. H. F. Bunn. The triumph of good over evil: protection by the sickle gene against malaria. *Blood*, 121(1):20–5, 2013.

385. E. E. Lattman and G. D. Rose. Protein folding–what's the question? *Proc. Natl. Acad. Sci. USA*, 90(2):439–41, 1993.

386. P. A. Alexander, Y. He, Y. Chen, J. Orban, and P. N. Bryan. The design and characterization of two proteins with 88% sequence identity but different structure and function. *Proc. Natl. Acad. Sci. USA*, 104(29):11963–8, 2007.

387. P. A. Alexander, Y. He, Y. Chen, J. Orban, and P. N. Bryan. A minimal sequence code for switching protein structure and function. *Proc. Natl. Acad. Sci. USA*, 106(50):21149–54, 2009.

388. B. W. Matthews. Structural and genetic analysis of the folding and function of T4 lysozyme. *FASEB J.*, 10(1):35–41, 1996.

389. R. Sasidharan and C. Chothia. The selection of acceptable protein mutations. *Proc. Natl. Acad. Sci. USA*, 104(24):10080–5, 2007.

390. J. A. Gerlt and P. C. Babbitt. Mechanistically diverse enzyme superfamilies: the importance of chemistry in the evolution of catalysis. *Curr. Opin. Chem. Biol.*, 2(5):607–12, 1998.

391. A. E. Todd, C. A. Orengo, and J. M. Thornton. Evolution of function in protein superfamilies, from a structural perspective. *J. Mol. Biol.*, 307(4):1113–43, 2001.

392. J. A. Gerlt, P. C. Babbitt, and I. Rayment. Divergent evolution in the enolase superfamily: the interplay of mechanism and specificity. *Arch. Biochem. Biophys.*, 433(1):59–70, 2005.

393. M. E. Glasner, J. A. Gerlt, and P. C. Babbitt. Evolution of enzyme superfamilies. *Curr. Opin. Chem. Biol.*, 10(5):492–7, 2006.

394. C. J. Jeffery, B. J. Bahnson, W. Chien, D. Ringe, and G. A. Petsko. Crystal structure of rabbit phosphoglucose isomerase, a glycolytic enzyme that moonlights as neuroleukin, autocrine motility factor, and differentiation mediator. *Biochemistry*, 39(5):955–64, 2000.

395. C. Spiess, A. Beil, and M. Ehrmann. A temperature-dependent switch from chaperone to protease in a widely conserved heat shock protein. *Cell*, 97(3):339–47, 1999.

396. S. C. Rison, S. A. Teichmann, and J. M. Thornton. Homology, pathway distance and chromosomal localization of the small molecule metabolism enzymes in *Escherichia coli*. *J. Mol. Biol.*, 318(3):911–32, 2002.

397. H. M. Wilks, K. W. Hart, R. Feeney, C. R. Dunn, H. Muirhead, W. N. Chia, D. A. Barstow, T. Atkinson, A. R. Clarke, and J. J. Holbrook. A specific, highly active malate dehydrogenase by redesign of a lactate dehydrogenase framework. *Science*, 242(4885):1541–4, 1988.

398. O. Khersonsky, C. Roodveldt, and D. S. Tawfik. Enzyme promiscuity: evolutionary and mechanistic aspects. *Curr. Opin. Chem. Biol.*, 10(5):498–508, 2006.

399. R. A. Jensen. Enzyme recruitment in evolution of new function. *Annu. Rev. Microbiol.*, 30:409–25, 1976.

400. E. Meyer. Internal water molecules and H-bonding in biological macromolecules: a review of structural features with functional implications. *Protein Sci.*, 1(12):1543–62, 1992.

401. Y. Levy and J. N. Onuchic. Water mediation in protein folding and molecular recognition. *Annu. Rev. Biophys. Biomol. Struct.*, 35:389–415, 2006.

402. V. P. Denisov, K. Venu, J. Peters, H. D. Hörlein, and B. Halle. Orientational Disorder and Entropy of Water in Protein Cavities. *J. Phys. Chem. B*, 101(45):9380–9389, 1997.

403. B. Halle. Water in biological systems: the NMR picture. In M. C. Bellisent-Funel, editor, *Hydration processes in biology: theoretical and Experimental approaches*, pages 221–232. IOS Press, 1999.

404. S. J. Hubbard and P. Argos. Cavities and packing at protein interfaces. *Protein Sci.*, 3(12):2194–206, 1994.

405. K. Rother, R. Preissner, A. Goede, and C. Frommel. Inhomogeneous molecular density: reference packing densities and distribution of cavities within proteins. *Bioinformatics*, 19(16):2112–21, 2003.

406. J. A. Ernst, R. T. Clubb, H. X. Zhou, A. M. Gronenborn, and G. M. Clore. Demonstration of positionally disordered water within a protein hydrophobic cavity by NMR. *Science*, 267(5205):1813–7, 1995.

407. S. Park and J. G. Saven. Statistical and molecular dynamics studies of buried waters in globular proteins. *Proteins: Struct., Funct., Bioinf.*, 60:450–63, 2005.

408. J. C. Covalt Jr, M. Roy, and P. A. Jennings. Core and surface mutations affect folding kinetics, stability and cooperativity in IL-1 beta: does alteration in buried water play a role? *J. Mol. Biol.*, 307(2):657–69, 2001.

409. D. H. Adamek, L. Guerrero, M. Blaber, and D. L. Caspar. Structural and energetic consequences of mutations in a solvated hydrophobic cavity. *J. Mol. Biol.*, 346(1):307–18, 2005.

410. N. Agmon. The Grotthuss mechanism. *Chem. Phys. Lett.*, 244:456–462, 1995.

411. F. Garczarek and K. Gerwert. Functional waters in intraprotein proton transfer monitored by FTIR difference spectroscopy. *Nature*, 439(7072):109–12, 2006.

412. W. Guo, S. Lampoudi, and J. E. Shea. Posttransition state desolvation of the hydrophobic core of the src-SH3 protein domain. *Biophys. J.*, 85(1):61–9, 2003.

413. L. D. Barron, L. Hecht, and G. Wilson. The lubricant of life: a proposal that solvent water promotes extremely fast conformational fluctuations in mobile heteropolypeptide structure. *Biochemistry*, 36(43):13143–7, 1997.

414. J. M. Zanotti, M. C. Bellissent-Funel, and J. Parello. Hydration-coupled dynamics in proteins studied by neutron scattering and NMR: the case of the typical EF-hand calcium-binding parvalbumin. *Biophys. J.*, 76(5):2390–411, 1999.

415. J. Fitter. The temperature dependence of internal molecular motions in hydrated and dry alpha-amylase: the role of hydration water in the dynamical transition of proteins. *Biophys. J.*, 76(2):1034–42, 1999.

416. J. Janin. Wet and dry interfaces: the role of solvent in protein-protein and protein-DNA recognition. *Structure*, 7(12):R277–9, 1999.

417. F. Rodier, R. P. Bahadur, P. Chakrabarti, and J. Janin. Hydration of protein-protein interfaces. *Proteins: Struct., Funct., Bioinf.*, 60(1):36–45, 2005.

418. M. Levitt and B. H. Park. Water: now you see it, now you don't. *Structure*, 1(4):223–6, 1993.

419. R. Loris, U. Langhorst, S. De Vos, K. Decanniere, J. Bouckaert, D. Maes, T. R. Transue, and J. Steyaert. Conserved water molecules in a large family of microbial ribonucleases. *Proteins*, 36(1):117–34, 1999.

420. H. R. Bairagya, B. P. Mukhopadhyay, and A. K. Bera. Conserved water mediated recognition and the dynamics of active site Cys-331 and Tyr-411 in hydrated structure of human IMPDH-II. *J. Mol. Recognit.*, 24(1):35–44, 2011.

421. H. Ponstingl, T. Kabir, D. Gorse, and J. M. Thornton. Morphological aspects of oligomeric protein structures. *Prog. Biophys. Mol. Biol.*, 89(1):9–35, 2005.

422. I. M. Klotz, D. W. Darnall, and N. R. Langerman. Quaternary structure of proteins. In H. Neurath and R. L. Hill, editors, *The Proteins*, volume 1, pages 293–411. Academic Press, New York, 1975.

423. D. S. Goodsell and A. J. Olson. Structural symmetry and protein function. *Annu. Rev. Biophys. Biomol. Struct.*, 29:105–53, 2000.

424. A. J. Venkatakrishnan, E. D. Levy, and S. A. Teichmann. Homomeric protein complexes: evolution and assembly. *Biochem. Soc. Trans.*, 38(4):879–882, 2010.

425. I. M. Nooren and J. M. Thornton. Diversity of protein-protein interactions. *EMBO J.*, 22(14):3486–92, 2003.

426. E. H. McConkey. Molecular evolution, intracellular organization, and the quinary structure of proteins. *Proc. Natl. Acad. Sci. USA*, 79(10):3236–40, 1982.

427. A. J. Wirth and M. Gruebele. Quinary protein structure and the consequences of crowding in living cells: leaving the test-tube behind. *BioEssays*, 35(11):984–93, 2013.

428. W. B. Monteith, R. D. Cohen, A. E. Smith, E. Guzman-Cisneros, and G. J. Pielak. Quinary structure modulates protein stability in cells. *Proc. Natl. Acad. Sci. USA*, 112(6):1739–42, 2015.

429. M. H. Ali and B. Imperiali. Protein oligomerization: how and why. *Bioorg. Med. Chem.*, 13(17):5013–20, 2005.

430. L. Lo Conte, C. Chothia, and J. Janin. The atomic structure of protein-protein recognition sites. *J. Mol. Biol.*, 285(5):2177–98, 1999.

431. A. A. Bogan and K. S. Thorn. Anatomy of hot spots in protein interfaces. *J. Mol. Biol.*, 280(1):1–9, 1998.

432. T. Clackson and J. A. Wells. A hot spot of binding energy in a hormone-receptor interface. *Science*, 267(5196):383–6, 1995.

433. B. Morgenstern and W. R. Atchley. Evolution of bHLH transcription factors: modular evolution by domain shuffling? *Mol. Biol. Evol.*, 16(12):1654–63, 1999.

434. T. L. Blundell and N. Srinivasan. Symmetry, stability, and dynamics of multidomain and multicomponent protein systems. *Proc. Natl. Acad. Sci. USA*, 93(25):14243–8, 1996.

435. W. K. Kim, A. Henschel, C. Winter, and M. Schroeder. The many faces of protein-protein interactions: A compendium of interface geometry. *PLoS Comput. Biol.*, 2(9):e124, 2006.

436. C. Chothia and J. Janin. Principles of protein-protein recognition. *Nature*, 256(5520):705–8, 1975.

437. M. J. Bennett, M. P. Schlunegger, and D. Eisenberg. 3D domain swapping: a mechanism for oligomer assembly. *Protein Sci.*, 4(12):2455–68, 1995.

438. Y. Liu and D. Eisenberg. 3D domain swapping: as domains continue to swap. *Protein Sci.*, 11(6):1285–99, 2002.

439. S. Posy, L. Shapiro, and B. Honig. Sequence and structural determinants of strand swapping in cadherin domains: do all cadherins bind through the same adhesive interface? *J. Mol. Biol.*, 378(4):954–68, 2008.

440. S. S. Chirala, W. Y. Huang, A. Jayakumar, K. Sakai, and S. J. Wakil. Animal fatty acid synthase: functional mapping and cloning and expression of the domain I constituent activities. *Proc. Natl. Acad. Sci. USA*, 94(11):5588–93, 1997.

441. M. Leibundgut, T. Maier, S. Jenni, and N. Ban. The multienzyme architecture of eukaryotic fatty acid synthases. *Curr. Opin. Struct. Biol.*, 18(6):714–25, 2008.

442. D. A. Johnson, P. Akamine, E. Radzio-Andzelm, M. Madhusudan, and S. S. Taylor. Dynamics of cAMP-dependent protein kinase. *Chem. Rev.*, 101(8):2243–70, 2001.

443. J. B. Shabb. Physiological substrates of cAMP-dependent protein kinase. *Chem. Rev.*, 101(8):2381–411, 2001.

444. H. Walden, G. S. Bell, R. J. Russell, B. Siebers, R. Hensel, and G. L. Taylor. Tiny TIM: a small, tetrameric, hyperthermostable triosephosphate isomerase. *J. Mol. Biol.*, 306(4):745–57, 2001.

445. S. Yooseph, G. Sutton, D. B. Rusch, A. L. Halpern, S. J. Williamson, et al. The Sorcerer II Global Ocean Sampling expedition: Expanding the universe of protein families. *PLoS Biol.*, 5:e16, 2007.

446. J. C. Venter, K. Remington, J. F. Heidelberg, A. L. Halpern, D. Rusch, J. A. Eisen, D. Wu, I. Paulsen, K. E. Nelson, W. Nelson, D. E. Fouts, S. Levy, A. H. Knap, M. W. Lomas, K. Nealson, O. White, J. Peterson, J. Hoffman, R. Parsons, H. Baden-Tillson, C. Pfannkoch, Y. H. Rogers, and H. O. Smith. Environmental genome shotgun sequencing of the Sargasso Sea. *Science*, 304(5667):66–74, 2004.

447. F. Wold. In vivo chemical modification of proteins (post-translational modification). *Annu. Rev. Biochem.*, 50:783–814, 1981.

448. C. T. Walsh. *Posttranslational Modification of Proteins: Expanding Nature's Inventory*. Roberts & Company Publishers, Greenwood Village, CO, 2005.

449. R. G. Krishna and F. Wold. Posttranslational Modifications. In R. H. Angeletti, editor, *Proteins: analysis and design*, pages 121–206. Academic Press, San Diego, CA, 1998.

450. M. R. Wilkins, E. Gasteiger, A. A. Gooley, B. R. Herbert, M. P. Molloy, P. A. Binz, K. Ou, J. C. Sanchez, A. Bairoch, K. L. Williams, and D. F. Hochstrasser. High-throughput mass spectrometric discovery of protein post-translational modifications. *J. Mol. Biol.*, 289(3):645–57, 1999.

451. J. Godovac-Zimmermann and L. R. Brown. Perspectives for mass spectrometry and functional proteomics. *Mass Spectrom. Rev.*, 20(1):1–57, 2001.

452. L. N. Johnson and R. J. Lewis. Structural basis for control by phosphorylation. *Chem. Rev.*, 101(8):2209–42, 2001.

453. M. Mann, S. E. Ong, M. Gronborg, H. Steen, O. N. Jensen, and A. Pandey. Analysis of protein phosphorylation using mass spectrometry: deciphering the phosphoproteome. *Trends Biotechnol.*, 20(6):261–8, 2002.

454. G. Manning, D. B. Whyte, R. Martinez, T. Hunter, and S. Sudarsanam. The protein kinase complement of the human genome. *Science*, 298(5600):1912–34, 2002.

455. H. Steen, M. Fernandez, S. Ghaffari, A. Pandey, and M. Mann. Phosphotyrosine mapping in Bcr/Abl oncoprotein using phosphotyrosine-specific immonium ion scanning. *Mol. Cell. Proteomics*, 2(3):138–45, 2003.

456. M. Landau, S. J. Fleishman, and N. Ben-Tal. A putative mechanism for downregulation of the catalytic activity of the EGF receptor via direct contact between its kinase and C-terminal domains. *Structure*, 12(12):2265–75, 2004.

457. M. Landau and N. Ben-Tal. Dynamic equilibrium between multiple active and inactive conformations explains regulation and oncogenic mutations in ErbB receptors. *Biochim. Biophys. Acta*, 1785(1):12–31, 2008.

458. B. F. Volkman, D. Lipson, D. E. Wemmer, and D. Kern. Two-state allosteric behavior in a single-domain signaling protein. *Science*, 291(5512):2429–33, 2001.

459. J. Lätzer, T. Shen, and P. G. Wolynes. Conformational Switching upon Phosphorylation: A Predictive Framework Based on Energy Landscape Principles. *Biochemistry*, 47(7):2110–2122, 2008.

460. J. Kitchen, R. E. Saunders, and J. Warwicker. Charge environments around phosphorylation sites in proteins. *BMC Struct. Biol.*, 8:19, 2008.

461. A. Narayanan and M. P. Jacobson. Computational studies of protein regulation by post-translational phosphorylation. *Curr. Opin. Struct. Biol.*, 19(2):156–63, 2009.

462. G. W. Hart, R. S. Haltiwanger, G. D. Holt, and W. G. Kelly. Glycosylation in The Nucleus and Cytoplasm. *Annu. Rev. Biochem.*, 58(1):841–874, 1989.

463. G. W. Hart, M. P. Housley, and C. Slawson. Cycling of O-linked β-N-acetylglucosamine on nucleocytoplasmic proteins. *Nature*, 446(7139):1017–1022, 2007.

464. H. Nothaft and C. M. Szymanski. Protein glycosylation in bacteria: sweeter than ever. *Nat. Rev. Microbiol.*, 8(11):765–78, 2010.

465. J. Eichler. Extreme sweetness: protein glycosylation in archaea. *Nat. Rev. Microbiol.*, 11(3):151–6, 2013.

466. Y. Mechref and M. V. Novotny. Structural investigations of glycoconjugates at high sensitivity. *Chem. Rev.*, 102(2):321–69, 2002.

467. R. A. Dwek. Glycobiology: Toward Understanding the Function of Sugars. *Chem. Rev.*, 96(2):683–720, 1996.

468. H. S. Lee, Y. Qi, and W. Im. Effects of N-glycosylation on protein conformation and dynamics: Protein Data Bank analysis and molecular dynamics simulation study. *Sci. Rep.*, 5:8926, 2015.

469. H. C. Joao and R. A. Dwek. Effects of glycosylation on protein structure and dynamics in ribonuclease B and some of its individual glycoforms. *Eur. J. Biochem.*, 218(1):239–44, 1993.

470. T. A. Gerken, K. J. Butenhof, and R. Shogren. Effects of glycosylation on the conformation and dynamics of O-linked glycoproteins: carbon-13 NMR studies of ovine submaxillary mucin. *Biochemistry*, 28(13):5536–43, 1989.

471. D. N. Hebert, L. Lamriben, E. T. Powers, and J. W. Kelly. The intrinsic and extrinsic effects of N-linked glycans on glycoproteostasis. *Nat. Chem. Biol.*, 10(11):902–10, 2014.

472. R. J. Sola, J. A. Rodriguez-Martinez, and K. Griebenow. Modulation of protein biophysical properties by chemical glycosylation: biochemical insights and biomedical implications. *Cell. Mol. Life Sci.*, 64(16):2133–52, 2007.

473. P. W. Fenimore, H. Frauenfelder, B. H. McMahon, and R. D. Young. Bulk-solvent and hydration-shell fluctuations, similar to alpha- and beta-fluctuations in glasses, control protein motions and functions. *Proc. Natl. Acad. Sci. USA*, 101(40):14408–13, 2004.

474. P. W. Fenimore, H. Frauenfelder, B. H. McMahon, and F. G. Parak. Slaving: solvent fluctuations dominate protein dynamics and functions. *Proc. Natl. Acad. Sci. USA*, 99(25):16047–51, 2002.

475. A. Rubinstein and S. Sherman. Influence of the solvent structure on the electrostatic interactions in proteins. *Biophys. J.*, 87(3):1544–57, 2004.

476. R. Affleck, C. A. Haynes, and D. S. Clark. Solvent dielectric effects on protein dynamics. *Proc. Natl. Acad. Sci. USA*, 89(11):5167–70, 1992.

477. R. J. Sola and K. Griebenow. Influence of modulated structural dynamics on the kinetics of alpha-chymotrypsin catalysis. Insights through chemical glycosylation, molecular dynamics and domain motion analysis. *FEBS J.*, 273(23):5303–19, 2006.

478. D. Shental-Bechor and Y. Levy. Effect of glycosylation on protein folding: a close look at thermodynamic stabilization. *Proc. Natl. Acad. Sci. USA*, 105(24):8256–61, 2008.

479. F. Xin and P. Radivojac. Post-translational modifications induce significant yet not extreme changes to protein structure. *Bioinformatics*, 28(22):2905–13, 2012.

480. R. J. Sola and K. Griebenow. Effects of glycosylation on the stability of protein pharmaceuticals. *J. Pharm. Sci.*, 98(4):1223–45, 2009.

481. M. Molinari. N-glycan structure dictates extension of protein folding or onset of disposal. *Nat. Chem. Biol.*, 3(6):313–320, 2007.

482. N. Mitra, S. Sinha, T. N. Ramya, and A. Surolia. N-linked oligosaccharides as outfitters for glycoprotein folding, form and function. *Trends Biochem. Sci.*, 31(3):156–63, 2006.

483. C. J. Bosques and B. Imperiali. The interplay of glycosylation and disulfide formation influences fibrillization in a prion protein fragment. *Proc. Natl. Acad. Sci. USA*, 100(13):7593–8, 2003.

484. G. Z. Lederkremer. Glycoprotein folding, quality control and ER-associated degradation. *Curr. Opin. Struct. Biol.*, 19(5):515–23, 2009.

485. M. Li, J. Luo, C. L. Brooks, and W. Gu. Acetylation of p53 inhibits its ubiquitination by Mdm2. *J. Biol. Chem.*, 277(52):50607–11, 2002.

486. C. Choudhary, C. Kumar, F. Gnad, M. L. Nielsen, M. Rehman, T. C. Walther, J. V. Olsen, and M. Mann. Lysine

acetylation targets protein complexes and co-regulates major cellular functions. *Science*, 325(5942):834–40, 2009.

487. B. M. Turner. Cellular memory and the histone code. *Cell*, 111(3):285–91, 2002.

488. D. P. Lane. Cancer. p53, guardian of the genome. *Nature*, 358(6381):15–6, 1992.

489. A. M. Bode and Z. Dong. Post-translational modification of p53 in tumorigenesis. *Nat. Rev. Cancer*, 4(10):793–805, 2004.

490. A. C. Joerger and A. R. Fersht. Structural biology of the tumor suppressor p53. *Annu. Rev. Biochem.*, 77:557–82, 2008.

491. D. R. Johnson, R. S. Bhatnagar, L. J. Knoll, and J. I. Gordon. Genetic and biochemical studies of protein N-myristoylation. *Annu. Rev. Biochem.*, 63:869–914, 1994.

492. S. McLaughlin and A. Aderem. The myristoyl-electrostatic switch: a modulator of reversible protein-membrane interactions. *Trends Biochem. Sci.*, 20(7):272–6, 1995.

493. M. A. Kostiuk, B. O. Keller, and L. G. Berthiaume. Palmitoylation of ketogenic enzyme HMGCS2 enhances its interaction with PPARalpha and transcription at the Hmgcs2 PPRE. *FASEB J.*, 24(6):1914–24, 2010.

494. M. D. Resh. Fatty acylation of proteins: new insights into membrane targeting of myristoylated and palmitoylated proteins. *Biochim. Biophys. Acta*, 1451(1):1–16, 1999.

495. S. S. Sanders, D. D. Martin, S. L. Butland, M. Lavallee-Adam, D. Calzolari, C. Kay, J. R. Yates 3rd, and M. R. Hayden. Curation of the Mammalian Palmitoylome Indicates a Pivotal Role for Palmitoylation in Diseases and Disorders of the Nervous System and Cancers. *PLoS Comput. Biol.*, 11(8):e1004405, 2015.

496. M. H. Glickman and A. Ciechanover. The Ubiquitin-Proteasome Proteolytic Pathway: Destruction for the Sake of Construction. *Physiol. Rev.*, 82:373–428, 2002.

497. M. Y. Sherman and A. L. Goldberg. Cellular defenses against unfolded proteins: a cell biologist thinks about neurodegenerative diseases. *Neuron*, 29(1):15–32, 2001.

498. C. M. Dobson. Protein folding and misfolding. *Nature*, 426(6968):884–90, 2003.

499. E. L. Eskelinen and P. Saftig. Autophagy: A lysosomal degradation pathway with a central role in health and disease. *Biochim. Biophys. Acta*, 1793(4):664–73, 2009.

500. H. Abriel, J. Loffing, J. F. Rebhun, J. H. Pratt, L. Schild, J. D. Horisberger, D. Rotin, and O. Staub. Defective regulation of the epithelial Na$^+$ channel by Nedd4 in Liddle's syndrome. *J. Clin. Invest.*, 103(5):667–73, 1999.

501. J. M. Younger, L. Chen, H. Y. Ren, M. F. Rosser, E. L. Turnbull, C. Y. Fan, C. Patterson, and D. M. Cyr. Sequential quality-control checkpoints triage misfolded cystic fibrosis transmembrane conductance regulator. *Cell*, 126(3):571–82, 2006.

502. A. M. Weissman. Themes and variations on ubiquitylation. *Nat. Rev. Mol. Cell Biol.*, 2(3):169–78, 2001.

503. K. A. Wilkinson and J. M. Henley. Mechanisms, regulation and consequences of protein SUMOylation. *Biochem. J.*, 428(2):133–145, 2010.

504. A. Flotho and F. Melchior. Sumoylation: A Regulatory Protein Modification in Health and Disease. *Annu. Rev. Biochem.*, 82(1):357–385, 2013.

505. A. Seifert, P. Schofield, G. J. Barton, and R. T. Hay. Proteotoxic stress reprograms the chromatin landscape of SUMO modification. *Sci. Signal.*, 8(384):rs7–rs7, 2015.

506. Y. Shi, F. Lan, C. Matson, P. Mulligan, J. R. Whetstine, P. A. Cole, and R. A. Casero. Histone demethylation mediated by the nuclear amine oxidase homolog LSD1. *Cell*, 119(7):941–53, 2004.

507. H. S. Gill, G. M. Pfluegl, and D. Eisenberg. Multicopy crystallographic refinement of a relaxed glutamine synthetase from *Mycobacterium tuberculosis* highlights flexible loops in the enzymatic mechanism and its regulation. *Biochemistry*, 41(31):9863–72, 2002.

508. C. W. Pugh and P. J. Ratcliffe. Regulation of angiogenesis by hypoxia: role of the HIF system. *Nat. Med.*, 9(6):677–84, 2003.

509. E. Chapman, M. D. Best, S. R. Hanson, and C. H. Wong. Sulfotransferases: structure, mechanism, biological activity, inhibition, and synthetic utility. *Angew. Chem. Int. Ed.*, 43(27):3526–48, 2004.

510. K. Chen and L. Kurgan. Investigation of atomic level patterns in protein–small ligand interactions. *PLoS One*, 4(2):e4473, 2009.

511. C. Andreini, I. Bertini, G. Cavallaro, G. L. Holliday, and J. M. Thornton. Metal ions in biological catalysis: from enzyme databases to general principles. *J. Biol. Inorg. Chem.*, 13(8):1205–1218, 2008.

512. Micronutrient Information Center. http://lpi.oregonstate.edu/mic/minerals.

513. Nutritional Disorders: Merck Manual. http://www.merckmanuals.com/professional/nutritional-disorders, 2017.

514. L. M. Plum, L. Rink, and H. Haase. The essential toxin: impact of zinc on human health. *Int. J. Environ. Res. Public. Health.*, 7(4):1342–65, 2010.

515. F. L. Zhang and P. J. Casey. Protein prenylation: molecular mechanisms and functional consequences. *Annu. Rev. Biochem.*, 65:241–69, 1996.

516. Q. Deng and J. T. Barbieri. Molecular mechanisms of the cytotoxicity of ADP-ribosylating toxins. *Annu. Rev. Microbiol.*, 62:271–88, 2008.

517. S. Baron. *Medical Microbiology*. University of Texas Medical Branch at Galveston, 4th edition, 1996.

518. K. Aktories and I. Just. *Bacterial Protein Toxins*. Springer, Berlin, 2000.

519. I. R. Vetter, F. Hofmann, S. Wohlgemuth, C. Herrmann, and I. Just. Structural consequences of mono-glucosylation of Ha-Ras by *Clostridium sordellii* lethal toxin. *J. Mol. Biol.*, 301(5):1091–5, 2000.

520. G. Schmidt, P. Sehr, M. Wilm, J. Selzer, M. Mann, and K. Aktories. Gln-63 of Rho is deamidated by *Escherichia coli* cytotoxic necrotizing factor-1. *Nature*, 387(6634):725–9, 1997.

521. R. N. Perham. Swinging arms and swinging domains in multifunctional enzymes: catalytic machines for multistep reactions. *Annu. Rev. Biochem.*, 69:961–1004, 2000.

522. K. E. Krueger and S. Srivastava. Posttranslational protein modifications: current implications for cancer detection, prevention, and therapeutics. *Mol. Cell. Proteomics*, 5(10):1799–810, 2006.

523. K. Giehl. Oncogenic Ras in tumour progression and metastasis. *Biol. Chem.*, 386(3):193–205, 2005.

524. C. Tsatsanis, A. Zafiropoulos, and D. A. Spandidos. Oncogenic Kinases in Cancer. In *eLS*. John Wiley & Sons, Ltd., 2010.

525. E. P. Booy, D. Johar, S. Maddika, H. Pirzada, M. M. Sahib, I. Gehrke, S. Loewen, S. F. Louis, K. Kadkhoda, M. Mowat, and M. Los. Monoclonal and bispecific antibodies as novel therapeutics. *Arch. Immunol. Ther. Exp.*, 54:85–1001, 2006.

526. D. Matei, D. D. Chang, and M. H. Jeng. Imatinib mesylate (Gleevec) inhibits ovarian cancer cell growth through a mechanism dependent on platelet-derived growth factor receptor alpha and Akt inactivation. *Clin. Cancer Res.*, 10:681–690, 2004.

527. S. Hakomori. Tumor-associated carbohydrate antigens defining tumor malignancy: basis for development of anti-cancer vaccines. *Adv. Exp. Med. Biol.*, 491:369–402, 2001.

528. M. M. Fuster, J. R. Brown, L. Wang, and J. D. Esko. A disaccharide precursor of sialyl Lewis X inhibits metastatic potential of tumor cells. *Cancer Res.*, 63(11):2775–81, 2003.

529. L. Zitvogel, A. Tesniere, and G. Kroemer. Cancer despite immunosurveillance: immunoselection and immunosubversion. *Nat. Rev. Immunol.*, 6(10):715–27, 2006.

530. L. M. Krug, G. Ragupathi, C. Hood, M. G. Kris, V. A. Miller, J. R. Allen, S. J. Keding, S. J. Danishefsky, J. Gomez, L. Tyson, B. Pizzo, V. Baez, and P. O. Livingston. Vaccination of patients with small-cell lung cancer with synthetic fucosyl GM-1 conjugated to keyhole limpet hemocyanin. *Clin. Cancer Res.*, 10(18 Pt 1):6094–100, 2004.

531. C. H. Westphal, M. A. Dipp, and L. Guarente. A therapeutic role for sirtuins in diseases of aging? *Trends Biochem. Sci.*, 32(12):555–60, 2007.

532. J. C. Milne and J. M. Denu. The Sirtuin family: therapeutic targets to treat diseases of aging. *Curr. Opin. Chem. Biol.*, 12:11–17, 2008.

533. M. Mann and O. N. Jensen. Proteomic analysis of post-translational modifications. *Nat. Biotechnol.*, 21(3):255–61, 2003.

534. P. Roepstorff. Mass spectrometry in protein studies from genome to function. *Curr. Opin. Biotechnol.*, 8(1):6–13, 1997.

535. E. van Hoffman and V. Strootbant. *Mass Spectrometry, Principles and Applications*. John Wiley & Sons, Inc., 2nd edition, 2001.

536. S. J. Fey and P. M. Larsen. 2D or not 2D. Two-dimensional gel electrophoresis. *Curr. Opin. Chem. Biol.*, 5(1):26–33, 2001.

537. T. Rabilloud. Two-dimensional gel electrophoresis in proteomics: old, old fashioned, but it still climbs up the mountains. *Proteomics*, 2(1):3–10, 2002.

538. J. Pogliano. The bacterial cytoskeleton. *Curr. Opin. Cell Biol.*, 20(1):19–27, 2008.

539. P. de Lanerolle, T. Johnson, and W. A. Hofmann. Actin and myosin I in the nucleus: what next? *Nat. Struct. Mol. Biol.*, 12(9):742–6, 2005.

540. I. Grummt. Actin and myosin as transcription factors. *Curr. Opin. Genet. Dev.*, 16(2):191–6, 2006.

541. M. R. K. Mofrad. Rheology of the Cytoskeleton. *Annu. Rev. Fluid Mech.*, 41:433–53, 2009.

542. D. S. Goodsell. Molecular Machinery: A Tour of the Protein Data Bank, 2008.

543. M. B. Omary, P. A. Coulombe, and W. H. McLean. Intermediate filament proteins and their associated diseases. *N. Engl. J. Med.*, 351(20):2087–100, 2004.

544. C. Guzman, S. Jeney, L. Kreplak, S. Kasas, A. J. Kulik, U. Aebi, and L. Forro. Exploring the mechanical

properties of single vimentin intermediate filaments by atomic force microscopy. *J. Mol. Biol.*, 360(3):623–30, 2006.

545. P. A. Janmey and C. A. McCulloch. Cell mechanics: integrating cell responses to mechanical stimuli. *Annu. Rev. Biomed. Eng.*, 9:1–34, 2007.

546. P. A. Coulombe and M. B. Omary. 'Hard' and 'soft' principles defining the structure, function and regulation of keratin intermediate filaments. *Curr. Opin. Cell Biol.*, 14(1):110–22, 2002.

547. L. H. Gu and P. A. Coulombe. Keratin function in skin epithelia: a broadening palette with surprising shades. *Curr. Opin. Cell Biol.*, 19(1):13–23, 2007.

548. M. Beil, A. Micoulet, G. von Wichert, S. Paschke, P. Walther, M. B. Omary, P. P. Van Veldhoven, U. Gern, E. Wolff-Hieber, J. Eggermann, J. Waltenberger, G. Adler, J. Spatz, and T. Seufferlein. Sphingosylphosphorylcholine regulates keratin network architecture and visco-elastic properties of human cancer cells. *Nat. Cell Biol.*, 5(9):803–11, 2003.

549. R. D. Goldman, B. Grin, M. G. Mendez, and E. R. Kuczmarski. Intermediate filaments: versatile building blocks of cell structure. *Curr. Opin. Cell Biol.*, 20:28–34, 2008.

550. J. Schmidt and C. Handsaker. Fluorescence microscopy of keratin intermediate filaments in epithelial cells. Wikipedia, the free encyclopedia. http://en.wikipedia.org/wiki/File:Keratinfl.jpg, 2004.

551. E. Fuchs and S. Raghavan. Getting under the skin of epidermal morphogenesis. *Nat. Rev. Genet.*, 3(3):199–209, 2002.

552. A. Z. Budzynski. Fibrinogen and fibrin: biochemistry and pathophysiology. *Crit. Rev. Oncol. Hematol.*, 6(2):97–146, 1986.

553. J. E. Sadler. K is for koagulation. *Nature*, 427:493–4, 2004.

554. M. D. Shoulders and R. T. Raines. Collagen Structure and Stability. *Annu. Rev. Biochem.*, 78:32.1–32.30, 2009.

555. J. Y. Exposito, C. Cluzel, R. Garrone, and C. Lethias. Evolution of collagens. *Anat. Rec.*, 268(3):302–16, 2002.

556. C. M. Kielty and M. E. Grant. The Collagen Family: Structure, Assembly, and Organization in the Extracellular Matrix. In P. M. Royce and B. Steinmann, editors, *Connective Tissue and Its Heritable Disorders: Molecular, Genetic, and Medical Aspects*, pages 159–222. Wiley-Liss, Inc., New York, 2nd edition, 2002.

557. S.-S. Sivan, E. Wachtel, E. Tsitron, N. Sakkee, F. van der Ham, J. DeGroot, S. Roberts, and A. Maroudas. Collagen turnover in normal and degenerate human intervertebral discs as determined by the racemization of aspartic acid. *J. Biol. Chem.*, 283(14):8796–8801, 2008.

558. N. Verzijl, J. DeGroot, S. R Thorpe, R. A Bank, J. N. Shaw, T. J. Lyons, J. W. J. Bijlsma, F. P. J. G. Lafeber, J. W. Baynes, and J. M. TeKoppele. Effect of collagen turnover on the accumulation of advanced glycation end products. *J. Biol. Chem.*, 275(50):39027–39031, 2000.

559. R. Berisio, L. Vitagliano, L. Mazzarella, and A. Zagari. Recent progress on collagen triple helix structure, stability and assembly. *Protein Pept. Lett.*, 9(2):107–16, 2002.

560. A. Bhattacharjee and M. Bansal. Collagen structure: the Madras triple helix and the current scenario. *IUBMB Life*, 57(3):161–72, 2005.

561. S. Viguet-Carrin, P. Garnero, and P. D. Delmas. The role of collagen in bone strength. *Osteoporos. Int.*, 17(3):319–36, 2006.

562. C. M. Nelson and M. J. Bissell. Of extracellular matrix, scaffolds, and signaling: tissue architecture regulates development, homeostasis, and cancer. *Annu. Rev. Cell. Dev. Biol.*, 22:287–309, 2006.

563. L. G. Griffith and M. A. Swartz. Capturing complex 3D tissue physiology in vitro. *Nat. Rev. Mol. Cell Biol.*, 7(3):211–24, 2006.

564. R. O. Hynes. Integrins: bidirectional, allosteric signaling machines. *Cell*, 110(6):673–87, 2002.

565. R. K. Assoian and X. Zhu. Cell anchorage and the cytoskeleton as partners in growth factor dependent cell cycle progression. *Curr. Opin. Cell Biol.*, 9(1):93–8, 1997.

566. A. E. Aplin, A. K. Howe, and R. L. Juliano. Cell adhesion molecules, signal transduction and cell growth. *Curr. Opin. Cell Biol.*, 11(6):737–44, 1999.

567. F. G. Giancotti and E. Ruoslahti. Integrin signaling. *Science*, 285(5430):1028–32, 1999.

568. I. Staniszewska, I. K. Sariyer, S. Lecht, M. C. Brown, E. M. Walsh, G. P. Tuszynski, M. Safak, P. Lazarovici, and C. Marcinkiewicz. Integrin $\alpha 9\beta 1$ is a receptor for nerve growth factor and other neurotrophins. *J. Cell Sci.*, 121(Pt 4):504–13, 2008.

569. D. E. Jaalouk and J. Lammerding. Mechanotransduction gone awry. *Nat. Rev. Mol. Cell Biol.*, 10(1):63–73, 2009.

570. M. J. Pallen and N. J. Matzke. From The Origin of Species to the origin of bacterial flagella. *Nat. Rev. Microbiol.*, 4(10):784–90, 2006.

571. J. L. Ross, M. Y. Ali, and D. M. Warshaw. Cargo transport: molecular motors navigate a complex cytoskeleton. *Curr. Opin. Cell Biol.*, 20(1):41–7, 2008.

572. E. Chevalier-Larsen and E. L. Holzbaur. Axonal transport and neurodegenerative disease. *Biochim. Biophys. Acta*, 1762(11–12):1094–108, 2006.

573. S. Inoue and E. D. Salmon. Force generation by microtubule assembly/disassembly in mitosis and related movements. *Mol. Biol. Cell*, 6(12):1619–40, 1995.

574. C. L. Craig and C. Riekel. Comparative architecture of silks, fibrous proteins and their encoding genes in insects and spiders. *Comp. Biochem. Physiol. B Biochem. Mol. Biol.*, 133(4):493–507, 2002.

575. Z. Shao and F. Vollrath. Surprising strength of silkworm silk. *Nature*, 418(6899):741, 2002.

576. E. Howell. Future Soldiers May Wear Bulletproof Spider Silk. *Live Science*, 2014.

577. A. Scott. Spider Silk Poised For Commercial Entry. *Chem. Eng. News*, 92(9):24–27, 2014.

578. X. Hu, K. Vasanthavada, K. Kohler, S. McNary, A. M. Moore, and C. A. Vierra. Molecular mechanisms of spider silk. *Cell. Mol. Life Sci.*, 63(17):1986–99, 2006.

579. R. Moll, W. W. Franke, D. L. Schiller, B. Geiger, and R. Krepler. The catalog of human cytokeratins: patterns of expression in normal epithelia, tumors and cultured cells. *Cell*, 31(1):11–24, 1982.

580. J. E. Plowman. The proteomics of keratin proteins. *J. Chromatogr. B. Analyt. Technol. Biomed. Life Sci.*, 849(1–2):181–9, 2007.

581. K. Van Den Bossche, J. M. Naeyaert, and J. Lambert. The quest for the mechanism of melanin transfer. *Traffic*, 7(7):769–78, 2006.

582. Y. Horiguchi, D. Sawamura, R. Mori, H. Nakamura, K. Takahashi, and H. Shimizu. Clinical heterogeneity of 1649delG mutation in the tail domain of keratin 5: a Japanese family with epidermolysis bullosa simplex with mottled pigmentation. *J. Invest. Dermatol.*, 125(1):83–5, 2005.

583. S. Nobuhara, T. Idea, Y. Miyachi, and K. Takahashi. The head domain of keratin 5 binds to a dynein light chain, the cytoplasmic cargo complex, and might be involved in the distribution of keratin filaments and melanosomes. *J. Invest. Dermatol.*, 121:498, 2003.

584. H. R. Byers, S. Maheshwary, D. M. Amodeo, and S. G. Dykstra. Role of cytoplasmic dynein in perinuclear aggregation of phagocytosed melanosomes and supranuclear melanin cap formation in human keratinocytes. *J. Invest. Dermatol.*, 121(4):813–20, 2003.

585. B. T. Helfand, L. Chang, and R. D. Goldman. Intermediate filaments are dynamic and motile elements of cellular architecture. *J. Cell Sci.*, 117(Pt 2):133–41, 2004.

586. L. Alonso and E. Fuchs. The hair cycle. *J. Cell Sci.*, 119(Pt 3):391–3, 2006.

587. K. M. McGowan, X. Tong, E. Colucci-Guyon, F. Langa, C. Babinet, and P. A. Coulombe. Keratin 17 null mice exhibit age- and strain-dependent alopecia. *Genes Dev.*, 16(11):1412–22, 2002.

588. X. Tong and P. A. Coulombe. Keratin 17 modulates hair follicle cycling in a TNF-alpha-dependent fashion. *Genes Dev.*, 20(10):1353–64, 2006.

589. D. DePianto and P. A. Coulombe. Intermediate filaments and tissue repair. *Exp. Cell Res.*, 301(1):68–76, 2004.

590. S. Kim, P. Wong, and P. A. Coulombe. A keratin cytoskeletal protein regulates protein synthesis and epithelial cell growth. *Nature*, 441(7091):362–5, 2006.

591. T. Scheibel. Protein fibers as performance proteins: new technologies and applications. *Curr. Opin. Biotechnol.*, 16(4):427–33, 2005.

592. H. Lodish, A. Berk, C. A. Kaiser, M. Krieger, M. P. Scott, A. Bretscher, H. Ploegh, and P. Matsudaira. *Molecular Cell Biology*. W. H. Freeman and Company, New York, 6th edition, 2007.

593. D. A. Parry, S. V. Strelkov, P. Burkhard, U. Aebi, and H. Herrmann. Towards a molecular description of intermediate filament structure and assembly. *Exp. Cell Res.*, 313(10):2204–16, 2007.

594. N. Becker, E. Oroudjev, S. Mutz, J. P. Cleveland, P. K. Hansma, C. Y. Hayashi, D. E. Makarov, and H. G. Hansma. Molecular nanosprings in spider capture-silk threads. *Nat. Mater.*, 2(4):278–83, 2003.

595. D. A. Parry, R. D. Fraser, and J. M. Squire. Fifty years of coiled-coils and alpha-helical bundles: a close relationship between sequence and structure. *J. Struct. Biol.*, 163(3):258–69, 2008.

596. C. Cohen and K. C. Holmes. X-ray diffraction evidence for alpha-helical coiled-coils in native muscle. *J. Mol. Biol.*, 6:423–32, 1963.

597. F. H. Crick. Is alpha-keratin a coiled coil? *Nature*, 170(4334):882–3, 1952.

598. F. H. C. Crick. The Fourier transform of a coiled-coil. *Acta Cryst.*, 6:685–689, 1953.

599. F. H. C. Crick. The packing of α-helices: simple coiled-coils. *Acta Cryst.*, 6:689–697, 1953.

600. J. F. Conway and D. A. Parry. Structural features in the heptad substructure and longer range repeats of two-stranded alpha-fibrous proteins. *Int. J. Biol. Macromol.*, 12(5):328–34, 1990.

601. R. A. Kammerer, T. Schulthess, R. Landwehr, A. Lustig, J. Engel, U. Aebi, and M. O. Steinmetz. An autonomous folding unit mediates the assembly of two-stranded coiled coils. *Proc. Natl. Acad. Sci. USA*, 95(23):13419–24, 1998.

602. K. C. Wu, J. T. Bryan, M. I. Morasso, S. I. Jang, J. H. Lee, J. M. Yang, L. N. Marekov, D. A. Parry, and P. M. Steinert. Coiled-coil trigger motifs in the 1B and 2B rod domain segments are required for the stability of keratin intermediate filaments. *Mol. Biol. Cell*, 11(10):3539–58, 2000.

603. L. Kreplak, H. Bar, J. F. Leterrier, H. Herrmann, and U. Aebi. Exploring the mechanical behavior of single intermediate filaments. *J. Mol. Biol.*, 354(3):569–77, 2005.

604. W. T. Astbury and F. O. Bell. Themolecular structure of the fibers of the collagen group. *Nature*, 145:421–22, 1940.

605. G. N. Ramachandran and G. Kartha. Structure of collagen. *Nature*, 174(4423):269–70, 1954.

606. A. Rich and F. H. Crick. The molecular structure of collagen. *J. Mol. Biol.*, 3:483–506, 1961.

607. P. M. Cowan, S. McGavin, and A. C. North. The polypeptide chain configuration of collagen. *Nature*, 176(4492):1062–4, 1955.

608. A. Rich and F. H. Crick. The structure of collagen. *Nature*, 176(4489):915–6, 1955.

609. J. Bella and H. M. Berman. Crystallographic evidence for $C\alpha-H\cdots O=C$ hydrogen bonds in a collagen triple helix. *J. Mol. Biol.*, 264(4):734–42, 1996.

610. J. Bella, M. Eaton, B. Brodsky, and H. M. Berman. Crystal and molecular structure of a collagen-like peptide at 1.9 Å resolution. *Science*, 266(5182):75–81, 1994.

611. J. Emsley, C. G. Knight, R. W. Farndale, M. J. Barnes, and R. C. Liddington. Structural basis of collagen recognition by integrin $\alpha2\beta1$. *Cell*, 101(1):47–56, 2000.

612. D. J. Hulmes and A. Miller. Quasi-hexagonal molecular packing in collagen fibrils. *Nature*, 282(5741):878–80, 1979.

613. S. Perumal, O. Antipova, and J. P. Orgel. Collagen fibril architecture, domain organization, and triple-helical conformation govern its proteolysis. *Proc. Natl. Acad. Sci. USA*, 105:2824–2829, 2008.

614. J. P. Orgel, T. C. Irving, A. Miller, and T. J. Wess. Microfibrillar structure of type I collagen in situ. *Proc. Natl. Acad. Sci. USA*, 103(24):9001–5, 2006.

615. E. Leikina, M. V. Mertts, N. Kuznetsova, and S. Leikin. Type I collagen is thermally unstable at body temperature. *Proc. Natl. Acad. Sci. USA*, 99(3):1314–8, 2002.

616. J. P. Orgel, A. Miller, T. C. Irving, R. F. Fischetti, A. P. Hammersley, and T. J. Wess. The in situ supermolecular structure of type I collagen. *Structure*, 9(11):1061–9, 2001.

617. J. P. Orgel, T. J. Wess, and A. Miller. The in situ conformation and axial location of the intermolecular cross-linked non-helical telopeptides of type I collagen. *Structure*, 8(2):137–42, 2000.

618. D. J. Cram. The design of molecular hosts, guests, and their complexes. *Science*, 240(4853):760–7, 1988.

619. R. A. Berg and D. J. Prockop. The thermal transition of a non-hydroxylated form of collagen. Evidence for a role for hydroxyproline in stabilizing the triple-helix of collagen. *Biochem. Biophys. Res. Commun.*, 52(1):115–20, 1973.

620. S. Jimenez, M. Harsch, and J. Rosenbloom. Hydroxyproline stabilizes the triple helix of chick tendon collagen. *Biochem. Biophys. Res. Commun.*, 52(1):106–14, 1973.

621. J. Myllyharju and K. I. Kivirikko. Collagens, modifying enzymes and their mutations in humans, flies and worms. *Trends Genet.*, 20(1):33–43, 2004.

622. G. N. Ramachandran, M. Bansal, and R. S. Bhatnagar. A hypothesis on the role of hydroxyproline in stabilizing collagen structure. *Biochim. Biophys. Acta*, 322(1):166–71, 1973.

623. G. N. Ramachandran and R. Chandrasekharan. Interchain hydrogen bonds via bound water molecules in the collagen triple helix. *Biopolymers*, 6(11):1649–58, 1968.

624. J. D. Dunitz and R. Taylor. Organic fluorine hardly ever accepts hydrogen bonds. *Chem. Eur. J.*, 3:89–98, 1997.

625. S. K. Holmgren, L. E. Bretscher, K. M. Taylor, and R. T. Raines. A hyperstable collagen mimic. *Chem. Biol.*, 6(2):63–70, 1999.

626. S. K. Holmgren, K. M. Taylor, L. E. Bretscher, and R. T. Raines. Code for collagen's stability deciphered. *Nature*, 392(6677):666–7, 1998.

627. L. E. Bretscher, C. L. Jenkins, K. M. Taylor, M. L. DeRider, and R. T. Raines. Conformational Stability of Collagen Relies on a Stereoelectronic Effect. *J. Am. Chem. Soc.*, 123(4):777–778, 2001.

628. O. P. Boryskina, T. V. Bolbukh, M. A. Semenov, A. I. Gasan, and V .Y. Maleev. Energies of peptide-peptide and peptide-water hydrogen bonds in collagen: evidences from infrared spectroscopy, quartz piezo-gravimetry, and differential scanning calorimetry. *J. Mol. Struct.*, 827:1–10, 2007.

629. K. Reiser, R. J. McCormick, and R. B. Rucker. Enzymatic and nonenzymatic cross-linking of collagen and elastin. *FASEB J.*, 6(7):2439–49, 1992.

630. M. J. Buehler. Nature designs tough collagen: explaining the nanostructure of collagen fibrils. *Proc. Natl. Acad. Sci. USA*, 103(33):12285–90, 2006.

631. K. Kar, P. Amin, M. A. Bryan, A. V. Persikov, A. Mohs, Y. H. Wang, and B. Brodsky. Self-association of collagen triple helic peptides into higher order structures. *J. Biol. Chem.*, 281(44):33283–90, 2006.

632. R. Nelson, M. R. Sawaya, M. Balbirnie, A. O. Madsen, C. Riekel, R. Grothe, and D. Eisenberg. Structure of the cross-beta spine of amyloid-like fibrils. *Nature*, 435(7043):773–8, 2005.

633. C. A. Kim and J. M. Berg. Thermodynamic beta-sheet propensies measured using a zinc-finger host peptide. *Nature*, 362(6417):267–70, 1993.

634. D. L. Minor Jr and P. S. Kim. Measurement of the beta-sheet-forming propensities of amino acids. *Nature*, 367(6464):660–3, 1994.

635. M. Meydani, L. Zubik, and F-Y. Tang. Vitamin C Deficiency. In *Encyclopedia of Life Sciences*. John Wiley & Sons, Ltd., 2003.

636. J. Lind. *A Treatise on the Scurvy*. Millar, A., London, 1753.

637. M. Levine. New concepts in the biology and biochemistry of ascorbic acid. *N. Engl. J. Med.*, 314(14):892–902, 1986.

638. H. E. Sauberlich. Pharmacology of vitamin C. *Annu. Rev. Nutr.*, 14:371–91, 1994.

639. A. Strohle and A. Hahn. Vitamin C and immune function. *Med. Monatsschr. Pharm.*, 32(2):49–54; quiz 55–6, 2009.

640. T. Hazell. Vitamin C has a key physiological role in facilitating the absorption of non-heme iron from the diet. *Hum. Nutr. Appl. Nutr.*, 41(4):286–7, 1987.

CHAPTER 3

Methods of Structure Determination and Prediction

3.1 INTRODUCTION

The deterministic approach of Western science is quite old, and has been in practice since well before scientists could actually study molecules in detail. With the emergence of technological advances in the first half of the 20th century, methods for the structural characterization of small molecules and later macromolecules began to appear. However, the first structures of proteins were determined only in the 1960s (see Chapter 2 for details). In contrast to looking through the lenses of a microscope, the determination of protein structures turned out to be a lengthy and quite difficult task. This difficulty led scientists to prioritize the structures to be determined, mostly according to practical considerations, such as medical relevance. As a result, the structure of each protein was determined only after a great deal of experimental data (genetic, biochemical, immunological) had already been collected on its function, tissue localization, binding partners, etc. This approach began to change around the turn of the millennium, with the sequencing of whole animal genomes (including the human genome) and the considerable improvement of structure determination techniques.

Accordingly, an ambitious project called 'structural genomics' was initiated, aimed at determining the structures of all proteins coded by the human genome, including those of unknown function [1]. The initiative has been carried out by several bodies, including the National Institutes of Health (NIH) in the US [2], in addition to institutes in Japan, Canada, Israel, and Europe. Determining the structures of all human proteins in a period of several years seems unrealistic, as the number of these proteins is huge and the process is long and costly (see more below). Therefore, it has been decided that, initially, the structures of a representative set of proteins will be determined. This set comprises proteins whose folds collectively represent the complete 'fold space' observed in nature. This decision was based on the assumption that the number of unique folds in nature is much smaller than the number of proteins. Although the exact number of folds is yet to be determined (see Chapter 2, Subsection 2.4.2.2.4 for estimates), it is known that proteins tend to converge to similar structures [3], which supports the assumption. Once the structures of all proteins included in the chosen set have been determined, it will be possible to predict the structures of the remaining structures computationally, on the basis of sequence similarity (see Subsection 3.4.3 below for details on homology-based modeling) [4]. Indeed, many structural biologists agree with the underlying assumption that the entire protein fold space can be represented by a set of distinct structures [5]. A relevant question is whether protein space is discrete [6,7], continuous [8–10], or features a combination of discrete and continuous regions [11]. Macromolecules in general can be studied using a variety of methods (Figure 3.1), some of which emerged as early as the beginning of the 20th century. However, there are only a handful of methods used for the determination of full (or major parts of) protein structures. These can be roughly separated into two groups; the first includes methods that are based on the diffraction or scattering of either subatomic particles or electromagnetic waves. The second group includes spectroscopic methods, which rely on changes in the energy states of protein atoms; these changes take place as a result of the atoms' interaction with electromagnetic radiation of different frequencies. The first part of this chapter reviews these methods. Since this is a very wide topic, the discussion focuses on the principles of the methods, as well as on their main advantages and disadvantages. The second part reviews computational methods, some of which are used for protein structure prediction, whereas others are used as a means of optimizing experimentally-determined structures.

3.2 DIFFRACTION AND SCATTERING METHODS

Knowledge about the structure of molecules can be gathered by observing the way beams are diffracted or scattered when hitting three-dimensional crystals of the molecule. This approach has been used to study small molecules since 1915, and was later adapted for proteins and other macromolecules [13,14]. In the following subsection we discuss three such methods, each of which is based on the diffraction or scattering of a different particle.

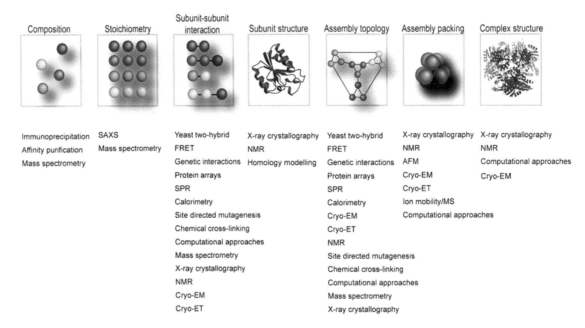

FIGURE 3.1 **Biophysical, biochemical and genetic methods used to characterize different levels of protein 3D structure.** AFM: atomic force microscopy; Cryo-ET: cryo-electron tomography; EM: electron microscopy; FRET: fluorescence resonance energy-transfer; NMR: nuclear magnetic resonance; SAXS: small-angle X-ray scattering; SPR: surface plasmon resonance. The image is taken from [12].

3.2.1   X-ray diffraction and scattering

3.2.1.1   Principles

X-ray diffraction is considered today the most accurate method for the structural determination of proteins. In this method, X-rays fired at a crystal of the investigated molecule are diffracted by the electron clouds of the atoms in the crystal, and form a unique pattern on a film or on another sensitive material (Figure 3.2a,b). The diffraction pattern is a collection of dots resembling the surface of a sieve (Figure 3.2b), but in fact it contains all the information needed for determining the structure of the molecule at high resolution. The conversion of the diffraction pattern into molecular structure is carried out mathematically, as well as by implementing physicochemical knowledge. A clear, yet detailed description of the method is given in Yaffe's excellent review paper [15]*1. In the following paragraphs we summarize the principles described in this review.

When we look at an object, we essentially see the light returned from its direction. The lenses of our eyes are capable of focusing the light beams, thus conferring sharpness to the image. However, this capability is limited to objects larger than a certain size. As a result, two points may look to us as one if the distance separating them is shorter than ~0.2 mm, i.e., beyond the *resolution* capability of our eyes. Therefore, in order to separate small objects, e.g., of micrometric size (such as biological cells), we use a microscope. The microscope contains lenses with a higher focusing capability than that of our eyes. In fact, given the appropriate lenses, the only factor limiting the resolution capability (r) is the wavelength (λ) of light (or any other type of electromagnetic radiation) reflected from the object we are observing:

$$r \sim \frac{\lambda}{2} \tag{3.1}$$

This explains why we cannot see separate molecules; the shortest wavelength of visible light (purple) is 4,000 Å, whereas the distance between atoms in molecules is 1 to 2 Å. In contrast, **X-rays have just the right wavelength for 'seeing' molecules** (λ = 0.1 to 100 Å). The use of X-rays solves the wavelength problem, but creates a new one; unlike visible light, X-rays cannot be focused by lenses. Consequently, there is no 'X-ray microscope' that can be used to view molecular structures directly. Instead, scientists use an indirect approach — they look at how the molecule under investigation diffracts X-rays. Another problem is that biomolecules tend to deflect X-rays rather weakly. This is because such molecules are made primarily of light atoms (C, N, O and H), which do not include many electrons. In principle, this problem could be solved by increasing the concentration of the molecule in solution, but this results in an incoherent signal, as the different individual molecules have different orientations in solution, and therefore scatter the rays in different directions. Thus, a real solution requires not only that there be many molecules in the sample, but also that they be arranged identically in three-dimensional space. This type of repetitive organization appears in crystals, which is why the method also called 'X-ray crystallography', requires the crystallization of the investigated molecule. It should be mentioned that low-resolution (10 to 12 nm) images of proteins can be obtained in solution by an X-ray-based method called *'small angle X-ray scattering'* (*SAXS*) [16], which is described in Subsection 3.2.1.5 below.

---

*1See also this video by the Diamond Light Source synchrotron facility:

3.2.1.2 Steps of procedure

1. **Isolation and purification.** The protein to be crystallized must be purified, for obvious reasons. The purification process is carried out gradually, using different biochemical methods. Often, the protein is modified prior to crystallization, in order to be more compatible with the process. Simple chemical modifications can be carried out post-translationally. For example, flexible segments of the polypeptide chain (e.g., loops) are often removed by proteases. Conversely, when the modification includes the replacement of an amino acid within the sequence of the protein, recombinant DNA techniques are used. Since crystallization requires large quantities of the protein, the latter is over-expressed using molecular techniques.

2. **Crystallization.** The crystallization process starts from a highly concentrated solution of the protein, and may last weeks to months until crystals of sufficient size (20 to 300 μm^3) are obtained. These crystals undergo treatment that reduces their sensitivity to the deleterious X-rays. For example, they may be soaked in solution and frozen to −140 °C.

3. **Irradiation and data collection.** An individual crystal is put in a rotating device, and the X-ray beam is turned on (Figure 3.2a). The diffracted beam hits a film or an X-ray-sensitive detector, producing a sieve-like pattern (Figure 3.2b).

4. **Diffraction pattern analysis.** Since the X-rays are diffracted by the outer electrons of the molecule, there is a relationship between the pattern of deflection and the density of these electrons. The relationship is described by a mathematical operation called *'Fourier transform'*. In other words, if we assume that the electron density around atoms is a mathematical function, the X-ray diffraction pattern is the Fourier transform of this function. This means that the electron density (and therefore the molecular structure) can be extracted from the diffraction pattern by inverting the Fourier transform. This, however, is where another problem appears, called *'the phase problem'* [17]. The rays diffracted by the molecule are electromagnetic waves, and as such they can be characterized by the combination of three quantities: *wavelength* (or frequency), *amplitude*, and *phase*. The physical meaning of these quantities can best be understood by analogy to the motion of a swing, which is also periodic and has a characteristic wavelength (or frequency). The motion can be described in terms of its amplitude, i.e., the (angular) length at the peak, and the time at the peak, i.e., the phase. Two swings may share the same amplitude but reach their peaks at different times if their phases are different. In the analysis of X-ray diffraction, the phase determines the fraction of the amplitude that the scattered wave marks on the detector. When the Fourier transform is inverted, **the amplitude of the waves provides information on the size of the atoms diffracting it, whereas the phase provides information on their relative location.** Since the X-rays are not recorded directly but rather indirectly (via the diffraction pattern), the collected data include only the intensity of the dots. **This intensity level is proportional to the square of the amplitude, but contains no information about its phase.** Fortunately, there are different ways to get the missing information. For example, in the *molecular replacement* method, the crystallographer relies on a known structure as a template to deduce the structure of the unknown protein. This method is particularly useful in pharmaceutical drug development, where

scientists look for the structure of a given protein in complex with different candidate drugs (see Chapter 8). In such cases, the already known structure of the apo-protein (i.e., the ligand-free structure) can be used to produce the phase for the protein's other structures, e.g., in complex with a compound. This approach has been used for the development of different drugs, including antibiotics and antihypertensives (drugs that are used to lower blood pressure). The obtained structural model of the protein is optimized in multiple iterations, until good agreement with the physical data is reached. In order to assess the agreement and make sure the process has not been overdone, the crystallographer calculates a parameter called an 'R-factor'. This parameter corresponds to the difference between the data measured and the data corresponding to the predicted structural model; a low R-factor indicates a good fit between the structural model and the physical data. In proteins, an R-factor range of 0.15 to 0.25 is considered satisfactory. The end result of the structural optimization process described above is an *electron density map* (Figure 3.2c). Simply put, this map presents graphically the locations in space where the density of electrons is high enough to indicate the presence of atoms. This information is important, albeit incomplete; in order to get a full description of the molecule, the crystallographer must also determine the atom type at each location. This is where the crystallographer's knowledge and experience come in handy. These allow him or her to interpret the electron density map in a way that yields a structural model of the molecule (Figure 3.2d). In the case of proteins, the crystallographer uses what is known about the properties of amino acids, as well as the sequence of the protein, to assign the right amino acid atom to each of the electron densities in the map. Also, to describe the atom locations and their bond properties accurately, the crystallographer uses energy- and geometry-based calculations. The chances of getting a correct structure depend on all of the aforementioned factors, but also on the protein's composition; ordered regions in the protein are much easier to determine compared to disordered or otherwise flexible regions (especially when loops are involved). The latter tend to have low electron densities, and in extreme cases do not register at all, and must therefore be defined as 'unresolved'.

3.2.1.3   Information obtained from crystallography

1. **Molecular structure.** The structure is usually deposited in a file, which specifies the type and location of each atom, in addition to the connectivity (see Section 3.6 below).

2. **Deviation.** As explained in Chapter 5, atoms are capable of absorbing heat energy from their environment and converting it into kinetic energy. This means that atoms are not static, but vibrate in space. The location obtained for each atom by the crystallography experiment is in fact its *average location*. In other words, it is the most popular location of the atom. However, the experiment also provides information on the vibration-induced deviations of each atom from its average location. This information is represented by the *B-factor*, also called the *'temperature factor'* [20,21]. Thus, the B-factor represents the uncertainty of atomic position per atom. B-factor values usually range between 20 and 80, where values lower than 45 indicate ordered atoms, and values higher than 60 indicate flexible atoms. The B-factor is often interpreted as a measure of atomic dynamic fluctuations, which is inaccurate, as it may also reflect

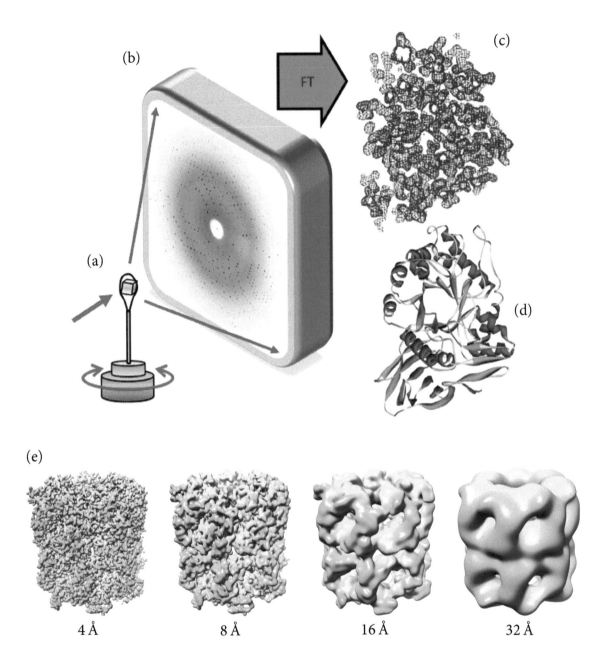

FIGURE 3.2 **Structure determination using X-ray diffraction.** (a) through (d) The structure determination process ([18]; the figure is reproduced from *Biomolecular Crystallography: Principles, Practice, and Application to Structural Biology* by Bernhard Rupp. Garland Science/Taylor & Francis LLC© 2011). The X-rays are passed through a rotating crystal (a) and the diffraction pattern is recorded (b). Since the diffraction images are not direct images of the molecule, the reciprocal space has to be back-transformed to direct space using the Fourier transform (FT). This, together with separately acquired phases for each diffraction spot, enable the crystallographer to reconstruct the electron density (blue grid) of the molecules self-assembled into the diffracting crystal (c). An atomic model of the structure is then built into the three-dimensional electron density (d). (e) Illustration of the structural features of proteins visible at resolution values between 4 Å and 32 Å. The structure shown is GroEL (PDB entry 1j4z). The image is taken from [19].

TABLE 3.1 **Protein structural features revealed at different resolutions.**

Resolution (Å)	Discernable Features
4.0–5.0	Protein shape General location of secondary structures
3.5	Outline (trace) of polypeptide chain
2.0–2.5	Amino acid side chains
1.0–1.2	Individual atoms, some hydrogen atoms (e.g., [26,27])
0.8	Hydrogen atoms[*a] Covalent bond valence

[*a]Hydrogen atom visibility can be used to determine the protonation states of ionizable residues, as well as the orientations of water molecules around (or inside) the protein.

static disorder in the crystal [22]. The latter represents an ensemble of conformational variations of the molecule, which exist in solution and are trapped in the crystal.

3. **Resolution.** In crystallography, this parameter refers to the smallest separation at which one can distinguish two separate atoms in the electron density map [15,23]. In proteins, the continuous range of resolutions can be separated into different groups, each of which reveals different features of the structure [23–25] (Table 3.1).

The features shown by lower resolutions are illustrated by Figure 3.2e. It should be noted, though, that in real structures these limits are not always as stringent. For example, it is common to find structures of transmembrane proteins in which individual amino acids are resolved despite the overall >2.5 Å resolution.

As of July 2017, only 11% of the structures in the Protein Data Bank (PDB)[*1] are high-resolution structures (≤1.5 Å[*2]). Moreover, less than 1% of the structures have a resolution of 1.0 Å or better, which allows for determination of the atomic locations of most hydrogen atoms in the protein. Knowing the location of these atoms is important, as it enables the protonation states of ionizable residues to be determined, as well as the orientations of protein-bound water molecules, some of which have catalytic or other structurally important roles. For structures with lower resolution, residues' protonation states can be predicted using either energy-based calculations or statistical data derived from pre-existing structures[*3] [25]. In the last decade, with the enhancement of crystallization techniques, as well as the development of more powerful sources of X-ray radiation, it has become possible to determine structures at very high resolution (e.g., [32–34]). Interestingly, inspection of some of the high-resolution structures shows that atoms' locations deviate by ~0.3 Å from their locations in the same structures, determined at a resolution of 2.0 Å. This value may seem negligible, but it may be significant for scientists working, e.g., in drug discovery. That is

---

[*1]The largest repository of protein structures (see Section 3.6 below).

[*2]Currently, the highest-resolution PDB structures are 3nir and 5d8v (0.48 Å in both cases).

[*3]Popular protonation-prediction web servers include MolProbity [28] (http://muscle.research.duhs.duke.edu/~rlab/), which uses the method Reduce [29], and PDB2PQR [30] (http://nbcr-222.ucsd.edu/pdb2pqr_2.0.0/), which uses the method PROPKA [31].

because this type of research focuses on protein-ligand interactions (see Chapter 8), which tend to be highly sensitive to the distance and angles between the interacting atoms [25].

When the crystallographic experiment provides a resolution of 0.8 Å or better, the data can be used to determine the distribution of electrons around atoms. This means that the chemical reactivity of each of the atoms in the molecule can be determined, at least in theory. This, in turn, should enable scientists to investigate chemical transformations involving protein atoms; until relatively recently, such studies were considered to be possible only for small molecules.

3.2.1.4 Problems of method

Crystallography is considered even today to be the most accurate way to determine the three-dimensional structures of macromolecules, including proteins. Yet, the method is associated with several significant difficulties and problems, which emerge at different stages of the process:

1. **Preparation of the protein.** The preparatory stage is perhaps the most difficult one. First, crystallization requires an accurate set of conditions, such as temperature, pH, salinity, etc. The optimal conditions change from one protein to another, and finding the right combination for a new system is usually carried out by trial and error, which takes time. Second, the need to overexpress the protein usually means working with organisms much simpler than those producing the protein naturally. As a consequence, many steps of the post-translational processing of the protein might be missing, which could result in misfolding, especially when the protein is not highly soluble. This problem is particularly pronounced in membrane-bound proteins, which undergo considerable modification. Third, before the protein is crystallized, its quality must be ascertained, which requires the use of more methods, e.g., *dynamic light scattering* [14]. Fourth, while any molecule should theoretically be able to undergo the crystallographic process [35], crystallization of large proteins is very difficult. Again, this problem is particularly pronounced in membrane proteins, which are surrounded by lipids. These molecules are flexible, and therefore do not tend to crystallize [36]. The solution is usually to replace the lipids with detergent molecules, but this creates other problems (see below). Nevertheless, various large proteins and protein complexes have been determined successfully in recent years using X-ray crystallography. These include the ribosome [37–40], ATP synthase [41–43], RNA polymerase II [44–47], and photosystem I [48–50]. Another trait of proteins making them difficult to crystallize is flexibility. The latter may result from structural factors, such as the presence of hinges between domains, or from chemical factors, such as the presence of numerous glycosyl or fatty groups attached to the surface of the protein [36].

2. **Data collection.** The inability to focus X-rays limits crystallographers to indirect data collection, i.e., measuring diffraction patterns. Since the collected data do not indicate the phase, this information must be obtained by using additional methods (see above). In addition, the high-energy X-rays tend to inflict damage on the investigated protein. The radiation damage can be reduced by combining cryogenic cooling of the crystals and the use of micron or sub-micron beams [51,52].

3. **Quality of the results.** One of the major problems of crystallographic structures (if not the biggest) is that they are determined under unnatural conditions. The protein is crystallized, which means it is no longer in solution, i.e., its natural environment. This is a potential problem, as biological macromolecules such as proteins have evolved to function in solution. The unnatural crystal environment may harm the structure in two ways. The first is compaction forces acting in the crystal, which may change the conformation of the protein to a non-native one. Indeed, while structures of the same proteins determined by X-ray and NMR are overall similar, there seem to be distinct differences between them, e.g., in contacts per residue and number of main chain hydrogen bonds [53,54]. The second problem is loss of the dynamic behavior characterizing the protein in solution (remember, the crystal structure is an average one). As we will see in Chapter 5, the inherent dynamics of a protein is highly important for its function. In membrane-bound proteins, the lipid environment is replaced by detergent molecules. Whereas currently-developed detergents are much better than older ones in their capacity to facilitate the extraction, purification, and crystallization of membrane proteins [51], the structural and chemical differences between a detergent micelle and the lipid bilayer may affect the native structure of the protein. Another type of problem regarding the reliability of crystal structures is the use of computational prediction methods. The resolution obtained by crystallography may be better than resolutions obtained by other methods, but it is still worse than $1.0\,\text{Å}$ in most cases. This means that crystallographers must use calculations to predict the locations of the atoms, bond valence, protonation state, and the orientation of water molecules. Such calculations are a potential source of errors.

As a result of the above factors, the overall success rate of structure determination by X-ray crystallography is only ~5% [55]. That is, if 20 proteins are selected for structure determination by X-ray crystallography, only one is expected to be successfully determined.

3.2.1.5 X-ray scattering

To obtain high-resolution structures of proteins, it is necessary to carry out X-ray experiments on crystals. However, when large protein complexes are studied, it is sometimes possible to construct a complex from high-resolution structures of its components, as long as their organization inside the complex is known (see Section 3.5 below). One method that can yield such information is SAXS, mentioned above [56–59]. In a typical SAXS experiment, the sample containing the protein in its natural medium (aqueous or lipid) is irradiated by a monochromatic beam of X-rays at a very low angle (up to 1°)[*1] (Figure 3.3a). The elastically scattered X-ray waves are registered by a detector, and their *intensities* (*I*) can be presented as a function of the *momentum transfer* (*q*)[*2] (Figure 3.3b, red circles). This characteristic profile is proportional to the scattering from a single particle averaged over all orientations, as well as to the solute's concentration [59]. In solution, the scattering is isotropic due to the random orientations of the protein particles. However, the use of two-dimensional detectors provides better statistical accuracy of the signal after radial averaging [60].

[*1] The scattering angle, 2θ, is the angle between the incident and scattered beams (Figure 3.3a). The experiment is also carried out on the blank solvent, and the resulting scattering is then subtracted from that recorded for the protein-containing solvent.

[*2] q, also called the magnitude of the *scattering vector*, is related directly to the scattering angle (2θ) by: $q = 4\pi \sin\theta/\lambda$ (where λ is the wavelength of the X-ray beam).

The resolution embodied in a SAXS profile is poor relative to that of an X-ray diffraction pattern or an electron microscopy density map (see below). Still, when the SAXS measurements are carried out on a mono-disperse solution of identical, non-interacting protein particles, the following parameters can be extracted directly from the scattering profile:

1. The molecular mass, volume, and specific surface area $(S)^{*1}$ of the hydrated protein. These can be used to estimate the possible oligomeric state of the protein and, in the case of a complex, its partial dissociation level.

2. The radius of gyration (R_g). This describes the mass distribution of a macromolecule around its center of gravity. An increase in this parameter is generally consistent with an opening of the macromolecule, whereas a decrease suggests compaction [57].

3. The overall shape of the protein at low resolution (~15 to 20 Å). The 'ab initio' three-dimensional shape is obtained in two principal steps. First, the scattering profile is converted into an approximate distribution of pairwise electron distances in the protein via a Fourier transform. Then, the shape is determined as the one that is consistent with the pair-distribution function [61]. Despite its low resolution, the obtained shape can be very useful; it can be used to validate high-resolution models of the protein by finding those that are consistent with it. In the case of a protein complex, the shape can also be used to guide the modeling of the complex's supra-molecular structure using high-resolution structures of the individual chains (see below).

Another way of using SAXS data to determine the structure of the protein is by comparing the measured scattering profile with the calculated (theoretical) profiles of different models [62]. That is, a theoretical SAXS profile is calculated for each of the existing models, and the one whose theoretical profile is most similar to the measured profile is chosen (Figure 3.3, black line). Since the theoretical profile can be calculated very quickly, this approach is useful for the screening of multiple concurrent models of the protein, particularly those that are predicted by computational methods (e.g., [63]).

Since the SAXS measurement is carried out on the non-crystallized protein in solution, the high degree of fluctuation in such media usually yields low-resolution data (10 to 50 Å). As mentioned above, at these resolutions, SAXS is only useful for finding the overall shapes of proteins or their assemblies, although such data may be combined with more structural data to obtain a high-resolution structure. It should be mentioned, however, that the accuracy of SAXS has been improving in recent years thanks to developments in both hardware (beam intensity, detectors) and data analysis software [58]. In addition, local regions of the protein can be better resolved by carrying out the measurements at higher angles ($\theta = 5°$ to $60°$, $q > 0.5\,\text{Å}^{-1}$), a method called 'wide-angle X-ray scattering' (WAXS). These experiments can provide important information on secondary structures in the protein. Also, WAXS is highly sensitive to small structural changes and can therefore be used for identifying structural similarities between proteins and for studying structural fluctuations (e.g., [66]).

In sum, SAXS offers the following important advantages [57,62]:

1. It does not require crystallization, and is therefore easy and fast. As such, SAXS can be used to guide computational structure prediction methods (see Subsection 3.5 below). Since crystallization is not required, SAXS can be used to characterize (i) highly flexible or unstructured regions in the protein that usually do not crystallize well [67], and (ii) protein assemblies that are too large for X-ray crystallography and NMR.

---

*1 S is the total surface area per unit of mass.

(a)

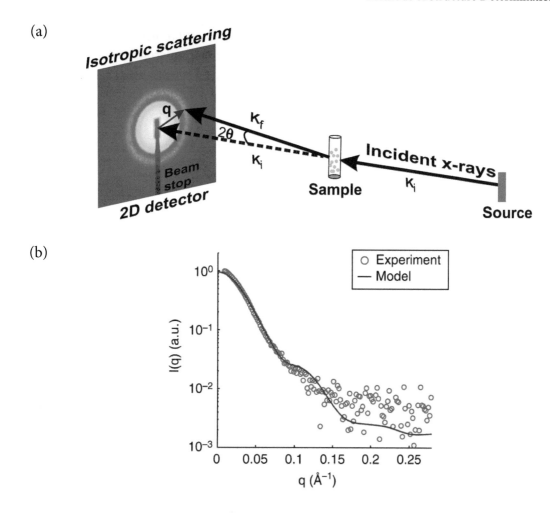

(b)

FIGURE 3.3 **Small-angle X-ray scattering (SAXS).** (a) In the SAXS experiment, 2θ is the angle between the incident and scattered beam, and q is the modulus of momentum transfer. Taken from Figure 3a in [64] (© IOP Publishing and Deutsche Physikalische Gesellschaft, CC BY-NC-SA). (b) A typical SAXS profile. Adapted from [65] (Figure 4d). The red circles are the measured intensities. At high q values, which correspond to smaller scattering angles, the measurements are particularly sensitive to structural details and changes in the sample. The black line is the calculated profile of a model structure.

2. It can be performed in the protein's natural environment, under a wide variety of conditions. Thus, it provides a more realistic view of the protein, which can be used to (*i*) identify artifacts in related X-ray crystal structures, (*ii*) determine the biologically relevant state(s) (e.g., [68]), (*iii*) characterize the conformational ensemble of the protein[*1], and (*iv*) follow discrete dynamic events in the protein, such as conformational changes (e.g., [69]), folding and unfolding (e.g., [70]), oligomerization, and fibril formation. These events are followed by employing *time-resolved SAXS*.

[*1]Characterization of the protein's conformational ensemble using SAXS data is far from being trivial, since a handful of conformations must be chosen out of thousands, and their fit to the measured data must be calculated. However, several approaches have been developed for solving these problems. For example, minimal ensemble search (MES) is one of the approaches used to choose the most probable ensemble, using a genetic algorithm.

3. The measurement is very rapid, taking between minutes and hours. This time range is even shorter than the time range required for cryo-electron microscopy, another popular method for obtaining low-resolution structures and for guiding structure prediction (see below).

4. It is sensitive to the macromolecule's hydration shell (e.g., [71]). This facilitates investigation of protein hydration in physiological solution, as well as investigation of the involvement of such hydration in stabilization and/or destabilization of the protein (e.g., by denaturants).

5. The experiment can be designed to track ligand-binding events (e.g., [72]).

6. The experiment requires minimal amounts of sample (15 μL at 0.1 to 1 mg/mL).

3.2.2 Neutron scattering

3.2.2.1 Principles

X-ray crystallography, which we discussed earlier, produces molecular structures of higher resolution than those determined by other methods. However, in most of these structures, the locations of hydrogen atoms are unresolved, owing to the very low electron density of this atom. As explained, the protonation states of amino acids, as well as the orientation of water molecules, may influence protein function significantly. Therefore, efforts have been made to develop methods for accurately determining the location of hydrogen atoms in the structure. Neutrons are just about right for this task, for three reasons. First, their wavelength (~0.1 to 10 Å) is of the same order as that of the atomic separation in molecules. Second, they penetrate deeply into the sample molecule. This is because, being electrically uncharged, these particles do not interact with the electron clouds occupying most of the space between atoms, but rather with the small nuclei [13,73–75]. Third, neutrons have low energy content and therefore do not damage the sample. But, most importantly, **unlike X-rays, neutrons interact with the nuclei of atoms, rather than with their electron shells.** As a result, their interaction with an atom does not depend strongly on the size of the atom, but rather on its nuclear mass, spin (see Subsection 3.3.1 below), and isotopic type [73]. This means that **a strong signal can be obtained even from the interaction of neutrons with small hydrogen atoms, especially when their common isotope *protium* (^1H) is replaced with the heavier *deuterium* (^2H) (see more below).** Structure determination using neutrons is similar to the process described for X-rays; it includes firing a particle stream at the sample and analyzing the scattering pattern. Two types of scattering may be observed, depending on the net spin of the nucleus concerned. *Coherent scattering* results from correlated interactions of neutrons with atomic nuclei, and provides information on the structure of the molecule. *Incoherent scattering* results from spatially isotropic interaction of the neutrons with each nucleus, and provides information on the dynamics of the molecule. The two types are described in the following subsections.

3.2.2.1.1 *Coherent scattering*

Coherent scattering may result from both elastic and inelastic collisions of neutrons with the nuclei of atoms. In elastic collisions the scattering occurs without changing the energy of the neutrons, whereas in inelastic collisions both the neutron's energy and its spin are affected. Furthermore, in elastic scattering the intensity of the scattering points to the relative location of the deflecting atom (albeit in a probabilistic manner) [76]. Thus, **by analyzing the scattering pattern computationally, the location of each atom in the molecule can be obtained**. Another advantage of neutron scattering is that it enables the protein structure to be determined in its natural environment. This feature, which yields a protein structure that is biologically more realistic than the crystal structure, results from the fact that scattering is sensitive to the type of isotope of the atom that is hit by the neutrons. For example, protium (^1H) scatters neutrons differently than the heavier deuterium (^2H, also marked as 'D') [13]. Changing the D_2O/H_2O ratio in the sample enables the contrast between the protein and its environment to be changed, thus allowing the scientist to focus on each separately. This technique is particularly useful in membrane-bound proteins, whose lipid environment makes them difficult to crystallize. Setting a D_2O/H_2O ratio of 10% to 14% renders the lipid component of the system invisible in the neutron scattering experiment, and a clear structure of the protein in its natural environment can be obtained [73]. In fact, this can be done even when the protein is surrounded by detergent [74], despite its highly disordered nature (e.g., [77]). The determination of hydrogen atoms' location also enables structural scientists to determine the solvent density at different distances from the protein (Figure 3.4), as well as the specific interactions between protein residues and membrane glycolipids (in membrane-bound proteins) [78]. The use of coherent neutron scattering to collect data on the structure of a protein can be done in two ways [73,76]:

1. Neutron diffraction. Like X-ray diffraction, this method uses crystals of the protein (individual, or powder crystals) to determine its structure at high resolution.

2. Small-angle neutron scattering (SANS). This method is used for structural determination of large proteins or protein complexes in solution [16], but at low resolution (10 to 12 nm). While it cannot provide accurate structures of single proteins, the method is very helpful in determining the relative locations of subunits within large complexes (e.g., the ribosome [79]). This is done by selectively labeling the subunits with deuterium.

3.2.2.1.2 Incoherent scattering

Incoherent scattering results from the interaction of neutrons with individual nuclei that are in motion. These collisions have varying degrees of elasticity, with each degree of elasticity corresponding to a different set of motions (see Chapter 5 for more details on types of motions in proteins) [73,76]:

1. Elastic (EINS) – results from atomic vibrations in the 10^{-13}–10^{-9} s time-range, particularly those of hydrogen atoms.

2. Quasi-elastic (QINS) – results from diffusive motions or shifts between near-native conformations.

3. Inelastic (IINS) – results from vibrational or rotational transitions between different energy levels.

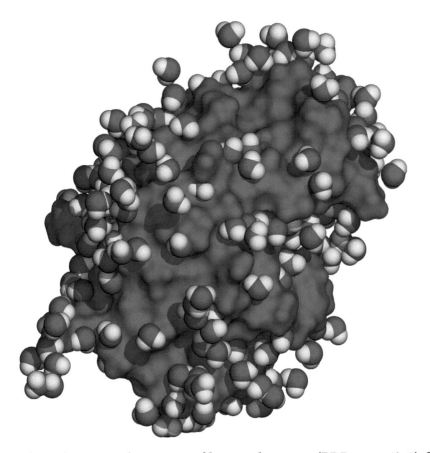

FIGURE 3.4 **Three-dimensional structure of hen egg lysozyme (PDB entry 1io5) determined by neutron quasi-Laue diffraction at 2 Å resolution.** The protein is shown in blue surface representation, and the surrounding (deuterated) water molecules are shown as spheres colored by atom type. Unlike X-ray diffraction, neutron diffraction is able to resolve the hydrogen atoms in the sample, thus revealing the location and orientation of the protein's hydration shell. The image shows the first hydration shell of the protein. These solvent molecules are very close to the protein surface, and display a behavior quite different from that of bulk solvent molecules (see Chapter 5 for details).

Evidently, the more dynamic the protein atoms, the less elastic the neutron-nucleus collision. Indeed, quasi-elastic neutron scattering and inelastic scattering are used often in the study of protein dynamics. In such neutron-nucleus interactions, both the energy and the momentum (spin) of the neutrons change, which allows scientists to follow the atomic motions in the protein.

3.2.2.2 Advantages and shortcomings

Earlier we discussed the main advantages of neutron scattering, including its capacity to determine the location of hydrogen atoms in proteins, to discern the protein from solvent, and to map the general location of protein components in large complexes. In addition, neutron radiation is relatively harmless to biological samples (compared to X-radiation) [80]. Finally, neutrons have a *magnetic moment*, and are therefore scattered also by magnetic fields within the protein, such as the field that emanates from unpaired electrons. This means that (elastic) **neutron scattering can also be used for studying the distribution of electrons in certain atoms.** The method has a few shortcomings. First, creating neutrons requires great effort; it is done in nuclear reactors or by initiated collisions between atomic nuclei and energetic radiation particles (*spallation*). In any case, the procedure requires larger installations than those used in the production of X-rays. Second, when neutron diffraction is used, the required crystals are larger than those required for X-ray diffraction, and the measurements require longer exposure time. Due to these problems, only ~100 structures in the protein data bank (PDB) have been solved by neutron scattering (0.08%).

3.2.3   Electron microscopy (EM)

3.2.3.1   Principles

Like X-rays and neutrons, electrons can be used to study the structure of molecules, including proteins and other macromolecules[*1]. Moreover, **the negatively charged electrons can be focused, using electromagnetic lenses, which allow scientists to observe them directly under the (electron) microscope**. This is a huge advantage of EM over X-ray and neutron diffraction, which can only be used to study molecules indirectly, by observing their diffraction patterns. Furthermore, **EM methods can image molecules directly as single particles**, which means that three-dimensional crystals of the sample protein are not required for the measurement [82–84]. In the measurements, which are done using a transmission electron microscope (TEM, see Figure 3.5a) in vacuum[*2], images of individual assemblies are either recorded on film or (more recently) captured digitally by direct detection device (DDD) cameras [86–88]. The images obtained by the electron microscope are two-dimensional and represent the sum of the density along the beam path [81]. In order to obtain the 3D structure of the sample molecule, 10^3 to 10^6 images of many particles in random orientations have to be recorded and then computationally aligned and merged [84]. This process also performs *averaging* of the images obtained for different particles. Such averaging decreases the noise and therefore improves the resolution of the final images.

Despite the advantages of direct EM measurements over measurements carried out with X-ray crystallography, until several years ago EM was able to produce only low-resolution structures, which ranged between 4 and 20 Å (Figure 3.5b[*3])[*4]. This limitation resulted from the following factors:

- The lower energy of electrons compared to X-rays

- The need to use low-dose beams to minimize radiation damage to the protein

- The movement of the sample within the electron beam

As a result, **EM has traditionally been used mostly for delineating the overall structures of proteins, which are then enhanced by using X-ray diffraction or NMR data to obtain**

[*1] In the transmission electron microscope, where electrons are accelerated, their wavelength is smaller than 0.04 Å [81]. Thus, electrons are, at least theoretically, suitable for investigating molecular structures at high resolution.

[*2] The high vacuum used in electron microscopes ensures that the electrons, which have little mass, are not scattered by air molecules [85].

[*3] See also Figure 3.2e for illustration of the structural features visible at this resolution range.

[*4] Note that in contrast to X-ray crystallography, where the accuracy of the density map can be assessed objectively (e.g., by the R-factor), in single-particle EM there is no such objective quality criterion that is simple and easy to use [83]. Instead, other, statistical measures are used [89]. As a result, determining the resolution of single-particle EM structures is a controversial issue.

higher-resolution models [90]. This approach was particularly useful in studying transmembrane proteins and large protein complexes, which could not be determined by more accurate methods (see Subsection 3.2.1 above). In the case of complexes, individual parts were determined separately at high resolution, and then attached to each other using the general constraints dictated by the EM data.

As we saw in X-ray crystallography, one way of improving image resolution is to crystallize the sample protein and measure the diffraction of particles by the crystals. This approach is based on the premise that crystals have an inherent periodicity that improves the signal. The same approach can be used in EM, and indeed, to obtain structures with better resolution, microscopists have turned to *electron crystallography* [91,92], obtaining images by measuring the diffraction of electrons from the crystallized protein. However, unlike X-ray crystallography, which uses highly ordered three-dimensional crystals, electron crystallography traditionally uses two-dimensional crystals of the sample protein. Two-dimensional crystals form much more readily than 3D crystals do, and despite the fact that they are relatively disordered and include only one or two layers, they can be used to obtain high-resolution images. Such crystals could not be used by X-ray crystallography because the X-rays, which interact with atoms about 10^4 times less strongly than electrons, would just pass through the crystals [93]. In an EM experiment, structural analysis of the sample is first carried out from a top view, which provides a two-dimensional map of the protein. Then, the data are collected again, with the sample tilted up to 60° to 70° to produce a three-dimensional map [94]. These raw data are Fourier-transformed to yield a structural map of the protein.

Until a few years ago, electron crystallography was virtually the only way to obtain high-resolution EM structures of proteins, reaching atomic resolutions (e.g., Figure 3.5cI). This approach was especially popular for solving structures of membrane proteins, since the 2D crystals readily form in the membrane environment [95]. Two famous 3D structures of membrane proteins solved with this method are those of *bacteriorhodopsin* (3.5 Å) [96] and the *light-harvesting complex II* (3.4 Å) [97]. In another structure, that of aquaporin-0, an atomic resolution of 1.9 Å was achieved, and the packing and interactions of the lipids surrounding the protein were described [98]. The electron crystallography field has recently advanced with the development of the *MicroED* technique, in which high-resolution structures are obtained by using three-dimensional micro-crystals [99,100] (e.g., Figure 3.5cI, left). As of July 2017, there are 65 structures in the PDB that have been determined by electron crystallography, 28 of which have a resolution that is equal to or better than 3 Å (e.g., [98,100,101]). Another repository of EM-determined structures is *EMDataBank*[*1], which also accepts low-resolution structures.

In recent years the exclusivity of electron crystallography in providing high-resolution EM structures of proteins has started to diminish, due to the development of better technologies for the direct recording and analysis of proteins in their natural environments [84]. These technologies include highly efficient electron detectors (DDD cameras [86–88]) and automated image processing tools [102], which correct for sample movements and classify images according to different structural states [86,103]. One known source of noise in EM measurements is beam-induced motion of the sample. The capacity of the new cameras to record the electron beam over time enables scientists to computationally correct for this effect by aligning the frames [87,104]. **These developments have enabled high-resolution structures**

[*1]http://emdatabank.org/

to be determined from direct, single-particle measurements [84,105]. Indeed, structures with atomic resolutions of 2.2 Å [106], 2.6 Å [107] (Figure 3.5cII), 2.8 Å [108], and 2.9 Å [109,110] are currently available (see also [103] for near-atomic resolution structures (3 to 4 Å)). Obviously, the added effect of direct EM is the largest for proteins that are difficult to crystallize. In addition, single-particle EM can use samples that are smaller and less pure than those required by crystallography [105].

As discussed earlier, one of the concerns in EM measurements is radiation damage to the sample, which results from inelastic electron scattering that causes breakage of chemical bonds. It is possible to reduce this damage by minimizing the electron dose used on the sample; however, doing so also increases the noise in the recorded micrographs. Another EM-associated problem in direct measurements is drying up of the sample due to the vacuum inside the microscope. To prevent both problems, it is necessary to treat the sample in a certain way before inspecting it under the microscope [111]. In the past, electron microscopists relied mainly on a treatment called *negative staining* [112,113], in which the sample was immersed in a layer of heavy metal, and the measurement was carried out on the metallic impression. Since the metal is more resilient to radiation than organic material, it was possible to use a stronger beam of electrons, producing better contrast. However, since the metal could not penetrate the protein fully, the template included only the surface of the protein. Today, microscopists commonly use a different approach, *cryogenics* (freezing). The corresponding method, called *cryo-electron microscopy*, involves quick freezing of the sample in liquid ethane, nitrogen, or helium [83,114–116]. The freezing process creates a layer of vitreous ice that is almost identical in structure to liquid water [81]. Thus, despite the very low temperature (~ − 200 °C), the environment of the protein in the sample is still considered to be more natural than in crystals. In order to prevent the formation of ice crystals, the microscopist usually adds sugar molecules (e.g., glucose) to the sample; these molecules replace the water molecules. In addition, the fragile sample must be supported by a surface that is strong enough yet penetrable by electrons (usually a carbon film). While cryo-EM provides a better resolution than negative staining, it has lower contrast, which is probably the reason why negative staining is still in use [105]. EM has additional uses in protein science besides determining three-dimensional structures. One such application is studying protonation and ionization states. This is possible with EM because the electrons are charged, and their scattering should therefore depend on the charge of the atoms they hit. Although this approach cannot currently be used on entire proteins, small regions can be targeted and analyzed. Another possible application of EM is studying protein dynamics, which is made possible by the fact that the investigated proteins are in solution, and therefore retain their inherent dynamic behavior. One way to study protein dynamics using EM is to freeze the sample in less than a millisecond, which enables intermediate states to be captured. It is possible to use this procedure to target functional states of the protein, by spraying the protein with its natural ligand shortly before freezing[*1]. A study of the ribosome is one known example of the use of EM to investigate conformational changes [117]. In this study, single-particle cryo-EM was used to follow the structures of the ribosome during the four stages of translation: initiation, elongation, termination, and recycling.

[*1] As explained in Chapter 5, folded proteins are in equilibrium between several different conformations. Ligand binding shifts the equilibrium toward certain conformations.

(a)

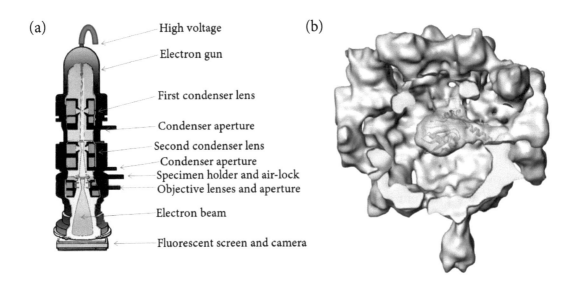

High voltage

Electron gun

First condenser lens

Condenser aperture

Second condenser lens

Condenser aperture

Specimen holder and air-lock

Objective lenses and aperture

Electron beam

Fluorescent screen and camera

(b)

(c) I

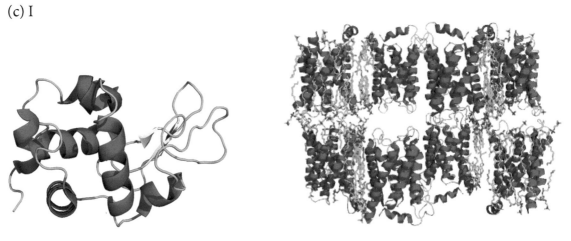

(c) II

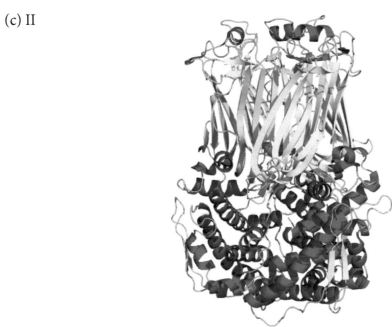

3.2.3.2 Advantages and shortcomings

In sum, EM offers some important advantages [111,120]:

- Unlike X-ray crystallography and neutron scattering, it enables the molecule to be observed directly. Direct observation is advantageous, as it provides both the amplitude and phase of the waves.

- It does not (necessarily) require the preparation of crystals, with the exception of two-dimensional crystals and three-dimensional micro-crystals used in electron crystallography [91,99,100].

- Recent progress in the field makes it possible to obtain structures at near-atomic- and even atomic resolution [83,84,103]. As will be explained in Section 3.5 below, near-atomic structures produced by EM can be further enhanced either through recognition of known folds of domains or motifs that fit the EM density maps, or through use of the latter as constraints that guide computational modeling tools [121].

- There is practically no upper limit size for EM[*1], such that microscopists can use the approach to study large protein complexes that are usually unavailable for X-ray crystallography.

- It does not require large quantities of the protein.

Some of the traditional problems of EM, such as image resolution and sample preparation, have been solved at least partially by the recent technological advances and automation techniques described above. However, major issues still remain. In particular, EM instrumentation is very costly and can reach many millions of dollars, far beyond the grasp of individual

FIGURE 3.5 **Electron microscopy.** (Opposite) (a) The transmission electron microscope. The electron beam generated by the electron gun is focused onto the specimen by electromagnetic lenses [85]. The deflected electrons hit a photographic film or an electron-sensitive camera, which creates an image. In protein structure-determination applications of EM, numerous images are recorded and processed (not shown). The figure was created by Graham Beards [118]. (b) Low-resolution EM structures of large protein complexes. The image shows a cutaway view through the density map of the spliceosomal protein p14 (gray). The yellow part in the middle of the structure is a peptide derived from the p14-associated U2 snRNP component SF3b155, for which an X-ray crystal structure also exists (see ribbon inside the density map). The figure was taken from [119]. (c) High-resolution structures obtained by two EM methods. (I) Electron crystallography. *Left*: A 2.5 Å structure of a lysozyme obtained by using the microED technique (PDB entry 5a3e [99]). *Right*: The 1.9 Å assembly of lens aquaporin-0 (AQP0) inside a lipid bilayer, obtained by using 2D crystals (PDB entry 2b6o [98]). In the assembly, packing interactions between AQP0 tetramers (shown as ribbons) are mediated by lipid molecules (shown as green wires). (II) Single-particle EM. The image shows the 2.6 Å structure of rotavirus VP6 protein (PDB entry 3j9s [107]).

[*1] In contrast to crystallography, which becomes easier the smaller the structure is, EM becomes easier the larger the structure is [81]. This is because in small structures, especially those that possess low symmetry, errors in the alignment of the projection images make the analysis more challenging. Theoretically speaking, it is possible to apply cryo-EM to any protein larger than ~100 kDa to obtain a high-resolution structure [93], but the technique is usually applied to proteins larger than ~200 kDa [59].

investigators, and most universities. The maintenance of EM microscopes is also expensive and demanding. One solution for these problems is the establishment of national and international facilities that share the costs. These are starting to emerge, e.g., in the US, UK, Germany, and Sweden.

3.3 SPECTROSCOPIC METHODS

Spectroscopy is a field that includes different analytic methods, all based on the interaction of a molecule with electromagnetic radiation, followed by changes in the energy levels of the former [122] (excitation, relaxation, etc.). These changes lead to emission of energy from the irradiated molecule, which can be recorded by the instrumentation. Since the magnitude and frequency of the emission usually depend on the local chemical environment of the emitting atoms, such data can be used to derive structural information related to the investigated molecule. This is particularly true in macromolecules, in which the chemical environment of each atom is diverse and can change dramatically as a result of biologically-relevant processes such as folding, ligand binding, etc.

A number of spectroscopic methods are available for the study of proteins; each is based on the use of electromagnetic radiation of a different frequency. For example, *nuclear magnetic resonance (NMR) spectroscopy* employs radio waves, *electron paramagnetic resonance (EPR) spectroscopy* employs microwaves, and *Fourier transform infrared (FTIR)* and *Raman spectroscopies* use infrared radiation. Although there are several methods in use to determine either the global or partial structures of proteins, NMR and EPR spectroscopies are the most common. We therefore focus on these methods. The principles of the two methods are described in the following subsections, with emphasis on NMR, the most efficient spectroscopic method for determination of global protein structure.

3.3.1 Nuclear magnetic resonance (NMR) spectroscopy

3.3.1.1 Principles

NMR is based on the behavior of atoms with magnetic properties, exposed to an externally applied magnetic field [123–125]. Certain atoms are said to have a *'nuclear spin'*, which results from the motion of charges in their nucleus. The motion creates a magnetic field around the nucleus that has a moment (μ) proportional to the spin. Not all nuclei have spin; those with an equal number of protons and neutrons usually have zero spin (e.g., the common isotopes: ^{12}C, ^{14}N and ^{16}O), whereas others have a spin of $\frac{1}{2}$ (^{1}H, ^{13}C, ^{15}N, ^{19}F and ^{31}P). When an atom with a nuclear spin is placed within a strong external magnetic field, it tends to realign spatially in one of two orientations. The first is parallel to the field lines and has low energy, whereas the second is 180° opposite (anti-parallel), and has higher energy. In a macromolecule such as a protein, most nuclei will be in the low-energy alignment. The imbalance between the parallel and anti-parallel magnetic moments will create a small polarization of the spins, resulting in a net macroscopic magnetization. It is possible to destabilize this equilibrium state by firing an energy pulse toward the nuclei[*1] (*excitation*); the magnitude of the pulse should be equal to the energy difference between the parallel and anti-parallel states. For most atoms this difference is smaller than 0.1 cal/mol, and has a

---

[*1] An oscillating magnetic field.

frequency within the radio range (20 to 900 MHz)[*1]. The energy pulse causes some of the nuclei to invert their spins from one energy state to the other. Since the inversion occurs in both directions, it is a resonance process. It is very fast (100 µs), and results in the return of the nuclei to their original states while emitting radio waves (*decay* or *relaxation*) that are sensed and amplified by the NMR instrument. The most basic NMR experiment, called '*1D NMR*', provides for each nucleus in the sample the frequency that was required for changing its spin. This information is shown in the output as a single peak per nucleus at the corresponding frequency (Figure 3.6a). In the case of macromolecules, the sample contains many nuclei, such that the NMR output contains numerous peaks, which form a jagged plot extending along the frequency scale.

Each nucleus has its own unique resonance spectrum. Therefore, given the strength of the external field, the excitation frequency, and the type of nucleus (e.g., ^1H), the signal emitted by a certain isolated nucleus can be identified on the measurement output. However, when the atom containing this nucleus is part of a molecule, i.e., surrounded by other nuclei, the signal it emits is slightly different from the one emitted in the isolated state. This phenomenon, called '*chemical shift*', results from the masking of the nucleus by the electron clouds of adjacent atoms. For example, two identical protons will emit signals of different frequencies if one is part of an amide group whereas the other is part of a methyl group. Such differences drive the capacity of the NMR method to determine the structure of molecules: By using the known chemical shifts of nuclei in different chemical environments, it is possible to decipher the structure of the protein from its NMR spectrum. The NMR peaks are assigned to the various protein nuclei on the basis of correlations identified between nuclei (*couplings*), which indicate their proximity to each other. Since ^1H has the strongest magnetic moment among all atoms having a spin value of $\frac{1}{2}$, it is customary to focus on its spectrum (*homonuclear NMR*). However in certain cases, ^{13}C or ^{15}N may be used after being introduced into the protein as isotopic labels, and the couplings between these nuclei and ^1H are inspected (*heteronuclear NMR*). NMR spectroscopists focus on two types of couplings between protein nuclei [124]:

1. J-coupling – This coupling exists between covalently bonded nuclei (up to three) and results from indirect interaction between the nuclei via polarization of bond electrons. The measurement used to locate this coupling (*COSY*) reveals the connectivity of atoms within the molecule.

2. Nuclear Overhauser effect (NOE) – this type of coupling occurs between proximal (less than 5 Å) non-bonded nuclei. It results from direct interaction between the magnetic dipoles of the nuclei. The measurements used to locate this coupling ('*NOESY*') reveal the tertiary structure of the molecule, which determines the proximity between non-bonded atoms. In fact, NOESY measurements do not provide a single location for each atom, but rather a range of possible locations. In other words, the measurements provide spatial constraints for each atom. Thus, **NMR does not provide a single structure, but rather an ensemble of structures that are consistent with the spatial constraints** (Figure 3.6b). A good ensemble samples the complete conformational space allowed by the constraints, **which makes it a potential source of information regarding the dynamics of the molecule**. In proteins, these data correspond to folded-state dynamics (see Chapter 5). Sometimes the ensemble is reported

---

[*1]The energy (E) is proportional to the frequency (v), through Planck's constant (h): $E = hv$.

along with a structure created by averaging the individual conformations composing it. Such average structures often contain deformations and improbable bond angles, as the averaging is carried out on each atom separately. One acceptable alternative is to report the conformation that is closest to the average.

As explained, the NMR measurements rely on the difference between the number of low-energy spins and the number of high-energy spins. Unfortunately, such differences are typically quite small, which leads to low sensitivity of the method. The method's sensitivity can be boosted in (at least) two ways:

1. Increasing the magnetic field.

2. Increasing the number of atoms in the sample. This is why NMR requires concentrated solutions of at least ~1 mM.

Still, the output of NMR measurements in macromolecules contains hundreds, sometimes thousands of resonance peaks, resulting from the numerous nuclei. In order to separate the peaks, the measurement can be carried out repetitively, where the time difference between the consecutive measurements allows adjacent nuclei to affect each other and form a correlation (*homonuclear 2D-COSY/NOESY* [124]). Indeed, this technique efficiently separates overlapping peaks in the 1D-NMR output, and provides good structure determination in small proteins (~10 kDa). In larger proteins, higher dimensions (3D and 4D-NMR) may be required. Although carrying out the measurements in several dimensions improves peak separation, it necessitates the use of complicated protocols that include series of radio pulses, as well as numerous repetitions. An alternative way to improve the measurements is to label the protein with ^{13}C or ^{15}N and measure the effect of these nuclei on ^{1}H nuclei (*heteronuclear 2D TOCSY/NOESY*). Finally, in proteins larger than 30 kDa it is customary to replace the ^{1}H atoms in the sample with ^{2}H.

3.3.1.2 Steps in protein structure determination by NMR spectroscopy

1. **Preparation of protein solution.** The solution on which the NMR experiment is carried out has to be prepared carefully to have the right pH, ionic strength, and temperature [124].

2. **NMR measurement.** The measurement may be conducted in several different ways. For example, in the *continuous wave* technique, the sample is put inside a glass tube and placed on a rotating surface between the two poles of the magnet. The system also contains two coils; one transmits radio waves at the right frequency towards the sample, whereas the other, which surrounds the glass tube, receives the energy emitted by the sample as a result of its excitation and relaxation. Also, the measurement may be conducted by changing the magnetic field within a narrow range while following the emitted radio signals, or by keeping the field constant and changing the radio pulses fired towards the sample.

3. **Signal assignment.** This is done according to the data obtained from the J-coupling analysis, and by using pre-existing knowledge about the protein sequence.

4. **Determining atomic distance constraints.** This is done using the NOE results.

(a)

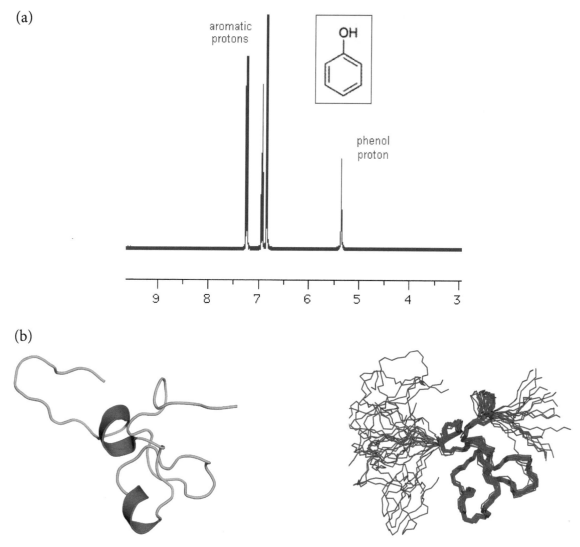

FIGURE 3.6   **Nuclear magnetic resonance spectroscopy.** (a) One-dimensional H-NMR spectrum map of phenol (taken from [126]). The different peaks in the spectrum correspond to different proton types in the molecule, as noted. For example, the peak of the hydroxyl proton is well separated from the one corresponding to the aromatic ring protons. (b) NMR structure. The figure shows the secondary structure of the somatomedin B domain of human vitronectin (PDB entry 1ssu) (left) and the corresponding 20-mer ensemble (right). All the structures of the ensemble are within the distance constraints indicated by the NMR measurements. Clearly, the central part of the structure is much more ordered than the flanking edges.

5. **Structure determination.** This is done by converting the constraints extracted directly from the NMR measurement into structural details. The procedure requires the use of calculations, which apply additional considerations to the experimentally derived data [124]. The calculations are as follows:

   (a) *Geometry-based calculations* – These calculations use pre-existing data regarding the size of atoms, as well as allowed distances and angles of the covalent bonds linking them, to determine the probability of the experimental structures. The calculations are carried out numerous times while the structures are modified, until they are consistent with the theoretical constraints.

(b) *Energy-based calculations* – The potential energy of the experimental structure is calculated, and is used to determine the structure's probability.

3.3.1.3 Advantages and shortcomings

The greatest advantage of NMR spectroscopy is that the protein structure is determined in its native environment. This is important for cytoplasmic proteins, as well as for proteins that exist in extracellular body fluids, such as blood, saliva, lymph, and GI fluids. The ability to study the protein in its natural environment provides the following advantages:

1. The structure of the protein can be determined even when it is difficult or impossible to crystallize, as in the case of intrinsically unstructured proteins and membrane-bound proteins. As yet, however, very few NMR structures of membrane proteins are available.

2. The natural dynamics of the protein can be studied at the nanosecond timescale (and shorter). As explained above, X-ray diffraction provides information that reflects the freedom of movement of different protein regions (i.e., the B-factor). However, it is often difficult to determine whether such X-ray diffraction data represent real dynamics or static disorder resulting from the crystallization process [22]. Furthermore, it is now possible to characterize the dynamics of proteins not only *in vitro* but also inside a living cell, by using *in-cell NMR* [127,128]. This development is important, since the inherent dynamics of proteins is affected by the molecular crowding in the cytoplasm (e.g., [129,130]). Thus, measurements inside living cells provide a more accurate view of the conformational freedom of proteins.

3. The protein's hydration layer can be studied both structurally and kinetically (e.g., [131]).

4. Protein binding processes can be analyzed, including enzyme-substrate, enzyme-cofactor, and protein-drug binding. The use of NMR for this purpose relies, among other qualities, on the method's high sensitivity to weak, short-range interactions. In fact, NMR is one of the methods used today for the development of new drugs [132]. The specific application used for this purpose (*$^{15}N-^{1}H$ heteronuclear single quantum coherence*) is used for the screening and optimization of the most suitable molecule, out of hundreds of thousands. The method involves labeling the protein with ^{15}N and measuring the NMR spectrum in the presence and absence of ligand. First, the binding strength of small individual chemical groups to relevant protein regions is determined. Next, the various groups are joined into a whole drug molecule, and the molecule is optimized. This NMR technique has been used in the development of an anti-cancer drug that inhibits the anti-apoptotic protein Bcl-2, and the drug's *in vivo* activity against solid tumors has been demonstrated [133].

5. The development of techniques such as liquid-crystal state NMR and solid state NMR (see below) has enabled scientists to track the insertion of proteins and peptides into membranes and to determine their orientation with respect to the membrane's vertical axis (e.g., [134–136]).

Another advantage of NMR spectroscopy is that it does not harm the sample, as the measurements use low-energy radio waves. Finally, a method called 'solid-state NMR' allows the measurements to be conducted on membrane proteins, or on proteins that form amyloid precipitates [137–139]. These proteins tend to acquire specific orientations in their respective environments, and NMR measurements can also be used for determining these orientations, which often play a central role in protein function [140,141]. Earlier we discussed one of the major problems of NMR spectroscopy, namely, the need to discern the magnetic properties of the numerous atoms present in large proteins. In such systems, the rapid nucleus relaxation produces wide-spectrum lines with low resolution. As a result of this problem, the method is limited to small proteins (40 kDa or less), although successful measurements have recently been obtained in larger proteins (~100 kDa), and protein complexes (see [22] for details). Scientists have succeeded in extending the method to very large proteins by using different techniques, such as labeling by ^2H and other isotopes, as well as by developing methods such as *TROSY*. The latter is an NMR technique that reduces the relaxation rate and therefore also the extent of signal loss [124]. Because of the difficulty in resolving spectrum lines, as well as the diversity of techniques in this field, NMR involves less automation compared with other methods, which is a problem unto itself.

3.3.2 Electron paramagnetic resonance (EPR) spectroscopy

The principles of EPR spectroscopy are very similar to those of NMR spectroscopy: subatomic particles possessing a magnetic spin, and which are located within the protein, are subjected to a strong, externally applied magnetic field. The spin of each particle aligns in either a parallel or anti-parallel orientation with respect to the field lines. An energy pulse fired at the particles allows them to switch momentarily between the two states, i.e., to resonate, and when they relax back to their original orientation they emit an energy signal that is recorded by the instrumentation. The emitted signal depends on the local environment of each particle, and when integrated, all the recorded signals provide structural information about the protein. So what are the differences between NMR and EPR? First, the subatomic particle in EPR is not the nucleus, but rather an unpaired electron with paramagnetic properties [142]. This naturally leads to the second difference; since proteins usually do not contain unpaired electrons (except for some metal-containing proteins), EPR requires that they be labeled with a *spin label*, i.e., a molecule that does. There are several such molecules; each is a small organic compound containing a *nitroxide group* with an unpaired electron. Some of these molecules bind to specific types of protein residues, whereas others are analogues of natural ligands, such as ATP [142]. Finally, electron spins during the EPR experiment undergo a larger energy transition compared with nuclear spins. As a result, the energy pulse used to excite the sample is in the microwave region of the electromagnetic scale, not the radio region [122].

In EPR, the signal emitted from the relaxing electrons depends on several factors [142]:

1. The orientation of the paramagnetic label with respect to the magnetic field

2. The mobility of the label

3. The presence of other unpaired electrons in the vicinity of the label, which gives rise to spin-spin interaction

The orientation and motion dependencies of EPR both result from the anisotropic interaction of the electron spin with the external magnetic field, which in turn results from the unpaired electron in the label occupying an asymmetric orbital. The second and third factors affecting the EPR signal mean that this method, in addition to providing structural information on proteins, can be used to study their dynamic behavior and interaction with ligands (see examples in [142]). Regarding dynamics, it is interesting to note that the motions that have the most observable effects on the EPR measurements are in the 10^{-11} to 10^{-9} s range, whereas the more biologically-relevant range (10^{-9} to 2×10^{-7} s) is harder to measure [142]. Motions slower than this range cannot be measured using basic EPR, and require a technique called 'saturation transfer EPR' (ST-EPR).

Being distance-dependent, the effect of spin-spin interactions on EPR output can be used for several purposes. First it enables scientists to measure distances in the range of 2 to 25 Å between two spin-labeled sites in the protein. Second, in membrane-bound proteins and peptides the effect can be used to determine the orientation of the peptide or protein segment with respect to the membrane's vertical axis, the distance of a specific residue from a lipid chemical group (e.g., [143,144]), or the aggregation of independent bioactive peptides (e.g., [145]). Similar measurements can be used to determine the solvent accessibility of specific residues within proteins, whether the solvent is the aqueous solution or the biological membrane (e.g., [146]).

Because of the requirement for spin labeling, EPR spectroscopy cannot be used for full structure determination of proteins. However, it can be used to provide local or qualitative information. For example, local structures, such as α-helices and β-sheets, display periodicity of 3.6 and 2, respectively, in the mobility and solvent accessibility of the side chains [142]. Also, both side chain mobility and solvent accessibility are diminished when the label is in a region of tertiary or quaternary contacts. The latter can also be followed by tracking the increased interaction between spin labels on different subunits.

3.3.3 Information derived from other methods

Classical spectroscopic methods, such as visible and UV absorbance and fluorescence, are extensively used by biochemists and molecular biologists, mainly because these methods are straightforward and are readily available [122]. More advanced spectroscopic methods also exist, such as Fourier transform infrared (FTIR) spectroscopy, Raman spectroscopy, circular dichroism, and mass spectroscopy. These require expertise and sophisticated instrumentation, but are invaluable in biochemical research. Although only some of these spectroscopic methods relate to the three-dimensional structures of molecules, and most of them cannot be used for the determination of complete protein structures, they may provide important information that is complementary to the data obtained in NMR and EPR studies. Below we briefly describe fluorescent and circular dichroism spectroscopies, as well as mass spectrometry.

3.3.3.1 Fluorescent spectroscopy

The simplest spectroscopic method for gaining information on the tertiary structure of proteins is probably fluorescence spectroscopy. When certain chemical groups (fluorophores) are electronically excited by photons of suitable wavelength (UV to blue-green), they relax by undergoing certain internal processes, one of which results in the emission of photons of a longer wavelength: fluorescence [147]. In proteins, fluorescence is mainly associated with aromatic residues. When these residues are exposed to the solvent, the fluorescence is recorded fully by the instruments. However, when they are inside the protein, at least some of the fluorescence is 'quenched'. Thus, the extent of recorded fluorescent emission may be used to learn about a protein's structure and the degree of folding. To avoid having to rely on naturally occurring aromatic residues in the investigated protein, the spectroscopist can chemically tag the protein with fluorescent dyes at different positions of the macromolecule. A sophisticated variation on this theme is *fluorescence resonance energy transfer (FRET)* [148], in which the emitted fluorescence energy is transferred non-radiatively between two fluorophores. That is, a pair of chromophores is used to tag the protein at specific locations. The pair is selected so that the emission spectrum of one fluorophore (the donor) overlaps with the absorption spectrum of the other fluorophore (the acceptor). The energy transfer decays with the sixth power of the distance between the two fluorophores, and can therefore be used to measure the distance between them.

3.3.3.2 Circular dichroism spectroscopy

Circular dichroism (CD) is a more sophisticated spectroscopy method, used mainly for determining secondary structures [149] but also for characterizing protein-ligand interactions and structural disorder [150]. The method relies on the differences in the absorption of left-handed versus right-handed polarized light by asymmetric molecules. In proteins, this asymmetry results from the dominance of the L configuration over the D configuration of amino acid residues. When residues appear in a repeating pattern, as occurs in secondary structures, their peptide bonds produce typical far-UV CD spectra (180 to 250 nm), revealing the existence of such structures. The method can reveal the percentage of residues in the protein occupying a certain secondary element, but not the identity of these residues. In addition to secondary structures, CD spectroscopy can also be used to learn about certain aspects of the tertiary structure. This is done by inspecting the near-UV CD spectra of the protein, which are produced by aromatic residues and disulfide bonds. The spectra are sensitive to changes in the three-dimensional structure of the protein, and processes that involve such changes, such as folding or ligand-binding, can therefore be detected by the method. For example, a misfolded protein comprising only secondary elements (see Chapter 5 for details) would only produce a significant far-UV spectrum.

In recent years, several advances in the field have made the use of CD spectroscopy more accurate. For example, synchrotron radiation CD (SRCD) spectroscopy is a relatively recent technique that extends the limits of conventional CD spectroscopy by broadening the spectral range, increasing the signal-to-noise ratio, and accelerating data acquisition in the presence of absorbing components (buffers, salts, etc.) [151,152]. As efficient as the data collection may be, the assignment of secondary structures from the CD spectra requires algorithms designed specifically for this task. The assignment is not trivial, especially in the case of β-sheet-rich proteins, due to their spectral variety and lower spectral amplitudes [151]. Here, too, much progress has been made, and the currently used algorithms achieve an overall good assignment of secondary structures [151,153,154]. These developments have led to a significant increase in the number of proteins characterized by CD spectroscopy. Today, measured CD spectra and related data are deposited in the Protein Circular Dichroism Data Bank (PCDDB) [155], which is fully available to the general public via the Internet[*1].

[*1]http://pcddb.cryst.bbk.ac.uk/home.php

3.3.3.3 Mass spectrometry

Mass spectrometry (MS) is a popular method for protein analysis, although historically it has mainly been used to identify small molecules for forensic purposes, environmental chemical analysis, and drug discovery. In a mass spectrometry measurement the elements and molecules in the sample are hit with an intense electron beam, which ionizes them by knocking away electrons [156,157]. The ions are then accelerated and deflected in vacuum by electric and magnetic fields, which separate them by mass and charge. Finally, the separated ions are detected by an electronic sensor, and the results are displayed as spectra of the relative abundance of the detected ions as a function of their *mass-to-charge ratio* (*m/z*, where *m* is in Dalton units and *z* is in charge units) (Figure 3.7a). By comparing the results to the known mass values of elements and molecules, it is possible to obtain the molecular composition of the original sample. However, the electron beam used in the measurement usually breaks the sample molecules into fragments, which greatly increases the number of chemical species that are detected, making direct identification very difficult. Therefore, chemical compounds are usually identified according to their typical *fingerprints*, i.e., their already known mass spectra patterns. This approach enables high-throughput experiments to be run with sample sizes as small as micrograms. Molecules analyzed by mass spectrometry first have to be volatilized (i.e., transferred into gas phase) and only then ionized. This is easy for small compounds, which are thermally stable. Larger molecules (e.g., most biomolecules) are less volatile and more thermally labile. Several techniques for ionizing such molecules have been developed in the past three decades [156]. The most commonly used among these are *electrospray ionization* (*ESI*) [158] and *matrix-assisted laser desorption/ionization* (*MALDI*) [159]. In ESI, the solution-based sample is passed through a narrow capillary (several micrometers wide[*1]) that is at a potential difference relative to a counter electrode, at voltages between 500 and 4,500 V. The electrostatic spraying of the sample creates an aerosol of charged droplets, and the sample ions make their way to the mass analyzer of the spectrometer [156].

In the study of biological molecules, ESI offers many advantages:

- It has virtually no size limit, such that it can be used with very large macromolecules.

- It is a 'soft' method, ionizing noncovalent complexes without degrading them. Thus, it can be used for studying protein-protein, protein-DNA, and protein-small molecule complexes.

- It can couple between mass spectrometry and liquid separation techniques such as HPLC. This type of coupled method is called '*liquid chromatography MS*' (*LC/MS*). This feature enables ESI to be used on complex biological samples that contain different molecules and macromolecules.

One of the problems of ESI is that the sample has to keep flowing and be consumed constantly, which leads to wastage. This problem is addressed in MALDI, where the ions are produced by pulsed-laser irradiation of the sample, which is co-crystallized with a solid matrix. The synchronization between ion formation and analysis also makes the method very sensitive. MALDI is also quick, which makes it suitable for high-throughput studies.

[*1]Because of the narrow diameter of the capillary, the method is also called '*microspray*' or '*nanospray*'. Mass spectrometry has many applications in protein characterization. Here we address some of the main applications. For a more detailed description see [12,160–166].

Finally, the method has high tolerance to salts and buffers, which makes it suitable for the analysis of physiological samples. One problem associated with MALDI is the noise generated when analyzing samples of small molecular weight (below 500 Da). With respect to biological samples, MALDI is more suitable for analyzing simple peptide mixtures, whereas more complex samples are usually analyzed by ESI (LC/MS).

1. **Protein identification, quantification, and sequencing**

As described above, current MS techniques can be used to identify whole proteins in a sample. The sample, which has been collected from a biological source, first needs to be purified; then, its constituent proteins are separated by chromatographic or electrophoretic methods [167,168]. Since MS of whole proteins is less sensitive than peptide MS, a different strategy is usually employed. The isolated proteins are enzymatically degraded to short peptides, which are then separated by chromatographic methods and identified using MS (Figure 3.7b). Knowledge of the composition of each peptide derived from each of the isolated proteins enables these proteins to be sequenced. When more accurate quantification is required, the proteins or peptides are labeled metabolically, enzymatically, or chemically with heavier isotopes and then subjected to MS. The above techniques can be used on a large scale, e.g., for proteomic analysis of biological organisms [167]. That is, samples can be collected from different organisms or different organelles inside a single organism, as well as from the same organism under different conditions, and analyzed in a high-throughput manner to construct a quantitative protein expression profile. Such analysis can also be carried out to differentiate between alternatively spliced or translated forms of a protein. Proteomics analysis can be carried out by different MS techniques. For example, LC/MS can be used, in which the collected samples are first purified by chromatographic and/or electrophoretic means and then subjected to MS (see above). An alternative technique is *tandem MS (MS/MS)* [169,170], which separates ions of specific m/z values from the original ion source by chemically reacting them [156]. With this technique, complex biological samples can be handled without first undergoing purification, and samples can be screened rapidly for certain compound types. When the sample proteins are first degraded in the MS protocol, the degradation products are usually analyzed in MS/MS by peptide mass fingerprinting.

There have been more than a few studies that demonstrate the power of MS in proteomics. In one of these studies, MHC class I-bound peptides were analyzed by LC-MS/MS [171], resulting in the quantification of 200 different peptide species. In two other studies, MS/MS was used to identify proteins associated with the human and mosquito stages of the malaria parasite [172,173]. These studies identified proteins that are specific to certain life-cycle stages of the parasite, and proposed some of them as candidate drug or vaccine targets. Finally, recent studies have presented an MS-based draft of the human proteome based on information assembled from different human tissues, cell lines and body fluids [174,175]. Proteomics studies focus on various aspects of biological systems that are beyond the mere types and quantities of cellular and tissue proteins. Two main aspects are the post-translational modifications that are applied to proteins under certain conditions, and the diverse patterns of interactions between individual proteins. These are described below.

2. Identification of post-translational modifications

Post-translational modifications (PTMs) of amino acids in proteins are of major importance in cells and tissues, as they assist in regulating and fine-tuning the activity of many enzymes and other proteins (see Chapter 2 for more details). Whether these modifications include the transfer of a chemical group, deletion of amino acids, or chemical alteration of amino acids in the target protein, they usually change the mass and/or electric charge of the protein. This means that such modifications can be identified and monitored by mass spectrometry.

Here, too, MS/MS is highly efficient in isolation and identification of specific products, where the sample proteins are usually degraded first to peptides, which are subsequently analyzed. When certain modifications are looked for, the MS experiment can be designed to scan the peptides specifically for these modifications. It should be noted that in the case of regulatory modifications that occur in only one or a few amino acids along the sequence (e.g., phosphorylation), the process is complicated by the low stoichiometry of the specific modification [167]. This makes it difficult to characterize regulatory modifications on the proteomic scale, although different techniques have been developed to cope with this problem, e.g., employing an affinity selection for the specific modification on which the study focuses [176].

3. Three-dimensional structure characterization and interaction mapping

When coupled with other methods, MS can be used to characterize the 3D structure of a protein [162]. The main MS-coupled methods include the following:

- **Hydrogen-deuterium exchange mass spectrometry (HDX-MS)** [177]. In this method, the exchange between ^1H and the heavier ^2H in the protein's amide group is monitored by MS and used to learn about the positions and secondary structures of different parts of the protein. This technique is based on the fact that under fixed pH and temperature, the $^1H/^2H$ exchange is slower in regions of the protein that are either buried or possess a secondary structure that involves hydrogen bonds (helices, sheets).

- **Hydroxyl-radical footprinting** [178]. This is another method that can be used to detect the solvent accessibility of different parts of the protein. In this method, side chain atoms are covalently oxidized by hydroxyl radicals. The oxidation rate depends on the reactivity of each side chain and also occurs more quickly in solvent-accessible parts of the protein. Since hydroxylation renders the modified residues heavier, it can be detected by MS as long as the mass shift is large enough (this is not the case for serine and threonine, for example). Identifying solvent-accessible regions in proteins is important not only for topological reasons, but also for guiding searches for interaction sites with other molecules.

- **Chemical crosslinking.** This method is used to detect parts inside proteins, or between protein subunits, that are close to each other in 3D space [160,161,179]. This is done by using chemical reagents called 'crosslinkers', which are able to covalently bind amino acids that are in close proximity to each other[*1]. After

---

[*1] The most popular crosslinkers are NHS esters that target amino groups of lysine side chains [180]. However, there are also crosslinkers that are specific to aspartate/glutamate, arginine, and other residues (see [160] for details).

the crosslinking is carried out, the protein is enzymatically degraded to short peptides, and the sequence of each peptide is determined by MS as described above[*1] (Figure 3.7c). This procedure is able to detect 'normal' (i.e., linear) peptides, as well as peptide fragments in which amino acids are crosslinked via their side chains. The latter provide the identity of residues that are adjacent to each other in the intact protein, which can be translated into distance constraints. When this procedure is carried out with crosslinkers of varying lengths and different residue specificities, a sufficient number of distance constraints can be determined to help learn about the folded structure of the protein, or the subunit composition of a protein complex (e.g., [181–183]). This is usually done in combination with protein structure prediction algorithms.

As mentioned above, certain MS techniques allow intact noncovalent complexes to be ionized. In these cases, MS can be used to characterize the quaternary structures of proteins (i.e., protein complexes), and other protein-protein interactions (see below). Furthermore, such complexes can be studied in their native environments and cellular locations [167]. To identify protein complexes, the scientist may 'tag' one of the proteins with an amino acid sequence that is targeted by an antibody for purification. Since many of the interactions are transient, only some of the complexes are usually identified. This problem is usually solved in one of the following ways: (*i*) carrying out the MS experiments iteratively, (*ii*) employing chemical crosslinking of the interacting proteins to prevent subunit dissociation, and (*iii*) combining the MS experiments with data derived from other methods, including bioinformatic approaches. MS studies of protein complexes can be carried out to identify and characterize protein-protein interactions on different biological levels:

(a) Individual proteins with quaternary structure. Different aspects of the structure can be characterized at different levels of details by using various MS techniques. These aspects include the shape of the complex, its composition and stoichiometry, the relative location of subunits (core versus peripheral), the existence of sub-complexes, and the structure's biogenesis [12].

(b) Large protein machines that act within cells. Such studies have been undertaken, e.g., for the spliceosome [184,185] and the nuclear pore complex [186].

(c) Local protein networks that form in certain cell locations and serve specific cellular functions. These include, for example, enzymes that act within the same metabolic pathway and form functional complexes to enhance metabolic efficiency (see Chapter 2, Subsection 2.5.1 for more details).

(d) Large cellular networks of stable or transient protein-protein interactions. Such proteomic studies (e.g., [187,188]) often reveal higher-order coordination between proteins and between small protein networks, which provide a unique view of cellular function and regulation.

[*1]The purification of the individual peptides is usually done by size-exclusion or ion exchange chromatography, coupled to the MS procedure.

(a)

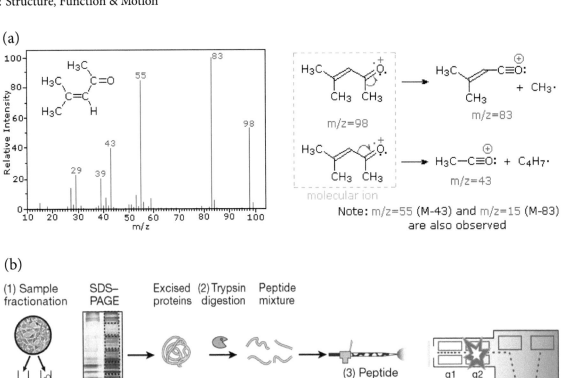

Note: m/z=55 (M-43) and m/z=15 (M-83) are also observed

(b)

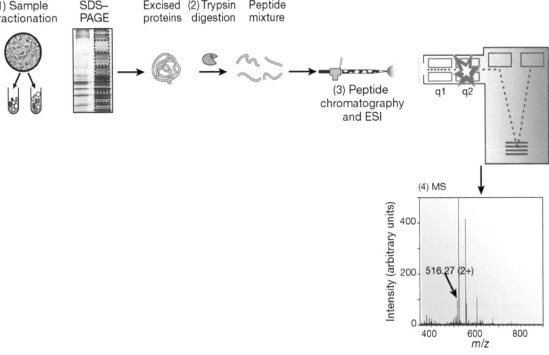

(1) Sample fractionation   SDS–PAGE   Excised proteins   (2) Trypsin digestion   Peptide mixture

(3) Peptide chromatography and ESI

(4) MS

516.27 (2+)

(c)

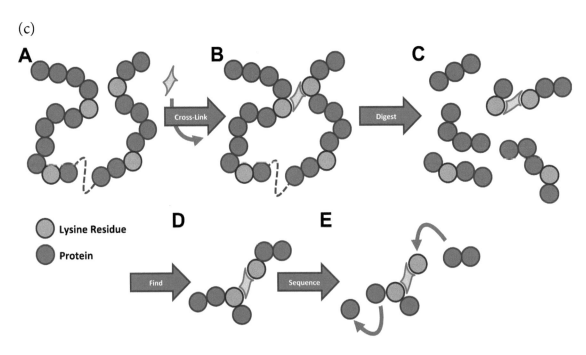

A   Cross-Link   B   Digest   C

Lysine Residue
Protein

D   Find   E   Sequence

3.4 COMPUTATIONAL METHODS FOR STRUCTURE PREDICTION

3.4.1 Introduction

Protein structure determination has progressed significantly thanks to the development of the methods discussed in the previous section. However, these methods are generally slow, and as a result the number of solved protein structures is only a small percentage of the number of all known sequences [189]. Furthermore, in spite of improvements in various structure-determination techniques, sequencing technologies have improved much more rapidly (and their costs have decreased considerably). Thus, the gap between the number of available protein sequences and structures keeps increasing. One possible solution for this problem is to use computational methods to predict protein structure on the basis of amino acid sequence. Computational methods are significantly faster and cheaper than experimental (i.e., lab) methods. The ability to predict the native structure of proteins based on their amino acid sequences alone has always been one of the most desirable goals of computational biology, as it would allow entire proteomes to be determined quickly and cheaply. This need is particularly felt in the pharmaceutical industry, where rapid prediction of protein structure is expected to save the years of laborious and expensive experiments that are currently required for the development of a single drug [190] (see Chapter 8).

The numerous computational methods that are available for protein structure prediction can be grouped into two general approaches [191,192]: (1) *ab initio* (a.k.a. *de novo*) methods, in which the structure is predicted from scratch using physical first principles, and (2) *template-based* methods, which use information from proteins of known 3D structure. The latter can be further divided into *homology (comparative) modeling* methods, which are based on sequence-sequence comparison between the predicted protein and the template, and *fold recognition* methods, which are based on other similarities between them. In the following subsection we review the principal aspects of the two approaches, their relative success rates, and the most common ways in which they are used. We will also briefly discuss an interesting approach that has recently been developed, which predicts the three-dimensional structure of a protein using information on correlated mutations in its sequence.

FIGURE 3.7 **Mass spectrometry.** (Opposite) (a) Mass spectra of 4-methyl-3-pentene-2-one. The spectra are shown on the left. The right image shows the ion of the intact molecule and the ions of two fragmentation products. The images were created by Prof. William Reusch and used with permission. (b) A generic mass spectrometry procedure for proteins. Step 1: The proteins to be analyzed are isolated from cell lysate or tissues by biochemical fractionation or affinity selection. This process often includes a final step of one-dimensional gel electrophoresis. Step 2: The proteins are degraded enzymatically to peptides, usually by trypsin. Stage 3: The peptides are separated chromatographically in very fine capillaries and eluted into an electrospray ion source, where they are nebulized in small, highly-charged droplets. After evaporation, the peptides enter the mass spectrometer. Stage 4: A mass spectrum of the peptides eluting at this time point is taken. The image is taken from [167]. (c) Protein analysis by crosslinking MS. Step 1: The protein is incubated with a residue-specific crosslinking reagent. Residues within the range of the crosslinking reagent are covalently bonded. Step 2: The protein is enzymatically digested to form peptides. Step 3: Data-dependent acquisition is used to identify peptides as they elute from an HPLC directly coupled to the mass spectrometer. Step 4: The identified peptides are then fragmented to provide sequence-specific information. The image is taken from [160].

3.4.2 *Ab initio* (physical) approach

The most obvious way to predict the native fold of a proteins is probably to follow nature itself, that is, to accurately characterize how physical forces drive the protein to fold, and to use this characterization to reproduce the folding process computationally for a protein of unknown structure. Since there are many possible folds (i.e., conformations) available for a given polypeptide chain, the prediction would have to rely on the thermodynamic premise that the native fold is the lowest-energy state of the protein [193], that is, the most stable three-dimensional organization of the protein's atoms. **A more accurate view of the native structure of a protein is that it is the conformation possessing the least energy, while retaining enough energy to maintain the level of dynamics required for the protein to execute its function.**

The 'total' energy of a system, also known as the *'free energy'* (see Chapter 4 for details), can be decomposed into different components: potential, thermal, kinetic, etc. Indeed, it is the free energy that determines the stability of the system. Purely energy-based predictions use only information on the types of atoms in the system, their relative locations in three-dimensional space, and their bonding and non-bonding interactions with other atoms. This information is used to calculate the energy content of the system and the forces acting on each atom. Such methods are therefore referred to as *'ab initio'*, which in Latin means 'from the beginning'[*1]. Although different methods are included in this approach, all are based on two basic abilities:

1. Calculating the energy content of the system in a single configuration[*2].

2. Sampling numerous configurations and finding one with the lowest energy (i.e., the most stable configuration).

The *ab initio* approach emerged in the 1960s with the work of a handful of people who implemented their knowledge in computational chemistry on macromolecular systems to explore their structure and dynamics [194,195]. This group included Shneior Lifson from the Weizmann Institute and his students and coworkers Arieh Warshel, Michael Levitt and Martin Karplus. The latter three won the 2013 Nobel Prize in Chemistry for their contribution to the field of computational molecular biophysics [196].

3.4.2.1 Calculating total potential energy of system

Protein-based systems include — in addition to the protein itself — atoms of the solvent and any other chemical species present (ions, cofactors, etc.). At any given moment, each of these atoms occupies a single point in space. During protein folding, numerous such configurations are sampled by the system. As explained in Chapter 1, a faithful characterization of all physical forces operating in the system between the different atoms would only be obtained by applying quantum-mechanical (QM) calculations. This is because the forces result from the spatial distribution of electrons around the atoms, requiring QM calculations. Unfortunately, such calculations are computationally costly, and the currently available computer power is insufficient for this type of rigorous characterization of a macromolecular system.

[*1] A configuration is the exact spatial distribution of all system atoms at a certain point in time.

[*2] The term *'ab initio'* as used here should not be confused with the same term used to describe quantum chemical (electronic structure) calculations of organic and inorganic compounds.

QM calculations carried out on a single conformation of a small protein (or even a single part of it) may take months [197]*1.

Therefore, scientists often describe the investigated system using approximations of the real physical forces in it, as these are much easier to calculate. Structure predictions relying on this approach emerged in the middle of the 1980s, first for short peptides and then for polypeptides (i.e., proteins) [202–204]. The field from which these calculations are taken is called 'molecular mechanics' (MM), since it approximates molecular systems by using expressions taken from Newton's classical mechanics. That is, all atoms of the system are considered*2, and the energy content is described using a mathematical equation (force field), which treats atoms and covalent bonds as balls and springs, respectively (see Box 3.1 below). Thus, any electronic motions in the atoms that would necessitate a QM description are ignored. The force field includes several separate expressions. Each of the expressions describes the potential energy*3 resulting from a different interaction between any two atoms in the system. For example:

$$U_{tot} = U_{bond} + U_{elec} + U_{np} + U_{vdw} \tag{3.2}$$

(where U_{tot} is the total potential energy, U_{bond} is the potential energy resulting from properties of covalent bonds, U_{elec} is the potential energy resulting from electrostatic interactions, U_{np} is the potential energy resulting from nonpolar interactions, and U_{vdw} is the potential energy resulting from van der Waals interactions).

Indeed, the approximated description provided by force fields allows the potential energy calculation of many macromolecular systems to be carried out in less than a second [197]. Obviously, the exact calculation time varies across different systems, and is dependent on the number of atoms and type of computer used. The first program for calculating the potential energy in proteins was developed in 1969 by Lifson and Levitt at the Weizmann Institute [205]. It employed a simple energy function, the consistent force field (CFF), which had been developed a year earlier by Lifson and Warshel from data on small crystalline molecules [206].

---

*1 It should be noted, though, that such point energy calculations can be completed faster using the QM/MM approach [198,199]. In this approach the QM calculations are usually carried out on a small region of the enzyme that contains the substrate and functionally important residues. The other regions of the enzyme are subjected to molecular mechanics (MM) calculations, which are approximate and therefore much faster. The QM/MM approach was introduced by the studies of Warshel, Levitt and Karplus [200,201], who, as mentioned above, also won the 2013 Nobel Prize in Chemistry for their contribution to computational biophysics [194,195].

*2 This type of representation is called 'explicit' (see Figure 3.8).

*3 A detailed description of the meaning of potential energy is provided in Chapter 4. For our current needs, we can describe it as the energy resulting from all covalent bonds and noncovalent interactions in a single configuration of the system.

(a) (b)

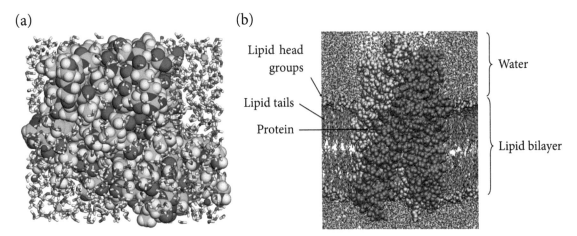

FIGURE 3.8 **Explicit descriptions in molecular mechanics calculations.** (a) A small globular protein (barnase, PDB entry 1b2x) soaked in solvent, which includes water molecules and ions (0.5 M NaCl). The protein is presented as atom spheres colored by atom type, whereas the surrounding water molecules are presented as sticks. Na^+ and Cl^- ions are presented as small yellow and magenta spheres, respectively. All atoms should be included in the simulation. (b) A molecular dynamics snapshot of the ABC transporter Sav1866 in a palmitoyl-oleoyl phosphatidylethanolamine (POPE) lipid bilayer, surrounded by aqueous solvent. The lipid head groups and the protein are shown in a space-fill representation, whereas the lipid tails and aqueous solvent are shown in a wire-frame representation. Lipid molecules are gray; water molecules are blue (oxygen) and white (hydrogen); and the two protein subunits are colored in red and orange, respectively. Courtesy of Drew Bennett and Peter Tieleman, University of Calgary.

3.4.2.2 Sampling configurational space of system

Assuming that the force field is accurate, a calculation of the potential energy of the system allows us to determine the stability of one configuration[*1]. In principle, finding the native fold of the protein requires one to sample the entire configurational space of the system efficiently. In other words, the scientist must consider all possible atomic locations in the system, calculate the potential energy in each case, and pick the one with the lowest energy. Despite the approximated nature of force-field-based calculations, this task is virtually impossible, since, in addition to the numerous atoms of the protein and associated molecules (ligand, cofactor, etc.) the system also includes numerous solvent molecules with many degrees of freedom. The total number of atomic combinations in such a system is difficult to fathom, let alone integrate into a calculation. To overcome this problem, **scientists use different configuration-searching methods, which try to reach the lowest energy configuration without sampling the entire configurational space of the system.** The simplest method, called 'energy minimization', starts from an initial, arbitrarily picked configuration, and applies the following steps:

1. The potential energy of the initial configuration is calculated. Derivation of this energy value at different locations in the system makes it possible to calculate the forces acting on each atom by the rest of the system [204] (see Equation (3.4) below).

[*1] Here, 'configuration' refers to the overall arrangement of all components of the system (protein, solvent, ions, membrane, etc.), i.e., the relative locations of all atoms (not to be confused with the *chemical configuration* of organic molecules; see Chapter 1). In the case of protein atoms the collective locations are referred to as a 'conformation'.

2. A small change is introduced in the location of each atom of the system in response to the forces applied to the atom by the remainder of the system, as calculated in the previous step.

3. The potential energy of the new configuration is calculated.

4. If the new configuration has lower energy than the previous one, the new configuration is adopted and subjected to steps 2 and 3. However, if it has higher energy it is rejected. Then, a new configuration is created and subjected to steps 2 and 3. The process continues until no new lower-energy configurations can be found.

Thus, the method shifts from one configuration to another down the gradient of the potential energy surface until it converges to the nearest local minimum [204].

Although different energy minimization procedures have been developed with the goal of efficiently searching for the lowest-energy configuration (i.e., the *global energy minimum*), they all tend to get 'stuck' in one of the many *local energy minima* on the way. That is, conformations that have relatively low energy, just not as low as the native conformation[*1] (see Chapter 5 for details on the energy landscapes of proteins). In order to shift from a local minimum conformation to the native one, the protein often needs to pass through another conformation, which may be geometrically similar to the previous, but has high energy due to improper atom-atom interactions (Figure 3.9). Such a high-energy conformation is termed an 'energy barrier'. Going back to the minimization procedure, when the algorithm 'walks' down the energy gradient potential, it reaches the local minimum configuration closest to the one from which it started the search. Since this local minimum is surrounded by energy barriers, and since the algorithm is designed to reject any configuration with energy higher than the previous configuration, it gets stuck in the local minimum conformation. This is similar to a ball going down a slope, but before it reaches the bottom it gets stuck in a pit on the way. The ball cannot leave the pit, since any movement makes it hit the walls of the pit. In order to get out of the pit, the ball needs to be pushed so as to overcome the local physical barrier, namely, the pit's walls.

The same solution is also valid in molecular systems; it is possible to 'push' the energy minimization algorithm out of the local minimum 'pit' by providing it with extra energy. This allows the algorithm to overcome the local energy barriers (i.e., high-energy configurations), get out of the local minimum, and keep searching for new configurations with even lower energy. One way of achieving this is by raising the temperature of the system, i.e., adding virtual heat energy. The added energy allows the atoms of the system to increase their motions and create new configurations that are outside the neighborhood of the local minimum. This method is called 'molecular dynamics' (MD) [204,207], and it focuses on calculating the time-dependent motions of the atoms in the system. The calculation is carried out according to the equations of motion defined in classical mechanics [204]. These couple the change in the motion of a body (i.e., the change in its location (∂r)) over time (t) to its mass (m) and the force acting on it (F):

$$F = ma = m \left(\frac{\partial v}{\partial t} \right) = m \left(\frac{\partial^2 r}{\partial t^2} \right) \tag{3.3}$$

where a is the body's acceleration and v is its velocity.

[*1]Note that we refer to the *conformation* of the protein, and not the entire configuration of the system (protein, solvent, and other molecules and atoms). We do that for the sake of simplicity; in reality, the potential energy results from the entire configuration of the system, not just the internal energy of the protein conformation.

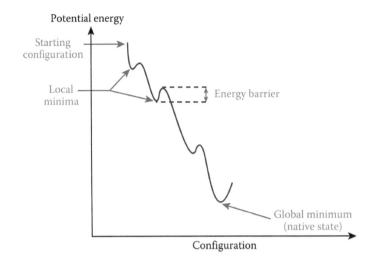

FIGURE 3.9 **Local energy minima and barriers in proteins.** The hypothetical energy landscape of a protein-solvent system is shown. In the plot, each distinct system configuration is characterized by a potential energy value. During a simulation, different configurations are sampled in a certain pattern. As the plot demonstrates, finding the global energy minimum (corresponding to the most stable 'native state' configuration) often requires the simulation to overcome energy barriers that separate different local energy minima.

The force acting on an atom i in the system (F_i) can be derived from the potential energy of the system (U_{tot}), which in turn is calculated using the force field:

$$F_i = -\frac{\partial U_{\text{tot}}}{\partial r_i} \tag{3.4}$$

Thus, the motion of each atom in the system is calculated according to its energy at any given moment. Indeed, in MD simulations the atoms are assigned initial velocity values (proportional to the simulation temperature), and they keep moving in space according to the corresponding changes in the system's potential energy. Again, the drive for the atomic motion is the heat energy supplied by raising the temperature. Therefore, MD simulations are often carried out in repetitive cycles of heating and cooling. The heating allows the system to acquire high-energy configurations, whereas the cooling (followed by energy minimization) allows it to relax into close low-energy configurations. This MD procedure is called 'simulated annealing' [204,208].

The first MD simulation of a protein was performed in 1977 by McCammon, Gelin, and Karplus [209], to study the folding dynamics of the 58-residue pancreatic trypsin inhibitor (BPTI) [195]. That study followed the pioneering work of Levitt and Warshel on the same protein [210]. Due to the poor computational resources at that time, both studies had to replace the all-atom representation of the investigated system with a reduced one. In Levitt and Warshel's study each amino acid of BPTI was represented by only two spheres [195] (a *coarse-grained model*). In the study of McCammon et al., the protein was fully (explicitly) represented, but it was simulated in vacuum. As explained below and in Box 3.1, reduced models have been, and are still, used extensively to simulate large proteins and/or long processes. One common way of making the MD search more efficient is to split it into two steps [211]. The first is a low-resolution search focusing on finding a collection of well-packed structures with nonpolar interactions. This approach is based on the notion that the folded protein features a hydrophobic core (see Chapter 5 for details). This type of search can employ a simple (*implicit*) description of the system (see more below), and does not require significant computational resources. The second step employs a high-resolution search for the native structure of the protein, among the structures isolated in the first step. This intensive search is based on an all-atom description of the system (Figure 3.8). The search emphasizes electrostatic and van der Waals interactions, because such interactions depend significantly on the packing of the atoms, and should be optimal in the native conformation, characterized by high packing efficiency.

3.4.2.3 Limitations and partial solutions

The physical approach to protein structure determination has several major problems, which have made many computational scientists wonder whether this approach should be used at all. A full discussion of these problems is beyond the scope of this book. Instead, we discuss two of the problems we feel to be most pressing, and then offer conclusions. The aim of this discussion is not to confirm or overrule the physical approach as a tool in structure prediction, but rather to provide the reader with a general sense of the factors that scientists must take into account when trying to determine how best to use the approach.

3.4.2.3.1 Finding lowest-energy configuration

Currently, the primary problem associated with MM/MD methods is the difficulty in using them to cover biologically relevant processes such as protein folding. What is the source of this difficulty? As explained above, a protein's native conformation is assumed to be the one that has the lowest content of free energy compared to all other possible conformations. Thus, any simulation that aims to identify a protein's native conformation must be able to calculate the free energy of the system. Unfortunately, force-field-based calculations typically provide only the potential energy of the system, which is just one of the two components of free energy (see Chapter 4). The missing component is *entropy*, an elusive physical quantity that is related to the number of ways in which the atoms of the system can be organized. The only way to estimate the entropy and free energy from potential energy calculations is to run these calculations on all possible configurations of the system and then integrate. And this is where the problem lies: MD simulations are indeed able to sample numerous configurations of the system, but since they rely on highly explicit models of the system, they cannot cover all possible configurations. Specifically, it is difficult for such simulations to consider all configurations of the aqueous solvent, as it includes numerous molecules [212]. As a result, MD simulations, even the most extensive ones, sample only a fraction of the configurational space, and when the calculated potential energies are integrated, the result is not the real free energy of the system, but rather a *potential of mean force (PMF)*. This problem is particularly pronounced in the calculation of electrostatic interactions between the protein and the solvent (see Chapter 1, Box 1.3). Thus, **MD simulations are not recommended for describing solvent effects**, although some force fields contain expressions that address solvent polarization (see Chapter 1 and Box 3.1). As explained in Box 3.1, one way to ease the burden of explicit MM/MD calculations is to use implicit descriptions of the system, or at least parts of it. In such an approach, the parts chosen to be described implicitly are represented by an average property, which is why this approach is called a '*mean field approach*'. The solvent is usually the least interesting element of the system from a biological viewpoint, and it is also the part of the system that is the most difficult to describe explicitly. Thus, it is customary to choose the solvent as an implicitly described component. In the implicit description, the solvent is treated as a homogeneous bulk, and the property used to describe the essence of the solvent is its dielectric (the *continuum-solvent* approach, Figure 3.10) [213,214]. The use of an average property enables scientists to calculate the free energy of the system without having to sample different system configurations. Unfortunately, the implicit description obscures specific aspects of the system. Indeed, this is the main problem of the mean field approach. For example, the interactions between protein atoms and the less mobile water molecules at the protein-water boundary are neglected[*1]. This problem is even more critical when the 'solvent' is not aqueous, but instead is the biological membrane. This is because of two aspects of the membrane. First, the lipid membrane is anisotropic, which makes the interactions of proteins with its cytoplasmic side different from their interactions with its exoplasmic side (see Chapter 7). Second, the lipids themselves are chemically diverse, and can therefore interact with proteins in various ways, all of which are obscured in the continuum-solvent approach.

It would therefore seem that the use of energy-based methods must always involve some kind of compromise; when the details of the system are most important, a fully explicit model of the system is used, and when the underlying energetics is important, mean-field models are preferable. Could there be a way to combine the best of the two model types? Indeed, recent efforts have produced different forms of *mixed force fields*, combining explicit calculations on the protein with mean-field calculations on the solvent [215]. Such a task is

(a)   (b)

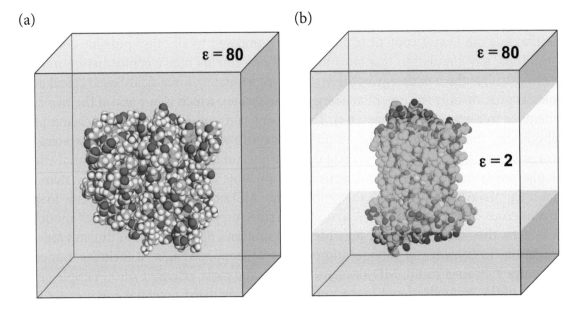

FIGURE 3.10 **Implicit (mean-field) descriptions of protein-based systems.** The figure shows a continuum-solvent model of the system, which is a type of mean-field treatment. In this model, the protein is described explicitly, whereas the solvent is described implicitly, using its dielectric properties. Two common continuum-solvent models are shown. (a) A globular protein in a high-dielectric environment representing the aqueous solution. (b) A membrane-bound protein in a heterogeneous environment, which includes a high-dielectric aqueous solvent and a low-dielectric slab of fixed dimensions, representing the hydrocarbon region of a biological lipid bilayer (in semi-transparent yellow).

[*1]It is noteworthy, though, that immobile water (or ligand) molecules that are considered important for the mechanism can be treated explicitly, as extra amino acids.

not as easy as it may sound, especially when the treatment of electrostatic interactions is considered. In mean-field models, it is customary to describe such interactions by solving the *Poisson-Boltzmann equation (PBE)*, which provides a more accurate solution to the electrostatic free energy than the well-known Coulomb's equation. As explained in Box 1.3, the PBE allows us to relate to the polarization effect of charges on their environment, which is highly important for describing the burial of charges within the protein core [211]. Unfortunately, an accurate solution comes with a price; the PBE can only be solved numerically, which takes time and computer resources. This would not pose a problem if only a single configuration of the system were being considered. However, when the PBE is integrated within a mixed force field, numerous configurations must be sampled, due to the MM-related expressions. To solve this problem, most mixed methods calculate the electrostatic free energy not by solving the PBE, but instead by solving a 'lighter' version of this equation, the *generalized Born (GB) formalism* [216]. The latter describes the protein as a simple geometric shape, and can therefore be solved analytically. As a result, the calculations can easily be carried out on many different configurations of the system, as has already been done in several studies (e.g., [217]). PB and GB calculations cover only the polar (electrostatic) component of the free energy. As explained in detail in Chapters 1 and 4, the nonpolar component of the energy correlates with the surface area of the molecule that is involved in these interactions (see Eq. 1.6). Thus, the entire free energy can be calculated by a single protocol that combines PB or GB calculations with those that rely on surface area (SA). These protocols are called *PBSA* and *GBSA* [218], respectively. Again, these protocols can be combined with molecular mechanics-based calculations to account also for specific interactions in the system. Such *MM-PBSA* or *MM-GBSA* calculations are widely used today to obtain energy values that are more accurate than those obtained from force-field-based calculations alone [219–221].

3.4.2.3.2   Reliability of force fields

 Another problem with force-field-based calculations is the need to avoid using overlapping expressions. For example, in one approach, the ionic interactions are calculated using Coulomb's equation, whereas hydrogen bonds are calculated separately using a geometry-dependent expression [222,223]. As both types of interactions are electrostatic, some interactions might be counted twice. A similar problem exists for van der Waals and nonpolar interactions. A third problem is the accuracy of the calculations. Generally speaking, force fields are only approximations of real interactions. Aside from the aspects described above concerning the calculation of electrostatic interactions and solvation effects, there are some concerns about the accuracy of other components as well. For example, the force field expression relating to the dihedral angles of the protein's backbone seems to need improvement, to address the balance between helical and extended conformations [224]. One 'trick' for reducing the effects of the above problems, which can easily be used when implementing a computational approach, is to design thermodynamic cycles in which most of the unknown (or difficult to calculate) interactions are canceled out [225]. However, this is not always possible, and in any case, does not constitute a real solution to the problem.

3.4.2.3.3   Conclusions

In principle, MD simulations are still heavily burdened by the explicit, all-atom description of the protein-based system. Consequently, most simulations can only cover a short period of time ($\sim 10^{-12}$ to 10^{-9} s). Since most proteins fold within $\sim 10^{-6}$ s or more, the timescale covered by the simulation is not enough to fully describe protein folding. This may very well change in the future. Indeed, simulations have been made longer and more efficient in recent years [226,227], thanks to a number of steps taken:

1. **Improvement of computational resources.** There are three main avenues used today for improving computational power: (*i*) Construction of more powerful, faster computers, which usually employ parallel computing (*supercomputers*). An interesting case in this category is *Anton*, a supercomputer developed by D.E. Shaw Research [228,229]. Anton was designed specifically to optimize MD simulations by using tightly interconnected *application-specific integrated circuits* (*ASICs*). The second generation of Anton is capable of covering 85 μs/day for a molecular system comprising ~23,600 atoms (~180 times faster than any other general-purpose computer) [229]. At these rates this machine can simulate millisecond-long folding processes of small proteins, as well as extract kinetic and thermodynamic quantities of fast-folding proteins (e.g., [229–231]). (*ii*) Employing CPU clusters, in which the computation is run in parallel by multiple processors in a controlled and scheduled manner. This option is commonly used in universities and other large institutions. GPU-accelerated computing has also become popular in recent years (e.g., [232]) (*iii*) Employing distributed (grid) computing [233], that is, using a large network of personal computers volunteered by people to run computational jobs of medical or biological importance. The jobs are run when the computers are idle, so their everyday functions are not interrupted. A well-known example of a distributed computing network is Folding@Home [234], which is operated by the Pande group at Stanford University and includes ~170,000 computers [235]. One of the major achievements of Folding@Home was simulating the entire folding process of the 86-residue acyl-coenzyme A binding protein (ACBP), which is known to require ~10 milliseconds to fold [236]. Moreover, the simulations, supported by experimental data, identified metastable folding intermediates that changed the view of this protein's folding kinetics. Another distributed computing initiative is Baker's Rosetta@Home [237] (~86,000 computers), which focuses on protein structure prediction and design rather than on the folding process.

2. **Statistical sampling.** In certain studies the calculation of the actual motion of atoms has been replaced by statistical sampling of their locations, which is less computationally costly. The most popular sampling method is based on the *Monte Carlo* (*MC*) approach. MC simulations sample conformations randomly. A new conformation with energy lower than the previously sampled conformation is always accepted (as in minimization), but high-energy conformations may also be accepted probabilistically [238].

3. **Protein fragmentation.** The calculations are carried out on separate, short fragments of the protein and then integrated again. One example of software that uses this strategy is QUARK [239]*1. The fragment-based approach is also widely used in methods that combine template-based prediction with *ab initio* assembly and refinement (e.g., [240,241]; see more in Subsection 3.4.4 below).

---

*1QUARK can be run via the following webserver: http://zhanglab.ccmb.med.umich.edu/QUARK/

4. **Replacing the force field with a knowledge-based scoring function.** Such scoring functions are often composed of expressions relating to the statistical tendency of any two atoms, chemical groups, or amino acids to interact with each other. Although this approach achieves good results in some cases (e.g., [222,239]), the reliability of the scoring function usually depends on the database from which it was derived. It should be mentioned that since methods using this approach rely heavily on statistical data, they cannot be regarded as pure '*ab initio*'. Instead, they are often included under a broader definition, as 'free modeling' methods.

5. **Integrating experimental (i.e., lab) data with predictions** [235]. Low-resolution methods for the determination of protein structures (e.g., electron microscopy) have recently been used for deriving geometric constraints, which can be applied along with computational methods to achieve better predictions. This issue is further discussed in Section 3.5 below.

In conclusion, **ab initio approaches are currently unable to predict the structures of most proteins on the basis of sequence alone** [191,192]. However, they are very efficient in doing so when the starting point of the prediction is a near-native structure. Therefore, these methods are widely used for enhancing raw structures obtained from structure-determination methods, such as X-ray diffraction and NMR spectroscopy [242]. In addition, **MD simulations provide invaluable information about protein dynamics, both during folding and in the folded state** (see Chapter 5). Indeed, MD simulations often provide millions of conformations of the investigated protein, and although they do not cover the entire configuration space of the system, they are still very meaningful in terms of revealing the dynamic behavior of the protein. Taken together, the advantages of MM/MD have led these methods to be used not only for gaining knowledge but also for various applications, such as protein engineering and drug discovery.

BOX 3.1 FORCE FIELDS

I. Overview

Molecular mechanics (MM) and molecular dynamics (MD) are used by biophysicists for different purposes, but they both share a common goal, which is to calculate the total potential energy of a molecular system. They do so by describing the different chemical bonds and physical interactions between atoms in the system. They require three types of input:

1. An explicit model of the system, that is, a description of the precise location in three-dimensional space of each of the atoms in the system. This type of input is provided by high-resolution experimental methods such as X-ray crystallography and NMR, in the form of coordinate files.

2. An atom-based description of the magnitude and direction of all covalent bonds and noncovalent interactions between any two types of atoms.

3. Parameters, such as radii and (partial) charges, relating to each of the atom types

in the system. These values are derived from experiments (or electronic structure calculations) with small molecules.

The magnitude and direction of covalent bonds and noncovalent interactions are provided in the form of a mathematical function called a 'force field' [243–247]. When applied to a given 'snapshot' of the system, which includes the coordinates of all atoms in a single static configuration, this function provides an estimate of the total potential energy of the system in that configuration. The typical force field is not a single function, but rather a sum of terms, each corresponding to a different type of chemical bond or interaction (Figure 3.1.1). More specifically, each term describes the dependency of the system's potential energy on certain qualities of specific bonds or interaction types. The qualities include bond length, atom charge, etc. The terms can be separated into two general types:

1. **Bonded (Figure 3.1.1a–d):** The expressions corresponding to this category refer to covalent bonds. In other words, they describe the dependency of the system's potential energy on the length and angle of the bonds. The expressions are approximations, adapted from Newton's classical mechanics. That is, the electronic distribution and motions are ignored; the atoms are described as simple spheres of typical radii and masses, and the bonds connecting them are described as harmonic springs that can be stretched or bent [207]. The force constants and other parameters used in these terms can be derived from small organic compounds that resemble the backbones and side chains of the amino acids. This can be achieved in two ways. The first involves using these compounds in calculations of physical properties, such as vibrational spectra, and adjusting them according to the measured values [197]. The second way entails fitting the parameters to highly accurate quantum chemistry calculations.

2. **Non-bonded (Figure 3.1.1e–f):** These terms relate to noncovalent interactions between atoms, typically electrostatic and van der Waals. The former is usually described by Coulomb's law, and the latter by the Lennard-Jones equation. Here too the physical expressions include parameters (e.g., atomic partial charges in Coulomb's equation) that have been produced by fitting to experimental data or quantum-chemical calculations ([248]). As discussed in Box 1.3 (Chapter 1), nonpolar interactions and the polarization component of electrostatic interactions are particularly difficult to calculate using explicit models, since they involve thousands of solvent molecules. Recent force fields have been designed to overcome this problem, as discussed below.

II. Types of force fields

Force fields for biological systems are relatively new. The first, called a 'consistent force field' (CFF), was developed by Lifson and Warshel in 1968 [206]. It relied on data obtained from small crystalline molecules, and applied the data to macromolecular systems. Other force fields started to appear around the mid-1970s [246]; these were based on CFF and other early functions that were developed for simple organic chemical systems (e.g., [249]). Since then, numerous force fields have been developed for biosystems.

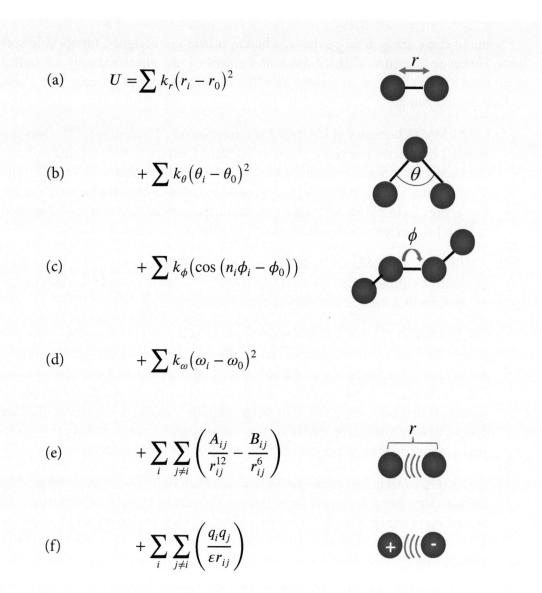

(a) $$U = \sum k_r(r_i - r_0)^2$$

(b) $$+ \sum k_\theta(\theta_i - \theta_0)^2$$

(c) $$+ \sum k_\phi(\cos(n_i\phi_i - \phi_0))$$

(d) $$+ \sum k_\omega(\omega_i - \omega_0)^2$$

(e) $$+ \sum_i \sum_{j \neq i} \left(\frac{A_{ij}}{r_{ij}^{12}} - \frac{B_{ij}}{r_{ij}^6} \right)$$

(f) $$+ \sum_i \sum_{j \neq i} \left(\frac{q_i q_j}{\varepsilon r_{ij}} \right)$$

FIGURE 3.1.1 **A simple force field.** (a) through (d) The dependency of the potential energy of the system on the following covalent bond factors: (a) Length, r_i. (b) Valence angle, θ_i. (c) Dihedral angle, ϕ_i. (d) Improper dihedral angle, ω_i, i.e., deviation from planarity. In all cases, the interaction is depicted as a comparison between the actual value found in the coordinate file (e.g., r_i and θ_i) and the theoretical equilibrium value (e.g., $r_{i,0}$ and $\theta_{i,0}$). The parameters $k_{r,i}$, $k_{\theta,i}$, $k_{\phi,i}$, $k_{\omega,i}$, are the force constants of interactions (a) through (d), respectively. (e) The dependency of the van der Waals potential energy on the distance (r_{ij}) between interacting atoms i and j. A_{ij} and B_{ij} are parameters associated with the van der Waals interaction. (f) The dependency of the electrostatic potential energy on the local dielectric (ε), and the distance (r_{ij}) between atoms i and j, with charges q_i and q_j.

Some of these are general-purpose, whereas others are designed for specific systems. However, in terms of distribution and extent of use, there are only a handful of popular force fields used in simulations that involve biological systems [250]. These include the following:

1. CHARMM (Chemistry at HARvard Macromolecular Mechanics) [251] – was developed by Martin Karplus's group at Harvard University for proteins and nucleic acids. Significant contributions to CHARMM have also been made by the Shneior Lifson and Harold Scheraga groups, as well as by Michael Levitt. The latest version, CHARMM36 [252], is a general-purpose force field that can be applied to proteins, nucleic acids, carbohydrates, lipids, and small molecules.

2. AMBER (Assisted Model Building with Energy Refinement) [253] – Developed originally by Peter Kollman's group at UCSF [254], this force field is used mainly for systems that include proteins and nucleic acids.

3. GROMOS/GROMACS (GROningen MAchine for Chemical Simulations) [255–257] – was developed originally by the University of Groningen for proteins, nucleotides and carbohydrates in aqueous or nonpolar solvent. GROMOS is a *united atom force field*. That is, of all the hydrogen atoms in the system, it describes explicitly only those that are polarized (i.e., those that are bound to electronegative atoms such as oxygen and nitrogen). This is done to make calculations or simulations based on the force field much faster.

4. OPLS (Optimized Potentials for Liquid Simulations) [258] – was developed by William Jorgensen, originally from Purdue University. OPLS was parameterized according to experimental properties of liquids, such as density and heat of vaporization. OPLS includes an all-atom form (OPLS-AA) and a united-atom form (OPLS-UA).

5. MM-family force fields (e.g., MM3 [259], MMFF94 [260–262]) – were designed for a wide range of molecules; the first versions were applicable mainly to small molecules, whereas more recent versions (MMFF94) are also applicable to macromolecules. The broad applicability of these force fields makes them particularly useful for calculations on protein-ligand systems, e.g., in drug design. Other force fields that cover a wide range of molecules include CFF [206], CVFF [263,264], UFF [265], Dreiding [266], and Tripos [267].

There are several similarities among the various force fields [204]:

1. Most are *class I force fields* that use similar basic descriptions of the various bond and interaction types: Bond lengths and angles are described using harmonic terms; bond torsion angles are described by Fourier series; van der Waals interactions are described by the Lennard-Jones dependency; and electrostatic interactions are described by Coulomb's law. Exceptions to this rule include *class II force fields* (e.g., MM3 [259]), which incorporate higher-order terms [268,269], that is, expressions in which the energy is proportional to the third, fourth, or higher power of the bond length or angle.

2. Most force fields are additive. That is, they assume that the energies corresponding to the different bonds and interactions do not affect each other, and can therefore be calculated separately and then summed up to yield the total potential energy. Again, class II force fields are an exception, due to their incorporation of cross (or coupling) terms or higher-order terms.

3. All force fields assume that the approximate nature of the expressions used to describe the various bond and interaction types can be corrected by using force constants and other parameters.

So, how do current force fields differ from one another? The two main differences are in the derivation of force constants and other parameters, and the use of statistically derived non-energy expressions for describing non-bonded interactions:

Parameter derivation. Though they are used on protein-based systems, many force fields rely on parameters derived for small organic molecules or fragments. The derivation process may rely on experiments or calculations that differ across force fields in many aspects, such as the type of solvent. Current force fields (e.g., CHARMM [251]) may contain different sets of parameters for different macromolecules (proteins, nucleic acids, lipids, carbohydrates).

Non-energy expressions. The determination of numerous protein structures in recent decades has enabled scientists to derive important statistical data reflecting different tendencies of protein residues. One of the uses found for these data was to turn them into energy-like expressions that describe the tendency of each residue to participate in certain non-bonded interactions. A well-known example involves hydrogen bonds. Whereas the original force fields described hydrogen bonds as Coulomb interactions between fixed dipoles, modern force fields often describe them as geometry-dependent, based on statistical data obtained from experimentally determined structures [222,223]. This representation is based on the recognition that the hydrogen bond's energy depends largely on the orientation of the groups involved (see Chapter 1, Subsection 1.3.1.3 for details).

Each of the major force fields mentioned above, including CHARMM [251], AMBER [253], OPLS [258] and GROMOS [255,257], is offered as part of a larger MM/MD package. The package corresponding to a given force field may contain features that were originally available only in other force fields. For example, the CHARMM package provides the option to consider higher-order energy terms, which were not included in the original CHARMM force field. The packages are updated every few years, with emphasis on the following aspects:

1. **Consideration of atoms and molecules that are not proteins or nucleic acids.** Many of the original force fields have been fitted to reproduce experiments involving either biological macromolecules or small organic molecules similar to their functional groups. However, metals and organic groups that can be found as protein prosthetic groups and enzyme cofactors often lack corresponding parameters in force fields. Class II force fields such as MMFF [260–262,270] are usually compatible with these *heteromolecules*. However, updates or additions to

the common force fields, such as the 'general AMBER force field (GAFF)' [271], also enable these non-protein molecules to be considered. This issue is probably most pressing in the field of drug design, as most drug molecules are foreign to biological systems, and are therefore not recognized by the common force fields. Current efforts focus on including parameters for such molecules in known force fields (e.g., OPLS-AA [272,273], and the general 'CGenFF' force field used by CHARMM [274-276]).

2. **Inclusion of solvent molecules.** Whereas some early force fields were parameterized to reproduce only molecular properties measured in vacuum, many of the current force fields are able to treat the aqueous solvent explicitly (e.g., the TIP3P model used by CHARMM, AMBER and OPLS; the TIP4P, which is also used by OPLS; and the SPC model used by GROMOS). Some are also parameterized to use nonpolar solvents (e.g., GROMOS).

3. **Inclusion of Quantum-Mechanical (QM) expressions (hybrid force fields [198,199,277,278]).** This enables the force field to address aspects of the system that involve changes in the distributions of electrons around atom nuclei, such as electronic polarization[*a] and the formation and breaking of covalent bonds [279]. Inclusion of this treatment creates a *polarizable force field*. Indeed, studies demonstrate that it is possible to substantially increase the accuracy of protein-ligand simulations by accounting for the polarization of the ligand due to the binding site's electric field [280,281]. The inclusion of electronic polarization in the calculations also enables scientists to inspect certain details of the system that would otherwise remain obscure. Such details include, for example, the effect of polarizability of the first and second hydration shells around the protein, interactions between aromatic groups, the effect of lone electron pairs on the geometry of hydrogen bonds, and ion selectivity in channels. However, the use of QM calculations is not without problems. One known problem of this treatment is its poor description of dispersion forces [197]. This can be addressed by describing the van der Waals component of the energy using a classical term, such as the Lennard-Jones expression. In CHARMM, different types of treatments have been added to account for the effects of polarization [281-284]. The latest treatment approximates polarization by using the classical Drude oscillator model [284,285]. The use of this model requires relatively little computing power, such that MD simulations that incorporate it can run in a time period of microseconds. The model has recently been shown to capture the folding cooperativity and temperature dependence of an α-helical peptide, more accurately than simple, additive force fields [286].

4. **Inclusion of terms that address solvent effects.** As explained above, such effects are very difficult to handle using explicit treatment, due to the large number of solvent molecules in the system. This problem can be resolved, at least partially, by using implicit (mean field) models [197] (see Chapter 1, Box 1.3). Thus, the

---

[*a]The redistribution of atomic electron charges, induced by changes in the electric fields to which the atoms are subjected [197].

Poisson-Boltzmann (PB) equation can be used to calculate the polarization component of electrostatic interactions, whereas a surface area (SA) expression can be used to describe the hydrophobic effect. This *PBSA* approach [218] works quite well, and can be incorporated into force fields employed in MM calculations (the combined approach is called '*MM-PBSA*' [219-221]). The MM calculations in this case are carried out in the absence of solvent molecules, and the PBSA terms are used instead for calculating solvent effects. As explained in the main text, MM calculations produce potential energy values, whereas PBSA calculations produce free energy values. In order to combine the two, the MM calculations are replaced by dynamic simulations (MD), in which numerous conformations are sampled to obtain free energy values. However, this creates a new problem; PB calculations take several minutes per one conformation of the protein, and therefore cannot be used in simulations that sample a huge number of such conformations. Fortunately, there is a simpler variation of the PB model, called the '*generalized Born (GB) model*', that approximates the shape of the protein as a collection of charged spheres, the electrostatic potential around which can be calculated analytically, thus allowing the calculations to be completed much faster. Indeed, the GBSA model [287] has already been successfully incorporated into force-field-based simulations (the approach is called '*MM-GBSA*'). In some systems, it is necessary to carefully assign protonation and tautomer states in order to obtain high accuracy [197].

3.4.3   Template-based (comparative) approach

3.4.3.1   Introduction

The proteins that exist in nature today developed through long evolutionary processes, progressing via random mutations and natural selection. The genetic revolution that started in the 1950s enabled scientists to determine the nucleotide sequences of many different species, and this information, in turn, was used to determine the amino acid sequences of their proteins. Around the beginning of the 1990s, a new field in biology called '*bioinformatics*' emerged, in which scientists sought to predict the characteristics of new proteins on the basis of properties of their sequences. For example, **one can learn about the 3D structure of a new protein by finding another protein with similar sequence properties and a known 3D structure (i.e., a template)** [288]. Such comparative, bioinformatic methods for protein structure prediction are termed '*template-based*', '*statistically-based*', or '*data-derived*'. Some template-based methods rely on sequence similarity (i.e., sequence homology) between the target protein and the template, whereas others rely on similarities in certain propensities that are determined by the sequence, such as formation of specific secondary structures. These propensities reflect physicochemical principles and evolutionary constraints, and they are usually extracted through statistical analysis of large protein databases [289,290]. For this reason, the field has benefited greatly from the growing number of fully sequenced genomes in recent years, including the human genome. In accordance with the above, template-based methods can be divided into two main groups based on the type of similarity used for comparison. Homology modeling methods rely on sequence similarity, whereas fold recognition methods rely on similarity of sequence-derived propensities or statistical tendencies. In the following subsections we describe each of these types.

3.4.3.2 Homology modeling

Homology modeling methods are based on the paradigm that sequence codes for structure. That is, **two highly similar amino acid sequences in two different (yet evolutionarily related) proteins should acquire the same local structure**[*1]. This assumption is not entirely accurate, as the local environments of the similar segments may be different in the two proteins, making them fold into different structures. However, if the common sequence in the two segments is evolutionarily conserved, it is reasonable to expect that the segments do share the same structure. This is because conserved sequence motifs are usually structurally and/or functionally important, and as we already know, they are often linked to important biological functions (e.g., substrate binding, catalysis, etc.). Thus, if a certain sequence of an unknown protein also appears in a protein of known structure, we can use it as a template to predict the structure of the former. As we will see in Chapter 5, proteins are dynamic entities; they undergo conformational changes due to the thermal energy in the solution surrounding them. Thus, each protein exists as an ensemble of conformations, and may even undergo larger conformational changes when its environment changes, when it binds a ligand, or when it is post-translationally modified [291]. Related proteins, sharing the same function, should alternate among similar conformations. Indeed, it has been found that proteins having one conformation in common also share other similar conformations [292]. To complete the structural picture, it is necessary to model all the relevant conformations of the query protein.

3.4.3.2.1 Principal steps

Homology modeling typically consists of the following steps [5,293]:

1. **Template search and selection.** This step entails finding at least one protein ('template') of known structure that has high sequence similarity to the unknown protein (i.e., the 'query'). This task is easy if the structure of a close homologue of the query protein has already been solved [293]. In some groups of proteins, however, such as membrane proteins, solved structures are scarce, making the task of identifying a template more difficult. There are several well-known accessible algorithms for finding templates, such as *PSI-BLAST* [294] and *HMM* (*Hidden Markov Models*) [295] (the HHpred method is commonly used [296,297]; see Subsection 3.4.3.3 below). Finding the right template(s) and characterizing their homology to the query protein is usually what determines the success of the entire prediction. In other words, when the sequence similarity between a query and a prospective template is too low, the prediction is worthless [298]. Fortunately, even when the *global sequence similarity* is low, certain segments in the query protein may be similar to specific segments in different templates (i.e., *local sequence similarity*) [299].

2. **Building a multiple sequence alignment (MSA) that includes the template and query proteins (Figure 3.11).** In this step, the modeler exploits evolutionary information to improve the alignment between the sequences of the template and query. Accordingly, it is difficult to establish an alignment between distant homologues, as in the case of a eukaryotic query and prokaryotic templates [293]. There are sev-

---

[*1]For example, Chothia and Lesk noted that proteins with 50% sequence identity have a corresponding structural r.m.s.d. of ~1 Å [3].

eral freely available tools for creating MSAs, such as *ClustalW* [300,301], *MAFFT* [302], *MUSCLE* [303], and *T-Coffee* [304]. Creating a good alignment is not a trivial task; even when the query and templates are globally similar, low-similarity regions within the proteins' sequence might skew the alignment and interfere with the prediction. One way of overcoming this problem is to use different MSA algorithms and compare the structural models that they produce (e.g., [305]). Another way is to incorporate structural information into the MSA, e.g., the tendency of certain sequences to acquire certain secondary structures (see Box 3.2 on secondary structure prediction and assignment).

3. **Assigning the spatial coordinates of the templates to the sequence of the target protein.** This is done for each segment according to its sequence similarity in the MSA. Well-known tools that are often used for this purpose include MODELLER [288], NEST [306], and SWISS-Model [307]. Because the alignment between target and template often includes insertions and deletions, and since in many cases several templates are used, the structure obtained for the target protein is usually deformed and may include suboptimal bond lengths and angles, and even overlapping atoms.

4. **Refinement of the model-structure.** This step is carried out by algorithms that compare geometric characteristics of the model-structure to those observed in proteins, or by energy-based calculations that detect unfavorable atomic configurations [308].

5. **Evaluation and validation of the resulting structure.** The reliability of the prediction can be assessed in different ways. First, certain qualities of the target protein can be compared to pre-determined statistical tendencies. For example, the Verify3D algorithm [309] ranks short segments of the model structure according to the known tendency of residues to appear in certain secondary structures or become buried inside the protein core. Second, the energy of the structure, which reflects the statistical qualities mentioned above, may be calculated [308]. This approach is particularly successful when the prediction yields several similar structures. Indeed, in such cases energy-based methods have proved to be efficient in selecting the right structure, as long as the structures do not differ significantly [310,311]. Third, the evolutionary conservation levels of the amino acids can be correlated with their buried or exposed status, where the anticipation is for the protein core to be conserved and for the periphery to be variable. This can easily be done visually by conducting ConSurf calculations [312]*1 with the model-structure. The ConQuass method [313] assigns a score to the model that reflects the degree to which it fulfills this anticipation. Finally, the model may be validated according to its consistency with experimentally derived data, such as low-resolution structures of the protein, and/or mutational data.

3.4.3.2.2 Shortcomings and overall efficiency

Generally speaking, homology modeling requires that the target and template proteins share at least 30% sequence identity (throughout their shared regions) [4]. However, studies indicate that reliable prediction requires at least 50% identity. In the latter cases, most homology-modeling software programs perform similarly well [315], with a backbone r.m.s.d.*2 of

---

*1 Server: http://consurf.tau.ac.il
*2 That is, the root mean square deviation between the backbone atoms of the predicted model and those of the real structure (see Chapter 2, Box 2.5).

(a) (b)

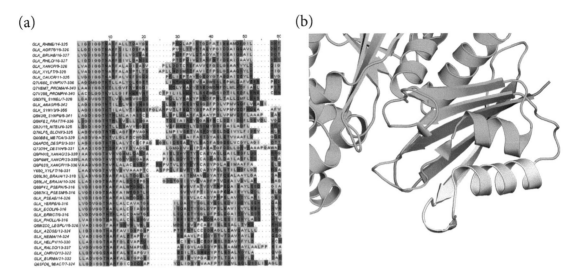

FIGURE 3.11 **Multiple sequence alignment.** (a) Part of a multiple sequence alignment of homologous glucokinase enzymes. For clarity, the different residue types (e.g., polar-neutral, nonpolar, charged, etc.) are colored differently. The alignment was taken from the Pfam database [314]. The sequence names are listed on the left, and the sequences are aligned to maximize the amino acid similarity in each column without opening too many sequence gaps. (b) The structural location of the first two sequence blocks found in the alignment. The structure shown is that of *E. coli* glucokinase (PDB entry 1sz2). The first block (residues 1 to 21) is colored in orange and the second block (residues 29 to 52) in yellow. The rest of the structure is colored in green. Both blocks correspond to organized secondary elements.

1 to 2 Å [4,191,316]*[1]. In the 30% to 50% sequence identity range, the corresponding r.m.s.d. value increases to 2 to 4 Å for most of the protein's core structure [317]. In any case, because of the reliance of comparative modeling on target-template similarity, a model structure produced by this approach is likely to be more similar to the template(s) than to the actual native structure of the target sequence [191]. As a result, it is necessary to efficiently refine the model by using energy or knowledge-based procedures. We should remember, though, that such procedures are not perfect (see Subsection 3.4.2.3). Therefore, their application may create a model structure that is even further away from the native one. As might be expected, most of the problems in homology-modeled structures tend to appear in regions of low similarity. These have mutated extensively during evolution and tend to organize as loops. Unfortunately, these loop regions are quite important when studying protein-ligand binding (e.g., in drug discovery; see Chapter 8), as they tend to be part of protein binding sites. The reliance of protein-ligand binding studies on homology models is also burdened by the fact that any deviation of the modeled protein from the native one tends to be larger for side chain atoms. This is because protein-ligand interactions depend significantly on the exact position of side chain atoms in the binding site.

Another problem with homology modeling is that there are many target proteins that do not have any homologues of known structure, or that contain elongated amino acid segments that are not covered by the template(s). As explained in Section 3.2, the latter are usually proteins that are hard to crystallize, such as membrane-bound proteins. In such cases homology modeling is useless, and other approaches must be used, usually with somewhat

---

*[1]The values refer to the regions of the query that are covered by the template(s).

limited success (see following section). Finally, similar sequences might not always share the same structure [318]. In spite of all these problems, the bottom line is that **homology modeling is currently the best computational method for predicting protein structures**, and its applicability is bound to increase with the determination of more protein structures that can serve as templates. A test of homology modeling efficiency carried out in 2007 has shown that in single-domain proteins comprising 90 residues or fewer, the structures predicted by this method differed from their corresponding native structures by 2 to 6 Å [190]. The challenge homology modeling faces is achieving an r.m.s.d. of 3 Å or better, particularly in large proteins and proteins with significant β structure content. Other challenges include modeling of multi-domain and membrane proteins.

Aside from structure prediction, homology modeling is also used in other fields [5]. In drug design, for example, the method is used to analyze structural differences between proteins that are targeted by the same drug, to find those that determine its binding specificity [319]. Another application is enzymatic catalysis, where the method can be used for analyzing catalytic mechanisms. For example, homology modeling has been used to predict a new peptide hydrolysis mechanism in a family of enzymes consisting of non-conserved catalytic residues [320].

BOX 3.2 SECONDARY STRUCTURE PREDICTION [321]

I. Overview

The problem of how to predict the complete three-dimensional (tertiary) structures of proteins is complex and has yet to be solved. As explained in Chapter 2, the complexity results from the numerous degrees of freedom in proteins and the difficulty of describing the interactions stabilizing this structure accurately. In contrast, secondary structures are simpler than the tertiary structure and are therefore easier to predict. Indeed, five decades of research have yielded efficient secondary structure prediction methods, with accuracy levels as high as 80%. This means that we can currently predict the secondary structures of four out of five residues in a protein from sequence alone.

Given that all important functions in proteins result from the three-dimensional structure, what does secondary structure prediction give us? In fact, quite a bit:

- Secondary structures tend to be more evolutionarily conserved than other chain arrangements. Therefore, incorporating knowledge on secondary structures often contributes toward predicting the complete 3D structure of a protein. For example, in template-based methods (see Subsection 3.4.3 in the main text) the inclusion of secondary structure information may help in two ways: (*i*) detection of protein homologues for the given sequence, and (*ii*) improving the alignment of the target protein's sequence with the sequences of the template proteins[*a].

[*a]PRALINE [322] (http://www.ibi.vu.nl/programs/pralinewww/) is an example of a method that iteratively optimizes multiple sequence alignments by incorporating secondary structure information. HHpred [296,297], which predicts protein structure on the basis of sequence profiles (see Subsection 3.4.3.3 in the main text), also uses secondary structure information in the prediction process.

This is especially true in *fold recognition* methods [323–325], which, for a given protein sequence, identify the most probable fold in a database (e.g., the PDB), based on certain properties of the sequence (see Subsection 3.4.3.3 below). Certain *ab initio* methods have also been demonstrated to improve by incorporating knowledge on secondary structures [326].

- The formation of the tertiary structure of a protein is often preceded by the formation of secondary structure elements (see Chapter 5). Thus, knowledge of the locations of secondary structures in the protein may be useful towards studying the folding process and topology of the folded structure. Indeed, such studies have revealed certain tendencies involving secondary structures, such as the strong tendency of β-α-β motifs to have a right-handed chirality, and the tendency of α/β proteins to contain core β-sheets covered by surrounding α-helices.

- In membrane-bound proteins the transmembrane segments are almost always α-helical or have a β-strand conformation. Thus, correct prediction of secondary structure locations in these proteins often provides the location of the transmembrane segments and overall topology of the protein.

II. Principles of secondary structure prediction

The two basic approaches to predicting the secondary structure of an amino acid sequence are as follows:

1. By using the statistical preference (propensity) of each amino acid in the sequence to form a specific type of secondary structure (see Chapter 2, Subsection 2.3.6). The propensities are calculated from proteins of known structures, e.g., those that are deposited in the PDB or other databases. This calculation relies on the assumption that the secondary structures are correctly assigned in the first place to the structures in the database. Such assignment is carried out by software such as DSSP [327,328] and STRIDE [329], based on backbone dihedral angles and/or hydrogen bonding patterns between backbone groups. In addition to α-helices and β-strands, both software programs are designed to recognize other types of secondary structures, such as 3_{10} and π helices, β-bridges, bends, and turns.

2. By using the evolutionary conservation of different amino acid stretches in the given sequence. The conservation levels can be extracted from multiple sequence alignments of the target sequence with its homologues.

The two approaches are not mutually exclusive and can be integrated within the same method.

Early methods — e.g., the Chou–Fasman method [330] — relied exclusively on the first approach (evaluation of statistical propensities of amino acids to form any of the three secondary structure types), since they only used single sequences. For this reason, and also because only a few protein structures were known at the time (the 1970s), the accuracy of these methods was 50% to 60% [331]. The propensities were calculated for short stretches over the sequence (typically 6 residues, sampled using a sliding window), where in some cases a score of short-range interactions between the residues was also included [332]. Development of the GOR method [333,334], which used a 17-residue window, increased the accuracy of the prediction to ~65%. Subsequent breakthroughs in the field, which eventually led to the current methods, emerged with the ability to 'train' the prediction algorithm to obtain better prediction rules [335–337]. Such training entailed applying machine learning algorithms to two key elements:

1. Multiple sequence alignments of the target sequence with its homologues. This enabled the prediction algorithm to incorporate signals regarding the evolutionary conservation levels of the residues and secondary structures.

2. Known 3D structures of proteins, whose number had increased considerably by that time. This enabled the prediction algorithm to incorporate structural information.

Neural networks constitute a popular and efficient machine-learning approach used by secondary structure prediction software [338]. The name is based on the similarity between the process by which these algorithms operate and the functioning of neurons in a nerve tissue. A neural network is constructed from interconnected layers of input and output units, where each unit receives information from other connected units, and generates outputs based on the input's weight. Neural network-based secondary structure prediction usually employs a sliding window for the training process (Figure 3.2.1).

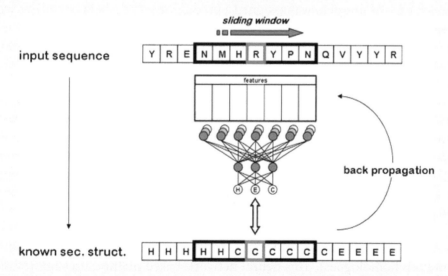

FIGURE 3.2.1 **Schematic representation of a sliding window approach used to train a neural network.** A window around the middle position, to be predicted, slides over the sequence. The trained neural network converts the window information (here 'NMHRYPN') into a prediction (here 'C', for coil) for the middle residue (here 'R'). The neural network depicted represents a prediction based on a single sequence. Modern methods typically use a multiple sequence alignment as input, which is then converted into a profile comprising a frequency table of the amino acids appearing at each alignment position. Taken from [321].

In addition to neural networks, other types of computational formalisms are used, such as *k-Nearest-Neighbor approaches*, *Hidden Markov Models* (*HMM*) methods and *Consensus* approaches. HMM is a probabilistic approach implemented in multiple fields, including speech recognition and weather forecasting. In bioinformatics it is often used to find homologues of a target protein [295], based on which a multiple sequence alignment can be built (e.g., see Subsection 3.4.3.3 on fold recognition). HMM is particularly useful in finding distant homologues for query proteins [339]. Consensus methods are based on the assumption that no single prediction method is better than the others in all cases. Therefore, to make a realistic prediction, consensus methods apply different prediction algorithms that employ different strategies, and they decide on the final prediction by using a simple majority voting scheme. Most widely used methods for secondary structure prediction employ combinations of neural networks, support vector machines, HMM, and consensus algorithms, as described in the following subsection.

III. Examples of secondary structure prediction software and methods

PHD (now PredictProtein): This was the first method to combine database search with neural network prediction based on multiple sequence alignment (MSA). The first version, developed by Rost and Sander [337], produced an accuracy level higher than 70%. In this version, the MSA was used to produce a sequence profile, which was fed to a three-layered neural network. Each layer produced a prediction more accurate than the one before it, until a final prediction emerged, in which each amino acid was assigned a prediction (alpha, beta or coil), as well as a reliability index. The most recent version of the software uses a better homology searching method (PSI-BLAST [294]), which identifies more distant homologues and therefore produces better MSA [340,341].
Web server: https://www.predictprotein.org/

PsiPred [342]: A neural network-based method that is popular because of its easy usage and high performance. Like PHD, this method also uses PSI-BLAST, but it feeds the position-specific scoring matrix (PSSM) profiles that are obtained from PSI-BLAST directly to the neural network. The latter contains only two layers, but its training protocol is effective.
Web server: http://bioinf.cs.ucl.ac.uk/psipred/

SSPro [343]: One of the most accurate algorithms (79% accuracy for proteins with no PDB homologues). This neural network-based method uses information on long-range interactions in the prediction process, and relies on several windows that slide from opposite sides. A recent version called SSPro8 replaces the standard three-class categorization of secondary structures (α-helix, β-strand, loop) with a full DSSP 8-class output classification (see above).
Web server: http://scratch.proteomics.ics.uci.edu/

SAM-T06: The SAM-T series of secondary structure prediction, which includes SAM-T99, SAM-T02 and SAM-T06, combines HMM for homologue search and neural networks for processing the resulting MSA. The current version (SAM-T06 [344]) has a refined protocol and also contains options for additional types of predictions.

Web server: https://compbio.soe.ucsc.edu/SAM_T06/T06-query.html

Jpred: A consensus method with accuracy of ~82%. Whereas the original algorithm [345] used several different secondary structure prediction methods (including PHD), later versions [346] used Jnet instead, an algorithm that relies on PHD-like strategies. The current version [347] uses PSI-BLAST-derived position-specific scoring matrices (PSSMs) and HMM profiles, which are passed on to Jnet for the prediction.

Web server: http://www.compbio.dundee.ac.uk/jpred4/index.html

IV. Accuracy assessment

The accuracy of secondary structure prediction can be assessed in absolute or relative terms. Absolute evaluation criteria can include, for example, the percentage of residues assigned the correct secondary structure by a given method. To be able to use such criteria, it is necessary for the structure of the benchmark protein to be known, and for the structure to be assigned the right secondary elements (see above for relevant software). The most extensive assessment of absolute prediction accuracy, both for secondary and tertiary structures, is the CASP contest, which is carried out every two years [348] (see Subsection 3.4.5 below for more details)[*a]. The prediction accuracy of a given method can also be assessed relative to other methods. Such comparison across methods is carried out by EVA [349], a web server[*b] that was launched in 2000 and has been continuously testing prediction algorithms since then. EVA is not limited to secondary structure prediction; it also tests algorithms of 3D structure prediction that employ homology modeling or fold recognition approaches. The reference data of EVA are updated daily from the PDB, and the new sequences are submitted to the different tested algorithms.

3.4.3.3 Fold recognition via threading

When the target sequence has no close homologues of known structure, homology modeling cannot be employed to predict its structure. However, we have already seen that in proteins structure is more conserved than sequence [3], as reflected by the fact that the number of protein folds in nature is much smaller than the number of sequences [350–352]. This means that proteins with different sequences may still form similar structures because of certain shared properties that are encoded in their sequences. If these properties or their statistical recurrence could somehow be identified, it would be possible to predict the structure of a

---

[*a]Interestingly, in the first rounds of CASP, secondary structure prediction was a separate category. This category was cancelled after the organizers noticed that the winners in this category used a somewhat circular approach. They predicted the 3D structure and used their model structure to decipher the secondary structure elements.

[*b]URL: http://pdg.cnb.uam.es/eva/

new protein based on a template that shares the same properties, even if the two have dissimilar sequences [353]. This is essentially what fold recognition methods do; they look for certain shared sequence properties or tendencies between the target protein and a protein of known structure (stored in the PDB), which would signify that the two have a similar fold or similar structural motifs [191]*1. To enable such a search to take place, the sequence properties of both the target protein and all proteins in the PDB must be represented in a simple manner. As will be further explained below, fold recognition methods represent sequence-encoded properties and tendencies either as a simple profile or as a pair potential [357,358] (see more below).

Structure prediction based on fold recognition usually includes two basic steps:

1. Identifying templates, i.e., proteins of known structure whose fold should be similar to that of the target protein. This can be done by *threading*, i.e., by aligning the sequence of the target protein with the structure of each fold in the PDB on the basis of *shared sequence properties*, and then choosing the best match. The sequence properties can be grouped into two types:

   (a) Properties that are explicitly related to the structure of amino acids [323–325]. For example, different amino acids have different tendencies to form certain secondary structures (see Chapter 2), to be exposed to the surrounding solvent, to have certain dihedral angles, to interact with certain amino acids, etc. Examples of software using these properties include GenTHREADER, which relies on solvation information, pGenTHREADER [359], which employs secondary structure information, and the SPARKS-X server [360], which takes into account secondary structure, dihedral angles, and solvent-accessible surface area. The latter has been found to contribute the most to prediction accuracy [360].

   (b) Properties that are not necessarily explained in terms of structure yet can be detected statistically by identifying amino acid frequencies in multiple sequence alignments. One way to represent these statistical tendencies is via a *profile*, which denotes amino acids that frequently appear in certain positions in the alignment. Such a profile is more informative than a simple amino acid sequence since it contains evolutionary information, i.e., the conservation levels of specific positions in the sequence, the chances of insertions or deletions, etc. The profile of the target protein is compared to corresponding profiles of all proteins with known folds (*profile-to-profile alignment*), and proteins that are identified as matches (i.e., that are sufficiently similar to the target protein) can be used as templates. Again, the fact that profiles incorporate additional information beyond simple amino acid sequences makes profile-profile alignments more sensitive for finding templates. A profile matching procedure is employed by HHPred [296,297], a popular, open-source software that can be used in a stand-alone mode or via the Internet (http://toolkit.tuebingen.mpg.de/hhpred).

---

*1 In many cases it is possible to find multiple proteins with shared properties (i.e., templates) in the PDB. But more importantly, the likelihood of finding at least one template for a given sequence is high, given estimates that most of the folds of natural globular proteins are already known [354–356]. This is not surprising considering the high conservation of folds and the high frequency of superfolds (see Chapter 2, Subsection 2.4.2.2.4). Note, however, that these estimates depend on how folds are defined and whether the fold space is discrete or continuous.

Specifically, HHPred employs pairwise comparison of profile hidden Markov models (HMM, see Box 3.2 above) between target and template proteins[*1].

In most current methods, both types of sequence properties are used to find the best template(s).

2. Constructing a model structure for the target protein based on the above alignment.

Fold recognition and threading software typically includes the following three components[192]:

1. A library of potential folds or structural templates (usually the PDB).

2. A scoring function for evaluating any particular placement of a target sequence into a given fold. As mentioned above, the function is usually constructed statistically by analyzing multiple sequence alignments, but often includes components derived from more structurally related amino acid propensities as well. The scoring function predicts how favorable each target protein-template amino acid replacement will be at each position of the sequence.

3. A method for searching the entire space of possible replacements between each sequence and each fold for the best set that gives the best total score.

There are many fold recognition methods, and most have comparable accuracy. Still, each method is a little different form the others, and may therefore provide more accurate predictions in certain cases. One way to obtain good results is to use meta-servers, that is, servers that use different threading methods on the input (target) protein and choose the most realistic model according to certain considerations. Examples of such meta-servers include 3D-Jury[361], LOMETS[362], Pcons.net[363], and TASSER[364]. It should be noted, though, that some individual methods have reached a level of accuracy that is comparable to and even better than that achieved by meta-servers. For example, HHPred is considered to be one of the best servers for template-based structure prediction[365], with an added bonus of being the fastest. SPARKS-X[360] is another example of an efficient server.

3.4.4 Integrative and fragment-based methods

Each of the approaches discussed above has advantages and disadvantages, and one might be more suitable than others for specific prediction tasks. For example, homology modeling is the best method when the target protein has at least one close homologue of known structure, whereas fold recognition is best when such homologues cannot be found. *Ab initio* methods cannot be used to predict the structure of an average protein starting from an open conformation, but they are usually very useful in refining models that are close to the global energy minimum of the system. Thus, an integrative method that uses several approaches and applies them according to their respective strengths is expected to have a clear advantage over any of the individual approaches. Indeed, such methods are used by most, if not all current protein structure prediction software programs and servers. For example, homology

---

[*1]Since HMM methods are more sensitive than others in finding remote sequence homologues, the HHPred server is also widely used for finding homologues for query proteins, e.g., in searching for homology modeling templates.

modeling methods always refine their initial models by using further calculations or simulations. These may be based on energy potentials, knowledge-based scoring functions, or both. Similarly, the use of profile-based methods to detect possible templates is not limited to fold recognition techniques. As mentioned in Subsection 3.4.2.3.3 above, many current methods employ *fragment-based calculations* to make predictions more feasible. In this approach, the prediction process starts with the identification of short protein fragments that are consistent with certain calculated properties of the input protein sequence. Then, the entire protein structure is assembled from the selected fragments in a process that involves the creation and scoring of numerous possible combinations of the individual fragments.

One tool that combines different approaches is *I-TASSER* [366]*1, developed by the Zhang group. The software is an extension of TASSER [364], and the procedure it employs includes the following three steps [366] (Figure 3.12):

1. Threading: The software identifies fold templates for the target protein by threading its sequence through a library of representative PDB files, using LOMETS [362]. The threading employs a sequence profile and secondary structure assignments, both of which are calculated from the sequence of the target protein. Each template is ranked by a variety of sequence-based and structure-based scores.

2. Structural assembly: Continuous fragments in threading alignments are excised from the template structures and used to assemble conformations of the aligned sections. The unaligned regions are constructed by replica-exchange MC simulations of a coarse model (no side chain atoms). The force field guiding the simulations includes both energy-based and knowledge-based components. The different resulting structures are clustered, and those with the lowest energy are selected.

3. Model selection and refinement: Further refinement of the resulting model structures is carried out by a second MC simulation, this time with spatial constraints taken from the threading alignments and the PDB. The missing side chain atoms of the lowest-energy structures are then added through the optimization of hydrogen-bonding networks.

I-TASSER received the highest rank for automated structure prediction in CASP 7-11.

Another popular method that integrates different approaches is *Rosetta* [240,241], which was developed by the Baker group. Rosetta can use both *ab initio* and comparative protocols for structure prediction. In the former case, the software uses fragments of unrelated proteins to predict the complete structure of the target protein (Figure 3.13). First, the software scans its database for fragments whose sequences are similar to local regions in the sequence of the target protein. These fragments are assumed to have a native-like structure, and in the next step they are assembled in different combinations to form the complete structure. Each of the resulting models is ranked by a scoring function, which includes various components, including physicochemical features such as radius of gyration, solvation, and residue pair interactions. The best structures are then refined using MC simulations. Thus, Rosetta employs both knowledge-based (statistical) and energy-related considerations in the search for the global structure of the target protein.

---

*1http://zhanglab.ccmb.med.umich.edu/I-TASSER/

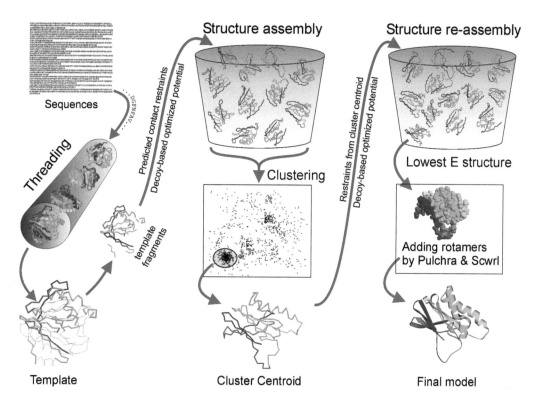

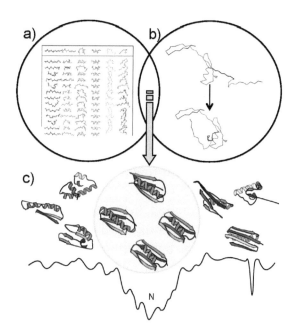

FIGURE 3.13 **Rosetta's *de novo* folding algorithm.** Rosetta starts from (a) fragment libraries with sequence-dependent (ϕ and ψ) angles that capture the local conformational space accessible to a sequence. (b) Combining different fragments from the libraries, the algorithm folds the protein through optimization of non-local contacts. (c) A low-resolution energy function smooths the rough energy surface, resulting in a deep, broad minimum for the native conformation. Metropolis Monte Carlo minimization drives the structure toward the global minimum. The image is taken from [368] (http://pubs.acs.org/doi/full/10.1021/bi902153g).

Finally, a relatively new approach for structure prediction relies on correlated mutations. In this approach, positions that are close to each other in the folded protein are predicted based on their linked tendencies to undergo mutations throughout evolution. This approach can be viewed as an intermediary between the *ab initio* and template-based approaches; it does not require any template with known structure, but it does require a sufficient number of homologous sequences to the target protein to carry out the evolutionary analysis. The approach, which can be used on its own or in combination with other methods, is discussed in Box 3.3.

BOX 3.3 STRUCTURE PREDICTION BASED ON EVOLUTIONARY SEQUENCE VARIATION

The appearance of mutations in protein sequences during their evolution depends on various aspects, both structural and functional (see Chapter 2). Interestingly, sequence analysis of protein families shows that certain positions tend to coevolve. That is, the appearance of a mutation in one position is accompanied by a mutation in another. Such linkage might occur in positions that are close in 3D space and interact with each other [369–371]. If a mutation in one position leads to disruption of its interaction with the adjacent position, a compensatory mutation of the latter may remedy the problem. For example, if two positions were originally polar and were involved in a favorable electrostatic interaction with each other, mutation of one position from polar to nonpolar would disrupt the interaction. However, if the adjacent position also mutates from polar to nonpolar, the original, favorable electrostatic interaction would be replaced with another favorable interaction, a nonpolar one. The above suggests that it is possible to identify positions that are located near each other in a protein's 3D folded structure by analyzing the protein's sequence variation throughout evolution and observing which positions have coevolved [372–375] (Figure 3.3.1a). Furthermore, if a sufficient number of such positions is found, distributed throughout the entire sequence of the protein, they can be used to predict its native 3D structure. This is done by mapping all predicted contacts (Figure 3.3.1b) and incorporating them into a structure prediction algorithm. The key step of the entire process is identifying the coevolved positions on the basis of sequence information. This is done by focusing on the target protein's evolutionary family, the members of which have the same fold (an isostructural family) [372]. Specifically, the sequence of the target protein is aligned with the sequences of the other family members, and a statistical analysis is carried out to find all positions whose evolutionary changes (i.e., mutations) are correlated. Each pair correlation found this way, if strong enough, is assumed to represent two positions that are close in 3D space in the folded protein.

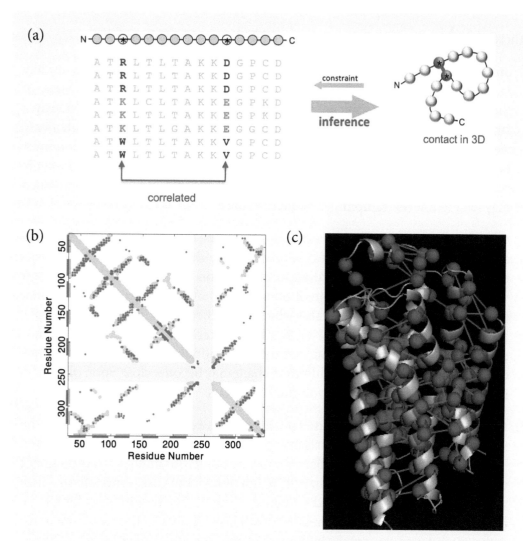

FIGURE 3.3.1 **Structure prediction based on identification of correlated mutations.** The process is based on the inference of contacts between nearby residues in the native structure of a protein from correlated mutations in its sequence. (a) The prediction process, demonstrated for clarity on two positions forming a single 3D contact. *Left*: Multiple sequence alignment of the target protein with the sequences of other family members, all of which have the same fold. The two correlated positions are marked. *Right*: A schematic representation of the 3D structure of the folded protein, showing the proximity of the two correlated positions. The images are taken from [372]. (b) A contact map of top-ranked evolutionary inferred couplings (EICs) (red) overlaid on contacts from a crystal structure of human β_2-adrenergic receptor (grey). The top 350 evolutionary couplings are shown, covering nearly the entire length of the protein. EICs between residues that are less than 5 Å away from each other are omitted; hence, there are no red marks along the long diagonal. The dark and light green bars on the axes show the locations of experimentally determined and predicted α-helices in the sequence, respectively; these are easily inferred from the high density of EICs along the short diagonals. The light blue areas correspond to a large flexible loop missing from the crystal structure. The image was adapted from the Evfold [373] website by Yana Gofman. (c) Top 100 EICs (red), projected onto the known 3D structure of human β_2-adrenergic receptor (PDB entry 2rh1) (grey ribbon). The image demonstrates how EICs can be used as constraints in the structure prediction process carried out by other algorithms (e.g., fold recognition, *ab initio*, etc.). The image is taken from the Evfold [373] website.

There are three main obstacles to the correlated mutation approach [372]:

1. **Statistical noise** – Sequence variation analysis within a protein family usually yields multiple positions that display different degrees of correlation. Many of these, however, are insignificant (i.e., noise), and should be ignored. Obviously, a strong correlation is more likely to reflect a real contact in 3D space than a very weak one. However, it is not simple to determine how strong a correlation must be in order to be considered significant. The statistical noise problem is especially common in cases where an insufficient number of sequences is sampled, which may lead to uneven sampling in sequence space.

2. **Correlation between distant residues** – Positions that are distant in the folded structure of a protein sometimes coevolve as well. This phenomenon, while making structure prediction according to coevolution difficult, may be important when studying protein function; it often reflects indirect interactions that link the two coevolving positions, and which are difficult to identify. For example, in allosteric proteins, an active or binding site residue may be remotely influenced by an allosteric site residue, via dynamic motions. Alternatively, in homo-oligomers, distant residues within each chain may become close upon oligomerization and therefore display evolutionary linkage.

3. **Insufficient number of correlated positions** – As explained above, to predict the structure of a protein on the basis of correlated mutations, the positions inferred by the prediction process as adjacent must be numerous and spread throughout the protein sequence. This, however, is not always the case, especially in cases where the number of compared sequences is small. The requirement for a large and highly dispersed set of correlated positions is similar to the requirement for sufficient distance constraints in the case of structure determination using NMR.

There are certain statistical approaches, such as *maximum entropy*, which help in addressing the above problems, at least partially, and identifying meaningful positional couplings. However, when the target protein has a small number of homologues, the evolutionary analysis and prediction process are usually highly inaccurate. One possible solution to this problem is to combine the evolutionary data with other predicted features of the protein, such as secondary structure and solvent accessibility (see below).

While the idea of using evolutionary covariation to find adjacent positions is not new, these methods have only recently become accurate enough to be used for structure prediction [374]. Today there are quite a few computational tools for analyzing correlated mutations and predicting 3D contacts; some of these tools can be used as web servers. Most of these tools, such as *DCA* [376], *MISTIC* [377], *CMAT* [378], *MetaPSICOV* [379], and *I-COMS* [380], provide the user with information on coevolved positions only[*a].

[*a]MetaPSICOV, which employs neural network calculations and three different coevolution methods, also predicts positions that are involved in long-range hydrogen bonding. Furthermore, when the available sequence alignment is poor, MetaPSICOV down-weights the coevolution calculations and instead uses other contributions, such as predicted secondary structure and solvent accessibility, through a process of machine learning.

However, a few tools also use the inferred contacts as spatial constraints for predicting the 3D structure of the target protein:

- **EVfold** [372,373]: This method incorporates evolutionary spatial constraints (as well as predicted secondary structure constraints) into distance geometry algorithms, which are usually employed for structure determination based on NMR spectroscopy. The initial 3D conformations of the target protein that are generated are then refined using simulated annealing. This method can handle both globular [372] and transmembrane [373] proteins, and is currently optimized for individual domains. A variation of EVfold called *EVcomplex* [381] uses a similar approach to predict positions in different proteins that interact with each other. It relies on the premise that two sequences in the same genome from nearby locations are likely to be co-expressed and physically interact. Both EVfold and EVcomplex are freely accessible as web servers.

- **FILM3** [375]: This tool was developed by the David Jones group to predict the 3D structures of membrane proteins. FILM3 uses evolutionary inferred contacts predicted by PSICOV [382], and the FILM structure prediction platform [383]. In fact, the original statistical potential of FILM has been replaced in FILM3 by a scoring potential that is based entirely on the estimated probabilities of predicted residue-residue contacts. The method has been tested on 28 membrane proteins with diverse topologies, and most of the models achieved a TM-score of > 0.5 following further refinement by MODELLER [375].

- **PconsFold** [384]: A pipeline for *ab initio* protein structure prediction of single-domain proteins. It uses predicted contacts from PconsC [385] and a Rosetta folding protocol. PconsFold seems to be more accurate than EVfold; its accuracy is attributed to its use of PconsC for contact prediction [384]. PconsFold is freely available for download.

- **Rosetta-GREMLIN** [386]: This method integrates the evolutionary constraints calculated by GREMLIN [387] with the Rosetta algorithm. Aside from the use of Rosetta, this tool offers two additional advantages over other tools:

  1. The distance constraints are assigned weights that are proportional to the strength of the coevolution signal.
  2. To avoid becoming trapped in wrong conformations during the folding simulation, the protocol first applies constraints between pairs of amino acids that are close in the sequence, and only then adds constraints between residues that are more distant from each other in the sequence.

This method was implemented in the CASP11 competition, where its predictions were found to be considerably more accurate than any previous predictions made in CASP's history for proteins comprising more than 100 amino acids and lacking homologues of known structure. The same method was then used to predict the structures of representative proteins belonging to 58 prokaryotic families, for which there are no known structures. The vast majority of these are membrane proteins.

3.4.5 Prediction assessment and verification

One of the indications of the importance of protein structure prediction is the effort invested in assessing and validating it. Since 1994, every two years, an international contest called CASP[*1] is held among scientists worldwide to find the best protein structure prediction algorithms and methods [348]. The contest starts with the release of sequences of proteins, whose structure is soon to be determined experimentally, and the participating research groups compete in predicting the native structures of the proteins until they are published. Since the prediction methods are diverse, the assessment of prediction efficiency is carried out in different categories. For example, *ab initio* methods belong to a separate category from comparative methods, and fully-automated predictions are distinguished from predictions obtained using human intervention. Results are analyzed on the basis of certain criteria, such as the number of residues whose locations have been predicted with a certain level of accuracy, successful identification of secondary structures, domain boundaries, side chain contacts, and disordered regions [388].

Inspection of recent CASP contests shows that most of the progress in recent years has occurred in the field of comparative prediction, mainly in the identification of remote sequence homologues and alignment of multiple sequences [224]. Despite the growth of computational power, the progress of the physical approach has been relatively moderate, and most of its prominent successes have been achieved with small proteins of the all-α type [298,389–391]. The reason for this seems to be the lack of significant innovation in the descriptions of physical interactions in the system by force fields, especially polar interactions that include the solvent. As a result, structural biologists who carry out predictions tend to rely on comparative methods for creating an overall model of the target protein, and then use energy-based methods only for refining the model.

The importance of CASP to structural biology is immense; in addition to creating an incentive for many researchers to develop new prediction tools, the analysis it provides and its publication accelerate the development process. In particular, we have seen the emergence of different automated methods for protein structure prediction. Some of these tools are accessible on the Internet as web servers, and they draw much public interest, as they do not require any pre-existing skills [392]. As a result, these servers can be used not only by bioinformatics experts but also by other scientists and students. The results of these assessment tests show that the best automated predictions are achieved either by consensus selection (see Subsection 3.4.3.3 above) or by using information on multiple templates to guide the physics-based assembly of the model (see Subsection 3.4.4 above) [191]. Both of these strategies are used by *meta-servers*, which are mentioned above. However, as CASP indicates, even such meta-servers are not as good as human researchers who rationally combine different tools, especially when the predicted protein has more than a single domain. Interestingly, automated and human predictions for single-domain proteins are almost identically efficient, implying that one of the problems of automated tools is identifying domains. Another known problem of such tools is the correct identification of the native (or best) structure among a group of decoys. The best solution so far has been based on consensus selection, which is described above.

[*1]Acronym: **C**ritical **A**ssessment of techniques for protein **S**tructure **P**rediction.

3.5 EXPERIMENTALLY GUIDED COMPUTATIONAL PREDICTION

3.5.1 Introduction

In Section 3.4 above, we discussed computational methods as a possible alternative for experimental methods in obtaining 3D structures of proteins. However, it is sometimes possible to combine the two approaches to determine a structure that is not amenable to either approach alone. We have seen that even when a protein can be crystallized and subjected to high-resolution diffraction experiments, the measurement provides only raw data, and the 3D structure is calculated using computational means. For example, in X-ray crystallography, computation is used both for obtaining the electron density map and for converting it into an atomic 3D model of the protein. Similarly, in NMR, different types of calculations are used to create an ensemble of atomic models that are consistent with the measured constraints. Methods such as EM or SAXS usually yield lower-resolution data than do X-ray crystallography and NMR. Therefore, EM and SAXS are mainly used for determining the general shapes of macromolecular complexes. However, recent technological developments, especially in single-particle cryo-EM, have given rise to structures at near-atomic-resolution (4 to 5 Å), at least for some proteins (see Section 3.2 above). Such data (either from EM or SAXS) can be used as spatial constraints that guide the search process employed by computational structure prediction [62,288,393,394]. Indeed, in recent years, several prediction software programs have been designed or adapted to integrate experimentally derived constraints into the structure prediction and modeling process [395] (see Subsection 3.5.2 below for details).

The integration of experimental and computational methods is also very useful for determining the supra-molecular structures of protein complexes. When the high-resolution X-ray or NMR structures of the individual complex subunits are known, the structure of the entire complex can often be determined by fitting these structures into the constraints provided by EM density maps or SAXS scattering profiles[*1]. In fact, this approach is often used even when the X-ray structure of the entire complex is available, as the oligomeric structure is often affected by crystallographic packing forces [396]. When EM data are used, the corresponding maps or models can also be used to obtain initial phases for determining the high-resolution structure [397]. Indeed, the combination of EM data and high-resolution X-ray and NMR structures has become popular in the determination of protein complexes due to the recent improvements in this method (see Subsection 3.2.3 above), as well as to the subsequent emergence of sub-nanometer-resolution EM structures of such complexes (e.g., [398–400]).

In many cases, however, a high-resolution structure of the individual components of the complex is unavailable. This is usually the case with membrane proteins, which are hard to produce and crystallize, though there are also numerous water-soluble proteins for which structures have yet to be determined. In such cases computational modeling may be the only way to obtain high-resolution structures, and the structures obtained can be combined with low-resolution data of the entire complex as described above. This is particularly true for membrane proteins, whose structures are constrained by the surrounding lipids (see Chapter 7) and are therefore easier to predict compared with water-soluble proteins. For example, Ben-Tal and coworkers combined low-resolution cryo-EM data with computationally

---

[*1]This can be done by rigid-body docking of the individual chains, followed by relaxation of the protein-protein interfacial atoms.

derived sequence conservation data to predict the topology of each of the transmembrane segments of EmrE, a prototypical small multidrug resistance antiporter [401]. The predicted model, which was the first to suggest that the two protein monomers have opposite orientations in the membrane, was later validated by X-ray crystallography [402], solution NMR dynamics experiments [403], and ample biochemical data (see [404] and references therein).

In many cases, data derived from other biophysical or biochemical methods are used to increase the accuracy of predicted structures. Such data may be taken from spectroscopic methods such as circular dichroism (CD), fluorescence resonance energy transfer (FRET) and electron paramagnetic resonance (EPR) spectroscopy, as well as biochemical methods such as site-directed mutagenesis and chemical crosslinking. One popular approach in this respect couples between different biochemical methods and mass spectrometry (see Subsection 3.3.3.3 above). The coupling enables specific attributes of the protein's 3D structure to be characterized (e.g., solvent accessibility and residue-residue distances), and this information can be converted into spatial constraints that guide structure predictions [162]. For example, crosslinking mass spectrometry uses chemical reagents to crosslink adjacent residues inside proteins or between protein subunits. After enzymatic degradation of the protein, the modified parts are identified by mass spectrometry and translated into distance constraints. By using crosslinkers with different lengths and residue specificities, it is possible to obtain a data set that can be used to refine structural predictions of both single proteins and complexes [161,405].

3.5.2 Applications and tools

In the last decade an increasing number of studies using combined experimental-computational approaches have been published (e.g., see reviews by [397,406,407]). One of the most notable is the study of Sali and coworkers [408,409], who used 10 different types of proteomic and biophysical data (including EM data) to obtain a low-resolution model of the yeast nuclear envelope (NPC), a huge protein complex that includes 456 subunits. This was a remarkable feat, as the NPC is both large in size and highly flexible, two traits that make its structure determination extremely difficult. In this study, the various data that were collected resulted in an ensemble of medium-resolution models. The models were then converted into a three-dimensional probability map, which was used to localize the 456 constituent monomers with an average precision of approximately 5 nm. Another interesting example of a combined study focused on the 2,500-residue protein talin [410]. In that study, the authors used a range of methods to study the following aspects of the protein's structure:

- The structure of the C-terminal part of the protein was determined using NMR and secondary structure prediction.

- Dimer formation was studied using X-ray crystallography and mutagenesis of the dimerization helix.

- The X-ray and NMR structures were fitted into a lower-resolution SAXS structure using molecular docking.

- EM data were used to find the stoichiometry of the complex and the modes of interaction between the individual chains.

Another large complex whose structure was determined using different methods is the 26S proteasome complex. Though the 20S core of the complex was solved by X-ray crystallography in the 1990s, the holo- (26S) complex could not be determined because of its conformational and compositional heterogeneity. The subunit architecture of the holo-complex was determined only in 2012 by two studies that employed low-resolution cryo-EM [411,412]. In one of these studies [412] the architecture of the huge (2.6 MDa) complex was determined by combining extensive data from different sources, both experimental and computational. These data included 8.5 Å cryo-EM maps of the complex, residue-specific chemical crosslinking data, proteomics data referring to the protein's subunit composition, and comparative structure models of the individual proteins. These studies were followed by others that described the complex at near-atomic resolution [413], as well as its conformational heterogeneity [414].

Studies such as the above required the development of computational tools capable of using different types of data to produce better structures. Popular examples of such tools include *Rosetta* [415–418] (see below), the *Integrative Modeling Platform (IMP)* [419], *EM-FOLD* [420], *BCL::EM-Fit* [421], and *FOLD-EM* [422]. Many of these software tools focus on predicting the structures of large macromolecular complexes. As described above, they use computational docking techniques to fit high-resolution structures of individual protein chains into the boundaries of the corresponding complex, which are determined by low-resolution methods. For example, the *MultiFit* server [423]*1 fits atomic structures into the EM density map of the complex, at resolutions as low as 25 Å. The fitting in this case is optimized by using a scoring function that includes the quality-of-fit of chains in the map, their protrusion from the map envelope, and the shape complementarity between pairs of chains. For more information on EM-fitting tools, see reviews [121,424].

It is also possible to use 1D SAXS profiles to fit high-resolution structures to macromolecular complexes. The *FoXSDock* server [425]*2, for example, uses SAXS data to guide the docking of atomic structures to form a complex. Similar tools include IMP [419] and CORAL [426]; the latter also account for missing fragments and symmetry of oligomeric structures. These and other similar software tools are reviewed in [58]. As in other cases, the fitting of high-resolution structures to low-resolution SAXS data is usually based on molecular docking procedures. In many cases, however, additional parameters are considered. For example, one modeling study achieved a flexible fitting protocol by incorporating a modified form of normal mode analysis (see Chapter 5) [427]. This algorithm allowed for large-scale conformational changes while maintaining secondary structures.

In addition to the above computational tools, there are others that do not settle for simple fitting or docking of two structure types; they directly integrate the experimentally derived constraints into the computational prediction process. Furthermore, some of these tools are not limited to the prediction of large protein complexes and can also be used to build or refine structures of individual chains, by relying on constraints derived from high-resolution data (e.g., NMR). One of the most popular tools in this group is *Rosetta* [415–418]. In the last decade a few Rosetta implementations have been developed for integrating experimental data. One such implementation is *CS-Rosetta* [418,428], which uses NMR chemical shift data relating to backbone atoms. We have seen earlier that Rosetta predicts protein structures by sampling and assembling numerous models from a library of short protein fragments. In CS-Rosetta, this procedure is guided by the chemical shift data. The advantage of this method

---

*1URL: http://modbase.compbio.ucsf.edu/multifit/
*2URL: http://modbase.compbio.ucsf.edu/foxsdock/

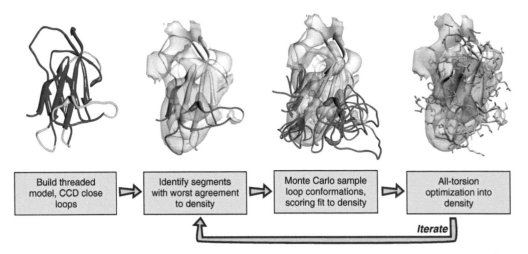

FIGURE 3.14 **Model building in which Rosetta's comparative modeling is guided by cryo-EM density maps.** A threaded model is built from some alignment (blue), using cyclic coordinate descent to close gaps in the alignment (cyan). This model is then docked into the EM density map, and regions with poor local agreement with the density data are identified (red). The conformations in these regions are aggressively resampled, and each potential conformation is scored by Rosetta's low-resolution energy function together with an agreement-to-density score. Finally, side chain rotamers are optimized, and all backbone and side chain torsions are minimized using Rosetta's high-resolution potential, also augmented with this agreement-to-density score. These three steps are iterated over until the lowest-energy models converge; at each iteration, the population of models is enriched with models characterized both by favorable Rosetta energy and by good fit to density. The image is taken from [417].

is that it avoids the bottleneck of traditional NMR structure determination, i.e., assigning the locations of all side chain atoms based on NOE data. In a recent study, CS-Rosetta was used with transmission electron microscopy (TEM) to model the structure of parallel Aβ (amyloid) fibrils [429]. Two other software packages that employ NMR chemical shifts in structure prediction are *CHESHIRE* [430] and *CS23D* [431].

Another Rosetta implementation uses cryo-EM density maps for model building and refinement. The original form of this implementation [417] used the EM data to rebuild missing regions of protein backbone (Figure 3.14). Specifically, the fit between the model and the EM density was used to guide the structure refinement process and to identify regions that are incompatible with the density. These were then targeted for extensive rebuilding. The new form of this implementation [432] relies on a more unified approach, involving the following:

- MC sampling with local optimization, guided by near-atomic-resolution EM maps

- Rosetta all-atom refinement

- Real-space B-factor fitting

Direct integration tools exist also for the use of SAXS data [62]. For example, *SAXTER* [433] integrates SAXS data with the MUSTER threading algorithm [434]. Specifically, MUSTER's fold recognition protocol is first used to identify templates for the target protein and generate model structures for the latter via threading. Then, the match between the target and template structures is scored by a combination of (*i*) MUSTER's Z-score, which is derived from the threading alignment, and (*ii*) the match between the SAXS profiles of the target and template structures. SAXSTER was tested on 412 proteins, and the results confirmed that the integration of SAXS data improved template selection compared to cases in which only MUSTER's threading score was used [433]. SAXSTER is freely available as a web server.

3.6 CONCLUSIONS

Protein structure determination still relies heavily on X-ray crystallography, which is considered to be the most accurate method [14], although cryo-EM structures of equal resolution are also emerging (e.g., [106]). Recent developments in cloning and protein expression, as well as in solving the phase problem, have accelerated the pace at which protein structures are being solved. The inherent problems of the method still exist, but today they can be overcome, at least partially, by using additional information that is obtained through other methods. For example, the locations of hydrogen atoms can be determined by using neutron scattering, and the size problem can be solved by using low-resolution methods such as electron microscopy and SAXS to provide structural constraints. Other problems that are related to the mandatory use of crystals can be solved by using NMR with the protein in solution. NMR not only provides the structure of the protein in its natural environment but also relates to its dynamics, and can be applied quickly. Indeed, NMR can be used to determine 'on the spot' the ability of different proteins to fold spontaneously [435]. Although the technique cannot provide three-dimensional structures, it does enable scientists to assess whether it is worthwhile to determine the structure of a given protein, even before starting the crystallization attempts. NMR spectroscopy suffers mainly from size limitations. Again, low-resolution methods such as electron microscopy can resolve this problem by providing structural constraints. When these are combined with high-resolution (monomeric) structures from X-ray crystallography, structural biologists are able to get a clear picture of large protein complexes. About 20 years ago, when the first structure prediction methods began to emerge, the goal of predicting the native structures of proteins on the basis of sequence alone seemed impossible. Today, the integration of different prediction approaches (physical and comparative) yields structures at resolution of a few angstroms. Moreover, many of these methods are currently accessible for non-computational scientists, in the form of web servers. Although the predictions provided by such servers are less accurate than those obtained by experts, they are good enough as a starting point for further characterization of protein structure. Finally, some of the prediction tools developed in previous decades have been enhanced substantially by the integration of experimental data that guide the prediction process. This integration improves prediction accuracy and enables such tools to assemble models of supra-molecular complexes. One should remember, though, that efficient protein structure prediction is just a means to an end; the real challenge of structural biology has always been (and still is) to deduce the functions of proteins on the basis of their sequences and/or structural data [190].

3.7 PROTEIN DATA BANK (PDB)

Protein structures that have been determined using the aforementioned methods (and others) are deposited in the *protein data bank (PDB)* [436]. The PDB was established in 1971 by the Brookhaven National Laboratory, and since 1998 it has been managed by the Research

Collaboratory for Structural Bioinformatics (RCSB). In 2003 the American RCSB, together with two other existing PDB centers, PDBe (Europe) and PDBj (Japan), created the Worldwide Protein Data Bank (wwPDB). The latter was established to ensure that standards are set and met for data representation and data quality [437]. Indeed, structures submitted to the PDB are carefully examined and must comply with certain standards in order to be accepted. Nevertheless, the PDB still suffers from several unresolved consistency problems.

A user who accesses the PDB can search for a desired structure via an electronic form offering many search options: name of the protein, name of the scientists determining the structure, time of deposition, method used to determine the structure, resolution, source organism, and more (Figure 3.15a). Such search forms can be found in the many PDB mirror sites that currently exist on the Internet (see examples below). A selected structure can then be downloaded as a file containing Cartesian coordinates representing the three-dimensional location of each atom in the structure (Figure 3.15b). The 3D structure can be viewed (remotely) by a built-in molecular graphics applet on the PDB website(s) or on other websites (e.g., FirstGlance in Jmol and NGL Viewer [438]), or (locally) by using molecular viewers that can easily be obtained online (e.g., PyMOL). The PDB file also provides information on the protein, such as its source, method of determination, resolution, amino acid sequence, location of secondary structures, and more. Recent entries also contain the electron density map, showing the extent to which the coordinates are derived from the data (as opposed to the force field used in the optimization). In structures solved by X-ray crystallography, the B-factor is specified as well. Many current PDB websites also provide functionalities that enable the user to view/identify protein-ligand interactions or to find sequence and/or structural homologues for a particular structure. Finally, the PDB websites also contain many references to other websites offering additional information about the various proteins and related biological and medical aspects, as well as options for viewing each structure, analyzing it, or manipulating it. Aside from proteins, the PDB also contains independent structures of nucleic acids, as well as structures of carbohydrates, lipids, and small organic molecules found in complex with proteins.

As of July 2017, the PDB contained about 131,500 structures, 93% of which are proteins. The number of new entries added to the database annually is about 10,000 [439]. This makes the PDB the most vast and important protein structural data set in existence. It should be noted, however, that the PDB has certain biases [440], mainly due to the following two aspects:

- Crystallization – Since ~90% of the structures have been determined by X-ray crystallography, the PDB is biased towards protein types and forms that are easy to crystallize. Thus, membrane-bound, intrinsically unstructured and complex proteins (oligomers, protein-nucleic acid and protein-carbohydrate complexes) are underrepresented in the PDB. Encouragingly, this bias has been getting smaller with the recent progress in methods such as cryo-EM and SAXS, which can be used to determine the structures of such proteins with increasing speed and accuracy (see Sections 3.2.1, 3.2.3 and 3.5 above).

  For example, since the inception of the PDB, the average molecular weight of the included structures has increased from less than 30,000 daltons to over 110,000 daltons [437]. Another bias that results from the prevalence of crystal structures in the PDB is that most structures present a static view of proteins and do not relate to their dynamic nature. NMR structures do provide insights into the molecular dynamics of the molecule, but they are relatively few and limited to small proteins.

- Scientific interest – Proteins that are studied more extensively than others (e.g., HIV protease, MHC proteins and MAP kinase) are over-represented in the PDB. One reason for the preference of particular proteins is medical relevance. Thus, proteins that serve as common drug targets (e.g., enzymes) can be found widely in the PDB, usually complexed with their targeting drugs. As a result, certain structural and functional features that are common in such proteins are also over-represented in the PDB. These include, e.g., metal-binding sites, which are common in certain enzyme groups [441].

Some folds, such as the TIM barrel and the Rossmann fold, are very common in the PDB, as are the supersecondary structures that build them. This, however, is not a bias, as they are also prevalent in the genomes of many organisms [440,442] (see Chapter 2, Subsection 2.4.2.2.4 for more information).

Protein structures predicted by computational methods can be found in various databases, such as the *Protein Model Portal* (*PMP*) at the National Institutes of Health [443] (see URL below).

Examples of relevant URLs:

- US site (RCSB): http://www.rcsb.org/pdb

- Worldwide site (wwPDB): http://www.wwpdb.org/

- European site (PDBe): http://www.ebi.ac.uk/pdbe

- Japanese site (PDBj): http://pdbj.org/

- Detailed search options: http://bip.weizmann.ac.il/oca-bin/ocamain

- PDBSum (summary for each structure): http://www.ebi.ac.uk/thornton-srv/databases/pdbsum/

- PDB statistics: http://www.pdb.org/pdb/statistics/holdings.do

- PDB documentation: http://www.wwpdb.org/docs.html

- Proteopedia (a structure-function-oriented database of proteins): http://www.proteopedia.org/wiki/index.php/Main_Page

- PMP (protein model portal): http://www.proteinmodelportal.org/

- Various tools for protein search, alignment, structure prediction and more: http://toolkit.tuebingen.mpg.de/

- FirstGlance in Jmol for viewing 3D structures on the Internet: http://FirstGlance.Jmol.Org

- PyMOL for viewing 3D structures locally: https://www.pymol.org/

(a)

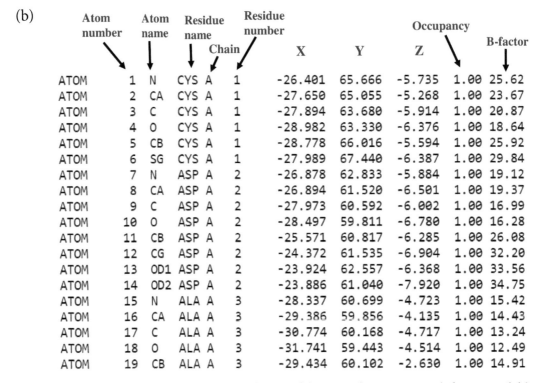

(b)

FIGURE 3.15 **The Protein Data Bank.** (a) One of the PDB electronic search forms available on the Internet. (b) The structural coordinates of a protein deposited in the PDB. Besides the coordinates, the PDB file contains additional structurally relevant parameters specified in the figure. Only the first two amino acids are presented.

3.8 SUMMARY

- Protein structure can be determined using various methods. These can be separated into two groups: diffraction methods, which rely on the diffraction or scattering of either subatomic particles or electromagnetic waves by the protein, and spectroscopic methods, which rely on the excitation and subsequent relaxation of protein atoms in response to electromagnetic radiation.

- The most common diffraction method is X-ray crystallography. It yields three-dimensional structures with the highest resolution, yet has some serious shortcomings, the most significant being the necessity to crystallize the investigated protein. Crystallization is a long and difficult process, and it also produces structures of proteins outside their environment. These structures are devoid of any dynamic properties, and they may sometimes be deformed; however, the structural similarities observed among homologous proteins, proteins that were determined in several crystal forms, and proteins that were determined using both crystallography and cryo-EM or NMR suggest that such deformations are rare.

- Small-angle X-ray scattering (SAXS) yields lower-resolution structures compared to X-ray crystallography. However, SAXS structures are very useful in setting positional constraints for large protein complexes, into which high-resolution X-ray structures can be 'molded'.

- Neutron scattering and electron crystallography are two methods used to provide scientists with complementary information to that obtained by X-ray crystallography. Neutron scattering is able to determine the locations of hydrogen atoms in the structure; such information is missing in most X-ray structures. It can therefore be used to determine protonation states of residues and the directions of bound water molecules within and around proteins.

- Until recently, electron crystallography yielded structures of lower resolution. However, as in the case of SAXS, these structures too can be used as constraints for placing high-resolution structures of single chains within large protein complexes. In addition, the method is easy to use, as it does not require the preparation of three-dimensional crystals. Recent developments in EM technology are showing great promise in producing atomic-resolution structures.

- The main spectroscopic method used for protein structure determination is NMR. This method relies on the excitation of atomic nuclei under a strong magnetic field and their subsequent structure-dependent relaxation. This approach enables scientists to deal with the protein in its natural environment (solution, membrane) and provides important information on its dynamics. EPR, a similar method, is based on the excitation and relaxation of electrons around the atoms of the protein, and requires tagging of protein residues with paramagnetic labels.

- The technological progress of recent decades has led to the development of several computational tools for the prediction of protein structure. Some of these methods, such as molecular dynamics, rely on mathematical characterization of the main physical forces acting on protein structures. These energy-based methods are currently unable to systematically predict the native structure of a protein from an unfolded conformation, but they are efficient in refining near-native structures and also provide important information about their dynamics. Their performance is often improved by the incorporation of knowledge-based expressions alongside the physical terms.

- The other approach of structure prediction is based on comparison to proteins of known structure (template-based approach). It includes two major methods. The first is homology modeling, which is currently the best method for protein structure prediction. This approach relies on the plethora of protein sequences and the growing number of protein structures present in current databases, and it uses them to predict the structure of a query protein on the basis of its sequence similarity to a known structural template. The second method is fold recognition via threading, which relies on other shared sequence properties between the target protein and the template(s), and is most useful when the compared sequences have no detectable homology.

- Current methods achieve the best results by using the following strategies: (*i*) integrating template-based prediction with physics-based refinement, (*ii*) using fragment-based calculations, and (*iii*) (where possible) using data from NMR, SAXS and EM experiments to guide the prediction algorithm.

- Protein structures determined by experimental methods are deposited in the Protein Data Bank after scrutiny. They are freely accessible to the general public via the Internet, and can be easily downloaded, viewed, and manipulated using molecular graphics software.

EXERCISES

3.1 Describe the main differences between X-ray diffraction and NMR spectroscopy. Refer to both the methodology and the quality of results.

3.2 Explain how neutron diffraction is used for finding the orientations of water molecules in proteins.

3.3 What uses are there for NMR spectroscopy, other than structure determination?

3.4 Explain how low-resolution images produced by electron microscopy are used for protein structure determination. Why are these used instead of low-resolution images produced by X-ray diffraction?

3.5 Explain the main differences between NMR and EPR spectroscopies.

3.6 Explain the main advantages and disadvantages of explicit and implicit descriptions employed by structure-prediction methods.

3.7 A. Why are energy-minimization methods unable to predict the native structures of proteins when starting from an unfolded state?

   B. What solutions have been developed for this problem, and how well do they work?

3.8 Describe the principles of the continuum-solvent model approach, its advantages and disadvantages in describing protein-related systems, and its current uses.

3.9 Why is homology modeling currently considered to be the best structure prediction approach?

3.10 Protein X has the following sequence:

```
MVFSDQQLFEKVVEILKPFDLSVVDYEEICDRMGESMRLGLQKSTNEKSSIKMFPSYVT
KTPNGTETGNFLALDLGGTNYRVLSVTLEGKGKSPRIQERTYCIPAEKMSGSGTELFKY
IAETLADFLENNGMKDKKFDLGFTFSFPCVQKGLTHATLVRWTKGFSADGVEGHNVAEL
LQTELDKRELNVKCVAVVNDTVGTLASCALEDPKCAVGLIVGTGTNVAYIEDSSKVELM
DGVKEPEVVINTEWGAFGEKGELDCWRTQFDKSMDIDSLHPGKQLYEKMVSGMYLGELV
RHIIVYLVEQKILFRGDLPERLKVRNSLLTRYLTDVERDPAHLLYNTHYMLTDDLHVPV
VEPIDNRIVRYACEMVVKRAAYLAGAGIACILRRINRSEVTVGVDGSLYKFHPKFCERM
TDMVDKLKPKNTRFCLRLSEDGSGKGAAAIAASCTRQN
```

A. Using the Internet resources mentioned in Chapters 1 through 3, find the name of the protein, its function, and the organism from which it was obtained.

B. Has the three-dimensional structure of the protein been determined experimentally? If so, answer the following questions:
  I. Which method was used to determine this structure?
  II. Does the structure include a ligand?
  III. Does the structure contain any hetero-atoms?
  IV. Which secondary structure does the structure include?

REFERENCES

1. O. Gileadi, S. Knapp, W. H. Lee, B. D. Marsden, S. Muller, F. H. Niesen, K. L. Kavanagh, L. J. Ball, F. von Delft, D. A. Doyle, U. C. Oppermann, and M. Sundstrom. The scientific impact of the Structural Genomics Consortium: a protein family and ligand-centered approach to medically-relevant human proteins. *J. Struct. Funct. Genomics*, 8(2–3):107–19, 2007.
2. PSI. The Protein Structure Initiative. https://www.nigms.nih.gov/Research/specificareas/PSI/Pages/default.aspx, 2015.
3. C. Chothia and A. M. Lesk. The relation between the divergence of sequence and structure in proteins. *EMBO J.*, 5(4):823–6, 1986.
4. D. Baker and A. Sali. Protein structure prediction and structural genomics. *Science*, 294(5540):93–6, 2001.
5. D. Petrey and B. Honig. Protein structure prediction: inroads to biology. *Mol. Cell*, 20(6):811–9, 2005.
6. A. G. Murzin, S. E. Brenner, T. Hubbard, and C. Chothia. SCOP: a structural classification of proteins database for the investigation of sequences and structures. *J. Mol. Biol.*, 247(4):536–40, 1995.
7. C. A. Orengo, A. D. Michie, S. Jones, D. T. Jones, M. B. Swindells, and J. M. Thornton. CATH–a hierarchic classification of protein domain structures. *Structure*, 5(8):1093–108, 1997.
8. I. N. Shindyalov and P. E. Bourne. An alternative view of protein fold space. *Proteins*, 38(3):247–60, 2000.
9. A. S. Yang and B. Honig. An integrated approach to the analysis and modeling of protein sequences and structures. III. A comparative study of sequence conservation in protein structural families using multiple structural alignments. *J. Mol. Biol.*, 301(3):691–711, 2000.
10. R. Kolodny, L. Pereyaslavets, A. O. Samson, and M. Levitt. On the universe of protein folds. *Annu. Rev. Biophys.*, 42:559–82, 2013.
11. S. Nepomnyachiy, N. Ben-Tal, and R. Kolodny. Global view of the protein universe. *Proc. Natl. Acad. Sci. USA*, 111(32):11691–6, 2014.

12. M. Sharon. How Far Can We Go with Structural Mass Spectrometry of Protein Complexes? *J. Am. Soc. Mass Spectrom.*, 21(4):487–500, 2010.

13. M. H. Koch, P. Vachette, and D. I. Svergun. Small-angle scattering: a view on the properties, structures and structural changes of biological macromolecules in solution. *Q. Rev. Biophys.*, 36(2):147–227, 2003.

14. H. L. Liu and J. P. Hsu. Recent developments in structural proteomics for protein structure determination. *Proteomics*, 5(8):2056–68, 2005.

15. M. B. Yaffe. X-ray crystallography and structural bioloby. *Crit. Care Med.*, 22(12 (suppl)):S435–S440, 2005.

16. M. V. Petoukhov and D. I. Svergun. Analysis of X-ray and neutron scattering from biomacromolecular solutions. *Curr. Opin. Struct. Biol.*, 17(5):562–71, 2007.

17. G. Taylor. The phase problem. *Acta Crystallogr. Sect. D*, 59(Pt 11):1881–90, 2003.

18. V. V. Krishnan and B. Rupp. Macromolecular Structure Determination: Comparison of X-ray Crystallography and NMR Spectroscopy. In *Encyclopedia of Life Sciences*. John Wiley & Sons, Ltd., 2001.

19. Vossman. Series of Resolutions for w:GroEL. Wikipedia, the free encyclopedia. http://en.wikipedia.org/wiki/File:GroEL_resolution_series.png, 2008.

20. S. Kundu, J. S. Melton, D. C. Sorensen, and G. N. Phillips Jr. Dynamics of proteins in crystals: comparison of experiment with simple models. *Biophys. J.*, 83(2):723–32, 2002.

21. H. Frauenfelder, G. A. Petsko, and D. Tsernoglou. Temperature-dependent X-ray diffraction as a probe of protein structural dynamics. *Nature*, 280(5723):558–63, 1979.

22. K. Henzler-Wildman and D. Kern. Dynamic personalities of proteins. *Nature*, 450(7172):964–72, 2007.

23. E. Martz, J. Sussman, W. Decatur, E. Hodis, Y. L. Jiang, and J. Prilusky. Proteopedia: life in 3D. In *Resolution*. Martz, E., 2014.

24. A. McPherson. *Introduction to Macromolecular Crystallography*. John Wiley & Sons, Inc., Hoboken, NJ, 2003.

25. R. E. Cachau and A. D. Podjarny. High-resolution crystallography and drug design. *J. Mol. Recognit.*, 18(3):196–202, 2005.

26. U. Mueller, H. Schubel, M. Sprinzl, and U. Heinemann. Crystal structure of acceptor stem of tRNA(Ala) from *Escherichia coli* shows unique G. U. wobble base pair at 1.16 Å resolution. *RNA*, 5(5):670–7, 1999.

27. U. Mueller, D. Perl, F. X. Schmid, and U. Heinemann. Thermal stability and atomic-resolution crystal structure of the *Bacillus caldolyticus* cold shock protein. *J. Mol. Biol.*, 297(4):975–88, 2000.

28. V. B. Chen, W. B. Arendall, J. J. Headd, D. A. Keedy, R. M. Immormino, G. J. Kapral, L. W. Murray, J. S. Richardson, and D. C. Richardson. MolProbity: all-atom structure validation for macromolecular crystallography. *Acta Crystallogr. Sect. D*, 66(1):12–21, 2010.

29. J. M. Word, S. C. Lovell, J. S. Richardson, and D. C. Richardson. Asparagine and glutamine: using hydrogen atom contacts in the choice of side-chain amide orientation. *J. Mol. Biol.*, 285(4):1735–1747, 1999.

30. T. J. Dolinsky, J. E. Nielsen, J. A. McCammon, and N. A. Baker. PDB2PQR: an automated pipeline for the setup of Poisson-Boltzmann electrostatics calculations. *Nucleic Acids Res.*, 32(suppl 2):W665–7, 2004.

31. C. R. Søndergaard, M. H. M. Olsson, M. Rostkowski, and J. H. Jensen. Improved treatment of ligands and coupling effects in empirical calculation and rationalization of pKa values. *J. Chem. Theory. Comput.*, 7(7):2284–2295, 2011.

32. C. Jelsch, M. M. Teeter, V. Lamzin, V. Pichon-Pesme, R. H. Blessing, and C. Lecomte. Accurate protein crystallography at ultra-high resolution: valence electron distribution in crambin. *Proc. Natl. Acad. Sci. USA*, 97(7):3171–6, 2000.

33. P. Kuhn, M. Knapp, S. M. Soltis, G. Ganshaw, M. Thoene, and R. Bott. The 0.78 Å structure of a serine protease: *Bacillus lentus* subtilisin. *Biochemistry*, 37(39):13446–52, 1998.

34. E. I. Howard, R. Sanishvili, R. E. Cachau, A. Mitschler, B. Chevrier, P. Barth, V. Lamour, M. Van Zandt, E. Sibley, C. Bon, D. Moras, T. R. Schneider, A. Joachimiak, and A. Podjarny. Ultrahigh resolution drug design I: details of interactions in human aldose reductase-inhibitor complex at 0.66 Å. *Proteins*, 55(4):792–804, 2004.

35. M. Mueller, S. Jenni, and N. Ban. Strategies for crystallization and structure determination of very large macromolecular assemblies. *Curr. Opin. Struct. Biol.*, 17(5):572–9, 2007.

36. M. G. Rossmann, M. C. Morais, P. G. Leiman, and W. Zhang. Combining X-ray crystallography and electron microscopy. *Structure*, 13(3):355–62, 2005.

37. B. S. Schuwirth, M. A. Borovinskaya, C. W. Hau, W. Zhang, A. Vila-Sanjurjo, J. M. Holton, and J. H. Cate. Structures of the bacterial ribosome at 3.5 Å resolution. *Science*, 310(5749):827–34, 2005.

38. M. Selmer, C. M. Dunham, F. V. th Murphy, A. Weixlbaumer, S. Petry, A. C. Kelley, J. R. Weir, and V. Ra-

makrishnan. Structure of the 70S ribosome complexed with mRNA and tRNA. *Science*, 313(5795):1935–42, 2006.

39. G. Blaha, G. Gurel, S. J. Schroeder, P. B. Moore, and T. A. Steitz. Mutations outside the anisomycin-binding site can make ribosomes drug-resistant. *J. Mol. Biol.*, 379(3):505–19, 2008.

40. J. Harms, F. Schluenzen, R. Zarivach, A. Bashan, S. Gat, I. Agmon, H. Bartels, F. Franceschi, and A. Yonath. High resolution structure of the large ribosomal subunit from a mesophilic eubacterium. *Cell*, 107(5):679–88, 2001.

41. C. Gibbons, M. G. Montgomery, A. G. Leslie, and J. E. Walker. The structure of the central stalk in bovine F(1)-ATPase at 2.4 Å resolution. *Nat. Struct. Biol.*, 7(11):1055–61, 2000.

42. V. K. Rastogi and M. E. Girvin. Structural changes linked to proton translocation by subunit c of the ATP synthase. *Nature*, 402(6759):263–8, 1999.

43. J. V. Bason, M. G. Montgomery, A. G. Leslie, and J. E. Walker. How release of phosphate from mammalian F1-ATPase generates a rotary substep. *Proc. Natl. Acad. Sci. USA*, 112(19):6009–14, 2015.

44. A. L. Gnatt, P. Cramer, J. Fu, D. A. Bushnell, and R. D. Kornberg. Structural basis of transcription: an RNA polymerase II elongation complex at 3.3 Å resolution. *Science*, 292(5523):1876–82, 2001.

45. C. Walmacq, L. Wang, J. Chong, K. Scibelli, L. Lubkowska, A. Gnatt, P. J. Brooks, D. Wang, and M. Kashlev. Mechanism of RNA polymerase II bypass of oxidative cyclopurine DNA lesions. *Proc. Natl. Acad. Sci. USA*, 112(5):E410–9, 2015.

46. S. Sainsbury, J. Niesser, and P. Cramer. Structure and function of the initially transcribing RNA polymerase II-TFIIB complex. *Nature*, 493(7432):437–440, 2013.

47. X. Liu, D. A. Bushnell, D. A. Silva, X. Huang, and R. D. Kornberg. Initiation complex structure and promoter proofreading. *Science*, 333(6042):633–7, 2011.

48. A. Ben-Shem, F. Frolow, and N. Nelson. Crystal structure of plant photosystem I. *Nature*, 426(6967):630–5, 2003.

49. H. N. Chapman, P. Fromme, A. Barty, T. A. White, R. A. Kirian, A. Aquila, M. S. Hunter, J. Schulz, D. P. DePonte, U. Weierstall, R. B. Doak, F. R. Maia, A. V. Martin, I. Schlichting, L. Lomb, N. Coppola, R. L. Shoeman, S. W. Epp, R. Hartmann, D. Rolles, A. Rudenko, L. Foucar, N. Kimmel, G. Weidenspointner, P. Holl, M. Liang, M. Barthelmess, C. Caleman, S. Boutet, M. J. Bogan, J. Krzywinski, C. Bostedt, S. Bajt, L. Gumprecht, B. Rudek, B. Erk, C. Schmidt, A. Homke, C. Reich, D. Pietschner, L. Struder, G. Hauser, H. Gorke, J. Ullrich, S. Herrmann, G. Schaller, F. Schopper, H. Soltau, K. U. Kuhnel, M. Messerschmidt, J. D. Bozek, S. P. Hau-Riege, M. Frank, C. Y. Hampton, R. G. Sierra, D. Starodub, G. J. Williams, J. Hajdu, N. Timneanu, M. M. Seibert, J. Andreasson, A. Rocker, O. Jonsson, M. Svenda, S. Stern, K. Nass, R. Andritschke, C. D. Schroter, F. Krasniqi, M. Bott, K. E. Schmidt, X. Wang, I. Grotjohann, J. M. Holton, T. R. Barends, R. Neutze, S. Marchesini, R. Fromme, S. Schorb, D. Rupp, M. Adolph, T. Gorkhover, I. Andersson, H. Hirsemann, G. Potdevin, H. Graafsma, B. Nilsson, and J. C. Spence. Femtosecond X-ray protein nanocrystallography. *Nature*, 470(7332):73–7, 2011.

50. Y. Mazor, A. Borovikova, and N. Nelson. The structure of plant photosystem I super-complex at 2.8 Å resolution. *eLife*, 4:e07433, 2015.

51. I. Moraes, G. Evans, J. Sanchez-Weatherby, S. Newstead, and P. D. Stewart. Membrane protein structure determination – the next generation. *Biochim. Biophys. Acta*, 1838(1 Pt A):78–87, 2014.

52. D. Axford, R. L. Owen, J. Aishima, J. Foadi, A. W. Morgan, J. I. Robinson, J. E. Nettleship, R. J. Owens, I. Moraes, E. E. Fry, J. M. Grimes, K. Harlos, A. Kotecha, J. Ren, G. Sutton, T. S. Walter, D. I. Stuart, and G. Evans. In situ macromolecular crystallography using microbeams. *Acta Crystallogr. Sect. D*, 68(Pt 5):592–600, 2012.

53. S. O. Garbuzynskiy, B. S. Melnik, M. Y. Lobanov, A. V. Finkelstein, and O. V. Galzitskaya. Comparison of X-ray and NMR structures: Is there a systematic difference in residue contacts between X-ray-and NMR-resolved protein structures? *Proteins*, 60(1):139–147, 2005.

54. M. Andrec, D. A. Snyder, Z. Zhou, J. Young, G. T. Montelione, and R. M. Levy. A large data set comparison of protein structures determined by crystallography and NMR: statistical test for structural differences and the effect of crystal packing. *Proteins: Struct., Funct., Bioinf.*, 69(3):449–465, 2007.

55. E. Martz. *Success Rates in Protein Crystallography*. University of Massachusetts, 2012.

56. H. D. Mertens and D. I. Svergun. Structural characterization of proteins and complexes using small-angle X-ray solution scattering. *J. Struct. Biol.*, 172(1):128–41, 2010.

57. R. P. Rambo and J. A. Tainer. Bridging the solution divide: comprehensive structural analyses of dynamic RNA, DNA, and protein assemblies by small-angle X-ray scattering. *Curr. Opin. Struct. Biol.*, 20(1):128–37, 2010.

58. M. A. Graewert and D. I. Svergun. Impact and progress in small and wide angle X-ray scattering (SAXS and WAXS). *Curr. Opin. Struct. Biol.*, 23(5):748–54, 2013.

59. M. V. Petoukhov and D. I. Svergun. Applications of small-angle X-ray scattering to biomacromolecular solutions. *Int. J. Biochem. Cell Biol.*, 45(2):429–37, 2013.

60. P. Bernado and D. I. Svergun. Structural analysis of intrinsically disordered proteins by small-angle X-ray scattering. *Mol. Biosyst.*, 8(1):151–67, 2012.

61. D. I. Svergun. Restoring low resolution structure of biological macromolecules from solution scattering using simulated annealing. *Biophys. J.*, 76(6):2879–86, 1999.

62. D. Schneidman-Duhovny, S. J. Kim, and A. Sali. Integrative structural modeling with small angle X-ray scattering profiles. *BMC Struct. Biol.*, 12:17, 2012.

63. C. Pons, M. D'Abramo, D. I. Svergun, M. Orozco, P. Bernado, and J. Fernandez-Recio. Structural characterization of protein-protein complexes by integrating computational docking with small-angle scattering data. *J. Mol. Biol.*, 403(2):217–30, 2010.

64. S. K. Ghosh, S. Castorph, O. Konovalov, R. Jahn, M. Holt, and T. Salditt. In vitro study of interaction of synaptic vesicles with lipid membranes. *New J. Phys.*, 12:105004, 2010.

65. D. Mizrachi, Y. Chen, J. Liu, H. M. Peng, A. Ke, L. Pollack, R. J. Turner, R. J. Auchus, and M. P. DeLisa. Making water-soluble integral membrane proteins in vivo using an amphipathic protein fusion strategy. *Nat. Commun.*, 6:6826, 2015.

66. L. Makowski, J. Bardhan, D. Gore, J. Lal, S. Mandava, S. Park, D. J. Rodi, N. T. Ho, C. Ho, and R. F. Fischetti. WAXS studies of the structural diversity of hemoglobin in solution. *J. Mol. Biol.*, 408(5):909–21, 2011.

67. R. P. Rambo and J. A. Tainer. Characterizing flexible and intrinsically unstructured biological macromolecules by SAS using the Porod-Debye law. *Biopolymers*, 95(8):559–71, 2011.

68. T. E. Williamson, B. A. Craig, E. Kondrashkina, C. Bailey-Kellogg, and A. M. Friedman. Analysis of self-associating proteins by singular value decomposition of solution scattering data. *Biophys. J.*, 94(12):4906–23, 2008.

69. E. M. Mandelkow, A. Harmsen, E. Mandelkow, and J. Bordas. X-ray kinetic studies of microtubule assembly using synchrotron radiation. *Nature*, 287(5783):595–9, 1980.

70. S. V. Kathuria, L. Guo, R. Graceffa, R. Barrea, R. P. Nobrega, C. R. Matthews, T. C. Irving, and O. Bilsel. Minireview: structural insights into early folding events using continuous-flow time-resolved small-angle X-ray scattering. *Biopolymers*, 95(8):550–8, 2011.

71. F. Poitevin, H. Orland, S. Doniach, P. Koehl, and M. Delarue. AquaSAXS: a web server for computation and fitting of SAXS profiles with non-uniformly hydrated atomic models. *Nucleic Acids Res.*, 39:W184–9, 2011.

72. H. Tidow, L. R. Poulsen, A. Andreeva, M. Knudsen, K. L. Hein, C. Wiuf, M. G. Palmgren, and P. Nissen. A bimodular mechanism of calcium control in eukaryotes. *Nature*, 491(7424):468–72, 2012.

73. O. Byron and R. J. Gilbert. Neutron scattering: good news for biotechnology. *Curr. Opin. Biotechnol.*, 11(1):72–80, 2000.

74. T. Gutberlet, U. Heinemann, and M. Steiner. Protein crystallography with neutrons–status and perspectives. *Acta Crystallogr. Sect. D*, 57(Pt 2):349–54, 2001.

75. B. P. Schoenborn and R. B. Knott. *Neutrons in Biology*. Plenum Press, New York, 1996.

76. R. Pynn. Neutron scattering: a primer. *Los Alamos Sci.*, 19:1–31, 1990.

77. S. Penel, E. Pebay-Peyroula, J. Rosenbusch, G. Rummel, T. Schirmer, and P. A. Timmins. Detergent binding in trigonal crystals of OmpF porin from *Escherichia coli*. *Biochimie*, 80(5–6):543–51, 1998.

78. M. Weik, H. Patzelt, G. Zaccai, and D. Oesterhelt. Localization of glycolipids in membranes by in vivo labeling and neutron diffraction. *Mol. Cell*, 1(3):411–9, 1998.

79. M. S. Capel, D. M. Engelman, B. R. Freeborn, M. Kjeldgaard, J. A. Langer, V. Ramakrishnan, D. G. Schindler, D. K. Schneider, B. P. Schoenborn, I. Y. Sillers, et al. A complete mapping of the proteins in the small ribosomal subunit of *Escherichia coli*. *Science*, 238(4832):1403–6, 1987.

80. B. Deme, T. Forsyth, G. Fragneto, M. Hartlein, R. May, P. A. Timmins, and G. Zaccai. Neutrons in biology: Institut Laue-Langevin, 4–7 September 2005. *Eur. Biophys. J.*, 35:549–550, 2005.

81. S. J. Ludtke and W. Chiu. Electron Cryomicroscopy and Three-dimensional Computer Reconstruction of Biological Molecules. In *Encyclopedia of Life Sciences*. John Wiley & Sons, Ltd., 2011.

82. W. Chiu, M. L. Baker, W. Jiang, M. Dougherty, and M. F. Schmid. Electron cryomicroscopy of biological machines at subnanometer resolution. *Structure*, 13(3):363–72, 2005.

83. Y. Cheng, N. Grigorieff, P. A. Penczek, and T. Walz. A primer to single-particle cryo-electron microscopy. *Cell*, 161(3):438–49, 2015.

84. A. Doerr. Single-particle electron cryomicroscopy. *Nat. Methods*, 11(1):30, 2014.

85. J. J. Bozzola. *Electron Microscopy*. John Wiley & Sons, Ltd., 2001.

86. A. R. Faruqi and G. McMullan. Electronic detectors for electron microscopy. *Q. Rev. Biophys.*, 44(3):357–90, 2011.

87. X. Li, P. Mooney, S. Zheng, C. R. Booth, M. B. Braunfeld, S. Gubbens, D. A. Agard, and Y. Cheng. Electron counting and beam-induced motion correction enable near-atomic-resolution single-particle cryo-EM. *Nat. Methods*, 10(6):584–90, 2013.

88. A. C. Milazzo, A. Cheng, A. Moeller, D. Lyumkis, E. Jacovetty, J. Polukas, M. H. Ellisman, N. H. Xuong, B. Carragher, and C. S. Potter. Initial evaluation of a direct detection device detector for single particle cryo-electron microscopy. *J. Struct. Biol.*, 176(3):404–8, 2011.

89. P. A. Penczek. Resolution measures in molecular electron microscopy. *Methods Enzymol.*, 482:73–100, 2010.

90. S. J. Fleishman, V. M. Unger, and N. Ben-Tal. Transmembrane protein structures without X-rays. *Trends Biochem. Sci.*, 31(2):106–13, 2006.

91. L. Renault, H. T. Chou, P. L. Chiu, R. M. Hill, X. Zeng, B. Gipson, Z. Y. Zhang, A. Cheng, V. Unger, and H. Stahlberg. Milestones in electron crystallography. *J. Comput. Aided Mol. Des.*, 20(7–8):519–27, 2006.

92. P. N. Unwin and R. Henderson. Molecular structure determination by electron microscopy of unstained crystalline specimens. *J. Mol. Biol.*, 94(3):425–40, 1975.

93. R. Henderson. The potential and limitations of neutrons, electrons and X-rays for atomic resolution microscopy of unstained biological molecules. *Q. Rev. Biophys.*, 28(2):171–93, 1995.

94. K. Grunewald, O. Medalia, A. Gross, A. C. Steven, and W. Baumeister. Prospects of electron cryotomography to visualize macromolecular complexes inside cellular compartments: implications of crowding. *Biophys. Chem.*, 100(1–3):577–91, 2003.

95. R. C. Ford and A. Holzenburg. Electron crystallography of biomolecules: mysterious membranes and missing cones. *Trends Biochem. Sci.*, 33(1):38–43, 2008.

96. R. Henderson, J. M. Baldwin, T. A. Ceska, F. Zemlin, E. Beckmann, and K. H. Downing. Model for the structure of bacteriorhodopsin based on high-resolution electron cryo-microscopy. *J. Mol. Biol.*, 213(4):899–929, 1990.

97. W. Kuhlbrandt, D. N. Wang, and Y. Fujiyoshi. Atomic model of plant light-harvesting complex by electron crystallography. *Nature*, 367(6464):614–21, 1994.

98. T. Gonen, Y. Cheng, P. Sliz, Y. Hiroaki, Y. Fujiyoshi, S. C. Harrison, and T. Walz. Lipid-protein interactions in double-layered two-dimensional AQP0 crystals. *Nature*, 438(7068):633–8, 2005.

99. B. L. Nannenga, D. Shi, A. G. Leslie, and T. Gonen. High-resolution structure determination by continuous-rotation data collection in MicroED. *Nat. Methods*, 11(9):927–30, 2014.

100. D. Shi, B. L. Nannenga, M. G. Iadanza, and T. Gonen. Three-dimensional electron crystallography of protein microcrystals. *eLife*, 2:e01345, 2013.

101. K. Tani, T. Mitsuma, Y. Hiroaki, A. Kamegawa, K. Nishikawa, Y. Tanimura, and Y. Fujiyoshi. Mechanism of aquaporin-4's fast and highly selective water conduction and proton exclusion. *J. Mol. Biol.*, 389(4):694–706, 2009.

102. D. Lyumkis, A. Moeller, A. Cheng, A. Herold, E. Hou, C. Irving, E. L. Jacovetty, P. W. Lau, A. M. Mulder, J. Pulokas, J. D. Quispe, N. R. Voss, C. S. Potter, and B. Carragher. Automation in single-particle electron microscopy connecting the pieces. *Methods Enzymol.*, 483:291–338, 2010.

103. X. C. Bai, G. McMullan, and S. H. W. Scheres. How cryo-EM is revolutionizing structural biology. *Trends Biochem. Sci.*, 40(1):49–57, 2015.

104. M. G. Campbell, A. Cheng, A. F. Brilot, A. Moeller, D. Lyumkis, D. Veesler, J. Pan, S. C. Harrison, C. S. Potter, B. Carragher, and N. Grigorieff. Movies of ice-embedded particles enhance resolution in electron cryo-microscopy. *Structure*, 20(11):1823–8, 2012.

105. E. J. Boekema, M. Folea, and R. Kouril. Single particle electron microscopy. *Photosynth. Res.*, 102(2–3):189–96, 2009.

106. A. Bartesaghi, A. Merk, S. Banerjee, D. Matthies, X. Wu, J. L. Milne, and S. Subramaniam. 2.2 Å resolution cryo-EM structure of beta-galactosidase in complex with a cell-permeant inhibitor. *Science*, 348(6239):1147–51, 2015.

107. T. Grant and N. Grigorieff. Measuring the optimal exposure for single particle cryo-EM using a 2.6 Å reconstruction of rotavirus VP6. *eLife*, 4:e06980, 2015.

108. M. G. Campbell, D. Veesler, A. Cheng, C. S. Potter, and B. Carragher. 2.8 Å resolution reconstruction of the *Thermoplasma acidophilum* 20S proteasome using cryo-electron microscopy. *eLife*, 4, 2015.

109. J. Jiang, B. L. Pentelute, R. J. Collier, and Z. H. Zhou. Atomic structure of anthrax protective antigen pore elucidates toxin translocation. *Nature*, 521(7553):545–549, 2015.

110. N. Fischer, P. Neumann, A. L. Konevega, L. V. Bock, R. Ficner, M. V. Rodnina, and H. Stark. Structure of the *E. coli* ribosome-EF-Tu complex at <3 Å resolution by C-corrected cryo-EM. *Nature*, 2015.

111. R. I. Koning. *Cryo-electron crystallography: from protein reconstitution to object reconstruction*. PhD thesis, University of Groningen, 2003.

112. R. M. Oliver. Negative stain electron microscopy of protein macromolecules. *Methods Enzymol.*, 27:616–72, 1973.

113. J. E. Mellema, E. F. J. van Bruggen, and M. Gruber. Uranyl oxalate as a negative stain for electron microscopy of proteins. *Biochim. Biophys. Acta*, 140:180–182, 1967.

114. J. Dubochet, M. Adrian, J. J. Chang, J. C. Homo, J. Lepault, A. W. McDowall, and P. Schultz. Cryo-electron microscopy of vitrified specimens. *Q. Rev. Biophys.*, 21(2):129–228, 1988.

115. J. Dubochet, F. P. Booy, R. Freeman, A. V. Jones, and C. A. Walter. Low temperature electron microscopy. *Annu. Rev. Biophys. Bio.*, 10:133–49, 1981.

116. W. Chiu. Electron microscopy of frozen, hydrated biological specimens. *Annu. Rev. Biophys. Biophys. Chem.*, 15:237–57, 1986.

117. K. Mitra and J. Frank. Ribosome dynamics: insights from atomic structure modeling into cryo-electron microscopy maps. *Annu. Rev. Biophys. Biomol. Struct.*, 35:299–317, 2006.

118. G. Beards. Electron Microscope. Wikipedia, the free encyclopedia. https://en.wikipedia.org/wiki/Electron_microscope#/media/File:Electron_Microscope.png, 2009.

119. M. J. Schellenberg, R. A. Edwards, D. B. Ritchie, O. A. Kent, M. M. Golas, H. Stark, R. Luhrmann, J. N. Glover, and A. M. MacMillan. Crystal structure of a core spliceosomal protein interface. *Proc. Natl. Acad. Sci. USA*, 103(5):1266–71, 2006.

120. N. Volkman and D. Hanein. Electron Microscopy. In P. E. Bourne and H. Weissig, editors, *Structural Bioinformatics*, volume 44 of *Methods of Biochemical Analysis*, pages 115–133. John Wiley & Sons, Inc., 2003.

121. A. Brown, F. Long, R. A. Nicholls, J. Toots, P. Emsley, and G. Murshudov. Tools for macromolecular model building and refinement into electron cryo-microscopy reconstructions. *Acta Crystallogr. Sect. D*, 71(Pt 1):136–53, 2015.

122. H. J. Dyson. Spectroscopic Techniques. In *Encyclopedia of Life Sciences*. John Wiley & Sons, Ltd., 2001.

123. T. L. James. Fundamentals of NMR. *Online Biophysics Textbook*, 1999.

124. G. Wider. Structure determination of biological macromolecules in solution using nuclear magnetic resonance spectroscopy. *Biotechniques*, 29(6):1278–82, 1284–90, 1292 passim, 2000.

125. J. Cavanagh, W. Fairbrother, A. G. Palmer III, M. Rance, and N. Skelton. *Protein NMR Spectroscopy: Principles and Practice*. Academic Press, 2007.

126. Cfromber. One-dimentional H-NMR spectrum map of phenol. Wikipedia, the free encyclopedia. http://en.wikibooks.org/wiki/File:Phenol.gif, 2006.

127. D. S. Burz and A. Shekhtman. Structural biology: Inside the living cell. *Nature*, 458(7234):37–8, 2009.

128. Z. Serber, A. T. Keatinge-Clay, R. Ledwidge, A. E. Kelly, S. M. Miller, and V. Dotsch. High-resolution macromolecular NMR spectroscopy inside living cells. *J. Am. Chem. Soc.*, 123(10):2446–7, 2001.

129. D. Sakakibara, A. Sasaki, T. Ikeya, J. Hamatsu, T. Hanashima, M. Mishima, M. Yoshimasu, N. Hayashi, T. Mikawa, M. Walchli, B. O. Smith, M. Shirakawa, P. Guntert, and Y. Ito. Protein structure determination in living cells by in-cell NMR spectroscopy. *Nature*, 458(7234):102–5, 2009.

130. K. Inomata, A. Ohno, H. Tochio, S. Isogai, T. Tenno, I. Nakase, T. Takeuchi, S. Futaki, Y. Ito, H. Hiroaki, and M. Shirakawa. High-resolution multi-dimensional NMR spectroscopy of proteins in human cells. *Nature*, 458(7234):106–9, 2009.

131. M. Bokor, V. Csizmok, D. Kovacs, P. Banki, P. Friedrich, P. Tompa, and K. Tompa. NMR relaxation studies on the hydrate layer of intrinsically unstructured proteins. *Biophys. J.*, 88(3):2030–7, 2005.

132. K. Takeuchi and G. Wagner. NMR studies of protein interactions. *Curr. Opin. Struct. Biol.*, 16(1):109–17, 2006.

133. T. Oltersdorf, S. W. Elmore, A. R. Shoemaker, R. C. Armstrong, D. J. Augeri, B. A. Belli, M. Bruncko, T. L. Deckwerth, J. Dinges, P. J. Hajduk, M. K. Joseph, S. Kitada, S. J. Korsmeyer, A. R. Kunzer, A. Letai, C. Li, M. J. Mitten, D. G. Nettesheim, S. Ng, P. M. Nimmer, J. M. O'Connor, A. Oleksijew, A. M. Petros, J. C. Reed, W. Shen, S. K. Tahir, C. B. Thompson, K. J. Tomaselli, B. Wang, M. D. Wendt, H. Zhang, S. W. Fesik, and S. H. Rosenberg. An inhibitor of Bcl-2 family proteins induces regression of solid tumours. *Nature*, 435(7042):677–81, 2005.

134. R. F. Epand, I. Martin, J. M. Ruysschaert, and R. M. Epand. Membrane orientation of the SIV fusion peptide

determines its effect on bilayer stability and ability to promote membrane fusion. *Biochem. Biophys. Res. Commun.*, 205(3):1938–43, 1994.

135. S. Kimura, A. Naito, S. Tuzi, and H. Saito. Dynamics and orientation of transmembrane peptide from bacteriorhodopsin incorporated into lipid bilayer as revealed by solid state (31)P and (13)C NMR spectroscopy. *Biopolymers*, 63(2):122–31, 2002.

136. S. Yamaguchi, D. Huster, A. Waring, R. I. Lehrer, W. Kearney, B. F. Tack, and M. Hong. Orientation and dynamics of an antimicrobial peptide in the lipid bilayer by solid-state NMR spectroscopy. *Biophys. J.*, 81(4):2203–14, 2001.

137. L. K. Thompson. Solid-state NMR studies of the structure and mechanisms of proteins. *Curr. Opin. Struct. Biol.*, 12(5):661–9, 2002.

138. D. A. Middleton. Solid-state NMR spectroscopy as a tool for drug design: from membrane-embedded targets to amyloid fibrils. *Biochem. Soc. Trans.*, 35(Pt 5):985–90, 2007.

139. S. J. Opella and F. M. Marassi. Structure determination of membrane proteins by NMR spectroscopy. *Chem. Rev.*, 104(8):3587–606, 2004.

140. M. Montal and S. J. Opella. The structure of the M2 channel-lining segment from the nicotinic acetylcholine receptor. *Biochim. Biophys. Acta*, 1565(2):287–93, 2002.

141. A. Kessel, D. S. Cafiso, and N. Ben-Tal. Continuum solvent model calculations of alamethicin-membrane interactions: thermodynamic aspects. *Biophys. J.*, 78(2):571–83, 2000.

142. P. G. Fajer. Electron Paramagnetic Resonance (EPR) and Spin-labelling. In *Encyclopedia of Life Sciences*. John Wiley & Sons, Ltd., 2002.

143. M. Barranger-Mathys and D. S. Cafiso. Membrane structure of voltage-gated channel forming peptides by site-directed spin-labeling. *Biochemistry*, 35(2):498–505, 1996.

144. E. S. Karp, J. J. Inbaraj, M. Laryukhin, and G. A. Lorigan. Electron paramagnetic resonance studies of an integral membrane peptide inserted into aligned phospholipid bilayer nanotube arrays. *J. Am. Chem. Soc.*, 128(37):12070–1, 2006.

145. F. Scarpelli, M. Drescher, T. Rutters-Meijneke, A. Holt, D. T. Rijkers, J. A. Killian, and M. Huber. Aggregation of Transmembrane Peptides Studied by Spin-Label EPR. *J. Phys. Chem. B*, 2009.

146. A. Volkov, C. Dockter, T. Bund, H. Paulsen, and G. Jeschke. Pulsed EPR determination of water accessibility to spin-labeled amino acid residues in LHCIIb. *Biophys. J.*, 96(3):1124–41, 2009.

147. P. T. C. So and C. Y. Dong. Fluorescence Spectrophotometry. In *Encyclopedia of Life Sciences*. John Wiley & Sons, Ltd., 2001.

148. D. Chhabra and C. G. dos Remedios. Fluorescence Resonance Energy Transfer. In *Encyclopedia of Life Sciences*. John Wiley & Sons, Ltd., 2005.

149. N. C. Price. Circular Dichroism: Studies of Proteins. In *Encyclopedia of Life Sciences*. John Wiley & Sons, Ltd., 2001.

150. B. A. Wallace. Protein characterisation by synchrotron radiation circular dichroism spectroscopy. *Q. Rev. Biophys.*, 42(4):317–70, 2009.

151. A. Micsonai, F. Wien, L. Kernya, Y. H. Lee, Y. Goto, M. Refregiers, and J. Kardos. Accurate secondary structure prediction and fold recognition for circular dichroism spectroscopy. *Proc. Natl. Acad. Sci. USA*, 112(24):E3095–103, 2015.

152. B. Wallace. Synchrotron radiation circular-dichroism spectroscopy as a tool for investigating protein structures. *J. Synchrotron Radiat.*, 7(5):289–295, 2000.

153. N. J. Greenfield. Using circular dichroism spectra to estimate protein secondary structure. *Nat. Protoc.*, 1(6):2876–2890, 2006.

154. B. Woollett, L. Whitmore, R. W. Janes, and B. A. Wallace. ValiDichro: a website for validating and quality control of protein circular dichroism spectra. *Nucleic Acids Res.*, 41:W417–21, 2013.

155. L. Whitmore, B. Woollett, A. J. Miles, D. P. Klose, R. W. Janes, and B. A. Wallace. PCDDB: the Protein Circular Dichroism Data Bank, a repository for circular dichroism spectral and metadata. *Nucleic Acids Res.*, 39(suppl 1):D480–6, 2011.

156. G. L. Glish and R. W. Vachet. The basics of mass spectrometry in the twenty-first century. *Nat. Rev. Drug Discov.*, 2(2):140–50, 2003.

157. E. van Hoffman and V. Strootbant. *Mass Spectrometry, Principles and Applications*. John Wiley & Sons, Inc., 2nd edition, 2001.

158. M. Yamashita and J. B. Fenn. Electrospray ion source. Another variation on the free-jet theme. *J. Phys. Chem.*, 88(20):4451–4459, 1984.

159. M. Karas, D. Bachmann, U. Bahr, and F. Hillenkamp. Matrix-assisted ultraviolet laser desorption of nonvolatile compounds. *Int. J. Mass. Spectrom. Ion. Process.*, 78:53–68, 1987.

160. A. N. Holding. XL-MS: Protein cross-linking coupled with mass spectrometry. *Methods*, 89:54–63, 2015.

161. B. Q. Tran, D. R. Goodlett, and Y. A. Goo. Advances in protein complex analysis by chemical cross-linking coupled with mass spectrometry (CXMS) and bioinformatics. *Biochim. Biophys. Acta*, 2015.

162. J. Pi and L. Sael. Mass Spectrometry Coupled Experiments and Protein Structure Modeling Methods. *Int. J. Mol. Sci.*, 14(10):20635, 2013.

163. S. W. Englander. Hydrogen exchange and mass spectrometry: A historical perspective. *J. Am. Soc. Mass Spectrom.*, 17(11):1481–9, 2006.

164. M. Mann, S. E. Ong, M. Gronborg, H. Steen, O. N. Jensen, and A. Pandey. Analysis of protein phosphorylation using mass spectrometry: deciphering the phosphoproteome. *Trends Biotechnol.*, 20(6):261–8, 2002.

165. J. R. Yates. Mass Spectrometry in Biology. In *Encyclopedia of Life Sciences*. John Wiley & Sons, Ltd., 2001.

166. H. Steen and A. Pandey. Mass Spectrometry in Protein Characterization. In *Encyclopedia of Life Sciences*. John Wiley & Sons, Ltd., 2001.

167. R. Aebersold and M. Mann. Mass spectrometry-based proteomics. *Nature*, 422(6928):198–207, 2003.

168. A. Bensimon, A. J. Heck, and R. Aebersold. Mass spectrometry-based proteomics and network biology. *Annu. Rev. Biochem.*, 81:379–405, 2012.

169. P. Hernandez, M. Muller, and R. D. Appel. Automated protein identification by tandem mass spectrometry: issues and strategies. *Mass Spectrom. Rev.*, 25(2):235–54, 2006.

170. F. M. Harris. Book review of Busch *et al.* (1989) Mass spectrometry/mass spectrometry: Techniques and applications of tandem mass spectrometry. *Rapid Commun. Mass Spectrom.*, 3(8):i–i, 1989.

171. D. F. Hunt, R. A. Henderson, J. Shabanowitz, K. Sakaguchi, H. Michel, N. Sevilir, A. L. Cox, E. Appella, and V. H. Engelhard. Characterization of peptides bound to the class I MHC molecule HLA-A2.1 by mass spectrometry. *Science*, 255(5049):1261–3, 1992.

172. E. Lasonder, Y. Ishihama, J. S. Andersen, A. M. Vermunt, A. Pain, R. W. Sauerwein, W. M. Eling, N. Hall, A. P. Waters, H. G. Stunnenberg, and M. Mann. Analysis of the *Plasmodium falciparum* proteome by high-accuracy mass spectrometry. *Nature*, 419(6906):537–42, 2002.

173. L. Florens, M. P. Washburn, J. D. Raine, R. M. Anthony, M. Grainger, J. D. Haynes, J. K. Moch, N. Muster, J. B. Sacci, D. L. Tabb, A. A. Witney, D. Wolters, Y. Wu, M. J. Gardner, A. A. Holder, R. E. Sinden, J. R. Yates, and D. J. Carucci. A proteomic view of the *Plasmodium falciparum* life cycle. *Nature*, 419(6906):520–6, 2002.

174. M. Wilhelm, J. Schlegl, H. Hahne, A. Moghaddas Gholami, M. Lieberenz, M. M. Savitski, E. Ziegler, L. Butzmann, S. Gessulat, H. Marx, T. Mathieson, S. Lemeer, K. Schnatbaum, U. Reimer, H. Wenschuh, M. Mollenhauer, J. Slotta-Huspenina, J. H. Boese, M. Bantscheff, A. Gerstmair, F. Faerber, and B. Kuster. Mass-spectrometry-based draft of the human proteome. *Nature*, 509(7502):582–7, 2014.

175. M. S. Kim, S. M. Pinto, D. Getnet, R. S. Nirujogi, S. S. Manda, R. Chaerkady, A. K. Madugundu, D. S. Kelkar, R. Isserlin, S. Jain, J. K. Thomas, B. Muthusamy, P. Leal-Rojas, P. Kumar, N. A. Sahasrabuddhe, L. Balakrishnan, J. Advani, B. George, S. Renuse, L. D. Selvan, A. H. Patil, V. Nanjappa, A. Radhakrishnan, S. Prasad, T. Subbannayya, R. Raju, M. Kumar, S. K. Sreenivasamurthy, A. Marimuthu, G. J. Sathe, S. Chavan, K. K. Datta, Y. Subbannayya, A. Sahu, S. D. Yelamanchi, S. Jayaram, P. Rajagopalan, J. Sharma, K. R. Murthy, N. Syed, R. Goel, A. A. Khan, S. Ahmad, G. Dey, K. Mudgal, A. Chatterjee, T. C. Huang, J. Zhong, X. Wu, P. G. Shaw, D. Freed, M. S. Zahari, K. K. Mukherjee, S. Shankar, A. Mahadevan, H. Lam, C. J. Mitchell, S. K. Shankar, P. Satishchandra, J. T. Schroeder, R. Sirdeshmukh, A. Maitra, S. D. Leach, C. G. Drake, M. K. Halushka, T. S. Prasad, R. H. Hruban, C. L. Kerr, G. D. Bader, C. A. Iacobuzio-Donahue, H. Gowda, and A. Pandey. A draft map of the human proteome. *Nature*, 509(7502):575–81, 2014.

176. A. Pandey, A. V. Podtelejnikov, B. Blagoev, X. R. Bustelo, M. Mann, and H. F. Lodish. Analysis of receptor signaling pathways by mass spectrometry: identification of vav-2 as a substrate of the epidermal and platelet-derived growth factor receptors. *Proc. Natl. Acad. Sci. USA*, 97(1):179–84, 2000.

177. L. Konermann, J. Pan, and Y.-H. Liu. Hydrogen exchange mass spectrometry for studying protein structure and dynamics. *Chem. Soc. Rev.*, 40(3):1224–1234, 2011.

178. J. G. Kiselar and M. R. Chance. Future directions of structural mass spectrometry using hydroxyl radical footprinting. *J. Mass Spectrom.*, 45(12):1373–82, 2010.

179. M. Fioramonte, A. M. dos Santos, S. McIlwain, W. S. Noble, K. G. Franchini, and F. C. Gozzo. Analysis of secondary structure in proteins by chemical cross-linking coupled to MS. *Proteomics*, 12(17):2746–2752, 2012.

180. A. J. J. Lomant and G. Fairbanks. Chemical probes of extended biological structures: synthesis and prop-

erties of the cleavable protein cross-linking reagent [^{35}S] dithiobis (succinimidyl propionate). *J. Mol. Biol.*, 104(1):243–261, 1976.

181. N. Jaya, V. Garcia, and E. Vierling. Substrate binding site flexibility of the small heat shock protein molecular chaperones. *Proc. Natl. Acad. Sci. USA*, 106(37):15604–15609, 2009.

182. M. Sharon, T. Taverner, X. I. Ambroggio, R. J. Deshaies, and C. V. Robinson. Structural organization of the 19S proteasome lid: insights from MS of intact complexes. *PLoS Biol.*, 4(8):e267, 2006.

183. S. Kang, A. M. Hawkridge, K. L. Johnson, D. C. Muddiman, and P. E. Prevelige Jr. Identification of subunit-subunit interactions in bacteriophage P22 procapsids by chemical cross-linking and mass spectrometry. *J. Proteome Res.*, 5(2):370–7, 2006.

184. G. Neubauer, A. Gottschalk, P. Fabrizio, B. Seraphin, R. Luhrmann, and M. Mann. Identification of the proteins of the yeast U1 small nuclear ribonucleoprotein complex by mass spectrometry. *Proc. Natl. Acad. Sci. USA*, 94(2):385–90, 1997.

185. G. Neubauer, A. King, J. Rappsilber, C. Calvio, M. Watson, P. Ajuh, J. Sleeman, A. Lamond, and M. Mann. Mass spectrometry and EST-database searching allows characterization of the multi-protein spliceosome complex. *Nat. Genet.*, 20(1):46–50, 1998.

186. M. P. Rout, J. D. Aitchison, A. Suprapto, K. Hjertaas, Y. Zhao, and B. T. Chait. The yeast nuclear pore complex: composition, architecture, and transport mechanism. *J. Cell Biol.*, 148(4):635–51, 2000.

187. A. C. Gavin, M. Bosche, R. Krause, P. Grandi, M. Marzioch, A. Bauer, J. Schultz, J. M. Rick, A. M. Michon, C. M. Cruciat, M. Remor, C. Hofert, M. Schelder, M. Brajenovic, H. Ruffner, A. Merino, K. Klein, M. Hudak, D. Dickson, T. Rudi, V. Gnau, A. Bauch, S. Bastuck, B. Huhse, C. Leutwein, M. A. Heurtier, R. R. Copley, A. Edelmann, E. Querfurth, V. Rybin, G. Drewes, M. Raida, T. Bouwmeester, P. Bork, B. Seraphin, B. Kuster, G. Neubauer, and G. Superti-Furga. Functional organization of the yeast proteome by systematic analysis of protein complexes. *Nature*, 415(6868):141–7, 2002.

188. Y. Ho, A. Gruhler, A. Heilbut, G. D. Bader, L. Moore, S. L. Adams, A. Millar, P. Taylor, K. Bennett, K. Boutilier, L. Yang, C. Wolting, I. Donaldson, S. Schandorff, J. Shewnarane, M. Vo, J. Taggart, M. Goudreault, B. Muskat, C. Alfarano, D. Dewar, Z. Lin, K. Michalickova, A. R. Willems, H. Sassi, P. A. Nielsen, K. J. Rasmussen, J. R. Andersen, L. E. Johansen, L. H. Hansen, H. Jespersen, A. Podtelejnikov, E. Nielsen, J. Crawford, V. Poulsen, B. D. Sorensen, J. Matthiesen, R. C. Hendrickson, F. Gleeson, T. Pawson, M. F. Moran, D. Durocher, M. Mann, C. W. Hogue, D. Figeys, and M. Tyers. Systematic identification of protein complexes in *Saccharomyces cerevisiae* by mass spectrometry. *Nature*, 415(6868):180–3, 2002.

189. P. Radivojac, L. M. Iakoucheva, C. J. Oldfield, Z. Obradovic, V. N. Uversky, and A. K. Dunker. Intrinsic disorder and functional proteomics. *Biophys. J.*, 92(5):1439–56, 2007.

190. K. A. Dill, S. Banu Ozkan, T. R. Weikl, J. D. Chodera, and V. A. Voelz. The protein folding problem: when will it be solved? *Curr. Opin. Struct. Biol.*, 17(3):342–346, 2007.

191. A. Roy and Y. Zhang. Protein Structure Prediction. In *Encyclopedia of Life Sciences*. John Wiley & Sons, Ltd., 2012.

192. M. Dorn, M. B. E Silva, L. S. Buriol, and L. C. Lamb. Three-dimensional protein structure prediction: Methods and computational strategies. *Comput. Biol. Chem.*, 53PB:251–276, 2014.

193. C. B. Anfinsen. Principles that Govern the Folding of Protein Chains. *Science*, 181(96):223–230, 1973.

194. A. R. Fersht. Profile of Martin Karplus, Michael Levitt, and Arieh Warshel, 2013 nobel laureates in chemistry. *Proc. Natl. Acad. Sci. USA*, 110(49):19656–7, 2013.

195. Y. Levy and J. Skolnick. Guest Editorial: Computational Molecular Biophysics: 40 Years of Achievements. *Isr. J. Chem.*, 54(8–9):1039–1041, 2014.

196. J. C. Smith and B. Roux. Eppur Si Muove! The 2013 Nobel Prize in Chemistry. *Structure*, 21(12):2102–2105, 2013.

197. M. K. Gilson and H. X. Zhou. Calculation of protein-ligand binding affinities. *Annu. Rev. Biophys. Biomol. Struct.*, 36:21–42, 2007.

198. H. M. Senn and W. Thiel. QM/MM studies of enzymes. *Curr. Opin. Chem. Biol.*, 11(2):182–7, 2007.

199. A. Warshel. Computer simulations of enzyme catalysis: methods, progress, and insights. *Annu. Rev. Biophys. Biomol. Struct.*, 32:425–43, 2003.

200. A. Warshel and M. Karplus. Calculation of ground and excited state potential surfaces of conjugated molecules. I. Formulation and parametrization. *J. Am. Chem. Soc.*, 94(16):5612–5625, 1972.

201. A. Warshel and M. Levitt. Theoretical studies of enzymic reactions: dielectric, electrostatic and steric stabilization of the carbonium ion in the reaction of lysozyme. *J. Mol. Biol.*, 103(2):227–249, 1976.

202. U. H. E. Hansmann and Y. Okamoto. Prediction of peptide conformation by multicanonical algorithm: new approach to the multiple-minima problem. *J. Comput. Chem.*, 14:1333–1338, 1993.

203. Z. Li and H. A. Scheraga. Monte Carlo-minimization approach to the multiple-minima problem in protein folding. *Proc. Natl. Acad. Sci. USA*, 84(19):6611–5, 1987.

204. S. A. Adcock and J. A. McCammon. Molecular dynamics: survey of methods for simulating the activity of proteins. *Chem. Rev.*, 106(5):1589–615, 2006.

205. M. Levitt and S. Lifson. Refinement of protein conformations using a macromolecular energy minimization procedure. *J. Mol. Biol.*, 46(2):269–79, 1969.

206. S. Lifson and A. Warshel. Consistent Force Field for Calculations of Conformations, Vibrational Spectra, and Enthalpies of Cycloalkane and n-Alkane Molecules. *J. Chem. Phys.*, 49(11):5116–5129, 1968.

207. M. Karplus and J. A. McCammon. Molecular dynamics simulations of biomolecules. *Nat. Struct. Biol.*, 9(9):646–52, 2002.

208. M. Nilges, A. M. Gronenborn, A. T. Brunger, and G. M. Clore. Determination of three-dimensional structures of proteins by simulated annealing with interproton distance restraints. Application to crambin, potato carboxypeptidase inhibitor and barley serine proteinase inhibitor 2. *Protein Eng.*, 2(1):27–38, 1988.

209. J. A. McCammon, B. R. Gelin, and M. Karplus. Dynamics of folded proteins. *Nature*, 267(5612):585–90, 1977.

210. M. Levitt and A. Warshel. Computer simulation of protein folding. *Nature*, 253(5494):694–698, 1975.

211. O. Schueler-Furman, C. Wang, P. Bradley, K. Misura, and D. Baker. Progress in Modeling of Protein Structures and Interactions. *Science*, 310(5748):638–642, 2005.

212. R. Unger and J. Moult. Finding the lowest free energy conformation of a protein is an NP-hard problem: proof and implications. *Bull. Math. Biol.*, 55(6):1183–98, 1993.

213. M. Feig and C. L. Brooks. Recent advances in the development and application of implicit solvent models in biomolecule simulations. *Curr. Opin. Struct. Biol.*, 14(2):217–224, 2004.

214. B. Honig and A. Nicholls. Classical electrostatics in biology and chemistry. *Science*, 268(5214):1144–1149, 1995.

215. J. Chen, C. L. Brooks, and J. Khandogin. Recent advances in implicit solvent-based methods for biomolecular simulations. *Curr. Opin. Struct. Biol.*, 18(2):140–8, 2008.

216. W. C. Still, A. Tempczyk, R. C. Hawley, and T. Hendrickson. Semianalytical treatment of solvation for molecular mechanics and dynamics. *J. Am. Chem. Soc.*, 112:6127–6129, 1990.

217. H. Lei and Y. Duan. Ab initio folding of albumin binding domain from all-atom molecular dynamics simulation. *J. Phys. Chem. B*, 111(19):5458–63, 2007.

218. D. Sitkoff, K. A. Sharp, and B. Honig. Accurate calculation of hydration free energies using macroscopic solvation models. *J. Phys. Chem.*, 98:1978–1988, 1994.

219. H. Sun, Y. Li, S. Tian, L. Xu, and T. Hou. Assessing the performance of MM/PBSA and MM/GBSA methods. 4. Accuracies of MM/PBSA and MM/GBSA methodologies evaluated by various simulation protocols using PDBbind data set. *Phys. Chem. Chem. Phys.*, 16(31):16719–29, 2014.

220. L. Xu, H. Sun, Y. Li, J. Wang, and T. Hou. Assessing the performance of MM/PBSA and MM/GBSA methods. 3. The impact of force fields and ligand charge models. *J. Phys. Chem. B*, 117(28):8408–21, 2013.

221. T. Hou, J. Wang, Y. Li, and W. Wang. Assessing the performance of the MM/PBSA and MM/GBSA methods. 1. The accuracy of binding free energy calculations based on molecular dynamics simulations. *J. Chem. Inf. Model.*, 51(1):69–82, 2011.

222. T. Kortemme, A. V. Morozov, and D. Baker. An orientation-dependent hydrogen bonding potential improves prediction of specificity and structure for proteins and protein-protein complexes. *J. Mol. Biol.*, 326(4):1239–59, 2003.

223. A. V. Morozov, T. Kortemme, K. Tsemekhman, and D. Baker. Close agreement between the orientation dependence of hydrogen bonds observed in protein structures and quantum mechanical calculations. *Proc. Natl. Acad. Sci. USA*, 101(18):6946–51, 2004.

224. K. A. Dill, S. B. Ozkan, M. S. Shell, and T. R. Weikl. The protein folding problem. *Annu. Rev. Biophys.*, 37:289–316, 2008.

225. V. G. Eijsink, A. Bjork, S. Gaseidnes, R. Sirevag, B. Synstad, B. van den Burg, and G. Vriend. Rational engineering of enzyme stability. *J. Biotechnol.*, 113(1–3):105–20, 2004.

226. A. Perez, J. A. Morrone, C. Simmerling, and K. A. Dill. Advances in free-energy-based simulations of protein folding and ligand binding. *Curr. Opin. Struct. Biol.*, 36:25–31, 2016.

227. E. A. Proctor and N. V. Dokholyan. Applications of Discrete Molecular Dynamics in biology and medicine. *Curr. Opin. Struct. Biol.*, 37:9–13, 2016.

228. D. E. Shaw, M. M. Deneroff, R. O. Dror, J. S. Kuskin, R. H. Larson, J. K. Salmon, C. Young, B. Batson, K. J.

Bowers, and J. C. Chao. Anton, a special-purpose machine for molecular dynamics simulation. *Commun. ACM*, 51(7):91–97, 2008.

229. D. E. Shaw, J. P. Grossman, J. A. Bank, B. Batson, J. A. Butts, J. C. Chao, M. M. Deneroff, R. O. Dror, A. Even, C. H. Fenton, A. Forte, J. Gagliardo, G. Gill, B. Greskamp, C. R. Ho, D. J. Ierardi, L. Iserovich, J. S. Kuskin, R. H. Larson, T. Layman, L.-S. Lee, A. K. Lerer, C. Li, D. Killebrew, K. M. Mackenzie, S. Y.-H. Mok, M. A. Moraes, R. Mueller, L. J. Nociolo, J. L. Peticolas, T. Quan, D. Ramot, J. K. Salmon, D. P. Scarpazza, U. B. Schafer, N. Siddique, C. W. Snyder, J. Spengler, P. T. P. Tang, M. Theobald, H. Toma, B. Towles, B. Vitale, S. C. Wang, and C. Young. Anton 2: raising the bar for performance and programmability in a special-purpose molecular dynamics supercomputer. In *Proceedings of the International Conference for High Performance Computing, Networking, Storage and Analysis*, pages 41–53. IEEE Press, 2014.

230. S. Piana, K. Lindorff-Larsen, and D. E. Shaw. Protein folding kinetics and thermodynamics from atomistic simulation. *Proc. Natl. Acad. Sci. USA*, 109(44):17845–17850, 2012.

231. S. Piana, K. Lindorff-Larsen, and D. E. Shaw. Atomic-level description of ubiquitin folding. *Proc. Natl. Acad. Sci. USA*, 110(15):5915–5920, 2013.

232. H. Nguyen, D. R. Roe, and C. Simmerling. Improved Generalized Born Solvent Model Parameters for Protein Simulations. *J. Chem. Theory Comput.*, 9(4):2020–2034, 2013.

233. O. S. Belden, S. C. Baker, and B. M. Baker. Citizens unite for computational immunology! *Trends Immunol.*, 36(7):385–387, 2015.

234. S. M. Larson, C. D. Snow, M. Shirts, and V. S. Pande. Folding@Home and Genome@Home: Using distributed computing to tackle previously intractable problems in computational biology. *arXiv preprint arXiv:0901.0866*, 2009.

235. R. F. Service. Problem solved* (*sort of). *Science*, 321(5890):784–6, 2008.

236. V. A. Voelz, M. Jager, S. Yao, Y. Chen, L. Zhu, S. A. Waldauer, G. R. Bowman, M. Friedrichs, O. Bakajin, L. J. Lapidus, S. Weiss, and V. S. Pande. Slow unfolded-state structuring in Acyl-CoA binding protein folding revealed by simulation and experiment. *J. Am. Chem. Soc.*, 134(30):12565–77, 2012.

237. R. Das, B. Qian, S. Raman, R. Vernon, J. Thompson, P. Bradley, S. Khare, M. D. Tyka, D. Bhat, D. Chivian, D. E. Kim, W. H. Sheffler, L. Malmström, A. M. Wollacott, C. Wang, I. Andre, and D. Baker. Structure prediction for CASP7 targets using extensive all-atom refinement with Rosetta@home. *Proteins: Struct., Funct., Bioinf.*, 69(S8):118–128, 2007.

238. U. H. E. Hansmann and Y. Okamoto. New Monte Carlo algorithms for protein folding. *Curr. Opin. Struct. Biol.*, 9(2):177–83, 1999.

239. S. Xu and Y. Zhang. Ab initio protein structure assembly using continuous structure fragments and optimized knowledge-based force field. *Proteins: Struct., Funct., Bioinf.*, 80(7):1715–1735, 2012.

240. K. T. Simons, R. Bonneau, I. Ruczinski, and D. Baker. Ab initio protein structure prediction of CASP III targets using ROSETTA. *Proteins*, Suppl 3:171–6, 1999.

241. C. A. Rohl, C. E. Strauss, K. M. Misura, and D. Baker. Protein structure prediction using Rosetta. *Methods Enzymol.*, 383:66–93, 2004.

242. W. F. van Gunsteren, J. Dolenc, and A. E. Mark. Molecular simulation as an aid to experimentalists. *Curr. Opin. Struct. Biol.*, 18(2):149–53, 2008.

243. W. L. Jorgensen and J. Tirado-Rives. Potential energy functions for atomic-level simulations of water and organic and biomolecular systems. *Proc. Natl. Acad. Sci. USA*, 102(19):6665–70, 2005.

244. C. M. Summa and M. Levitt. Near-native structure refinement using in vacuo energy minimization. *Proc. Natl. Acad. Sci. USA*, 104(9):3177–82, 2007.

245. F. E. Boas and P. B. Harbury. Potential energy functions for protein design. *Curr. Opin. Struct. Biol.*, 17(2):199–204, 2007.

246. J. W. Ponder and D. A. Case. Force fields for protein simulations. *Adv. Protein Chem.*, 66:27–85, 2003.

247. M. Bixon and S. Lifson. Potential functions and conformations in cycloalkanes. *Tetrahedron*, 23:769–84, 1967.

248. S. Emma and R. Ulf. Comparison of methods for deriving atomic charges from the electrostatic potential and moments. *J. Comput. Chem.*, 19(4):377–395, 1998.

249. A. T. Hagler and S. Lifson. Energy functions for peptides and proteins. II. The amide hydrogen bond and calculation of amide crystal properties. *J. Am. Chem. Soc.*, 96(17):5327–35, 1974.

250. A. D. Mackerell Jr. Empirical force fields for biological macromolecules: overview and issues. *J. Comput. Chem.*, 25(13):1584–604, 2004.

251. B. R. Brooks, R. E. Bruccoleri, B. D. Olafson, D. J. States, S. Swaminathan, and M. Karplus. CHARMM: A

Program for Macromolecular Energy, Minimization, and Dynamics Calculations. *J. Comput. Chem.*, 4:187–217, 1983.

252. R. B. Best, X. Zhu, J. Shim, P. E. Lopes, J. Mittal, M. Feig, and A. D. Mackerell Jr. Optimization of the additive CHARMM all-atom protein force field targeting improved sampling of the backbone ϕ, ψ and side-chain χ_1 and χ_2 dihedral angles. *J. Chem. Theory Comput.*, 8(9):3257–3273, 2012.

253. P. K. Weiner and P. A. Kollman. AMBER: assisted model building with energy refinement. A general program for modeling molecules and their interactions. *J. Comput. Chem.*, 2:287–303, 1983.

254. S. J. Weiner, P. A. Kollman, D. A. Case, U. c. Singh, C. Ghio, G. Alagona, S. Profeta Jr, and P. Weiner. A new force field for molecular mechanical simulation of nucleic acids and proteins. *J. Am. Chem. Soc.*, 106:765–784, 1984.

255. W. R. P. Scott, P. H. Hunenberger, I. G. Tironi, A. E. Mark, S. R. Billeter, J. Fennen, A. E. Torda, T. Huber, P. Kruger, and W. F. van Gunsteren. The Gromos Biomolecular Simulation Program Package. *J. Phys. Chem. A*, 103:3596–3607, 1999.

256. D. Van Der Spoel, E. Lindahl, B. Hess, G. Groenhof, A. E. Mark, and H. J. Berendsen. GROMACS: fast, flexible, and free. *J. Comput. Chem.*, 26(16):1701–18, 2005.

257. M. Christen, P. H. Hunenberger, D. Bakowies, R. Baron, R. Burgi, D. P. Geerke, T. N. Heinz, M. A. Kastenholz, V. Krautler, C. Oostenbrink, C. Peter, D. Trzesniak, and W. F. van Gunsteren. The GROMOS software for biomolecular simulation: GROMOS05. *J. Comput. Chem.*, 26(16):1719–51, 2005.

258. W. L. Jorgensen and J. Tirado-Rives. The OPLS Potential Functions for Proteins. Energy Minimizations for Crystals of Cyclic Peptides and Crambin. *J. Am. Chem. Soc.*, 110:1657–1666, 1988.

259. B. P. Hay, L. Yang, J-H. Lii, and N. L. Allinger. An extended MM3(96) force field for complexes of the group 1A and 2A cations with ligands bearing conjugated ether donor groups. *J. Mol. Struct. THEOCHEM*, 428:203–219, 1998.

260. T. A. Halgren. Merck Molecular Force Field. I. Basis, form, scope, parameterization, and performance of MMFF94. *J. Comput. Chem.*, 17(5–6):490–519, 1996.

261. T. A. Halgren. Merck Molecular Force Field. II. MMFF94 van der Waals and Electrostatic Parameters for Intermolecular Interactions. *J. Comput. Chem.*, 17:520–552, 1996.

262. T. A. Halgren. Merck Molecular Force Field. III. Molecular Geometrics and Vibrational Frequencies for MMFF94. *J. Comput. Chem.*, 17:553–586, 1996.

263. P. Dauber-Osguthorpe, V. A. Roberts, D. J. Osguthorpe, J. Wolff, M. Genest, and A. T. Hagler. Structure and energetics of ligand binding to proteins: *Escherichia coli* dihydrofolate reductase-trimethoprim, a drug-receptor system. *Proteins*, 4(1):31–47, 1988.

264. K. Gaedt and H.-D. Höltje. Consistent valence force-field parameterization of bond lengths and angles with quantum chemical ab initio methods applied to some heterocyclic dopamine D3-receptor agonists. *J. Comput. Chem.*, 19(8):935–946, 1998.

265. A. K. Rappe, C. J. Casewit, K. S. Colwell, W. A. Goddard, and W. M. Skiff. UFF, a full periodic table force field for molecular mechanics and molecular dynamics simulations. *J. Am. Chem. Soc.*, 114(25):10024–10035, 1992.

266. S. L. Mayo, B. D. Olafson, and W. A. Goddard. DREIDING: a generic force field for molecular simulations. *J. Phys. Chem.*, 94(26):8897–8909, 1990.

267. M. Clark, R. D. Cramer, and N. Van Opdenbosch. Validation of the general purpose tripos 5.2 force field. *J. Comput. Chem.*, 10(8):982–1012, 1989.

268. C. S. Ewig, R. Berry, U. Dinur, J. R. Hill, M. J. Hwang, H. Li, C. Liang, J. Maple, Z. Peng, T. P. Stockfisch, T. S. Thacher, L. Yan, X. Ni, and A. T. Hagler. Derivation of class II force fields. VIII. Derivation of a general quantum mechanical force field for organic compounds. *J. Comput. Chem.*, 22(15):1782–1800, 2001.

269. K. Palmo, B. Mannfors, N. G. Mirkin, and S. Krimm. Potential energy functions: from consistent force fields to spectroscopically determined polarizable force fields. *Biopolymers*, 68(3):383–94, 2003.

270. T. A. Halgren. Merck Molecular Force Field. IV. Conformational Energies and Geometries. *J. Comput. Chem.*, 17:587–615, 1996.

271. J. Wang, R. M. Wolf, J. W. Caldwell, P. A. Kollman, and D. A. Case. Development and testing of a general amber force field. *J. Comput. Chem.*, 25(9):1157–74, 2004.

272. S. W. I. Siu, K. Pluhackova, and R. A. Böckmann. Optimization of the OPLS-AA Force Field for Long Hydrocarbons. *J. Chem. Theory. Comput.*, 8(4):1459–1470, 2012.

273. C. E. S. Bernardes and A. Joseph. Evaluation of the OPLS-AA Force Field for the Study of Structural and Energetic Aspects of Molecular Organic Crystals. *J. Phys. Chem. A*, 119(12):3023–3034, 2015.

274. K. Vanommeslaeghe, E. Hatcher, C. Acharya, S. Kundu, S. Zhong, J. Shim, E. Darian, O. Guvench, P. Lopes,

I. Vorobyov, and A. D. Mackerell Jr. CHARMM general force field: A force field for drug-like molecules compatible with the CHARMM all-atom additive biological force fields. *J. Comput. Chem.*, 2009.

275. K. Vanommeslaeghe and A. D. MacKerell Jr. Automation of the CHARMM General Force Field (CGenFF) I: bond perception and atom typing. *J. Chem. Inf. Model.*, 52(12):3144–54, 2012.

276. K. Vanommeslaeghe, E. P. Raman, and A. D. MacKerell Jr. Automation of the CHARMM General Force Field (CGenFF) II: assignment of bonded parameters and partial atomic charges. *J. Chem. Inf. Model.*, 52(12):3155–68, 2012.

277. J. Gao. Hybrid Quantum and Molecular Mechanical Simulations: An Alternative Avenue to Solvent Effects in Organic Chemistry. *Acc. Chem. Res.*, 29(6):298–305, 1996.

278. W. Wang, O. Donini, C. M. Reyes, and P. A. Kollman. Biomolecular simulations: recent developments in force fields, simulations of enzyme catalysis, protein-ligand, protein-protein, and protein-nucleic acid non-covalent interactions. *Annu. Rev. Biophys. Biomol. Struct.*, 30:211–43, 2001.

279. A. Warshel, P. K. Sharma, M. Kato, and W. W. Parson. Modeling electrostatic effects in proteins. *Biochim. Biophys. Acta*, 1764(11):1647–76, 2006.

280. A. E. Cho, V. Guallar, B. J. Berne, and R. Friesner. Importance of accurate charges in molecular docking: quantum mechanical/molecular mechanical (QM/MM) approach. *J. Comput. Chem.*, 26:915–931, 2005.

281. H. Li, V. Ngo, M. C. Da Silva, D. R. Salahub, K. Callahan, B. Roux, and S. Y. Noskov. Representation of Ion–Protein Interactions Using the Drude Polarizable Force-Field. *J. Phys. Chem. B*, 2015.

282. S. Patel and C. L. Brooks. CHARMM fluctuating charge force field for proteins: I parameterization and application to bulk organic liquid simulations. *J. Comput. Chem.*, 25(1):1–15, 2004.

283. S. Patel, A. D. Mackerell Jr, and C. L. Brooks. CHARMM fluctuating charge force field for proteins: II protein/solvent properties from molecular dynamics simulations using a nonadditive electrostatic model. *J. Comput. Chem.*, 25(12):1504–14, 2004.

284. P. E. M. Lopes, J. Huang, J. Shim, Y. Luo, H. Li, B. Roux, and A. D. MacKerell. Polarizable Force Field for Peptides and Proteins Based on the Classical Drude Oscillator. *J. Chem. Theory. Comput.*, 9(12):5430–5449, 2013.

285. K. Vanommeslaeghe and A. D. MacKerell Jr. CHARMM additive and polarizable force fields for biophysics and computer-aided drug design. *Biochim. Biophys. Acta*, 1850(5):861–871, 2015.

286. J. Huang and A. D. MacKerell Jr. Induction of peptide bond dipoles drives cooperative helix formation in the (AAQAA)3 peptide. *Biophys. J.*, 107(4):991–7, 2014.

287. D. Qiu, P. S. Shenkin, F. P. Hollinger, and W. C. Still. The GB/SA continuum model for solvation: a fast analytical method for the calculation of approximate Born radii. *J. Phys. Chem.*, 101:3005–14, 1997.

288. A. Sali and T. L. Blundell. Comparative protein modelling by satisfaction of spatial restraints. *J. Mol. Biol.*, 234(3):779–815, 1993.

289. D. F. Feng and R. F. Doolittle. Progressive sequence alignment as a prerequisite to correct phylogenetic trees. *J. Mol. Evol.*, 25(4):351–60, 1987.

290. I. M. Wallace, G. Blackshields, and D. G. Higgins. Multiple sequence alignments. *Curr. Opin. Struct. Biol.*, 15(3):261–6, 2005.

291. M. Kosloff and R. Kolodny. Sequence-similar, structure-dissimilar protein pairs in the PDB. *Proteins*, 71(2):891–902, 2008.

292. A. Narunsky, S. Nepomnyachiy, H. Ashkenazi, R. Kolodny, and N. Ben-Tal. ConTemplate suggests possible alternative conformations for a query protein of known structure. *Structure*, 23:1–9, 2015.

293. M. Schushan and N. Ben-Tal. Modeling and Validation of Transmembrane Protein Structures. In *Introduction to Protein Structure Prediction: Methods and Algorithms*, pages 369–401. John Wiley & Sons, Inc., 2010.

294. S. F. Altschul, T. L. Madden, A. A. Schaffer, J. Zhang, Z. Zhang, W. Miller, and D. J. Lipman. Gapped BLAST and PSI-BLAST: a new generation of protein database search programs. *Nucleic Acids Res.*, 25(17):3389–402, 1997.

295. S. R. Eddy. Profile hidden Markov models. *Bioinformatics*, 14(9):755–63, 1998.

296. J. Soding, M. Remmert, and A. Biegert. HHrep: de novo protein repeat detection and the origin of TIM barrels. *Nucleic Acids Res.*, 34(suppl 2):W137–42, 2006.

297. J. Söding, A. Biegert, and A. N. Lupas. The HHpred interactive server for protein homology detection and structure prediction. *Nucleic Acids Res.*, 33(suppl 2):W244–W248, 2005.

298. A. N. Lupas. The long coming of computational structural biology. *J. Struct. Biol.*, 2008.

299. C. Bystroff and D. Baker. Blind predictions of local protein structure in CASP2 targets using the I-sites library. *Proteins*, Suppl 1:167–71, 1997.

300. M. A. Larkin, G. Blackshields, N. P. Brown, R. Chenna, P. A. McGettigan, H. McWilliam, F. Valentin, I. M. Wallace, A. Wilm, R. Lopez, J. D. Thompson, T. J. Gibson, and D. G. Higgins. Clustal W and Clustal X version 2.0. *Bioinformatics*, 23(21):2947–8, 2007.

301. J. D. Thompson, D. G. Higgins, and T. J. Gibson. CLUSTAL W: improving the sensitivity of progressive multiple sequence alignment through sequence weighting, position-specific gap penalties and weight matrix choice. *Nucleic Acids Res.*, 22(22):4673–80, 1994.

302. K. Katoh, K. Misawa, K. Kuma, and T. Miyata. MAFFT: a novel method for rapid multiple sequence alignment based on fast Fourier transform. *Nucleic Acids Res.*, 30(14):3059–66, 2002.

303. R. C. Edgar. MUSCLE: a multiple sequence alignment method with reduced time and space complexity. *BMC Bioinformatics*, 5:113, 2004.

304. C. Notredame, D. G. Higgins, and J. Heringa. T-Coffee: A novel method for fast and accurate multiple sequence alignment. *J. Mol. Biol.*, 302(1):205–17, 2000.

305. D. Fischer. 3D-SHOTGUN: a novel, cooperative, fold-recognition meta-predictor. *Proteins*, 51(3):434–41, 2003.

306. D. Petrey, Z. Xiang, C. L. Tang, L. Xie, M. Gimpelev, T. Mitros, C. S. Soto, S. Goldsmith-Fischman, A. Kernytsky, A. Schlessinger, I. Y. Koh, E. Alexov, and B. Honig. Using multiple structure alignments, fast model building, and energetic analysis in fold recognition and homology modeling. *Proteins*, 53 Suppl 6:430–5, 2003.

307. T. Schwede, J. Kopp, N. Guex, and M. C. Peitsch. SWISS-MODEL: An automated protein homology-modeling server. *Nucleic Acids Res.*, 31(13):3381–5, 2003.

308. T. Lazaridis and M. Karplus. Effective energy functions for protein structure prediction. *Curr. Opin. Struct. Biol.*, 10(2):139–45, 2000.

309. D. Eisenberg, R. Luthy, and J. U. Bowie. VERIFY3D: assessment of protein models with three-dimensional profiles. *Methods Enzymol.*, 277:396–404, 1997.

310. F. Fogolari and S. C. Tosatto. Application of MM/PBSA colony free energy to loop decoy discrimination: toward correlation between energy and root mean square deviation. *Protein Sci.*, 14(4):889–901, 2005.

311. D. Petrey and B. Honig. Free energy determinants of tertiary structure and the evaluation of protein models. *Protein Sci.*, 9(11):2181–91, 2000.

312. A. Armon, D. Graur, and N. Ben-Tal. ConSurf: an algorithmic tool for the identification of functional regions in proteins by surface mapping of phylogenetic information. *J. Mol. Biol.*, 307(1):447–63, 2001.

313. M. Kalman and N. Ben-Tal. Quality assessment of protein model-structures using evolutionary conservation. *Bioinformatics*, 26(10):1299–307, 2010.

314. R. D. Finn, J. Mistry, B. Schuster-Bockler, S. Griffiths-Jones, V. Hollich, T. Lassmann, S. Moxon, M. Marshall, A. Khanna, R. Durbin, S. R. Eddy, E. L. Sonnhammer, and A. Bateman. Pfam: clans, web tools and services. *Nucleic Acids Res.*, 34(suppl 1):D247–51, 2006.

315. M. A. Dolan, J. W. Noah, and D. Hurt. Comparison of common homology modeling algorithms: application of user-defined alignments. *Methods Mol. Biol.*, 857:399–414, 2012.

316. A. Sali, L. Potterton, F. Yuan, H. van Vlijmen, and M. Karplus. Evaluation of comparative protein modeling by MODELLER. *Proteins*, 23(3):318–26, 1995.

317. R. Jauch, H. C. Yeo, P. R. Kolatkar, and N. D. Clarke. Assessment of CASP7 structure predictions for template free targets. *Proteins*, 69 Suppl 8:57–67, 2007.

318. I. Friedberg and A. Godzik. Connecting the protein structure universe by using sparse recurring fragments. *Structure*, 13(8):1213–24, 2005.

319. M. Jacobson and A. Sali. Comparative Protein Structure Modeling and its Applications to Drug Discovery. *Annu. Rep. Med. Chem.*, 39:259–276, 2004.

320. S. Bjelic and J. Aqvist. Computational prediction of structure, substrate binding mode, mechanism, and rate for a malaria protease with a novel type of active site. *Biochemistry*, 43(46):14521–8, 2004.

321. W. Pirovano and J. Heringa. Protein Secondary Structure Prediction. In O. Carugo and F. Eisenhaber, editors, *Data Mining Techniques for the Life Sciences*, volume 609 of *Methods in Molecular Biology*, chapter 19, pages 327–348. Humana Press, 2010.

322. W. Pirovano, K. A. Feenstra, and J. Heringa. PRALINETM: a strategy for improved multiple alignment of transmembrane proteins. *Bioinformatics*, 24(4):492–7, 2008.

323. R. B. Russell, R. R. Copley, and G. J. Barton. Protein fold recognition by mapping predicted secondary structures. *J. Mol. Biol.*, 259(3):349–65, 1996.

324. K. K. Koretke, R. B. Russell, R. R. Copley, and A. N. Lupas. Fold recognition using sequence and secondary structure information. *Proteins*, Suppl 3:141–8, 1999.

325. H. Zhou and Y. Zhou. Single-body residue-level knowledge-based energy score combined with sequence-profile and secondary structure information for fold recognition. *Proteins*, 55(4):1005–13, 2004.

326. J. Skolnick, A. Kolinski, and A. R. Ortiz. MONSSTER: a method for folding globular proteins with a small number of distance restraints. *J. Mol. Biol.*, 265(2):217–41, 1997.

327. W. Kabsch and C. Sander. Dictionary of protein secondary structure: Pattern recognition of hydrogen-bonded and geometrical features. *Biopolymers*, 22(12):2577–2637, 1983.

328. C. A. Andersen, A. G. Palmer, S. Brunak, and B. Rost. Continuum secondary structure captures protein flexibility. *Structure*, 10(2):175–84, 2002.

329. M. Heinig and D. Frishman. STRIDE: a web server for secondary structure assignment from known atomic coordinates of proteins. *Nucleic Acids Res.*, 32(suppl 2):W500–2, 2004.

330. P. Y. Chou and G. D. Fasman. Conformational parameters for amino acids in helical, beta-sheet, and random coil regions calculated from proteins. *Biochemistry*, 13(2):211–22, 1974.

331. W. Kabsch and C. Sander. How good are predictions of protein secondary structure? *FEBS Lett.*, 155(2):179–82, 1983.

332. K. Nagano. Logical analysis of the mechanism of protein folding. I. Predictions of helices, loops and beta-structures from primary structure. *J. Mol. Biol.*, 75(2):401–20, 1973.

333. J. Garnier, D. J. Osguthorpe, and B. Robson. Analysis of the accuracy and implications of simple methods for predicting the secondary structure of globular proteins. *J. Mol. Biol.*, 120(1):97–120, 1978.

334. J. Garnier, J. F. Gibrat, and B. Robson. GOR method for predicting protein secondary structure from amino acid sequence. *Methods Enzymol.*, 266:540–53, 1996.

335. M. J. Zvelebil, G. J. Barton, W. R. Taylor, and M. J. Sternberg. Prediction of protein secondary structure and active sites using the alignment of homologous sequences. *J. Mol. Biol.*, 195(4):957–61, 1987.

336. J. M. Levin, S. Pascarella, P. Argos, and J. Garnier. Quantification of secondary structure prediction improvement using multiple alignments. *Protein Eng.*, 6(8):849–54, 1993.

337. B. Rost and C. Sander. Prediction of protein secondary structure at better than 70% accuracy. *J. Mol. Biol.*, 232(2):584–99, 1993.

338. D. E. Rumelhart, G. E. Hinton, and R. J. Williams. Learning representations by back-propagating errors. *Nature*, 323(6088):533–536, 1986.

339. K. Karplus, C. Barrett, and R. Hughey. Hidden Markov models for detecting remote protein homologies. *Bioinformatics*, 14(10):846–56, 1998.

340. D. Przybylski and B. Rost. Alignments grow, secondary structure prediction improves. *Proteins*, 46(2):197–205, 2002.

341. B. Rost, G. Yachdav, and J. Liu. The PredictProtein server. *Nucleic Acids Res.*, 32(suppl 2):W321–W326, 2004.

342. D. T. Jones. Protein secondary structure prediction based on position-specific scoring matrices. *J. Mol. Biol.*, 292(2):195–202, 1999.

343. P. Baldi, S. Brunak, P. Frasconi, G. Soda, and G. Pollastri. Exploiting the past and the future in protein secondary structure prediction. *Bioinformatics*, 15(11):937–46, 1999.

344. G. Shackelford and K. Karplus. Contact prediction using mutual information and neural nets. *Proteins*, 69 Suppl 8:159–64, 2007.

345. J. A. Cuff, M. E. Clamp, A. S. Siddiqui, M. Finlay, and G. J. Barton. JPred: a consensus secondary structure prediction server. *Bioinformatics*, 14(10):892–3, 1998.

346. J. A. Cuff and G. J. Barton. Application of multiple sequence alignment profiles to improve protein secondary structure prediction. *Proteins*, 40(3):502–11, 2000.

347. C. Cole, J. D. Barber, and G. J. Barton. The Jpred 3 secondary structure prediction server. *Nucleic Acids Res.*, 36(suppl 2):W197–201, 2008.

348. J. Moult. A decade of CASP: progress, bottlenecks and prognosis in protein structure prediction. *Curr. Opin. Struct. Biol.*, 15(3):285–9, 2005.

349. I. Y. Koh, V. A. Eyrich, M. A. Marti-Renom, D. Przybylski, M. S. Madhusudhan, N. Eswar, O. Grana, F. Pazos, A. Valencia, A. Sali, and B. Rost. EVA: Evaluation of protein structure prediction servers. *Nucleic Acids Res.*, 31(13):3311–5, 2003.

350. J. S. Richardson. The anatomy and taxonomy of protein structure. *Adv. Protein Chem.*, 34:167–339, 1981.

351. H. Li, R. Helling, C. Tang, and N. Wingreen. Emergence of preferred structures in a simple model of protein folding. *Science*, 273(5275):666–9, 1996.

352. Z. X. Wang. A re-estimation for the total numbers of protein folds and superfamilies. *Protein Eng.*, 11(8):621–6, 1998.

353. R. B. Russell, M. A. Saqi, P. A. Bates, R. A. Sayle, and M. J. Sternberg. Recognition of analogous and homologous protein folds: assessment of prediction success and associated alignment accuracy using empirical substitution matrices. *Protein Eng.*, 11(1):1–9, 1998.

354. A. Godzik. Metagenomics and the protein universe. *Curr. Opin. Struct. Biol.*, 21(3):398–403, 2011.

355. T. Sikosek and H. S. Chan. Biophysics of protein evolution and evolutionary protein biophysics. *J. R. Soc. Interface*, 11(100):20140419, 2014.

356. Y. Zhang, I. A. Hubner, A. K. Arakaki, E. Shakhnovich, and J. Skolnick. On the origin and highly likely completeness of single-domain protein structures. *Proc. Natl. Acad. Sci. USA*, 103(8):2605–10, 2006.

357. J. U. Bowie, R. Luthy, and D. Eisenberg. A method to identify protein sequences that fold into a known three-dimensional structure. *Science*, 253(5016):164–70, 1991.

358. D. T. Jones, W. R. Taylor, and J. M. Thornton. A new approach to protein fold recognition. *Nature*, 358(6381):86–9, 1992.

359. A. Lobley, M. I. Sadowski, and D. T. Jones. pGenTHREADER and pDomTHREADER: new methods for improved protein fold recognition and superfamily discrimination. *Bioinformatics*, 25(14):1761–7, 2009.

360. Y. Yang, E. Faraggi, H. Zhao, and Y. Zhou. Improving protein fold recognition and template-based modeling by employing probabilistic-based matching between predicted one-dimensional structural properties of query and corresponding native properties of templates. *Bioinformatics*, 27(15):2076–82, 2011.

361. K. Ginalski, A. Elofsson, D. Fischer, and L. Rychlewski. 3D-Jury: a simple approach to improve protein structure predictions. *Bioinformatics*, 19(8):1015–8, 2003.

362. S. Wu and Y. Zhang. LOMETS: a local meta-threading-server for protein structure prediction. *Nucleic Acids Res.*, 35(10):3375–82, 2007.

363. B. Wallner, P. Larsson, and A. Elofsson. Pcons.net: protein structure prediction meta server. *Nucleic Acids Res.*, 35(suppl 2):W369–74, 2007.

364. H. Zhou and J. Skolnick. Ab initio protein structure prediction using chunk-TASSER. *Biophys. J.*, 93(5):1510–8, 2007.

365. V. Mariani, F. Kiefer, T. Schmidt, J. Haas, and T. Schwede. Assessment of template based protein structure predictions in CASP9. *Proteins*, 79 Suppl 10:37–58, 2011.

366. A. Roy, A. Kucukural, and Y. Zhang. I-TASSER: a unified platform for automated protein structure and function prediction. *Nat. Protoc.*, 5(4):725–38, 2010.

367. S. Wu, J. Skolnick, and Y. Zhang. Ab initio modeling of small proteins by iterative TASSER simulations. *BMC Biol.*, 5:17, 2007.

368. K. W. Kaufmann, G. H. Lemmon, S. L. Deluca, J. H. Sheehan, and J. Meiler. Practically useful: what the Rosetta protein modeling suite can do for you. *Biochemistry*, 49(14):2987–98, 2010.

369. D. Altschuh, A. M. Lesk, A. C. Bloomer, and A. Klug. Correlation of co-ordinated amino acid substitutions with function in viruses related to tobacco mosaic virus. *J. Mol. Biol.*, 193(4):693–707, 1987.

370. E. Neher. How frequent are correlated changes in families of protein sequences? *Proc. Natl. Acad. Sci. USA*, 91(1):98–102, 1994.

371. A. Poon and L. Chao. The rate of compensatory mutation in the DNA bacteriophage phiX174. *Genetics*, 170(3):989–99, 2005.

372. D. S. Marks, L. J. Colwell, R. Sheridan, T. A. Hopf, A. Pagnani, R. Zecchina, and C. Sander. Protein 3D structure computed from evolutionary sequence variation. *PLoS One*, 6(12):e28766, 2011.

373. T. A. Hopf, L. J. Colwell, R. Sheridan, B. Rost, C. Sander, and D. S. Marks. Three-dimensional structures of membrane proteins from genomic sequencing. *Cell*, 149(7):1607–21, 2012.

374. D. S. Marks, T. A. Hopf, and C. Sander. Protein structure prediction from sequence variation. *Nat. Biotechnol.*, 30(11):1072–80, 2012.

375. T. Nugent and D. T. Jones. Accurate de novo structure prediction of large transmembrane protein domains using fragment-assembly and correlated mutation analysis. *Proc. Natl. Acad. Sci. USA*, 109(24):E1540–7, 2012.

376. F. Morcos, A. Pagnani, B. Lunt, A. Bertolino, D. S. Marks, C. Sander, R. Zecchina, J. N. Onuchic, T. Hwa, and M. Weigt. Direct-coupling analysis of residue coevolution captures native contacts across many protein families. *Proc. Natl. Acad. Sci. USA*, 108(49):E1293–301, 2011.

377. F. L. Simonetti, E. Teppa, A. Chernomoretz, M. Nielsen, and C. Marino Buslje. MISTIC: Mutual information server to infer coevolution. *Nucleic Acids Res.*, 41(W1):W8–14, 2013.

378. C. S. Jeong and D. Kim. Reliable and robust detection of coevolving protein residues. *Protein Eng. Des. Sel.*, 25(11):705–13, 2012.

379. D. T. Jones, T. Singh, T. Kosciolek, and S. Tetchner. MetaPSICOV: combining coevolution methods for

accurate prediction of contacts and long range hydrogen bonding in proteins. *Bioinformatics*, 31(7):999–1006, 2015.

380. J. Iserte, F. L. Simonetti, D. J. Zea, E. Teppa, and C. Marino-Buslje. I-COMS: Interprotein-COrrelated Mutations Server. *Nucleic Acids Res.*, 43(W1):W320–5, 2015.

381. T. A. Hopf, C. P. Scharfe, J. P. Rodrigues, A. G. Green, O. Kohlbacher, C. Sander, A. M. Bonvin, and D. S. Marks. Sequence co-evolution gives 3D contacts and structures of protein complexes. *eLife*, 3, 2014.

382. D. T. Jones, D. W. Buchan, D. Cozzetto, and M. Pontil. PSICOV: precise structural contact prediction using sparse inverse covariance estimation on large multiple sequence alignments. *Bioinformatics*, 28(2):184–90, 2012.

383. M. Pellegrini-Calace, A. Carotti, and D. T. Jones. Folding in lipid membranes (FILM): a novel method for the prediction of small membrane protein 3D structures. *Proteins*, 50(4):537–45, 2003.

384. M. Michel, S. Hayat, M. J. Skwark, C. Sander, D. S. Marks, and A. Elofsson. PconsFold: improved contact predictions improve protein models. *Bioinformatics*, 30(17):i482–8, 2014.

385. M. J. Skwark, A. Abdel-Rehim, and A. Elofsson. PconsC: combination of direct information methods and alignments improves contact prediction. *Bioinformatics*, 29(14):1815–6, 2013.

386. S. Ovchinnikov, L. Kinch, H. Park, Y. Liao, J. Pei, D. E. Kim, H. Kamisetty, N. V. Grishin, and D. Baker. Large scale determination of previously unsolved protein structures using evolutionary information. *eLife*, 4, 2015.

387. H. Kamisetty, S. Ovchinnikov, and D. Baker. Assessing the utility of coevolution-based residue-residue contact predictions in a sequence- and structure-rich era. *Proc. Natl. Acad. Sci. USA*, 110(39):15674–9, 2013.

388. A. Zemla, Venclovas, J. Moult, and K. Fidelis. Processing and evaluation of predictions in CASP4. *Proteins*, Suppl 5:13–21, 2001.

389. P. Bradley, K. M. Misura, and D. Baker. Toward high-resolution de novo structure prediction for small proteins. *Science*, 309(5742):1868–71, 2005.

390. H. Lei, C. Wu, H. Liu, and Y. Duan. Folding free-energy landscape of villin headpiece subdomain from molecular dynamics simulations. *Proc. Natl. Acad. Sci. USA*, 104(12):4925–30, 2007.

391. J. Moult, K. Fidelis, A. Kryshtafovych, T. Schwede, and A. Tramontano. Critical assessment of methods of protein structure prediction (CASP)–round x. *Proteins*, 82 Suppl 2:1–6, 2014.

392. D. Fischer. Servers for protein structure prediction. *Curr. Opin. Struct. Biol.*, 16(2):178–82, 2006.

393. W. Zheng and S. Doniach. Protein structure prediction constrained by solution X-ray scattering data and structural homology identification. *J. Mol. Biol.*, 316(1):173–87, 2002.

394. W. Zheng and S. Doniach. Fold recognition aided by constraints from small angle X-ray scattering data. *Protein Eng. Des. Sel.*, 18(5):209–19, 2005.

395. P. D. Adams, D. Baker, A. T. Brunger, R. Das, F. DiMaio, R. J. Read, D. C. Richardson, J. S. Richardson, and T. C. Terwilliger. Advances, interactions, and future developments in the CNS, Phenix, and Rosetta structural biology software systems. *Annu. Rev. Biophys.*, 42:265–87, 2013.

396. D. I. Svergun, M. V. Petoukhov, M. H. Koch, and S. Konig. Crystal versus solution structures of thiamine diphosphate-dependent enzymes. *J. Biol. Chem.*, 275(1):297–302, 2000.

397. M. D. Purdy, B. C. Bennett, W. E. McIntire, A. K. Khan, P. M. Kasson, and M. Yeager. Function and dynamics of macromolecular complexes explored by integrative structural and computational biology. *Curr. Opin. Struct. Biol.*, 27:138–48, 2014.

398. D. Lyumkis, J. P. Julien, N. de Val, A. Cupo, C. S. Potter, P. J. Klasse, D. R. Burton, R. W. Sanders, J. P. Moore, B. Carragher, I. A. Wilson, and A. B. Ward. Cryo-EM structure of a fully glycosylated soluble cleaved HIV-1 envelope trimer. *Science*, 342(6165):1484–90, 2013.

399. M. Liao, E. Cao, D. Julius, and Y. Cheng. Structure of the TRPV1 ion channel determined by electron cryo-microscopy. *Nature*, 504(7478):107–12, 2013.

100. W. Wong, X. C. Bai, A. Brown, I. S. Fernandez, E. Hanssen, M. Condron, Y. H. Tan, J. Baum, and S. H. Scheres. Cryo-EM structure of the *Plasmodium falciparum* 80S ribosome bound to the anti-protozoan drug emetine. *eLife*, 3, 2014.

401. S. J. Fleishman, S. E. Harrington, A. Enosh, D. Halperin, C. G. Tate, and N. Ben-Tal. Quasi-symmetry in the cryo-EM structure of EmrE provides the key to modeling its transmembrane domain. *J. Mol. Biol.*, 364(1):54–67, 2006.

402. Y. J. Chen, O. Pornillos, S. Lieu, C. Ma, A. P. Chen, and G. Chang. X-ray structure of EmrE supports dual topology model. *Proc. Natl. Acad. Sci. USA*, 104(48):18999–9004, 2007.

403. E. A. Morrison, G. T. DeKoster, S. Dutta, R. Vafabakhsh, M. W. Clarkson, A. Bahl, D. Kern, T. Ha, and

K. A. Henzler-Wildman. Antiparallel EmrE exports drugs by exchanging between asymmetric structures. *Nature*, 481(7379):45–50, 2012.

404. P. Lloris-Garcerá, J. S. G. Slusky, S. Seppälä, M. Prieß, L. V. Schäfer, and G. von Heijne. In Vivo Trp Scanning of the Small Multidrug Resistance Protein EmrE Confirms 3D Structure Models. *J. Mol. Biol.*, 425(22):4642–4651, 2013.

405. F. Stengel, R. Aebersold, and C. V. Robinson. Joining forces: integrating proteomics and cross-linking with the mass spectrometry of intact complexes. *Mol. Cell. Proteomics*, 11(3):R111 014027, 2012.

406. N. P. Cowieson, B. Kobe, and J. L. Martin. United we stand: combining structural methods. *Curr. Opin. Struct. Biol.*, 18(5):617–22, 2008.

407. A. B. Ward, A. Sali, and I. A. Wilson. Biochemistry. Integrative structural biology. *Science*, 339(6122):913–5, 2013.

408. F. Alber, S. Dokudovskaya, L. M. Veenhoff, W. Zhang, J. Kipper, D. Devos, A. Suprapto, O. Karni-Schmidt, R. Williams, B. T. Chait, M. P. Rout, and A. Sali. Determining the architectures of macromolecular assemblies. *Nature*, 450(7170):683–694, 2007.

409. F. Alber, S. Dokudovskaya, L. M. Veenhoff, W. Zhang, J. Kipper, D. Devos, A. Suprapto, O. Karni-Schmidt, R. Williams, B. T. Chait, A. Sali, and M. P. Rout. The molecular architecture of the nuclear pore complex. *Nature*, 450(7170):695–701, 2007.

410. A. R. Gingras, N. Bate, B. T. Goult, L. Hazelwood, I. Canestrelli, J. G. Grossmann, H. Liu, N. S. Putz, G. C. Roberts, N. Volkmann, D. Hanein, I. L. Barsukov, and D. R. Critchley. The structure of the C-terminal actin-binding domain of talin. *EMBO J.*, 27(2):458–69, 2008.

411. G. C. Lander, E. Estrin, M. E. Matyskiela, C. Bashore, E. Nogales, and A. Martin. Complete subunit architecture of the proteasome regulatory particle. *Nature*, 482(7384):186–91, 2012.

412. K. Lasker, F. Forster, S. Bohn, T. Walzthoeni, E. Villa, P. Unverdorben, F. Beck, R. Aebersold, A. Sali, and W. Baumeister. Molecular architecture of the 26S proteasome holocomplex determined by an integrative approach. *Proc. Natl. Acad. Sci. USA*, 109(5):1380–7, 2012.

413. F. Beck, P. Unverdorben, S. Bohn, A. Schweitzer, G. Pfeifer, E. Sakata, S. Nickell, J. M. Plitzko, E. Villa, W. Baumeister, and F. Forster. Near-atomic resolution structural model of the yeast 26S proteasome. *Proc. Natl. Acad. Sci. USA*, 109(37):14870–5, 2012.

414. P. Unverdorben, F. Beck, P. Sledz, A. Schweitzer, G. Pfeifer, J. M. Plitzko, W. Baumeister, and F. Forster. Deep classification of a large cryo-EM dataset defines the conformational landscape of the 26S proteasome. *Proc. Natl. Acad. Sci. USA*, 111(15):5544–9, 2014.

415. S. J. Hirst, N. Alexander, H. S. McHaourab, and J. Meiler. RosettaEPR: an integrated tool for protein structure determination from sparse EPR data. *J. Struct. Biol.*, 173(3):506–14, 2011.

416. P. Rossi, L. Shi, G. Liu, C. M. Barbieri, H. W. Lee, T. D. Grant, J. R. Luft, R. Xiao, T. B. Acton, E. H. Snell, G. T. Montelione, D. Baker, O. F. Lange, and N. G. Sgourakis. A hybrid NMR/SAXS-based approach for discriminating oligomeric protein interfaces using Rosetta. *Proteins*, 83(2):309–17, 2015.

417. F. DiMaio, M. D. Tyka, M. L. Baker, W. Chiu, and D. Baker. Refinement of protein structures into low-resolution density maps using Rosetta. *J. Mol. Biol.*, 392(1):181–90, 2009.

418. Y. Shen, O. Lange, F. Delaglio, P. Rossi, J. M. Aramini, G. Liu, A. Eletsky, Y. Wu, K. K. Singarapu, A. Lemak, A. Ignatchenko, C. H. Arrowsmith, T. Szyperski, G. T. Montelione, D. Baker, and A. Bax. Consistent blind protein structure generation from NMR chemical shift data. *Proc. Natl. Acad. Sci. USA*, 105(12):4685–90, 2008.

419. D. Russel, K. Lasker, B. Webb, J. Velazquez-Muriel, E. Tjioe, D. Schneidman-Duhovny, B. Peterson, and A. Sali. Putting the pieces together: integrative modeling platform software for structure determination of macromolecular assemblies. *PLoS Biol.*, 10(1):e1001244, 2012.

420. S. Lindert, N. Alexander, N. Wötzel, M. Karakaş, P. Stewart, and J. Meiler. EM-Fold: De Novo Atomic-Detail Protein Structure Determination from Medium-Resolution Density Maps. *Structure*, 20(3):464–478, 2012.

421. N. Woetzel, S. Lindert, P. L. Stewart, and J. Meiler. BCL::EM-Fit: rigid body fitting of atomic structures into density maps using geometric hashing and real space refinement. *J. Struct. Biol.*, 175(3):264–76, 2011.

422. M. Saha and M. C. Morais. FOLD-EM: automated fold recognition in medium- and low-resolution (4–15 Å) electron density maps. *Bioinformatics*, 28(24):3265–73, 2012.

423. E. Tjioe, K. Lasker, B. Webb, H. J. Wolfson, and A. Sali. MultiFit: a web server for fitting multiple protein structures into their electron microscopy density map. *Nucleic Acids Res.*, 39(suppl 2):W167–70, 2011.

424. E. Villa and K. Lasker. Finding the right fit: chiseling structures out of cryo-electron microscopy maps. *Curr. Opin. Struct. Biol.*, 25:118–25, 2014.

425. D. Schneidman-Duhovny, M. Hammel, and A. Sali. Macromolecular docking restrained by a small angle X-ray scattering profile. *J. Struct. Biol.*, 173(3):461–71, 2011.

426. M. V. Petoukhov, D. Franke, A. V. Shkumatov, G. Tria, A. G. Kikhney, M. Gajda, C. Gorba, H. D. Mertens, P. V. Konarev, and D. I. Svergun. New developments in the program package for small-angle scattering data analysis. *J. Appl. Crystallogr.*, 45(Pt 2):342–350, 2012.

427. W. Zheng and M. Tekpinar. Accurate flexible fitting of high-resolution protein structures to small-angle x-ray scattering data using a coarse-grained model with implicit hydration shell. *Biophys. J.*, 101(12):2981–91, 2011.

428. G. van der Schot, Z. Zhang, R. Vernon, Y. Shen, W. F. Vranken, D. Baker, A. M. Bonvin, and O. F. Lange. Improving 3D structure prediction from chemical shift data. *J. Biomol. NMR*, 57(1):27–35, 2013.

429. N. G. Sgourakis, W. M. Yau, and W. Qiang. Modeling an in-register, parallel "iowa" abeta fibril structure using solid-state NMR data from labeled samples with Rosetta. *Structure*, 23(1):216–27, 2015.

430. A. Cavalli, X. Salvatella, C. M. Dobson, and M. Vendruscolo. Protein structure determination from NMR chemical shifts. *Proc. Natl. Acad. Sci. USA*, 104(23):9615–20, 2007.

431. D. S. Wishart, D. Arndt, M. Berjanskii, P. Tang, J. Zhou, and G. Lin. CS23D: a web server for rapid protein structure generation using NMR chemical shifts and sequence data. *Nucleic Acids Res.*, 36(suppl 2):W496–502, 2008.

432. F. DiMaio, Y. Song, X. Li, M. J. Brunner, C. Xu, V. Conticello, E. Egelman, T. C. Marlovits, Y. Cheng, and D. Baker. Atomic-accuracy models from 4.5-Å cryo-electron microscopy data with density-guided iterative local refinement. *Nat. Methods*, 12(4):361–5, 2015.

433. M. A. dos Reis, R. Aparicio, and Y. Zhang. Improving protein template recognition by using small-angle x-ray scattering profiles. *Biophys. J.*, 101(11):2770–81, 2011.

434. S. Wu and Y. Zhang. MUSTER: Improving protein sequence profile-profile alignments by using multiple sources of structure information. *Proteins*, 72(2):547–56, 2008.

435. G. T. Montelione, D. Zheng, Y. J. Huang, K. C. Gunsalus, and T. Szyperski. Protein NMR spectroscopy in structural genomics. *Nat. Struct. Biol.*, 7 Suppl:982–5, 2000.

436. J. L. Sussman, D. Lin, J. Jiang, N. O. Manning, J. Prilusky, O. Ritter, and E. E. Abola. Protein Data Bank (PDB): database of three-dimensional structural information of biological macromolecules. *Acta Crystallogr. Sect. D*, 54(Pt 6 Pt 1):1078–84, 1998.

437. H. M. Berman, B. Coimbatore Narayanan, L. Di Costanzo, S. Dutta, S. Ghosh, B. P. Hudson, C. L. Lawson, E. Peisach, A. Prlic, P. W. Rose, C. Shao, H. Yang, J. Young, and C. Zardecki. Trendspotting in the Protein Data Bank. *FEBS Lett.*, 587(8):1036–45, 2013.

438. A. S. Rose and P. W. Hildebrand. NGL Viewer: a web application for molecular visualization. *Nucleic Acids Res.*, 43(W1):W576, 2015.

439. N. Savage. Plenty of proteins. *Commun. ACM*, 58(6):12–14, 2015.

440. R. A. Goldstein. The structure of protein evolution and the evolution of protein structure. *Curr. Opin. Struct. Biol.*, 18(2):170–177, 2008.

441. K. Peng, Z. Obradovic, and S. Vucetic. Exploring bias in the Protein Data Bank using contrast classifiers. *Pac. Symp. Biocomput.*, pages 435–46, 2004.

442. J. Söding and A. N. Lupas. More than the sum of their parts: on the evolution of proteins from peptides. *BioEssays*, 25(9):837–846, 2003.

443. J. Haas, S. Roth, K. Arnold, F. Kiefer, T. Schmidt, L. Bordoli, and T. Schwede. The Protein Model Portal: a comprehensive resource for protein structure and model information. *Database (Oxford)*, 2013:bat031, 2013.

CHAPTER 4

Energetics and Protein Stability

4.1 BASIC PRINCIPLES OF THERMODYNAMICS

4.1.1 Introduction

In Chapter 2 we discussed in detail the multi-level architecture of proteins. Like all other molecules, proteins are physical entities that are subjected to the physical forces that dominate our universe. The interplay between proteins and these forces results in folding and formation of the unique architectures discussed in Chapter 2. In accordance with the central structural-biological paradigm, by determining the structure of a protein, the physical forces that act on the protein also determine its biological function. It therefore stands to reason that a true understanding of proteins requires their characterization in terms of forces and energies. This approach is the basis of a specific field in life sciences termed 'Structural Biophysics'. The field not only contributes to the understanding of protein structure but is also used in various applied areas, such as protein engineering and rational drug design (see Chapter 8). Structural biophysics deals with various physical aspects of proteins, but can be separated into two major fields:

1. *Energetics* studies the principal forces that affect protein folding and stability. This is the subject of the present chapter, focusing on globular, water-soluble proteins. The energetics of fibrous and membrane-bound proteins is covered in Chapters 6 and 7, respectively.

2. *Dynamics* studies the conformational changes of the polypeptide chain during folding, and also those that occur in the folded (native) state. These issues are covered in Chapter 5.

The biophysical approach requires the characterization of the physical forces acting on the system. It also requires determination of the directions of the processes involving these forces, that is, folding and unfolding of a protein. These goals can be achieved, to a significant extent, by studying the thermodynamics of the system. In fact, protein folding alone has been characterized thermodynamically for over 70 years [1,2]. The basic principles of this field and their application in the study of protein folding will be summarized in the following subsections.

4.1.2 Free energy and spontaneous processes

Thermodynamics is a veteran field in physics. It enables different states in nature to be characterized, by studying the dependency of some of their general properties (e.g., energy) on temperature, pressure, volume, and chemical concentration. **Most importantly, thermodynamics allows us to use macroscopic characterization, namely, the total energy content of a system in different states, to predict the direction and probability of a process without delving into the microscopic details**. This constitutes an important tool for studying different natural systems. Indeed, the rules of thermodynamics are applied today in many fields, including engineering, chemistry and biology. When the studied system is well characterized, as in many biological systems, the general measured quantities can be assigned to specific elements and processes of the systems. In such cases, thermodynamics is invaluable for obtaining data that cannot be produced by other, sometimes more sophisticated, methods.

The direction of a process (e.g., folding or unfolding of a protein) has originally been predicted by measuring *entropy*, a quantity related to the number of possible configurations of the system (see next subsection for a detailed description). This approach is based on the second law of thermodynamics, stating that in an isolated system[*1], spontaneous processes approach equilibrium by increasing the entropy of the system [3]. Biological systems are not isolated. That is, they exist in a state of constant temperature and pressure, not volume and energy. In such systems it is more convenient to use a different thermodynamic quantity for determining the direction of a given process. This quantity, which is called the Gibbs (after Josiah Willard Gibbs) free energy or simply the free energy, is denoted by G. It represents the 'useful' energy[*2] in systems under constant temperature and pressure.

How does free energy relate to the direction of a process? Spontaneous processes always proceed towards equilibrium by decreasing the free energy of the system to a minimum (Figure 4.1). At equilibrium, the free energy change is zero, and the change in the standard free energy (ΔG^0)[*3] is related to the equilibrium constant (K_{eq})[*4]:

$$\Delta G^0 = -RT \ln K_{eq} \tag{4.1}$$

(where R is the universal gas constant $(1.989 \, \text{cal/mol K})$[*5] and T is the temperature, in K. At room temperature (298 K), $RT \approx 0.6 \, \text{kcal/mol}$).

Standard conditions are unrealistic, and it is therefore more convenient to use the value of ΔG, which can be calculated from ΔG^0, provided that the concentrations of the initial and final states of the process are known [5]:

$$\Delta G = \Delta G^0 + RT \ln \frac{f \, [\text{products}]}{f \, [\text{reactants}]} \tag{4.2}$$

(where f signifies the function of the ratio of actual product and reactant concentrations, the exact mathematical form of which depends on the stoichiometry).

[*1] A system that cannot exchange matter or energy with its surroundings, and therefore has constant energy and volume.

[*2] Energy that can be used for doing non-expansion work, e.g., execution of a chemical process or changing molecular conformation or configuration.

[*3] The free energy change under standard conditions, i.e., concentration of 1 molar (M) and pressure of 1 atmosphere.

[*4] K_{eq} is a measurable quantity. In molecular systems it is often determined by spectroscopy [4].

[*5] R can be replaced by the Boltzmann constant $(k_B = 3.3 \times 10^{-24} \, \text{cal/K})$ to obtain the free energy per molecule instead of per mole.

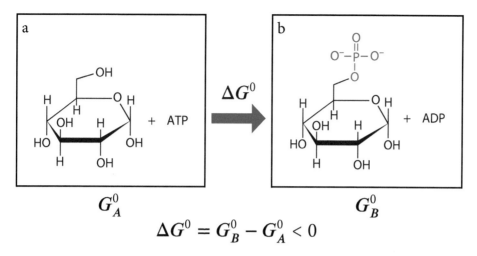

$$\Delta G^0 = G_B^0 - G_A^0 < 0$$

FIGURE 4.1 **Spontaneous processes involve a decrease in the standard free energy of the system.** The figure depicts the process of glucose oxidation to glucose 6-phosphate, an enzyme-catalyzed spontaneous process (the phosphate group is in red). The final state of the system has lower free energy (G) compared to the initial state (i.e., it is more stable). Therefore, the standard free energy change accompanying the process (ΔG^0) is negative.

For example, let us examine the hydrolysis of adenosine triphosphate (ATP) to adenosine diphosphate (ADP) and inorganic phosphate (P$_i$), a chemical reaction very familiar to students of life sciences:

$$ATP \longleftrightarrow ADP + P_i$$

The standard free energy of this reaction (ΔG^0) is −7.3 kcal/mol. However, since the cellular concentration of ATP is much higher than that of ADP ($\dfrac{ATP}{ADP} \approx 10\text{–}100$ [6,7]), the free energy of the reaction (ΔG) is even more negative than ΔG^0:

$$\Delta G = \Delta G^0 + RT \ln \frac{[ADP][P_i]}{[ATP]}$$
$$= -7.3\,\text{kcal/mol} + RT \ln \frac{0.04\,\text{mM} \cdot 4.15\,\text{mM}}{3.75\,\text{mM}}$$
$$= -9.2\,\text{kcal/mol}^{*1*2}$$

Thus, by measuring the free energy change of any process, one is able to determine whether or not it is spontaneous. Moreover, by rearranging Eq. (4.1), the value of the free energy change can be used to calculate the equilibrium populations of the initial and final states of the process:

$$e^{-\frac{\Delta G}{RT}} = K_{eq} = \frac{\#\,\text{final}}{\#\,\text{initial}} \tag{4.3a}$$

[*1] Values taken from [6].

[*2] In cells, the hydrolysis reaction involves other chemical species, such as Mg^{2+} ions. The concentrations of these further lower the free energy of hydrolysis to a value of ~12 kcal/mol [7].

For example, in a chemical reaction turning substance A into substance B, the expression $\frac{\text{\# final}}{\text{\# initial}}$ will signify the equilibrium ratio between the number of molecules that underwent the transformation to the number of molecules that did not, i.e., $\frac{\text{\# B}}{\text{\# A}}$.

Since there is a correlation between $\frac{\text{\# final}}{\text{\# initial}}$ and the probability (P) of the process, the latter can be estimated from the change in free energy [8][*1]:

$$P \propto e^{-\frac{\Delta G}{RT}} \tag{4.3b}$$

Note that the probability depends exponentially on the free energy change. This means that even a small difference in the free energy of the two states of the system can be manifested as a large change in its probability to shift from the less stable to the more stable state. The free energy is a state function: its magnitude depends only on the end points of the process and is independent of the specific path taken by the system to connect them.

4.1.3   Enthalpy, entropy, and molecular thermodynamics

Knowing the free energy of a system is very helpful in predicting the relative stability of states and the directions of processes, including protein folding, protein-ligand binding, and enzyme-catalyzed processes (see Box 4.1). However, biologists are ultimately interested in the specifics of the system, which the total free energy values cannot provide. Fortunately, the free energy change itself can be broken down into contributions from changes in two other quantities, **enthalpy** (H) and **entropy** (S). At constant temperature:

$$\Delta G = \Delta H - T\Delta S \tag{4.4}$$

H and S are also general quantities, but unlike the free energy, they can be associated with specifics of the molecular system. As explained in the following paragraphs, these quantities represent the general tendency to minimize energy and maximize disorder.

4.1.3.1   Enthalpy

Enthalpy (H) is a thermodynamic quantity related to the system's internal energy (E), pressure (P), and volume (V):

$$H = E + PV \tag{4.5}$$

Thus, changes in enthalpy must result from changes in the internal energy of the system, or from pressure-volume exchange with the environment. Biological systems are based on solids and liquids, and therefore do not usually experience changes in pressure or volume. Therefore, enthalpy changes (ΔH) in such systems result, in essence, from changes in the internal energy (ΔE) alone:

$$\Delta H \approx \Delta E \tag{4.6}$$

But what is the meaning of the internal energy in molecular terms? The internal energy can be broken down into two different contributions, potential energy (U) and kinetic energy (K):

$$E = U + K \tag{4.7}$$

[*1]This expression draws from statistical thermodynamics, a field that describes the microscopic behaviors of thermodynamic systems using probability.

Potential energy is defined as 'energy of position'. In broad terms, it is the energy of an object that is located within some kind of field (gravitational, electric, magnetic, etc.). In molecules, atoms create electric fields that affect other atoms in their vicinity. These fields result from either fully or partially charged atoms. If we sum all the atom-field effects in the molecular system, we obtain its potential energy, which is position-dependent. The atom-field effects can also be presented as covalent bonds and noncovalent interactions. In the former, the short distance between atoms leads to electron sharing. In the latter, the atoms are too distant to share electrons, but may form ionic, hydrogen, or van der Waals interactions, depending on the charge intensity and whether it is fixed or induced. Thus, the potential energy can be defined as the type of energy resulting from all covalent and most noncovalent[*1] interactions in the molecular system [9]. The other component of the internal energy, K, is the result of thermally induced atomic motions in the molecule.

In sum, the ΔH component of ΔG involves the following molecular phenomena:

A. Formation or breaking of covalent bonds

B. Changes in electrostatic or van der Waals interactions

C. Changes in thermally-induced atomic motions

A common property of all of these changes is that under constant pressure they involve heat transfer ($q_{(P)}$) between the system and its environment at constant pressure. Equation (4.6) can therefore be rewritten as:

$$q_{(P)} = \Delta H \approx \Delta E \tag{4.8}$$

This is an important property, since it enables the structural biophysicist to track enthalpy changes in the system by measuring the heat transfer to and from the system[*2]. The formation of an energetically favorable bond or noncovalent interaction leads to the decrease of ΔH by releasing heat from the system to its surroundings. Biological systems involve different processes in which such occurrences take place (see Box 4.1). In systems based on macromolecules such as proteins, whose structure results from the folding of a long polymeric chain, enthalpy changes involve mainly noncovalent (electrostatic and van der Waals) interactions [9,12].

Heat transfer, following the formation of chemical bonds or noncovalent interactions, can be measured in the lab using calorimetric methods, mainly differential scanning calorimetry (DSC) and isothermal calorimetry (ITC) [13]. Calorimetry measures a quantity called 'heat capacity' (C_P), which is defined as the amount of heat ($q_{(P)}^{rev}$) required for elevating the temperature of 1 mol of substance by 1 K, in a reversible process (at constant pressure) [14]. Since $q_{(P)}^{rev} = \Delta H$, the heat capacity can be defined as a change of energy per change of temperature:

$$C_P = \frac{q_{(P)}^{rev}}{\Delta T} = \frac{\Delta H}{\Delta T} \tag{4.9}$$

Thus, ΔH can be calculated from the measurable value of the heat capacity.

In a typical DSC experiment, C_P is measured as a function of temperature, and ΔH is extracted as the area under the transition curve [5] (Figure 4.2).

[*1] This does not include nonpolar interactions, which involve mainly solvent entropy changes.

[*2] ΔH can also be measured using the Van't Hoff relation between K_{eq} and T, but studies show this approach to be less accurate [10,11].

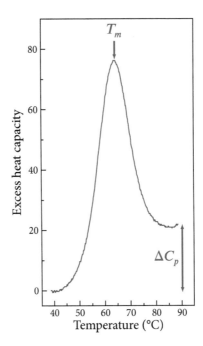

FIGURE 4.2 **DSC measurement of the thermal denaturation of the protein ubiquitin.** The curve depicts C_P as a function of temperature, where T_m is the transition temperature (adapted from [5]).

The relation between the heat capacity and the enthalpy change is apparent when the molecular significance of the latter is considered [12]. Positive enthalpy changes in macromolecules ($\Delta H > 0$) are attributed to the breaking of favorable noncovalent interactions[*1]. In fact, the system uses the heat energy to drive the endergonic process of breaking the interactions. For this reason, heating a sample containing a macromolecule is not accompanied by a temperature increase (as would normally be expected[*2]), as long as there are intact interactions left within the macromolecule. The higher the number of interactions in a macromolecule, the greater the energy needed to increase the temperature; namely, the C_P of the macromolecule is higher. Thus, the heat capacity can be viewed as a measure of the number of noncovalent interactions in the system. This is similar to the process of ice melting (ice is not a macromolecule, but contains numerous hydrogen bonds, similarly to proteins). The process also happens during the thermal denaturation of proteins[*3], with one important difference. As in the case of ice, the heating of the folded polypeptide chain involves the disruption of noncovalent interactions[*4]. However, unlike ice (considered in this case as a solute), the protein is surrounded by solvent, i.e. water molecules, some of which hydrogen-bond to the protein's polar groups, whereas the others participate in intra-water bonds. The measurable changes in C_P during protein denaturation result from changes in intra-protein interactions, as well as in protein-solvent and intra-solvent interactions. Most studies suggest that the two latter dominate, although this matter is still under debate [14].

[*1] In studying the folding or unfolding of macromolecules, it is assumed that covalent bonds remain unchanged, and that any enthalpy changes measured result from changes in noncovalent interactions. Indeed, formation or breaking of a covalent bond would result in energies on a much larger scale.

[*2] In 'normal' cases the heat energy is converted into kinetic energy of atoms in the molecule.

[*3] For experimental convenience, measurements usually follow protein denaturation, which is the thermodynamic reversal of folding.

[*4] In the case of ice, the noncovalent interactions include only hydrogen bonds, whereas in the protein they also include ionic and van der Waals interactions.

BOX 4.1 FREE ENERGY CHANGES IN BIOLOGICAL PROCESSES

Any natural state can be characterized by its free energy content. The free energy emanates from the overall properties of the system in the specific state. In a molecular system these properties include the number and types of atoms, their location in 3-D space, the nature of chemical bonds and interactions connecting them to each other, their freedom of movement, etc. The higher the free energy of the system, the less stable the system is. That is, a system capable of existing in two alternative states with different free energy values will spontaneously shift from the high-energy state to the low-energy one. These spontaneous processes are called 'exergonic'. Biological processes are often chemical, and therefore involve a change in the molecular properties of the system. This is particularly true for metabolic processes, in which complex molecules are degraded and oxidized to fuel endergonic (energy-demanding) processes. The latter include the transport of ions across the plasma membrane and the construction of complex cellular building blocks, such as proteins, DNA, oligosaccharides and complex lipids. A central exergonic process in metabolism is the complete degradation-oxidation of glucose to carbon dioxide and water (Figure 4.1.1). Although it is spontaneous ($\Delta G = -686$ kcal/mol), this process is carried out in many steps, and over two different subcellular compartments. This is done in order to make the energy available for any endergonic process in the cell. The small amounts of energy produced along the oxidative path are chemically stored in the form of ATP molecules; each contains 7.3 kcal/mol of free energy. Considering the total amount of energy released in the process, the number of ATP molecules produced should be $\dfrac{686}{7.3} = 94$. In fact, only 38 or so molecules are produced, revealing a ~40% efficiency of the biological energy-extracting and storing mechanism.

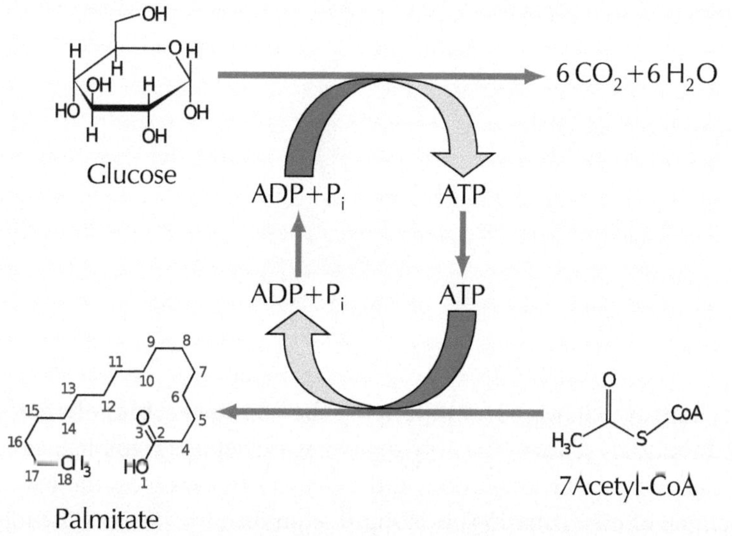

FIGURE 4.1.1 **Metabolic thermodynamics.** Exergonic processes in the cell (e.g., glucose oxidation to CO_2 and water) drive endergonic processes (e.g., fatty acid production from acetyl-CoA). The energy obtained from the former is temporarily stored chemically, as ATP.

Free energy changes in biological systems need not necessarily involve chemical transformations. In fact, many biological processes lead to significant energy changes without the formation or degradation of chemical bonds. Such processes may involve changes in noncovalent interactions, or in the degree of order of the system. Noncovalent interactions may change when a molecule changes conformation. The chemical composition of the system remains the same, but since the spatial location of some atoms changes, so do the interactions between them. This effect is particularly significant in macromolecules, which may include tens of thousands of atoms. In principle, electrostatic and van der Waals interactions in molecules are favorable, and involve a decrease in the free energy of the system. In macromolecules, hydrogen bonds are very common and help stabilize the native structure of the molecule. Changes in noncovalent interactions also occur during molecular binding, a key process in living organisms.

Ordering and disordering processes are very common in biological systems. These processes take place, for example, during the active transport of ions or other solutes across biological membranes, and also during the folding of biopolymers, such as proteins (Figure 4.1.2). In the latter case, ordering occurs, because in the folded state the atoms of the polymer are more confined than in the unfolded state, and therefore have a smaller number of configurations they may acquire. Ordering events tend to increase the free energy of the system, i.e., destabilize it. However, such events are common in bio-systems, because they are often over-compensated energetically by other factors. For example, in the folding of proteins, the ordering of the polypeptide chain is accompanied by the disordering of water around the protein, as well as by the formation of favorable noncovalent interactions.

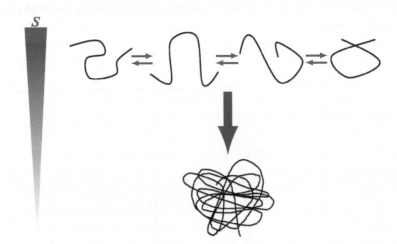

FIGURE 4.1.2 **Entropy changes accompanying the folding of a biopolymer.** The unfolded (top) and folded (bottom) states of the polymer chain are depicted simplistically. The entropy (S) change of the polypeptide chain during the process is shown on the left. The unfolded state is an ensemble of conformations existing in equilibrium, whereas the folded state has only one average conformation[*a]. Thus, the entropy change (of the polypeptide) of the folding process is negative.

[*a] As explained in detail in Chapter 5, the native state is itself an ensemble of conformations, which exist in equilibrium. However, it is much smaller than the unfolded state's ensemble. Thus, a significant drop in the entropy of the polypeptide chain takes place upon folding.

The effect of the phenomena mentioned above on the free energy of the system is expressed in the central thermodynamic equation: $\Delta G = \Delta H - T\Delta S$, where ΔH represents free energy changes resulting from chemical bonding or changes in noncovalent interactions, and $-T\Delta S$ represents free energy changes resulting from changes in the system's degree of order (see details in the main text).

4.1.3.2 Entropy

The other component of the free energy, $-T\Delta S$, involves a different thermodynamic quantity, entropy (S). The microscopic meaning of this quantity is provided by statistical thermodynamics, which relates entropy to the number of possible configurational states of the system (Ω) [8]:

$$S = k_B \ln \Omega \tag{4.10}$$

The number of possible states is inversely related to the order in the system; a highly ordered system can only access a few different states, whereas a disordered system can sample many. Thus, entropy can also be considered as a measure of **disorder** in the system. In molecular systems, entropy changes (ΔS) usually represent changes in the freedom of movement of atoms belonging to both the solute and the solvent.

A good example for this type of process is the folding of a macromolecule, e.g., a protein. During this process, the atoms of the polymeric chain become more confined, and the entropy decreases [15,16]. However, in cases where folding involves burial of nonpolar atoms inside the macromolecule (as in the case of proteins), the hydrophobic effect driving this process [9] is accompanied by an increase in the entropy of the aqueous solvent. The increase results from dismantling the 'cage' of water formed around the nonpolar regions of the unfolded macromolecule (see Chapter 1 for details). As it turns out, the increase in solvent entropy over-compensates for the loss of entropy of the macromolecule, thus leading to an overall (favorable) increase in the entropy of the whole system.

In classical thermodynamics, at constant temperature, the entropy is related to the enthalpy as:

$$\Delta S = \frac{\Delta H}{T} = \frac{q_P^{\text{rev}}}{T} \tag{4.11}$$

This relation suggests another definition of ΔS: in an exothermic (heat-releasing) process, ΔS expresses the number of molecular states of the surroundings, over which the heat energy can be dispersed, provided that this energy is not used for doing work (e.g., expansion or chemical work) [17]. Using Equation (4.9), Equation (4.11) can be rearranged to express the relation between the entropy and the heat capacity (at constant temperature)[*1]:

$$\Delta S = \frac{C_P}{T} \Delta T \tag{4.12}$$

This enables the structural biophysicist to obtain ΔS experimentally by performing simple heat capacity measurements. The $\Delta S - C_P$ relation is consistent with the molecular interpretations given above; when a system contains a high number of configurational states over which heat can be dispersed, it will take more heat to produce a temperature change, or in other words, the heat capacity of the system will be higher.

[*1]Because $\Delta G = \Delta H - T\Delta S$ and both entropy and enthalpy change in the same direction, free energy has a much weaker dependence on temperature than either of its components [14].

4.1.3.3 Computational approaches focus on individual interactions

The experimental thermodynamic approach described above has contributed significantly to our understanding of protein folding and stability. Recently, the classical calorimetric measurements have been supplemented with site-directed mutagenesis, allowing scientists to study the effects of single residues on protein stability. However, interpreting general enthalpy and entropy measurements in molecular terms is sometimes difficult and leads to inaccuracies [18], particularly in the case of interactions that involve both enthalpic and entropic contributions. In such cases it is difficult to focus on one type of interaction using heat capacity measurements alone. For example, the electrostatic interaction between two charges is usually regarded as a basically enthalpic phenomenon. As explained in Chapter 1, at least one component of this interaction, the polarization energy, is partly entropic. This is because the polarization of solvent molecules involves their three-dimensional reorganization into a more ordered form. Another example is the hydrophobic effect, which is considered an entropic effect, but includes an enthalpic contribution as well. The latter results from the breaking of hydrogen bonds between water molecules surrounding nonpolar parts of the solute (see previous subsection for further detail). As explained, this endothermic effect, measured as a change in the heat capacity of the system, unsuccessfully opposes the entropy-driven clustering of nonpolar groups in the solute[*1].

One way to correctly assign measured thermodynamic parameters to specific factors in the system is by using a computational approach. These methods are not limited to the general thermodynamic quantities, and can refer directly to the specific forces acting on the system. The total free energy change in the system (ΔG_{tot}) is decomposed into separate contributions, each resulting from a different noncovalent interaction. For example:

$$\Delta G_{tot} = \Delta G_{elec} + \Delta G_{np} + \Delta G_{vdW} + \Delta G_{ent} \tag{4.13}$$

where the free energy contributions on the right-hand side of the equation represent the following interactions, from left to right: electrostatic, nonpolar, and van der Waals, respectively. The last component represents free energy changes resulting from entropic effects related to the macromolecule.

ΔG_{tot} is calculated based on the molecular structure of the macromolecule (and sometimes the solvent as well), and mathematical expressions describing the different energy contributions. These depend on the model used to describe the physical interactions acting on the system (see Chapter 3 for further details). For example, a molecular mechanics (MM) type of model uses an explicit description of the solute-solvent system, and a set of equations, which are collectively called a 'force field', to describe the energy (see Box 3.1). The force field contains different expressions, each referring to a different type of atom-atom interaction. However, force fields describe only the *potential energy* of the system, whereas the free energy in biological systems also includes kinetic and entropic contributions (see Equations (4.4), (4.5) and (4.7)). The kinetic energy is a function of the temperature and can be included by allowing the atoms in the system to move (see Chapter 3 for a description of the molecular dynamics approach).

Calculating the entropic contribution to the free energy is somewhat more difficult, as it requires rigorous sampling of the system's configurational space, and integrating the potential energy values over all configurations. Thanks to significant advances in computing

---

[*1]Such a phenomenon is called '*enthalpy-entropy compensation*' [17,19].

power and algorithms, molecular dynamics simulations have reached a point where they are able to sample numerous configurations of the macromolecular system. Still, when a large protein is involved, and the solvent is described explicitly, current computational means do not allow for rigorous sampling. Recent developments in force field composition and sampling techniques have enabled computational biophysicists to derive average energy values from simulations (e.g., [20,21]). These *potentials of mean force* are indeed closer to the free energy compared with the potentials obtained from static molecular mechanics calculations. Alternatively, implicit models (e.g., the continuum-solvent model [22]) can be used for calculating free energy values directly [23] (see Box 1.3 and Chapter 3).

4.1.4 Thermodynamics and protein structure

Thermodynamic characterization of proteins may highlight various aspects. These include the formation and stability of protein structure, protein-ligand binding, and enzyme catalysis. The latter is discussed in Chapter 9. Protein-ligand binding is discussed in Chapter 8, with emphasis on thermodynamic aspects. The structural energetics of proteins is discussed in the next section, following the thermodynamic approach presented in the previous section. Obviously, the basic question in this respect is, 'How stable are proteins?' This matter is discussed by referring to the process of protein folding (Figure 2.23). Although important, the matter of stability *per se* is meaningless without a thorough characterization of the identity, direction, and magnitude of the forces and interactions affecting it. Accordingly, our discussion focuses on the factors that stabilize or destabilize the native structure of a protein. We note that the current quantitative (and in some cases qualitative) understanding of the forces stabilizing such complex systems is still lacking [18], and conclusions must therefore be drawn with the appropriate caution.

In the third section we discuss two topics that are related to protein thermodynamics: denaturation of proteins under extreme conditions, and the manner in which certain organisms adapt to living under such conditions, namely, by introducing changes in the sequences of their proteins. In the last section of the chapter we discuss how the insights derived from studies about the thermodynamics of protein structure are applied in the industry, through a relatively new approach termed 'protein engineering'.

4.2 PROTEIN STABILITY AND FORCES INVOLVED

4.2.1 How stable are proteins?

In thermodynamic terms, the structural stability of proteins is reflected by the free energy difference between the folded and unfolded states[*1] (Figure 2.23). Since the free energy is a state function, the dynamic and kinetic details of protein folding are unimportant for this determination. It is, however, important to define the meaning of the term 'unfolded', as it constitutes the reference state for the folding process. The term 'unfolded' refers to two possible states of protein structure:

1. The state of the protein prior to its folding, i.e., just after the polypeptide chain is constructed by the cellular machinery.

[*1]The native structure is considered the thermodynamically most stable state, as demonstrated by Anfinsen [24].

2. The state resulting from complete protein denaturation, an occurrence that may be induced in cells by extreme changes in environmental conditions.

Again, from a thermodynamic point of view, it is unimportant which of the two is considered. We are concerned only with the structural features of this state. The 'unfolded' state is often called 'random coil', to denote the dynamic nature of the polypeptide chain, and its ability to shift randomly between numerous conformations of similar energy. This definition, however, is inaccurate. As explained earlier, the unfolded chain often includes stable remnants of regular local structures (see Chapter 2). In this sense, the random coil is more of a hypothetical state than an actual entity. In addition, 'random coil' is often the name used for describing parts of the folded chain lacking regular secondary structures. While such parts are more flexible than those with stable secondary structures, they are not necessarily random, and may even be evolutionarily conserved [12].

Determining the free energy difference between the folded and unfolded states of the protein can be achieved by denaturing the polypeptide chain. This is done by placing the protein under conditions that enable the chain to open in a controlled manner. When done correctly, the equilibrium constant and free energy difference of the process can be determined (Equation (4.1)). Surprisingly, such experiments reveal protein stability to be marginal [25,26]. That is to say, despite the thousands of noncovalent interactions existing in the structures of folded proteins [27], they are merely 5 to 15 kcal/mol more stable than the unfolded chain [28–30]. For comparison, the dissociation of a single covalent bond requires ~65 to 175 kcal/mol [3]. The low stability of proteins can be explained in two ways. First, even in the lower stability limit (−5 kcal/mol) the folded state is over 99.9% occupied [31]. Thus, it is plausible that there has been no impetus for the evolutionary process to create proteins, which are orders of magnitude more stable than what we see today. The second reason for the marginal stability of proteins has to do with their dynamics. The numerous noncovalent bonds that are present in a protein help the protein to maintain its native structure while allowing it to keep a certain degree of flexibility. As discussed in Chapter 5, this structural flexibility is crucial for protein activity. Such flexibility would not be possible in a structure stabilized by covalent bonds[*1]. Interestingly, the 15 kcal/mol limit is kept even in proteins that are artificially engineered for increased stability (see Section 4.4 below), provided that the stability is measured at temperatures in which these proteins are active [32]. For example, proteins that are engineered to be active at high temperatures may possess stabilities exceeding 15 kcal/mol at room temperature. However, they are not fully active under these conditions. Under the high temperatures required for these proteins' activity, the increased dynamics of the protein chain (due to the added heat energy) lowers the net stability below 15 kcal/mol. The same is true for natural proteins in *hyperthermophiles*, i.e., organisms that live in very hot environments (see more below).

4.2.2   Dominant driving forces

The noncovalent interactions formed upon protein folding have been reviewed in the previous chapters. They include electrostatic, nonpolar and van der Waals[*2] interactions. Protein

---

[*1]S—S bonds may appear in proteins, but they are relatively scarce, and do not rigidify the entire structure.

[*2]Van der Waals interactions are also electrostatic in nature, but, as described in Chapter 1, they are usually described differently from interactions involving charges and/or fixed dipoles, and are therefore treated separately.

stability is also influenced by effects related to the protein's configurational entropy. The free energies corresponding to these interactions and effects are as described by Equation (4.13). Following Equation (4.4), the last contribution (ΔG_{ent}) can be written as $T\Delta S_{con}$, and thus Equation (4.13) takes the following form:

$$\Delta G_{folding} = \Delta G_{elec} + \Delta G_{np} + \Delta G_{vdW} - T\Delta S_{con} \tag{4.14}$$

The marginal stability of proteins suggests that some interactions may contribute more than others to stability. Alternatively, some interactions may counteract others, as initially postulated by Brandts [33,34]. In order to determine the contribution of each interaction type to $\Delta G_{folding}$, the effect of each on both folded and unfolded states must be examined. This, of course, must be done quantitatively, as will be reviewed below. Nevertheless, some general conclusions can be drawn using a simple qualitative 'inventory check':

The unfolded state: In this state, the polypeptide chain is organized such that most of its parts are accessible to the aqueous solvent. This allows all solvent-accessible atoms in the protein to interact in a van der Waals manner with the surrounding water molecules of the solvent. Polar atoms in the protein can also interact electrostatically with the surrounding water dipoles. In addition to these classical interactions, the unfolded chain is also stabilized by its high configurational entropy, which reflects its ability to shift between different conformations by movements of backbone and side chain atoms [15,16].

The folded state: This state is stabilized by the same types of noncovalent interactions that are found in the unfolded state. However many of the interactions occur between atoms inside the protein core, whereas some still occur between surface atoms and the water molecules of the solvent (as in the unfolded state). In addition, the folded state is stabilized by nonpolar interactions (i.e., the hydrophobic effect).

This overview alone reveals factors that are unique to each of the two states: Specifically, nonpolar interactions stabilize only the folded state, whereas protein configurational entropy stabilizes only the unfolded state. Based on this, we can assume that the former drives folding (as already predicted in 1953 by Crick [35]), and the latter opposes it. The following subsections discuss this assumption, and the role of the other types of interactions.

4.2.2.1 Nonpolar interactions (ΔG_{np})

The dominance of nonpolar interactions as a driving force of protein folding[*1] has been demonstrated by a wide range of studies, from simple observations of protein denaturation by organic solvent, to complex thermodynamic and spectroscopic measurements [25,30,37]. This dominance is so profound that, at least for some proteins, the overall structure of a protein can be maintained even after its sequence is randomized, as long as the original hydrophobic-hydrophilic pattern is kept [38–40]. While virtually all structural biophysicists agree that the hydrophobic effect (i.e., nonpolar interactions) is the major driving force of protein folding[*2], they disagree on the magnitude of this effect. One way to investigate this is

---

[*1] As originally suggested by Kauzmann [36] and later by Dill [30].

[*2] Although backbone hydrogen bonds have also been suggested to play a major part in folding energetics [41].

to mutate specific residues participating in nonpolar interactions, and to see how the free energy of denaturation[*1] changes compared to that of the original protein (e.g., [42,43], see more below). Although accurate, such measurements are often controversial due to the molecular interpretation of the measured values. For example, the residues chosen for mutation may participate in more than one interaction, e.g., van der Waals packing interactions [44,45].

A simpler way to obtain ΔG_{np} is to use a model that can focus on nonpolar interactions alone (reviewed by [46]). This has been done following Kauzmann's suggestions [36]; the model included single amino acids, which were transferred from water into a nonpolar medium, such as liquid alkane. When carried out for nonpolar amino acids, the process represents the hydrophobic effect. This is because it mimics the transfer of nonpolar residues from an aqueous to a nonpolar environment, similar to what happens in protein folding. Thus, measuring the free energy change accompanying the transfer of different nonpolar amino acids enables the researcher to evaluate the strength of the hydrophobic effect, namely, ΔG_{np}[*2]. Several studies have been carried out using this model, most notably by the research groups of Tanford [47–49], and later Privalov [28]. These studies provided important data, but did not agree on the magnitude of the hydrophobic effect, mainly because of the different experimental conditions employed by different researchers (e.g., the type of nonpolar medium).

Significant progress was achieved when the researchers observed that ΔG_{np} obtained from the transfer experiments of nonpolar amino acids correlated with their surface area (ΔSA) [50,51]:

$$\Delta G_{np} \approx -B\Delta\text{SA} \qquad (4.15)$$

(where B is the correlation constant).

As expected, different research groups obtained different correlation constants, due to the differences in nonpolar media and methods used to calculate the surface areas. However, the different values revolved around ~25 cal/mol for each 1 Å^2 buried in the protein core upon folding [51,52] (a comprehensive review is provided by Baldwin et al. [46]). Interestingly, a nearly identical value was obtained by molecular dynamics simulations of hydrocarbons in water [53]. Once the correlation constant was known, the contribution of the hydrophobic effect to the folding of the entire protein could be calculated according to its surface area[*3].

While the surface area is the quantity that is most commonly used for calculating ΔG_{np}, other characteristics have also been suggested. These include the atom-atom contact area, and the number of methyl ($-CH_3$) or methylene ($-CH_2-$) groups that get buried inside the protein core during folding. Although all describe the same general property (dimensions), researchers disagree regarding which quantity is best correlated with the free energy of transfer. Alternative experimental approaches have also been suggested for obtaining the correlation constant. One method, for example, involves measuring the change in the free energy of folding or denaturation following replacement of bulky residues with smaller ones (e.g., Ile → Ala). Such studies have produced free energy values of ~−1.5 kcal/mol for each methyl or methylene group buried inside the protein core [43,56–58]. Considering the

---

[*1]Namely, the change in free energy associated with protein denaturation.
[*2]Although van der Waals interactions may also be involved.
[*3]The quantity used for this purpose is usually the solvent-accessible surface area. It can be easily calculated (if the protein structure is known) by computationally 'rolling' a water-sized probe around the structure [54,55] (see Figure 2.4.5 in Box 2.4).

surface area of a methyl group to be ~80 Å² [53], this value is in agreement with the value of ~−25 cal/(mol Å²) mentioned above.

Another quantity suggested for predicting ΔG_{np} is the volume reduced by each residue that gets buried inside the protein [58]. This quantity is somewhat different from the dimensions of the residue itself, for technical reasons related to the way it was measured in the study. The nonpolar free energy value measured in these studies was ~22 cal/mol per 1 Å⁻³ of reduced space[*1] [58–60]. This value differs from the values mentioned above by an order of magnitude. How can this be explained? Li and coworkers encountered a similar problem when studying the influence of the hydrophobic effect on protein-protein interactions [61]. The values they measured produced good correlations in some proteins, but not in others. On the basis of the heterogeneity of protein-protein interfaces, the authors suggested that the nonpolar free energy depends not only on the dimensions of the buried residue, but also on its local environment. The protein core is more homogeneous than protein-protein interfaces, and includes primarily nonpolar residues. Nevertheless, it, too, may include clusters of polar residues, and even water molecules. Thus, nonpolar interactions within proteins may differ in strength, depending on the local environment of the interacting residues.

4.2.2.2 Configurational entropy effect ($-T\Delta S_{con}$)

The loss of configurational entropy upon protein folding (ΔS_{con}) includes two components: the decrease in the number of accessible conformers[*2], and the decrease in the free movement of atoms within the energy well corresponding to each conformer [62]. The second component is much smaller than the first, and is therefore neglected. Still, it is difficult to quantify ΔS_{con} and the corresponding free energy change ($-T\Delta S_{con}$). Equation (4.10) relates entropy to the number of possible configurational states of the system, which in the case of proteins means the number of allowed conformations. Thus, by measuring or calculating the decrease in the accessible conformational space upon folding, the magnitude of the entropy change can be estimated. Studies using these methods have estimated the penalty on decreasing the entropy to be around 4 to 6 cal/(molK)) per residue (reviewed by [12,46]). At physiological temperatures (37 °C = 310 K), these values correspond to free energy changes of 1.2 to 1.9 kcal/mol per residue. Similar values were obtained by spectroscopic studies (e.g. [63]), which are more direct, owing to their ability to track the allowed conformations in both folded and unfolded states[*3]. Is this range of values to be expected? Assuming that the surface area of an 'average' residue is ~100 Å² and the strength of the hydrophobic effect is −25 cal/(mol Å²), then the free energy gain due to burying one residue is similar to, and slightly higher in magnitude than, the penalty paid due to the loss of entropy of the protein chain. This coarse evaluation is in agreement with the suggestion made earlier, namely, that the marginal stability of proteins results mainly from two opposing effects, which nearly balance each other, with slight superiority of the folding-driving force[*4].

[*1]Obviously, Equation (4.15) was not used in these measurements, as they refer to volume, rather than the surface area of amino acids.

[*2]Conformers are inter-convertible conformations that correspond to an energy minimum.

[*3]Entropy changes in proteins can also be estimated experimentally based on the extent of sub-nanosecond motions in the molecule [64,65], e.g., by using NMR spectroscopy [65,66].

[*4]Brandts reached a similar conclusion in 1964, following his studies on chymotrypsinogen [33,34]. However, he concluded that the two opposable effects are entropy and enthalpy, when in fact the effects are both entropic in nature.

Another interesting question regarding the entropy changes of folding is to what extent the main chain of the protein and the amino acid side chains contribute, respectively, to these changes. It is difficult to determine these relative contributions experimentally, since doing so requires the decomposition of the measured values into separate contributions. As in other cases mentioned in this book, such problems can be solved, at least partially, by using a computational approach. Certain computational models allow us to evaluate the degrees of freedom of protein atoms before and after folding. Using statistical thermodynamics (Equation (4.10)), such data can be converted into entropy values. Studies using these methods show that the loss of entropy following protein folding is more significant for the backbone than for side chains [46,67]. This is indeed an unexpected finding, as the backbone should inherently have fewer degrees of freedom to lose compared with the side chains[*1]. One possible explanation for this finding is that the side chain atoms somehow retain significant freedom of movement after folding. This is indeed the case for surface residues, and to some extent for residues at subsurface locations (Figure 4.3). The atomic packing in this region is far less tight than in the core, and the entropy change accompanying folding is minimal. Core residues lose a significant portion of their movement capability, but the overall change is moderated by surface and subsurface residues. In addition, the backbone tends to acquire an ordered secondary structure after folding, and therefore the loss of its entropy during folding is significant.

Electrostatic interactions (ionic, hydrogen bonds) and van der Waals interactions exist in both the folded and unfolded states of proteins. This suggests that such interactions are not the dominant factors affecting protein folding and its stability. However, since their magnitude is not necessarily equal in both states, they may still affect folding to some extent. The following subsections discuss our current knowledge regarding the role of these interactions.

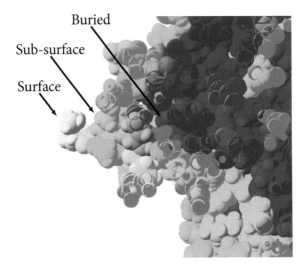

FIGURE 4.3 **Solvent accessibility of protein residues.** The protein acetylcholinesterase (PDB entry 1acj) is colored by solvent accessibility, from yellow (most accessible) to dark blue (least accessible). The front of the protein has been removed, to make the core visible. Surface, subsurface, and core locations in the protein are marked.

[*1]This is because most side chains contain several single bonds, the movement around which is free.

4.2.3 Electrostatic interactions (ΔG_{elec})

In the protein's unfolded state, electrostatic interactions occur between polar residues in the polypeptide chain and the surrounding water. In the folded state, there are three types of electrostatic interactions:

1. Interactions between polar residues at the protein surface and the surrounding water molecules

2. Interactions between polar residues at the protein surface

3. Interactions between polar residues inside the protein core

The first type of interaction should not affect protein stability, as it also exists in the unfolded state. Its main contribution is making the protein soluble in the cytoplasm or extracellular fluid. The second type of electrostatic interaction is expected to have only a marginal influence on protein stability, due to the strong screening effect of the aqueous solvent in that region[*1].

The third type, namely, electrostatic interactions within the protein core, is the subject of a long dispute. The issue has been investigated using different methods and approaches, with emphasis on salt bridges and hydrogen bonds. Some studies have suggested that these interactions stabilize the folded state [69–73], whereas others have demonstrated destabilization [74–78]. The major difference between the folded and unfolded states with respect to electrostatic interactions is the polarity of the medium in which the interacting charges reside; the unfolded chain is completely exposed to the high-dielectric solvent, whereas in the folded state the atoms are desolvated and exposed to a mostly nonpolar environment. The notion that electrostatic interactions involving opposite charges can be destabilizing seems ridiculous at first; as any first year student knows, opposite charges interact favorably with each other, and the interaction just becomes more favorable with the drop in the dielectric of their environment, following Coulomb's law. However, as explained in Chapter 1, this is only part of the story. Electrostatic interactions include *two* different contributions: pairwise (Coulomb) interactions between the solute's charges, and polarization (Born) interactions between solute charges and those of the medium (Figure 1.13). The effect of medium polarity on the two types of interactions is completely opposite: pairwise (Coulomb) interactions are more favorable in nonpolar media, whereas polarization interactions are more favorable in polar media [79], due to the lack of screening [80]. Since protein folding involves desolvation of some charges, this process should involve a favorable change in pairwise energy (ΔG_{Coul}), and an unfavorable change in polarization energy (ΔG_{pol}). Thus, the total effect of the electrostatic interactions on protein folding depends on the magnitude of the ΔG_{pol} vs. ΔG_{Coul} components.

Hendsch and Tidor addressed this problem directly by using the continuum-solvent (CS) model [22,81,82] (see Box 1.3) [77]*2. To determine the effect of ionic interactions on the stability of folded proteins, the authors replaced specific pairs of buried salt bridges with uncharged residues of similar size. Then, the electrostatic free energy difference (ΔG_{elec}) between folded and unfolded states was calculated for both the wild-type (i.e., original) and

---

*1Although some challenge this notion [68].

*2As explained in Box 1.3, the CS model enables ΔG_{pol} to be calculated, thanks to the use of the dielectric constant (Box 1.1), which is an average property of the medium. In comparison, explicit models of the protein-water system require rigorous sampling of the system's configurational space to obtain free energy values.

mutated proteins. The results showed that the ΔG_{pol} of folding is, on average, 3.5 kcal/mol larger than the Coulomb energy. The conclusion was, therefore, that salt bridges are unfavorable to the folding of the polypeptide chain, due to the severe penalty on the desolvation of protein charges. Similar results were recently obtained by García-Moreno and coworkers using an artificial (engineered) ion pair [83].

The study of Hendsch and Tidor (and others mentioned above) seems to be in stark contrast to studies demonstrating (*i*) stabilization of the folded state by electrostatic interactions (see below), (*ii*) the prevalence of polar groups (neutral and charged) inside the protein core [72,84], and (*iii*) hydrogen bonds being shorter [85] and stronger [86] as their environment in the protein becomes less polar. How can these seemingly contradictory studies be reconciled? The answer may have to do with the number and spatial organization of charges within local areas of the protein core. As explained in Chapter 2, the transfer of a single charge from water to a nonpolar environment of dielectric 4 leads to an increase of ~25 kcal/mol in the polarization (desolvation) free energy. However, the presence of more than one charge inside the protein core may change that (Figure 4.4). That is, a few adjacent charges screening each other *via* salt bridges or hydrogen bonds may reduce the unfavorable ΔG_{pol} to values that can be overcompensated for by the favorable ΔG_{Coul}. The presence of water molecules in the region would decrease ΔG_{pol} further. Clusters of polar groups in a confined region of the protein interior may be viewed as a high-dielectric microenvironment, in which the desolvation penalty is significantly lower. In fact, Schutz and Warshel [71] estimated that such microenvironments may have dielectrics of up to 20, a value that is 5 to 10 times higher than those normally assigned to protein cores [87,88]. Finally, the organization of polar groups in clusters within the protein core may also strengthen the favorable ΔG_{Coul} by adding more pairwise interactions.

These conclusions are supported by the study of Kumar and Nussinov [89], in which the effects of more than 200 salt bridges on protein stability were investigated using the same methodology used by Hendsch and Tidor. Kumar and Nussinov found that most of the salt bridges were organized in clusters and stabilized the folded state of the protein. Other studies reached similar findings (e.g., [90,91]). This is probably true for hydrogen bonds as well, which are highly common inside proteins [92–94] and tend to be bifurcated (i.e., have more than one donor or acceptor) in certain instances [95]. Furthermore, hydrogen bonds tend to be cooperative [96–98] and therefore stronger[*1]. In the study of Kumar and Nussinov, the average stabilization energy of a salt bridge was found to be −3.7 kcal/mol, but the total energy range was between 0 and −7.5 kcal/mol. Other studies reached different values (e.g., [90,101]), all within this range. Similar findings were reached by Pace and coworkers for hydrogen bonds; the stabilization energy per bond was highly context-dependent (also see [86]), and while the average value was about −1 kcal/mol, it ranged between 0 and −3.6 kcal/mol [72][*2*3]. The favorable contribution to stability was found to be similar for backbone and side chain groups.

To conclude, it seems that **while most electrostatic interactions inside proteins tend**

[*1]Cooperativity of hydrogen bonds means that two or more such interactions that are close to each other have a mutual strengthening effect. This phenomenon is not entirely understood and seems to result from various effects, such as bond polarization and resonance ([99,100]).

[*2]Similar, yet smaller ranges were found by the research groups of Kelly (0 to −1.2 kcal/mol) [86] and Schreiber (−1 to −1.5 kcal/mol) [90].

[*3]Note that charged hydrogen bonds are on average ~2 kcal/mol stronger than neutral ones, and their energy can reach −7 kcal/mol [102].

to be stabilizing, the degree of stabilization is highly variable and context-dependent. It depends on the identity of the atoms or groups involved in the interaction, their relative geometry, and the nature of their microenvironment. To make electrostatic interactions as stabilizing as possible, evolution created clusters of polar residues that form interaction networks, thus reducing the desolvation penalty for burying the charges. Another way proteins may increase the stabilizing effect of electrostatic interactions is via long-range effects. A recent study shows that such effects may be induced by partially solvent-exposed charged residues [103]. This allows the protein to strengthen Coulomb interactions without paying the full penalty for desolvation.

The data presented above suggest that electrostatic interactions may be favorable to protein folding, provided that the interacting groups 'enjoy' a certain microenvironment within the protein core. Still, at least some of the numerous electrostatic interactions inside proteins must be surrounded by a nonpolar environment, which means they are unfavorable to folding. If so, why do these interactions appear at all inside protein cores [93,104]? There seem to be two major reasons:

1. **Functional importance**: Polar residues are involved in key cellular processes executed by proteins, such as enzyme-mediated catalysis, protein folding, and protein-ligand binding [105–109]. In enzyme-mediated catalysis, electrostatic interactions have been directly implicated in decreasing the activation energy barrier, which is the hallmark of the catalytic process [110]. They do so, e.g., by electrostatically stabilizing the substrate in its transition state, which is often charged. In fact, Schultz and coworkers

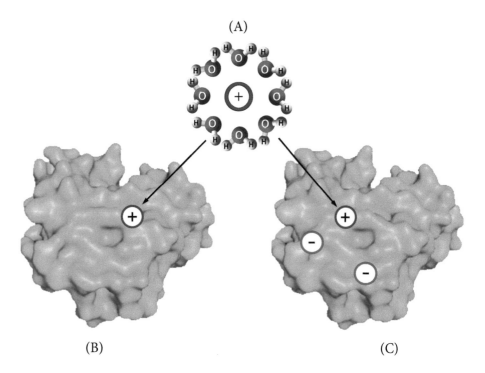

FIGURE 4.4 **Two views of charge desolvation during protein folding.** In the unfolded state, protein charges (partial or full) are electrostatically screened by the surrounding water molecules (A). Theoretically, during folding, these charges should be desolvated inside the protein core, losing the screening of each of the charges (B). However, in reality, buried charges are almost always surrounded by other charges, which provide at least partial screening (C).

have already demonstrated this phenomenon in catalytic antibodies [111]. In enzymes, the contribution of electrostatics to catalysis may even involve polar residues that are not in proximity to the substrate [112]. These usually act indirectly, by affecting catalytic residues in their vicinity. For instance, they may change the pKa of the catalytic residues, thus changing their ionization state.

The most popular example for this mechanism is probably the catalytic cycle of serine proteases, described in Chapters 1 and 2. Briefly, these enzymes rely on a serine residue as an initiator of a nucleophilic attack on the substrate, despite the fact that Ser is a weak nucleophile [113,114]. The catalysis is made possible thanks to a nearby His, which, by acting as a general base, decreases the pKa of the catalytic serine's hydroxyl group, thereby allowing it to deprotonate and become a strong nucleophile. Another pKa-related mechanism operates in the enzyme *aspartic protease*; a key Asp at the active site must function as a general acid (i.e., donate a proton), despite the fact that the local pH is higher than its pKa. This problem is solved by an adjacent (negatively charged) Asp, which induces an increase of the first Asp's pKa, thus rendering it neutral [115]. These examples demonstrate an important phenomenon in proteins: structural stability may be partially sacrificed for functional reasons. This conclusion is supported by many studies featuring enzymes.[116] For example, by using site-directed mutagenesis, Shoichet *et al.* showed that the catalytic residues themselves are frequently responsible for the decreased stability [117]. It seems that in the long run, such sacrifices of stability for functionality are worthwhile.

2. **Structural importance**: The primary driving force for protein folding is the hydrophobic effect, which also drives the aggregation of lipids into oil drops in soup. However, hydrophobicity provides low structural specificity. That is, there are a number of folded conformations that have the same nonpolar free energy, and the protein may shift among these conformations instead of being committed to one [118,119]. In contrast, electrostatic interactions are highly specific, due to their strong dependence on the local dielectric [104,105,120]. In other words, any change in conformation resulting in the breaking of electrostatic interactions (e.g., a hydrogen bond) leads to the energetically unfavorable exposure of polar groups to the low-dielectric environment of the protein core. Although possible, such an event is highly improbable (see Equation (4.3b)), which means it will rarely be observed. As a result, the native structure of a protein, which is also the biologically active state, is maintained within a limited range of allowed conformations. The importance of electrostatic interactions in promoting the specific fold of the native protein is supported by the fact that artificial perturbation of these interactions does not lead to complete denaturation of the protein, but rather to formation of a partially folded conformation[*1], which has proper topology, yet improper tertiary contacts [105]. It is also supported by the high conservation of buried hydrogen-bonding polar residues inside proteins [121]. Furthermore, the conservation of these residues is higher than that of non-hydrogen-bonding (buried) residues, and much higher than that of nonpolar residues.

Salt bridges should be particularly efficient in conferring structural specificity, due to the charged groups constituting them. This may explain the high evolutionary conservation of

---

[*1]This conformation, termed '*molten globule*', is considered to be the penultimate step of folding, just before the native fold is acquired (see Chapter 5 for more detail).

salt bridges within protein cores [105]. Nevertheless, specificity can also be conferred by other electrostatic interactions, which are highly dependent on geometry. These include primarily hydrogen bonds and, to a lesser extent, weaker interactions such as those involving the sulfur atom of Met/Cys [122,123], π-π interactions between aromatic residues [124], and interactions between aromatic residues and X−H groups (where X is S, N, O, and even C) [125].

4.2.4 van der Waals interactions (ΔG_{vdW})

Like electrostatic interactions, van der Waals interactions exist in both the folded and un- folded states of the protein. In the former state they occur mostly between protein atoms, whereas in the latter they occur between protein atoms and those of the solvent. However, these interactions are expected to be stronger in the folded state, because they are extremely short-ranged, and thus optimized in the tightly-packed cores of proteins [44,126,127]. This con- clusion is supported by experimental studies (e.g. [128]). It is difficult to determine quanti- tatively the van der Waals free energy difference between the folded and unfolded states, since the measured values correspond to packing forces, which also include nonpolar com- ponents. In addition, being extremely short-ranged, van der Waals interactions change dra- matically between different regions of the protein, so the measured value is only an average, with a large standard deviation. Nevertheless, several studies have used calculations to gen- erate estimates of the van der Waals energy. For example, Karplus and coworkers calculated the corresponding interaction energy of a methylene ($-CH_2-$) group in the protein interior as -3.1 kcal/mol [127], although this value is probably too high[*1]. Despite the uncertainty re- garding the real van der Waals energy in proteins, most structural biophysicists consider these interactions to be secondary to the hydrophobic effect as a driving force of folding [30].

4.2.5 Summary and conclusions

Numerous studies have provided us with invaluable knowledge about the key physical forces and energetics in proteins, and their effect on protein folding. However, accurate determi- nation of the magnitude of those forces, even in relative terms, has always been lacking. Measurements and calculations show that protein folding is driven primarily by the hy- drophobic effect, and to a lesser extent, by van der Waals forces. These two are responsible for the dense packing of protein cores, a hallmark of globular proteins. Electrostatic interac- tions may either stabilize or destabilize the folded state of proteins, depending on the spatial organization of charges, and on their local dielectric. In any case, electrostatic interactions play important roles in proteins, and are largely responsible for their specific native confor- mations; they make the native conformation(s) of a given protein stand out in comparison to alternative, similar packing interactions.

Protein folding can therefore be described as a process that minimizes the exposure of nonpolar regions to the surrounding aqueous solvent, while optimizing packing interac- tions and hydrogen bonds. The native structure of a protein is the balance of those forces and interactions. The main opponent to protein folding is the loss of configurational entropy. The magnitude of this effect has not yet been accurately determined, but since the overall stability of proteins is marginal, it is assumed to be comparable to that of the hydrophobic

---

[*1] As we saw earlier, the nonpolar 'interaction' energy corresponding to the same group has been estimated as ~-1.5 kcal/mol by mutational studies [43,56–58]. The van der Waals energy is most likely smaller.

effect. In that sense, it should be noted that the penalty due to the loss of configurational entropy might be lower than initially assumed. This conclusion emerges from studies demonstrating that the 'unfolded' chain contains a certain degree of secondary structure [129] (see also [130–132]). A similar phenomenon is found in intrinsically unstructured proteins (see Chapter 6), in which the polypeptide chain lacks organized tertiary structure, but still contains certain secondary structures. These include mainly the polyproline II (PPII) helix [133], which may comprise up to ~50% of the amino acid sequence [132–134].

It is yet to be determined why the PPII helix is so common in unfolded proteins. Especially confusing is the fact that this structure is not limited at all to proline-rich sequences. The 'popularity' of the PPII helix may have to do with its intrinsically high configurational entropy. Another suggestion is that the helix creates only a minor disruption of the solvent's structure (relative to other structures), and is therefore energetically preferred [131,135].

Interestingly, the value of the ϕ angle of PPII is similar to that of the α-helix, whereas the value of the ψ angle is similar to that of the β conformation. Thus, PPII is, in a way, a hybrid of the two most common secondary structures in proteins, which may also be the reason for the relative ease with which α-helices and β-strands change into PPII during unfolding.

In any case, the above suggests that instead of a completely unorganized chain, the process of unfolding produces a partially organized state, which involves the inter-conversion of alternative secondary structures. This idea is also supported by the high crowding of the cytoplasm [136]. Indeed, the intracellular environment contains very high concentrations of molecules (~40% of it is dry material, and the concentration of macromolecules is 300 to 400 g/L) [137,138]). Generally speaking, the crowding should stabilize the folded structure *via quinary interactions* [139–141] (see Chapter 2), especially inside intracellular organelles, whose volume can be as small as 10^{-14} μl. However, the degree to which crowding affects the preference of the folded structure in cells is still under debate [142]. There is also evidence that at least some proteins are destabilized inside living cells, presumably by sequence- and context-specific interactions with certain cellular components that are yet to be identified [143].

It is interesting to note that whereas the overall (thermodynamic) stability of most proteins is within a relatively small range, their *kinetic stability*, which refers to their rate of unfolding, can be quite diverse. Indeed, many proteins unfold within minutes or hours, whereas some are 'trapped' in their native conformation and have unfolding half-lives of weeks to billions of years [144–149]*1. This phenomenon results from high-energy transition state(s) that separate the folded from the unfolded state (see Chapter 5), and which create a virtually insurmountable barrier. This barrier does not affect the thermodynamic stability of the protein, as the latter results only from the energy difference between the folded and unfolded states. Kinetic and thermodynamic stabilities often correlate, but not always. For example, when a certain part of α-lytic protease is removed, its thermodynamic stability decreases significantly, yet it remains highly kinetically stable [150]. The trapping of kinetically stable proteins in the native state reduces their ability to sample unfolded or partially folded conformations, and therefore protects them from chemical denaturation (see following section) and proteolysis by cellular enzymes, which act on partially folded proteins [151]. This testifies to the importance of kinetic stability and suggests that this trait was selected by

---

*1 The subject of kinetic stability is more suitable for discussion in Chapter 5, which focuses on protein folding and dynamics. However, since kinetic and thermodynamic stabilities are often confused we chose to discuss this topic here.

evolution to maintain proper activity of cellular proteins, as well as regulating it[*1]. A study of kinetically stable proteins shows that they tend to have a rigid structure [145,152]. This may result, e.g., from a large β-sheet content [145] or other structural characteristics that involve long-range, distributed interactions [153]. Rigidity may increase kinetic stability by preventing the solvent water molecules from accessing the weak points on the protein's surface, thus inducing partial (local) unfolding [145]. Indeed, kinetically stable proteins were also found to have 'reinforced' surfaces as a result of stabilizing disulfide bonds and metal-involving interactions [145]. Proteins that are not sufficiently rigid may still be rendered kinetically stable by oligomerization [145], post-translational modifications (see below), and, in the case of membrane proteins, by interactions with the confining membrane. One example of a common post-translational modification that affects the energy landscape in membrane-bound and secreted proteins, as well as the balance between the forces driving folding and those opposing it, is glycosylation.

Although it is agreed that glycosylation stabilizes the structure of proteins, the exact reason is not completely understood. In this respect, it is worth mentioning two very different mechanisms of stabilization that have been proposed. The first is based on the stabilization of the folded structure, resulting from glycan-induced repulsion of the solvent from the protein surface [154]. The second mechanism is based on enthalpic destabilization of the unfolded state [155]. That is, the bulky glycan residues interfere with the formation of favorable interactions between protein residues in the unfolded state alone, thus increasing the magnitude of the folding free energy.

As discussed later on in Chapter 8, the same interactions that affect protein folding also affect the binding between different proteins, and between different subunits of the same protein. Nevertheless, there are some differences between the two cases. First, in the case of binding, the loss of configurational entropy is supplemented by the loss of translations and rotations of the binding partners [62]. However, these are much smaller in magnitude compared to the loss of configurational entropy accompanying protein folding [156]. Second, the geometric complementarity of atoms within the protein core, and therefore the optimization of van der Waals and nonpolar interactions, is better than in protein-protein or subunit-subunit interfaces. Third, the binding interfaces are more polar than the core, which affects the energetics of the system in two opposite ways: on one hand, the nonpolar driving force for protein-protein complex formation is smaller than that of protein folding. However on the other hand, the desolvation penalty is lower in the former case, due to better electrostatic screening [157].

4.3 PROTEIN DENATURATION AND ADAPTATION TO EXTREME CONDITIONS

4.3.1 Denaturation as experimental tool

Protein stability has been studied for many years, using thermodynamics, spectroscopy, and other approaches. In many of these, protein denaturation, rather than folding, is used for assessing the different interactions stabilizing the native structure. Denaturation is defined as a process in which proteins lose their activity due to changes in environmental condi-

---

[*1]It is possible that evolution selected kinetically stable proteins for roles requiring long periods of activity, whereas less kinetically stable proteins were selected for roles requiring fast protein turnover (e.g., in signal transduction networks).

tions. However, this process always involves unfolding of the protein to some degree, in accordance with the central structure-function dogma. Therefore, denaturation is usually considered in structural studies as a reversal of protein folding. This assumption is inaccurate as far as kinetics is concerned; although denaturation may ultimately lead to complete unfolding, denaturation and unfolding do not necessarily take the same route. However, since stability studies deal with free energy (a state function), the approximation is considered justified. Denaturation can be induced in the lab by changing some of the basic environmental conditions, such as temperature, pH, ionic concentration, and pressure, to extreme values that are incompatible with the folded structure of the protein. Such procedures must be conducted carefully, to prevent the protein from undergoing harsh chemical (covalent) changes.

4.3.1.1 Temperature-dependent denaturation

Changes in the environmental conditions of proteins may induce denaturation via different paths. The most studied form of denaturation, i.e. thermal, takes place when the temperature is elevated to ~45 °C or higher [158]. The increased heat absorbed by the protein is converted into kinetic energy, which in turn increases atomic vibrations and other motions within the protein. This weakens the noncovalent interactions stabilizing the protein, until eventually the native structure is lost. A more thermodynamically-oriented explanation would relate to the entropy difference between the folded and unfolded states. This difference is the main free energy component opposing folding, but it is usually not large enough to fully counteract the other, pro-folding components (mainly, the hydrophobic effect). In thermal denaturation, the added kinetic energy increases the configurational entropy of the unfolded state much more than it increases that of the folded state. Given enough heat, the entropy difference between the two states becomes large enough to overwhelm the pro-folding components, resulting in unfolding.

Interestingly, nonpolar interactions, which are not inter-atomic interactions in the classical sense, also exhibit temperature-dependent behavior; at moderate temperatures they result from changes in the entropy of the aqueous solvent (ΔS_{water}) [28], as stated in Chapter 1. However, at high temperatures they are driven by changes in the enthalpy of the solvent (ΔH_{water}) (although ΔS_{water} is still positive) [159,160]. This shift from an entropy- to an enthalpy-driven process has to do with the decrease in ΔS_{water}, induced by the heating; at high temperatures, the kinetic energy of the water molecules surrounding the protein is significantly large. This makes ΔS_{water} too small to drive the hydrophobic effect [30]. The latter still drives folding, just by a different mechanism.

Denaturation may also occur at low temperatures ($T < 5$ °C), although in these cases it is usually reversible. The loss of heat, which is the source of kinetic energy, is expressed as a decrease in internal dynamics (vibrations, motions). This in turn leads directly to loss of activity (see Chapter 5 on the role of dynamics in protein function). At freezing temperatures ($T < 0$ °C), the increase in intra-water hydrogen bonds characterizing the formation of ice, comes at the expense of protein-water bonds. This may lead to a loss of solubility and, as a result, to aggregation and sedimentation of the protein [158]. This is why in cryo-EM methods the protein is frozen very fast, to prevent unfolding and aggregation. In addition, the increase of salt concentration in the leftover liquid water may induce denaturation, e.g., via pH changes (see the following subsection).

4.3.1.2 pH-dependent denaturation

The activity of every protein reaches its optimum at a certain pH range. Outside this range, changes occurring in the protein structure may lead to loss of function. The structural change results from the influence of the pH on the ionization state of protein residues, as explained in Chapter 1. Briefly, when the solvent pH is lower than the pKa of a residue, the residue tends to be protonated. This renders the acidic residues Glu and Asp electrically neutral, and the basic residues His, Arg, and Lys positively charged. When the pH of the solvent is higher than the pKa of a residue, the residue tends to be deprotonated. This renders the acidic residues Glu and Asp negatively charged, and the basic residues His, Arg, and Lys electrically neutral. The loss of activity following pH changes may be direct, as a result of changes in the ionization state of catalytic residues, or indirect, resulting from structural changes. The latter involves an unfavorable change in the electrostatic free energy of the protein's folded state. For example, protonation or deprotonation of a charged residue may disrupt a salt bridge in which the residue is involved, by rendering the residue electrically neutral. Such a process would have two unfavorable consequences:

1. Weakening the Coulomb interaction between the two residues involved in the bond would make the associated free energy (ΔG_{Coul}) less negative.

2. The lack of electrostatic screening for the other residue involved in the salt bridge, which is still charged and exposed to the nonpolar environment, would make the associated free energy (ΔG_{pol}) more positive.

It should be noted that the pH change must be large enough to create an unfavorable ionization state. This is because of the pKa shift that is often observed in protein residues, which occurs to prevent the unfavorable state [12] (see details in Chapter 2).

4.3.1.3 Pressure-induced denaturation

Elevating the hydrostatic pressure of the protein environment increases the density of the bulk water surrounding the protein and 'pushes' the water towards the protein. What happens next is a sort of reversal of protein folding; water molecules close to the protein surface start to penetrate it and disrupt water-excluded cavities in its core [161]. This leads first to partial exposure of the hydrophobic core [162] and then to full unfolding. Pressure-induced denaturation seldom happens in 'real life'. Although some organisms live in deep underwater regions of the ocean where the pressure is high, they are adapted to such conditions. Thus, pressure-induced denaturation is mainly a lab tool for studying proteins.

4.3.1.4 Chemical denaturation

Certain chemicals may cause denaturation in proteins. These can be separated into two general groups. The first comprises organic solvents, which interact with nonpolar residues inside the protein. These interactions stabilize the unfolded state of the protein and thus increase the probability of unfolding. The second group of chemical denaturants includes polar molecules, such as urea and guanidinium chloride. The mechanism by which these chemicals cause denaturation is not entirely understood [12]. Being polar, they are expected to interact with polar atoms in the protein, thus interfering with intra-protein polar interactions [162–164]. However, their effect may be indirect, by disrupting the structure of the aqueous solvent (e.g. [165]). Such an event is expected to weaken the hydrophobic effect, which is responsible for protein stability. The latter mechanism requires high concentration of the denaturing chemical, which is indeed the requirement in the case of urea.

Some polar chemicals have the opposite effect on protein structure, i.e., they stabilize it. These molecules, called *organic osmolytes*, include sugars, amino acids, poly-alcohols, and methylamines. In fact, the biological significance of organic osmolytes far exceeds the protection of protein structure; they are used by certain organisms living under extreme environmental conditions, such as high temperature, pressure and dryness, as a means of maintaining cellular pressure, neutralizing toxins and oxidative damages, reducing metabolites, and protecting the plasma membrane [166]. Some of these chemicals, e.g., the extensively-studied *trimethylamine N-oxide* (*TMAO*; $(CH_3)_3N^+O^-$), are also used for protecting proteins against denaturation. The amphipathic structure of TMAO enables it to hydrogen-bond with water, as well as to self-associate. Indeed, TMAO has been found to remain detached from the protein, separated from it by a high-order layer of water. Although it is not entirely understood how TMAO stabilizes the protein, a mechanism based on the observed ordering of the protein hydration shell has been suggested [166]; this action is expected to minimize favorable protein-water interactions, forcing the protein to acquire a more compact (and thus more stable) structure. Other osmolytes, e.g., diglycerol phosphate, act more directly, by interacting with the protein electrostatically.

On the basis of the above, a unified mechanism has been suggested by Bolen and Rose for the action of both denaturing and protecting osmolytes [41]. This mechanism revolves around the ability of the protein backbone to form intramolecular hydrogen bonds. The authors suggest that chemical denaturants such as urea (see above) interact favorably with backbone polar groups, thus shifting the folding equilibrium backwards. In contrast, chemical protectors such as TMAO reduce the solvent's ability to hydrogen-bond with the protein backbone, thus shifting the folding equilibrium forward. This model has been supported by several subsequent studies carried out by different groups. One study, which used infrared spectroscopy, demonstrated that TMAO's stabilization effect has both enthalpic and entropic contributions [167]. The enthalpic stabilization results from the decreased water-protein hydrogen bonds, which results in increased hydrogen bonds between protein backbone groups. The entropic stabilization results from the crowding effect of TMAO on the protein, which results in decreased conformational entropy of the latter.

4.3.2 Adaptation of proteins to extreme environments

In complex multicellular organisms, homeostasis enables proteins to exist in a virtually unchanged environment, in which fluctuations of environmental conditions are rare [168]. Consequently, proteins in these organisms are relatively protected from denaturation. This is not the case for unicellular and simple multicellular organisms devoid of homeostatic capabilities. Their proteins are under a constant threat of denaturation, which may be irreversible when followed by aggregation and sedimentation. This is a real threat for the entire cell; studies show that sedimentation of 28% of the proteins in a cell leads to its instant death [169]. Fortunately, cells contain protective measures, such as molecular chaperones and heat-shock proteins, which help denatured proteins refold before the process becomes irreversible.

Certain organisms, many of which belong to the *Archaea* kingdom, live constantly under extreme environmental conditions [170]. They can be found in areas with high temperatures (e.g., geysers), high salt concentration (e.g., the Dead Sea), or high pressure (e.g., the ocean floor). Since the proteins of these organisms are exposed constantly to extreme conditions, their cellular protection mechanisms are insufficient to prevent denaturation completely. The solution adopted by evolution was to change the sequence of these proteins, thus creating three-dimensional structures that can withstand harsh conditions without losing their function. By comparing these proteins with those taken from *mesophiles* (organisms that live under moderate conditions), scientists are able to learn about the molecular determinants of stability, without the need to mutate proteins or denature them. Adaptation to extreme temperatures is the most extensively studied form of adaptation to extreme environments, and will be the focus of the following paragraphs.

Organisms living at temperatures of 45 to 80 °C are called 'thermophiles', whereas those living at higher temperatures (80 to 113 °C) are called 'hyper-thermophiles' [171]. Beyond 115 °C, life is not expected to exist, due to ATP degradation, although certain proteins have been known to maintain their structures under higher temperatures. Indeed, the most thermostable protein known, CutA1 from the hyper-thermophile *Pyrococcus horikoshii*, has a melting temperature of 150 °C at pH 7 [172]. Protein stability studies usually reach similar conclusions for thermophiles and for hyper-thermophiles; therefore, in what follows we will refer to both as thermophiles. Such organisms live in warm springs, openings of volcanoes, and hydrothermal vents on the ocean floor. As noted above, high temperatures increase the structural dynamics inside proteins, eventually leading to the disruption of noncovalent interactions, and hydrogen bonds in particular. Surprisingly, comparison of thermophilic proteins to mesophilic proteins shows very few sequence-related differences. However, the few differences that are observed have implications that can be interpreted as mechanisms for increasing stability under high temperatures [173,174]. Although different thermophiles use different strategies [175,176], there seem to be a few general trends:

1. **Structural rigidification** is achieved by increasing the percentage of Pro in the sequence (or decreasing that of Gly), and by adding aromatic residues in positions that allow stacking interactions to take place [177]. Such changes act primarily by reducing the configurational entropy of the unfolded chain [160]. This makes the unfavorable entropy loss during folding smaller, and increases the fraction of folded proteins.

2. **Strengthening the protein core** is achieved by increasing the percentage of nonpolar residues (primarily I, V, L, Y, W [178]) in core locations, as well as residues that support the formation of secondary structures [179,180]. Secondary-structure formation is beneficial for two reasons. First, it reduces local chain motions. Second, it allows the pairing of polar backbone groups in hydrogen bonds. As described in Chapter 2, this reduces the exposure of these groups to the nonpolar core of the protein, thus stabilizing it.

3. **Strengthening electrostatic interactions inside the protein** is achieved by increasing the number of salt bridges [180,181]. As described above, the net contribution of buried salt bridges to protein stability is under debate. The stabilizing effect of these interactions in thermophiles may result from the fact that the desolvation penalty associated with burying charged groups in a low-dielectric environment is smaller at high temperatures, due to changes in water properties [182,183]. The desolvation penalty may further be decreased by formation of salt bridge networks, which optimizes the electrostatic masking of the charged groups [184].

4. **Strengthening protein-solvent interactions** is achieved by increasing the percentage of polar and charged residues at the protein periphery [176,181,185].

It should be mentioned that with respect to thermophiles, the term 'adaptation' is somewhat misleading; geological and genetic evidence suggests that the first forms of life on earth were similar to today's thermophiles [171]. If this is true, then the original proteins were similar to those currently present in thermophiles, and the adaptation taking place during evolution involved getting used to living under moderate temperatures, instead of vice versa.

The adaptation of enzymes to high temperature is not just an academic issue. Indeed, enzymes are used today commercially as catalysts in various industrial fields, including textiles, food, and pharmaceuticals [186–193]. While enzymes are usually used for their ability to function under moderate conditions, some of the industrial processes involving enzymes are carried out under harsh conditions, including high temperatures and alkaline pH. In such cases, further 'engineering' is required to adapt the enzymes to their new environment. In most cases, this involves increasing the enzymes' thermostability. This issue is discussed in Section 4.4 below.

Organisms living at low temperatures ($T < 15\,°C$) are termed *psychrophiles*. These include fish[*1], insects, plants and microorganisms. As explained earlier, the proteins of such organisms face the danger of losing their internal dynamics, leading to the loss of function. In cases of solvent freezing, the proteins also face a decrease in solubility, due to the loss of electrostatic interactions with the aqueous solvent. The adaptations observed in these proteins include the following:

1. **Increasing the internal dynamics** is achieved by increasing the percentage of polar residues at the expense of nonpolar residues at the protein core.

2. **Strengthening the interaction with the solvent** is achieved by increasing the percentage of hydrogen-bonding residues at the protein's periphery [194]. Interestingly, a similar strategy is also used by psychrophiles to protect themselves on the physiological level [195,196]. That is, these organisms produce *antifreeze proteins* (*AFPs*) that inhibit ice formation in their bodily fluids by lowering the freezing point and raising the melting point of ice (the *thermal hysteresis gap*). This is done through the *Gibbs-Thomson effect*, whereby the adsorption of AFPs causes an increase in the micro-curvature of the ice [197]. A recent molecular dynamics study of an insect AFP shows that the adsorption of the AFP to the ice is stereo-specific, virtually irreversible, and indirect; it is mediated by a linear array of ordered water molecules that are structurally distinct from ice [198].

4.3.3   Conclusions

Thermophile adaptation to high temperatures provides us with important insights regarding the factors contributing to protein stability. The mechanisms of adaptation seem to support the principles emanating from thermodynamic measurements and calculations. These studies show that the hydrophobic effect and chain entropy are the major free energy components supporting and opposing protein structure, respectively. Indeed, these two factors are the main targets of thermophile adaptation. The adaptation of psychrophiles also addresses these factors, although oppositely. Psychrophilic proteins face the danger of over-stability due to low temperature, and their adaptations act to reduce stability by weakening nonpolar interactions and strengthening polar interactions. Moreover, both thermophilic

---

[*1] These fish are usually found on the ocean floor, e.g., at the Earth's poles.

and psychrophilic adaptations act to strengthen protein solubility via polar interactions. This issue is unrelated to protein stability, but still affects protein function. The issue of solubility is important in thermophiles and psychrophiles, for different reasons. In thermophiles, the added kinetic energy increases the dynamics of both protein and solvent atoms, thus decreasing the likelihood of long-lasting hydrogen bonds between the two. In psychrophiles, the reduced water dynamics increases the likelihood of long-lasting intrawater hydrogen bonds, thus decreasing the chances of such bonds being established with the protein.

4.4 STABILITY ENHANCEMENT OF INDUSTRIAL ENZYMES USING PROTEIN ENGINEERING

4.4.1 Enzymes in industry

While enzymes are usually discussed with respect to their physiological roles in cells and tissues, they also play a role in industry, as catalysts [186–193]. This topic is discussed in detail in Chapter 9. Briefly, enzymes are used in different industries, such as the food (and feed), textile, and pharmaceutical industries. For example, in the food industry enzymes are used, among other things, for baking bread and making beer. In the textile industry they are used for processing fabrics or as laundry detergents. In the pharmaceutical industry they may be used for the synthesis of drugs, or as a drug itself. The industrial use of enzymes started a few decades ago and gained popularity around the end of the 1980s, with the emergence of large-scale fermentation and recombinant DNA technologies. The combination of these methods enabled industrial scientists to produce large quantities of desired enzymes in large containers termed 'fermentors'.

The extensive use of enzymes relies on several advantages of these macromolecules over simple catalysts, such as metals. First, enzymes display selectivity towards their respective substrates and/or reactions (see Chapters 1 and 9). Second, they can be produced in large quantities and in an unlimited fashion. Third, they are easily degraded. Fourth, they function under moderate temperatures and pH values. Fifth, their use does not involve the production of contaminants. Sixth, as proteins, enzymes can be changed to increase stability, efficiency, selectivity, etc. Although enzymes are still the exception rather than the rule in industrial catalyses, there are many reactions in which the traditional simple catalysts have been replaced by enzymes. Examples include the following [186]:

1. In laundry detergents, phosphates have been replaced with enzymes such as proteases and cellulases (e.g., [189]). These are very efficient in degrading protein-based stains, which are caused, for example, by blood, milk, egg, and sauces. In fact, *alkali proteases* used as detergents account for over a third of the industrial enzyme market. These primarily include the bacterial serine proteases of the *subtilisin* family. The extensive use of subtilisins is driven by their high stability, low substrate specificity (an advantage in this application), and the ease with which they can be isolated.

2. In bread baking, lipases have replaced chemical emulsifiers.

3. In textile processing, amylases and pectinases have replaced sodium hydroxide.

4.4.2 Enzyme engineering

Unlike cells and tissues, industrial fermentors often work under non-physiological conditions. These include elevated temperatures, alkaline or acidic pH, and exposure to lipophilic media (e.g., organic solvents). For example, laundry detergents usually work at pH 9 or 10, in temperatures of 50 to 60 °C, and in the presence of oxidants that are derived from bleaching agents [186]. Some enzymes, such as those isolated from extremophiles, are able to work under some of these conditions. However, **most other enzymes exhibit decreased catalytic efficiency (e.g., due to oxidation of catalytic residues), decreased stability and decreased solubility when exposed to harsh conditions.** In such cases, it is necessary to modify the enzyme to increase its tolerance. Engineering may also be used in enzymes working under moderate conditions, in order to improve their basal activity.

Current technology enables scientists to modify proteins very easily by introducing mutations into the appropriate genes. Indeed, many industrial enzymes today are *engineered* for different purposes. For example, almost all enzymes used as detergents are engineered variants of the bacterial enzyme *subtilisin*. Whereas introducing the mutations is technologically straightforward, it is highly challenging to determine which modifications are required in order to achieve a certain function. Two general approaches are used to solve this problem. The first is *rational*. That is, scientists use their own experience and the extensive data collected on the functional outcomes of different mutations in order to design specific changes in the enzyme [18]. This approach works best when a small number of point mutations suffices to create the desired effect. A classic example is the use of a single residue replacement to protect subtilisin in laundry detergents from oxidation [199]. As explained above, detergent enzymes are often inactivated by oxidative damage cause by the bleaching agent (e.g., H_2O_2). Specifically, the oxidation tends to damage certain residues, such as Met, which is oxidized to sulfoxide. In the cited case, Met 222, which is adjacent to the functionally important residue Ser 221, was replaced by residues that are not oxidized by the bleaching agent (Ser, Ala, and Leu). This rendered the enzyme resistant to oxidation-related inactivation.

The second approach is to randomly introduce mutations in the enzyme, and to apply selective pressure to isolate the fittest mutants. This *'directed evolution'* [186,200–203] approach has become very popular in the last decade, especially with the development of high-throughput selection methods. Directed evolution is recommended for cases in which the desired function can only be obtained by introducing more than just a few mutations. In these cases, the interplay between the numerous mutations is difficult to predict using the rational approach, and the desired combination is more likely to be discovered through the blunt force of random mutagenesis. In most applications, however, the best approach is to combine the rational approach with directed evolution (a *semi-rational* approach). In line with the structure-function view of this book, we focus on the rational design of enzymes.

4.4.3 Rational engineering of enzymes for increased stability

Increasing the thermostability of enzymes[*1] is a general goal of industrial scientists, regardless of the specific role intended for the enzyme. This is because more stable enzymes have a lower turnover rate, which means a longer shelf life [18]. In addition, enzymes intended for processes involving high temperatures must possess higher thermostability compared

---

[*1]Usually expressed as T_{50}, which is the temperature required for reducing the enzyme's activity by 50%.

with the average cellular enzyme. Elevated temperatures may be necessary for several reasons, such as increasing substrate solubility or reducing the risk of microbial contamination (particularly in the feed industry).

Rational protein engineering has benefited greatly from two types of studies:

1. Mutational studies on enzymes such as subtilisin [204]. These studies have analyzed different aspects of enzyme structure and function, including stability.

2. Studies comparing enzymes from mesophiles with their thermophilic homologues, in terms of both sequence and structure.

Both types of studies have reached similar conclusions about the changes proteins may undergo to increase their thermostability, as described in Subsection 4.3.2 above. These conclusions point to general approaches that are likely to increase protein stability without inducing conformational changes that might lead to loss of function (Figure 4.5a): (1) reducing the entropy of the unfolded state (rigidification) by introducing Pro residues, eliminating Gly residues, introducing aromatic residues, or creating disulfide bonds, (2) strengthening the protein's core by introducing nonpolar residues, (3) strengthening electrostatic interactions inside the protein by increasing the number of salt bridges, (4) introducing residues that promote α-helical formation, and (5) strengthening protein-solvent interactions. The last approach suggests that the protein surface, mainly in terms of electrostatic interactions, is important for protein stability. This suggestion is indeed supported by different studies [68,205–208][*1]. This may seem counter-intuitive at first. It does, however, make more sense when considering the following. Many enzymes, especially when put in industrial fermentors, do not unfold reversibly. Instead, they undergo an unfolding step followed by an irreversible process, such as aggregation or autolysis [18]. Studies show that the first unfolding step is partial, and suggest that it involves the protein surface (e.g., [209–211]). In such cases, it is worthwhile to focus on this region of the protein when attempting to increase stability by introducing mutations. Indeed, the same conclusions were reached through mutagenesis studies of extensively investigated enzymes, such as thermolysin, thermolysin-like protease (TLP), subtilisin, and α-amylase (see [18] for references) (Figure 4.5b).

Although the effect of the protein surface on stability results mainly from electrostatic interactions that are either introduced or eliminated by mutations, nonpolar effects also seem to be important. For example, near-surface nonpolar clusters have been found to be particularly stabilizing when appearing in β-sheets [98,212,213]. Presumably, the stabilization results from the strengthening of interactions between adjacent strands, by these clusters [18]. Nonpolar surface residues may stabilize the protein even when adjacent to polar residues, provided that the latter have large enough aliphatic or aromatic side chains, with which the nonpolar residues may interact favorably (e.g., [214]). Finally, stability may also be affected by ion binding; in some Ca^{2+}-dependent proteases, such as TLP [215,216] and subtilisin [204], mutations that strengthen the Ca^{2+}-binding capability of the enzyme also stabilize it (Figure 4.5b).

While the insights outlined above have improved our understanding of protein stability and greatly improved the success of rational protein design, some aspects of stability remained unclear. For example, some studies demonstrate the cumulative contribution of

---

[*1]This was reflected in the > 1 kcal/mol increase in the stability of the investigated protein upon introduction of single mutations that created or optimized pairwise Coulomb interactions.

many point mutations to the stability of different proteins, whereas other studies show that a small number of mutations (or even a single one) may be sufficient to produce very large stability differences (e.g., [217]). Another aspect that complicates the design of stabilizing mutations is the tendency of enzymes in an industrial environment to unfold in two steps: partial (local) unfolding, followed by aggregation or autolysis. As a result, mutations designed to affect the overall (thermodynamic) stability of the enzyme are likely to affect only its kinetic stability, which is associated with the first step (unfolding). For protein engineers, this means that instead of focusing on globally stabilizing mutations, it is necessary to target local regions that participate in the first step (and therefore promote the second). These regions include, e.g., surface regions that unfold and externalize nonpolar residues, and therefore promote aggregation [18]. The partial unfolding of the protein also creates a problem of non-additivity. That is, introducing two stabilizing mutations does not necessarily mean that the effect of both will be the sum of the effect of the individual mutants. The reason is that the two mutations may reside in different regions of the protein, each participating in the unfolding process to a different extent.

Despite these issues, protein engineering is considered a success, especially in conferring increased stability to proteins used by industry. Several groups have taken the concept of engineering a step further and developed automated computational methods for protein redesign (see reviews in [218–223]). Their approach can be viewed as the inverse of protein structure prediction (see Chapter 3). Whereas the latter aims to find the most probable structure for a given sequence, protein redesign methods look for the 'best' (e.g., the most stable) sequence yielding a required structure.

(a)   (b)

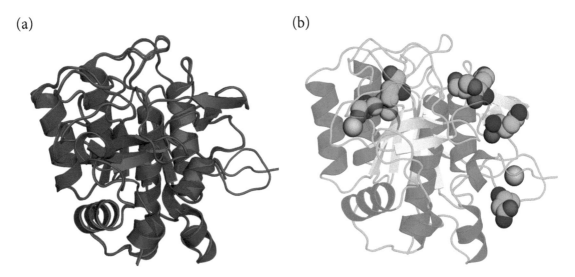

FIGURE 4.5 **Stabilizing mutations in subtilisin.** (a) The structures of both wild-type (magenta) and mutated (blue) subtilisin BPN′ from *Bacillus amyloliquefaciens* are shown in a ribbon representation. As can be clearly seen, no significant structural changes are induced by the mutations. (b) The mutated subtilisin BPN. The six mutations shown (as atom-type colored spheres) have been found to confer the highest degree of stabilization to the protein: N218S, G169A, Y217K, M50F, Q206C, and N76D. The change of stability is represented by the free energy change of unfolding, as measured by differential scanning calorimetry [224]. Each individual mutation increased stability by 0.3 to 1.3 kcal/mol, and together they increased stability by 3.8 kcal/mol. The image demonstrates the surface location of all the mutations save one, and the proximity of some to the two Ca^{2+} binding sites (the two cations are colored gray).

4.5 SUMMARY

- To understand how proteins fold and maintain stability, it is necessary to consider the ways in which they respond to the physical forces acting on their atoms. This energy-centered view underlies the field of structural biophysics.

- The biophysical approach to studying protein structure and function requires not only the characterization of the physical forces acting on the system, but also their direction and probability. This is done using thermodynamics, which describes the free energy content of the system in different states and under different conditions. Spontaneous processes, such as protein folding, always involve a drop in the free energy of the system, which is measurable.

- The free energy of a system has two major components: enthalpy, which results from the formation and breaking of chemical bonds or physical interactions, and entropy, which reflects the disorder in the system. Referring to these components helps understanding some of the changes occurring in a system when a protein folds or binds to a ligand. However, in order to understand the molecular details of these processes, the total free energy changes that accompany them must be interpreted in terms of distinct physical forces.

- Numerous studies have been carried out with the goal of deciphering the molecular and energy changes accompanying protein folding. These studies used both experimental and computational approaches, and offered different interpretations for the measured and calculated thermodynamic quantities they obtained. All of these studies indicate that proteins have marginal stability, which allows them to maintain both their native fold and dynamic nature, both of which are important for function. The studies also agree that protein folding is driven primarily by the hydrophobic effect, and to a lesser extent by van der Waals interactions.

- However, the exact magnitude of these effects is still unclear. Furthermore, the effect of electrostatic interactions on the folding and stability of proteins is under intense dispute. Not only is their magnitude unknown, but even their direction is a matter of disagreement. These problems account, at least partially, for the inability of current energy-based algorithms to predict the native folds of most proteins on the basis of their sequence.

- One of the phenomena that best illustrates the different aspects of protein stability is the adaptation of certain biological organisms to environments characterized by extreme conditions (temperature, pressure, salinity, pH). Such organisms, termed 'extremophiles', have undergone evolutionary processes that changed the sequences of their proteins to render them resistant to these conditions. By comparing the sequences of such proteins to those of their moderate homologues, scientists have gained important insights regarding the factors that affect protein stability.

- Enzymes' catalytic efficiency and selectivity have made them desired targets for numerous industrial applications. Indeed, enzymes are used today in the textile, food, and pharmaceutical industries, as well as in other industrial fields. In many cases, these enzymes must function under less-than-favorable conditions. To enable them to do so, scientists apply the knowledge accumulated on protein stability and engineer the enzymes using recombinant DNA technology. In recent years, this exciting field of protein science has taken a step forward in the form of automated algorithms used for enhancing enzyme stability and catalytic power.

EXERCISES

4.1 A room-temperature chemical reaction in which substrate A is turned into the product B at atmospheric pressure involves a change of −5 kcal/mol in the standard free energy. When the reaction reaches equilibrium, the concentration of A is 0.01 mM. Calculate the equilibrium concentration of B considering a 1:1 reaction stoichiometry.

4.2 A chemical reaction was carried out at 25 °C (room temperature) and under constant pressure. At equilibrium, the change in free energy was measured as −3 kcal/mol, and the reaction released 5 kcal/mol of heat, as measured by calorimetry. Calculate the change in entropy that accompanied the reaction.

4.3 Protein folding is a favorable process. Yet, it decreases the entropy of the polypeptide chain, in contrast to Nature's tendency to increase entropy. Explain how this seeming contradiction is possible.

4.4 What is the source of disagreement between scientists regarding the favorability of electrostatic interactions between protein core charges?

4.5 Is the loss of entropy upon protein folding equal among all parts of the protein? Explain.

4.6 Explain the mechanism through which high temperature disrupts the folded structure of a protein. How does Nature solve this problem for hyper-thermophilic organisms?

4.7 A. List a few industrial uses of enzymes.

B. Pick one example and elaborate.

4.8 Suggest ways to stabilize the structure of a mesophilic protein in acidic environments.

REFERENCES

1. H. Wu. Studies on denaturation of proteins. XIII. A theory of denaturation. *Chin. J. Physiol.*, V:321–344, 1931.
2. A. E. Mirsky and L. Pauling. On the Structure of Native, Denatured, and Coagulated Proteins. *Proc. Natl. Acad. Sci. USA*, 22(7).439–447, 1936.
3. K. E. van Holde, W. C. Johnson, and P. S. Ho. *Principles of Physical Biochemistry*. Pearson Education, New Jersey, US, 2nd edition, 2006.
4. A. D. Robertson and K. P. Murphy. Protein Structure and the Energetics of Protein Stability. *Chem. Rev.*, 97(5):1251–1268, 1997.
5. J. B. Chaires. Calorimetry and thermodynamics in drug design. *Annu. Rev. Biophys.*, 37:135–51, 2008.
6. B. E. Corkey, J. T. Deeney, M. C. Glennon, F. M. Matschinsky, and M. Prentki. Regulation of steady-state free Ca_{2+} levels by the ATP/ADP ratio and orthophosphate in permeabilized RINm5F insulinoma cells. *J. Biol. Chem.*, 263(9):4247–53, 1988.

7. H. Qian. Phosphorylation energy hypothesis: open chemical systems and their biological functions. *Annu. Rev. Phys. Chem.*, 58:113–42, 2007.

8. D. Chandler. *Introduction to Modern Statistical Mechanics*. Oxford University Press, New York, 1987.

9. K. A. Dill and S. Bromberg. *Molecular Driving Forces*. Garland Science, New York, 2003.

10. J. R. Horn, D. Russell, E. A. Lewis, and K. P. Murphy. Van't Hoff and calorimetric enthalpies from isothermal titration calorimetry: are there significant discrepancies? *Biochemistry*, 40(6):1774–8, 2001.

11. J. B. Chaires. Possible origin of differences between van't Hoff and calorimetric enthalpy estimates. *Biophys. Chem.*, 64(1–3):15–23, 1997.

12. A. Cooper. Thermodynamics of Protein Folding and Stability. In *Protein: A Comprehensive Treatise*, volume 2, pages 217–270. JAI Press Inc., Stamford, CT, 1999.

13. A. Cooper and C. M. Johnson. Introduction to microcalorimetry and biomolecular energetics. *Methods Mol. Biol.*, 22:109–24, 1994.

14. N. V. Prabhu and K. A. Sharp. Heat capacity in proteins. *Annu. Rev. Phys. Chem.*, 56:521–48, 2005.

15. C. Tanford. Protein denaturation. *Adv. Protein Chem.*, 23:121–282, 1968.

16. P. J. Flory. *Statistical mechanics of chain molecules*. Wiley, New York, 1969.

17. A. Cooper. Thermodynamic analysis of biomolecular interactions. *Curr. Opin. Chem. Biol.*, 3(5):557–63, 1999.

18. V. G. Eijsink, A. Bjork, S. Gaseidnes, R. Sirevag, B. Synstad, B. van den Burg, and G. Vriend. Rational engineering of enzyme stability. *J. Biotechnol.*, 113(1–3):105–20, 2004.

19. J. D. Dunitz. Win some, lose some: enthalpy-entropy compensation in weak intermolecular interactions. *Chem. Biol.*, 2(11):709–12, 1995.

20. C. L. Wee, D. Gavaghan, and M. S. Sansom. Lipid Bilayer Deformation and the Free Energy of Interaction of a Kv Channel Gating-Modifier Toxin. *Biophys. J.*, 95:3816–26, 2008.

21. M. Liu, T. Sun, J. Hu, W. Chen, and C. Wang. Study on the mechanism of the BtuF periplasmic-binding protein for vitamin B12. *Biophys. Chem.*, 135(1–3):19–24, 2008.

22. B. Honig and A. Nicholls. Classical electrostatics in biology and chemistry. *Science*, 268(5214):1144–1149, 1995.

23. M. Feig and C. L. Brooks. Recent advances in the development and application of implicit solvent models in biomolecule simulations. *Curr. Opin. Struct. Biol.*, 14(2):217–224, 2004.

24. C. B. Anfinsen. Principles that Govern the Folding of Protein Chains. *Science*, 181(96):223–230, 1973.

25. W. Guo, J. E. Shea, and R. S. Berry. The physics of the interactions governing folding and association of proteins. *Ann. N. Y. Acad. Sci.*, 1066:34–53, 2005.

26. C. N. Pace, S. Trevino, E. Prabhakaran, and J. M. Scholtz. Protein structure, stability and solubility in water and other solvents. *Philos. Trans. R. Soc. Lond. B: Biol. Sci.*, 359(1448):1225–34; discussion 1234–5, 2004.

27. A. R. Fersht, A. Matouschek, and L. Serrano. The folding of an enzyme. I. Theory of protein engineering analysis of stability and pathway of protein folding. *J. Mol. Biol.*, 224(3):771–82, 1992.

28. P. L. Privalov and S. J. Gill. Stability of protein structure and hydrophobic interaction. *Adv. Protein Chem.*, 39:191–234, 1988.

29. C. N. Pace. Conformational stability of globular proteins. *Trends Biochem. Sci.*, 15(1):14–7, 1990.

30. K. A. Dill. Dominant forces in protein folding. *Biochemistry*, 29(31):7133–55, 1990.

31. S. Warszawski, R. Netzer, D. S. Tawfik, and S. J. Fleishman. A "fuzzy"-logic language for encoding multiple physical traits in biomolecules. *J. Mol. Biol.*, 426(24):4125–38, 2014.

32. T. Sikosek and H. S. Chan. Biophysics of protein evolution and evolutionary protein biophysics. *J. R. Soc. Interface*, 11(100):20140419, 2014.

33. J. F. Brandts. The Thermodynamics of Protein Denaturation. I. The Denaturation of Chymotrypsinogen. *J. Am. Chem. Soc.*, 86:4291–301, 1964.

34. J. F. Brandts. The Thermodynamics of Protein Denaturation. II. A Model of Reversible Denaturation and Interpretations Regarding the Stability of Chymotrypsinogen. *J. Am. Chem. Soc.*, 86:4302–14, 1964.

35. F. H. C. Crick. The packing of α-helices: simple coiled-coils. *Acta Cryst.*, 6:689–697, 1953.

36. W. Kauzmann. Some factors in the interpretation of protein denaturation. *Adv. Protein Chem.*, 14:1–63, 1959.

37. K. A. Dill, S. Banu Ozkan, T. R. Weikl, J. D. Chodera, and V. A. Voelz. The protein folding problem: when will it be solved? *Curr. Opin. Struct. Biol.*, 17(3):342–346, 2007.

38. M. H. Hecht, A. Das, A. Go, L. H. Bradley, and Y. Wei. De novo proteins from designed combinatorial libraries. *Protein Sci.*, 13(7):1711–23, 2004.

39. S. Kamtekar, J. M. Schiffer, H. Xiong, J. M. Babik, and M. H. Hecht. Protein design by binary patterning of polar and nonpolar amino acids. *Science*, 262(5140):1680–5, 1993.

40. D. E. Kim, H. Gu, and D. Baker. The sequences of small proteins are not extensively optimized for rapid folding by natural selection. *Proc. Natl. Acad. Sci. USA*, 95(9):4982–6, 1998.

41. D. W. Bolen and G. D. Rose. Structure and energetics of the hydrogen-bonded backbone in protein folding. *Annu. Rev. Biochem.*, 77:339–62, 2008.

42. P. L. Privalov and N. N. Khechinashvili. A thermodynamic approach to the problem of stabilization of globular protein structure: a calorimetric study. *J. Mol. Biol.*, 86(3):665–84, 1974.

43. C. N. Pace, H. Fu, K. L. Fryar, J. Landua, S. R. Trevino, B. A. Shirley, M. M. Hendricks, S. Iimura, K. Gajiwala, J. M. Scholtz, and G. R. Grimsley. Contribution of hydrophobic interactions to protein stability. *J. Mol. Biol.*, 408(3):514–28, 2011.

44. J. Chen and W. E. Stites. Packing is a key selection factor in the evolution of protein hydrophobic cores. *Biochemistry*, 40(50):15280–9, 2001.

45. W. A. Baase, L. Liu, D. E. Tronrud, and B. W. Matthews. Lessons from the lysozyme of phage T4. *Protein Sci.*, 19(4):631–41, 2010.

46. R. L. Baldwin. Energetics of protein folding. *J. Mol. Biol.*, 371(2):283–301, 2007.

47. C. Tanford. Contribution of Hydrophobic Interactions to the Stability of the Globular Conformation of Proteins. *J. Am. Chem. Soc.*, 84(22):4240–4247, 1962.

48. C. Tanford. Isothermal Unfolding of Globular Proteins in Aqueous Urea Solutions. *J. Am. Chem. Soc.*, 86(10):2050–2059, 1964.

49. Y. Nozaki and C. Tanford. The Solubility of Amino Acids and Two Glycine Peptides in Aqueous Ethanol and Dioxane Solutions: Establishment Of A Hydrophobicity Scale. *J. Biol. Chem.*, 246(7):2211–2217, 1971.

50. R. B. Hermann. Theory of hydrophobic bonding. II. The correlation of hydrocarbon solubility in water with cavity surface area. *J. Phys. Chem.*, 76:2754–2759, 1972.

51. C. Chothia. Hydrophobic bonding and accessible surface area in proteins. *Nature*, 248(446):338–339, 1974.

52. W. C. Wimley, T. P. Creamer, and S. H. White. Solvation energies of amino acid side chains and backbone in a family of host-guest pentapeptides. *Biochemistry*, 35(16):5109–5124, 1996.

53. T. M. Raschke, J. Tsai, and M. Levitt. Quantification of the hydrophobic interaction by simulations of the aggregation of small hydrophobic solutes in water. *Proc. Natl. Acad. Sci. USA*, 98(11):5965–9, 2001.

54. K. A. Sharp, A. Nicholls, R. F. Fine, and B. Honig. Reconciling the magnitude of the microscopic and macroscopic hydrophobic effects. *Science*, 252(5002):106–9, 1991.

55. S. A Sridharan, A. Nicholls, and B. Honig. A new vertex algorithm to calculate solvent accessible surface area. *Biophys. J.*, 61:A174, 1992.

56. J. T. Kellis Jr, K. Nyberg, and A. R. Fersht. Energetics of complementary side-chain packing in a protein hydrophobic core. *Biochemistry*, 28(11):4914–22, 1989.

57. C. N. Pace. Contribution of the hydrophobic effect to globular protein stability. *J. Mol. Biol.*, 226(1):29–35, 1992.

58. M. Bueno, L. A. Campos, J. Estrada, and J. Sancho. Energetics of aliphatic deletions in protein cores. *Protein Sci.*, 15(8):1858–72, 2006.

59. J. Xu, W. A. Baase, E. Baldwin, and B. W. Matthews. The response of T4 lysozyme to large-to-small substitutions within the core and its relation to the hydrophobic effect. *Protein Sci.*, 7(1):158–77, 1998.

60. K. Takano, K. Ogasahara, H. Kaneda, Y. Yamagata, S. Fujii, E. Kanaya, M. Kikuchi, M. Oobatake, and K. Yutani. Contribution of hydrophobic residues to the stability of human lysozyme: calorimetric studies and X-ray structural analysis of the five isoleucine to valine mutants. *J. Mol. Biol.*, 254(1):62–76, 1995.

61. Y. Li, Y. Huang, C. P. Swaminathan, S. J. Smith-Gill, and R. A. Mariuzza. Magnitude of the hydrophobic effect at central versus peripheral sites in protein-protein interfaces. *Structure*, 13(2):297–307, 2005.

62. L. M. Amzel. Calculation of entropy changes in biological processes: folding, binding, and oligomerization. *Methods Enzymol.*, 323:167–77, 2000.

63. J. Fitter. A measure of conformational entropy change during thermal protein unfolding using neutron spectroscopy. *Biophys. J.*, 84(6):3924–30, 2003.

64. M. Karplus, T. Ichiye, and B. M. Pettitt. Configurational entropy of native proteins. *Biophys. J.*, 52(6):1083–5, 1987.

65. T. I. Igumenova, K. K. Frederick, and A. J. Wand. Characterization of the fast dynamics of protein amino acid side chains using NMR relaxation in solution. *Chem. Rev.*, 106(5):1672–99, 2006.

66. K. K. Frederick, M. S. Marlow, K. G. Valentine, and A. J. Wand. Conformational entropy in molecular recognition by proteins. *Nature*, 448(7151):325–330, 2007.

67. G. P. Brady and K. A. Sharp. Entropy in protein folding and in protein-protein interactions. *Curr. Opin. Struct. Biol.*, 7(2):215–21, 1997.

68. S. S. Strickler, A. V. Gribenko, T. R. Keiffer, J. Tomlinson, T. Reihle, V. V. Loladze, and G. I. Makhatadze. Protein stability and surface electrostatics: a charged relationship. *Biochemistry*, 45(9):2761–6, 2006.

69. S. Marqusee and R. T. Sauer. Contributions of a hydrogen bond/salt bridge network to the stability of secondary and tertiary structure in λ repressor. *Protein Sci.*, 3(12):2217–25, 1994.

70. V. Lounnas and R. C. Wade. Exceptionally stable salt bridges in cytochrome P450cam have functional roles. *Biochemistry*, 36(18):5402–17, 1997.

71. C. N. Schutz and A. Warshel. What are the dielectric "constants" of proteins and how to validate electrostatic models? *Proteins*, 44(4):400–17, 2001.

72. C. N. Pace, H. Fu, K. Lee Fryar, J. Landua, S. R. Trevino, D. Schell, R. L. Thurlkill, S. Imura, J. M. Scholtz, K. Gajiwala, J. Sevcik, L. Urbanikova, J. K. Myers, K. Takano, E. J. Hebert, B. A. Shirley, and G. R. Grimsley. Contribution of hydrogen bonds to protein stability. *Protein Sci.*, 23(5):652–661, 2014.

73. A. R. Fersht, J. P. Shi, J. Knill-Jones, D. M. Lowe, A. J. Wilkinson, D. M. Blow, P. Brick, P. Carter, M. M. Waye, and G. Winter. Hydrogen bonding and biological specificity analysed by protein engineering. *Nature*, 314(6008):235–8, 1985.

74. L. A. Campos, S. Cuesta-Lopez, J. Lopez-Llano, F. Falo, and J. Sancho. A double-deletion method to quantifying incremental binding energies in proteins from experiment: example of a destabilizing hydrogen bonding pair. *Biophys. J.*, 88(2):1311–21, 2005.

75. N. Ben-Tal, D. Sitkoff, I. A. Topol, A. S. Yang, S. K. Burt, and B. Honig. Free energy of amide hydrogen bond formation in vacuum, in water, and in liquid alkane solution. *J. Phys. Chem. B*, 101(3):450–457, 1997.

76. B. H. Honig and W. L. Hubbell. Stability of "salt bridges" in membrane proteins. *Proc. Natl. Acad. Sci. USA*, 81(17):5412–6, 1984.

77. Z. S. Hendsch and B. Tidor. Do salt bridges stabilize proteins? A continuum electrostatic analysis. *Protein Sci.*, 3(2):211–26, 1994.

78. C. D. Waldburger, J. F. Schildbach, and R. T. Sauer. Are buried salt bridges important for protein stability and conformational specificity? *Nat. Struct. Biol.*, 2(2):122–8, 1995.

79. C. Tanford and J. G. Kirkwood. Theory of Protein Titration Curves. I. General Equations for Impenetrable Spheres. *J. Am. Chem. Soc.*, 79:5333, 1957.

80. K. A. Sharp and B. Honig. Electrostatic interactions in macromolecules: theory and applications. *Annu. Rev. Biophys. Biophys. Chem.*, 19(1):301–332, 1990.

81. B. Roux and T. Simonson. Implicit solvent models. *Biophys. Chem.*, 78(1–2):1–20, 1999.

82. N. A. Baker. Improving implicit solvent simulations: a Poisson-centric view. *Curr. Opin. Struct. Biol.*, 15(2):137–143, 2005.

83. A. C. Robinson, C. A. Castaneda, J. L. Schlessman, and E. B. Garcia-Moreno. Structural and thermodynamic consequences of burial of an artificial ion pair in the hydrophobic interior of a protein. *Proc. Natl. Acad. Sci. USA*, 111(32):11685–90, 2014.

84. G. J. Lesser and G. D. Rose. Hydrophobicity of amino acid subgroups in proteins. *Proteins*, 8(1):6–13, 1990.

85. N. H. Joh, A. Min, S. Faham, J. P. Whitelegge, D. Yang, V. L. Woods, and J. U. Bowie. Modest stabilization by most hydrogen-bonded side-chain interactions in membrane proteins. *Nature*, 453(7199):1266–70, 2008.

86. J. Gao, D. A. Bosco, E. T. Powers, and J. W. Kelly. Localized Thermodynamic Coupling between Hydrogen Bonding and Microenvironment Polarity Substantially Stabilizes Proteins. *Nat. Struct. Mol. Biol.*, 16(7):684–690, 2009.

87. S. Bone and R. Pethig. Dielectric studies of the binding of water to lysozyme. *J. Mol. Biol.*, 157(3):571–5, 1982.

88. M. K. Gilson and B. H. Honig. The dielectric constant of a folded protein. *Biopolymers*, 25(11):2097–119, 1986.

89. S. Kumar and R. Nussinov. Salt bridge stability in monomeric proteins. *J. Mol. Biol.*, 293(5):1241–55, 1999.

90. S. Albeck, R. Unger, and G. Schreiber. Evaluation of direct and cooperative contributions towards the strength of buried hydrogen bonds and salt bridges. *J. Mol. Biol.*, 298(3):503–20, 2000.

91. D. Lee, J. Lee, and C. Seok. What stabilizes close arginine pairing in proteins? *Phys. Chem. Chem. Phys.*, 15(16):5844–53, 2013.

92. D. F. Sticke, L. G. Presta, K. A. Dill, and G. D. Rose. Hydrogen bonding in globular proteins. *J. Mol. Biol.*, 226(4):1143–1159, 1992.

93. I. K. McDonald and J. M. Thornton. Satisfying hydrogen bonding potential in proteins. *J. Mol. Biol.*, 238(5):777–793, 1994.

94. P. J. Fleming and G. D. Rose. Do all backbone polar groups in proteins form hydrogen bonds? *Protein Sci.*, 14(7):1911–1917, 2005.

95. E. S. Feldblum and I. T. Arkin. Strength of a bifurcated H bond. *Proc. Natl. Acad. Sci. USA*, 111(11):4085–90, 2014.

96. N. Kobko, L. Paraskevas, E. del Rio, and J. J. Dannenberg. Cooperativity in amide hydrogen bonding chains: implications for protein-folding models. *J. Am. Chem. Soc.*, 123(18):4348–9, 2001.

97. R. J. Kennedy, K.-Y. Tsang, and D. S. Kemp. Consistent Helicities from CD and Template t/c Data for N-Templated Polyalanines: Progress toward Resolution of the Alanine Helicity Problem. *J. Am. Chem. Soc.*, 124(6):934–944, 2002.

98. J. F. Espinosa, V. Munoz, and S. H. Gellman. Interplay between hydrophobic cluster and loop propensity in β-hairpin formation. *J. Mol. Biol.*, 306(3):397–402, 2001.

99. P. Gilli, V. Bertolasi, L. Pretto, V. Ferretti, and G. Gilli. Covalent versus electrostatic nature of the strong hydrogen bond: discrimination among single, double, and asymmetric single-well hydrogen bonds by variable-temperature X-ray crystallographic methods in β-diketone enol RAHB systems. *J. Am. Chem. Soc.*, 126(12):3845–55, 2004.

100. J. J. Dannenberg, Laury Haskamp, and Artëm Masunov. Are Hydrogen Bonds Covalent or Electrostatic? A Molecular Orbital Comparison of Molecules in Electric Fields and H-Bonding Environments. *J. Phys. Chem. A*, 103(35):7083–7086, 1999.

101. D. E. Anderson, W. J. Becktel, and F. W. Dahlquist. pH-induced denaturation of proteins: a single salt bridge contributes 3–5 kcal/mol to the free energy of folding of T4 lysozyme. *Biochemistry*, 29(9):2403–8, 1990.

102. Z. Cao and J. U. Bowie. An energetic scale for equilibrium H/D fractionation factors illuminates hydrogen bond free energies in proteins. *Protein Sci.*, 23(5):566–75, 2014.

103. B. A. Joughin, D. F. Green, and B. Tidor. Action-at-a-distance interactions enhance protein binding affinity. *Protein Sci.*, 14(5):1363–9, 2005.

104. D. N. Bolon and S. L. Mayo. Polar residues in the protein core of *Escherichia coli* thioredoxin are important for fold specificity. *Biochemistry*, 40(34):10047–53, 2001.

105. S. Kumar and R. Nussinov. Close-range electrostatic interactions in proteins. *ChemBioChem*, 3(7):604–617, 2002.

106. M. F. Perutz. Stereochemistry of cooperative effects in haemoglobin. *Nature*, 228(5273):726–39, 1970.

107. A. R. Fersht. Conformational equilibria in α- and δ-chymotrypsin: The energetics and importance of the salt bridge. *J. Mol. Biol.*, 64(2):497–509, 1972.

108. D. J. Barlow and J. M. Thornton. Ion-pairs in proteins. *J. Mol. Biol.*, 168(4):867–85, 1983.

109. T. K. Harris and G. J. Turner. Structural basis of perturbed pKa values of catalytic groups in enzyme active sites. *IUBMB Life*, 53(2):85–98, 2002.

110. A. Warshel. Electrostatic origin of the catalytic power of enzymes and the role of preorganized active sites. *J. Biol. Chem.*, 273(42):27035–8, 1998.

111. P. G. Schultz. The interplay between chemistry and biology in the design of enzymatic catalysts. *Science*, 240(4851):426–33, 1988.

112. A. Gutteridge and J. M. Thornton. Understanding nature's catalytic toolkit. *Trends Biochem. Sci.*, 30(11):622–629, 2005.

113. J. Kraut. Serine proteases: structure and mechanism of catalysis. *Annu. Rev. Biochem.*, 46:331–58, 1977.

114. L. Hedstrom. Serine protease mechanism and specificity. *Chem. Rev.*, 102(12):4501–24, 2002.

115. C. Frazao, I. Bento, J. Costa, C. M. Soares, P. Verissimo, C. Faro, E. Pires, J. Cooper, and M. A. Carrondo. Crystal structure of cardosin A, a glycosylated and Arg-Gly-Asp-containing aspartic proteinase from the flowers of *Cynara cardunculus L. J. Biol. Chem.*, 274(39):27694–701, 1999.

116. B. H. Dessailly, M. F. Lensink, and S. J. Wodak. Relating destabilizing regions to known functional sites in proteins. *BMC Bioinformatics*, 8(1):141, 2007.

117. B. K. Shoichet, W. A. Baase, R. Kuroki, and B. W. Matthews. A relationship between protein stability and protein function. *Proc. Natl. Acad. Sci. USA*, 92(2):452–6, 1995.

118. K. B. Wong and V. Daggett. Barstar has a highly dynamic hydrophobic core: evidence from molecular dynamics simulations and nuclear magnetic resonance relaxation data. *Biochemistry*, 37(32):11182–92, 1998.

119. K. Lindorff-Larsen, R. B. Best, M. A. Depristo, C. M. Dobson, and M. Vendruscolo. Simultaneous determination of protein structure and dynamics. *Nature*, 433(7022):128–32, 2005.

120. K. J. Lumb and P. S. Kim. A buried polar interaction imparts structural uniqueness in a designed heterodimeric coiled coil. *Biochemistry*, 34(27):8642–8, 1995.

121. C. L. Worth and T. L. Blundell. Satisfaction of hydrogen-bonding potential influences the conservation of polar sidechains. *Proteins*, 75(2):413–29, 2009.

122. M. Iwaoka, S. Takemoto, M. Okada, and S. Tomoda. Weak Nonbonded S···X (X=O, N, and S) Interactions in Proteins. Statistical and Theoretical Studies. *Bull. Chem. Soc. Jpn.*, 75(7):1611–1625, 2002.

123. M. Iwaoka and N. Isozumi. Possible roles of S···O and S···N interactions in the functions and evolution of phospholipase A$_2$. *Biophysics*, 2:23–34, 2006.

124. C. A. Hunter, J. Singh, and J. M. Thornton. π-π interactions: The geometry and energetics of phenylalanine-phenylalanine interactions in proteins. *J. Mol. Biol.*, 218(4):837–846, 1991.

125. M. Levitt and M. F. Perutz. Aromatic rings act as hydrogen bond acceptors. *J. Mol. Biol.*, 201(4):751–754, 1988.

126. I. D. Kuntz and G. M. Crippen. Protein densities. *Int. J. Pept. Protein Res.*, 13(2):223–8, 1979.

127. T. Lazaridis, G. Archontis, and M. Karplus. Enthalpic contribution to protein stability: insights from atom-based calculations and statistical mechanics. *Adv. Protein Chem.*, 47:231–306, 1995.

128. V. V. Loladze, D. N. Ermolenko, and G. I. Makhatadze. Thermodynamic consequences of burial of polar and nonpolar amino acid residues in the protein interior. *J. Mol. Biol.*, 320(2):343–57, 2002.

129. J. Klein-Seetharaman, M. Oikawa, S. B. Grimshaw, J. Wirmer, E. Duchardt, T. Ueda, T. Imoto, L. J. Smith, C. M. Dobson, and H. Schwalbe. Long-range interactions within a nonnative protein. *Science*, 295(5560):1719–22, 2002.

130. N. Prabhu and K. Sharp. Protein-solvent interactions. *Chem. Rev.*, 106(5):1616–23, 2006.

131. N. C. Fitzkee, P. J. Fleming, H. Gong, N. Panasik Jr, T. O. Street, and G. D. Rose. Are proteins made from a limited parts list? *Trends Biochem. Sci.*, 30(2):73–80, 2005.

132. Z. Shi, K. Chen, Z. Liu, and N. R. Kallenbach. Conformation of the backbone in unfolded proteins. *Chem. Rev.*, 106(5):1877–97, 2006.

133. A. Rath, A. R. Davidson, and C. M. Deber. The structure of "unstructured" regions in peptides and proteins: role of the polyproline II helix in protein folding and recognition. *Biopolymers*, 80(2–3):179–85, 2005.

134. R. K. Dukor and T. A. Keiderling. Reassessment of the random coil conformation: vibrational CD study of proline oligopeptides and related polypeptides. *Biopolymers*, 31(14):1747–61, 1991.

135. J. C. Ferreon and V. J. Hilser. The effect of the polyproline II (PPII) conformation on the denatured state entropy. *Protein Sci.*, 12(3):447–57, 2003.

136. A. P. Minton. Influence of macromolecular crowding upon the stability and state of association of proteins: predictions and observations. *J. Pharm. Sci.*, 94(8):1668–75, 2005.

137. S. B. Zimmerman and S. O. Trach. Estimation of macromolecule concentrations and excluded volume effects for the cytoplasm of *Escherichia coli*. *J. Mol. Biol.*, 222(3):599–620, 1991.

138. F.-X. Theillet, A. Binolfi, T. Frembgen-Kesner, K. Hingorani, M. Sarkar, C. Kyne, C. Li, P. B. Crowley, L. Gierasch, G. J. Pielak, A. H. Elcock, A. Gershenson, and P. Selenko. Physicochemical properties of cells and their effects on intrinsically disordered proteins (IDPs). *Chem. Rev.*, 114(13):6661–6714, 2014.

139. E. H. McConkey. Molecular evolution, intracellular organization, and the quinary structure of proteins. *Proc. Natl. Acad. Sci. USA*, 79(10):3236–40, 1982.

140. A. J. Wirth and M. Gruebele. Quinary protein structure and the consequences of crowding in living cells: leaving the test-tube behind. *BioEssays*, 35(11):984–93, 2013.

141. W. B. Monteith, R. D. Cohen, A. E. Smith, E. Guzman-Cisneros, and G. J. Pielak. Quinary structure modulates protein stability in cells. *Proc. Natl. Acad. Sci. USA*, 112(6):1739–42, 2015.

142. A. Politou and P. A. Temussi. Revisiting a dogma: the effect of volume exclusion in molecular crowding. *Curr. Opin. Struct. Biol.*, 30C:1–6, 2014.

143. J. Danielsson, X. Mu, L. Lang, H. Wang, A. Binolfi, F. X. Theillet, B. Bekei, D. T. Logan, P. Selenko, H. Wennerstrom, and M. Oliveberg. Thermodynamics of protein destabilization in live cells. *Proc. Natl. Acad. Sci. USA*, 112(40):12402–7, 2015.

144. C. M. Dobson. Protein folding and misfolding. *Nature*, 426(6968):884–90, 2003.

145. M. Manning and W. Colon. Structural basis of protein kinetic stability: resistance to sodium dodecyl sulfate suggests a central role for rigidity and a bias toward beta-sheet structure. *Biochemistry*, 43(35):11248–54, 2004.

146. C. Puorger, O. Eidam, G. Capitani, D. Erilov, M. G. Grutter, and R. Glockshuber. Infinite kinetic stability against dissociation of supramolecular protein complexes through donor strand complementation. *Structure*, 16(4):631–42, 2008.

147. R. E. Jefferson, T. M. Blois, and J. U. Bowie. Membrane Proteins Can Have High Kinetic Stability. *J. Am. Chem. Soc.*, 135(40):15183–15190, 2013.

148. J. P. Schlebach and C. R. Sanders. The safety dance: biophysics of membrane protein folding and misfolding in a cellular context. *Q. Rev. Biophys.*, 48(1):1–34, 2015.

149. D. Baker and D. A. Agard. Kinetics versus thermodynamics in protein folding. *Biochemistry*, 33(24):7505–9, 1994.

150. S. S. Jaswal, J. L. Sohl, J. H. Davis, and D. A. Agard. Energetic landscape of alpha-lytic protease optimizes longevity through kinetic stability. *Nature*, 415(6869):343–6, 2002.

151. C. Park, S. Zhou, J. Gilmore, and S. Marqusee. Energetics-based protein profiling on a proteomic scale: identification of proteins resistant to proteolysis. *J. Mol. Biol.*, 368(5):1426–37, 2007.

152. K. Xia, M. Manning, H. Hesham, Q. Lin, C. Bystroff, and W. Colón. Identifying the subproteome of kinetically stable proteins via diagonal 2D SDS/PAGE. *Proc. Natl. Acad. Sci. USA*, 104(44):17329–17334, 2007.

153. A. Broom, S. M. Ma, K. Xia, H. Rafalia, K. Trainor, W. Colon, S. Gosavi, and E. M. Meiering. Designed protein reveals structural determinants of extreme kinetic stability. *Proc. Natl. Acad. Sci. USA*, 112(47):14605–10, 2015.

154. R. J. Sola, J. A. Rodriguez-Martinez, and K. Griebenow. Modulation of protein biophysical properties by chemical glycosylation: biochemical insights and biomedical implications. *Cell. Mol. Life Sci.*, 64(16):2133–52, 2007.

155. D. Shental-Bechor and Y. Levy. Effect of glycosylation on protein folding: a close look at thermodynamic stabilization. *Proc. Natl. Acad. Sci. USA*, 105(24):8256–61, 2008.

156. F. B. Sheinerman, R. Norel, and B. Honig. Electrostatic aspects of protein-protein interactions. *Curr. Opin. Struct. Biol.*, 10(2):153–9, 2000.

157. D. Xu, S. L. Lin, and R. Nussinov. Protein binding versus protein folding: the role of hydrophilic bridges in protein associations. *J. Mol. Biol.*, 265(1):68–84, 1997.

158. J. C. Bischof and X. He. Thermal stability of proteins. *Ann. N. Y. Acad. Sci.*, 1066:12–33, 2005.

159. B. Lee. Solvent reorganization contribution to the transfer thermodynamics of small nonpolar molecules. *Biopolymers*, 31(8):993–1008, 1991.

160. A. M. Lesk. Hydrophobicity: getting into hot water. *Biophys. Chem.*, 105(2–3):179–82, 2003.

161. J. Roche, J. A. Caro, D. R. Norberto, P. Barthe, C. Roumestand, J. L. Schlessman, A. E. Garcia, B. E. Garcia-Moreno, and C. A. Royer. Cavities determine the pressure unfolding of proteins. *Proc. Natl. Acad. Sci. USA*, 109(18):6945–50, 2012.

162. G. A. de Oliveira and J. L. Silva. A hypothesis to reconcile the physical and chemical unfolding of proteins. *Proc. Natl. Acad. Sci. USA*, 2015.

163. W. K. Lim, J. Rösgen, and S. W. Englander. Urea, but not guanidinium, destabilizes proteins by forming hydrogen bonds to the peptide group. *Proc. Natl. Acad. Sci. USA*, 106(8):2595–2600, 2009.

164. G. D. Rose, P. J. Fleming, J. R. Banavar, and A. Maritan. A backbone-based theory of protein folding. *Proc. Natl. Acad. Sci. USA*, 103(45):16623–33, 2006.

165. X. Hoccart and G. Turrell. Raman spectroscopic investigation of the dynamics of urea–water complexes. *J. Chem. Phys.*, 99(11):8498–8503, 1993.

166. P. H. Yancey. Organic osmolytes as compatible, metabolic and counteracting cytoprotectants in high osmolarity and other stresses. *J. Exp. Biol.*, 208(Pt 15):2819–30, 2005.

167. J. Ma, I. M. Pazos, and F. Gai. Microscopic insights into the protein-stabilizing effect of trimethylamine N-oxide (TMAO). *Proc. Natl. Acad. Sci. USA*, 111(23):8476–81, 2014.

168. E. R. Andrews. Multicellular Organs and Organisms. *eLS*, 2001.

169. X. He, W. F. Wolkers, J. H. Crowe, D. J. Swanlund, and J. C. Bischof. In situ thermal denaturation of proteins in dunning AT-1 prostate cancer cells: implication for hyperthermic cell injury. *Ann. Biomed. Eng.*, 32(10):1384–98, 2004.

170. G. Antranikian. Extremophiles, in Encyclopedia of Life Sciences, 2001.

171. J. Wiegel and M. W. W. Adams. *Thermophiles: The keys to molecular evolution and the origin of life?* Taylor and Francis, London, 1998.

172. M. Sawano, H. Yamamoto, K. Ogasahara, S. Kidokoro, S. Katoh, T. Ohnuma, E. Katoh, S. Yokoyama, and K. Yutani. Thermodynamic basis for the stabilities of three CutA1s from *Pyrococcus horikoshii*, *Thermus thermophilus*, and *Oryza sativa*, with unusually high denaturation temperatures. *Biochemistry*, 47(2):721–30, 2008.

173. S. Kumar, C. J. Tsai, and R. Nussinov. Thermodynamic differences among homologous thermophilic and mesophilic proteins. *Biochemistry*, 40(47):14152–65, 2001.

174. R. B. Greaves and J. Warwicker. Mechanisms for stabilisation and the maintenance of solubility in proteins from thermophiles. *BMC Struct. Biol.*, 7:18, 2007.

175. M. Sadeghi, H. Naderi-Manesh, M. Zarrabi, and B. Ranjbar. Effective factors in thermostability of thermophilic proteins. *Biophys. Chem.*, 119(3):256–70, 2006.

176. G. A. Petsko. Structural basis of thermostability in hyperthermophilic proteins, or "there's more than one way to skin a cat". *Methods Enzymol.*, 334:469–78, 2001.

177. K. A. Henzler-Wildman, M. Lei, V. Thai, S. J. Kerns, M. Karplus, and D. Kern. A hierarchy of timescales in protein dynamics is linked to enzyme catalysis. *Nature*, 450(7171):913–6, 2007.

178. K. B. Zeldovich, I. N. Berezovsky, and E. I. Shakhnovich. Protein and DNA sequence determinants of thermophilic adaptation. *PLoS Comput. Biol.*, 3(1):e5, 2007.

179. S. Kumar, C. J. Tsai, and R. Nussinov. Factors enhancing protein thermostability. *Protein Eng.*, 13(3):179–91, 2000.

180. S. Kumar and R. Nussinov. How do thermophilic proteins deal with heat? *Cell. Mol. Life Sci.*, 58(9):1216–33, 2001.

181. R. Das and M. Gerstein. The stability of thermophilic proteins: a study based on comprehensive genome comparison. *Funct. Integr. Genomics*, 1(1):76–88, 2000.

182. A. H. Elcock and J. A. McCammon. Continuum Solvation Model for Studying Protein Hydration Thermodynamics at High Temperatures. *J. Phys. Chem. B*, 101(46):9624–9634, 1997.

183. A. H. Elcock. The stability of salt bridges at high temperatures: implications for hyperthermophilic proteins. *J. Mol. Biol.*, 284(2):489–502, 1998.

184. L. Xiao and B. Honig. Electrostatic contributions to the stability of hyperthermophilic proteins. *J. Mol. Biol.*, 289(5):1435–44, 1999.

185. C. N. Pace. Single surface stabilizer. *Nat. Struct. Biol.*, 7(5):345–6, 2000.

186. J. R. Cherry and A. L. Fidantsef. Directed evolution of industrial enzymes: an update. *Curr. Opin. Biotechnol.*, 14(4):438–43, 2003.

187. L. De Maria, J. Vind, K. M. Oxenboll, A. Svendsen, and S. Patkar. Phospholipases and their industrial applications. *Appl. Microbiol. Biotechnol.*, 74(2):290–300, 2007.

188. A. Kilara and K. M. Shahani. The use of immobilized enzymes in the food industry: a review. *CRC Crit. Rev. Food Sci. Nutr.*, 12(2):161–98, 1979.

189. K. H. Maurer. Detergent proteases. *Curr. Opin. Biotechnol.*, 15(4):330–4, 2004.

190. A. Schmid, F. Hollmann, J. B. Park, and B. Bühler. The use of enzymes in the chemical industry in Europe. *Curr. Opin. Biotechnol.*, 13(4):359–366, 2002.

191. S. M. Thomas, R. DiCosimo, and V. Nagarajan. Biocatalysis: applications and potentials for the chemical industry. *Trends Biotechnol.*, 20(6):238–42, 2002.

192. J. B. van Beilen and Z. Li. Enzyme technology: an overview. *Curr. Opin. Biotechnol.*, 13(4):338–44, 2002.

193. G. A. Walsh, R. F. Power, and D. R. Headon. Enzymes in the animal-feed industry. *Trends Biotechnol.*, 11(10):424–30, 1993.

194. G. N. Somero. Protein adaptations to temperature and pressure: complementary roles of adaptive changes in amino acid sequence and internal milieu. *Comp. Biochem. Physiol. B Biochem. Mol. Biol.*, 136(4):577–91, 2003.

195. J. A. Raymond and A. L. DeVries. Adsorption inhibition as a mechanism of freezing resistance in polar fishes. *Proc. Natl. Acad. Sci. USA*, 74(6):2589–2593, 1977.

196. P. L. Davies and C. L. Hew. Biochemistry of fish antifreeze proteins. *FASEB J.*, 4(8):2460–8, 1990.

197. Y. Yeh and R. E. Feeney. Antifreeze Proteins: Structures and Mechanisms of Function. *Chem. Rev.*, 96(2):601–618, 1996.

198. M. J. Kuiper, C. J. Morton, S. E. Abraham, and A. Gray-Weale. The biological function of an insect antifreeze protein simulated by molecular dynamics. *eLife*, 4, 2015.

199. D. A. Estell, T. P. Graycar, and J. A. Wells. Engineering an enzyme by site-directed mutagenesis to be resistant to chemical oxidation. *J. Biol. Chem.*, 260(11):6518–21, 1985.

200. L. G. Otten and W. J. Quax. Directed evolution: selecting today's biocatalysts. *Biomol. Eng.*, 22(1–3):1–9, 2005.

201. W. P. Stemmer. Rapid evolution of a protein in vitro by DNA shuffling. *Nature*, 370(6488):389–91, 1994.

202. N. J. Turner. Directed evolution drives the next generation of biocatalysts. *Nat. Chem. Biol.*, 5(8):567–73, 2009.

203. C. Jackel, P. Kast, and D. Hilvert. Protein design by directed evolution. *Annu. Rev. Biophys.*, 37:153–73, 2008.

204. P. N. Bryan. Protein engineering of subtilisin. *Biochim. Biophys. Acta*, 1543(2):203–222, 2000.

205. K. L. Schweiker, A. Zarrine-Afsar, A. R. Davidson, and G. I. Makhatadze. Computational design of the Fyn

SH3 domain with increased stability through optimization of surface charge charge interactions. *Protein Sci.*, 16(12):2694–702, 2007.

206. Y. Bi, J. H. Cho, E. Y. Kim, B. Shan, H. Schindelin, and D. P. Raleigh. Rational design, structural and thermodynamic characterization of a hyperstable variant of the villin headpiece helical subdomain. *Biochemistry*, 46(25):7497–505, 2007.

207. J. M. Schwehm, C. A. Fitch, B. N. Dang, E. B. Garcia-Moreno, and W. E. Stites. Changes in stability upon charge reversal and neutralization substitution in Staphylococcal nuclease are dominated by favorable electrostatic effects. *Biochemistry*, 42(4):1118–28, 2003.

208. G. R. Grimsley, K. L. Shaw, L. R. Fee, R. W. Alston, B. M. Huyghues-Despointes, R. L. Thurlkill, J. M. Scholtz, and C. N. Pace. Increasing protein stability by altering long-range Coulombic interactions. *Protein Sci.*, 8(9):1843–9, 1999.

209. V. G. Eijsink, G. Vriend, B. van der Vinne, B. Hazes, B. van den Burg, and G. Venema. Effects of changing the interaction between subdomains on the thermostability of Bacillus neutral proteases. *Proteins*, 14(2):224–36, 1992.

210. A. L. Fink. Protein aggregation: folding aggregates, inclusion bodies and amyloid. *Fold. Des.*, 3(1):R9–23, 1998.

211. J. M. Finke, M. Roy, B. H. Zimm, and P. A. Jennings. Aggregation events occur prior to stable intermediate formation during refolding of interleukin 1β. *Biochemistry*, 39(3):575–83, 2000.

212. M. A. Ceruso, A. Amadei, and A. Di Nola. Mechanics and dynamics of B1 domain of protein G: role of packing and surface hydrophobic residues. *Protein Sci.*, 8(1):147–60, 1999.

213. L. C. Tisi and P. A. Evans. Conserved structural features on protein surfaces: small exterior hydrophobic clusters. *J. Mol. Biol.*, 249(2):251–8, 1995.

214. B. Van den Burg, B. W. Dijkstra, G. Vriend, B. Van der Vinne, G. Venema, and V. G. Eijsink. Protein stabilization by hydrophobic interactions at the surface. *Eur. J. Biochem.*, 220(3):981–5, 1994.

215. O. R. Veltman, G. Vriend, H. J. Berendsen, B. Van den Burg, G. Venema, and V. G. Eijsink. A single calcium binding site is crucial for the calcium-dependent thermal stability of thermolysin-like proteases. *Biochemistry*, 37(15):5312–9, 1998.

216. O. R. Veltman, G. Vriend, F. Hardy, J. Mansfeld, B. van den Burg, G. Venema, and V. G. Eijsink. Mutational analysis of a surface area that is critical for the thermal stability of thermolysin-like proteases. *Eur. J. Biochem.*, 248(2):433–40, 1997.

217. M. Sandgren, P. J. Gualfetti, A. Shaw, L. S. Gross, M. Saldajeno, A. G. Day, T. A. Jones, and C. Mitchinson. Comparison of family 12 glycoside hydrolases and recruited substitutions important for thermal stability. *Protein Sci.*, 12(4):848–60, 2003.

218. A. Jaramillo, L. Wernisch, S. Hery, and S. J. Wodak. Automatic procedures for protein design. *Comb. Chem. High Throughput Screen*, 4(8):643–59, 2001.

219. S. M. Lippow and B. Tidor. Progress in computational protein design. *Curr. Opin. Biotechnol.*, 18(4):305–11, 2007.

220. H. Madaoui, E. Becker, and R. Guerois. Sequence search methods and scoring functions for the design of protein structures. *Methods Mol. Biol.*, 340:183–206, 2006.

221. S. Park, X. Yang, and J. G. Saven. Advances in computational protein design. *Curr. Opin. Struct. Biol.*, 14(4):487–94, 2004.

222. M. Rosenberg and A. Goldblum. Computational protein design: a novel path to future protein drugs. *Curr. Pharm. Des.*, 12(31):3973–97, 2006.

223. P. S. Shah, G. K. Hom, S. A. Ross, J. K. Lassila, K. A. Crowhurst, and S. L. Mayo. Full-sequence computational design and solution structure of a thermostable protein variant. *J. Mol. Biol.*, 372(1):1–6, 2007.

224. M. W. Pantoliano, M. Whitlow, J. F. Wood, S. W. Dodd, K. D. Hardman, M. L. Rollence, and P. N. Bryan. Large increases in general stability for subtilisin BPN' through incremental changes in the free energy of unfolding. *Biochemistry*, 28(18):7205–13, 1989.

CHAPTER 5

Protein Dynamics

5.1 INTRODUCTION

From the end of the 19th century until the middle of the 20th century, scientists perceived proteins as rigid entities, following Emil Fischer's 'lock and key' model [1]. The model originated in Fischer's attempt to explain the substrate specificity of proteins acting as enzymes. He posited that specificity is the result of rigid geometric complementarity between the three-dimensional structures of the enzyme and of its substrate. The static view of proteins changed when Daniel Koshland, who also studied enzyme specificity, proposed the 'induced fit' model [2]. Koshland suggested that an enzyme undergoes limited conformational changes to create a better fit between its structure and that of its substrate[*1]. Although the current view of protein behavior is more complex than Koshland's, numerous studies have demonstrated that different structural changes take place in proteins [4-6], thus establishing these macromolecules as dynamic entities. Theoretically speaking, this realization is hardly surprising; the basic polymeric structure of proteins consists of many potential hinges of motion. These are mainly backbone and side chain single bonds, the movement around which is limited only by steric hindrance. Indeed, protein dynamics is characterized by a diverse set of atom and group motions, within a large range of time (10^{-15} to 10^4 sec), amplitude (0.01 to 100 Å), and energy (0.1 to 100 kcal/mol) [6-9] (Table 5.1).

Most importantly, many of the inherent motions in proteins occur on the timescales of central biochemical processes [8-10]:

1. Vibrations and local motions (10^{-15} to 10^{-10} sec) – correspond to chemical events that occur during enzyme-mediated catalysis. These events include:

   - The making or breaking of covalent bonds (10^{-14} to 10^{-10} sec)

   - Formation of hydrogen bonds (10^{-12} to 10^{-10} sec)

   - Transfer of electrons, protons or hydride ions between chemical groups (10^{-12} sec = 1 ps)

2. Motions of large side chains, secondary elements, and domains (10^{-9} to 10^{-3} sec) – correspond to the following:

   - Proton transport (10^{-9} to 10^{-4} sec)

[*1]In fact, the adaptability of proteins to their bound ligands via conformational changes was already suggested in 1950 by Karush, for serum albumin [3], but the exact mechanism underlying this adaptability was slightly different from the mechanism proposed, as described below.

- Electron tunneling (10^{-9} to 10^{-4} sec)
- Water structure reorganization (10^{-8} sec)
- Ligand binding (10^{-8} to 10^1 sec)
- Local denaturation (10^{-5} to 10^1 sec)
- Allostery (10^{-5} to 1 sec)[*1]

The compatibility between the timescales of protein motions and biological processes implies that, as most protein scientists today agree, **the inherent dynamics of proteins is important for their function** [11,12]. Indeed, when the dynamics of a protein is slowed down through temperature reduction, the protein's activity diminishes as well. Most of the more detailed evidence connecting protein dynamics and function is based on observations carried out on relatively long timescales. For example, in many enzymes the rate of catalysis is set by conformational changes needed for the correct positioning of catalytic residues with respect to the substrate, where the changes occur at the secondary structure level [10]. Another example is substrate and ligand-binding (or releasing) processes, which require conformational changes leading to the opening or closing of protein segments acting as gates. Indeed, many enzymes tend to undergo dynamic changes that close their binding sites upon substrate binding [13]. This closure completely or partially prevents the solvent from accessing the binding site. In doing so, it strengthens protein-ligand electrostatic interactions that may be important for catalysis, and also reduces the solvent's ability to interact with the substrate at the expense of binding site groups (competing interactions). Although all of the above represent long-range dynamics, short motions in proteins, from concerted vibrations of covalent bonds to motions of small side chains, are also known to contribute to the overall dynamics and function of the protein.

TABLE 5.1 **Timescales of motions in proteins.** The data are taken from [8–10].

Type of Movement	Example	Timescale (sec)
Local motions	• Overall	10^{-15}–10^{-6}
	• Bond vibration	10^{-15}–10^{-13}
	• Elastic vibration of globular region	10^{-12}–10^{-11}
	• Methyl group rotation around connecting bond to molecule	10^{-12}–10^{-9}
	• Rotation of surface side chains	10^{-11}–10^{-10}
	• Hinge bending at domain interfaces	10^{-11}–10^{-7}
	• Loop movement FV	10^{-9}–10^{-6}
	• α-helix formation	10^{-8}–10^{-7}
Rigid body motions	• Helix, domain, subunit	10^{-9}
Motions of large domains		10^{-6}–10^{-3}
Protein folding		10^{-6}–10^{-4}

[*1] See details in Subsection 5.3.2.1 below.

The *types* of motions that can be found in proteins are also diverse, and include hinge motions, rotations, translations, and even the folding and unfolding of secondary and super-secondary elements. Hinge motions (Figure 5.1), which occur on the ns (10^{-9} sec) timescale, are particularly common in proteins, and usually involve proline or glycine residues (e.g. [14]). Moreover, in many cases, this type of motion can be assigned a specific function. For example, in *motor proteins* such as myosin, the hinge motion is utilized for the conversion of chemical energy (ATP) into mechanical force. In many enzymes, this type of motion is used for the transduction of allosteric changes, e.g., by opening or closing active site gates, thus changing the site's availability to the substrate [15]. Indeed, **protein hinges have been shown to be evolutionarily conserved, which confirms their general functional importance** [16]. It is difficult to demonstrate all protein-related motions in the 'static' format of this book. For a more extensive and visually clear presentation of protein motions, we recommend the *molmovdb database* [17], which can be easily accessed via the Internet[*1]. The database provides information on the various motions in their biological context, as well as animations illustrating the motions using experimentally-determined structures of proteins. A more focused view can be found in the accompanying paper [4].

The field of protein dynamics focuses on two main topics: folding kinetics and the dynamics of the folded state. The interest in a protein's folded state (or states) is obvious, as the folded state is the biologically active state. Studies in this field focus on dynamics in the μs–ms (10^{-6} to 10^{-3} sec) range, for two reasons. First, this range corresponds to biological processes such as enzyme-mediated catalysis, protein-protein interactions, and signal transduction [6,18]. Second, motions within this range can be studied in a straightforward manner with existing methods [6,18] (see below). Folding kinetics may not relate to protein structure

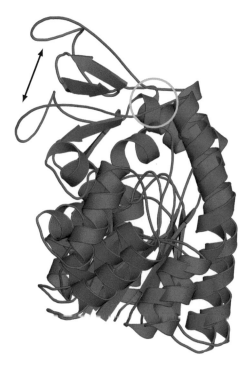

FIGURE 5.1 **Hinge motion in adenylate kinase.** The motion of the loop in the upper-left region is illustrated by superimposition of the two conformations of the proteins (PDB entries 2eck, 4ake). The hinge region is circled.

[*1]http://www.molmovdb.org. The database is maintained by Mark Gerstein's group at Yale University.

or function[*1]; however, an understanding of this topic can provide further insights into the behavior of proteins, and can also contribute towards improving the performance of computational algorithms attempting to predict protein structure. In addition, characterization of the folding process may assist in the understanding (and perhaps even the treatment) of pathologies that are related to protein misfolding (see Box 5.1).

5.2 PROTEIN FOLDING

5.2.1 Kinetic aspects

5.2.1.1 Levinthal's paradox and energy landscape theory

Proteins fold at different rates, and these rates are distributed over a range of eight orders of magnitude [20]. Studies show that the folding rate correlates not only with protein size [21]*2, but also with the topology of the protein's native fold [22]. That is, fast-folding proteins, i.e., proteins with a folding rate on a nanosecond timescale, tend to have larger proportions of local secondary elements (α-helices, β-turns) compared with slower-folding proteins, whereas slow-folding proteins tend to have larger proportions of global elements, namely β-sheets [22]*3. Nevertheless, proteins with the same topology may substantially differ from each other in their folding rates [23]. This implies that the rate is affected by additional factors, such as the connectivity of secondary elements.

The thermodynamic approach discussed in Chapter 4 requires a simple consideration of protein folding as a two-state process (folded and unfolded). In many proteins, especially the small ones, such a simplistic view may seem justifiable, mainly since protein folding is a highly cooperative process. However, in others the folding may be much more gradual [24]. This is particularly true in proteins of 100 residues or more, which constitute ~90% of all cellular proteins [25]. Theoretically, the gradual nature of protein folding is hardly surprising, as the folding process includes numerous intermediate states, resulting from the large number of degrees of freedom in the protein. This very large number of degrees of freedom seems to be incompatible with the extremely high folding rate measured in most proteins. This problem was presented in 1968 by Cyrus Levinthal, and was named 'Levinthal's Paradox' [26]. The paradox can be presented as follows.

Assuming that the protein folding process involves the free sampling of all possible conformations of the protein (i.e., of each residue independently), and that each residue has at least 3 states, then the folding of a 100-residue protein is expected to sample $3^{100} = 5 \times 10^{47}$ conformations. Now, if we assume that it takes the protein 1 ps to sample a single conformation, then the time it takes to sample all possible conformations in order to find the right one should be $3^{100} \times 10^{-12}$ sec $= 5 \times 10^{35}$ sec $= 1.6 \times 10^{28}$ years. This period of time is about 10^{18} times longer than the age of the universe, and therefore cannot be reconciled with the extremely short period of time it takes most proteins to fold in real life, i.e., between one ms and a second [27-29]. Hence, the paradox.

[*1] As Anfinsen has demonstrated, the structure of the native state essentially depends on the sequence of the protein, and not on its folding process [19].

[*2] It has been shown that the folding time increases exponentially with the square root of the number of amino acids [21].

[*3] Helices and turns are considered to be local because they typically extend along a short segment of amino acids. Conversely, β-sheets are more global because they bring together segments that are separated from each other in the sequence.

So, how can this paradox be settled? The obvious answer is that **proteins do not actually sample all possible conformations to get to the right one. Rather, they fold in a cooperative manner, in which each step further limits the folding possibilities of subsequent steps.** Indeed, studies carried out in the 1980s and onward by Dill, Wolynes, Bryngelson and others confirm that the folding process, which involves the formation of residue-residue interactions and compact secondary structures [30]*1, considerably lowers the number of conformations available to the protein, as it proceeds [31]. Furthermore, studies of protein-protein complexes show that folding cooperativity depends not only on backbone connectivity, but also on tertiary contacts [32]. Protein folding cooperativity is depicted graphically by the widely accepted [33–37] *energy landscape theory*, which describes the folding process as an energy-entropy funnel (Figure 5.2). Each of the conformations involved in the folding process is represented as a single point on a hyper-plane, plotted according to its energy and entropy values. Together, the conformations form the funnel shape. The plot can be presented in two or three dimensions, depending on the amount of information available. In any case, the funnel shape calls attention to the following key aspects of folding:

1. **The folding process involves a decrease of both energy and entropy of the protein.** The folding is completed when the lowest energy conformation of the protein is reached. Thus, the protein's native structure constitutes the global minimum of the landscape. This is because the native structure is able, for the most part, to avoid energy 'frustrations', which result from unfavorable atom-atom interactions and other energy conflicts*2.

2. **The folding process involves many local energy minima separated by high-energy barriers*3.** This property is reflected in the ragged surface of the funnel. Each intermediate constitutes a barrier that has to be overcome in order for the folding process to continue and reach completion (i.e., the global energy minimum). The barriers constitute a kinetic obstacle; they do not change the overall free energy difference between the folded and unfolded states, but rather the time needed for the protein to fold. The inverse dependency of the folding rate on the height of the barriers (Ea) is exponential, as described by the Arrhenius equation that is mentioned in Chapter 1 (Equation (1.1)). Each of the local minima in the energy landscape involves favorable, yet non-native interactions. As such, these minima constitute potential *kinetic traps* for the folding process. Thus, the folding process is burdened by two needs: (*i*) overcoming the energy barriers, and (*ii*) avoiding getting trapped in local minima. It seems that evolution has selected protein sequences that have relatively smooth energy landscapes, which allow them to fold in a physiologically relevant amount of time [30].

3. **The folding process may take different paths**; this is also reflected in the ragged form of the funnel.

---

*1 As explained in Chapter 2, the formation of secondary structures creates intramolecular hydrogen bonds that are energetically more favorable than the backbone-water hydrogen bonds dominating the unfolded state. The favorable residue-residue interactions formed during folding act more locally to optimize the structure within the limitations already set by the secondary elements.

*2 This is following Bryngelson and Wolynes' *principle of minimal frustration* [34].

*3 See Chapter 3 for definition of energy minima and barriers. Here, high-energy barriers mean barriers that are over $1kT$ (~0.6 kcal/mol).

(a)

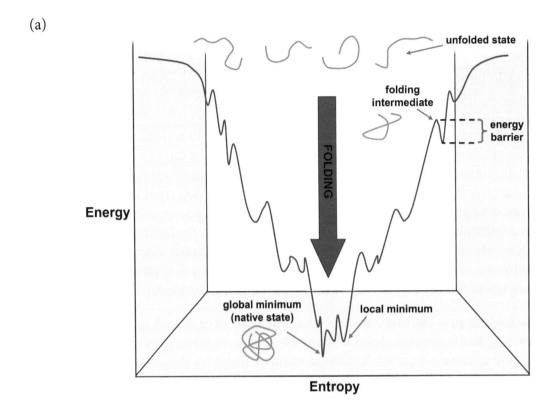

(b)

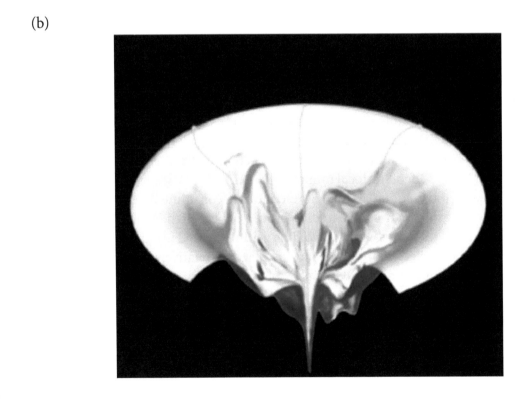

FIGURE 5.2 **The folding energy landscape in two (a) and three (b) dimensions.** The green shapes in (a) represent the polypeptide chain of the folding protein. The funnel shape of the curve reflects the simultaneous decrease in energy and entropy of the protein during its folding. (b) Adapted from [38].

5.2.1.2 Folding models and mechanisms

The energy landscape theory provides a general solution to the protein folding paradox. In order to understand this solution thoroughly, one has to address the *kinetics* of the process, that is, to follow the time-dependent change in protein structure, including the high-energy intermediate conformations, while the protein folds. This has always been a problem for protein scientists, as many of the intermediate conformations, being of similar energy, inter-convert very quickly, i.e., appear for only very short periods of time. Classical experimental methods usually provide time-averaged information rather than detailed data on individual fluctuations, especially when the latter are extremely short-lived. In recent decades new and sophisticated methods have been developed, providing partial or complete solutions to this problem. These include spectroscopic methods, such as nuclear magnetic resonance (NMR) spectroscopy, electron paramagnetic resonance (EPR) spectroscopy, and laser-based methods, as well as others [20] (see Section 5.4 below). In addition, scientists learned how to regulate the conditions of the measurements so as to slow down the folding rate, to an extent that enabled them to trap individual high-energy intermediates. Taken together, these developments have enabled scientists to gain ample knowledge about folding, although not to the extent of having a single description of the process. Instead, several models have been proposed to describe the kinetics of folding [20,39] (Figure 5.3). For example, the *'hydrophobic collapse'* model is based on the idea that the hydrophobic effect is the major driver of protein folding. The model proposes that proteins fold in two steps. In the first, the hydrophobic effect induces the collapse of nonpolar parts of the protein to form a compact, partially folded conformation. In the second step, secondary structures are formed, and the partially folded structure samples the limited conformational space available to it, which allows it to complete the formation of the proper tertiary contacts that characterize the native fold. Electrostatic interactions are particularly important in this step, as they determine the specific conformation of the native state. Supporting this idea is the observation that when these interactions are perturbed by lowering the pH, the protein tends to remain in the partially folded state [40]. During the protein folding process, different thermodynamic intermediates are formed. One of these, which is often mentioned in folding studies and seems to appear towards the end of the process, is called the *'molten globule'* [41]. This is a compact conformation with a secondary structure similar to that of the native fold. However, it lacks proper tertiary organization, to the extent of exposure of nonpolar residues to the aqueous solvent. In particular, the side chains tend to be loosely packed and therefore have a high degree of freedom [42]. These abnormalities make the molten globule state less stable than the native state, and it therefore tends to change into the latter at the end of the folding process.

One of the supporting studies of the hydrophobic collapse model comes from molecular dynamics (MD) simulations, carried out so as to follow the folding of a 36-residue sequence taken from the protein villin [43]. The simulation managed to cover 0.3 ms of the folding process, during which the unfolded structure collapsed quickly (after 20 ns) into a more compact form. After the collapse, the resulting structure underwent fine-tuning for about 5 μs, until the most stable conformation was achieved. Interestingly, the rate-limiting step of the folding process was the formation of an interaction network between aromatic residues in the core of the protein. Compared with simulation results, experimental support for the hydrophobic collapse model is much more difficult to obtain, as the collapsed intermediate is hard to isolate.

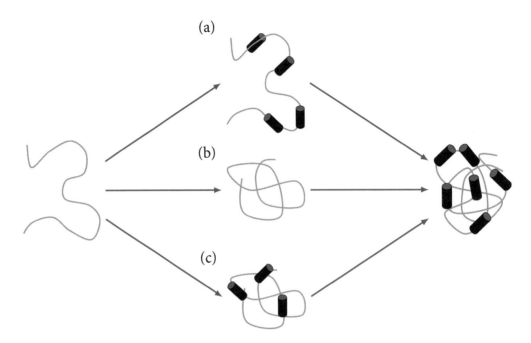

FIGURE 5.3 **Proposed models for protein folding.** (a) The framework model. (b) The hydrophobic collapse model. (c) The nucleation–condensation model. The green shapes represent the polypeptide chains whereas the black cylinders are α-helices, representing secondary structures.

Unlike the hydrophobic collapse model, the *'framework'* (a.k.a. *diffusion-collision* [44]) model proposes a more gradual process of folding, following the hierarchy of protein structure [20,45,46]. That is, the local (secondary) structures are formed first in the different regions of the chain, then complex folds, domains, and finally the complete tertiary fold. The tertiary contacts between residues stabilize the secondary elements, which are of marginal stability in isolation [20]. The framework model is attractive in the sense that, by allowing secondary elements to be formed first, it achieves the following:

1. Enthalpic optimization by creating local, energetically-favorable interactions.

2. A chance for the protein to 'pay' the entropic price of folding in two 'installments'.

3. A plausible solution to Levinthal's paradox: initial formation of local structures eliminates the need to sample irrelevant conformations later. When the second step commences (i.e., global folding), the protein has a much smaller conformational space to sample.

This model has been used to describe the folding of some proteins, such as myoglobin [47,48], barnase [49], and the λ repressor fragment [50]. In addition, an NMR study of the α-helical protein BBL shows that when the protein is heated, the first interactions to be lost are tertiary contacts between side chains, whereas those stabilizing the secondary elements in the protein are lost only later [51]. The latter observation provides strong support to the framework model, and also implies that there is no strong coupling between tertiary and secondary contacts [24]. Another finding that supports the framework model comes from the aforementioned studies of Plaxco, Simons and Baker, who demonstrated a correlation between the number of local contacts in proteins, such as those formed in α-helices and β-turns, and their overall folding rates [22].

Finally, the *nucleation–condensation* model [52] proposes that both hydrophobic collapse and secondary-element formation take place simultaneously during folding [53]. As in the diffusion–collision model, here too the secondary elements are stabilized by the formation of the protein's nonpolar core and tertiary contacts. However, in the nucleation–condensation model, stabilization occurs while the secondary elements are being formed, and not later. Considering the marginal stability of secondary elements in isolation, it indeed seems more realistic that they are stabilized gradually rather than after they are formed in solution. The nucleation–condensation model has been used to describe the folding of some proteins, such as chymotrypsin inhibitor 2 (CI2) [54].

The fact that each of the different models mentioned above can be observed in some, but not all, proteins implies that **there is no single, universal folding mechanism, but rather a collection of possible mechanisms that may be used.** In fact, the folding mechanism can vary even across proteins that are members of a single group. For example, in the *homeodomain superfamily* of DNA-binding proteins, both nucleation–condensation and diffusion–collision mechanisms have been observed [55,56]. To make this problem even more complex, some proteins seem to fold via hybrid mechanisms, as in the case of the small $\alpha\beta$ protein domain PTP-BL PDZ2 (reviewed in [57]).

The 'preference' of a protein to use a certain folding mechanism may depend on various qualities. One quality seems to be the secondary structure of the protein. For example, in proteins dominated by α-helices the folding is often hierarchical, whereas proteins containing more global folds may use other folding mechanisms [58]. Another quality is the structural context of the folding unit. Historically, most studies of protein folding have been carried out on single-domain proteins, yet most proteins (over 70% in eukaryotes) contain more than one domain. Indeed, interactions between the folding units may affect both the mechanism and rate of the folding. This issue has been investigated through comparison of thermodynamic and kinetic data from different multi-domain proteins [58]. The results suggest that in such proteins the folding rate is enhanced by favorable nonpolar interactions between the domains, and that this effect is more pronounced in domains that are larger and more tightly packed [59]. The topology of the protein is also important; when the domains are separated by short limiting linkers, their interactions with each other are not completely satisfied, which makes their effect on the folding rate smaller [60].

In conclusion, protein folding exhibits a wide range of paths, mechanisms, and rates, which depend on multiple parameters, such as protein composition and the folding conditions [61]. However, regardless of the exact mechanism, the folding process always follows the energy landscape theory. That is, folding is always accompanied by a decrease in the number of conformations that must be sampled by the protein, thus allowing the system to avoid Levinthal's paradox.

5.2.2   *In vivo* folding

5.2.2.1   *In vivo* factors that complicate folding

The rate of protein translation in biological cells is two to eight amino acids per second [62], which means that folding occurs while the polypeptide chain is being built. Small proteins can fold independently, following the Anfinsen model. Large proteins, however, particularly those having a complex structure stabilized by long-range interactions or a large number of domains, may encounter difficulties during their folding [63]. Other factors that make *in vivo* folding even more difficult include the following [63]:

1. The dense cytoplasm – The excluded volume might lead to non-specific interactions between unfolded proteins, which in turn might associate into aggregates of different types (Figure 5.4) [63]. The association is mediated by interactions between nonpolar residues of adjacent, misfolded, polypeptide chains. In correctly folded proteins, the same nonpolar residues are involved in intramolecular interactions, which prevent them from interacting with other chains. The aggregates may be energetically more stable than the folded monomers (Figure 5.4). However, they are not as active (if at all) and may even be toxic (see Box 5.1).

2. The translation process – The internal organization of the ribosome does not allow folding to proceed beyond the formation of α-helices [64,65], and this prevents the last 40 to 60 residues of the chain from participating in long-range interactions [66]. Thus, proper folding is made possible only after the entire protein (or domain) is completely translated. This would not pose a problem if the translation process were very fast. However, since it is rather slow (see above), the nascent chain spends a considerable amount of time in partially folded states, which increases the chances of both misfolding and aggregation.

A similar problem may arise in folded proteins, which, according to studies, undergo multiple and frequent folding-unfolding episodes during their lifetimes [67].

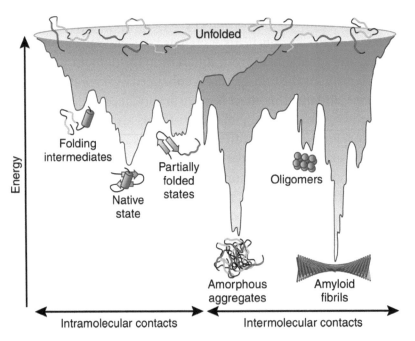

FIGURE 5.4 **The energy landscape of protein folding and aggregation.** The purple surface represents the folding process in which single-chain proteins, via intramolecular contacts, form their native structure. It is equivalent to the scheme shown in Figure 5.2. The pink area shows a similar process, but one that involves intermolecular contacts between different polypeptide chains, and results in amorphous aggregates, oligomers, or amyloid fibrils (see Box 5.1). Note that the two energy landscapes overlap in certain areas. That is, aggregates and amyloid fibrils may be formed during *de novo* folding, but also by destabilization of the native state into partially folded states. Cell-toxic oligomers may occur as off-pathway intermediates of amyloid fibril formation. The figure (and parts of the caption) are taken from [63].

BOX 5.1 PATHOLOGIES THAT RESULT FROM PROTEIN MISFOLDING

I. Protein misfolding leads to disease

Proteins, being dynamic entities, normally undergo reversible structural changes. These changes are usually within a limited range, but from time to time may amount to partial unfolding, which is followed by rapid refolding. As this unfolding-refolding process is reversible, in most cases it does not pose a threat to either the existence or the function of the protein. However, some factors may lead proteins to undergo a permanent loss of structure, and when this occurs in a large number of proteins or protein copies, the entire organism may be in danger. The loss of structure in a protein might harm the cell, tissue or organism in two different ways [68]. The first is straightforward; loss of structure leads to loss of function, following the Anfinsen paradigm [19]. The more central the unfolding protein to the function of the cell, the harsher the damage to the organism. The well-known disease *cystic fibrosis*, for example, results from the misfolding of CFTR, a protein functioning as a chloride (Cl^-) channel. The loss of CFTR function interferes with the body's ability to efficiently secrete fluids and salts, resulting in various physiological impairments, such as blockage of pancreatic secretions, accumulation of mucus in respiratory pathways, and a decrease in salt absorption from sweat [69].

Loss of protein structure may also harm the organism in another, less obvious way. Proteins undergoing this process tend to have increased hydrophobicity, which makes them aggregate and precipitate. The precipitates are toxic to cells, although the exact mechanism of the toxicity is not entirely clear. The aggregation-induced toxicity can be referred to as a 'gain of function' type of misfolding impairment. In recent decades a heterogeneous group of seemingly unrelated pathologies, which damage different organs and have different clinical manifestations, have all turned out to involve protein aggregation in the form of fine fibers (i.e., fibrils, Figure 5.1.1), followed by precipitation of these fibers [70–72]. Today, more than 40 such pathological conditions are known in humans [70]. The first scientists studying this phenomenon were under the wrong impression that the precipitates were made of starch, and therefore referred to them as 'amyloids'*a. As in the case of many other historic names, this one has also remained in use, even after it was found that the precipitates are in fact made of protein. Accordingly, amyloid-related pathologies are referred to, even today, as 'amyloidoses'. It should be mentioned, though, that the precipitates may also contain other components, such as metal ions and sugar derivates [73].

---

*a*Amylum* is the Latin word for starch.

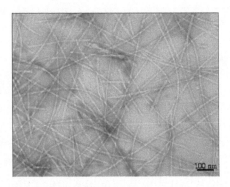

FIGURE 5.1.1 **A transmission electron micrograph of amyloid fibrils.** The figure was taken from [74].

Despite the great variance between the different amyloidoses, they can be separated into two general groups:

1. **Neuropathic** [75]. This group includes the following neurodegenerative diseases:

   I. *Alzheimer's disease (AD)* is the most common form of progressive dementia in the elderly, and of neurodegenerative disease in general [76]. AD is characterized by the aggregation and precipitation of two protein species [77]: (1) a peptide called *'amyloid-β' (Aβ)*, which precipitates in the spaces between neurons. Aβ is the degradation product of the transmembrane amyloid-α protein; (2) *Tau*, a microtubule-associated protein that in AD forms neurofibrillary tangles within neurons [78]. Another protein associated with AD is *apolipoprotein E*, although it does not seem to cause the disease [79]. Rather, its ε4 allele is a risk factor for developing late-onset AD.

   II. *Parkinson's disease (PD)* is a motor disorder that is common among the elderly (but can also hurt young people), resulting from damage in a brain region called the *'substantia nigra'* [80]. The damage leads to gradual loss of nerve cells that secrete the neurotransmitter *dopamine*. PD is characterized by the aggregation and precipitation of the protein *α-synuclein*.

   III. *Huntington's disease (HD)* is a genetic disease characterized by involuntary movements, dementia, and emotional problems. HD is characterized by the aggregation and precipitation of the protein *huntingtin*, which in its pathological form includes a large number of glutamine residues.

   IV. *Amyotrophic lateral sclerosis (ALS)* is a genetic disease characterized by the death of motor nerve cells in the brain, brainstem, and spinal cord, resulting in fatal paralysis. ALS is characterized by the aggregation and precipitation of the protein superoxide dismutase (SOD), normally involved in the antioxidation of harmful oxygen radicals.

   V. *Creutzfeldt-Jakob disease (CJD)* – An either congenital or acquired disease, manifesting as a variety of disorders, including progressive motor dysfunction, cognitive impairment, and cerebral ataxia. CJD is characterized by the aggregation and precipitation of the protein *prion*.

2. **Non-neuropathic.** This group includes many diseases, such as *cataract*, in which the aggregating protein is *γ-crystalline*, and *type II diabetes*, which involves the protein *amylin*. In such diseases, the aggregating protein may harm the cells or the tissue directly (i.e., as a result of its toxicity), or indirectly, by obstruction. For example, accumulation of the protein precipitate in the tongue may lead to difficulties in swallowing, whereas accumulation in the joints is likely to cause pain during motion.

The protein precipitates in amyloidoses may appear in two major forms. In the first, the protein precipitates accumulate in the extracellular environment, in which case they are referred to as 'plaques'. This happens, e.g., with Aβ peptides in Alzheimer's disease. In the second form the precipitates accumulate inside the cell, in which case they are often called 'inclusions' [81]. This happens with α-synuclein in Parkinson's disease, in which case the precipitates are referred to as 'Lewy bodies' [80] (Figure 5.1.2c).

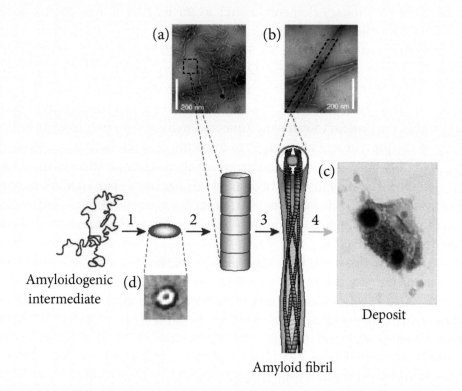

FIGURE 5.1.2 **A schematic representation of the general mechanism of aggregation to form amyloid fibrils (taken from [82]).** Unfolded or partially unfolded proteins associate with each other to form small, soluble aggregates that undergo further assembly into protofibrils or protofilaments (a) and then mature fibrils (b) (electron micrographs with 200 nm scale bars; taken from [83]). The fibrils often accumulate in plaques or other structures, such as the Lewy bodies associated with Parkinson's disease (c). Some of the early aggregates seem to be amorphous or micellar in nature, although others form ring-shaped species with diameters of approximately 10 nm. (d) Taken from [84]. Thus, the entire process of plaque formation includes the following steps; each is marked by a numbered arrow: (1) aggregation of the amyloidogenic protein to form a ring-shaped aggregate; (2) assembly of the aggregate into a protofibril; (3) further assembly of the protofibril into protofilament; and (4) accumulation of the protofilaments as plaques, Lewy bodies, or other structures.

II. The structures of amyloid aggregates

Amyloid precipitates were originally identified using light microscopy; this was possible because of the tendency of the protein precipitates to interact with dyes such as *Congo red*. This tendency also allowed scientists to find to cellular or tissue localization of the precipitates. Later, when transmission electron microscopy was developed, detailed characterization of the amyloid fibril itself became possible. In contrast to other cellular aggregates characterized by amorphous structures, amyloid precipitates were found to have an organized structure (Figure 5.1.2): each fibril consisted of 2 to 6 *protofilaments*, of diameter 20 to 50 Å each [85]. The next step, i.e., structural determination of the proteins in the amyloid fibril, required higher-resolution methods, such as X-ray diffraction and solid-state NMR. These have revealed that while protein fibrils associated with different pathologies differ from one another; the compositions of their secondary elements, as well as the lengths of their chains, have similar characteristics, the most pronounced of which is the cross-β structure at the core of each protofilament[*a] [74] (Figure 5.1.3a). A high-resolution structure of the cross-β structure was solved only in 2005, by David Eisenberg's group [86]. All these data led to the realization that the formation of amyloid fibrils results from *misfolding*, i.e., the partial loss of the native structure [71]. Misfolding leads to externalization of nonpolar residues within the protein towards the aqueous environment, thus lowering the water solubility of the protein and allowing it to form the characteristic aggregates. Again, the aggregates organize as protofilaments, which then assemble to form the final fibril structure [72,82,87]. The cellular toxicity of amyloids has been associated for a long time with the insoluble fibrils. However, evidence that has accumulated in recent years suggests that the toxicity may result from soluble oligomers of the proteins [88], which are released from the fibrils [89].

Amyloids occur also in bacteria, although, in contrast to eukaryotic amyloids, bacterial amyloids are functional and do not pose a threat to the cell in which they are produced [90]. The functions carried out by bacterial amyloids are diverse. For example, in *Staphylococcus aureus*, amyloids formed by phenol-soluble modulins (PSMs) play a part in the virulence of the bacterium [91]. The atomic structure of bacterial amyloid fibrils has until recently been unknown. Meytal Landau and co-workers used X-ray crystallography to determine the high-resolution structure of fibrils formed by the 22-peptide PSMα3, the most cytotoxic member of the PSM family [92]. Surprisingly, the fibrils had a cross-α, rather than a cross-β, structure (Figure 5.1.3b). Still, the helices stacked as amphipathic sheets along the fibril axis, similar to the stacking of the cross-β elements in the eukaryotic amyloids. Mutations confirmed that the cross-α fibrillar form of PSMα3 is involved in cytotoxicity. It is still unclear whether the cross-α structure appears also in some of the eukaryotic, disease-associated amyloid fibrils. However, several such proteins contain α-helices in their soluble, monomeric or prefibrillar intermediary states (e.g. [93]). This supports the suggestion mentioned above, that the soluble states, rather than the fibrillar state, are responsible for cytotoxicity in eukaryotic amyloids.

[*a]The β structure is perpendicular to the fibril's axis.

(a)

(b)

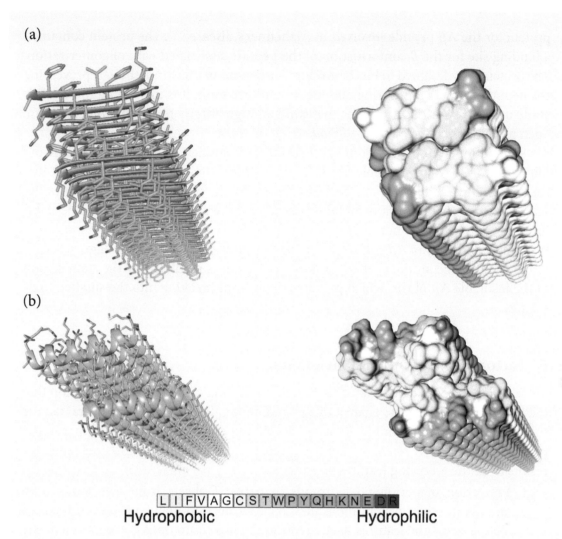

L I F V A G C S T W P Y Q H K N E D R
Hydrophobic   Hydrophilic

FIGURE 5.1.3 **The structures of amyloid fibrils** (courtesy of Meytal Landau and Einav Tayeb-Fligelman). (a) The cross-β structure of the amyloid forming peptide KLVFFA from amyloid β. In the structure, the β strands are positioned perpendicularly to the fibril axis (PDB entry 3ow9). The image on the left shows the supersecondary structure of the fiber, whereas the right image shows the surface of the structure, colored by the Kessel-Ben-Tal Scale of hydrophobicity [94]. (b) The cross-α structure of PSMα3 from *Staphylococcus aureus* [92] (PDB entry 5i55) with the α-helices also being perpendicular to the fibril axis, reminiscent of the classical cross-β architecture. The supersecondary structure and surface properties of the cross-α fiber are shown as in (a), demonstrating the amphipathic nature of the structure.

III. Therapeutic implications

The establishment of the misfolding theory had important implications, one of which was the possibility of treating related diseases by stabilizing the native structures of the corresponding proteins. One of the strategies proposed in this respect was to use antibodies that are specific to the native structure of the aggregating protein, thus stabilizing it [95]. A slightly different strategy is to stabilize the monomeric form of the protein, regardless of whether it has a native structure, in order to prevent aggregation. Such a strategy was used by Hoyer and colleagues, who utilized a specific binding protein for the Aβ peptide involved in Alzheimer's disease [96]. The protein contained a binding site for the β conformation of the peptide, associated with oligomerization. The protein was designed to bind the nonpolar regions of this form, thereby preventing the peptide from oligomerizing, despite its conformation. It is interesting to note this study used non-antibody proteins that still had high affinity to the monomeric form. Such proteins are referred to as 'affibodies'.

As mentioned in the previous subsection, some studies implicate soluble amyloid oligomers rather than insoluble fibrils in amyloid toxicity. Their findings suggest that the therapeutic strategy for amyloidoses should be the opposite of the strategies described above. That is, the amyloid fibers instead of their components should be stabilized. In support of this notion, Eisenberg and coworkers discovered small molecules that reduced toxicity of Aβ without reducing the number of its insoluble aggregates [97]. It appears that the antitoxic effect of these molecules is mediated by the stabilization of the insoluble Aβ fibrils, which prevents them from breaking into the smaller, toxic entities.

IV. Factors causing protein misfolding

The long list of pathologies related to protein misfolding may leave the impression that the human body is helpless against this phenomenon. In fact, there are several factors that should prevent misfolding. First, we should remember that to begin with, cellular proteins have a strong tendency not to aggregate [98]. Studies show that this tendency results from the following features of proteins:

1. Protection of peripheral β-strands in native β-sheets from interacting with strands in other protein molecules. Such protection may be achieved through simple structural features, such as formation of a continuous β-sheet, which may fold into a β-barrel. However, analyses of multiple protein structures (e.g., [99]) reveal more profound solutions, at the sequence level — that is, evolutionary selection of sequences that distort peripheral β-strands, and thereby reduce their propensity to interact with other strands. Such sequences may include inward-pointing charges, Pro residues, β-bulge*a-forming residues, and Gly residues, which promote bends and twists. This type of solution is called 'negative design' [99], because it is not dictated by the desired final structure, but by the need to avoid a different structure (i.e., the aggregate).

2. The flanking of all aggregation-prone segments in the protein with *gatekeeper residues* [103], that is, residues such as Pro, Lys, Arg, Glu, and Asp, which have very low propensity to aggregate, mainly due to their polarity and low β-sheet propensity [98].

3. Selection against sequences containing alternating polar and nonpolar residues [104]. This pattern may form amphipathic β-sheets that are efficient in packing surface residues. However, the pattern is also known to promote the formation of amyloid aggregates [105]. When sequences with such a pattern do appear in proteins, they are usually buried and can therefore do no harm.

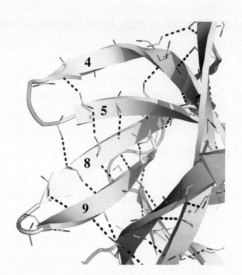

FIGURE 5.1.4 **The β-bulge.** The bulge is shown in interleukin-1β (PDB entry 1i1b) between the consecutive β-strands 4 and 5. The distortion formed by the bulge is clearly seen, as well as the missing backbone hydrogen bonds between the strands (backbone atoms are shown as lines, colored according to atom type; hydrogen bonds are shown as dotted lines). For comparison, the consecutive β-strands 8 and 9, which do not contain a bulge, are characterized by a regular hydrogen bonding pattern.

Cells also contain several built-in mechanisms that serve as a second layer of protection against protein misfolding. These mechanisms include molecular chaperones, which give misfolded proteins 'a second chance' to fold, as well as degradation mechanisms (proteasomal, lysosomal) for those proteins that are beyond repair. Indeed, strengthening of these mechanisms lowers the frequency of misfolding diseases [106], whereas weakening them promotes the diseases [107]. The mechanisms make sure that occasional misfolding in our cells is dealt with quickly before it can cause further damage to the organism.

Why, then, do misfolding diseases occur at all? Certain factors seem to induce extensive protein misfolding, beyond the normal repair and/or degradation capabilities of the body. In such cases, the natural systems are overwhelmed, and the excess of misfolded proteins eventually leads to disease. The factors associated with this process are as follows:

- Environmental – Changes in pH, temperature, etc. (see Chapter 4). This factor is mostly relevant to single-cell organisms, which cannot maintain constant environmental conditions.

- Genetic – Some mutations destabilize the native structures of proteins, or induce aggregation in other ways. As in other genetic diseases, the mutations may be congenital (as in the familial amyloidoses) or acquired.

---

*aA *β-bulge* is a region appearing in β-sheets (primarily anti-parallel), in which one strand includes two or more residues opposite a single residue on the other strand [100,101] (Figure 5.1.4). This type of local structure disturbs the typical backbone hydrogen-bonding pattern of the sheet. β-bulges appear on average twice per protein [102].

- Infectious – Protein aggregation may be the result of infection, as in the case of CJD or mad cow disease.

Of the three factors associated with misfolding, only the latter two are relevant to multicellular organisms, and we will therefore focus on them.

IV.I. Mutation-induced misfolding

Since at least some of the amyloidoses have a genetic background, scientists have tried to find sequence-related characteristics that might induce misfolding and aggregation. Aromatic amino acids were first implicated in amyloidogenesis by Gazit [108], on the basis of the following. First, aromatic amino acids have been known for a while to participate in self-assembly processes of biological macromolecules [109,110], including protein-protein interactions [111]. Second, they are common in amyloid-related proteins, and tend to appear on the surfaces of aggregating proteins at higher frequency than they appear on the surfaces of normal proteins. Third, the π-π interaction between aromatic residues is geometry-dependent, and therefore provides directionality to the assembly process.

Of the two aromatic residues, Phe seems to be most responsible for aggregation, as it is the most hydrophobic, is capable of π-π interactions, and tends to appear in β-sheets. In fact, Phe (and also Met, which is not aromatic) fits the average distance between β-sheets in amyloid fibrils (10 Å), whereas the indole side chain of Trp is too large for this space. In an *in vitro* study, Dobson and coworkers tried to characterize the effect of mutations on the tendency of proteins to aggregate, with an emphasis on the proteins' physicochemical properties [112]. This study demonstrated that the largest pro-aggregation effects resulted from mutations leading to the following changes in the protein:

1. Increasing the hydrophobicity of the protein

2. Promoting an $\alpha \rightarrow \beta$ conformational change

3. Decreasing the electric charge on the protein surface

Similar conclusions were also reached by *in vivo* studies. Thus, in conclusion, it seems that despite the fact that both environmental and physiological[*a] factors contribute to a protein's tendency to aggregate, aggregation is primarily determined by the protein's physicochemical properties.

[*a]For example, in Parkinson's disease the proteins aggregate and precipitate in only one location in the brain.

IV.II. Infection-induced misfolding: prion diseases

Protein misfolding that results from infection has been found in a group of diseases called 'transmissible spongiform encephalopathies' (TSEs)*a, which include CJD in humans, mad cow disease (BSE) in bovines, and scrapie in sheep [113]. Once it was realized that TSEs are caused by infection, viruses were naturally suspected. However, when the infectious agent turned out to be resistant to nucleases*b and ultra-violet radiation, while at the same time sensitive to protein-neutralizing agents, scientists realized they were dealing with an infectious protein. This was a true revelation, as all infectious agents encountered up to that point had been based on nucleic acids (DNA or RNA). The surprise grew even larger when it turned out that the infectious protein, called a 'prion'*c (abbreviated as PrP), was identical in sequence to a normal, membrane-bound glycoprotein. Further studies have demonstrated that the only difference between the benign form of the protein (PrP^C) and its harmful form (PrP^{Sc}) lies in their conformation: the former includes mainly α-helices, whereas the latter is based primarily on β-sheets, which tend to form amyloid fibrils and to deposit as precipitate. According to the accepted model, PrP^{Sc} acts as a template for normal PrP^C molecules, thus promoting their conversion from the α conformation into the β conformation, which renders them pathogenic. How exactly this is done is probably the most important and intriguing question related to prions, yet it is not entirely clear despite extensive research [115,116]. Another surprising fact discovered about prions is that they are resistant to gastric acids, which allows them to be transmitted, unharmed, via ingestion of contaminated tissue. Non-infectious forms of prion diseases are also known. In such cases, mutations, which may be either hereditary or randomly acquired, convert the prion protein from its harmless form into its harmful (amyloid) form.

Another cause for amyloidoses may be old age. The constant attrition associated with aging involves, among many other things, the decreased function of mechanisms that identify and repair misfolded proteins [117].

V. Amyloid mechanism of toxicity

The exact mechanism through which amyloid fibrils cause misfolding-related pathologies is not entirely clear, and it may very well be that different diseases are characterized by different mechanisms. While scientists agree that the protein precipitates are toxic to the cells, and different biochemical, cellular, and physiological effects of these precipitates have been documented in recent years, it is still premature to draw final conclusions. One intriguing possibility emerging from these studies is that the amyloid fibrils themselves may not be the toxic element, but rather the oligomeric state(s) preceding them [70,115]. In this respect, it has been proposed that these oligomers may hurt the cells by accumulating on the plasma membrane, and increasing its permeation

---

*aThis is because the brain of an affected individual is filled with holes, like a sponge. The holes are the result of neuronal cell death.

*bEnzymes that cleave nucleic acids.

*cPrion is short for 'proteinaceous infectious particle', a term coined by Stanley Prusiner [114].

either non-specifically or via the formation of ion channels (see details in [118,119]). If this is true, amyloid precipitates would be sharing their toxicity mechanisms with antibacterial toxins, i.e., proteins and peptides that are produced by many organisms and act as bio-weapons against invading bacteria [120–122] (see Chapter 7).

VI. Are amyloid proteins all bad?

The catastrophic consequences of amyloidosis are undeniable, and a great deal of research is carried out to find ways to prevent it. Still, proteins with increased propensity for aggregation exist [123], which may imply that such proteins are associated with an evolutionary advantage [124], similar to the one responsible for the endurance of the sickle hemoglobin isoform in third-world countries. A particularly interesting hypothesis was made by Eric Kandel[*a], who investigated a neuronal member of the CPEB family (cytoplasmic poly-adenylation element binding protein) [125]. CPEB regulates mRNA translation in nerve cells, and has also been implicated in the regulation of long-term memory [126], a process that is known to involve protein synthesis.

Kandel and coworkers found that the N-terminus of the CPEB isoform in *Aplysia* confers prion-like behavior to CPEB, i.e., induces aggregation. And yet, despite being aggregation-prone, CPEB is a functioning and necessary protein in neurons. Considering this, as well as CPEB's role in the formation of long-term memory in neurons, Kandel hypothesized that CPEB's aggregation capability was related to its involvement in long-term memory formation. The idea behind this suggestion was that amyloid-like aggregation is a stable form of the protein, which makes such aggregation a long-term process. Thus, when CPEB switches from its water-soluble state to its aggregated state, it can serve as a long-term marker directly, or via the accompanying change in its ability to regulate mRNA translation.

If true, Kandel's hypothesis has implications beyond the maintenance of memories in neurons; it means that protein aggregation might be a benevolent phenomenon, selected by evolution because it achieves specific purposes. The pathological consequences of protein misfolding might then be blamed upon the failure of cellular or physiological regulatory mechanisms to prevent such an occurrence, rather than on the aggregation propensity itself. Indeed, most misfolding diseases appear in the late stages of life, when body regulation and repair mechanisms start to fail.

5.2.2.2 Assisted folding

Since protein unfolding and misfolding might endanger the cell and even the entire organism, it is imperative that such cases be dealt with immediately and efficiently. Indeed, cells have built-in mechanisms designed to deal with the emergence of misfolded proteins [127], the primary mechanism being *molecular chaperones* [128,129]. Chaperones are proteins; they interact with unfolded or misfolded proteins in a way that helps the latter regain their proper folds before they aggregate non-specifically. Chaperones also have other cellular roles, such

---

[*a]Kandel, winner of the 2000 Nobel Prize in Physiology or Medicine, studies molecular mechanisms in the nervous system.

as facilitating the unfolding of proteins that are to be transported into a cellular compartment. Although many chaperones are expressed constitutively, they are also referred to as *'heat-shock proteins'*, because their levels rise sharply when the cell is exposed to heat or to other stress-related conditions, which increase the chance of protein denaturation [130] (i.e., the *heat-shock response* [131]). These conditions include extreme pH levels or salt concentration, elevated atmospheric pressure, exposure to heavy metals or toxins, inflammation, osmotic changes, certain hormonal changes, and even old age. The latter is associated, among other things, with chronic oxidative stress that leads to the gradual disintegration of cellular and physiological systems.

Although there are different types of chaperones, they can be divided into three basic groups according to their mechanisms of action:

1. **Stabilizers**

   These bind either to free proteins that have undergone misfolding or to the (unfolded) nascent chains of new proteins, while they are being translated by ribosomes. The unfolded or misfolded chains are recognized according to their solvent-exposed nonpolar residues, which in folded proteins are buried inside the core. The binding of the stabilizers to the polypeptide chains prevents the latter from interacting non-specifically with other proteins via their exposed nonpolar groups, i.e., aggregating.

   One well-known group of stabilizers includes the common, *small* (12 to 43 kDa) *heat-shock proteins (sHSPs)* [130]. The binding of sHSPs to their target proteins does not require an input of energy, which makes them perfectly suited for stressful conditions[*1]. The emergency action of sHSPs continues until energy reserves are restored, at which point ATP-dependent *chaperonins* (a.k.a. *HSP60s*, see below) replace the sHSPs and allow the target protein to refold. The sHSPs also act under normal conditions in monitoring the general folding status of cellular proteins. Interestingly, *α*-crystallin, the protein responsible for the transparence of our eyes, and which has always been considered to be inert, is also a sHSP. As such, its expression is not limited to the eyes, but also occurs in other tissues, primarily the heart, kidneys, and striated muscles.

   Another protein family whose members function as stabilizers is *Hsp70*. These proteins have more diverse roles in assisted folding, such as helping other proteins to pass through intracellular membranes (endoplasmic reticulum (ER), mitochondria, chloroplasts), as well as oligomerization. Another difference between Hsp70 proteins and sHSPs is that the former depend on ATP-derived energy to detach from their target protein. Finally, Hsp70 chaperones also act as central organizers of the chaperone network [63]. In this capacity, they distribute subsets of proteins to downstream chaperones, such as the chaperonins (see below) and the Hsp90 system, which will not be discussed here.

2. **Chaperonins**

   These are large protein complexes (~800 kDa) that function routinely in assisting the refolding process of proteins [132]. Like stabilizers, chaperonins recognize unfolded and misfolded proteins via their solvent-exposed nonpolar residues, and prevent them from interacting with other molecules. However, they act downstream to stabilizers [63]. That is, they do not recognize ribosome-bound polypeptide chains. Furthermore, their mechanism of action requires the input of ATP-derived energy, and is far

---

[*1]Such conditions are usually accompanied by depletion of energy reserves.

more complicated than the one employed by stabilizers. Briefly, the action of chaperonins involves induced unfolding of the misfolded protein and either its release to the solution to refold, or encapsulation of the protein and promotion of folding through comprehensive conformational changes that alter the polarity around the misfolded protein. A more detailed description is given in Box 5.2.

3. **Assisting enzymes**
These include *protein disulfide isomerase (PDI)*, which catalyzes the formation and reorganization of disulfide bonds inside the ER lumen, and *peptidyl prolyl cis-trans isomerase (PPI)*, which catalyzes the *cis-trans* shift of the peptide bonds of proline residues in the protein.

Membrane-bound and secreted proteins are also assisted by chaperones, which reside inside the ER lumen [127]. This is because these proteins must unfold in order to be inserted into the ER membrane or lumen. The chaperones unique to the ER[*1] help the inserted protein to refold, and prevent it from leaving the ER before the process is completed. In cases in which the refolding process fails, ER stress induces the *unfolded protein response* [127], which delivers the unfolded protein back to the cytosol, and induces its degradation by either the ubiquitin-proteasome pathway [82,133,134] or autophagy [135]. In some extreme cases the folding of ER proteins is altogether inhibited. In such a case, the appropriate signal is sent, and the cell responds by undergoing apoptosis.

BOX 5.2 CHAPERONINS

Chaperonins are ubiquitous[*a], large protein complexes (~800 kDa) that function routinely in assisting the refolding of unfolded or misfolded proteins [137–139][*b]. A major aspect of their function is their ability to encapsulate unfolded substrate proteins in a cage-like structure. By doing so, they allow the misfolded protein to fold or refold, unperturbed by other 'sticky' molecules. The entire process is active, i.e., it requires ATP-derived energy. There are two groups of chaperonins. Group I comprises chaperonins that are found in the bacterial cytoplasm and in the eukaryotic mitochondria and chloroplasts. Chaperonins in this group mediate protein folding activity in complex with a second protein component, the co-chaperonin. Specifically, the chaperonin (*GroEL* in bacteria and *Hsp60 or cpn60* in mitochondria and chloroplasts) forms the cavity in which the misfolded protein is trapped. The structure may appear in two forms, open and closed. Closure of the cage occurs in the presence of ATP (bound to the chaperonin), by a smaller 'lid' co-chaperonin protein (*GroES* in bacteria and *Hsp10 and cpn10* in mitochondria and chloroplast). Group II chaperonins occur in Archaea (*thermosomes*) and in the eukaryotic cytosol (*TRiC/CCT*). They too have a cage-like structure and require ATP to function, but their structure is different from that of Group I chaperonins. Specifically, their lid component is built-in, such that they do not require an Hsp10-like partner.

[*1] For example, BiP, which belongs to the Hsp70 protein family.

Most of what we know about chaperonins comes from studies carried out on the bacterial *GroEL-GroES* complex [137,140]. The importance of this complex for protein folding is reflected in the fact that deletion of either GroEL or GroES genes is lethal for bacteria. This observation is not surprising because, under normal conditions, out of the ~2,400 cytoplasmic proteins expressed in *E. coli*, 250 (~10%) are assisted by the complex, and ~85 (~3.5%) completely depend on it to fold [141]. These are mainly proteins that weigh between 20 kDa and 50 kDa, and are characterized by complex $\alpha\beta$ topologies, especially the TIM barrel fold [63,142]. The reason for this preference is that the architecture of such proteins includes numerous long-range interactions, and they are therefore likely to get trapped in local minima during folding [63,142,143].

The GroEL-GroES complex is formed by two homo-oligomeric proteins: GroEL is a large 14-mer cylinder that is arranged in two heptameric rings, and GroES consists of seven subunits that are arranged in a dome-like structure (Figure 5.2.1). The structure of a GroEL subunit is divided into three regions:

1. Equatorial domain, which has ATPase activity.

2. Apical domain, which binds the substrate.

3. Intermediate domain, which links the equatorial and apical domains to each other.

The complex can cycle between a few distinct states by undergoing the following steps [137] (Figure 5.2.2):

1. Cooperative binding of up to seven ATP molecules to the equatorial region of one of the GroEL rings (marked as *cis* in the image). This induces conformational changes in the apical domains of the ring [144], which lead to the subsequent step (see below). The other (*trans*) ring remains unbound to ATP due to negative cooperativity in ATP binding between the two rings [144].

2. Binding of the lid protein (GroES) to the apical region of the *cis* ring.

3. Hydrolysis of the bound ATP molecules after ~15 seconds and release of P_i molecules. This results in another conformational change in the *cis* ring and decreases the negative influence on the *trans* ring.

4. Binding of ATP molecules to the *trans* ring leads to release of ADP and GroES from the *cis* ring (due to the negative cooperativity).

5. Occurrence of steps 2 through 4 in the *trans* ring.

Note that due to the negative cooperativity between the two rings, only one ring tends to bind ATP and GroES at any given time, resulting in an asymmetric structure [144]. The positive cooperativity within each ring is also important; the binding of ATP to one site in the ring promotes ATP binding to all apical domains, which is imperative for GroES binding.

---

*a In human (HeLa) cells they constitute 2% (wt/vol) of the total mass of intracellular proteins [136].

*b For brevity, we will refer to such proteins as 'misfolded'.

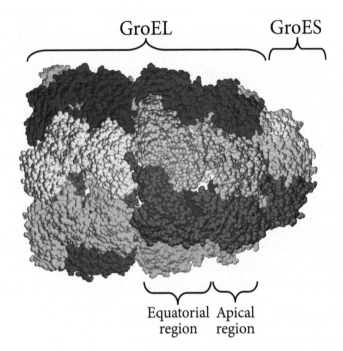

FIGURE 5.2.1 **The structure of the bacterial GroEL-GroES-ADP chaperonin complex.** The multi-subunit structure (PDB entry 1aon) is shown in a sphere representation, with each chain colored differently. The various parts of the structure are marked.

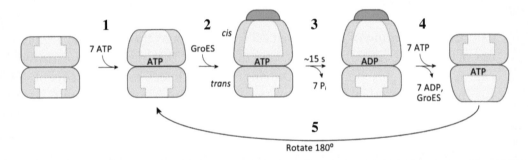

FIGURE 5.2.2 **Main molecular states of the GroEL-GroES chaperonin.** See main text for details. Taken from [137].

Based on the steps described in Figure 5.2.2, the following model has been suggested to describe the assisted folding of misfolded proteins by the GroEL-GroES complex (Figure 5.2.3):

1. Binding of a misfolded substrate protein to nonpolar residues lining the inner walls of GroEL's apical domains (Figure 5.2.4), followed by the binding of up to seven ATP molecules to the *cis* ring of apo GroEL. The residues attract misfolded proteins in a non-specific manner, as the latter have a greater fraction of solvent-exposed nonpolar residues compared with folded proteins. Binding to GroEL induces transient expansion and stretching motions in the substrate protein [145,146] (see more below).

2. GroES binds to the top of GroEL's apical domains, and this binding displaces the substrate protein inside the cage, trapping it there, where the protein has only a few seconds to unfold and then refold into its native structure[*a] (see more below).

3. Following ATP binding to the *trans* ring, GroEL undergoes a conformational change, which results in release of both GroES and the folded substrate protein into solution. If the protein is still unfolded or misfolded, it may rebind to GroEL and go through the entire process all over again.

The encapsulation of the protein serves an important role: it prevents the exposure of the protein to other unfolded proteins, which might adhere to it and form an insoluble, dysfunctional aggregate. Indeed, the main problem with misfolded proteins in the cytosol or in the lumen of organelles is not the fact that they are incapable of refolding, but rather that they risk being scavenged by other molecules before they have the chance to refold properly. It should be noted that the GroEL-GroES complex has a size limit and can handle misfolded proteins of 60 kDa or less. It is unclear how the folding of larger proteins is assisted, although a few suggestions have been made. For example, these proteins may fold with the assistance of the Hsp70 system [148], or undergo a repetitive process of binding and unbinding to GroEL (without entering the cage), until they refold [149].

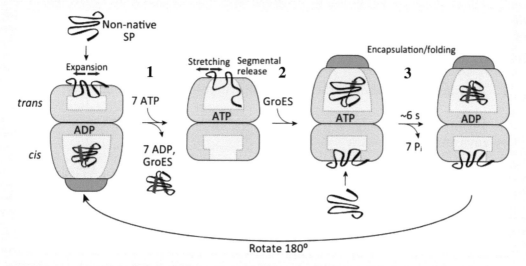

FIGURE 5.2.3 **The asymmetrical model for chaperonin-assisted protein folding.** See main text for details. SP – substrate protein. Taken from [137].

[*a]The time limit results from the rate of ATP hydrolysis by GroEL, which is about three times faster in the presence of the bound substrate protein than in its absence [147].

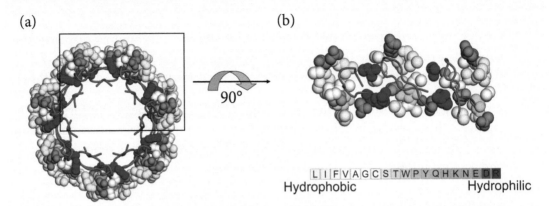

FIGURE 5.2.4 **The nonpolar lining of GroEL's inner walls.** The structure used (PDB entry 1mnf) is that of GroEL bound to GroES (not shown), with each of the apical substrate-binding subunits bound to a short unfolded peptide. (a) A top view of the peptides (red backbone only), along with the GroEL residues closest to them (spheres). The residues are colored according to the Kessel-Ben-Tal hydrophobicity scale [94], with the color index presented at the bottom of the figure. (b) A blowup of the peptide-binding residues from three GroEL subunits, viewed from the center of the lumen. This view was obtained by a 90° rotation of (a). The image demonstrates the nonpolar patch (yellow), which attracts misfolded proteins to GroEL. Also, highly polar residues (blue) seem to be placed between subunits, probably to prevent the bound protein from shifting from one to another.

As explained above, the binding of the misfolded substrate protein to the apical domains of GroEL (Figure 5.2.3, step 1) involves interaction between nonpolar residues in both. The interaction is important for attracting the protein to GroEL and holding it there, but it actually prevents it from refolding, for two reasons:

1. The interaction stabilizes the misfolded structure of the protein.

2. The folding process is driven by the hydrophobic effect, which requires a polar environment.

So how does the protein refold after all? As it turns out, the binding of ATP and GroES to GroEL in the next step induces a significant conformational change in GroEL (Figure 5.2.5a), which involves the relocation of the nonpolar residues away from the caged protein and renders the inner walls of the apical domain polar (Figure 5.2.5b). **The change of polarity not only releases the misfolded protein into the lumen of the cage, but it also drives it to refold.**

(a) (b)

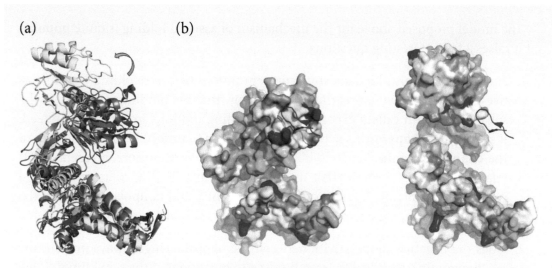

FIGURE 5.2.5 **Conformational and polarity changes in GroEL upon nucleotide binding.** (a) The conformational change. The image shows a superimposition of the apo (nucleotide unbound) form of GroEL (PDB entry 1mnf, colored in blue) and the ADP-bound form (PDB entry 1aon, colored in yellow). In both cases a single subunit is shown, containing the apical (top), intermediat (middle) and equatorial (bottom) domains. The green arrow shows the hinge motion resulting from the conformational change. The substrate peptide, taken from the nucleotide-unbound form) is shown as a red ribbon. (b) The corresponding polarity change. The image shows a surface representation of the two forms of GroEL, colored according to polarity as in Figure 5.2.3. In the peptide-bound form (left image) nonpolar residues of the apical domain face the bound peptide. The hinge motion that follows nucleotide binding moves the nonpolar residues upwards, leaving a larger and more polar cavity in the vicinity of the substrate peptide (right image).

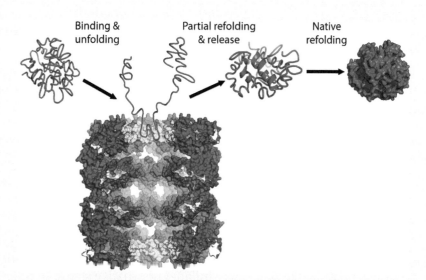

FIGURE 5.2.6 **Scheme of apo GroEL acting as an unfolding facilitator.** Apo-GroEL (blue) mediates iterative cycles of binding to the upper cavity (yellow), unfolding, release, and out-of-cage refolding, thereby converting high-affinity misfolded polypeptide substrates (left) into partially unfolded intermediates (center) that fold spontaneously in solution into low-affinity native products (right). The figure is taken from [150].

The model proposed above for the mechanism of assisted folding is quite popular, but it raises a few interesting questions:

1. How does the misfolded substrate protein unfold before refolding into its native structure? As we saw earlier, studies show that the binding of GroEL to the substrate protein causes expansion or stretching motions in the latter. Indeed, **it seems that chaperonins actively help proteins to unfold before giving them the chance to refold** [145,146,151–154] (Figure 5.2.6). Furthermore, a recent NMR relaxation study [153] shows that this activity is intrinsic (i.e., cofactor-free) and relies on hydrophobic interactions between GroEL and nonpolar regions in the substrate protein, which are exposed during the unfolding process.

2. Does the negative cooperativity between the opposite rings always prevent the co-chaperonin (e.g., GroES) from binding to both rings at the same time? Some studies suggest this is not necessarily true, and that a symmetric 'football'-shaped complex may be formed regardless of whether the misfolded substrate protein is bound or not [155–157] (see Figure 5.2.7 for the mitochondrial Hsp60-Hsp10 complex). In the absence of a substrate protein, the asymmetric form is considered to be dominant. However, the studies cited above and others suggest that, in the presence of the protein, the symmetric complex may be dominant, with both rings of the chaperonin acting simultaneously. Such a model of assisted folding is shown in Figure 5.2.8. The process starts from the asymmetric form of the complex, where only one of the rings binds ATP molecules, the substrate protein, and the co-chaperonin. The following steps then take place. (1) Binding of additional ATP molecules, a second substrate protein, and the co-chaperonin to the opposite ring, creating the symmetric form of the complex. The protein refolds inside the opposite ring. (2) ATP hydrolysis and P_i release from the first ring. (3) Release of co-chaperonin and folded substrate protein from the first ring. (4) Binding of a new misfolded substrate protein to the first ring. Steps 2 through 4 then take place in the *trans* ring. It should be noted that this model is highly debated [137] and further studies are required to confirm it. Also, it is inconsistent with the negative cooperativity known to exist between the two rings.

3. Do chaperones accelerate the folding process? The above model suggests that chaperones, despite being impressively complex structures, 'merely' mediate passive protein folding [158]. While there is agreement that the tendency to fold into the native structure is inherent to the proteins assisted by chaperones, some studies suggest that the latter accelerate the process up to ten times [151,159,160] (compared to the spontaneous process in the absence of aggregation). This acceleration is attributed mainly to the confinement of the misfolded protein, which favors compact native-like states over inflated non-native states. In other words, confining the unfolded chain to the limited space of the chaperonin cage reduces the entropy penalty of folding significantly in comparison to the penalty in bulk solution.

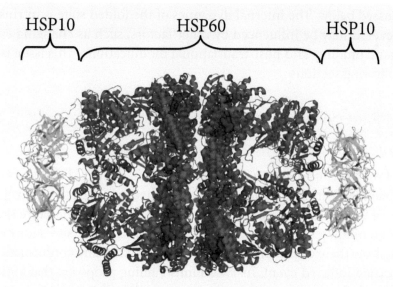

FIGURE 5.2.7 **The structure of the human mitochondrial Hsp60-Hsp10-ADP complex, demonstrating a football shape (PDB entry 4pj1** [161]**).** Hsp60 is colored in red, Hsp10 in green, and ADP in blue (spheres).

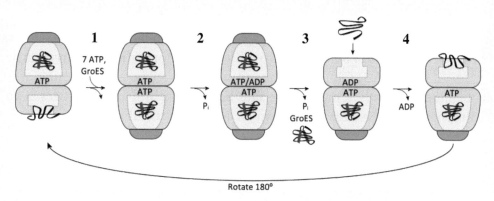

FIGURE 5.2.8 **A symmetrical model for chaperonin-assisted folding.** In this model, both rings are active in protein binding and refolding. See main text for details. Taken from [137].

5.3 FOLDED STATE DYNAMICS

Our discussions in Chapters 2 and 4 emphasized the uniqueness of the protein's native structure, as well as its functional importance. This emphasis might have created the impression that a protein has a single functional conformation — yet, in fact, the conformation considered to correspond to a given protein's native structure is only the *time-averaged* one. Although the native structure contains considerably fewer degrees of freedom compared to the unfolded chain, it still undergoes constant internal motions within a certain limit. Moreover, the internal dynamics of proteins is important for their function, whether catalysis, binding of small ligands, or formation of large macromolecular complexes [18,162–168]. For example, the internal dynamics helps to fine-tune the exact positioning of binding and catalytic residues, and to create a pathway leading the substrate to the catalytic site. This issue is further discussed below. The internal dynamics of the folded state is intrinsic and spontaneous. However, it may be influenced by other factors, such as changing environmental conditions, ligand binding, and post-translational modifications. This issue is discussed in detail in the following sections.

5.3.1 Spontaneous dynamics

5.3.1.1 Proteins are conformational ensembles

We have seen how the consideration of proteins as static entities changed around the middle of the 20th century, with Koshland's 'induced fit' theory [2]. The theory originally related to the creation of better enzyme-substrate complementarity (and therefore specificity), following the initial binding of the substrate to the enzyme's active site. However, many scientists projected the theory onto proteins in general, making any conformational change a ligand- or substrate-induced event. In 1940 Linus Pauling proposed that antibodies might acquire different configurations with identical energies [169]. This implied that proteins could undergo conformational changes in the absence of bound ligand or substrate. Unfortunately, Pauling did not turn his idea into a detailed model, so it had to wait for acceptance until 1965, when Monod and coworkers suggested the *Monod-Wyman-Changeux (MWC) model* [170]. The model suggested that ligand-free proteins*1 might exist in two different equilibrium conformations (the *pre-existing equilibrium* theory). Over time, these ideas have been combined and transformed into the current model [171], which assumes that **proteins are in equilibrium between many different conformations, referred to as 'substates'***2 [172-181] (Figure 5.5). Thus, each protein can be described as an *ensemble* of conformations that are constantly sampled. The shift between the different substates is possible because the energy barriers separating them are low (see next subsection).

5.3.1.2 Statistical-thermodynamic view of protein dynamics

The different substates constituting the protein's ensemble share the same fold and usually also the same secondary structure, but differ in their detailed atomic coordinates [182]. The latter differences may reflect differences in the position of one atom, or in the positions of multiple atoms — including groups of atoms corresponding to entire regions of the protein. This is why the movements involved in the shift between one substate and another are of different timescales. To illustrate, imagine the motions involving the side chain of the amino acid isoleucine. In the few nanoseconds it takes this side chain to move as a whole, one of its methyl groups may change its position a few thousand times (Table 5.1). This means that the number of substates available to each protein is enormous, and can only be described statistically [183]. The probability of the protein to acquire a certain substate (P) depends inversely and exponentially on its free energy (ΔG) [184], following Equation (4.3b) described in Chapter 4. In other words, the protein is expected to spend most of the time occupying those substates that have the least energy. Of these, **the conformation we refer to as 'native', and which appears, e.g., in crystallographic structures, is considered to be the most**

---

*1 The ligand-free and ligand-bound forms of the protein are referred to as *'apoprotein'* and *'holoprotein'*, respectively.
*2 Also known as 'microstates'.

FIGURE 5.5 **Backbone dynamics of folded proteins.** The conformational dynamics of chemotaxis Y protein is illustrated by superimposing 46 allowed conformations of the single protein, solved by NMR (PDB entry 1cey). The structures are colored according to secondary structure (helices are red, sheets are yellow, and loops are green).

stable one, i.e., have the lowest energy. The other conformations sampled by the protein are also of low energy, just not as low as the latter.

If the distribution of substates is statistical in nature, does this mean that any substate can be spontaneously acquired by the protein? To answer this we must first understand the forces that allow proteins to change conformation. In the absence of any other force, the 'engine' driving the conformational sampling by the protein is the thermal energy drawn from its environment. As explained in Chapter 1, the thermal energy exists in any temperature above the absolute zero, and has a value of RT (0.6 kcal/mol at room temperature). This energy is absorbed by the protein and converted into kinetic energy, in the form of atomic motions and bond vibrations. The dynamics resulting directly from this energy is in the picosecond range, but since the protein's atoms are connected to each other, the small movements are cooperative along the polypeptide chain [185,186]. Thus, the limited, thermally induced movements are translated into larger ones. This issue is further discussed in Subsection 5.3.1.4 below.

The fact that the spontaneous sampling of conformations is driven by the thermal energy implies that only conformations that are within +0.6 kcal/mol of the native conformation can be sampled. Indeed, over 20 years ago, Elber and Karplus demonstrated, using MD simulations, that the potential energy surface of myoglobin includes a large number of local minima within the thermal accessibility of the global minimum (i.e., the native structure) [179]. However, the assumption of RT being an absolute limit is inaccurate; although the conformations within the *RT* range are most likely to be sampled, others also have a chance to be acquired, just a lower one. **In other words, the protein spends most of the time in the native state or in other conformations whose energy content is no more than 0.6 kcal/mol higher than that of the native conformation** [179,187]. However, given enough time, other conformations with higher energy are expected to appear briefly. Obviously, conformations of very high energy have near-zero probability to be sampled during the lifetime of the protein. For example, conformations with steric clashes, i.e., two atoms occupying the same space, are not expected to appear at all.

5.3.1.3 Dynamics of disordered proteins

The type of dynamics described above characterizes well-defined regions within globular proteins. Some proteins include intrinsically disordered regions, and in some cases the entire protein is disordered (see Chapter 6). The dynamics of these proteins is different [188]; the atomic fluctuations are large and do not revolve around equilibrium values, as in the case of folded regions and proteins. Rather, the fluctuations are non-cooperative, and therefore random. For this reason, the disorder characterizing such proteins does not rule out the formation of secondary elements. These are as likely to appear as disordered backbone conformations, but unlike their counterparts in globular proteins, the secondary elements in disordered proteins tend to change or disappear constantly, due to the lack of stabilizing tertiary interactions.

5.3.1.4 Biological significance of thermally induced conformational changes

Despite the fact that most of the conformations sampled by a given protein are within the *RT* energy range, the shifting between them may still affect the protein significantly at the functional level. The following paragraphs describe some of the recorded effects of protein dynamics on function.

1. **Changing active site properties**

   Near-native conformations that a particular protein samples routinely are expected to be geometrically very similar to the native one. Nevertheless, they may differ in other physical traits that have functional implications, especially when these manifest in the binding or active site. For example, the electrostatic potential of the binding site may change significantly even upon a relatively small conformational change (Figure 5.6). This is due to the high sensitivity of the potential to the dielectric environment, which in turn depends on the exact conformation of the protein. In such cases, **the change in the electrostatic properties of the binding site may be translated into a change in the affinity of the protein to its substrate or ligand.**

2. **Facilitating substrate diffusion into a buried active site**

   Binding sites for ligands are often depicted as depressions on the protein's surface, whereas in reality they might be buried deep inside the core. For example, proteins that bind nonpolar gases (O_2, H_2), such as hemoglobin and myoglobin, often have such binding sites. In such cases, it is not always clear how the substrate might access the binding site form the solvent. This question has been studied in hemoglobin and myoglobin. The investigators found pre-existing voids inside these proteins, which seemed to outline the path of the substrate into the catalytic site [190,191]. In addition,

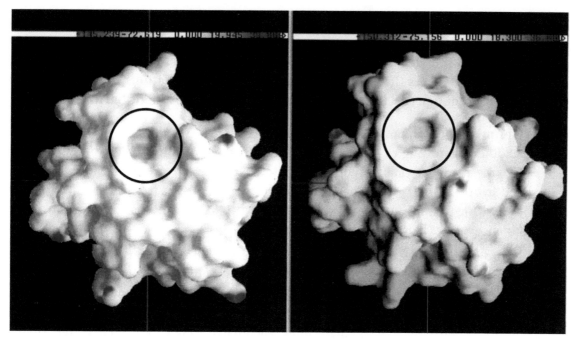

FIGURE 5.6 **Electrostatic differences between two similar microstates.** The figure shows the electrostatic potential of two microstates of cytochrome c, generated using MD simulations. Although highly similar in conformation, the potential of certain regions may differ significantly (e.g., see circled region). The figure was taken from [189].

certain dynamic changes in the proteins have been implicated in facilitating the diffusion to the site. The same aspects were also studied in the enzyme *cpI hydrogenase*, which catalyzes the reduction of protons to H_2, as well as the reverse process. In this case, too, no visible path was initially observed between the buried active site and the surface of the enzyme. The investigators used MD simulations in order to characterize the transient conformational changes that transpire, and which cannot be tracked by observing the static structure [192]. The simulations suggested that within the picosecond (ps) to nanosecond (ns) timescale, spontaneous dynamic changes create transient nonpolar voids inside the protein core. Similarly to the pre-existing voids found in hemoglobin and myoglobin, the transient voids in cpI hydrogenase seem to outline the path of H_2 to the catalytic site, or from it to the surface. This type of dynamics is sometimes referred to as *'breathing'*, due to its fast and repetitive nature.

3. **Translating short thermal fluctuations into wide-amplitude conformational changes**

Observations of complex structures such as proteins suggest that, in many cases, short (ps to ns) fluctuations from all regions of the molecule act cooperatively so as to create a long range effect (on a much larger timescale), which affects protein function. This resembles the so-called 'butterfly effect', i.e., a butterfly flapping its wings in one part of the world may cause a hurricane elsewhere. Henzler-Wildman and colleagues directly demonstrated the cooperation of atomic fluctuations in the enzyme *adenylate kinase* [186]. The investigators combined NMR measurements and MD simulations in order to cover a wide range of fluctuations. Previous studies had shown that the catalytic rate of the enzyme is limited by the 'slow' dynamics of the μs to ms timescale, which is attributed to the opening of the substrate-binding site (needed for product release). Henzler-Wildman and colleagues demonstrated that the μs to ms fluctuations are in fact a direct result of cooperative thermal fluctuations on the ps to ns timescale, occurring in key residues in that area. Thus, **proteins are somehow able to act as transducers that convert short, low-energy fluctuations into motions that are large enough and have enough energy to overcome kinetic barriers.**

4. **Optimization of quantum tunneling events**

   Another interesting example of the effect short fluctuations and vibrations have on enzyme-mediated catalysis comes from studies on the *quantum tunneling* of hydrogen and hydrogen-like particles[*1]. The name 'tunneling' comes from the realization that the hydrogen atom 'tunnels' through the energy barrier instead of crossing it using its kinetic energy (Figure 5.7a).

   Hydrogen tunneling may occur in cases where the hydrogen transfer process involves an insurmountable energy barrier [193–199][*2]. Although the hydrogen does not possess sufficient kinetic energy to pass the barrier, it can sometimes 'tunnel' through it. This is possible because of the quantum nature of very small particles (electrons, protons, etc.), described by quantum mechanics. Such particles have a dual character. That is, they behave both as particles and as waves. As waves, they can be described using a function that provides the probabilities of finding them in different locations. Thus, the entire organization of electrons around the nucleus of an atom can be described by solving their wave function. Again, the solution does not provide the absolute location of each electron, but instead the probabilities of finding it at various locations. This forms the basis for the way we perceive atoms today, i.e., as nuclei surrounded by 'shells'; each shell is a collection of points in space where electrons have high probability to be found.

   How does this relate to the tunneling reaction? Quantum mechanics shows that the lighter the particle, the larger its characteristic wavelength is. Hydrogen-like particles are light, and therefore have large wavelengths (and electrons are lighter and have even larger wavelengths). In the case of hydrogen transfer discussed above, the X−H bond (X is the donor) does not have sufficient energy to get completely over the energy barrier of the transfer reaction; it can only 'climb' part of the way. However, at some point along the way, it reaches a position where the hydrogen's wavelength exceeds the barrier width, which allows it to transfer to the acceptor without having reached the energy of the transition state [197] (Figure 5.7b). This illustrates one of the main oversights of the classical treatment of catalysis (i.e., the transition-state theory), which considers the reactants merely as particles that have only one possible location in space. The quantum-mechanical (QM) treatment acknowledges that each reactant has a most probable location, but also less probable ones, which may even be on the other side of the reaction's energy barrier.

   The QM model described above is more accurate than the basic transition-state theory model, but it still does not accurately describe the tunneling process. A more realistic model must relate to the energies and wave functions (i.e., probabilities) of both

---

[*1]I.e., protons or hydride ions.

[*2]One such case is enzymatic catalysis. The importance of hydrogen tunneling to a common catalytic mechanism of enzymes called 'acid-base catalysis' is further discussed in Chapter 9.

reactants and products of the tunneling reaction[*1]. This is because efficient hydrogen tunneling requires the probability of finding the hydrogen particle at the reactant's energy well to overlap with the probability of finding it at the product's energy well (Figure 5.7c) [197]. The overlap depends on the width and *degeneracy* of the two energy wells[*2]. In the structural sense, this is translated into several quantities, such as the donor-acceptor distance, their orientation, and the electrostatic properties of their immediate environment.

In many enzymes, the evolutionarily-selected (native) structure already contains donor-acceptor couples that are suitable for hydrogen tunneling [196,199]. This is often achieved by multiple noncovalent interactions between bulky enzyme residues and the substrate, which keep the reacting atoms of both enzyme and substrate in place. Nevertheless, theoretical considerations suggest that the presence of these couples is insufficient for the tunneling process to transpire, and that protein dynamics must play a role as well. For example, dynamic fluctuations are required for the enzyme to avoid energy degeneracy of the reactants and products, an occurrence that would obviously make the hydrogen particle tunnel back and forth with no net reaction actually taking place. Indeed, computational studies suggest that the fast (fs–ps) thermal vibrations in enzymes act to increase the rate of hydrogen tunneling by optimizing the geometric-electrostatic factors contributing to the reactant-product probability overlap [200,201]. Here, too, the fast vibrations are coupled, and lead to much slower conformational changes, which optimize the tunneling-contributing factors. Furthermore, the importance of these vibrations may explain the negative impact of certain mutations far from the active site on the tunneling process [200]. That is, mutations whose roles have so far eluded scientists, due to their distant locations and their non-additive effects on tunneling, have been found to act by disrupting the network of coupled vibrations that optimize the tunneling process. This insight is expected to have implications for protein engineering and drug design, which rely heavily on site-directed mutagenesis [200].

In addition to hydrogen tunneling, other quantum effects have been implicated in protein function [203,204]. For example, *electron tunneling* has been shown to facilitate the transfer of electrons between protein redox centers that are separated by 15 to 30 Å (e.g., in cellular respiration) [205,206]. Another quantum effect, called *coherence*, has been implicated in photosynthesis. As explained in Chapter 1, the first step of photosynthesis involves the highly efficient harvesting of light energy by antenna arrays made of chlorophyll molecules and other pigments. This process involves the conversion of electromagnetic (solar) energy into electronic excitation waves in the pigment molecules. These waves are channeled to the *photosynthetic reaction center*, where their energy facilitates electron delocalization and subsequent redox reactions. The reason for the efficiency of light harvesting in photosynthetic systems is not entirely clear, especially when the propagation of the excitation waves through the antenna arrays has always been considered as random, and therefore wasteful. However, studies carried in 2007 and later suggest that these waves are in fact coherent [207–210], and can therefore find the most efficient way to the photosynthetic reaction center by exploring numerous paths simultaneously. These findings are far from being trivial;

[*1]Here, the reactant and product are the donor-hydrogen and acceptor-hydrogen bond states, respectively.
[*2]*Degeneracy* means that both energy wells have identical minima, like the ones shown in Figure 5.7c.

quantum effects are usually observed in a vacuum under low temperature, and quantum coherence should theoretically be destroyed by the ambient temperatures and chaotic environment of biological organisms. Computer simulations suggest that not only can the coherence endure the environment of the photosynthetic center, but in fact the random noise in this environment actually helps the propagation of the excitation waves [211]. Specifically, the noise helps waves to propagate by preventing them from being trapped in sites along the photosynthetic chain. Quantum coherence has also been suggested to facilitate the extremely fast and specific photoisomerization of the *retinal* molecule in the protein rhodopsin, which is responsible for sight (see [204] and references therein).

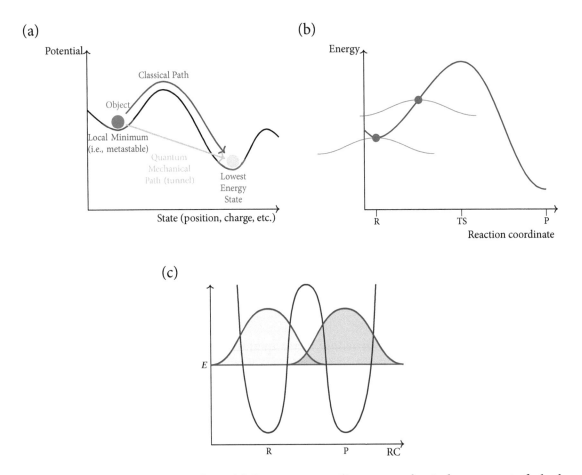

FIGURE 5.7 **Quantum tunneling.** (a) Quantum tunneling versus classical movement of a hydrogen atom between a hypothetical protein donor (X) and an acceptor (Y), depicted as the crossing of an energy barrier. The figure is adapted from [202]. (b) Explanation of quantum tunneling based on the wave properties of the reactant (i.e., the transferred hydrogen). The potential energy of a hydrogen reaction is depicted along the reaction coordinate (R – reactants, TS – transition state, P – products). The mean, most probable location of the hydrogen particle along the reaction coordinate is specified by the red sphere, and its probability distribution is represented by the green curve. As the plot shows, even when the hydrogen particle has only enough energy for 'climbing' halfway to the top of the energy barrier, it has a finite probability to appear at the other side of the barrier and complete the transfer. (c) Tunneling requires an overlap between the wave functions of the reactants and those of the products. The tunneling efficiency increases with the extent of the overlap. The figure was adapted from [197].

5.3.1.5   Effects of solvents on protein dynamics

Water, as we learned earlier, is the universal solvent in biological systems and is essential for the ability of biomolecules to stay mobile and interact with each other without forming non-specific aggregates. On the single-molecule level, it has been found that proteins may retain some function even after partial dehydration, but they require at least one layer of water molecules around them to remain mobile and active [212] (see more below). One of the most important means by which the aqueous solvent keeps proteins active is by affecting their inherent dynamics, and it does so in different ways [213].

The role of the solvent in conducting the thermal energy to the protein has already been mentioned above. This role, however, does not depend on the behavior of specific solvent molecules. An additional effect exists, one that depends on the dynamic fluctuations of the individual water molecules. The fluctuations are particularly strong in bulk water, but are also transmitted to water molecules that form the protein's *hydration shell*. These water molecules have fewer degrees of freedom, and are considered by many biophysicists as part of the protein. **When the atoms of these molecules are in physical contact with those of the protein, the fluctuations of the former may affect those of the latter, provided that both water and protein atoms have similar frequencies** [214]. This effect is not just mechanical; due to the physical properties of water molecules, their mechanical fluctuations are accompanied by fluctuations of their dipole moment, which manifests as constant changes in the electric field surrounding them. Polar residues on the surface of the protein are within the range of these changing electric fields, and this induces similar fluctuations in them [215–217]. The resulting dynamics of the surface polar residues is referred to as 'slaved', because it is induced by water fluctuations [216]. This phenomenon has been shown to be complex in nature [183]: on the one hand, it weakens the inherent electrostatic interactions between protein surface residues, whereas on the other hand it may facilitate biological processes such as enzyme-mediated catalysis, which are carried out by the protein and depend on its dynamics (see Chapter 9 for details). In this respect there are two additional issues that are interesting to note. First, it seems that the induced fluctuations are not limited to the protein surface, and can propagate into the core. Second, even large conformational changes seem to be influenced by the water fluctuations. Specifically, although they cannot drive such changes, water fluctuations seem to be able to limit their rate.

Interestingly, the effects describe above are mutual, and the dynamics of the hydration shell is also affected by the protein. In fact, it has been found that the hydration shell dynamics is heterogeneous and is strongly influenced by the local geometry of the protein surface [218]. That is, water molecules bound to a certain region of the protein surface experience completely different dynamics from those bound to other surface regions. Thus, the hydration shell around the protein should probably be viewed as a collection of independent water clusters, each with its own dynamics and surface affinity. Such an approach is expected to be helpful in protein design studies, or when searching for pharmaceutical drugs that target specific protein binding sites [219].

5.3.2 External effects on protein dynamics

The previous sections demonstrated that folded proteins are conformational ensembles, constantly shifting between different substates due to the thermal energy in their environment. Although thermal energy is what drives conformational sampling, other factors can influence the process, including ligand binding, post-translational modifications [220,221], and environmental changes (pH, temperature, etc.; see Subsection 5.3.2.2 below). Understanding this influence is important, as it often has significant implications for the protein's activity (see below), stability, and solubility [222]. The extensive research carried out in this field shows that the effects of all these factors are based on the same physical principle. That is, they all bias the conformational equilibrium towards certain conformations by changing the energy landscape of the protein [223,224]. The exact means by which this takes place varies across the different factors. The following subsections elaborate mainly on the effect of ligand binding, which has been the focus of most of the research carried out in this field. Other aspects of protein-ligand interactions are described in Chapter 8. The effects of phosphorylation and glycosylation on protein structure and dynamics are discussed in Chapter 2, Section 2.6.

It should be pointed out that whereas the conformational changes induced by the above factors are usually small-to-moderate (backbone r.m.s.d. $\leq$ ~2.5 Å [225,226])[*1*2], they may sometimes be quite large. This is the case with *metamorphic proteins*, which are able to alternate between (at least) two very different, yet interconvertible, conformations [228,229]. In fact, the difference between the alternate conformations may be so large that they do not even share the same fold, as is the case with lymphotactin [230] (Figure 5.8). As described above, the pre-existing equilibrium between the different conformations is biased by external factors such as ligand binding, temperature, salt concentration, and even the redox state of the environment [230,231]. For example, in lymphotactin, one conformation (Ltn10) is predominant at 10 °C and salt concentration of 200 mM, whereas the other conformation (Ltn40) is predominant at 40 °C and low salt concentrations; the latter is further stabilized by dimerization of the protein. Because the two conformations are very different, the transition between them is unlikely to happen via a unique intermediate structure. Instead, it has been suggested that such interconversion requires global unfolding of the chain [232]. Another example of a protein that undergoes a large conformational change when influenced by external factors is calmodulin, which we have already encountered in Chapter 2. In this case, binding of Ca^{2+} ion and a target ligand to the protein induces a dramatic change, from an open to a closed conformation [5,233].

5.3.2.1 Ligand-induced dynamics and allostery

5.3.2.1.1 *Allostery*

The general notion that ligand binding can affect protein structure is not new; it was proposed by Wyman at the end of the 1940s for hemoglobin [234,235], and since then such an

---

[*1] Note, however, that the small backbone changes are usually accompanied by large side chain changes, and that these changes may have significant implications for the binding and/or catalytic properties of the protein (see following subsections).

[*2] A survey of the *CoDNaS database* of protein conformational diversity [227] shows that mutations have a larger (average) effect on protein conformation compared with ligand binding, oligomeric state, and post-translational modifications [226].

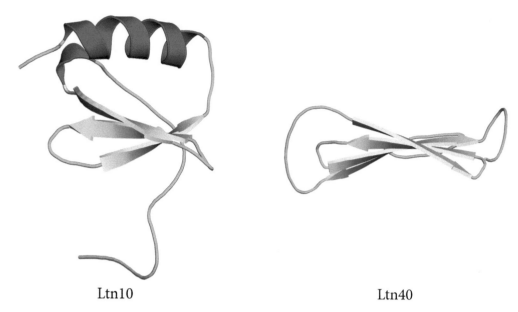

Ltn10   Ltn40

FIGURE 5.8 **NMR structures of the two stable conformations of human lymphotactin (Ltn).** *Left*: Ltn10 (PDB entry 1jt0). For clarity, the C-terminal tail (residues 71 through 93) is not shown. *Right*: Ltn40 (PDB entry 2jp1). This conformation is predominant and exists as a dimer (here only one monomer is shown).

effect has been demonstrated in other proteins as well [5,236,237]. Although the effect is structural in essence, it may have functional implications if the new chosen conformation has a higher or lower activity than the original. Within cells, ligand binding usually serves some specific purpose: enzymes bind their substrates, receptors bind their cognate hormones or messengers, antibodies their antigens, etc. In enzyme-substrate binding, the substrate participates in the chemical reaction. Conversely, in binding of a receptor to its cognate hormone or messenger, the latter is not changed by the binding, but rather activates the receptor's biological function. In both enzymes and receptors, the inherent function of the protein (catalysis, binding), which occurs at a certain location (e.g., the active site in enzymes), can be modulated by binding of a ligand at a different location. This phenomenon is called *'allostery'*[1], and the site to which the modulating ligand binds is called *'allosteric site'*. The term *allostery* was originally used by Monod and Jacob [238] to describe non-competitive enzyme inhibition, in which a molecule that differs from the substrate inhibits an enzyme by binding to an allosteric site [171][2] (see Chapter 9). Today the term is used for all proteins whose activity is modulated by allosteric molecules. As we will see later, the allosteric effect is mediated by conformational changes in the protein. Allostery is one of the most important phenomena in the biological world, since it allows cells to accurately regulate central processes, such as metabolism, division, and defense mechanisms. Specifically, allostery affects enzyme-mediated catalysis, signal transduction, transport, and more. Indeed, Monod, who discovered allostery, referred to it as 'the second secret of life' (with DNA structure being the first) [241]. Although allostery is only one of the many manifestations of protein dynamics,

---

[1]The term *'allosteric'* means 'other solid or object', from the Greek *'allos'* (ἄλλος – other) and *'stereos'* (στερεός – solid or object).

[2]While the term *allostery* was coined in 1961 by Monod and Jacob [238], the idea behind it was already proposed in 1935 by Pauling [239], who suggested a mechanism for the positive cooperativity in ligand binding observed in hemoglobin (see Box 5.3). This idea was later adopted by Koshland in his famous KNF model [240].

its functional implications have led to it being the most studied aspect of dynamics. Studies of allostery have yielded both qualitative and quantitative models, which are described in Box 5.3, and which have contributed to our general understanding of protein dynamics.

Cells rely heavily on allostery to regulate their multi-step biochemical pathways [242]. In most cases, only key enzymes in the pathway are regulated, that is, enzymes that commit the cell to the specific process. In some cases, the allosteric ligand is a cellular messenger produced by a certain signal transduction cascade, in response to a hormone or growth factor. This type of regulation, which is referred to as 'heterotropic', is particularly common in multicellular organisms, where it facilitates the systemic modulation of metabolic processes in certain cells, tissues, and organs. In other cases, the allosteric ligand is itself part of the biochemical process it regulates, i.e., it is the substrate (*homotropic* regulation), product, or intermediate. This is a more local type of regulation, which allows biochemical pathways to modulate their own rates automatically. For example, in *product inhibition*, the product of a biosynthetic pathway acts as an allosteric inhibitor, which shuts down its own production, via negative feedback, when its levels become sufficient for cellular needs. The opposite type of regulation, i.e., substrate activation, enhances the efficiency of proteins by turning substrate binding into a cooperative process. In any case, substrate- or product-mediated regulation demonstrates that, contrary to what was originally postulated, allostery is not limited to ligands that do not participate in the reactions. It should also be mentioned that although allostery usually refers to noncovalent ligand binding, it can also include the covalent addition of chemical groups, such as phosphorylation [166]. This is because some of the covalent modifications are reversible inside cells due to the action of enzymes, such as phosphatases.

As mentioned above, allostery works by coupling two different sites in the protein via conformational changes (see, e.g., [243–246]). Such changes need not be large. For example, in the aspartate receptor, a 1-Å conformational change in the regulation site leads to a significant increase of activity at a site 100 Å away! [247]. In order to understand how this is possible, one must first understand the physical basis for the effect of ligand binding on the protein. Note that this effect, which leads to conformational changes in the protein, exists in any type of ligand binding, not just an allosteric ligand. **Indeed, in both enzymes and receptors, the binding of the primary ligand (substrate in enzymes, hormone or messenger in receptors) may also induce conformational changes in the protein.** This phenomenon is especially prominent in receptors, in which the conformational changes induced by the binding of the primary ligand to its target site (termed *'orthosteric site'*) activate the protein. Thus, whereas the following discussion on the physical effect of ligand binding is general, its goal is to provide a better understanding of allostery. The effects induced by the binding of the primary ligand are discussed in Chapter 7.

5.3.2.1.2 *Physical effect of ligand binding*

Protein folding, like all spontaneous processes in nature, involves a decrease in free energy. This energy has different components; some are internal, whereas others result from the environmental conditions, such as pH, temperature, pressure, etc. The native conformation of the protein is therefore the one that under physiological conditions has the lowest free energy. When the protein binds a certain ligand, a new physicochemical entity is introduced into the native conformation, which changes the free energy. This change is a direct consequence of the physicochemical interactions between the ligand and the protein (as well as the effects of ligand binding on the environment). The change in free energy may be either negative or positive. The former means further stabilization of the native conformation, thus making it more highly populated. Conversely, a positive change in the free energy upon ligand binding means destabilization of the native conformation, making it less populated. If the free energy increase of the native conformation is large enough, one of the near-native conformations of the isolated protein may dominate the population upon ligand addition. In such cases, the binding of the ligand results in a new native conformation. The above description implies that, following ligand binding, a conformational change takes place that is consistent with the pre-existing equilibrium known to characterize proteins. In other words, conformational shifts happen spontaneously, whether or not the protein binds its ligand. Accordingly, **the ligand is believed to act by biasing the pre-existing equilibrium, i.e., preferentially binding the conformation that has the least energy when bound to it** [6,164,181,223,224,248,249] (Figure 5.9a)[*1]. This notion emerges from the *conformational selection model* [181,250,251], which is described below and in Box 5.3.

The biasing effect of the ligand can be demonstrated by using the simple model shown in Figure 5.9b [252]. The model refers to two different conformations of the protein (T and R). Both are capable of ligand binding, but with different affinities. The R conformation is better suited for the ligand than the T conformation; it binds the ligand more strongly. The differences in affinities towards the ligand manifest in the different equilibrium constants of the binding process. Since ΔG is a state function, the shift from the ligand-free T state (T_0) to the ligand-bound R state (R_1) can conveniently be described in two ways; the first corresponds to the orange path in Figure 5.9b, whereas the second corresponds to the magenta path:

$$\Delta G_{(T_0 \to R_1)} = \Delta G_T + \Delta G_1 = \Delta G_0 + \Delta G_R \tag{5.1}$$

In terms of equilibrium constants, the above can be described as:

$$\Delta G_{(T_0 \to R_1)} = -RT(\ln(K_T) + \ln(K_1)) = -RT(\ln(K_0) + \ln(K_R)), \tag{5.2}$$

which means that:

$$\frac{K_T}{K_R} = \frac{K_0}{K_1} \tag{5.3}$$

Put simply, **if the ligand-binding affinity of one conformation is higher than that of the other by a given factor, then the equilibrium constant for the R $\to$ T transition in the ligand-free and ligand-bound states must be related by the same factor. This means that ligand binding changes the R $\to$ T transition, which is exactly what happens in allostery.** Of course, in reality the protein exists in multiple conformations, but the bottom line is the same as explained here.

The notion of pre-existing conformational equilibrium underlying allostery is easy to imagine when the conformations are overall similar. It is, however, less imaginable in cases where the protein has two or more conformations that, despite having similar energy, possess very different architectures. How can such a protein interconvert spontaneously and quickly between the different conformations? We saw earlier that in extreme cases, such as that of the metamorphic protein lymphotactin, the interconversion requires the protein to

---

[*1]Although this view is considered to be relatively new, a similar notion was already suggested in 1950 by Karush, for serum albumin [3]. Specifically, Karush suggested that binding sites in serum albumin exist in different equilibrium conformations of nearly equal energy, and the ability of the protein to shift between these conformations allows it to bind different ligands.

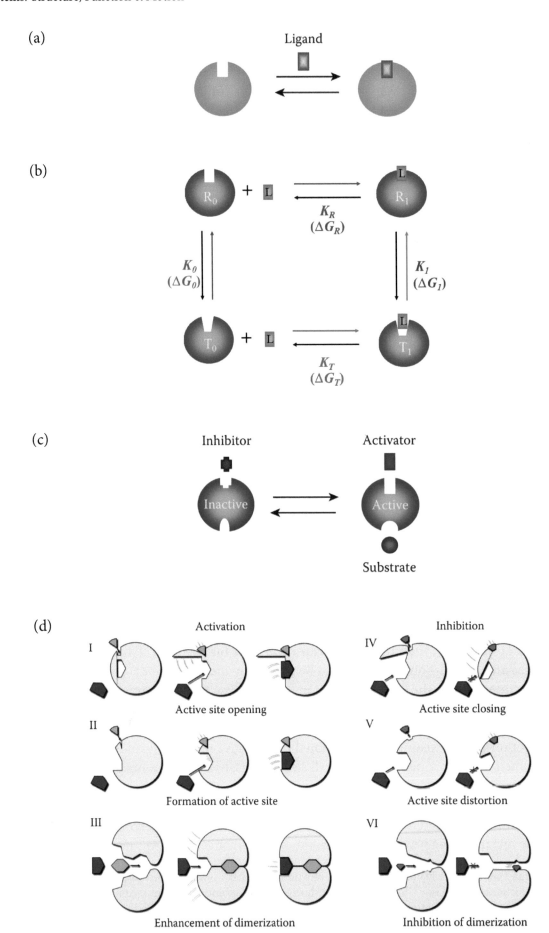

(a)

Ligand

(b)

(c)

Inhibitor
Activator
Inactive
Active
Substrate

(d)

Activation

Inhibition

I

Active site opening

IV

Active site closing

II

Formation of active site

V

Active site distortion

III

Enhancement of dimerization

VI

Inhibition of dimerization

FIGURE 5.9 **Schematic illustration of the mechanism of allosteric ligands.** (Opposite) (a) The general mechanism of allosteric ligands. The ligand binds to the conformation that is best suited to it stereochemically, and biases the natural conformational equilibrium of the protein towards this conformation. (b) A thermodynamic cycle (following [252]) describing the path from the free low-affinity conformation (T_0) to the ligand-bound, high-affinity conformation (R_1) (orange path). R and T are the low-affinity and high-affinity conformations, respectively. L is the ligand. 0 and 1 denote the ligand-free and ligand-bound states, respectively. The path may proceed via the ligand-bound low-affinity conformation (T_1) (orange path) or the free high-affinity conformation (R_1) (magenta path). The equilibrium constants (K) and related free energy changes (ΔG) associated with each step of the cycle are noted in different colors. (c) The effects of positive and negative allosteric ligands. Both types of ligands act as in (a), but they bind to different conformations. The first is inactive due to a substrate-incompatible binding site, and the second is active due to a substrate-compatible binding site. (d) Specific mechanisms of allosteric activators and inhibitors, affecting active site accessibility, active site shape, and oligomerization capability. Adapted from [253].

partially unfold and then re-fold into the new conformation. This is, however, not necessarily true for other metamorphic proteins. For example, it has been suggested that such interconversions can be made possible by the following factors:

1. **Subunit symmetry** [254] – Different conformations of an oligomeric protein can be achieved by alternative packing of the same subunits in a number of symmetry-equivalent ways having similar energies.

2. **Local frustrations** [255] – As explained above, native structures of proteins are stable because they avoid energy conflicts (frustrations), resulting, for example, from unfavorable atomic interactions. However, at certain key points in the protein chain, local frustrations may act as hinges that promote large conformational changes.

Although conformational selection is believed to be the model that best describes the effect of ligands on protein structure and function [256], conformational changes following ligand binding (e.g., induced fit) are also known to happen in some cases (e.g., [257]). This phenomenon demonstrates once again that reality is much more complex than the models we use to describe it. A more detailed discussion of the different theories concerning the physical basis of ligand binding is given in Chapter 8. A focus on models of allostery is given in Box 5.3. Finally, another effect of ligand binding is rigidification of the binding site region, which makes additional conformational changes even less probable.

BOX 5.3 MODELS OF ALLOSTERY

Allostery, a phenomenon known for decades, plays a key role in the function of numerous, perhaps even most proteins found in nature. One of the main questions occupying biochemists has always been whether a single model can describe both qualitatively and quantitatively all aspects of allostery, such as cooperativity, homotropy versus heterotropy, activation versus inhibition, etc. The first researchers to propose a detailed model for allostery (in enzymes) were Jacques Monod and coworkers [170] (Figure 5.3.1). Their suggestion, referred to as the *Monod-Wyman-Changeux (MWC) model*, addressed two main aspects of allostery in enzymes: substrate-binding (positive) cooperativity and product inhibition (see main text for definitions). The MWC model posited the following:

1. **An allosteric enzyme includes more than one polypeptide chain** (called subunits), arranged symmetrically. Although a generalization [258], this assumption has been confirmed in many allosteric enzymes, as well as in membrane-bound and nuclear receptors [259].

2. **An allosteric enzyme exists in two naturally occurring conformations.** The first, termed 'R' ('relaxed') has high affinity for the ligand, whereas the other, termed 'T' ('tense') has low affinity to it. The T/R assumption is important not only for allostery but also for understanding protein dynamics in general, since it implies that proteins are capable of shifting spontaneously between different conformations. As mentioned in the main text, this is also the current understanding of proteins [260].

3. **The inner symmetry of the subunits' organization in the enzyme must be maintained, even when the enzyme shifts between the R and T conformations.** This implies that a conformational change in one subunit must also occur in the other, so as to maintain the symmetry. That is, any conformational change, even in a small region of the protein, must induce a total change in the quaternary structure of the entire protein. In reality, this posit does not hold for all proteins undergoing conformational changes, but it is true in many enzymes displaying substrate-binding cooperativity.

FIGURE 5.3.1 **Jacques Monod (1910–1976), Nobel Prize laureate in Physiology and Medicine, 1965.** The Nobel Foundation [261].

Monod and coworkers addressed positive cooperativity, in which initial substrate binding to the enzyme increases the affinity of the latter to substrate molecules that it subsequently encounters. They concluded that in such cases the substrate acts as a *positive allosteric regulator*, i.e., an *activator*. That is, **the substrate increases the affinity of the enzyme by shifting its R⟷T equilibrium towards the R conformation.**

Since R is also the active conformation, substrate binding is expected to render the protein more active. Again, the conformational change starts in one subunit, but since the MWC model stipulates that symmetry must remain, the same change must also take place simultaneously in the other subunit(s)[*a], even though they do not bind the substrate. Thus, substrate binding results in all subunits shifting from the low-affinity T conformation to the high-affinity R conformation. **The net effect of ligand binding to one subunit is therefore the increased affinity of all the other subunits to it**. This is the essence of positive cooperativity. This is also where the MWC model differs from another popular model of the time, the KNF model of Koshland and coworkers [240]. The KNF model also assumes that the allosteric protein is a homo-oligomer, which can be in either an R or a T state. However, this model does not require symmetry to be maintained during the conformational change, but rather describes cooperativity as a sequential process[*b]. That is, substrate binding to one subunit involves an induced-fit type of conformational change from the T to the R state. This change, in turn, induces slight conformational changes in the adjacent subunits and increases their affinity to the substrate, with no T→R change. Although the two models coexisted at first, the MWC model prevailed when later studies carried out on several proteins demonstrated a change in the quaternary structure of the investigated proteins following substrate binding.

One of the best benchmarks for testing the MWC model and studying allostery mechanisms in general is vertebrate hemoglobin. This is mainly because hemoglobin has a quaternary structure and exhibits cooperativity in binding its substrate, oxygen (O_2). Studies extending over 40 years on hemoglobin have confirmed most of the stipulations of the MWC model, at least with respect to cooperativity in this protein. New findings published in recent years have shed new light on the model, compelling scientists to extend it.

I. Allostery in hemoglobin

Hemoglobin is one of the most extensively studied proteins in the history of biology and chemistry. Studies on this protein were published a century ago, whereas research on allostery appeared only about 50 years later. For this reason, as well as for the reasons mentioned above, hemoglobin is considered to be a good model for allosteric processes. Hemoglobin, which is the primary protein of red blood cells, is responsible in mammals, birds and reptiles for the highly regulated transport of molecular oxygen (dioxygen; O_2) from the lungs to peripheral tissues, as well as for the reverse transport of carbon dioxide (CO_2), a metabolic waste product, back to the lungs. This process involves binding oxygen in the arteries of the lungs, exchanging it for carbon dioxide in the vicinity of peripheral cells, and returning the latter to the lungs via venous blood, where it can be unloaded and exhaled. This is the reason why the color of arterial blood is different from that of venous blood; the bright red color of the former results from

---

[*a]This is why the model is also referred to as the 'concerted model' of allostery.

[*b]Accordingly, this model is referred to as the 'sequential model' of allostery.

its high oxygen content, whereas the cyan color of the latter results from its low oxygen content. In other words, arterial blood is oxidized, whereas venous blood is reduced.

The description above reflects the *ambivalent* nature of hemoglobin, i.e., its capacity to both bind and release the same substance (O_2 or CO_2), depending on its location in the body. This capacity implies that the structure of hemoglobin is sufficiently complex to support this functional sophistication. The assumption of a complex structure seems even more justified considering that oxygen binding to hemoglobin exhibits *cooperativity*. That is, the affinity of hemoglobin to oxygen increases with the concentration of the latter[*a]. Such cooperativity can be clearly seen in hemoglobin's binding curve (Figure 5.3.2); instead of the hyperbolic curve expected in normal binding, the protein exhibits a sigmoid curve, which is a sign of cooperativity.

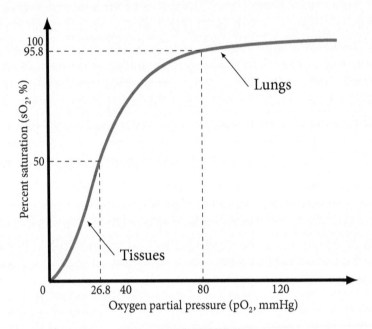

FIGURE 5.3.2 **Hemoglobin-oxygen binding curve.** The curve describes the dependence of the binding (represented by the percent saturation) on the oxygen's concentration (represented by its partial pressure). In the lungs, high oxygen concentration leads to strong binding, whereas in peripheral tissues the binding is weak due to low oxygen concentrations. Furthermore, the sigmoid shape of the curve indicates positive cooperativity in oxygen binding. The image is adapted from [262].

I.I. Structure of hemoglobin

In our previous encounters with hemoglobin in these pages, we witnessed its quaternary structure (Figure 2.34). This includes four α-helical subunits (α_1, β_1, α_2, β_2) of similar structures but dissimilar sequences, each containing an oxygen-binding heme group (Figure 5.3.3a). The subunits are arranged so the most significant noncovalent

---

[*a]In gases, the chemical concentration is usually expressed as *partial pressure*.

interactions are between different types $(\alpha\beta)$, which makes hemoglobin a 'dimer of dimers': $\alpha_1\beta_1$, and $\alpha_2\beta_2$.

The heme is a planar *porphyrin* group, which consistes of four interconnected pyrrole rings[*a], with an iron cation (Fe^{2+}) in the middle (Figure 5.3.3b). The latter is coordinated to the four nitrogen atoms of the surrounding rings in the heme plane, which leaves Fe^{2+} two more potential bonds, one above and one below the plane of the heme group. The first of these bonds is formed between Fe^{2+} and a histidine residue on one side of the heme (Figure 5.3.3c). This residue is referred to as '*proximal*', and it is part of the subunit's F helix[*b]. The second potential bond is usually formed between Fe^{2+} and O_2. The *ferrous* form of the iron cation (Fe^{2+}) sometimes becomes oxidized to the *ferric* form (Fe^{3+}). Fe^{3+} is unable to bind O_2, and any hemoglobin molecule undergoing such oxidation (called '*methemoglobin*'), is therefore useless as an oxygen carrier[*c]. A small percentage of hemoglobin in our body exists as methemoglobin. The percentage does not normally increase, thanks to enzymatic reduction of the molecule back to its active state. However, certain factors, which are either genetic (e.g., glucose-6-phosphate dehydrogenase (G6PD) deficiency) or environmental (e.g., pharmaceutical drugs such as nitrates and sulfonamides), might expose the body to oxidative stress and increase the fraction of methemoglobin in the blood. This condition, called *methemoglobinemia*, is potentially hazardous, as it causes tissue hypoxia [264].

On the other side of the heme group, beyond the oxygen, lies another histidine residue referred to as '*distal*', which is part of the E helix of the protein. This residue stabilizes the heme-bound oxygen by hydrogen bonding [265]. Moreover, the distal histidine is important for reducing the risk of carbon monoxide (CO) poisoning. This compound is formed by incomplete oxidation of hydrocarbons, as well as by routine physiological processes in our body, such as the degradation of heme groups. CO is highly toxic due to its high affinity to the Fe^{2+} cation of the heme. As a matter of fact, the affinity of CO to Fe^{2+} in isolated heme is about 2×10^4 times higher than the affinity of oxygen, which means that CO could essentially block all free hemoglobin molecules in our body and prevent oxygen supply to the tissues. However, the affinity of CO to the heme group drops 100-fold when the latter is bound to hemoglobin. Again, this is because of the distal histidine, although the exact mechanism responsible for the phenomenon is not entirely clear. Originally, it was suggested that the histidine residue, via steric hindrance, compels any ligand bound to the heme to tilt, which in the case of

---

[*a]Porphyrins are very common in biochemistry, participating in photosynthesis, cellular respiration, methane biosynthesis, redox reactions involving nitrite and sulfite, and the detoxification of hazardous chemicals [263]. The porphyrin group always contains a transition metal cation in the middle. For example, iron-containing porphyrins (hemes) participate in oxygen binding and electron transport; manganese-containing porphyrin (chlorophyll) participates in photosynthesis; and cobalt-containing porophyrin (cobalamine, derived from vitamin B_{12}, see Chapter 9) participates in nucleotide biosynthesis. The prevalence of porphyrins in such a large variety of processes is attributed to the transition metal and the conjugated aromatic rings, which render the molecule photo-reactive (important for photosynthesis), and provide it with multiple redox levels (important for electron-transfer reactions).

[*b]This is the only covalent bond between the heme group and the protein.

[*c]In partially oxidized hemoglobin (one to three heme groups are in the ferric form), the heme groups that are in the ferrous form bind oxygen with increased affinity. Thus, the release of oxygen to the target tissues by such hemoglobin molecules is reduced. This, in addition to the inability of ferric heme groups to bind oxygen, results in tissue hypoxia [264].

CO significantly lowers its affinity to the heme. However, later studies raised doubts as to the capability of steric hindrance alone to lower the affinity of CO to that extent [266]. Today, it is believed that the effect of the distal histidine results from several factors, including electrostatic stabilization of the bound oxygen. In any case, the action of the distal histidine protects us from the endogenous levels of CO, and limits poisoning to rare cases of exposure to external sources of the gas. Indeed, CO levels over 0.1% (of air) are hazardous.

(a)

(b)

(c)

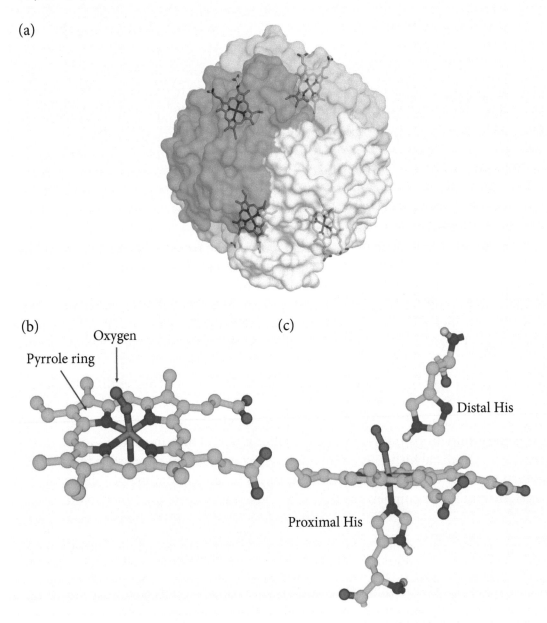

Oxygen

Pyrrole ring

Distal His

Proximal His

FIGURE 5.3.3 **Hemoglobin structure.** (a) The four subunits are marked with different colors, and the heme groups are shown in bond-stick models (PDB entry 1hho). (b) Structure of the heme group. Atoms are colored according to type, with the Fe^{2+} ion colored in gray. The location of the bound oxygen molecule is marked. (c) Proximal and distal histidine residues in hemoglobin.

In addition to the important role described above, the distal histidine fulfills another role, that of creating passage for the oxygen from the solvent to the heme groups [190]. The binding of oxygen to the heme groups has always been a mystery, as the latter are buried inside the protein, and the experimentally determined structures of hemoglobin do not show any passage leading from the solvent to the heme and back. By combining different methods, researchers have found that such a passage is formed by permanent voids within the core of the protein, and that the diffusion of oxygen between the voids is facilitated by conformational changes that originate in the distal histidine [190,191].

I.II. Cooperativity and allostery in hemoglobin

Hemoglobin has been long known to exist in two functional states. The first was termed 'deoxy-hemoglobin', as it has low affinity towards oxygen, and is therefore unbound to it most of the time. The second was termed 'oxy-hemoglobin', because its oxygen affinity is 150 to 300 times higher than that of *deoxy*-hemoglobin. Max Perutz (Figure 5.3.4), who determined the structure of hemoglobin by using X-ray diffraction [267], found that the *oxy* and *deoxy* states of hemoglobin have different quaternary structures (in accordance with Monod's MWC model [170]), and that the *deoxy* conformation is stabilized by salt bridges between subunits $\alpha_1\alpha_2$, $\alpha_1\beta_2$, and $\alpha_2\beta_1$. These salt bridges break when the conformation changes from *deoxy* to *oxy* (e.g., see Figure 5.3.5). Oxygen binding by hemoglobin was known at that time to depend cooperatively on the partial pressure of oxygen. Perutz realized that the cooperativity is induced by oxygen itself, which in addition to being the substrate also acts as an allosteric activator that shifts the protein from the *deoxy* state to the *oxy* state [268]*a. Drawing from the similarity between the MWC model, which dictates the principles of allostery in enzymes, and Perutz's own observations on hemoglobin (a non-enzyme protein), Perutz formulated his 'stereochemical' model for describing both allostery and cooperativity in this protein. He considered the *oxy* and *deoxy* states to be the hemoglobin equivalents of Monod's R and T enzyme states (respectively), i.e., to differ in quaternary structure [271]. This idea, combined with the assumption that oxygen serves as an allosteric activator, explained the duality of hemoglobin: In the lungs, where the partial pressure of oxygen is high (100 mmHg [272]) (Figure 5.3.2), initial oxygen binding induces a change in hemoglobin quaternary structure from the T conformation (*deoxy*-hemoglobin) to the R conformation (*oxy*-hemoglobin). Since the R conformation has higher oxygen affinity, the binding of the other three O_2 molecules to hemoglobin is cooperative. When the hemoglobin-containing blood cells move towards the peripheral tissues, the partial pressure of oxygen drops to 40 mmHg, and the hemoglobin starts to shift back to the T state while releasing oxygen for the use of the cells.

---

*a The tendency of the hemoglobin tetramer to change into the *oxy* state, following the binding of four oxygen molecules, increases by 10^9 times compared to its tendency to do so in the absence of bound oxygen molecules [269,270].

FIGURE 5.3.4   **Max Ferdinand Perutz (1914–2002), Nobel Prize laureate in Chemistry, 1962.** The Nobel Foundation [273].

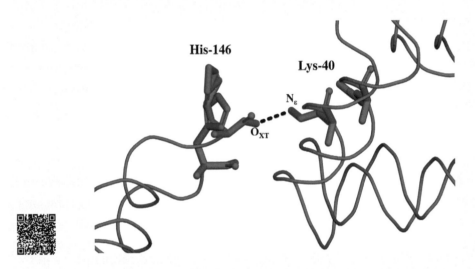

FIGURE 5.3.5   **A salt bridge that stabilizes the T state of hemoglobin** [268].

Perutz's model successfully rationalized both the duality and the cooperativity of hemoglobin. Yet, it was unclear how exactly oxygen induces the conformational change between the states. Studies carried out since then have shown that the binding of oxygen to the heme groups in hemoglobin involves structural changes at different levels. At the most basic level, the binding changes the position of the Fe^{2+} cation with respect to the heme ring. In the absence of oxygen, Fe^{2+} slightly protrudes from the ring plane towards the proximal histidine (Figure 5.3.6a). This is because the oxygen-free Fe^{2+} is in a high-spin state, with a radius too large to fit within the plane of the heme group [13]. However, when O_2 binds to Fe^{2+}, the ion is converted into a distorted-octahedral, low-spin state, whose radius is smaller than that of the oxygen-free cation, and can therefore slide into the middle of the ring plane. Fe^{2+} is also bound to the proximal histidine residue, and when it slides into the ring plane it pulls the histidine slightly (0.4 Å [13]) in that direction (Figure 5.3.6a). Since the histidine is part of the F helix in hemoglobin, its motion causes the entire helix to move as well (Figure 5.3.6b). Thus, the dynamics, which began with oxygen binding, spreads to the other parts of the subunit, making it shift from the T conformation to the R conformation within less than 1 μs [274]. The conformational change does not stop at the single subunit level; the subunit's interactions with the other subunits allow it to affect their conformations as well, making the entire protein shift to the R state [275–277] (Figure 5.3.6c). The global movement that transpires in hemoglobin following this shift is significant, manifesting in a 15° tilt between the $\alpha_1\beta_1$ dimer and the $\alpha_2\beta_2$ dimer. The propagation and amplification of the 0.4 Å movement of the Fe cation into a large-scale conformational change is one of the most interesting and least understood phenomena in proteins. Obviously, it results from the high flexibility of protein structure, and is probably related to the specific structure of hemoglobin, which can be described as a system of interconnected springs (i.e., the helices). However, a true biophysical characterization of the force or energy transmission yielding this amplification has yet to be carried out.

Any model describing the behavior of hemoglobin is expected also to address the well-known *Bohr effect* [278–280], which is named after the Danish physician Christian Bohr. This effect refers to the lower affinity of hemoglobin to O_2 in a low-pH environment[*a]. As mentioned earlier, the T conformation contains certain salt bridges that are missing in the R conformation (Figure 5.3.5). Some of these bridges exist in the interfaces between subunits, whereas others occur within chains, and they may involve backbone or side chain atoms. Some of these salt bridges are formed between positively charged histidine side chains and negatively charged glutamate or aspartate side chains. Whereas the Glu or Asp side chains are almost always charged under physiological pH, those of His are only charged part of the time, because their pKa is close to the neutral pH. In the lungs, the pH is 7.6, which renders most of the His side chains deprotonated, i.e., uncharged. However, in peripheral tissues the pH drops to 7.2 due to high CO_2 levels[*b], which renders most of the His side chains protonated, i.e., positively charged. Thus, the His-Glu or Asp salt bridges that stabilize the T conformation of hemoglobin tend to form in peripheral blood rather than in the lungs. In other words, the R→T conformational shift initiated by the drop in oxygen concentration following the transfer of hemoglobin from the lungs to peripheral tissues is aided by the simultaneous drop in pH. One of the prominent salt bridges, which studies show to have the largest contribution to the Bohr effect, is formed between the side chains of His-146 and Asp-94 [275,281] (Figure 5.3.7). In addition to this effect, the ability of hemoglobin to bind protons in peripheral blood also makes it a good buffer, which prevents sharp changes in blood pH. The stabilization of the T conformation in peripheral tissues is also aided by two other factors. The first is the covalent (yet reversible) binding of CO_2 to N-terminal amino groups, which forms carbamates. Like the His side chains, these too participate in salt bridges that are unique to the T state [13]. The second factor is phosphorylated molecules that bind to the central cavity of the T state via electrostatic interactions and hydrogen bonds to His and Arg side chains in the β subunits. The phosphorylated molecules include 2,3 bis-phosphoglycerate (2,3 BPG) in mammals and inositol hexaphosphate (IHP) in birds and reptiles [13].

[*a]In fact, Bohr did not mention pH at all. He originally observed that when blood CO_2 levels increase, the affinity of hemoglobin to oxygen decreases. The two, however, are linked; high CO_2 levels in blood decrease its pH, since the CO_2 turns into carbonic acid (H_2CO_3).

[*b]CO_2 is a metabolic byproduct of cells.

Since the publication of Perutz's model in 1970, other studies on cooperativity and allostery in hemoglobin have been conducted, using advanced spectroscopic methods. These studies contributed important data on those mechanisms (especially kinetic data on the conformational changes), and even succeeded in quantifying some of them. Some studies tried to refute Perutz's model, but the evidence in most cases was inconclusive. One of the claims made by the latter studies was that cooperativity in hemoglobin is too complicated to be explained by quaternary changes alone. This claim was supported by experiments showing that, in the presence of certain allosteric inhibitors, the oxygen affinity of the T state changes without any changes in quaternary structure [282]. Moreover, it was demonstrated that a single quaternary structure can harbor different tertiary conformations [283]. Following these revelations the MWC model was extended to include tertiary changes as well [259], although quaternary changes remained a requirement. Thus, the 'new and improved' model may be viewed as combination of the MWC model with some of the ideas incorporated into the KNF model.

Finally, one may wonder why hemoglobin appears as a hetero-tetramer and not as a homo-tetramer. Surely, it would be much easier to synthesize and use a homo-tetramer. Insights into this issue have been provided by research on a hemoglobin-related disease called α-thalassemia[*a]. In healthy people, the four hemoglobin chains are produced in matching quantities, which promotes the formation of a hetero-tetramer [285]. However, in people who have α-thalassemia, increasing loss of hemoglobin α genes during different stages of the disease leads to shortage in the production of α-chains. This gives rise to two distinct homo-tetrameric forms of hemoglobin: *HbH*, which consists of four γ chains (γ_4)[*b], and *homeglobin Bart's*, which consists of four β chains (β_4). Studies have shown that HbH binds oxygen with greater affinity than does wild-type, $\alpha_2\beta_2$ hemoglobin (HbA) [286], but fails to display the Bohr effect or cooperativity [287]. The same has been found for Hb Bart's [288].

The reason for these differences between HbH, Hb Bart's and HbA became clear when the structures of the two homo-tetrameric hemoglobins were solved [288–290]; both were found to exist constantly in the R state, such that they can bind oxygen efficiently, but are precluded from undergoing the R→T conformational change. Thus, it seems likely that the hetero- form of hemoglobin was selected during evolution due to its unique ability to respond to changing environmental conditions by changing conformation, an ability that allowed hemoglobin to extend its physiological function from simple oxygen and CO_2 binding, to oxygen and CO_2 transport. As explained above, such a role requires the protein to load its ligand at one location in the body and unload it in a different location, which can only be achieved if the protein can shift between two different states, each having a different affinity to the ligand.

[*a]Thalassemias are a group of inherited disorders characterized by absence or markedly decreased accumulation of one of the globin subunits of hemoglobin [284]. The prefix (α or β) denotes the identity of the chain which is absent or insufficiently produced.

[*b]The γ chain normally appears in fetal hemoglobin. After birth it is replaced by the β chain.

(a)

(b)

(c)

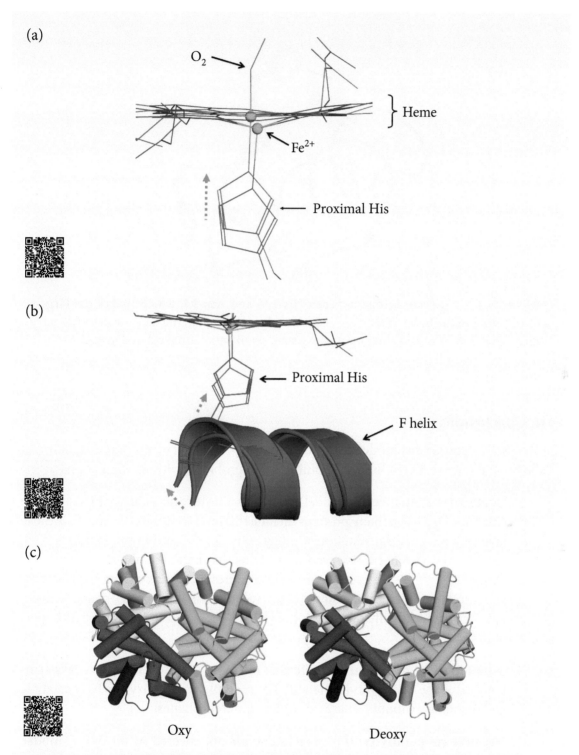

FIGURE 5.3.6 **Hemoglobin dynamics.** The figure shows the effects of oxygen binding to hemoglobin on various regions of the molecule. The induced motions are demonstrated by superimposition of the oxy (red) and deoxy (blue) states. The dashed green arrows mark the directions of the motions. (a) Motion of the iron atom (green) from below the heme plane into the middle of the heme plane, and of the proximal histidine in the same direction. (b) The motion of the F helix. (c) Motions at the level of the entire molecule. Here the motions are complex, so we present the two structures and refer the reader to the global change in conformation. For example, note the change in the size of the central cavity between the two structures.

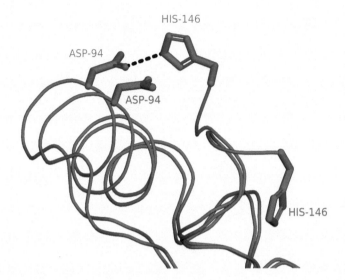

FIGURE 5.3.7 **The salt bridge between His-146 and Asp-94, which makes the largest contribution to the Bohr effect** [281]. The salt bridge is present in the deoxy state (PDB entry 2dn2, colored in blue) and absent in the oxy state (PDB entry 2dn3, colored in red).

II. Conclusions

The MWC (concerted) model constitutes the basis for our current understanding of allostery. Yet, more recent findings point to some problems with its stipulations [171]:

1. **Quaternary structure:** the model conditions allostery on a change in quaternary structure. Although many proteins include this level of structure, especially those involved in signal transduction [259] (e.g., GPCRs [291] and protein kinases [292]), allostery has also been observed in proteins that include only one polypeptide chain [166,258,293,294]. In fact, even when the protein does contain a quaternary structure it may exhibit allostery that is accompanied by tertiary changes alone, as in the case of shellfish hemoglobin (HbI) [295,296] and in the bacterial chaperonin GroEL [297,298]*a.

2. **Symmetry:** the quaternary structure of oligomeric proteins is not always symmetrical as the MWC model dictates. For example, many enzymes have catalytic and regulatory subunits characterized by different shapes.

3. **Negative cooperativity:** this phenomenon is not covered by the MWC model. Describing it requires the incorporation of other models.

4. **Atomistic and mechanistic details:** the MWC (and KNF) model does not provide a detailed description of the principles of allostery. Such a description is provided by the current *conformational selection model* of allostery [181,250,251]*b.

---

*a As a result, Eaton and coworkers extended the MWC model to the *tertiary two-state (TTS) model*, which accounts for tertiary changes as well [277].

*b See also [6,164,180,248,249,299,300].

Like the MWC model, the conformational selection model also assumes a dynamic equilibrium between different conformations in the protein (referred to as *pre-equilibrium*) [3], where one of the conformations is stabilized by the allosteric regulator. However, the conformational selection model assumes multiple conformations instead of just two. Positive regulators (i.e., activators) stabilize the active conformation(s), whereas negative regulators (i.e., inhibitors) stabilize the inactive conformation(s). This is an extension of the MWC model, which provides a description of both *homotropic* and *heterotropic* types of regulation, instead of just the first. Since the allosteric regulator merely changes the statistical distribution of the different available conformations of the folded state, it has been proposed that any protein has the potential to be allosteric [248]. This proposal is supported by studies in which non-allosteric proteins were rendered allosteric by simple point mutations or chemical modifications, or just by using certain ligands [301–304].

III. Partial agonists

So far, we have encountered two types of allosteric ligands — activators and inhibitors — each stabilizing different conformations with different functionalities. This general description applies to different protein-ligand systems, such as protein-substrate, receptor-hormone, channel or transporter-solute, etc. One of the most extensively studied protein-ligand systems involves membrane-bound receptors, acting as ion channels. In such systems, activators and inhibitors are referred to as *agonists* and *antagonists*, respectively, and they are assumed to act as described above, i.e., by shifting the protein's conformational equilibrium. However, in these systems there is also a third type of allosteric regulator, referred to as a *partial agonist*. As the name implies, this type of ligand binds to the receptor and partially activates it. Studies of the acetylcholine-activated channel have demonstrated that this protein shifts among several closed conformations of the channel [305,306]. This observation prompted the following model for partial agonism. At least one of the closed conformations acts as an 'activated-closed' conformation. Both partial and 'full' agonists bind with the same (high) affinity to the open conformation of the channel. They differ only in their affinity to the activated-closed conformation; the affinity is high in partial agonist binding, and very low in full agonist binding. In other words, the partial agonist can bind to (and stabilize) both open and activated-closed conformations, with a preference for the latter, whereas the full agonist binds only to the open conformation. This explains why the partial agonist can induce the opening of the channel part of the time. It also explains how partial agonists differ from antagonists, which bind only to the closed conformation, and therefore never induce channel opening. This model has been supported by a study conducted on acetylcholine and glycine channels, both of which belong to the nicotinic channel superfamily [307].

5.3.2.1.3 Functional effects of ligand binding

In our earlier discussion of spontaneous protein dynamics, we saw how conformational changes may affect protein function by changing the physicochemical characteristics of the binding or active site, to which the protein's primary ligands bind. The same is true for conformational changes that are induced by allosteric ligands. In fact, these may alter protein function more effectively than spontaneous changes, as they are less affected by the *RT* (thermal) limit, and are therefore more pronounced (i.e., have a larger amplitude). Different allosteric ligands induce different types of changes. Positive regulators stabilize active conformations of the protein, whose primary binding sites are geometrically and physicochemically compatible with the protein's substrate or activating ligand, whereas negative regulators stabilize conformations that are not, and which are inactive (Figure 5.9c). The primary binding sites formed by the active conformation are expected to differ from those of the inactive conformations in terms of the locations of binding or catalytic residues, and sometimes in their overall electrostatic properties. A survey carried out by Thornton and coworkers [253] used known structures of apo- and holoenzymes to identify the structural and functional outcomes of allosteric changes. The survey demonstrates three major outcomes (Figure 5.9d, examples are given in the paper):

1. Active site opening/closing (Figure 5.9d I/IV) – This category includes all types of structural changes that influence the availability of the enzyme's binding site to the substrate. The structural changes may differ significantly in range; dramatic cases involve high-amplitude motions of rigid elements around hinges, e.g., in the boundary of two domains, whereas more subtle changes may involve, e.g., rotation of a single side chain.

2. Active site formation or distortion (Figure 5.9d II/V) – This includes minor modifications that are still significant enough to change the binding affinity of the substrate.

3. Enhancement or inhibition of oligomerization (Figure 5.9d III/VI) – This includes cases in which binding of the ligand either facilitates or interferes with the formation of a biologically active oligomer.

Other studies have also shown that the apo and holo forms of some allosteric proteins differ in the presence of specific water molecules that bridge residue-residue or residue-substrate interactions. Again, the differences need not necessarily be large, as the sensitivity of binding and catalysis to the exact positioning of binding and catalytic residues (respectively) is high.

The long-range effects in allosteric regulation and the sufficiency of limited motions for creating a significant conformational change are both nicely demonstrated in the regulation of the enzyme *dihydrofolate reductase (DHFR)* [7]. DHFR catalyzes the reduction of dihydrofolate (DHF) to tetrahydrofolate (THF) using the reducing agent NADPH. The THF product is required for biological methylation reactions, such as those that occur during the biosynthesis of the nucleotide thymine. For this reason, DHFR activity is particularly important in dividing cells (which constantly replicate DNA), and inhibition of DHFR by drugs such as *methotrexate* is a successful anti-cancer therapy. Studies combining NMR, single-point mutations, and MD simulations demonstrate that the binding of DHF and NADPH to DHFR induces conformational changes in different regions of the enzyme. Surprisingly, it was found that the largest conformational change occurs in a loop called Met-20, which

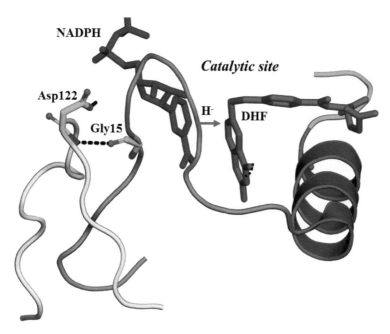

FIGURE 5.10 **Internal motions in the enzyme DHFR from *E. coli* (PDB entry 7dfr).** The figure shows the catalytic site region of DHFR. For clarity, only some of the secondary structures and residues are shown. DHFR catalysis involves hydride (H^-) transfer from NADPH (magenta) to DHF (cyan). DHF binding leads to closure of the Met-20 loop (orange) around it and to acceleration of catalysis. Interestingly, NMR studies show that this motion involves creation of a hydrogen bond between Gly-115 of the Met-20 loop and Asp-122 of the βF-βG loop (yellow), a region that does not include catalytic residues.

does not contain any catalytic residues (Figure 5.10). The new conformation is stabilized by a hydrogen bond between Gly-15 and Asp-122, which is about 8 Å from the active site of the enzyme. The conformational change in the Met-20 loop and in an adjacent loop is 'transmitted' to the active site by small (0.5 Å), concerted motions of residues in this region, and eventually leads to the formation of an active site that has high affinity for the substrate's transition state [7]. As explained in Chapters 1 and 9, transition-state stabilization is the principal strategy that enzymes use to catalyze reactions. The importance of the non-catalytic residues on the distant loops of DHFR is reflected in their high evolutionary conservation. Additional examples of long-range dynamics in other enzymes can be found in [7].

5.3.2.1.4 Allostery without conformational change [171,308]

The original definition of allostery relates to ligand-induced change of activity in proteins. As explained earlier, such change was originally attributed to conformational changes, sometimes considerably large ones, which were observed in the studied proteins (e.g., in metamorphic proteins [228,229]). More recent studies, however, have revealed that some proteins undergo changes in activity with no accompanying conformational changes, or with small ones. Proteins in which this curious phenomenon has been detected include calbindin D9k [260], eglin C [309], catabolite activator protein [310], methionine repressor [309], and the PDZ domain [224]. In these proteins and in others displaying this phenomenon, NMR measurements[*1] demonstrated changes in bond vibrations as a result of ligand binding. In cases where calorimetric measurements were also conducted, entropy changes were observed, despite the absence of conformational changes. Together, these results seem to be in line with the suggestions made by Cooper and Dryden over 20 years ago, regarding allostery involving dynamic, yet non-conformational changes in proteins [311]. According to this model, **ligand binding may either increase or decrease the entropy of the protein by changing the frequency and/or amplitude of its thermal fluctuations**[*2]. Specifically, such changes to a protein's thermal fluctuations result in a new array of vibrations, ranging from highly correlated low-frequency harmonic fluctuations to local anharmonic ones of individual atoms or chemical groups. Although the exact underlying mechanism is not entirely clear, these new vibrations can yield two possible outcomes: rigidification of the polypeptide chain (i.e., reduction of its entropy), or enhanced flexibility of the chain (i.e., an increase in its entropy).

How do the allosteric-induced changes in entropy that are described above affect ligand binding? The binding of the protein's primary ligand (substrate, hormone, or messenger) to its cognate binding site leads to a decrease in the entropy of the protein, which (unsuccessfully) opposes the binding[*3]. However, in cases in which the entropy of the free protein has already been partially decreased by binding to the allosteric ligand, the second binding (to the primary ligand) has little added effect, which makes the second binding more plausible and increases the activity of the protein. In such cases, the allosteric ligand acts as an activator. In other cases, the allosteric ligand may act as an inhibitor, by increasing the entropy of the free protein, thus making the binding to the natural ligand entropically more costly.

The new model changes the previous definition of allostery; it postulates that ligand binding still brings about a change in protein activity, but not necessarily via conformational changes, which are enthalpic in essence[*4]. The model suggests three cases of allosteric regulation:

1. Regulation involving only entropy changes (i.e., only vibrational changes)

2. Regulation involving only enthalpy changes (i.e., only conformational changes)

3. Regulation involving both entropy and enthalpy changes (i.e., both vibrational and conformational changes)

In other words, binding of a positive regulator to an allosteric site in a protein may increase the activity at the protein's primary site by any one of the following ways, or by both:

1. Decreasing the entropy cost of binding the primary ligand (i.e., rigidification of the primary site)

2. Increasing the enthalpy gain of the binding (i.e., changing the conformation of primary site)

A negative regulator does the opposite.

[*1]NMR is particularly suitable for this type of measurement, since it covers a wide range of motions, from ps to ms (see Section 5.4 below).

[*2]That is, covalent bond vibrations. In contrast to atomic translations, which characterize conformational changes, vibrations behave like harmonic oscillators, i.e., they vibrate around the same conformation in a spring-like manner.

[*3]This is because a reduction in entropy is in essence an unfavorable process (second law of thermodynamics; see Chapter 4).

[*4]This is because they change noncovalent interactions in the protein, and create new ones between protein and primary ligand.

Although the entropic model constitutes a refreshing innovation of an old theory, it has yet to gain wide acceptance in the scientific community. For example, Cui and Karplus have argued that vibrational changes alone cannot explain the significant change of free energy that allostery involves [171]. This is because the vibrations, which are calculated on the basis of measurements of low-frequency normal modes[*1], are collective, which makes them insensitive to the level of detail of the method used for their calculation [315]. In addition, Nussinov and Tsai [316] have contended that conformational changes may have been present in the aforementioned studies, even if they were not observed. Failure to observe such conformational changes could have happened for various reasons, including the following:

- Crystallization conditions and crystal effects

- One of the conformational states was disordered

- The structural comparisons disregarded the quaternary protein structure

- Synergy effects among allosteric effectors were overlooked

5.3.2.1.5 Applications

The discussion above demonstrates why an understanding of the workings of cells requires a deep understanding of allostery. However, an understanding of allostery is also required for other, more applicative fields [253]. One such field is drug design. As discussed in Chapter 8, the design of new drugs, which are in most cases small organic molecules, is based mainly on the physicochemical adaptation of the drug molecule to the active site of the target protein. In allosteric enzymes and other proteins, the binding site for the allosteric regulator is an additional potential target for the design process. Nevertheless, allosteric sites are seldom involved in the design process, because it is much more difficult to predict how ligand binding to such a site affects protein activity. As potential targets, allosteric sites should be even more attractive than the active sites, since they are under less evolutionary pressure, and are therefore more species-specific [317,318]. One case in which drugs were designed to target an allosteric site is that of p38 MAP kinase [294].

Another applicative field that could benefit from the consideration of allosteric mechanisms is protein design, that is, the rational design of either completely new proteins or improved versions of existing proteins. Protein design is carried out for specific goals, such as improving protein stability and/or activity, changing substrate and/or reaction specificity, and developing new chemical reactions. Again, designing an allosteric site is far more challenging than modifying an active site, because of the difficulty of predicting long-range effects. One case in which this has been done successfully is the engineering of a new allosteric site in tyrosine phosphatase (PTP) [319] (see details in [253]).

[*1]Normal modes (NM) are measurable vectors representing concerted atomic vibrations in molecules [312]. By describing the interatomic bonds in proteins as harmonic oscillating springs, one can calculate the frequency of a periodic motion associated with each NM [313] (see Section 5.4 below). Since NM can explain all the collective motions within the protein [314], analysis of NM provides a way to find correlations between different parts of the protein.

5.3.2.2   Dynamics induced by environmental changes

As explained in Subsection 5.3.2.1.2 above, the native conformation of the protein is that of lowest energy under the specific environmental conditions of the protein (pH, temperature, pressure, and salts concentrations). Thus, changes in these conditions may favor a different conformation over the native one. In other words, **changing the environmental conditions of a protein may induce conformational changes, with accompanying changes in function.** The changes may be extreme, as in thermal denaturation, caused by a heating-induced increase in protein dynamics up to the point of hydrogen-bond disruption and structural unfolding (see Chapter 4). Milder environmental changes may induce moderate conformational or dynamic changes that do not endanger the structure and/or function of the protein, and these are sometimes used for regulatory purposes. The role of mild environmental changes is nicely demonstrated in the case of hemoglobin; a slight drop in pH changes the function of hemoglobin, in accordance with its physiological role (see Box 5.3). The pH drop changes the ionization of certain histidine residues, and this leads to the formation or breaking of noncovalent interactions. Since such interactions either stabilize or destabilize protein structure, changing them may lead to conformational changes.

5.3.2.3   Enzyme-mediated protein dynamics

The two types of external factors mentioned in the previous sections, i.e., ligand binding and environmental changes, influence the energetic aspect of protein dynamics. There is at least one dynamic change in the polypeptide chain that requires kinetic assistance, i.e., the help of an enzyme. This is the *cis-trans* isomerization of the peptide bond, located between a proline residue and the residue preceding it [320]. As explained in Chapter 2, this process occurs in proline alone, because in other amino acids the *trans* configuration is overwhelmingly more stable than the *cis*. In the case of proline, the energy barrier between the two configurations is much lower, so despite the low rate of the isomerization process [321], it is still observed in proteins. The enzymes catalyzing the process are called *peptidyl prolyl cis-trans isomerases* (*PPIases*). Their action has been investigated mainly in the context of protein folding, but some studies focused on the folded state as well. In most cases the process seems to involve proline residues on outwardly facing loops, and it affects the conformation of the loop. Although the conformational change is usually local, there are documented cases of long-range changes. In such cases, the proline isomerization change seems to act as a molecular switch, which turns the biological function of the protein on or off.

Such is the case with the SH2 regulatory domain of *interleukin-2 tyrosine kinase* (*Itk*), functioning in the activation of T-lymphocytes during the immune response. Within this domain, the peptide bond between Asn-286 and Pro-287 undergoes *cis-trans* isomerization, which leads to conformational changes that extend far beyond the specific loop in which this peptide bond resides (Figure 5.11). These changes redesign the surface of the SH2 domain, and as a result change its ligand preference; in the *trans* configuration the domain binds to phosphotyrosine residues on peptides, whereas in the *cis* configuration it binds the SH3 domain of an adjacent Itk protein [322]. Again, similar cases have been documented in other proteins as well, and it seems that, compared with the *trans* configuration, the *cis* configuration provides the proline-containing segment with greater capacity to create contacts with other parts of the protein. It is possible that these contacts stabilize the *cis* configuration, thus making it easier for the protein to overcome the energy barrier separating the two configurations.

(a)

(b)

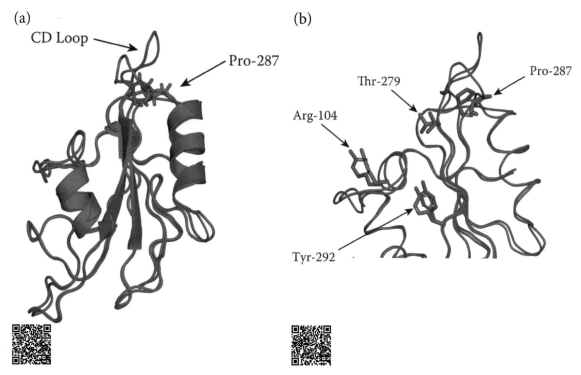

FIGURE 5.11 **Conformational changes in the SH2 domain of interleukin-2 tyrosine kinase (Itk), following isomerization of Pro-287 from *trans* to *cis*.** (a) The two conformations (PDB entries 1luk and 1lun) are colored differently and superimposed. The changes occur in different regions of the domain, but are most pronounced in the CD loop, which contains the Pro-287. (b) Long-range effects of the isomerization. The effects are illustrated via three residues along the phosphotyrosine binding site in the SH2 domain: Thr-279, Arg-104, and Tyr-292. Although the latter two are distant from Pro-287, they too display structural differences between the *cis* and *trans* conformations.

Another case in which an enzyme-assisted reaction affects protein dynamics is the formation of disulfide bonds. First, such bonds rigidify the polypeptide chain, thus reducing its inherent dynamics. Moreover, the formation of such bonds at one region of the protein may trigger a conformational change affecting another region, similarly to the workings of allosteric ligands [253,323]. For example, in botulinum neurotoxin type A, two chains are held together by a single disulfide bond, such that the active site in the catalytic chain is blocked by the other chain. When the toxin invades the cell, either the reducing environment in the cytoplasm or the acidity of the endosome reduces the S−S bond, thus removing the blockage and activating the catalytic chain [324]. Other examples of disulfide-involving protein dynamics can be found in [253]. Interestingly, analysis of 'allosteric' disulfide bonds in different proteins shows that these bonds all have conformations of high potential energy [325]. In other words, such disulfides are less stable than others, and are therefore much easier to cleave.

5.4 METHODS FOR STUDYING PROTEIN DYNAMICS

In Chapter 3 we saw how certain biophysical methods for structure determination can also be used for studying the dynamic behavior of proteins; these methods include NMR, EPR, incoherent neutron scattering, and certain applications of electron microscopy. Generally speaking, the study of protein dynamics is somewhat problematic, as many of the substates sampled by proteins are short-lived. It is therefore very difficult to characterize these substates at high resolution, as such characterization is usually carried out by X-ray crystallography, which requires static and homogeneous structures [6]. It could be argued that the dynamics of such structures is embodied within the B-factors specified along with the structure coordinates, but, as explained in Chapter 3, these could also represent static disorder in the crystal [326]. Another method that is based on diffraction of particles but that does provide dynamics-related data is neutron diffraction. This method uses collisions of different elasticity to determine atomic motions in molecules, from fast vibrations to diffusion. Some of the other tools in use are described in the following subsections.

5.4.1   Tools for studying slow (ms–sec) to intermediate (ns–μs) motions

One solution to the problem of multiple substates comes in the form of biochemical methods and approaches (e.g., [327]) that allow biophysicists to 'trap' individual substates and characterize them. Another common method used for studying motions in the ms to sec range is *hydrogen ^1H-^2H exchange* [178,328]. In this method, the protein is placed in an aqueous solution of 'heavy' hydrogen isotopes (deuterium, ^2H), and the exchange of the light hydrogen isotopes (protium, ^1H) in backbone amide groups with the heavy isotopes in solution is observed. The exchange is possible only for those hydrogen atoms that are exposed to the solvent and that are not involved in chemical bonds. Thus, conformational changes in the protein that involve the exposure or burial of hydrogen atoms, or the creation or breaking of hydrogen bonds, are expected to affect the exchange, and therefore provide information about the changes. The location of the heavy hydrogen isotopes has been measured in the past by NMR alone. Today, such measurements can also be made using *mass spectrometry*, which detects the change in weight following the exchange, although on a slower timescale that mainly corresponds to folding or unfolding and ligand binding.

The most common method for following long-range dynamics with high resolution, under physiological conditions, and without the need for metabolites to trap substates, is NMR [329–331]. The principles of NMR as a means of structure determination are explained in Chapter 3. In studies of protein dynamics, this method entails exciting the nuclei of atoms in the sample and extracting relevant data during their relaxation[*1]. In the case of very rapid motions the informative capacity of solution NMR is limited, but the range can be stretched from ps to s by using solid-state NMR, in which a dry powder of the protein is used [6]. It is possible to render the measurement location-specific by tagging certain regions in the protein with detectable isotopes of hydrogen (^1H, ^2H), carbon (^{13}C), or nitrogen (^{15}N). Moreover, the measurements are able to provide not only the conformations involved in the dynamic process, but also the probabilities of their sampling, and the corresponding rates of the shifts between them. Finally, thanks to the extensive substate-sampling capacity of NMR measurements, it is possible to determine a protein's entropy changes that characterize certain biologically relevant processes, such as ligand binding [333].

It should be noted that most of the NMR techniques developed thus far employ diluted protein preparations. The environment of the proteins in such preparations is very different from the environment in living cells, where the macromolecular concentration is roughly

---

[*1]NMR techniques for characterization of internal protein dynamics include heteronuclear Overhauser effect spectroscopy (HOESY), relaxation dispersion, paramagnetic relaxation enhancement (PRE), and residual dipolar coupling (RDC) [332].

300 to 400 g/L. This difference is likely to influence both the structure and the dynamics of the proteins. To overcome this problem, new NMR techniques have been developed, which measure protein dynamics either inside cells (*in-cell NMR*) or in cell lysates [332]. Since NMR requires mM concentrations of the investigated protein, and the cellular concentration of most cellular proteins is sub-μM (and ~100 μM at the most), the measurements usually involve protein overexpression and isotopic enrichment/labeling.

When NMR is carried out in solution or inside a cell, where a large number of protein molecules exist, the measurements are limited to synchronized motions. In order to measure non-synchronized motions, the biophysicist is compelled to use a single protein molecule. This means that a method must be used in which the individual protein is tagged. Such methods are often based on fluorescent tagging of the protein [334]. That is, the protein is attached to a tag that emits fluorescent radiation upon excitation (i.e., fluorophore), and this emission is measured[*1]. The emission is affected by the immediate environment of the fluorophore. Thus, conformational changes around the fluorophore are expected to change the emission pattern, and provide data on protein dynamics in that region. The high sensitivity of fluorescent methods allows scientists not only to detect single protein molecules, but also to follow their dynamics within the wide range of 10^{-15} to 1 sec [6]. One of the most efficient variations of this approach, called *fluorescence resonance energy transfer (FRET)* [335], uses two fluorophores, and measures the transfer of radiation energy between them[*2]. In this application of the method, fluorescence is measured for a period of time sufficient to provide data regarding the dynamics in that region (e.g., see [336]). It should be mentioned, though, that this information is limited[*3], which is why biophysicists often integrate it with high-resolution data of the protein.

In addition to the lab methods mentioned above (and others), computational methods are also used to learn about protein dynamics, with MD simulations being the primary tool [337–339]. The principles of such computational methods have been described in Chapter 3 in relation to structure prediction. As explained, MD simulations are (currently) inefficient in describing the entire folding of the typical protein, mainly because the calculations are too heavy to describe processes that are longer than the μs scale[*4]. However, this approach can be used for describing folding intermediates [340,342] or dynamic processes occurring in the folded state (e.g., ligand-induced fit of an active site or allosterically-induced conformational changes). The computational approach has a few important advantages. First, it is faster and cheaper compared with laboratory methods. Second, methods such as MD provide atomically detailed descriptions of the simulated systems. Third, all the intermediary states are available to the biophysicist, provided they are within the timescale covered by the simulation. Finally, in addition to the atomic locations and movements, the calculations also provide the corresponding energies. Although the absolute energy values are considered inaccurate, the relative values are usually good enough for drawing general conclusions. The data extracted from MD simulations are particularly useful when combined

---

[*1]Fluorescence techniques are used because of their higher sensitivity compared to NMR.

[*2]See Chapter 3, Subsection 3.3.3.1 in for a brief description of the method.

[*3]FRET has a lower spatial resolution than, e.g., NMR. In addition, the fluorescent tag may sometimes affect the inherent dynamics of the protein.

[*4]Note, however, that recent progress in supercomputing (e.g., the Anton machine) and in distributed computing (e.g., Folding@Home) has enabled researchers to simulate milliseconds-long folding processes of small proteins, and to extract thermodynamic and kinetic quantities [340,341] (see Chapter 3 for further details).

with data from other methods. Indeed, MD simulations can bridge the gaps between the structural data provided by X-ray crystallography and the kinetic data provided by NMR.

As mentioned above, cooperative thermal fluctuations in proteins are believed to result in large equilibrium motions that have functional significance. Examples of such an equilibrium motion include the motion of a loop, which uncovers a ligand binding site in a receptor or an enzyme, and the relative motions of subunits in a membrane-bound transporter, which create a path for the transported ion or molecule. Often, such motions mediate allosteric regulation of the protein per cellular needs. As in the case of folding, these large motions are usually too slow to be fully covered by MD simulations. In such cases, a different method called *'normal mode analysis'* (*NMA*) [182,343] is often used to capture the collective modes of motion that underlie the equilibrium dynamics of macromolecules. In this method, proteins are usually described using *elastic network models* (*ENM*) [344], in which each protein is represented by a network of nodes connected by harmonic oscillators (i.e., springs). The springs represent the protein's interacting virtual bonds, and the nodes are the atoms connected by these bonds. Since large motions in proteins tend to be insensitive to atomic details, the atoms need not necessarily be described explicitly. For example, in the popular *Gaussian network model* (*GNM*) [345,346] (the simplest form of ENM), each node is occupied by a single mass, which accounts for an entire residue. By accounting for the collective motions of the springs, NMA is able to calculate the resulting lower-frequency motions (*normal modes*), which, as mentioned above, are usually of biological interest. In each of these modes, all parts of the protein move with the same frequency. The use of GNM and other coarse-grained models (e.g., the *anisotropic network model*, or *ANM*) allow normal mode analysis to be implemented on large protein systems, such as the ribosome [347,348] or the viral capsid [349,350]. There are several freely accessible web servers for running NMA calculations, including the following:

- **ANM 2.0:** http://anm.csb.pitt.edu/cgi-bin/anm2/anm2.cgi

- **WEBnm@:** http://apps.cbu.uib.no/webnma/home

- **FlexServ:** http://mmb.pcb.ub.es/FlexServ/input.php

- **HingeProt:** http://www.prc.boun.edu.tr/appserv/prc/hingeprot/

- **DynaFace:** http://safir.prc.boun.edu.tr/dynaface/

5.4.1.1 Tools for studying rapid motions (fs–ps)

Some of the tools mentioned in the previous section are also capable of tracking rapid protein motions. Still, covering vibrations in the fs to ps range (e.g., when studying catalytic processes) requires other methods. One of the methods developed specifically for this purpose is based on ultra-fast laser pulses that initiate catalytic processes in photoactivated enzymes [8]. Zewail's group from the California Institute of Technology extended this idea to a research method called *'4D ultra-fast electron diffraction, crystallography, and microscopy'* [351]. In this approach, electrons are shot very fast and in a timed manner at the protein, and the time-dependent change in their scattering is detected by a charge-coupled device (CCD) sensor. The measurements can be carried out in an aqueous sample using diffraction or microscopy, or in a protein crystal, using crystallography. Since electrons are used, no tagging is needed, and the data collected relate to the entire protein. The method is sufficiently fast to study catalytic events in enzymes, in addition to the slower conformational changes. Another biophysical method used for studying fast dynamics in proteins is *incoherent neutron scattering*, which we encountered in Chapter 3. The greatest advantage of this method is probably that it can apply to both protein and solvent dynamics. However, it requires equipment that is not readily accessible, and the sample often has to be tagged with a heavy isotope to boost the signal.

Computational methods have also evolved to deal with the very short timescales of atomic vibrations. MD simulations are capable of describing molecular events up to the ps timescale, such as proton transfer of electron tunneling. Dynamic events on the fs timescale involve rearrangements of electrons, such as in the formation and breaking of covalent bonds [8], which can only be described by quantum-mechanical (QM) calculations [339,352]. As mentioned in Chapter 3, QM calculations require immense computer resources, and therefore cannot be applied to macromolecules. However, in recent years methods have been developed for the integration of QM calculations with molecular mechanical and dynamic (MM/MD) calculations in a manner that increases the accuracy of the latter [353,354]. In this approach, most parts of the protein are described by MM models, whereas those parts in which ultra-fast events occur (electron transfer, formation of reaction intermediates, electronic excitation, etc.) are described by the QM model. The main challenge in this approach is usually the coupling of the two different protein regions, and the description of the boundary regions between them.

5.5 SUMMARY

- Proteins are dynamic entities; they exhibit numerous different motions that vary in timescale, amplitude, and energy.

- Protein dynamics is important for function.

- Proteins may fold in different ways and at different rates, but they all do so while minimizing both their energy and entropy. The folding process involves numerous intermediary conformations of high energy, which act as kinetic barriers that have to be overcome to complete the process.

- Inside cells, folding of large proteins is often assisted by molecular chaperones.

- The folded state itself is dynamic and consists of an ensemble of conformations that are constantly sampled by the protein. The native conformation is the one of least energy. The other conformations are sampled using thermal energy (RT), drawn from the environment and converted into kinetic energy. As a result, these conformations all have similar energies that, at room temperature, are at most 0.6 kcal/mol (RT) higher than the native conformation.

- The natural conformational equilibrium of the folded protein can be affected by various factors, including ligand binding, post-translational modification, and change in environmental conditions (e.g., pH). In each case, the influencing factor affects the conformational equilibrium in the protein. For example, a ligand binds preferentially to the conformation that has the least energy when bound to it. When this conformation is not the native one, the net effect of the binding is a conformational change. This may in turn lead to functional changes, e.g., when the conformation chosen by the ligand has an increased or decreased affinity to the natural substrate or ligand of the protein. This phenomenon, called allostery, is one of the most important regulatory tools used by cells to modulate the rate of their metabolic processes. Another strategy is to phosphorylate protein residues; this strategy is based on a similar premise, stabilizing a conformation that is different from the native one.

• Different methods, both experimental and computational, have been developed to describe the dynamic events in proteins. Experimental approaches include biochemical methods used to trap reaction intermediates, as well as NMR, incoherent neutron scattering, and biophysical methods relying on fast lasers or highly sensitive fluorescence tags. Computational methods mainly include normal mode analysis and MD simulations, which are usually combined with structural and kinetic data obtained experimentally. Ultra-fast events require quantum-mechanical calculations, which, due to their complexity, are integrated with other, more computationally economic models.

EXERCISES

5.1 Describe protein folding in terms of conformation, free energy, entropy, and kinetic barriers.

5.2 Explain the main differences between the framework, hydrophobic collapse, and nucleation–condensation models of protein folding.

5.3 Is the native structure the most stable form of a protein? Explain your answer.

5.4 Describe the process of amyloid fibril formation.

    A. List the main types of molecular chaperones.

    B. What are the unique features of chaperonins such as the GroEL–GroES complex?

5.5 Explain how small structural changes in the folded conformation of a protein can lead to dramatic changes in its activity.

5.6 List the three main factors that may affect the folded-state dynamics of proteins.

5.7 A. Explain how ligand binding can allosterically affect protein activity. Base your answer on the MWC, KNF, and conformational selection models.

    B. Do the models explain the function of both protein activators and inhibitors? How?

5.8 A. How does Max Perutz's 'stereochemical' model explain the well-known phenomenon of positive cooperativity in hemoglobin action?

    B. Briefly explain the main structural features underlying this phenomenon.

    C. Does the model also explain the well-known Bohr effect of hemoglobin?

REFERENCES

1. E. Fischer. Einfluss der Configuration auf die Wirkung der Enzyme. *Ber. Dtsch. Chem. Ges.*, 27(3):2985–2993, 1894.
2. D. E. Koshland Jr. Enzyme flexibility and enzyme action. *J. Cell. Comp. Physiol.*, 54:245–58, 1959.
3. F. Karush. Heterogeneity of the Binding Sites of Bovine Serum Albumin1. *J. Am. Chem. Soc.*, 72(6):2705–2713, 1950.
4. M. Gerstein and N. Echols. Exploring the range of protein flexibility, from a structural proteomics perspective. *Curr. Opin. Chem. Biol.*, 8(1):14–9, 2004.
5. J. Evenas, S. Forsen, A. Malmendal, and M. Akke. Backbone dynamics and energetics of a calmodulin domain mutant exchanging between closed and open conformations. *J. Mol. Biol.*, 289(3):603–17, 1999.
6. K. Henzler-Wildman and D. Kern. Dynamic personalities of proteins. *Nature*, 450(7172):964–72, 2007.
7. A. Tousignant and J. N. Pelletier. Protein motions promote catalysis. *Chem. Biol.*, 11(8):1037–42, 2004.
8. D. Zhong. Ultrafast catalytic processes in enzymes. *Curr. Opin. Chem. Biol.*, 11(2):174–81, 2007.
9. S. J. Benkovic and S. Hammes-Schiffer. A perspective on enzyme catalysis. *Science*, 301(5637):1196–202, 2003.
10. M. Akke. NMR methods for characterizing microsecond to millisecond dynamics in recognition and catalysis. *Curr. Opin. Struct. Biol.*, 12(5):642–7, 2002.
11. K. Teilum, J. G. Olsen, and B. B. Kragelund. Functional aspects of protein flexibility. *Cell. Mol. Life Sci.*, 66(14):2231–47, 2009.
12. P. Agarwal. Enzymes: An integrated view of structure, dynamics and function. *Microb. Cell Fact.*, 5(1):2, 2006.
13. M. F. Dunn. Protein–Ligand Interactions: General Description. In *Encyclopedia of Life Sciences.* John Wiley & Sons, Ltd., 2001.
14. A. F. Neuwald. The glycine brace: a component of Rab, Rho, and Ran GTPases associated with hinge regions of guanine- and phosphate-binding loops. *BMC Struct. Biol.*, 9:11–28, 2009.
15. G. Bao. Mechanics of biomolecules. *J. Mech. Phys. Solids*, 50:2237–2274, 2002.
16. Y. Liu and I. Bahar. Sequence evolution correlates with structural dynamics. *Mol. Biol. Evol.*, 29(9):2253–63, 2012.
17. N. Echols, D. Milburn, and M. Gerstein. MolMovDB: analysis and visualization of conformational change and structural flexibility. *Nucleic Acids Res.*, 31(1):478–82, 2003.
18. E. Z. Eisenmesser, O. Millet, W. Labeikovsky, D. M. Korzhnev, M. Wolf-Watz, D. A. Bosco, J. J. Skalicky, L. E. Kay, and D. Kern. Intrinsic dynamics of an enzyme underlies catalysis. *Nature*, 438(7064):117–21, 2005.
19. C. B. Anfinsen. Principles that Govern the Folding of Protein Chains. *Science*, 181(96):223–230, 1973.
20. K. A. Dill, S. B. Ozkan, M. S. Shell, and T. R. Weikl. The protein folding problem. *Annu. Rev. Biophys.*, 37:289–316, 2008.
21. A. N. Naganathan and V. Munoz. Scaling of folding times with protein size. *J. Am. Chem. Soc.*, 127(2):480–1, 2005.
22. K. W. Plaxco, K. T. Simons, and D. Baker. Contact order, transition state placement and the refolding rates of single domain proteins. *J. Mol. Biol.*, 277(4):985–94, 1998.
23. A. R. Fersht. Transition-state structure as a unifying basis in protein-folding mechanisms: contact order, chain topology, stability, and the extended nucleus mechanism. *Proc. Natl. Acad. Sci. USA*, 97(4):1525–9, 2000.
24. J. W. Kelly. Structural biology: proteins downhill all the way. *Nature*, 442(7100):255–6, 2006.
25. D. J. Brockwell and S. E. Radford. Intermediates: ubiquitous species on folding energy landscapes? *Curr. Opin. Struct. Biol.*, 17(1):30–7, 2007.
26. C. Levinthal. Are there pathways for protein folding? *J. Chim. Phys. Phys.- Chim. Biol.*, 65:44–49, 1968.
27. J. Kubelka, T. K. Chiu, D. R. Davies, W. A. Eaton, and J. Hofrichter. Sub-microsecond protein folding. *J. Mol. Biol.*, 359(3):546–53, 2006.
28. J. Kubelka, J. Hofrichter, and W. A. Eaton. The protein folding 'speed limit'. *Curr. Opin. Struct. Biol.*, 14(1):76–88, 2004.
29. Y. Xu, P. Purkayastha, and F. Gai. Nanosecond folding dynamics of a three-stranded beta-sheet. *J. Am. Chem. Soc.*, 128(49):15836–42, 2006.
30. G. D. Rose, P. J. Fleming, J. R. Banavar, and A. Maritan. A backbone-based theory of protein folding. *Proc. Natl. Acad. Sci. USA*, 103(45):16623–33, 2006.
31. R. F. Service. Problem solved* (*sort of). *Science*, 321(5890):784–6, 2008.

32. C. J. Tsai and R. Nussinov. Hydrophobic folding units at protein-protein interfaces: implications to protein folding and to protein-protein association. *Protein Sci.*, 6(7):1426–37, 1997.

33. K. A. Dill. Theory for the folding and stability of globular proteins. *Biochemistry*, 24(6):1501–9, 1985.

34. J. D. Bryngelson and P. G. Wolynes. Spin glasses and the statistical mechanics of protein folding. *Proc. Natl. Acad. Sci. USA*, 84(21):7524–8, 1987.

35. P. G. Wolynes, J. N. Onuchic, and D. Thirumalai. Navigating the folding routes. *Science*, 267(5204):1619–20, 1995.

36. J. D. Bryngelson, J. N. Onuchic, N. D. Socci, and P. G. Wolynes. Funnels, pathways, and the energy landscape of protein folding: a synthesis. *Proteins*, 21(3):167–95, 1995.

37. C. M. Dobson, A. Sali, and M. Karplus. Protein folding: a perspective from theory and experiment. *Angew. Chem. Int. Ed.*, 37(7):868–893, 1998.

38. R. Kapon, R. Nevo, and Z. Reich. Protein energy landscape roughness. *Biochem. Soc. Trans.*, 36(Pt 6):1404–8, 2008.

39. V. Daggett and A. R. Fershet. Is there a unifying mechanism for protein folding? *Trends Biochem. Sci.*, 28(1):18–25, 2003.

40. S. Kumar and R. Nussinov. Close-range electrostatic interactions in proteins. *ChemBioChem*, 3(7):604–617, 2002.

41. M. Ohgushi and A. Wada. 'Molten-globule state': a compact form of globular proteins with mobile side-chains. *FEBS Lett.*, 164(1):21–4, 1983.

42. F. Gabel, D. Bicout, U. Lehnert, M. Tehei, M. Weik, and G. Zaccai. Protein dynamics studied by neutron scattering. *Q. Rev. Biophys.*, 35(4):327–67, 2002.

43. B. Zagrovic, C. D. Snow, M. R. Shirts, and V. S. Pande. Simulation of folding of a small alpha-helical protein in atomistic detail using worldwide-distributed computing. *J. Mol. Biol.*, 323(5):927–37, 2002.

44. M. Karplus and D. L. Weaver. Protein folding dynamics: the diffusion-collision model and experimental data. *Protein Sci.*, 3(4):650–68, 1994.

45. R. L. Baldwin and G. D. Rose. Is protein folding hierarchic? II. Folding intermediates and transition states. *Trends Biochem. Sci.*, 24(2):77–83, 1999.

46. R. L. Baldwin and G. D. Rose. Is protein folding hierarchic? I. Local structure and peptide folding. *Trends Biochem. Sci.*, 24(1):26–33, 1999.

47. D. Bashford, F. E. Cohen, M. Karplus, I. D. Kuntz, and D. L. Weaver. Diffusion-collision model for the folding kinetics of myoglobin. *Proteins*, 4(3):211–27, 1988.

48. C. Nishimura, S. Prytulla, H. Jane Dyson, and P. E. Wright. Conservation of folding pathways in evolutionarily distant globin sequences. *Nat. Struct. Biol.*, 7(8):679–86, 2000.

49. K. B. Wong, J. Clarke, C. J. Bond, J. L. Neira, S. M. Freund, A. R. Fersht, and V. Daggett. Towards a complete description of the structural and dynamic properties of the denatured state of barnase and the role of residual structure in folding. *J. Mol. Biol.*, 296(5):1257–82, 2000.

50. J. K. Myers and T. G. Oas. Contribution of a buried hydrogen bond to lambda repressor folding kinetics. *Biochemistry*, 38(21):6761–8, 1999.

51. M. Sadqi, D. Fushman, and V. Munoz. Atom-by-atom analysis of global downhill protein folding. *Nature*, 442(7100):317–21, 2006.

52. V. I. Abkevich, A. M. Gutin, and E. I. Shakhnovich. Specific nucleus as the transition state for protein folding: evidence from the lattice model. *Biochemistry*, 33(33):10026–36, 1994.

53. A. R. Fersht. Nucleation mechanisms in protein folding. *Curr. Opin. Struct. Biol.*, 7(1):3–9, 1997.

54. L. S. Itzhaki, D. E. Otzen, and A. R. Fersht. The structure of the transition state for folding of chymotrypsin inhibitor 2 analysed by protein engineering methods: evidence for a nucleation–condensation mechanism for protein folding. *J. Mol. Biol.*, 254(2):260–88, 1995.

55. S. Gianni, N. R. Guydosh, F. Khan, T. D. Caldas, U. Mayor, G. W. White, M. L. DeMarco, V. Daggett, and A. R. Fersht. Unifying features in protein-folding mechanisms. *Proc. Natl. Acad. Sci. USA*, 100(23):13286–91, 2003.

56. G. W. White, S. Gianni, J. G. Grossmann, P. Jemth, A. R. Fersht, and V. Daggett. Simulation and experiment conspire to reveal cryptic intermediates and a slide from the nucleation–condensation to framework mechanism of folding. *J. Mol. Biol.*, 350(4):757–75, 2005.

57. C. Travaglini-Allocatelli, Y. Ivarsson, P. Jemth, and S. Gianni. Folding and stability of globular proteins and implications for function. *Curr. Opin. Struct. Biol.*, 19(1):3–7, 2009.

58. J. H. Han, S. Batey, A. A. Nickson, S. A. Teichmann, and J. Clarke. The folding and evolution of multidomain proteins. *Nat. Rev. Mol. Cell Biol.*, 8(4):319–30, 2007.

59. G. A. Papoian and P. G. Wolynes. The physics and bioinformatics of binding and folding-an energy landscape perspective. *Biopolymers*, 68(3):333–49, 2003.

60. K. Itoh and M. Sasai. Cooperativity, connectivity, and folding pathways of multidomain proteins. *Proc. Natl. Acad. Sci. USA*, 105(37):13865–70, 2008.

61. K. Sridevi, G. S. Lakshmikanth, G. Krishnamoorthy, and J. B. Udgaonkar. Increasing stability reduces conformational heterogeneity in a protein folding intermediate ensemble. *J. Mol. Biol.*, 337(3):699–711, 2004.

62. R. D. Palmiter. Quantitation of parameters that determine the rate of ovalbumin synthesis. *Cell*, 4(3):189–197, 1975.

63. F. U. Hartl and M. Hayer-Hartl. Converging concepts of protein folding in vitro and in vivo. *Nat. Struct. Mol. Biol.*, 16(6):574–81, 2009.

64. J. Lu and C. Deutsch. Folding zones inside the ribosomal exit tunnel. *Nat. Struct. Mol. Biol.*, 12(12):1123–9, 2005.

65. C. A. Woolhead, P. J. McCormick, and A. E. Johnson. Nascent membrane and secretory proteins differ in FRET-detected folding far inside the ribosome and in their exposure to ribosomal proteins. *Cell*, 116(5):725–36, 2004.

66. G. Kramer, D. Boehringer, N. Ban, and B. Bukau. The ribosome as a platform for co-translational processing, folding and targeting of newly synthesized proteins. *Nat. Struct. Mol. Biol.*, 16(6):589–97, 2009.

67. S. E. Jackson. How do small single-domain proteins fold? *Fold. Des.*, 3(4):R81–91, 1998.

68. L. M. Luheshi, D. C. Crowther, and C. M. Dobson. Protein misfolding and disease: from the test tube to the organism. *Curr. Opin. Chem. Biol.*, 12(1):25–31, 2008.

69. OMIM (Online Mendelian Inheritance in Man) database. Cystic Fibrosis (entry No. 219700). http://www.ncbi.nlm.nih.gov/entrez/dispomim.cgi?id=219700, 2008.

70. F. Chiti and C. M. Dobson. Protein misfolding, functional amyloid, and human disease. *Annu. Rev. Biochem.*, 75:333–66, 2006.

71. J. W. Kelly. The alternative conformations of amyloidogenic proteins and their multi-step assembly pathways. *Curr. Opin. Struct. Biol.*, 8(1):101–6, 1998.

72. J. C. Rochet and P. T. Lansbury Jr. Amyloid fibrillogenesis: themes and variations. *Curr. Opin. Struct. Biol.*, 10(1):60–8, 2000.

73. A. T. Alexandrescu. Amyloid accomplices and enforcers. *Protein Sci.*, 14(1):1–12, 2005.

74. T. Stromer and L. C. Serpell. Structure and morphology of the Alzheimer's amyloid fibril. *Microsc. Res. Tech.*, 67(3–4):210–7, 2005.

75. G. G. Kovacs. Molecular Pathological Classification of Neurodegenerative Diseases: Turning towards Precision Medicine. *Int. J. Mol. Sci.*, 17(2), 2016.

76. D. J. Selkoe. Alzheimer's disease. In the beginning. *Nature*, 354(6353):432–3, 1991.

77. G. S. Bloom. Amyloid-beta and tau: the trigger and bullet in Alzheimer disease pathogenesis. *JAMA Neurol.*, 71(4):505–8, 2014.

78. H. Zempel and E. Mandelkow. Lost after translation: missorting of Tau protein and consequences for Alzheimer disease. *Trends Neurosci.*, 37(12):721–32, 2014.

79. A. S. Carlo. Sortilin, a novel APOE receptor implicated in Alzheimer disease. *Prion*, 7(5):378–82, 2013.

80. M. G. Spillantini, M. L. Schmidt, V. M. Lee, J. Q. Trojanowski, R. Jakes, and M. Goedert. Alpha-synuclein in Lewy bodies. *Nature*, 388(6645):839–40, 1997.

81. P. Westermark, M. D. Benson, J. N. Buxbaum, A. S. Cohen, B. Frangione, S. Ikeda, C. L. Masters, G. Merlini, M. J. Saraiva, and J. D. Sipe. Amyloid: toward terminology clarification. Report from the Nomenclature Committee of the International Society of Amyloidosis. *Amyloid*, 12(1):1–4, 2005.

82. C. M. Dobson. Protein folding and misfolding. *Nature*, 426(6968):884–90, 2003.

83. M. Fandrich and C. M. Dobson. The behaviour of polyamino acids reveals an inverse side chain effect in amyloid structure formation. *EMBO J.*, 21(21):5682–90, 2002.

84. H. A. Lashuel, D. Hartley, B. M. Petre, T. Walz, and P. T. Lansbury Jr. Neurodegenerative disease: amyloid pores from pathogenic mutations. *Nature*, 418(6895):291, 2002.

85. L. C. Serpell, M. Sunde, M. D. Benson, G. A. Tennent, M. B. Pepys, and P. E. Fraser. The protofilament substructure of amyloid fibrils. *J. Mol. Biol.*, 300(5):1033–9, 2000.

86. R. Nelson, M. R. Sawaya, M. Balbirnie, A. O. Madsen, C. Riekel, R. Grothe, and D. Eisenberg. Structure of the cross-beta spine of amyloid-like fibrils. *Nature*, 435(7043):773–8, 2005.

87. M. Stefani and C. M. Dobson. Protein aggregation and aggregate toxicity: new insights into protein folding, misfolding diseases and biological evolution. *J. Mol. Med.*, 81(11):678–99, 2003.

88. J. P. Cleary, D. M. Walsh, J. J. Hofmeister, G. M. Shankar, M. A. Kuskowski, D. J. Selkoe, and K. H. Ashe. Nat-

ural oligomers of the amyloid-beta protein specifically disrupt cognitive function. *Nat. Neurosci.*, 8(1):79–84, 2005.

89. R. Krishnan, J. L. Goodman, S. Mukhopadhyay, C. D. Pacheco, E. A. Lemke, A. A. Deniz, and S. Lindquist. Conserved features of intermediates in amyloid assembly determine their benign or toxic states. *Proc. Natl. Acad. Sci. USA*, 109(28):11172–11177, 2012.

90. W. H. DePas and M. R. Chapman. Microbial manipulation of the amyloid fold. *Res. Microbiol.*, 163(9–10):592–606, 2012.

91. G. Y. Cheung, D. Kretschmer, S. Y. Queck, H. S. Joo, R. Wang, A. C. Duong, T. H. Nguyen, T. H. Bach, A. R. Porter, F. R. DeLeo, A. Peschel, and M. Otto. Insight into structure-function relationship in phenol-soluble modulins using an alanine screen of the phenol-soluble modulin (PSM) α3 peptide. *FASEB J.*, 28(1):153–61, 2014.

92. E. Tayeb-Fligelman, O. Tabachnikov, A. Moshe, O. Goldshmidt-Tran, M. R. Sawaya, N. Coquelle, J.-P. Colletier, and M. Landau. The cytotoxic *Staphylococcus aureus* PSMα3 reveals a cross-α amyloid-like fibril. *Science*, 355(6327):831–833, 2017.

93. D. Ghosh, P. K. Singh, S. Sahay, N. N. Jha, R. S. Jacob, S. Sen, A. Kumar, R. Riek, and S. K. Maji. Structure based aggregation studies reveal the presence of helix-rich intermediate during α-Synuclein aggregation. *Sci. Rep.*, 5:9228, 2015.

94. A. Kessel and N. Ben-Tal. Free energy determinants of peptide association with lipid bilayers. *Curr. Top. Membr.*, 52:205–253, 2002.

95. J. C. Sacchettini and J. W. Kelly. Therapeutic strategies for human amyloid diseases. *Nat. Rev. Drug Discov.*, 1(4):267–75, 2002.

96. W. Hoyer, C. Gronwall, A. Jonsson, S. Stahl, and T. Hard. Stabilization of a β-hairpin in monomeric Alzheimer's amyloid-β peptide inhibits amyloid formation. *Proc. Natl. Acad. Sci. USA*, 105:5099–5104, 2008.

97. L. Jiang, C. Liu, D. Leibly, M. Landau, M. Zhao, M. P. Hughes, and D. S. Eisenberg. Structure-based discovery of fiber-binding compounds that reduce the cytotoxicity of amyloid beta. *eLife*, 2:e00857, 2013.

98. E. Monsellier and F. Chiti. Prevention of amyloid-like aggregation as a driving force of protein evolution. *EMBO Rep.*, 8(8):737–42, 2007.

99. J. S. Richardson and D. C. Richardson. Natural beta-sheet proteins use negative design to avoid edge-to-edge aggregation. *Proc. Natl. Acad. Sci. USA*, 99(5):2754–9, 2002.

100. B. Offmann, M. Tyagi, and A. G. de Brevern. Local protein structures. *Curr. Bioinform.*, 2(3):165–202, 2007.

101. J. S. Richardson, E. D. Getzoff, and D. C. Richardson. The beta bulge: a common small unit of nonrepetitive protein structure. *Proc. Natl. Acad. Sci. USA*, 75(6):2574–8, 1978.

102. A. W. Chan, E. G. Hutchinson, D. Harris, and J. M. Thornton. Identification, classification, and analysis of beta-bulges in proteins. *Protein Sci.*, 2(10):1574–90, 1993.

103. F. Rousseau, L. Serrano, and J. W. Schymkowitz. How evolutionary pressure against protein aggregation shaped chaperone specificity. *J. Mol. Biol.*, 355(5):1037–47, 2006.

104. B. M. Broome and M. H. Hecht. Nature disfavors sequences of alternating polar and nonpolar amino acids: implications for amyloidogenesis. *J. Mol. Biol.*, 296(4):961–8, 2000.

105. M. W. West, W. Wang, J. Patterson, J. D. Mancias, J. R. Beasley, and M. H. Hecht. De novo amyloid proteins from designed combinatorial libraries. *Proc. Natl. Acad. Sci. USA*, 96(20):11211–6, 1999.

106. P. K. Auluck, H. Y. Chan, J. Q. Trojanowski, V. M. Lee, and N. M. Bonini. Chaperone suppression of alpha-synuclein toxicity in a Drosophila model for Parkinson's disease. *Science*, 295(5556):865–8, 2002.

107. E. A. Nollen, S. M. Garcia, G. van Haaften, S. Kim, A. Chavez, R. I. Morimoto, and R. H. Plasterk. Genome-wide RNA interference screen identifies previously undescribed regulators of polyglutamine aggregation. *Proc. Natl. Acad. Sci. USA*, 101(17):6403–8, 2004.

108. E. Gazit. A possible role for pi-stacking in self-assembly of amyloid fibrils. *FASEB J.*, 16:77–83, 2002.

109. S. K. Burley and G. A. Petsko. Aromatic-aromatic interaction: a mechanism of protein structure stabilization. *Science*, 229(4708):23–8, 1985.

110. C. G. Claessens and J. F. Stoddart. π-π interactions in self-assembly. *J. Phys. Org. Chem. Biol.*, 10:254–272, 1997.

111. B. Ma and R. Nussinov. Trp/Met/Phe hot spots in protein-protein interactions: potential targets in drug design. *Curr. Top. Med. Chem.*, 7(10):999–1005, 2007.

112. F. Chiti, M. Stefani, N. Taddei, G. Ramponi, and C. M. Dobson. Rationalization of the effects of mutations on peptide and protein aggregation rates. *Nature*, 424(6950):805–8, 2003.

113. J. Shorter and S. Lindquist. Prions as adaptive conduits of memory and inheritance. *Nat. Rev. Genet.*, 6(6):435–50, 2005.

114. S. B. Prusiner. Novel proteinaceous infectious particles cause scrapie. *Science*, 216(4542):136–44, 1982.

115. N. J. Cobb and W. K. Surewicz. Prion Diseases and Their Biochemical Mechanisms (dagger). *Biochemistry*, 48:2574–2585, 2009.

116. P. Saá, D. A. Harris, and L. Cervenakova. Mechanisms of prion-induced neurodegeneration. *Expert Rev. Mol. Med.*, 18:e5, 2016.

117. E. Cohen, J. Bieschke, R. M. Perciavalle, J. W. Kelly, and A. Dillin. Opposing activities protect against age-onset proteotoxicity. *Science*, 313(5793):1604–10, 2006.

118. H. Jang, J. Zheng, R. Lal, and R. Nussinov. New structures help the modeling of toxic amyloid beta ion channels. *Trends Biochem. Sci.*, 33(2):91–100, 2008.

119. H. A. Lashuel and P. T. Lansbury Jr. Are amyloid diseases caused by protein aggregates that mimic bacterial pore-forming toxins? *Q. Rev. Biophys.*, 39(2):167–201, 2006.

120. K. A. Brogden. Antimicrobial peptides: pore formers or metabolic inhibitors in bacteria? *Nat. Rev. Microbiol.*, 3(3):238–50, 2005.

121. R. E. Hancock and G. Diamond. The role of cationic antimicrobial peptides in innate host defences. *Trends Microbiol.*, 8(9):402–10, 2000.

122. M. Zasloff. Antimicrobial peptides of multicellular organisms. *Nature*, 415(6870):389–95, 2002.

123. E. Monsellier, M. Ramazzotti, N. Taddei, and F. Chiti. Aggregation propensity of the human proteome. *PLoS Comput. Biol.*, 4(10):e1000199, 2008.

124. P. Hunter. Shedding a negative image. Research into their mechanism of infectivity reveals that prions might have important biological roles. *EMBO Rep.*, 7(12):1196–8, 2006.

125. K. Si, S. Lindquist, and E. R. Kandel. A neuronal isoform of the aplysia CPEB has prion-like properties. *Cell*, 115(7):879–91, 2003.

126. J. D. Richter. Think globally, translate locally: what mitotic spindles and neuronal synapses have in common. *Proc. Natl. Acad. Sci. USA*, 98(13):7069–71, 2001.

127. M. Schroder and R. J. Kaufman. The mammalian unfolded protein response. *Annu. Rev. Biochem.*, 74:739–89, 2005.

128. R. J. Ellis. Chaperone function: the orthodox view. In B. Henderson and G. Pockley, editors, *Molecular Chaperones and Cell Signalling.* Cambridge University Press, Cambridge, UK, 2005.

129. J. C. Young, V. R. Agashe, K. Siegers, and F. U. Hartl. Pathways of chaperone-mediated protein folding in the cytosol. *Nat. Rev. Mol. Cell Biol.*, 5(10):781–91, 2004.

130. T. M. Treweek, A. M. Morris, and J. A. Carver. Intracellular protein unfolding and aggregation: the role of small heat-shock chaperone proteins. *Aust. J. Chem.*, 56:357–367, 2003.

131. S. D. Westerheide and R. I. Morimoto. Heat shock response modulators as therapeutic tools for diseases of protein conformation. *J. Biol. Chem.*, 280(39):33097–100, 2005.

132. R. U. Mattoo and P. Goloubinoff. Molecular chaperones are nanomachines that catalytically unfold misfolded and alternatively folded proteins. *Cell. Mol. Life Sci.*, 71(17):3311–25, 2014.

133. M. H. Glickman and A. Ciechanover. The Ubiquitin-Proteasome Proteolytic Pathway: Destruction for the Sake of Construction. *Physiol. Rev.*, 82:373–428, 2002.

134. M. Y. Sherman and A. L. Goldberg. Cellular defenses against unfolded proteins: a cell biologist thinks about neurodegenerative diseases. *Neuron*, 29(1):15–32, 2001.

135. E. L. Eskelinen and P. Saftig. Autophagy: A lysosomal degradation pathway with a central role in health and disease. *Biochim. Biophys. Acta*, 1793(4):664–73, 2009.

136. A. Finka and P. Goloubinoff. Proteomic data from human cell cultures refine mechanisms of chaperone-mediated protein homeostasis. *Cell Stress Chaperones*, 18(5):591–605, 2013.

137. M. Hayer-Hartl, A. Bracher, and F. U. Hartl. The GroEL-GroES Chaperonin Machine: A Nano-Cage for Protein Folding. *Trends Biochem. Sci.*, 41(1):62–76, 2016.

138. L. Skjaerven, J. Cuellar, A. Martinez, and J. M. Valpuesta. Dynamics, flexibility, and allostery in molecular chaperonins. *FEBS Lett.*, 589(19 Pt A):2522–32, 2015.

139. A. Leitner, L. A. Joachimiak, A. Bracher, L. Monkemeyer, T. Walzthoeni, B. Chen, S. Pechmann, S. Holmes, Y. Cong, B. Ma, S. Ludtke, W. Chiu, F. U. Hartl, R. Aebersold, and J. Frydman. The molecular architecture of the eukaryotic chaperonin TRiC/CCT. *Structure*, 20(5):814–25, 2012.

140. D. Thirumalai and G. H. Lorimer. Chaperonin-mediated protein folding. *Annu. Rev. Biophys. Biomol. Struct.*, 30:245–69, 2001.

141. M. J. Kerner, D. J. Naylor, Y. Ishihama, T. Maier, H. C. Chang, A. P. Stines, C. Georgopoulos, D. Frishman,

M. Hayer-Hartl, M. Mann, and F. U. Hartl. Proteome-wide analysis of chaperonin-dependent protein fold-ing in *Escherichia coli*. *Cell*, 122(2):209–20, 2005.

142. A. Azia, R. Unger, and A. Horovitz. What distinguishes GroEL substrates from other *Escherichia coli* pro-teins? *FEBS J.*, 279(4):543–50, 2012.

143. M. M. Gromiha and S. Selvaraj. Inter-residue interactions in protein folding and stability. *Prog. Biophys. Mol. Biol.*, 86(2):235–77, 2004.

144. R. Gruber and A. Horovitz. Allosteric Mechanisms in Chaperonin Machines. *Chem. Rev.*, 2016.

145. S. Sharma, K. Chakraborty, B. K. Müller, N. Astola, Y.-C. Tang, D. C. Lamb, M. Hayer-Hartl, and F. U. Hartl. Monitoring Protein Conformation along the Pathway of Chaperonin-Assisted Folding. *Cell*, 133(1):142–153, 2008.

146. Z. Lin, D. Madan, and H. S. Rye. GroEL stimulates protein folding through forced unfolding. *Nat. Struct. Mol. Biol.*, 15(3):303–11, 2008.

147. A. J. Gupta, S. Haldar, G. Milicic, F. U. Hartl, and M. Hayer-Hartl. Active cage mechanism of chaperonin-assisted protein folding demonstrated at single-molecule level. *J. Mol. Biol.*, 426(15):2739–54, 2014.

148. V. R. Agashe, S. Guha, H. C. Chang, P. Genevaux, M. Hayer-Hartl, M. Stemp, C. Georgopoulos, F. U. Hartl, and J. M. Barral. Function of trigger factor and DnaK in multidomain protein folding: increase in yield at the expense of folding speed. *Cell*, 117(2):199–209, 2004.

149. T. K. Chaudhuri, G. W. Farr, W. A. Fenton, S. Rospert, and A. L. Horwich. GroEL/GroES-mediated folding of a protein too large to be encapsulated. *Cell*, 107(2):235–46, 2001.

150. S. Priya, S. K. Sharma, V. Sood, R. U. Mattoo, A. Finka, A. Azem, P. De Los Rios, and P. Goloubinoff. GroEL and CCT are catalytic unfoldases mediating out-of-cage polypeptide refolding without ATP. *Proc. Natl. Acad. Sci. USA*, 110(18):7199–204, 2013.

151. Z. Lin and H. S. Rye. Expansion and compression of a protein folding intermediate by GroEL. *Mol. Cell*, 16(1):23–34, 2004.

152. Z. Lin and H. S. Rye. GroEL-mediated protein folding: making the impossible, possible. *Crit. Rev. Biochem. Mol. Biol.*, 41(4):211–39, 2006.

153. D. S. Libich, V. Tugarinov, and G. M. Clore. Intrinsic unfoldase/foldase activity of the chaperonin GroEL directly demonstrated using multinuclear relaxation-based NMR. *Proc. Natl. Acad. Sci. USA*, 112(29):8817–8823, 2015.

154. M. J. Todd, G. H. Lorimer, and D. Thirumalai. Chaperonin-facilitated protein folding: optimization of rate and yield by an iterative annealing mechanism. *Proc. Natl. Acad. Sci. USA*, 93(9):4030–5, 1996.

155. D. Yang, X. Ye, and G. H. Lorimer. Symmetric GroEL:GroES2 complexes are the protein-folding functional form of the chaperonin nanomachine. *Proc. Natl. Acad. Sci. USA*, 110(46):E4298–305, 2013.

156. A. Azem, S. Diamant, M. Kessel, C. Weiss, and P. Goloubinoff. The protein-folding activity of chap-eronins correlates with the symmetric GroEL14(GroES7)2 heterooligomer. *Proc. Natl. Acad. Sci. USA*, 92(26):12021–5, 1995.

157. A. Koike-Takeshita, M. Yoshida, and H. Taguchi. Revisiting the GroEL-GroES reaction cycle via the sym-metric intermediate implied by novel aspects of the GroEL(D398A) mutant. *J. Biol. Chem.*, 283(35):23774–81, 2008.

158. A. C. Apetri and A. L. Horwich. Chaperonin chamber accelerates protein folding through passive action of preventing aggregation. *Proc. Natl. Acad. Sci. USA*, 105(45):17351–5, 2008.

159. Y. C. Tang, H. C. Chang, A. Roeben, D. Wischnewski, N. Wischnewski, M. J. Kerner, F. U. Hartl, and M. Hayer-Hartl. Structural features of the GroEL-GroES nano-cage required for rapid folding of encapsu-lated protein. *Cell*, 125(5):903–14, 2006.

160. A. Brinker, G. Pfeifer, M. J. Kerner, D. J. Naylor, F. U. Hartl, and M. Hayer-Hartl. Dual function of protein confinement in chaperonin-assisted protein folding. *Cell*, 107(2):223–33, 2001.

161. S. Nisemblat, O. Yaniv, A. Parnas, F. Frolow, and A. Azem. Crystal structure of the human mitochondrial chaperonin symmetrical football complex. *Proc. Natl. Acad. Sci. USA*, 112(19):6044–9, 2015.

162. D. D. Boehr, D. McElheny, H. J. Dyson, and P. E. Wright. The dynamic energy landscape of dihydrofolate reductase catalysis. *Science*, 313(5793):1638–42, 2006.

163. M. Karplus and J. Kuriyan. Molecular dynamics and protein function. *Proc. Natl. Acad. Sci. USA*, 102(19):6679–85, 2005.

164. M. Vendruscolo and C. M. Dobson. Structural biology. Dynamic visions of enzymatic reactions. *Science*, 313(5793):1586–7, 2006.

165. A. Mittermaier and L. E. Kay. New tools provide new insights in NMR studies of protein dynamics. *Science*, 312(5771):224–8, 2006.

166. B. F. Volkman, D. Lipson, D. E. Wemmer, and D. Kern. Two-state allosteric behavior in a single-domain signaling protein. *Science*, 291(5512):2429–33, 2001.

167. R. J. Sola and K. Griebenow. Influence of modulated structural dynamics on the kinetics of alpha-chymotrypsin catalysis. Insights through chemical glycosylation, molecular dynamics and domain motion analysis. *FEBS J.*, 273(23):5303–19, 2006.

168. G. G. Hammes. Multiple conformational changes in enzyme catalysis. *Biochemistry*, 41(26):8221–8, 2002.

169. L. Pauling. A theory of the structure and process of formation of antibodies. *J. Am. Chem. Soc.*, 62:2643–2657, 1940.

170. J. Monod, J. Wyman, and J. P. Changeux. On the nature of allosteric transitions: a plausible model. *J. Mol. Biol.*, 12:88–118, 1965.

171. Q. Cui and M. Karplus. Allostery and cooperativity revisited. *Protein Sci.*, 17(8):1295–307, 2008.

172. H. Frauenfelder, S. G. Sligar, and P. G. Wolynes. The energy landscapes and motions of proteins. *Science*, 254(5038):1598–603, 1991.

173. H. Taketomi, Y. Ueda, and N. Go. Studies on protein folding, unfolding and fluctuations by computer simulation. I. The effect of specific amino acid sequence represented by specific inter-unit interactions. *Int. J. Pept. Protein Res.*, 7(6):445–59, 1975.

174. A. Kitao, S. Hayward, and N. Go. Energy landscape of a native protein: jumping-among-minima model. *Proteins*, 33(4):496–517, 1998.

175. J. A. McCammon, B. R. Gelin, and M. Karplus. Dynamics of folded proteins. *Nature*, 267(5612):585–90, 1977.

176. G. A. Petsko and D. Ringe. Fluctuations in protein structure from X-ray diffraction. *Annu. Rev. Biophys. Bio.*, 13:331–71, 1984.

177. H. Frauenfelder, F. Parak, and R. D. Young. Conformational substates in proteins. *Annu. Rev. Biophys. Biophys. Chem.*, 17:451–79, 1988.

178. Y. Bai, T. R. Sosnick, L. Mayne, and S. W. Englander. Protein folding intermediates: native-state hydrogen exchange. *Science*, 269(5221):192–7, 1995.

179. R. Elber and M. Karplus. Multiple conformational states of proteins: a molecular dynamics analysis of myoglobin. *Science*, 235(4786):318–21, 1987.

180. L. C. James and D. S. Tawfik. Conformational diversity and protein evolution – a 60-year-old hypothesis revisited. *Trends Biochem. Sci.*, 28(7):361–8, 2003.

181. G. Weber. Ligand binding and internal equilibrium in proteins. *Biochemistry*, 11(5):864–878, 1972.

182. I. Bahar, T. R. Lezon, A. Bakan, and I. H. Shrivastava. Normal mode analysis of biomolecular structures: functional mechanisms of membrane proteins. *Chem. Rev.*, 110(3):1463–97, 2010.

183. H. Frauenfelder, P. W. Fenimore, and R. D. Young. Protein dynamics and function: insights from the energy landscape and solvent slaving. *IUBMB Life*, 59(8–9):506–12, 2007.

184. D. Chandler. *Introduction to Modern Statistical Mechanics*. Oxford University Press, New York, 1987.

185. B. Halle. Flexibility and packing in proteins. *Proc. Natl. Acad. Sci. USA*, 99(3):1274–9, 2002.

186. K. A. Henzler-Wildman, M. Lei, V. Thai, S. J. Kerns, M. Karplus, and D. Kern. A hierarchy of timescales in protein dynamics is linked to enzyme catalysis. *Nature*, 450(7171):913–6, 2007.

187. D. T. Leeson and D. A. Wiersma. The Energy Landscape of Myoglobin: An Optical Study. *J. Chem. Phys. B*, 101:6331–6340, 1997.

188. P. Radivojac, L. M. Iakoucheva, C. J. Oldfield, Z. Obradovic, V. N. Uversky, and A. K. Dunker. Intrinsic disorder and functional proteomics. *Biophys. J.*, 92(5):1439–56, 2007.

189. M. Laberge, J. M. Vanderkooi, and K. A. Sharp. The effect of a protein electric field on the CO-stretch frequency. Finite Difference Poisson-Boltzmann calculations on carbonmonoxy cytochrome c. *J. Phys. Chem.*, 100:10793–10801, 1996.

190. M. Brunori and Q. H. Gibson. Cavities and packing defects in the structural dynamics of myoglobin. *EMBO Rep.*, 2(8):671–9, 2001.

191. B. Vallone and M. Brunori. Roles for holes: are cavities in proteins mere packing defects? *Ital. J. Biochem.*, 53(1):46–52, 2004.

192. J. Cohen, K. Kim, P. King, M. Seibert, and K. Schulten. Finding gas diffusion pathways in proteins: application to O_2 and H_2 transport in CpI [FeFe]-hydrogenase and the role of packing defects. *Structure*, 13(9):1321–9, 2005.

193. B. J. Bahnson and J. P. Klinman. Hydrogen tunneling in enzyme catalysis. *Methods Enzymol.*, 249:373–97, 1995.

194. Y. Cha, C. J. Murray, and J. P. Klinman. Hydrogen tunneling in enzyme reactions. *Science*, 243(4896):1325–1330, 1989.

195. Z. X. Liang and J. P. Klinman. Structural bases of hydrogen tunneling in enzymes: progress and puzzles. *Curr. Opin. Struct. Biol.*, 14(6):648–55, 2004.

196. J. P. Klinman. Linking protein structure and dynamics to catalysis: the role of hydrogen tunnelling. *Philos. Trans. R. Soc. Lond. B: Biol. Sci.*, 361(1472):1323–31, 2006.

197. A. Kohen and J. P. Klinman. Hydrogen tunneling in biology. *Chem. Biol.*, 6(7):R191–8, 1999.

198. J. P. Layfield and S. Hammes-Schiffer. Hydrogen tunneling in enzymes and biomimetic models. *Chem. Rev.*, 114(7):3466–94, 2014.

199. J. P. Klinman and A. Kohen. Hydrogen tunneling links protein dynamics to enzyme catalysis. *Annu. Rev. Biochem.*, 82:471–96, 2013.

200. S. Hammes-Schiffer. Hydrogen tunneling and protein motion in enzyme reactions. *Acc. Chem. Res.*, 39(2):93–100, 2006.

201. M. P. Meyer, D. R. Tomchick, and J. P. Klinman. Enzyme structure and dynamics affect hydrogen tunneling: the impact of a remote side chain (I553) in soybean lipoxygenase-1. *Proc. Natl. Acad. Sci. USA*, 105(4):1146–51, 2008.

202. Cranberry. A (simplified) diagram of Quantum Tunneling. Wikipedia, the free encyclopedia. http://en.wikipedia.org/wiki/File:QuantumTunnel.jpg, 2008.

203. P. Ball. Physics of life: The dawn of quantum biology. *Nature*, 474(7351):272–4, 2011.

204. N. Lambert, Y.-N. Chen, Y.-C. Cheng, C.-M. Li, G.-Y. Chen, and F. Nori. Quantum biology. *Nat. Phys.*, 9(1):10–18, 2013.

205. A. A. Stuchebrukhov. Long-Distance Electron Tunneling in Proteins: A New Challenge for Time-Resolved Spectroscopy. *Laser Phys.*, 20(1):125–138, 2010.

206. H. B. Gray and J. R. Winkler. Electron Flow through Proteins. *Chem. Phys. Lett.*, 483(1–3):1–9, 2009.

207. G. S. Engel, T. R. Calhoun, E. L. Read, T. K. Ahn, T. Mancal, Y. C. Cheng, R. E. Blankenship, and G. R. Fleming. Evidence for wavelike energy transfer through quantum coherence in photosynthetic systems. *Nature*, 446(7137):782–6, 2007.

208. H. Lee, Y. C. Cheng, and G. R. Fleming. Coherence dynamics in photosynthesis: protein protection of excitonic coherence. *Science*, 316(5830):1462–5, 2007.

209. G. Panitchayangkoon, D. Hayes, K. A. Fransted, J. R. Caram, E. Harel, J. Wen, R. E. Blankenship, and G. S. Engel. Long-lived quantum coherence in photosynthetic complexes at physiological temperature. *Proc. Natl. Acad. Sci. USA*, 107(29):12766–70, 2010.

210. H. B. Chen, J. Y. Lien, C. C. Hwang, and Y. N. Chen. Long-lived quantum coherence and non-Markovianity of photosynthetic complexes. *Phys. Rev. E: Stat. Nonlin. Soft. Matter Phys.*, 89(4):042147, 2014.

211. M. Mohseni, P. Rebentrost, S. Lloyd, and A. Aspuru-Guzik. Environment-assisted quantum walks in photosynthetic energy transfer. *J. Chem. Phys.*, 129(17):174106, 2008.

212. G. Careri. Collective effects in hydrated proteins. In M. C. Bellisent-Funel, editor, *Hydration processes in biology: theoretical and Experimental approaches*, pages 143–155. IOS Press, 1999.

213. C. Mattos. Protein-water interactions in a dynamic world. *Trends Biochem. Sci.*, 27(4):203–8, 2002.

214. C. L. Brooks and M. Karplus. Solvent effects on protein motion and protein effects on solvent motion. Dynamics of the active site region of lysozyme. *J. Mol. Biol.*, 208(1):159–81, 1989.

215. P. W. Fenimore, H. Frauenfelder, B. H. McMahon, and R. D. Young. Bulk-solvent and hydration-shell fluctuations, similar to alpha- and beta-fluctuations in glasses, control protein motions and functions. *Proc. Natl. Acad. Sci. USA*, 101(40):14408–13, 2004.

216. P. W. Fenimore, H. Frauenfelder, B. H. McMahon, and F. G. Parak. Slaving: solvent fluctuations dominate protein dynamics and functions. *Proc. Natl. Acad. Sci. USA*, 99(25):16047–51, 2002.

217. A. Rubinstein and S. Sherman. Influence of the solvent structure on the electrostatic interactions in proteins. *Biophys. J.*, 87(3):1544–57, 2004.

218. N. V. Nucci, M. S. Pometun, and A. J. Wand. Site-resolved measurement of water-protein interactions by solution NMR. *Nat. Struct. Mol. Biol.*, 18(2):245–9, 2011.

219. V. J. Hilser. Structural biology: Finding the wet spots. *Nature*, 469(7329):166–7, 2011.

220. F. Xin and P. Radivojac. Post-translational modifications induce significant yet not extreme changes to protein structure. *Bioinformatics*, 28(22):2905–13, 2012.

221. H. S. Lee, Y. Qi, and W. Im. Effects of N-glycosylation on protein conformation and dynamics: Protein Data Bank analysis and molecular dynamics simulation study. *Sci. Rep.*, 5:8926, 2015.

222. A. De Simone, A. Dhulesia, G. Soldi, M. Vendruscolo, S.-T. D. Hsu, F. Chiti, and C. M. Dobson. Experi-

mental free energy surfaces reveal the mechanisms of maintenance of protein solubility. *Proc. Natl. Acad. Sci. USA*, 108(52):21057–21062, 2011.

223. A. del Sol, C. J. Tsai, B. Ma, and R. Nussinov. The origin of allosteric functional modulation: multiple pre-existing pathways. *Structure*, 17(8):1042–50, 2009.

224. J. F. Swain and L. M. Gierasch. The changing landscape of protein allostery. *Curr. Opin. Struct. Biol.*, 16(1):102–8, 2006.

225. A. Gutteridge and J. M. Thornton. Conformational changes observed in enzyme crystal structures upon substrate binding. *J. Mol. Biol.*, 346(1):21–8, 2005.

226. G. Parisi, D. J. Zea, A. M. Monzon, and C. Marino-Buslje. Conformational diversity and the emergence of sequence signatures during evolution. *Curr. Opin. Struct. Biol.*, 32:58–65, 2015.

227. A. M. Monzon, E. Juritz, M. S. Fornasari, and G. Parisi. CoDNaS: a database of conformational diversity in the native state of proteins. *Bioinformatics*, 29(19):2512–4, 2013.

228. A. G. Murzin. Biochemistry. Metamorphic proteins. *Science*, 320(5884):1725–6, 2008.

229. P. N. Bryan and J. Orban. Proteins that switch folds. *Curr. Opin. Struct. Biol.*, 20(4):482–8, 2010.

230. R. L. Tuinstra, F. C. Peterson, S. Kutlesa, E. S. Elgin, M. A. Kron, and B. F. Volkman. Interconversion between two unrelated protein folds in the lymphotactin native state. *Proc. Natl. Acad. Sci. USA*, 105(13):5057–5062, 2008.

231. D. R. Littler, S. J. Harrop, W. D. Fairlie, L. J. Brown, G. J. Pankhurst, S. Pankhurst, M. Z. DeMaere, T. J. Campbell, A. R. Bauskin, R. Tonini, M. Mazzanti, S. N. Breit, and P. M. Curmi. The intracellular chloride ion channel protein CLIC1 undergoes a redox-controlled structural transition. *J. Biol. Chem.*, 279(10):9298–305, 2004.

232. R. C. Tyler, N. J. Murray, F. C. Peterson, and B. F. Volkman. Native state interconversion of a metamorphic protein requires global unfolding. *Biochemistry*, 50(33):7077–7079, 2011.

233. J. Gsponer, J. Christodoulou, A. Cavalli, J. M. Bui, B. Richter, C. M. Dobson, and M. Vendruscolo. A coupled equilibrium shift mechanism in calmodulin-mediated signal transduction. *Structure*, 16(5):736–46, 2008.

234. J. Wyman Jr and D. W. Allen. The problem of the heme interactions in hemoglobin and the basis of the Bohr effect. *J. Polym. Sci., Part A: Polym. Chem.*, 7(5):499–518, 1951.

235. J. Wyman. Heme proteins. *Adv. Protein Chem.*, 4:407–531, 1948.

236. K. K. Frederick, M. S. Marlow, K. G. Valentine, and A. J. Wand. Conformational entropy in molecular recognition by proteins. *Nature*, 448(7151):325–330, 2007.

237. J. A. Hanson, K. Duderstadt, L. P. Watkins, S. Bhattacharyya, J. Brokaw, J. W. Chu, and H. Yang. Illuminating the mechanistic roles of enzyme conformational dynamics. *Proc. Natl. Acad. Sci. USA*, 104:18055–18060, 2007.

238. J. Monod and F. Jacob. Teleonomic mechanisms in cellular metabolism, growth, and differentiation. *Cold Spring Harb. Symp. Quant. Biol.*, 26:389–401, 1961.

239. L. Pauling. The Oxygen Equilibrium of Hemoglobin and Its Structural Interpretation. *Proc. Natl. Acad. Sci. USA*, 21(4):186–91, 1935.

240. D. E. Koshland Jr, G. Nemethy, and D. Filmer. Comparison of experimental binding data and theoretical models in proteins containing subunits. *Biochemistry*, 5(1):365–85, 1966.

241. J. Monod, J. P. Changeux, and F. Jacob. Allosteric proteins and cellular control systems. *J. Mol. Biol.*, 6:306–29, 1963.

242. A. B. Pardee. Regulatory molecular biology. *Cell Cycle*, 5(8):846–52, 2006.

243. E. L. Roberts, N. Shu, M. J. Howard, R. W. Broadhurst, A. Chapman-Smith, J. C. Wallace, T. Morris, J. E. Cronan Jr, and R. N. Perham. Solution structures of apo and holo biotinyl domains from acetyl coenzyme A carboxylase of *Escherichia coli* determined by triple-resonance nuclear magnetic resonance spectroscopy. *Biochemistry*, 38(16):5045–53, 1999.

244. D. Zhao, C. H. Arrowsmith, X. Jia, and O. Jardetzky. Refined solution structures of the *Escherichia coli* trp holo- and aporepressor. *J. Mol. Biol.*, 229(3):735–46, 1993

245. K. Gunasekaran, B. Ma, B. Ramakrishnan, P. K. Qasba, and R. Nussinov. Interdependence of backbone flexibility, residue conservation, and enzyme function: a case study on beta1,4-galactosyltransferase-I. *Biochemistry*, 42(13):3674–87, 2003.

246. B. Ramakrishnan, P. V. Balaji, and P. K. Qasba. Crystal structure of beta1,4-galactosyltransferase complex with UDP-Gal reveals an oligosaccharide acceptor binding site. *J. Mol. Biol.*, 318(2):491–502, 2002.

247. K. M. Ottemann, W. Xiao, Y. K. Shin, and D. E. Koshland Jr. A piston model for transmembrane signaling of the aspartate receptor. *Science*, 285(5434):1751–4, 1999.

248. K. Gunasekaran, B. Ma, and R. Nussinov. Is allostery an intrinsic property of all dynamic proteins? *Proteins*, 57(3):433–43, 2004.

249. I. Bahar, C. Chennubhotla, and D. Tobi. Intrinsic dynamics of enzymes in the unbound state and relation to allosteric regulation. *Curr. Opin. Struct. Biol.*, 17(6):633–40, 2007.

250. B. Ma, S. Kumar, C. J. Tsai, and R. Nussinov. Folding funnels and binding mechanisms. *Protein Eng.*, 12(9):713–20, 1999.

251. H. R. Bosshard. Molecular recognition by induced fit: how fit is the concept? *News Physiol. Sci.*, 16:171–3, 2001.

252. M. B. Jackson. Allosteric Mechanisms in the Activation of Ligand-Gated Channels. In L. De Flice, editor, *Biophysics Textbook Online*, pages 1–48. Biophysical Society, Bethesda, MD, 2002.

253. R. A. Laskowski, F. Gerick, and J. M. Thornton. The structural basis of allosteric regulation in proteins. *FEBS Lett.*, 583:1692–8, 2009.

254. P. G. Wolynes. Symmetry and the energy landscapes of biomolecules. *Proc. Natl. Acad. Sci. USA*, 93(25):14249–55, 1996.

255. D. U. Ferreiro, J. A. Hegler, E. A. Komives, and P. G. Wolynes. On the role of frustration in the energy landscapes of allosteric proteins. *Proc. Natl. Acad. Sci. USA*, 108(9):3499–503, 2011.

256. A. D. Vogt, N. Pozzi, Z. Chen, and E. Di Cera. Essential role of conformational selection in ligand binding. *Biophys. Chem.*, 186:13–21, 2014.

257. J. P. Junker, F. Ziegler, and M. Rief. Ligand-dependent equilibrium fluctuations of single calmodulin molecules. *Science*, 323(5914):633–7, 2009.

258. P. Ascenzi and M. Fasano. Allostery in a monomeric protein: the case of human serum albumin. *Biophys. Chem.*, 148(1–3):16–22, 2010.

259. J. P. Changeux and S. J. Edelstein. Allosteric mechanisms of signal transduction. *Science*, 308(5727):1424–8, 2005.

260. D. Kern and E. R. Zuiderweg. The role of dynamics in allosteric regulation. *Curr. Opin. Struct. Biol.*, 13(6):748–57, 2003.

261. Nobel Foundation. The Nobel Prize in Physiology or Medicine 1965. http://nobelprize.org/nobel_prizes/medicine/laureates/1965/, 2009.

262. OldakQuill. Hb saturation curve. Wikipedia, the free encyclopedia. http://en.wikipedia.org/wiki/File:Hb_saturation_curve.png, 2005.

263. D. C. Mauzerall. Evolution of porphyrins. *Clin. Dermatol.*, 16(2):195–201, 1998.

264. T. S. do Nascimento, R. O. Pereira, H. L. de Mello, and J. Costa. Methemoglobinemia: from diagnosis to treatment. *Rev. Bras. Anestesiol.*, 58(6):651–64, 2008.

265. S. E. Phillips and B. P. Schoenborn. Neutron diffraction reveals oxygen-histidine hydrogen bond in oxymyoglobin. *Nature*, 292(5818):81–2, 1981.

266. S. Borman. A mechanism essential to life. *Chem. Eng. News*, 77:31–36, 1999.

267. M. F. Perutz, M. G. Rossmann, A. F. Cullis, H. Muirhead, G. Will, and A. C. T. North. Structure of haemoglobin: A three-dimensional Fourier synthesis at 5.5-Å resolution, obtained by X-ray analysis. *Nature*, 185:416–422, 1960.

268. M. F. Perutz, A. J. Wilkinson, M. Paoli, and G. G. Dodson. The stereochemical mechanism of the cooperative effects in hemoglobin revisited. *Annu. Rev. Biophys. Biomol. Struct.*, 27:1–34, 1998.

269. C. A. Sawicki and Q. H. Gibson. Quaternary conformational changes in human hemoglobin studies by laser photolysis of carboxyhemoglobin. *J. Biol. Chem.*, 251:1533–1542, 1976.

270. C. M. Jones, A. Ansari, E. R. Henry, G. W. Christoph, J. Hofrichter, and W. A. Eaton. Speed of intersubunit communication in proteins. *Biochemistry*, 31(29):6692–702, 1992.

271. M. F. Perutz, G. Fermi, B. Luisi, B. Shaanan, and R. C. Liddington. Stereochemistry of Cooperative Mechanisms in Hemoglobin. *Acc. Chem. Res.*, 20(9):309–321, 1987.

272. A. C. Guyton and J. E. Hall. *Textbook of medical physiology*. W. B. Saunders Company, Philadelphia, 9th edition, 1996.

273. Nobel Foundation. The Nobel Prize in Chemistry 1962. http://nobelprize.org/nobel_prizes/chemistry/laureates/1962/, 2009.

274. S. Adachi, S. Y. Park, J. R. Tame, Y. Shiro, and N. Shibayama. Direct observation of photolysis-induced tertiary structural changes in hemoglobin. *Proc. Natl. Acad. Sci. USA*, 100(12):7039–44, 2003.

275. M. F. Perutz. Stereochemistry of cooperative effects in haemoglobin. *Nature*, 228(5273):726–39, 1970.

276. A. W. Lee and M. Karplus. Structure-specific model of hemoglobin cooperativity. *Proc. Natl. Acad. Sci. USA*, 80(23):7055–9, 1983.

277. E. R. Henry, S. Bettati, J. Hofrichter, and W. A. Eaton. A tertiary two-state allosteric model for hemoglobin. *Biophys. Chem.*, 98(1–2):149–64, 2002.

278. C. Bohr. Theoretische Behandlung der quantitativen Verhaltnisse der Kohlensaurebindung des Hamoglobins. *Zentralblatt Für Physiologie*, XVII:713–716, 1903.

279. C. Bohr, K. Hasselbalch, and A. Krogh. About a new biological relation of high importance that the blood carbonic acid tension exercises on its oxygen binding. *Skand. Arch. Physiol.*, 16:404–4012, 1904.

280. F. B. Jensen. Red blood cell pH, the Bohr effect, and other oxygenation-linked phenomena in blood O_2 and CO_2 transport. *Acta Physiol. Scand.*, 182(3):215–27, 2004.

281. G. Zheng, M. Schaefer, and M. Karplus. Hemoglobin Bohr effects: atomic origin of the histidine residue contributions. *Biochemistry*, 52(47):8539–55, 2013.

282. T. Yonetani, S. I. Park, A. Tsuneshige, K. Imai, and K. Kanaori. Global allostery model of hemoglobin. Modulation of O(2) affinity, cooperativity, and Bohr effect by heterotropic allosteric effectors. *J. Biol. Chem.*, 277(37):34508–20, 2002.

283. C. Viappiani, S. Bettati, S. Bruno, L. Ronda, S. Abbruzzetti, A. Mozzarelli, and W. A. Eaton. New insights into allosteric mechanisms from trapping unstable protein conformations in silica gels. *Proc. Natl. Acad. Sci. USA*, 101(40):14414–9, 2004.

284. B. G. Forget and H. F. Bunn. Classification of the Disorders of Hemoglobin. *Cold Spring Harb. Perspect. Med.*, 3(2):a011684, 2013.

285. R. Valdes Jr and G. K. Ackers. Thermodynamic studies on subunit assembly in human hemoglobin. Self-association of oxygenated chains (alphaSH and betaSH): determination of stoichiometries and equilibrium constants as a function of temperature. *J. Biol. Chem.*, 252(1):74–81, 1977.

286. R. Benesch and R. E. Benesch. Homos and heteros among the hemos. *Science*, 185(4155):905–8, 1974.

287. A. Kurtz, H. S. Rollema, and C. Bauer. Heterotropic interactions in monomeric beta SH chains from human hemoglobin. *Arch. Biochem. Biophys.*, 210(1):200–3, 1981.

288. R. D. Kidd, H. M. Baker, A. J. Mathews, T. Brittain, and E. N. Baker. Oligomerization and ligand binding in a homotetrameric hemoglobin: two high-resolution crystal structures of hemoglobin Bart's ($\gamma(4)$), a marker for alpha-thalassemia. *Protein Sci.*, 10(9):1739–49, 2001.

289. G. E. Borgstahl, P. H. Rogers, and A. Arnone. The 1.9 Å structure of deoxy beta 4 hemoglobin. Analysis of the partitioning of quaternary-associated and ligand-induced changes in tertiary structure. *J. Mol. Biol.*, 236(3):831–43, 1994.

290. G. E. Borgstahl, P. H. Rogers, and A. Arnone. The 1.8 Å structure of carbonmonoxy-β 4 hemoglobin. Analysis of a homotetramer with the R quaternary structure of liganded $\alpha_2\beta_2$ hemoglobin. *J. Mol. Biol.*, 236(3):817–30, 1994.

291. D. Wootten, A. Christopoulos, and P. M. Sexton. Emerging paradigms in GPCR allostery: implications for drug discovery. *Nat. Rev. Drug Discov.*, 12(8):630–44, 2013.

292. H. S. Meharena, P. Chang, M. M. Keshwani, K. Oruganty, A. K. Nene, N. Kannan, S. S. Taylor, and A. P. Kornev. Deciphering the structural basis of eukaryotic protein kinase regulation. *PLoS Biol.*, 11(10):e1001680, 2013.

293. H. Frauenfelder, B. H. McMahon, R. H. Austin, K. Chu, and J. T. Groves. The role of structure, energy landscape, dynamics, and allostery in the enzymatic function of myoglobin. *Proc. Natl. Acad. Sci. USA*, 98(5):2370–4, 2001.

294. C. Pargellis, L. Tong, L. Churchill, P. F. Cirillo, T. Gilmore, A. G. Graham, P. M. Grob, E. R. Hickey, N. Moss, S. Pav, and J. Regan. Inhibition of p38 MAP kinase by utilizing a novel allosteric binding site. *Nat. Struct. Biol.*, 9(4):268–72, 2002.

295. W. E. Royer Jr. High-resolution crystallographic analysis of a co-operative dimeric hemoglobin. *J. Mol. Biol.*, 235(2):657–81, 1994.

296. W. E. Royer Jr, W. A. Hendrickson, and E. Chiancone. Structural transitions upon ligand binding in a co-operative dimeric hemoglobin. *Science*, 249(4968):518–21, 1990.

297. A. Horovitz and K. R. Willison. Allosteric regulation of chaperonins. *Curr. Opin. Struct. Biol.*, 15(6):646–51, 2005.

298. A. van der Vaart, J. Ma, and M. Karplus. The unfolding action of GroEL on a protein substrate. *Biophys. J.*, 87(1):562–73, 2004.

299. S. Kumar, B. Ma, C. J. Tsai, N. Sinha, and R. Nussinov. Folding and binding cascades: dynamic landscapes and population shifts. *Protein Sci.*, 9(1):10–9, 2000.

300. D. D. Boehr, R. Nussinov, and P. E. Wright. The role of dynamic conformational ensembles in biomolecular recognition. *Nat. Chem. Biol.*, 5(11):789–96, 2009.

301. X. Wang and R. G. Kemp. Reaction path of phosphofructo-1-kinase is altered by mutagenesis and alternative substrates. *Biochemistry*, 40(13):3938–42, 2001.

302. Y. Ikeda, N. Taniguchi, and T. Noguchi. Dominant negative role of the glutamic acid residue conserved in the pyruvate kinase M(1) isozyme in the heterotropic allosteric effect involving fructose-1,6-bisphosphate. *J. Biol. Chem.*, 275(13):9150–6, 2000.

303. W. A. Lim. The modular logic of signaling proteins: building allosteric switches from simple binding domains. *Curr. Opin. Struct. Biol.*, 12(1):61–8, 2002.

304. C. M. Falcon and K. S. Matthews. Engineered disulfide linking the hinge regions within lactose repressor dimer increases operator affinity, decreases sequence selectivity, and alters allostery. *Biochemistry*, 40(51):15650–9, 2001.

305. C. Grosman, M. Zhou, and A. Auerbach. Mapping the conformational wave of acetylcholine receptor channel gating. *Nature*, 403(6771):773–6, 2000.

306. P. Purohit, A. Mitra, and A. Auerbach. A stepwise mechanism for acetylcholine receptor channel gating. *Nature*, 446(7138):930–3, 2007.

307. R. Lape, D. Colquhoun, and L. G. Sivilotti. On the nature of partial agonism in the nicotinic receptor superfamily. *Nature*, pages 1–7, 2008.

308. C. J. Tsai, A. del Sol, and R. Nussinov. Allostery: absence of a change in shape does not imply that allostery is not at play. *J. Mol. Biol.*, 378(1):1–11, 2008.

309. M. W. Clarkson, S. A. Gilmore, M. H. Edgell, and A. L. Lee. Dynamic coupling and allosteric behavior in a nonallosteric protein. *Biochemistry*, 45(25):7693–9, 2006.

310. N. Popovych, S. Sun, R. H. Ebright, and C. G. Kalodimos. Dynamically driven protein allostery. *Nat. Struct. Mol. Biol.*, 13(9):831–8, 2006.

311. A. Cooper and D. T. Dryden. Allostery without conformational change. A plausible model. *Eur. Biophys. J.*, 11(2):103–9, 1984.

312. B. Brooks and M. Karplus. Harmonic dynamics of proteins: normal modes and fluctuations in bovine pancreatic trypsin inhibitor. *Proc. Natl. Acad. Sci. USA*, 80(21):6571–5, 1983.

313. M. Delarue and P. Dumas. On the use of low-frequency normal modes to enforce collective movements in refining macromolecular structural models. *Proc. Natl. Acad. Sci. USA*, 101(18):6957–62, 2004.

314. I. Bahar, T. R. Lezon, L. W. Yang, and E. Eyal. Global dynamics of proteins: bridging between structure and function. *Annu. Rev. Biophys.*, 39:23–42, 2010.

315. D. A. Kondrashov, A. W. Van Wynsberghe, R. M. Bannen, Q. Cui, and G. N. Phillips Jr. Protein structural variation in computational models and crystallographic data. *Structure*, 15(2):169–77, 2007.

316. R. Nussinov and C. J. Tsai. Allostery without a conformational change? Revisiting the paradigm. *Curr. Opin. Struct. Biol.*, 30:17–24, 2015.

317. G. S. Salvesen and S. J. Riedl. Caspase inhibition, specifically. *Structure*, 15(5):513–4, 2007.

318. A. Schweizer, H. Roschitzki-Voser, P. Amstutz, C. Briand, M. Gulotti-Georgieva, E. Prenosil, H. K. Binz, G. Capitani, A. Baici, A. Pluckthun, and M. G. Grutter. Inhibition of caspase-2 by a designed ankyrin repeat protein: specificity, structure, and inhibition mechanism. *Structure*, 15(5):625–36, 2007.

319. X. Y. Zhang and A. C. Bishop. Site-specific incorporation of allosteric-inhibition sites in a protein tyrosine phosphatase. *J. Am. Chem. Soc.*, 129(13):3812–3, 2007.

320. A. H. Andreotti. Native state proline isomerization: an intrinsic molecular switch. *Biochemistry*, 42(32):9515–24, 2003.

321. C. Grathwohl and K. Wüthrich. Nmr studies of the rates of proline cis-trans isomerization in oligopeptides. *Biopolymers*, 20:2623–2633, 1981.

322. R. J. Mallis, K. N. Brazin, D. B. Fulton, and A. H. Andreotti. Structural characterization of a proline-driven conformational switch within the Itk SH2 domain. *Nat. Struct. Biol.*, 9(12):900–5, 2002.

323. P. J. Hogg. Disulfide bonds as switches for protein function. *Trends Biochem. Sci.*, 28(4):210–4, 2003.

324. D. B. Lacy, W. Tepp, A. C. Cohen, B. R. DasGupta, and R. C. Stevens. Crystal structure of botulinum neurotoxin type A and implications for toxicity. *Nat. Struct. Biol.*, 5(10):898–902, 1998.

325. B. Schmidt, L. Ho, and P. J. Hogg. Allosteric disulfide bonds. *Biochemistry*, 45(24):7429–33, 2006.

326. H. Frauenfelder, G. A. Petsko, and D. Tsernoglou. Temperature-dependent X-ray diffraction as a probe of protein structural dynamics. *Nature*, 280(5723):558–63, 1979.

327. I. Schlichting, J. Berendzen, K. Chu, A. M. Stock, S. A. Maves, D. E. Benson, R. M. Sweet, D. Ringe, G. A. Petsko, and S. G. Sligar. The catalytic pathway of cytochrome p450cam at atomic resolution. *Science*, 287(5458):1615–22, 2000.

328. S. W. Englander. Hydrogen exchange and mass spectrometry: A historical perspective. *J. Am. Soc. Mass Spectrom.*, 17(11):1481–9, 2006.

329. J. Cavanagh, W. Fairbrother, A. G. Palmer III, M. Rance, and N. Skelton. *Protein NMR Spectroscopy: Principles and Practice*. Academic Press, 2007.

330. T. L. James. Fundamentals of NMR. *Online Biophysics Textbook*, 1999.

331. G. Wider. Structure determination of biological macromolecules in solution using nuclear magnetic resonance spectroscopy. *Biotechniques*, 29(6):1278–82, 1284–90, 1292 passim, 2000.

332. C. Li and M. Liu. Protein dynamics in living cells studied by in-cell NMR spectroscopy. *FEBS Lett.*, 587(8):1008–11, 2013.

333. V. A. Jarymowycz and M. J. Stone. Fast time scale dynamics of protein backbones: NMR relaxation methods, applications, and functional consequences. *Chem. Rev.*, 106(5):1624–71, 2006.

334. X. Michalet, S. Weiss, and M. Jager. Single-molecule fluorescence studies of protein folding and conformational dynamics. *Chem. Rev.*, 106(5):1785–813, 2006.

335. D. Chhabra and C. G. dos Remedios. Fluorescence Resonance Energy Transfer. In *Encyclopedia of Life Sciences*. John Wiley & Sons, Ltd., 2005.

336. M. Diez, B. Zimmermann, M. Borsch, M. Konig, E. Schweinberger, S. Steigmiller, R. Reuter, S. Felekyan, V. Kudryavtsev, C. A. Seidel, and P. Graber. Proton-powered subunit rotation in single membrane-bound F0F1-ATP synthase. *Nat. Struct. Mol. Biol.*, 11(2):135–41, 2004.

337. K. A. Dill, S. Banu Ozkan, T. R. Weikl, J. D. Chodera, and V. A. Voelz. The protein folding problem: when will it be solved? *Curr. Opin. Struct. Biol.*, 17(3):342–346, 2007.

338. S. A. Adcock and J. A. McCammon. Molecular dynamics: survey of methods for simulating the activity of proteins. *Chem. Rev.*, 106(5):1589–615, 2006.

339. H. A. Scheraga, M. Khalili, and A. Liwo. Protein-folding dynamics: overview of molecular simulation techniques. *Annu. Rev. Phys. Chem.*, 58:57–83, 2007.

340. V. A. Voelz, M. Jager, S. Yao, Y. Chen, L. Zhu, S. A. Waldauer, G. R. Bowman, M. Friedrichs, O. Bakajin, L. J. Lapidus, S. Weiss, and V. S. Pande. Slow unfolded-state structuring in Acyl-CoA binding protein folding revealed by simulation and experiment. *J. Am. Chem. Soc.*, 134(30):12565–77, 2012.

341. D. E. Shaw, J. P. Grossman, J. A. Bank, B. Batson, J. A. Butts, J. C. Chao, M. M. Deneroff, R. O. Dror, A. Even, C. H. Fenton, A. Forte, J. Gagliardo, G. Gill, B. Greskamp, C. R. Ho, D. J. Ierardi, L. Iserovich, J. S. Kuskin, R. H. Larson, T. Layman, L.-S. Lee, A. K. Lerer, C. Li, D. Killebrew, K. M. Mackenzie, S. Y.-H. Mok, M. A. Moraes, R. Mueller, L. J. Nociolo, J. L. Peticolas, T. Quan, D. Ramot, J. K. Salmon, D. P. Scarpazza, U. B. Schafer, N. Siddique, C. W. Snyder, J. Spengler, P. T. P. Tang, M. Theobald, H. Toma, B. Towles, B. Vitale, S. C. Wang, and C. Young. Anton 2: raising the bar for performance and programmability in a special-purpose molecular dynamics supercomputer. In *Proceedings of the International Conference for High Performance Computing, Networking, Storage and Analysis*, pages 41–53. IEEE Press, 2014.

342. S. L. Kazmirski and V. Daggett. Non-native interactions in protein folding intermediates: molecular dynamics simulations of hen lysozyme. *J. Mol. Biol.*, 284(3):793–806, 1998.

343. Q. Cui and I. Bahar. *Normal Mode Analysis: Theory and Applications to Biological and Chemical Systems*. Mathematical and Computational Biology. Chapman & Hall/CRC, 2006.

344. I. Bahar and A. J. Rader. Coarse-grained normal mode analysis in structural biology. *Curr. Opin. Struct. Biol.*, 15(5):586–92, 2005.

345. I. Bahar, A. R. Atilgan, and B. Erman. Direct evaluation of thermal fluctuations in proteins using a single-parameter harmonic potential. *Fold. Des.*, 2(3):173–81, 1997.

346. T. Haliloglu, I. Bahar, and B. Erman. Gaussian Dynamics of Folded Proteins. *Phys. Rev. Lett.*, 79(16):3090–3093, 1997.

347. F. Tama, M. Valle, J. Frank, and C. L. Brooks. Dynamic reorganization of the functionally active ribosome explored by normal mode analysis and cryo-electron microscopy. *Proc. Natl. Acad. Sci. USA*, 100(16):9319–23, 2003.

348. Y. Wang, A. J. Rader, I. Bahar, and R. L. Jernigan. Global ribosome motions revealed with elastic network model. *J. Struct. Biol.*, 147(3):302–14, 2004.

349. H. W. van Vlijmen and M. Karplus. Normal mode calculations of icosahedral viruses with full dihedral flexibility by use of molecular symmetry. *J. Mol. Biol.*, 350(3):528–42, 2005.

350. F. Tama and C. L. Brooks. Diversity and identity of mechanical properties of icosahedral viral capsids studied with elastic network normal mode analysis. *J. Mol. Biol.*, 345(2):299–314, 2005.

351. A. H. Zewail. 4D ultrafast electron diffraction, crystallography, and microscopy. *Annu. Rev. Phys. Chem.*, 57:65–103, 2006.

352. M. H. Olsson, W. W. Parson, and A. Warshel. Dynamical contributions to enzyme catalysis: critical tests of a popular hypothesis. *Chem. Rev.*, 106(5):1737–56, 2006.

353. H. M. Senn and W. Thiel. QM/MM studies of enzymes. *Curr. Opin. Chem. Biol.*, 11(2):182–7, 2007.

354. A. Warshel. Computer simulations of enzyme catalysis: methods, progress, and insights. *Annu. Rev. Biophys. Biomol. Struct.*, 32:425–43, 2003.

CHAPTER 6

Intrinsically Unstructured Proteins

6.1 INTRODUCTION

According to the central paradigm of structural biology, the structure of a protein directly determines its function. This paradigm — supported by the well-accepted models of Pauling and Fischer [1,2], as well as by the first experimentally-determined structures [3,4] (and by most of those following them) and by Anfinsen's denaturation experiments [5] — is particularly prominent in enzymes. Specifically, the three-dimensional globular structures of enzymes facilitate the organization of binding and active sites that fit physically and chemically to the enzymes' natural substrates, and enable catalytic residues to be precisely aligned such that highly efficient catalysis can take place. However, studies in the 1990s began to identify disordered regions in proteins (e.g., Figure 6.1), thereby calling the direct structure–function relationship into question [6]. The size of these so-called *'intrinsically disordered regions'* (*IDRs*) was found to vary considerably across proteins, ranging from short segments of the polypeptide chain to the entire protein [7–9]. Proteins that are entirely disordered are referred to in the literature as *'intrinsically unstructured proteins'* (IUPs), *'intrinsically disordered proteins'* (IDPs) or *'natively unfolded proteins'* (NUPs). Herein, we will refer to these proteins as IUPs. Interestingly, ~14% of the domains in the Pfam database, which are defined by sequence alone, have been predicted to have more than 50% of their residues in disordered state [10]. Examples of such domains include the KID domain of Cdk inhibitors and the Wiskott–Aldrich syndrome protein (WASP)-homology domain 2 (WH2) of actin-binding proteins [10]. Moreover, disordered regions have been found in proteins with different functions, such as DNA packing and repair, ion transport, nuclear trafficking of proteins, and regulation of key cellular processes (the cell cycle, transcription, and splicing), as well as in proteins involved in common pathologies, such as cancer [11] and cardiovascular diseases (see review in [12]).

It is difficult to establish the full extent to which IUPs and IDRs are involved in cellular functions, because the main methods in use for determining protein structure (mainly X-ray crystallography) almost exclusively identify ordered structures. In contrast, disordered regions within proteins exist as ensembles of conformations that interconvert rapidly [13,14] due to the flat energy landscape of the protein [15], and are therefore largely unresolved in X-ray structures. Thus, the main database of protein structures (PDB) under-represents both IUPs and protein IDRs. Fortunately, other lab methods such as NMR, EPR, CD and fluorescence spectroscopies, high-speed atomic force microscopy, Raman optical activity, and time-resolved small angle X-ray scattering (TR-SAXS) are able to track disorder in proteins. Of these, NMR is particularly useful. First, it provides data on protein structure and dynamics at the residue level [16]. Second, it provides data on various aspects of disorder, such as the location of IDRs and binding-folding coupling (see below) [10]. Single-molecule FRET is also very useful in studying disorder, as it can be employed to characterize different states of IUPs, such as the unbound ensemble and the various folding intermediates formed during binding of the IUP to other proteins (e.g., [17,18]).

The data obtained by the above lab methods are often supplemented by computational methods that can predict disorder with some degree of success [8,9,19]. Methods that predict disorder on a genome-wide scale identify sequences with certain properties (e.g., IUPred [20]), employ machine-learning algorithms on training sets (e.g., DISO-PRED [21]), or integrate predictions from several successful algorithms (e.g., metaPr-DOS [22]) [10]. Characterization of disorder and the conformational ensemble of IDRs in individual proteins facilitates the use of more computationally exhaustive methods such as molecular dynamics simulations [23,24].

Bioinformatic methods suggest that **the extent to which IUPs are prevalent in the genome of an organism correlates with the organism's complexity** [25], implying that these proteins play complex roles [26]. According to bioinformatic studies carried on whole proteomes, it is estimated that for mammals [27]:

1. About 25% of all proteins are IUPs.

2. About 75% of all signaling proteins, which make up about 50% of the entire proteome, contain long disordered regions.

To account for IUPs in a manner that reflects their prevalence, several IUP databases have been constructed and are freely available over the Internet. For example:

DisProt [28] contains experimentally obtained data on IUPs/IDRs, their functions, and the location of the disordered regions. URL: http://www.disprot.org

IDEAL [29,30] contains experimentally obtained data on IUPs/IDPs as well, but focuses on those that undergo coupled binding-folding. It also predicts putative IDRs that are expected to undergo induced folding upon binding. URL: http://www.ideal.force.cs.is.nagoya-u.ac.jp/IDEAL

MoBiDB [31,32] contains IDR information obtained from various sources, including the above databases and the protein data bank. URL: http://MoBiDB.bio.unipd.it

pE-DB [33] contains structural ensembles of IUPs, predicted by computational algorithms. URL: http://pedb.vib.be/

D^{2}P^2 [34] contains predictions of protein disorder and of functionally different IDRs, such as MoRFs (see below), post-translational sites, and domains. URL: http://d2p2.pro

IDRs appear in many proteins of different functions, but they are particularly common in those regulating complex biological processes such as communication and signal transduction [35], cell cycle (e.g., see [36]), and gene expression. Accordingly, diseases such as cancer[*1], which are associated with such complex functions, are also tightly related to protein disorder [37]. Thus, IDRs seem to complement the functional repertoire of globular proteins, which mainly includes enzyme-mediated catalysis and transport [12]. This is also the

---

[*1]Many cancer markers and oncogenes such as p53, cMyc, and BRCA1 manifest high levels of disorder [37].

(a)

(b)

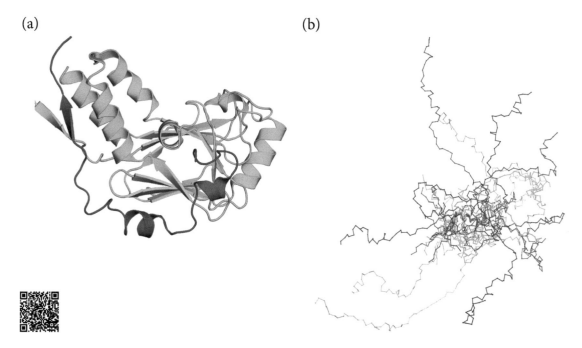

FIGURE 6.1 **Protein disorder.** (a) Disorder of a relatively short stretch of a polypeptide chain (PDB entry 1dev). The image shows the disordered Smad-binding domain (SBD) domain (pink) of SARA (Smad Anchor for Receptor Activation), bound to the Smad2 MH2 domain (green). The SBD domain is one of the relatively few cases in which binding of an IUP to its target does not induce folding into a compact structure. Although the domain contains three short segments of helical or strand conformations, most of the length of the polypeptide chain is unstructured and extended. (b) Large-scale disorder. The image shows 20 NMR structures of the intrinsically-unstructured thylakoid soluble phosphoprotein (PDB entry 2fft) (each structure is colored differently), illustrating the very large conformational freedom of the unstructured chain.

conclusion reached by Uversky and coworkers, who conducted a survey of SwissProt annotations for both structured and unstructured proteins [38–40]. The possible reasons for this functional dichotomy are discussed below. It should be mentioned, though, that this functional distinction is not absolute, as enzymes themselves catalyze diverse processes, which include regulation and signal transduction. Moreover, even within enzymes, many parts directly involved in regulation are now known to be disordered [41]. Surveys of IUPs with known functions have assigned 28 different functions to these proteins, which can be divided into a few major groups; these groups are described in the following subsections. A comprehensive review of the different functions is provided in [10]. In all cases, the function results from the disorder of the protein segment involved, whether it is of limited length or extends over the entire protein. Thus, in the following descriptions we ignore the distinction between IUPs and IDRs, and treat them both as 'disordered'.

6.1.1 Molecular recognition

Proteins that belong to this group mainly function in protein-protein recognition during signal transduction, but they are also involved in other processes requiring the binding of nucleic acids, membranes, and small molecules [9,10] (e.g., see Box 6.1). **The binding is usually accompanied by the folding of the disordered segment (Figure 6.2), although the IUP or IDR may retain significant flexibility even after binding** [42,43] (see more below).

Binding processes in these proteins can be further classified as either permanent or transient. IDRs involved in permanent binding belong to the following functional subgroups:

1. *Assemblers* function in the assembly of large protein complexes, such as the ribosome, cytoskeleton, and chromatin. The flexibility of the IDRs participating in this function is sometimes crucial, as some complexes cannot be assembled from rigid components due to topological limitations [26]. Moreover, since assemblers tend to remain partly unfolded even when interacting with their binding partners [44], each IDR has a large interface that is able to interact with multiple proteins. It should be noted that assemblers not only bring complex components together by providing a mutual binding interface, but also constitute a scaffold that directs the correct spatial assembly of the components.

2. *Effectors* regulate the activity of other proteins or other regions of the same protein. Examples of disordered proteins that regulate activity in other proteins include p21 and p27, which participate in the regulation of the mammalian cell cycle [36]. IDRs that regulate activity within their own proteins are observed, e.g., in the Wiskott-Aldrich syndrome protein, which is inhibited by its own GBD domain [45]. This inhibition occurs when the GBD domain folds onto other parts of the protein, thus blocking the capacity of these regions to interact with cytoskeletal actin.

3. *Scavengers* store and/or neutralize small ligands. One well-known example of a scavenger is the phosphoprotein *casein*, which prevents the precipitation of calcium phosphate crystals in milk by scavenging calcium [46]. Other phosphoproteins such as osteopontin and fetuin act similarly in other body fluids and tissues. Scavengers bind their ligands with high affinity, in the nM range [9].

IDRs involved in transient binding belong to the following functional subgroups:

1. *Display sites* create target sites in their respective proteins for post-translational modifications (PTMs), such as phosphorylation, methylation, acylation, ADP-ribosylation, ubiquitin attachment, and proteolytic cleavage [12]. Since PTMs affect the stability, turnover rate, interaction potential, and localization of many cellular proteins, the function of disordered display sites has important regulative implications [10].

2. *Chaperones* assist in the folding of proteins and RNA molecules. The latter group seems to require a particularly high degree of disorder: over half the residues in RNA chaperones have been found to be disordered, compared to only about a third in protein chaperones [47]. In addition to binding RNA, these IDRs also seem to loosen the structures of misfolded proteins and RNA. In contrast to the activity of large chaperones, discussed in Chapter 5 (e.g., Gro-EL and Gro ES), chaperone activity based on disordered regions does not require ATP [10]. Since this activity involves concomitant (partial) folding of the disordered region and unfolding of the substrate protein, its mechanism may be based on some type of entropy exchange.

As might be expected, IDRs involved in transient binding bind their partners with low to moderate affinity ($Kd =$ mM to μM, respectively).

As mentioned above, IUPs and IDRs involved in binding and recognition may remain

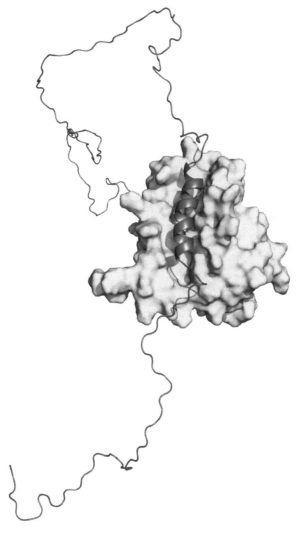

FIGURE 6.2 **An IUP-protein complex involving partial folding.** The image shows the binding of the anti-apoptotic protein Bcl-xL (yellow surface) to the unstructured protein Bad (blue ribbon). Upon binding, a 25-residue segment from the death promoting region of Bad acquires an α-helical conformation, whereas the rest of the protein remains unstructured. The complex between Bcl-XL and the 25-residue segment of Bad is taken from PDB entry 1g5j, whereas the rest of Bad is modelled as a random coil.

significantly flexible even after becoming bound. This phenomenon, often referred to as 'fuzziness' [42], may have several functions [10]. For example, different conformations of the IDR in the bound state may be used to induce different structures of the binding partner or different binding affinities [48,49]. The disorder in fuzzy complexes may manifest as multiple stable conformations (*static disorder*) or as a constantly fluctuating ensemble of conformations (*dynamic disorder*). In extreme cases of fuzziness (e.g., in elastin fibers [50]) the bound IUP or IDR is only slightly more ordered than the unbound one, and the binding is mediated largely by linear sequence motifs in the IDR. Examples of fuzzy complexes with different degrees of disorder are given in [42,48].

6.1.2 Entropic chain activity

Proteins that belong to this group serve in roles that require freedom of movement of the polypeptide chain. The need for freedom of movement explains why disordered proteins are most suitable for this job, and also why they remain disordered during their function, in contrast to many of the proteins belonging to the first group. The function of proteins involved in entropic chain activity can generally be described as either creating a force that opposes structural changes, or affecting the orientation or localization of attached domains within the protein. The proteins found to function in these ways can be separated into the following groups:

1. *Springs* generate 'passive force', i.e., force that does not require the use of external energy. For example, in titin, a disordered sequence called *PEVK*[*1] creates passive force in muscle fibers, helping them to restore their relaxed length when over-stretched [51,52].

2. *Bristles* keep adjacent segments of the chain separated. For example, the C' tails of the neurofilament units NF-M and NF-H maintain the space between adjacent filaments [53].

3. *Linkers* link different regions within the protein, such as domains, thus allowing them to acquire different orientations with respect to each other.

BOX 6.1 INTRINSIC DISORDER IN NUCLEAR PORE COMPLEX [54]

Regular transport of molecules from the cytoplasm to the nucleoplasm (the inner environment of the cell nucleus), as well as in the opposite direction, is imperative for the correct functioning of the cell. For example, transcription factors, required for the regulated expression of genes, are imported from the cytoplasm into the nucleoplasm. Likewise, newly-transcribed messenger RNA molecules and ribosomal precursors that are formed inside the nucleus are exported into the cytoplasm, where genetic information is translated into protein molecules. Proteins to be transported (*cargo*) contain short targeting sequences[*a], which are recognized by specific protein factors[*b]. The latter bind to the cargo and allow it to undergo the transport process, which is active and regulated by the small GTPase *ran*. RNA molecules either are recognized directly by the protein factors or bind first to adaptor proteins that are subsequently recognized. Although scientists have gained some understanding of the transport process in recent years (see below), the process is quite complex and is still under intense investigation [55,56].

The transport process is carried out by the *nuclear pore complex* (*NPC*), a very large body (50 to 110 MDa) that includes ~30 distinct proteins (*nucleoproteins*) and an over-

---

[*1]The sequence is rich in the amino acids proline, glutamate, valine, and lysine (PEVK), hence the name.

[*a]'*Nuclear localization signals*' (*NLS*) in the case of imported proteins, and '*nuclear export signals*' (*NES*) is the case of exported proteins.

[*b]These are called '*importins*' and '*exportins*', for cargo import and export, respectively. These proteins are also named more generally '*karyopherins*' and '*transportins*'.

all number of ~456 individual protein molecules (depending on the organism) [57]. The complex has two modes of transport. Ions and small molecules up to ~26 Å in size are transported by passive diffusion, whereas larger cargo (up to ~380 Å in size) is transported actively, using the RanGTP-GDP cycle [58]. The sheer size of the NPC has challenged scientists in their attempts to determine the three-dimensional structure of the complex accurately. To explore the structure of the entire NPC and its numerous components, a plethora of data has been collected using various methods, such as X-ray diffraction, NMR, atomic force microscopy, electron microscopy, SAXS, and different biochemical methods (see reviews in [55,59]).

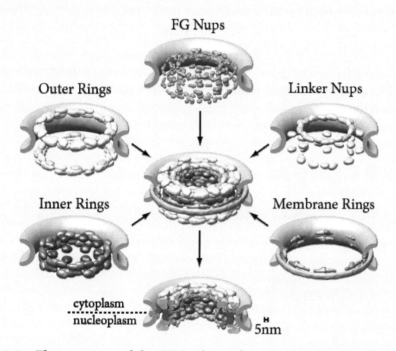

FIGURE 6.1.1 **The structure of the NPC.** The nucleoporins composing the NPC are colored according to their classification into five distinct substructures on the basis of their location and functional properties: the outer rings in yellow, the inner rings in purple, the membrane rings in brown, the linker nucleoporins in blue and pink, and the FG nucleoporins (Nups) in green. The pore membrane is shown in grey. The central structure contains all rings, whereas the structures surrounding it show each ring separately, for clarity. The bottom structure shows a cross section through the NPC. A 5-nm size scale is shown at the bottom of the image. The figure is based on the structure of Alber et al. [54], and is adapted from [60].

A nice example of the integrative strategy needed to study the structure and organization of the NPC is given by the combined experimental and computational work of Alber et al. on the *Saccharomyces cerevisiae* NPC [61] (other examples are given in [62–66]). The experimental data, obtained by methods such as electron microscopy, mass spectrometry, and quantitative immunoblotting, were used to produce spatial constraints for the complex. Then, computational methods were used to produce a collection of structural solutions that satisfied the constraints. The averaged structure produced by this procedure is shown in Figure 6.1.1. It is ~980 Å in diameter and ~370 Å in height. The inner pore has a diameter of ~380 Å, corresponding to the maximal size of macromolecular cargo that can go through the NPC [67]. Other studies that came later added more details on the structure of the individual components of the NPC, their oligomerization states, and their overall organization (e.g., [60,62–66,68,69]). Most of the individual nucleoproteins composing the NPC join together to form five coaxial rings (see Figure 6.1.1 for details), whereas others link the rings. Although each ring is unique, the five rings can be separated into two groups. The first includes four of the five rings. All of these rings are made from coat nucleoproteins that have an ordered three-dimensional structure. Furthermore, their folds and the pattern in which they are organized are similar to those of vesicle-coating proteins (e.g., clathrin). This similarity should not come as a surprise, as both types of complexes have a similar function, that is, to form coating scaffolds that curve membranes. The dominant folds of these proteins are the β-propeller and the α-solenoid (Figures 6.1.2b and 6.1.2c). Whereas the former creates a tough core, the latter seems to be almost designed to confer conformational flexibility. In the NPC, such flexibility is required to accommodate the large cargo molecules that go through the pore [54].

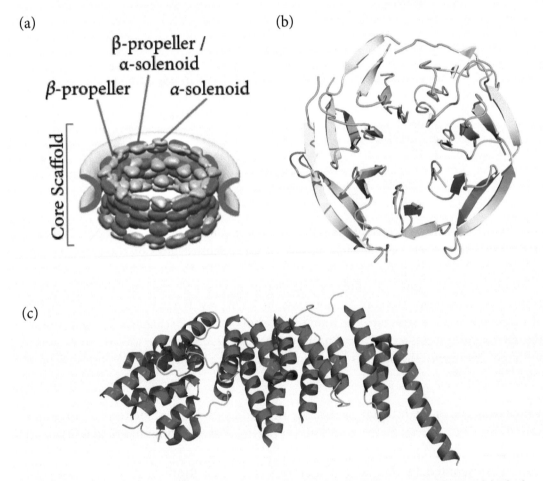

FIGURE 6.1.2 **The β-propeller and the α-solenoid folds of NPC coat proteins.** (a) The locations of the scaffold proteins containing these folds in the NPC (adapted from [60]). (b) The β-propeller fold of Seh1 (PDB entry 3ewe, chain A). (c) The α-solenoid fold of Nup85 (PDB entry 3ewe, chain B).

The second group includes the FG ring[*a], which lines the innermost region of the NPC and constitutes a ~300-Å-thick permeability barrier of the central channel. Each of the nucleoproteins composing this ring contains a small structured domain anchoring it to the other rings, and a large disordered region facing the pore [70]. Being disordered, this region may acquire multiple conformations [71] and, as a result, appears in the averaged structure as a low-density cloud surrounding the structurally-resolved region [54] (Figure 6.1.3). The shape and distribution of the cloud may be suggestive of a few interesting aspects of NPC transport:

1. The cloud extends to both the nucleoplasm and the cytoplasm, which means that the FG nucleoproteins are accessible to molecules approaching either side of the NPC. This property points to the possibility that FG nucleoproteins may serve as the NPC component that mediates transport. Indeed, it has been suggested that FG nucleoproteins constitute the NPC's docking sites for cargo-factor complexes [61,70,72–75], provided that the latter contain FG repeat binding sites, e.g., hydrophobic pockets in the factor proteins, which bind to phenylalanine residues in the FG nucleoproteins [76]. Moreover, mutational studies demonstrate that different FG nucleoproteins have different functions and handle different cargo molecules [76,77]. These functions are affected by post-translational modifications of FG sequences, such as glycosylation and phosphorylation [77–79].

2. The disordered parts of the FG domains partially fold upon cargo binding [58]. This folding process is accompanied by an unfavorable decrease in chain entropy, suggesting that the FG domains may participate in the gating of the transport process by serving as 'repulsive gating barriers' [58].

3. The cloud leaves a ~10-Å space at the very center of the channel, which is roughly the maximal size of molecules that can freely diffuse through the pore.

4. The long reach of the individual FG nucleoproteins, reflected in the significant overlap in the cloud, may be interpreted as a transport mechanism. That is, it may suggest a mechanism in which the cargo protein shifts quickly between different FG nucleoproteins, such that it is able to traverse the pore in minimal time (~5 ms). As explained in the main text, IUPs are indeed characterized by the capacity to exchange binding partners rapidly, and this capacity is based on the relative weakness of the interactions between the proteins and their partners. In the case of FG nucleoproteins, such interactions seem to be based on the short FG repeats, each of which binds to the target protein with low affinity [56].

5. A small number of FG docking sites also face the pore membrane, suggesting that they may be involved in the transport of membrane-bound proteins.

[*a]The abbreviation 'FG' comprises the one-letter codes for phenylalanine and glycine, respectively. The ring is named so because the nucleoproteins composing it are in most part made of phenylalanine-glycine repeats, organized as motifs (e.g., FXFG and GLFG). The 5- to 30-residue sequences linking these motifs are enriched in polar amino acids [55].

(a)

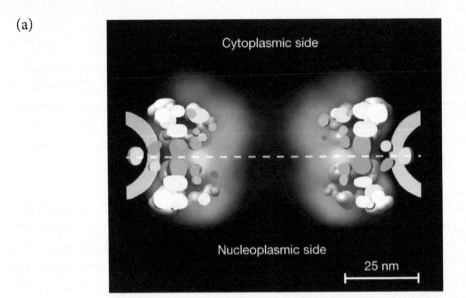

(b)

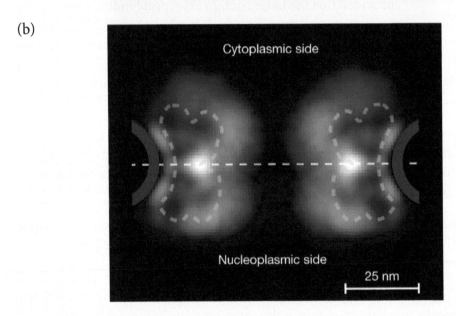

FIGURE 6.1.3 **Distribution of the disordered FG-repeat regions in the NPC.** (a) Slice through the NPC, showing the structured domains of all nucleoporins, colored according to their classification (see Figure 6.1.1). Also shown are the localization probabilities of the unstructured regions of all FG nucleoporins (green cloud). (b) Projection of the localization probabilities of the FG-repeat regions from all the FG nucleoporins is shown by a density plot, sampled in a plane perpendicular to the central Z-axis from $X = -50$ Å to $X = +50$ Å. Projections from the FG-repeat regions belonging to FG nucleoporins anchored mainly or exclusively on one side of the NPC are indicated: red for those that are cytoplasmically disposed, blue for those nucleoplasmically disposed, and white for those that are present to equal degrees on both sides. The equatorial plane of the NPC is indicated by a dashed white line, the position of the NPC density is indicated by a dashed grey line, and the position of the pore membrane is shown in purple. A scale bar of 25 nm (250 Å) is shown. The figure was taken from [54].

6.2 SEQUENCE AND STRUCTURAL ORGANIZATION OF IUPs AND IDRs

Although IUPs and IDRs are defined as disordered, measurements show that **they may contain secondary elements** [80] **such as α-helices and β-sheets (comprising 10% to 20% of the protein residues** [26]**), as well as non-α helices.** However, these elements are unstable, and tend to change often and in a random manner. This instability is significantly reduced upon binding of the protein to its target, thus allowing the secondary elements to remain stable without the help of tertiary contacts (as in globular proteins). Among the secondary elements found in IUPs, the PPII helix is most common, and CD measurements show that 20% to 25% (and up to 50%) of the residues in IUPs tend to occupy this conformation. As mentioned in Chapter 4, the PPII helix was originally identified in proline-rich sequences, but is not limited to these. It tends to appear in disordered segments, whether they belong to IUPs or denatured globular proteins. The functional significance of the PPII helix in IUPs is discussed in Section 6.3 below. The presence of secondary elements with no tertiary structure makes the structure of some IUPs reminiscent of the molten globule conformation, mentioned in Chapter 5 as a folding intermediate in globular proteins. Still, other IUPs may look very different, with a 'random coil' structure characterized by the almost total absence of tertiary contacts and a very high degree of freedom of movement.

In addition to being structurally unique, disordered protein segments also tend to have unique amino acid composition. As expected, they are enriched with proline; however, they are also enriched in the polar amino acids Met, Lys, Arg, Ser, Gln, and Glu. This should come as no surprise, as **the polarity of IDR sequences ensures that a nonpolar, compact, and ordered core cannot be stabilized.** For the same reason, IDRs tend to contain very few Trp, Tyr, Ile, Phe, Val, and Leu residues [41], all of which are capable of nonpolar interactions. Cysteine is also under-represented in IDRs, as it tends to form disulfide bonds that stabilize the ordered tertiary and quaternary structures of proteins. Another characteristic sequence-related property of IDRs is low evolutionary conservation (compared to structured segments) [81]. This is to be expected, as IDRs do not contain buried residues, and therefore lack many of the structural constraints on sequence that are present in structured segments [27]. The above sequence-related properties of IDRs have been used successfully to automatically predict disorder in proteins [82].

Finally, IDR sequences are enriched with short linear motifs (comprising 3 to 10 residues) termed 'LMs', 'SLiMs' or 'MiniMotifs' [83–86]. These motifs regulate low-affinity interactions of the protein with its binding partners in ways that may lead to different outcomes: a change in stability, formation of a complex, subcellular localization, or recruitment of post-translational modifying enzymes [10]. SLiMs can be categorized according their mechanisms of action: those in one category promote modifications in the protein (structural, proteolytic, group addition or removal), whereas those in the other category act as ligands (for complex formation, docking to other proteins, or for targeting/trafficking) [10] (Figure 6.3). For example, members of the *complex-promoting* group often act as scaffolds for components of protein complexes. This group includes, e.g., the tyrosine-containing motif recognized by SH_2 domains upon phosphorylation, as well as the PxxP motifs recognized by SH_3 domains [87]. Members of the *docking* group act by increasing the specificity and/or efficiency of modification events by providing additional binding surface [10]. For example, *degrons* affect protein stability by recruiting members of the ubiquitin-proteasome system [88,89].

(a)

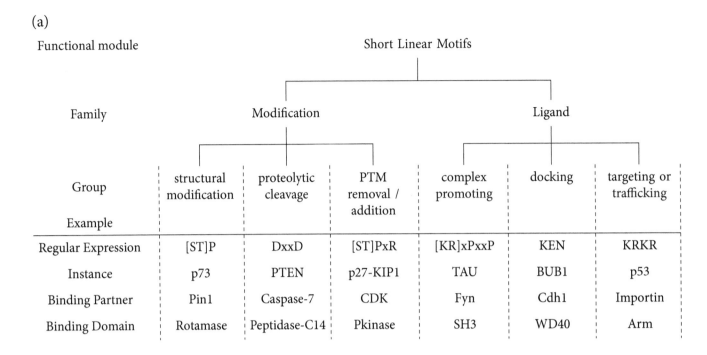

Functional module	Short Linear Motifs					
Family	Modification			Ligand		
Group	structural modification	proteolytic cleavage	PTM removal / addition	complex promoting	docking	targeting or trafficking
Example						
Regular Expression	[ST]P	DxxD	[ST]PxR	[KR]xPxxP	KEN	KRKR
Instance	p73	PTEN	p27-KIP1	TAU	BUB1	p53
Binding Partner	Pin1	Caspase-7	CDK	Fyn	Cdh1	Importin
Binding Domain	Rotamase	Peptidase-C14	Pkinase	SH3	WD40	Arm

(b)

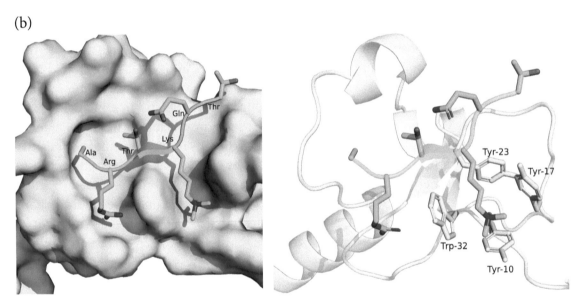

FIGURE 6.3 **Short linear motifs (SLiMs).** (a) Functional classification of SLiMs. All SLiMs can be divided into two major groups: (*i*) those that act as modification sites, i.e., act as recognition sites for the active sites of modifying enzymes, and (*ii*) those that act as ligands, i.e., they are recognized by the binding surface of a protein partner. These two main families can be further divided into groups according to the specific function. In the regular expressions, x corresponds to any amino acid, while other letters represent single-letter codes of amino acids; letters within square brackets mean either residue is allowed in that position. The image is taken from [10] (http://pubs.acs.org/doi/full/10.1021/cr400525m). (b) *Left*: A methylated SLiM in a histone H3-derived peptide (shown as sticks), bound to human BPTF (yellow surface, PDB entry 2f6j). The histone tail corresponding to the peptide is disordered in the unbound form, and acquires a short β conformation upon binding. The addition of three methyl groups to Lys-4 enables this residue to fit well with the BPTF binding site, where it interacts with three tyrosine residues and one tryptophan. *Right*: BPTF is shown as sticks with the partially transparent backbone for clarity. Histone lysine methylation is associated with the demarcation of transcriptionally active sites in eukaryotic genes.

6.3   STRUCTURE-FUNCTION RELATIONSHIP

6.3.1   IUP binding to target proteins

6.3.1.1   IUPs are designed for fast protein binding and release

IUPs tend to participate in the regulation of biological processes, and their function in this capacity usually involves binding and/or recognition of other proteins. In fact, it has been found that the more binding targets an IUP has, the more disordered it is [12]. This implies that **disorder is inherently important for binding and recognition roles**. Although the precise reasons for this are not fully understood, disorder does seem to confer a few advantages in binding, especially when it takes place within large functional or regulatory networks. First, **freedom of conformation of the polypeptide chain (which results from disorder) confers plasticity of binding (e.g., Figure 6.4)**. This is a huge advantage for hub proteins that interact with multiple partners in large networks. For example, the p21^{Cip1} protein is able to bind different cell-cycle regulatory proteins, depending on cellular conditions [26]. Another well-known example is p53, whose N- and C-terminal domains are disordered, and are involved in interactions with numerous binding partners that modulate the protein's activity [90]. As we saw earlier, in some IUPs forming 'fuzzy' complexes, the IDRs are able to sample different conformations even when bound to the same protein, which allows the IUP to modulate the affinity of binding to its partner [48,49].

Secondly, IUPs are particularly sensitive to proteolytic cleavage, because of the high solvent exposure of their polypeptide chains. This sensitivity makes it much easier for the cell to

(a)   (b)

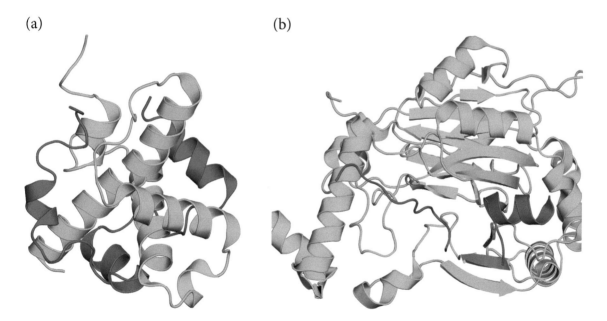

FIGURE 6.4   **Different conformations of HIF-1α when bound to different targets.** The figure shows the C-terminal activation domain of the hypoxia-inducible factor HIF-1α (red) in complex with: (a) the transcription activation zinc finger (Taz1) domain of CREB-binding protein (CBP) (PDB entry 1l8c, green); (b) the inhibiting factor FIH (PDB entry 1h2k, green). As is clearly shown, HIF-1α has an α-helical conformation in the first complex, and a largely coiled conformation in the second. In the latter complex, six residues of HIF are missing (i.e., unresolved in the experiment), reflecting the disordered nature of this protein.

regulate the activity of IUPs by controlling their turnover rate, which, again, is advantageous in proteins that regulate complex processes. Third, and perhaps most importantly, **IUPs excel in forming weak, yet specific interactions with their target proteins**. This ability is highly important, as proteins participating in regulatory and signal transduction processes need to bind and release their targets rapidly to facilitate rapid signal transfer. The properties of IUPs that facilitate this type of binding are as follows:

1. **Low IUP-target affinity results from binding-folding coupling**
   The ability of IUPs to bind their targets weakly is a consequence of their tendency to (partially) fold upon binding. This is because folding involves entropy loss. In fact, any intermolecular binding involves a loss of entropy, even when both binding partners are already folded. However, in the latter case the decrease is only of the *translational and rotational entropy* of the proteins ($\Delta S_{trans/rot}$). When an IUP binds to its target, $\Delta S_{trans/rot}$ is supplemented by the loss of *configurational entropy* (ΔS_{con}), resulting from the confinement of protein structure by the folding. Binding still occurs, in both globular proteins and IUPs, because the enthalpy gained from binding interactions (ΔH_{bind}) over-compensates for the loss of entropy, thus making the total binding energy (ΔG_{bind}) favorable (e.g., [91]):

$$\Delta G_{bind} = \Delta H_{bind} - T\left(\Delta S_{trans/rot} + \Delta S_{con}\right) \qquad (6.1)$$

(at constant temperature).

However, assuming that IUP binding and globular protein binding are associated with similar values of ΔH_{bind}, the larger entropy loss in IUP binding means that the ΔG_{bind} of these interactions is smaller than it is in globular protein binding. This difference is manifested in lower affinity of the IUP for its target. It should be mentioned, though, that in most cases the folding of the IUP is partial [92]: First, it often involves 10 to 70 residues, amphipathic sequences called '*molecular recognition elements or features*' (*MoREs or MoRFs*) [93–95], rather than the whole disordered domain or protein (see more below). Second, even those parts that fold upon binding may retain significant flexibility [42,43]. Thus, ΔS_{con} is expected to be moderate.

The assumption that ΔH_{bind} is similar in both IUPs and globular proteins is not always accurate. In many IUPs, the loss of configurational entropy is so large that ΔH_{bind} needs to be especially strong to achieve *any* binding affinity. The strong binding interactions in such cases are achieved by at least three properties possessed by IUP binding interfaces:

(a) The average interaction surface area per residue is larger in IUP interfaces than in globular protein interfaces [96]. Thus, despite the fact that the total surface area of the binding interface is similar in the two protein types, IUPs achieve stronger interactions.

(b) The binding interface of an IUP is constructed from one or two segments of the polypeptide chain, instead of many short segments that are separated in sequence, as is the case in globular proteins. IUP interfaces are much easier to construct following binding to the target, since they allow the protein to minimize the entropic cost down to a necessary minimum.

(c) The binding interfaces of IUPs are enriched with nonpolar residues, compared with those of globular proteins [96], thus increasing the interaction affinity of the former. This is intriguing, as IDRs generally tend to be polar. This implies that the folding process is not random, but rather intended to create particularly strong binding interfaces. Globular proteins tend to do the opposite, i.e., to bury their nonpolar residues instead of using them for constructing binding interfaces.

The extra energy needed to compensate for the folding of IUPs may be one reason why none of these proteins has been found so far to have catalytic activity. **Enzymes use their substrate-binding interactions (ΔH_{bind}) to drive catalysis. IUPs waste at least some of this energy on folding, and the rest may not suffice for driving catalysis** [41]. Thus, although the functions of IUPs and globular proteins both involve binding, the former bind their ligands in the context of signal transduction processes, whereas the latter are involved also in catalysis.

2. **IUP-target specificity involves polar groups**

   IUPs bind to their targets specifically and selectively, as is often required in biological systems, especially in complex functional networks. This property seems to be in disagreement with the nature of IUPs; the fact that they are unfolded prior to binding suggests low specificity, due to conformational flexibility. How is it done, then? We have already seen that IUPs tend to form binding interfaces with a larger-than-usual binding surface per residue. The large number of contact points formed in such surfaces provides, at least theoretically, a higher potential for accurate contacts. However, do all IUP-target contacts contribute equally to specificity? This issue was studied in depth for the interactions between the Smad2 protein and a disordered segment of the TGF$_\beta$ receptor [97]. The results demonstrated a functional division among the different interactions present between the interacting proteins: nonpolar interactions contribute to the affinity of binding, whereas electrostatic interactions contribute to its specificity. As discussed in Chapter 4, the same functional division also exists amongst noncovalent interactions within folded proteins, and is explained by the higher specificity of electrostatic interactions in the nonpolar environment, compared with nonpolar interactions. How can this be reconciled with the high plasticity that IUPs have towards their targets? As a matter of fact, there is no real conflict between the two; plasticity results from changing environmental conditions (as in the aforementioned case of p21^{Cip1} binding to cell-cycle proteins). This means that **under fixed conditions the IUP is expected to bind to a single target protein**.

   The polar groups conferring specificity to IUP-target binding are diverse. Most belong to the residue's side chains, whereas others are backbone groups (C=O, N−H). The folding that accompanies binding is supposed to pair those backbone groups in hydrogen bonds, thus reducing the energetic penalty associated with their exposure to the nonpolar interface. This is the exact process occurring in secondary structure formation in globular proteins (see Chapter 2). In IUPs, however, the local structures formed are not necessarily canonical (i.e., α-helical or β-sheeted), which means at least some of the polar backbone groups remain unpaired. **The exposure of these groups to the low dielectric interface involves some destabilization, but at the same time provides the interface with additional hydrogen bond donors and acceptors, thus rendering the binding more specific.** This phenomenon has been confirmed in *PPII helices*, which are highly common in IUPs, but has to be further studied in other non-α and non-β conformations. Finally, another factor contributing to the binding specificity of IUPs and IDRs is the high prevalence of MoREs and MoRFs [93–95] in disordered regions; these are linear sequences that are characterized by high evolutionary conservation and tend to fold upon binding to their partners[*1] (see above). Functional analysis of MoREs/MoRFs implicates many of them in signal transduction and alternative splicing, and it seems that their activity is mainly regulated by phosphorylation [94]. Earlier, we encountered SLiMs, which are another type of linear sequence that tends to be disordered. While MoREs and MoRFs are distinguished from SLiMs, which are much shorter, the two share certain features besides linearity and enrichment in IDRs; they both tend to fold upon binding (all MoRFs and ~60% of SLiMs) and to promote complex formation [10].

6.3.1.2 Mechanism and kinetics of binding-folding coupling in IUPs

Binding-folding coupling is a key aspect of IUP action. As mentioned in the previous subsection, the reason for the coupling is the need to reduce the energy penalty on the exposure of polar backbone groups to the hydrophobic interface, by pairing them in hydrogen bonds. However, the mechanism through which this is done is not necessarily that simple. Generally speaking, there are two basic mechanisms through which the coupling can work [92]. In the first, the folding is induced by the binding, similarly to Koshland's 'induced fit' theory, but with a more extensive change (see below). Such a process involves quick binding followed by slower folding (as in [99], for example). On the basis of their studies, Wolynes and coworkers proposed a two-step process called *'fly casting'* [100], which supports the induced folding mechanism. First, the IUP binds weakly to its target, forming an *encounter complex*. Then, it partially folds into the conformation that provides the strongest interaction with its target protein. This mechanism has two important advantages:

1. Since the initial binding is weak, the IUP can sample many conformations rapidly, which increases its chances of finding the best possible one (Figure 6.5). Globular proteins, being already folded, can only undergo a limited change of conformation (i.e., induced fit), which does not necessarily create the best possible binding site. This may explain the fact that in IUPs, the average contact surface area per residue is larger than in globular proteins [96].

2. Since the sampling occurs while the IUP is already bound to its target, there is no need for the two to undergo diffusion and rotation in order to find each other, as occurs in globular proteins.

The second mechanism through which binding-folding in IUPs may occur involves selection of a pre-existing conformation in the unbound IUP, following the 'conformational selection' model explained in Chapter 5. So, which of the two mechanisms is right? Apparently, both are [92]. Some studies support the induced folding mechanism (e.g., [101–104]), and some support the conformational selection mechanism (e.g., [105]). Interestingly, Kataoka

---

[*1]Although the conformational equilibrium of the unbound state tends to be biased toward the bound conformation, which is why MoREs and MoRFs are also called pre-structured motifs (PreSMos) [98].

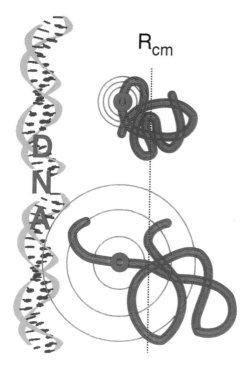

FIGURE 6.5 **Folding speed enhancement by the 'fly casting' mechanism.** The figure shows a DNA-binding protein (right) interacting with its target (left). At an approach distance R_{cm}, the partially folded ensemble is already able to interact weakly with its binding site. In contrast, the folded structure remains out of range because of the smaller fluctuations in the folded state. The weak interaction between the unfolded ensemble and the target molecule allows the former to search for and find its specific binding site within its bound partner, while completing the folding process. The figure was taken from [100].

and coworkers have found that the *same IUP* (staphylococcal nuclease in this case) can employ both mechanisms [106]. Clearly, it is necessary to carry out much more research on this matter in order to draw definite conclusions.

6.3.1.3 Significance of PPII helix in IUPs

The PPII conformation is particularly common in IUPs, implying that it may be at least partially responsible for their binding efficiency. One of the salient characteristics of this helical conformation is the absence of any intramolecular interactions. **Backbone and side chain groups are therefore free to interact with the target protein** [107]. Many of the side chain groups are nonpolar, and responsible for creating the characteristic nonpolar interfaces. As mentioned earlier, the high hydrophobicity of these interfaces is responsible for the larger-than-usual contact surface area per residue, which compensates for the otherwise weak binding affinity associated with IUPs.

PPII helices also appear in IDRs *within* globular proteins, where, as in IUPs, they tend to participate in protein-protein binding. However, the amino acid composition of PPII helices in globular proteins differs from that in IUPs. Specifically, in globular proteins, PPII helices are as polar as other protein segments, and they tend to be enriched with proline residues. Therefore, their ability to participate in binding must rely on other properties than those described above for IUPs. As this issue is related to the function of common binding domains in proteins, it will be further discussed in Chapter 8, which deals with protein-ligand interactions.

6.3.1.4 Disorder can be used for regulation

We have seen how the weak binding of IUPs to their targets can be advantageous in complex functional networks that require rapid association and dissociation of the protein from its target. Another advantage of disorder in such processes is that it makes the protein much easier to regulate. Tight regulation is crucial in signal transduction, which mediates some of the most dramatic outcomes in cells: division, growth, death, and mass production and/or release of bioactive chemicals. **Disorder promotes regulation mainly by conferring an extended conformation to the polypeptide chain, which makes most residues solvent-exposed, and therefore available to modifying enzymes** [36]. Indeed, post-translational modification (PTM) sites in proteins are often clustered within IDRs [39,108]. The types of PTM used for protein regulation include phosphorylation, acetylation, ubiquitination, hydroxylation, and methylation (see Chapter 2). For example, the first three are known to change the function of the tumor-suppressing protein p53, as well as its localization and turnover [109]. Most of the sites for these p53 PTMs are located within IDRs.

Of the above-mentioned PTMs used for the regulation of protein activity, phosphorylation is by far the most common. As explained in Chapter 2, phosphorylation may act in different ways, one of which is inducing a conformational change in the protein, which leads to a change in its activity [110,111]. The inherent disorder of IUPs provides these proteins with a wider range of conformational changes compared with globular proteins. Therefore, it also provides (at least theoretically) more ways for phosphorylation to change the activity of the protein [112]. This is demonstrated by the Ets-1 transcriptional activator, which binds in its disordered form to DNA. Following phosphorylation at multiple sites, Ets-1 gradually acquires an ordered conformation [113]. Whereas the disordered form has high affinity to DNA, the ordered form does not. Thus, phosphorylation is used to control the activity of Ets-1 by inducing folding. This is a nice example of how cells combine the inherent disorder of IUPs with existing means of regulation (in this case phosphorylation) to fine-tune key processes, such as gene expression.

6.3.2 Entropy assistance-related roles

The functioning of IUPs as entropy assistants is obvious, due to their characteristically large freedom of movement. This property is particularly important for IDRs separating domains, for different reasons. For example, in enzymes carrying out different catalytic steps consecutively, using different domains, there is a real advantage in the capability of the domains to sample different orientations with respect to each other, so as to be able to pass on the substrate between them efficiently. In enzymes that have both regulatory and catalytic domains, the flexible linking IDR allows the domains to interact efficiently with each other upon the binding of an activator or inhibitor to the regulatory domain. Flexible linkers may also allow the domains they are linking to undergo a more extensive conformational change needed for interacting with another protein. A well-known example is calmodulin (CaM), in which the two globular domains containing the Ca^{2+} helix-loop-helix (HLH) motif are linked by a short sequence of 4 to 8 residues, which is helical in the crystal structure but disordered in solution [27]. As described in Chapter 2, the binding of Ca^{2+} to the HLH motif in the globular domains induces an extensive conformational change in the entire molecule, allowing it to wrap around the target protein segment, which is usually in a helical conformation (Figure 2.24). The flexibility of the disordered linker, as well as of the methionine side chains exposed during this change, allows CaM not only to undergo the conformational change, but also to bind different targets [114]. Finally, in nucleoporins, which reside in the nuclear lamina, the flexible linker also acts as a gate, which either opens or closes the entry point of cellular molecules into the nucleus [54].

6.4 IUPs *IN VIVO*

In the sections above we have seen how the lack of structure in IUPs can be advantageous for different functions, even though this idea seems to be at odds with the central structure-function paradigm of structural and molecular biology. Many of the studies demonstrating the properties of IUPs, as well as their activity, have been carried out using *in vitro* experiments [92]. Although the general use of the *in vitro* approach is not in question, its use for studying IUPs is problematic, for two main reasons. **First, the crowding effect resulting from the highly dense cytoplasm [115,116] is expected to induce folding in at least some of the proteins considered to be IUPs.** NMR studies, which have been employed to investigate this issue, have so far yielded contradictory results for different proteins (e.g., [117,118]).

Another open question concerning IUPs inside cells is how they avoid chaperones. As explained in the previous chapter, molecular chaperones reside inside the cytoplasm and certain organelles, where they 'hunt' misfolded proteins and provide them with a protected environment in which they can refold. While this is advantageous for globular proteins gone awry, IUPs, being incapable of folding (at least in their free state), might be incapacitated by the action of chaperones. A recent *in vivo* study demonstrates low tendency of IUPs to bind to chaperones, despite their unfolded nature [119]. This may be explained by the highly polar nature of IUPs. Indeed, chaperones recognize misfolded proteins by detecting their exposed nonpolar residues. **Being polar and even charged, IUPs are not expected to be recognized by chaperones.** Another reason might be structural; whereas IUPs primarily have a PPII helical conformation, misfolded globular proteins tend to have some type of order, usually in the form of a β conformation [120]. A final reason might be the ability of IUPs to bind to their natural cellular targets very quickly. This binding, which often involves folding, would 'hide' the IUP from chaperones in its environment.

6.5 SUMMARY

- Some proteins, referred to as IUPs, may be completely devoid of regular three-dimensional structure or contain unstructured regions. Although they are disordered, IUPs participate in complex biological processes such as signal transduction and regulation, where they function mainly in molecular recognition, but also in assisting in the assembly of large protein complexes, scavenging small molecules, and even inducing conformational changes in other proteins or assisting their folding.

- IUPs are devoid of tertiary and quaternary structure, but may contain unstable secondary elements, with the PPII helix being the most common. These elements are often stabilized when the IUP binds to its target protein. The absence of tertiary structure in unbound

IUPs is ensured by their highly polar composition, which prevents the formation of a stable nonpolar core.

- The disordered nature of IUPs confers plasticity of binding. Still, under fixed conditions a given IUP tends to bind one ligand in a specific manner. The need to fold upon binding weakens the overall affinity of IUPs to their targets by increasing the entropy cost of the process. The weak binding allows IUPs to participate in complex biological processes, which require rapid binding and releasing of ligands. However, the 'waste' of energy on the folding process is thought to prevent IUPs from functioning as enzymes, which require all the energy produced by substrate binding to be channeled towards driving catalysis.

- Despite the overall weak affinity of IUPs towards their target proteins, their interfaces optimize binding interactions. This property, as well as their specificity towards their targets, is largely achieved by the availability of many polar groups, which in folded (globular) proteins are mostly paired in hydrogen bonds and are therefore unavailable for binding. Specificity is also achieved through binding-folding coupling, which allows IUPs to keep searching for the best binding conformation even after the formation of the initial encounter complex ('fly casting' mechanism).

EXERCISES

6.1 Explain how intrinsically unstructured proteins (IUPs) can be so common and yet so few of them are present in the Protein Data Bank.

6.2 Use the example of the nuclear pore complex to explain how IUPs may carry out certain functional roles in cells more efficiently than structured proteins.

6.3 Which local structure(s) are common in IUPs? Explain their compatibility with the roles assigned to IUPs.

6.4 Explain the principles of the 'fly casting' mechanism.

REFERENCES

1. E. Fischer. Einfluss der Configuration auf die Wirkung der Enzyme. *Ber. Dtsch. Chem. Ges.*, 27(3):2985–2993, 1894.
2. L. Pauling, R. B. Corey, and H. R. Branson. The structure of proteins; two hydrogen-bonded helical configurations of the polypeptide chain. *Proc. Natl. Acad. Sci. USA*, 37(4):205–11, 1951.
3. C. C. Blake, D. F. Koenig, G. A. Mair, A. C. North, D. C. Phillips, and V. R. Sarma. Structure of hen egg-white lysozyme. A three-dimensional Fourier synthesis at 2 Angstrom resolution. *Nature*, 206(986):757–61, 1965.
4. J. C. Kendrew, R. E. Dickerson, B. E. Strandberg, R. G. Hart, D. D. Davis, D. C. Phillips, and V. C. Shore. The three dimensional structure of myoglobin. *Nature*, 185:422–427, 1960.
5. C. B. Anfinsen. Principles that Govern the Folding of Protein Chains. *Science*, 181(96):223–230, 1973.
6. V. J. Hilser and E. B. Thompson. Intrinsic disorder as a mechanism to optimize allosteric coupling in proteins. *Proc. Natl. Acad. Sci. USA*, 104(20):8311–5, 2007.
7. H. J. Dyson and P. E. Wright. Intrinsically unstructured proteins and their functions. *Nat. Rev. Mol. Cell Biol.*, 6(3):197–208, 2005.
8. C. J. Oldfield and A. K. Dunker. Intrinsically disordered proteins and intrinsically disordered protein regions. *Annu. Rev. Biochem.*, 83:553–84, 2014.

9. P. Tompa, E. Schad, A. Tantos, and L. Kalmar. Intrinsically disordered proteins: emerging interaction specialists. *Curr. Opin. Struct. Biol.*, 35:49–59, 2015.

10. R. van der Lee, M. Buljan, B. Lang, R. J. Weatheritt, G. W. Daughdrill, A. K. Dunker, M. Fuxreiter, J. Gough, J. Gsponer, D. T. Jones, P. M. Kim, R. W. Kriwacki, C. J. Oldfield, R. V. Pappu, P. Tompa, V. N. Uversky, P. E. Wright, and M. M. Babu. Classification of Intrinsically Disordered Regions and Proteins. *Chem. Rev.*, 114(13):6589–6631, 2014.

11. L. M. Iakoucheva, C. J. Brown, J. D. Lawson, Z. Obradovic, and A. K. Dunker. Intrinsic disorder in cell-signaling and cancer-associated proteins. *J. Mol. Biol.*, 323(3):573–84, 2002.

12. P. Radivojac, L. M. Iakoucheva, C. J. Oldfield, Z. Obradovic, V. N. Uversky, and A. K. Dunker. Intrinsic disorder and functional proteomics. *Biophys. J.*, 92(5):1439–56, 2007.

13. H. J. Dyson and P. E. Wright. Equilibrium NMR studies of unfolded and partially folded proteins. *Nat. Struct. Biol.*, 5 Suppl:499–503, 1998.

14. S. Mukhopadhyay, R. Krishnan, E. A. Lemke, S. Lindquist, and A. A. Deniz. A natively unfolded yeast prion monomer adopts an ensemble of collapsed and rapidly fluctuating structures. *Proc. Natl. Acad. Sci. USA*, 104(8):2649–54, 2007.

15. C. K. Fisher and C. M. Stultz. Constructing ensembles for intrinsically disordered proteins. *Curr. Opin. Struct. Biol.*, 21(3):426–31, 2011.

16. H. J. Dyson and P. E. Wright. Unfolded proteins and protein folding studied by NMR. *Chem. Rev.*, 104(8):3607–22, 2004.

17. B. Schuler and H. Hofmann. Single-molecule spectroscopy of protein folding dynamics: expanding scope and timescales. *Curr. Opin. Struct. Biol.*, 23(1):36–47, 2013.

18. P. R. Banerjee and A. A. Deniz. Shedding light on protein folding landscapes by single-molecule fluorescence. *Chem. Soc. Rev.*, 43(4):1172–88, 2014.

19. D. Eliezer. Biophysical characterization of intrinsically disordered proteins. *Curr. Opin. Struct. Biol.*, 19(1):23–30, 2009.

20. Z. Dosztanyi, V. Csizmok, P. Tompa, and I. Simon. IUPred: web server for the prediction of intrinsically unstructured regions of proteins based on estimated energy content. *Bioinformatics*, 21(16):3433–4, 2005.

21. J. J. Ward, J. S. Sodhi, L. J. McGuffin, B. F. Buxton, and D. T. Jones. Prediction and functional analysis of native disorder in proteins from the three kingdoms of life. *J. Mol. Biol.*, 337(3):635–45, 2004.

22. T. Ishida and K. Kinoshita. Prediction of disordered regions in proteins based on the meta approach. *Bioinformatics*, 24(11):1344–8, 2008.

23. A. H. Mao, N. Lyle, and R. V. Pappu. Describing sequence-ensemble relationships for intrinsically disordered proteins. *Biochem. J.*, 449(2):307–18, 2013.

24. C. M. Baker and R. B. Best. Insights into the binding of intrinsically disordered proteins from molecular dynamics simulation. *Wiley Interdiscip. Rev. Comput. Mol. Sci.*, 4(3):182–198, 2014.

25. A. K. Dunker, Z. Obradovic, P. Romero, E. C. Garner, and C. J. Brown. Intrinsic protein disorder in complete genomes. *Genome Inform. Ser. Workshop Genome Inform.*, 11:161–71, 2000.

26. P. Tompa. The interplay between structure and function in intrinsically unstructured proteins. *FEBS Lett.*, 579(15):3346–54, 2005.

27. A. K. Dunker, I. Silman, V. N. Uversky, and J. L. Sussman. Function and structure of inherently disordered proteins. *Curr. Opin. Struct. Biol.*, 18(6):756–64, 2008.

28. M. Sickmeier, J. A. Hamilton, T. LeGall, V. Vacic, M. S. Cortese, A. Tantos, B. Szabo, P. Tompa, J. Chen, V. N. Uversky, Z. Obradovic, and A. K. Dunker. DisProt: the Database of Disordered Proteins. *Nucleic Acids Res.*, 35(suppl 1):D786–93, 2007.

29. S. Fukuchi, T. Amemiya, S. Sakamoto, Y. Nobe, K. Hosoda, Y. Kado, S. D. Murakami, R. Koike, H. Hiroaki, and M. Ota. IDEAL in 2014 illustrates interaction networks composed of intrinsically disordered proteins and their binding partners. *Nucleic Acids Res.*, 42(D1):D320–5, 2014.

30. S. Fukuchi, S. Sakamoto, Y. Nobe, S. D. Murakami, T. Amemiya, K. Hosoda, R. Koike, H. Hiroaki, and M. Ota. IDEAL: Intrinsically Disordered proteins with Extensive Annotations and Literature. *Nucleic Acids Res.*, 40(D1):D507–11, 2012.

31. E. Potenza, T. Di Domenico, I. Walsh, and S. C. Tosatto. MobiDB 2.0: an improved database of intrinsically disordered and mobile proteins. *Nucleic Acids Res.*, 43(D1):D315–20, 2015.

32. T. Di Domenico, I. Walsh, A. J. Martin, and S. C. Tosatto. MobiDB: a comprehensive database of intrinsic protein disorder annotations. *Bioinformatics*, 28(15):2080–1, 2012.

33. M. Varadi, S. Kosol, P. Lebrun, E. Valentini, M. Blackledge, A. K. Dunker, I. C. Felli, J. D. Forman-Kay, R. W. Kriwacki, R. Pierattelli, J. Sussman, D. I. Svergun, V. N. Uversky, M. Vendruscolo, D. Wishart, P. E.

Wright, and P. Tompa. pE-DB: a database of structural ensembles of intrinsically disordered and of unfolded proteins. *Nucleic Acids Res.*, 42(D1):D326–35, 2014.

34. M. E. Oates, P. Romero, T. Ishida, M. Ghalwash, M. J. Mizianty, B. Xue, Z. Dosztanyi, V. N. Uversky, Z. Obradovic, L. Kurgan, A. K. Dunker, and J. Gough. D^2P^2: database of disordered protein predictions. *Nucleic Acids Res.*, 41(D1):D508–16, 2013.

35. P. E. Wright and H. J. Dyson. Intrinsically disordered proteins in cellular signalling and regulation. *Nat. Rev. Mol. Cell Biol.*, 16(1):18–29, 2015.

36. C. A. Galea, Y. Wang, S. G. Sivakolundu, and R. W. Kriwacki. Regulation of cell division by intrinsically unstructured proteins: intrinsic flexibility, modularity, and signaling conduits. *Biochemistry*, 47(29):7598–609, 2008.

37. V. N. Uversky, C. J. Oldfield, and A. K. Dunker. Intrinsically disordered proteins in human diseases: introducing the D^2 concept. *Annu. Rev. Biophys.*, 37:215–46, 2008.

38. S. Vucetic, H. Xie, L. M. Iakoucheva, C. J. Oldfield, A. K. Dunker, Z. Obradovic, and V. N. Uversky. Functional anthology of intrinsic disorder. 2. Cellular components, domains, technical terms, developmental processes, and coding sequence diversities correlated with long disordered regions. *J. Proteome Res.*, 6(5):1899–916, 2007.

39. H. Xie, S. Vucetic, L. M. Iakoucheva, C. J. Oldfield, A. K. Dunker, Z. Obradovic, and V. N. Uversky. Functional anthology of intrinsic disorder. 3. Ligands, post-translational modifications, and diseases associated with intrinsically disordered proteins. *J. Proteome Res.*, 6(5):1917–32, 2007.

40. H. Xie, S. Vucetic, L. M. Iakoucheva, C. J. Oldfield, A. K. Dunker, V. N. Uversky, and Z. Obradovic. Functional anthology of intrinsic disorder. 1. Biological processes and functions of proteins with long disordered regions. *J. Proteome Res.*, 6(5):1882–98, 2007.

41. A. K. Dunker, C. J. Brown, J. D. Lawson, L. M. Iakoucheva, and Z. Obradovic. Intrinsic disorder and protein function. *Biochemistry*, 41(21):6573–82, 2002.

42. P. Tompa and M. Fuxreiter. Fuzzy complexes: polymorphism and structural disorder in protein-protein interactions. *Trends Biochem. Sci.*, 33(1):2–8, 2008.

43. T. Mittag, L. E. Kay, and J. D. Forman-Kay. Protein dynamics and conformational disorder in molecular recognition. *J. Mol. Recognit.*, 23(2):105–16, 2010.

44. M. S. Cortese, V. N. Uversky, and A. K. Dunker. Intrinsic disorder in scaffold proteins: getting more from less. *Prog. Biophys. Mol. Biol.*, 98(1):85–106, 2008.

45. A. S. Kim, L. T. Kakalis, N. Abdul-Manan, G. A. Liu, and M. K. Rosen. Autoinhibition and activation mechanisms of the Wiskott-Aldrich syndrome protein. *Nature*, 404(6774):151–8, 2000.

46. C. Holt. Unfolded phosphopolypeptides enable soft and hard tissues to coexist in the same organism with relative ease. *Curr. Opin. Struct. Biol.*, 23(3):420–5, 2013.

47. R. Ivanyi-Nagy, L. Davidovic, E. W. Khandjian, and J. L. Darlix. Disordered RNA chaperone proteins: from functions to disease. *Cell. Mol. Life Sci.*, 62(13):1409–17, 2005.

48. M. Fuxreiter. Fuzziness: linking regulation to protein dynamics. *Mol. Biosyst.*, 8(1):168–77, 2012.

49. L. Renault, B. Bugyi, and M. F. Carlier. Spire and Cordon-bleu: multifunctional regulators of actin dynamics. *Trends Cell Biol.*, 18(10):494–504, 2008.

50. M. S. Pometun, E. Y. Chekmenev, and R. J. Wittebort. Quantitative observation of backbone disorder in native elastin. *J. Biol. Chem.*, 279(9):7982–7, 2004.

51. A. Nagy, L. Grama, T. Huber, P. Bianco, K. Trombitas, H. L. Granzier, and M. S. Kellermayer. Hierarchical extensibility in the PEVK domain of skeletal-muscle titin. *Biophys. J.*, 89(1):329–36, 2005.

52. L. Tskhovrebova and J. Trinick. Titin: properties and family relationships. *Nat. Rev. Mol. Cell Biol.*, 4(9):679–89, 2003.

53. J. H. Hoh. Functional protein domains from the thermally driven motion of polypeptide chains: a proposal. *Proteins*, 32(2):223–8, 1998.

54. F. Alber, S. Dokudovskaya, L. M. Veenhoff, W. Zhang, J. Kipper, D. Devos, A. Suprapto, O. Karni-Schmidt, R. Williams, B. T. Chait, A. Sali, and M. P. Rout. The molecular architecture of the nuclear pore complex. *Nature*, 450(7170):695–701, 2007.

55. R. L. Adams and S. R. Wente. Uncovering nuclear pore complexity with innovation. *Cell*, 152(6):1218–21, 2013.

56. S. Milles, D. Mercadante, I. V. Aramburu, M. R. Jensen, N. Banterle, C. Koehler, S. Tyagi, J. Clarke, S. L. Shammas, M. Blackledge, F. Gräter, and E. A. Lemke. Plasticity of an Ultrafast Interaction between Nucleoporins and Nuclear Transport Receptors. *Cell*, 163(3):734–745, 2015.

57. J. M. Cronshaw, A. N. Krutchinsky, W. Zhang, B. T. Chait, and M. J. Matunis. Proteomic analysis of the mammalian nuclear pore complex. *J. Cell Biol.*, 158(5):915–27, 2002.

58. M. Eibauer, M. Pellanda, Y. Turgay, A. Dubrovsky, A. Wild, and O. Medalia. Structure and gating of the nuclear pore complex. *Nat. Commun.*, 6:7532, 2015.

59. E. Hurt and M. Beck. Towards understanding nuclear pore complex architecture and dynamics in the age of integrative structural analysis. *Curr. Opin. Cell Biol.*, 34:31–8, 2015.

60. J. D. Aitchison and M. P. Rout. The Yeast Nuclear Pore Complex and Transport Through It. *Genetics*, 190(3):855–883, 2012.

61. F. Alber, S. Dokudovskaya, L. M. Veenhoff, W. Zhang, J. Kipper, D. Devos, A. Suprapto, O. Karni-Schmidt, R. Williams, B. T. Chait, M. P. Rout, and A. Sali. Determining the architectures of macromolecular assemblies. *Nature*, 450(7170):683–694, 2007.

62. K. H. Bui, A. von Appen, A. L. DiGuilio, A. Ori, L. Sparks, M. T. Mackmull, T. Bock, W. Hagen, A. Andres-Pons, J. S. Glavy, and M. Beck. Integrated structural analysis of the human nuclear pore complex scaffold. *Cell*, 155(6):1233–43, 2013.

63. Y. Shi, J. Fernandez-Martinez, E. Tjioe, R. Pellarin, S. J. Kim, R. Williams, D. Schneidman-Duhovny, A. Sali, M. P. Rout, and B. T. Chait. Structural characterization by cross-linking reveals the detailed architecture of a coatomer-related heptameric module from the nuclear pore complex. *Mol. Cell. Proteomics*, 13(11):2927–43, 2014.

64. K. Thierbach, A. von Appen, M. Thoms, M. Beck, D. Flemming, and E. Hurt. Protein interfaces of the conserved Nup84 complex from *Chaetomium thermophilum* shown by crosslinking mass spectrometry and electron microscopy. *Structure*, 21(9):1672–82, 2013.

65. A. von Appen, J. Kosinski, L. Sparks, A. Ori, A. L. DiGuilio, B. Vollmer, M. T. Mackmull, N. Banterle, L. Parca, P. Kastritis, K. Buczak, S. Mosalaganti, W. Hagen, A. Andres-Pons, E. A. Lemke, P. Bork, W. Antonin, J. S. Glavy, K. H. Bui, and M. Beck. In situ structural analysis of the human nuclear pore complex. *Nature*, 526(7571):140–3, 2015.

66. T. Stuwe, A. R. Correia, D. H. Lin, M. Paduch, V. T. Lu, A. A. Kossiakoff, and A. Hoelz. Nuclear pores. Architecture of the nuclear pore complex coat. *Science*, 347(6226):1148–52, 2015.

67. N. Pante and M. Kann. Nuclear pore complex is able to transport macromolecules with diameters of about 39 nm. *Mol. Biol. Cell*, 13(2):425–34, 2002.

68. S. Bilokapic and T. U. Schwartz. 3D ultrastructure of the nuclear pore complex. *Curr. Opin. Cell Biol.*, 24(1):86–91, 2012.

69. T. Maimon, N. Elad, I. Dahan, and O. Medalia. The human nuclear pore complex as revealed by cryo-electron tomography. *Structure*, 20(6):998–1006, 2012.

70. D. P. Denning, S. S. Patel, V. Uversky, A. L. Fink, and M. Rexach. Disorder in the nuclear pore complex: the FG repeat regions of nucleoporins are natively unfolded. *Proc. Natl. Acad. Sci. USA*, 100(5):2450–5, 2003.

71. B. Fahrenkrog, B. Maco, A. M. Fager, J. Köser, U. Sauder, K. S. Ullman, and U. Aebi. Domain-specific antibodies reveal multiple-site topology of Nup153 within the nuclear pore complex. *J. Struct. Biol.*, 140(1–3):254–267, 2002.

72. R. Y. Lim and B. Fahrenkrog. The nuclear pore complex up close. *Curr. Opin. Cell Biol.*, 18(3):342–7, 2006.

73. L. A. Strawn, T. Shen, N. Shulga, D. S. Goldfarb, and S. R. Wente. Minimal nuclear pore complexes define FG repeat domains essential for transport. *Nat. Cell Biol.*, 6(3):197–206, 2004.

74. M. P. Rout, J. D. Aitchison, M. O. Magnasco, and B. T. Chait. Virtual gating and nuclear transport: the hole picture. *Trends Cell Biol.*, 13(12):622–8, 2003.

75. S. M. Liu and M. Stewart. Structural basis for the high-affinity binding of nucleoporin Nup1p to the *Saccharomyces cerevisiae* importin-beta homologue, Kap95p. *J. Mol. Biol.*, 349(3):515–25, 2005.

76. L. J. Terry and S. R. Wente. Flexible Gates: Dynamic Topologies and Functions for FG Nucleoporins in Nucleocytoplasmic Transport. *Eukaryotic Cell*, 8(12):1814–1827, 2009.

77. B. B. Hülsmann, A. A. Labokha, and D. Görlich. The Permeability of Reconstituted Nuclear Pores Provides Direct Evidence for the Selective Phase Model. *Cell*, 150(4):738–751, 2012.

78. A. A. Labokha, S. Gradmann, S. Frey, B. B. Hülsmann, H. Urlaub, M. Baldus, and D. Görlich. Systematic analysis of barrier-forming FG hydrogels from Xenopus nuclear pore complexes. *EMBO J.*, 32(2):204–218, 2012.

79. E. Laurell, K. Beck, K. Krupina, G. Theerthagiri, B. Bodenmiller, P. Horvath, R. Aebersold, W. Antonin, and U. Kutay. Phosphorylation of Nup98 by Multiple Kinases Is Crucial for NPC Disassembly during Mitotic Entry. *Cell*, 144(4):539–550, 2011.

80. J. Klein-Seetharaman, M. Oikawa, S. B. Grimshaw, J. Wirmer, E. Duchardt, T. Ueda, T. Imoto, L. J.

Smith, C. M. Dobson, and H. Schwalbe. Long-range interactions within a nonnative protein. *Science*, 295(5560):1719–22, 2002.

81. C. J. Brown, S. Takayama, A. M. Campen, P. Vise, T. W. Marshall, C. J. Oldfield, C. J. Williams, and A. K. Dunker. Evolutionary rate heterogeneity in proteins with long disordered regions. *J. Mol. Evol.*, 55(1):104–10, 2002.

82. Z. Dosztanyi and P. Tompa. Prediction of protein disorder. *Methods Mol. Biol.*, 426:103–15, 2008.

83. H. Dinkel, K. Van Roey, S. Michael, N. E. Davey, R. J. Weatheritt, D. Born, T. Speck, D. Kruger, G. Grebnev, M. Kuban, M. Strumillo, B. Uyar, A. Budd, B. Altenberg, M. Seiler, L. B. Chemes, J. Glavina, I. E. Sanchez, F. Diella, and T. J. Gibson. The eukaryotic linear motif resource ELM: 10 years and counting. *Nucleic Acids Res.*, 42:D259–66, 2014.

84. F. Diella, N. Haslam, C. Chica, A. Budd, S. Michael, N. P. Brown, G. Trave, and T. J. Gibson. Understanding eukaryotic linear motifs and their role in cell signaling and regulation. *Front. Biosci.*, 13:6580–603, 2008.

85. N. E. Davey, K. Van Roey, R. J. Weatheritt, G. Toedt, B. Uyar, B. Altenberg, A. Budd, F. Diella, H. Dinkel, and T. J. Gibson. Attributes of short linear motifs. *Mol. Biosyst.*, 8(1):268–81, 2012.

86. M. Fuxreiter, I. Simon, P. Friedrich, and P. Tompa. Preformed structural elements feature in partner recognition by intrinsically unstructured proteins. *J. Mol. Biol.*, 338(5):1015–26, 2004.

87. T. Pawson and P. Nash. Assembly of cell regulatory systems through protein interaction domains. *Science*, 300(5618):445–52, 2003.

88. C. M. Pfleger and M. W. Kirschner. The KEN box: an APC recognition signal distinct from the D box targeted by Cdh1. *Genes Dev.*, 14(6):655–65, 2000.

89. J. He, W. C. Chao, Z. Zhang, J. Yang, N. Cronin, and D. Barford. Insights into degron recognition by APC/C coactivators from the structure of an Acm1-Cdh1 complex. *Mol. Cell*, 50(5):649–60, 2013.

90. A. C. Joerger and A. R. Fersht. Structural biology of the tumor suppressor p53. *Annu. Rev. Biochem.*, 77:557–82, 2008.

91. H. J. Choi, A. H. Huber, and W. I. Weis. Thermodynamics of beta-catenin-ligand interactions: the roles of the N- and C-terminal tails in modulating binding affinity. *J. Biol. Chem.*, 281(2):1027–38, 2006.

92. P. E. Wright and H. J. Dyson. Linking folding and binding. *Curr. Opin. Struct. Biol.*, 19(1):31–8, 2009.

93. C. J. Oldfield, Y. Cheng, M. S. Cortese, P. Romero, V. N. Uversky, and A. K. Dunker. Coupled folding and binding with alpha-helix-forming molecular recognition elements. *Biochemistry*, 44(37):12454–70, 2005.

94. A. Mohan, C. J. Oldfield, P. Radivojac, V. Vacic, M. S. Cortese, A. K. Dunker, and V. N. Uversky. Analysis of molecular recognition features (MoRFs). *J. Mol. Biol.*, 362(5):1043–59, 2006.

95. V. Vacic, C. J. Oldfield, A. Mohan, P. Radivojac, M. S. Cortese, V. N. Uversky, and A. K. Dunker. Characterization of molecular recognition features, MoRFs, and their binding partners. *J. Proteome Res.*, 6(6):2351–66, 2007.

96. B. Meszaros, P. Tompa, I. Simon, and Z. Dosztanyi. Molecular Principles of the Interactions of Disordered Proteins. *J. Mol. Biol.*, 372(2):549–61, 2007.

97. G. Wu, Y. G. Chen, B. Ozdamar, C. A. Gyuricza, P. A. Chong, J. L. Wrana, J. Massague, and Y. Shi. Structural basis of Smad2 recognition by the Smad anchor for receptor activation. *Science*, 287(5450):92–7, 2000.

98. S. H. Lee, D. H. Kim, J. J. Han, E. J. Cha, J. E. Lim, Y. J. Cho, C. Lee, and K. H. Han. Understanding pre-structured motifs (PreSMos) in intrinsically unfolded proteins. *Curr. Protein Pept. Sci.*, 13(1):34–54, 2012.

99. K. Sugase, H. J. Dyson, and P. E. Wright. Mechanism of coupled folding and binding of an intrinsically disordered protein. *Nature*, 447(7147):1021–5, 2007.

100. B. A. Shoemaker, J. J. Portman, and P. G. Wolynes. Speeding molecular recognition by using the folding funnel: the fly-casting mechanism. *Proc. Natl. Acad. Sci. USA*, 97(16):8868–73, 2000.

101. Q. Lu, H. P. Lu, and J. Wang. Exploring the mechanism of flexible biomolecular recognition with single molecule dynamics. *Phys. Rev. Lett.*, 98(12):128105, 2007.

102. M. E. Ferreira, S. Hermann, P. Prochasson, J. L. Workman, K. D. Berndt, and A. P. Wright. Mechanism of transcription factor recruitment by acidic activators. *J. Biol. Chem.*, 280(23):21779–84, 2005.

103. R. Narayanan, O. K. Ganesh, A. S. Edison, and S. J. Hagen. Kinetics of folding and binding of an intrinsically disordered protein: the inhibitor of yeast aspartic proteinase YPrA. *J. Am. Chem. Soc.*, 130(34):11477–85, 2008.

104. Y. Levy, J. N. Onuchic, and P. G. Wolynes. Fly-casting in protein-DNA binding: frustration between protein folding and electrostatics facilitates target recognition. *J. Am. Chem. Soc.*, 129(4):738–9, 2007.

105. J. Song, L. W. Guo, H. Muradov, N. O. Artemyev, A. E. Ruoho, and J. L. Markley. Intrinsically disordered gamma-subunit of cGMP phosphodiesterase encodes functionally relevant transient secondary and tertiary structure. *Proc. Natl. Acad. Sci. USA*, 105(5):1505–10, 2008.

106. M. Onitsuka, H. Kamikubo, Y. Yamazaki, and M. Kataoka. Mechanism of induced folding: Both folding before binding and binding before folding can be realized in staphylococcal nuclease mutants. *Proteins*, 72(3):837–47, 2008.

107. A. Rath, A. R. Davidson, and C. M. Deber. The structure of "unstructured" regions in peptides and proteins: role of the polyproline II helix in protein folding and recognition. *Biopolymers*, 80(2–3):179–85, 2005.

108. L. M. Iakoucheva, P. Radivojac, C. J. Brown, T. R. O'Connor, J. G. Sikes, Z. Obradovic, and A. K. Dunker. The importance of intrinsic disorder for protein phosphorylation. *Nucleic Acids Res.*, 32(3):1037–49, 2004.

109. A. M. Bode and Z. Dong. Post-translational modification of p53 in tumorigenesis. *Nat. Rev. Cancer*, 4(10):793–805, 2004.

110. J. Lätzer, T. Shen, and P. G. Wolynes. Conformational Switching upon Phosphorylation: A Predictive Framework Based on Energy Landscape Principles. *Biochemistry*, 47(7):2110–2122, 2008.

111. L. N. Johnson and R. J. Lewis. Structural basis for control by phosphorylation. *Chem. Rev.*, 101(8):2209–42, 2001.

112. B. Ma and R. Nussinov. Regulating highly dynamic unstructured proteins and their coding mRNAs. *Genome Biol.*, 10(1):204, 2009.

113. M. A. Pufall, G. M. Lee, M. L. Nelson, H. S. Kang, A. Velyvis, L. E. Kay, L. P. McIntosh, and B. J. Graves. Variable control of Ets-1 DNA binding by multiple phosphates in an unstructured region. *Science*, 309(5731):142–5, 2005.

114. G. Fiorin, R. R. Biekofsky, A. Pastore, and P. Carloni. Unwinding the helical linker of calcium-loaded calmodulin: a molecular dynamics study. *Proteins*, 61(4):829–39, 2005.

115. S. B. Zimmerman and S. O. Trach. Estimation of macromolecule concentrations and excluded volume effects for the cytoplasm of *Escherichia coli*. *J. Mol. Biol.*, 222(3):599–620, 1991.

116. F.-X. Theillet, A. Binolfi, T. Frembgen-Kesner, K. Hingorani, M. Sarkar, C. Kyne, C. Li, P. B. Crowley, L. Gierasch, G. J. Pielak, A. H. Elcock, A. Gershenson, and P. Selenko. Physicochemical properties of cells and their effects on intrinsically disordered proteins (IDPs). *Chem. Rev.*, 114(13):6661–6714, 2014.

117. M. M. Dedmon, C. N. Patel, G. B. Young, and G. J. Pielak. FlgM gains structure in living cells. *Proc. Natl. Acad. Sci. USA*, 99(20):12681–4, 2002.

118. C. Li, L. M. Charlton, A. Lakkavaram, C. Seagle, G. Wang, G. B. Young, J. M. Macdonald, and G. J. Pielak. Differential dynamical effects of macromolecular crowding on an intrinsically disordered protein and a globular protein: implications for in-cell NMR spectroscopy. *J. Am. Chem. Soc.*, 130(20):6310–1, 2008.

119. H. Hegyi and P. Tompa. Intrinsically disordered proteins display no preference for chaperone binding in vivo. *PLoS Comput. Biol.*, 4(3):e1000017, 2008.

120. F. Zhu, J. Kapitan, G. E. Tranter, P. D. Pudney, N. W. Isaacs, L. Hecht, and L. D. Barron. Residual structure in disordered peptides and unfolded proteins from multivariate analysis and ab initio simulation of Raman optical activity data. *Proteins*, 70(3):823–33, 2008.

CHAPTER 7

Membrane-Bound Proteins

7.1 INTRODUCTION

The plasma membrane is a lipid body containing proteins and carbohydrate units. It defines the shape of the cell and acts as a physical barrier between the cytoplasm and the external environment (*exoplasm*). Since these two environments are very different in terms of their chemical composition, the formation of the first membranes signified the evolutionary emergence of the first biological organisms; this event took place between 3.2 and 3.8 billion years ago [1-4]. The cells of all organisms are enveloped by plasma membranes, and some cells contain other membranes as well. Gram-negative bacteria, for example, also have an outer membrane, which has its own set of characteristic proteins, and is rich in polysaccharides. Eukaryotic cells, which evolved about 1.5 billion years ago, contain inner membranes, which define the various organelles.

The primary role of the plasma membrane is to maintain the unique chemical environment of the cell, which includes the following:

1. **Ions.** Intracellular ions are chemically diverse. Some are elements such as Na^+, K^+, Cl^-, Mg^{2+}, Mn^{2+}, Cu^{2+}, Zn^+, Co^{2+} and $Fe^{2+/3+}$, whereas others are molecules of various sizes (e.g., PO_2^{3-}). Maintaining a constant concentration of these ions inside the cell is crucial for the routine execution of numerous cellular and physiological processes. First, by stabilizing charged groups in the active sites of enzymes, or participating in the binding of atoms and molecules, ions enable routine biochemical pathways to take place [5,6]. Second, the concentration gradient of Na^+, K^+ and Cl^- across the plasma membrane is responsible for its electric potential, which is used to drive processes such as cellular transport, neural transmission, and muscle contraction. Finally, the physiological ionic balance is important for the regulation of body hydration.

2. **Small metabolites.** Small organic molecules, such as ATP, amino acids, monosaccharides and disaccharides, nucleotides, pyruvate, and others, participate in, and are formed by, metabolic processes.

3. **Macromolecules.** Proteins, carbohydrates, lipids, and nucleic acids are the functional units of cells and tissues, as explained in Chapter 1.

Many of these chemicals are polar (neutral or charged), which explains why the type of barrier chosen by evolution to keep them inside the cell is based on a lipid structure, whose permeability to such molecules is extremely low. In order for a polar compound to cross the membrane, it must first transfer from the aqueous environment (cytoplasm or exoplasm) into the membrane itself. This process is highly unlikely, due to the desolvation of the polar compound. The transfer energy can be estimated using the Born equation of solvation (Equation (2.2)), as follows.

Assuming that ion transfer into the membrane is determined by electrostatic contributions, the transfer free energy is:

$$\Delta G_{\text{transfer}} = G_m - G_w = 166 \left(\frac{q}{r}\right) \left(\frac{1}{\varepsilon_m} - \frac{1}{\varepsilon_w}\right) \qquad (7.1)$$

(where m and w are the membrane and aqueous environments, respectively; q is the net charge of the molecule; r is its effective radius; and ε is the dielectric. See Chapter 2 for more details.).

For simplicity, let us consider a simple spherical ion of radius 1 Å and charge +1. Taking the dielectrics of the membrane and aqueous environment to be 2 and 80, respectively, we obtain $\Delta G_{\text{transfer}} = +80\,\text{kcal/mol}$. Thus, the fraction of ions that transfer into the membrane (Equation (4.3b)) is: $P \propto e^{(-\Delta G/RT)} = e^{(-80/0.6)} \approx 1 \times 10^{-58}$. **Thus, in essence, at a physiological ion concentration of about 150 mM, no cation will partition into the membrane.** Computational [7] and experimental [8] studies show that this is also true for large ions, such as the side chains of charged amino acids.

Since most of the compounds inside cells are produced and utilized constantly by metabolic processes, it is not enough to prevent them from leaving the cell; keeping their concentrations fixed requires the constant import of some compounds and the export of others. In addition, the entire process must be carefully controlled, so as to avoid the loss of important metabolites or the internalization of wastes and/or toxic chemicals. As in the other cases we have encountered, here too evolution has assigned the job to proteins. There are two types of *transport proteins*: *channels* and *transporters*. A channel crosses the entire length of the membrane, and contains a water annulus, which enables polar chemicals (often ions) to move across the membrane, down their electrochemical gradient. A transporter binds polar chemicals on one side of the membrane and releases them on the other side. Some transporters span the entire width of the membrane, whereas others (*carriers*) do not, and have to diffuse from one side to the other in order to release the 'substrate' into the right compartment. While some transporters transfer molecules down their electrochemical gradient, others (*pumps*) do so in the opposite direction, by using an external source of energy, which can be direct (ATP), or indirect (the electrochemical gradient of another molecule). The transport process is controlled in both channels and transporters; channels only let in molecules that are small enough to enter the water annulus, and can also open or close in response to different signals, such as ligand binding, change in cross-membrane voltage, or application of mechanical pressure. Transporters can bind or release their 'substrates' when the latter are present, but in many cases they do so pending the binding of the right regulatory ligand. The function of transport proteins is so basic that in microorganisms they constitute 40% to 50% of all membrane proteins [9].

In addition to transport, membrane proteins play other important roles, most of which are mentioned in Chapter 1:

1. **Communication and signal transduction.** Numerous cellular proteins are membrane-bound receptors. These act as antennae, and pass communication signals arriving

from the external environment into the cell. Most of these receptors respond to chemical messengers, such as hormones, neurotransmitters, pheromones, odorants, and local mediators, whereas others respond to other types of signals, such as electromagnetic radiation (light) or mechanical pressure. Thus, the proper functioning of these proteins is crucial not only to individual cells, but also to entire physiological systems, specifically, the nervous, endocrine, and immune systems. Receptors span the entire width of the membrane, with their extracellular side designed to bind or respond to the external messenger, and their intracellular side interacting with different cytoplasmic proteins. The latter transmit messages into the cell, often by catalyzing enzymatic reactions. For example, *growth factor receptors* relay the message 'grow' or 'divide' into the cell by promoting phosphorylation of various cellular components. While most of the cytoplasmic proteins receiving this message are water-soluble, some, such as *G-proteins* or the enzymes *protein kinase A and C* are membrane-bound, at least part of the time.

2. **Cell-cell and cell-ECM recognition.** Certain membrane-bound proteins, such as *integrins* or *cadherins*, bind proteins or other elements that reside either on other cells or in the extracellular matrix (ECM). Such interactions are important, e.g., for the cell's ability to recognize its neighbors or become anchored to its biological tissue.

3. **Energy production and photosynthesis.** A number of proteins that reside inside the inner mitochondrial membrane (or the plasma membrane of bacteria) participate in the extraction of chemical energy from foodstuff, and its storage in a freely available form, ATP. Most of these proteins act as electron carriers and proton pumps, whereas others construct the ATP-producing component of this system. A similar system functions in the thylakoid membrane of plants and algae (or the plasma membrane of photosynthetic bacteria), in the assimilation of solar energy, i.e., photosynthesis.

4. **Defense.** Certain membrane-bound proteins participate in the defense of the cell or the entire body against invading pathogens, i.e., bacteria, viruses, and parasites. These proteins may fulfill different functions, most of which involve recognition of pathogen-related molecules. A well-known example is the *T-cell receptor*, which spans the membrane of a T lymphocyte, and recognizes peptides that have been taken from degraded pathogens.

5. **Cellular trafficking.** Membrane proteins often serve as attachment points for other proteins. This function enables cells to concentrate metabolic enzymes or signal transduction proteins in certain locations. The trafficking of vesicles carrying lipids and proteins between cellular compartments is also affected by certain membrane-bound proteins.

The significance of membrane proteins is reflected not only in their functions but also in their prevalence; it is estimated that 20% to 30% of any genome codes for membrane proteins [10,11], and a recent, extensive survey of the human proteome indicates a similar value, 23% [12]. Defects in membrane proteins manifest as various pathologies, including neural or cardiovascular disorders, depression, obesity, and cancer. Accordingly, it is estimated that ~60% of approved pharmaceutical drugs act on membrane proteins [13,14], most of which are G protein-coupled receptors (GPCRs; see below) [15,16].

We discussed earlier how proteins are structurally, thermodynamically, and functionally affected by their environments. It therefore stands to reason that in order to understand the behavior of membrane proteins, one must first understand the nature of biological membranes. This conclusion is reinforced by the fact that membranes are by nature much more complex, both chemically and physically, than the aqueous solution constituting the cytoplasm. In the following section we review the key characteristics of the biological membrane, and in Section 7.4 we focus on the effects of these properties on the behavior of membrane proteins, and *vice versa*.

7.2 STRUCTURE AND ORGANIZATION OF BIOLOGICAL MEMBRANES

7.2.1 General structure and properties

In 1972, Singer and Nicolson proposed their well-known *fluid mosaic (FM) model* to describe the structures and characteristics of biological membranes [17]. The model depicts the membrane as a structure made of two layers of lipid molecules (*lipid bilayer*), in which various proteins reside (Figure 7.1a). These proteins are separated into two general types:

1. **Integral proteins** reside inside the lipid bilayer, with one or more segments of their polypeptide chain crossing the full width of the bilayer (*transmembrane (TM) domain*). Isolating these proteins requires the disruption of the bilayer structure by detergents.

2. **Peripheral proteins** are loosely attached to one of the lipid monolayers, or to an integral protein. Isolating such proteins does not require membrane disruption; a mild treatment, e.g., elevating the salt concentration, is sufficient.

In addition to lipids and proteins, membranes also contain different types of carbohydrates, in the form of long and branched chains. These are attached to both lipid and protein molecules on the extracellular side of the membrane. The entire carbohydrate coat of the membrane is referred to as the 'glycocalyx'. Contrary to its depiction in old biochemistry and cell biology books, the glycocalyx is of formidable size, and is visible to the electron microscope. The carbohydrate chains provide physical protection to the membrane, but also participate in molecular recognition processes. These can be between cells, or between the cell and a water-soluble molecule within the body.

The bilayer structure, on which the entire membrane is based, is made of numerous lipid molecules packed tightly against one another (Figure 7.1b). Nevertheless, the FM model posits that since each lipid molecule is inherently dynamic, the bilayer is only mildly viscous (hence the term 'fluid' in the name of the model). To test this posit, early studies focused on the protein-to-lipid ratio in different biological membranes. They concluded that although most biological membranes have a weight ratio of ~0.5, some membranes differ considerably in this parameter. For example, in the myelin membrane, which surrounds the axons of nerve cells, the ratio is ~0.2, whereas in the inner mitochondrial membrane it is ~0.8. A study carried out on red blood cells took a slightly different approach by considering a different parameter, the proportion of the membrane surface occupied by the protein component [18]. The results demonstrated that proteins occupy at least 23% of the membrane surface, i.e., a much higher value than expected based on the protein-to-lipid ratio. This high value was attributed to the fact that many integral membrane proteins have large extramembrane domains, and also to the fact that such proteins tend to form large oligomers. These results suggested that, contrary to the fluid depiction put forward by the FM model and studies carried out in pure lipid bilayers — an approach that had dominated the scientific view up until that point — biological membranes have a certain rigidity. Today, the membrane is considered to have intermediary properties between fluid and gel, which enable it to block the free movement of polar solutes, but at the same time retain its flexibility, which is highly important for its function. For example, flexibility is important for the formation of transport vesicles, which carry protein and lipid cargo between intracellular membranes and the plasma membrane. The cargo molecules may reside in the membrane or be secreted in a process of exocytosis [19,20]. Transport vesicles also enable polar solutes to be internalized by the cell through endocytosis.

(a)

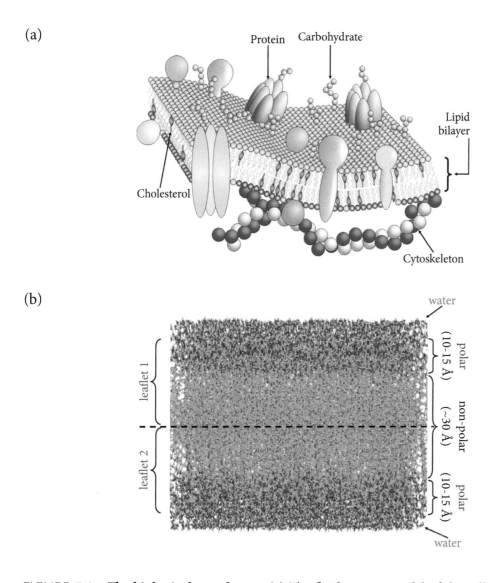

(b)

FIGURE 7.1 **The biological membrane.** (a) The fluid mosaic model of the cell membrane. The image was adapted from [21]. (b) An atomistic representation of the hydrated lipid bilayer. The polar and nonpolar regions are noted, as well as their lengths (from [22]). The dashed line marks the border between two leaflets. For clarity, the lipid bilayer is shown in its ordered phase. In reality, the lipid chains are disordered and dynamic.

7.2.2 Composition of lipid bilayer

The biological membrane contains numerous lipid molecules of different types [23,24]. Even red blood cells, which are considered to be highly simple, contain in their plasma membranes over 200 different types of lipids [25]. The following subsections summarize the structures and properties of the main types.

7.2.2.1 Glycerophospholipids

Glycerophospholipids are the most abundant type of lipids in biological membranes [26]. Their name alludes to their chemical structure; each contains two fatty acids esterified to a glycerol backbone, with the third glycerol carbon attached to a negatively charged phosphate group (Figure 7.2a). In most phospholipids, the phosphate group is attached on its other side to an alcohol group, which can be *serine, choline, ethanolamine, inositol 4,5-bisphosphate (IP$_2$)*, or even *glycerol* [27] (Figure 7.2b). The name of each phospholipid includes the prefix *'phosphatidyl'*, followed by the name of the alcohol group it contains. Since the various phospholipids differ only in the identity of their respective alcohol groups, it is the alcohol group that determines the overall physicochemical uniqueness of each phospholipid, including its size and electric charge (Table 7.1). *Cardiolipin* differs substantially from other phospholipids in its shape. Its alcohol group is an entire phosphatidylglycerol group, which means it contains two phosphate groups, two glycerol groups, and four acyl chains[*1] [28]. The acyl chain in the first glycerol position *(sn-1)* tends to be either saturated[*2] or monounsaturated[*3], whereas that in the second position *(sn-2)* tends to be polyunsaturated[*4] [29]. Most acyl chains in biological membranes contain 18 carbons [30], which creates an average hydrophobic width of ~30 Å [31].

TABLE 7.1 **Glycerophospholipids composing biological membranes.**

Full Name	Abbreviation	Acyl Chains	Alcoholic Head Group	Net Charge
Phosphatidyl**choline**	PC	2	trimethylammonium	0
Phosphatidyl**ethanolamine**	PE	2	amino	0
Phosphatidyl**serine**	PS	2	amino/carboxyl	−1
Phosphatidyl**inositol bisphosphate**	PIP$_2$	2	hydroxyl/phosphate	−5
Phosphatidyl**glycerol**	PG	2	hydroxyl	−1
Diphosphatidyl**glycerol** (cardiolipin)	CL	4	hydroxyl/diacyl	−2

7.2.2.2 Sphingolipids

Sphingolipids have similar properties to glycerophospholipids, except for the following:

1. The backbone of the molecule contains *dihydrosphingosine* instead of glycerol (Figure 7.3a).

[*1] When attached to another molecule or group, the fatty acids are called 'acyl chains'.
[*2] That is, devoid of any double bonds.
[*3] That is, containing one double bond.
[*4] That is, containing several double bonds.

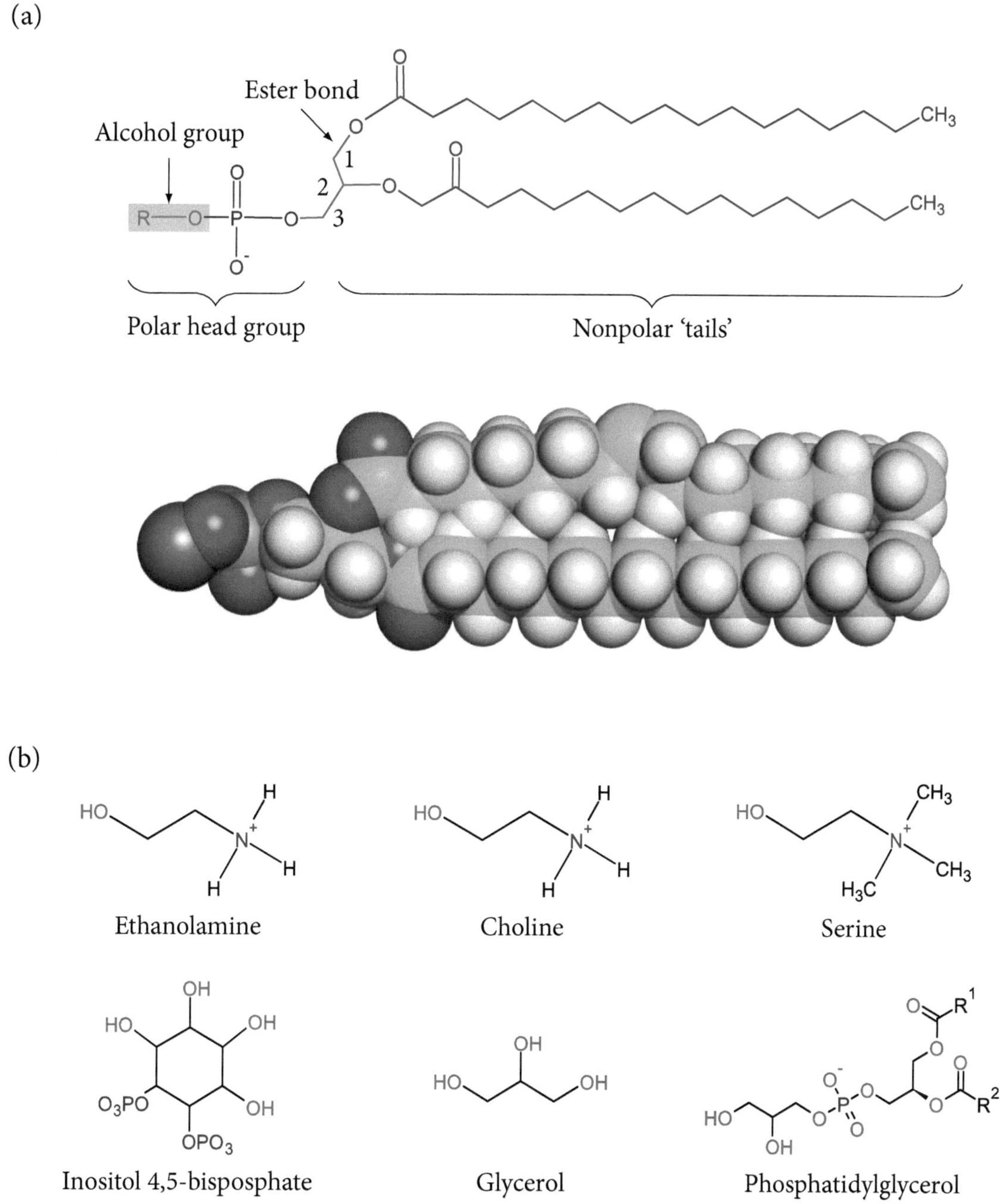

(a)

Ester bond

Alcohol group

Polar head group Nonpolar 'tails'

(b)

Ethanolamine Choline Serine

Inositol 4,5-bisposphate Glycerol Phosphatidylglycerol

FIGURE 7.2 **Glycerophospholipids.** (a) General structure. *Top*: Chemical structure. The glycerol backbone is colored in red and numbered; the acyl chains are colored in blue; and the alcohol group (R−OH) is surrounded by a green box. The nonpolar and polar parts of the phospholipid define the corresponding regions of the lipid bilayer (see Figure 7.1). *Bottom*: Three-dimensional structure. The atoms are colored by atom type, with the R moiety colored in grey. (b) Common types of alcohol groups that appear in phospholipids.

(a)

Acyl chain

Alcohol group

Sphingosine

(b)

Acyl chain

Phosphocholine   Sphingosine

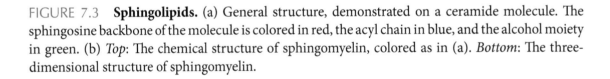

FIGURE 7.3 **Sphingolipids.** (a) General structure, demonstrated on a ceramide molecule. The sphingosine backbone of the molecule is colored in red, the acyl chain in blue, and the alcohol moiety in green. (b) *Top*: The chemical structure of sphingomyelin, colored as in (a). *Bottom*: The three-dimensional structure of sphingomyelin.

2. Only one acyl chain is present, attached to the sphingosine backbone (this conjugate is called 'ceramide'). However, since the structure of sphingosine itself includes a long hydrocarbon chain that resembles an acyl chain, the general shape of the lipid molecule is still similar to that of glycerol phospholipids.

3. Although in many cases the carbon at the third position of sphingosine is attached to a phosphocholine group (this molecule is called *sphingomyelin* [32]; Figure 7.3b), in other cases the phosphate group may be replaced by a large carbohydrate group. Such complex molecules are referred to as 'glycosphingolipids' (*GSLs*). Some GSLs contain a *sialic acid* group (*N-acetylneuraminic acid*) covalently attached to the sugar moiety. These GSLs are called 'gangliosides', and are particularly prevalent in neuronal membranes, where they constitute 2% to 10% of the total lipid component [33].

GSLs are ubiquitous components of animal cell membranes [33] and constitute a particularly interesting category of sphingolipids. The complex carbohydrate patterns in GSLs, and the fact that most GSLs reside on the outer leaflet of the lipid bilayer, make these molecules highly suitable for molecular recognition processes. Indeed, GSLs are known to interact with an extensive set of extracellular ligands, such as lectins, toxins, hormones, and viruses. The membrane composition of GSLs is carefully regulated, and is known to depend on the developmental condition of the cell. In addition, this composition has been found to change dramatically in some abnormal events, such as neurological diseases and cancerous transformation of cells.

7.2.2.3 Sterols

In eukaryotes, a third type of lipid, the *sterol*, can be found. Sterols have a characteristic structure of four fused rings with a hydroxyl group at one end of the molecule and a lipid 'tail' at the other end (Figure 7.4a). The specific type of sterol in the membrane depends on the type of organism: Plants, fungi and animals contain *stigmasterol, ergosterol*, and *cholesterol*, respectively [34] (Figure 7.4b–d). Compared to the slender-flexible phospholipids, sterols are bulky and rigid. These two properties of sterols have an important effect on the properties of the entire membrane, as explained below. The general importance of cholesterol in the mammalian membrane is reflected in its narrow concentration range in the membrane [34,35]. This range is actively monitored by the cell.

7.2.2.4 Ethers

Archaeans are among the most ancient organisms on Earth. Not surprisingly, they tend to live in niches such as the hydrothermal vents at the bottom of the ocean, which have extreme conditions resembling those that dominated our planet ~3.5 billion years ago. Though they are considered prokaryotes, Archaeans have several characteristics that distinguish them from eubacteria ('modern' bacteria), as well as from eukaryotes. One of these differences lies in the chemistry of their membrane lipids. While in eukaryotes and eubacteria most membrane lipids include fatty acids esterified to glycerol backbones, in Archaeans the lipid chain in the first position is attached to the glycerol via an *ether bond* (e.g., in *plasmalogen*) (Figure 7.5). It is possible that the ether bond, which is more stable than an ester, confers an important advantage at the extreme conditions these organisms live in.

7.2.2.5 Variability

Cells possess different mechanisms that enable them to control the lipid composition in their membranes [36–38]. Although the compositions of most membranes share several general characteristics (e.g., the dominance of phospholipids), substantial variability is observed across membranes from different origins, as follows:

1. **Different groups of organisms.** In eukaryotes the major phospholipid is phosphatidylcholine (PC; ~20% of the lipids in the rat liver plasma membrane [39]), whereas in most bacteria it is phosphatidylethanolamine (PE) or phosphatidylglycerol (PG) [40]. Conversely, in the mycobacteria *M. tuberculosis* the dominant phospholipid is cardiolipin (CL) [41].

2. **Different tissues within the same organism.** For example, the intestinal brush border membrane has no CL, whereas membranes in the nervous system that are rich in

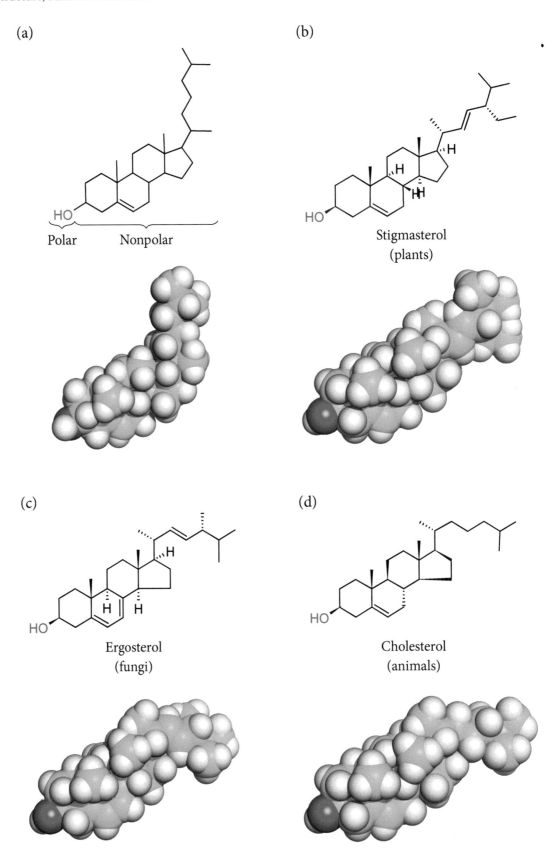

FIGURE 7.4 **Sterols.** (a) The general sterol structure, containing four fused rings, a hydrophobic tail, and a hydroxyl group on the other side. (b) Stigmasterol. (c) Ergosterol. (d) Cholesterol. *Top*: The chemical structures of the lipids. *Bottom*: The three-dimensional structures of the lipids.

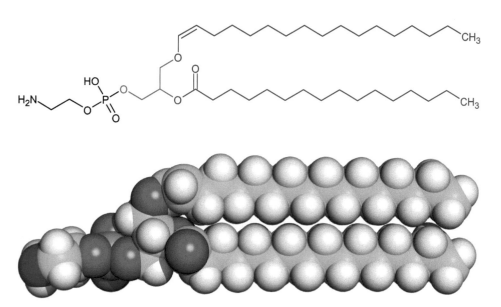

FIGURE 7.5 **Ether-linked lipids.** *Top*: The chemical structure of these lipids is exemplified by plasmalogen. The glycerol and acyl chains are shown as in Figure 7.2. *Bottom*: The three-dimensional structure of plasmalogen.

cholinergic receptors have very little sphingomyelin (SM) and no phosphatidylinositol (PI) [42].

3. **Plasma versus inner membranes in eukaryotes.** For example, CL constitutes ~20% of lipids in the mitochondrial inner membrane, whereas it is virtually absent in the ER and plasma membranes [43]. In addition, animal cholesterol resides mainly in the plasma membrane, and is present only in negligible amounts in the ER membrane [44–46]. Finally, most of the SM in the cell is concentrated in the plasma membrane and the lysosomal membrane [42].

4. **Cytoplasmic versus exoplasmic leaflets of the plasma membrane.** In eukaryotic membranes, the exoplasmic leaflet contains mainly choline phospholipids (PC and SM), whereas the cytoplasmic leaflet contains mainly amino phospholipids (PS and PE), as well as PI, in much smaller quantities [47–50]. Since both PC and SM are electrically neutral, whereas phosphatidylserine (PS) and PI are negatively charged, the lipid asymmetry leads to a charge difference between the two leaflets. That is, **the cytoplasmic leaflet is negative compared to the exoplasmic leaflet.** In the bacterial inner membrane the exoplasmic leaflet is enriched with PG, whereas the cytoplasmic leaflet is enriched with PE and PI [50].

5. **Different regions of the same membrane.** Certain lipid molecules of similar characteristics tend to gather at defined regions of the membrane, called *microdomains*, or *rafts* [51,52]. The formation of microdomains is one of the results of lipid-protein interactions, and usually has functional implications. For example, PIP_2 microdomains are important for certain signal transduction processes (see Subsection 7.4.1.2.2 below for details).

7.2.3 Lipid property effects on membranes

7.2.3.1 Amphipathicity

Despite their marked differences, lipid molecules in biological membranes share one common characteristic, namely, *amphipathicity*. That is, each membrane lipid includes a polar region and a nonpolar region. For example, in glycerophospholipids the polar region includes the ester-glycerol-phosphate-alcohol (or carbohydrate) groups, whereas the nonpolar region includes the acyl chains. These two regions are often referred to as the *'polar head'* and *'nonpolar tails'*, respectively. In cholesterol, the fused ring structure and attached hydrophobic tail constitute the nonpolar region, and the hydroxyl group constitutes the polar region. In the aqueous environment typical to biological systems, the hydrophobic effect and amphipathic nature of these lipids drive them to form larger structures, in which the polar regions face the aqueous medium and the nonpolar regions face each other. One such stable structure is the lipid bilayer. As described earlier, this structure is organized so the nonpolar tails of all lipids create a ~30-Å hydrophobic core[*1], and their head groups form two ~10 to 15-Å polar layers facing the external aqueous environment[53] (Figure 7.1b). This structural organization is fundamental to the lipid bilayer's most important trait, i.e., *impermeability* to most polar solutes. Again, because of the membrane's impermeability, the cell can tightly regulate the concentration of its metabolites by using specific transport proteins as the sole means of entry into and exit from the cytoplasm.

7.2.3.2 Asymmetry

As explained in Subsection 7.2.2.5 above, the membrane is asymmetric in terms of its lipid distribution. For example, in eukaryotic membranes the exoplasmic leaflet contains mainly the choline-containing lipids PC and SM, as well as glycolipids, whereas the cytoplasmic leaflet contains mainly the amino lipids PS and PE[33,47–50,54]. Since phospholipids can change sides in a matter of hours, the asymmetry must be maintained by an active mechanism: namely, membrane-bound enzymes that transfer lipids from one side of the bilayer to the other, using ATP as an energy source[54,55]. In particular, there are two enzymes working in opposite directions:

1. **Flippase** (aminophospholipid translocase) transfers the amino peptides PS and PE from the exoplasmic side of the bilayer to the cytoplasmic side.

2. **Floppase** transfers PC and cholesterol (in some tissues) from the cytoplasmic side to the exoplasmic side.

Membrane lipid asymmetry is diminished by certain processes, such as programmed or accidental cell death (apoptosis and necrosis, respectively), as well as cancerous transformation of cells[49]. This reduction of asymmetry happens as a result of either a decrease in the activity of flippase or activation of another enzyme, *scramblase*, which transfers phospholipids equally to both sides of the bilayer. The loss of asymmetry may in turn affect the cell and tissue, at least in the case of PS[49]. Specifically, the presence of PS in the exoplasmic leaflet has been found to mediate several physiological processes that involve cellular recognition:

[*1] The width of the hydrophobic core is measured between the glycerol groups of the two opposite layers.

1. **Recognition of apoptotic cells by macrophages.** As explained earlier, macrophages are phagocytes that are able to engulf and internalize a variety of entities, from single proteins to entire cells. In doing this, they play a double role. First, they kill invading bacteria that may harm the body. Second, they assist in disposing of dead cells from tissues. The latter role is important not only for cleaning purposes, but also for preventing the development of a harmful inflammatory response in the tissue following the apoptosis of cells. Conversely, when cells die by necrosis, which is not 'planned' by the body but rather inflicted by some kind of trauma, inflammation ensues rapidly. PS on the surface of apoptotic cells has been implicated in macrophages' capacity to recognize these cells, and therefore has a role in the prevention of inflammation.

2. **Recognition of activated endothelial cells by T-lymphocytes.** One of the roles of the immune system is to detect tissues invaded by pathogens and to act quickly to eradicate the invaders. The problem is that lymphocytes are normally on the move inside circulating blood and lymph, and do not linger in one place. Thus, when pathogens are detected in a certain tissue, it is necessary to prevent lymphocytes in the vicinity of this tissue from moving elsewhere. This is done by nearby endothelial cells, i.e., cells that line the blood vessels at the vicinity of the invaded tissue. These undergo a process that exposes their PS to the extracellular environment, and the latter is recognized and bound by nearby T-lymphocytes.

3. **Recognition of bacteria by the complement system.** The complement system includes several proteins that normally exist in an inert state. However, when activated during pathogenic invasion, they form a complex that attacks the invading cells. The attack involves damaging both the membrane and the cytoplasmic components of those cells. In most cases, the complement system is activated against invading bacteria, already recognized by antibodies. However, alternative activation pathways also exist, and they seem to involve PS on the exoplasmic leaflet of the invading bacteria. If this is indeed the case, it is likely that cancer cells are also recognized this way, as such cells are known to have lower membrane lipid asymmetry compared with healthy cells.

7.2.3.3   Degree of order and thickness

Though the lipid bilayer is commonly depicted as being overall fluid (in accordance with the fluid mosaic model), its fluidity may vary within a certain range. This variation is determined by the degree of order of the individual lipids, and specifically their hydrophobic tails. Linear tails are tightly packed within the bilayer structure, making it more viscous [56]. Such a structure is referred to as 'liquid ordered' (l_o). Conversely, bent tails form a less tightly packed and more fluid structure, referred to as 'liquid disordered' (l_d). Under physiological conditions the two types (or *phases*) coexist within the lipid bilayer, each contributing to its biological properties: The l_o phase enhances the bilayer's capacity to serve as a physical barrier for polar solutes, whereas the l_d phase provides it with a certain degree of dynamics. Indeed, bilayer regions that assume the l_d phase allow individual lipids not only to diffuse along the surface of the bilayer (*lateral diffusion*), but also to 'flip' from one leaflet to another (*transverse diffusion*). Moreover, the l_d phase allows the membrane to undergo structural organization that is needed for certain biological processes, such as the formation, budding, and fusion of transport vesicles.

The packing tightness of individual lipids is determined by two main factors:

1. **Degree of lipid saturation.** Fully saturated lipids have linear hydrophobic tails, which tend to form an l_o-type of bilayer. Conversely, lipids containing one or more double bonds have bent tails that pack in an l_d-type of structure.

2. **Presence of sterols in the bilayer.** Sterols have a dual effect on the properties of the lipid bilayer [34,57]. On the one hand, the sterol acts as a plug that prevents the free passage of solutes through the cavities between phospholipids. On the other hand, the rigid and bent structure of the sterol molecule creates a spatial disturbance in the tight phospholipid packing of the lipid bilayer, which prevents the bilayer from solidifying. This is important for the biological function of the membrane, which requires the bilayer to remain dynamic. **Thus, sterols allow biological membranes to remain dynamic without losing their basic function as a physical barrier.**

The cell can modulate the two factors to achieve the right balance between different properties. For example, the ER membrane contains small quantities of sterols [44–46], but retains flexibility due to large quantities of unsaturated phospholipids. Studies evaluating lipid composition, sensitivity to detergents, and biophysical measurements of lipid motions suggest that **extensive regions in biological membranes exist in the l_o phase** [51], and that **the l_d phase is usually restricted to those regions involved in dynamic activity**, such as the formation of transport vesicles.

An important trait of the lipid bilayer that is derived from its degree of order is its thickness. Measurements in pure lipid bilayers indicate that the bilayer has an average thickness of ~50 to 60 Å, with a ~30-Å nonpolar core and two polar lipid-water interfaces measuring ~10 to 15 Å each [53] (Figure 7.1b). These values vary (within a certain range) across different organisms, and even among different compartments of the same cell. In liver cells, for example, the *apical membrane*, which faces the lumen, is ~5 Å thicker than the ER membrane, and ~7 Å thicker than the *basolateral membrane*, which faces other cells [57]. When a region of the lipid bilayer changes from the l_o to the l_d phase, it becomes thinner due to a decrease in the length of the acyl chains. This difference affects the interactions between the phospholipids and the integral membrane proteins in their vicinity. Since such effects also influence the stability of the latter, proteins tend to concentrate in regions of the membrane in which only one of the phases exists, and, as a result, groups of membrane proteins tend to be physically separated from one another. The concentration of proteins in certain membrane regions is often used to enhance signal transduction pathways, which require proximity of their components. Finally, changes in protein-lipid interactions following phase changes may directly affect the activity of the former. These issues are further discussed in Section 7.4 below.

7.2.3.4 Curvature

Although the lipid bilayer is traditionally depicted as planar, it may curve temporarily in certain regions [58]. This phenomenon facilitates various processes, such as the vesicular transport of proteins and lipids among the ER, Golgi apparatus, and plasma membrane. Vesicular transport begins with the gradual curving of the source membrane until the vesicle is formed, continues with the separation of the vesicle from the membrane (*budding*) and its diffusion towards the target membrane, and finally ends with fusion of the two [19]

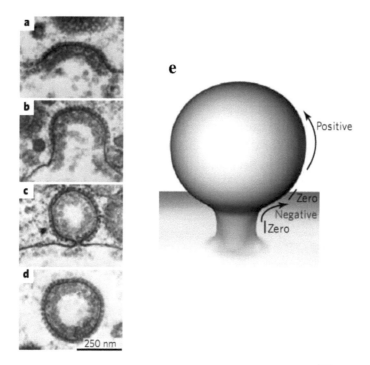

FIGURE 7.6 **Curvature changes in the budding of transport vesicles** [58]. (a) through (d) Stages in vesicle budding. (e) Positive and negative curvatures in the transport vesicle.

(Figure 7.6). Exocytosis, endocytosis [20], and inter-organelle exchange all involve this process.

Membrane curvature is affected by lipid and protein composition. The effect of lipids involves the ratio between the effective cross-section area of the lipids' head groups and that of their tails. When the ratio is ~1, the lipids arrange side-by-side in parallel, forming a roughly planar bilayer structure. Conversely, when there is a mismatch between the area of the head group region and that of the tail group region, the lipids form a curved membrane [59]. There are two such cases:

1. **Positive membrane curvature** forms when the head group section is wider than the tail section, as is the case with choline-containing lipids (PC and SM), as well as with PG. The leaflet formed by these lipids has an inherent tendency to curve convexly (Figure 7.7a).

2. **Negative membrane curvature** forms when the head group section is narrower than the tail section, as is the case with PE (Figure 7.7b). The leaflet formed by these lipids has an inherent preference to form a concave curvature. Lipids that induce negative curvature reduce the stability of the bilayer membrane, which might ultimately lead to bilayer disintegration.

Biological membranes feature a mixture of lipids of different curvature preferences, as well as proteins, and it is not always easy to predict the exact shape that will arise from a certain lipid composition. For example, the bacterial plasma membrane remains, in essence, planar, despite the fact that PE constitutes 70% of its lipids [40]. This is because the remaining 30% are PGs, which induce a compensational positive curvature. In fact, studies show that as long as the concentration of negative curvature-inducing lipids is less than ~20%, the membrane will remain planar and whole even in the absence of compensatory lipids [60]. Nevertheless, the presence of a mix of lipids with different curvature-inducing properties does create mechanical frustration within the membrane. It has been suggested that cells use this so-called '*curvature frustration*' to render the membrane metastable, which is advantageous in cases where the membrane's biological function requires frequent curvature changes (e.g., in intracellular transport) [61]. Thus, the effect of lipid shape on bilayer curvature is present, yet weak. Lipid shape makes the ER and Golgi membranes, for example, slightly curved. Integral membrane proteins exert a much stronger effect on membrane curvature; these are responsible for dramatic changes such as formation of transport vesicles. This issue is further discussed in Section 7.4.2 below.

(a)   (b)

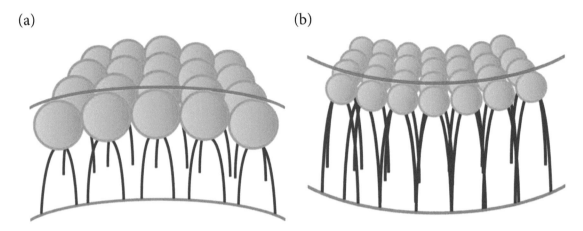

FIGURE 7.7 **Effects of different lipids on membrane curvature.** The figure shows a highly schematic illustration of the following: (a) Formation of positive curvature and convex membrane by lipids with large head group sections and small tails. (b) Formation of negative curvature and concaved membrane by lipids with a small head group sections and large tail sections. In both images, only one leaflet of the lipid bilayer is shown. The shape of the other leaflet depends on its lipid composition.

7.3 PRINCIPLES OF MEMBRANE PROTEIN STRUCTURE

7.3.1 Overview

Membrane-bound proteins can be separated into two major groups: integral proteins and peripheral proteins. **The membrane-spanning region of an integral protein may appear in two forms. The first includes α-helical segments (Figure 7.8a), whereas the second is structured as a β-barrel (Figure 7.8b)** (see details in Subsection 7.3.2 below). Certain antibiotic peptides such as gramicidin have alternating D and L amino acids, which allow them to create a third type of structure, the β-helix (Figure 7.8c). The β-helix is wider than the α-helix, and can therefore function as a channel, transferring monovalent ions through the membrane. In this section we focus primarily on helical membrane proteins, which constitute the vast majority of integral membrane proteins [62]. A discussion of the properties of β-barrel membrane proteins is provided in Section 7.3.2.2.2 below. Helical membrane proteins may be separated into subgroups according to the number of membrane-crossing segments they contain. *Bitopic* membrane proteins contain a single transmembrane segment (Figure 7.8d), whereas *polytopic* membrane proteins contain several such segments (Figure 7.8e). Comparison among different organisms suggests that in unicellular organisms, integral membrane proteins containing 6 or 12 transmembrane segments are more common than others, whereas in higher organisms (*Caenorhabditis elegans* and *Homo sapiens*) there is a weak preference for membrane proteins containing seven transmembrane segments each [10]. GPCRs are a well-known example of the latter type of protein; these proteins play a central role in animal physiology and constitute a major target for pharmaceutical drugs [63,64]. GPCRs are the focus of the last section of this chapter.

Integral membrane proteins constitute most of the membrane protein population, and have diverse roles. Monotopic[*1] and bitopic proteins tend to function as recognition and/or adhesion molecules, as well as receptors to growth-factor-like messengers. Their extracellular region is responsible for binding the chemical messenger, whereas their cytoplasmic region passes the signal into the cell by binding soluble elements or cytoskeletal proteins. Polytopic proteins usually function as receptors or transporters. For example, GPCRs, mentioned above, respond to a variety of messengers, including hormones, neurotransmitters, odorants, pheromones, and even electromagnetic radiation (i.e., light) [65,66]. Peripheral membrane proteins are anchored to membrane lipids or integral proteins on either side of the membrane. Lipid attachment may be direct or mediated by carbohydrate moieties.

As integral membrane proteins are surrounded by the lipid bilayer, their structure (more specifically, the structure of their transmembrane domains) is determined by rules quite different from those corresponding to water-soluble proteins. Therefore, our discussion will focus primarily on the structure of integral membrane proteins, whereas peripheral proteins, which are mostly surrounded by a water-based environment, will be discussed mainly with respect to their membrane anchoring.

7.3.2   Structures of integral membrane proteins

Integral membrane proteins are considered to be globular, like their cytoplasmic counterparts. However, their presence in an environment so different from the aqueous cytoplasm suggests that the energy determinants of their structural stability might differ from those affecting the structure of water-soluble proteins. Understanding these determinants requires analysis of numerous structures, as has been done in the last decades for water-soluble proteins. As explained in Chapter 3, determining the structure of a membrane protein is challenging, due to the difficulty to overexpress, extract and purify such proteins, as well as to crystallize them [67]. The crystallization problem is usually addressed by replacing the surrounding lipids with detergent molecules, though the new environment might change the structure of the protein, making findings irrelevant. In the last few years, researchers have made impressive progress in the experimental determination of membrane protein structure [67]. This progress includes the development and perfection of methods such as electron cryomicroscopy (cryo-EM), circular dichroism (CD), and small-angle X-ray scattering (SAXS) (see Chapter 3), which have provided valuable information on hard-to-crystallize membrane proteins, as well as on the supra-molecular assemblies they form. In addition, X-ray crystallography has advanced substantially in various aspects, including (*i*) the capacity to overexpress proteins in different hosts; (*ii*) the development of new detergents and lipids for more efficient solubilization and crystallization; (*iii*) protein stabilization via mutations, fusion with other proteins, or binding to monoclonal antibodies; (*iv*) hardware-related methods for optimizing the crystallization process; and (*v*) developments in beam-line and synchrotron radiation (see recent review in [67]). And yet, despite all this progress, the membrane proteins whose structures have been experimentally determined constitute only ~3.5% of all known protein structures (as of Dec 2017). Fortunately, while the lipid bi-

---

[*1]Proteins that are anchored to the membrane from one side.

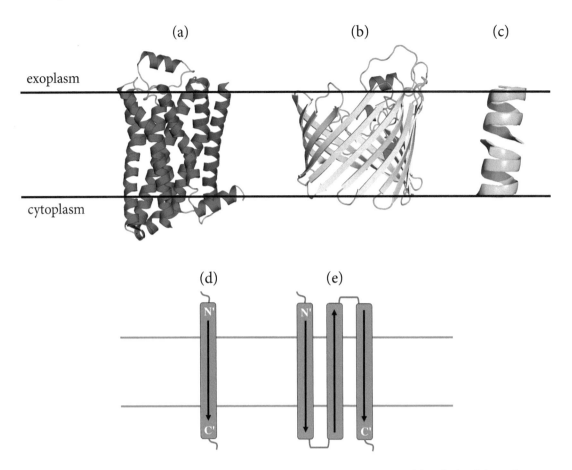

FIGURE 7.8 **General classification of integral membrane proteins.** (a) α-helical (β_1-adrenergic receptor; PDB entry 2vt4). (b) β-barrel (bacterial porin; PDB entry 2por). (c) β-helical (head-to-head gramicidin dimer; PDB entry 1grm). (d) Bitopic (single-pass). The transmembrane segment of each protein is represented by a grey cylinder, with the termini and the direction of the polypeptide chain marked. (e) Polytopic (multi-pass). The extramembrane connections between the transmembrane segments are shown in red.

layer surrounding membrane proteins impedes crystallization, it makes the understanding and even the prediction of their structure easier than in water-soluble proteins. This is because of the anisotropic and chemically complex nature of the lipid bilayer, which imposes constraints on the structure of resident proteins [68]. As a result, the general architecture of membrane proteins is relatively simple, and fewer structures are needed for understanding the basic principles determining that architecture [69].

The main determinant of integral membrane protein structure is the energetic cost of burying the protein's polar peptide bonds inside the hydrophobic hydrocarbon core of the lipid bilayer [70,71]. To compensate for this cost, the sequences of the transmembrane segments are highly hydrophobic [72], and have a strong tendency to form organized secondary structures [9,53]. Additional determinants exist, with secondary, yet important influence on membrane protein structure. In the following subsections we review the principles determining membrane protein structure, as we understand them today, according to the structural hierarchy used for water-soluble proteins. For further details, we recommend the reviews written by von Heijne [73,74], White [75], Engelman [76], and Bowie [77,78].

7.3.2.1 Primary structure

7.3.2.1.1 Polarity and length

The polypeptide chain of an integral membrane protein crosses the lipid bilayer at least once. The hydrocarbon core of the bilayer is highly hydrophobic, which requires the transmembrane domains of the proteins to be hydrophobic as well [8,53,72] (Figure 7.9a). Indeed, **the most pronounced trait of integral membrane proteins is their low polarity compared to water-soluble proteins, particularly in their transmembrane segments.** Though all types of nonpolar residues are common in transmembrane segments, Leu, Ile, Val, and Phe are particularly highly enriched in integral membrane proteins in comparison to water-soluble proteins [72]. **Polar residues also appear in transmembrane domains, but they are less common, especially in single-pass proteins, where they constitute in total only ~20% of the sequence [79]. In multi-pass membrane proteins polar residues are usually buried in the core (especially if they are charged) rather than facing the membrane, which is more hydrophobic.** As in the cores of water-soluble proteins, here too the presence of polar residues in a highly hydrophobic environment serves a specific function, justifying the unavoidable structural destabilization [80]*¹. The destabilization is mitigated to some extent by the fact that the buried polar residue is surrounded by water molecules, other polar residues [81,82], or both (e.g., in the voltage-sensing K^+ channel [83–85]). Integral membrane proteins are inserted into the ER membrane co-translationally, via the translocon machinery [75]*². How, then, is the translocon able to scan the nested polypeptide chain and detect transmembrane segments? Structural studies show that, in addition to the main channel pore that accommodates the nested polypeptide chain, the translocon structure contains a 'side gate', which opens to the lipid bilayer at a certain frequency, thus exposing the sequences inside it [86].

Another characteristic trait of transmembrane segments is their length. **Statistical analyses demonstrate that though helical transmembrane segments may include 15 to 39 residues, the 'average' transmembrane helix includes 21 to 26 residues, and there is strong preference for helices with over 20 residues [72,87,88].** Again, this is a result of the restrictions imposed by the membrane environment combined with the structural properties of α-helices. That is, **because of the characteristic 1.5 Å rise per residue along the helix axis, a 20-residue-long α-helical transmembrane segment would correspond to a length of ~30 Å, matching the average thickness of the hydrocarbon core of the lipid bilayer.** Obviously, to cross the membrane, the helix should be hydrophobic enough. As we will see later, longer helices usually tilt to maximize their nonpolar interactions with the membrane's core (see Section 7.4 below). For comparison, in water-soluble proteins, whose environment does not impose the restrictions observed in the bilayer, α-helices tend to be shorter on average, with a broader length distribution (15 ± 9 residues [72]).

The significance of the low polarity and characteristic length of transmembrane segments is demonstrated by the fact that these characteristics can be successfully used to detect membrane proteins automatically in whole genomes and to predict the number of their transmembrane segments, according to sequence alone. Algorithms that are used for this purpose are discussed in Box 7.1.

---

*¹This also explains the evolutionary conservation of polar residues in transmembrane proteins.

*²This enables cells to prevent non-specific aggregation of the highly nonpolar membrane proteins in the aqueous environment of the cytoplasm.

(a)

(b)

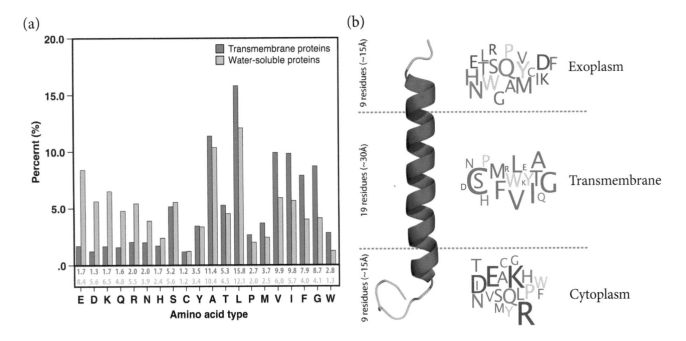

FIGURE 7.9 **Amino acid preferences in water-soluble and transmembrane proteins.** (a) Amino acid type distributions from 792 transmembrane and 7,348 water-soluble helices from a set of non-redundant proteins of known structure. The distribution for transmembrane helices is in blue, and the distribution for water-soluble helices is in orange. (b) Amino acid location prevalence in a membrane. Letter size is proportional to the relative prevalence of a given amino acid in the corresponding region in the membrane. Colors: red – charged amino acids (KRED), orange – polar-uncharged amino acids (QHN), green – aromatic amino acids plus Pro (PYW), blue – other amino acids (CMTSGVFAIL). The images are taken from [72].

7.3.2.1.2 Pro and Gly

Transmembrane segments arranged as α-helices often include Pro and Gly, as well as β-branched residues (Figure 7.9b). This is highly unexpected, as such residues rarely appear in α-helices of globular proteins (see Chapter 2, Section 2.3.6.1). Proline is particularly common in helical transmembrane segments, and is usually adjacent to Ser or Thr [89]. Structural analysis shows why these residues are so important in membrane proteins; this is discussed in Section 7.3.2.2.1 below.

7.3.2.1.3 Aromatic residues

Transmembrane segments tend to have 'aromatic belts' near the boundaries of the hydrocarbon region of the lipid bilayer [90-92] (Figure 7.10a). Such a belt includes the aromatic residues Trp and Tyr, which are normally rare in proteins. This is intriguing, especially in the case of Trp, whose frequency in membrane proteins is three times higher than in water-soluble proteins [91,93]. The location of the aromatic residues is near the termini of transmembrane helices, placing them at the interface between the nonpolar tails and the polar head group region of the lipid bilayer. There, they can participate in complex interactions with both parts of the lipid (see Subsection 7.4 below). According to the accepted theory, these interactions are used to anchor the transmembrane segments to the bilayer, thus preventing them from 'sliding' into the cell or out of it [94,95]. The affinity of Trp and Tyr to the interface area may be explained by their amphipathic nature, which allows them to interact with the amphipathic membrane interface. That is, their polar NH and OH groups hydrogen-bond with the polar lipid head groups of the interface, whereas their large surface area allows them to interact with the nonpolar lipid tails. In addition, the rigid, bulky side chains of Trp and Tyr are expected to disfavor insertion into the highly disordered acyl chains of the membrane core.

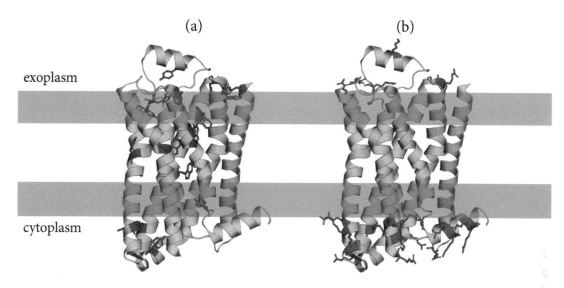

FIGURE 7.10   **Locations of aromatic (Trp and Tyr; (a)) and basic (Arg and Lys; (b)) residues in the β_1-adrenergic receptor (PDB entry 2vt4).** The residues are colored purple, and the polar head group regions of the bilayer are in light blue. As the figure demonstrates, most aromatic residues are concentrated near the acyl-head group interface, and some are buried in the protein. Most of the basic residues are positioned on the cytoplasmic side of the membrane, in accordance with the 'positive inside' rule (see also Figure 7.9b).

7.3.2.1.4   Basic residues

Transmembrane segments tend to include the basic residues Lys and Arg in their cytoplasmic regions [72,96,97] **(Figure 7.9b and 7.10b). This tendency was discovered by von Heijne, who referred to it as 'the positive inside rule'** [96]. Histidine also displays such a preference, though the prevalence of His in these regions is half that of Lys or Arg [97]. This makes sense, considering that the His side chain has almost equal probabilities of being positively charged or electrically neutral at physiological pH. There are several possible explanations for the positive inside tendency. For example, it may have to do with the inherent phospholipid asymmetry of lipid bilayers. We have seen earlier that eukaryotic membranes place electrically neutral phospholipids (PC and SM) at the exoplasmic leaflet of the bilayer, and negatively charged phospholipids (PS, and PI) at the cytoplasmic leaflet (see Subsection 7.2.3.2 above). In inner bacterial membranes both the exoplasmic and cytoplasmic leaflets contain negatively charged lipids (PG and PI, respectively). Thus, in both prokaryotic and eukaryotic membranes, transmembrane segments that have basic residues at their cytoplasmic regions could form salt bridges with these negatively charged lipids, and stabilize the protein-membrane system. The opposite, i.e., presence of acidic residues on the exoplasmic side of the bilayer, does not occur, as the lipid bilayer does not include any positively charged lipids.

Other reasons for the positive inside rule may be the 'membrane potential', i.e., the electric potential across biological membranes, where the cytoplasm is more negative than the periplasm. Alternatively, the positive inside rule could reflect bias in the translocon machinery. Finally, the compatibility of Lys and Arg with the membrane interface probably has to do with their side chains, which each include a polar group at the end of a long nonpolar chain. Thus, the polar group can interact favorably with phospholipid head groups even when the residue is positioned deeper inside the hydrocarbon core of the bilayer. This phenomenon is referred to as 'snorkeling' [98]. Interestingly, a recent study suggests that only Arg significantly stabilizes the cytoplasmic side of the membrane, although its preference for this region is similar to that of Lys [97]. This may have to do with the electronic delocalization on the side chain of Arg (which is not present in Lys), which spreads the stabilizing positive charge over a larger area. Also, compared with that of Lys, the side chain of Arg can participate in more hydrogen bonds with phospholipid head groups.

In the β-barrel proteins of the Gram-negative bacterial membrane (see below), the distribution of positive residues is the opposite of that in other membrane proteins [68]. That is, basic residues appear mainly in the outside-facing loops of the protein ('positive outside') [99]. This distribution serves a purpose; in contrast to other cellular membranes, the bacterial outer membrane is highly negatively charged on its exoplasmic side due to the abundance of lipopolysaccharides (LPS). The basic residues of membrane proteins in this region stabilize the negatively charged LPS through ionic interactions. On its opposite side, a bacterial outer-membrane protein is enriched in negatively charged residues; these interact with periplasmic cationic chaperones, such as Skp, which assist in the insertion and folding of the β-barrel proteins into the outer membrane [100].

7.3.2.1.5 Small residues

Small residues such as Gly, Ala, and Ser are common in transmembrane segments. These residues tend to appear in α-helices, and allow adjacent helices to optimize their van der Waals interactions, as well as their hydrogen bonds. This topic is further discussed below.

BOX 7.1 PREDICTING LOCATIONS AND MEMBRANE TOPOLOGIES OF TRANSMEMBRANE SEGMENTS IN AMINO ACID SEQUENCES

Predicting the three-dimensional structures of membrane proteins has been an important goal of computational-structural biologists for decades, particularly considering the lack of experimentally determined structures. Paradoxically, because of the lack of structures, progress towards this goal has been slow, due to the difficulty in understanding the basic rules governing membrane proteins at the atomic level. And yet, different prediction methods have emerged (see Section 7.3.2.3.4 below for details). This process happened gradually; early trials began with relatively humble goals, such as locating transmembrane segments of membrane proteins, or trying to predict their overall topology, i.e., the sidedness of their termini in the membrane. These tasks relied on simple rules at the sequence level, and were therefore a good starting point for the prediction process. In the following paragraphs we give a short description of these methods.

I. Locating transmembrane segments

I.I. Hydrophobicity scales and hydropathy plots

The transmembrane segments of an integral membrane protein contain mainly nonpolar residues, whose number in each segment tends to fit roughly the thickness of the hydrophobic core of the membrane (see main text for details). In α-helical proteins, which constitute the bulk of integral membrane proteins, this number is ~20 residues. This understanding has prompted scientists to devise computer algorithms that can locate transmembrane segments within genomes, based on these tendencies [71,101]. The first attempt was carried out by Kyte and Doolittle [102]. Their general idea was to use a (virtual) sliding window covering ~20 amino acid positions, to locate transmembrane segments along the protein sequence. For each position of the window along the sequence, either the overall or average hydrophobicity of the 20-amino acid-long sequence was calculated. In those places where the calculated value exceeded a certain threshold, the segment covered by the window was considered to be a potential transmembrane segment. The results of this procedure were presented as a *hydropathy plot* (Figure 7.1.1), which represents the probability of each consecutive 20-residue segment in the sequence to be a transmembrane segment.

(a) (b)

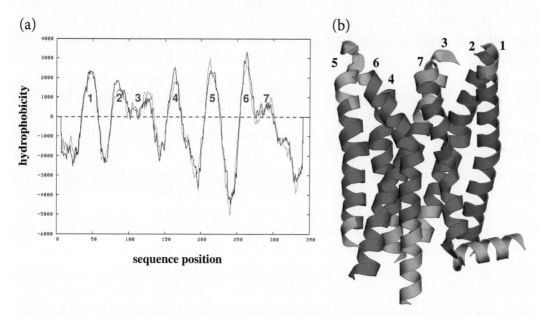

FIGURE 7.1.1 **Hydropathy plot.** (a) A hydropathy plot calculated for the GPCR bovine rhodopsin using TMpred [103,104]*a. The calculation was based on Kyte and Doolittle's scale. The red dashed line marks the threshold, above which the sequence is considered to be hydrophobic enough to span the membrane. The seven peaks in the figure correspond to the seven transmembrane segments of rhodopsin. (b) The three-dimensional structure of bovine rhodopsin with the predicted transmembrane segments colored in red. The image demonstrates the main problem of such prediction algorithms: Although they often provide a rough indication of the locations of the transmembrane segments, they fail in identifying their exact boundaries.

*a http://www.ch.embnet.org/software/TMPRED_form.html

The critical component of this method was (and still is) the calculation of the hydrophobicity of each candidate sequence. This was carried out by using a *hydrophobicity scale*, in which each of the 20 natural amino acid types was assigned a hydrophobicity value. The construction of this scale may seem trivial at first, but it is (still) a matter of controversy. First, there is the matter of selecting the physical quantity that can be used to represent hydrophobicity. The first quantity that comes to mind is the polarity of the molecule. However, despite the fact that polarity (inversely) affects hydrophobicity, it does not necessarily account for it fully. This is because polarity results merely from the geometric distribution of electronegative atoms, whereas hydrophobicity reflects all the qualities contributing to the molecule's tendency to prefer a nonpolar medium over a polar one. Thus, as explained in Chapter 1, a large residue such as tyrosine, which is considered to be polar due to its side chain OH group, might still turn out to be hydrophobic if its large phenyl group can produce strong enough nonpolar interactions. Accordingly, the Kyte-Doolittle (KD) scale [102] was produced empirically, based on the partition of the amino acids between polar (water) and nonpolar (vacuum) media (Figure 7.1.2). The result were converted into an energy-like value using Equation (4.1) (see Chapter 4):

$$\Delta G^0 = -RT \ln K_p$$

(where K_p is the equilibrium constant of partitioning). However, the empirical values were not used 'as is', but rather were normalized, in some cases using arbitrary considerations.

This leads us to the second problem associated with producing a hydrophobicity scale, namely, the types of media used in the measurement of hydrophobicity. In order for such measurement to be efficient, the polar and nonpolar media chosen for measuring the hydrophobicity values must be as close as possible to what they represent, i.e., the cytoplasm and the lipid bilayer, respectively. Whereas water has always been accepted as representative of the cytoplasm, disagreement has arisen concerning the medium that should be chosen to represent the lipid bilayer [96,102,105–107]. Vacuum, which was used by Kyte and Doolittle, is indeed nonpolar ($\varepsilon = 1$; a value even lower than that of the hydrocarbon region of the membrane); however it is not amphipathic, and therefore cannot faithfully represent the biological membrane. To take this important property of the membrane into account, other scales have been produced by using octanol (e.g., [108]), and even real lipid bilayers (e.g., the Goldman, Engelman, and Steitz (GES) scale [106]) as the nonpolar medium. White, von Heijne and coworkers took another step in making the hydrophobicity values more accurate [109–111]; instead of measuring the spontaneous, yet artificial partitioning of residues between simple polar and nonpolar media, they used a reconstructed system containing all the biological components involved in the insertion of transmembrane segments into the membrane in real cells, including the ribosome-translocon complex. The hydrophobicity scale they produced is perhaps more realistic than the scales produced by transferring amino acids and/or peptides between simple media. However, the amino acid transfer energies obtained by White, von Heijne, and coworkers were significantly lower in magnitude than those obtained by the above studies [112], which raised doubts as to their accuracy. It was suggested that these low energies might have resulted from several approximations that were made in the study [113] and/or interactions of the inserted peptides with other membrane components [114]. Notably, a recent study of amino acids transfer energetics, which, like the study of White, von Heijne and coworkers, used a real biological membrane, yielded similar transfer energies to those obtained in the studies that used simple media [97]. These results support the reliability of the latter studies and suggest that simple organic solvents faithfully represent the hydrophobicity at the core of biological membranes.

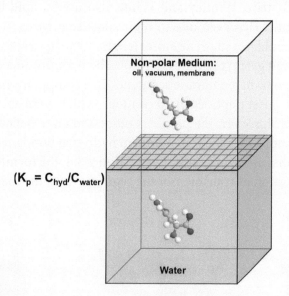

FIGURE 7.1.2 **Partitioning between polar and nonpolar media.** The amino acid residue is put in one of the compartments, and its equilibrium concentrations in each (C, representing molar concentrations) are measured to produce the partitioning constant K_p, from which the free energy of transfer is derived.

The use of knowledge-based scales circumvents the two problems discussed above, i.e., selecting the physical quantity that should be used to construct the hydropathy scales and the type of media that should be used for the amino acid transfer experiments [92,115,116]. These relatively new scales replace the physically meaningful hydrophobicity with the non-physical probability of a residue to appear inside a transmembrane segment. The probability is calculated on the basis of statistical data collected from membrane proteins of known 3D structure. Such data were not available until recently, due to the lack of membrane protein structures, but with the growth in the number of such structures over the past decades it is now possible to extract this information. It should be noted that knowledge-based scales are biased by functional constraints. This bias is especially prominent in the case of charged residues, whose real transfer energies into nonpolar media are highly unfavorable; yet their statistical tendency to appear in transmembrane domains is higher than suggested by these energies because they are needed for functional reasons (binding, catalysis, etc.).

The third controversial issue associated with producing hydrophobicity scales is determining how to represent the residues in the transmembrane segment. The KD and GES scales use individual amino acids, despite the fact that in reality the amino acids are interconnected by peptide bonds. The bonds reduce the full electric charges on the α-amino and α-carboxyl groups of the amino acids into mere partial dipoles. Since the partitioning between polar and nonpolar media depends considerably on the magnitude of the charge on the molecule, it has been suggested that the use of individual amino acids is methodologically flawed. To solve this problem, Wimley and White [105] used whole peptides in producing their own hydrophobicity scale, following the *'host-guest'* approach. The peptides they used in the various measurements were identical except for one position, which contained a different residue in each case. Like Goldman, Engelman, and Steitz, Wimley and White also used a lipid bilayer instead of a simple nonpolar mimic. However, due to their short sequences (5 or 6 residues), the host-guest peptides could not span the entire length of the lipid bilayer, and partitioned only into the polar head group region. As a result, the scale produced by this procedure could not be applied to transmembrane segments. In a computational study, Kessel and Ben-Tal [8] used host-guest peptides of 20 residues, which were able to span the entire length of the bilayer. Moreover, the peptides possessed an α-helical conformation, in which the polar backbone groups were paired in hydrogen bonds, as in real transmembrane segments. As discussed in Chapter 2, Section 2.3.4, the formation of such bonds is crucial for the insertion of transmembrane segments into the membrane.

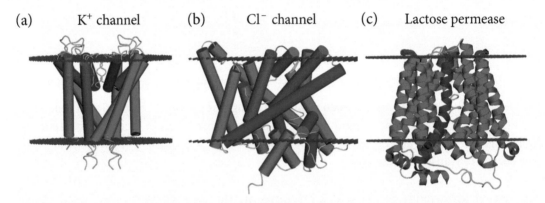

(a)   K⁺ channel   (b)   Cl⁻ channel   (c)   Lactose permease

FIGURE 7.1.3 **Examples of non-canonical transmembrane α-helices.** (a) Four short helices within the 're-entrant loops' of the potassium ion channel (PDB entry 1bl8), which are too short to span the entire thickness of the lipid bilayer. As a result, one polar terminus of each of these four 'half-helices' resides roughly in the bilayer midplane. The termini are shielded from the lipid tails because of their location in the protein core. The red and blue planes mark the predicted boundaries of the membrane respectively (the OPM database [117]). The polar head group regions of the bilayer are in light green. Note that in reality, the membrane probably deforms to match its thickness to the hydrophobic lengths of the transmembrane segments of the proteins. (b) A transmembrane α-helix in the chloride channel (PDB entry 1otu) whose hydrophobic length far exceeds the thickness of the lipid bilayer core. Water exposure of nonpolar groups is reduced by the tilting of the helix. (c) Core exposure of polar groups in the middle of lactose permease (PDB entry 1pv6), due to a helix-distorting kink (circled). The figure was prepared following [118].

I.II. Shortcomings and future leads

The first algorithms for identifying transmembrane segments were designed and used at a time when only a few 3D structures of membrane proteins existed. As structures started to emerge, such as those of the K$^+$ channel [119], it became clear that the length and hydrophobicity values of some of the transmembrane segments deviated from the basic tendencies assumed by the prediction algorithms [101,120]. First, certain transmembrane segments were buried inside the core of the membrane, despite the fact that they were too short to span its entire thickness (Figure 7.1.3a). Nevertheless, the unpaired polar groups in the termini of these segments were not exposed to the hydrophobic core, as they were electrostatically masked by polar groups on adjacent segments, or by water molecules filling the intramembrane pore. Other transmembrane segments were found to have hydrophobic lengths that exceeded the hydrophobic thickness of the membrane (Figure 7.1.3b). These segments usually acquired a tilted orientation with respect to the membrane normal, in order to allow as many nonpolar residues as possible to interact with the hydrocarbon region of the membrane. Finally, some segments appeared to be kinked in a way that distorted the helical structure (Figure 7.1.3c). These observations have made it clear that, in order to be efficient, predictions of transmembrane segments must incorporate data beyond sequence tendencies.

II. Predicting topologies of transmembrane segments

The search for rules of thumb describing transmembrane segments has led scientists to the issue of topology, i.e., predicting which regions of each segment face the cytoplasmic or exoplasmic sides of the membrane. This type of prediction constitutes a critical preliminary step in the prediction of the overall structure of an integral membrane protein, as it limits the number of ways in which the transmembrane segments can be spatially organized with respect to each other. The first algorithm designed for this purpose, *TopPred* [96], combined the general tendency of transmembrane segments for hydrophobicity with the aforementioned 'positive inside' rule. Algorithms that were designed later, such as *MEMSAT* [121] and *TMHMM* [122], mainly relied on statistical data concerning the locations of transmembrane segments in proteins of known structure. Today, numerous prediction methods and algorithms are fully accessible to the general public, and anyone can use them to produce a good starting model. For example, the current TMHMM algorithm [123], which is accessible via server[*a], has been demonstrated to achieve 80% success in predicting the topology of transmembrane segments in bacterial membrane proteins [124].

Similarly, the MEMSAT-SVM, which integrates both signal peptide predictions and re-entrant helix predictions, can achieve an accuracy level of 89% [125]. This method is also freely available, via the PsiPred server[*b]. Again, the accuracy of these tools usually decreases considerably in the case of short-buried helices (half-helices, re-entrant helices) and those that possess kinks.

[*a]URL: http://www.cbs.dtu.dk/services/TMHMM/
[*b]URL: http://bioinf.cs.ucl.ac.uk/psipred/

A recent graph-based method called *'TopGraph'*[*a] represents an approach that differs from the aforementioned methods, in that it relies on extensive empirical data. Top-Graph uses apparent insertion free energies of host peptides into membranes by the bacterial TOXCAT-β-lactamase system [126]. It is especially useful for predicting the topologies of membrane proteins with low similarities to known structures, as it does not rely on data that were derived for specific structures. It also allows constraints to be added on the basis of prior knowledge regarding the query protein (e.g., an amino acid known to be on the cytoplasmic side of the membrane). Finally, there are prediction methods that focus on β-barrels, e.g., BOCTOPUS [127]*b (see additional methods in [128]). However, most of these tools have been trained on proteins of the bacterial outer membrane, and can therefore deal only with single-chain β-barrels [128].

7.3.2.2 Secondary structure

One of the most prominent characteristics of integral membrane proteins is the substantial extent to which α-helical (and to a lesser extent the β-strand) conformations occupy their transmembrane segments [70,86] (Figure 7.10a and b). In Chapter 2 we saw that one of the main purposes of the α and β conformations in water-soluble proteins is to pair backbone polar groups in hydrogen bonds, thus lowering the energetic penalty associated with their exposure to the nonpolar protein core during folding (see Chapter 2, Section 2.3.4). The reason for the high prevalence of these conformations in membrane proteins is essentially the same, and even more salient; the transmembrane segments of such proteins are exposed to the cores of both the protein and the lipid bilayer, with the latter being extremely nonpolar. Electrostatic masking of backbone polar groups is therefore even more critical in membrane proteins than in their water-soluble counterparts [53]. The tendency of transmembrane segments to have extensive α-helical and β-strand/sheet content has been exploited in algorithms for structural prediction of membrane proteins. In particular, specific algorithms have been developed for GPCRs (see details in [71,101]).

7.3.2.2.1 α-Helical proteins

Most transmembrane segments in integral membrane proteins are organized as α-helices. As mentioned above, these helices have a high content of Pro residues [129], despite the 'helix-breaking' properties of this residue (Figure 7.11a, Box 7.1) and its low prevalence in the cores of helical segments of water-soluble proteins [130]. On the basis of their MD simulations, Sansom and coworkers proposed that **Pro residues act as hinges of motion in transmembrane segments, thus allowing better adjustment of helices' orientations, as well as mediating functionally-important conformational changes** [131]. As we will see later, conformational changes are highly important to the function of integral membrane proteins, e.g., for facilitating gating in channels, and transitions between inside- and outside-open states in transporters, or between active and inactive states in receptors and enzymes. These changes include a range of motions in the protein, from hinge bending or displacement of

---

*aURL: http://topgraph.weizmann.ac.il/
*bURL: http://boctopus.cbr.su.se/

individual helices in screw or pivot motions, to positional changes of whole domains and subunits. As mentioned above, Pro is known to create kinks in helices and confer rigidity to the polypeptide chain. Therefore, the importance of Pro to conformational changes in transmembrane proteins may seem surprising at first. However, Pro residues in transmembrane segments tend to appear within certain sequence motifs, which also contain Ser and Thr, such as *(S/T)P, (S/T)AP, PAA(S/T)* [89]. Simulations demonstrate that these residues compensate for the structural distortion created by the Pro, and suggest that these structural effects allow the helices to undergo the required conformational changes [89]. Experiments support this proposition, showing that replacement of Ser and Thr within these motifs changes the activity of the protein [132–134]. Still, one may wonder why these phenomena are observed almost exclusively in integral membrane proteins rather than in all proteins. This may have to do with the restrictions imposed on membrane protein motions by the lipid bilayer structure. That is, since integral membrane proteins are more restricted by their environment than water-soluble proteins, special characteristics such as inclusion of Pro residues inside secondary structures may have developed as a means of providing these proteins with the same degree of flexibility that normally exists in their water-soluble counterparts. Thus, in the absence of the surrounding membrane, many of the advantages conferred by Pro are likely to become liabilities. This proposition raises the interesting possibility that Pro may also be important to membrane proteins through so-called 'negative design'. Specifically, it prevents membrane proteins from folding outside the membrane, in which case they would be unstable and subject to degradation. A different study has shown that the mere presence of Pro residues in certain positions on transmembrane segments can protect membrane proteins from misfolding, by disfavoring non-native structures that contain an array of β-strands [135].

Transmembrane segments may contain other structural distortions as well, such as tight turns of 3_{10} helices and wide turns of π-helices [76]. The latter two irregularities have been studied less extensively than Pro-induced kinks, so it is still unclear whether they play a functional role. One of the advantages conferred by such distortions is that they allow proximity between polar groups in adjacent helices, thereby facilitating better electrostatic masking than that achieved by intrahelical interactions alone. Indeed, nearly 40% of the transmembrane helices in membrane proteins are distorted, compared to only 19% of the helices in water-soluble proteins [87]. In addition, transmembrane helices may be discontinuous or penetrate halfway into the membrane core, forming re-entrant loops [68] (Figure 7.11b, Box 7.1). The latter are particularly common in channels such as aquaporins and K^+-channels, where the exposed residues in the re-entrant loops usually serve as binding sites for ions or other substrates (see Figure 7.15b below).

7.3.2.2.2 β-Sheet proteins

Integral membrane proteins with extended (β) conformations [136,137] are less common than those having α-helical conformations, and are estimated to constitute only a small percentage of the total proteome [62]. These structures, called β-barrels, usually consist of a single chain but can also comprise multiple chains. Most β-barrels belong to the porin superfamily of small molecule channels, which reside in the membranes of Gram-negative bacteria [138,139], as well as in the outer membranes of mitochondria and chloroplasts [140–143]. In bacteria, the exposure of porins to the external environment turns many of them into attachment sites for phages and bacterial toxins [9]. In fact, some of the toxins (e.g., *α-hemolysin*

(a)   (b)

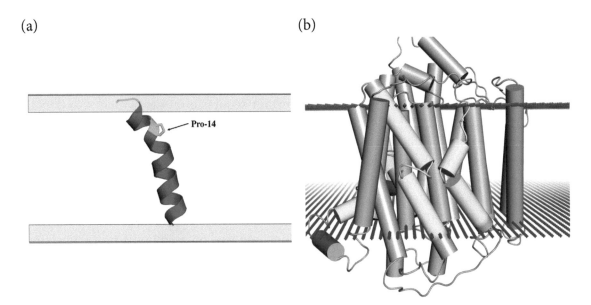

FIGURE 7.11 **Irregularities in transmembrane helices.** (a) A kink-inducing Pro in the transmembrane peptide alamethicin. The peptide is represented as a ribbon, with Pro-14 shown also as sticks (PDB entry 1amt). The polar head group regions of the bilayer are in light green. (b) Distorted helices, half-helices and membrane-exposed loops in one of the subunits of the membrane domain of respiratory complex I from *E. coli* (PDB entry 3rko). For clarity, the helices are presented as cylinders. The protein is colored in grey with irregular helices colored in yellow. The red and blue planes mark the predicted boundaries of the membrane, respectively (the OPM database [117]).

from *Staphylococcus aureus* [144]) create a *β*-barrel structure in the host membrane. In the barrel structure, the strands are anti-parallel, connected by short loops at the periplasmic side, and long loops at the external side of the cell or organelle [145] (Figure 7.8b). Based on this structure, it has been suggested that the *β*-hairpin motif is the principal evolutionary unit of all *β*-barrel proteins [146]. In accordance with the role of porins, the barrel structure is amphipathic[*1], with a water-filled center and nonpolar exterior. Moreover, the large width of the barrel is associated with low selectivity. As a result, porins are able to transport a larger variety of polar molecules, compared with the channels that reside in the plasma membrane or inner mitochondrial membrane, all of which are made up of *α*-helical bundles. Finally, porins tend to oligomerize within the membrane [147]. Porins should not be confused with aquaporins, which are *α*-helical channels that belong to the *major intrinsic protein* (*MIP*) superfamily (see Subsection 7.3.2.3.3 below).

7.3.2.3 Tertiary structure

7.3.2.3.1 *Key characteristics*

Integral membrane proteins exist within a lipid environment, which explains why they have nonpolar exteriors, as well as the fact that many of their polar residues tend to face the protein interior. On the basis of these observations, it was initially proposed that these proteins are '*inside-out*' versions of water-soluble proteins. However, with the structural characterization of different membrane-bound proteins, this assumption has turned out to be an oversimplification [75,148]. Rather, the structure of integral membrane proteins is similar in

---

[*1]Contains polar residues on one face and nonpolar residues on the opposite face.

certain aspects to structures of water-soluble proteins [86,149]. First, in both protein types, the core is tightly packed, contains mainly nonpolar residues with few functionally important polar residues, and is evolutionarily more conserved than the surface of the protein [150]. The high conservation of the core probably results from the fact that inter-residue packing is tighter than residue-lipid packing, such that the residues in the core are more structurally constrained [71]. Second, the loops in the structures of both protein types serve similar roles in ligand binding and signal transduction.

As mentioned above, the transmembrane segments of integral membrane proteins tend to include small residues such as Ala, Ser, and Gly, which facilitate tight packing of helices. The packing is important not only for helices within polytopic proteins, but also for bitopic proteins, which tend to dimerize or oligomerize within the membrane [151]. Moreover, the distribution of the small residues among larger residues produces *grooves and ridges* (respectively) along the helix, which creates geometric complementarity between adjacent helices ('*knobs-into-holes' packing*) (Figure 7.12a,b). Since these ridges and grooves are not geometrically parallel, but rather curl around each helix, the best fit between adjacent helices requires them to tilt across the membrane. Indeed, **though transmembrane helices in membrane proteins may have a 5° to 35° tilt** [77], **their average tilt is ~20°, which seems to be optimal for interhelical packing** [88,152].

One of the most common models used for studying helix-helix interactions in membrane proteins is *glycophorin A*, which forms an α-helical dimer when solubilized in detergent or lipid bilayers. The protein contains the sequence motif *LIxxGVxxGVxxT* (x is any residue), located within the helix-helix interface. When these residues are replaced with mutations, the protein dimer separates into two monomers. The motif includes a smaller motif, *GxxxG*, which is over-represented in transmembrane segments [129,153], and also appears in many interacting helices of water-soluble proteins. As explained in Chapter 2, the two Gly residues are located on the same face of the helix, a turn away from each other, and allow the two interacting helices to become separated by only 6 Å. This short distance optimizes van der Waals interactions and allows the C_α−H group of one helix to hydrogen-bond to a backbone carbonyl group (C=O) in the other [154]. Many cases have been observed, however, in which the small residues Ser and Ala appear instead of Gly residues, thus extending the *GxxxG* motif into a '*GxxxG-like (or GAS) motif*' [129,155,156]. Other motifs suggested to mediate tertiary interactions in membrane proteins include the *leucine-isoleucine zipper* [157]*1 (Figure 7.12c), the heptad *serine zipper* (e.g., *SxxLxxx*) [158], and the *GxxxGxxxG glycine zipper* [159] motifs. These and other linear motifs that mediate helix-helix and protein-protein interactions can be found in the MeMotif database [160], which can also identify such motifs in specific sequences*2.

Finally, the charged residues Glu, Asp, Lys, and Arg within transmembrane helices have been implicated as mediators of helix-helix interactions [82,151] (Figure 7.12d). The energetic implications of such interactions are discussed in the following subsection. As in water-soluble proteins (see Chapter 4), polar interactions inside the protein have an added benefit; they are more specific than nonpolar interactions that rely on steric complementarity alone.

Certain structural arrangements found in integral membrane proteins tend to recur [161]. **The most common is by far the α-helical bundle, which reappears in different**

---

*1 As we saw in Chapter 2, the leucine zipper motif mediates interhelical interactions in coiled coil-forming water-soluble and fibrous proteins as well. In membrane proteins, however, this motif often appears coincidental due to the high prevalence of Leu, Ile, and Val in transmembrane segments.

*2 URL: http://projects.biotec.tu-dresden.de/memotif/en/Special:Search

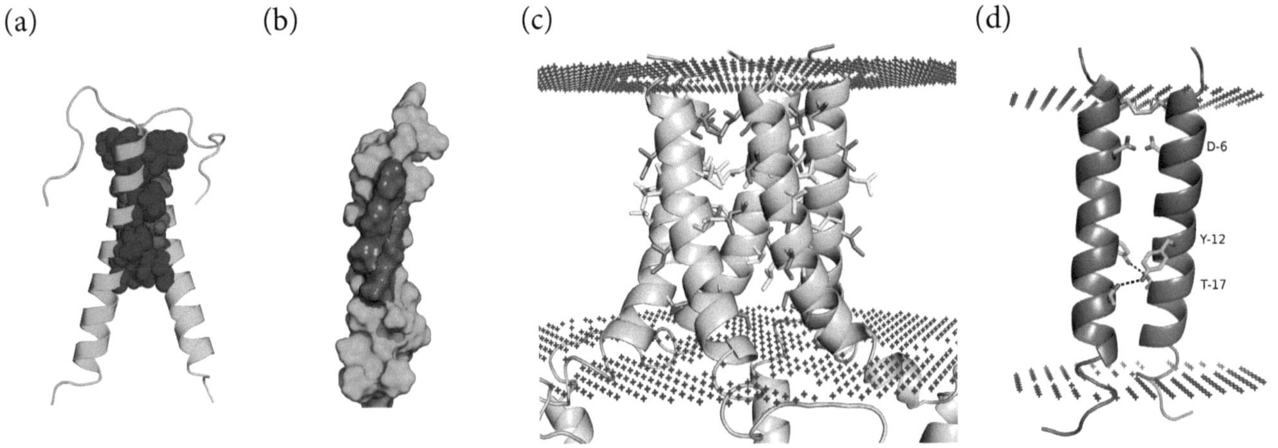

(a) (b) (c) (d)

FIGURE 7.12 **Packing motifs and interhelical interactions in transmembrane segments.**
(a) and (b) Ridges and grooves. (a) The dimeric structure of glycophorin A in detergent micelles
(PDB entry 1afo), prepared after Figure 12 in [53]. The backbone is shown as a ribbon, and the side
chains of the helix-helix interface are shown as spheres. Valine and isoleucine residues, forming the
ridges, are colored in blue. Glycine and threonine residues, forming the grooves, are colored in red.
(b) The ridges and grooves in the interface of one of the chains. (c) Leucine and isoleucine zipper in
the pentameric structure of phospholamban (PDB entry 1zll). The backbone is shown as a ribbon,
and the side chains of the helix-helix interface are shown as sticks. Leucine and isoleucine residues
are colored in yellow and orange, respectively. The red and blue dashed lines mark the predicted
boundaries of the membrane, respectively (the OPM database [117]). (d) Polar interactions. The image shows hydrogen bonds involving two Asp residues and two Tyr-Thr pairs in the TCR-ζ chain
dimer (PDB entry 2hac). The close proximity of the two Asp residues suggests that one of them is
protonated. Furthermore, NMR studies indicate that these residues are stabilized by an extensive
hydrogen bond network with a buried water molecule (absent in the presented structure) and other
residues. The two chains are also connected by a disulfide bond.

forms. For example, GPCRs include a characteristic seven-helical bundle, some transporter
groups include 12 or 14 helical bundles, and so on. A less common structural motif is the β-
barrel, which characterizes channels with low selectivity in the outer membranes of bacteria
and eukaryotic organelles of bacterial origin [136,137] (see Subsection 7.3.2.2.2 above).

7.3.2.3.2 Energetics

Integral membrane proteins reach their final active state in the membrane through a complex process, during which each transmembrane segment undergoes the following major
steps [74–76,162,163]:

1. Translation by the ribosome

2. Insertion by the translocon complex into the membrane

3. Acquiring secondary structure

4. Assembly with the other transmembrane segments into the mature protein, usually
   in the form of an α-helical bundle

Since membrane insertion often depends on secondary structure formation in transmembrane segments [105,164–166] steps 2 and 3 are usually referred to as a single coupled step. In

any case, the entire four-step process is complex, and although recent structural studies have clarified some of the complexities [75], it is still not entirely understood. One alternative for studying this process, with all the biological components included therein, is to use model systems that are based either on whole proteins [167] or on isolated peptides, and to focus on the energetics of the four key steps instead of getting into the numerous kinetic barriers and minima that the complex translocon system involves. Studies with such systems, as well as with more realistic setups [109], have produced estimates of the energies associated with the key steps, especially the first three (see [8,53] and references therein). Generally speaking, the energetics of transfer of transmembrane segments from the aqueous phase into the membrane is dominated by the following [75]:

1. The free energy penalty associated with partitioning of the (polar) peptide backbone into the lipid environment. This penalty has been calculated as +2.1 kcal/mol for C=O and N−H backbone groups that are hydrogen-bonded to each other, and +6.4 kcal/mol when the hydrogen bond is not satisfied [168,169].

2. The favorable free energy contribution due to the hydrophobic effect. On the basis of simple partition experiments, this free energy value has been estimated at $\sim -25 \pm 30\%$ cal/mol per Å^2 of the protein involved in nonpolar interactions [108,170–173]. Experiments using a more realistic system, in which peptides were inserted into the ER membrane by the Sec61 translocon [174], produced a smaller value of -10 cal/mol per Å^2.

The favorable free energy must therefore compensate for the free energy penalty, as well as for the cost of inserting polar side chains, in order for the net membrane-partitioning free energy to be favorable (i.e., negative).

Inside the membrane, the nonpolar environment induces secondary structure formation in the inserted segments [105,164–166]. The energetics of this step is essentially the same as the energetics of the induction of secondary structure in water-soluble proteins upon folding (see Chapter 2). **Atomic force microscopy measurements carried out on bacteriorhodopsin indicate that the free energy of the coupled insertion-folding process of a single α-helical transmembrane segment is -1.3 kcal/mol per residue** [175]. Estimates concerning the insertion of each of the transmembrane segments into the membrane can be presented in the form of a scale. In the scale, each of the naturally occurring amino acids is assigned a value describing the free energy of its insertion into the polar head group region [105,108] or the hydrocarbon core [8] (see also Box 7.1).

In the final step of membrane protein formation, each of the folded transmembrane segments interacts inside the membrane with its neighbors to form the fully folded protein [75]*1. The net energy of this process has recently been measured for bacteriorhodopsin as ~ -11 kcal/mol [178], which is on the scale of the energy of water-soluble protein folding. However, compared with the earlier stages of membrane insertion and acquisition of secondary structure of each transmembrane segments, this stage is associated with much more controversy with regard to its energy components and their magnitude [86]. We have seen

---

*1Note that this also pertains to dimerization and oligomerization of (helical) membrane proteins; here, too, the process involves interhelical packing, with the only exception being that the interacting transmembrane segments are not connected by loops. Dimerization and oligomerization are common in membrane proteins [176], with 50% to 70% of the complexes containing the same chains (homodimers and oligomers) [177].

that in water-soluble proteins the driving force for folding is the hydrophobic effect (non-polar interactions). Integral membrane proteins are surrounded by a lipid medium, which means the driving force for their folding must be different. Studies show that the stability of membrane-bound proteins correlates with the amount of surface area of the protein that becomes buried during folding [179]. In the absence of the hydrophobic effect, this correlation should reflect mainly van der Waals interactions. Indeed, **van der Waals interactions are expected to play a more significant role in driving the folding of membrane proteins compared to their water-soluble counterparts.** This expectation stems mainly from the fact that water molecules, being smaller than lipid molecules, are better at rearranging around the unfolded protein and forming a tight interaction shell. As a result, the van der Waals interactions between the unfolded protein and the molecules of its environment should be stronger in the case of water-soluble proteins than in the case of membrane proteins. This means that the increase in strength of these interactions during folding is more pronounced in membrane proteins. In addition, the aforementioned motifs in membrane proteins facilitate especially tight packing of transmembrane segments, which, in turn, optimizes the van der Waals interactions between them.

Aside from van der Waals interactions, the stability-surface area correlation may also reflect another effect, which is entropic in nature, and is reminiscent of the hydrophobic effect in water-soluble proteins [180]. The lipids that make up the membrane have a certain freedom of movement, which decreases when a protein is inserted into the membrane. This restriction of movement pertains mostly to lipids that are in direct physical contact with protein residues. During protein folding, some of those lipids are released into the bulk, and as a result their freedom of movement increases. In other words, **folding decreases the entropy of the polypeptide chain but increases the entropy of the lipid bilayer. Since this process is favorable, it can be considered as a driving force of folding.** This process is analogous to the hydrophobic effect in water-soluble proteins, because in both cases folding of the protein releases 'solvent' molecules (lipids or water), thereby increasing the overall entropy. Quantitatively speaking, this effect is probably weaker than the hydrophobic effect, as the change in the freedom of movement of lipids is expected to be smaller than in the case of the much smaller water molecules.

As in the case of water-soluble proteins, the exact effect of electrostatic interactions on membrane protein folding is also not entirely clear [86]. We saw earlier that various motifs known to promote helix-helix packing in membrane proteins include polar residues, suggesting that polar interactions are overall favorable in this setting. Indeed, studies that focused on hydrogen bonds between transmembrane segments suggested that these bonds are favorable and drive the assembly of the protein [181–183]. For example, DeGrado and coworkers studied this issue using helical model peptides that dimerize [183]. When hydrogen bonds involving the amino acid residues Asn, Gln, Asp and Glu were disrupted by mutagenesis, dimerization did not occur. Thermodynamically, it makes sense that hydrogen bonds contribute favorably to the assembly of the helices into a folded protein, since in their isolated state (inside the lipid bilayer) the potential hydrogen bond donors and acceptors are surrounded by a less polar environment than in the folded state (Figure 7.13). Again, the fact that polar residues such as Ser and Thr are more prevalent inside membrane-bound proteins than inside water-soluble proteins supports this suggestion [152].

Assuming that transmembrane hydrogen bonds do stabilize the folded state of membrane proteins, how significant is the stabilization? A study by Bowie and coworkers [184] investigated this issue using bacteriorhodopsin, a bacterial light-activated proton pump of known structure, which is often used as a model for membrane proteins. They mutated hydrogen-bonding pairs to non-hydrogen-bonding residues that were highly similar to the original residues (in this case, alanine) in all other aspects, and measured the resulting change in protein stability. Their results suggested that the average contribution of hydrogen bonds to stability is quite modest, on the scale of $1k_BT$ (i.e., 0.6 kcal/mol). What could be the evolutionary reason for maintaining such marginally stabilizing forces? First, the additive stabilization achieved with multiple hydrogen bonds between transmembrane segments that contain several polar residues can be substantial. Second, Bowie and coworkers speculated that, as in the case of Pro-related hinges, these weak interactions might just be what membrane proteins need to remain highly flexible and form the helical distortions characterizing their structure. Finally (as controversies go), some studies suggest that hydrogen bonds destabilize membrane proteins [185,186].

The effects of ionic interactions like salt bridges on membrane protein stability are more complicated, as the exposure of charged residues to the core of the protein involves a large desolvation penalty. As discussed at length in Chapter 4, studies carried out on salt bridges in the cores of water-soluble proteins indicate that such interactions may be overall stabilizing if they are optimized spatially and constitute part of a larger network of electrostatic interactions that include also hydrogen bonds [187–189]. In membrane proteins the salt bridges may also be partially exposed to the lipid environment, which should make the desolvation penalty even larger than in water-soluble proteins. However, the fact that salt bridges are found in transmembrane segments, where they are organized as motifs that promote helix-helix interactions, suggests that these interactions are overall stabilizing in membrane proteins.

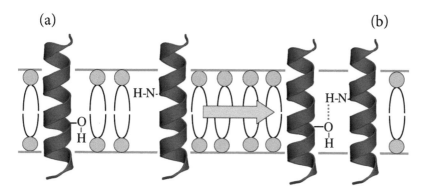

(a) (b)

FIGURE 7.13 **Polar interactions in integral membrane proteins.** (a) Polar groups in unfolded membrane proteins are exposed to the hydrophobic region of the membrane, which is energetically unfavorable. (b) Assembly of transmembrane segments allows these polar groups to hydrogen-bond and mask each other in the slightly less hydrophobic environment of the protein core.

7.3.2.3.3 Architecture

As mentioned above, the constraints imposed on membrane proteins by the chemically complex and highly anisotropic lipid bilayer have led to a collection of structures that share similar characteristics. For example, transmembrane segments are almost always arranged as α-helical bundles, and to a lesser extent as β-barrels. Nevertheless, the shared general architecture of membrane proteins may manifest in different forms, which are suited to the proteins' specific biological functions. Several important examples of such architectural themes and their functional significance are reviewed by von Heijne [9] and by Gouaux and McKinnon [190]. Here we focus on proteins that are involved in the transport of polar molecules across biological membranes, i.e., channels and transporters. In Section 7.5 below we describe the structure-function relationship in the other major category of membrane proteins, ligand-activated receptors.

1. **Aquaporins**

   Aquaporins are ancient channels that can be found in a wide range of organisms, where they facilitate the passive movement of water in cells and tissues [191]. In animals, for example, they participate in the re-absorption of water from urine in the kidneys. Certain aquaporins, termed 'aqua-glyceroporins', can also transport small solutes like glycerol, ammonia, CO_2, and O_2. Aquaporins are part of the major intrinsic protein superfamily and exist as tetramers in which each monomer is an independent pore. The pores created by aquaporins have an hourglass-shaped structure that includes six transmembrane segments and two half-helices (Figure 7.14a). The selectivity of the channel against molecules larger than water results from the narrow region at the center of the channel (2.8 Å). In fact, even the water molecules themselves can pass through the narrow channel only in single file, though the transport rate is extremely high (~10^9 molecules/sec). The narrow part of this region is formed from side chains that face the inner side of the pore, and which belong to two types of motifs:

   - Two conserved *Asn-Pro-Ala motifs* (*NPA*, yellow region in Figure 7.14a), which reside in the two half-helices (see more below).

   - An *ar/R motif* ('ar' for aromatic and 'R' for arginine), which resides ~8 Å away from the NPA motif, towards the extracellular side. This motif is considered to be the major barrier for large uncharged solutes. Indeed, in aqua-glyceroporins the ar/R residues create a wider opening, allowing larger solutes to be transported [192] (Figure 7.14b).

   As mentioned above, the *NPA* motif plays a secondary role in the selectivity of the channel against uncharged solutes. On the other hand, it is crucial for the selectivity of aquaporins against protons, which are smaller than water molecules. Protons are positively charged, and the selectivity mechanism is, predictably, electrostatic. Each water molecule that passes through the center of the channel reorients such that its partially negative oxygen atom can hydrogen-bond with the *NPA* asparagine residues surrounding it (Figure 7.14c). Protons are positively charged and therefore cannot form these interactions. The exclusion of protons is further assisted by the dipoles of the half-helices flanking the *NPA* motif. Thus, protons cannot pass through the channel, nor can they 'hop' between the single-file water molecules that reside in the pore [193]. Finally, the partial charges of the *NPA* asparagine have also been suggested to contribute to the fast passage of the water through the channel [194].

2. **Ion channels**

   Ion channels are involved in numerous physiological functions, such as neural transmission, molecular transport, muscle contraction, energy production, and more. Accordingly, malfunction of these proteins leads to various pathologies (*channelopathies*), which include cystic fibrosis, Bartter syndrome, and paralysis [195]. The

(a)   (b)   (c)

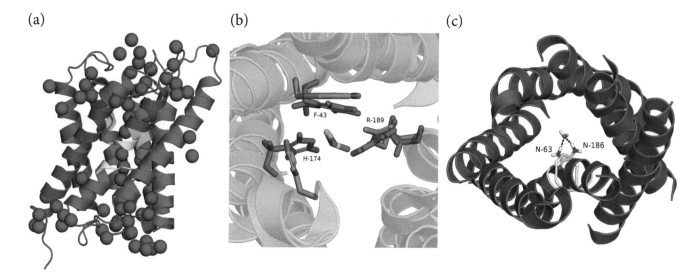

FIGURE 7.14 **Structural characteristics of aquaporins.** (a) A general view of the channel (PDB entry 1rc2). The image shows the single-file distribution of water molecules along the channel (oxygen atoms as red spheres), and the location of the two motifs that create the central constriction: the NPA (yellow), and the ar/R (green) motifs. (b) Superimposition of aquaporin (blue) and aqua-glyceroporin (orange, PDB entry 1fx8). The residues forming the motif in aquaporin are noted. A glycerol molecule in the aqua-glyceroporin is shown as sticks. (c) Hydrogen bonds between asparagine residues of the NPA motif and a water molecule (as sticks) inside the channel.

structure of an ion channel is arranged around a central water-filled pore that traverses the lipid bilayer, and in which ions can dissolve without having to pay the energy cost of exposure to the hydrophobic environment of the bilayer core. One of the popular models for such channels is the pH-dependent bacterial K^+ channel (KcsA)[*1]. K^+ channels have a characteristic structure that includes a transmembrane domain and a long (35-residue) cytoplasmic domain, both of which are α-helical (Figure 7.15a). The channel is a tetramer, where each monomer contributes two transmembrane helices (TM1 and TM2) and one cytoplasmic helix at the C-terminus. The transmembrane domain has an inverted teepee structure, and includes a wide, water-filled pore that extends over 2/3 of the lipid bilayer. Thus, the potassium ion can diffuse freely most of its way across the membrane (Figure 7.15b). This property, as well as the electric repulsion between adjacent K^+ ions, leads to a very high K^+ passage rate (~10^8 ions per second). The cytoplasmic domain is an extension of the four TM2 helices that form the inner part of the channel. This domain influences different functional aspects of the transport (permeation, gating), stabilizes the channel, and allows it to interact with regulatory elements.

Despite the high variability of channels, in terms of shape, size, and diameter, virtually all possess some degree of specificity towards the ions they transport. The specificity can be broad, e.g., allowing all elemental ions of a certain charge to pass, or narrow, e.g., enabling only Ca^{2+} ions to pass. In most cases, the selection seems to require the 'candidate' ion to lose its solvation shell and to bind directly to channel residues. This mechanism enables the channel to assess the suitability of the ion. In the KcsA chan-

[*1]For solving the structure of the KcsA channel's transmembrane domain and elucidating the selectivity mechanism [119], Roderick MacKinnon received the 2003 Nobel Prize in Chemistry.

nel the selectivity-related residues appear at the end of the ion's path, in a loop-based substructure called the *'selectivity filter'* (Figure 7.15bI). The loop composing the selectivity filter in each chain includes four residues (the *TVGYG* signature sequence), forming four evenly-spaced K^+ binding sites termed S1 through S4 (Figure 7.15bII). The sites coordinate the four K^+ atoms via four backbone carbonyl groups and one side chain hydroxyl group. The filter determines whether or not the desolvated ion can keep moving all the way to the other side. The biological importance of the filter structure is reflected, for example, in the fact that it is targeted by toxins, such as *kaliotoxin*, produced by scorpions [196]. Whereas some ion channels use a simple size cutoff for selecting the right ion(s), others may be more sophisticated. This seems to be the case with the KcsA channel, which selects K^+ over Na^+, despite the smaller size of the latter. The selection mechanism can be referred to as *'molecular mimicry'*. That is, **the polar uncharged oxygen atoms that coordinate each K^+ ion in the selectivity filter are positioned in a way that precisely mimics the arrangement of water-derived oxygen atoms around the K^+ ion in bulk solution (Figure 7.15c)**. Thus, the K^+ ion encounters virtually no energy barriers during its transfer from solution to the selectivity filter (i.e., desolvation). Being smaller, Na^+ ions can enter the selectivity filter, but they are too far from the filter's oxygen atoms to achieve efficient electrostatic masking by the latter. Therefore, they tend to remain in bulk solution, i.e., outside the channel. The molecular mimicry mechanism, which was proposed by MacKinnon in his Nobel Prize-winning work, is elegant and explains the 1,000-fold preference of the KcsA channel for K^+ over Na^+. However, data from later studies suggest that the selectivity mechanism is probably more complicated. For example, NMR measurements have demonstrated increased protein flexibility in the selectivity filter region, suggesting that this region should be able to adapt to ions smaller than K^+ [197]. Similarly, molecular dynamics (MD) simulations of the KcsA channel in a lipid membrane demonstrate fluctuations of the selectivity filter over a range of ion–carbonyl distances, sufficient to coordinate either ion [198]. The authors of the latter study suggested that selectivity is controlled by the intrinsic electrostatic properties of the coordinating carbonyl groups and not by the average size of the pore. Finally, a study combining electrophysiology, X-ray crystallography, and MD suggested that smaller ions like Na^+ and Li^+ can bind easily to the selectivity filter, but find it difficult to reach the filter from the intracellular side due to a K^+-dependent energy barrier [199]. Again, these studies suggest that the selectivity of ion channels is more complex than previously assumed. Interestingly, despite the fact that small elemental ions such as K^+ occupy very little space, their binding often involves considerable changes in channel conformation, as demonstrated for the K^+, Na^+ and Ca^{2+} channels [190].

As in the case of many cellular proteins, the transport carried out by ion channels must be regulated, i.e., it must occur only in response to the right signal. For this purpose, most channels have a gate that prevents the passage of ions in the resting state. The signal activating the channel may be electric, chemical or mechanical in nature. Accordingly, the gate always includes a region of amino acids that acts as a sensor and responds to an activating or inactivating signal. For example, in voltage-gated K^+ (K_v) channels, the sensor includes positively charged amino acids (Lys and Arg) that reside on S4 helices [200,201]. In the channel's resting state the abundance of Na^+ ions on the extracellular side of the membrane repels the positively charged S4 helices, keeping the channel closed. However, when the membrane depolarizes during neuronal signaling, the significant decrease in the number of Na^+ ions at the extracellular side reduces the electric repulsion, enabling the S4 helices to be displaced, such that the channel opens. This process can be viewed as a conversion of electric energy into mechanical energy, which induces motion. As we will see later, energy conversions happen also in transporters that function as ion pumps, where the chemical energy stored in ATP, or the electrochemical energy stored in an ion gradient, is converted into motion, and vice versa.

The gating mechanism of the KcsA channel is rather complex and relies on two separate gates [202,203]:

(a) H^+-activated gate, which resides at the region covering the intracellular side of the transmembrane domain and the beginning of the cytoplasmic domain. This gate is closed at high pH levels and opens when the pH decreases.

(b) H^+-independent gate, which resides at the selectivity filter.

Understandably, most studies of KcsA gating have focused on the first gate, since it is the one that responds to a change in pH. Activation of this gate leads to conformational changes in the TM2 helices, in both the transmembrane and cytoplasmic domains. In the transmembrane domain, the helices undergo a ~15° hinge-bending motion around Gly-104 [204] (Figure 7.15dI). In the cytoplasmic domain, the conformational change is focused around a region in which the TM2 helices create a bulge, near Val-115 (Figure 7.15dII). There, at the narrowest point for K^+ permeation, each TM2 helix shifts ~4 Å outwardly upon activation, resulting in an overall ~20-Å opening of the channel[*1] [204,206]. The bulging region also seems to be the place where the pH sensing takes place; it contains three residues, Arg-117, Glu-118, and Glu-120, which have been implicated as the pH sensors of the KcsA channel (Figure 7.15dII) [207,208]. Although the exact mechanism of this sensing is not entirely clear, it has been suggested that a pH drop beyond the channel's pKa (~4.2) leads to protonation of the two glutamate residues (118 and 120), thus rendering them electrically neutral. As a result, these residues are no longer able to mask and stabilize the positive charges on Arg-117. The electrostatic repulsion that ensues between Arg-117 from adjacent monomers induces displacement of helices, and opening of the channel. As mentioned above, the channel also contains a second gate, at the selectivity filter. It is thought that the conformational change described for the H^+-activated gate affects the second gate allosterically, which leads the channel to assume a fully open state [205,209].

It addition to the active and resting states, ion channels can also exist in an inactivated state. Channels normally become inactivated after a period of activity, to attenuate the cell's response. There are two major types of inactivation in K^+ channels [210]. The first, called N-type inactivation, is observed in voltage-gated K channels. This is a fast process that involves physical blockage of the pore by an electrically charged N-terminal

---

[*1] Interestingly, when the large cytoplasmic domain is truncated, activation results in a ~32-Å opening of the transmembrane domain [205]. The ~20-Å opening obtained in the full-length structure [204] is probably a lower bound, as the large C-terminal domain was further stabilized by bound antibody fragments. Thus, the real opening of the channel is most likely somewhere between 20 Å and 32 Å.

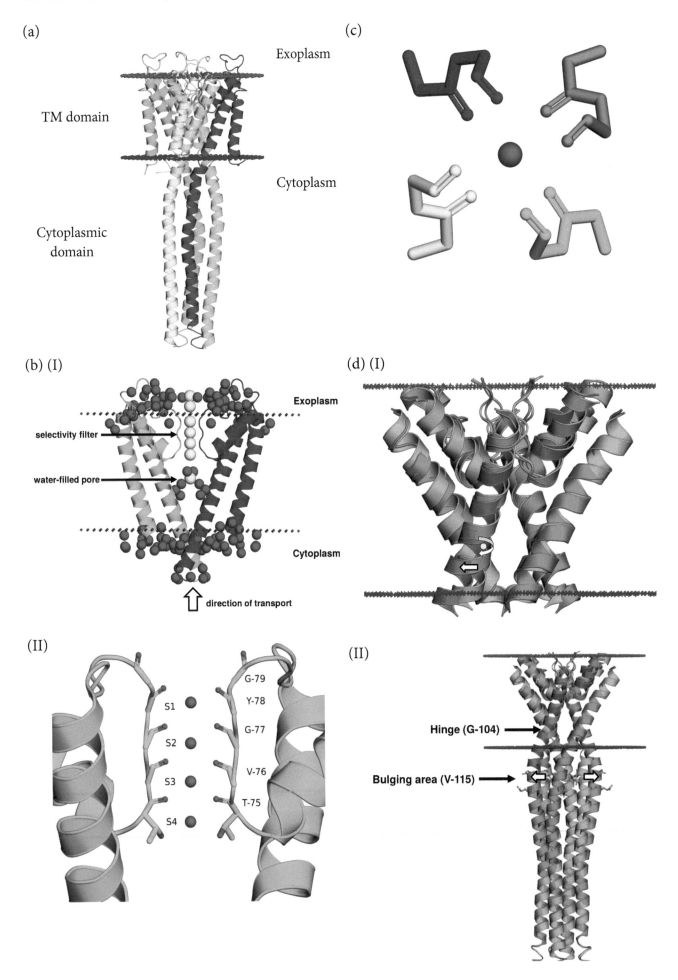

segment of the channel. The second type of inactivation, called *C*-type inactivation, is observed in the KcsA channel, and is much slower than *N*-type inactivation. *C*-type inactivation is thought to involve the H^+-independent gate at the selectivity filter, through reorientation of the backbone carbonyls and subsequent destabilization of K^+ ions inside the filter [203]. Interestingly, though the large cytoplasmic domain limits the degree of opening of the transmembrane domain upon activation (see above), it also seems to slow the rate of channel inactivation [204].

3. **Transporters**

Like ion channels, some transporters facilitate the passive transfer of ions down their electrochemical gradients. However, many other transporters transfer larger solutes (amino acids, sugars, etc.), and some transporters are even capable of transferring solutes against their electrochemical gradients, or in other words, 'pump' them. In any case, it is obvious that transporters must use a more sophisticated transport mechanism than the one used by channels. Indeed, **transporters do not form simple water-filled structures, but rather exist in (at least) two different states; one allows the ligand to enter the transporter on one side of the membrane, whereas the other releases it on the other side.** This model, abstractly introduced by Oleg Jardetzky, is commonly called 'the alternating access mechanism' [211]. Structurally, it requires each transporter to possess at least two distinct conformations, corresponding to the states mentioned above (Figure 7.16). Such a mechanism is observed in the *SemiSweet transporter* [212], which facilitates passive diffusion of sugars into bacteria. The function of

FIGURE 7.15 **Structural characteristics of the KcsA K^+ channel.** (Opposite) (a) A general structure of the tetrameric channel in the closed state (PDB entry 3eff), colored by chain. The transmembrane and cytoplasmic domains are noted. The red and blue planes mark the predicted boundaries of the membrane, respectively (the OPM database [117]). (b) (I) The different parts of the transmembrane domain in the conductive channel at high K^+ concentration (PDB entry 1k4c). For clarity, only two of the four chains are shown, and are colored differently. K^+ ions are presented as yellow spheres, and the oxygen atoms of water molecules are shown as red spheres. Most of the transmembrane domain, from the cytoplasmic side, forms a water-filled pore (a hydrated K^+ ion is shown). The top third of the domain forms a selectivity filter that can accommodate four dehydrated K^+ ions. The last part of the domain is just outside the membrane, where the ions are hydrated again. The filter is made of the re-entrant loops interfacing with the K^+ ions. (II) The selectivity filter. The filter is composed of the signature *TVGYG* sequence, which binds the K^+ ions via four backbone carbonyls and one side chain hydroxyl per chain. Each K^+ ion is coordinated by oxygen atoms from two layers, and thus, the filter contains four K^+ binding sites, termed S1 through S4. (c) A view from above showing interactions between one of the K^+ ions (blue sphere, reduced size for clarity) and two layers of backbone carbonyl oxygen atoms in the selectivity filter (eight carbonyls altogether, shown as sticks, colored by chain). (d) Conformational changes in the pH-activated gate of the KcsA channel. (I) The hinge motion of the TM2 helices. The image shows a superposition of the closed-state (grey, PDB entry 3eff) and open-state (orange, PDB entry 3pjs) structures, with a close-up on the transmembrane domain. For clarity, the TM1 helices are not shown. The Gly-104 hinge is marked by the yellow circle, and the hinge motion is delineated by the curved arrow. The resulting shift of the cytoplasmic part of one of the TM2 helices is noted by the thick arrow. (II) The conformational change in the cytoplasmic domain, focused around the bulging area (V-115). The 4-Å shifts in two helices are noted by the thick arrows. The residues constituting the pH sensor are shown as sticks.

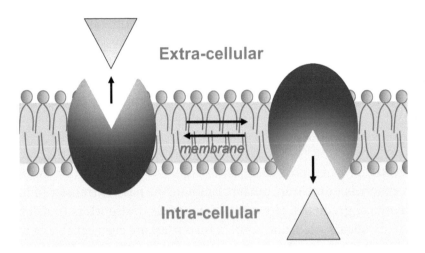

FIGURE 7.16 **The alternating access mechanism for substrate transfer in transporters.** A highly schematic illustration of the mechanism. The purple shapes symbolize two different equilibrium conformations of the transporter inside the membrane (cyan). In each conformation the substrate-binding site faces a different cellular compartment, where it binds or releases the substrate (yellow triangle). The figure demonstrates an import process, where the left 'conformation' binds the substrate in the exoplasm, and then changes into the right conformation to release the substrate into the cytoplasm. The shapes symbolizing the transporter, substrate, and membrane are grossly disproportionate to emphasize the main aspects of the transport process.

this transporter is analogous to that of the GLUT and SGLT transporters in humans, which are key elements in sugar metabolism. As Figure 7.17a shows, in the outward-open state of the transporter (left), the inner side is blocked by a cluster of aromatic and nonpolar residues (intracellular (IC) gate). The switch from the outward-open state to the inward-open state involves several conformational changes. First, a hinge motion of the cytoplasmic half of TM1 (TM1b), around the highly conserved Pro-21, tilts TM1b 30° with respect to the exoplasmic half of TM1 (TM1a). Second, a rotational movement of TM1a, TM2, TM3, and TM1b of the other monomer, leads to an overall 'binder clip' movement. That is, the exoplasmic parts of the transmembrane helices become closer to each other while their cytoplasmic parts are pushed away from each other. This motion leads to the simultaneous closure of the extracellular gate (EC gate) and opening of the cytoplasmic gate. The EC gate is composed of Tyr-53, Arg-57 and Asp-59. When the gate is closed, these residues form polar interactions with the equivalent residues in the adjacent monomer, thus stabilizing the closure.

An active transporter, or 'pump', transfers solutes against its electrochemical gradient. This process requires a source of energy, either ATP or the electrochemical gradient of a common cellular ion. Like passive transporters, active transporters switch between outward- and inward-facing conformations. However, in an active transporter the dominant conformation at any given time is determined by the binding of ATP/ADP/P_i or ions to a specific site in the protein[*1]. This is because the binding of each of these species to the protein stabilizes a different conformation. Thus, in an

---

[*1] The binding of the transported solute and possibly of other chemical species (common ions, lipids, etc.) is likely to stabilize certain transporter conformations as well, but the effects of these species are not necessarily unique to active transporters.

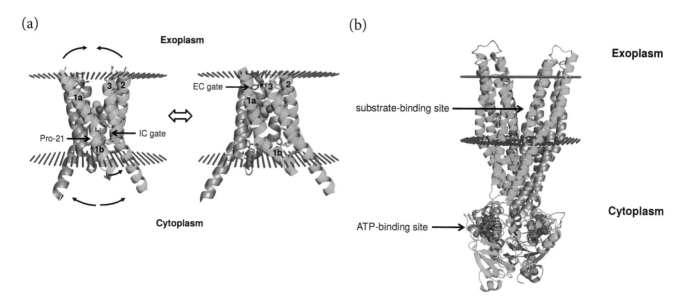

FIGURE 7.17   **Conformational changes in transporters.** (a) Outward (left) and inward (right) conformations of the bacterial SemiSweet transporter (PDB entry 4x5n). The dimeric structure is colored by chain. The glycerol moiety of 1-oleoyl-R-glycerol (PDB entry 4x5m), which mimics the transported sugar, is shown as spheres in the putative substrate-binding site between the two monomers. Residues belonging to the transporter's extracellular gate (EC gate) and intracellular gate (IC gate) are shown as sticks. The curved arrows show the 'binder clip' motion in which the protein shifts from its outward-open state into the inward-open state. (b) The ATP-bound state of the Sav1866 ABC transporter (PDB entry 2hyd), with its substrate-binding site open to the extracellular side of the membrane. The dimeric transporter is colored by chain, and the bound ADP molecules are shown as magenta spheres. The red and blue planes mark the predicted boundaries of the membrane, respectively (the OPM database [117]).

ATP-dependent transporter, a cycle that includes (1) ATP binding, (2) ATP hydrolysis to ADP and P_i, and (3) release of one or two of the latter, involves switching of the protein between different conformations that face different sides of the membrane. Such a mechanism is exemplified by the bacterial multidrug transporter Sav1866, which belongs to the group of *ATP-binding cassette (ABC) transporters*. These proteins appear in both prokaryotes and eukaryotes, where they pump small molecules from (or to) the cell. The action of these transporters is also medically important; some ABC transporters in bacteria pump out antibiotics, leading to resistance. Similarly, tumor cells use such transporters to pump out chemotherapeutic agents, which reduces the efficiency of the treatment. The solute export cycle of Sav1866 begins when the protein is in a monomeric state and its transmembrane domain is exposed to the cell's interior. This conformation enables the targeted solute to bind the transmembrane domain. ATP binding to the nucleotide-binding domains of two Sav1866 monomers induces dimerization and an outward-facing conformation (Figure 7.17b) [213,214]. ATP hydrolysis is thought to induce additional conformational changes that bring the transporter back to an inward-facing state, ready to bind a new solute molecule. It should be noted that the exact nature of the coupling between ATP binding or hydrolysis and the transport of solutes in ABC transporters is not always as simple as depicted above, and may sometimes (and in some variants) be very weak. For example, the bacterial vitamin B12 importer $BtuC_2D_2$ hydrolyzes many ATP molecules for each substrate, and hydrolysis takes place even in the absence of the substrate [215].

The effects of nucleotide or ion binding on the shifts between inward- and outward-facing conformations in active transporters do not really explain how these transporters can draw a solute from a low-concentration compartment and release it into a high-concentration compartment. This capability has to do with the inherent *affinity* of each conformation to the solute. In passive transporters the affinity of the inward-facing conformation may be similar to that of the outward-facing conformation, since the binding and release of the solute are governed by its concentration in the two opposite compartments. In active transporters, the conformation facing the compartment that has low solute concentration must have a sufficiently high affinity to the solute to bind it, whereas the conformation facing the compartment with high solute concentration must have a sufficiently low affinity in order to release the solute [216,217]. By selectively stabilizing the two different conformation types, binding of nucleotides or ions indirectly determines changes in affinity during the transport cycle. For example, in the ABC transporters described above, ATP binding stabilizes the outward-facing conformation, which has a low affinity to the solute, thus enabling the solute to be released into the cell's exterior (see [217] and references therein). Some transporters are assisted by binding proteins that scavenge the solute and then bind to a certain domain in the transporter [213].

In conclusion, ATP hydrolysis and electrochemical gradient dissipation are able to fuel active transport, not by a direct release of energy, but rather through sequential binding of nucleotides or ions that stabilize distinct conformations of the transporter. These conformations differ from one another in (1) the exposure of the transporter's transmembrane domains to the inner or outer sides of the membrane, and (2) the affinity of the transmembrane domains to the transported solute.

The principles of active transport that are described above exist in *ATP-dependent ion transporters*. These proteins, which constitute key elements in any organism, are divided into the following groups [218]:

P-ATPases (E_1E_2-ATPases). These transporters are found in bacteria, in fungi, and in eukaryotic plasma membranes and organelles. They transport a variety of different ions across membranes, including H^+, Na^+, K^+, and Ca^{2+}. This category includes some of the cell's key primary active transport proteins, such as the Na^+/K^+-ATPase (Figure 7.18), H^+/K^+-ATPase, and the Ca^{2+}-ATPase (Figure 7.19). The electrochemical gradients formed by these proteins are used for different purposes. For example, bacteria use the proton gradient to drive processes such as chemotaxis and secondary active transport. P-ATPases contain cytoplasmic and transmembrane domains. The cytoplasmic domain includes phosphorylation (P)[*1], nucleotide-binding (N), and actuator (A) subdomains [219]. The transmembrane domain includes six helices, through which the membrane transport takes place. Most P-ATPases, however, also include additional transmembrane helices. For example, in the Na^+/K^+-ATPase and Ca^{2+}-ATPase the transmembrane domain includes 10 helices. These helices also contain cation-binding sites, and are thought to contribute to the ion selectivity of the transporter. Finally, some P-ATPases, such as the Na^+/K^+-ATPase and the H^+/K^+-ATPase,

---

[*1]The mechanism of P-ATPases includes a phosphorylated intermediate (aspartate in a *DKTG* motif, see more below), hence the name.

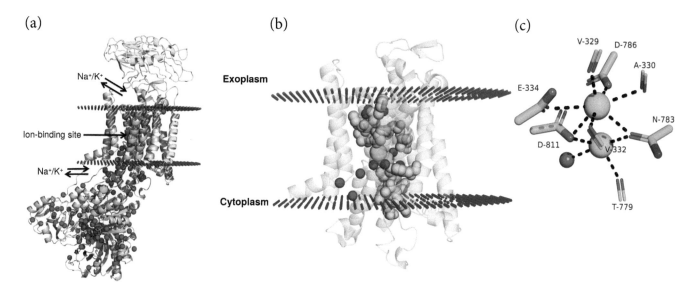

FIGURE 7.18 **Shark Na⁺/K⁺-ATPase bound to K⁺ [221] (PDB entry 2zxe).** (a) A global view of the protein and its position inside the membrane. The α subunit (grey) is responsible for the ion transport and ATPase activity. The β subunit (yellow) is involved in the assembly and trafficking of the protein, and also in the binding of K⁺ [221]. Potassium ions are colored in cyan, and water molecules (oxygen atoms only) in red. The entry and exit points of the ions with respect to the transmembrane domain are noted, as well as a transmembrane K⁺ binding site. (b) Experimentally-determined K⁺-binding sites in the transmembrane domain of the Na⁺/K⁺-ATPase [221,222]. The residues constituting the binding sites are shown as light-orange spheres. For clarity, the transmembrane helices are semi-transparent, and other parts of the protein have been removed. (c) Polar interactions (black dashed lines) between the two K⁺ ions located in the transmembrane domain and protein groups. The ionization states of the acidic residues were predicted by MolProbity [223].

also include an extracellular β subunit that is important for the proper trafficking of the transporter to the plasma membrane [220], and that also affects other functional aspects of the transport.

F-ATPases (F₁F₀-ATPases). These H⁺-ATPases reside in mitochondria, chloroplasts, and bacterial plasma membranes. F-ATPases operate in the opposite direction to P-ATPases; in F-ATPases, proton transport is used to fuel ATP synthesis, instead of the other way around. Specifically, F-ATPases use the H⁺ electrochemical gradient that is created during cellular respiration (in mitochondria) or photosynthesis (in chloroplasts) to drive the synthesis of ATP. Thus, these proteins are usually referred to as *'ATP synthases'* rather than ATPases. The structure and mechanism of the mitochondrial F-ATPase are described in Chapter 9, Section 9.1.5.3.

V-ATPases (V₁V₀-ATPases). These H⁺-ATPases are primarily found in the membranes engulfing eukaryotic organelles (e.g., vacuoles and lysosomes). In the latter, V-ATPases function in acidifying the organelle by pumping protons into it. In addition, in segments of the kidneys they contribute to the net excretion of acid into urine. They have a complex structure, containing more than 10 subunits (Figure 7.20).

A-ATPases (A₁A₀-ATPases) are found in Archaea and function similarly to F-ATPases.

E-ATPases are cell-surface enzymes that hydrolyze a range of nucleoside triphosphates (NTPs), including extracellular ATP.

(a)

(b)

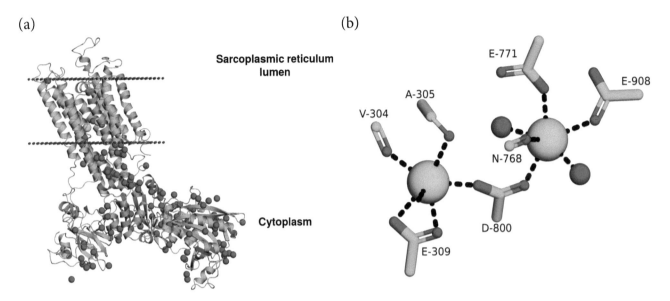

FIGURE 7.19 **Sarcoplasmic reticulum calcium ATPase (SERCA) bound to Ca^{2+} [224] (PDB entry 1su4).** (a) A global view of the protein and its position inside the membrane. The Ca^{2+} ion and water molecules are colored as in panel (b) Polar interactions between the two Ca^{2+} ions located in the transmembrane domain of the transporter and protein groups. The ionization states of the acidic residues were predicted by MolProbity [223].

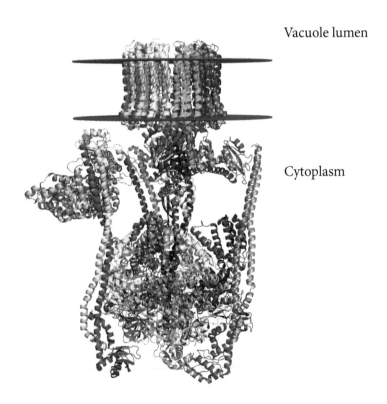

FIGURE 7.20 **A low-resolution electron microscopy structure of the complete V-ATPase pump from yeast [225] (PDB entry 3j9t).** The protein is colored by chain.

Like other transporters, ion pumps work by cycling among different conformations, which 'pick up' the solute on one side of the membrane and release it on the other side. Evidence of such a cycle can be seen in the three-dimensional structures of the Na^+/K^+-ATPase (Figure 7.18a) and the Ca^{2+}-ATPase (Figure 7.19a). Both structures demonstrate that in order to cross the membrane, the ions must 'hop' between amino acids inside the transporter, which serve as transient binding sites (see Figure 7.18b for the Na^+/K^+-ATPase). As Figures 7.18c and 7.19b show, the cation-binding sites in the Na^+/K^+-ATPase and the Ca^{2+}-ATPase include the negatively charged side chains of glutamate and aspartate residues, but also the partial charges of main chain or side chain carbonyl and hydroxyl groups, as well as individual molecules. **The involvement of partial charges that transiently bind desolvated ions is similar to what we have seen in the KcsA (potassium) channel. There too, the use of partial charges ensures that ions can move through the protein instead of getting stuck in one place.** The absence of a water-filled path in transporters may seem disadvantageous, but it is in fact one of the factors preventing the unwanted (simultaneous) exposure of the transported ion to both sides of the membrane, an event that would allow the ion to go back to the compartment from which it was taken.

We have seen earlier that in active ATP-dependent transporters (e.g., the ABC transporter Sav1866), the conformational changes that underlie the transport process are driven by ATP binding and hydrolysis. This is true for all ATPases, including the ion pumps described in the previous paragraphs. The mechanism through which ATPases translate these events into conformational changes involves the effects of ATP and of its hydrolysis products (ADP and P_i) on the transporter's energy landscape [219]. That is, the binding of ATP, ADP, or P_i to the transporter changes its energy, induces a conformational change, and brings it back to its energy minimum. Since ATP, ADP, and P_i have different effects on the energy of the transporter, the binding of each leads to a different conformation. The evolution of ATPases enabled them to cycle between conformations in accordance with the binding of ATP, ADP, and P_i, and this cycling results in ion transport. In P-ATPases such as the Na^+/K^+-, Ca^{2+}-, and H^+-ATPases, the conformational changes begin at the cytoplasmic domain with the movements of the A, N and P subdomains relative to each other, and propagate to the transmembrane domain through linkers and tertiary contacts. To illustrate the overall transport cycle we will look at the Na^+/K^+-ATPase, which transports three Na^+ ions from the cytoplasm to the cell's exterior and two K^+ ions in the reverse direction. The transport cycle includes the following steps (Figure 7.21):

- The ATP-bound form of the transporter, termed E1, has a high affinity to Na^+ ions, and its Na^+-binding site faces the cytoplasm. These properties result in the binding of three cytosolic Na^+ ions to the protein and occlusion of the binding site (Figure 7.21, step 1).

- The transporter has an inherent ATPase activity, and the hydrolysis of ATP is coupled to the transfer of the γ-phosphate to a conserved aspartate in the protein's P domain (Figure 7.21, step 2). This event, which is accompanied by ADP release, has two effects. First, it considerably reduces the affinity of the transporter to Na^+, and second, it promotes a conformational change that exposes the Na^+-binding site to the cytoplasm (the E2 state). These changes result in the release of the three Na^+ ions to the extracellular side of the membrane and binding of two K^+ ions, as well as protonation of the protein (Figure 7.21, steps 3 and 4).

- Binding of the K$^+$ ions promotes dephosphorylation of the protein's conserved aspartate and occlusion of the binding site (Figure 7.21, step 5), release of P$_i$ (Figure 7.21, step 6), and binding of new ATP (Figure 7.21, step 7).

- The ATP-bound protein reverts to the E1 form, resulting in the release of the bound K$^+$ ions to the cytoplasm (Figure 7.21, step 8).

In P-ATPases that transport only one type of ion (e.g., Ca^{2+}-ATPase), the transport cycle does not include the steps involving binding and release of the second ion type (steps 4 and 8 in Figure 7.21, respectively).

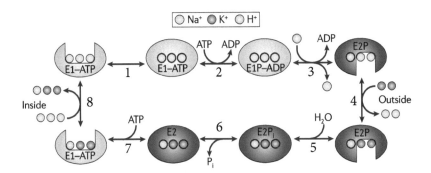

FIGURE 7.21 **A schematic representation of the transport process employed by the Na$^+$/K$^+$-ATPase.** Step 1 – occlusion of the Na$^+$-binding site. Step 2 – ATP hydrolysis and phosphorylation of the conserved aspartate in the protein. Step 3 – protonation of the transporter is accompanied by a conformational change of the protein from the E1 state to the low-affinity E2 state (binding site faces cell's exterior). The change also involves the release of ADP and one Na$^+$ atom. Step 4 – exchange of the two remaining Na$^+$ ions with two K$^+$ ions from the cell's exterior. Step 5 – dephosphorylation of the protein and occlusion of the binding site. Step 6 – release of P$_i$. Step 7 – ATP binding to the protein changes it back to the E1 form, in which the binding site is exposed to the cytoplasm and the affinity to Na$^+$ is high. Step 8 – deprotonation and release of the two K$^+$ ions from the protein to the cytoplasm, and binding of 3 new Na$^+$ ions. The image is taken from [219].

7.3.2.3.4 Structure prediction

We have seen how the strong sequence-related tendencies of transmembrane segments enable scientists to identify such segments in the amino acid sequence and to predict their topology with respect to the other parts of the protein. The two remaining steps for successful prediction of the entire structure relate to the orientations of helices and the conformation of the protein backbone (in cases in which the helices are distorted) and side chains. Thus, one way to classify structural prediction methods of membrane proteins is according to the properties they predict: secondary structure, topology, tertiary structure, etc. (see [128] for details). Another way of classifying structure prediction methods is by the information they use to carry out their predictions. According to the latter criterion, methods can be separated into three main groups [71,101]. The first group includes *ab initio* methods, which rely solely on physicochemical principles characteristic to membrane proteins, such as length, hydrophobicity, etc. PREDICT [226] is an example of such a method, one that was developed specifically for GPCRs. The greatest advantage of such methods is that they do not require any additional information regarding the protein, aside from the amino acid sequence. Their main disadvantage is their requirement of massive conformational sampling, which is computationally demanding [227]. The second group includes methods that rely on statistical tendencies of residues to appear in certain regions of the protein. One example is kPROT [228], a scale of the propensities of amino acids to face either the lipid bilayer or the protein core; this scale was developed on the basis of statistical data extracted from bitopic and polytopic proteins, respectively. Three other statistical propensities successfully used in structure prediction algorithms are evolutionary conservation [229,230], correlated mutations [231] (see Box 3.3), and tight packing against other residues [232]. The third group of structural prediction methods of membrane proteins includes methods that rely on sequence-based similarity between the query protein and membrane proteins whose structure is already known (*template-based methods*). When the query and template proteins are homologous (sequence identity > ~30%), the preferred method is *homology modeling* (see Chapter 3, Subsection 3.4.3). However, this method, which is highly successful in water-soluble proteins, is less efficient in membrane proteins because of the limited number of homologues with known structure that can be used as templates. In such cases, *fold recognition (threading)* methods may be used (see Chapter 3, Subsection 3.4.3.3). In such methods, the template is chosen not by the similarity of its sequence to that of the query protein, but rather by similarity of sequence properties. Some of these properties are extracted statistically from multiple sequence alignments, whereas others are physically meaningful, e.g., propensity to form a certain secondary structure, to be exposed to the surrounding medium, to have certain dihedral angles, to interact with certain amino acids, etc.

Current prediction algorithms are often hybrid, i.e., they are based on a scoring function that includes expressions adopted from different approaches [233] (see Chapter 3, Subsection 3.4.4). For example, some of the expressions may rely on physicochemical considerations, whereas others rely heavily on statistically derived tendencies of certain amino acids to form certain structures or to be involved in certain physical interactions. Two well-known methods, *Rosetta* [234,235]*1 and *I-TASSER* [237]*2, offer a unified modeling framework in which different approaches are used for different modeling challenges [233]. Indeed, both methods have been tested on different proteins and yielded good predictions [236,239,240] (see more details in [233]).

Finally, a relatively new approach has been adopted that uses experimental data in order to make predictions more efficient [118] (see Chapter 3, Section 3.5). These data are derived from experimental methods that provide low-resolution structures of membrane proteins; such methods include mainly cryo-EM, SAXS, and NMR, but also CD and FRET. The general idea is to use the extracted data as spatial constraints that narrow down the conformational space searched by the prediction methods, and thus significantly increase their chances of providing the native protein structure [241–244]. Cryo-EM looks especially promising in this sense; EM methods have traditionally been used to provide data concerning the number, tilt, and overall locations of transmembrane helices in query structures [233]. However, recent technological developments in single-particle cryo-EM have facilitated the production of near-atomic-resolution structures (4 or 5 Å), at least for some proteins. Indeed, in recent years, several software programs have been designed or adapted to integrate experimentally derived constraints into the structure prediction and modeling process [245]. These programs include *Rosetta* [246–250] (mentioned above), *CHESHIRE* [251], *CS23D* [252],

---

*1Rosetta has a version designed specifically for membrane proteins, called RosettaMembrane [234,236].

*2I-TASSER has a version designed specifically for GPCRs, called GPCR-I-TASSER [238] (server: http://zhanglab.ccmb.med.umich.edu/GPCR-I-TASSER/).

and *SAXTER* [253]. Biochemical methods also provide experimental data that can be used to guide prediction tools; such methods include (1) proteolytic cleavage of extramembrane regions of the proteins or their identification using antibodies, (2) point mutations of different residues of the protein (conserved residues often face the core [150]), and (3) chemical crosslinking. One major problem in the use of such methods for structure prediction is that they are carried out on samples containing large numbers of molecules. As a result, they might provide data corresponding to different substates of the protein, thereby complicating the prediction process [233].

7.3.3 Peripheral membrane proteins

Peripheral membrane proteins are attached to either the exoplasmic side or the cytoplasmic side of the membrane, with the bulk of their surface in the aqueous solution (extracellular matrix or cytoplasm, respectively). Thus, the principles determining their structures are similar to those of water-soluble proteins. Accordingly, our discussion of peripheral membrane proteins focuses on their attachment to the lipid bilayer. This attachment can take place through the following three mechanisms:

1. **Electrostatic binding (Figure 7.22a).** Certain membrane proteins contain a binding site that is geometrically suited for binding a certain membrane phospholipid. In many cases the latter is negatively charged (PS, PG, PIP_2), and the binding site contains basic residues that are capable of interacting with it favorably (e.g., in the pleckstrin homology domain, Figure 7.22b). Other proteins, such as MARCKS, contain a *patch* of basic amino acids, which renders the entire region positively charged. This charge allows the protein to adhere non-specifically to areas of the lipid bilayer containing microdomains of negatively charged phospholipids [254–256]; in eukaryotes, such areas are present on the cytoplasmic side of the membrane. In these cases, the protein is positioned about 3 Å away from the phospholipid head group, a distance that is electrostatically optimal. That is, the Coulomb attraction between the two charged entities at this distance is maximal, and over-compensates for the unfavorable Born repulsion, which is minimal since a water layer separates the protein and lipid membrane. A well-known example of the two forms of electrostatic binding described here is given in Subsection 7.4.1.2 below, which discusses the interaction between proteins and membrane PIP_2.

2. **Covalent binding (Figure 7.22a).** Some proteins undergo post-translational modifications that enable them to bind covalently to membranes. The modifications include N'-myristoylation, S-palmitoylation, and S-prenylation (see also Chapter 2, Section 2.6). Such binding can be observed in certain key signal transduction proteins such as ras and src, which bind to lipid chains from the cytoplasmic side of the membrane.

3. **Integrated binding of amphipathic helices (Figure 7.22c).** Some membrane proteins contain amphipathic helices that enable them to partially penetrate one of the bilayer leaflets. In this mode of protein-membrane binding, the nonpolar residues of the hydrophobic face of the helix interact with the hydrocarbon core of the bilayer, and the polar residues on the opposite face interact with the lipid head groups (e.g. [257,258]). This form of binding is also observed in peptides [259,260], e.g., antimicrobial peptides attaching to bacterial membranes.

The last form of association is particularly interesting. First, it is observed both in membrane proteins and in peptides. Second, it affects the structure of the lipid bilayer, as further discussed in Section 7.4 below. Interestingly, none of the three forms of binding seems to be sufficient on its own, as demonstrated in the case of the signal transduction protein MARCKS. This protein uses the first two forms of binding, and possibly the third as well, to remain attached to the membrane. Upon phosphorylation or binding to calmodulin [261], the electrostatic component of the binding is nullified [262], which leads to the release of the protein into the cytoplasm.

7.4 PROTEIN-MEMBRANE INTERACTION

7.4.1 Lipid bilayer effects on membrane proteins

The lipid bilayer is a chemically complex medium, and as such it has diverse effects on its resident proteins [30]. Still, these effects, which result from the physicochemical interactions between the lipids and proteins, can be separated into two types. The first type results from the physical properties of the bilayer as a bulk, which can be considered as a complex solvent. Indeed, studies show that membrane proteins are affected by general properties of the lipid bilayer, such as topology [70,263], degree of order [264], viscosity [265], hydrophobic thickness [266], curvature [61], degree of acyl chain packing [267], free volume [268], rigidity [269], and more. The second type of bilayer effect is mediated via specific interactions between the proteins and certain lipid molecules. The two types of effects have been studied in the last decades using different techniques. The main findings are reviewed in the following subsections.

7.4.1.1 Effects of general bilayer properties

7.4.1.1.1 Topology

Transmembrane segments of membrane proteins tend to acquire an ordered secondary structure (mainly α-helical) to avoid the unfavorable exposure of their backbone polar groups to the hydrophobic core of the lipid bilayer. A similar phenomenon also occurs when amphipathic peptides or protein segments bind to the bilayer interface. That is, the amphipathic peptides or segments fold and acquire an ordered helical structure [70,263]. The folding in this case, however, is driven in part by the need to preserve the amphipathic nature of the peptide or segment, which allows it to interact favorably with the lipid bilayer (see Subsection 7.3.3 above). Thus, the basic polar/nonpolar topology of the lipid bilayer can affect the structure of proteins inside it.

7.4.1.1.2 Degree of order and thickness

Membrane proteins differ in their preference for regions in the membrane of particular degrees of lipid order. Studies show that many integral proteins prefer l_d-phase regions, yet some integral proteins have been shown to prefer l_o-phase regions [270,271]. The different preferences lead to *lateral segregation* of proteins within the membrane, and to the formation of microdomains that have their own unique lipid and protein compositions. In some cases the preference results from direct protein-lipid interactions, as in the case of proteins that are covalently attached to the bilayer via a fatty acid or other types of hydrocarbon chains. In such cases, the preference has to do with the degree of saturation of the acyl chain. For

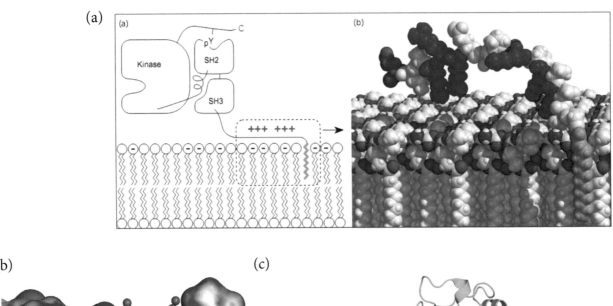

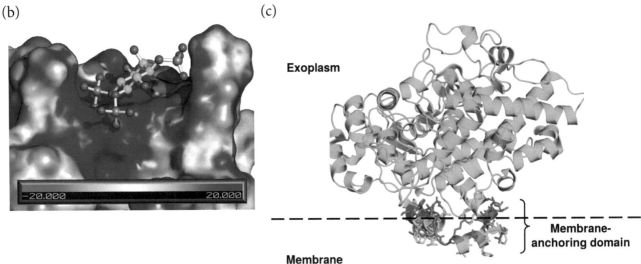

FIGURE 7.22 **Common interactions of peripheral proteins with biomembranes.** (a) Electrostatic adhesion and covalent binding to an acyl chain. The structure of c-Src is shown. The cartoon on the left shows the domain composition of the protein and suggests its orientation with respect to the membrane. The right image shows a blowup view of how the N' of the protein could interact electrostatically with the polar head groups of the lipid bilayer, and covalently with an acyl chain. Basic residues in c-Src are blue; acidic residues are red; and the acyl chain (myristate) bound to the protein is green. The membrane is represented here by a 2:1 PC:PS bilayer, with the acidic lipid, PS, identified by its exposed nitrogen, colored blue. The figures are taken from [254]. (b) Electrostatic and geometric compatibility of the pleckstrin homology (PH) domain binding site to inositol 1,4,5-trisphosphate. The image shows the structure of the PH domain of Arhgap9 (PDB entry 2p0d). The protein surface is colored according to electrostatic potential, in the range specified by the scale at the bottom of the figure (in k_BT/e units). Inositol 1,4,5-trisphosphate is shown as sticks and balls. The latter binds into a cleft on the surface of the protein, which is characterized by a strong positive potential, created by Arg and Lys residues. The positive potential of the protein cleft matches the negative potential of the ligand, which results from the three phosphate groups in the latter. (c) A schematic representation of the electrostatic and nonpolar interactions between the amphipathic helices of the enzyme cyclooxygenase-1 (PDB entry 1eqg) and the lipid bilayer. The membrane-anchoring region of the protein extends between positions 73 and 116 (shown as sticks). Polar residues are colored in orange, whereas nonpolar residues are colored in cyan. The dashed line marks the approximate location of the (exoplasmic) membrane core boundary.

example, proteins that are covalently attached to GPI (glycophosphatidylinositol) prefer ordered regions of the bilayer (l_o), because the GPI's acyl chain is saturated [272]. This is also true for src and other kinases of that family, which are myristoylated or palmitoylated [273]. Conversely, GTPases of the ras family, which are attached to the unsaturated prenyl chain, prefer disordered (l_d) regions of the bilayer [274].

The bilayer's degree of order also affects proteins indirectly, by influencing the membrane's thickness. Saturated acyl chains are more extended than unsaturated ones, and therefore tend to form more ordered bilayers. The proteins are affected not by the overall thickness of the bilayer but rather by the thickness of the hydrocarbon region. As described earlier, nonpolar residues in transmembrane segments usually extend over a length that roughly matches the hydrophobic thickness of the lipid bilayer. Still, there are cases where transmembrane segments *hydrophobically mismatch* [266,267] the bilayer core. As we will see in Section 7.4.2 below, such a mismatch is likely to affect the shape of the lipid bilayer. Still, the protein may also undergo some changes in order to minimize the mismatch. *Positive hydrophobic mismatch* occurs when the thickness of the bilayer's core is smaller than the length of the nonpolar stretch in the transmembrane segment (Figure 7.23a). This prevents the nonpolar residues of the transmembrane segment from interacting fully with the core of the bilayer. In order to overcome this problem, the transmembrane segment may tilt with respect to the bilayer's vertical axis (Figure 7.23c). The opposite situation, i.e., when the thickness of the bilayer's core is greater than the length of the nonpolar stretch in the transmembrane segment (*negative hydrophobic mismatch*, Figure 7.23b), is much less favorable energetically. The problem is the partitioning of polar amino acid residues of the protein to the nonpolar environment of the bilayer core, which, as we have seen earlier, may seriously destabilize the system. There are several ways to minimize negative mismatch or its effects [264]:

1. **Lateral diffusion (Figure 7.23d).** The protein may move along the plane of the lipid bilayer towards regions with lower degrees of order and smaller hydrophobic thickness. This creates microdomains in the lipid bilayer. Signal transduction is an example of a cellular process that is highly dependent on the presence of such microdomains [52,270]. Specifically, signal transduction requires a concentration of certain protein and lipid elements in one confined region of the membrane. An example for such a requirement is given by PIP_2-dependent signal transduction processes (see Subsection 7.4.1.2.3 below). Another cellular process affected by microdomains is the trafficking of proteins inside the cell [275]. That is, a protein sent to a certain cellular compartment must hydrophobically match the membrane of that compartment[*1].

2. **Conformational changes in the protein (Figure 7.23e).** Membrane proteins that are large enough may undergo conformational changes in order to reduce hydrophobic mismatch. Such changes usually include screw or slide motions of structural protein units, such as helices or domains. Though they solve the mismatch problem, such conformational changes might create new problems by reducing the activity of the protein [276–279]. Such a reduction in activity occurs, e.g., in the enzyme Ca^{2+}-ATPase within the *sarcoplasmic reticulum (SR)*[*2] membranes of muscle cells, when the thickness of the latter becomes different from that of the plasma membrane [30,280]. The

---

[*1]The membranes of the different organelles have different lipid compositions, and therefore different thicknesses.

[*2]The equivalent of the ER in muscle cells.

implications of this reduction are not necessarily negative, as **it provides a means of regulating the activity of the enzyme under different conditions that change the thickness of the SR membrane.**

3. **Oligomerization (Figure 7.23f).** When two separate transmembrane segments interact unfavorably with the lipid bilayer (e.g., due to hydrophobic mismatch), the system can reduce the number of unfavorable interactions by replacing some of them with favorable protein-protein interactions. That is, the two transmembrane segments associate. In some cases association may facilitate activation. This is observed in the antibiotic protein gramicidin [281]. The association in this case allows two short segments to dimerize into a transmembrane segment long enough to span the membrane and function as an ion channel [282,283]. **This may reflect an evolutionary mechanism using protein-membrane (hydrophobic) mismatches to activate certain proteins and peptides.**

All the above-mentioned membrane-induced effects may lead to changes in the activity of the protein. Indeed, studies have already proved this to be the case in many proteins, such as Na/K-ATPase [278,284], cytochrome c oxidase [280], Ca^{2+}-ATPase [30,280], melibiose permease [285], and diacylglycerol kinase [279].

7.4.1.1.3 *Viscosity*

The activity of virtually all proteins requires them to dynamically shift between different conformations (see Chapter 5 for details). Most globular proteins are surrounded by water molecules, which can adjust rapidly to any new conformation the protein acquires. Such adjustment makes it easy for the protein to undergo structural changes, although it does involve friction due to water-water interactions (van der Waals, hydrogen bonds). Membrane proteins, in contrast, are surrounded by lipids, which are less mobile than water, and limited in their capacity to reorganize in response to conformational changes. As a result, conformational changes of integral membrane proteins involve significant friction with neighboring lipids, particularly with the acyl chains [30]. On the one hand, such friction opposes the change, but on the other hand it may make the change 'smoother' by inhibiting post-change vibrations. The importance of having constant viscosity in biological membranes is revealed in prokaryotes, which are exposed to changing environmental conditions. In these organisms membrane viscosity is a *homeostatic* property, i.e., it is kept constant; this homeostasis is achieved through the lipid composition of the bilayer, which can change in response to changes in environmental conditions [286]. For example, a bacterium will respond to elevation of the external temperature by increasing the percentage of long saturated phospholipids, which oppose the heat-induced rise in membrane dynamics.

7.4.1.1.4 *Curvature*

The capacity of the lipid bilayer to acquire positive or negative curvature in specific regions has been found to affect the activity of integral proteins in these regions. Many of the studies investigating this issue have focused on processes that create negative curvature in the bilayer. This is because in extreme cases such processes may lead to loss of planarity of the bilayer, and create, e.g., an inverted hexagonal phase [59]. The results demonstrate a complex situation, in which enrichment of PE, a lipid known to induce negative curvature, increases the activity of some proteins (e.g., see [287]) while decreasing the activity of others (e.g., see [288]).

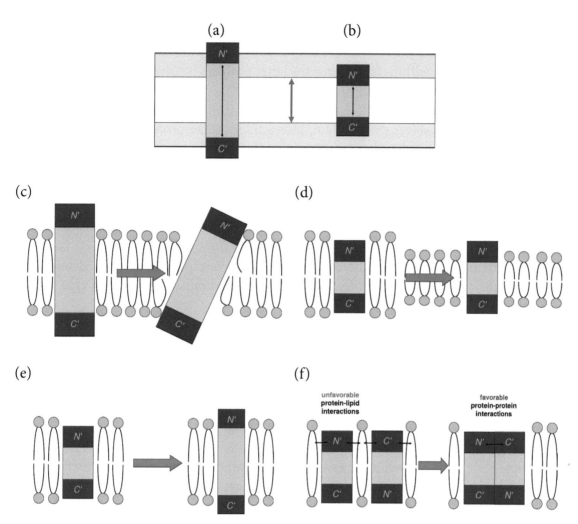

FIGURE 7.23 **Hydrophobic mismatch.** The figure illustrates (a) positive and (b) negative mismatches between the hydrophobic length of the transmembrane helix (or protein) and the thickness of the hydrocarbon region of the lipid bilayer. The transmembrane segments are depicted as rectangles, with their hydrophobic regions colored in grey, and their polar termini in purple. The hydrophobic lengths of the transmembrane segments are marked by the black arrows. The lipid bilayer is depicted as in Figure 7.11, with its hydrophobic thickness marked by the red arrow. (c) Tilting of the transmembrane segment reduces the positive mismatch. The transmembrane segments are depicted as in (a) and (b), and representative lipids of the bilayer are depicted schematically. (d) through (f) Adaptations of proteins to negative mismatch: (d) Lateral diffusion to a thinner area of the lipid bilayer; (e) Conformational changes that exclude the polar termini from the nonpolar bilayer core; (f) Oligomerization, which replaces unfavorable interactions between the polar termini of the transmembrane segments and the nonpolar core of the bilayer with favorable interactions between the polar termini of the two transmembrane segments. (The interactions are favorable because the two segments are positioned in an anti-parallel topology, allowing their partially positive N' to interact with their partially negative C').

7.4.1.1.5 Mechanical pressure

In prokaryotes exposed to environmental changes, a sudden drop in the solute concentration of the external environment leads to massive entry of water into the cell. This is a dangerous situation, because aside from the resulting drop in intracellular solute concentration, the stretching of the plasma membrane may lead to its rupture due to the immense pressure applied to it from within. In bacteria, the latter problem is solved by certain membrane proteins that function as mechanosensitive sensors [289]. Upon stretching the membrane, these proteins allow a rare event to happen: the massive efflux of cellular solutes. This leads to water efflux, which solves the problem. Interestingly, these proteins have been found to respond to membrane stretching only when the increase in membrane surface area reaches 4%, which happens to be the threshold for membrane rupture[*1] [289].

7.4.1.2 Effects of specific bilayer lipids

7.4.1.2.1 General lipid types

Biological membranes include different kinds of lipids, as explained above in detail. However, in terms of their interactions with proteins, these lipids, regardless of type, can be divided into three basic categories [290]:

1. **Bulk lipids**, include all lipids that are not engaged in specific contacts with proteins, and whose diffusion is therefore determined by their interaction with neighboring lipids.

2. **Annular lipids**, include all lipids that form a contact layer around a protein but still move constantly between this layer and the bulk [291]. EPR measurements show this movement to be about one order of magnitude slower than the diffusion rate of bulk lipids [292]. This contact layer assists in positioning integral proteins vertically in the lipid bilayer, and seals the protein-lipid interface. Annular lipids are not confined to the periphery of the protein and can also be found in large spaces within multimeric proteins, as can be seen clearly in the structure of the V-Type Na^+-ATPase [293].

3. **Bound lipids**, include lipid molecules that interact strongly with the protein. The interaction inhibits the motions of these lipids considerably, allowing them to co-crystallize with the protein (Figure 7.24). They can therefore be seen in structures obtained by X-ray diffraction. Bound lipids may be found in clefts on the protein surface (usually located in inter-subunit interfaces), or buried within the protein [290]. Lipids that are deeply buried within a protein or complex are also called 'integral'. The lipid-binding site often contains evolutionarily conserved residues [294] (for example, the cholesterol-binding motif in G-proteins; see Section 7.5 below). The bound lipid molecule tends to acquire a conformation that provides the best possible interaction with the protein, and this often leads to distortion of the lipid molecule. Such distortion is known to happen even when the lipid is saturated, and may result in its translation inward, towards the bilayer core. The lipid distortion may even lead to situations in which the lipid polar head group is positioned below the normal location of the phosphoester groups of the bilayer, or its acyl chains curve and wrap around α-helices of the protein [295].

[*1]The low tolerance of the membrane to stretching probably results from the exposure of its nonpolar core to the aqueous solvent [30].

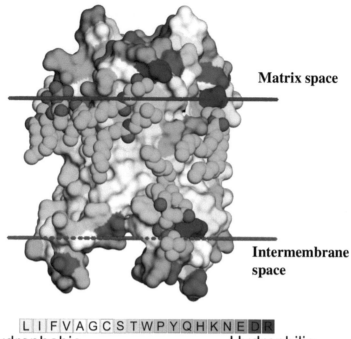

Matrix space

Intermembrane space

L I F V A G C S T W P Y Q H K N E D R
Hydrophobic   Hydrophilic

FIGURE 7.24 **Protein-bound lipids.** The structure of the mitochondrial ADP/ATP carrier complex with bound lipids is shown (PDB entry 1okc). The surface of the protein is colored by hydrophobicity using the Kessel-Ben-Tal Scale [8] (see Figure 2.7 for details), and the annular lipids are depicted as atom type-colored spheres. As the image clearly shows, some of the lipids are distorted significantly, which prevents the highly unfavorable interaction between the nonpolar tails and the polar regions of the protein. The red and blue lines mark the predicted boundaries of the membrane (the OPM database [117]).

7.4.1.2.2 Phospholipid-protein interaction

Specific protein-lipid interactions are mostly noncovalent, and include electrostatic interactions involving the lipid's polar head group [30], in addition to nonpolar and van der Waals interactions involving the acyl chains [290]. As mentioned earlier, some of the most favorable interactions involving lipid head groups are between basic protein residues and acidic head groups on the electronegative side of the membrane (the mitochondrial matrix, the stroma of chloroplasts, or the cytoplasmic side of the plasma membrane) [295,296]. While interactions involving the alcoholic groups of lipids are diverse [295], those that involve the phosphodiester group, common to all phospholipids, are mediated by several two-residue combinations of basic and polar-neutral residues [290]:

KT, KW, KY, RS, RW, RY, RN, HS, HW, HY

The primary interaction with the phosphodiester group involves the following residues, ranked by the number of occurrences:

Arg > Lys > Tyr > His > Trp, Ser, Asn

Other favorable interactions involve Thr and Gln. The interactions at the other, less electronegative side of the bilayer also involve recurring residues, most of which are polar-neutral, and very few of which are basic. In the case of PC, the main phospholipid on that side of the membrane, the interactions do not involve Lys and Arg at all, probably because of the presence of the positively charged trimethylamino group. They do, however, involve a combination of His and Ser, or His, Ser, and Thr, separately. Another special case is that of cardiolipin (CL), which contains two phosphodiester groups. These interact with a three-residue motif, in which the first two are basic and the third is polar-neutral.

As mentioned above, the aromatic residues Tyr and Trp in transmembrane segments tend to appear near the polar-nonpolar interface, where they function as anchors preventing the segment from sliding out or in (see Subsection 7.3.2.1.3 above). Much of the anchoring effect results from complex interactions between these residues and adjacent phospholipids (Figure 7.25). These interactions rely on two properties of Tyr/Trp [297]:

1. **A ring structure.** This allows the residue to have significant van der Waals and non-polar interactions with the acyl chains of the phospholipid. The planarity of the ring makes these interactions geometry-dependent.

2. **Polarity.** This property results both from the chemical composition of the residue (OH group in Tyr and NH in Trp) and from its aromaticity (i.e., the delocalized π electrons). It allows the residue to interact electrostatically with polar head groups of adjacent lipids. Due to the location of the aromatic residue, the interaction primarily includes hydrogen bonds with phospholipid carbonyl groups, although a recent NMR study suggests that these are not important for membrane anchoring [298].

7.4.1.2.3 Effects on membrane proteins

Specific lipid molecules within the bilayer may affect the stability, folding, assembly, and activity of integral membrane proteins [290]. Evolutionary 'forces' made these effects beneficial in most cases, and **some proteins are already known to be active only when surrounded with certain lipids** [299]. For example, the activity of the metabolic proteins NADH dehydrogenase, ADP/ATP carriers, cytochrome c oxidase, ATP synthase, and cytochrome bc1 depends on cardiolipin [300–302], which is abundant in the inner mitochondrial membrane. In such cases, the lipid molecule is considered to be a cofactor. In some cases the three-dimensional structure of the protein and bound lipid sheds light on the molecular basis for the functional dependency. This is the case with the light-harvesting complex of photosystem II (LHC-II) in plants. The LHC is a trimer, whose formation (and therefore activity) depends on PG [303]. Close inspection of the structure of LHC shows that the phospholipid molecule is located at the subunit interface, where one of its acyl chains is positioned inside the trimer [304,305] (Figure 7.26). Thus, PG assists in stabilizing the oligomeric structure of LHC. Furthermore, PG also interacts with other lipids in the vicinity of LHC, such as chlorophyll and carotenoids, which help stabilize the loosely packed and marginally hydrophobic α-helices in the LHC structure [306]. Similarly, in cytochrome c oxidase, two cardiolipin molecules that face the mitochondrial intermembrane side of the protein seem to stabilize the dimeric structure of the protein, whereas two other cardiolipin molecules that face the matrix side seem to function as proton traps, thus facilitating proton translocation along the protein's surface [307]. Other examples of lipid molecules that can be attributed specific functional roles are given in [68,294]. Nevertheless, it is not always easy to deduce the molecular basis for the lipid dependency of the protein from its three-dimensional structure. This is the case with the tetrameric KcsA channel, the activity of which depends on PG. The

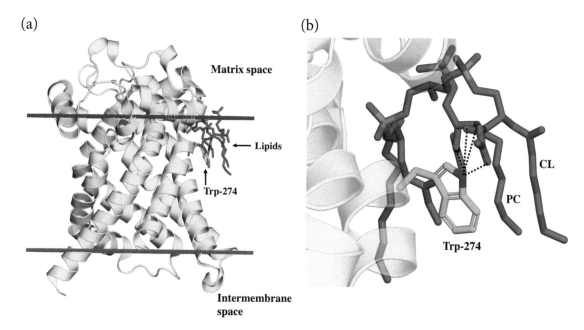

FIGURE 7.25 **Interaction between aromatic residues and lipids in biological membranes.** (a) The structure of the mitochondrial ADP/ATP carrier complex, colored in yellow (PDB entry 1okc). Trp-274 is shown as green sticks, and the adjacent lipids are shown as blue sticks. The red and blue lines mark the predicted boundaries of the membrane (the OPM database [117]). (b) A blowup of the region containing Trp-274, showing the interactions between this residue and two lipids, cardiolipin (CL) and phosphatidylcholine (PC). As explained in the main text, the large ring structure of Trp enables van der Waals and nonpolar interactions to take place between the residue and the nonpolar acyl chains of the lipids, whereas the polar NH group interacts electrostatically with the lipids' ester oxygen atoms and phosphate groups (black dashed lines). For clarity the orientation of the protein is different from that in (a).

phospholipid molecule binds to two Arg residues; each is contributed by a different subunit. This suggests that the role of PG is to help stabilize the quaternary structure of the protein. However, other studies show that KcsA does not require PG specifically, but settles with any negatively charged phospholipid [308]. Thus, it seems that PG's role in this system is to reduce the electrostatic repulsion between positive charges in the subunit interface, rather than to form a specific interaction.

One of the most extensively studied examples of specific protein-lipid interactions is that of PIP_2, a phospholipid that appears in minute quantities in the plasma membrane, almost exclusively on its cytoplasmic side [262,309]. Despite its rarity, PIP_2 is involved in important cellular processes, including endocytosis, exocytosis, phagocytosis, and vesicle transport within the cell. The interest in this molecule began when it was found to be a substrate for the cytoplasmic enzyme *phospholipase C (PLC$_\gamma$)*. PLC$_\gamma$ splits PIP_2 into *inositol 1,4,5-trisphosphate (IP$_3$)* and *diacylglycerol (DAG)*, two prominent second messengers in signal transduction pathways. Their combined action leads to the activation of the enzyme *protein kinase C (PKC)*, which phosphorylates numerous targets within the cell and induces significant biological responses. Today we know that PIP_2 is in fact the source for three second messengers, and that its influence on the aforementioned processes is carried out mainly through them [310]. However, it seems that PIP_2 may also directly affect membrane proteins, primarily ion channels [309,311] and transporters [312,313]. What advantage does the PIP_2 dependency confer to these proteins? At least two come to mind, both related to regula-

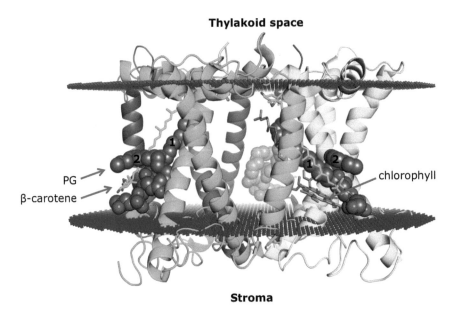

Thylakoid space

PG

β-carotene

chlorophyll

Stroma

FIGURE 7.26 **Interactions between the light-harvesting complex of photosystem II (LHC-II) and phosphatidylglycerol (PG) in plants.** The backbone structure of the trimeric LHC is shown, with each subunit colored differently. The red and blue lines mark the predicted boundaries of the membrane (the OPM database [117]). The space between any two adjacent monomers contains a PG molecule (blue spheres) with one of its fatty acid chains penetrating into the trimer (1) and the other (2) pointing towards the lipid bilayer. Each PG molecule interacts with protein residues, as well as with a β-carotene pigment molecule (green sticks) that runs parallel to one of PG's chains, and with a chlorophyll molecule (orange sticks) that lies below this chain.

tion [309]. The first is simple; the dependency on PIP_2 ensures that these proteins are inactive unless they are attached to the membrane. Indeed, there are many cases in which the plasma membrane acts as a 'meeting place' for signaling proteins. This makes it easier for the different components of the pathway to interact with each other and propagate the signal. The second advantage of the PIP_2 dependency of proteins is indirect; since the levels of PIP_2 are affected by external signals (via the activation of PLC_γ), the dependency of certain proteins on this phospholipid subjects them as well to the same external signals.

Proteins bind PIP_2 in two main forms:

1. **Specifically.** This is carried out via a geometrically and chemically compatible binding site that recognizes PIP_2 using basic and other residues. There are several known binding sites of this kind; the best known is the *pleckstrin homology (PH) domain* [315]. This domain contains over 100 residues and has already been found to be present in about 250 human proteins. It is composed of a seven-strand and one α-helix β-sandwich (Figure 7.27). The binding to PIP_2 is carried out using basic residues that form salt bridges with the lipid's phosphate groups, and also via hydrogen bonds to other forming residues [316]. Different PH domains bind different phosphoinositides ($PI(3,4)P_2$, $PI(3,4,5)P_3$), where the spatial arrangement of the residues in the binding site determines the specificity. Other PIP_2 binding sites can be found in other domains, such as *FYVE, PX,* and *ENTH* [315].

2. **Non-specifically.** This is carried out via disordered protein segments, which adhere electrostatically and non-specifically to clusters of negatively charged PIP_2 in certain

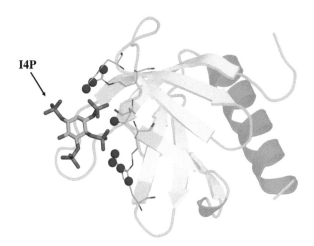

FIGURE 7.27 **Specific binding by the PH domain.** The structure of the PH domain of DAPP1 bound to inositol 1,3,4,5-tetraphosphate (I4P) is shown (PDB entry 1fao). The protein is shown using a ribbon representation. The Arg and Lys residues, interacting electrostatically with I4P, are shown as lines, with the N_ζ of Lys and $N_{\eta 1}$, $N_{\eta 2}$ of Arg shown as small spheres. For clarity, the backbone is rendered partially transparent. The basic residues in the figure are positioned such that they interact optimally and specifically with the phosphate groups of the ligand.

regions of the membrane. The adhesion is made possible by a number of basic residues at proximal positions within the sequence, which provide their corresponding segments with a general positive charge. The proximity of the basic residues to each other is also the cause for the disordered nature of such segments. One frequently used example of this type of binding is the aforementioned MARCKS, a 331-residue protein that is overall acidic, except for a 24-residue stretch, in which 13 residues are basic [261,317]. The positively charged part of this segment allows it to bind PIP_2 molecules and cluster them into a microdomain (Figure 7.28a). Moreover, the binding of MARCKS competes with PIP_2 interactions with other basic cytoplasmic proteins, and it seems that this is used as a regulatory mechanism of PIP_2-dependent signal transduction processes. Indeed, the IP_3-mediated rise in cytoplasmic Ca^{2+} levels (see above) leads to the binding of calmodulin (Ca^{2+}/CaM) to MARCKS, turning the positive electrostatic potential of MARCKS negative, thus inducing the departure of MARCKS from the PIP_2-rich region of the membrane [262] (Figure 7.28b–d). This makes PIP_2 available to other cytoplasmic signaling proteins, thereby allowing the signal to propagate. It should be mentioned that the electric field on MARCKS is also changed by PKC-induced phosphorylation of the protein. As explained above, PKC is activated by the same signal pathway that activates Ca^{2+}/CaM.

The two forms of PIP_2 binding described above differ both in binding specificity and in the structural requirements of the binding site. Although both forms exist, it seems that specific binding is more prominent. A mechanistic characterization of PIP_2 dependency is often not straightforward, although some studies have yielded interesting insights. For example, it has been suggested (based on models) that PIP_2 contributes to the gating mechanism of tetrameric K^+ channel-like proteins, by applying an electrostatic force to parts of the protein, which, as a result, shift and open the channel [309]. Other models that have been suggested are described in [311].

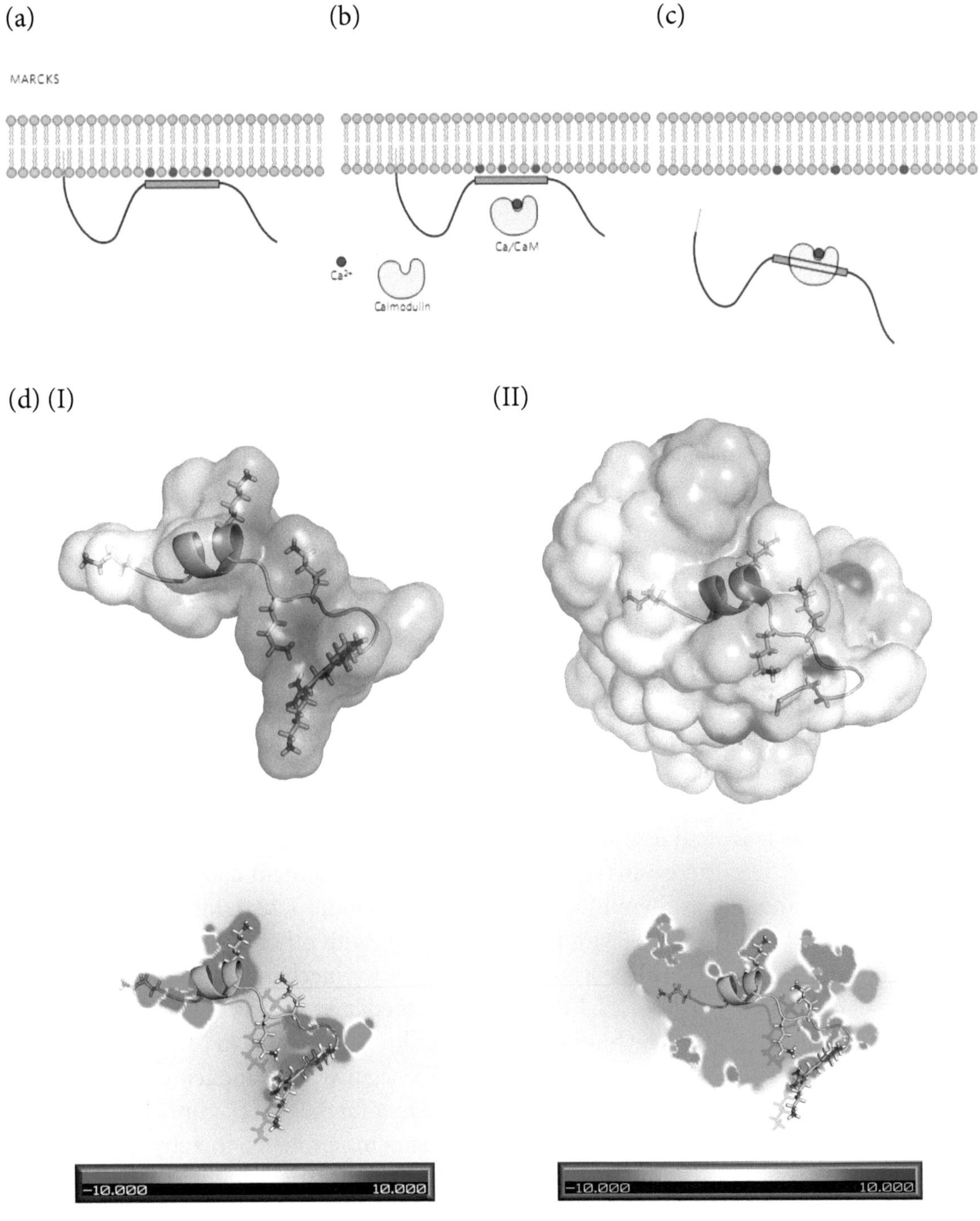

(a)   (b)   (c)

(d) (I)   (II)

FIGURE 7.28 **Modulation of PIP$_2$ availability during signal transduction by MARCKS and Ca^{2+}/CaM.** (a) through (c) A schematic depiction. (a) MARCKS binding to PIP$_2$ lipid (red) via electrostatic interactions, during the resting state. (b) Ca^{2+} binding to CaM following the right signal, and the migration of the complex to the lipid bilayer. (c) Binding of Ca^{2+}/CaM to MARCKS induces detachment of MARCKS from the membrane. The figure was taken from [262]. (d) Electrostatic potential mapped on the surface (top) and on a two-dimensional slice (bottom) of (I) the MARCKS-derived 19-residue basic segment, known to bind to both PIP$_2$ and calmodulin (CaM), and (II) the complex between the same segment and CaM (PDB entry 1iwq). Negative potentials ($0k_BT/e > \Phi > -10k_BT/e$) are red; positive potentials ($0k_BT/e < \Phi < 10k_BT/e$) are blue; and neutral potentials are white (see color code at the bottom). The electrostatic potential was calculated using APBS [314]. As can clearly be seen, the free MARCKS segment has a strong positive potential, which is reversed upon binding to CaM. As explained in the main text, a similar effect on MARCKS' potential is achieved by PKC-induced phosphorylation of the segment (not shown).

7.4.2 Effects of membrane proteins on lipid bilayer properties

Hydrophobic inclusions such as proteins inside a membrane perturb the lipid order. The perturbation works at several levels, and may lead to different reactions of the bilayer's lipids. In the following subsections we summarize the main effects.

7.4.2.1 Decrease in mobility

Membrane lipids are dynamic, with motions that range from limited vibrations to large-scale movements such as lateral diffusion or flipping between bilayer leaflets. The mere presence of a hydrophobic inclusion in the lipid bilayer reduces the dynamics of lipids bordering the rigid inclusion, and this translates to loss of entropy. Statistical-thermodynamic models show that the insertion of even a single helix into the lipid bilayer reduces the entropy, with a corresponding free energy penalty of +2 kcal/mol [267,318].

7.4.2.2 Deformation and curvature changes

We have seen earlier that, because transmembrane segments vary in length, hydrophobic mismatch may arise between the length of a given transmembrane segment and the thickness of the hydrocarbon region of the lipid bilayer. Such mismatch may prompt changes not only in the structure and/or orientation of the transmembrane segment (see Subsection 7.4.1.1.2 above) but also in the membrane lipids. The primary response of the lipid bilayer to such mismatch is deformation, that is, stretching or compression of the acyl chains around the transmembrane segment, in order to compensate for negative or positive mismatch, respectively [259,266,319] (Figure 7.29a,c). The deformation leads to a local change in the curvature of the lipid bilayer, at the protein-lipid interface. This is made possible by the soft nature of the lipid bilayer, whose compressibility is 10^9 to 10^8 N/m^2 [320,321]. Recent measurements have shown that the effect of protein-induced deformation on membrane thickness is five times more significant than the effect of cholesterol, which was considered for a long time to be the prominent factor determining membrane thickness in higher eukaryotes [57]. Lipid-induced changes in curvature seem to be involved primarily in concentrating certain signaling proteins within the same membrane region [58]. The deformation of the membrane involves an energy cost, which has been assessed at 0.4 kcal/mol for a 4-Å reduction in membrane thickness [259] (see also [322] for other estimates). Interestingly, computational studies suggest that even when the length of the transmembrane segment is equal to or shorter than the hydrophobic thickness of the membrane, the segment may still tilt at least 10° from the membrane's normal, to increase its *entropy of procession* [323–325] (Figure 7.29b). This requires the membrane to deform inwardly, but the deformation penalty is compensated for by the entropy gain. In fact, the deformation penalty is associated with another form of entropy, that of the lipid chains. This is yet another example of two entropy-based terms that balance each other, in this case in the determination of the optimal tilt angle of the transmembrane segment in the lipid bilayer.

Protein shape has also been found to affect membrane curvature. That is, integral proteins with asymmetric profiles, i.e., proteins whose extracellular regions are smaller or wider than the intracellular regions, create either a positive or negative curvature (depending on the location of the wide region), especially when they oligomerize or aggregate [326] (Figure 7.30a).

(a) Positive mismatch   (b) Perfect match   (c) Negative mismatch

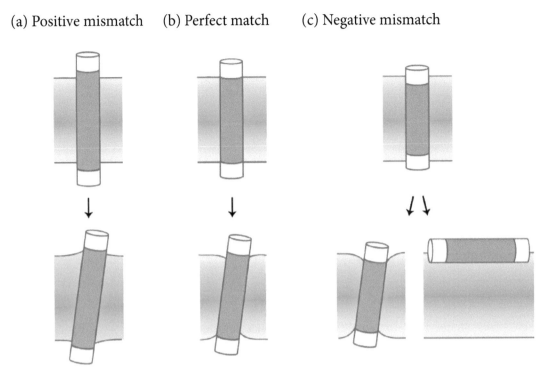

FIGURE 7.29 **Membrane deformation resulting from hydrophobic mismatch between the protein and lipid bilayer.** (a) Positive hydrophobic mismatch. (b) Perfect match. (c) Negative hydrophobic mismatch. The helix is represented as a cylinder, with the hydrophobic core in purple and the hydrophilic termini in white. (a) At positive mismatch, the transmembrane helix tilts, and the membrane expands to match the helix hydrophobic core. (b) At perfect match, the helix tilts because of the favorable increase in precession entropy, and the membrane thins so that the polar helix termini can remain in the lipid head group region rather than partition into the hydrocarbon region of the membrane. (c) At slight negative mismatch (lower left panel), the transmembrane helix tilts and the membrane thins locally as in perfect match. In cases of excessive mismatch, the helix adopts a surface orientation instead of forcing the membrane to thin beyond its elastic limit (lower right panel). The image is taken from [325] (http://pubs.acs.org/doi/full/10.1021/ct300128x).

In addition to these general properties of proteins, some specific cases are known in which the behavior of certain proteins has marked effects on membrane curvature:

1. **Actin polymerization.** The ability of the cytoskeletal protein actin to polymerize in response to certain signals is directly linked to changes in the plasma membrane, including curvature changes. Specifically, the polymerization creates mechanical pressure on the membrane [327], to the point of inducing curvature. This effect is important for several cellular processes, such as the formation of pseudopodia, phagocytic cups, endocytic invaginations, and even axonal growth cones (formed during the creation of the neural synapse) [328,329].

2. **Vesicle formation by coat proteins.** The formation of transport vesicles inside cells is carried out by *coat proteins* such as *clathrin, caveolin, COPI* and *COPII*, which are bound to the membrane peripherally [330–332]. Their activity leads to the application of mechanical pressure to the bilayer, and this pressure gradually increases the bilayer's curvature until the transport vesicle is formed. It was once assumed that the polymerization of these proteins was responsible for creating pressure on the membrane [333]. However, recent data have shown that clathrin, COPI and COPII do not form direct

(a)

(b)

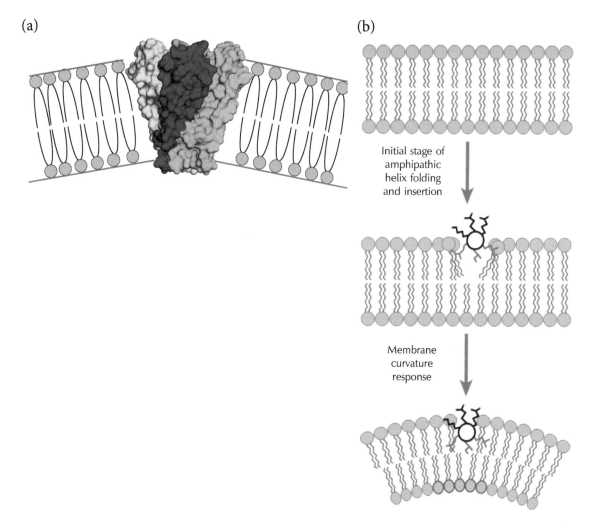

Initial stage of
amphipathic
helix folding
and insertion

Membrane
curvature
response

FIGURE 7.30 **Membrane deformation resulting from (a) protein shape and (b) mode of binding.** The protein in (a) is the KcsA voltage-gated potassium channel (PDB entry 1j95). With its teepee-like shape it has an uneven profile along the membrane normal, and is thus expected to cause membrane deformation. Each of the four chains of the channel is in a different color. (b) A model illustrating the bending effect of an epsin-like amphipathic helix, which partially immerses inside the bilayer and may ultimately induce membrane curvature. The figure was taken from [58].

contacts with the membrane, but rather use other proteins (e.g., epsins), which create the mechanical pressure [334]. The latter usually contain amphipathic α-helices that are partially immersed inside the bilayer with their nonpolar side facing the hydrophobic core and their polar side stuck like a wedge between the lipid head groups. The lipids in contact with such α-helices change conformation to compensate for the membrane distortion, and that creates positive curvature in the bilayer [58]. What, then, is the role of clathrin, COPI and COPII? It seems that these proteins are responsible for concentrating the pressure-creating proteins in a particular region of the membrane, and after the vesicle has formed they polymerize to form a scaffold around it.

Formation of transport vesicles is an example of a dramatic, yet physiologically relevant effect of proteins on membrane curvature.

7.5   G PROTEIN-COUPLED RECEPTORS

Despite the limited number of experimentally determined structures of membrane proteins, the extensive research carried out has yielded numerous insights regarding the structural and sequence-related requirements for the function of these proteins. In this subsection we discuss structure-function relationships in G protein-coupled receptors (GPCRs), which play a central role in many physiological processes and related disease, and whose action involves structural complexity, such as changing conformation and binding multiple partners.

7.5.1   Introduction

Single-cell organisms and cells in multicellular organisms communicate with their environments via highly complex signal transduction systems (see Chapter 1, Subsection 1.1.3.5). Such systems contain multiple components, starting from the cell surface and ending in intracellular proteins and small molecules (Figure 7.31a). The first cellular component responding to the incoming message is the membrane-bound receptor. Cells contain a multitude of receptors that respond to different types of messengers. These include external messengers, absorbed from the organism's environment (odorants, pheromones, tastes, etc.), and internal ones (hormones, neurotransmitters, local modulators). When activated, cell-surface receptors can relay the signal into the cell, to a diverse set of enzymes and small molecules or elemental ions. Most of these species function as *transducers-amplifiers*, since they pass on the message while amplifying it by acting on multiple cellular targets. Others act as *end effectors*, i.e., proteins whose activation (or inhibition) leads to an end result. This result may range from relatively small cellular responses, such as the production and/or release of a chemical compound, to more dramatic responses, such as cellular division and even suicide.

Membrane-bound receptors can be classified according to their types of responses to ligand binding:

1. Ion channels

2. Tyrosine kinases

3. Serine and threonine kinases

4. Guanylate cyclases

5. Cytokine receptors (defined by ligand type)

6. G protein-coupled receptors (GPCRs)

GPCRs are by far the largest and most common family of membrane receptors. They are widely represented in most life forms[*1]; in vertebrates they constitute 1% to 5% of the entire genome [336–338], and in the human genome they are encoded by more than 800 genes [339].

[*1]GPCRs seem to be missing in plants, although this matter is controversial. G-proteins do exist in plants, but it has been claimed that they are activated by receptor-like kinases (RLKs) rather than by GPCRs [335].

Another impressive trait of GPCRs is their ability to respond to a huge variety of external messengers, including proteins, peptides, small organic molecules, elemental ions, and even photons of light. These messengers may function as hormones, neurotransmitters, local mediators, pheromones, or environmental factors. Accordingly, GPCRs participate in numerous physiological processes [340], and are involved in many diseases and pathological syndromes, in which they are either inactive or overactive[*1]. These diseases include hypertension, congestive heart failure, stroke, cancer, thyroid dysfunction, congenital bowel obstruction, abnormal bone development, night blindness, and neonatal hyperparathyroidism [343]. Clearly, GPCRs are promising drug targets, and indeed, it is estimated that 30% to 50% of clinically prescribed drugs act by binding to GPCRs and changing their activity [343–346].

7.5.2   GPCR signaling

7.5.2.1   General view

As their name implies, GPCRs relay signals into cells primarily via large GDP/GTP-binding proteins, called *G-proteins*. Once activated, G-proteins may activate different effector proteins in a process resembling a cascade. That is, each molecule activates a number of effector proteins, and the number of activated proteins grows as the signal advances downstream of the pathway. Thus, the end result of GPCR signaling usually involves either the activation or inhibition of a large number of effector proteins, which include enzymes, ion channels, proteins associated with transport vesicles, and others. **The types of proteins activated in a given GPCR pathway depend on the messenger molecule, GPCR, and G-protein that are activated**. For example, in the *cAMP-PKA pathway* (Figure 7.31b), signaling via certain GPCR and G-protein types leads to activation of *adenylyl cyclase* (AC) and the production of the second messenger *cAMP*. The latter activates *protein kinase A* (PKA), which in turn phosphorylates numerous cytoplasmic proteins. The phosphorylation activates some of the proteins, while inactivating others. In another universal signaling pathway, the G-protein activates *phospholipase C* (PLC) instead of adenylyl cyclase. PLC hydrolyzes the membrane lipid *phosphatidylinositol 4,5-bisphosphate* to two second messengers, *diacylglycerol* (DAG) and *inositol 1,4,5-trisphosphate* (IP$_3$). The combined action of both messengers leads, through a massive, yet short-lived Ca^{2+} influx into the cytoplasm, to the activation of *protein kinase C* (PKC), which phosphorylates different proteins than does its PKA counterpart, thus leading to different outcomes.

Because it involves multiple components, GPCR signaling is inherently complex. This complexity is compounded by the following properties:

1. A single GPCR may activate different G-proteins, and even certain non-G proteins [349]. For example, the β_2-adrenergic receptor is known to activate the *MAP kinase* pathway.

2. Most GPCRs tend to have some degree of baseline activity [350]. That is, they are active to some extent even when not binding their activating ligands (*agonists*).

[*1]For example, overactive GPCRs may affect the formation and spreading of tumors by trans-activating cancer-related receptors such as the epidermal growth factor receptor (EGFR) [341], and by promoting cell migration during metastasis [342].

(a)

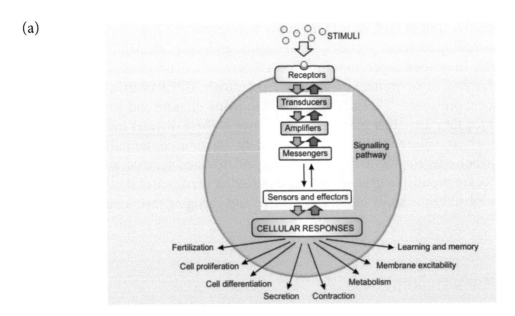

(b)

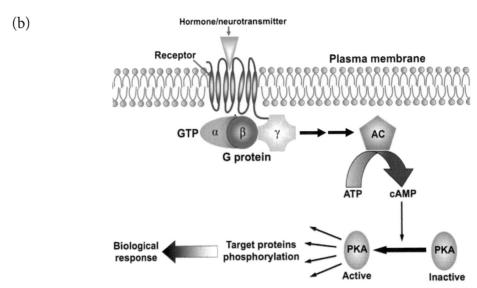

FIGURE 7.31 **The main players in a typical signal transduction cascade.** (a) A general scheme showing the principal components in a signal transduction cascade. Stimuli (e.g., hormones, neurotransmitters or growth factors) act on cell-surface receptors, which activate transducers to relay the signal into the cell. The transducers use amplifiers to generate internal messengers, which either act locally or diffuse throughout the cell. These messengers then engage sensors that are coupled to the effectors responsible for activating cellular responses. Note that the order of some of the components may differ somewhat across different signaling pathways. For example, messengers may be used to activate amplifiers instead of being produced by them (see panel (b)). The green arrows indicate ON mechanisms, which enable information to flow down the pathway, and the red arrows indicate opposing OFF mechanisms, which switch off the different steps of the signaling pathway. Virtually all of the components mentioned above may be proteins. The image is taken from [347]. (b) The cAMP-PKA cascade. Binding of an external chemical messenger (hormone, neurotransmitter, etc.) to a membrane-bound protein receptor induces the activation of an enzyme called a G-protein, which acts as a transducer. The activation of the G-protein leads to the activation of adenylyl cyclase (AC), which catalyzes the conversion of ATP into cyclic AMP. The latter acts as an intracellular messenger. It binds to and activates the enzyme amplifier PKA, which, in turn, phosphorylates a large set of cytoplasmic proteins. The phosphorylated proteins may activate other cellular components, or perform a certain function (that is, they may act as sensors and/or effectors). In any case, this signal transduction eventually leads to changes in the cell's behavior, i.e., to a biological response.

3. A single GPCR may respond to different types of ligands, each eliciting a different outcome [351] (Figure 7.32):

- *Full agonists* induce maximal receptor activity by stabilizing an active conformation.

- *Inverse agonists* decrease the baseline (constitutive) activity of the receptor by stabilizing an inactive conformation.

- *Partial agonists* induce partial activity since they have some affinity to both active and inactive conformations.

- *Antagonists* prevent other ligands from binding to the GPCR and activating it.

These observations are in line with two currently accepted models of protein dynamics [350]. The first, referred to as the *'pre-existing equilibrium' model* [352–362] (see Chapter 5), stipulates that even in the absence of a bound agonist, the active conformation of the protein is sampled sufficiently to yield some degree of baseline activity [363]. The second is the so-called *'conformational selection' model* [364–370] (see Chapter 8), which stipulates that each ligand binds and stabilizes a different conformation of the GPCR, which has a different intrinsic activity. Since a single GPCR can modulate different pathways, the same ligand may have opposing effects on two different pathways that are modulated by the same GPCR, by stabilizing a conformation of the GPCR that is compatible with only one of the pathways [371]. This phenomenon is often used in the design of drugs acting on GPCRs.

4. The activity of GPCRs may be affected by their oligomerization (see below), by their localization to certain membrane compartments, or by the lipid composition of the membrane [371].

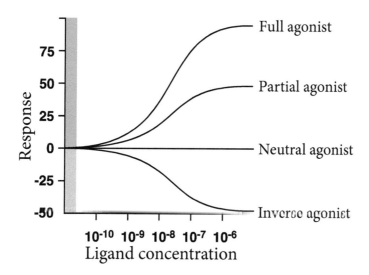

FIGURE 7.32 **Idealized dose response curves of a cellular receptor to a full agonist, partial agonist, neutral antagonist, and inverse agonist.** The constitutive activity of the receptor is assumed to be zero. Note, however, that many receptors have a baseline activity even in the absence of an agonist. Adapted from [348].

7.5.2.2   G-protein mechanisms and regulation

As explained above, GPCRs relay external signals into the cell via G-proteins. Each of these proteins contains three subunits, α, β, and γ (or Gα, Gβ and Gγ, respectively), which may appear in different forms; so far, 23 genes have been found to code for Gα, 5 for Gβ, and 12 for Gγ [338]. Based on the Gα types, G-proteins have been classified into four different families, each of which tends to activate or inhibit specific targets [372]:

- $G\alpha s$ – activates adenylyl cyclase → cAMP-PKA signaling cascade. This family is also over-activated by the *cholera toxin* via covalent modification.
- $G\alpha i_{/o}$ – inhibits adenylyl cyclase and activates c-Src tyrosine kinases; covalently inactivated by the *pertussis toxin*.
- $G\alpha q_{/11}$ – activates PLC$_\beta$ → IP$_3$-PKC signaling cascade.
- $G\alpha_{12/13}$ – leads to Rho activation.
- $G\alpha_{transducin}$ – activates cyclic GMP (cGMP) phosphodiesterase (transducin) in the retina.

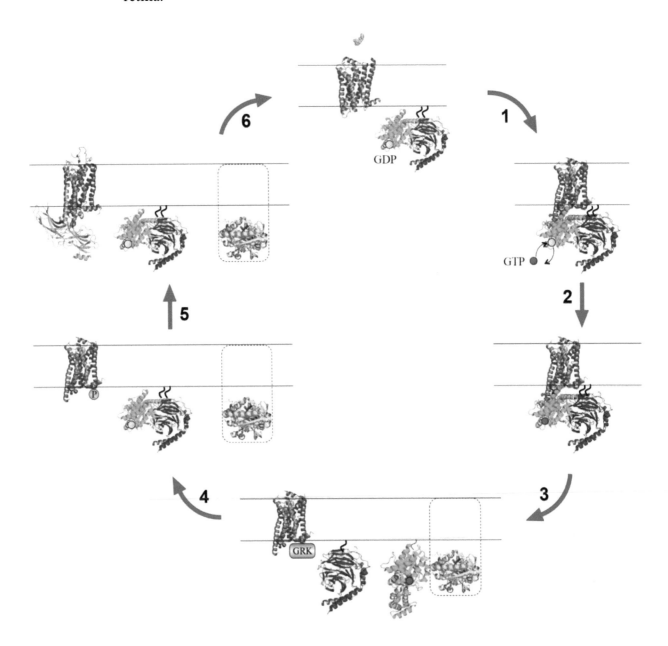

In the G-protein's resting state, its three subunits, Gα, Gβ and Gγ, are bound tightly to each other, and Gα binds the guanine nucleotide GDP (Figure 7.33). The G-protein is attached to the membrane via covalently-bound lipid chains; a palmitoyl or myristoyl chain bound to Gα's N-terminus, and an isoprenyl chain bound to the C-terminal *CAAX* motif of Gγ (not shown). Gβ is tightly bound to Gγ via nonpolar interactions and therefore need not be attached to the membrane covalently. The 'resting' G-protein may bind to an active (agonist-bound) or inactive GPCR, or just drift in the membrane. However, when it binds to an active GPCR, it switches from a resting state to an active state. Specifically, **the activated, agonist-induced GPCR conformation induces a conformational change in the G-protein's α subunit (Gα), leading to the exchange of GDP with GTP (Figure 7.33, step 2).** Thus, the immediate role of GPCRs is to act as *GDP/GTP exchange factors (GEF)*[*1]. The binding of GTP to Gα leads to the departure of the latter from the G$\beta\gamma$ complex, and the binding of Gα to an effector protein (Figure 7.33, step 3). This binding activates the effector protein. The G$\beta\gamma$ complex also has its effectors, which may be different from or identical to those of Gα (e.g., PLC [376]). Gα possesses GTPase activity, and after a short while hydrolyzes its bound GTP to GDP. This returns Gα to its original conformation, allowing it to reattach to the $\beta\gamma$ subunits (Figure 7.33, step 4). The resting state is restored, and the entire system is ready for another activation cycle. Interestingly, the intrinsic GTPase activity of Gα, though present, is too slow for cellular requirements. G-proteins must therefore use the assistance of proteins referred to as *'regulators of G protein signaling' (RGSs)* to accelerate GTP hydrolysis to the required speed [377]. The GPCR cycle also includes regulative steps that ensure the inactivation and internalization of the GPCR (Figure 7.33, steps 4 through 6). The steps are discussed below.

FIGURE 7.33 **The GPCR signaling cycle.** (Opposite) The cycle includes six basic steps: (1) binding of agonist (aquamarine spheres) to the extracellular domain of the inactive GPCR (orange ribbon), followed by the binding of G-protein to the intracellular part of the GPCR. The GPCR structures shown here are those of the β_2-adrenergic receptor: the inactive structure is taken from PDB entry 2rh1 and the active structure is taken from entry 3sn6. In the G-protein's resting state, the Gα subunit, represented by the green ribbon, is attached to the $\beta\gamma$ subunits, represented by the purple and blue ribbons, respectively. Gα in this state is bound to GDP (yellow sphere). The Gα subunit and the Gγ subunit are covalently attached to the membrane (short wavy curves). (2) and (3) Activation of the G-protein, which involves exchange of GDP with GTP (red sphered) in Gα's nucleotide binding site (2), separation of Gα from G$\beta\gamma$ and binding of Gα to an effector protein (3). The Gα-effector complex shown here is taken from PDB entry 1cul. The effector in this complex is the enzyme *adenylyl cyclase*, and only the catalytic subunits are shown (yellow ribbon). The entire region of adenylyl cyclase, which also consists of a transmembrane domain, is marked by the dashed line. The binding of the G$\beta\gamma$ complex to its effector protein is not shown. (4) after a certain amount of time Gα hydrolyses its bound GTP molecule, which causes Gα to separate from its effector protein and re-bind to the G$\beta\gamma$ complex. Following the activation of the GPCR and G-protein, the GPCR undergoes deactivation to terminate the signal. This process includes binding of a G protein-coupled receptor kinase (GRK) to the intracellular domain of the GPCR (3), phosphorylation of the domain by the GRK (4), and binding of arrestin (cyan) to the phosphorylated domain (5, PDB entry 4zwj). The binding prevents the GPCR from re-binding to the G-protein and also induces the internalization and recycling of the GPCR (see Subsection 7.5.5). The entire activation-deactivation cycle is brought to completion with arrestin's departure from the GPCR and return of the system to its resting state (6).

[*1]Note that while GPCRs can bind several types of G-proteins, each GPCR has a strong preference for one G-protein [375].

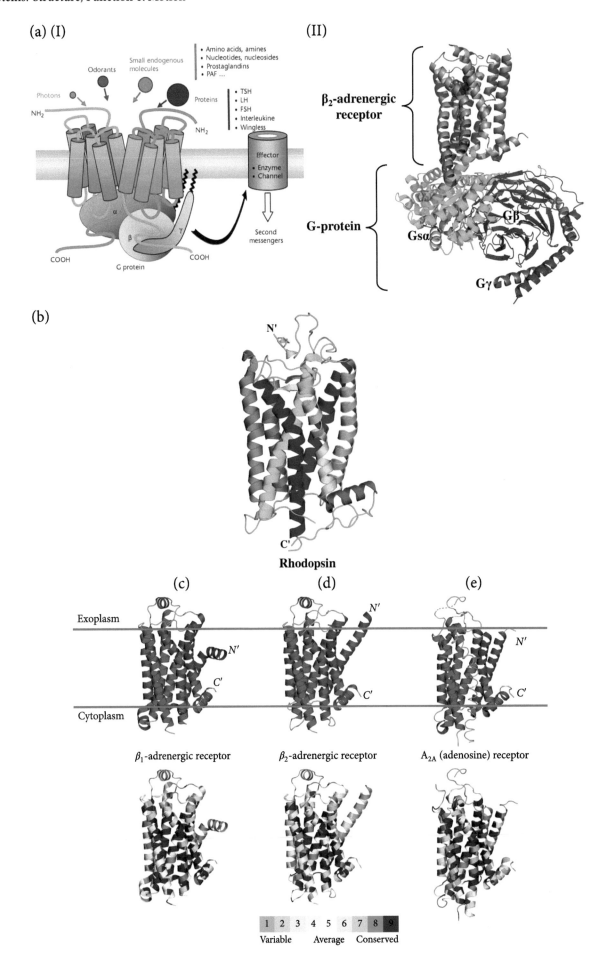

Rhodopsin

β_1-adrenergic receptor

β_2-adrenergic receptor

A_{2A} (adenosine) receptor

7.5.3   GPCR structure

7.5.3.1   General features

As in the case of other membrane-bound proteins, GPCRs' structures have also been late to arrive due to technical difficulties. However, in recent years the difficulties have been successfully tackled, and numerous structures of GPCRs have emerged, although most of these belong to class A[*1]. Many of these structures have been solved by Brian Kobilka's group. Kobilka and Robert Lefkowitz received the 2012 Nobel Prize in Chemistry for their work on GPCRs, starting from the detection of the β-adrenergic receptor and identification of the gene that codes for it. As of the end of 2017, the PDB contained 220 GPCR structures that have been experimentally determined. These correspond to ~25 different receptors (without counting subtypes), including those for adrenaline, adenosine, dopamine, histamine, acetylcholine, serotonin, glutamate, opioids, sphingosine, chemokines, neurotensin, purine nucleosides, and free fatty acids (see reviews in [378,379]). **These studies indicate that GPCRs share several common structural characteristics, the primary characteristic being a transmembrane core containing seven α-helical segments, termed TM1 through TM7, and arranged in a counter-clockwise configuration (when viewed from the extracellular side; Figure 7.34a–e)** [336,372]. Although not all seven-transmembrane receptors are GPCRs, most are. The seven helices are preceded by the extracellular N', and followed by the intracellular C'. The transmembrane segments are interconnected by loops of different lengths at both the extracellular and intracellular sides (these loops are termed ECL1-3 and ICL1-3, respectively). Spectroscopic studies focusing on the refolding of bacteriorhodopsin (a seven-

FIGURE 7.34   **G protein-coupled receptors (GPCRs).** (Opposite) (a) The overall structure of the GPCR–G-protein system. (I) A schematic depiction of GPCRs, taken from [336]. The left side of the image shows a GPCR dimer, bound to its cognate G-protein. The upper-left side shows the different ligands that may bind to the extracellular side of the GPCR and activate it, and the upper-right side shows examples of two of the possible ligands, i.e., proteins and small molecules. The effector molecule, i.e., the enzyme or channel that may be affected by the activated GPCR, and as a result induce the action of a second messenger, is shown on the right side of the figure. (II) The three-dimensional structure of the β_2-adrenergic receptor in complex with Gs-protein (PDB entry 3sn6). The GPCR is shown in orange, with the agonist presented as grey spheres. The α, β, and γ subunits of the G-protein are colored in green, magenta, and blue, respectively. (b) The crystal structure of inactive bovine rhodopsin (PDB entry 1hzx). The N' of the polypeptide chain faces the extracellular side of the membrane. Each of the seven transmembrane helices in the core of the protein is colored differently. (c) through (e) Other GPCRs of known structure positioned similarly to rhodopsin in (b), colored by secondary structure (top) and evolutionary conservation level (bottom). Conservation levels (cyan – lowest, maroon – highest; see color code in figure) are estimated using ConSurf (http://consurf.tau.ac.il) [373,374]). (c) The β_1-adrenergic receptor bound to the antagonist cyanopindolol (PDB entry 2vt4). The structure contains several point mutations in the transmembrane region, as well as deletions in the C' region, which were introduced to increase its stability. (d) The β_2-adrenergic receptor bound to the inverse agonist carazolol (PDB entry 2rh1). (e) The adenosine (A_{2A}) receptor bound to the antagonist ZM241385 (PDB entry 3eml). It is noteworthy that one of the transmembrane spans forms a banana-like helix.

[*1]For a list of all known GPCR structures, including annotations on co-crystallized ligands and their functional effects, see the GPCRDB database (URL: http://gpcrdb.org/).

transmembrane protein that does not interact with a G-protein) in lipid vesicles imply that the common seven-transmembrane core of GPCRs is not accidental, as seven is the minimal number of transmembrane helices required for preserving the environment of the ligand. On the basis of that observation, White speculated that seven helices can "*provide ample space for ligands, through relatively minor helix distortions and reorientations, without the need to increase or decrease the number of transmembrane helices*" [161]. **Besides the number of transmembrane helices and loops, most GPCRs also share certain sequence motifs (see below), and a disulfide bond between a cysteine residue at the extracellular tip of TM3 and another cysteine residue in ECL2**[*1]. The disulfide bond is important in shaping the entrance to the ligand-binding pocket. As we will see below, the extracellular domain (ECD) of GPCRs is the principal ligand-binding site, although in many GPCRs the ligand may also interact with parts of the transmembrane domain (TMD) [380].

Despite sharing several structural characteristics, GPCRs differ in several ways, and these differences give rise to diversity. Most of the differences are localized at the intra- and extracellular regions of the protein, but some are in the transmembrane region. For example, a GPCR may contain additional helices beyond the seven that span the membrane. This is the case in the β-adrenergic receptors, which contain an eighth helix, positioned along the extracellular membrane plane (Figure 7.34c–d). There have also been reports of a μ-opioid receptor variant that has only six transmembrane helices [381,382], as well as of a GPCR-like sequence with a predicted transmembrane domain comprising only five helices [383], but these predictions have not been confirmed structurally. Phylogenetic analysis shows that GPCRs can be grouped into six classes, based on their similarity (see [336] for further detail):

- **Class A or 1 (rhodopsin class).** This class includes most GPCRs (~85%); members of this class respond to both endogenous and exogenous (odorants, pheromones) ligands. Proteins in this class can be further grouped into the following subclasses:

  Subclass I includes receptors that respond to small ligands (e.g., neurotransmitters or even light photons), and whose ligand-binding sites reside within the transmembrane region. Clustering of GPCRs belonging to this subclass [384] suggests that they can be further divided into the following groups: *amine, opsin, melatonin, prostaglandin,* and *MECA* (melatonin, EDG, cannabinoid and adenosine).

  Subclass II includes receptors for peptides. The ligand-binding site of each member of this subclass is built from several segments on the protein's extracellular side.

  Subclass III includes receptors for glycoprotein hormones. The binding site of each protein in this subclass resides primarily in the protein's very large extracellular domain.

- **Class B or 2 (secretin class).** These proteins are similar to class AIII despite a lack of sequence similarity. They respond to large protein and peptide hormones such as:

  Gastrointestinal hormones and factors: glucagon, secretin, vasoactive intestinal peptide (VIP), glucagon-like peptide 1 (GLP-1), glucose-dependent insulinotropic polypeptide (GIP), and others.

[*1] Many GPCRs contain multiple disulfide bonds in the extracellular domain; these bonds stabilize the protein's structure.

Other hormones and factors: corticotropin-releasing factor (CRF), growth hormone-releasing factors (GRF), calcitonin, pituitary adenylate cyclase activating polypeptide (PACAP), and parathyroid hormone (PTH).

Class B GPCRs are also targeted by *α-latrotoxin*, the toxin produced by the black widow spider.

- **Class C or 3 (glutamate class).** This class includes metabotropic glutamate receptors (mGluRs), Ca^{2+}-sensing receptors (CaSRs) of the parathyroid, kidney, and brain [385], $GABA_B$ receptors, pheromone receptors, sweet and amino acid taste receptors (TAS1R), and odorant receptors in fish [386].

- **Class D or 4** includes fungal mating pheromone receptors.

- **Class E or 5** includes *cAMP* receptors in *Dictyostelium discoideum* (slime mold). They are involved in developmental control of the organism.

- **Class F or 6 (Frizzled/Smoothened class).** Members of this class are involved in many cellular and physiological processes (e.g., embryonic development). They activate key signaling pathways, such as the *Wnt pathway*.

Note that all classes also include *orphan receptors*, i.e., proteins that share structural characteristics with known receptors, but are activated by natural ligands that are yet to be found [387].

In the following subsections we will focus on class A GPCRs, for which we have ample structural knowledge. In Subsection 7.5.6 we discuss GPCRs of classes B, C, and F.

GPCRs were initially assumed to work as single polypeptide chains, in contrast to receptor tyrosine kinases, which are known to dimerize. Today, however, it is recognized that **many GPCRs dimerize or oligomerize within biological membranes** [336,388–390]**, and such events are believed to have a role in GPCR signal transduction and crosstalk between different signaling pathways** [391,392]. Much of the information gathered on GPCR dimerization comes from class C receptors, which are active only in their dimeric form [393,394]. Whereas some of these, such as mGluRs, are homodimers, others, such as the $GABA_B$ receptor, are heterodimers. In the case of the $GABA_B$ receptor, one polypeptide chain binds the ligand whereas the other binds the G-protein [395], and is also needed for bringing the entire receptor to the cell surface. Dimerization and oligomerization, with a preference for homodimers, have also been observed in class A GPCRs [396–399] (Figure 7.35). Interestingly, in this class the monomeric form is also active [397,400], and the role of dimerization (or oligomerization) is probably regulatory for the most part. That is, dimerization facilitates better regulation of GPCR activity through cooperativity, crosstalk between different GPCRs in the same heterodimer, etc. [389] Dimerization is usually mediated by the transmembrane region of the GPCRs, but other regions may contribute as well [379]. For example, in the β_1-adrenergic receptor the dimer has two interfaces, involving residues from transmembrane helices, as well as from extracellular and intracellular loops [401]. Moreover, membrane lipids may also contribute to dimerization, as has been suggested in the case of the β_2-adrenergic [402], μ-opioid [403], and mGlu [404] receptors.

(a)　　　　　　　　　　　　　　　　　(b)

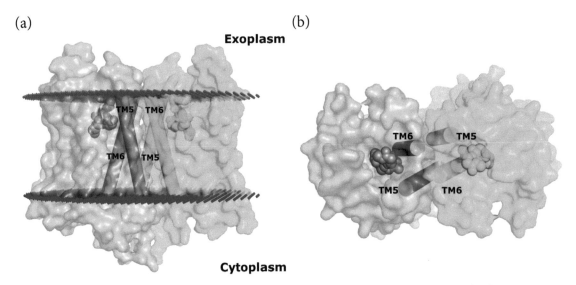

FIGURE 7.35 **A two-fold symmetric dimer formed by the *μ*-opioid receptor** [396]. The receptor (PDB entry 4dkl) is shown in a surface representation from (a) the side and (b) the top (extracellular). Each monomer is colored differently. The two bound ligands (one in each monomer) are shown as spheres. The red and blue lines mark the predicted boundaries of the membrane (the OPM database [117]). The dimerization is mediated by a four-helix bundle motif, formed by TM5 and TM6 (shown as cylinders). On their opposite faces TMs 5 and 6 line the ligand-binding sites, which may be related to the role of dimerization in regulating receptor activity.

7.5.3.2　Structural variations among GPCRs

Since the year 2000, bovine rhodopsin in its inactive state has been the only source for accurate structural analysis of GPCRs. Rhodopsin is a light-activated protein in retinal rod cells, whose function is to convert visual input into neuronal signals that can be transmitted to the brain (via the optic nerves) for processing. Although it has a highly specialized function, rhodopsin has always been a model for the structure of class A GPCRs, mostly because of its overwhelming abundance[*1] and stability; in complete darkness, it assumes only a single (inactive) conformation. These properties enabled scientists to characterize the structure of rhodopsin much earlier compared with other GPCRs, and to study the protein extensively using various approaches [405]. *β*-adrenergic receptors, which respond to *catecholamines* (see Box 7.2), have also served as popular models of class A GPCRs. As in the case of rhodopsin, their popularity stems from historical reasons; the *β*-adrenergic receptors were the first GPCRs to be sequenced and cloned after rhodopsin [406], and their three-dimensional structures were also solved quite early. Compared with rhodopsin, however, adrenergic receptors and many other GPCRs are considerably less stable, and their crystallization requires the use of stabilization methods. These usually include mutations or truncation of unstable parts (e.g., the third intracellular loop, ICL3) and attachment of these parts to an antibody fragment or to another protein (e.g., T4-lysozyme). Lower stability is often related to increased flexibility, which suggests that even when the GPCR is agonist-free and inactive, it can still sample other conformations, including the active one. Indeed, as noted above, most GPCRs are characterized by baseline activity even in the absence of an activating signal, whereas rhodopsin has no baseline activity.

[*1]It constitutes 90% of the protein in purified retinal rod outer segments.

As explained above, GPCRs share very similar structures (especially within each class), which is remarkable considering the low sequence identity among members of the group. For example, rhodopsin and the β_2-adrenergic receptor share only 21% identical amino acids, yet the r.m.s.d. between their structures is merely 2.3 Å, and in the transmembrane region it is 1.6 Å. Nevertheless, some differences do exist between GPCRs, and these merit analysis. In the following paragraphs we review the main conclusions obtained thus far for class A GPCRs, by addressing the three different regions of GPCRs (i.e., extracellular, transmembrane and intracellular) separately.

1. **The extracellular (EC) region**
   As expected, **most of the differences between GPCRs are localized to the extracellular domain (ECD), and particularly to the ligand-binding site and adjacent loops** [371]. ECL2 is particularly important in distinguishing among GPCR structures, as it is the largest loop and can therefore assume different conformations. In contrast, ECLs 1 and 3 are much shorter and usually do not have a distinct secondary structure. Here are a few examples:

   Rhodopsin (Figure 7.36a): The EC region has a compact, rigid tertiary structure, which significantly restricts the access of solvent and other molecules to the ligand-binding pocket [407]. This is not surprising; first, rhodopsin's 'ligand' is light (i.e., photons), which does not require a wide entrance. Second, solvent access to the retinal cofactor (see Subsection 7.5.4 below) would result in hydrolysis of the bond connecting the retinal to the polypeptide chain [371]. The main structural element preventing the solvent from reaching the retinal is ECL2, which forms a short β-sheet. It also blocks the main entrance to the ligand-binding pocket and prevents movement of the transmembrane helices [405].

   β-adrenergic receptors (Figure 7.36b): These represent the opposite case, with an open EC region. In particular, ECL3 does not interact with any of the other loops, ECL2 forms a very short helix, and the N' is altogether disordered [371,407].

   Adenosine (A_{2A}) receptor (Figure 7.36c): ECL2 lacks secondary structure, yet remains rigid due to disulfide bridges that stabilize it and the entire extracellular region of the receptor. Polar and van der Waals interactions involving the three loops also contribute to stabilization. It has been suggested that the role of the disulfide bridges is to constrain segments of the EC regions that are involved in ligand binding [407].

   Neurotensin receptor (Figure 7.36d) and other peptide-binding GPCRs: The extracellular region is more open than that in non-peptide-binding GPCRs, and ECL2 forms a hairpin structure.

   Sphingosine 1-phosphate receptor (Figure 7.36e) and other lipid-activated GPCRs: The extracellular region is capped by the N' (organized as a helix) and ECL1, which block the entrance of the ligand to its binding pocket. In these GPCRs, the highly hydrophobic ligands are thought to gain access to their binding pockets through the membrane.

2. **The transmembrane region and ligand-binding pocket**
   This region is where GPCRs vary the least, especially those that belong to the same

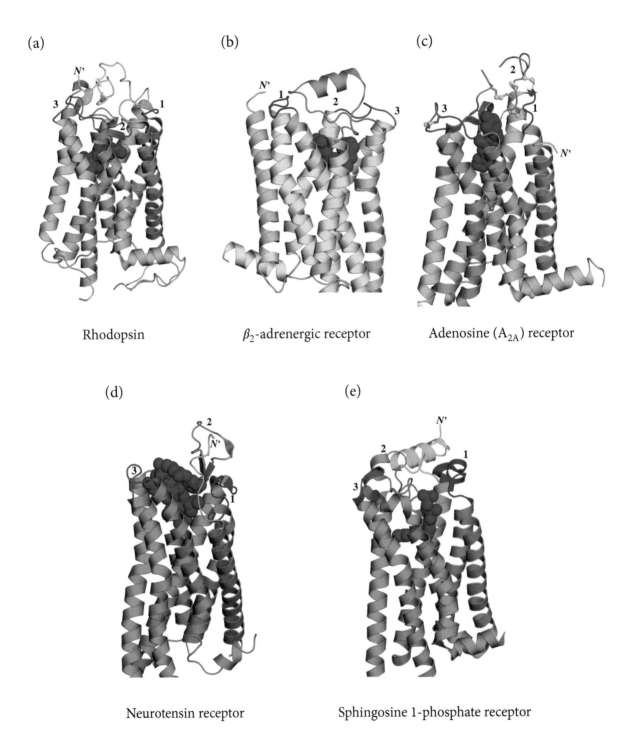

(a) Rhodopsin

(b) β_2-adrenergic receptor

(c) Adenosine (A_{2A}) receptor

(d) Neurotensin receptor

(e) Sphingosine 1-phosphate receptor

FIGURE 7.36 **The extracellular regions of different class A GPCRs.** (a) Bovine rhodopsin (PDB entry 1hzx). The first, second and third extracellular loops are in blue, orange, and red (respectively), whereas the N-terminus is in green (the loop numbers are also noted in the image). The retinal cofactor is presented as magenta spheres, within the transmembrane region (in grey). (b) Human β_2-adrenergic receptor (PDB entry 2rh1). The ligand, carazolol, is presented as magenta spheres. (c) Human adenosine (A_{2A}) receptor (PDB entry 3eml). Sulfur atoms of loop-related disulfide bonds are shown as small cyan spheres, and the ligand as magenta spheres. (d) Rat neurotensin receptor (PDB entry 4xee). (e) Human sphingosine 1-phosphate receptor (PDB entry 3v2w).

class. Still, even this region contains parts that are structurally more similar than others. This is evident in a structural alignment between the various structures, showing a core of 97 residues that have a C_α-r.m.s.d. of only 1.3 Å [407]. The other, less structurally similar residues, are expected to be involved in those functions that make GPCRs distinct from each other, i.e., ligand- and G-protein-binding. One of the common structural characteristics of the GPCRs listed above is the chemical environment of the highly conserved *NPxxY motif*, located at the intracellular end of TM7 [371]. This region is involved in key conformational changes occurring during GPCR activation (see Subsection 7.5.4 below). Interestingly, the helix in all the above GPCR structures is distorted in this region due to the presence of Pro in the sequence motif. The helical distortion is, however, stabilized electrostatically by hydrogen bonds to other residues or adjacent water molecules (Figure 7.37).

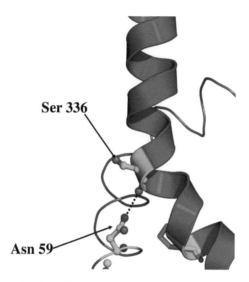

FIGURE 7.37 **Stabilization of helical kink by hydrogen bond.** TM7 of the β_2-adrenergic receptor is shown as a red ribbon. The kink of the helix is seen clearly. The proline residue that induces the kink is presented as sticks. The kink breaks the backbone hydrogen bond involving the carbonyl oxygen of Ser-336. However, the latter is stabilized by Asn-59 of TM1 (blue ribbon), which hydrogen-bonds to Ser-336 (dotted black line).

The structure of the β_2-adrenergic receptor [402,408] provides interesting insights regarding GPCR-lipid interactions. The structure contains a cholesterol-binding site between transmembrane helices 2 through 4, which comprises evolutionarily conserved residues. These include Trp-158 and Ile-154 on TM4, and Ser-74 on TM2 (Figure 7.38). Trp-158, which is highly conserved, is geometrically compatible with the ring(s) of the bound cholesterol. This compatibility optimizes the nonpolar, van der Waals and CH-π interactions between Trp-158 and cholesterol. It has been suggested that the cholesterol-binding residues constitute an allosteric site of GPCRs [407], in accordance with the proposed role of cholesterol as a modulator of the function of membrane proteins in general [409,410], and of GPCRs in particular [411–413]. As mentioned above, cholesterol and other membrane lipids are also thought to contribute to the dimerization of GPCRs, which itself is believed to participate in the regulation of GPCR signaling.

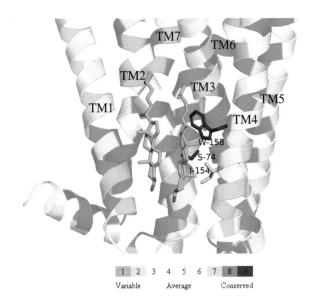

FIGURE 7.38 **Cholesterol binding to the β_2-adrenergic receptor (PDB entry 3d4s).** The cholesterol molecules are shown as sticks (colored by atom type), whereas the receptor is represented as ribbons, colored according to evolutionary conservation level (cyan – lowest, maroon – highest; see color code in figure). The conservation levels are calculated using ConSurf (http://consurf.tau.ac. il) [373,374]. The evolutionarily conserved residues Trp-158, Ile-154 and Ser-74 are shown as sticks. These residues are adjacent to the site of many therapeutic agents that act on class A GPCRs [407], and their location is an attractive target for future drugs.

Though the ligands of all class A GPCRs interact with both the transmembrane and EC domains, the position of the binding pocket varies significantly across the different receptors, although not in all cases. For example, the pockets of rhodopsin and the β-adrenergic receptors are quite similar to each other, with the ligand extending between TMs 3 and 7 and the interface between TMs 5 and 6 [407]. However, in rhodopsin the bound retinal cofactor extends further and is in physical proximity with Trp-265 on TM6, which is part of the conserved *CWxP motif* (Figure 7.39a). This residue, together with Phe-208, is part of a mechanism called the *'transmission switch'* [414], which participates in the propagation of the signal to the intracellular part of rhodopsin upon activation (see Subsection 7.5.4 below). In the β-adrenergic receptors the inverse agonist is separated from Trp-286 (the equivalent of Trp-265) by aromatic residues. In contrast to rhodopsin and the β-adrenergic receptors, the binding pocket of the adenosine (A_{2A}) receptor has a very different location. First, it is located closer to the interface between TM6 and TM7, where the ligand can interact with the second extracellular loop, ECL2. Second, the ligand extends perpendicularly to the membrane plane, and seems to be shifted towards the membrane-solvent interface, where part of it is solvent-exposed [117].

Peptide-binding GPCRs such as the chemokine, neurotensin, and opioid receptors have to accommodate ligands that are larger than biogenic amines like adrenaline, or nucleotides like adenosine. As a result, the binding sites in these GPCRs tend to be shallower than those in the other class A GPCRs, and the peptide ligands tend to bind closer to the extracellular domain (Figure 7.39b). Still, as in GPCRs that bind small molecules, there is variability in ligand locations and interactions within the peptide-binding GPCRs [378].

(a)

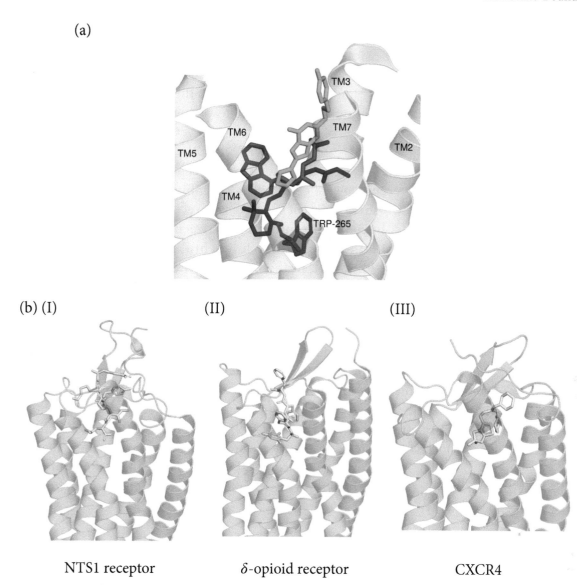

(b) (I)   (II)   (III)

NTS1 receptor   δ-opioid receptor   CXCR4

FIGURE 7.39 **Ligand binding sites in class A GPCRs.** (a) The sites of rhodopsin, the β_2-adrenergic receptor, and the A_{2A} receptors are represented by their respective ligands, as blue, red, and green sticks. Rhodopsin's Trp-265 is shown in magenta. For clarity, only rhodopsin's helices are shown, in grey. (b) Peptide binding to class A GPCRs. (I) the neurotensin 1 (NTS1) receptor (PDB entry 4xee), (II) the δ-opioid receptor (PDB entry 4rwa), and (III) CXCR4 (the human chemokine receptor 4, PDB entry 3odu). As the image demonstrates, in peptide-binding receptors, the peptides, shown as yellow sticks, tend to bind closer to the extracellular domain of the GPCR than compared to the small ligands of other class A GPCRs.

3. **The intracellular region**

The structure of the intracellular region of GPCRs is relatively conserved, probably because of the limited diversity of potential binding partners, which include G-proteins, arrestins, and G protein-coupled receptor kinases (GRKs; see Subsection 7.5.5 below). Members of class A GPCRs have several distinct structural characteristics, one of which is the conserved sequence motif *D/ERY* at the intracellular side of TM3 [336]. The three-dimensional structure of rhodopsin, which was the first GPCR structure to be determined, demonstrated an electrostatic interaction, referred to as the '*ionic lock*', between Arg-135 of the motif and the adjacent Glu-247 on TM6 (Figure 7.40). The sequence involved in the ionic lock was conserved, suggesting that the interaction was important. Furthermore, it was found that activation of rhodopsin leads to the disruption of the ionic lock (see below). On the basis of these data, in addition to data obtained from mutational studies, it was suggested that the ionic lock is important for the stabilization of the inactive state of GPCRs, and possibly for their activation or signaling as well [371,407]. However, when the three-dimensional structures of other inactive GPCRs (e.g., the β-adrenergic receptors, the A_{2A} receptor, and the muscarinic (M_2) receptor) were determined, the ionic lock was found to be absent in these structures. **Other studies of β-adrenergic receptors have indicated that the ionic lock is in fact in constant equilibrium between two conformations, one in which it is formed and the other in which it is broken [415,416]. The partial or complete absence of the ionic lock in the inactive forms of β and A_{2A} receptors may explain why these receptors have measurable baseline activity in their inactive states, whereas rhodopsin does not [417].**

Another interesting point emerges from comparing the structures of vertebrate and invertebrate rhodopsins, represented by bovine [418] and squid [419,420] rhodopsins, respectively. As far as their function is concerned, the two proteins respond to the same agonist and differ only in their G-protein specificity. Intriguingly, this difference seems sufficient to create detectible topological differences in the intracellular regions of the two structures [407]. The largest difference is in the third intracellular loop, ICL3, which is attributed to the longer sequence of squid rhodopsin in that region [421]. Indeed, ICL3 is considered to confer specificity to the intracellular binding partner; swapping this part between GPCRs results in switching the receptors' G-protein selectivity [422].

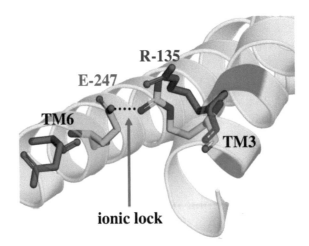

FIGURE 7.40 **The D/ERY motif.** The figure shows the Arg residue of the *D/ERY* motif in the intracellular domain of rhodopsin (colored by atom type) and the β_2-adrenergic receptor (red). In rhodopsin, this residue forms an ionic interaction (dashed line) with an adjacent Glu residue (colored by atom type). However, in the β_2-adrenergic receptor it is too far from its respective Glu (red) to be engaged in a salt bridge.

BOX 7.2 ADRENERGIC RECEPTORS: FIGHT YOUR ASTHMA AND YOUR ENEMIES

Adrenergic receptors mediate the physiological effects of adrenaline and noradrenaline in the animal body [423] (Figure 7.2.1)[*a]. Adrenaline and noradrenaline belong to a group of hormones and neurotransmitters called 'catecholamines', which are produced from the amino acid tyrosine [424–426]. There are two sources of catecholamines in the animal body: the nervous system and the medulla (core) or the adrenal glands. The nervous system implicates catecholamines in *neurotransmission*[*b], whereas the medulla implicates them — and particularly adrenaline — in *hormonal action*. The two catecholamines have a very important function; they prepare the body for a situation called *'fight-or-flight'* [428] (Figure 7.2.2). It is easy to understand this situation when considering the lives of wild animals and prehistoric man. In both cases, life in the wild is full of danger for the individual. The danger can appear abruptly, in the form of a predator, competitor, and even a routine, yet catastrophic, act of nature. In such cases, the difference between life and death often lies in the ability of the individual to respond very quickly to the danger. This response involves perceiving the danger and taking appropriate action.

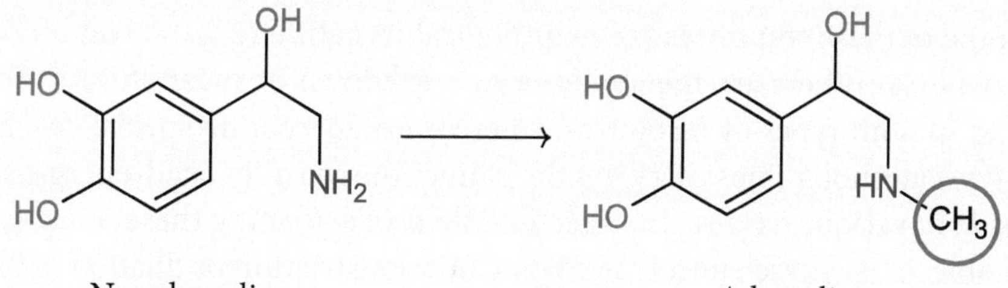

FIGURE 7.2.1 **The molecular structures of adrenaline and noradrenaline.** Adrenaline is formed by *N*-methylation of noradrenaline.

Indeed, upon detection of danger, the brain, which collects input from sensory organs, innervates other organs (via sympathetic nerves) in order to get them ready for action. Innervation of the two adrenal glands, located on top of the kidneys, stimulates their neurosecretory cells to secrete adrenaline (and to a lesser extent noradrenaline) into circulation, which further enhances the body's readiness to handle the danger.

As the term 'fight-or-flight' implies, the immediate response of the body to danger, following the action of adrenaline and noradrenaline, may be reduced to two simple options: fighting or fleeing, depending on the situation. Both types of responses require top performance of several body systems. First, sensory organs, such as the eyes, must collect as much information as possible from the environment, so as to be able to alert the body to danger. Second, skeletal muscles must be ready to respond quickly and

---

[*a]Adrenaline and noradrenaline are also called epinephrine and norepinephrine, respectively.

[*b]Although both adrenaline and noradrenaline are used as neurotransmitters in the central nervous system (CNS), and the *sympathetic* branch of the *autonomic nervous system* (ANS), noradrenaline is much more common in this capacity [427].

powerfully. This response involves the functions of several organs; the heart and lungs must supply ample oxygen to the skeletal muscles, and the body's sugar stores in the liver and muscles must be broken down quickly to supply the muscles with available fuel. In addition, bodily functions that usually take place in the resting state and require the functioning of certain organs (e.g., digestion by the gastrointestinal (GI) system), must be inhibited, so as to allow peripheral blood to reach the muscles in high volumes.

FIGURE 7.2.2 **A wolf in a fight-or-flight posture.** The image is taken from [429].

Clearly, some of these responses are neurological in nature (e.g., activation of skeletal muscles), whereas others are metabolic (e.g., breakdown of sugar stores). Adrenaline participates in both types of responses, whereas noradrenaline specializes in the former. The two catecholamines carry out their functions via a set of adrenergic receptors, which reside in various tissues. The specific effects of activating these receptors are detailed in Table 7.2.1. These effects mainly include constriction or dilation of blood vessels (depending on the organ), as well as contraction of some muscles and relaxation of others. Together, these effects create the following physiological outcomes [424]:

1. Increased heart rate and stroke volume[*a].

2. Diversion of blood from the skin and GI tract to the heart, lungs, brain, and skeletal muscles.

3. Increased blood pressure.

4. Pupil dilation.

5. Windpipe dilation.

6. Relaxation of GI smooth muscle.

7. Increased blood clotting rate → less danger of hemorrhages following injury.

8. Increased sweating → cooling down of overworked body.

9. Increased metabolic rate, resulting from the breakdown of both liver lipid stores and muscle glycogen stores.

[*a]The amount of blood the heart can pump out in a single beat.

TABLE 7.2.1 **Some of the physiological effects of adrenergic receptors.**

Receptor Subtype	Location	Effect
α_1	Blood vessels	Smooth muscle contraction $\longrightarrow$ Vasoconstriction
	Heart	Increased cardiac contraction and rate
α_2	Blood vessels	Smooth muscle relaxation $\longrightarrow$ Vasodilation
β_1	Heart	Increased cardiac contraction and rate
	Kidney	Activation of renin-angiotensin-aldosterone system $\longrightarrow$ Na$^+$–K$^+$ exchange $\longrightarrow$ Na$^+$ retention
β_2	Blood Vessels	Smooth muscle relaxation $\longrightarrow$ Vasodilation
	Heart	Increased heart rate and output
	GI tract	Smooth muscle relaxation
	Pancreas	Glucagon secretion $\longrightarrow$ glycogen breakdown and gluconeogenesis
β_3	Liver	Lipid breakdown

The adrenal glands act not only in response to fight-or-flight threats, which are immediate and potentially life-threatening, but also to more prolonged types of threat, called *stress* [430]. In the latter case, however, the adrenal glands are activated hormonally by the brain; the *hypothalamus*, which is the hormonal control center of the brain, stimulates the pituitary *adrenocorticotropic hormone (ACTH)*. This physiologically-active peptide is secreted from the anterior pituitary gland into circulation under orders from the brain's central metabolic coordinator, the hypothalamus. This mode of stimulation is referred to as the *'hypothalamic-pituitary-adrenal (HPA) axis'*; it also stimulates the cortex of this gland to secrete large amounts of *glucocorticoid hormones*, primarily *cortisol*, into the circulation. Cortisol is implicated in the *stress response*, which is less immediate than the fight-or-flight response. Indeed, like many other steroid hormones, cortisol acts more slowly than adrenaline and noradrenaline, but has extensive effects on the body. For example, its metabolic effects are anabolic in the liver and catabolic in muscle and fat cells, and act to increase blood glucose levels. Cortisol also affects other systems, such as the cardiovascular system, the central nervous system, the immune system, and the kidneys. Most importantly, cortisol inhibits the inflammatory response, which can be potentially hazardous following injury.

The effects of the adrenaline-noradrenaline system on multiple organs have made it a target for many medicinal drugs [425]. These can be separated into the following groups:

1. **Adrenergic agonists.** β_2 receptors can be bound in the trachea (windpipe) and bronchi. Therefore, β_2 agonists, such as albuterol, are used to treat asthma. The α receptors can be found on smooth muscles, such as those controlling the diameter of blood vessels. Whereas the activation of α_1 receptors leads to smooth muscle contraction and the consequent constriction of blood vessels (*vasoconstriction*), α_2 receptors function in regulating their α_1 counterparts, and their activation leads to the opposite effect (*vasodilation*). Thus, α_2 agonists, such as

clonidine, are used as *antihypertensive* drugs, i.e., to treat high blood pressure. High blood pressure has been termed 'the silent killer' because of its devastating effects on untreated individuals and the relative absence of symptoms.

2. **Adrenergic antagonists (blockers).** β_1 receptors reside primarily in the heart, and their antagonists (e.g., atenolol) are used for treating angina pectoris, hypertension, and some arrhythmias.

3. **Reuptake inhibitors.** The action of adrenaline and noradrenaline in the nervous system is stopped primarily by their uptake or reuptake away from the synapse and into their secreting cells. These processes are carried out by transporters, which have become targets for drugs termed 'reuptake inhibitors'. These drugs block transporters in order to elevate the levels of these neurotransmitters in the synapse, thereby intensifying their action. The primary medical use for adrenaline and noradrenaline reuptake inhibitors is fighting depression. Some of the older-generation antidepressants, such as the *tricyclics* (e.g., *desipramine*) inhibit the reuptake of both catecholamines and indoleamines (e.g., serotonin). Antidepressants of newer generations (SSRIs) are more specific; they are designed to boost only serotonin levels. Yet, a relatively new class of antidepressants (SNRIs) elevate the levels of both serotonin and noradrenaline. The catecholamine reuptake system is also a target for certain types of drugs of abuse, i.e., *amphetamines* (*cocaine, MDMA*). These drugs work similarly to the adrenaline-noradrenaline reuptake inhibitors mentioned above, but with much higher intensity. Amphetamines are addictive, and have serious adverse effects on the cardiovascular system.

4. **Monoamine oxidase (MAO) inhibitors.** MAO is an enzyme that catalyzes the oxidative deamination of catecholamines and of indoleamines. Its inhibition therefore increases adrenergic and serotonergic effects. MAO inhibitors (e.g., *selegiline*) are used primarily as antidepressants.

In addition to the drugs listed above, there are other drugs that achieve similar results by acting on the opposite branch of the autonomic nervous system, i.e., the parasympathetic system. For example, an antagonist of the receptor for acetylcholine (the principal neurotransmitter in the parasympathetic system), such as atropine, induces some of the physiological effects of agonists of adrenergic receptors.

7.5.4   GPCR and G-protein activation

Much of what we know today about the changes that GPCRs undergo following activation comes from extensive biochemical and biophysical studies performed on class A receptors [336]. Studying the activation process requires knowledge of GPCRs in both inactive and active states. There are currently many structures of GPCRs, which have been crystallized in complex with an agonist. However, such structures are only partly activated; studies show that in order to assume a fully active conformation, the GPCR must also bind a G-protein or a protein mimicking it (e.g., part of an antibody) on its intracellular side [431–433].

To date, only three GPCRs, all of which belong to class A, have been crystallized in fully active conformations:

- **Rhodopsin.** Two active structures have been determined, one bound on its intracellular side to an 11-amino acid fragment representing the C-terminus of $G\alpha$ [434], and the other bound to an antibody.

- **The β_2-adrenergic receptor.** Two active structures have been determined in complex with an agonist, one bound to an entire G-protein molecule [435] on its intracellular side, and the other bound to a nanobody (the heavy chain of an antibody) [436].

- **The muscarinic (M_2) receptor.** Two active structures have been determined; each is in complex with an agonist and bound to a nanobody on its intracellular side [437]. One of these structures is also bound to an allosteric activator.

In our discussion of GPCR activation below, we focus on these three GPCRs, but also refer to some of the other GPCRs, such as the A_{2A} receptor, for which a partly active structure is known. In the case of the β_2-adrenergic receptor, we refer only to the active structure bound to a G-protein molecule, as it has been found to be very similar to the nanobody-bound structure. For a more detailed account of the data obtained from known GPCR structures, please see the review by Shonberg et el. [378].

7.5.4.1 Structural changes in GPCRs upon activation

Rhodopsin was the first GPCR for which 3D structures of both the active and inactive states were obtained (Figure 7.41a), and it was therefore the first source of knowledge about the activation process. Rhodopsin is a photoactivated protein residing in the membranes of retinal rod cells; it enables these cells to relay visual input to the brain. Since polypeptide chains are not well suited for responding to electromagnetic photons, rhodopsin uses an organic cofactor called '11-cis-retinal' (a *vitamin A* derivate, Figure 7.41b, left) as the photoreactive element. This molecule is covalently, yet reversibly bound to the polypeptide chain via a Schiff base to Lys-296 (on TM7). In the inactive state, the protonated (i.e., positively charged) Schiff base is stabilized by Glu-113 on TM3 (the '3 to 7 lock'). When hit by light, the retinal responds by undergoing isomerization. That is, it changes from an 11-*cis* to an *all-trans* configuration (Figure 7.41b, right)[*1]. This, in turn, induces conformational changes in the protein, which create a binding site for rhodopsin's cognate G-proteins[*2]. The α subunit of the G-protein binds to the intracellular side of rhodopsin, and stabilizes its active conformation[*3]. After activation, rhodopsin is *bleached*. That is, the Schiff base binding the retinal is hydrolyzed, allowing the retinal to leave the receptor and render the latter inactive, for about 30 minutes. The retinal-free polypeptide chain that remains, which also represents the active form of rhodopsin, is referred to as '*opsin*'.

In 2011, two structures of fully active rhodopsin were determined in the presence of all-*trans* retinal, one of the structures also bound to a Gs-derived 11-amino-acid fragment in the intracellular domain [434]. A first glance at the superimposed structures of inactive and active

---

[*1]Thus, 11-*cis* retinal is in fact a covalently-bound inverse agonist.

[*2]*Transducin* is the main rhodopsin-related G-protein.

[*3]This means that there are two factors stabilizing the active GPCR: its agonist, and its cognate G-protein [405].

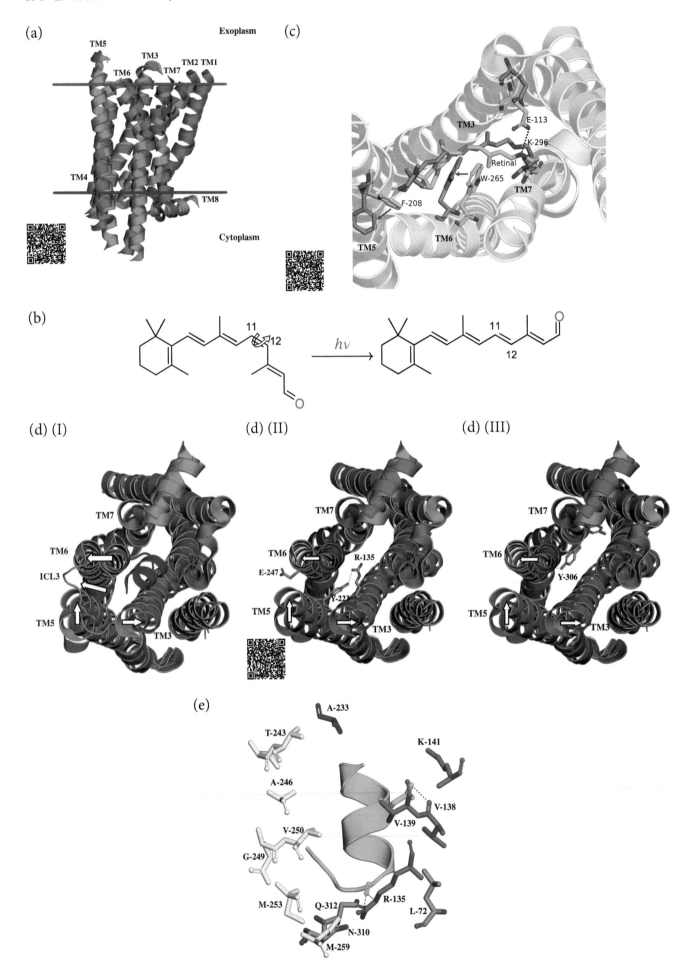

states of rhodopsin reveals only modest changes in conformation, which mostly involve helices 5, 6, and 7 [405] (Figure 7.41a). A closer look at the retinal pocket seems to confirm the first impression, with relatively subtle changes observed in that region. For example:

- Certain side chain movements create room and promote the *cis*-to-*trans* change of the retinal cofactor. These movements involve the side chain of Phe-208 on TM5, and that of Trp-265 on TM6, which moves into a space formerly occupied by the β-ionone ring of the retinal cofactor (Figure 7.41c).

- A distance is created between TM3 and TM7, mainly due to a shift in TM7. This increases the distance between Glu-113 and the Schiff base, from 3.5 Å to 5.3 Å, which disrupts the salt bridge between them (i.e., the 3 to 7 lock). However, the conformational change also strengthens the existing salt bridge between the Schiff base and Glu-181, which replaces Glu-113 as the main stabilizer of the Schiff base.

NMR studies also suggest a disruption of hydrogen bonds between ECL2 and the extracellular parts of helices 4, 5, and 6, occurring just before the dissociation of the retinal cofactor [438].

FIGURE 7.41 **Structural changes in rhodopsin following its activation.** (Opposite) (a) General view of activation-induced shifts of helices. The two structures presented here are of inactive (grey; PDB 1gzm) and active rhodopsin (orange; PDB 3pqr). For clarity, the loops have been removed. The red and blue lines mark the predicted boundaries of the membrane (the OPM database [117]). (b) The retinal cofactor. The figure shows in one step the activation process in which 11-*cis* retinal (left) transitions into all-*trans* retinal (right) using the electromagnetic energy of a photon (*hv*). Carbon atoms 11 and 12 are noted, and the rotation around the bond connecting them is represented by the circular arrow. (c) Conformational changes in the retinal-binding pocket. The backbone of the inactive state (PDB 1gzm) is shown from the extracellular side as grey ribbons. Residue conformations corresponding to the inactive state are colored by atom type, whereas those corresponding to the active state (PDB 3pqr) are in orange. Movements of specific residues are marked by blue arrows. The movements of Trp-265 and Phe-208 are clearly seen, as well as the breaking of the salt bridge between Glu-113 and Lys-296 (dashed line). (d) Changes on the intracellular side of rhodopsin, allowing the binding of transducin. The structures and representation of the active and inactive forms of rhodopsin are the same as in panel (a), but the view is from the intracellular side. (I) Transducin binding. The *C*-terminal 11 amino acids of transducin are presented as a purple ribbon. The white arrows denote the direction of movement of TMs 3, 5, and 6, as well as of ICL3, when rhodopsin shifts from the inactive state to the active state. This movement clearly creates space for the transducin fragment. (II) The conformational change disrupts the inactive state-related ionic lock, which involves Glu-247 of TM6 and Arg-135 of TM3's *D/ERY* motif. In its new position, Arg-135 is stabilized by a hydrogen bond with Tyr-223 (dashed line). (III) The conformational change of rhodopsin also includes a large movement of Tyr-306 of the *NPxxY* motif, which stabilizes the active conformation. A hydrogen bond between Tyr-306 and TM6 is shown. (e) Stabilization of rhodopsin's active conformation by transducin. Transducin is presented as a green ribbon. Residues of rhodopsin that interact with transducin are shown as sticks, colored in magenta (helix 5) cyan (helix 6), and orange (other secondary elements). The hydrogen bonds of transducin with Arg-135 and Val-138 are presented as dashed lines.

However, inspection of the intracellular side of rhodopsin's TMD reveals larger changes in TMs 3, 5, 6, and 7. **The shifts of TMs 3 and 6**[*1] **away from each other and from the center of the protein create a space for transducin binding (Figure 7.41dI). These conformational changes result from the local movements of Trp-265 (the CWxP motif) and Phe-208 in the retinal-binding pocket (i.e., the transmission switch)** [414]. **The space created between TMs 3 and 6 leads to disruption of the 'ionic lock' between Glu-247 and Arg-135 of the D/ERY motif (Figure 7.41dII).** The loss of this interaction is partly compensated for by new interactions formed between Arg-135 and some of the residues of TM5 (e.g., Tyr-223) and TM6, which become closer due to the shifts. As explained above, in the other structurally determined GPCRs the ionic lock is absent, in at least some cases, which has been proposed to account for the differences in baseline activity between the GPCRs. Indeed, whereas rhodopsin is completely inactive in the dark, the other three GPCRs (like many others) retain some activity even when they do not bind their agonists. This activity can be inhibited by inverse agonists, such as carazolol, which acts on the β_2-adrenergic receptor [440].

As mentioned above, the *NPxxY* motif in TM7 differs from the D/ERY motif in that it is involved in the activation of GPCRs, rather than in stabilizing the inactive state. Indeed, Tyr-306 of the motif inserts into a space previously occupied by TM6 (Figure 7.41dIII), which stabilizes the active conformation of rhodopsin. Similar movements are observed also in the equivalent positions of Tyr-306 in the β_1- and β_2-adrenergic receptors, the muscarinic (M_2) receptor, and the A_{2A} receptor upon activation, and it is believed that in these GPCRs, too, the movements participate in the activation process. In the β_2-adrenergic receptor, stabilization of the active site, conferred by the Tyr-306-equivalent position, relies in part on a water-mediated hydrogen-bond with the Tyr-223-equivalent position on TM5 [441]. The active structure of the M_2 receptor does not contain water molecules, but the same water-mediated interaction between the two tyrosine residues is thought to happen there too [437]. The conservation of this interaction and the similar positions of the two tyrosine residues in the three activated GPCRs suggests that the interaction is a hallmark of GPCR activation.

In conclusion, the changes described above, though they involve different parts of rhodopsin, are overall small, i.e., within 2 to 6 Å [421]. **Nevertheless, they serve their purpose, in creating a space between transmembrane helices 3, 5, 6, and 7, which serves as a binding site for transducin** [371,405]. **As we will see later, the β_2-adrenergic receptor and the M_2 receptor display similar changes in their overall conformations upon activation.** The experimentally determined structures of the rhodopsin-transducin complex (PDB entries 3cap and 3pqr) show multiple interactions, both polar and nonpolar, between residues of transducin and those of rhodopsin, which stabilize the active conformation of the latter (Figure 7.41e). The direct interaction between Arg-135 of the *D/ERY* motif and a backbone group in transducin may seem important, considering the high conservation level of Arg-135. However, this interaction is absent in the β_2-adrenergic receptor, which has been crystallized in complex with a complete G-protein molecule (see below). Rhodopsin-transducin binding seems to be driven by nonpolar interactions, with hydrogen bonds rendering the orientation of the transducin-derived peptide specific [371].

[*1] In agreement with spectroscopic data showing a 5 Å outward rotation of TM6 [439].

7.5.4.2 Agonist effect and G-protein activation

In the previous subsection we described the conformational changes that GPCRs undergo when activated. The ultimate goal of GPCR research, however, is to understand how these changes are induced by ligand binding, and how they lead to the activation of the G-protein. This aspect also has important pharmacological implications, since different ligands designed to bind to the same pocket in the GPCR may induce different responses; some may act as agonists, whereas others may act as antagonists or inverse agonists. These questions cannot be answered by the aforementioned structures of rhodopsin, since this GPCR has a non-diffusible agonist (i.e., the covalently-bound all-*trans* retinal)[*1], and the corresponding structure of the activated receptor is bound to a small fragment of the Gs-protein. Fortunately, the fully active structures of both the β_2-adrenergic receptor and the muscarinic (M$_2$) receptor have been determined.

The active structure of the β_2-adrenergic receptor was crystallized in 2011, bound to a high-affinity agonist (BI-167107) and an entire Gs-protein (see above for details)[435]. The G-protein in the crystallized structure is nucleotide-free, which means it was captured in the middle of the GDP-GTP exchange process. Comparison between the active structure of the β_2-adrenergic receptor and its inactive structure, bound to the inverse agonist carazolol, reveals the following[442]. As in rhodopsin, activation of the β_2-adrenergic receptor induces only small structural changes on the protein's extracellular side and larger changes on its intracellular side (Figure 7.42a). The bound agonist forms three hydrogen bonds with residues in TM5: two with Ser-203 and one with Ser-207. These interactions seem to pull TM5 slightly inward (2 Å at position 207, Figure 7.42b). This small movement leads to rearrangement of the hydrophobic interaction network formed between Phe-208 (TM5), Pro-211 (TM5), Ile-221 (TM3), and Phe-282 (TM6)[443]. As a result, TM6 (and to a lesser extent TM5) undergoes a hinge movement that pushes its intracellular tip outward (Figure 7.42c); this movement is accompanied by much smaller inward movements of TMs 3 and 7 (not shown). **The movements of TMs 5 and 6 create sufficient room for accommodating the carboxyl end of the G-protein's α5 helix, which is pushed to a partial extent into the transmembrane core of the receptor**[*2]. This process leads to a large displacement of one of Gα's domains with respect to the other, both of which hold the GDP cofactor in the inactive G-protein (Figure 7.42d), and this displacement is suggested to promote the exchange of GDP with GTP during activation[444]. Note that the overall conformational change of the β_2-adrenergic receptor is relatively similar to the changes observed in rhodopsin, the β_1-adrenergic receptor, and the muscarinic (M$_2$) receptor (Figure 7.42e, see also below). Interestingly, the displacement of TM6 in the β_1-adrenergic receptor is much smaller than that observed in the β_2-adrenergic receptor. This difference may have to do with the fact that of the two structures, only that of the β_2-adrenergic receptor was crystallized when bound to a G-protein, which was likely to have induced larger conformational changes. Small displacements of helices are also observed in the adenosine (A$_{2A}$) receptor, whose active structure, like the active structure of the β_1-adrenergic receptor, was crystallized in the absence of a G-protein[445,446]. Indeed, **Gα binding has been shown by NMR**

[*1]Although the all-*trans* and *cis* retinal can be considered as an agonist and an inverse agonist, respectively.

[*2]This is also made possible by conformational changes in the second intracellular loop, ICL2. These changes involve rearrangement of the interactions between Asp-130 of the *D/ERY* motif, Asn-68, and Tyr-141, resulting in displacement of the latter from the space, now occupied by the α5 helix. In its new position, the α5 helix interacts with different residues in TMs 3, 5, and 6, as well as in ICL2.

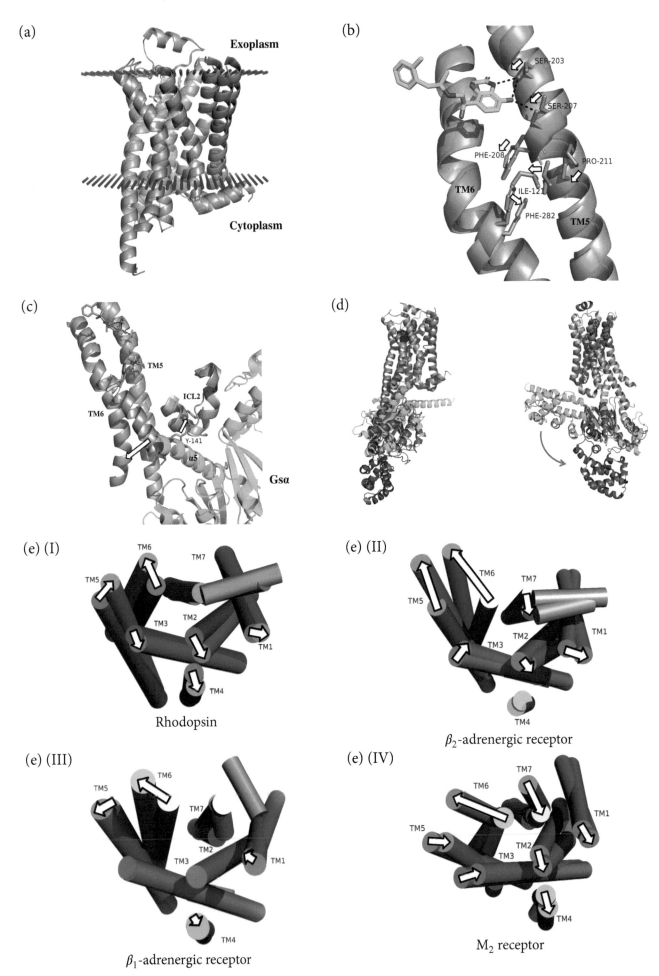

(a)

(b)

(c)

(d)

(e) (I) Rhodopsin

(e) (II) β_2-adrenergic receptor

(e) (III) β_1-adrenergic receptor

(e) (IV) M_2 receptor

and molecular dynamics studies to be required for fully stabilizing the conformational changes induced by the agonist [431–433].

We have seen that the chain of events leading to the large conformational changes on the intracellular side of the β_2-adrenergic receptor starts with the hydrogen bonds formed between the agonist and serines 203 and 207 in TM5. These interactions pull TM5 slightly inward, causing rearrangement of other, more downstream interactions, in addition to movement of TM6. Interestingly, whereas Phe-208 is involved in this rearrangement, the other residue included in the transmission switch, Trp-286, does not seem to play a part in the activation of the β-adrenergic receptor. The involvement of serines 203 and 207 in the activation is not surprising considering their high conservation levels in aminergic receptors [447]. Carazolol, which is bound to the inactive structure of the β_2-adrenergic receptor, also interacts with TM5. However, it forms only a single hydrogen bond with Ser-203, and it seems that this interaction, which is weaker than that of BI-167107 in the active structure, is insufficient to induce the aforementioned structural changes. This may explain why carazolol acts as an inverse agonist, whereas BI-167107 acts as an agonist. The same differential interactions with TM5 are observed in the β_1-adrenergic receptor; in this case, they involve the agonist isoprenaline and the antagonist cyanopindolol, although the resulting conformational changes observed in this GPCR are much smaller. In contrast, in the A_{2A} receptor, TM2 and TM7 are the helices that interact differently with the agonist

FIGURE 7.42 **Structural changes in the β_2-adrenergic receptor following its activation.** (Opposite) (a) Superimposition of the active and inactive structures of the β_2-adrenergic receptor. The inactive structure (grey; PDB entry 2rh1) is bound to the inverse agonist carazolol, and the active structure (orange; PDB entry 3sn6) is bound to the agonist BI-167107. The red and blue lines mark the predicted boundaries of the membrane (the OPM database [117]). (b) Agonist-induced movements of TMs 5 and 6, and of key residues that act as molecular switches. The agonist is shown as sticks, colored by atom type. The hydrogen bonds between the agonist and residues in TM5 are shown as dashed lines. The resulting movement of TM5 leads to local movements of the downstream nonpolar residues Phe-208, Pro-211, Ile-121, and Phe-282, and the hydrophobic contacts between them. The movement of each of the above residues is marked by arrows. (c) Conformational changes on the intracellular side of the receptor upon agonist binding. The changes described in (b) induce a large hinge movement (curved arrow) in TM6, which creates room on the intracellular side for the $\alpha 5$ helix of G_α (marked). In addition, ICL2 shifts away from TM6, with a large change in the position of Tyr-141. Smaller changes in the positions of TM3 and TM7 are not shown. As can be clearly seen, the position of TM6 in the inactive state clashes with the helix of $G\alpha$, which prevents the binding of the latter to the GPCR. (d) Superimposition of the structure of $G\alpha$ in complex with the β_2-adrenergic receptor (PDB entry 3sn6; $G\alpha$ is colored in green and the β_2-adrenergic receptor is colored in orange) and free $G\alpha$ bound to 5′-guanosine-diphosphate-monothiophosphate (GTPγS) (PDB entry 1gia; $G\alpha$ is colored in magenta). The superimposed structures are shown from two angles, rotated 90° from each other. The superimposition shows that the activation of $G\alpha$ involves a substantial displacement of one of its domains with respect to the other (red arrow). (e) Comparison between movements of helices in rhodopsin and those in the β_2-adrenergic receptor, β_1-adrenergic receptor (where the inactive and active structures correspond to PDB entries 2vt4 and 2y03, respectively) and muscarinic (M_2) receptor. The inactive and active conformations of all three GPCRs are colored in grey and orange, respectively. The arrows mark movements of the intracellular sides of the helices, where the length of each arrow is proportional to the degree of movement. For clarity, the helices are shown as cylinders, and the loops are not shown.

(a)

(b)

(c)

(d)

(NECA) as compared to the inverse agonist (ZM241385), and the conformational changes are transmitted to the intracellular side by TMs 3, 6, and 7, rather than TM5 [445]. Thus, **it seems that whereas the interactions of antagonists and inverse agonists with their respective ligand-binding pockets generally differ from the interactions of agonists, the exact mechanisms through which these interactions inhibit or activate the GPCR may differ across GPCRs, and involve different molecular switches and/or transmembrane segments.**

Muscarinic receptors belong to the same class as rhodopsin and the β-adrenergic receptors (class Aα). These receptors, which mediate cholinergic transmission, include five subtypes. They modulate a variety of physiological functions and are targeted by drugs for treating different diseases (e.g., Alzheimer's disease, Parkinson's disease, and schizophrenia). In 2012 the antagonist-bound inactive structures of the M_2 [448] and M_3 [449] receptor subtypes were determined, and a year later the fully active structure of the M_2 receptor was determined, bound to the agonist iperoxo and to a nanobody on its intracellular side [437]. Comparison between the inactive and active structures of the M_2 receptor shows conformational changes at the ligand-binding site that involve TMs 5, 6, and 7, and which are larger than those seen in rhodopsin and the β_2-adrenergic receptor. These changes lead to the complete burial of the agonist inside the pocket, occluded from the solvent (Figure 7.43a, top). This occlusion is created by an 'aromatic lid' above the agonist, composed of Tyr-104,

FIGURE 7.43 **Activation of M_2 receptor and allostery.** (Opposite) (a) Conformational changes in the ligand-binding site of the M_2 receptor. *Top*: A cross-section through the binding site of the inactive (grey; PDB entry 3uon) and active (orange; PDB entry 4mqs) structures of the M_2 receptor. The inactive structure is bound to the antagonist QNB, and the active structure is bound to the agonist iperoxo. The aromatic lid residues above the ligands are colored in yellow. Activation induces conformational changes that lead to the complete occlusion of the agonist from the solvent. *Bottom*: Activation-induced closure of the aromatic lid (yellow sticks). (b) Propagation of the conformational change in TM6 from the ligand-binding site to the intracellular site of the receptor. The image shows a superimposition of the active and inactive structures of the M_2 receptor. For clarity, only TMs 3 and 6 are shown. In the inactive state the antagonist is hydrogen-bonded to Asn-404 on TM6 (black dashed line). Activation of the GPCR induces a small movement of TM6 towards TM3 and allows Asn-404 to remain hydrogen-bonded to the agonist (orange dashed line), which is located further away from TM6 than the antagonist is. The pivot motion of TM6 around Thr-399 (curved white arrow) converts the small movement of the helix near the ligand-binding site into a large movement at the intracellular site, away from TM3. A coordinated movement of TM3 away from TM6 allows a hydrogen bond to be formed between Asp-120 and Asn-58. The activation also involves a conformational change in TM7, which changes the orientation of Tyr-440 (the *NPxxY* motif), thus allowing it to interact with Tyr-206. Note that although the interaction is shown here as a direct hydrogen bond (black dashed line), in reality it is mediated by a water molecule, which is missing in the solved structure. (c) Binding of the allosteric activator LY2119620 to the agonist-bound M_2 receptor (PDB entry 4mqt). LY2119620 binds above the agonist pocket, where it is separated from the agonist iperoxo by the aromatic lid residues (in yellow). (d) Membrane-facing position of an allosteric site in the $P2Y_1$ receptor (PDB entry 4xnv). The allosteric site is occupied by the negative modulator BPTU (with blue carbon spheres). The orthosteric site, which resides in the extracellular loop region of the receptor, is shown with a bound antagonist (orange spheres; PDB entry 4xnw). Co-crystallized lipid molecules are also shown, in pink. The red and blue lines mark the predicted boundaries of the membrane (the OPM database [117]).

Tyr-403, and Tyr-426 (yellow patch in Figure 7.43a, top). In the receptor's inactive state the lid is partly open, but upon activation, movements of these residues, especially Tyr-403 and Tyr-426, allow them to hydrogen-bond with each other and close the lid (Figure 7.43a, bottom).

On the muscarinic receptor's intracellular side, the conformational changes that follow activation are overall similar to those observed in rhodopsin and in the β_2-adrenergic receptor (Figure 7.42eIV). In particular, the outward movements of TMs 6 and 2, and the accompanying inward movements of TMs 7 and 3, are very similar to the movements seen in the β_2-adrenergic receptor, and create the G-protein binding site in a similar manner. As in the other GPCRs, the large conformational change of TM6 begins in the ligand-binding pocket and propagates to the intracellular side. In the ligand-binding pocket both the antagonist and the agonist hydrogen-bond with Asn-404. However, the agonist, which is smaller than the antagonist, is further away from TM6 and closer to TM3 than the antagonist is. The interaction between the agonist and Asn-404 therefore pulls TM6 closer to TM3 in this region (Figure 7.43b). TM6 undergoes a pivot motion around Thr-399, and its intracellular side is displaced farther away from TM3. In light of the above, in addition to the results of mutational studies [437,450], Asn-404 is thought to be a key residue in the activation of the M_2 receptor. The distance created between TMs 3 and 6 upon activation is also the result of a slight displacement of TM3, which involves the formation of a stabilizing hydrogen bond between Asp-120 of the *D/ERY* motif and Asn-58 (the equivalent of Asn-68 in the β-adrenergic receptor). We have already encountered this interaction in the active β_2-adrenergic receptor, suggesting that it plays a general role in GPCR activation, and refuting the association of the *D/ERY* motif with stabilization of the inactive state. The second motif implicated in GPCR activation, *NPxxY*, also seems to be important for the activation of the M_2 receptor. Specifically, Tyr-440 of the motif (the equivalent of rhodopsin's Tyr-306) on TM7 becomes closer to Tyr-206 (the equivalent of rhodopsin's Tyr-223) on TM5, and the water-mediated interaction that is thought to occur between them is expected to stabilize the active state of the receptor.

The study described above also investigated allostery in the M_2 receptor [437]. The protein was crystallized in complex with the agonist iperoxo and the positive allosteric modulator LY2119620. The allosteric activator binds directly above the agonist (Figure 7.43c). **Excluding a few small adjustments of the GPCR structure to the activator, mainly involving residues that interact with the latter, the two structures (i.e., bound and unbound to the activator) are very similar, indicating that the allosteric site is pre-formed by the agonist. This idea is in line with our current view of allostery, which posits that allosteric activators stabilize an active conformation of the protein, whereas allosteric inhibitors stabilize an inactive conformation (see Chapter 5 for more details).** Allosteric modulators have been recognized in many class A GPCRs, including the adenosine, dopamine, histamine, serotonin and chemokine receptors, as well as in class C GPCRs [451]. The allosteric sites identified in these GPCRs reside in the extracellular or transmembrane domains [379]. The modulators are chemically diverse and include lipids (e.g., fatty acids, phospholipids, and cholesterol), amino acids, ions (e.g., Na^+), and various small molecules. In fact, the G-proteins and other intracellular binding partners of GPCRs (e.g., *GRK* and *arrestins*, see following subsection) can also be regarded as allosteric modulators, as each binds preferentially to an active or inactive conformation of the receptor and stabilizes it. As discussed in Section 7.5.7 below, allosteric modulators are highly sought-after by the pharmaceutical industry, for various reasons. Allosteric sites in GPCRs may appear in different locations of the protein, including the outer, membrane-facing surface. Such a position of an allosteric site is observed, e.g., in the P2Y$_1$ (purine) receptor [452] (Figure 7.43d).

To conclude, the studies discussed above, as well as many others carried out in recent years, have produced the following insights about the activation of class A GPCRs:

- The activation process usually involves relatively small conformational changes in the extracellular and ligand-binding domains. One exception is the P2Y$_{12}$ receptor, where 5 to 10 Å shifts are observed on the extracellular sides of TM6 and TM7.

- The activation results in the transmission of small, local conformational changes from the ligand-binding site of the GPCR to its intracellular side by a variety of molecular triggers, which include ionic locks and transmission switches (see [453] for a more detailed description). Although these triggers do not act identically in all GPCRs, they involve physically similar mechanisms and structurally equivalent positions, which are included in highly conserved motifs, such as *D/ERY* and *NPxxY*.

- The local conformational changes are amplified as they propagate towards the intracellular side, resulting in relatively large shifts of transmembrane helices, especially TM6, and to a lesser extent TMs 3, 5, and 7, which create a binding site for the G-protein.

- The rearrangements of helices on the intracellular side are overall similar in both light-activated rhodopsin and agonist-activated GPCRs.

- Gα undergoes large conformational changes upon binding to its receptor, which in turn promotes nucleotide exchange and activation of downstream signaling.

Nevertheless, there are aspects of the activation process that are yet to be clarified. These include the following:

- The generality of the structural changes observed in rhodopsin and in the β-adrenergic receptors. Evaluating generality would require the determination of additional fully-active GPCR structures, i.e., bound simultaneously to an agonist and to an intracellular binding partner. It is particularly important to gain such information for GPCRs of other classes beyond A, for which no fully active structures are available.

- The GDP-GTP exchange. The β_2-adrenergic receptor was determined in complex with nucleotide-free Gs. To obtain a complete view of the activation process, it is necessary to carry out additional studies on the GTP-bound form of GPCRs, as well as on the intermediates that may exist between the two states. It should be noted that crystallizing the active state of the GPCR with GTP-bound G-protein is not trivial, as the presence of GTP promotes dissociation of the G-protein from the receptor.

- Ligand selectivity. Additional studies are required to understand the mechanistic effects of ligands with different functionality (agonists, antagonists, inverse agonists) and receptor subtype specificity, as well as the effects of *biased ligands*, that is, ligands whose binding to the GPCR leads to activation of specific signaling pathways [454,455]. Moreover, GPCRs may also be affected by allosteric ligands, and it will be interesting to characterize the mechanisms of activation or inhibition employed by such ligands.

- At the system level it is intriguing that GPCRs are an order of magnitude more diverse than their G-proteins (~600 versus ~20). From a signal processing view one may wonder what benefit is obtained from having such a diverse 'sensing end' (GPCRs) that eventually reduces to the limited G-protein repertoire.

7.5.5   GPCR desensitization

Part of the regulation of GPCR action involves GPCR inhibition following short or prolonged activation, so as to prevent over-stimulation of the signaling system [456]. The first type, i.e., downregulation that occurs very soon after the activation of the GPCR, is called *desensitization* [336,457,458]. There are two types of desensitization:

1. **Homologous** desensitization acts on the activated receptor and is mediated by phosphorylation of Ser and Thr residues in ICL3 or the C' of the GPCR (Figure 7.33, step 4). The phosphorylation is carried out by a specific group of Ser/Thr kinases called *GRKs* [459], which act only on the agonist-bound conformation of the receptor. The phosphorylation serves to increase the affinity of the GPCR to proteins of the *arrestin family*, and facilitate their binding to the receptor (Figure 7.33, step 5). The binding to arrestin has two outcomes: first, it prevents the GPCR from interacting with its cognate G-protein, thus stopping the signaling. Second, it recruits *clathrin* and its adaptor protein AP-2, which induces the internalization of the GPCR into the cell via *clathrin-coated vesicles* [460], after which the receptor is recycled or degraded [461]. Interestingly, class A GPCRs lose the clathrin coat and become dephosphorylated following internalization; as a result, it is thought that they may be able to continue signaling even when inside the endosome [462]. In such a case, however, the GPCR and its effectors are closer to the cell's nucleus compared to their initial location in the plasma membrane, which might make the activation of the transcriptional pathway more efficient [463]. In contrast, class B GPCRs remain bound to arrestin following internalization, which leads to their ubiquitylation and degradation. It should be noted that, according to recent studies, arrestins' involvement in cellular signaling is much more complicated than that mentioned above [464]. Specifically, it seems that arrestins are involved in biased agonism: by binding to GPCRs they stabilize certain conformations that block certain signaling pathways and promote other pathways [455,464]. Thus, arrestins should be viewed as multifunctional adapter proteins rather than signal terminators [464].

2. **Heterologous** desensitization acts on other receptors and is mediated by second messenger-activated kinases, such as PKA or PKC.

There are additional downregulation mechanisms that act on activated GPCRs following prolonged stimulation of the receptor. These may act at several levels, including gene transcription and translation [461].

In 2015 the structure of active rhodopsin bound to visual arrestin [465] was determined (Figure 7.44a). Interestingly, the structure showed that arrestin binds to rhodopsin asymmetrically, which should allow arrestin's conserved hydrophobic residues (Phe-197, Phe-198, Met-199, Phe-339, and Leu-343, Figure 7.44a, pink area) to touch, or even insert into the nonpolar region of the membrane. In contrast to G-proteins and GRKs, arrestins are not attached to hydrophobic chains (palmitoyl or prenyl) that anchor them to the membrane. Thus, the conserved hydrophobic patch may be the only means by which arrestins can become anchored to the membrane, and this anchoring may in turn stabilize arrestin's interaction with the GPCR [465]. Indeed, mutation of any of these residues to alanine affects the binding of arrestin to rhodopsin [466]. The highly asymmetric shape of the rhodopsin-arrestin complex has also been suggested to affect the curvature of the membrane, perhaps as part of arrestin's role in initiating rhodopsin's endocytosis [465].

From the arrestin end, binding to rhodopsin is mediated by several elements. A short helical segment of arrestin inserts into the intracellular side of rhodopsin, in a process similar to the insertion of the $G\alpha$ subunit of rhodopsin's G-protein, transducin (Figure 7.44b). The helical segment of arrestin interacts with the carboxy-terminus of TM7 and with the amino terminus of helix 8. Indeed, both of these elements have been implicated in previous studies as important for arrestin binding [467,468]. Thus, arrestin directly competes with transducin. Another interesting interaction occurs between arrestin and the second intracellular loop of rhodopsin (ICL2, Figure 7.44b left). In its apo form (detached from rhodopsin), arrestin assumes a closed conformation in which the interaction region is inaccessible (Figure 7.44c, left). Upon binding it opens up, to accommodate rhodopsin's ICL2 (Figure 7.44c, right). In arrestin-bound rhodopsin, ICL2 adopts a helical conformation, whereas in arrestin-free, active rhodopsin it is organized as a loop.

As mentioned above, arrestin binds with high affinity to activated rhodopsin only after the latter has been phosphorylated by a GRK. In the rhodopsin-arrestin structure described above, rhodopsin is not phosphorylated and the binding was made possible by the introduction of mutations into the two binding partners. Thus, the structure of arrestin in that case represents a pre-activated state of the protein. Determining the structure of phosphorylated rhodopsin bound to activated arrestin will enable us to understand the process of arrestin activation and also, hopefully, the initiation of signaling pathways associated with arrestin binding.

7.5.6   GPCRs of other classes

The number of structures determined for GPCRs outside class A is much smaller than the number of class A structures. Known non-class-A structures include the following:

Class B – glucagon and corticotropin-releasing factor (CRF) receptors

Class C – mGlu1, mGlu5, and $GABA_B$ receptors

Class F – Smoothened protein (SMO, a GPCR-like receptor)

7.5.6.1   Class B GPCRs

Class B GPCRs bind large peptide hormones, and therefore constitute attractive targets for therapeutic drugs used to treat diseases associated with glucose metabolism (e.g., diabetes), the stress response, cardiovascular regulation, etc. Unfortunately, the structural data on these GPCRs is relatively scarce, with only two members of the group characterized structurally: the corticotropin-releasing factor (CRF) [469] and glucagon [470] receptors. Moreover, stabilization of the structures for crystallization requires, among other things, the removal of large portions of the N- and C-termini. Thus, the structures represent mainly the transmembrane domains of the receptors. The structure of CRF receptor subtype 1 (CRFR1) was determined in complex with the non-peptide antagonist CP-376395. There are three striking differences between the structure of CRFR1 and those of the class A GPCRs discussed above:

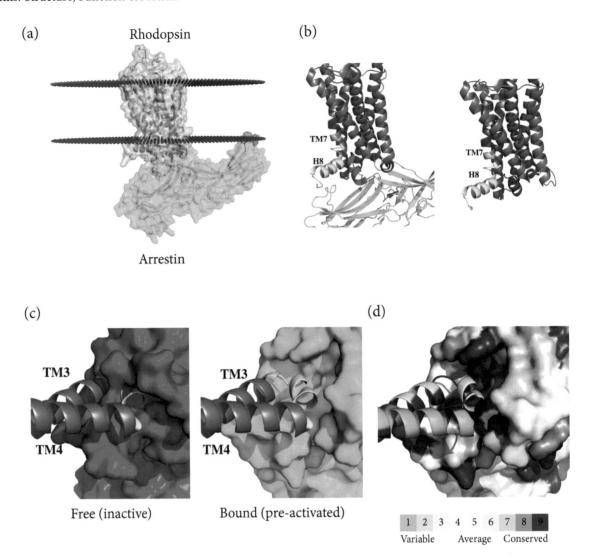

FIGURE 7.44 **Binding of arrestin to active rhodopsin.** (a) The overall structure of the complex (PDB entry 4zwj), where rhodopsin is in gold and arrestin is in green. The conserved hydrophobic patch in arrestin is colored in pink. The red and blue lines mark the predicted boundaries of the membrane (the OPM database [117]). (b) *Left*: The interaction of arrestin's short helical segment with rhodopsin's TM7 and helix 8, where rhodopsin is in blue, the transducin segment is in magenta, and the TM7-H8 binding site is in yellow. *Right*: The structure of active rhodopsin bound to a short segment of the Gα subunit of transducin (PDB entry 3pqr), where the transducin segment is in magenta. The segment binds to the same intracellular pocket of rhodopsin that binds arrestin's short segment, although the orientations of the two segments within the pocket are different. (c) Interaction between arrestin and rhodopsin's ICL2 (colored yellow). *Left*: Free arrestin in its basal (inactive) state (red, PDB entry 1cf1), superimposed on the bound arrestin. *Right*: The rhodopsin-arrestin complex. The rhodopsin is not phosphorylated, and arrestin is therefore said to be in its pre-activated state (see main text). The bound arrestin has an open conformation that accommodates ICL2. For clarity, only TM3, TM4 and ICL2 are shown in rhodopsin. The ICL2-interacting pocket is close, and sterically clashes with the short helix of ICL2. (d) Evolutionary conservation of the rhodopsin-arrestin interface. The image is identical to the one shown on the right side of (c), but both rhodopsin and arrestin are colored according to conservation levels (cyan – lowest, maroon – highest; see color code in figure). The conservation levels are calculated by the ConSurf web server (http://consurf.tau.ac.il) [373,374]. As can clearly be seen, the residues in both proteins that form the interface are evolutionarily conserved.

1. **CRFR1 is V-shaped and has a large cavity** (Figure 7.45a).

2. The antagonist's binding site is located near the intracellular side. Note, however, that the native (peptide) ligands of class B GPCRs bind to both the extracellular and transmembrane domains of the receptor (see more below).

3. TM7 has a sharp kink above the midpoint of the transmembrane domain, around Gly-356. This residue is part of a conserved *QGxxV* motif in class B GPCRs. Gly-356 allows TM7 to acquire the kink, similarly to the *NPxxY* proline in class A GPCRs, which also induces a distortion in TM7. The intracellular side of the transmembrane domain, on the other hand, is similar in shape to that of class A GPCRs (despite the lack of ICL2), indicating that this structure is likely to be able to bind the cognate G-protein.

The glucagon receptor (GCGR), which was crystallized in the same year as CRFR1, was also shown to include a large cavity (Figure 7.45b), but since the ligand could not be resolved, the exact binding location is unknown. The structure of the glucagon receptor does not have the pronounced V shape seen in the CRFR1 structure. Moreover, GCGR seems to be more similar in structure to class A GPCRs than CRFR1 is, in terms of the orientations and positions of the transmembrane helices. One clear difference between the GCGR structure and the structures of class A GPCRs is TM1, which is significantly longer in the former. This region is thought to be involved in glucagon binding and in the positioning of the ECD with respect to the TMD.

The structures described above lack portions of the ECD and a bound peptide. This prevents us from understanding the spatial relation of the ECD to the TMD, how the natural peptide agonist binds to the receptor, and how the receptor is activated. There are, however, biochemical data implicating a *GWGxP* motif in a functionally important network of interactions. These interactions are seen in the CRFR1 structure, but since the active structure of the receptor is unknown, their exact role is yet to be understood. As mentioned above, the natural peptide ligands of class B GPCRs bind to both the ECD and the TMD. Specifically, the *C*-terminus of the peptide agonist binds primarily to the ECD, whereas the *N*-terminus binds to the TMD. This pattern has prompted studies of the isolated ECD of class B GPCRs in complex with a peptide agonist (see review by Parthier and coworkers [471]). The studies indicate that members of this group have similar ECDs, all of which are quite large. This is consistent with the large ligands of class B GPCRs, i.e., peptides comprising ~30 amino acids [472]. The peptide-binding site is composed of two small β-sheets and an adjacent α-helix, and is stabilized by three disulfide bonds (Figure 7.46a). **The binding of the peptide ligands seems to be coupled to their folding (Figure 7.46b), similarly to what we have seen in intrinsically unstructured proteins (Chapter 6).** Upon binding, the C' of the peptide ligand is squeezed between the two β-sheets of the ECD (Figure 7.46b), where it forms mainly nonpolar but also polar contacts with complementary ECD residues of loops 2 and 4, and of the C' of the GPCR (Figure 7.46c). On the other side, the N' of the peptide ligand is held close to the TMD (Figure 7.46a). In this orientation, the N' of the ligand can interact with extracellular loops and transmembrane helices of the GPCR. The binding is also accompanied by conformational changes in ECD loops, which may be transmitted to the TMD, and therefore constitute part of the activation process [471].

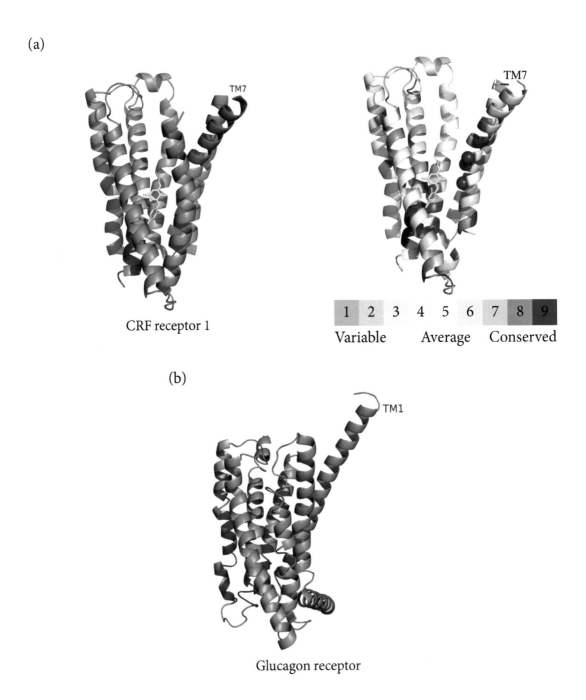

(a)

CRF receptor 1

| 1 | 2 | 3 | 4 | 5 | 6 | 7 | 8 | 9 |

Variable · · · · Average · · · Conserved

(b)

Glucagon receptor

FIGURE 7.45 **The transmembrane domain of class B GPCRs.** (a) The inactive corticotropin-releasing factor receptor 1 (CRFR1) bound to the non-peptide antagonist CP-376395 (PDB entry 4k5y). *Left*: The full structure of the receptor (blue ribbons), with the bound ligand shown as yellow sticks. TM7, which has an unusual kink, is marked. *Right*: The same structure, colored according to evolutionary conservation level (cyan – lowest, maroon – highest; see color code in figure). The Cα atoms of the highly-conserved Gln-355, Gly-356, and Val-359 (the *QGxxV* motif) are shown as spheres. TM5 is removed for clarity. Conservation levels are calculated by the ConSurf web server (http://consurf.tau.ac.il) [373,374]. (b) The inactive glucagon receptor (PDB entry 4l6r). The receptor was crystallized in the presence of the antagonist NNC0640, but the latter was not resolved in the structure.

(a)

(b)

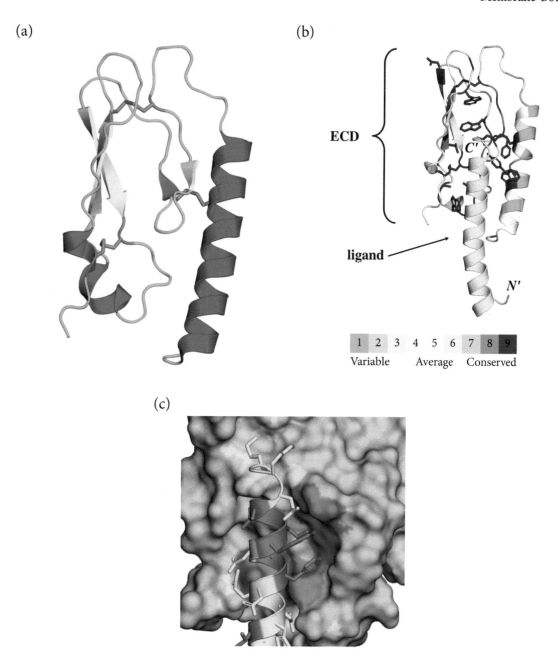

FIGURE 7.46 **Ligand binding in class B GPCRs.** The extracellular domain (ECD) of the human gastric inhibitory peptide (GIP) receptor, in complex with the hormone GIP (PDB entry 2qkh). (a) The structure of the ECD, colored by secondary structure. The disulfide bonds are represented as orange sticks. (b) General view of the ECD-GIP complex. The ECD is colored according to conservation level (cyan – lowest, maroon – highest; see color code in figure), and the peptide ligand is in yellow, with its termini noted. The side chains of the highly conserved residues in the ECD are shown as sticks. As would be expected, most of the highly-conserved residues in the ECD face the peptide ligand. Conservation levels are calculated by the ConSurf web server (http://consurf.tau.ac.il) [373,374]. (c) ECD-ligand interactions. The surface of the receptor's ECD is shown in grey. The ligand peptide is shown as a ribbon, with its side chains represented as sticks. In both the ligand and the ECD, polar residues involved in protein-ligand interactions are in yellow, and nonpolar interacting residues are in orange. The interactions are similar across different class B GPCRs, and the ECD loops 2 and 4, which mediate these interactions, are consistently the parts undergoing the most significant conformational changes following activation.

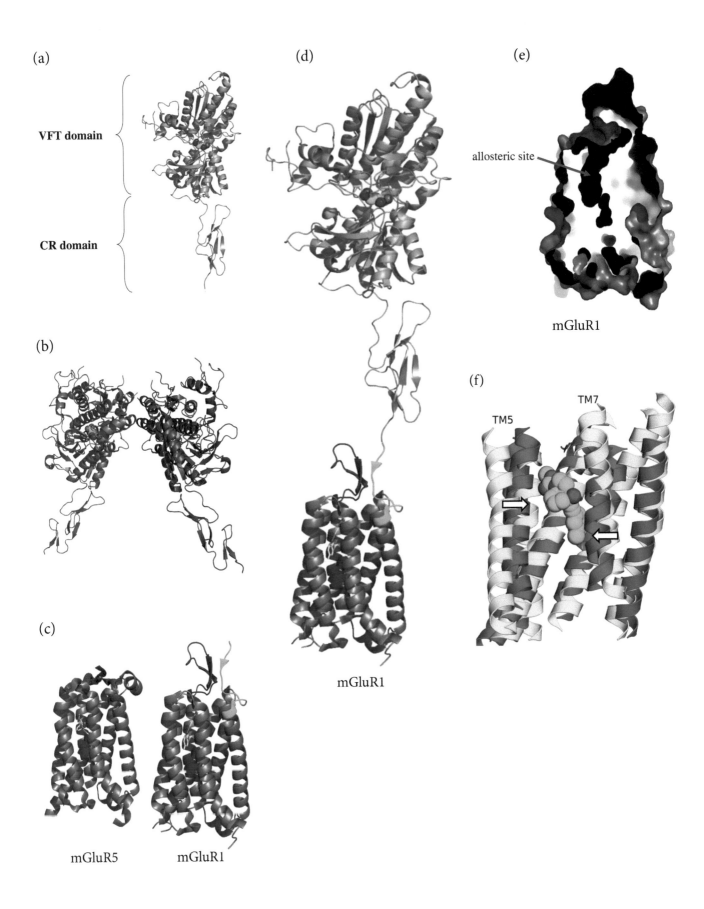

(a)

VFT domain

CR domain

(b)

(c)

mGluR5   mGluR1

(d)

mGluR1

(e)

allosteric site

mGluR1

(f)

TM5

TM7

7.5.6.2   Class C GPCRs

Studies on class C GPCRs have concentrated mainly on the metabotropic glutamate receptors (mGluR), which are classified into three subgroups based on sequence similarity, agonist selectivity, and effector:

>   **I** – mGluR subtypes 1 and 5

>   **II** – mGluR subtypes 2 and 3

>   **III** – mGluR subtypes 4, 6, 7, and 8

The focus on mGlu receptors is not surprising considering their physiological, medical and therapeutic importance. Glutamate, the native agonist of these receptors, is the principal excitatory neurotransmitter in the central and peripheral nervous systems, and it is involved in numerous neurological functions, including memory and learning, sensory and motor functions, emotions, etc. Thus, mGluR malfunctioning leads to diseases such as epilepsy, neurodegeneration, chronic pain, schizophrenia, anxiety, and autism [474]. Accordingly, mGlu receptors are promising targets for different neurological and psychiatric drugs.

The structure of the ECD of mGlu receptors has been determined [473] and shown to consist of a *'Venus fly trap' (VFT) domain*, which binds the glutamate agonist, and a *cysteine-rich (CR) domain*, which links the VFT domain to the TMD (Figure 7.47a). Thus, in contrast to GPCRs of classes A and B, class C GPCRs bind their ligands only through the ECD. The VFT domain also mediates the dimerization of mGluR[*1], and its crystallized structures demonstrate substantial conformational flexibility, which is stabilized by agonist binding. In 2014, the structure of the TMD of mGluR5 was determined in complex with the negative allosteric modulator mavoglurant [475] (Figure 7.47c, left). This structure was of great importance for two reasons. First, allosteric modulation of mGluR5 was postulated to be effective in treating anxiety disorders (negative modulation), as well as schizophrenia and disorders of cognitive function (positive modulation) [476]. Second, as explained earlier, drugs acting on allosteric

FIGURE 7.47   **The structure of the metabotropic glutamate receptor (mGluR).** (Opposite) (a) The extracellular domain of mGlu3 monomer in complex with glutamate [473] (PDB entry 2e4u). The Venus fly trap (VFT) and cysteine-rich (CR) domains are noted. (b) The dimeric structure of the EC domain of mGlu3. Each monomer is in a different color. The monomers interact via the VFT domain. The disulfide bond between the monomers is located in a disordered region that was not resolved in the structure. (c) The transmembrane domains of mGluR5 (PDB entry 4oo9) and mGluR1 (PDB entry 4or2), in complex with the negative allosteric modulators mavoglurant and FITM (respectively), shown as yellow sticks. ECL2, which caps the entrance to the allosteric site, is colored in magenta. The region in mGluR1 linking the TMD to the ECD is colored in green. (d) The postulated structure of the entire monomeric mGluR (for a more elaborate structure and domain interactions, see [404]). The structure was produced by joining the TMD and ECD through the linker regions. (e) A cross-section through mGluR1, showing the narrow allosteric site. (f) Superposition of mGluR5 and rhodopsin (PDB entry 1gzm), showing the differences in the relative orientations of TMs 5 and 7 between them. For clarity, only transmembrane helices are shown.

[*1]mGlu and calcium-sensing receptors create disulfide-linked homodimers (Figure 7.47b; note that the region forming the inter-chain disulfide link is missing in the structure). In contrast, the GABA$_B$ and taste 1 (TAS1) receptors form non-crosslinked heterodimers.

sites are more likely to be subtype-specific, since allosteric sites are less conserved than or-thosteric (agonist-binding) sites. As in the case of the class B receptors described above, large flexible portions of the extracellular and intracellular domains of mGluR5 had to be removed for crystallization, so the exact positioning of the ECD with respect to the TMD is still unknown. The structure of the TMD of mGluR1 was determined in the same year, bound to the negative allosteric modulator FITM [404] (Figure 7.47c, right). In the case of mGluR1, however, the linker region to the ECD was resolved, allowing us to postulate the structure of the entire monomeric receptor (Figure 7.47d). Also, the structure was solved as a dimer, with six cholesterol molecules residing at the interface between the monomers (not shown). **These observations support the suggestion made earlier, that cholesterol molecules inside the membrane are involved in GPCR dimerization.**

The TMDs of mGluR1 and mGluR5 are overall similar to the TMDs of class A and class B GPCRs, especially on the intracellular side. As already explained above, this makes sense, considering that there are only a few intracellular binding partners for all GPCRs, such that minimal structural diversity is required in the intracellular region. When the different classes are compared in terms of the conformation of each helix in the TMD, most of the differences seem to be located in TMs 5 through 7 (see more below). As Figure 7.47c shows, in both mGluR structures, the allosteric binding site is deeper than the average class A binding site, yet is not as deep as the (class B) CRF receptor's binding site for the antagonist CP-376395 (see above). Still, FITM extends further towards the extracellular side of the transmembrane domain. The different locations and interactions of the two allosteric modulators with the TMD highlight the aforementioned potential of such modulators to serve as subtype-specific drugs. The entrance to the allosteric site in both structures is occluded by ECL2 (Figure 7.47c), and it is quite narrow (Figure 7.47e), mainly due to TMs 5 and 7, which have a more inward orientation compared to their counterparts in class A and class B GPCRs (Figure 7.47f). The capping of the TMD by ECL2 is consistent with the fact that, unlike in class A GPCRs, the native ligands of mGlu receptors bind to the ECD, so there is no need for a wide entrance to the TMD. Finally, the mGlu receptors include a set of ionic locks and other interactions[*1], which, as we have seen in class A GPCRs, are important for structural stabilization and the activation process. Although the positions involved in these interactions differ among the GPCR classes, their functional mechanisms are similar. A more detailed comparison between the functional motifs of class A GPCRs and those in class C (mGlu5R) is given in [475].

7.5.6.3 Class F GPCRs

Smoothened (SMO) is a GPCR-like protein that constitutes a part of the *hedgehog (Hh) signaling pathway*, which regulates animal embryonic development [477]. Malfunctioning of this pathway leads to embryonic malformations, and sometimes to cancer in adults. Several structures of SMO have been determined since 2013, in complex with different ligands [478–480]. In the first structure, of human SMO in complex with the antagonist LY2940680 [478], the architecture of the TMD was similar to that of class A GPCRs, de-

---

[*1]For example, in mGluR5 the ionic lock between Lys-665 (the equivalent of rhodopsin's Arg-135) and Glu-770 contributes to the stabilization of the inactive state. Indeed, disrupting the lock by mutating these residues to alanine leads to constitutive activity of the receptor [475].

spite low sequence identity ($< 10\%$) between the two types of proteins (Figure 7.48a)[*1]. The differences in this domain are mostly in TMs 5 through 7, similar to what we have seen in the glutamate (mGlu) receptors. The ligand resides in the interface between the transmembrane and complex extracellular domains, and interacts mainly with ECL2 and ECL3 (Figure 7.48b)[*2]. As in the mGlu receptors, the ligand-binding cavity in SMO is narrow, partly because of the inward positioning of TM5. In the case of SMO, however, ECL2 is located inside the TMD.

In 2014 the antagonist-bound and agonist-bound structures of SMO were determined [479] (Figure 7.48c). One of the most pronounced differences between the two structures involves an ionic lock in the ligand-binding site, between Arg-400 (TM5) and Asp-473 (TM6). This interaction exists in the antagonist-bound structure but is eliminated upon agonist binding, due to a conformational change in Arg-400 (Figure 7.48c). In its new position, Arg-400 hydrogen-bonds with Asn-477, which is also part of TM6. This 'remodeling' of interactions is likely to serve as a molecular switch in the activation of SMO, although the current structures do not reveal the underlying mechanism. On the intracellular side of the transmembrane domain, the only significant difference between the two structures seems to be in the orientation of TM5 (Figure 7.48d), in contrast to the case of class A GPCRS, whose activation primarily involves changes in TM6. This difference, however, does not necessarily constitute a fundamental distinction between class A and class F GPCRs; the agonist-bound SMO lacks an intracellular binding partner[*3], which means it is not fully activated. Complete activation of the receptor may induce changes more reminiscent of those we observed in rhodopsin and in the β_2-adrenergic receptor.

7.5.7 GPCR-targeting drugs

As mentioned above, it is estimated that 30% to 50% of clinically prescribed drugs target GPCRs [343–346]. This is not surprising considering the numerous physiological processes that are regulated by GPCRs and the high accessibility of these cell-surface receptors. Therefore, the pharmaceutical industry constantly attempts to find new drugs that act on GPCRs [482]. GPCR-acting drugs are currently used to treat a plethora of pathological conditions and disorders (see Table 7.2). These drugs may be grouped as follows:

1. **Directly-acting drugs** affect the activity of the GPCR by binding to it. They can be further classified according to binding site type:

   • **Orthosteric drugs** constitute most of the GPCR-acting drugs. This type of compound binds to the orthosteric site of the GPCR and acts as an agonist, antagonist, or inverse agonist. The latter type is particularly interesting, as demonstrated by the antipsychotic drug *aripiprazole*. Psychosis is associated with overactivation of D_2 (dopamine) receptors in certain brain areas, which is why many antipsychotic drugs, e.g., *haloperidol*, act by blocking these receptors. Unfortu-

---

[*1]The low sequence identity results in the absence of most of the conserved motifs of class A GPCRs, including *D/ERY* (TM3), *CWxP* (TM6) and *NPxxY* (TM7).

[*2]In another antagonist-binding structure the ligand is bound deeper than LY2940680, but it still interacts with both the TMD and ECD [479], a hallmark of class F GPCRs.

[*3]SMO has been shown to activate G-proteins [481], but in the hedgehog pathway it may act (also) by activating other binding partners. For example, in fruit flies, SMO binds the kinesin-like protein *Costal-2 (Cos2)*. The exact way in which SMO activates the mammalian pathway is less understood, and may involve G-proteins.

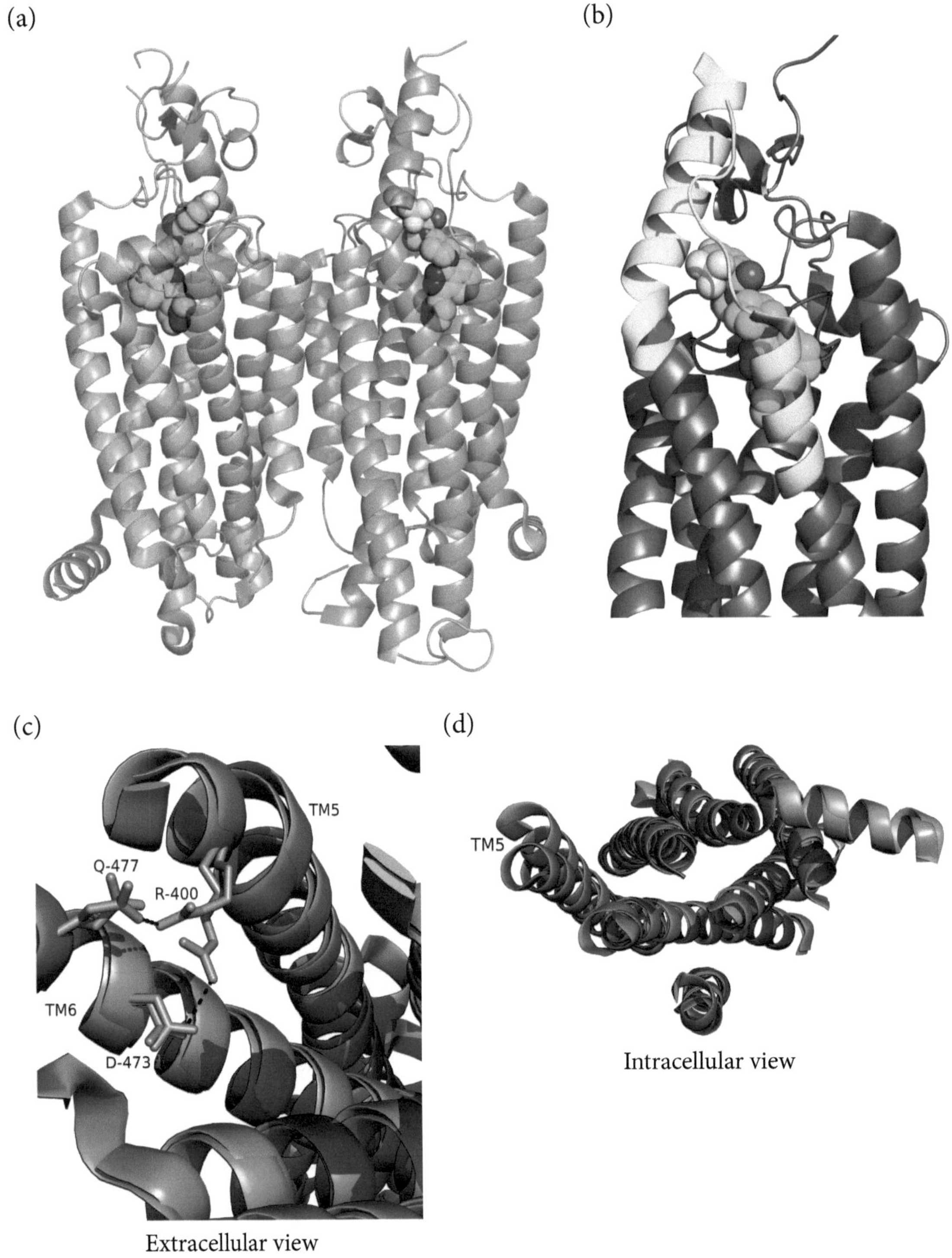

(a)

(b)

(c)

(d)

FIGURE 7.48 **The smoothened protein (SMO).** (a) The structure of the dimeric human SMO, bound to the antagonist LY2940680 (PDB entry 4jkv). Each subunit is colored differently, and the ligand is shown as spheres, colored by atom type. (b) The location of the ligand at the interface between the TMD and the ECD allows the ligand to interact extensively with ECL2 (red) and ECL3 (yellow). (c) and (d) Superposition of an antagonist-bound structure and an agonist-bound structure of SMO (PDB entries 4n4w and 4qin, respectively). For clarity, ECD segments and intracellular loops have been removed. (c) An extracellular view of the superposed structures, showing differences in noncovalent interactions that involve Arg-400, Asp-473, and Asn-477. (d) An intracellular view of the superposed structures showing conformational changes in the orientation of transmembrane helices.

nately, this blockage is a double-edged sword, which also leads to neurological side effects such as *tardive dyskinesia* (i.e., tremors). Like haloperidol and other antipsychotics, aripiprazole also acts on D_2 receptors. However, instead of blocking these receptors, it acts as a partial agonist. Not only does this action of aripiprazole lead to alleviation of psychotic symptoms, but it apparently also results in fewer side effects [483]. Finally, some of the directly acting drugs act as *bivalent ligands*, i.e., they bind in a way that promotes dimerization of GPCR [484].

- **Allosteric drugs** bind to an allosteric site and modulate the activity of the GPCR directly, or by changing its affinity to the natural ligand. *Maraviroc* is an example of an allosterically acting therapeutic agent. It is an antiretroviral drug used for the treatment of HIV infection. The drug targets CCR5 (the C-C chemokine receptor type 5), which is a co-receptor of HIV on certain white blood cells. By allosterically stabilizing an inactive conformation of the receptor [485], maraviroc prevents the binding of the virus to CCR5, thus blocking HIV entry into the host cell. The 3D structure of the CCR5-maraviroc complex (Figure 7.49) suggests that the drug stabilizes the inactive conformation via nonpolar interactions with the transmission switch Trp-248 (the equivalent of rhodopsin's Trp-265). This should prevent movement of the tryptophan residue, which, as we saw earlier, is part of the activation process of certain GPCRs. Allosteric drugs are of great interest to the pharmaceutical industry [482], for at least two reasons. First, they constitute an alternative to the veteran orthosteric drugs, which have already been studied extensively. Second, allosteric drugs enable pharmaceutical scientists to control the activity of target GPCRs more accurately (see Chapter 9 for details), and, since allosteric sites are less conserved than orthosteric sites, allosteric drugs tend to be more subtype-selective.

2. **Indirectly-acting drugs** change the activity of GPCRs by affecting other proteins. SSRIs are well-known examples of such drugs; these antidepressants elevate the levels of serotonin in the brain by inhibiting its reuptake from the synapse [486,487] (see Chapter 1). Although SSRIs directly target serotonin transporters rather than GPCRs, the resulting elevation of serotonin levels leads to increased action of this neurotransmitter on its cognate receptor, which is a GPCR [378].

Academic and industrial labs worldwide are continuously searching for GPCR-acting drugs. The drug discovery process is similar to that corresponding to other protein targets, as described in Chapter 8. Briefly, this process relies mostly on the screening of many known molecules with potential to bind to the target protein and change its activity. In the past the screening process was conducted solely in the lab using various binding and functional assays. Such assays were, however, costly and time-consuming, and could only be carried out effectively by large organizations. The subsequent development of computational methods for small molecules enabled pharmaceutical scientists to conduct *structure-activity relationship (SAR)* studies, in which the structural and chemical properties of a desired drug were deduced on the basis of the activity of other, yet similar, known drugs. Later, more sophisticated computational methods were developed, in which, given a 3D structure of a protein and a ligand, it became possible to characterize the physicochemical interaction between the two molecules (*receptor-based approach*). The determination of numerous high-resolution structures of GPCRs in various states and with various bound ligands has enabled scientists to use the receptor-based approach to predict the binding of many millions of potential compounds to a target GPCR, in a relatively short time (*virtual screening*) [482]. The top-ranking compounds are then tested in the lab to confirm the binding and characterize their pharmacological activity.

As mentioned above, there is a growing effort to develop allosteric drugs that will facilitate more accurate modulation of GPCR activity and complement the arsenal of existing drugs, most of which act as agonists or antagonists on orthosteric sites. However, discovering allosteric drugs is not an easy task, as it requires knowledge of active and inactive conformations of the protein, which usually requires crystallization of the GPCR with various ligands or under different conditions. Additional desired goals in GPCR drug discovery include the following drug types [482]:

- **Biased drugs** that activate specific signaling pathways [454,455]. Such drugs render treatment more specific, and their use is therefore expected to result in fewer side effects. This is particularly important in the case of GPCRs, which, as explained above, are involved in virtually all physiological processes in the body. Furthermore, the use of biased drugs is also expected to allow the targeting of the same GPCR for treating different diseases [482]. In such cases, each biased drug inhibits or amplifies a different signaling route that is initiated by the same GPCR.

- **Dual-acting drugs** that can bind to different receptors or other proteins. Such drugs

TABLE 7.2 **Examples of GPCR-targeting drugs.**

Type of Disease or Disorder	Disease or Disorder	Drug	Target GPCR
Psychiatric	Depression or anxiety	Buspirone	5-HT_{1A} and D_2 receptors
	Schizophrenia	Aripiprazole	D_2 (dopamine) receptor
	Insomnia	Suvorexant	Orexin receptor
Cardiovascular	Hypertension	Valsartan	Angiotensin receptor
	Congestive heart failure		
	Thrombosis	Clopidogrel	$P2Y_{12}$ receptor
Neurological	Pain	Oxycodone	Opioid receptors
	Migraine	Sumatriptan	5-HT_1 receptor
	Vomiting or nausea	Dolasetron	5-HT_3 receptor
Respiratory	Asthma or chronic obstructive pulmonary disease (COPD)	Salmeterol	β_2-adrenergic receptor
Metabolic	Diabetes	Albiglutide	GLP-1 receptor
Hormonal	Hypothyroidism	Parathyroid hormone	PTH receptor
	Acromegaly	Octreotide	Somatostatin receptor
Gastrointestinal	Gastric ulcers	Ranitidine	H_2 receptor
Cancer	Prostate cancer	Leuprolide	GNRH receptor

(a) (b)

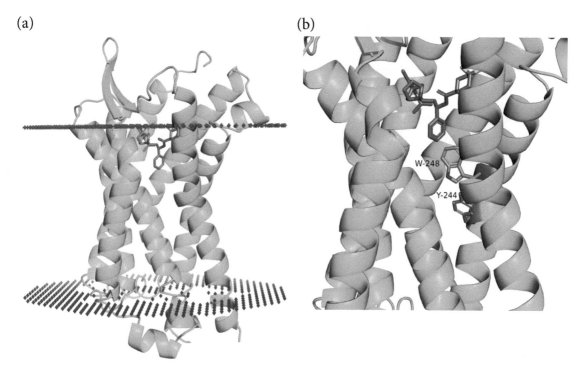

FIGURE 7.49 **Binding of the drug maraviroc to CCR5.** (a) An overall view of the receptor. The receptor is shown as a green ribbon, and maraviroc, which is bound to the ECD-TMD interface, is shown as blue sticks. (b) A magnification of the maraviroc binding site. The drug forms nonpolar interactions with Trp-248, which is thought to prevent the protein from undergoing activation-induced conformational changes. Trp-248 and Tyr-244 constitute a switch that relays the ligand-induced activation process to the intracellular domain. Thus, preventing Trp-248 from moving is expected to inhibit CCR5 activation, which in turn prevents CCR5-HIV binding and viral entry into the cell.

can couple the activity of different GPCRs, or the activity of a GPCR to that of another protein that complements it. For example, the small molecule *donecopride* can both activate 5-HT$_4$ (serotonin) receptors and inhibit the enzyme acetylcholinesterase. The combined effect of the two activities is potentially beneficial for treating Alzheimer's disease [488].

- **Monoclonal antibodies** that bind to desired GPCRs in a highly specific manner and modulate their activity (e.g., [489]). However, the administration and use of such drugs is much more complicated than in the case of small molecules.

- **Drugs that affect the trafficking or desensitization of the GPCR** (e.g., GRKs and arrestins [490]).

7.6 SUMMARY

- Membrane-bound proteins constitute 20% to 30% of the human genome and are involved in numerous cellular and physiological processes. As such, they are also involved in different pathologies, and therefore constitute prime targets for drugs.

- Membrane proteins reside in a two-layered lipid body referred to as the lipid bilayer.

They are either immersed inside it or anchored to it peripherally. The highly anisotropic lipid bilayer has complex physicochemical properties, such as amphipathicity, asymmetry, quasi-fluidity, and curvature.

- These properties, in addition to specific interactions between membrane proteins and lipids, affect the properties of both. Importantly, the properties of the lipid bilayer significantly constrain the structural space available for membrane proteins. Therefore, despite the limited number of experimentally solved structures of membrane proteins, the principles governing their structure can be deduced.

- At the primary structure level, the transmembrane regions of membrane proteins tend to be highly nonpolar and contain on average 21 to 26 residues. These strong tendencies are the basis for current computer algorithms that trace membrane proteins in raw genomic sequences.

- Transmembrane segments of membrane proteins have an extremely high tendency to form secondary structures, mostly α-helices. This property is mainly driven by a need for lipid-exposed polar backbone groups to be paired in hydrogen bonds.

- At the tertiary structure level, transmembrane domains are mostly arranged as α-helical bundles, and in some cases as β-barrels. The near total dominance of the former structure probably results from its high adaptability. Indeed, helical bundles can form a range of structures, each fulfilling a different role. For example, they can form water-filled ion channels through which ions can diffuse across the membrane, or water-free transporters that undergo considerable conformational changes in order to 'pick up' a ligand on one side of the membrane and release it on the other side. Both of these structures contain domains (or elements) responsible for transport, as well as domains (or elements) that regulate transport according to cellular needs.

- Some membrane proteins have complex structures and are involved in equally complex physiological processes. Examples of such proteins are G protein-coupled receptors (GPCRs), which respond to a variety of hormones, factors, and even light photons. Most GPCRs have a similar seven-transmembrane structures, and they differ mainly in their extracellular domains. The response of a GPCR to its ligand involves conformational changes, which allow the protein to bind and activate its cognate G-protein(s) in the cytoplasm. Thus, an entire signal transduction system is activated, with results that are often dramatic to the cell. The large number of GPCR structures determined in recent years demonstrates that the activation process involves various molecular triggers and switches, which relay the ligand-binding signal to the intracellular side of the protein, thus creating a binding site for the G-protein. Moreover, the binding induces conformational changes in, and activation of, the G-protein. Although the molecular triggers and switches differ to some extent between different classes of GPCRs, many of them are associated with shared conserved motifs, and work in similar ways.

EXERCISES

7.1 Explain why the interest in membrane proteins far exceeds their relative proportion in a cell.

7.2 The lipids found in biological membranes differ significantly from each other in chemical structure and composition. What makes them all suitable as membrane building blocks?

7.3 Unlike other membrane lipids, cholesterol has a bulky, rigid structure. In your opinion, how would this structure affect membrane properties?

7.4 Suggest reasons why membranes of different organelles, cells, and organs differ so significantly in their lipid composition.

7.5 A. Estimate how many encounters between a divalent cation and the plasma membrane are needed in order for the cation to cross the membrane successfully. Assume that the radius of the cation is 1 Å, and that the dielectrics of the membrane and cytoplasm or extracellular matrix are 2 and 80, respectively.

   B. If each unsuccessful cation-membrane encounter lasts 10^{-12} s, how much time may be needed for the system to achieve a successful encounter?

7.6 Following the study of Elazar et al. (2016) (*Elife* **5**:e12125), state the expected general locations of the following amino acids in an integral membrane protein: Val, Trp, Arg, Pro, and Asp. Explain your prediction.

7.7 What are the main differences between the driving forces for the folding of globular proteins and those of integral membrane proteins?

7.8 Explain the following observations regarding membrane proteins that function in the transport of ions:

   I. Channels contain a water annulus, yet make the passing ions lose their solvation shell for a short part of the way.

   II. Carriers do not contain a constant water annulus linking the bulk solvent on both sides of the membrane.

7.9 Peripheral membrane proteins require more than a single type of noncovalent interaction to bind to the membrane. Explain the underlying advantage(s).

7.10 List the different means that the protein-membrane system can use to ameliorate the energy cost of positive hydrophobic mismatch.

7.11 Explain how activation of adrenergic receptors serves the 'fight-or-flight' response in animals.

7.12 Unlike other class A GPCRs, rhodopsin has no baseline activity in the absence of an agonist. Which structural features of GPCRs have been proposed to explain this phenomenon?

7.13 List the structural features of GPCR activation that have emerged from the study of rhodopsin's active and inactive structures.

REFERENCES

1. J. W. Schopf. Microfossils of the Early Archean Apex Chert: new evidence of the antiquity of life. *Science*, 260:640–6, 1993.

2. J. W. Schopf and B. M. Packer. Early Archean (3.3-billion to 3.5-billion-year-old) microfossils from Warrawoona Group, Australia. *Science*, 237:70–3, 1987.

3. S. J. Mojzsis, G. Arrhenius, K. D. McKeegan, T. M. Harrison, A. P. Nutman, and C. R. Friend. Evidence for life on Earth before 3,800 million years ago. *Nature*, 384(6604):55–9, 1996.

4. L. J. Ducka, M. Gliksona, S. D. Goldinga, and R. D. Webb. Microbial remains and other carbonaceous forms from the 3.24 Ga Sulphur Springs black smoker deposit, Western Australia. *Precambrian Res.*, 154(3–4):205–220, 2007.

5. I. Bertini, A. Sigel, and H. Sigel. *Handbook on metalloproteins*. Marcel Dekker, New York, 2001.

6. J. J. R. Frausto da Silva and R. J. P. Williams. *The biological chemistry of the elements: the inorganic chemistry of life*. Oxford University Press, New York, 2001.

7. S. Dorairaj and T. W. Allen. On the thermodynamic stability of a charged arginine side chain in a transmembrane helix. *Proc. Natl. Acad. Sci. USA*, 104(12):4943–8, 2007.

8. A. Kessel and N. Ben-Tal. Free energy determinants of peptide association with lipid bilayers. *Curr. Top. Membr.*, 52:205–253, 2002.

9. G. von Heijne. The membrane protein universe: what's out there and why bother? *J. Intern. Med.*, 261:543–557, 2007.

10. E. Wallin and G. von Heijne. Genome-wide analysis of integral membrane proteins from eubacterial, archaean, and eukaryotic organisms. *Protein Sci.*, 7(4):1029–38, 1998.

11. J. D. Bendtsen, T. T. Binnewies, P. F. Hallin, and D. W. Ussery. Genome update: prediction of membrane proteins in prokaryotic genomes. *Microbiology*, 151(Pt 7):2119–21, 2005.

12. M. Uhlen, L. Fagerberg, B. M. Hallstrom, C. Lindskog, P. Oksvold, A. Mardinoglu, A. Sivertsson, C. Kampf, E. Sjostedt, A. Asplund, I. Olsson, K. Edlund, E. Lundberg, S. Navani, C. A. Szigyarto, J. Odeberg, D. Djureinovic, J. O. Takanen, S. Hober, T. Alm, P. H. Edqvist, H. Berling, H. Tegel, J. Mulder, J. Rockberg, P. Nilsson, J. M. Schwenk, M. Hamsten, K. von Feilitzen, M. Forsberg, L. Persson, F. Johansson, M. Zwahlen, G. von Heijne, J. Nielsen, and F. Ponten. Proteomics. Tissue-based map of the human proteome. *Science*, 347(6220):1260419, 2015.

13. M. A. Yildirim, K.-I. Goh, M. E. Cusick, A.-L. Barabasi, and M. Vidal. Drug–target network. *Nat. Biotechnol.*, 25(10):1119–1126, 2007.

14. D. Salom and K. Palczewski. Structural Biology of Membrane Proteins. In A. S. Robinson, editor, *Production of Membrane Proteins: Strategies for Expression and Isolation*, chapter 9, pages 249–273. Wiley-Vch, 2011.

15. A. L. Hopkins and C. R. Groom. The druggable genome. *Nat. Rev. Drug Discov.*, 1(9):727–30, 2002.

16. A. P. Russ and S. Lampel. The druggable genome: an update. *Drug Discov. Today*, 10(23–24):1607–10, 2005.

17. S. J. Singer and G. L. Nicolson. The fluid mosaic model of the structure of cell membranes. *Science*, 175(23):720–31, 1972.

18. A. D. Dupuy and D. M. Engelman. Protein area occupancy at the center of the red blood cell membrane. *Proc. Natl. Acad. Sci. USA*, 105:2848–2852, 2008.

19. J. E. Rothman and L. Orci. Budding vesicles in living cells. *Sci. Am.*, 274(3):70–5, 1996.

20. N. A. Bright, M. J. Gratian, and J. P. Luzio. Endocytic delivery to lysosomes mediated by concurrent fusion and kissing events in living cells. *Curr. Biol.*, 15(4):360–5, 2005.

21. J. Pietzsch. Mind the membrane. In *Horizon Symposia: Living Frontier*, volume 1. Nature Publishing Group, 2004.

22. S. H. White and G. von Heijne. How translocons select transmembrane helices. *Annu. Rev. Biophys.*, 37:23–42, 2008.

23. E. K. Fridriksson, P. A. Shipkova, E. D. Sheets, D. Holowka, B. Baird, and F. W. McLafferty. Quantitative analysis of phospholipids in functionally important membrane domains from RBL-2H3 mast cells using tandem high-resolution mass spectrometry. *Biochemistry*, 38(25):8056–63, 1999.

24. B. Brugger, G. Erben, R. Sandhoff, F. T. Wieland, and W. D. Lehmann. Quantitative analysis of biological membrane lipids at the low picomole level by nano-electrospray ionization tandem mass spectrometry. *Proc. Natl. Acad. Sci. USA*, 94(6):2339–44, 1997.

25. J. J. Myher, A. Kuksis, and S. Pind. Molecular species of glycerophospholipids and sphingomyelins of human erythrocytes: improved method of analysis. *Lipids*, 24(5):396–407, 1989.

26. J. I. MacDonald and H. Sprecher. Phospholipid fatty acid remodeling in mammalian cells. *Biochim. Biophys. Acta*, 1084(2):105–21, 1991.

27. D. L. Nelson and M. M. Cox. *Lehninger Principles of Biochemistry*. W. H. Freeman and Company, 2004.

28. J. Lecocq and C. E. Ballou. On the Structure of Cardiolipin. *Biochemistry*, 3:976–80, 1964.

29. D. Hishikawa, H. Shindou, S. Kobayashi, H. Nakanishi, R. Taguchi, and T. Shimizu. Discovery of a lysophospholipid acyltransferase family essential for membrane asymmetry and diversity. *Proc. Natl. Acad. Sci. USA*, 105:2830–2835, 2008.

30. A. G. Lee. How lipids affect the activities of integral membrane proteins. *Biochim. Biophys. Acta*, 1666(1–2):62–87, 2004.

31. B. A. Lewis and D. M. Engelman. Lipid bilayer thickness varies linearly with acyl chain length in fluid phosphatidylcholine vesicles. *J. Mol. Biol.*, 166(2):211–7, 1983.

32. F. M. Goni and A. Alonso. Biophysics of sphingolipids I. Membrane properties of sphingosine, ceramides and other simple sphingolipids. *Biochim. Biophys. Acta*, 1758(12):1902–21, 2006.

33. B. Maggio, M. L. Fanani, C. M. Rosetti, and N. Wilke. Biophysics of sphingolipids II. Glycosphingolipids: an assortment of multiple structural information transducers at the membrane surface. *Biochim. Biophys. Acta*, 1758(12):1922–44, 2006.

34. F. R. Maxfield and I. Tabas. Role of cholesterol and lipid organization in disease. *Nature*, 438(7068):612–21, 2005.

35. J. L. Goldstein and M. S. Brown. Molecular medicine. The cholesterol quartet. *Science*, 292(5520):1310–2, 2001.

36. R. Wood and R. D. Harlow. Structural analyses of rat liver phosphoglycerides. *Arch. Biochem. Biophys.*, 135(1):272–81, 1969.

37. M. Schlame, S. Brody, and K. Y. Hostetler. Mitochondrial cardiolipin in diverse eukaryotes. Comparison of biosynthetic reactions and molecular acyl species. *Eur. J. Biochem.*, 212(3):727–35, 1993.

38. A. Yamashita, T. Sugiura, and K. Waku. Acyltransferases and transacylases involved in fatty acid remodeling of phospholipids and metabolism of bioactive lipids in mammalian cells. *J. Biochem.*, 122(1):1–16, 1997.

39. T. K. Ray, V. P. Skipski, M. Barclay, E. Essner, and F. M. Archibald. Lipid composition of rat liver plasma membranes. *J. Biol. Chem.*, 244(20):5528–36, 1969.

40. S. Morein, A. Andersson, L. Rilfors, and G. Lindblom. Wild-type *Escherichia coli* cells regulate the membrane lipid composition in a "window" between gel and non-lamellar structures. *J. Biol. Chem.*, 271(12):6801–9, 1996.

41. M. B. Coren. Biosynthesis and structures of phospholipids and sulfatides. In G. P. Kubica and L. G. Wayne, editors, *The Mycobacteria*, pages 379–415. Marcel Dekker, New York, 1984.

42. P. L. Yeagle. *The structure of biological membranes*. CRC Press, Boca Raton, New York, 2nd edition, 2005.

43. C. Tanford. *The hydrophobic effect*. Wiley, New-York, 1973.

44. T. W. Keenan and D. J. Morre. Phospholipid class and fatty acid composition of Golgi apparatus isolated from rat liver and comparison with other cell fractions. *Biochemistry*, 9(1):19–25, 1970.

45. L. Orci, R. Montesano, P. Meda, F. Malaisse-Lagae, D. Brown, A. Perrelet, and P. Vassalli. Heterogeneous distribution of filipin–cholesterol complexes across the cisternae of the Golgi apparatus. *Proc. Natl. Acad. Sci. USA*, 78(1):293–7, 1981.

46. P. F. Pimenta and W. de Souza. Localization of filipin-sterol complexes in cell membranes of eosinophils. *Histochem. Cell Biol.*, 80(6):563–7, 1984.

47. L. D. Bergelson and L. I. Barsukov. Topological asymmetry of phospholipids in membranes. *Science*, 197(4300):224–30, 1977.

48. J. A. Op den Kamp. Lipid asymmetry in membranes. *Annu. Rev. Biochem.*, 48:47–71, 1979.

49. R. F. Zwaal and A. J. Schroit. Pathophysiologic implications of membrane phospholipid asymmetry in blood cells. *Blood*, 89(4):1121–32, 1997.

50. D. Marquardt, B. Geier, and G. Pabst. Asymmetric lipid membranes: towards more realistic model systems. *Membranes*, 5(2):180–196, 2015.

51. S. Mukherjee and F. R. Maxfield. Membrane domains. *Annu. Rev. Cell. Dev. Biol.*, 20:839–66, 2004.

52. K. Simons and W. L. Vaz. Model systems, lipid rafts, and cell membranes. *Annu. Rev. Biophys. Biomol. Struct.*, 33:269–95, 2004.

53. S. H. White and W. C. Wimley. Membrane protein folding and stability: physical principles. *Annu. Rev. Biophys. Biomol. Struct.*, 28:319–65, 1999.

54. J. C. Holthuis and T. P. Levine. Lipid traffic: floppy drives and a superhighway. *Nat. Rev. Mol. Cell Biol.*, 6(3):209–20, 2005.

55. D. L. Daleke. Regulation of transbilayer plasma membrane phospholipid asymmetry. *J. Lipid Res.*, 44(2):233–42, 2003.

56. D. A. Brown and E. London. Structure and origin of ordered lipid domains in biological membranes. *J. Membr. Biol.*, 164(2):103–14, 1998.

57. K. Mitra, I. Ubarretxena-Belandia, T. Taguchi, G. Warren, and D. M. Engelman. Modulation of the bilayer thickness of exocytic pathway membranes by membrane proteins rather than cholesterol. *Proc. Natl. Acad. Sci. USA*, 101(12):4083–8, 2004.

58. H. T. McMahon and J. L. Gallop. Membrane curvature and mechanisms of dynamic cell membrane remodelling. *Nature*, 438(7068):590–6, 2005.

59. J. M. Seddon. Structure of the inverted hexagonal (HII) phase, and non-lamellar phase transitions of lipids. *Biochim. Biophys. Acta*, 1031(1):1–69, 1990.

60. L. T. Boni and S. W. Hui. Polymorphic phase behaviour of dilinoleoylphosphatidylethanolamine and palmitoyloleoylphosphatidylcholine mixtures. Structural changes between hexagonal, cubic and bilayer phases. *Biochim. Biophys. Acta*, 731(2):177–85, 1983.

61. S. M. Gruner. Intrinsic curvature hypothesis for biomembrane lipid composition: a role for nonbilayer lipids. *Proc. Natl. Acad. Sci. USA*, 82(11):3665–9, 1985.

62. W. C. Wimley. Toward genomic identification of beta-barrel membrane proteins: composition and architecture of known structures. *Protein Sci.*, 11:301–312, 2002.

63. S. Watson and S. Arkinstall. *The C-Protein Linked Receptor-Facts Book*. Academic Press, London, 1994.

64. F. Horn, E. Bettler, L. Oliveira, F. Campagne, F. E. Cohen, and G. Vriend. GPCRDB information system for G protein-coupled receptors. *Nucleic Acids Res.*, 31(1):294–7, 2003.

65. T. M. Bridges and C. W. Lindsley. G-Protein-Coupled Receptors: From Classical Modes of Modulation to Allosteric Mechanisms. *ACS Chem. Biol.*, 3:530–41, 2008.

66. T. H. Ji, M. Grossmann, and I. Ji. G protein-coupled receptors. I. Diversity of receptor-ligand interactions. *J. Biol. Chem.*, 273(28):17299–302, 1998.

67. I. Moraes, G. Evans, J. Sanchez-Weatherby, S. Newstead, and P. D. Stewart. Membrane protein structure determination: the next generation. *Biochim. Biophys. Acta*, 1838(1 Pt A):78–87, 2014.

68. I. D. Pogozheva, H. I. Mosberg, and A. L. Lomize. Life at the border: Adaptation of proteins to anisotropic membrane environment. *Protein Sci.*, 23(9):1165–1196, 2014.

69. A. Oberai, Y. Ihm, S. Kim, and J. U. Bowie. A limited universe of membrane protein families and folds. *Protein Sci.*, 15(7):1723–1734, 2006.

70. S. H. White, A. S. Ladokhin, S. Jayasinghe, and K. Hristova. How membranes shape protein structure. *J. Biol. Chem.*, 276(35):32395–8, 2001.

71. N. Hurwitz, M. Pellegrini-Calace, and D. T. Jones. Towards genome-scale structure prediction for transmembrane proteins. *Philos. Trans. R. Soc. Lond. B: Biol. Sci.*, 361(1467):465–75, 2006.

72. C. Baeza-Delgado, M. A. Marti-Renom, and I. Mingarro. Structure-based statistical analysis of transmembrane helices. *Eur. Biophys. J.*, 42(2–3):199–207, 2013.

73. G. von Heijne. Recent advances in the understanding of membrane protein assembly and structure. *Q. Rev. Biophys.*, 32(4):285–307, 1999.

74. G. von Heijne. Membrane-protein topology. *Nat. Rev. Mol. Cell Biol.*, 7(12):909–18, 2006.

75. F. Cymer, G. von Heijne, and S. H. White. Mechanisms of integral membrane protein insertion and folding. *J. Mol. Biol.*, 427(5):999–1022, 2015.

76. J.-L. Popot and D. M. Engelman. Helical membrane protein folding, stability, and evolution. *Annu. Rev. Biochem.*, 69:881–922, 2000.

77. J. U. Bowie. Helix packing in membrane proteins. *J. Mol. Biol.*, 272(5):780–9, 1997.

78. J. U. Bowie. Understanding membrane protein structure by design. *Nat. Struct. Biol.*, 7(2):91–4, 2000.

79. R. Worch, C. Bökel, S. Höfinger, P. Schwille, and T. Weidemann. Focus on composition and interaction potential of single-pass transmembrane domains. *Proteomics*, 10(23):4196–4208, 2010.

80. K. Illergård, A. Kauko, and A. Elofsson. Why are polar residues within the membrane core evolutionary conserved? *Proteins*, 79(1):79–91, 2011.

81. M. Bañó-Polo, C. Baeza-Delgado, M. Orzáez, M. A. Marti-Renom, C. Abad, and I. Mingarro. Polar/Ionizable residues in transmembrane segments: effects on helix-helix packing. *PLoS One*, 7(9):e44263, 2012.

82. T. H. Walther and A. S. Ulrich. Transmembrane helix assembly and the role of salt bridges. *Curr. Opin. Struct. Biol.*, 27:63–68, 2014.

83. S. Chakrapani, L. G. Cuello, D. M. Cortes, and E. Perozo. Structural dynamics of an isolated voltage-sensor domain in a lipid bilayer. *Structure*, 16(3):398–409, 2008.

84. D. Krepkiy, M. Mihailescu, J. A. Freites, E. V. Schow, D. L. Worcester, K. Gawrisch, D. J. Tobias, S. H. White, and K. J. Swartz. Structure and hydration of membranes embedded with voltage-sensing domains. *Nature*, 462(7272):473–479, 2009.

85. A. G. Lee. Structural biology: Highly charged meetings. *Nature*, 462(7272):420–421, 2009.

86. J. U. Bowie. Solving the membrane protein folding problem. *Nature*, 438(7068):581–9, 2005.

87. M. Gimpelev, L. R. Forrest, D. Murray, and B. Honig. Helical packing patterns in membrane and soluble proteins. *Biophys. J.*, 87(6):4075–4086, 2004.

88. T. A. Eyre, L. Partridge, and J. M. Thornton. Computational analysis of α-helical membrane protein structure: implications for the prediction of 3D structural models. *Protein Eng. Des. Sel.*, 17(8):613–624, 2004.

89. X. Deupi, M. Olivella, C. Govaerts, J. A. Ballesteros, M. Campillo, and L. Pardo. Ser and Thr residues modulate the conformation of pro-kinked transmembrane alpha-helices. *Biophys. J.*, 86(1 Pt 1):105–15, 2004.

90. E. Granseth, G. von Heijne, and A. Elofsson. A study of the membrane-water interface region of membrane proteins. *J. Mol. Biol.*, 346(1):377–85, 2005.

91. C. Landolt-Marticorena, K. A. Williams, C. M. Deber, and R. A. Reithmeier. Non-random distribution of amino acids in the transmembrane segments of human type I single span membrane proteins. *J. Mol. Biol.*, 229(3):602–8, 1993.

92. M. B. Ulmschneider, M. S. Sansom, and A. Di Nola. Properties of integral membrane protein structures: derivation of an implicit membrane potential. *Proteins*, 59(2):252–65, 2005.

93. G. V. Heijne. The distribution of positively charged residues in bacterial inner membrane proteins correlates with the trans-membrane topology. *EMBO J.*, 5(11):3021–3027, 1986.

94. M. Schiffer, C. H. Chang, and F. J. Stevens. The functions of tryptophan residues in membrane proteins. *Protein Eng.*, 5(3):213–4, 1992.

95. A. N. Ridder, S. Morein, J. G. Stam, A. Kuhn, B. de Kruijff, and J. A. Killian. Analysis of the role of interfacial tryptophan residues in controlling the topology of membrane proteins. *Biochemistry*, 39(21):6521–8, 2000.

96. G. von Heijne. Membrane protein structure prediction. Hydrophobicity analysis and the positive-inside rule. *J. Mol. Biol.*, 225(2):487–94, 1992.

97. A. Elazar, J. Weinstein, I. Biran, Y. Fridman, E. Bibi, and S. J. Fleishman. Mutational scanning reveals the determinants of protein insertion and association energetics in the plasma membrane. *eLife*, 5:e12125, 2016.

98. J. P. Segrest, H. De Loof, J. G. Dohlman, C. G. Brouillette, and G. M. Anantharamaiah. Amphipathic helix motif: classes and properties. *Proteins*, 8(2):103–17, 1990.

99. R. Jackups and J. Liang. Interstrand pairing patterns in β-barrel membrane proteins: the positive-outside rule, aromatic rescue, and strand registration prediction. *J. Mol. Biol.*, 354(4):979–993, 2005.

100. J. Qu, S. Behrens-Kneip, O. Holst, and J. H. Kleinschmidt. Binding regions of outer membrane protein A in complexes with the periplasmic chaperone Skp. A site-directed fluorescence study. *Biochemistry*, 48(22):4926–4936, 2009.

101. S. J. Fleishman and N. Ben-Tal. Progress in structure prediction of alpha-helical membrane proteins. *Curr. Opin. Struct. Biol.*, 16(4):496–504, 2006.

102. J. Kyte and R. F. Doolittle. A simple method for displaying the hydropathic character of a protein. *J. Mol. Biol.*, 157(1):105–32, 1982.

103. K. P. Hofmann and W. Stoffel. TMbase: A database of membrane spanning proteins segments. *Biol. Chem.*, 374:166, 1993.

104. TMpred. http://www.ch.embnet.org/software/TMPRED_form.html, 2017.

105. W. C. Wimley and S. H. White. Experimentally determined hydrophobicity scale for proteins at membrane interfaces. *Nat. Struct. Biol.*, 3(10):842–8, 1996.

106. D. M. Engelman, T. A. Steitz, and A. Goldman. Identifying nonpolar transbilayer helices in amino acid sequences of membrane proteins. *Annu. Rev. Biophys. Biophys. Chem.*, 15:321–53, 1986.

107. D. Eisenberg, W. Wilcox, and A. D. McLachlan. Hydrophobicity and amphiphilicity in protein structure. *J. Cell Biochem.*, 31(1):11–7, 1986.

108. W. C. Wimley, T. P. Creamer, and S. H. White. Solvation energies of amino acid side chains and backbone in a family of host-guest pentapeptides. *Biochemistry*, 35(16):5109–5124, 1996.

109. T. Hessa, H. Kim, K. Bihlmaier, C. Lundin, J. Boekel, H. Andersson, I. Nilsson, S. H. White, and G. von Heijne. Recognition of transmembrane helices by the endoplasmic reticulum translocon. *Nature*, 433(7024):377–81, 2005.

110. T. Hessa, N. M. Meindl-Beinker, A. Bernsel, H. Kim, Y. Sato, M. Lerch-Bader, I. Nilsson, S. H. White,

and G. Von Heijne. Molecular code for transmembrane-helix recognition by the Sec61 translocon. *Nature*, 450(7172):1026–1030, 2007.

111. K. Öjemalm, S. C. Botelho, C. Stüdle, and G. von Heijne. Quantitative analysis of SecYEG-mediated insertion of transmembrane α-helices into the bacterial inner membrane. *J. Mol. Biol.*, 425(15):2813–2822, 2013.

112. A. Bernsel, H. Viklund, J. Falk, E. Lindahl, G. von Heijne, and A. Elofsson. Prediction of membrane-protein topology from first principles. *Proc. Natl. Acad. Sci. USA*, 105(20):7177–7181, 2008.

113. D. Shental-Bechor, S. J. Fleishman, and N. Ben-Tal. Has the code for protein translocation been broken? *Trends Biochem. Sci.*, 31(4):192–6, 2006.

114. A. C. V. Johansson and E. Lindahl. Protein contents in biological membranes can explain abnormal solvation of charged and polar residues. *Proc. Natl. Acad. Sci. USA*, 106(37):15684–15689, 2009.

115. A. Senes, D. C. Chadi, P. B. Law, R. F. S. Walters, V. Nanda, and W. F. DeGrado. E(z), a depth-dependent potential for assessing the energies of insertion of amino acid side-chains into membranes: derivation and applications to determining the orientation of transmembrane and interfacial helices. *J. Mol. Biol.*, 366(2):436–448, 2007.

116. C. A. Schramm, B. T. Hannigan, J. E. Donald, C. Keasar, J. G. Saven, W. F. DeGrado, and I. Samish. Knowledge-based potential for positioning membrane-associated structures and assessing residue-specific energetic contributions. *Structure*, 20(5):924–935, 2012.

117. M. A. Lomize, A. L. Lomize, I. D. Pogozheva, and H. I. Mosberg. OPM: orientations of proteins in membranes database. *Bioinformatics*, 22:623–625, 2006.

118. S. J. Fleishman, V. M. Unger, and N. Ben-Tal. Transmembrane protein structures without X-rays. *Trends Biochem. Sci.*, 31(2):106–13, 2006.

119. D. A. Doyle, J. Morais Cabral, R. A. Pfuetzner, A. Kuo, J. M. Gulbis, S. L. Cohen, B. T. Chait, and R. MacKinnon. The structure of the potassium channel: molecular basis of K$^+$ conduction and selectivity. *Science*, 280(5360):69–77, 1998.

120. J. M. Cuthbertson, D. A. Doyle, and M. S. Sansom. Transmembrane helix prediction: a comparative evaluation and analysis. *Protein Eng. Des. Sel.*, 18(6):295–308, 2005.

121. D. T. Jones, W. R. Taylor, and J. M. Thornton. A model recognition approach to the prediction of all-helical membrane protein structure and topology. *Biochemistry*, 33(10):3038–49, 1994.

122. E. L. Sonnhammer, G. von Heijne, and A. Krogh. A hidden Markov model for predicting transmembrane helices in protein sequences. *Proc. Int. Conf. Intell. Syst. Mol. Biol.*, 6:175–82, 1998.

123. A. Krogh, B. Larsson, G. von Heijne, and E. L. Sonnhammer. Predicting transmembrane protein topology with hidden Markov model: application to complete genomes. *J. Mol. Biol.*, 305(3):567–80, 2001.

124. D. O. Daley, M. Rapp, E. Granseth, K. Melen, D. Drew, and G. von Heijne. Global topology analysis of the *Escherichia coli* inner membrane proteome. *Science*, 308(5726):1321–3, 2005.

125. T. Nugent and D. T. Jones. Transmembrane protein topology prediction using support vector machines. *BMC Bioinformatics*, 10:1–11, 2009.

126. A. Elazar, J. J. Weinstein, J. Prilusky, and S. J. Fleishman. Interplay between hydrophobicity and the positive-inside rule in determining membrane-protein topology. *Proc. Natl. Acad. Sci. USA*, pages 10340–5, 2016.

127. S. Hayat and A. Elofsson. BOCTOPUS: improved topology prediction of transmembrane β barrel proteins. *Bioinformatics*, 28(4):516–522, 2012.

128. J. Koehler Leman, M. B. Ulmschneider, and J. J. Gray. Computational modeling of membrane proteins. *Proteins: Struct., Funct., Bioinf.*, 83(1):1–24, 2015.

129. A. Senes, M. Gerstein, and D. M. Engelman. Statistical analysis of amino acid patterns in transmembrane helices: the GxxxG motif occurs frequently and in association with beta-branched residues at neighboring positions. *J. Mol. Biol.*, 296(3):921–36, 2000.

130. K. T. O'Neil and W. F. DeGrado. A thermodynamic scale for the helix-forming tendencies of the commonly occurring amino acids. *Science*, 250(4981):646–51, 1990.

131. M. S. Sansom and H. Weinstein. Hinges, swivels and switches: the role of prolines in signalling via transmembrane alpha-helices. *Trends Pharmacol. Sci.*, 21(11):445–51, 2000.

132. C. Govaerts, C. Blanpain, X. Deupi, S. Ballet, J. A. Ballesteros, S. J. Wodak, G. Vassart, L. Pardo, and M. Parmentier. The TXP motif in the second transmembrane helix of CCR5. A structural determinant of chemokine-induced activation. *J. Biol. Chem.*, 276(16):13217–25, 2001.

133. A. Peralvarez, R. Barnadas, M. Sabes, E. Querol, and E. Padros. Thr-90 is a key residue of the bacteriorhodopsin proton pumping mechanism. *FEBS Lett.*, 508(3):399–402, 2001.

134. Y. Ri, J. A. Ballesteros, C. K. Abrams, S. Oh, V. K. Verselis, H. Weinstein, and T. A. Bargiello. The role of a

conserved proline residue in mediating conformational changes associated with voltage gating of Cx32 gap junctions. *Biophys. J.*, 76(6):2887–98, 1999.

135. W. C. Wigley, M. J. Corboy, T. D. Cutler, P. H. Thibodeau, J. Oldan, M. G. Lee, J. Rizo, J. F. Hunt, and P. J. Thomas. A protein sequence that can encode native structure by disfavoring alternate conformations. *Nat. Struct. Mol. Biol.*, 9(5):381–388, 2002.

136. S. K. Buchanan. Beta-Barrel proteins from bacterial outer membranes: Structure, function and refolding. *Curr. Opin. Struct. Biol.*, 9:455–461, 1999.

137. G. E. Schulz. Beta-barrel membrane proteins. *Curr. Opin. Struct. Biol.*, 10(4):443–7, 2000.

138. R. Benz and K. Bauer. Permeation of hydrophilic molecules through the outer membrane of Gram-negative bacteria. Review on bacterial porins. *Eur. J. Biochem.*, 176(1):1–19, 1988.

139. T. Schirmer. General and specific porins from bacterial outer membranes. *J. Struct. Biol.*, 121(2):101–9, 1998.

140. R. Benz. Permeation of hydrophilic solutes through mitochondrial outer membranes: review on mitochondrial porins. *Biochim. Biophys. Acta*, 1197(2):167–96, 1994.

141. D. C. Bay and D. A. Court. Origami in the outer membrane: the transmembrane arrangement of mitochondrial porins. *Biochem. Cell Biol.*, 80(5):551–62, 2002.

142. D. Duy, J. Soll, and K. Philippar. Solute channels of the outer membrane: from bacteria to chloroplasts. *Biol. Chem.*, 388(9):879–89, 2007.

143. B. Bolter and J. Soll. Ion channels in the outer membranes of chloroplasts and mitochondria: open doors or regulated gates? *EMBO J.*, 20(5):935–40, 2001.

144. S. Bhakdi and J. Tranum-Jensen. Alpha-toxin of *Staphylococcus aureus*. *Microbiol. Rev.*, 55(4):733–751, 1991.

145. G. von Heijne. Principles of membrane protein assembly and structure. *Prog. Biophys. Mol. Biol.*, 66(2):113–39, 1996.

146. K. Zeth and M. Thein. Porins in prokaryotes and eukaryotes: common themes and variations. *Biochem. J.*, 431(1):13–22, 2010.

147. K. Seshadri, R. Garemyr, E. Wallin, G. von Heijne, and A. Elofsson. Architecture of beta-barrel membrane proteins: analysis of trimeric porins. *Protein Sci.*, 7(9):2026–32, 1998.

148. T. J. Stevens and I. T. Arkin. Are membrane proteins "inside-out" proteins? *Proteins*, 36(1):135–43, 1999.

149. D. C. Rees, L. DeAntonio, and D. Eisenberg. Hydrophobic organization of membrane proteins. *Science*, 245(4917):510–3, 1989.

150. D. Donnelly, J. P. Overington, S. V. Ruffle, J. H. Nugent, and T. L. Blundell. Modeling alpha-helical transmembrane domains: the calculation and use of substitution tables for lipid-facing residues. *Protein Sci.*, 2(1):55–70, 1993.

151. D. T. Moore, B. W. Berger, and W. F. DeGrado. Protein-protein interactions in the membrane: sequence, structural, and biological motifs. *Structure*, 16(7):991–1001, 2008.

152. M. Eilers, A. B. Patel, W. Liu, and S. O. Smith. Comparison of helix interactions in membrane and soluble alpha-bundle proteins. *Biophys. J.*, 82(5):2720–36, 2002.

153. W. P. Russ and D. M. Engelman. The GxxxG motif: a framework for transmembrane helix-helix association. *J. Mol. Biol.*, 296(3):911–9, 2000.

154. B. K. Mueller, S. Subramaniam, and A. Senes. A frequent, GxxxG-mediated, transmembrane association motif is optimized for the formation of interhelical Cα−H hydrogen bonds. *Proc. Natl. Acad. Sci. USA*, 111(10):E888–E895, 2014.

155. A. Senes, D. E. Engel, and W. F. DeGrado. Folding of helical membrane proteins: the role of polar, GxxxG-like and proline motifs. *Curr. Opin. Struct. Biol.*, 14(4):465–479, 2004.

156. R. F. S. Walters and W. F. DeGrado. Helix-packing motifs in membrane proteins. *Proc. Natl. Acad. Sci. USA*, 103(37):13658–13663, 2006.

157. N. A. Noordeen, F. Carafoli, E. Hohenester, M. A. Horton, and B. Leitinger. A transmembrane leucine zipper is required for activation of the dimeric receptor tyrosine kinase DDR1. *J. Biol. Chem.*, 281(32):22744–22751, 2006.

158. B. North, L. Cristian, X. Fu Stowell, J. D. Lear, JG. Saven, and W. F. Degrado. Characterization of a membrane protein folding motif, the Ser zipper, using designed peptides. *J. Mol. Biol.*, 359:930–939, 2006.

159. S. Kim, T.-J. Jeon, A. Oberai, D. Yang, J. J. Schmidt, and J. U. Bowie. Transmembrane glycine zippers: physiological and pathological roles in membrane proteins. *Proc. Natl. Acad. Sci. USA*, 102(40):14278–14283, 2005.

160. A. Marsico, K. Scheubert, A. Tuukkanen, A. Henschel, C. Winter, R. Winnenburg, and M. Schroeder. MeM-

otif: a database of linear motifs in α-helical transmembrane proteins. *Nucleic Acids Res.*, 38:D181–D189, 2010.

161. S. H. White. Biophysical dissection of membrane proteins. *Nature*, 459:344–6, 2009.

162. J.-L. Popot, S.-E. Gerchman, and D. M. Engelman. Refolding of bacteriorhodopsin in lipid bilayers: a thermodynamically controlled two-stage process. *J. Mol. Biol.*, 198(4):655–676, 1987.

163. R. E. Jacobs and S. H. White. The nature of the hydrophobic binding of small peptides at the bilayer interface: implications for the insertion of transbilayer helices. *Biochemistry*, 28(8):3421–37, 1989.

164. D. M. Engelman and T. A. Steitz. The spontaneous insertion of proteins into and across membranes: the helical hairpin hypothesis. *Cell*, 23(2):411–422, 1981.

165. C. M. Deber and N. K. Goto. Folding proteins into membranes. *Nat. Struct. Mol. Biol.*, 3(10):815–818, 1996.

166. J.-L. Popot. Integral membrane protein structure: transmembrane α-helices as autonomous folding domains. *Curr. Opin. Struct. Biol.*, 3(4):532–540, 1993.

167. K. G. Fleming. Energetics of membrane protein folding. *Annu. Rev. Biophys.*, 43:233–255, 2014.

168. N. Ben-Tal, A. Ben-Shaul, A. Nicholls, and B. Honig. Free-energy determinants of alpha-helix insertion into lipid bilayers. *Biophys. J.*, 70(4):1803–12, 1996.

169. N. Ben-Tal, D. Sitkoff, I. A. Topol, A. S. Yang, S. K. Burt, and B. Honig. Free energy of amide hydrogen bond formation in vacuum, in water, and in liquid alkane solution. *J. Phys. Chem. B*, 101(3):450–457, 1997.

170. C. Chothia. Hydrophobic bonding and accessible surface area in proteins. *Nature*, 248(446):338–339, 1974.

171. J. A. Reynolds, D. B. Gilbert, and C. Tanford. Empirical Correlation Between Hydrophobic Free Energy and Aqueous Cavity Surface Area. *Proc. Natl. Acad. Sci. USA*, 71(8):2925–2927, 1974.

172. S. Vajda, Z. Weng, and C. DeLisi. Extracting hydrophobicity parameters from solute partition and protein mutation/unfolding experiments. *Protein Eng.*, 8(11):1081–1092, 1995.

173. P. Andrew Karplus. Hydrophobicity regained. *Protein Sci.*, 6(6):1302–1307, 1997.

174. K. Öjemalm, T. Higuchi, Y. Jiang, Ü. Langel, I. Nilsson, S. H. White, H. Suga, and G. von Heijne. Apolar surface area determines the efficiency of translocon-mediated membrane-protein integration into the endoplasmic reticulum. *Proc. Natl. Acad. Sci. USA*, 108(31):E359–E364, 2011.

175. D. J. Müller, M. Kessler, F. Oesterhelt, C. Möller, D. Oesterhelt, and H. Gaub. Stability of bacteriorhodopsin α-helices and loops analyzed by single-molecule force spectroscopy. *Biophys. J.*, 83(6):3578–3588, 2002.

176. F. Cymer and D. Schneider. Oligomerization of polytopic α-helical membrane proteins: causes and consequences. *Biol. Chem.*, 393:1215–30, 2012.

177. A. J. Venkatakrishnan, E. D. Levy, and S. A. Teichmann. Homomeric protein complexes: evolution and assembly. *Biochem. Soc. Trans.*, 38(4):879–882, 2010.

178. Y.-C. Chang and J. U. Bowie. Measuring membrane protein stability under native conditions. *Proc. Natl. Acad. Sci. USA*, 111(1):219–224, 2014.

179. S. Faham, D. Yang, E. Bare, S. Yohannan, J. P. Whitelegge, and J. U. Bowie. Side-chain contributions to membrane protein structure and stability. *J. Mol. Biol.*, 335(1):297–305, 2004.

180. V. Helms. Attraction within the membrane. Forces behind transmembrane protein folding and supramolecular complex assembly. *EMBO Rep.*, 3(12):1133–8, 2002.

181. F. X. Zhou, M. J. Cocco, W. P. Russ, A. T. Brunger, and D. M. Engelman. Interhelical hydrogen bonding drives strong interactions in membrane proteins. *Nat. Struct. Biol.*, 7(2):154–60, 2000.

182. C. Choma, H. Gratkowski, J. D. Lear, and W. F. DeGrado. Asparagine-mediated self-association of a model transmembrane helix. *Nat. Struct. Biol.*, 7(2):161–6, 2000.

183. W. F. DeGrado, H. Gratkowski, and J. D. Lear. How do helix-helix interactions help determine the folds of membrane proteins? Perspectives from the study of homo-oligomeric helical bundles. *Protein Sci.*, 12(4):647–65, 2003.

184. N. H. Joh, A. Min, S. Faham, J. P. Whitelegge, D. Yang, V. L. Woods, and J. U. Bowie. Modest stabilization by most hydrogen-bonded side-chain interactions in membrane proteins. *Nature*, 453(7199):1266–70, 2008.

185. E. Arbely and I. T. Arkin. Experimental measurement of the strength of a Cα–H$\cdots$O bond in a lipid bilayer. *J. Am. Chem. Soc.*, 126:5362–5363, 2004.

186. S. Yohannan, S. Faham, D. Yang, D. Grosfeld, A. K. Chamberlain, and J. U. Bowie. A Cα–H$\cdots$O hydrogen bond in a membrane protein is not stabilizing. *J. Am. Chem. Soc.*, 126:2284–2285, 2004.

187. S. Kumar and R. Nussinov. Salt bridge stability in monomeric proteins. *J. Mol. Biol.*, 293(5):1241–55, 1999.

188. S. Albeck, R. Unger, and G. Schreiber. Evaluation of direct and cooperative contributions towards the strength of buried hydrogen bonds and salt bridges. *J. Mol. Biol.*, 298(3):503–20, 2000.

189. D. Lee, J. Lee, and C. Seok. What stabilizes close arginine pairing in proteins? *Phys. Chem. Chem. Phys.*, 15(16):5844–53, 2013.

190. E. Gouaux and R. Mackinnon. Principles of selective ion transport in channels and pumps. *Science*, 310(5753):1461–5, 2005.

191. U. G. Hacke and J. Laur. Aquaporins: Channels for the Molecule of Life. In *Encyclopedia of Life Sciences*. John Wiley & Sons, Ltd., 2016.

192. J. S. Hub and B. L. De Groot. Mechanism of selectivity in aquaporins and aquaglyceroporins. *Proc. Natl. Acad. Sci. USA*, 105(4):1198–1203, 2008.

193. Y. Fujiyoshi, K. Mitsuoka, B. L. de Groot, A. Philippsen, H. Grubmuller, P. Agre, and A. Engel. Structure and function of water channels. *Curr. Opin. Struct. Biol.*, 12(4):509–15, 2002.

194. A. Barati Farimani, N. R. Aluru, and E. Tajkhorshid. Thermodynamic insight into spontaneous hydration and rapid water permeation in aquaporins. *Appl. Phys. Lett.*, 105(8):083702, 2014.

195. Y. Arinaminpathy, E. Khurana, D. M. Engelman, and M. B. Gerstein. Computational analysis of membrane proteins: the largest class of drug targets. *Drug Discov. Today*, 14(23–24):1130–5, 2009.

196. A. Lange, K. Giller, S. Hornig, M. F. Martin-Eauclaire, O. Pongs, S. Becker, and M. Baldus. Toxin-induced conformational changes in a potassium channel revealed by solid-state NMR. *Nature*, 440(7086):959–62, 2006.

197. J. H. Chill, J. M. Louis, J. L. Baber, and A. Bax. Measurement of 15N relaxation in the detergent-solubilized tetrameric KcsA potassium channel. *J. Biomol. NMR*, 36(2):123–36, 2006.

198. S. Y. Noskov, S. Berneche, and B. Roux. Control of ion selectivity in potassium channels by electrostatic and dynamic properties of carbonyl ligands. *Nature*, 431(7010):830–4, 2004.

199. A. N. Thompson, I. Kim, T. D. Panosian, T. M. Iverson, T. W. Allen, and C. M. Nimigean. Mechanism of potassium-channel selectivity revealed by Na^+ and Li^+ binding sites within the KcsA pore. *Nat. Struct. Mol. Biol.*, 16(12):1317–1324, 2009.

200. Q. Kuang, P. Purhonen, and H. Hebert. Structure of potassium channels. *Cell. Mol. Life Sci.*, 72(19):3677–3693, 2015.

201. D. M. Kim and C. M. Nimigean. Voltage-gated potassium channels: a structural examination of selectivity and gating. *Cold Spring Harb. Perspect. Biol.*, 8(5):a029231, 2016.

202. S. Chakrapani, J. F. Cordero-Morales, and E. Perozo. A quantitative description of KcsA gating I: macroscopic currents. *J. Gen. Physiol.*, 130(5):465–478, 2007.

203. S. Imai, M. Osawa, K. Takeuchi, and I. Shimada. Structural basis underlying the dual gate properties of KcsA. *Proc. Natl. Acad. Sci. USA*, 107(14):6216–6221, 2010.

204. S. Uysal, L. G. Cuello, D. M. Cortes, S. Koide, A. A. Kossiakoff, and E. Perozo. Mechanism of activation gating in the full-length KcsA K^+ channel. *Proc. Natl. Acad. Sci. USA*, 108(29):11896–11899, 2011.

205. L. G. Cuello, V. Jogini, D. M. Cortes, and E. Perozo. Structural mechanism of C-type inactivation in K^+ channels. *Nature*, 466(7303):203–208, 2010.

206. Y.-S. Liu, P. Sompornpisut, and E. Perozo. Structure of the KcsA channel intracellular gate in the open state. *Nat. Struct. Mol. Biol.*, 8(10):883–887, 2001.

207. L. G. Cuello, D. M. Cortes, V. Jogini, A. Sompornpisut, and E. Perozo. A molecular mechanism for proton-dependent gating in KcsA. *FEBS Lett.*, 584(6):1126–1132, 2010.

208. L. G. Cuello, V. Jogini, D. M. Cortes, A. Sompornpisut, M. D. Purdy, M. C. Wiener, and E. Perozo. Design and characterization of a constitutively open KcsA. *FEBS Lett.*, 584(6):1133–1138, 2010.

209. L. G. Cuello, V. Jogini, D. M. Cortes, A. C. Pan, D. G. Gagnon, O. Dalmas, J. F. Cordero-Morales, S. Chakrapani, B. Roux, and E. Perozo. Structural basis for the coupling between activation and inactivation gates in K^+ channels. *Nature*, 466(7303):272–275, 2010.

210. T. Hoshi, W. N. Zagotta, and R. W. Aldrich. Two types of inactivation in Shaker K^+ channels: Effects of alterations in the carboxy-terminal region. *Neuron*, 7(4):547–556, 1991.

211. O. Jardetzky. Simple Allosteric Model for Membrane Pumps. *Nature*, 211(5052):969–970, 1966.

212. Y. Lee, T. Nishizawa, K. Yamashita, R. Ishitani, and O. Nureki. Structural basis for the facilitative diffusion mechanism by SemiSWEET transporter. *Nat. Commun.*, 6:6112, 2015.

213. S. Wilkens. Structure and mechanism of ABC transporters. *F1000Prime Rep*, 7, 2015.

214. R. J. P. Dawson and K. P. Locher. Structure of a bacterial multidrug ABC transporter. *Nature*, 443(7108):180–185, 2006.

215. N. Livnat-Levanon, A. I. Gilson, N. Ben-Tal, and O. Lewinson. The uncoupled ATPase activity of the ABC transporter BtuC2D2 leads to a hysteretic conformational change, conformational memory, and improved activity. *Sci. Rep.*, 6:21696, 2016.

216. P. A. Mitchell. General theory of membrane transport from studies of bacteria. *Nature*, 180:134–136, 1957.

217. C. F. Higgins and K. J. Linton. The ATP switch model for ABC transporters. *Nat. Struct. Mol. Biol.*, 11(10):918–926, 2004.

218. L. Reuss. Ion Transport across Nonexcitable Membranes. In *Encyclopedia of Life Sciences*. John Wiley & Sons, Ltd., 2001.

219. J. P. Morth, B. P. Pedersen, M. J. Buch-Pedersen, J. P. Andersen, B. Vilsen, M. G. Palmgren, and P. Nissen. A structural overview of the plasma membrane Na^+,K^+-ATPase and H^+-ATPase ion pumps. *Nat. Rev. Mol. Cell Biol.*, 12(1):60–70, 2011.

220. U. Hasler, G. Crambert, J.-D. Horisberger, and K. Geering. Structural and Functional Features of the Transmembrane Domain of the Na,K-ATPase β Subunit Revealed by Tryptophan Scanning. *J. Biol. Chem.*, 276(19):16356–16364, 2001.

221. T. Shinoda, H. Ogawa, F. Cornelius, and C. Toyoshima. Crystal structure of the sodium-potassium pump at 2.4 Å resolution. *Nature*, 459(7245):446–450, 2009.

222. J. P. Morth, B. P. Pedersen, M. S. Toustrup-Jensen, T. L. M. Sorensen, J. Petersen, J. P. Andersen, B. Vilsen, and P. Nissen. Crystal structure of the sodium-potassium pump. *Nature*, 450(7172):1043–1049, 2007.

223. V. B. Chen, W. B. Arendall, J. J. Headd, D. A. Keedy, R. M. Immormino, G. J. Kapral, L. W. Murray, J. S. Richardson, and D. C. Richardson. MolProbity: all-atom structure validation for macromolecular crystallography. *Acta Crystallogr. Sect. D*, 66(1):12–21, 2010.

224. C. Toyoshima, M. Nakasako, H. Nomura, and H. Ogawa. Crystal structure of the calcium pump of sarcoplasmic reticulum at 2.6 Å resolution. *Nature*, 405(6787):647–655, 2000.

225. J. Zhao, S. Benlekbir, and J. L. Rubinstein. Electron cryomicroscopy observation of rotational states in a eukaryotic V-ATPase. *Nature*, 521(7551):241–245, 2015.

226. O. M. Becker, Y. Marantz, S. Shacham, B. Inbal, A. Heifetz, O. Kalid, S. Bar-Haim, D. Warshaviak, M. Fichman, and S. Noiman. G protein-coupled receptors: in silico drug discovery in 3D. *Proc. Natl. Acad. Sci. USA*, 101(31):11304–9, 2004.

227. O. Schueler-Furman, C. Wang, P. Bradley, K. Misura, and D. Baker. Progress in Modeling of Protein Structures and Interactions. *Science*, 310(5748):638–642, 2005.

228. Y. Pilpel, N. Ben-Tal, and D. Lancet. kPROT: a knowledge-based scale for the propensity of residue orientation in transmembrane segments. Application to membrane protein structure prediction. *J. Mol. Biol.*, 294(4):921–35, 1999.

229. W. R. Taylor, D. T. Jones, and N. M. Green. A method for alpha-helical integral membrane protein fold prediction. *Proteins*, 18(3):281–94, 1994.

230. S. J. Fleishman, S. Harrington, R. A. Friesner, B. Honig, and N. Ben-Tal. An automatic method for predicting transmembrane protein structures using cryo-EM and evolutionary data. *Biophys. J.*, 87(5):3448–59, 2004.

231. T. A. Hopf, L. J. Colwell, R. Sheridan, B. Rost, C. Sander, and D. S. Marks. Three-dimensional structures of membrane proteins from genomic sequencing. *Cell*, 149(7):1607–21, 2012.

232. S. J. Fleishman and N. Ben-Tal. A novel scoring function for predicting the conformations of tightly packed pairs of transmembrane alpha-helices. *J. Mol. Biol.*, 321(2):363–78, 2002.

233. M. Schushan and N. Ben-Tal. Modeling and Validation of Transmembrane Protein Structures. In *Introduction to Protein Structure Prediction: Methods and Algorithms*, pages 369–401. John Wiley & Sons, Inc., 2010.

234. V. Yarov-Yarovoy, J. Schonbrun, and D. Baker. Multipass membrane protein structure prediction using Rosetta. *Proteins*, 62(4):1010–25, 2006.

235. R. Das and D. Baker. Macromolecular modeling with Rosetta. *Annu. Rev. Biochem.*, 77:363–82, 2008.

236. P. Barth, J. Schonbrun, and D. Baker. Toward high-resolution prediction and design of transmembrane helical protein structures. *Proc. Natl. Acad. Sci. USA*, 104(40):15682–7, 2007.

237. A. Roy, A. Kucukural, and Y. Zhang. I-TASSER: a unified platform for automated protein structure and function prediction. *Nat. Protoc.*, 5(4):725–38, 2010.

238. J. Zhang, J. Yang, R. Jang, and Y. Zhang. GPCR-I-TASSER: A hybrid approach to G protein-coupled receptor structure modeling and the application to the human genome. *Structure*, 23(8):1538–49, 2015.

239. P. Barth, B. Wallner, and D. Baker. Prediction of membrane protein structures with complex topologies using limited constraints. *Proc. Natl. Acad. Sci. USA*, 106(5):1409–14, 2009.

240. Y. Zhang, M. E. Devries, and J. Skolnick. Structure modeling of all identified G protein-coupled receptors in the human genome. *PLoS Comput. Biol.*, 2(2):e13, 2006.

241. A. Sali and T. L. Blundell. Comparative protein modelling by satisfaction of spatial restraints. *J. Mol. Biol.*, 234(3):779–815, 1993.

242. W. Zheng and S. Doniach. Protein structure prediction constrained by solution X-ray scattering data and structural homology identification. *J. Mol. Biol.*, 316(1):173–87, 2002.

243. W. Zheng and S. Doniach. Fold recognition aided by constraints from small angle X-ray scattering data. *Protein Eng. Des. Sel.*, 18(5):209–19, 2005.

244. D. Schneidman-Duhovny, S. J. Kim, and A. Sali. Integrative structural modeling with small angle X-ray scattering profiles. *BMC Struct. Biol.*, 12:17, 2012.

245. P. D. Adams, D. Baker, A. T. Brunger, R. Das, F. DiMaio, R. J. Read, D. C. Richardson, J. S. Richardson, and T. C. Terwilliger. Advances, interactions, and future developments in the CNS, Phenix, and Rosetta structural biology software systems. *Annu. Rev. Biophys.*, 42:265–87, 2013.

246. S. J. Hirst, N. Alexander, H. S. McHaourab, and J. Meiler. RosettaEPR: an integrated tool for protein structure determination from sparse EPR data. *J. Struct. Biol.*, 173(3):506–14, 2011.

247. P. Rossi, L. Shi, G. Liu, C. M. Barbieri, H. W. Lee, T. D. Grant, J. R. Luft, R. Xiao, T. B. Acton, E. H. Snell, G. T. Montelione, D. Baker, O. F. Lange, and N. G. Sgourakis. A hybrid NMR/SAXS-based approach for discriminating oligomeric protein interfaces using Rosetta. *Proteins*, 83(2):309–17, 2015.

248. F. DiMaio, M. D. Tyka, M. L. Baker, W. Chiu, and D. Baker. Refinement of protein structures into low-resolution density maps using Rosetta. *J. Mol. Biol.*, 392(1):181–90, 2009.

249. Y. Shen, O. Lange, F. Delaglio, P. Rossi, J. M. Aramini, G. Liu, A. Eletsky, Y. Wu, K. K. Singarapu, A. Lemak, A. Ignatchenko, C. H. Arrowsmith, T. Szyperski, G. T. Montelione, D. Baker, and A. Bax. Consistent blind protein structure generation from NMR chemical shift data. *Proc. Natl. Acad. Sci. USA*, 105(12):4685–90, 2008.

250. F. DiMaio, Y. Song, X. Li, M. J. Brunner, C. Xu, V. Conticello, E. Egelman, T. C. Marlovits, Y. Cheng, and D. Baker. Atomic-accuracy models from 4.5-Å cryo-electron microscopy data with density-guided iterative local refinement. *Nat. Methods*, 12(4):361–5, 2015.

251. A. Cavalli, X. Salvatella, C. M. Dobson, and M. Vendruscolo. Protein structure determination from NMR chemical shifts. *Proc. Natl. Acad. Sci. USA*, 104(23):9615–20, 2007.

252. D. S. Wishart, D. Arndt, M. Berjanskii, P. Tang, J. Zhou, and G. Lin. CS23D: a web server for rapid protein structure generation using NMR chemical shifts and sequence data. *Nucleic Acids Res.*, 36(suppl 2):W496–502, 2008.

253. M. A. dos Reis, R. Aparicio, and Y. Zhang. Improving protein template recognition by using small-angle x-ray scattering profiles. *Biophys. J.*, 101(11):2770–81, 2011.

254. D. Murray, N. Ben-Tal, B. Honig, and S. McLaughlin. Electrostatic interaction of myristoylated proteins with membranes: simple physics, complicated biology. *Structure*, 5(8):985–9, 1997.

255. D. Murray and B. Honig. Electrostatic control of the membrane targeting of C2 domains. *Mol. Cell*, 9(1):145–54, 2002.

256. N. Ben-Tal, B. Honig, R. M. Peitzsch, G. Denisov, and S. McLaughlin. Binding of small basic peptides to membranes containing acidic lipids: theoretical models and experimental results. *Biophys. J.*, 71(2):561–75, 1996.

257. S. J. Dunne, R. B. Cornell, J. E. Johnson, N. R. Glover, and A. S. Tracey. Structure of the Membrane Binding Domain of CTP:Phosphocholine Cytidylyltransferase†. *Biochemistry*, 35(37):11975–11984, 1996.

258. B. Antonny, S. Beraud-Dufour, P. Chardin, and M. Chabre. N-Terminal Hydrophobic Residues of the G-Protein ADP-Ribosylation Factor-1 Insert into Membrane Phospholipids upon GDP to GTP Exchange. *Biochemistry*, 36(15):4675–4684, 1997.

259. A. Kessel, D. S. Cafiso, and N. Ben-Tal. Continuum solvent model calculations of alamethicin-membrane interactions: thermodynamic aspects. *Biophys. J.*, 78(2):571–83, 2000.

260. K. V. Damodaran, K. M. Merz Jr, and B. P. Gaber. Interaction of small peptides with lipid bilayers. *Biophys. J.*, 69(4):1299–308, 1995.

261. E. Yamauchi, T. Nakatsu, M. Matsubara, H. Kato, and H. Taniguchi. Crystal structure of a MARCKS peptide containing the calmodulin-binding domain in complex with Ca^{2+}-calmodulin. *Nat. Struct. Biol.*, 10(3):226–31, 2003.

262. S. McLaughlin and D. Murray. Plasma membrane phosphoinositide organization by protein electrostatics. *Nature*, 438(7068):605–11, 2005.

263. A. Kessel, D. Shental-Bechor, T. Haliloglu, and N. Ben-Tal. Interactions of hydrophobic peptides with lipid bilayers: Monte Carlo simulations with M2delta. *Biophys. J.*, 85(6):3431–44, 2003.

264. O. S. Andersen and R. E. Koeppe 2nd. Bilayer thickness and membrane protein function: an energetic perspective. *Annu. Rev. Biophys. Biomol. Struct.*, 36:107–30, 2007.

265. M. P. Sheetz and S. J. Singer. Biological membranes as bilayer couples. A molecular mechanism of drug-erythrocyte interactions. *Proc. Natl. Acad. Sci. USA*, 71(11):4457–61, 1974.

266. O. G. Mouritsen and M. Bloom. Mattress model of lipid-protein interactions in membranes. *Biophys. J.*, 46(2):141–53, 1984.

267. D. R. Fattal and A. Ben-Shaul. A molecular model for lipid-protein interaction in membranes: the role of hydrophobic mismatch. *Biophys. J.*, 65(5):1795–809, 1993.

268. D. C. Mitchell, M. Straume, J. L. Miller, and B. J. Litman. Modulation of metarhodopsin formation by cholesterol-induced ordering of bilayer lipids. *Biochemistry*, 29(39):9143–9, 1990.

269. J. A. Lundbaek, P. Birn, S. E. Tape, G. E. Toombes, R. Sogaard, R. E. Koeppe 2nd, S. M. Gruner, A. J. Hansen, and O. S. Andersen. Capsaicin regulates voltage-dependent sodium channels by altering lipid bilayer elasticity. *Mol. Pharmacol.*, 68(3):680–9, 2005.

270. K. Simons and E. Ikonen. Functional rafts in cell membranes. *Nature*, 387(6633):569–72, 1997.

271. H. Sprong, P. van der Sluijs, and G. van Meer. How proteins move lipids and lipids move proteins. *Nat. Rev. Mol. Cell Biol.*, 2(7):504–13, 2001.

272. R. J. Schroeder, S. N. Ahmed, Y. Zhu, E. London, and D. A. Brown. Cholesterol and sphingolipid enhance the Triton X-100 insolubility of glycosylphosphatidylinositol-anchored proteins by promoting the formation of detergent-insoluble ordered membrane domains. *J. Biol. Chem.*, 273(2):1150–7, 1998.

273. M. D. Resh. Membrane targeting of lipid modified signal transduction proteins. *Subcell. Biochem.*, 37:217–32, 2004.

274. T. Y. Wang, R. Leventis, and J. R. Silvius. Partitioning of lipidated peptide sequences into liquid-ordered lipid domains in model and biological membranes. *Biochemistry*, 40(43):13031–40, 2001.

275. S. Munro. An investigation of the role of transmembrane domains in Golgi protein retention. *EMBO J.*, 14(19):4695–704, 1995.

276. A. P. Starling, J. M. East, and A. G. Lee. Effects of phosphatidylcholine fatty acyl chain length on calcium binding and other functions of the $(Ca^{2+}-Mg^{2+})$-ATPase. *Biochemistry*, 32(6):1593–600, 1993.

277. P. A. Baldwin and W. L. Hubbell. Effects of lipid environment on the light-induced conformational changes of rhodopsin. 2. Roles of lipid chain length, unsaturation, and phase state. *Biochemistry*, 24(11):2633–9, 1985.

278. F. Cornelius. Modulation of Na,K-ATPase and Na-ATPase activity by phospholipids and cholesterol. I. Steady-state kinetics. *Biochemistry*, 40(30):8842–51, 2001.

279. J. D. Pilot, J. M. East, and A. G. Lee. Effects of bilayer thickness on the activity of diacylglycerol kinase of *Escherichia coli*. *Biochemistry*, 40(28):8188–95, 2001.

280. C. Montecucco, G. A. Smith, F. Dabbeni-sala, A. Johannsson, Y. M. Galante, and R. Bisson. Bilayer thickness and enzymatic activity in the mitochondrial cytochrome c oxidase and ATPase complex. *FEBS Lett.*, 144(1):145–8, 1982.

281. O. S. Andersen, R. E. Koeppe 2nd, and B. Roux. Gramicidin channels. *IEEE Trans. Nanobioscience*, 4(1):10–20, 2005.

282. A. M. O'Connell, R. E. Koeppe 2nd, and O. S. Andersen. Kinetics of gramicidin channel formation in lipid bilayers: transmembrane monomer association. *Science*, 250(4985):1256–9, 1990.

283. S. Bransburg-Zabary, A. Kessel, M. Gutman, and N. Ben-Tal. Stability of an ion channel in lipid bilayers: implicit solvent model calculations with gramicidin. *Biochemistry*, 41(22):6946–54, 2002.

284. A. Johannsson, G. A. Smith, and J. C. Metcalfe. The effect of bilayer thickness on the activity of $(Na^+ + K^+)$-ATPase. *Biochim. Biophys. Acta*, 641(2):416–21, 1981.

285. F. Dumas, J. F. Tocanne, G. Leblanc, and M. C. Lebrun. Consequences of hydrophobic mismatch between lipids and melibiose permease on melibiose transport. *Biochemistry*, 39(16):4846–54, 2000.

286. M. Sinensky. Homeoviscous adaptation: a homeostatic process that regulates the viscosity of membrane lipids in *Escherichia coli*. *Proc. Natl. Acad. Sci. USA*, 71(2):522–5, 1974.

287. M. F. Brown. Influence of nonlamellar-forming lipids on rhodopsin. *Curr. Top. Membr.*, 44:285–356, 1997.

288. A. P. Starling, K. A. Dalton, J. M. East, S. Oliver, and A. G. Lee. Effects of phosphatidylethanolamines on the activity of the Ca^{2+}-ATPase of sarcoplasmic reticulum. *Biochem. J.*, 320 (Pt 1):309–14, 1996.

289. O. P. Hamill and B. Martinac. Molecular basis of mechanotransduction in living cells. *Physiol. Rev.*, 81(2):685–740, 2001.

290. C. Hunte. Specific protein-lipid interactions in membrane proteins. *Biochem. Soc. Trans.*, 33(Pt 5):938–42, 2005.

291. A. C. Simmonds, J. M. East, O. T. Jones, E. K. Rooney, J. McWhirter, and A. G. Lee. Annular and non-annular binding sites on the $(Ca^{2+} + Mg^{2+})$-ATPase. *Biochim. Biophys. Acta*, 693(2):398–406, 1982.

292. P. F. Knowles, A. Watts, and D. Marsh. Spin-label studies of lipid immobilization in dimyristoylphosphatidylcholine-substituted cytochrome oxidase. *Biochemistry*, 18(21):4480–7, 1979.

293. T. Murata, I. Yamato, Y. Kakinuma, A. G. Leslie, and J. E. Walker. Structure of the rotor of the V-Type Na$^+$-ATPase from *Enterococcus hirae*. *Science*, 308(5722):654–9, 2005.

294. H. Palsdottir and C. Hunte. Lipids in membrane protein structures. *Biochim. Biophys. Acta*, 1666(1–2):2–18, 2004.

295. C. Lange, J. H. Nett, B. L. Trumpower, and C. Hunte. Specific roles of protein-phospholipid interactions in the yeast cytochrome bc1 complex structure. *EMBO J.*, 20(23):6591–600, 2001.

296. T. Tsukihara, K. Shimokata, Y. Katayama, H. Shimada, K. Muramoto, H. Aoyama, M. Mochizuki, K. Shinzawa-Itoh, E. Yamashita, M. Yao, Y. Ishimura, and S. Yoshikawa. The low-spin heme of cytochrome c oxidase as the driving element of the proton-pumping process. *Proc. Natl. Acad. Sci. USA*, 100(26):15304–9, 2003.

297. S. S. Deol, P. J. Bond, C. Domene, and M. S. Sansom. Lipid-protein interactions of integral membrane proteins: a comparative simulation study. *Biophys. J.*, 87(6):3737–49, 2004.

298. P. C. van der Wel, N. D. Reed, D. V. Greathouse, and R. E. Koeppe 2nd. Orientation and motion of tryptophan interfacial anchors in membrane-spanning peptides. *Biochemistry*, 46(25):7514–24, 2007.

299. K. Boesze-Battaglia and R. Schimmel. Cell membrane lipid composition and distribution: implications for cell function and lessons learned from photoreceptors and platelets. *J. Exp. Biol.*, 200(Pt 23):2927–36, 1997.

300. F. Jiang, M. T. Ryan, M. Schlame, M. Zhao, Z. Gu, M. Klingenberg, N. Pfanner, and M. L. Greenberg. Absence of cardiolipin in the crd1 null mutant results in decreased mitochondrial membrane potential and reduced mitochondrial function. *J. Biol. Chem.*, 275(29):22387–94, 2000.

301. S. Heimpel, G. Basset, S. Odoy, and M. Klingenberg. Expression of the mitochondrial ADP/ATP carrier in *Escherichia coli*. Renaturation, reconstitution, and the effect of mutations on 10 positive residues. *J. Biol. Chem.*, 276(15):11499–506, 2001.

302. B. Hoffmann, A. Stockl, M. Schlame, K. Beyer, and M. Klingenberg. The reconstituted ADP/ATP carrier activity has an absolute requirement for cardiolipin as shown in cysteine mutants. *J. Biol. Chem.*, 269(3):1940–4, 1994.

303. S. Nussberger, K. Dorr, D. N. Wang, and W. Kuhlbrandt. Lipid-protein interactions in crystals of plant light-harvesting complex. *J. Mol. Biol.*, 234(2):347–56, 1993.

304. J. Standfuss, A. C. Terwisscha van Scheltinga, M. Lamborghini, and W. Kuhlbrandt. Mechanisms of photoprotection and nonphotochemical quenching in pea light-harvesting complex at 2.5 Å resolution. *EMBO J.*, 24(5):919–28, 2005.

305. Z. Liu, H. Yan, K. Wang, T. Kuang, J. Zhang, L. Gui, X. An, and W. Chang. Crystal structure of spinach major light-harvesting complex at 2.72 Å resolution. *Nature*, 428(6980):287–92, 2004.

306. R. Horn and H. Paulsen. Folding In vitro of Light-harvesting Chlorophyll a/b Protein is Coupled with Pigment Binding. *J. Mol. Biol.*, 318(2):547–556, 2002.

307. C. Arnarez, S. J. Marrink, and X. Periole. Identification of cardiolipin binding sites on cytochrome c oxidase at the entrance of proton channels. *Sci. Rep.*, 3:1263, 2013.

308. L. Heginbotham, L. Kolmakova-Partensky, and C. Miller. Functional reconstitution of a prokaryotic K$^+$ channel. *J. Gen. Physiol.*, 111(6):741–9, 1998.

309. B. C. Suh and B. Hille. PIP2 is a necessary cofactor for ion channel function: how and why? *Annu. Rev. Biophys.*, 37:175–95, 2008.

310. M. J. Berridge and R. F. Irvine. Inositol phosphates and cell signalling. *Nature*, 341(6239):197–205, 1989.

311. M. A. Zaydman and J. Cui. PIP(2) regulation of KCNQ channels: biophysical and molecular mechanisms for lipid modulation of voltage-dependent gating. *Front. Physiol.*, 5, 2014.

312. P. J. Hamilton, A. N. Belovich, G. Khelashvili, C. Saunders, K. Erreger, J. A. Javitch, H. H. Sitte, H. Weinstein, H. J. Matthies, and A. Galli. PIP2 regulates psychostimulant behaviors through its interaction with a membrane protein. *Nat. Chem. Biol.*, 10(7):582–9, 2014.

313. F. Buchmayer, K. Schicker, T. Steinkellner, P. Geier, G. Stübiger, P. J. Hamilton, A. Jurik, T. tockner, J.-W. Yang, T. Montgomery, M. Holy, T. Hofmaier, O. Kudlacek, H. J. G. Matthies, V. Ecker, G. F. nd Bochkov, A. Galli, S. Boehm, and H. H. Sitte. Amphetamine actions at the serotonin transporter rely on the availability of phosphatidylinositol 4,5-bisphosphate. *Proc. Natl. Acad. Sci. USA*, 110(28):11642–11647, 2013.

314. N. A. Baker, D. Sept, S. Joseph, M. J. Holst, and J. A. McCammon. Electrostatics of nanosystems: Application to microtubules and the ribosome. *Proc. Natl. Acad. Sci. USA*, 98(18):10037–10041, 2001.

315. M. A. Lemmon. Phosphoinositide recognition domains. *Traffic*, 4(4):201–13, 2003.

316. K. M. Ferguson, J. M. Kavran, V. G. Sankaran, E. Fournier, S. J. Isakoff, E. Y. Skolnik, and M. A. Lem-

mon. Structural basis for discrimination of 3-phosphoinositides by pleckstrin homology domains. *Mol. Cell*, 6(2):373–84, 2000.

317. H. Tapp, I. M. Al-Naggar, E. G. Yarmola, A. Harrison, G. Shaw, A. S. Edison, and M. R. Bubb. MARCKS is a natively unfolded protein with an inaccessible actin-binding site: evidence for long-range intramolecular interactions. *J. Biol. Chem.*, 280(11):9946–56, 2005.

318. A. Ben-Shaul, N. Ben-Tal, and B. Honig. Statistical thermodynamic analysis of peptide and protein insertion into lipid membranes. *Biophys. J.*, 71(1):130–7, 1996.

319. J. A. Killian. Hydrophobic mismatch between proteins and lipids in membranes. *Biochim. Biophys. Acta*, 1376(3):401–15, 1998.

320. N. I. Liu and R. L. Kay. Redetermination of the pressure dependence of the lipid bilayer phase transition. *Biochemistry*, 16(15):3484–6, 1977.

321. E. A. Evans and R. M. Hochmuth. Mechanochemical properties of membranes. *Curr. Top. Membr. Transp.*, 10:1–64, 1978.

322. D. Marsh. Energetics of Hydrophobic Matching in Lipid-Protein Interactions. *Biophys. J.*, 94(10):3996–4013, 2008.

323. T. Kim and W. Im. Revisiting hydrophobic mismatch with free energy simulation studies of transmembrane helix tilt and rotation. *Biophys. J.*, 99(1):175–83, 2010.

324. S. K. Kandasamy and R. G. Larson. Molecular dynamics simulations of model trans-membrane peptides in lipid bilayers: a systematic investigation of hydrophobic mismatch. *Biophys. J.*, 90(7):2326–43, 2006.

325. Y. Gofman, T. Haliloglu, and N. Ben-Tal. The Transmembrane Helix Tilt May Be Determined by the Balance between Precession Entropy and Lipid Perturbation. *J. Chem. Theory. Comput.*, 8(8):2896–2904, 2012.

326. D. P. Siegel, V. Cherezov, D. V. Greathouse, R. E. Koeppe 2nd, J. A. Killian, and M. Caffrey. Transmembrane peptides stabilize inverted cubic phases in a biphasic length-dependent manner: implications for protein-induced membrane fusion. *Biophys. J.*, 90(1):200–11, 2006.

327. D. Raucher and M. P. Sheetz. Cell spreading and lamellipodial extension rate is regulated by membrane tension. *J. Cell Biol.*, 148(1):127–36, 2000.

328. M. D. Ledesma and C. G. Dotti. Membrane and cytoskeleton dynamics during axonal elongation and stabilization. *Int. Rev. Cytol.*, 227:183–219, 2003.

329. M. P. Sheetz. Cell control by membrane-cytoskeleton adhesion. *Nat. Rev. Mol. Cell Biol.*, 2(5):392–6, 2001.

330. B. Antonny, P. Gounon, R. Schekman, and L. Orci. Self-assembly of minimal COPII cages. *EMBO Rep.*, 4(4):419–24, 2003.

331. R. Nossal. Energetics of clathrin basket assembly. *Traffic*, 2(2):138–47, 2001.

332. B. Razani and M. P. Lisanti. Caveolins and caveolae: molecular and functional relationships. *Exp. Cell Res.*, 271(1):36–44, 2001.

333. R. J. Mashl and R. F. Bruinsma. Spontaneous-curvature theory of clathrin-coated membranes. *Biophys. J.*, 74(6):2862–75, 1998.

334. M. G. Ford, I. G. Mills, B. J. Peter, Y. Vallis, G. J. Praefcke, P. R. Evans, and H. T. McMahon. Curvature of clathrin-coated pits driven by epsin. *Nature*, 419(6905):361–6, 2002.

335. M. N. Aranda-Sicilia, Y. Trusov, N. Maruta, D. Chakravorty, Y. Zhang, and J. R. Botella. Heterotrimeric G proteins interact with defense-related receptor-like kinases in Arabidopsis. *J. Plant Physiol.*, 188:44–48, 2015.

336. J. Bockaert. G Protein-coupled Receptors. In *Encyclopedia of Life Sciences*. John Wiley & Sons, Ltd., 2009.

337. R. Fredriksson, M. C. Lagerstrom, L. G. Lundin, and H. B. Schioth. The G-protein-coupled receptors in the human genome form five main families. Phylogenetic analysis, paralogon groups, and fingerprints. *Mol. Pharmacol.*, 63:1256–1272, 2003.

338. C. D. Hanlon and D. J. Andrew. Outside-in signaling: a brief review of GPCR signaling with a focus on the Drosophila GPCR family. *J. Cell Sci.*, 128(19):3533–3542, 2015.

339. T. K. Bjarnadóttir, D. E. Gloriam, S. H. Hellstrand, H. Kristiansson, R. Fredriksson, and H. B. Schiöth. Comprehensive repertoire and phylogenetic analysis of the G protein-coupled receptors in human and mouse. *Genomics*, 88(3):263–273, 2006.

340. N. Wettschureck and S. Offermanns. Mammalian G Proteins and Their Cell Type Specific Functions. *Physiol. Rev.*, 85(4):1159–1204, 2005.

341. N. E. Bhola and J. R. Grandis. Crosstalk between G-protein-coupled receptors and epidermal growth factor receptor in cancer. *Front. Biosci. J. Virtual Library*, 13:1857–1865, 2008.

342. A. Madeo and M. Maggiolini. Nuclear Alternate Estrogen Receptor GPR30 Mediates 17β-Estradiol-

Induced Gene Expression and Migration in Breast Cancer–Associated Fibroblasts. *Cancer Res.*, 70(14):6036–6046, 2010.

343. G. Sliwoski, S. Kothiwale, J. Meiler, and E. W. Lowe Jr. Computational methods in drug discovery. *Pharmacol. Rev.*, 66(1):334–95, 2014.

344. Y. Fang, T. Kenakin, and C. Liu. Editorial: Orphan GPCRs As Emerging Drug Targets. *Front. Pharmacol.*, 6:295, 2015.

345. J. A. Salon, D. T. Lodowski, and K. Palczewski. The Significance of G Protein-Coupled Receptor Crystallography for Drug Discovery. *Pharmacol. Rev.*, 63(4):901–937, 2011.

346. S. L. Garland. Are GPCRs Still a Source of New Targets? *J. Biomol. Screen.*, 18(9):947–966, 2013.

347. M. J. Berridge. Introduction. In *Cell Signalling Biology*, chapter 1, pages 1–69. Portland Press, 2014.

348. Boghog. Dose response curves of a full agonist, partial agonist, neutral antagonist, and inverse agonist. Wikipedia, the free encyclopedia. https://en.wikipedia.org/wiki/Inverse_agonist#/media/File:Inverse_agonist_3.svg, 2014.

349. J. Bockaert, L. Fagni, A. Dumuis, and P. Marin. GPCR interacting proteins (GIP). *Pharmacol. Ther.*, 103(3):203–21, 2004.

350. W. I. Weis and B. Kobilka. Structural insights into G-protein-coupled receptor activation. *Curr. Opin. Struct. Biol.*, 18:734–740, 2008.

351. B. K. Kobilka. G protein coupled receptor structure and activation. *Biochim. Biophys. Acta – Biomembranes*, 1768(4):794–807, 2007.

352. A. Kitao, S. Hayward, and N. Go. Energy landscape of a native protein: jumping-among-minima model. *Proteins*, 33(4):496–517, 1998.

353. J. A. McCammon, B. R. Gelin, and M. Karplus. Dynamics of folded proteins. *Nature*, 267(5612):585–90, 1977.

354. R. H. Austin, K. W. Beeson, L. Eisenstein, H. Frauenfelder, and I. C. Gunsalus. Dynamics of ligand binding to myoglobin. *Biochemistry*, 14(24):5355–73, 1975.

355. G. A. Petsko and D. Ringe. Fluctuations in protein structure from X-ray diffraction. *Annu. Rev. Biophys. Bio.*, 13:331–71, 1984.

356. H. Frauenfelder, F. Parak, and R. D. Young. Conformational substates in proteins. *Annu. Rev. Biophys. Biophys. Chem.*, 17:451–79, 1988.

357. K. S. Kim and C. Woodward. Protein internal flexibility and global stability: effect of urea on hydrogen exchange rates of bovine pancreatic trypsin inhibitor. *Biochemistry*, 32(37):9609–13, 1993.

358. Y. Bai, T. R. Sosnick, L. Mayne, and S. W. Englander. Protein folding intermediates: native-state hydrogen exchange. *Science*, 269(5221):192–7, 1995.

359. R. Elber and M. Karplus. Multiple conformational states of proteins: a molecular dynamics analysis of myoglobin. *Science*, 235(4786):318–21, 1987.

360. L. Fetler, E. R. Kantrowitz, and P. Vachette. Direct observation in solution of a preexisting structural equilibrium for a mutant of the allosteric aspartate transcarbamoylase. *Proc. Natl. Acad. Sci. USA*, 104(2):495–500, 2007.

361. L. C. James and D. S. Tawfik. Conformational diversity and protein evolution: a 60-year-old hypothesis revisited. *Trends Biochem. Sci.*, 28(7):361–8, 2003.

362. A. Malmendal, J. Evenas, S. Forsen, and M. Akke. Structural dynamics in the C-terminal domain of calmodulin at low calcium levels. *J. Mol. Biol.*, 293(4):883–99, 1999.

363. B. K. Kobilka and X. Deupi. Conformational complexity of G-protein-coupled receptors. *Trends Pharmacol. Sci.*, 28(8):397–406, 2007.

364. K. Henzler-Wildman and D. Kern. Dynamic personalities of proteins. *Nature*, 450(7172):964–72, 2007.

365. M. Vendruscolo and C. M. Dobson. Dynamic Visions of Enzymatic Reactions. *Science*, 313(5793):1586–1587, 2006.

366. K. Gunasekaran, B. Ma, and R. Nussinov. Is allostery an intrinsic property of all dynamic proteins? *Proteins*, 57(3):433–43, 2004.

367. G. Weber. Ligand binding and internal equilibrium in proteins. *Biochemistry*, 11(5):864–878, 1972.

368. I. Bahar, C. Chennubhotla, and D. Tobi. Intrinsic dynamics of enzymes in the unbound state and relation to allosteric regulation. *Curr. Opin. Struct. Biol.*, 17(6):633–40, 2007.

369. O. F. Lange, N. A. Lakomek, C. Fares, G. F. Schroder, K. F. Walter, S. Becker, J. Meiler, H. Grubmuller, C. Griesinger, and B. L. de Groot. Recognition dynamics up to microseconds revealed from an RDC-derived ubiquitin ensemble in solution. *Science*, 320(5882):1471–5, 2008.

370. J. Gsponer, J. Christodoulou, A. Cavalli, J. M. Bui, B. Richter, C. M. Dobson, and M. Vendruscolo. A coupled equilibrium shift mechanism in calmodulin-mediated signal transduction. *Structure*, 16(5):736–46, 2008.

371. D. M. Rosenbaum, S. G. F. Rasmussen, and B. K. Kobilka. The structure and function of G-protein-coupled receptors. *Nature*, 459:356–363, 2009.

372. K. L. Pierce, R. T. Premont, and R. J. Lefkowitz. Seven-transmembrane receptors. *Nat. Rev. Mol. Cell Biol.*, 3(9):639–50, 2002.

373. H. Ashkenazy, S. Abadi, E. Martz, O. Chay, I. Mayrose, T. Pupko, and N. Ben-Tal. ConSurf 2016: an improved methodology to estimate and visualize evolutionary conservation in macromolecules. *Nucleic Acids Res.*, 44(W1):W344–W350, 2016.

374. F. Glaser, T. Pupko, I. Paz, R. E. Bell, D. Bechor-Shental, E. Martz, and N. Ben-Tal. ConSurf: identification of functional regions in proteins by surface-mapping of phylogenetic information. *Bioinformatics*, 19(1):163–4, 2003.

375. R. A. Cerione, C. Staniszewski, J. L. Benovic, R. J. Lefkowitz, M. G. Caron, P. Gierschik, R. Somers, A. M. Spiegel, J. Codina, and L. Birnbaumer. Specificity of the functional interactions of the beta-adrenergic receptor and rhodopsin with guanine nucleotide regulatory proteins reconstituted in phospholipid vesicles. *J. Biol. Chem.*, 260(3):1493–500, 1985.

376. W. W. I. Lau, A. S. L. Chan, L. S. W. Poon, J. Zhu, and Y. H. Wong. G$\beta\gamma$-mediated activation of protein kinase D exhibits subunit specificity and requires G$\beta\gamma$-responsive phospholipase Cβ isoforms. *Cell Commun. Signal.*, 11:22, 2013.

377. H. G. Dohlman and J. Thorner. RGS proteins and signaling by heterotrimeric G proteins. *J. Biol. Chem.*, 272:3871–3874, 1997.

378. J. Shonberg, R. C. Kling, P. Gmeiner, and S. Löber. GPCR crystal structures: Medicinal chemistry in the pocket. *Bioorg. Med. Chem.*, 23(14):3880–3906, 2015.

379. D. Zhang, Q. Zhao, and B. Wu. Structural Studies of G Protein-Coupled Receptors. *Mol. Cells*, 38(10):836–842, 2015.

380. A. J. Venkatakrishnan, Xavier Deupi, Guillaume Lebon, Christopher G. Tate, Gebhard F. Schertler, and M. Madan Babu. Molecular signatures of G-protein-coupled receptors. *Nature*, 494(7436):185–194, 2013.

381. S. A. Shabalina, D. V. Zaykin, P. Gris, A. Y. Ogurtsov, J. Gauthier, K. Shibata, I. E. Tchivileva, I. Belfer, B. Mishra, C. Kiselycznyk, M. R. Wallace, R. Staud, N. A. Spiridonov, M. B. Max, D. Goldman, R. B. Fillingim, W. Maixner, and L. Diatchenko. Expansion of the human μ-opioid receptor gene architecture: novel functional variants. *Hum. Mol. Genet.*, 18(6):1037, 2009.

382. M. Convertino, A. Samoshkin, J. Gauthier, M. S. Gold, W. Maixner, N. V. Dokholyan, and L. Diatchenko. μ-Opioid receptor 6-transmembrane isoform: A potential therapeutic target for new effective opioids. *Prog. Neuropsychopharmacol. Biol. Psychiatry*, 62:61–67, 2015.

383. N. Kamesh, Gopala K. Aradhyam, and Narayanan Manoj. The repertoire of G protein-coupled receptors in the sea squirt *Ciona intestinalis*. *BMC Evol. Biol.*, 8(1):129, 2008.

384. J. A. Hanson, K. Duderstadt, L. P. Watkins, S. Bhattacharyya, J. Brokaw, J. W. Chu, and H. Yang. Illuminating the mechanistic roles of enzyme conformational dynamics. *Proc. Natl. Acad. Sci. USA*, 104:18055–18060, 2007.

385. J. Bockaert, S. Claeysen, C. Bécamel, S. Pinloche, and A. Dumuis. G protein-coupled receptors: dominant players in cell-cell communication. *Int. Rev. Cytol.*, 212:63–132, 2002.

386. L. Chun, W.-h. Zhang, and J.-f. Liu. Structure and ligand recognition of class C GPCRs. *Acta Pharm. Sin.*, 33(3):312–323, 2012.

387. H. P. Nothacker. Orphan Receptors. In S. Offermanns and W. Rosenthal, editors, *Encyclopedia of Molecular Pharmacology*, pages 914–917. Springer, Heidelberg, 2008.

388. S. P. Lee, B. F. O'Dowd, and S. R. George. Homo- and hetero-oligomerization of G protein-coupled receptors. *Life Sci.*, 74(2–3):173–180, 2003.

389. S. Ferré, V. Casadó, L. A. Devi, M. Filizola, R. Jockers, M. J. Lohse, G. Milligan, J.-P. Pin, and X. Guitart. G Protein–Coupled Receptor Oligomerization Revisited: Functional and Pharmacological Perspectives. *Pharmacol. Rev.*, 66(2):413–434, 2014.

390. S. Ferré. The GPCR heterotetramer: challenging classical pharmacology. *Trends Pharmacol. Sci.*, 36(3):145–152, 2015.

391. M. J. Lohse. Dimerization in GPCR mobility and signaling. *Curr. Opin. Pharmacol.*, 10(1):53–58, 2010.

392. G. Milligan. G protein-coupled receptor hetero-dimerization: contribution to pharmacology and function. *Br. J. Pharmacol.*, 158(1):5–14, 2009.

393. J. Kniazeff, L. Prézeau, P. Rondard, J.-P. Pin, and C. Goudet. Dimers and beyond: The functional puzzles of class C GPCRs. *Pharmacol. Ther.*, 130(1):9–25, 2011.

394. X. C. Zhang, J. Liu, and D. Jiang. Why is dimerization essential for class-C GPCR function? New insights from mGluR1 crystal structure analysis. *Protein Cell*, 5(7):492–495, 2014.

395. J. P. Pin, J. Kniazeff, V. Binet, J. Liu, D Maurel, T. Galvez, B. Duthey, M. Havlickova, J. Blahos, L. Prézeau, and P. Rondard. Activation mechanism of the heterodimeric GABA(B) receptor. *Biochem. Pharmacol.*, 68(8):1565–1572, 2004.

396. A. Manglik, A. C. Kruse, T. S. Kobilka, F. S. Thian, J. M. Mathiesen, R. K. Sunahara, L. Pardo, W. I. Weis, B. K. Kobilka, and S. Granier. Crystal structure of the micro-opioid receptor bound to a morphinan antagonist. *Nature*, 485(7398):321–326, 2012.

397. R. Franco, E. Martínez-Pinilla, J. L. Lanciego, and G. Navarro. Basic Pharmacological and Structural Evidence for Class A G-Protein-Coupled Receptor Heteromerization. *Front. Pharmacol.*, 7:76, 2016.

398. B. Wu, E. Y. T. Chien, C. D. Mol, G. Fenalti, W. Liu, V. Katritch, R. Abagyan, A. Brooun, P. Wells, F. C. Bi, D. J. Hamel, P. Kuhn, T. M. Handel, V. Cherezov, and R. C. Stevens. Structures of the CXCR4 Chemokine GPCR with Small-Molecule and Cyclic Peptide Antagonists. *Science*, 330(6007):1066–1071, 2010.

399. H. Wu, D. Wacker, V. Katritch, M. Mileni, G. W. Han, E. Vardy, W. Liu, A. A. Thompson, X.-P. Huang, F. I. Carroll, S. W. Mascarella, R. B. Westkaemper, P. D. Mosier, B. L. Roth, V. Cherezov, and R. C. Stevens. Structure of the human kappa opioid receptor in complex with JDTic. *Nature*, 485(7398):327–332, 2012.

400. M. R. Whorton, B. Jastrzebska, P. S. Park, D. Fotiadis, A. Engel, K. Palczewski, and R. K. Sunahara. Efficient coupling of transducin to monomeric rhodopsin in a phospholipid bilayer. *J. Biol. Chem.*, 283(74387–4394), 2008.

401. J. Huang, S. Chen, J. J. Zhang, and X.-Y. Huang. Crystal structure of oligomeric β1-adrenergic G protein-coupled receptors in ligand-free basal state. *Nat. Struct. Mol. Biol.*, 20(4):419–425, 2013.

402. V. Cherezov, D. M. Rosenbaum, M. A. Hanson, S. G. F. Rasmussen, F. S. Thian, T. S. Kobilka, H. J. Choi, P. Kuhn, W. I. Weis, B. K. Kobilka, and R. C. Stevens. High-resolution crystal structure of an engineered human beta2-adrenergic G protein-coupled receptor. *Science*, 318(5854):1258–65, 2007.

403. H. Zheng, E. A. Pearsall, D. P. Hurst, Y. Zhang, J. Chu, Y. Zhou, P. H. Reggio, H. H. Loh, and P.-Y. Law. Palmitoylation and membrane cholesterol stabilize μ-opioid receptor homodimerization and G protein coupling. *BMC Cell Biology*, 13(1):6, 2012.

404. H. Wu, C. Wang, K. J. Gregory, G. W. Han, H. P. Cho, Y. Xia, C. M. Niswender, V. Katritch, J. Meiler, V. Cherezov, P. J. Conn, and R. C. Stevens. Structure of a Class C GPCR Metabotropic Glutamate Receptor 1 Bound to an Allosteric Modulator. *Science*, 344(6179):58–64, 2014.

405. T. W. Schwartz and W. L. Hubbell. Structural biology: A moving story of receptors. *Nature*, 455(7212):473–474, 2008.

406. R. J. Lefkowitz. A Brief History of G-Protein Coupled Receptors (Nobel Lecture). *Angew. Chem. Int. Ed.*, 52(25):6366–6378, 2013.

407. M. A. Hanson and R. C. Stevens. Discovery of New GPCR Biology: One Receptor Structure at a Time. *Structure*, 17:8–14, 2009.

408. S. G. F. Rasmussen, H. J. Choi, D. M. Rosenbaum, T. S. Kobilka, F. S. Thian, P. C. Edwards, M. Burghammer, V. R. Ratnala, R. Sanishvili, R. F. Fischetti, G. F. Schertler, W. I. Weis, and B. K. Kobilka. Crystal structure of the human beta2 adrenergic G-protein-coupled receptor. *Nature*, 450(7168):383–7, 2007.

409. E. M. Bastiaanse, K. M. Hold, and A. Van der Laarse. The effect of membrane cholesterol content on ion transport processes in plasma membranes. *Cardiovasc. Res.*, 33(2):272–83, 1997.

410. K. Burger, G. Gimpl, and F. Fahrenholz. Regulation of receptor function by cholesterol. *Cell. Mol. Life Sci.*, 57(11):1577–92, 2000.

411. T. J. Pucadyil and A. Chattopadhyay. Role of cholesterol in the function and organization of G-protein coupled receptors. *Prog. Lipid Res.*, 45(4):295–333, 2006.

412. J. Oates and A. Watts. Uncovering the intimate relationship between lipids, cholesterol and GPCR activation. *Curr. Opin. Struct. Biol.*, 21(6):802–807, 2011.

413. Y. D. Paila and A. Chattopadhyay. The function of G-protein coupled receptors and membrane cholesterol: specific or general interaction? *Glycoconj. J.*, 26(6):711, 2009.

414. X. Deupi and J. Standfuss. Structural insights into agonist-induced activation of G-protein-coupled receptors. *Curr. Opin. Struct. Biol.*, 21(4):541–551, 2011.

415. R. Moukhametzianov, T. Warne, P. C. Edwards, M. J. Serrano-Vega, A. G. W. Leslie, C. G. Tate, and G. F. X. Schertler. Two distinct conformations of helix 6 observed in antagonist-bound structures of a β1-adrenergic receptor. *Proc. Natl. Acad. Sci. USA*, 108(20):8228–8232, 2011.

416. R. O. Dror, D. H. Arlow, D. W. Borhani, M. Ø. Jensen, S. Piana, and D. E. Shaw. Identification of two distinct inactive conformations of the β2-adrenergic receptor reconciles structural and biochemical observations. *Proc. Natl. Acad. Sci. USA*, 106(12):4689–4694, 2009.

417. R. A. Bond and A. P. Ijzerman. Recent developments in constitutive receptor activity and inverse agonism, and their potential for GPCR drug discovery. *Trends Pharmacol. Sci.*, 27:92–96, 2006.

418. K. Palczewski, T. Kumasaka, T. Hori, C. A. Behnke, H. Motoshima, B. A. Fox, I. Le Trong, D. C. Teller, T. Okada, R. E. Stenkamp, M. Yamamoto, and M. Miyano. Crystal structure of rhodopsin: A G protein-coupled receptor. *Science*, 289(5480):739–45, 2000.

419. M. Murakami and T. Kouyama. Crystal structure of squid rhodopsin. *Nature*, 453(7193):363–7, 2008.

420. T. Shimamura, K. Hiraki, N. Takahashi, T. Hori, H. Ago, K. Masuda, K. Takio, M. Ishiguro, and M. Miyano. Crystal structure of squid rhodopsin with intracellularly extended cytoplasmic region. *J. Biol. Chem.*, 283(26):17753–6, 2008.

421. D. Mustafi and K. Palczewski. Topology of class A G protein-coupled receptors: insights gained from crystal structures of rhodopsins, adrenergic and adenosine receptors. *Mol. Pharmacol.*, 75(1):1–12, 2009.

422. B. K. Kobilka, T. S. Kobilka, K. Daniel, J. W. Regan, M. G. Caron, and R. J. Lefkowitz. Chimeric alpha 2-,beta 2-adrenergic receptors: delineation of domains involved in effector coupling and ligand binding specificity. *Science*, 240(4857):1310–1316, 1988.

423. C. C. Malbon and H. Y. Wang. Adrenergic Receptors. In *Encyclopedia of Life Sciences*. John Wiley & Sons, Ltd., 2005.

424. S. Nussey and S. Whitehead. *Endocrinology: An Integrated Approach*. BIOS Scientific Publishers Ltd, Oxford, England, 2001.

425. D. S. Goldstein. Adrenaline and noradrenaline. In *Encyclopedia of Life Sciences*. John Wiley & Sons, Ltd., 2001.

426. S. C. Stanford. Adrenaline and Noradrenaline: Introduction. In *Encyclopedia of Life Sciences*. John Wiley & Sons, Ltd., 2009.

427. J. H. Schwartz. Neurotransmitters. In *Encyclopedia of Life Sciences*. John Wiley & Sons, Ltd., 2001.

428. W. B. Cannon. *Bodily Changes in Pain, Hunger, Fear and Rage: An Account of Recent Research Into the Function of Emotional Excitement*. Appleton-Century-Crofts, New York, 1929.

429. A. Hachmon. An Arabian woof bristling its hair. Wikipedia, the free encyclopedia. http://en.wikipedia.org/wiki/File:Canis_lupus_arabic.JPG, 2008.

430. E. Charmandari, C. Tsigos, and G. Chrousos. Endocrinology of the stress response. *Annu. Rev. Physiol.*, 67:259–84, 2005.

431. D. M. Rosenbaum, C. Zhang, J. A. Lyons, R. Holl, D. Aragao, D. H. Arlow, S. G. F. Rasmussen, H.-J. Choi, B. T. DeVree, R. K. Sunahara, P. S. Chae, S. H. Gellman, R. O. Dror, D. E. Shaw, W. I. Weis, M. Caffrey, P. Gmeiner, and B. K. Kobilka. Structure and function of an irreversible agonist-β2 adrenoceptor complex. *Nature*, 469(7329):236–240, 2011.

432. R. Nygaard, Y. Zou, R. O. Dror, T. J. Mildorf, D. H. Arlow, A. Manglik, A. C. Pan, C. W. Liu, J. J. Fung, M. P. Bokoch, F. S. Thian, T. S. Kobilka, D. E. Shaw, L. Mueller, R. S. Prosser, and B. K. Kobilka. The Dynamic Process of β2-Adrenergic Receptor Activation. *Cell*, 152(3):532–542, 2013.

433. A. Manglik, T. H. Kim, M. Masureel, C. Altenbach, Z. Yang, D. Hilger, M. T. Lerch, T. S. Kobilka, F. S. Thian, W. L. Hubbell, R. S. Prosser, and B. K. Kobilka. Structural Insights into the Dynamic Process of β2-Adrenergic Receptor Signaling. *Cell*, 161(5):1101–1111, 2015.

434. H.-W. Choe, Y. J. Kim, J. H. Park, T. Morizumi, E. F. Pai, N. Krausz, K. P. Hofmann, P. Scheerer, and O. P. Ernst. Crystal structure of metarhodopsin II. *Nature*, 471(7340):651–655, 2011.

435. S. G. F. Rasmussen, B. T. DeVree, Y. Zou, A. C. Kruse, K. Y. Chung, T. S. Kobilka, F. S. Thian, P. S. Chae, E. Pardon, D. Calinski, J. M. Mathiesen, S. T. A. Shah, J. A. Lyons, M. Caffrey, S. H. Gellman, J. Steyaert, G. Skiniotis, W. I. Weis, R. K. Sunahara, and B. K. Kobilka. Crystal structure of the β2 adrenergic receptor-Gs protein complex. *Nature*, 477(7366):549–555, 2011.

436. S. G. F. Rasmussen, H.-J. Choi, J. J. Fung, E. Pardon, P. Casarosa, P. S. Chae, B. T. DeVree, D. M. Rosenbaum, F. S. Thian, T. S. Kobilka, A. Schnapp, I. Konetzki, R. K. Sunahara, S. H. Gellman, A. Pautsch, J. Steyaert, W. I. Weis, and B. K. Kobilka. Structure of a nanobody-stabilized active state of the β2 adrenoceptor. *Nature*, 469(7329):175–180, 2011.

437. A. C. Kruse, A. M. Ring, A. Manglik, J. Hu, K. Hu, K. Eitel, H. Hubner, E. Pardon, C. Valant, P. M. Sexton, A. Christopoulos, C. C. Felder, P. Gmeiner, J. Steyaert, W. I. Weis, K. C. Garcia, J. Wess, and B. K. Kobilka. Activation and allosteric modulation of a muscarinic acetylcholine receptor. *Nature*, 504(7478):101–106, 2013.

438. S. Ahuja, V. Hornak, E. C. Yan, N. Syrett, J. A. Goncalves, A. Hirshfeld, M. Ziliox, T. P. Sakmar, M. Sheves, P. J. Reeves, S. O. Smith, and M. Eilers. Helix movement is coupled to displacement of the second extracellular loop in rhodopsin activation. *Nat. Struct. Mol. Biol.*, 16:168–175, 2009.

439. C. Altenbach, A. K. Kusnetzow, O. P. Ernst, K. P. Hofmann, and W. L. Hubbell. High-resolution distance mapping in rhodopsin reveals the pattern of helix movement due to activation. *Proc. Natl. Acad. Sci. USA*, 105:7439–7444, 2008.

440. S. R. Sprang. A receptor unlocked. *Nature*, 450:355–6, 2007.

441. A. M. Ring, A. Manglik, A. C. Kruse, M. D. Enos, W. I. Weis, K. C. Garcia, and B. K. Kobilka. Adrenaline-activated structure of β2-adrenoceptor stabilized by an engineered nanobody. *Nature*, 502(7472):575–579, 2013.

442. I. Bang and H.-J. Choi. Structural Features of β2 Adrenergic Receptor: Crystal Structures and Beyond. *Mol. Cells*, 38(2):105–111, 2015.

443. R. O. Dror, D. H. Arlow, P. Maragakis, T. J. Mildorf, A. C. Pan, H. Xu, D. W. Borhani, and D. E. Shaw. Activation mechanism of the β2-adrenergic receptor. *Proc. Natl. Acad. Sci. USA*, 108(46):18684–18689, 2011.

444. K. Y. Chung, S. G. F. Rasmussen, T. Liu, S. Li, B. T. DeVree, P. S. Chae, D. Calinski, B. K. Kobilka, V. L. Woods, and R. K. Sunahara. Conformational changes in the G protein Gs induced by the β2 adrenergic receptor. *Nature*, 477(7366):611–615, 2011.

445. G. Lebon, T. Warne, P. C. Edwards, K. Bennett, C. J. Langmead, A. G. W. Leslie, and C. G. Tate. Agonist-bound adenosine A_{2A} receptor structures reveal common features of GPCR activation. *Nature*, 474(7352):521–525, 2011.

446. F. Xu, H. Wu, V. Katritch, G. W. Han, K. A. Jacobson, Z.-G. Gao, V. Cherezov, and R. C. Stevens. Structure of an Agonist-Bound Human A_{2A} Adenosine Receptor. *Science*, 332(6027):322–327, 2011.

447. G. Liapakis, J. A. Ballesteros, S. Papachristou, W. C. Chan, X. Chen, and J. A. Javitch. The forgotten serine. A critical role for Ser-2035.42 in ligand binding to and activation of the β2-adrenergic receptor. *J. Biol. Chem.*, 275(48):37779–37788, 2000.

448. K. Haga, A. C. Kruse, H. Asada, T. Yurugi-Kobayashi, M. Shiroishi, C. Zhang, W. I. Weis, T. Okada, B. K. Kobilka, T. Haga, and T. Kobayashi. Structure of the human M2 muscarinic acetylcholine receptor bound to an antagonist. *Nature*, 482(7386):547–551, 2012.

449. A. C. Kruse, J. Hu, A. C. Pan, D. H. Arlow, D. M. Rosenbaum, E. Rosemond, H. F. Green, T. Liu, P. S. Chae, R. O. Dror, D. E. Shaw, W. I. Weis, J. Wess, and B. K. Kobilka. Structure and dynamics of the M3 muscarinic acetylcholine receptor. *Nature*, 482(7386):552–556, 2012.

450. F. Heitz, J. A. Holzwarth, J.-P. Gies, R. M. Pruss, S. Trumpp-Kallmeyer, M. F. Hibert, and C. Guenet. Site-directed mutagenesis of the putative human muscarinic M2 receptor binding site. *Eur. J. Pharmacol.*, 380(2–3):183–195, 1999.

451. P. R. Gentry, P. M. Sexton, and A. Christopoulos. Novel Allosteric Modulators of G Protein-coupled Receptors. *J. Biol. Chem.*, 290(32):19478–19488, 2015.

452. D. Zhang, Z.-G. Gao, K. Zhang, E. Kiselev, S. Crane, J. Wang, S. Paoletta, C. Yi, L. Ma, W. Zhang, G. W. Han, H. Liu, V. Cherezov, V. Katritch, H. Jiang, R. C. Stevens, K. A. Jacobson, Q. Zhao, and B. Wu. Two disparate ligand-binding sites in the human P2Y1 receptor. *Nature*, 520(7547):317–321, 2015.

453. B. Trzaskowski, D. Latek, S. Yuan, U. Ghoshdastider, A. Debinski, and S. Filipek. Action of molecular switches in GPCRs–theoretical and experimental studies. *Curr. Med. Chem.*, 19(8):1090–109, 2012.

454. J. Shonberg, L. Lopez, P. J. Scammells, A. Christopoulos, B. Capuano, and J. R. Lane. Biased Agonism at G Protein-Coupled Receptors: The Promise and the Challenges: A Medicinal Chemistry Perspective. *Med. Res. Rev.*, 34(6):1286–1330, 2014.

455. E. Reiter, S. Ahn, A. K. Shukla, and R. J. Lefkowitz. Molecular Mechanism of β-Arrestin-Biased Agonism at Seven-Transmembrane Receptors. *Annu. Rev. Pharmacol. Toxicol.*, 52(1):179–197, 2012.

456. R. Tsao and M. von Zastrow. Downregulation of G protein-coupled receptors. *Curr. Opin. Neurobiol.*, 10(3):365–9, 2000.

457. S. K. Bohm, E. F. Grady, and N. W. Bunnett. Regulatory mechanisms that modulate signalling by G-protein-coupled receptors. *Biochem. J.*, 322(part 1):1–18, 1997.

458. S. S. Ferguson. Evolving concepts in G protein-coupled receptor endocytosis: the role in receptor desensitization and signaling. *Pharmacol. Rev.*, 53:1–24, 2001.

459. J. A. Pitcher, N. J. Freedman, and R. J. Lefkowitz. G protein-coupled receptor kinases. *Annu. Rev. Biochem.*, 67:653–92, 1998.

460. O. B. Goodman Jr, J. G. Krupnick, F. Santini, V. V. Gurevich, R. B. Penn, A. W. Gagnon, J. H. Keen, and J. L.

Benovic. Beta-arrestin acts as a clathrin adaptor in endocytosis of the beta2-adrenergic receptor. *Nature*, 383(6599):447–50, 1996.

461. S. Danner and M. J. Lohse. Regulation of beta-adrenergic receptor responsiveness modulation of receptor gene expression. *Rev. Physiol., Biochem. Pharmacol.*, 136:183–223, 1999.

462. F. Mullershausen, F. Zecri, C. Cetin, A. Billich, D. Guerini, and K. Seuwen. Persistent signaling induced by FTY720-phosphate is mediated by internalized S1P1 receptors. *Nat. Chem. Biol.*, 5(6):428–434, 2009.

463. N. G. Tsvetanova and M. von Zastrow. Spatial encoding of cyclic AMP signaling specificity by GPCR endocytosis. *Nat. Chem. Biol.*, 10(12):1061–1065, 2014.

464. J. S. Smith and S. Rajagopal. The β-Arrestins: Multifunctional regulators of G protein-coupled receptors. *J. Biol. Chem.*, 291(17):8969–8977, 2016.

465. Y. Kang, X. E. Zhou, X. Gao, Y. He, W. Liu, A. Ishchenko, A. Barty, T. A. White, O. Yefanov, G. Won Han, Q. Xu, P. W. de Waal, J. Ke, M. H. E. Tan, C. Zhang, A. Moeller, G. M. West, B. D. Pascal, N. Van Eps, L. N. Caro, S. A. Vishnivetskiy, R. J. Lee, K. M. Suino-Powell, X. Gu, K. Pal, J. Ma, X. Zhi, S. Boutet, G. J. Williams, M. Messerschmidt, C. Gati, N. A. Zatsepin, D. Wang, D. James, S. Basu, S. Roy-Chowdhury, C. E. Conrad, J. Coe, H. Liu, S. Lisova, C. Kupitz, I. Grotjohann, R. Fromme, Y. Jiang, M. Tan, H. Yang, J. Li, M. Wang, Z. Zheng, D. Li, N. Howe, Y. Zhao, J. Standfuss, K. Diederichs, Y. Dong, C. S. Potter, B. Carragher, M. Caffrey, H. Jiang, H. N. Chapman, J. C. H. Spence, P. Fromme, U. Weierstall, O. P. Ernst, V. Katritch, V. V. Gurevich, P. R. Griffin, W. L. Hubbell, R. C. Stevens, V. Cherezov, K. Melcher, and H. E. Xu. Crystal structure of rhodopsin bound to arrestin by femtosecond X-ray laser. *Nature*, 523(7562):561–567, 2015.

466. M. K. Ostermaier, C. Peterhans, R. Jaussi, X. Deupi, and J. Standfuss. Functional map of arrestin-1 at single amino acid resolution. *Proc. Natl. Acad. Sci. USA*, 111(5):1825–1830, 2014.

467. J. J. Liu, R. Horst, V. Katritch, R. C. Stevens, and K. Wuthrich. Biased Signaling Pathways in 2-Adrenergic Receptor Characterized by 19F-NMR. *Science*, 335(6072):1106–1110, 2012.

468. K. Kirchberg, T.-Y. Kim, M. Möller, D. Skegro, G. Dasara Raju, J. Granzin, G. Büldt, R. Schlesinger, and U. Alexiev. Conformational dynamics of helix 8 in the GPCR rhodopsin controls arrestin activation in the desensitization process. *Proc. Natl. Acad. Sci. USA*, 108(46):18690–18695, 2011.

469. K. Hollenstein, J. Kean, A. Bortolato, R. K. Y. Cheng, A. S. Doré, A. Jazayeri, R. M. Cooke, M. Weir, and F. H. Marshall. Structure of class B GPCR corticotropin-releasing factor receptor 1. *Nature*, 499(7459):438–443, 2013.

470. F. Y. Siu, M. He, C. de Graaf, G. W. Han, D. Yang, Z. Zhang, C. Zhou, Q. Xu, D. Wacker, J. S. Joseph, W. Liu, J. Lau, V. Cherezov, V. Katritch, M.-W. Wang, and R. C. Stevens. Structure of the human glucagon class B G-protein-coupled receptor. *Nature*, 499(7459):444–449, 2013.

471. C. Parthier, S. Reedtz-Runge, R. Rudolph, and M. T. Stubbs. Passing the baton in class B GPCRs: peptide hormone activation via helix induction? *Trends Biochem. Sci.*, 34:303–10, 2009.

472. H. A. Watkins, M. Au, and D. L. Hay. The structure of secretin family GPCR peptide ligands: Implications for receptor pharmacology and drug development. *Drug Discov. Today*, 17(17–18):1006–1014, 2012.

473. T. Muto, D. Tsuchiya, K. Morikawa, and H. Jingami. Structures of the extracellular regions of the group II/III metabotropic glutamate receptors. *Proc. Natl. Acad. Sci. USA*, 104(10):3759–3764, 2007.

474. C. M. Niswender and P. J. Conn. Metabotropic Glutamate Receptors: Physiology, Pharmacology, and Disease. *Annu. Rev. Pharmacol. Toxicol.*, 50(1):295–322, 2010.

475. A. S. Doré, K. Okrasa, J. C. Patel, M. Serrano-Vega, K. Bennett, R. M. Cooke, J. C. Errey, A. Jazayeri, S. Khan, B. Tehan, M. Weir, G. R. Wiggin, and F. H. Marshall. Structure of class C GPCR metabotropic glutamate receptor 5 transmembrane domain. *Nature*, 511(7511):557–562, 2014.

476. P. J. Conn, A. Christopoulos, and C. W. Lindsley. Allosteric modulators of GPCRs: a novel approach for the treatment of CNS disorders. *Nat. Rev. Drug Discov.*, 8(1):41–54, 2009.

477. A. Ruiz-Gómez, C. Molnar, H. Holguín, F. Mayor, and J. F. de Celis. The cell biology of Smo signalling and its relationships with GPCRs. *Biochim. Biophys. Acta Biomembranes*, 1768(4):901–912, 2007.

478. C. Wang, H. Wu, V. Katritch, G. W. Han, X.-P. Huang, W. Liu, F. Y. Siu, B. L. Roth, V. Cherezov, and R. C. Stevens. Structure of the human smoothened receptor 7TM bound to an antitumor agent. *Nature*, 497(7449):338–343, 2013.

479. C. Wang, H. Wu, T. Evron, E. Vardy, G. W. Han, X.-P. Huang, S. J. Hufeisen, T. J. Mangano, D. J. Urban, V. Katritch, V. Cherezov, M. G. Caron, B. L. Roth, and R. C. Stevens. Structural basis for Smoothened receptor modulation and chemoresistance to anticancer drugs. *Nat. Commun.*, 5, 2014.

480. U. Weierstall, D. James, C. Wang, T. A. White, D. Wang, W. Liu, J. C. H. Spence, B. R. Doak, G. Nelson, P. Fromme, R. Fromme, I. Grotjohann, C. Kupitz, N. A. Zatsepin, H. Liu, S. Basu, D. Wacker, G. Won Han, V. Katritch, S. Boutet, M. Messerschmidt, G. J. Williams, J. E. Koglin, M. Marvin Seibert, M. Klinker, C. Gati,

R. L. Shoeman, A. Barty, H. N. Chapman, R. A. Kirian, K. R. Beyerlein, R. C. Stevens, D. Li, S. T. A. Shah, N. Howe, M. Caffrey, and V. Cherezov. Lipidic cubic phase injector facilitates membrane protein serial femtosecond crystallography. *Nat. Commun.*, 5(3309), 2014.

481. N. A. Riobo, B. Saucy, C. Dilizio, and D. R. Manning. Activation of heterotrimeric G proteins by Smoothened. *Proc. Natl. Acad. Sci. USA*, 103(33):12607–12612, 2006.

482. K. A. Jacobson. New paradigms in GPCR drug discovery. *Biochem. Pharmacol.*, 98(4):541–555, 2015.

483. N. M. Urs, P. J. Nicholls, and M. G. Caron. Integrated approaches to understanding antipsychotic drug action at GPCRs. *Curr. Opin. Cell. Biol.*, 27(1):56–62, 2014.

484. J. Shonberg, P. J. Scammells, and B. Capuano. Design strategies for bivalent ligands targeting GPCRs. *ChemMedChem*, 6(6):963–974, 2011.

485. Q. Tan, Y. Zhu, J. Li, Z. Chen, G. W. Han, I. Kufareva, T. Li, L. Ma, G. Fenalti, J. Li, W. Zhang, X. Xie, H. Yang, H. Jiang, V. Cherezov, H. Liu, R. C. Stevens, Q. Zhao, and B. Wu. Structure of the CCR5 Chemokine Receptor-HIV Entry Inhibitor Maraviroc Complex. *Science*, 341(6152):1387–1390, 2013.

486. G. J. Siegel. *Basic neurochemistry: molecular, cellular and medical aspects.* Lippincott Williams & Wilkins, Philadelphia, 6th edition, 1999.

487. R. B. Russell and D. S. Eggleston. New roles for structure in biology and drug discovery. *Nat. Struct. Biol.*, 7 Suppl:928–30, 2000.

488. C. Rochais, C. Lecoutey, F. Gaven, P. Giannoni, K. Hamidouche, D. Hedou, E. Dubost, D. Genest, S. Yahiaoui, T. Freret, V. Bouet, F. Dauphin, J. S. De Oliveira Santos, C. Ballandonne, S. Corvaisier, A. Malzert-Fréon, R. Legay, M. Boulouard, S. Claeysen, and P. Dallemagne. Novel multitarget-directed ligands (MTDLs) with acetylcholinesterase (AChE) inhibitory and serotonergic subtype 4 receptor (5-HT$_4$R) agonist activities as potential agents against Alzheimer's disease: the design of donecopride. *J. Med. Chem.*, 58(7):3172–3187, 2015.

489. G. L. Harris, M. B. Creason, G. B. Brulte, and D. R. Herr. In vitro and in vivo antagonism of a G protein-coupled receptor (S1P3) with a novel blocking monoclonal antibody. *PLoS One*, 7(4):e35129, 2012.

490. S. M. Schumacher, E. Gao, W. Zhu, X. Chen, J. K. Chuprun, A. M. Feldman, J. J. G. Tesmer, and W. J. Koch. Paroxetine-mediated GRK2 inhibition reverses cardiac dysfunction and remodeling after myocardial infarction. *Sci. Transl. Med.*, 7(277):277ra31–277ra31, 2015.

CHAPTER 8

Protein-Ligand Interactions

8.1 INTRODUCTION

In the previous chapters we witnessed the large and diverse set of functions fulfilled by proteins in cells and tissues. Together, these functions enable the most basic, as well the most sophisticated processes to take place in living organisms. Protein functions result from the inherent structural complexity of these molecules. Still, all of these functions are based, at least partially, on the capability of each protein to bind a certain ligand (or multiple specific ligands). Such ligands might be small organic molecules (~600 Da or less), macromolecules or even elemental ions. The roles played by the ligands are equally diverse, and include the following:

1. **Catalysis.** Enzymes act on substrates, turning them into products. Both substrates and products can be considered as ligands, and may be small molecules, peptides, or macromolecules. Some substrates also act as cofactors. For example, NADH and FADH$_2$ serve as cofactors in multiple redox reactions that are catalyzed by enzymes called *oxidoreductases* (Figure 8.1a). These issues are discussed in Chapter 9, which focuses on enzyme catalysis.

2. **Regulation.** Many small organic molecules are routinely used by cells to regulate the activity of metabolic enzymes (see Chapter 9, Section 9.5), signal transduction proteins, or other key proteins (Figure 8.1b). The regulation may be simple, as in the case of product inhibition used in metabolic pathways, or more sophisticated, as in the case of hormone-activated control over key cellular processes. An example of regulation via product inhibition is the use of ATP as an inhibitor of a number of metabolic enzymes (e.g., *phosphofructokinase*). Sophisticated regulation usually involves complex signal transduction pathways that activate or inhibit numerous targets.

3. **Communication.** Ligands may participate at different points along cellular communication pathways: first messenger (hormone, neurotransmitter, or local mediator), second messenger (e.g., *cAMP*, *IP$_3$*), and downstream regulator. The action of such ligands may lead to different outcomes, such as cellular growth, division, biosynthesis of metabolites, activation of a defensive function, and so on.

4. **Protein trafficking.** Certain ligands serve as means by which organelles or other macromolecules are recognized by proteins. For example, by binding to the cytoplasmic enzyme *protein kinase C (PKC)*, the small ligand *diacylglycerol (DAG)* allows the

enzyme to attach to the plasma membrane, become activated, and participate in a major signal transduction process. Similarly, a nucleotide sequence at the beginning of a gene (i.e., the *promoter*) allows proteins functioning as transcription factors to recognize this location and activate the expression of that specific gene. Other nucleotide sequences are recognized by proteins that participate in DNA replication and RNA processing.

5. **Prosthetic groups.** Certain ligands bind tightly to proteins and help them execute certain functions. For example, the iron-containing heme group binds oxygen in *hemoglobin* (see Figure 5.2.3 in Box 5.2) and myoglobin, but in cytochromes binds electrons. Another prosthetic group, the retinal, allows the proteins rhodopsin and bacteriorhodopsin to become activated by light.

6. **Defense and offense.** Certain ligands act as toxins that attack other cells. In bacteria, they are secreted and can be used either as a defense against other bacteria, or as an offense against the host. Toxins are also produced by higher organisms (e.g., plants, insects, snakes), in which case they may be used for deterring predators or for catching prey.

In accordance with our general approach to biological processes, here too we have chosen to elaborate on basic aspects that we believe to be crucial for the understanding of these processes: structure, energetics, and dynamics. These are covered by the first part of the chapter. The most complex cellular processes involve protein-ligand interactions, in which the ligand itself is a protein. This type of interaction is also the most challenging for structural biologists, as it involves two binding partners of equal complexity. We have therefore chosen to focus in the second part of this chapter on protein-protein interactions.

The last part of the chapter is dedicated to protein-ligand interactions in drug development and design. Pharmaceutical drugs are most often small organic molecules, which bind to proteins that have gone awry as a result of disease. The binding inhibits the abnormal activity of the protein, and therefore has an overall positive influence on the body. In other cases, the drug does not bind to the abnormal protein directly, but rather to another protein, whose activity opposes that of the former. Thus, by activating the other protein, the drug can diminish the harmful effects of the abnormal protein. In any case, the drug's mechanism of action is based on its protein binding, which is why drug development requires scientists to harness the accumulated knowledge on protein-ligand interactions for the construction of suitable drug molecules. The large demand for pharmaceutical drugs and the resulting investment in their development have led to the emergence of state-of-the-art tools (both experimental and computational) intended to increase the efficiency of the drug discovery process to the extent that current technology allows. In our discussion we describe some of these tools briefly, as well as the main approaches for characterizing protein-drug interactions.

8.2 THEORIES ON PROTEIN-LIGAND BINDING AND DYNAMICS

The binding of various ligands to proteins has been studied for over a century. During the first decades of this period, research in this field focused on enzyme-substrate interactions. A central aspect of these studies was the ability of enzymes to bind selectively to their substrate(s). This selectivity is not trivial, as the cytoplasm is highly dense [1] and is expected to contain at least one molecule resembling the substrate. As explained in Chapter 5, specific

(a)

(b)

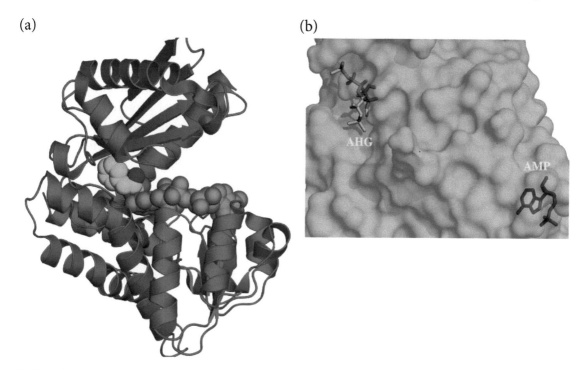

FIGURE 8.1 **Examples of biologically important protein-ligand interactions.** (a) Enzyme substrate-cofactor binding. Here, *L-phenylalanine dehydrogenase* is the enzyme (in blue ribbon representation), bound to both substrate (colored by atom type) and cofactor (NADH, colored in orange) (PDB entry 1c1d). (b) Allosteric regulators. The figure shows the protein fructose-1,6-bisphosphatase with 2,5-anhydro-D-glucitol-1,6-bisphosphate (AHG, an analogue of its substrate), and AMP (an allosteric inhibitor) (PDB entry 1fpd). As can be clearly seen, the allosteric ligand occupies a distinct binding site, which is quite distant from the catalytic site.

enzyme-substrate binding was initially explained by the *'lock and key' theory* [2], proposed by Emil Fischer in the 19th century (Figure 8.2a). This theory stipulated that the binding sites of enzymes are rigid and pre-adjusted geometrically to the natural substrate. The theory became widely accepted within the scientific community, and was even used to support another widely accepted idea concerning the *a priori* compatibility of proteins with their biological functions. In subsequent years, accumulating evidence began to indicate that in many cases the binding sites of enzymes do not match the substrate perfectly, and that the binding process is accompanied by conformational changes in the enzyme [3–7] It therefore became clear that Fischer's model was in need of a revision, which came in the form of Daniel Koshland's *'induced fit' theory* [8], whose validity has since been demonstrated in numerous proteins [9] (Figure 8.2b).

The induced fit theory suggested that enzymes do indeed match their substrates geometrically, but that this match is far from being perfect. Therefore, following the initial enzyme-substrate binding, certain conformational changes are required in the binding site to improve the match. In principle, both the aforementioned theories are applicable for describing the binding of any protein to its natural ligand, and are not confined to enzyme-substrate binding. The *Monod-Wyman-Changeux (MWC) model* [10], which appeared a few years after the induced fit theory (see Chapter 5, Subsection 5.3.2), contended that proteins are able to shift spontaneously between (at least) two different conformations, even in the absence of a ligand [10]. Indeed, this phenomenon has been demonstrated experimentally in many studies (e.g., [11]). The MWC model could also explain allostery, a well-known phenomenon in which the binding of the substrate to the catalytic site of the enzyme is affected by ligand binding to a different site (e.g., [12–15]). As explained in Chapter 5, the theory underlying the MWC model has undergone some changes since it was first suggested, and the current thinking is that proteins shift spontaneously among multiple conformations, called 'substates' [16–21]. This significant change in the perception of proteins has naturally led to corresponding changes in theories concerning protein-ligand binding. The currently accepted *conformational selection model* [22–24] posits that the ligand binds preferentially to one of the conformations sampled spontaneously by the protein, thus stabilizing it (Figure 8.2c; see also Chapter 5). In other words, by changing the protein's energy landscape, the ligand turns a previously less favorable conformation into the most favorable (and probable) one [25].

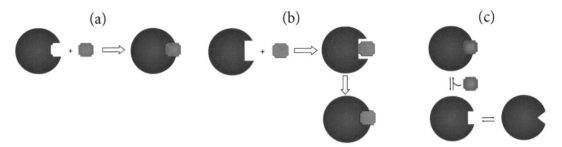

FIGURE 8.2 **Schematic representation of three popular protein-ligand binding theories.** (a) The 'lock and key' theory. The protein is represented as a blue circle and the ligand as an orange shape. The ligand-binding site of the protein matches the ligand perfectly. (b) The 'induced fit' theory. The binding site generally fits the ligand. However, the fit is significantly improved following the binding, due to ligand-induced conformational changes in the protein. (c) The 'conformational selection' theory. The protein undergoes constant conformational changes, with at least one of its conformations matching the ligand. Ligand binding selects this conformation.

The conformational selection theory does not necessarily refute Koshland's induced fit theory, and **in many cases even the best-fitting conformation of the ensemble does not match the ligand perfectly** [26]. **In such cases, the strain applied by the ligand to the binding site is expected to induce some changes in the latter, which further stabilize this conformation** [27–30]. In fact, both mechanisms may coexist in the same system, depending on circumstances. For example, a study on the effect of Ca^{2+} on *calmodulin* shows that the mechanism depends on the concentration of this ion [31]. That is, at low Ca^{2+} concentrations, Ca^{2+} acts mainly via the stabilization of existing conformations, whereas at high concentrations it induces conformational changes in calmodulin. In fact, even in its Ca^{2+}-bound form, calmodulin is highly dynamic [32,33], and it adopts distinct conformations only upon binding to its target protein. Interestingly, during binding, both calmodulin and its target seem to undergo a conformational search for their natively bound conformations [34]. This type of mechanism is referred to as *'mutually induced fit'* [35]. Induced fit and conformational selection are probably most prevalent in intrinsically unstructured proteins (IUPs), due to the inherent dynamics of these proteins (see Chapter 6).

As explained in Chapter 5, the conformational changes induced by ligand binding are usually small to moderate (backbone r.m.s.d. $\leq$ ~2.5 Å [36,37]), although in some cases they may be quite large [38,39]. We saw that ligand-induced conformational changes may have functional implications for the protein, through allostery. In addition, the inherent dynamics of the protein may also change as a result of ligand binding [40,41], and this change has implications for the energetics of the binding, as discussed in Subsection 8.3.2 below.

8.3   PROTEIN-LIGAND BINDING ENERGETICS

8.3.1   Total binding free energy

8.3.1.1   Protein-ligand binding displays diverse affinities

The diversity of protein ligand types naturally leads to a corresponding diversity in binding strength (affinity). Indeed, **binding energies may range between ~−2 and −22 kcal/mol** [42], corresponding to a range spanning 11 orders of magnitude in the value of the dissociation constant (Kd) [43]. To understand the factors determining these diverse affinities, one should first be able to determine the free energy change upon binding. This energy can be measured experimentally, via the equilibrium constant of the binding (K_{eq})[*1]. Methods for measuring K_{eq} are often based on the following two-step procedure [44]:

1. Phase separation between the protein-bound and free forms of the ligand.

2. Spectroscopic quantification of the free ligand's concentration. When such measurement is difficult, e.g., due to very high protein-ligand affinity, K_{eq} can be measured by quantifying the binding-induced changes in the spectroscopic properties of the protein, such as absorbance, fluorescence, or anisotropy (fluorescence polarization).

The binding free energy can also be calculated in many cases (see following subsection). **The binding strength displayed by a given protein matches the biological purpose of that specific binding.** For example, proteins that are involved in signal transduction networks tend to bind their ligands weakly, enabling these proteins to switch binding partners easily, thereby transmitting the signal. Conversely, cofactor binding by enzymes, which has to be reversible but also strong, is characterized by free energies of −5.5 to −9.5 kcal/mol [42] (see also below), comparable to the free energy of protein folding. Finally, some proteins display exceptionally strong binding. For example, the free energy of binding of the protein *avidin* to the small molecule *biotin* (vitamin H) is roughly −20 kcal/mol [45]. Avidin is a common glycoprotein in the egg whites of birds, reptiles and amphibians. Its reason for binding biotin is not entirely understood, but the high affinity between the avidin and biotin has been utilized extensively in biochemical lab assays and purification procedures [46]. Strong affinity is also observed in the binding of enzymes to their natural inhibitors. For example, the binding of *trypsin* to its inhibitor has a Kd of 60 fM [47], which corresponds to a binding energy of ~−18 kcal/mol. This strong affinity can be rationalized by the biological need to quickly neutralize proteases before they irreversibly damage cells or tissues [43].

An early survey of protein-ligand complexes carried out by Kuntz and coworkers suggested a maximal binding energy of ~1.5 kcal/mol per ligand atom[*2] [51]. However, this value was not constant, and found to decrease when the number of ligand atoms exceeded ~15. This decrease is probably driven by the fact that larger ligands need to optimize a larger number of contacts within the binding site, which might lead to structural compromises and, therefore, to reduced affinity [52]. The authors also found the highest-energy binding

---

[*1]As explained in Chapter 4, $\Delta G_0 = -RT \ln K_{eq}$ (Equation (4.1)), where $R = 1.989$ cal/(molK), and $RT \approx 0.6$ kcal/mol at room temperature.

[*2]A slightly higher value of −1.75 kcal/mol per atom was found by Carlson and coworkers [48], based on ~4,700 protein-ligand complexes from the Binding MOAD database [49,50] (URL: http://www.bindingmoad.org/). However, for 95% of the complexes the maximal binding energy was much lower, −0.83 kcal/mol per atom.

interactions (per atom) to be associated with metals, small anions, and (obviously) ligands that form covalent bonds. This observation is in agreement with a later study of Carlson and coworkers, who found that the most efficient binders tend to be small, charged ligands that are in contact with charged binding site residues or cofactors, and that form interaction networks [48] (see more about the contribution of polar networks to binding in Subsection 8.4.3 below). The strong binding of metals[*1] is to be expected considering their biological roles (helping catalysis and stabilizing protein substructures), most of which require the metal to remain bound to the protein. Similarly, covalently bound ligands, which include mostly prosthetic groups, are required to remain bound to their corresponding proteins because of their important roles in these proteins' biological function. Certain transition states of substrates formed during enzyme-mediated catalysis also become covalently bound to their related enzymes (see Chapter 9, Subsection 9.3.3.3). However, the catalytic mechanisms involving these transition states also include chemical species that break the bond very soon after it is formed. In fact, the high affinity of enzymes towards the transition states of their substrates (see following paragraph and Chapter 9, Subsection 9.3.3.2) is mainly the result of noncovalent interactions, and is part of what makes enzymatic catalysis both specific and highly efficient.

A continuation of the work done by Kuntz and coworkers was carried out a few years later by Zhang and Houk [54], who conducted a survey of 1,600 proteins. Their results and those of others showed a correlation between the different affinities of proteins towards their ligands and the general types of complexes formed (Table 8.1).

TABLE 8.1 **Binding energies and corresponding dissociation constants of different protein-ligand complexes.**

Protein	Ligand	Binding Energy (kcal/mol)	Dissociation Constant (Kd) (Molar)
Antibody	Antigen	−6 to −11 [54]*a	$\sim 10^{-2}$ to 10^{-8}
Receptor	Hormone	$\sim$−12 [42]	$\sim 10^{-9}$
Enzyme	Substrate	−4 to −8 [55]	$\sim 10^{-3}$ to 10^{-6}
	Cofactor	−5.5 to −9.5 [42]	$\sim 10^{-4}$ to 10^{-7}
	Inhibitor	−10 to −15 [54]*b	$\sim 10^{-7}$ to 10^{-11}
	Transition state	−16 to −27 [54]*c	$\sim 10^{-12}$ to 10^{-20}

*a A higher value (−14 kcal/mol ($Kd \approx 10^{-10}$)) was obtained for antibodies that underwent a process called 'maturation'. In this process, B-cell lymphocytes adapt to the invading pathogen by producing antibodies with better compatibility with the antigens produced by the pathogen.

*b A similar, yet slightly higher value of $\approx$ −18 kcal/mol ($Kd = 60$ fM) was measured for the binding of trypsin to pancreatic trypsin inhibitor, which is itself a protein [47].

*c Higher values have also been suggested. For example, Schramm [56] indicated an enzyme-transition state dissociation constant (Kd) ranging from 10^{-14} to 10^{-23} mol L^{-1}. This corresponds roughly to a binding free energy of −19 to −31 kcal/mol.

[*1] As explained in Chapter 2, cationic metals such as Fe^{2+}, Fe^{3+}, Zn^{2+}, Cu^{2+}, Cu^+, Mg^{2+}, Mn^{2+}, Mn^{3+}, Mo^{3+}, Mo^{4+}, Mo^{6+}, Co^{2+}, and nickel (Ni^+) are usually coordinated to the side chains of cysteine, histidine, and/or glutamate or aspartate, with the latter two also stabilizing the metal electrostatically [53]. These form coordination complexes around the metal cation, which normally include 2 to 4 residues. In other cases, metals may form clusters with non-metallic elements (e.g., Fe with S in cytochromes) or be bound to an organic structure, as in the protein cofactors heme and cobalamin (B_{12} cofactor).

These results seem to agree with some of the most basic and widely accepted notions regarding protein-ligand interactions:

1. The suggestion of Kuntz and coworkers [51] that the upper limit for protein-ligand affinity is 13 kcal/mol. This value matches the affinity values obtained for the first four complex types. Protein-inhibitor complexes are characterized by somewhat higher binding affinities, in line with their biological role (see above). Still, **most protein-ligand complexes do not exceed a binding energy of 15 kcal/mol, probably to avoid clearance problems**. The very high affinities of enzymes to their substrates' transition states are in line with the suggestion of Pauling [57] and others that **the binding site of an enzyme matches the transition state of the corresponding substrate much better than it does the ground state** (see Chapter 9). The very strong binding in this case does not pose a clearance problem, since the transition state is chemically and rapidly converted into another form (the reaction product), which has a much lower affinity to the binding site and therefore dissociates quickly from the enzyme.

2. The suggestion that **protein-ligand interactions are (in principle) noncovalent**. Again, enzyme-transition state complexes seem to be an exception, at least partially; certain enzymes are known to form covalent bonds with their transition states as part of the catalytic mechanism (see Chapter 9 for details). These bonds include those that are formed during nucleophilic catalysis (e.g., the formation of acyl-enzyme intermediates in serine proteases, or Schiff bases in aminotransferases), cofactor-mediated bonds (e.g., the cofactor TPP in pyruvate dehydrogenase), and 'semi-covalent bonds' such as low-barrier hydrogen bonds (LBHBs; see Chapter 1). In Zhang and Houk's study, 97% of the enzyme-transition state complexes were found to have a binding energy that could only be explained by the formation of covalent bonds.

8.3.1.2 Calculating absolute binding free energy

In a variety of situations, it might be necessary to use computational methods to calculate the energy of binding of proteins to their ligands. Such methods are often used, for example, when technical problems make it difficult or impossible to measure the binding free energy values directly. This is commonly the case in drug discovery (see below), which involves searching for the most appropriate drug molecule among an immense number of candidates. In these cases, it is often unrealistic to measure the energy of the binding of a given protein to each of the numerous candidate molecules, and this is where computational methods usually come into play. Another important reason for using calculations is that the computational models used often enable each of the components of the binding energy to be characterized separately, thus providing a more detailed picture of the binding process (see below).

There are several popular methods used to calculate binding free energy [58–60]:

1. **Alchemical methods (a.k.a. 'free energy perturbation')** [61,62]: The binding is described by introducing a series of non-physical intermediates between the bound and unbound states, and the binding energy is calculated by integrating them. The calculation requires the use of all-atom molecular dynamics (MD) simulations (FEP/MD[*1]).

[*1]FEP/MD calculations can be run using the free internet server *Ligand Binder* [63] (URL: http://www.charmm-gui.org/?doc=input/gbinding).

As explained in Chapter 3, the force field-based calculations used in MD simulations provide only the potential energy of the system. Therefore, adequate sampling of each state is required in order to account for the entropic component of the free energy. As a result, the FEP/MD method is considered rigorous and computationally expensive, although certain protocols have been devised to alleviate this problem [63,64]. Alchemical methods can also be used for calculating the relative binding energy, e.g., by artificially converting one ligand into another (see the following subsection).

2. **Steered methods** [65,66]: All-atom MD simulations are carried out under a constant force that pulls the ligand away from the binding site in a physically consistent manner, and the free energy required for separating the two is calculated. Such methods are computationally expensive due to the need to sample a sufficient number of configurations of the system (see more below).

3. **Linear interaction energy (LIE) method** [67,68]: The binding free energy is calculated as the difference between the ligand-bound and ligand-free states of the protein (endpoint calculation), where in both cases the system is described explicitly, and MD or Monte Carlo simulations are used to sample the two states. Since only the two endpoints are sampled, the calculation is less expensive than in steered MD or alchemical methods. The name of the method comes from the linear response assumption that is used to calculate the electrostatic interactions between the ligand and its environment.

4. **MM/PB(GB)SA** [69–72]: This, too, is an endpoint method, which calculates the binding energy as the difference between the energies of the ligand-bound and ligand-free states of the protein. Unlike LIE, however, MM/PB(GB)SA uses implicit solvent models of the protein-ligand system (PB or GB, respectively) for calculating solvation effects (see Chapter 3 for details). This makes the method less computationally expensive than LIE, as well as other methods that rely on all-atom simulations (see more below).

Generally speaking, calculating the change in absolute free energy accompanying protein-ligand binding is a challenging task, due to the following:

1. *The free energy of binding is small compared to its components.*
   The free energy of binding (ΔG_{bind}) has a small magnitude, since it is usually a small difference between large numbers [73]. This occurs for two reasons. First, the energy is in fact a difference between large energies, one of the complex (G_{PL}), and the other of the unbound partners ($G_P + G_L$):

$$P + L \rightarrow PL \tag{8.1}$$

$$\Delta G_{bind} = G_{PL} - \left(G_P + G_L\right) \tag{8.2}$$

Second, the free energy of binding results mainly from two large and opposite forces: the binding enthalpy (ΔH_{bind}) and the loss of entropy (of both protein and ligand) (ΔS_{bind}), which nearly cancel each other out ('enthalpy-entropy compensation' [74–78], Figure 8.3):

$$\Delta G_{bind} = \Delta H_{bind} - T\Delta S_{bind} \tag{8.3}$$

The values of the binding free energy, enthalpy, and entropy are collectively referred to as the *thermodynamic signature* of the binding. Two protein-ligand pairs may have the same binding energy, yet different thermodynamic signatures, due to differences in their binding enthalpies and/or entropies (e.g., [79]). Knowing these components not only helps in distinguishing between binding pairs but often provides information on the molecular interactions and the dynamic changes in each system. **Thus, when studying a binding process, it is often helpful to characterize its complete thermodynamic signature, not just the binding free energy.** As explained in Chapter 4, binding enthalpy can be measured directly by calorimetric methods such as isothermal calorimetry (ITC) [80], and the entropy component can be derived arithmetically when the binding energy and enthalpy are known. However, the binding entropy itself has sub-components, and it is often challenging to characterize their relative contributions (see Subsection 8.3.2 below). Moreover, both binding enthalpy and entropy involve the protein, ligand, and solvent, which makes their characterization at the molecular level even more challenging.

2. *The free energy of binding is highly conformation-dependent.*
   Although the conformation dependency of binding free energy does not directly influence the complexity of calculating this energy value, it does raise doubts about the validity of such calculations. That is, if the protein conformation to which the calculation is applied is not completely accurate, the results may be considerably different from what they should be. The conformation dependency results from the fact that both short- and long-range interactions take place in this system, as well as from the effect of the dielectric on electrostatic interactions (see Box 1.1).

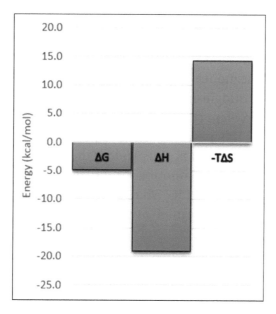

FIGURE 8.3 **Enthalpy-entropy compensation.** The image shows the free energy (ΔG) of binding of streptavidin to the peptide *HDHPQNL*, as well as the binding enthalpy (ΔH) and entropy ($-T\Delta S$) components. The values were measured by isothermal titration calorimetry (taken from the SCORPIO database [81]). Note that the enthalpy and entropy components are large and opposite in sign. The resulting binding free energy value is therefore relatively small.

3. *Accurate calculations require explicit treatment of the system.*

As explained above, some of the methods for calculating binding free energies use explicit models of the system in MD simulations. Generally speaking, explicit treatment is possible whenever a three-dimensional, high-resolution structure of the system is available. However, we have seen earlier that MD simulations of large, protein-based systems are usually unable to produce a statistically meaningful configurational ensemble from which the calculated potential energies can be translated into free energy values. This problem is particularly pronounced in binding events, which involve the exclusion of numerous water molecules from the protein-ligand interface. The advances achieved in recent years in computational resources, sampling methods, and accuracy of scoring functions have enabled MD simulations to become longer and more efficient [82]. Hopefully, these developments will soon enable scientists to employ such simulations for calculating binding free energies accurately and in a reasonable period of time. Until then, an alternative solution (albeit a partial one) is to use *mean field models*, in which the solvent is described implicitly, on the basis of its dielectric properties [83] (see Box 1.3 and Chapter 3). In such cases, solvation effects are usually calculated using the Poisson-Boltzmann (PB) [84] or generalized Born (GB) [85] models, whereas the hydrophobic effect is accounted for by using surface area (SA) models (see Chapters 3 and 4). Calculations using these formalisms are called *PBSA* [86] and *GBSA* [86], respectively [73]. Several studies (e.g., [87,88]) have shown that solvation effects are significant in protein-ligand binding, illustrating the need for adequate treatment of this energy component [89]. Neglecting this effect would make all protein-ligand electrostatic interactions appear energetically favorable, due to Coulombic attraction (see Chapter 1, Subsection 1.3.1.2). Inaccurate treatment of this effect might also lead to incorrect protonation states of residues like His, Glu, Asp and Cys, whose pKa often changes upon ligand binding [90–92] (see Chapter 2, Subsection 2.2.1.3.3.3 for a detailed explanation).

The two main advantages of mean field models are that they are fast, and that the calculated values correspond to the free energy of the system. However, the approximations used by these models might obscure important aspects of the system. For example, they do not distinguish between bulk and protein-bound water molecules [93], which may pose a problem considering that the latter are considerably less dynamic than the former [94]. This problem has been partially addressed in mean field models through the use of a distance-dependent dielectric [73], but this solution is still not as accurate as actually accounting for the different types of water molecules explicitly. Today, integration of the two different approaches (explicit and mean field) has become possible in the form of 'mixed force fields' [73], as well as *MM/PB(GB)SA* calculations [69–71] (see Chapter 3, Subsection 3.4.2 for details). As explained above, the latter are often used as a 'lighter' alternative to alchemical and MD methods for calculating the protein-ligand binding energy. The endpoint calculations used in such methods consider only the ligand-free and ligand bound systems. In some implementations of MM/PB(GB)SA, the two states are sampled, and snapshots of the produced ensemble are used in the calculation. In other implementations, the starting protein-ligand structure is energy minimized and then subjected to a single MM/PB(GB)SA calculation [72].

8.3.1.3 Calculating relative binding energies

Many proteins are capable of binding multiple different ligands, in some cases by using the same binding site (see Subsection 8.4.4 below). When the different ligands corresponding to a given protein are similar, their binding induces only small changes (if any) in the conformation of the protein (e.g., [95])[*1]. When the ligands are structurally and/or chemically different, they may target very different conformations of the protein. In either case, the protein is expected to have a different affinity towards each ligand (Figure 8.4, horizontal equilibria). When the structures of protein and ligand are available, the accurate way to calculate the associated free energy differences is, in principle, to use an explicit model of the entire system, including solvent, and apply force field-based calculations [83]. For convenience, let us consider the simplest case, in which the protein has only two possible ligands (L and L′). The two related affinities (ΔG_{PL} and $\Delta G_{PL'}$ in Figure 8.4) are obtained by subtracting the free energy of the protein-ligand complex from that of the unbound state:

$$\Delta G_{PL} = G_{PL} - G_P - G_L \tag{8.4a}$$

$$\Delta G_{PL'} = G_{PL'} - G_P - G_{L'} \tag{8.4b}$$

The difference between the two affinities ($\Delta\Delta G$) is defined as:

$$\Delta\Delta G = \Delta G_{PL'} - \Delta G_{PL} \tag{8.5}$$

As explained in Box 1.3, Chapter 3, and the subsections above, if an explicit model is used, it is not enough to consider a single configuration of the system. Instead, numerous configurations of all components must be rigorously sampled, in order to turn the calculated potential energies into free energy values. Since this task is virtually infeasible considering the large number of water molecules in the system, a different approach must be used. In the subsections above we discussed mean-field alternatives to the explicit model approach. Here, since we aim to compare the free energy values of the two binding processes instead of calculating the absolute binding free energy of each, a much easier solution is available. That is, the two binding processes can be turned into a full thermodynamic cycle (Figure 8.4). As free energy is a state function, the free energy differences accompanying the binding can also be described as:

$$\Delta\Delta G = \Delta G_{PL \to PL'} - \Delta G_{L \to L'} \tag{8.6}$$

$\Delta G_{PL \to PL'}$ and $\Delta G_{L \to L'}$ do not relate directly to a binding process, which means that the explicit simulations need not sample the rearrangements of numerous water molecules following their exclusion from the protein-ligand interface. To make the calculations more efficient, it is possible to apply this general approach using a step-wise perturbation procedure, as is used in the aforementioned alchemical methods [62,97]. In such cases, the 'alchemical changes' usually involve the conversion of one ligand into another, or alternatively, the two ligands' respective interactions with the protein.

Other solutions to the configurational sampling problem of dynamic simulations are reviewed in [98].

[*1]Although a recent study shows that even similar ligands may sometimes induce substantial changes in the conformation of the binding site [96].

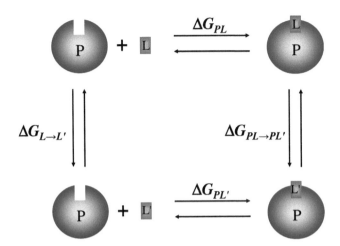

FIGURE 8.4 **A thermodynamic cycle describing protein-ligand binding and associated free energies (following** [98]**).** The cycle describes the binding of a protein (P) to two different ligands, L and L', with the corresponding free energies of binding ΔG_{PL} and $\Delta G_{PL'}$. Assuming that both ligands induce similar conformational changes in the protein upon binding, the free and bound states of the protein-ligand system can be connected by two non-physical processes: the first converts L into L' (with a corresponding free energy change $\Delta G_{L\to L'}$) and the second converts PL into PL' (with a corresponding free energy change $\Delta G_{PL\to PL'}$). The free energy of binding can then be calculated as the difference between $\Delta G_{PL\to PL'}$ and $\Delta G_{L\to L'}$, following Equation (8.6).

8.3.2 Thermodynamic determinants of binding energy

Thermodynamically speaking, the binding of a protein to its ligand resembles the process of protein folding in several ways. First, in both cases chemical groups are transferred from the aqueous solvent into a less polar environment, in which they interact with one another (Figure 8.5). Second, **both folding and binding involve a favorable enthalpy change and an unfavorable decrease in the entropy of the protein and the ligand.** Indeed, studies of protein-ligand model systems show that the unfavorable entropic contribution is rather significant, to the point of nearly nullifying the favorable enthalpy contribution [99,100]. This explains why the net binding free energy (like the folding free energy) has a relatively small value. While there is no doubt regarding the significance of the entropy loss to the net binding free energy, there is a debate about its exact source [101]. Traditionally, it is believed that most of the entropy loss originates from the restriction of movement around rotational bonds in the ligand, i.e., from conformational restrictions. However, binding of the ligand also restricts its ability to move as a whole. That is, the free ligand has three translational (x, y and z directions) and three orientational degrees of freedom relative to the receptor, which become significantly restricted upon binding [89,102]:

$$\Delta S_{\text{bind}} = \Delta S_{\text{trans}} + \Delta S_{\text{orient}} + \Delta S_{\text{conf}} \tag{8.7}$$

(where ΔS_{trans}, ΔS_{orient}, and ΔS_{conf} are the translational, rotational, and conformational changes in ligand entropy, respectively)

Some studies indicate that the correlation between the total entropy loss of the ligand and its number of rotatable bonds is not as high as originally thought [99,103,104], indicating that the related energy penalty ($-T\Delta S_{\text{conf}}$) might not be the dominant component. Interestingly, in some cases, despite the significant loss of entropy, ligands can still assume different orientations within the binding site [89].

In protein-protein and protein-peptide binding, the decrease in entropy may result from the different entropic components in both binding partners, not just the ligand. Therefore, the uncertainty about the relative contribution of ΔS_{conf} to ΔS_{bind} is even greater in such cases. Wand and coworkers developed a methodology based on nuclear magnetic resonance (NMR) relaxation measurements to assess the contribution of ΔS_{conf} to ΔS_{bind}, and implemented it on *calmodulin*-peptide systems [40,41]. The results suggest that the contribution of ΔS_{conf}, which corresponds to sub-nanosecond motions, is indeed significant, at least in protein-peptide binding (see also [101]). This finding is important towards understanding protein behavior better, and it also has implications for studies that focus on the engineering of protein-ligand interactions for more practical purposes. For example, in the drug industry, scientists design active molecules in pharmaceutical drugs to match their target proteins by modulating these molecules' physicochemical properties (see Section 8.6 below). This ensures optimization of potential interactions between drug and protein, thus making the binding stronger. If, as explained above, ligand-induced changes in the conformational entropy of the protein lead to significant changes in the binding energy, such changes must be taken into account. The neglect of this factor by most drug design procedures may explain their limited success so far. Another practical field that may benefit from the consideration of entropic aspects in assessing the binding energy is enzyme engineering, in which enzymes are mutated in strategic sites to increase their affinity to a given substrate (see Chapter 9).

A third point of resemblance between protein folding and ligand binding is that both processes are driven by the hydrophobic effect [105,106] and by van der Waals interactions, whereas electrostatic interactions render them specific[*1] [107]. The latter results from the strong dependency of the interaction energy on the local dielectric, and from the highly geometry-dependent nature of hydrogen bonds and other types of electrostatic interactions (see Subsection 8.4.3 below for further details). As explained in Chapter 1, the hydrophobic effect is an entropy-driven process associated with the ligand-induced removal of water molecules from the protein's binding site and their transfer to bulk water (not including 'structural' water molecules, which remain and play important roles; see Chapter 2, Subsection 2.4.4 for details). However, several studies have suggested that this process also has a favorable enthalpy contribution (see [108,109] and references therein). This is because the water molecules in the binding site are often unable to satisfy their full hydrogen-bonding potential (four bonds), but they can do so in bulk water. Therefore, their transfer to bulk water is enthalpically favorable, as it reduces the 'frustration' of these individual water molecules and allows them to form additional hydrogen bonds.

The role of the hydrophobic effect as a driving force in protein-ligand binding is reflected, e.g., in the fact that protein-protein binding interfaces tend to be more hydrophobic than surfaces that do not participate in binding [110] (see more below). This is not restricted to protein-protein interactions; studies on enzymes demonstrate the importance of the hydrophobic effect in binding of non-protein substrates, as well [54]. The dominance of the hydrophobic effect in driving protein-ligand binding requires both binding partners to be surrounded by water prior to binding. This does not necessarily mean that the protein's binding site must be filled with water, although that would increase the energy gain accompanying the process, particularly in the case of deep binding sites that are suited for the binding of small, non-protein ligands (see more below). Indeed, **almost all protein-binding**

[*1]Protein folding is a specific process since it leads to the formation of a specific 3D structure.

sites studied so far at a sufficient level of detail have been found to contain at least one water molecule, as expected, considering that the protein is surrounded by water. Exceptions primarily include proteins with hydrophobic binding sites that bind small nonpolar ligands [111–113]. β-lactoglobulin is an interesting example of such a protein [114]. Its binding site, adapted for hydrophobic ligands such as palmitic acid, is completely 'dry', despite its large size (315 Å³). Since there are no (or few) water molecules to expel in such cases, the thermodynamics of ligand binding to dry binding sites is expected to rely very little on the hydrophobic effect and more on van der Waals interactions, as well as on electrostatic interactions augmented by the low-dielectric environment.

During protein-ligand binding, both partners may undergo conformational changes involving a decrease in entropy, which opposes the binding. As mentioned above, this entropy decrease is considerable, at least in the case of protein-peptide binding [40,41]. In proteins that bind multiple ligands, the loss of entropy is gradual, as the first ligand induces some rigidification of the protein, thus making it easier for the second ligand to bind [115]. A well-known example of this phenomenon is hemoglobin-oxygen interactions; as explained in Chapter 5, the binding of the first substrate molecule to the protein stabilizes it in the R configuration, which raises the affinity of the protein to the other three oxygen molecules. Interestingly, in proteins that bind other proteins with low specificity, it seems that the entropy might even increase during binding, which consequentially increases the binding affinity while keeping the selectivity low [116] (see Subsection 8.5.3.1.2 below for details).

8.4 LIGAND-BINDING SITES

8.4.1 Overview

Evolution has equipped proteins with binding sites that match their natural ligands almost perfectly, thus optimizing the binding interactions. As mentioned above, proteins bind a diverse set of ligands, and it is therefore not surprising that binding sites are equally diverse. **The binding site-ligand match is based on two main features: geometry and electrostatics** (see details below). The former has traditionally been believed to be the dominant factor in ligand compatibility. However, as a recent study has demonstrated, different protein binding sites for the same ligand exhibit greater geometric variability than can be accounted for by the conformational variability of the ligand [119]. This suggests that geometry alone is insufficient to enable a binding site to recognize the correct ligand.

Since a given geometric or electrostatic pattern can be achieved using different binding site architectures, there is no single structural motif (i.e., loops, turns, helices, and sheets) dedicated to the binding function. That is, each of these motifs may be adapted to complement the ligand.

8.4.2 Geometric complementarity

The binding site is designed to geometrically complement the ligand's three-dimensional structure (Figure 8.6). Hence, binding sites for small molecules tend to be shaped as small and deep depressions [120] (Figure 8.6a), whereas those for peptides and proteins are larger and flatter [121,122] (although in some cases the site and ligand are intertwined; Figure 8.6b). The geometrical match optimizes all noncovalent interactions that mediate the binding, particularly the short-range van der Waals interactions. The match, however, is not perfect, and

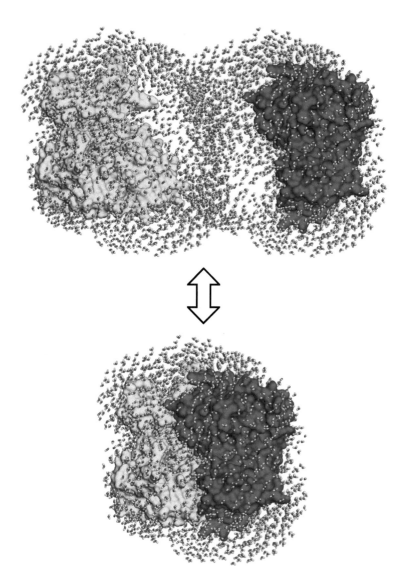

FIGURE 8.5 **A schematic representation of protein-ligand binding.** The figure illustrates protein-protein binding, but the principle is the same for non-protein ligands, as well. Ligand binding involves the exclusion of water surrounding the two binding surfaces, and formation of (mostly) noncovalent interactions between the desolvated atoms. Water molecules were added computationally using the 'PDB_hydro' server [117], by the Delarue group, Institut Pasteur [118].

as explained above, ligand-induced fit of the binding site occurs in many cases in order to optimize binding interactions. Yet, even this does not create perfect complementarity, and cases have been recorded in which binding creates strain in the protein, ligand, and both [89]. Such strain involves an energy cost that is often associated with deformation in one or two of the binding partners. This may be beneficial, for example, it may help enzymes to force certain geometries on their substrates, which in turn makes it easier for the latter to turn into products. Evidently, the binding interactions are sufficient to overcompensate for the strain energy (or else the complex would not form). **Thus, the unique capability of the native (ligand-bound) structure of the protein is not necessarily its capacity to maximize favorable interactions, but rather its capacity to do so while balancing the resulting unfavorable effects, such as strain, which are required for the protein's function [89,123].**

(a) (b)

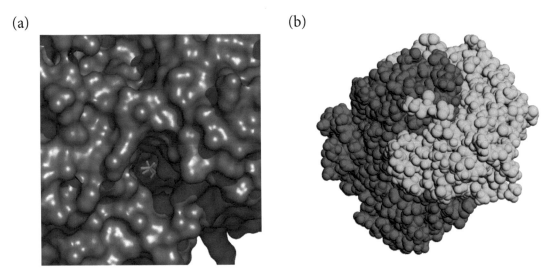

FIGURE 8.6 **Geometric match at the binding site.** (a) The structure of acetylcholinesterase (AChE; blue surface) in complex with its substrate, acetylcholine (Ach; yellow sticks; PDB entry 2ha4). (b) The intertwined subunit-subunit interface in the Bence Jones protein (immunoglobulin light chain dimer; PDB entry 1bjm). The protein atoms are represented as spheres, with each of the subunits in a different color.

8.4.3 Electrostatic complementarity

When a ligand carries an electric charge in its binding region, the corresponding protein binding site tends to carry the opposite charge. This electrostatic match increases the binding specificity, and since electrostatic forces are long-ranged, the match may also decrease the diffusion time of the ligand to the binding site [124]. As explained in Chapter 1, electrostatic interactions involve both full and partial electric charges, leading to a spectrum of electrostatic interactions: ionic interactions and salt bridges, hydrogen bonds, interactions involving π electrons (π-π, π-cation), halogen-involving X-bonds, and more (see Chapter 1 for details). As in protein folding (see Chapter 4), the specificity provided to protein-ligand binding by electrostatic interactions results from two qualities:

1. The strong dependence of these interactions on the local dielectric. While a protein and its ligand may form complexes of different conformations, only the right one pairs all the right polar groups in the two binding partners. This pairing compensates for the desolvation of these groups due to the drop in dielectric upon binding. Indeed, studies show that polar and charged species inside binding interfaces [48], as well as in the protein core [125-129], usually appear in clusters and form interaction networks to reduce the desolvation penalty and render the interactions favorable. These interactions often involve buried water molecules as well.

2. The geometric dependence of hydrogen bonds, π-π interactions, and other interactions that involve molecular orbitals (e.g., n $\rightarrow$ $\pi*$ interactions, X-bonds, and interactions involving the low-lying $\sigma*$ orbitals of the sulfur atom). Again, correct binding between a protein and its ligand is 'designed' to optimize these interactions. Other, non-native forms of binding are therefore less favorable energetically. Hydrogen bonds are particularly important for conferring specificity to protein-ligand interactions, as they are both stronger and more common than the other types mentioned

above. Indeed, **a survey of protein complexes of known structure shows that hydrogen bonds mediate recognition in 67% of the studied complexes of proteins and small organic ligands** [53]. The same study also demonstrates that hydrogen bonds are most likely to connect an NH group of a residue (donor) and the oxygen atom of the ligand (acceptor).

A well-known example of a protein-ligand electrostatic match is observed in the enzyme *acetylcholinesterase (AChE)*, which resides in cholinergic synapses of the nervous system. Its role is to put a time limit on the action of the neurotransmitter *acetylcholine (ACh)*, and it does so by hydrolyzing it to choline and acetic acid [130]. The importance of this enzyme is revealed by the grave consequences of its inhibition; the resultant buildup of acetylcholine at the synapses leads to overstimulation of the *parasympathetic* branch of the autonomic nervous system, and this might lead to paralysis and death (see Box 8.1). ACh carries a positive charge on its trimethylammonium group (Figure 8.7a). Calculations show that the substrate binding site of AChE has a strong negative electrostatic potential (primarily due to Asp-72), which draws ACh to the surface of the enzyme [84] (Figure 8.7b,c). However, the catalytic site of AChE is buried inside the protein core, about 20 Å away from the surface. How, then, is the substrate able to reach the catalytic site from its initial attachment point on the surface? Experimentally determined structures of the enzyme reveal a narrow pathway leading to the catalytic site. The distribution of residues along the path, together with the low dielectric in that region, creates an *electrostatic gradient* that leads the substrate to the bottom of the path, where the catalytic site resides [131–136]. This process is termed *'electrostatic steering'*[1]. At the bottom of the pathway, Trp-84 draws the trimethylammonium group of ACh so the entire substrate is optimally positioned for catalysis (Figure 8.7d). Surprisingly, the residues forming the negative electrostatic gradient are aromatic (including Trp-279, Tyr-121 and Phe-330), and therefore the narrow path containing them is called the *'aromatic gorge'*. The composition of the gorge is not accidental; aromatic residues have weak electrostatic potential, which results from the π electrons in their ring structures. **The weak potential is an important advantage in the AChE gorge, as it allows the charged substrate to move on quickly along the path** [131]. Fully charged residues (Asp, Glu) deep in the low-dielectric core of the enzyme would probably attract the substrate too strongly and hinder its movement into the catalytic site (or out of it when the catalysis is over). If this suggestion is true, it may explain why AChE is one of the fastest enzymes in nature, with a rate approaching the diffusion limit ($\sim 10^8$/(meter second)).

As mentioned above, there is no single structural motif dedicated to binding. What about individual residues? **Studies suggest that aromatic residues are more prevalent in protein-protein interfaces than in other interfaces** [138,139]. What makes these residues so well-adapted for binding? Olson and coworkers studied antibody-antigen binding, and suggested that the following properties provide amino acids an advantage in binding [140]:

1. Amphipathicity, which allows the amino acid to tolerate the drop in polarity upon binding.

2. Large size, which allows the amino acid to form extensive noncovalent interactions with chemical groups in the ligand.

3. Flexibility, which allows the amino acid to adapt to the antigen's shape, thus optimizing geometric complementarity.

---

[1]Electrostatic steering is responsible also for the rapid binding of proteins to one another, as has been demonstrated by Schreiber and Fersht in the barnase-barstar system [137].

All three aromatic residues are large, and Tyr and Trp are also amphipathic. Moreover, the aromaticity of these amino acids is advantageous for binding. Aromatic rings are capable of electrostatic π-π interactions, even if they do not include polar groups. **The combination of such interactions and the planarity of the rings makes the interactions geometry-dependent, which in turn confers specificity.**

Of the three aromatic residues, Tyr seems to possess the characteristics listed above (except for flexibility) to the greatest extent. Indeed, Tyr and its chemical derivatives play important roles in different molecular recognition processes. These include protein-protein interactions [139], in addition to interactions of proteins with small ligands. For example the *catecholamine* neurotransmitters *epinephrine, norepinephrine,* and *dopamine* (see Box 7.2), which act by binding to their cognate receptors in the synapses of the nervous system, all originate from Tyr and resemble it structurally. That is, they contain the *catechol* (dihydroxybenzene) ring, which is identical to Tyr's side chain, with an additional hydroxyl group. Another example is *acetaminophen*, the active ingredient of the painkilling and fever-reducing drug *Tylenol®*. This chemical compound includes the phenol ring of Tyr, and although it is artificial, it also acts by binding to proteins. Specifically, acetaminophen and related drugs termed *'non-steroidal anti-inflammatory drugs'* (NSAIDs) [141–143] bind to *cyclooxygenases* (COXs), enzymes that are involved in the mediation of inflammation, pain and fever (see Subsection 8.6.2.1.2 below for further details).

8.4.4 Binding specificity and promiscuity

The binding of proteins to their natural (cognate) ligands is generally considered to be specific. The binding specificity results from the geometric and electrostatic match between the protein's binding or active site and the ligand. As discussed above, the match involves several noncovalent interactions between the protein and the ligand, although the importance of each contribution to specificity is different; Electrostatic interactions (particularly hydrogen bonds and other geometry-dependent interactions) and van der Waals repulsion have a larger contribution to the binding specificity than nonpolar interactions and van der Waals attraction. Binding specificity also correlates with the number of interactions and their spatial distribution. The latter means that the binding is expected to be more specific as the ligand's interacting groups are more evenly distributed over the entire area of the ligand.

Despite the overall tendency of proteins to bind their natural ligands specifically, it seems that many proteins are capable of binding other ligands as well, a phenomenon termed *'binding promiscuity'* [107,144]. This type of binding has been studied in different protein-based systems, often by using chemical and thermodynamic analysis. Levels of binding promiscuity vary across proteins; some binding sites are capable of binding multiple ligands of the same type (e.g., nonpolar amino acids, sterols, etc.), whereas others are even less 'picky' and bind different types of ligands. Promiscuous binding may involve interactions of the different ligands with the same binding site residues, albeit with different affinity; alternatively, different ligands may interact with different residues [145], or both types of interactions may occur (Figure 8.8). In addition, in both cases different ligands may bind to different conformations of the binding site, which differ in certain properties such as residue accessibility or protonation state [145]. As mentioned above, the specificity of a binding site to its ligand(s) correlates with the number of interactions between them, especially when the interactions are geometry-dependent. The binding affinity, on the other hand, may depend on a few strong interactions. Thus, a promiscuous binding site may bind one ligand more specifically, yet less strongly than it binds another ligand.

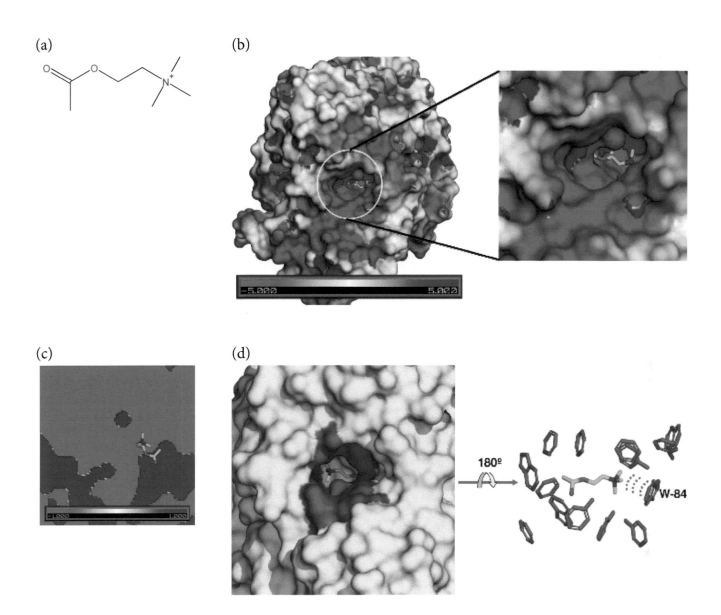

FIGURE 8.7 **Electrostatic complementarity between the protein's binding site and its ligand.**
(a) The chemical structure and electric charge of ACh. (b) The strong negative potential of the binding site in AChE. The figure shows the same structure as in Figure 8.6a. The enzyme is represented as a surface colored according to electrostatic potential, with positive, neutral, and negative potentials represented by blue, white, and red, respectively. The electrostatic potential ranges between −5 and $5k_BT/e$. *Left*: The entire enzyme. *Right*: A magnification of the binding site. (c) A two-dimensional slice of the electrostatic potential presented in (b), traversing AChE at the location of the binding site. The strong negative potential is seen clearly around the choline group of ACh. (d) The aromatic gorge of AChE. The structure shown in the figure is that of AChE complexed with ACh in its transition state (PDB entry 2ace). *Left*: The aromatic gorge (colored blue) when viewed from outside the enzyme (colored yellow). *Right*: The 14 conserved aromatic residues forming the gorge. These include F-120, F-288, F-290, F-330, F-331, W-84, W-233, W-279, W-432, Y-70, Y-121, Y-130, Y-334, and Y-442. The positively charged choline group of ACh interacts favorably with some of these residues (e.g., W-84) via cation-π interactions (red dashed lines).

(a)

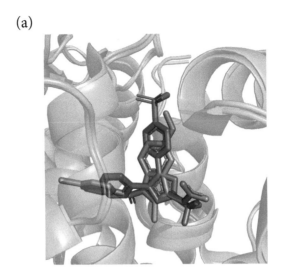

(b)   (c)

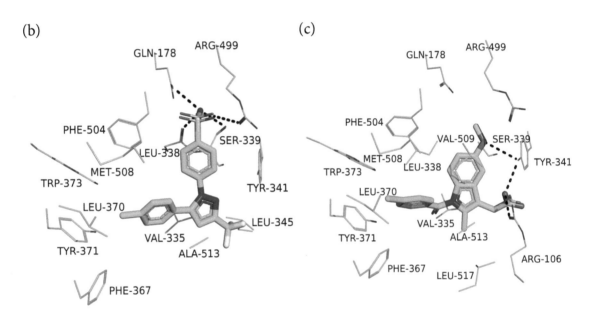

FIGURE 8.8 **Binding of two ligands to the same binding site in cyclooxygenase-2 (COX-2) (adapted from [144]).** (a) Superposition of two COX-2 structures bound to two different inhibitors. The first binds celecoxib (PDB entry 3ln1, blue), and the second binds indomethacin (PDB entry 4cox, orange). The binding site of COX-2 is shown as ribbons, and the inhibitors are shown as sticks. As can be clearly seen, the binding sites in the two structures share the same conformation, in essence, which means that the binding of the two different inhibitors is not based primarily on conformational selection. (b) and (c) The noncovalent interactions of celecoxib (b) and indomethacin (c) with binding site residues. In both cases the inhibitor is shown as sticks, colored according to atom type, and the binding site residues are noted and shown as lines. Most of the residues on the left side of the image stabilize both inhibitors via nonpolar and π-π interactions. In contrast, the polar residues on the upper-right side of the image stabilize one inhibitor or the other selectively, via hydrogen bonds (dashed lines); Q178, R499 and S339 form hydrogen bonds only with celecoxib, whereas Y341 forms hydrogen bonds only with indomethacin.

In enzymes, the binding of different substrates can also be characterized by using activity-related (kinetic) parameters, such as the *specificity factor* $\left(\dfrac{k_{cat}}{K_m}\right)$ and the rate-enhancement factor $\left(\dfrac{k_{cat}}{K_m k_2}\right)$ [146] (see Chapter 9 for details). Studies employing these parameters indicate that while a given enzyme may bind substrates other than its natural (biologically relevant) one, using the same binding site, the catalytic process that follows is considerably less efficient than in the case of the natural substrate. This makes sense considering that enzymes have been evolutionarily optimized to act on their natural substrates [107]. Such optimization is manifested in the orientation of catalytic residues, their pKa, structural dynamics, etc. As a result, the secondary (non-biological) reaction may involve a mechanism similar to that of the biological one, but differ in the catalytic residues involved or their relative contributions (see [107] and references therein). The subjects of enzyme binding specificity and promiscuity are further discussed in Chapter 9 (Subsection 9.3.2).

What makes protein binding sites promiscuous? Several factors seem to be involved. At the most basic level, promiscuity emerges from the inherently high reactivity of protein binding and active sites. This reactivity is supplemented by post-translational modifications of residues (e.g., [147]) and the use of organic or metallic cofactors (e.g., [148]), both of which extend the ability of binding and active sites to act on substrates. Finally, the flexibility of binding and active sites, resulting from proteins' inherent dynamic nature, allows these sites to adapt to the binding of non-natural ligands and substrates [119,149]. Examples of the latter property can be seen clearly in cytochrome P450 and in glutathione *S*-transferase [150,151].

The prevalence of binding promiscuity among proteins is in line with studies of protein-ligand complexes in the PDB, which demonstrate the *degeneracy* of binding sites [144,152]. That is, many different protein binding sites have similar shapes and chemical properties. As a result, the number of binding site types is much lower than the number of biological ligands. This suggests that at least some of these binding sites are able to bind multiple different ligands, but also that many ligands may bind to different binding sites that share certain properties. The latter phenomenon, termed *ligand promiscuity*, is not only interesting academically but is also of great importance to the pharmaceutical industry. As we will see later in this chapter, pharmaceutical drugs act by binding to molecular targets in our body, mainly proteins, and changing the activity or properties of these targets. To treat a certain illness or condition, a given pharmaceutical drug is supposed to bind to one target, affecting one molecular function. If, however, the drug binds to other targets as well (*off-target binding* [153]), it may influence other molecular functions, which leads to *side effects* that are highly undesirable and potentially dangerous. As explained in Subsection 8.6.2 below, one of the main reasons for drug side effects is the fact that biological molecules, which are mimicked by the drug, are evolutionarily selected to bind multiple biological targets[*1]. Thus, the drug molecule too might bind multiple targets, which leads to side effects. To prevent this occurrence, drugs are often designed by pharmaceutical scientists to interact with the target binding sites even more specifically than the mimicked biological molecules. This lowers the likelihood of binding of the drug to proteins other that the intended one. Ligand promiscuity is not all bad, however; if a single drug molecule can bind to multiple protein targets, it might have the potential to be used to treat other illnesses or conditions, aside from the one it was designed for [154–157]. Indeed, such *drug re-purposing* is one of the goals of current pharmaceutical research.

[*1]This is an important evolutionary feature; it provides biochemical robustness to the cell and allows it to prioritize molecular mechanisms via the modular activation of different proteins by the same ligand at a particular concentration.

BOX 8.1 ACETYLCHOLINESTERASE INHIBITION IN WAR AND PEACE

I. Acetylcholine as a neurotransmitter

Acetylcholine (ACh) is perhaps the most common neurotransmitter in the animal nervous system. It mediates neural transmission in both the central and peripheral nervous systems. In the latter, the activity of ACh can be found in the following two subsystems:

1. **The somatic (motor) system** – activates skeletal muscles at the *motor endplate*. In this capacity the cholinergic transmission is responsible for voluntary motor activities, such as locomotion, eye blinking, facial expressions, etc. [158]

2. **The autonomic system** – mediates transmission in the *pre-ganglionic synapses* of both sympathetic and parasympathetic branches of the system. In the latter, ACh also mediates *post-ganglionic transmission*, i.e., activates target involuntary (smooth) muscles and glands, in addition to inhibiting the myocardium (heart muscle). Thus, cholinergic transmission in that system affects homeostatic functions such as breathing (diaphragm contraction), blood pressure (relaxation of endothelial smooth muscle), digestion and peristalsis (gut contraction), excretory functions (excretion of saliva and tears), heart beating (myocardium contraction), thermoregulation (sweating), etc. [158]

There are two general types of acetylcholine receptors; each is activated by a different plant alkaloid, in addition to its response to acetylcholine. The *nicotinic acetylcholine receptor* (responds to nicotine) functions as a cation channel. Nicotinic receptors reside in pre-ganglionic spinal cord neurons of the autonomic nervous system, as well as in skeletal muscles and some neurons of the brain. *Muscarinic acetylcholine receptors* (respond to *muscarine*) function as GPCRs (see Chapter 7). These receptors reside in post-ganglionic neurons of the parasympathetic system, as well as in some neurons of the brain.

The involvement of ACh in numerous physiological processes naturally leads to its involvement in diseases. Many of these diseases are related to skeletal muscles, and result from dysfunction of cholinergic transmission at the motor endplate. For example, *myasthenia gravis*, a known disease resulting in muscle weakness, is the consequence of a significant decrease in the number of post-synaptic ACh receptors, due to autoimmune attack [158]. *Amyotrophic lateral sclerosis (ALS)* is another disease resulting in muscle weakness, but this condition is progressive and also leads to atrophy and death. It is a consequence of degeneration of motor neurons of the spinal cord. Some ACh-related neurodegenerative diseases result from changes in brain neurons. For example, Alzheimer's disease is thought to be the result of neuronal degeneration at the basal forebrain. Parkinson's disease and Huntington's disease are also associated with dysfunction of brain cholinergic neurotransmission.

II. Acetylcholinesterase

As is the case for all other neurotransmitters, the activity of ACh must be terminated quickly after it has been released to the synapse, to avoid neuronal overstimulation. In contrast to the *monoamine transmitters* (*epinephrine, norepinephrine, dopamine, serotonin*), whose activity is terminated by reuptake and oxidation, the activity of ACh is terminated primarily by its hydrolysis, and to a lesser extent by reuptake. ACh hydrolysis to acetic acid and choline is carried out by the enzyme *acetylcholinesterase* (*AChE, EC 3.1.1.7; Figure 8.1.1*). AChE belongs to the *serine esterase enzyme group*. As such, it uses an activated serine residue (Ser-200) to carry out a nucleophilic attack on the substrate (Figure 8.1.1b). This creates a covalent (tetrahedral) *acyl-enzyme intermediate*, followed by the cleavage of the ester bond (see Chapter 9 for further details). The nucleophilic properties of Ser-200 result from its deprotonation, which in turn is facilitated by His-440 and Glu-327. With Ser-200, these functionally coupled residues are called a '*catalytic triad*'. In short, the proximity of these residues allows Glu-327 to polarize His-440 and thus improve its capability to function as a general base. As a result, His-440 abstracts the proton from Ser-200, making it a good-enough nucleophile to attack the substrate.

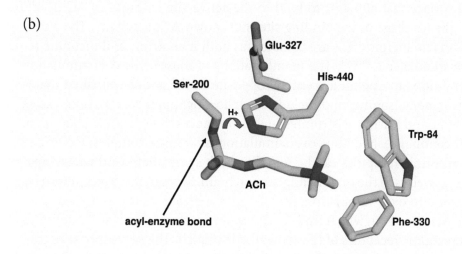

FIGURE 8.1.1 **AChE.** (a) The chemical reaction catalyzed by AChE. (b) The active site of AChE. The structure shown is that of the AChE-ACh transition state, in which ACh is covalently attached to Ser-200 and has a tetrahedral configuration (PDB entry 2ace). The acyl-enzyme intermediate results from deprotonation of Ser-200 (curved red arrow), followed by a nucleophilic attack of this residue on acetylcholine's carbonyl group. The proton of Ser-200 is transferred to His-440 thanks to a decrease in the latter's p*Ka* by Glu-327-induced polarization (see details in Chapter 9 Figure 9.26). The active site is a subsite of the aromatic gorge shown in Figure 8.10. It includes the catalytic triad, Trp-84, and Phe-330. Both residues stabilize the positively charged choline group of ACh through cation-π interactions.

The central role of ACh in the nervous system has led to the evolutionary development of natural compounds capable of targeting most of the components of the *cholinergic system*. These toxins include bacterial compounds, plant alkaloids, and components of animal and insect venoms. Some of these toxins act by reducing the effect of cholinergic transmission. For example, the bacterial *botulinum toxin (Botox)* inhibits the release of ACh at the neuromuscular junction [159]; the plant alkaloid *atropine* blocks muscarinic ACh receptors; and *tubocurarine*, which also originates from plants and is used by South American indigenous people as a paralytic agent in hunting, blocks nicotinic ACh receptors. In mild cases of toxin exposure, the resulting inhibition of parasympathetic regulation manifests as dryness of the mouth, skin, and eyes (due to decreased glandular secretions); blurred vision (due to pupil dilation); a decrease in digestive function; an increase in heart rate; and urinary retention. In more severe cases, muscle weakness manifests as difficulty in swallowing and speaking, and later spreads to lower parts of the body, causing *flaccid paralysis*. Interestingly, when used carefully, atropine is a precious antidote used against nerve agent poisoning (see below). This application is a result of atropine's capacity to prevent overstimulation of muscarinic receptors.

Other toxins do the exact opposite of atropine and botox; i.e., they overstimulate cholinergic transmission (*cholinergic crisis*). This is done by inhibiting AChE, which leads to accumulation of ACh and to prolonging ACh action on cholinergic receptors. For example, the *green mamba* toxin *fasciculin-2*, which is a small protein, has a positively charged surface that allows it to bind to the active-site entrance of AChE [132] (Figure 8.1.2). The blockage of the binding site shuts down AChE activity. The physiological results of cholinergic crisis are complex, as both muscarinic and nicotinic receptors are overstimulated [158,160]. The manifestations of muscarinic overstimulation in the peripheral nervous system are exactly opposite to the above-described manifestations of atropine poisoning, and include the following: pupil constriction, excess salivation, sweating and tearing, gastrointestinal disturbances (diarrhea and abdominal cramps), and incontinence. Nicotinic overstimulation is far more complex, for several reasons. First, nicotinic receptors can be found in both sympathetic and parasympathetic branches, as well as in the motor endplate and brain. Second, these receptors have a biphasic response to over stimulation; initially they increase their activity, but later they undergo desensitization, which leads to the opposite effect. Third, desensitization occurs only in nicotinic receptors of the sympathetic branch. The net response of prolonged nicotinic overstimulation is flaccid paralysis due to a decrease in cholinergic transmission at the motor endplate, bronchial spasm, and the aforementioned muscarinic effects, which are due to both the maintained activity of the parasympathetic branch and the failure of the sympathetic branch. Brain effects, which result from both nicotinic and muscarinic overstimulation, may include giddiness, anxiety, restlessness, headache, tremor, confusion, failure to concentrate, and in some cases even convulsions and respiratory depression [158]. In severe cases of AChE inhibition, such as in nerve gas poisoning (see below), death results from respiratory failure — due to paralysis of the diaphragm and intercostal (chest) muscles and inhibition of the respiratory center of the brain — and from heart failure.

As in many cases, man has learned from nature and succeeded in replicating natural mechanisms of interest in the lab. Furthermore, the ability to synthesize bioactive compounds has also provided an opportunity to enhance these compounds. In the following section we review some of the most popular uses for man-made AChE inhibitors. As we will see, these chemical agents often have dramatic effects on the host, and can be used for both benevolent and malevolent causes. Following this line of thought, we separate the following discussion of 'war' and 'peace' uses of AChE inhibitors.

(a)   (b)

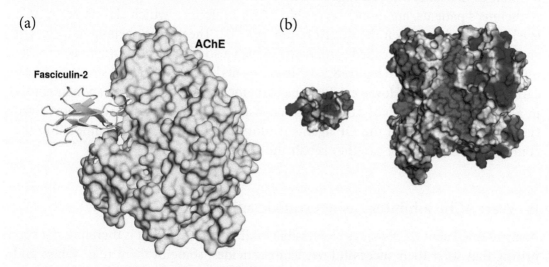

FIGURE 8.1.2 **AChE-fasciculin-2 complex.** (a) A side view of the complex, illustrating the geometric complementarity of the two interacting proteins. AChE is presented as a yellow surface and *fasciculin-2* as a blue ribbon. (b) A front view of both interacting proteins, presented separately as surfaces colored according to electrostatic potential. Negative potential ($0k_BT/e > \Phi > -60k_BT/e$) is red, positive potential ($0k_BT/e < \Phi < 60k_BT/e$) is blue, and neutral potential is white. To create this view, both proteins were rotated 90° compared to their position in (a), AChE to the right and fasciculin to the left. The electrostatic compatibility between the two proteins is clear; the positively charged part of fasciculin matches the entrance to AChE's binding site, which is negatively charged. The electrostatic potential was calculated using APBS [161] and presented using PyMOL [162].

III. Beneficial uses of AChE inhibitors

The most common use of AChE inhibitors is probably as insecticides. In fact, this was the first application assigned to a group of AChE inhibitors called organophosphates (OPs) (Figure 8.1.3a,b). OPs are phosphoesters that irreversibly phosphorylate the (catalytic) Ser-200 residue of AChE, thus blocking its action completely [163] (Figure 8.1.3c). Thus, the action of OPs is an irreversible form of competitive inhibition. OPs win the competition because of their structure; like ACh, they contain an ester bond that can be positioned proximally to Ser-200. However, the phosphate group within an OP has a tetrahedral configuration, much like ACh in its transition state structure (Figure 8.1.3d). Since the binding site of an enzyme geometrically and electrostatically complements the structure of the substrate's transition state rather than that of the ground state, AChE prefers to bind OPs rather than to bind its own substrate, ACh [164]*a. As mentioned above, relatively weak OPs such as *parathion* and *malathion* have been developed specifically to be used as insecticides.

---

*aSuch inhibitors are called *'transition state analogues'* [165,166].

Another beneficial use of AChE inhibitors is in medicine. They are used to treat 'peripheral' diseases such as glaucoma, myasthenia gravis, and gastrointestinal disturbances, as well as 'central' diseases like Alzheimer's disease [167]. The latter is a form of progressive dementia, which starts as reduced synaptic function in hippocampal regions of the brain, which create new memories, and progresses to massive cellular death in multiple regions of the brain [168]. As a result, the cognitive functions of the patient deteriorate until death ensues. Drugs such as *rivastigmine* and *galanthamine*, used to treat Alzheimer's patients, are in fact reversible AChE inhibitors, which elevate brain ACh levels, thus slowing down the deterioration rate [169]. These drugs belong to a class of cholinesterase inhibitors called *carbamates*, which also include *physostigmine, neostigmine*, and *pyridostigmine* (Figure 8.1.4). These agents block the catalytic site of AChE by carbamoylating it [160]. However, since the carbamoyl group hydrolyzes spontaneously in a few hours, the action of carbamates is considered to be reversible and short-lived (i.e., safe), unlike that of the OP agents. Unfortunately, these drugs cannot stop the course of the disease; nor can they repair the associated damage.

IV. War: AChE inhibitors as neurotoxic agents

As mentioned above, OPs act as irreversible inhibitors of AChE. It is therefore not surprising that, after their successful use as insecticides, some of them (e.g., tabun and sarin) were also developed as warfare agents (Figure 8.1.3b). This trend started in Germany just before World War II [170]. Indeed, both tabun and sarin were developed by the German war machine, although they were not actually used during the war. A much more efficient and particularly dangerous OP, VX, was developed during the 1950s and 1960s both by the United Kingdom and the United States [170]. Other highly toxic agents were developed during the 1980s by the Soviet Union. The devastating effects of chemical warfare became public knowledge following several instances when these agents were used [170]; one example is Iraq's use of chemical weapons on its own Kurdish citizens in 1988, which killed about 5,000 people. Today, the use of nerve agents is mainly associated with terrorism. The best-known examples are two instances of nerve agent usage by terrorists (1994 and 1995), both occurring in Japan [170].

One property that makes nerve agents so much more hazardous than OPs used as insecticides is the ability of some (e.g., soman and sarin) to undergo de-alkylation after binding to the enzyme, a process called 'aging' [160] (Figure 8.1.5a). Why is this process a problem? The answer has to do with recovery of affected AChE. Regular OPs can be removed from AChE with the aid of *oximes* (e.g., *pralidoxime*), positively charged compounds that initiate a nucleophilic attack on the OP-AChE bond and hydrolyze it (Figure 8.1.5b). However, aged enzymes do not respond to oximes, and therefore rule out their target AChE for future use.

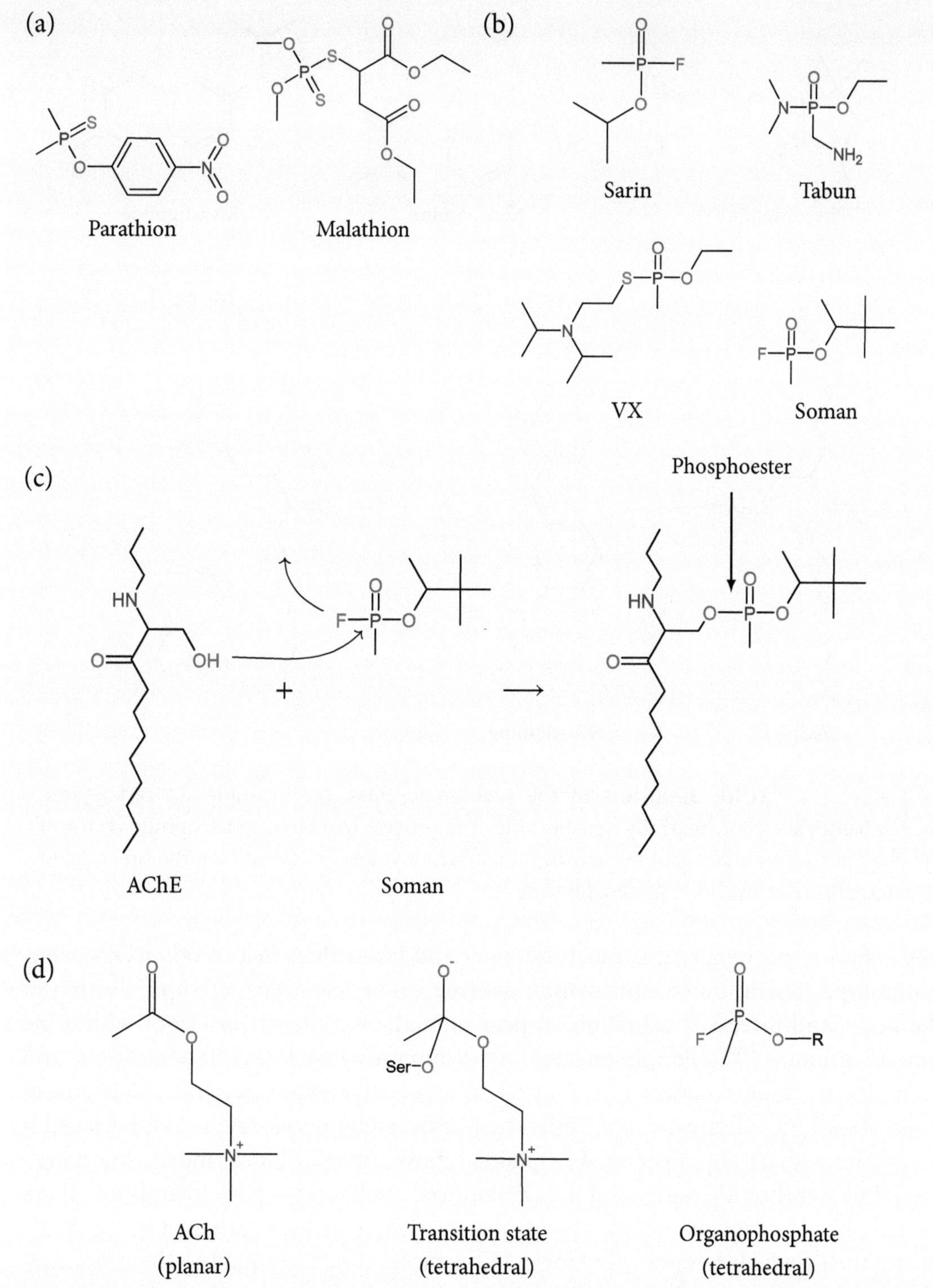

FIGURE 8.1.3 **Organophosphates (OPs).** (a) Weak OPs usually used as insecticides. (b) Strong OPs used as nerve agents. (c) Phosphorylation of AChE's serine residue by soman. The covalent bond is created following a nucleophilic attack of the enzyme's serine side chain on the phosphorus atom of soman, and the concomitant leaving of the latter's fluoride atom (see Chapter 9, Section 9.5 for more details). (d) The similarity between the configuration of acetylcholine (ACh) in its transition state and the configuration of the phosphate group in organophosphate.

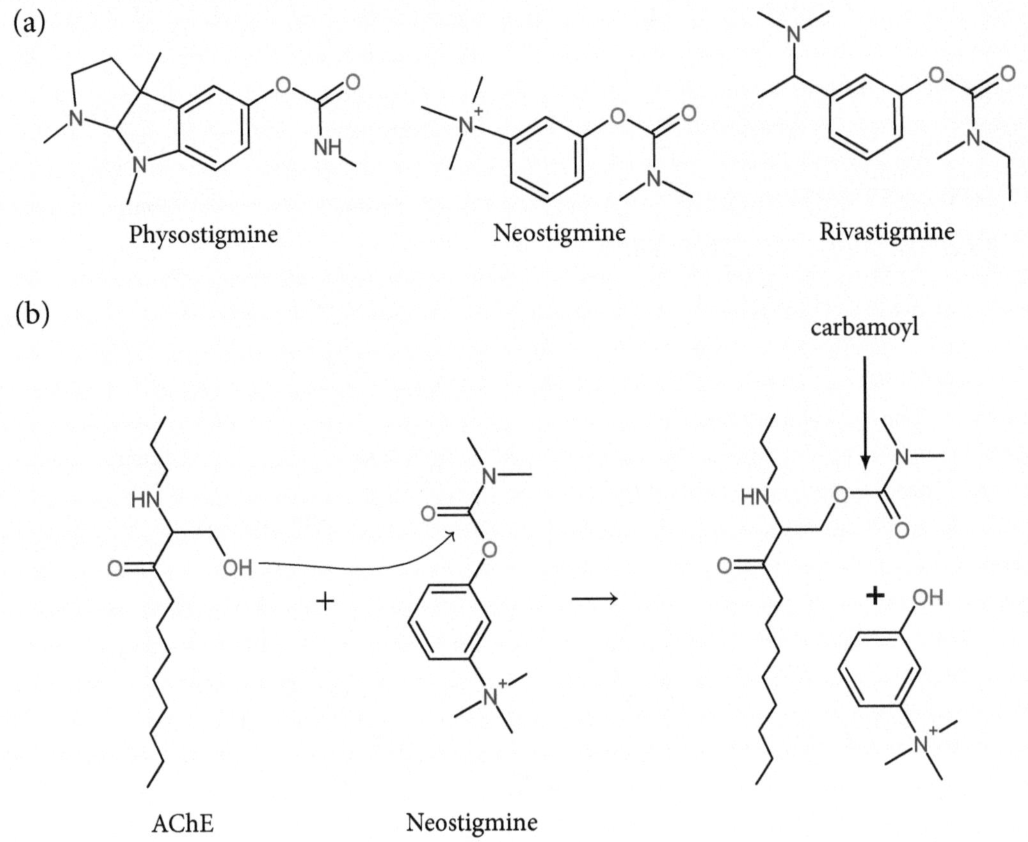

FIGURE 8.1.4 **AChE inhibitors of the carbamate class.** (a) Examples of carbamates. (b) Carbamoylation of AChE by neostigmine. The process involves a nucleophilic attack of the enzyme's serine side chain on neostigmine's carbonyl group. It results in the breaking of an internal carbamoyl bond in neostigmine.

The action of nerve agents is much stronger and faster than that of other OPs. Central and peripheral muscarinic symptoms (increased secretions, cramps, slowing of the heart, and increased urination) appear immediately after exposure or within the first 15 minutes [160]. Peripheral-nicotinic symptoms (weakness, fasciculation, and convulsions) appear between 20 and 60 minutes after exposure, and death ensues 30 to 60 minutes after exposure. Individuals affected by nerve agents can be saved if treated immediately by atropine. As explained above, atropine blocks muscarinic receptors, thus significantly mitigating the OP-induced cholinergic crisis. In addition, there are prophylactics against attacks by nerve agents. These include carbamates (e.g., pyridostigmine), which temporarily block AChE molecules, thus making them unavailable for the nerve agent. When the carbamoyl group dissolves, the previously inhibited enzyme molecules are reactivated. Another medical use for carbamates is in treating atropine poisoning. By inhibiting AChE using carbamates, doctors can artificially increase the synaptic concentrations of ACh, which in turn removes atropine molecules from muscarinic receptors by a simple competitive mechanism.

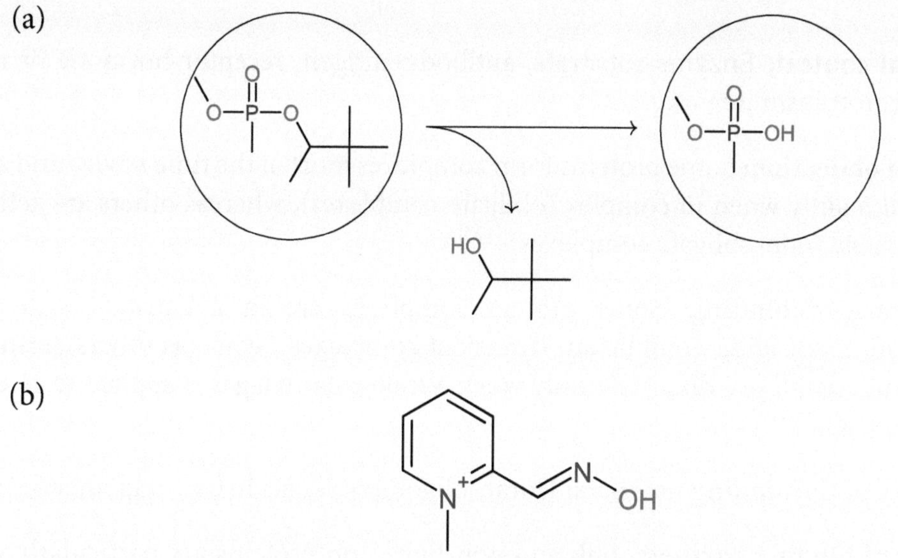

(a)

(b)

Pralidoxime

FIGURE 8.1.5 **OP aging.** (a) The aging process. AChE is schematically symbolized by the circle with the modified Ser-200 inside it. (b) Pralidoxime, used to recover AChE from non-aging OPs.

8.5   PROTEIN-PROTEIN INTERACTIONS

8.5.1   Overview

Protein complexes are the basis for key cellular processes such as signal transduction and communication, enzyme-mediated catalysis, the immune response, cellular division, programmed cell death, cell-cell recognition, and viral action. Moreover, the binding of proteins to one another allows them to participate in metabolic and genetic regulatory networks, which are assumed to have played a key role in the evolution of organisms. Today we know that complex organisms do not differ greatly from simple organisms in terms of the number of genes they possess. We have seen in Chapter 2 how post-translational modifications confer diversity to proteins that are encoded by limited sets of genes, and that the extent of such modifications correlates with the complexity of the organism. Another way for an organism to achieve complexity is to use diverse regulation on gene expression. This is done by *transcription factors*, which activate or inhibit gene expression. The functional diversity of transcription factors results from their ability to respond to various environmental changes and to the organism's own physiological state. Indeed, the percentage of transcription factors in organisms correlates with their complexity [171]. The tight regulation on the activity of transcription factors involves interactions among various proteins that relay activation or inhibition messages to the transcription machinery.

The extent to which two proteins are likely to be involved in a complex with each other depends on their ability to recognize and bind to each other *reversibly* within seconds [138]. The binding may involve conformational changes in one or both binding partners, which leads to functional changes (i.e., allostery). The entire population of protein complexes can be classified according to different criteria [172].

For example:

1. **Biological context:** Enzyme-substrate, antibody-antigen, receptor-hormone or receptor-neurotransmitter, etc.

2. **Degree of obligation:** Some proteins form complexes most of the time *in vivo* and are functional mainly when in complex (obligate complexes), whereas others are active also separately (non-obligate complexes) [173].

3. **Permanence of binding:** Some protein complexes are in a constant state of association-dissociation equilibrium (transient complexes), whereas others, termed 'permanent complexes', dissociate only when a molecular trigger is applied (e.g., extracellular signal) [173].

4. **Similarity of the binding partners:** Homo-oligomers versus hetero-oligomers.

5. **Number of binding partners:** hub and non-hub. Hub proteins are particularly interesting; a single hub protein is able to bind multiple partners [174,175] (tens in some cases), not necessarily at the same time. Consequently, hub proteins can participate in complex cellular networks, such as metabolic or signal transduction pathways and the cell cycle. Thus, hub proteins seem to refute the common perception that proteins have evolved to optimize their binding affinity and specificity. Instead, their characteristics suggest that each protein has evolved to serve its biological purpose, which may in some cases involve weak binding with low specificity. Weak binding interactions that create transient complexes in subcellular networks are sometimes called *quinary interactions* [176–178].

Some of the above categories may partially overlap. For example, homodimers are frequently permanent-obligate complexes, whereas transient complexes tend to include heteropartners with certain biological roles, such as enzyme inhibitors and hormone receptors [172]. Interestingly, the two types of interfaces seem to differ in their evolutionary behavior as well; permanent-obligate complexes tend to evolve slowly, allowing the partner proteins to coevolve, whereas transient complexes tend to evolve much faster, with no mutational correlation between the partners [179]. As explained in Chapter 2, in the case of heteromers it is sometimes difficult to determine whether the interacting partners are different individual proteins or different subunits of the same protein. In our discussion here we ignore the distinction, as the physicochemical aspects of binding are the same regardless of the exact type of complex.

8.5.2 Protein-protein binding domains

Thus far, several domains have been found to mediate protein-protein interactions. Each recognizes a different type of sequence motif, thus conferring some specificity to the binding. Here are some of the known domains:

1. **Src-homology 2 (SH2)** – Found initially in the signal transduction protein src, this domain binds phosphorylated Tyr residues (pY) in the target protein, particularly in auto-phosphorylated growth factor receptors. The bound sequence typically has an extended (β) conformation, and the pY interacts with conserved basic residues of the

domain. Other, more variable residues of the SH2 domain recognize the few residues flanking the pY in the bound peptide (mostly C' to the pY), which confers specificity to the binding [180,181].

2. **Src-homology 3 (SH3)** – Also found in src, this domain binds proline-rich sequences in target proteins, which contain the x-Pro-x-x-Pro sequence motif (where x represents any residue) [182–184]. It appears in many proteins that are involved in signal transduction, cytoskeletal organization, and receptor internalization. The bound sequence is organized as a PPII helix (see Subsection 8.5.3.2 below for a detailed description of the binding).

3. **WW** – This domain binds the following sequence motifs: Pro-Pro-x-Tyr, Pro-Pro-Leu-Pro, as well as other Pro-rich sequences. Its name comes from the appearance of two Trp residues in the domain, separated by 20 to 22 positions. It functions in signal transduction proteins, some of which have been associated with diseases such as Alzheimer's and Huntington's.

4. **Enabled VASP homology 1 (EVH1)** – This domain is found in proteins that regulate the cytoskeleton. It binds to the following sequence motifs: (Asp/Glu)-Phe-Pro-Pro-Pro-Pro, Pro-Pro-X-Phe.

A highly-connected protein (hub) is likely to include a repetitive sequence of a certain binding domain or a smaller binding motif, which allows the protein to create multiple interactions using combinatorial contacts [172].

8.5.3 Structure-function relationships

8.5.3.1 Protein-protein interface

8.5.3.1.1 *Interface properties depend on type of complex*

The ability of proteins to form biologically active complexes depends to a large extent on the properties of their binding surfaces [185]. This is why scientists put considerable effort into characterizing biological interfaces, looking for common properties. Studies in this field have used statistical [122,186] and physical [187,188] analyses of proteins of known structure, and examined the interfaces of both different proteins and different subunits of individual proteins. The analyses focused on certain properties of the interface, such as size, geometric compatibility, chemical composition, polarity, atom packing efficiency, hydrogen bond or salt bridge frequency, number of buried water molecules, interaction energy, residue conservation, and types of secondary structures [172]. The results suggest that interfaces in biologically active complexes tend to have the following characteristics:

1. Surface area of ~700–2,000 Å [122,185].

2. A shape flatter than that of other protein-ligand interfaces [121,122].

3. Tighter atom packing [110], lower polarity [110,138], and higher conservation [189] compared to non-binding protein surfaces. The high conservation is especially pronounced in residues located at the center of the interface [190].

It should be noted, however, that the exact values of the above parameters differed considerably across the different interfaces examined, and it seems that common characteristics could only be ascribed to subgroups of interfaces (e.g., those that are mentioned in Subsection 8.4.1 above) [122,186,187]. In particular, protein-protein interfaces were found to differ from subunit-subunit interfaces[*1]; the latter, which on average generate stronger binding, were found to be typically less planar and more hydrophobic (~66% of the residues), and to possess better geometric and electrostatic complementarity compared with the former [172,191,192] (Figure 8.9). These differences were particularly pronounced in homodimers. Indeed, in terms of hydrophobicity and complementarity, homodimeric interfaces resemble protein cores [193,194], whereas heterodimeric interfaces are more like the regular (nonbinding) protein surface [195][*2].

Most of the structures used by scientists for analyses such as those mentioned above have been produced by X-ray diffraction. This method, which yields high-resolution structures, requires the protein to appear in a crystal form. As explained in Chapter 3, the crystallization process creates artificial packing forces, which might lead to non-biological aggregation of polypeptide chains. As a result, biological complexes must be distinguished from artificial ones. To solve this problem, scientists have scrutinized different complexes in search of properties that might assist in differentiation. Recent work shows that while some properties are shared by both types of complexes[*3], the biological interface seems to be larger and less polar, and to contain fewer salt bridges compared with the artificial interface [110].

8.5.3.1.2 Division of labor among interfacial residues

Protein-protein interfaces are heterogeneous and contain residues that play different roles during the binding process. In our discussion of protein folding energetics (Chapter 4), we saw a basic 'division of labor' among protein residues; nonpolar residues are involved in driving folding via the hydrophobic effect, whereas polar residues confer specificity to the native fold via electrostatic interactions. **Such a functional distinction should also appear in protein-protein interfaces, as interprotein interactions, like intraprotein interactions, involve interactions between amino acid residues.** Therefore, the only significant difference between interprotein and intraprotein interactions should be the lack of chain connectivity in the former. In addition to affinity and specificity, some residues seem to contribute to the dynamics of the protein more than others. The three aspects of protein-protein binding are discussed in the subsections below.

8.5.3.1.2.1. Specificity

As in other types of protein-ligand interactions, specificity in protein-protein binding is conferred by electrostatic interactions [125,196] (see Subsection 8.4.3 above). **Binding specificity is highly important for the biological function of many proteins, particularly anti-**

[*1] As explained in Chapter 2, the distinction between protein-protein and subunit-subunit complexes can be rather confusing. Here, the former term is mainly used to refer to complexes consisting of different proteins that may also function in their uncomplexed state, whereas the latter term refers mainly to homomers.

[*2] Although such interfaces are less polar and tend to contain fewer charged residues compared with nonbinding protein surfaces.

[*3] For example, a hydrogen bond frequency of $\sim \dfrac{1}{100 \text{ Å surface area}}$.

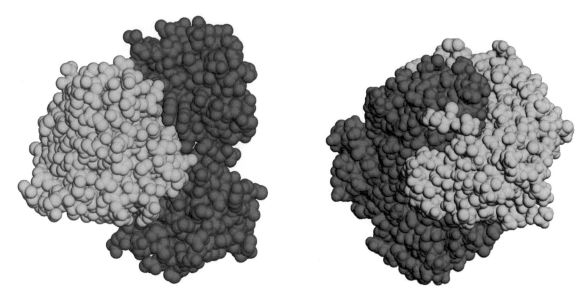

FIGURE 8.9 **Differences in shape and complementarity of protein-protein and subunit-subunit interfaces.** The former is represented by the Ras-RalGDS complex (left, PDB entry 1fld), whereas the latter is represented by the Bence-Jones protein (right, PDB entry 1bjm; see Figure 8.5c for more details). In both cases, the two complexed parties are colored differently. The figure demonstrates one of the observed differences between the two interface types; protein-protein interfaces tend to be more planar than subunit-subunit interfaces. In the latter the interacting subunits are often intertwined, which creates better geometric and electrostatic complementarity.

bodies, hormone-activated receptors and many enzymes. In addition, it prevents protein aggregation, which is mediated by non-specific (nonpolar) interactions. A statistical survey of numerous protein-protein interfaces found that ionic interactions (including ionic hydrogen bonds) involve Arg, rather than Lys [110]. This could be the result of the following:

1. The ability of Arg to participate in more hydrogen bonds than Lys. A higher number of hydrogen bonds leads to greater specificity and creates better masking of the positive charge. Interestingly, the geometry of both hydrogen bonds and salt bridges in protein-protein interfaces has been found to be less optimal than that in the protein core [125]. This problem is often solved by water molecules, which mediate polar interactions. This is probably why protein-protein interfaces contain a relatively high number of water molecules.

2. The lower flexibility of Arg's side chain compared to Lys'. As a result, forcing Arg's side chain into the rotamer corresponding to the bound state involves less entropy loss than in the case of Lys.

3. The higher pKa of Arg compared to Lys, which allows the former (but not the latter) to remain positively charged in virtually any state, and therefore to participate in ionic interactions.

Aromatic residues, which are enriched in protein-protein interfaces [138,197–200], are largely responsible for the specificity of protein-protein interactions. As explained above, these residues are involved in specific interactions, due to their unique geometry and electronic configuration. The specific interactions in which they are involved are diverse and include hydrophobic, van der Waals, and electrostatic contributions. The latter involve the π electrons of the aromatic residues, which are capable of participating in π-π and π-cation interactions, as well as in weak hydrogen bonding, all of which are highly geometry-dependent and therefore specific [201]. Interestingly, a survey of 111 antibody-protein complexes suggests that, in such interfaces, aromatic residues, especially Tyr, contribute primarily to the binding energy (i.e., affinity), whereas the binding specificity is conferred by short-range electrostatic interactions involving the short-chain polar residues Ser, Thr, Asp, and Asn [200].

Proteins may generate specificity in different ways. One example, namely, the *immunoglobulin domain* (*IgD*) in proteins of the immune system, has already been discussed in previous chapters. As explained, the strategy in these proteins is to use a fixed folding frame (*β*-sandwich) and generate specificity by changing the loop structures. This allows the Ig superfamily to avoid the need to use a different protein or domain just to generate a new antigen-binding site. Specific recognition in protein-protein binding is a complex issue, and not all of its aspects are understood. For example, in some cases the binding specificity is conferred by multiple residues in the binding interface [202,203] whereas in other cases it involves only a single residue [204], or even a mere conformational change [205].

Specific binding capabilities are not always an advantage in biological systems. Some proteins, especially those that function in signal transduction networks, are selected according to their ability to bind their partners with low specificity. For example, the binding domain (D/D) of *protein kinase A* (*PKA*) can bind multiple different proteins with high affinity, and this property enables PKA to attach to different cellular locations where its activity is required [206]. The binding interface of the D/D domain is overall hydrophobic, and MD simulations suggest that its ability to bind different proteins results from the high percentage of nonpolar residues serving as potential interaction sites[*1] [116]. Thus, **different proteins can bind to the same interface, each interacting with a different cluster of nonpolar residues**. The presence of different nonpolar binding sites in the same interface should also allow the protein ligand to sample different conformations, each interacting with a different arrangement of nonpolar residues. This, in turn, increases the entropy of the ligand, as prior to the binding its nonpolar residues interact with one another in a more static arrangement. This is a unique phenomenon, since **in all other known cases, binding between two molecules decreases their entropy**. The entropy increase in the case of the D/D domain should strengthen the affinity of the interacting partners via the $-T\Delta S$ component of the binding free energy, without affecting the low specificity.

As mentioned above, a hub protein may bind to different partners using the same binding interface but different nonpolar clusters. Some studies, however (see [172] and references therein), suggest that, in fact, such proteins use the same conserved interactions to bind different partners, and that the low binding specificity results from interface properties that confer high flexibility to the interface [172]. These properties are: small size, loose atomic packing, low geometric complementarity, and a high content of water molecules.

[*1]The involvement of nonpolar residues in non-specific binding has also been suggested in studies of antibody-antigen interactions [207,208], and enzyme-substrate interactions [107] (see also Section 8.4.4 above).

8.5.3.1.2.2. Affinity

Factors affecting affinity

As mentioned above, protein-protein binding is driven by nonpolar interactions (the hydrophobic effect) [192,209–211]. These interactions, however, are expected to be weaker than in protein folding, as the complementarity between interacting residues is lower in the interface than in the core. Electrostatic interactions are expected to be mildly destabilizing in protein-protein binding, but again, to a lesser extent than in protein folding. This is because of the higher polarity of protein-protein interfaces compared to the protein core, which should lower the desolvation penalty of buried electrostatic interactions in the former.

Proteins display a variety of affinities towards their ligands [204], with Kd in the 10^{-12} to 10^{-3} molar range [172,185]. This range corresponds to a free energy range of -6 to -19 kcal/mol [172]. The differences in affinity are driven not only by differences across binding domains and proteins but also by other factors, such as the following:

1. **Biological context.** The affinity of binding domains is not always optimal, and it seems that some proteins have been selected during evolution to form low-affinity complexes [212]. The capacity to form such complexes is common among proteins involved in signal transduction [212,213] and electron transfer [214], which are required to bind and release their ligands quickly. In addition, cell-surface proteins that are involved in cell-cell recognition tend to display lower affinity to their ligands than do hormone- or cytokine-binding receptors, such as those belonging to the Ig superfamily [215] (see Chapter 2). This is probably because cell-cell recognition proteins often appear in arrays on the cell surface, and despite the low binding affinity of individuals, together they strongly bind their cognate proteins on the surface of the other cell.

2. **Biological response.** There is at least one example of a system in which proteins change their affinity to their ligands as a result of a biological response. This happens in the immune system; antibodies and T-cell receptors are known to increase affinity to their ligands (by up to 1,500 times), as the immune response progresses [216–218]. This process, called 'maturation', happens at the genetic level (somatic mutations), and manifests as changes in the binding interfaces of these proteins (Figure 8.10). Surprisingly, the corresponding changes in the number of hydrogen bonds, interacting surface area, and geometric compatibility are much smaller than would be expected given the dramatic change of affinity [219].

Hot spots

The protein-protein binding interface is heterogeneous. With regard to affinity, this heterogeneity manifests as different contributions of different interface residues to the binding strength. That is, the affinity seems to result from a relatively small number of interface residues [210,211,222]. These 'hot spots' are defined as residues whose replacement by alanine leads to an affinity change of over 2 kcal/mol [222]*1. Studies employing this approach show that aromatic residues (and histidine), leucine, isoleucine, methionine and arginine

---

*1This method is called 'alanine scanning' [223]. It has been applied to numerous protein-protein interfaces, and the results can be found in databases such as ASEdb [224] and SKEMPI [225]; both are fully accessible on the Internet.

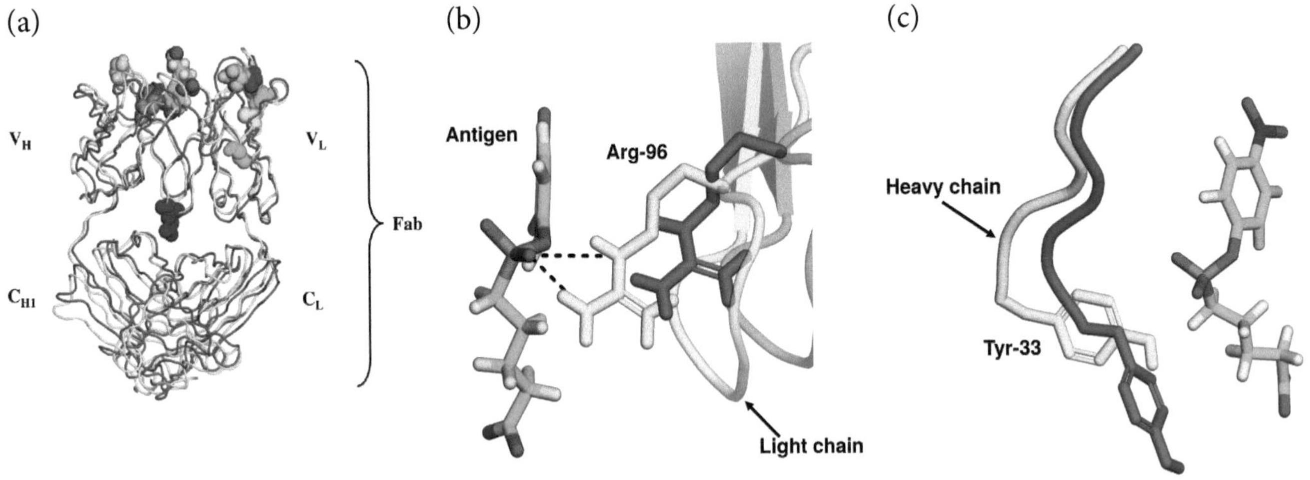

(a)

(b)

(c)

FIGURE 8.10 **Maturation of the 48G7 antibody.** The effects of the maturation process on antibody structure were investigated by comparing the germline (PDB entry 2rcs) and mature (PDB entry 1gaf) forms of the antigen-binding region of the antibody (Fab) [220,221]. (a) Superposition of the entire Fab of both forms of the antibody. The variable (V_H, V_L) and constant (C_{H1}, C_L) regions are marked. The germline form is colored in slate blue, and the mature form is colored in yellow. The antigen (5-(para-nitrophenyl phosphonate)-pentanoic acid) is shown as spheres, colored according to atom type. The maturation process involves somatic mutations of amino acids in the protein, shown here as spheres, each colored differently. Interestingly, only one of the nine mutations is in direct contact with the antigen. The rest of the mutations seem to act indirectly, by inducing global conformational changes in the antibody. The changes increase the affinity of the antibody to its antigen by increasing the number of interaction sites. (b) and (c) Two residues that do not interact with the antigen in the germline form. Following the maturation process, the induced conformational changes allow these residues to form electrostatic interactions with the phosphonate group of the antigen. (b) Arg-96 of the Fab light chain. In the new conformation the side chain of Arg-96 forms a hydrogen bond with the phosphate group of the ligand (c) Tyr-33 of the Fab heavy chain. The new conformation does allow hydrogen bond formation between the residue and the ligand, but the reduced distance between the two and the change in positioning of Tyr-33's phenol group may promote electrostatic stabilization of the negatively charged phosphate by the partially positive charges of the aromatic phenol group. For clarity, the backbone is shown (as ribbon) only in the region around the discussed residues.

are prevalent as hot spots [138,197–199]. The dominant effect of hot spots was recently demonstrated, quite dramatically, when scientists succeeded in maintaining the native affinity of a growth hormone receptor to its ligand, despite simultaneously replacing over 50% of the receptor's non-hot spot binding site residues [226]. Hot spots tend to be evolutionarily conserved (e.g. [227]), which enables them to be detected in the binding interfaces of proteins. Their conservation, in addition to their large contribution to the stability of their respective complexes, can be explained by the fact that they tend to appear in tightly packed regions of the interface [228].

The prevalence of the aforementioned residues as hot spots in protein-protein interfaces can be rationalized by considering the following properties of these residues:

- **Size:** The aromatic residues, as well as leucine, isoleucine, methionine and arginine, all have large side chains. This property allows each of these residues to interact with more than one partner residue and thus strengthens the binding. Indeed, Trp, which has the largest side chain, is known to form extensive interaction networks inside protein cores or interfaces [229]. Thus, despite the general scarcity of Trp in proteins, it is enriched in proteins that participate in binding, and in many cases it is even conserved in their binding surfaces [138,197,198,230]. As mentioned above, antibody-protein interfaces use Tyr residues for establishing strong binding [200]. The binding energy results from interactions between the Tyr residues and both main chain atoms and side chain carbons in the epitope of the bound protein.

- **Length:** The long side chains of Phe, Trp, Met, and Arg enable these residues to contribute to binding, even when they are partially buried away from the surface [43].

- **Aromaticity:** The planar, aromatic rings of Phe, Tyr, His, and Trp enable these residues to form complex, geometry-dependent interactions (see Chapter 2), which contribute to binding specificity.

- **Polarity:** Being moderately polar, Phe, Tyr, and Trp can prevent non-specific aggregation, but without the large desolvation penalty that accompanies the burial of polar-charged residues inside the interface. Arg is highly polar, but can form interaction networks with acidic and other residues to reduce its desolvation penalty.

Certain studies, carried out using synthetic antibodies and other binding proteins ([139] and references therein), suggest that Ser and Gly have important roles in binding; these roles do not involve interacting with residues on the other side of the interface, but rather are indirect. That is, Ser and Gly seem to be particularly important for allowing the two interacting proteins to get close enough to each other[*1]; Ala might share the same quality, but hot spot analysis would fail to reveal it, by definition. Gly has also been implicated in providing the polypeptide chain with the flexibility required to position the interacting residues in the appropriate orientation for binding.

In order to characterize the distribution of hot spots in binding interfaces, Bogan and Thorn examined the effects of over 2,300 mutations in interfacial regions on the binding free energy [211]. They found that hot spots tend to appear as clusters in the interface (Figure 8.11). There, nonpolar interactions are two times stronger than at the periphery [231]. The clustering of hot spots seems to result from solvation issues; strong binding requires exclusion of water molecules between the interacting surfaces, and in the densely packed regions populating the hot spots, water molecules are least prevalent [232]. In line with this theory, it has been suggested that the residues peripheral to the hot spot clusters act to exclude interfacial water molecules from the center, thus allowing hot spots to interact optimally (the 'O-ring' theory) [211]. Hence, despite the relatively small involvement of the peripheral residues in the binding itself, they facilitate binding by 'drying out' the interface. **If true, this theory may explain the failure of studies thus far to determine the strength of nonpolar interactions in protein-protein binding (and possibly in protein folding). Specifically, this theory implies that the interaction energy depends not only on the interacting residues, but also on their chemical-structural environments** [43,233].

In addition to determining the overall affinity of binding, hot spots also seem to be able to fine-tune it. This is clearly seen in the binding between the proteins TEM1 and BLIP [185,234]. A close examination of the binding interface shows that the hot spot residues are arranged as clusters, and that the residues within each cluster (from both sides of the

---

[*1]As in the case of interacting transmembrane segments of membrane-bound proteins (see Chapter 7).

interface) interact cooperatively. Conversely, the interactions between residues of different clusters are additive, i.e., their overall strength is the sum of the strengths of the individual interactions. This could have been a general evolutionary mechanism, i.e., to modulate the binding affinity of different protein complexes by assigning a different number of clusters of interacting residues to each interface [235,236].

8.5.3.1.2.3. Kinetics and Dynamics

The binding sites of proteins match their ligands both geometrically and electrostatically. At the same time, they remain dynamic, a property manifested in constant conformational changes. These characteristics should also apply to protein-protein interactions, yet are expected to be manifested in a more complicated way, as both binding partners have complex structures, and therefore complex dynamics. Accordingly, one would expect the formation of protein-protein complexes to take a considerably long time, which is (presumably) required for both proteins to sample multiple conformations in search of the one most suitable for binding. Since proteins bind each other very quickly, researchers have investigated the kinetics of the binding process, hoping to find an answer to this apparent paradox. Camacho and coworkers [238] have done so using MD simulations, which allowed them to follow changes in the interface during the binding process. The simulations, carried out on 39 different protein complexes, suggested that binding involves two steps:

1. Rapid formation of an initial complex ('encounter complex' [239]). This step involves interactions that confer specificity to the binding. The specific interactions are mediated by one to three 'anchor residues'. Interestingly, the simulations suggest that even before binding takes place, the anchor residues are already in their 'bound' conformation, which is in line with Fischer's 'lock and key' model. Other studies draw a more complex picture that is in line with the currently accepted *substate* and *conformational selection theories*: the two binding partners change their conformations constantly, with the dominant one maintaining latch residues (see below) in their 'bound' conformation [240,241].

2. Slow conformational sampling aimed at optimizing the binding affinity. The chosen conformations (of both partners) bind optimally, thanks to *latch residues*, which reside at the periphery of the interface. This type of binding seems to be in line with Koshland's *induced fit model*.

8.5.3.1.2.4. Interfacial water molecules

Interprotein association involves the exclusion of many water molecules from the interface, although some remain (Figure 8.12). In fact, these molecules seem to constitute (on average) ~30% of the interface [242,243]. The caged water molecules form hydrogen bonds with various polar groups of the interface, with preference for backbone carbonyl groups and the side chains of Glu, Asp and Arg. Surprisingly, it seems that the number of these hydrogen bonds is equal to the number of interfacial hydrogen bonds that do not involve water [244,245]. Such findings have drawn attention to interfacial water molecules, previously considered to be of minor importance. Our knowledge of protein folding suggests that water-mediated hydrogen bonds are unlikely to contribute significantly to the strength of protein-protein binding. Instead, their role is probably to minimize unfavorable desolvation effects associated with complex formation.

(a)

(b)

(c)

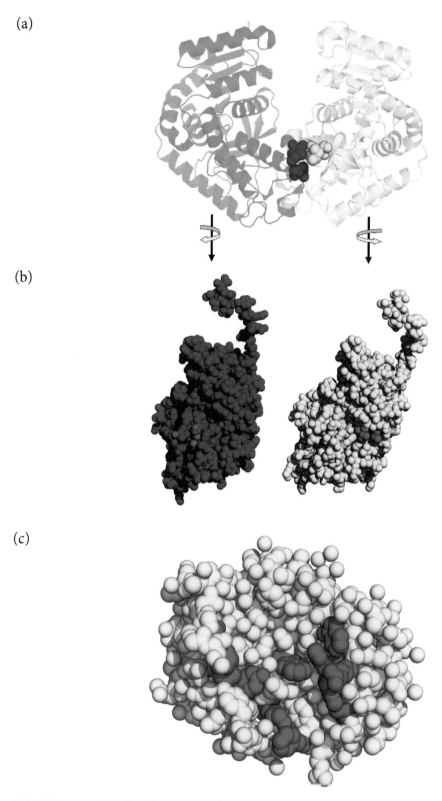

FIGURE 8.11 **Binding hot spots often appear as clusters.** (a) and (b) Hot spots on the interface between chains A (yellow) and C (magenta) of L-lactate dehydrogenase (PDB entry 1i10) (data taken from [172]). (a) The location of the hot spot residues (spheres) on the dimeric enzyme (ribbon, with the subunits in different colors). (b) An 'open sandwich' view of the same interface, where the two subunits are rotated 90° with respect to their orientations in panel (a). The proteins are presented using an atoms and spheres model, and the hot spot residues are marked in red. (c) Multiple hot spot clusters in the interface between β-lactamase (TEM) and its inhibitor (BLIP) (PDB entry 1jtg) (data taken from [237]). The image shows only the hot spots on BLIP, in red.

8.5.3.2 PPII helices in protein-protein interactions [246]

The importance of polyproline type II (PPII) helices in conferring ligand-binding capabilities to intrinsically unstructured proteins (IUPs) has been addressed in detail in Chapter 6. In fact, PPII helices mediate protein-protein interactions in folded proteins, as well[*1]. They are recognized by specialized domains, such as SH3, WW, and EVH-1. In an IUP, the nonpolar side chains of the PPII helix are free to interact with their counterparts in the binding partner, due to the unstructured nature of the protein. Such interactions can also occur in folded proteins, owing to the surface location of PPII helices, as well as to the tendency of these helices to appear in disordered regions. In addition, due to the relatively extended conformation of the PPII helix (compared to α-helices), backbone groups cannot form hydrogen bonds with each other, which makes them available for interprotein interactions. There are also other characteristics that render the PPII conformation suitable for mediating protein-protein interactions. This can be clearly seen in the PPII helix-SH3 domain complex (Figure 8.13). Within the entire PPII sequence, the part interacting with the SH3 domain includes seven to nine residues and the motif **x-P-x-x-P** [182–184]. The motif interacts with three binding pockets in the SH3 domain, with two of these pockets interacting directly with the **x-P** parts of the motif (Figure 8.13a,b). In this interaction, the side chains of the two Pro residues face the binding pockets. The pockets' residues interacting with the

(a)   (b)

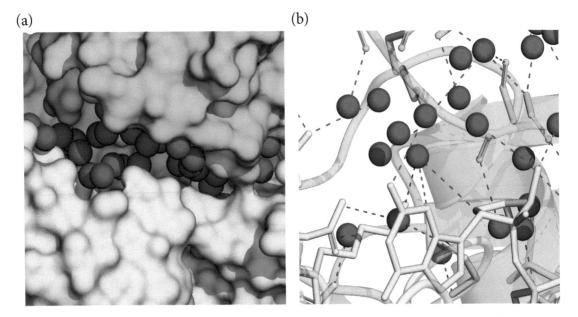

FIGURE 8.12 **Water molecules within the Ras-RalGDS interface (PDB entry 1fld).** (a) Burial of water molecules within the interface. The apoprotein is represented as a surface, with each chain colored differently, and the caged water molecules are represented as red spheres (around the oxygen centers). The image demonstrates the clear presence of water inside interfaces. The caging of the water molecules is unfavorable entropically, and is probably due to partial desolvation as well. It therefore seems that such molecules play specific roles in the interface, which justify their caging (see main text). (b) Hydrogen bonds (red dashed lines) involving some of the caged water molecules. The protein interface is presented as ribbons. Residues involved in the hydrogen bonds are presented as sticks. The negative enthalpy change resulting from these interactions compensates, at least partially, for the entropy penalty of caging the water molecule inside the protein.

[*1]Although even in folded proteins these helices reside in disordered regions.

x-P units are aromatic and highly evolutionarily conserved (Figure 8.13a): Tyr-8 and Tyr-45 (pocket 1), and Tyr-10 and Trp-36 (pocket 2). The two Pro residues are separated by 9 Å, which is the exact distance separating the two binding pockets. **Hence, the PPII helix conformation is required in order to position the two Pro residues at the correct distance to fit into the SH3 binding pockets.**

Lim and coworkers [247] have suggested an explanation for the exact relationship between the geometry of the PPII helix and the conservation of its residues. Briefly, the PPII helix has a triple symmetry, in which three consecutive residues face different directions (Figure 8.13c). The first position of the motif (X_1 in the figure) interacts with the SH3 via its C_α and C_β atoms. Since most amino acids contain these atoms, there is no clear preference for a certain amino acid in this position. Conversely, the second position of the motif interacts with the SH3 domain via its C_δ atom, which has to be covalently attached to the backbone's nitrogen atom in order to attain the correct position. Since the only amino acid that can fulfill this requirement is Pro, the second position of the unit is always populated by this residue. Finally, the amino acid populating the third position of the motif (X_2 in the figure) has its side chain pointing away from the domain. Therefore, this position may contain any residue. Indeed, this position has the lowest conservation in the motif.

8.5.4 Effect of molecular crowding on protein-protein interactions

Studies of protein-ligand interactions are usually carried out *in vitro*, in dilute buffer solutions. In contrast to these solutions, the cytoplasm of living cells is very dense; it has a macromolecular concentration of 300 to 400 g/L, which occupies ~40% of its volume [1,250]. This crowding is expected to promote the formation of molecular complexes, because binding partners occupy less volume in complex than in their unbound states [251–253]*1. As a result, complex formation increases the volume available for the free distribution of the surrounding molecules, which in turn increases their configurational entropy. Since complex formation is an equilibrium process, crowding affects it by tilting the equilibrium forward, following the law of mass action. This *crowding effect* should be particularly significant for complexes involving macromolecules like proteins, which occupy much volume and can form high-order oligomers. Indeed, it has been estimated that the crowding effect may increase the equilibrium constants of macromolecular association by as much as two or three orders of magnitude [254], although some studies suggest these values to be too high [255,256].

To consider the crowding effect, studies of protein-ligand interactions often use supplemented compounds such as polyethylene glycol (PEG), Ficoll, and dextran, which mimic the cellular macromolecules (e.g., [255]). This approach, however, is not trivial; the added compound must exert its effect only via steric repulsion, not specific physicochemical interactions. Still, such studies almost unanimously confirm the complex-promoting effect of crowding in various biological systems [257], albeit to different degrees (e.g., [255]). They also suggest that cells may somehow use the crowding effect to control binding processes, formation of protein oligomers, and even biochemical reactions.

---

*1We refer specifically to the *excluded volume* around proteins, named so because it is unavailable to other solutes due to geometric considerations. The excluded volume around any protein complex is smaller than the sum of excluded volumes around its free components.

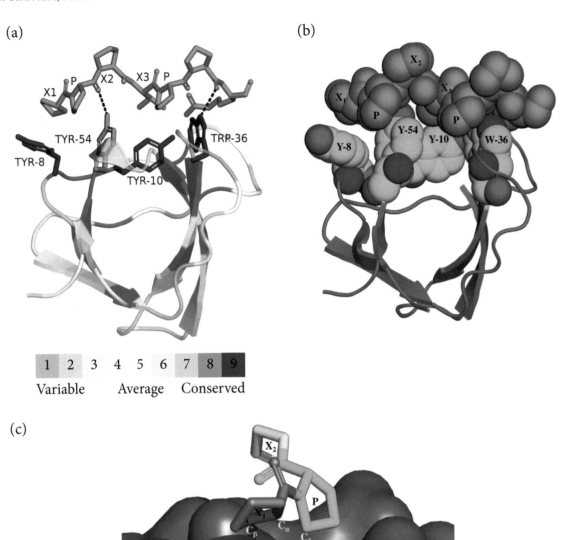

FIGURE 8.13 **The interaction of an SH3 domain with the mSos-derived polyproline peptide that has a PPII conformation.** The SH3 domain is from the actin-binding protein Abp 1P (PDB entry 1jo8). It is superimposed on the SH3 domain of the protein 1sem, which is bound to the mSos-derived peptide with the amino acid sequence PPPVPPRRR (PDB entry 1sem). (a) A front view. The SH3 domain is presented as ribbon with the residues interacting with the peptide presented as sticks. The protein is colored according to evolutionary conservation (bottom, cyan – lowest, maroon – highest; see color code in figure). The conservation is calculated by the ConSurf web server (http://consurf.tau.ac.il) [248,249]. The peptide is depicted as orange sticks, with the locations of the conserved motif positions noted. Hydrogen bonds between peptide carbonyl groups and residues in the SH3 domains are shown as dashed black lines. (b) As in (a), with the peptide and peptide-binding atoms in the SH3 domain represented as spheres, to illustrate their geometrical match. (c) A view of (a), where the entire complex is turned 90° to the right; the SH3 domain is depicted as a blue surface, and the peptide is depicted as sticks. The first three positions of the substrate are shown in different colors, and atoms interacting with the domain are noted.

8.6 PROTEIN-LIGAND INTERACTIONS IN DRUG ACTION AND DESIGN

8.6.1 Involvement of proteins in disease

The emergence of disease in the animal body may result from different factors. In some cases disease is due to an invading pathogen, which attacks the cells of the host or secretes toxins that induce a radical physiological response. In other cases, disease appears when external toxic compounds, industrial or others, enter the body via air, water, food, or injury, and chemically attack the body. Finally, disease might result from spontaneous processes inside the body, in which endogenous components of cells and tissues malfunction as a result of genetic, environmental, or age-related causes. This form of disease includes cancer, metabolic disorders, autoimmune diseases, hormonal imbalances, and more. Regardless of the specific reason for the emergence of disease, proteins are almost always involved. For example, disease can result from the following changes in proteins:

1. **Loss of enzymatic activity.** Many metabolic illnesses, such as *galactosemia, phenylketonuria (PKU)*, and liver storage diseases, result from genetic defects that cause either the biosynthesis of a malfunctioning enzyme, or the complete absence of a given enzyme. The loss of enzymes that are part of central metabolic pathways decreases the body's ability to metabolize or biosynthesize key metabolites. In addition, intermediate products of these pathways may accumulate, causing severe problems ranging from blockage of organs to chemotoxic effects.

2. **Receptor overstimulation.** Certain genetic defects lead to the formation of cell surface receptors that initiate signal transduction even in the absence of a bound agonist. When this happens in growth factor receptors, for example, the result may be loss of control over cellular growth and/or division, which might cause cells to undergo cancerous transformation.

3. **Misfolding and aggregation.** Natural proteins in our cells may, under certain circumstances, lose their native folds and aggregate non-specifically (see Chapter 5, Box 5.1). Such a process is potentially harmful; aside from the problems associated with the loss of the function of these proteins, aggregation of these proteins often leads to their precipitation, which is in most cases toxic to the tissue. Alzheimer's disease, Huntington's disease, and mad cow disease result from such processes.

4. **Autoimmune response.** In some people the immune system responds to 'self' components of the body in the same way that it responds to 'non-self' ones (pathogen-related antigens). The powerful action of the immune response makes such autoimmune conditions extremely dangerous, and people who suffer from them often require medical intervention for prolonged periods of time. Numerous proteins participate in such autoimmune responses, including 'self' proteins that come under attack, proteins that recognize these components (antibodies and cell-surface receptors), intracellular and intercellular proteins that transduce the information (cytokines, enzymes activated by second messengers), and proteins that act on it (T-cell-secreted perforins, the proteins of the complement system, etc.).

Proteins may be the cause of the disease. For example, some viral and bacterial toxins have toxic effects on the host cell. Bacterial toxins that are cell-surface proteins are released into the host's blood stream or tissues upon the degradation of the bacterium by the host's defense system. The damage inflicted by such toxins is double; not only do they hurt nearby cells by incapacitating some of their key processes, but they might also overstimulate the immune system, causing a dramatic and potentially life-threatening inflammatory or allergic response.

The extensive involvement of proteins in both the emergence and course of disease has made them prime targets for pharmaceutical drugs. Indeed, proteins constitute ~80% of all drug targets [258,259]. Other targets include mainly DNA and RNA molecules.

8.6.2 How pharmaceutical drugs work

8.6.2.1 Principal modes of action

A pharmaceutical drug is a molecule[*1] that elicits a cellular or physiological change in the body by binding to a molecular target and affecting its function [261]. With the exception of infection-fighting drugs, which act on enzymes, receptors, and other molecular targets in the pathogen, all other drug types bind to endogenous molecular targets, thereby causing a change in the body's physiology. The physiological change elicited by a drug is meant to overcome an existing disease or alleviate its symptoms. Endogenous drug targets can be grouped into six major types, most of which are proteins[*2]: enzymes, cell-surface receptors, nuclear hormone receptors, ion channels, transporters, and nucleic acids (DNA, RNA) [258,262]. Protein-targeting drugs may work directly, by inhibiting a malfunctioning protein, or indirectly, by modulating the activity of a different protein, so as to compensate for the abnormal activity of the malfunctioning protein. The two most common protein targets are cell-surface receptors and enzymes. Of the former, G protein-coupled receptors (GPCRs) are by far the proteins most commonly targeted by drugs[*3], due to their involvement in numerous diseases (see Chapter 7 for details). The most common enzymes targeted by drugs are protein kinases, which are also involved in a variety of diseases, including cancer, inflammatory diseases, hypertension and Parkinson's disease [265]. A list of common enzymes serving as drug targets is given in Table 8.2.

8.6.2.1.1 Drug effects on cell-surface receptors

Cell-surface receptors constitute the largest group of drug targets (44% of human targets) [266]. Most drugs acting on such proteins have antihypertensive (blood pressure-lowering) or anti-allergic activity. A drug that raises the activity of a protein receptor is called an 'agonist', whereas a drug that lowers the receptor's activity is called an 'antagonist' (see Chapter 7, Subsection 7.5.2.1 for details). For example, β-adrenergic receptors (see Chapter 7) can be activated by the agonist adrenaline (a.k.a. epinephrine), and inhibited by the non-selective antagonist alprenolol. In most cases, both agonists and antagonist act using 'molecular mimicry'. This means that the drug resembles the endogenous hormone or transmitter acting on the receptor, which enables it to bind to the receptor's binding site and execute its effect. Some examples are given in Figure 8.14.

[*1]Most drugs (~87% [260]) are small molecules such as acetaminophen and aspirin, and to a much lesser extent larger molecules such as peptides and proteins.

[*2]In humans, these proteins are the products of 618 genes [260].

[*3]It is estimated that 30% to 50% of the clinically prescribed drugs act by binding to GPCRs and changing their activity [59,263,264].

TABLE 8.2   **Some common enzymes used as drug targets.**

Enzyme	Physiological Role	Drug Family	Representative	Effect
Cyclooxygenase (COX)	Producing bioactive compounds that mediate pain, fever, inflammation, and blood clotting	Non-steroidal anti-inflammatory drugs (NSAIDs)	Aspirin	Analgesic, antipyretic, anti-inflammatory, and blood thinning
Angiotensin converting enzyme (ACE)	Increase blood pressure	ACE inhibitors	Enalapril	Vasodilation $\longrightarrow$ decreasing blood pressure
Serotonin transporter	Terminating serotonin signaling by reuptake	SSRIs	Fluoxetine (Prozac®)	Anti-depression, anxiolytic, increasing appetite, reducing obsessive-compulsive behavior
Monoamine oxidase (MAO)	Terminating catecholamine signaling by oxidative hydrolysis	MAO inhibitors	Pheniprazine	Anti-depression
Acetylcholinesterase (AChE)	Terminating ACh signaling by hydrolysis	Para-sympathomimetics	Rivastigmine	Symptomatic treatment of Alzheimer's disease
Bacterial ribosome	Protein synthesis	Tetracyclines	Tetracycline	Antibacterial action

Although agonists and antagonists have opposite effects, the chemical-structural difference between the two types of molecules may not be large. In the example given above, both adrenaline and alprenolol share the same aromatic ring and a similar substitution 2-(methylamino)-ethanol group, and differ in three other substitutions around the ring (Figure 8.15). As explained in Chapter 7, small molecules (natural or pharmaceutical) that affect cellular receptors may also act as *partial agonists* (induce partial activity), or as *inverse agonists* (decrease the baseline activity of the receptor) [267]. In contrast to antagonists, which prevent the activation of the receptor by other ligands, inverse agonists decrease the constitutive activity of the ligand-free receptor, provided that it has such activity. They do so by stabilizing an inactive conformation (or several inactive conformations) of the receptor. In that sense, their activity is opposite to that of agonists, which stabilize the active conformation.

Some drugs acting on hormone-activated receptors use allostery, which is a noncompetitive (and usually a noncovalent) mode of action [268,269]. These drugs bind to a different site from the one binding the hormone, and by inducing a conformational change in the site, they change its activity. Since such drugs do not compete with endogenous ligands, they need not resemble them structurally or chemically. Developing allosteric drugs is challenging because it requires pre-existing knowledge of the allosteric site, as well as the capacity to predict the consequences of binding to this site, which are not straightforward.

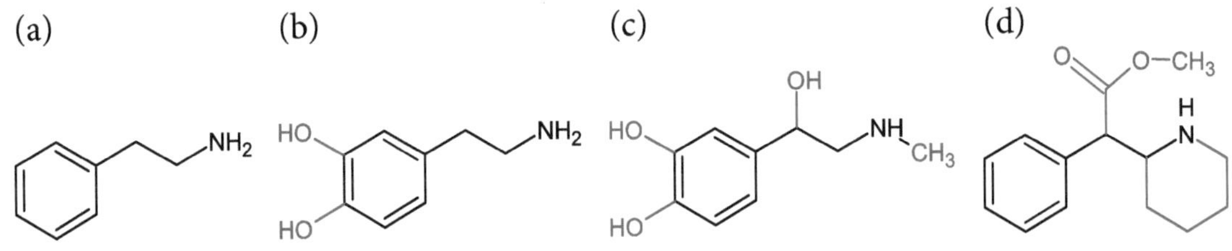

FIGURE 8.14 **Molecular mimicry in pharmaceutical drugs.** The figure shows the example of the catecholamine reuptake inhibitor *methylphenidate (Ritalin®)*, which mimics the catecholamines adrenaline and dopamine. All catecholamines contain the phenylethylamine (PEA) group within their structures (a). The structures of dopamine (b), adrenaline (c) and methylphenidate (d) are shown, with the PEA group in black in all molecules, and other, additional chemical groups in red.

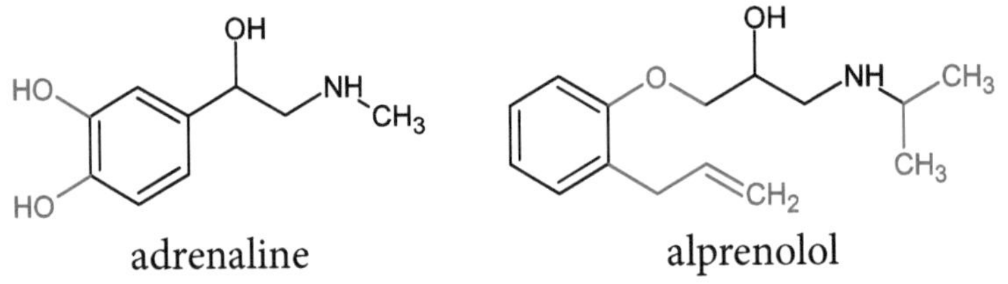

FIGURE 8.15 **Structural similarities between a β-adrenergic receptors' agonist (adrenaline) and antagonist (alprenolol).** The aromatic ring and 2-(methyl-amino)-ethanol group, both common to the two molecules, are colored in black, and the other chemical groups are in red.

Still, such drugs have advantages over orthosteric drugs (i.e., drugs that act on the active site). These advantages include the following [268]:

1. They tend to be more specific than orthosteric drugs, as allosteric sites are less conserved within protein families than active sites are. This means that the use of allosteric drugs is associated with fewer side effects.

2. They allow for modulation of protein activity (positively or negatively) rather than its complete elimination.

3. Since an allosteric drug generally binds to its corresponding protein when the latter is bound to the endogenous ligand, it acts when cellular conditions require the protein to work.

4. In the case of receptor-acting drugs, in addition to affecting the activity of each receptor molecule, these drugs may also affect the formation of large complexes that function as signaling units. This increases researchers' ability to fine-tune the reactions of specific cells and/or tissues to certain signals.

Well-known allosteric drugs include members of the *benzodiazepines* drug family (*Valium®*, *Xanax®*, etc.), which activate GABA receptors [270–272]. These drugs are used mainly as anxiolytic (tranquilizing) and hypnotic (sleep-inducing) agents, as well as muscle relaxants. Their binding to brain $GABA_A$ receptors stabilizes a conformation that has higher affinity to the natural agonist, GABA, thus upregulating the receptor's activity. Activation of the $GABA_A$ receptor opens a Cl^- channel within the protein, which leads to membrane hyper-polarization. In other words, benzodiazepines increase the frequency at which the

Cl⁻ channel inside the GABA receptor opens [270–272]. Since hyper-polarization of post-synaptic membranes is inhibitory in nature, overstimulation of brain GABA receptors leads to inhibition of brain functions, which accounts for some of the side effects of benzodiazepines: sleepiness, drowsiness, etc. Interestingly, the GABA receptor also contains another, yet different, allosteric site for *barbiturates* (e.g., *phenobarbital*). These constitute a different group of drugs, whose effects are similar to those of benzodiazepines. The binding of a barbiturate to a $GABA_A$ receptor prolongs the duration of Cl⁻ channel opening [270,273]. In the past, barbiturates were used for the same purposes as benzodiazepines, but over time they have been replaced by the latter drugs, which are safer; currently, barbiturates are more commonly used as anti-epileptic agents [271]. Other examples of allosteric drugs that act on cellular receptors include *cinacalcet*, a positive regulator at the Ca^{2+}-sensing receptor; and *maraviroc*, a negative modulator of the chemokine CCR5 [274].

8.6.2.1.2   Drug effects on enzymes

Enzymes constitute the second-largest group of drug targets; ~30% of all human targets [266]. The drugs targeting them usually have anti-inflammatory or anti-neoplastic (cancer-fighting) activity. Many of these drugs inhibit their enzyme targets by competitively displacing the natural substrate from the active site [275]. Other drugs may act in a non-competitive, uncompetitive, or even allosteric manner (see also Chapter 9, Section 9.5). For example, *ibuprofen* (*Advil®*), an NSAID, targets the enzyme *cyclooxygenase* (*COX*). COX is a key enzyme in animal physiology; it turns the polyunsaturated fatty acid *arachidonate* (a.k.a. arachidonic acid) into *prostaglandin* (*PG*) G_2, a bioactive substance (Figure 8.16a). The latter can be converted into other types of *prostaglandins*, as well as into *prostacyclins* and *thromboxanes*. These are three groups of powerful chemicals, which act as local mediators of numerous physiological effects. For example, prostaglandins mediate pain and inflammation. Indeed, NSAIDS, which inhibit the production of these compounds, have anti-inflammatory, antipyretic (fever-reducing), and analgesic (painkilling) effects. Ibuprofen acts by competing with arachidonate for the passageway into COX's catalytic site. As is to be expected, both ibuprofen and arachidonate have similar chemical structures, with a carboxylic acid group attached to a hydrophobic chain with a ring-shaped structure in the middle (Figure 8.16b). Interestingly, aspirin, which is the 'classic' NSAID, acts differently from the other NSAIDs; it acylates a serine residue in COX, which results in irreversible binding and inactivation of the enzyme [276].

Competitive enzyme-inhibiting drugs may bind either reversibly or irreversibly to their target enzymes. Reversible drugs are relatively easy to design but have a major drawback; they are overmatched by the natural substrate of the enzyme once the latter reaches high enough concentrations [261]. The only way for such drugs to inhibit their targets efficiently is to be administered in high concentrations, which usually leads to side effects (see the following subsection). Irreversible drugs bind so strongly to their enzyme targets that they effectively inactivate them for good. As a result, such drugs tend to have high toxicity. Compromises between these two options are 'suicide inhibitors' (a.k.a. 'mechanism-based inhibitors'). These are irreversible drugs, but they are administered in their unreactive form, and become activated only upon binding to their target enzymes. As a result, they react specifically with their targets and are therefore less toxic than other irreversible drugs. As in the case of cell-surface receptor drugs, there are also drugs that act allosterically on enzymes [261] (e.g., *imatinib* and *nevirapine*). That is, they stabilize a conformation of the en-

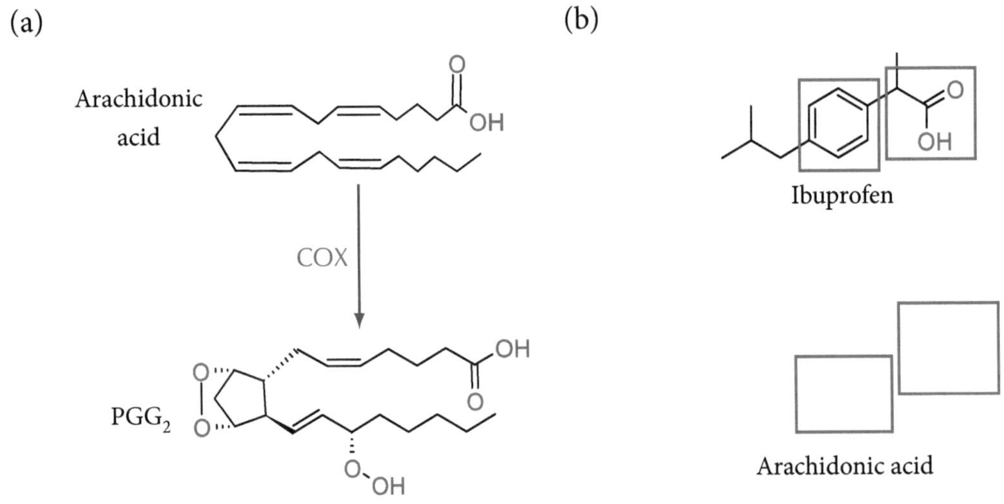

(a)

(b)

FIGURE 8.16 **Ibuprofen's mode of action.** (a) The conversion of arachidonic acid into PGG$_2$. (b) The structures of arachidonic acid and ibuprofen. The red rectangles mark the similar chemical groups in the two molecules.

zyme that has a different affinity to the substrate and/or a different activity. Currently, allosteric drugs are mostly used as inhibitors. However, in cases in which the target is a key cellular protein (e.g., a protein kinase) there is usually also interest in drugs that act by activation [277]. Like other drugs, allosteric drugs may bind their target enzymes either covalently or noncovalently [268]. Covalent allosteric enzyme inhibitors are relatively new, including, for example, those that act on caspases (proteolytic enzymes that play a central role in programmed cell death). The various means by which small molecules may inhibit enzymes are discussed in detail in Chapter 9, Section 9.5.

8.6.2.2 Selectivity and side effects

In Section 8.4 we encountered the phenomenon of ligand promiscuity, that is, the ability of a given ligand to bind to multiple different protein binding sites, which share certain geometric and/or chemical properties. As explained, this phenomenon is of great importance to the pharmaceutical industry, as it allows drugs to bind proteins other than those they were intended to, leading to potentially dangerous side effects. One of the main causes for drug promiscuity is the resemblance of many drug molecules to endogenous compounds, which, ironically, is what makes them biologically potent in the first place. The resemblance is a problem in this sense because the biological molecules themselves often act on several targets. For example, acetylcholine acts on both *nicotinic* and *muscarinic* receptors, despite the fact that they mediate different physiological processes. Similarly, noradrenaline (a.k.a. norepinephrine) acts on both α and β-adrenergic receptors; serotonin acts on at least six subtypes of the 5-HT receptor; and so on. Again, the same goes for drugs. A very well-known example is *aspirin* (see Table 8.1). As explained above, this drug belongs to the NSAID drug family, which also includes ibuprofen. As an NSAID, aspirin is an efficient analgesic, antipyretic, and anti-inflammatory agent. Since thromboxane A$_2$, which is formed within platelets by aspirin-inhibited COX also promotes blood clotting, aspirin is also used as a blood thinner, given to people with a high risk of developing vascular infarcts. The most common isoform of COX (COX-1) also resides in the stomach, synthesizing prostaglandins that help protect the lining of the stomach from the corrosive effects of stomach acid. Unfortunately, when aspirin and other NSAIDs inhibit COX-1, they also contribute to the disruption of the stomach lining. As a result, use of these drugs might cause stomach bleeding and ulcers. The solution to this problem has been solved in the form of COX-2-selective NSAIDs, such as *Celebrex®* and *Vioxx®*. COX-2, a much less common isoform of the enzyme, is absent in the stomach, and its inhibition is therefore relatively harmless, at least with respect to gastric problems.

8.6.3   Drug development and design

8.6.3.1   General sources of pharmaceutical drugs

Many drugs are naturally produced compounds that can be found in sources such as bacteria, fungi, plants, and animals. Known examples include antibiotics such as *penicillin* and *tetracycline*; drugs controlling cholesterol levels, like *lovastatin*; and anti-cancer drugs like *taxol* [278]. The medicinal effects of the sources of these compounds have been known since ancient times, when they were used either in their original forms or as extracts [279]. Use of natural sources for their medicinal content is problematic, however; these sources typically contain thousands of different compounds, some of which might elicit unwanted side effects, and others might even counteract the effect of the desirable ingredient itself. The development of technologies for the purification and analysis of chemical compounds has enabled medicinal scientists to isolate desirable ingredients from their natural sources, and to incorporate them into standardized preparations (pills, solutions, etc.) [280]. Such technologies also facilitate production of optimal concentrations of this ingredient, concentrations that are rarely achieved when using a natural source. Further biochemical research with isolated compounds has revealed the molecular basis of their medicinal effects, i.e., their ability to bind to macromolecules (usually proteins) within the animal body and modulate their activity[*1]. Another cornerstone in the history of pharmaceutical drugs was the discovery of penicillin and other antibiotics; the microbial source of these compounds prompted many drug companies to grow microorganisms industrially, in order to mass-produce these important pharmaceuticals.

As rich as nature may be, the number of molecules that can theoretically serve as potential drugs is considerably higher than the number of molecules found in living organisms. This is because non-biological compounds may also contain chemical groups that are absent in biomolecules, and thus, their chances of interacting with chemical groups in the binding sites of target proteins are much higher. Indeed, the technological advances of the last decades in the field of chemical synthesis and modification have significantly increased the numbers of both potential drug molecules and those that are in actual use. In addition, many synthetic compounds lack some of the 'drug-unfriendly' properties that natural compounds tend to have (see details in [278]). Like their biologically produced counterparts, these artificial drugs resemble endogenous compounds. However, since they are much more diverse structurally and chemically, their suitability for their respective biological targets can be designed to surpass that of naturally-produced drugs.

[*1] Actually, the general idea was proposed by Paul Ehrlich in the 19th century, on the basis of his observations of industrial dyes. However, actual molecular targets of small molecules were found only in the 20th century (see review by Drews [280]).

8.6.3.2 Drug development process

In the early days of drug development, natural substances with known therapeutic effects were collected from all over the globe (mainly in the form of plant extracts), and the activity of their key ingredients was tested using different assays. Such tests would only be the beginning of a long process of drug development [281], extending over ~15 years and including the following additional steps:

1. Purification and isolation of the active ingredient.

2. Testing its activity, selectivity, and toxicity on cell cultures and isolated tissues.

3. Repeating the tests on animal models, to address systemic effects, as well as drug stability, delivery, selectivity, and certain processes the drug may undergo within the animal body (degradation, detoxification, clearance, etc). These tests are termed '*pre-clinical studies*'.

4. Repeating the tests on human volunteers ('*clinical trials*').

Only drug candidates passing all tests (typically one in a few thousands) were cleared for use.

As mentioned above, development of methods for chemical synthesis enabled scientists to artificially create numerous versions of bioactive molecules, by changing their chemical groups systematically [282]. These developments constituted a significant step in pharmaceutical science, as they freed scientists from their total dependence on the large, yet finite reservoir of natural compounds, and allowed them (at least in principle) to create drugs with specific properties. Focal properties included characteristics related to the binding of the drug to its target protein, as well as to other clinically significant parameters, such as drug absorption, distribution, metabolism, excretion and toxicity (*ADMET* properties).

The new power acquired by pharmaceutical scientists was accompanied by an inherent problem [283]: the vast number of chemically feasible molecules that could (in principle) serve as drugs, estimated to be in the range of 10^{60} to 10^{100} [284–286]. Needless to say, the available technological means were insufficient to handle this quantity of molecules, let alone to repeat the analysis for each new drug design attempt. Thus, a more realistic approach, termed '*rational drug design*', was developed, in which the huge chemical space was reduced to manageable size. Indeed, by implementing physicochemical knowledge, as well as statistical data, scientists learned how to narrow down the options and focus only on those molecules that had better chances of fulfilling the desired function of the drug. Although this type of prediction relies, at least partially, on human knowledge and insights, its implementation requires complex calculations, and has therefore been improved considerably by the emergence of powerful computers and sophisticated algorithms. This '*in-silico*' approach is termed '*computer-aided drug design*' (*CADD*) [59,287–290].

Today, rational approaches to drug design carried out using computer calculations and simulations are incorporated into the drug development process, which makes it more time-efficient, as well as cheaper [291]. Rational methods are helpful mainly for the first stages of the process, i.e., for predicting the relative binding affinity and specificity of candidate molecules to the target protein. Predicting the physiological compatibility of the candidate drug is more difficult, and is often carried out by an approach termed '*quantitative structure-activity relationship*' (*QSAR*) [292]. Briefly, this approach uses statistically derived data relating certain chemical-structural properties of a given drug molecule to its physiological adaptability. For example, the potency (P) of a certain drug may be found to depend on the surface area (SA) of the group occupying position X in the drug, as well as on the dipole moment (μ) of the chemical group occupying position Y, in the following way:

$$P = 1.2SA - 0.4\mu \tag{8.8}$$

The disadvantage of this approach is that the data are limited, and are insufficient to enable the scientist to infer the reasons for the dependency. Thus, it is difficult to increase this type of knowledge without collecting new data. Because of the difficulty in predicting these properties of candidate drugs, it is often necessary to test candidates in long trials, which is one of the reasons why the drug development process is still long and expensive. In the following subsection we focus on the general principles of the first stages of the rational drug design process, with emphasis on computer-aided methods. In the last section we describe a famous case study of rational drug design, not because it demonstrates the importance of computations in this field, but rather because of the opposite; it illustrates how great results can sometimes be obtained solely on the basis of insightful thinking of scientists and their experience.

8.6.3.3 Principal steps in rational drug design

8.6.3.3.1 Overview

De novo drug design is the most challenging form of drug design confronted by pharmaceutical scientists, as they must generate an active molecule without any pre-existing knowledge, short of the structure of the target protein and other ligands or drugs that may bind to it. As mentioned above, the only way to do so is to reduce the nearly infinite number of candidate molecules into a defined group of prototypical molecules (called 'leads'), which bind to the target protein and can be further evaluated as drugs. This is done by establishing *constraints*, i.e., certain limitations on the number of possible chemical groups that may appear in certain regions of the molecule. These limitations embody all the knowledge the scientist has about the specific interactions between the molecule and the target protein. This information is formulated as 'rules of thumb', which determine the general properties of the different regions of the molecule (Figures 8.14 and 8.15). For example, the constraints might state that a certain region of the molecule must contain a hydrogen-bond donor group in order to interact with the target protein optimally. Or, they might state that another region of the molecule must contain an aromatic group, in order to avoid toxicity. There are various ways of deriving constraints on a drug candidate. For example, the scientist may use physicochemical 'rules' regarding the compatibility of certain chemical groups of the drug to potential interacting residues in the target protein. Alternatively, statistical data may be used for the same purpose. Two types of constraints might be established, which relate to different properties of the drug [283]. *Primary target constraints* relate to the protein-binding capabilities of the lead molecule. Conversely, *secondary target constraints* relate to the AD-MET properties of the lead (absorption, distribution, metabolism, excretion and toxicity), which are of clinical importance and are necessary in order to assess the compatibility of the lead as a possible drug.

(a)

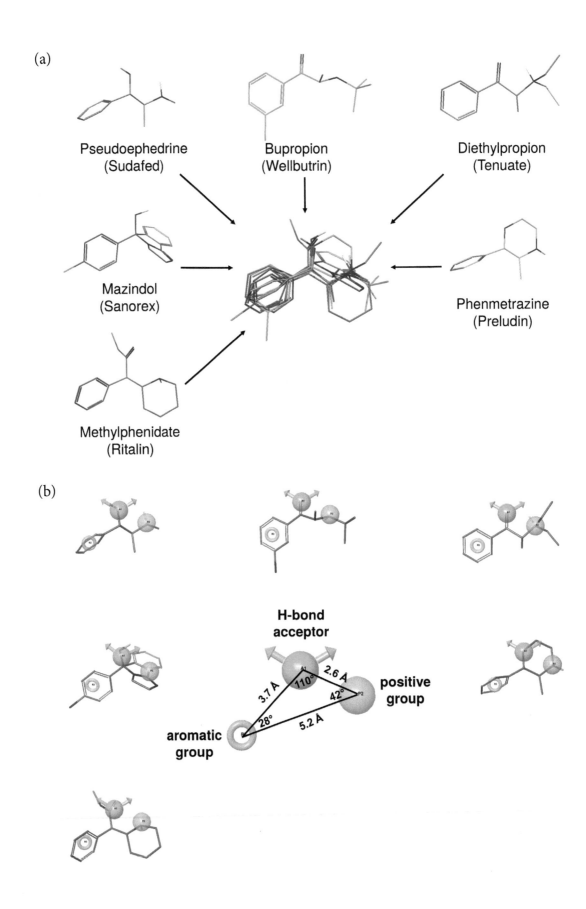

Pseudoephedrine
(Sudafed)

Bupropion
(Wellbutrin)

Diethylpropion
(Tenuate)

Mazindol
(Sanorex)

Phenmetrazine
(Preludin)

Methylphenidate
(Ritalin)

(b)

**H-bond
acceptor**

3.7 Å   110°   2.6 Å

42°

28°   5.2 Å

**positive
group**

**aromatic
group**

8.6.3.3.2 Establishing primary target constraints

There are basically two approaches that can be used to assign the proper primary target constraints, and the choice between them depends on the data that are available. One case is when there are several known ligands of the target protein, which bind to it similarly and have the same general effect planned for the designed drug. In such a case, the popular approach is the *ligand-based approach*. That is, the known ligands are treated as models for the drug, and used for the building of a rough 'template' of the drug; this template is referred to as a *pharmacophore* [293–296]. The IUPAC definition of a pharmacophore is *'an ensemble of steric and electronic features that is necessary to ensure the optimal supramolecular interactions with a specific biological target and to trigger (or block) its biological response'* [297]. In simpler terms, the pharmacophore is a reduced representation of the drug, which includes only those properties that are important for the desirable effect on the target protein (Figure 8.17). In the following subsection we briefly describe how pharmacophores are built.

A different starting point is when there are no known ligands that can be used for building a pharmacophore, but the target protein, its structure, and the location of the binding site are known. In this case, the scientist has to take the *receptor-based approach*[*1] for extracting primary target constraints. This approach entails mapping key characteristics of the binding site, such as geometry, electrostatic properties, and individual groups, which are likely to participate in ligand binding (*'hypothetical interaction sites'*) (Figure 8.18). Integration of all these characteristics yields a *'property map'* of the binding site, according to which the candidate drug can be designed. In some cases, the target protein and its structure are known, but the location of the catalytic site, which is often the desired target for drugs, is not. In such a case, the scientist first has to locate the binding site, and only then can he or she construct a property map. Finding the binding or catalytic site might prove to be a difficult task. Currently, there are several algorithms that are able to scan the surface of a protein and, on the basis of certain geometric, electrostatic, evolutionary conservation, and other properties, suggest putative locations for the binding or catalytic site.

FIGURE 8.17 **Generation of a simple, two-dimensional pharmacophore hypothesis using the ligand-based approach.** (Opposite) (a) Superimposition of six drug molecules that are used for different diseases, but that all inhibit dopamine uptake by the dopamine transporter (brand names and names of the active ingredients are noted). These include *methylphenidate*, which is used to treat attention deficit disorders (ADD); *pseudoephedrine*, which is used to treat various conditions (e.g., asthma and rhinitis); *diethylpropion* and *phenmetrazine*, which are used to treat obesity; *bupropion*, which is an antidepressant, and *mazindol*; which is used to treat Duchenne muscular dystrophy. The superimposed structures are shown in the center. (b) A two-dimensional pharmacophore hypothesis that best fits the six drug molecules listed in (a) (calculated by Elon Yariv using Phase [298] (Schrödinger, Inc.)). The pharmacophore summarizes the important groups in the drug as three centers, each having a different physicochemical property. The first center, which is represented here by the orange doughnut like shape, is aromatic. The second center is a hydrogen-bond acceptor. It is represented as a magenta sphere with two arrows, indicating the direction of the accepted hydrogen bonds. The third center, represented by the blue sphere, is a positively charged group. The distances and angles between the centers are noted. The individual drug molecules used for calculating the pharmacophore are shown around it, with the three centers specified in each. As can be seen, in all of the molecules, the second center of the pharmacophore (a hydrogen-bond acceptor) is populated by an oxygen-containing species: a ketone, hydroxyl, ester or ether group. Also, in all molecules, the third center (a positively charged group) is populated by an amino group.

[*1]Also referred to as *'structure-guided drug design'*.

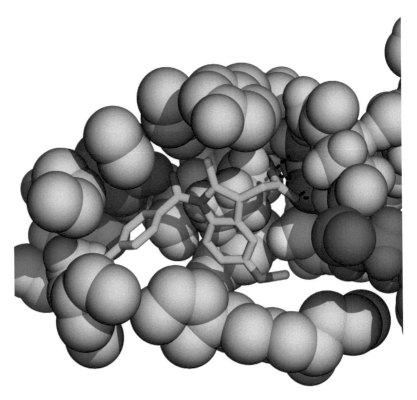

FIGURE 8.18 **Interaction sites in the target protein.** The figure shows the anti-inflammatory drug indomethacin (sticks colored according to atom type) within the binding site of D2 11-ketoreductase (spheres) (PDB entry 1s2a). Atoms involved in nonpolar interactions with the drug are colored in gray. Basic residues that interact electrostatically with electronegative atoms in the drug are colored in blue. Atoms involved in hydrogen bonds with the drug (black dashed lines) are colored in magenta. For clarity, only the residues mentioned above are shown.

Finally, our scientist might encounter a particularly dismal situation, in which the identity of the target protein is known, but its structure has not yet been determined, and there are no known ligands. In this case, the scientist can still use the receptor-based approach, provided that he or she can construct a model of the protein, and use it as a basis for the design of the drug, as described above. Unfortunately, modeling techniques often provide only approximate protein structures, whereas drug design requires accurate information about the three-dimensional locations of catalytic and/or ligand-binding residues. Still, if proteins of known structure that have high sequence similarity to the target protein can be found, homology-modeling techniques, combined with energy-based optimization, may prove to be sufficient for generating a good model. In the remainder of this section we elaborate briefly on the two main approaches described above for applying primary constraints to a designed drug.

8.6.3.3.2.1. Ligand-based approach: building a pharmacophore

As explained above, a pharmacophore is built according to known ligands of the target protein that are assumed to share the same binding mode [296]. This is done by aligning the ligands' structures in a way that reveals regions of common chemical properties, which are likely to play a role in receptor binding (Figure 8.17a). The common regions most often complement the binding site of the target protein geometrically and/or electrostatically. The pharmacophore is usually presented as a general scaffold, in the form of distances, angles, and sometimes full or partial charges (Figure 8.17b). Note that the pharmacophore generation procedure involves multiple fitting and optimization steps, and usually results in different possible models, each fitting the input molecules to different degrees. Therefore, the chosen scaffold is usually referred to as a *pharmacophore hypothesis*. As a rule, the more diverse the ligands are, the higher the generalization quality of the pharmacophore model. There are various software, and even Internet servers, that can construct pharmacophores quickly and automatically. However, since drug molecules are relatively small, the construction process is sometimes carried out manually, especially when the template ligands are of similar structure. Although the use of pharmacophores is frequently helpful in the design of new drugs, it should be noted that they are highly generalized representations, and may be unreliable. Problems associated with this method include the following [275]:

1. Finding the active ligand conformation. Calculation of a pharmacophore relies on knowing the most likely conformation of each of the input ligands. If all of the ligands have a rigid cyclic structure, the task is straightforward. However, if a ligand has a flexible structure, it exists as a conformational ensemble, with each of the conformations in this ensemble positioning the binding groups of the molecule differently relative to each other. The problem is that, for each ligand, only one of these conformations (called the *active conformation*) is important, as it is the conformation acquired when the ligand binds to its target protein. The active conformation can be identified by crystallization, by NMR spectroscopy or by running binding assays with rigid analogues of each ligand[*1]. If these approaches are unavailable and the active conformation of each ligand remains unknown, pharmacophore calculation must take into account numerous conformations and try to fit them into a decent hypothesis. This makes the calculation significantly less reliable.

2. Emphasis on 'functional' groups. Calculation of a pharmacophore considers only discrete functional groups (especially polar ones) that presumably contribute to the binding of the drug to the protein target. However, the entire skeleton of the drug is also involved in the binding, through van der Waals and nonpolar interactions. These may have a substantial and even crucial role in determining whether or not the drug will bind to the target protein.

3. Ignoring the size of the drug. A ligand can fit perfectly to a pharmacophore hypothesis, in the sense that it has all the right functional groups positioned in the right way, but may be too big to fit in the protein's binding site.

4. Ignoring alternative interactions. Candidate molecules that form alternative interactions with the target protein's binding site via groups that do not appear in the pharmacophore hypothesis will be unjustifiably rejected.

8.6.3.3.2.2. Receptor-based approach: mapping binding sites

When there are no known ligands that bind to the target protein and exert the same effect as the planned drug, the scientist has to work 'blindly', i.e., attempt to determine the important chemical groups in the drug molecule and the geometric relations among them according to the properties of the binding site. In principle, this can be done manually, by using certain

---

[*1]When the active conformation is known, a related approach called *'scaffold hopping'* [299] can be used, in which this conformation is used for obtaining a 'hypothetical receptor', which is utilized in turn for constructing leads.

physical-chemical 'rules of thumb' that help identify which chemical groups in the binding site are most likely to participate in protein-drug interactions. For example, aromatic rings would stand out due to their bulkiness and planarity, whereas alcohols, amines, carbonyls and organic acids would stand out as potential hydrogen bond donors and/or acceptors. Organic acids would be implicated in salt bridge formation, and so on. In practice, however, identification of the hypothetical interaction sites of a protein's binding cleft is usually carried out computationally. A description of past and current algorithms capable of conducting an automated characterization of a protein's binding site is given by Schneider and Fechner [283]. Briefly, the first algorithms mainly addressed hydrogen bond donors and acceptors, due to their strongly directional nature. With time, other interaction types were incorporated into the search, such as nonpolar interactions, covalent bonds, metal coordination, and complex hydrogen bonds.

'Rule-based' algorithms are used to search for discrete chemical groups in a target protein that are capable of participating in meaningful atom-atom interactions. These algorithms constitute a quick means of assessing the interaction capabilities of the binding site, but they may neglect certain aspects of the interactions, in particular, aspects that depend on chemical context (see Box 8.2) and the strong effects of the immediate environment (e.g., the effect of the dielectric on electrostatic interactions). These potential problems are addressed by other algorithms that use a *grid-based approach*. That is, the binding site is mapped onto a 3D grid, and different probe atoms or fragments are placed at each grid position so the interaction energy can be calculated. The type of energy calculated depends on the type of probe. For example, probes capable of participating in hydrogen bonding yield hydrogen-bond energy, and so on. A third type of algorithm uses *multiple copy simultaneous search (MCSS)*. In this approach, multiple copies of functional groups are randomly positioned inside the binding site and energy-minimized, and those configurations with low enough energy are adopted. Thus, the MCSS approach not only characterizes the binding site but also produces candidate combinations of chemical fragments, which may later become leads.

As explained above, the search for leads includes not only the generation of different molecular permutations, but also their evaluation, with binding site affinity being the primary evaluated factor. Algorithms may use different scoring functions to rank the different molecules generated. Functions that were developed early on evaluate the geometric compatibility of the molecule to the binding site. More sophisticated functions rely on explicit force fields, empirical functions, or knowledge-based functions [283]. Force field-based functions are described in detail in Box 3.1. Being explicit, these functions are the slowest means of ranking lead candidates. *Empirical scoring functions* may look like simplified force fields in the sense that they include individual expressions, each describing a different ligand–receptor interaction type. However, the expressions need not represent real energies, but rather a dependency produced by fitting the expressions to experimentally derived values. This dependency is expressed mainly via the weight assigned to each expression. Such scoring functions can be implemented rapidly, but since they rely on empirical databases, they are often biased. *Knowledge-based scoring functions* rely on statistical analysis of experimentally determined ligand–receptor structures. That is, the frequencies at which any two possible atoms interact with each other are extracted from the available structures. Frequent interactions are considered attractive, whereas infrequent interactions repulsive. Since binding energy values are not used by such functions, they rely on larger datasets than empirical functions do, and are therefore less biased.

BOX 8.2 CHEMICAL CONSIDERATIONS IN IDENTIFYING INTERACTING GROUPS [300,301]

I. General considerations

In many cases, the drug design process requires the scientist to identify and characterize chemical groups that are likely to participate in physical-chemical interactions. The groups may be part of the drug molecule, or of the amino acids building the binding site of the target protein. The interactions considered for this task include all noncovalent forces known from natural systems, and in some cases even covalent bonds. The former, which form the bulk of interactions observed in binding sites, can be separated into six general types (see Chapter 1 for details), which typically involve specific chemical groups, as follows:

1. **Hydrogen bonds** – alcohols, thiols, amines, carbonyls, organic acids, some heterocyclic groups

2. **Salt bridges** – carboxylate, phenolate, phosphate, sulfate

3. **Nonpolar interactions** – aliphatic and aromatic compounds

4. π **interactions** – aromatic compounds, double bond-containing compounds, metals and other positively charged species

5. **n→ $\pi*$ interactions** – groups containing oxygen, nitrogen, and sulfur

6. **X-bonds** – halogens

In this context, electrostatic interactions are most important, as they determine not only the strength of the binding but also its specificity. To identify which groups might participate in electrostatic interactions, the scientist must relate to their polarity, hydrogen-bonding capability, and tendency to undergo protonation or deprotonation. As drug molecules can be constructed from numerous combinations of chemical groups, this task is far from being trivial.

II. Further considerations: importance of chemical context

The functional groups mentioned above are very familiar to biologists, as they tend to appear in numerous biomolecules. Accordingly, their chemical properties are also known. However, in biomolecules these groups appear in a relatively limited chemical context. Artificial molecules, such as those serving as drugs, are not limited by evolutionary considerations, and may appear in many more chemical configurations. For scientists engaged in pharmacophore construction, this means greater diversity in physicochemical properties, which should be taken into account when considering the compatibility of these groups to specific protein-drug interactions. In the following paragraphs we review some examples.

II.I. Alcohol acidity

Aliphatic alcohols are known to be polar enough to participate in hydrogen bonds, but not enough to behave as acids, i.e., to undergo deprotonation. When the alcohol's hydroxyl group is attached to an aromatic ring, as in *phenol*, its pKa decreases to the extent that it can deprotonate under physiological pH, i.e., it becomes an acid (Figure 8.2.1a). This is because the phenyl group is an electron-withdrawing group, and as such it polarizes the hydroxyl group. The extent of this polarization can be modulated by attaching a third group to the phenyl ring; when the attached group is an electron-withdrawing group (e.g., nitro), the polarization of the hydroxyl is stronger (Figure 8.2.1b), and when it is an electron-releasing group (e.g., ethyl) the polarization is weaker (Figure 8.2.1c). The same is also true for the carboxylate group, which is acidic in its pure form, and its acidity can be altered by the attachment of different substituents.

(a) (b) (c)

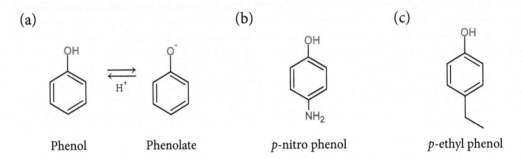

Phenol   Phenolate   *p*-nitro phenol   *p*-ethyl phenol

FIGURE 8.2.1 **Acidity of aromatic alcohols.** (a) The deprotonation of phenol. (b) Increased acidity of phenol by attachment of a *p*-nitro group. (c) Decreased acidity of phenol by attachment of a *p*-ethyl group.

II.II. Amino groups as acids and bases

Amino groups are known in biology to function mainly as weak bases. However, when they appear in particular chemical-structural contexts, they might display completely different behavior, as follows:

1. Whereas primary, secondary and tertiary amines behave as bases, quaternary amines do not. They are unable to do so since their lone electron pair is occupied in a covalent bond with one of the substituents (Figure 8.2.2a), and therefore cannot bind H$^+$ ions.

2. Aromatic amines (e.g., *aniline*) tend to behave as acids, because the aromatic ring acts as an '*electron sink*' that polarizes the amino group and enhances deprotonation (Figure 8.2.2b). Amino group deprotonation is even more favorable when the aromatic ring is bound on its other end to an electronegative group. This is the case with *sulfonamides*, a group of antibiotic drugs in which the aromatic amine is attached to a *sulfone group* (Figure 8.2.2c). Another type of compound in which the amino group is acidic is an *imide*. In this case the electrons of the group are drawn by a ketone group (or by several ketone groups) (Figure 8.2.2d).

3. *Pyrrole rings* contain a nitrogen atom, whose lone electron pair joins the other 4π electrons of the ring to render it aromatic (Figure 8.2.2e). This nitrogen cannot act as a base since the binding of H^+ would disrupt the aromaticity of the ring, a highly unfavorable process. A similar case is *amides*, in which the nitrogen's lone pair participates in *keto-enol tautomerization*, which makes it unavailable (Figure 8.2.2f). This group is therefore neither a base nor an acid.

4. In *amidines*, the NH group is basic, but much more so than an ordinary amino group. This is because protonation enables the highly favorable delocalization of the positive charge (Figure 8.2.2g).

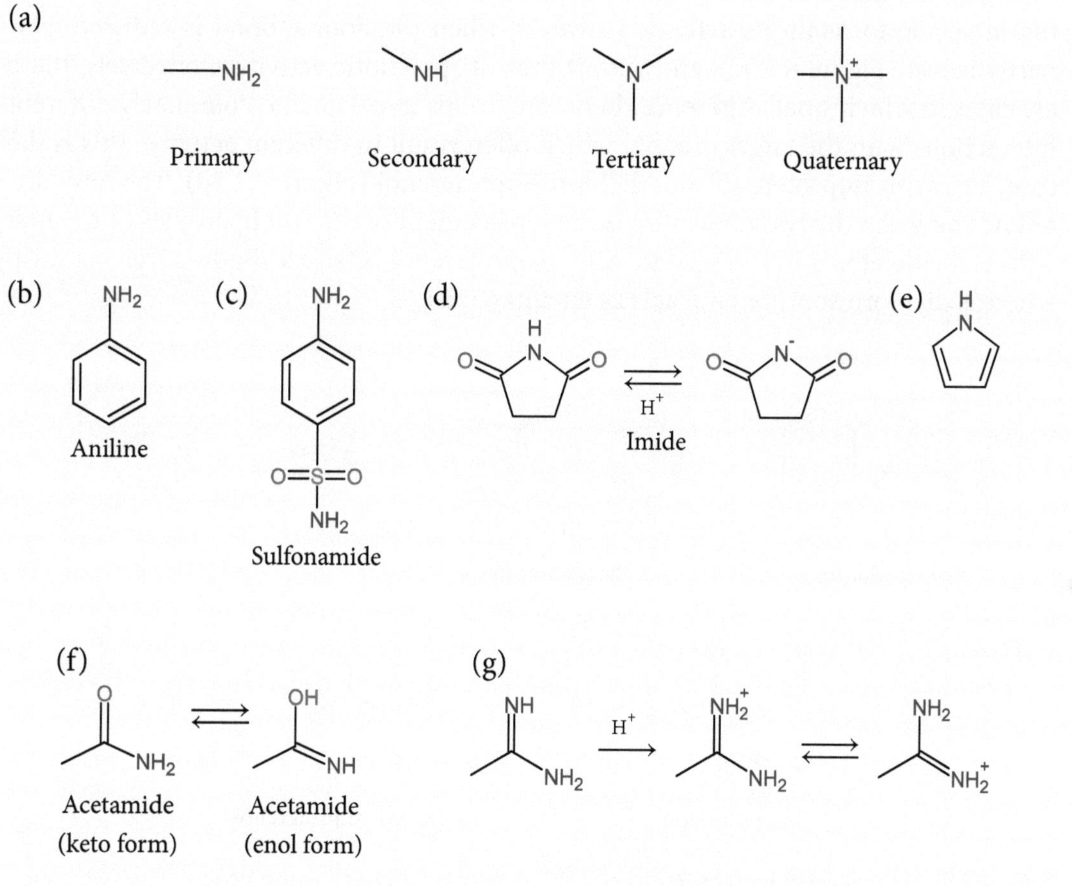

(a)

Primary   Secondary   Tertiary   Quaternary

(b) Aniline   (c) Sulfonamide   (d) Imide   (e)

(f) Acetamide (keto form) ⇌ Acetamide (enol form)   (g)

FIGURE 8.2.2 **Amines.** (a) Aliphatic amines. (b) Aniline. The aromatic ring changes the behavior of the amino group from base to acid. (c) Sulfonamide. The sulfonamide increases the acidity of the amino group even further. (d) An imide. The two keto groups act as electron-withdrawing groups, increasing the acidity of the amino group. (e) Pyrrole ring. The nitrogen's lone pair is occupied and therefore cannot bind further protons. (f) Keto-enol tautomerization in acetamide. (g) Charge delocalization in imidine.

III. Stereoelectronic properties of ring systems

There are additional factors that complicate the choice of chemical groups. For example, different groups may interact similarly with a given protein if their stereoelectronic properties are similar. This phenomenon is observed, for example, in the *phenothiazine* ring system, which is common among anti-psychotic drugs such as *chlorpromazine*. This group contains two phenyl rings flanking a heterocyclic ring with a sulfur atom (Figure 8.2.3a). The latter has four π electrons, a property that renders the middle ring aromatic, and therefore planar. The planarity of the phenothiazine rings seems to be important for the activity of the drug and must therefore be kept in any of its chemical derivates. In fact, even relatively large modifications may retain anti-psychotic activity, as long as the planarity is retained. For example when the sulfur atom is replaced by a double bond, the resulting dibenzazepine (Figure 8.2.3b) remains planar, which allows the molecule to retain its activity. However, when the double bond is reduced, planarity is lost (Figure 8.2.3c), and with it the anti-psychotic activity. Conversely, there are cases in which small differences between groups translate into completely different interactions with the target protein, which often result in different activity. This is the case, e.g., with isoproterenol and dichloroisoproterenol (Figure 8.2.3d). The only difference between the two molecules is the replacement of the two hydroxyls of the first with chloride atoms in the second. Still, isoproterenol acts as a β-adrenergic agonist, whereas dichloroisoproterenol acts as an antagonist.

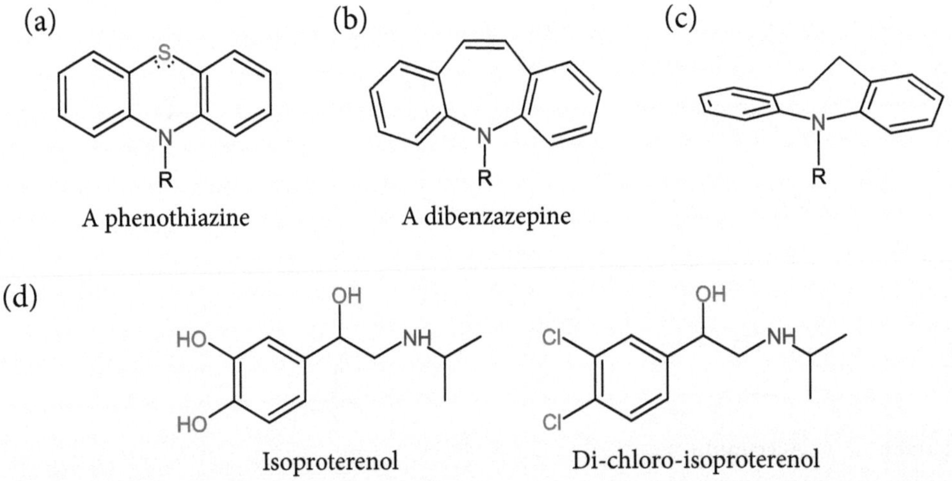

A phenothiazine · A dibenzazepine

Isoproterenol · Di-chloro-isoproterenol

FIGURE 8.2.3 **Stereoelectronic effects.** (a) A planar phenothiazine group. The two lone pairs of the sulfur atom are noted. (b) A planar dibenzazepine group, produced by replacement of the sulfur of phenothiazine with a double bond. (c) Loss of planarity upon reduction of the double bond of dibenzazepine. (d) Changing an agonist (isoproterenol) into an antagonist (dichloroisoproterenol) by replacement of two hydroxyls with chlorides.

IV. Stereoisomerism

Biomolecules are synthesized by stereospecific machineries (the ribosome, enzymes), and therefore a given biomolecule is likely to appear almost exclusively as a single isomer (*S* or *R*). In contrast, drug molecules are usually synthesized chemically in a process that yields the two isomers, often in identical ratios (*racemic mixture*). Unfortunately, the targets of these drug molecules are also stereospecific, which means that only one isomer of the drug must be used. In many cases the other isomer does not even bind to the target protein, whereas in others it may bind but exert a weaker or different effect. For example, *levorphanol* and *dextromethorphan* are two isomers of the same molecule (Figure 8.2.4). Both have an antitussive (cough-suppressing) effect, but whereas the former is also a highly addictive analgesic, the latter is not.

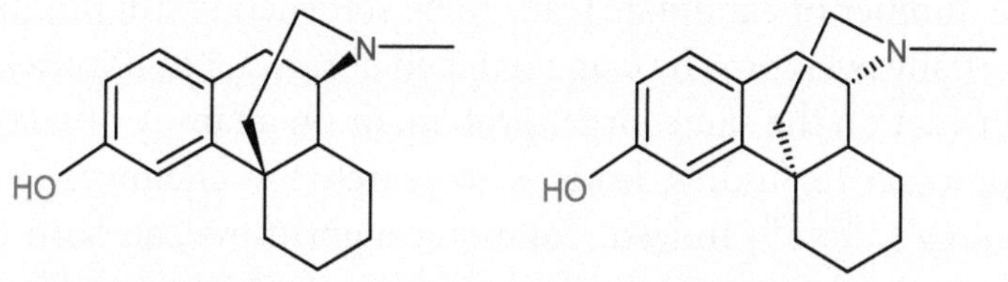

Levorphanol   Dextromethorphan

FIGURE 8.2.4 **Importance of using the right enantiomer.** Levorphanol and dextromethorphan differ only by absolute configuration, yet differ in their physiological effects.

8.6.3.3.3 Assigning secondary target constraints

Secondary constraints refer to properties of the drug that are usually of clinical importance (e.g., ADMET properties). Such constraints are usually established towards the end of the lead optimization process (see below). The constraints may be formulated separately, or as weights assigned to the overall score of the evaluated molecule [283]. As a rule, computational methods do less well in predicting the molecular characteristics that correspond to secondary constraints, as compared with predicting the characteristics corresponding to primary constraints. Thus, validation by biological assays is especially important. Still, some secondary properties are easier to predict than others. For example, the oral bioavailability of a drug has been found to correlate with Lipinski's '*rule of five*'[*1] [302]. Moreover, certain chemical substructures are known to reduce the stability of drug molecules.

8.6.3.3.4 Choosing and designing a lead compound

 After the constraints relating to a drug are assigned, the missing parts, i.e., the real chemical groups, can be added to generate a complete molecule. Since the constraints are non-specific, many molecules are likely to be able to fulfill these requirements. The scientist must try to generate and explore as many of these molecules as possible, and assess their compatibility as leads. The first task, i.e., the synthesis of the different molecular derivates, can be done by using methods of combinatorial chemistry [303]. That is, an automated system is used to synthesize the numerous candidate lead molecules by creating different combina-

---

[*1]A rule of thumb stating that orally active drugs should not violate more than one of the following criteria:
1. ≤ 5 hydrogen bond donors
2. ≤ 10 hydrogen bond acceptors
3. Molecular mass ≤ 500 Da
4. log P < 5
(Note that all the above numbers are multiples of five)

tions of the same (or similar) molecular fragments. The second task, i.e., to assay the vast number of molecules produced by this procedure ('*screening*'), is much more difficult to accomplish. Not only are there numerous molecules to be tested, but also several tests are often required for each molecule, such as measuring the binding to the target protein, quantifying the biochemical effect of the candidate drug, etc. In the last two decades, automation approaches have enabled high-throughput methods to be developed for screening chemical libraries [304–306]. These methods have accelerated the testing stage considerably, and they are commonly used in the pharmaceutical industry, with proven success (e.g., [307]). However, in many cases, the number of candidate leads to be screened is still prohibitively high. This problem can be partially overcome by using reduced libraries, i.e., libraries that include drug molecules known to act on the same target protein, or on a family of related targets [308].

A different approach to finding leads is to search the chemical space computationally ('*virtual screening*') [309,310]. Indeed, computer algorithms can scan huge databases of molecules, fragments, and even atoms, in search of the right combination of chemical groups that can translate a set of previously assigned constraints into a real, viable lead[*1]. This field arose in the 1970s [312,313] and has gained much popularity in the last decade. The advantage of this approach is obvious; virtual scanning of the chemical space is much faster than actual scanning, which means that a larger part of the space can be covered by the search (up to millions of molecules). The search for the best chemical match to a given set of constraints is usually carried out at the level of molecular fragments [283]. That is, individual fragments are combinatorically integrated into the molecular scaffold, and the binding affinity of the resulting molecule to the target protein is assessed using a scoring function[*2]. In contrast to the methods described in Subsection 8.3.1.2 above, which use rigorous, explicit calculations in order to obtain an accurate value of the protein-ligand binding energy, the goal in virtual screening is to rank the relative affinity of each of the sampled molecules to the target protein. Achieving such a ranking does not require very accurate calculations, but rather fast ones, which are able to rank as many molecules as possible in a reasonable amount of time. As a result, scoring functions for lead discovery tend to contain simplified terms, which are in many cases knowledge-based, rather than expressions describing real physical interactions.

This is not to say that the calculations are easy; they still have to consider two factors that affect the binding energy. The first is the conformational flexibility in the system (both the protein's and the lead's), and the second is the different orientations the lead may acquire with respect to the protein's binding site. Both of these considerations make the screening process computationally costly. The sampling of conformations and orientations is often

---

[*1]Current computational resources enable libraries containing $\sim 10^7$ molecules to be screened over several days or weeks [310]. There are several popular libraries that are used by both academic and industrial researchers. Many of these libraries contain molecules that have drug-like features, and that are commercially available. For example, the ZINC library [311] (http://zinc15.docking.org) contains over 10^8 such molecules and molecular fragments.

[*2]The following animation shows an example for screening of 17 small drug-like molecules (presented as grey balls-and-sticks) for binding to a protein binding site.

 The binding site is between two subunits, shown as cyan and dark green helices. The residues interacting with the drug-like molecules are shown as lines. Each drug-like molecule forms unique interactions with the protein, which contribute to the total binding energy. In the animation, hydrogen bonds and π-interactions, which contribute to the binding specificity, are shown as black and orange dashed lines, respectively. Created by Elon Yariv.

carried out by a computational procedure called *'molecular docking'*[*1] [310,314,315]. To avoid *combinatorial explosion*, many docking algorithms only change the orientation and/or conformation of the candidate lead and do not change those of the target protein (an approach called *'rigid docking'*). Thanks to advances in artificial intelligence and machine-learning technologies, the chemical-conformational space that needs to be scanned has been considerably reduced [73]. For example, the sampling techniques, which used to be systematic, may be conducted by using *stochastic patterns*. Stochastic patterns enable screening algorithms to identify regions in the chemical-conformational space that are more likely than others to contain the desirable solution. Thus, the search can focus on these regions alone instead of considering the entire space. Sampling the conformational space of both the protein and the lead is called *'flexible docking'*. Despite the advances in sampling methods mentioned above, the number of conformations that must be sampled to achieve 'full' flexible docking is too large to be computationally feasible. Thus, flexibility is often conferred only to a subset of protein residues, typically those that participate in ligand binding [83]. Since the selection of such residues is carried out by the scientist, it is a potential source for errors.

Even when conformational sampling of the protein is restricted to the binding site, docking can still be computationally demanding, especially when large numbers of potential drug molecules are screened. Indeed, in a typical docking procedure, 10^2 to 10^4 orientations are sampled for each ligand tested, and for each orientation, 10^2 conformations are sampled [310]. Thus, if a library of 10^7 ligands is used, the number of configurations sampled by the docking procedure can reach 10^{13}! One way to sample conformations in the binding site is to use Monte Carlo simulations, which are less computationally costly than MD simulations (see Chapter 3 for details). An alternative approach is to represent the continuous conformational space of the protein's binding site using an ensemble of discrete conformations. Such an ensemble can be predicted using methods such as normal mode analysis (see Chapter 3), as is done by the MRC web server [316][*2]. After the conformations are generated, each can be used in simple rigid docking simulations (*'cross-docking'*) to find the most favorable binding mode.

Finally, the scoring functions themselves are potential sources for problems. As explained above, they typically rely on either energy calculations or statistical tendencies of atoms or groups to appear in certain molecular contexts. Despite recent improvements, scoring functions are still far from reproducing binding affinities accurately, which makes their use in lead generation problematic. In particular, they are usually unable to reliably rank high-scoring ligands [310,314]. However, all things considered, computational tools with all their inherent problems make the search for leads and drug-related molecules much more efficient (see [283,290,310] for examples), and are often combined with lab tools to achieve the best results.

[*1]The following animation shows a 'blind' docking simulation, in which the ligand (spheres colored by atom types) is positioned at numerous different orientations with respect to the protein (white surface), and in each orientation the protein-ligand binding energy is calculated to find the most probable binding mode. The simulation sampled different conformations of both protein and ligand. As explained in the main text, most docking simulations sample only the ligand's conformations.

[*2]http://wwww-ablab.ucsd.edu/MRC/index.cgi

8.6.3.3.5 Lead optimization

The hit rate of virtual screening via docking is around 10%. That is, about one-tenth of the ligands that the screening process identifies as prospective leads actually test as such. Moreover, these ligands need to be further processed. Whereas most drugs display nM binding affinities to their target proteins, the first round of *de novo* design tends to yield leads with µM affinities [283]. This means that further binding optimization must be carried out to reach the desired drug. Consequently, the binding energy between the protein and the lead must be calculated using methods that are more accurate than those used in docking. As explained in Subsection 8.3.1.2 above, several such methods are available [58–60]. Alchemical (free energy perturbation) methods are probably the most accurate [61,63,64]; however, they require extensive sampling of various non-physical states separating the ligand-bound form of the protein from the ligand-free form, and therefore tend to be too computationally demanding for lead optimization. This problem can be partially alleviated by using a less rigorous protocol for calculating the *relative* binding energy, in which one ligand is artificially transformed into another [61]. The popular alternatives to alchemical methods are endpoint methods, such as LIE [67,68] and MM/PB(GB)SA [69–72], which focus only the bound and unbound states of the protein, and are therefore less computationally costly[*1]. These methods are generally less accurate than alchemical methods, yet more accurate than docking simulations. Other features of the leads that may require optimization are their physicochemical, pharmaceutical, ADMET, and pharmacokinetic properties [290]. For example, molecules suggested in the first round of screening may turn out to be toxic, difficult to synthesize, or to have non-drug-like properties (too large, too lipophilic, etc.) [308]. To avoid such features, and particularly toxicity, scientists typically use QSAR [290], that is, pre-existing knowledge pertaining to the relation of certain chemical groups with unwanted properties. When the three-dimensional structure of the target protein is known, a receptor-based optimization process may also help to address some of the clinical aspects. For example, by increasing the compatibility between the lead and the target's binding site, the lead's selectivity can be increased, which means reduced toxicity. A fine example of such an approach is the design of a novel inhibitor drug for the enzyme *adenosine deaminase*, which is significantly less toxic than its predecessors [317].

8.6.3.4 Rational drug design case study: ACE inhibitors

8.6.3.4.1 Introduction

Rational drug design, especially the structure-guided type, is a relatively new field, and there have already been a few successful cases. Two of the most well-known successes are as follows:

1. **COX-2 inhibitors** [318] – NSAIDs may act on the two types of cyclooxygenases in the human body. The first type, COX-1, can be found throughout the body. In the stomach, the activity of COX-1 protects the mucosa cells lining the stomach. This is why COX-1 inhibition by NSAIDs may cause gastric problems, such as bleeding and ulcers. The second type of COX enzyme, COX-2, is found almost exclusively in inflammatory cells, like macrophages. Rational drug design has given rise to the development of COX-2-selective NSAIDs, which can be used for fighting inflammation without hurting the stomach. Examples include *celecoxib* (*Celebrex®*).

[*1]This is particularly true for MM/PB(GB)SA implementations that do not use sampling of the endpoint states, but rather start from energy-minimized structures.

2. **Peripheral H$_1$ receptor antagonists** [319] – Histamine receptors ('H receptors') are involved in physiological processes but also in allergic reactions. There are several types of H receptors; some reside in the brain, whereas others can be found in the peripheral nervous system. Anti-histamines are drugs designed to block H receptors. They are widely used to treat people suffering from allergy. Old-generation anti-histamines like *diphenhydramine* (*Benadryl®*) can cross the *blood-brain barrier* (*BBB*) that protects the brain, and block central H receptors. Unfortunately, they also tend to block *muscarinic* (acetylcholine) receptors, which leads to '*anti-cholinergic side effects*', such as sedation, sleepiness, dizziness, and dry mouth. Through rational drug design, peripheral H$_1$ receptor anti-histamines have been developed; these drugs are specific to one subtype of histamine receptor and are also unable to cross the BBB, and therefore do not have central anti-cholinergic side effects. Examples include *loratadine* (*Claritin®*).

Another successful drug design endeavor was the development of inhibitors for *angiotensin-converting enzyme* (*ACE*), as a treatment of hypertension. In fact, the development of first-generation ACE inhibitors is considered to be one of the first true rational drug design efforts [320]. Below we describe the main steps of this development process, and also discuss the main rationale behind new-generation inhibitors. Before doing so, we briefly describe the physiological system within which ACE operates.

8.6.3.4.2 Hypertension and renin-angiotensin-aldosterone system

8.6.3.4.2.1. Overview

Chronic hypertension is one of the most common medical conditions in the developed world [321], affecting ~26% of the world's adult population [322]. It is potentially life-threatening, due to the resulting cardiovascular complications, such as kidney failure, heart failure, and stroke [320]. Accordingly, a variety of pharmaceutical drugs have been developed for treating hypertension, including diuretics, β-blockers, ACE inhibitors, calcium channel antagonists, angiotensin receptor blockers, and α-blockers [323]. Three of these drug types act on the *renin-angiotensin-aldosterone (RAA) system*, which is responsible for regulating blood pressure and water-electrolyte homeostasis in the mammalian body [324]. The system is complex and includes numerous different hormones and enzymes. We will therefore avoid getting into the full details of the system, and instead discuss its general principles (Figure 8.19).

When blood pressure drops, an occurrence that might result from dehydration, allergic reaction, or a hemorrhaging injury, the body responds immediately to raise blood pressure back to normal values. This response is part of the body's cardiovascular homeostasis, but it is also designed to prevent a life-threatening situation, i.e., a significant drop in blood pressure. Such a drop makes the heart work harder and faster, and if it is not treated in time, might even lead to cardiac arrest and death. The body's response to the decrease in blood pressure begins with the release of the enzyme *renin* from the kidneys into circulation. Renin is a protease that degrades *angiotensinogen*, a liver-derived 55-kDa plasma protein. One of the degradation products is the 10-amino acid peptide *angiotensin I* (Figure 8.20), which has no significant physiological effect. However, angiotensin I can be degraded by ACE, and the remaining 8-amino acid peptide (*angiotensin II*) is a potent *vasoconstrictor*. That is, angiotensin II binds to its cognate receptor on the smooth muscles of arterioles, which makes the muscles contract and the arterioles constrict. This constriction raises blood pressure back to its normal values, and prevents the aforementioned symptoms.

Angiotensin II also acts indirectly to increase blood pressure, by inducing the release into circulation of two other blood pressure-regulating hormones. These are the steroid hormone *aldosterone*, released from the adrenal cortex, and the peptide hormone *vasopressin (anti-diuretic hormone, or ADH)*, released from the pituitary gland. Aldosterone acts on the distal tubules and collecting ducts of the kidneys. There, it increases the reabsorption of Na^+ into the blood and the secretion of K^+ into urine. Since water follows the Na^+ ions, the increased reabsorption of the latter also means pumping water back into the blood. This has two results. First, water loss is diminished, which also helps in cases of dehydration. Second, the resulting increase in blood volume leads to an increase in blood pressure. Like aldosterone, vasopressin acts on the kidney to preserve water and increase blood pressure. However, unlike aldosterone, it acts directly, by increasing the water permeability of the distal tubules and the collecting ducts. Finally, ACE too may raise blood pressure indirectly, by degrading the 9-amino acid peptide *bradykinin* (RPPGFSPFR), which is a known vasodilator.

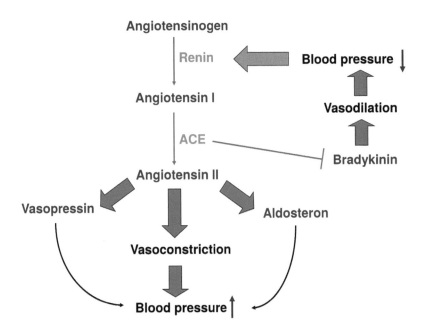

FIGURE 8.19 **A schematic description of the renin-angiotensin-aldosterone (RAA) system.** Proteins, peptides, or small molecules involved in the system are in blue, and enzymes are in green. Biochemical reactions are presented as red arrows. The increase or decrease of blood pressure is marked by the short purple arrows. Wide filled arrows lead from a condition to its effect, or mark hormonal induction (except for the vasopressin or aldosterone effects on blood pressure, which are marked by the curved black arrows). Finally, the red shape represents the degradation of bradykinin by angiotensin-converting enzyme (ACE). The RAA system includes a main biochemical path leading from the large protein angiotensinogen to the vasoconstrictor angiotensin II. The path is facilitated by two enzymes, renin and ACE. Angiotensin II reduces blood pressure directly, by inducing vasoconstriction, or indirectly, by inducing the release of the hormones vasopressin and aldosterone. ACE contributes to the decrease of blood pressure by creating angiotensin II, and also by degrading bradykinin, a peptide with vasodilating activity.

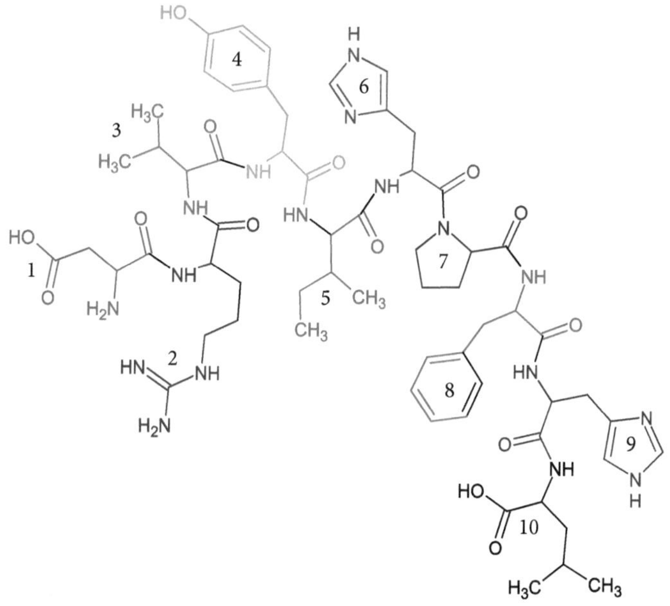

$^+H_3N-Asp-1-Arg-2-Val-3-Tyr-4-Ile-5-His-6-Pro-7-Phe-8-His-9-Leu-10-COO^-$

FIGURE 8.20 **Angiotensin I.** The structure (top) and sequence (bottom) of the 10-amino acid peptide are shown. Each of the residues in the structure is colored differently and numbered. The last two residues that are cleaved off the peptide by ACE are colored pink in the sequence.

8.6.3.4.2.2. Angiotensin-converting enzyme (ACE)

ACE, also known as *peptidyl dipeptidase A* (EC 3.4.15.1), is a heavily glycosylated, zinc-dependent carboxypeptidase. It resides in the plasma membrane of endothelial, intestinal brush border, and renal proximal tubule cells [325–327]. The exoplasmic region of endothelial ACE (a.k.a. 'somatic ACE'; sACE) faces blood, which allows the enzyme to act on the correct peptide substrates in circulation. *In vitro*, ACE may act on a variety of peptides, catalyzing the hydrolytic cleavage of their two C-terminal amino acids. It has low specificity: The penultimate peptide position can be occupied by any residue except Pro, and the ultimate position can be any residue except Asp or Glu. *In vivo*, sACE acts on several bioactive peptides, including *substance P, luteinizing hormone-releasing hormone (LHRH)*, and *neurotensin* [328]. **However, as ACE is a regulator of cardiovascular homeostasis, its primary substrates are angiotensin I and bradykinin** (see below).

Human ACE is one of the rare examples of a single polypeptide chain containing two homologous active sites. Indeed, ACE has two 55% identical catalytic domains (called *N* and *C*), which are thought to be the result of gene duplication [325,329]. Each active site contains a zinc cation (Zn^{2+}) bound by the *HEMGH* sequence, a known motif of zinc-dependent proteases. The two domains differ in some aspects. For example, each can act on a range of peptide sequences, but they do so with different efficiency. The *C*-domain is more active *in vivo* on angiotensin I than the *N*-domain [330–332], but is also more Cl^--dependent [329].

In addition to sACE, other forms of the enzyme can also be found:

1. Soluble ACE – is formed by the proteolytic cleavage of sACE's exoplasmic region. Although this form of the enzyme is active, its role in cardiovascular homeostasis is minor compared with that of membrane-bound ACE.

(a)

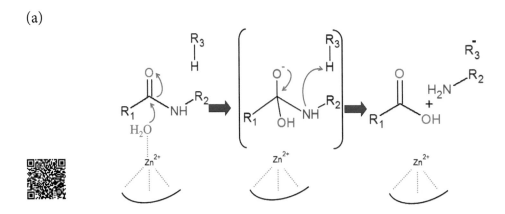

(b)

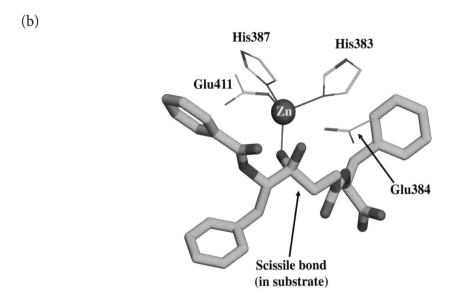

His387

His383

Glu411

Zn

Glu384

**Scissile bond
(in substrate)**

(c)

Zn

S_1'

P_1'

NH

P_2

P_1

N

S_2

HO

O

CH$_3$

P_2'

CH$_3$

S_2'

S_1

2. tACE – an alternate transcription product of ACE, found in the adult testis. It is identical to the C' domain of somatic ACE, except for the first 36 residues [333].

3. ACE2 (a.k.a. ACEH) – an ACE homologue found in the heart, kidney and testis of humans and rodents. It is active towards peptides with C-terminal hydrophobic or basic residues, but like ACE it acts on angiotensin I and regulates blood pressure and cardiac function [320]. Interestingly, it is unaffected by the classical ACE inhibitors (see below).

The mechanism of ACE, as in other zinc-dependent metalloproteases, is thought to include the following steps [335,336] (Figure 8.21a):

1. Polarization of a water molecule by the zinc cation, followed by its deprotonation. The resulting hydroxide anion is stabilized by the positive charge on the zinc.

2. Nucleophilic attack of the scissile carbonyl carbon by the hydroxide anion, followed by the formation of a tetrahedral transition state (Figure 8.21b).

3. Protonation of the scissile amide nitrogen, followed by peptide bond cleavage.

Although the exact involvement of specific residues in the mechanism is not entirely known, some are implicated in the catalytic process. For example, Glu-384 has been proposed to shuttle a proton between the water nucleophile and the leaving amide.

The orientation of the peptide with respect to the active site of ACE is usually described in relation to the zinc cation. Thus, the first and second residues that are N-terminal to the scissile bond's carbonyl group that binds to the zinc are termed P_1 and P_2, respectively (Figure 8.21c). The first and second residues that are C-terminal to the carbonyl are called P'_1 and P'_2, respectively. The subsites in ACE's active site that interact with these residues are termed, respectively, S_1, S_2, and S'_1, S'_2.

FIGURE 8.21 **The mechanism of ACE.** (Opposite) (a) A schematic representation of the ACE mechanism of action. Red arrows mark electronic rearrangements, green dotted lines represent coordinate bonds, and the curved line below the zinc cation represents the enzyme binding pocket in which the ion resides. The catalyzed reaction includes zinc-induced polarization and deprotonation of a water molecule, which then acts as a nucleophile. A tetrahedral transition state is formed (in parentheses), followed by protonation of the scissile amide nitrogen and peptide bond cleavage. The process is depicted in general, i.e., no specific ACE residues are noted. (b) A likely structure and orientation of the substrate's transition state in the active site of ACE. The structure shown is that of human testicular ACE (tACE) in complex with the inhibitor kAF (PDB entry 1bkk) [334]. kAF (shown as sticks) has a ketone group where the substrate's carbonyl is located. It is therefore able to undergo the first part of the catalyzed reaction inside ACE's active site, i.e., the nucleophilic attack. As a result, it transitions from its ground state into a tetrahedral transition state, representing that of the real substrate. However, the scissile amide nitrogen of the substrate is replaced in kAF with a carbon, so the bond cannot be cleaved. This property enabled Watermeyer and coworkers [334] to crystallize both enzyme and 'substrate'. The structure demonstrates how the *gem*-diol group of the transition state, in which the two oxygen atoms have a partial negative charge, chelates the zinc cation. The latter is also coordinated to three other residues, which are noted in the figure. (c) The four positions in the substrate and inhibitor (P_2, P_1, P'_1 and P'_2), noted in relation to the zinc-coordinated ketone group, and their corresponding binding subsites in the active site of ACE (S_2, S_1, S'_1 and S'_2). The scissors symbol marks the ester bond (in the substrate) cleaved by the enzyme.

(a) (b)

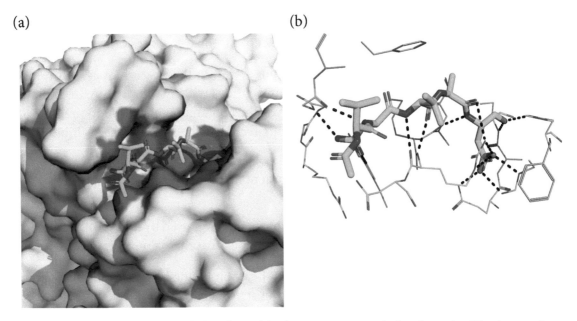

FIGURE 8.22 **Protein-peptide binding.** (a) The protein-peptide binding site. The image shows *caspase-2* in complex with the peptide *Acetyl-Val-Asp-Val-Ala-Asp-CHO* (PDB entry 3r6l). The surface of the protein is shown in yellow with ligand-interacting residues colored in red. The peptide is shown as sticks, colored according to atom type. (b) Protein-peptide interactions. The same binding site is shown, with the peptide-interacting residues presented as lines. Polar interactions between the enzyme and the peptide are shown as dashed lines. Hydrophobic residues involved in nonpolar interactions are colored orange. As demonstrated by the image, the high number of interacting groups in peptides results in many stabilizing interactions. Therefore, small molecules are usually poor competitive inhibitors when the target is a protein-peptide interface.

8.6.3.4.3 Design of ACE inhibitors [320]

8.6.3.4.3.1. First-generation inhibitors

The starting point for the rational design of the first ACE inhibitors by Cushman and Ondetti was the early discovery of a group of ACE-inhibiting peptides in the venom of the South American pit viper *Bothrops jararaca*. The inhibitory effects of the peptides were compared, and the tripeptide *Phe-Ala-Pro* was found to have the optimal effect (Figure 8.23a) [337]. **Peptides are advantageous as drugs that disrupt protein-protein or protein-peptide interactions, as they can mimic these interactions much better than small molecule drugs** [338]*1. **However, since peptides are digested in the stomach by proteases, they usually cannot be administered orally, which significantly reduces their 'druggability' potential.** Thus, the researchers in the ACE study turned to the design of non-peptide inhibitors with similar properties to those of the *Phe-Ala-Pro* peptide.

Assuming that the *Phe-Ala-Pro* peptide acted by competitive inhibition, the researchers looked for a feature of the binding site that could be efficiently targeted by the new inhibitor. The feature selected for this purpose was the metal zinc on which the activity of

---

*1 Peptides are elongated molecules with functional groups extending along an axis. As a result, they are perfect for interacting with protein-protein or protein-peptide interfaces, which are usually flat and elongated [121,122] (Figure 8.22; see also Figure 8.13a,b). Small molecules have functional groups concentrated in a small region and extending to different directions. They are therefore more suitable for interacting with deep, small cavities in the protein (see Figures 8.5a and 8.6).

ACE depended. The selection of zinc was inspired by studies carried out on carboxypeptidase A (CPA), another zinc-dependent enzyme, for which a three-dimensional structure already existed in the databases. Earlier studies of CPA had found that the enzyme was inhibited by *benzyl-succinic acid* [339] (Figure 8.23b). Cushman and Ondetti predicted that the inhibition had at least something to do with the coordination of the zinc by the carboxyl group of the inhibitor. Since the researchers believed that CPA and ACE shared the same mechanism, they looked for a succinyl-amino acid derivative that would do the same in the active site of ACE. On the basis of the inhibitory effect of the *Phe-Ala-Pro* peptide, they constructed the succinic acid derivative *methyl-succinyl-Pro* (Figure 8.23c left), which is analogous to carboxy-Ala-Pro. Indeed, the inhibitor was found to act specifically on ACE, with $IC_{50} = 22\,\mu M$[*1]. To improve the inhibition even further, the researchers replaced the carboxyl group with a thiol, a better metal cation chelator (Figure 8.23c right). The result was *captopril*, whose IC_{50} value was ~1,000-fold lower than that of its carboxyl counterpart [340,341], and which became the first anti-hypertensive ACE-acting drug. When the structure of ACE bound to captopril was solved in 2004 [342], it confirmed the vision of Cushman and Ondetti; the inhibitor was placed inside the active site with its thiol group coordinating the zinc ion, the methyl group occupying the S_1' subsite, and the prolyl group occupying the S_2' subsite (Figure 8.23d).

While captopril was efficient in inhibiting ACE, it had some unpleasant side effects (e.g., rash), which prompted Patchett and coworkers to look for a non-thiol inhibitor. Since non-thiol groups yielded weaker zinc binding, the researchers decided to target additional inhibitor-binding groups in the enzyme's active site. These included the S_1 pocket (P_1) and the groups forming hydrogen bonds with the amide nitrogen of the substrate's scissile bond. To this end, the researchers used different variations of *N-carboxy-alkyl dipeptides* [343] (Figure 8.24a), based on their previously characterized inhibition of *thermolysin*. Two variations were found to have better activity than the others. In both, a benzyl-methylene group was attached to the *N*-carboxyl group, at the location intended for the P_1 site in ACE. The difference between the two inhibitors resulted from the identity of the dipeptide group; in the first inhibitor, *enalaprilat*, the dipeptide was Ala-Pro (Figure 8.24b), whereas in the second inhibitor, *lisinopril*, it was Lys-Pro (Figure 8.24c). The activity of enalaprilat was not surprising, as it was very similar to the Phe-Ala-Pro tripeptide, which was already known to be an efficient ACE inhibitor (see above). However, the efficiency of lisinopril was unexpected, as the side chain of Lys is very different from that of Ala. The structure of the ACE-lisinopril complex, published in 2003 [344], confirmed that the long pentylamine side chain of lisinopril fits snugly into the deep S_1 subsite (Figure 8.24d), where it is stabilized electrostatically by Glu-162 (Figure 8.24e). This probably accounts for the greater affinity of ACE to lisinopril over enalaprilat [345]. In addition, lisinopril forms polar interactions via its amide nitrogen and two carboxylate groups, and uses the carboxy-alkyl group to coordinate the Zn^{2+} cation (Figure 8.24e).

8.6.3.4.3.2. C-domain selective inhibitors

The three molecules described above served as a basis for the design of others. Together, they constituted the first generation of ACE inhibitors. **The emphasis in the design of these inhibitors was conferring strong zinc-binding abilities, while trying to minimize side ef-**

[*1]The half maximal inhibitory concentration, which is a common measure of an inhibitor's effectiveness.

(a)

(b)

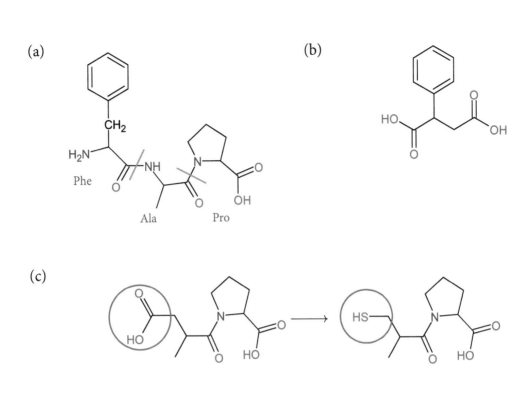

(c)

(d)

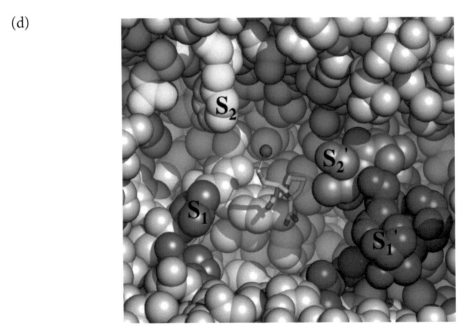

FIGURE 8.23 **Steps in the development of the first ACE inhibitor captopril.** (a) The *Phe-Ala-Pro* peptide from the venom of the South American pit viper *Bothrops jararaca*. The boundaries between amino acids are marked by the green lines. (b) Benzyl-succinic acid, the carboxypeptidase A by-product inhibitor. (c) Methyl-succinyl-Pro (left), and thiol derivative captopril (right). The two inhibitors differ only in the single encircled substitution. (d) The structure and orientation of captopril inside the active site of human tACE (PDB entry 1uzf). Captopril is shown as sticks, colored according to atom type, and the enzyme is shown as grey spheres. The S_2, S_1, S'_1 and S'_2 subsites in ACE are colored differently and noted. The coordination of the zinc cation (pink sphere) by the thiol group of captopril is clearly seen. For clarity, some atoms of the enzyme have been omitted.

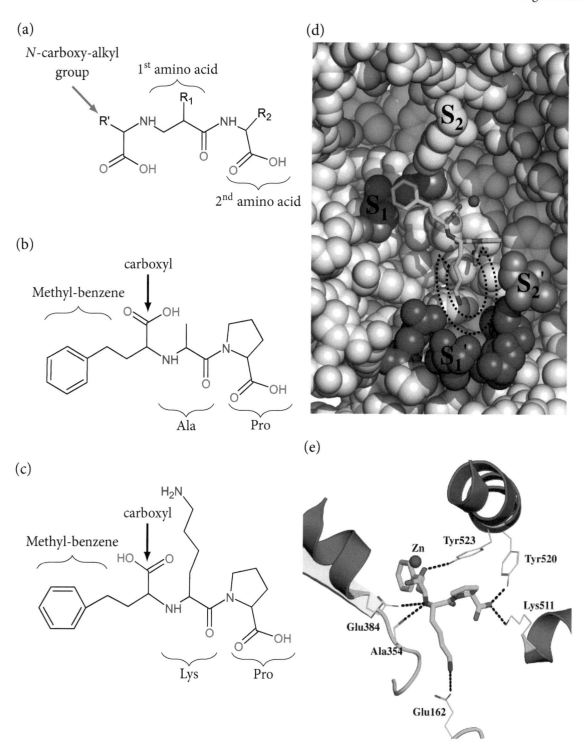

FIGURE 8.24 **Steps in the development of the ACE inhibitor lisinopril.** (a) The *N*-carboxy-alkyl dipeptide scaffold used for lisinopril. R′ denotes the chemical group substituted for the *N*-carboxy group, whereas R$_1$ and R$_2$ denote the side chains of the first and second amino acids, respectively. (b) Enalaprilat and (c) lisinopril, both designed using the *N*-carboxy-alkyl dipeptide scaffold. In both, a methylbenzene group substitutes for the *N*-carboxy group, but they differ in the identity of the first amino acid (alanine in enalaprilat and lysine in lisinopril). (d) The structure and orientation of lisinopril inside the active site of human tACE (PDB entry 1o86). Lisinopril and the enzyme are presented as in Figure 8.23d. For clarity, some atoms of the enzyme are omitted. The general shape of the S′$_1$ pocket is delineated by the dashed lines. (e) Some of the polar contacts formed between lisinopril and ACE residues. Lisinopril is shown as sticks, and the ACE residues are shown as lines. The polar contacts are shown as dashed black lines.

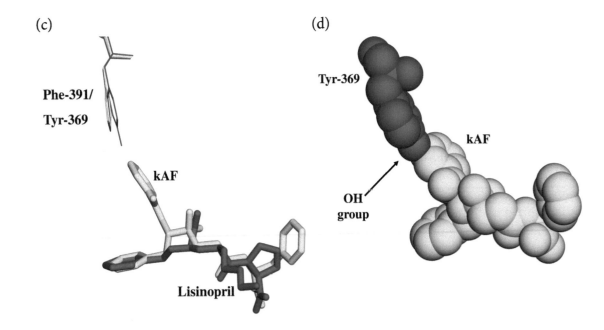

fects. This led to the construction of inhibitors in which the zinc binding group included different chemical species, such as carboxylate [343], ketone [346], phosphinic acid [347], hydroxamic acid [348] and silanediol [349]. Structure-activity relationship studies of the numerous inhibitors that were designed also yielded insights regarding the positions flanking the zinc-binding group [328]. For example, the P_2' position was found to tolerate a wide range of substituents, but to 'prefer' aromatic groups or other ring structures, which increase potency. In addition, nonpolar groups in the P_1 position were shown to increase activity via participation in nonpolar interactions with ACE active site residues.

At some point it was realized that some of the side effects, such as persistent cough and *angioedema*, which accompanied the use of first-generation inhibitors, resulted from the build-up of bradykinin [350,351]. Since first-generation inhibitors acted on both of ACE's domains, and since both were capable of cleaving bradykinin, it was suggested to reduce the side effects by inhibiting only the *C*-domain [352,353], that is, to design domain-selective inhibitors. *C*-domain inhibition would reduce blood pressure, whereas the still active *N*-domain would be free to keep degrading excess bradykinin. Some of the first-generation inhibitors already exhibited *C*-domain-selective inhibition, mainly due to the use of zinc-binding groups that were weaker than the thiol group of captopril. Indeed, there seemed to be a trade-off between *C*-domain selectivity and zinc binding affinity (see possible explanation below). Still, the degree of selectivity obtainable by changing the zinc-binding group was rather limited, and efforts were made to increase it by addressing other inhibitor properties. In the following paragraphs we provide a short discussion of the structure-function relationship in *ketone-based inhibitors*, which were used as part of this effort.

Keto-ACE (a.k.a. *kAP*) is a ketomethylene analogue of the ACE inhibitor benzyl-Phe-Gly-Pro (Figure 8.25a). The ketomethylene group has already been used extensively in the inhibition of other proteases [354]. kAP has been known since 1980 to inhibit ACE, and it displays modest (26 to 34-fold) *C*-domain selectivity [355,356]. Ketomethylene inhibitors dif-

FIGURE 8.25 **Ketomethylene inhibitors.** (Opposite) (a) The differences between ketomethylene inhibitors (represented by kAP) and the *N*-carboxy-alkyl dipeptides (represented by lisinopril). Both inhibitors include a dipeptide unit. However, the amide nitrogen of the *N*-carboxy-alkyl dipeptides is substituted by a carbon atom (blue rectangles), and the carboxyl group is substituted by a ketone group (red rectangles). In addition, the ketomethylene inhibitors also include an S_2 group, which is missing in the *N*-carboxy-alkyl dipeptides. (b) The three ketomethylene inhibitors used by Watermeyer and coworkers. The differences reside in the S_2' group (right), which includes Pro in kAP, Trp in kAW, and Phe in kAF. (c) and (d) Effect of the Phe-391→Tyr change in the *N*-domain of ACE on the accommodation of kAF in the active site. (c) The figure shows three ACE-inhibitor complexes: *C*-domain-lisinopril (PDB entry 1o86, red), *N*-domain-lisinopril (PDB entry 2c6n, blue), and *C*-domain-kAF (PDB entry 3bkk, yellow). The inhibitors of the latter two structures are shown, represented as sticks. The residues occupying position 391 (369 in the *N*-domain) in each of the three structures are in wireframe representation. For clarity, the other residues and the backbone of the structures are not shown. As can be clearly seen in the figure, the added benzoyl group at the S_2 position of kAF can be accommodated to a much greater extent in the *C*-domain (red and yellow) than in the *N*-domain. The same should be true for kAP and kAW, as the three ketomethylene inhibitors share the same benzoyl group at the S_2 position. (d) The same as in (c), except that both kAF and Tyr-369 of the *N*-domain are represented as spheres, to emphasize the steric clash between them. The atoms responsible for the clash are those of the Tyr-OH group, marked by the arrow.

fer from *N*-carboxy-alkyl dipeptides, such as lisinopril, in several structural aspects (Figure 8.25a):

1. The amide nitrogen of the *N*-carboxy-alkyl dipeptides is replaced by a carbon atom.

2. The zinc-binding carboxyl group is replaced by a ketone group, which displays weaker binding of zinc.

3. Unlike the *N*-carboxy-alkyl dipeptides and the first-generation inhibitors, ketomethylene inhibitors contain a P_2 group.

Thus, ketomethylene inhibitors do not rely on strong zinc binding, but rather on the binding of additional groups in the enzyme. This makes sense considering the need for domain selectivity. That is, the differential binding affinity of an inhibitor to the two domains is bound to result from differences in specific residues rather than from the zinc atom, which is the same in both. This may explain the aforementioned observation, that selectivity often appears when the zinc binding affinity is reduced. **Reducing the affinity component that results from zinc binding increases the relative contribution (to the overall affinity) of residue-specific components. Since the latter are domain-specific, this also increases the chances for domain-selective binding.**

Watermeyer and coworkers [334] used kAP and two of its derivatives, kAF and kAW, to find the molecular determinants of *C*-domain selectivity. kAF and kAW are analogues of the peptides Phe-Gly-Phe and Phe-Gly-Trp, respectively (Figure 8.25b), and display high *C*-domain selectivity. Since kAP has no P_1' group, and since the P_2' Pro showed no preference for one of the domains [357], the researchers concluded that **the P_1' and P_2' groups could not be responsible for kAP's 30-fold *C*-domain preference.** This left the P_1 and P_2 groups. In order to locate the specific residues involved in the preference, the researchers had to compare the structures of the *C* and *N* domains. While structures of the *C*-domain in complex with different inhibitors have been available for quite some time, the structure of the *N*-domain was solved only in 2006 [358], in complex with lisinopril.

When the two structures were compared, the researchers found only four S_1 or S_2 residues in proximity to the inhibitors, which differed between the two domains. They focused on three of the four residues, which had nonpolar contacts with the inhibitors: Ser-516 and Val-518 of S_1, and Phe-391 of S_2. Mutational studies indicated that Val-518 and Phe-391 were important for binding selectivity, especially the former. The question remained, however, of how these residues were involved in selectivity. **The authors suggested that the Val-518→Thr and Phe-391→Tyr changes made the *N*-domain less accommodating for kAP**, as follows:

1. Val-518 (S_1 subsite): The change of Val to Thr conferred polarity to this region of the *N*-domain's S_1 subsite. Assuming that the binding of kAP's bulky phenyl group to the S_1 subsite involves the latter's desolvation, the binding should be less favorable in the *N*-domain than in the *C* domain. This logic can also explain the moderate *C*-domain selectivity of both lisinopril and enalaprilat, which each have a P_1 phenyl group.

2. Phe-391 (S_2 subsite): The change of Phe to Tyr added a hydroxyl group, which made it difficult for the P_2 benzoyl group of kAP to be accommodated inside the S_2 subsite of the *N*-domain (Figure 8.25c and d).

Indeed, when Watermeyer and coworkers replaced the two residues in the *C*-domain with their *N*-domain counterparts, the binding of kAP, kAF and kAW was affected dramatically.

8.7 SUMMARY

- Protein-ligand interactions form the basis of virtually all biological processes: enzyme catalysis, hormonal action, neurotransmission, signal transduction, intracellular trafficking, and more. Accordingly, the chemical species recognized and bound by proteins are highly diverse, ranging from elemental ions to macromolecules. The interactions are also diverse, and may be covalent or noncovalent.

- Protein-ligand binding has been addressed both qualitatively and quantitatively since the 19th century. Early models assumed pre-existing shape complementarity between the ligand and the protein's binding site, whereas later models suggested a more dynamic interaction. The latter are also able to explain complex phenomena that are related to these interactions, such as cooperativity and different modes of inhibition.

- As in protein folding, protein-ligand interactions are relatively weak (< ~15 kcal/mol), driven mainly by the hydrophobic effect, and rendered specific by electrostatic interactions. Still, the interaction energies observed between different proteins and their ligands vary considerably and are evolutionarily dictated by the biological role of the binding.

- The relatively low protein-ligand interaction energy results, among other things, from the mutual compensation between the energy's enthalpic and entropic components. The weak interaction serves one of the biological imperatives of proteins, i.e., to be able to bind their target ligands reversibly, as often happens in protein-based regulatory and communication networks. Exceptions to this rule include metals and prosthetic groups, which are both required to stay inside the protein for much longer than are other types of ligands.

- Protein binding sites complement their natural ligands both geometrically and electrostatically. This fit is, however, not perfect, and proteins sometimes use the strain resulting from such imperfection to execute their functions. Protein-ligand complementarity is not the result of a specific structural motif, and can be achieved using a variety of structural combinations. Certain amino acids do, however, tend to function in binding, primarily the aromatic ones. The unique properties of these amino acids, especially Tyr, contribute to both the energetics and specificity of the binding.

- Interactions between polypeptide chains are highly diverse and include complex energetic and dynamic aspects. As a result, evolution has led to the development of protein domains specializing in the binding of specific types of proteins and peptides.

- Part of the inherent complexity of protein-protein interactions manifests in the 'division of labor' among different residues within the binding interface. Whereas certain residues at the center of the interface ('hot spots') interact directly with residues of the binding partner, it seems that some peripheral residues act indirectly to strengthen the binding, by removing water molecules from the center.

- Another expression of binding complexity in protein-protein interactions is the dependence of the binding specificity on the type of complex and its biological role. Thus, while antibodies, hormone-activated receptors, and many enzymes are designed for highly specific binding, signal transduction proteins often switch between different binding partners, as required by their biological role. Another biological aspect that affects protein-protein interactions is cellular crowding. This effect, however, is often neglected in *in vitro* studies.

● The accumulated knowledge on protein-ligand interactions is applied extensively in the design of medicinal drugs by the pharmaceutical industry. This is because pharmaceutical drugs act by binding to malfunctioning proteins within the body and changing their abnormal activity, or by manipulating proteins in pathogens. Indeed, a plethora of experimental and computational tools have been devised to facilitate the prediction and synthesis of specific molecules capable of targeting disease-related proteins. These methods try to determine the capacity of candidate molecules to bind to the target protein, exert the desired effect, and do so without having adverse effects on other body organs.

EXERCISES

8.1 List four specific biological processes that rely on protein-ligand binding, and explain for each how the binding serves its biological role.

8.2 Describe the three basic models proposed for protein-ligand binding and suggest which one is the most realistic.

8.3 Explain the phenomenon known as 'electrostatic steering'.

8.4 Which branches of the animal nervous system include the neurotransmitter acetylcholine?

    a. Central (brain and spinal cord)

    b. Somatic

    c. Sympathetic

    d. All of the above

8.5 Organophosphates are chemical agents that inactivate the enzyme acetylcholinesterase by binding covalently to its catalytic serine residue. Why, then, are some organophosphates (e.g., sarin) far more dangerous than others (e.g., parathion)?

8.6 A. List some of the different criteria used to classify protein-protein complexes.

    B. Are there significant differences between the interfaces of the different protein-protein complexes?

8.7 Explain the phenomenon known as 'antibody maturation'.

8.8 Elaborate on the differences in qualitative and quantitative contribution to binding, between different residues in protein-protein interfaces.

8.9 Morphine, heroin, methadone and papaverine are members of the opiate group of drugs:

| Morphine | Heroin | Methadone | Papaverine |

In your estimation, which chemical-structural features of these drugs are required for their activity? What type of analysis would you use to find these features?

8.10 Explain the main differences between N-carboxy-alkyl dipeptides and keto-based ACE inhibitors.

REFERENCES

1. S. B. Zimmerman and S. O. Trach. Estimation of macromolecule concentrations and excluded volume effects for the cytoplasm of *Escherichia coli*. *J. Mol. Biol.*, 222(3):599–620, 1991.
2. E. Fischer. Einfluss der Configuration auf die Wirkung der Enzyme. *Ber. Dtsch. Chem. Ges.*, 27(3):2985–2993, 1894.
3. J. R. Schnell, H. J. Dyson, and P. E. Wright. Structure, dynamics, and catalytic function of dihydrofolate reductase. *Annu. Rev. Biophys. Biomol. Struct.*, 33:119–40, 2004.
4. Y. Savir and T. Tlusty. Conformational proofreading: the impact of conformational changes on the specificity of molecular recognition. *PLoS One*, 2(5):e468, 2007.
5. C. A. Sotriffer, O. Kramer, and G. Klebe. Probing flexibility and "induced-fit" phenomena in aldose reductase by comparative crystal structure analysis and molecular dynamics simulations. *Proteins*, 56(1):52–66, 2004.
6. R. Brenk, M. T. Stubbs, A. Heine, K. Reuter, and G. Klebe. Flexible adaptations in the structure of the tRNA-modifying enzyme tRNA-guanine transglycosylase and their implications for substrate selectivity, reaction mechanism and structure-based drug design. *ChemBioChem*, 4(10):1066–77, 2003.
7. M. Kosloff and R. Kolodny. Sequence-similar, structure-dissimilar protein pairs in the PDB. *Proteins*, 71(2):891–902, 2008.
8. D. E. Koshland Jr. Enzyme flexibility and enzyme action. *J. Cell. Comp. Physiol.*, 54:245–58, 1959.
9. M. Gerstein and N. Echols. Exploring the range of protein flexibility, from a structural proteomics perspective. *Curr. Opin. Chem. Biol.*, 8(1):14–9, 2004.
10. J. Monod, J. Wyman, and J. P. Changeux. On the nature of allosteric transitions: a plausible model. *J. Mol. Biol.*, 12:88–118, 1965.
11. L. Fetler, E. R. Kantrowitz, and P. Vachette. Direct observation in solution of a preexisting structural equilibrium for a mutant of the allosteric aspartate transcarbamoylase. *Proc. Natl. Acad. Sci. USA*, 104(2):495–500, 2007.
12. E. L. Roberts, N. Shu, M. J. Howard, R. W. Broadhurst, A. Chapman-Smith, J. C. Wallace, T. Morris, J. E. Cronan Jr, and R. N. Perham. Solution structures of apo and holo biotinyl domains from acetyl coenzyme A carboxylase of *Escherichia coli* determined by triple-resonance nuclear magnetic resonance spectroscopy. *Biochemistry*, 38(16):5045–53, 1999.
13. D. Zhao, C. H. Arrowsmith, X. Jia, and O. Jardetzky. Refined solution structures of the *Escherichia coli* trp holo- and aporepressor. *J. Mol. Biol.*, 229(3):735–46, 1993.
14. K. Gunasekaran, B. Ma, B. Ramakrishnan, P. K. Qasba, and R. Nussinov. Interdependence of backbone flexibility, residue conservation, and enzyme function: a case study on beta1,4-galactosyltransferase-I. *Biochemistry*, 42(13):3674–87, 2003.

15. B. Ramakrishnan, P. V. Balaji, and P. K. Qasba. Crystal structure of beta1,4-galactosyltransferase complex with UDP-Gal reveals an oligosaccharide acceptor binding site. *J. Mol. Biol.*, 318(2):491–502, 2002.

16. A. Kitao, S. Hayward, and N. Go. Energy landscape of a native protein: jumping-among-minima model. *Proteins*, 33(4):496–517, 1998.

17. J. A. McCammon, B. R. Gelin, and M. Karplus. Dynamics of folded proteins. *Nature*, 267(5612):585–90, 1977.

18. R. H. Austin, K. W. Beeson, L. Eisenstein, H. Frauenfelder, and I. C. Gunsalus. Dynamics of ligand binding to myoglobin. *Biochemistry*, 14(24):5355–73, 1975.

19. G. A. Petsko and D. Ringe. Fluctuations in protein structure from X-ray diffraction. *Annu. Rev. Biophys. Bio.*, 13:331–71, 1984.

20. H. Frauenfelder, F. Parak, and R. D. Young. Conformational substates in proteins. *Annu. Rev. Biophys. Biophys. Chem.*, 17:451–79, 1988.

21. A. Malmendal, J. Evenas, S. Forsen, and M. Akke. Structural dynamics in the C-terminal domain of calmodulin at low calcium levels. *J. Mol. Biol.*, 293(4):883–99, 1999.

22. G. Weber. Ligand binding and internal equilibrium in proteins. *Biochemistry*, 11(5):864–878, 1972.

23. B. Ma, S. Kumar, C. J. Tsai, and R. Nussinov. Folding funnels and binding mechanisms. *Protein Eng.*, 12(9):713–20, 1999.

24. H. R. Bosshard. Molecular recognition by induced fit: how fit is the concept? *News Physiol. Sci.*, 16:171–3, 2001.

25. S. Kumar, B. Ma, C. J. Tsai, N. Sinha, and R. Nussinov. Folding and binding cascades: dynamic landscapes and population shifts. *Protein Sci.*, 9(1):10–9, 2000.

26. D. D. Boehr and P. E. Wright. Biochemistry. How do proteins interact? *Science*, 320(5882):1429–30, 2008.

27. L. C. James, P. Roversi, and D. S. Tawfik. Antibody multispecificity mediated by conformational diversity. *Science*, 299(5611):1362–7, 2003.

28. C. Berger, S. Weber-Bornhauser, J. Eggenberger, J. Hanes, A. Pluckthun, and H. R. Bosshard. Antigen recognition by conformational selection. *FEBS Lett.*, 450(1–2):149–53, 1999.

29. J. Foote and C. Milstein. Conformational isomerism and the diversity of antibodies. *Proc. Natl. Acad. Sci. USA*, 91(22):10370–4, 1994.

30. R. Grunberg, J. Leckner, and M. Nilges. Complementarity of structure ensembles in protein-protein binding. *Structure*, 12(12):2125–36, 2004.

31. J. P. Junker, F. Ziegler, and M. Rief. Ligand-dependent equilibrium fluctuations of single calmodulin molecules. *Science*, 323(5914):633–7, 2009.

32. Q. Wang, K.-C. Liang, A. Czader, M. N. Waxham, and M. S. Cheung. The Effect of Macromolecular Crowding, Ionic Strength and Calcium Binding on Calmodulin Dynamics. *PLoS Comput. Biol.*, 7(7):e1002114, 2011.

33. N. J. Anthis, M. Doucleff, and G. M. Clore. Transient, Sparsely Populated Compact States of Apo and Calcium-Loaded Calmodulin Probed by Paramagnetic Relaxation Enhancement: Interplay of Conformational Selection and Induced Fit. *J. Am. Chem. Soc.*, 133(46):18966–18974, 2011.

34. Q. Wang, P. Zhang, L. Hoffman, S. Tripathi, D. Homouz, Y. Liu, M. N. Waxham, and M. S. Cheung. Protein recognition and selection through conformational and mutually induced fit. *Proc. Natl. Acad. Sci. USA*, 110(51):20545–50, 2013.

35. J. R. Williamson. Induced fit in RNA-protein recognition. *Nat. Struct. Biol.*, 7(10):834–7, 2000.

36. A. Gutteridge and J. M. Thornton. Conformational changes observed in enzyme crystal structures upon substrate binding. *J. Mol. Biol.*, 346(1):21–8, 2005.

37. G. Parisi, D. J. Zea, A. M. Monzon, and C. Marino-Buslje. Conformational diversity and the emergence of sequence signatures during evolution. *Curr. Opin. Struct. Biol.*, 32:58–65, 2015.

38. A. G. Murzin. Biochemistry. Metamorphic proteins. *Science*, 320(5884):1725–6, 2008.

39. P. N. Bryan and J. Orban. Proteins that switch folds. *Curr. Opin. Struct. Biol.*, 20(4):482–8, 2010.

40. K. K. Frederick, M. S. Marlow, K. G. Valentine, and A. J. Wand. Conformational entropy in molecular recognition by proteins. *Nature*, 448(7151):325–330, 2007.

41. M. S. Marlow, J. Dogan, K. K. Frederick, K. G. Valentine, and A. J. Wand. The role of conformational entropy in molecular recognition by calmodulin. *Nat. Chem. Biol.*, 6(5):352–8, 2010.

42. M. F. Dunn. Protein–Ligand Interactions: General Description. In *Encyclopedia of Life Sciences*. John Wiley & Sons, Ltd., 2000.

43. P. L. Kastritis and A. M. Bonvin. Molecular origins of binding affinity: seeking the Archimedean point. *Curr. Opin. Struct. Biol.*, 23(6):868–77, 2013.

44. D. J. Winzor. Binding Constants: Measurement and Biological Range. In *Encyclopedia of Life Sciences*. John Wiley & Sons, Ltd., 2001.

45. N. M. Green. Avidin. *Adv. Protein Chem.*, 29:85–133, 1975.

46. F. Nau, C. Guérin-Dubiard, and T. Croguennec. Avidin. In R. Huopalahti, M. Anton, R. López-Fandiño, and R. Schade, editors, *Bioactive egg compounds (part I)*, pages 75–80. Springer, Berlin, 2007.

47. J. P. Vincent and M. Lazdunski. Trypsin-pancreatic trypsin inhibitor association. Dynamics of the interaction and role of disulfide bridges. *Biochemistry*, 11(16):2967–77, 1972.

48. R. D. Smith, A. L. Engdahl, J. B. Dunbar Jr, and H. A. Carlson. Biophysical limits of protein-ligand binding. *J. Chem. Inf. Model.*, 52(8):2098–106, 2012.

49. A. Ahmed, R. D. Smith, J. J. Clark, J. B. Dunbar, and H. A. Carlson. Recent improvements to Binding MOAD: a resource for protein–ligand binding affinities and structures. *Nucleic Acids Res.*, 43(D1):D465–D469, 2015.

50. L. Hu, M. L. Benson, R. D. Smith, M. G. Lerner, and H. A. Carlson. Binding MOAD (Mother Of All Databases). *Proteins: Struct., Funct., Bioinf.*, 60(3):333–340, 2005.

51. I. D. Kuntz, K. Chen, K. A. Sharp, and P. A. Kollman. The maximal affinity of ligands. *Proc. Natl. Acad. Sci. USA*, 96(18):9997–10002, 1999.

52. C. H. Reynolds, B. A. Tounge, and S. D. Bembenek. Ligand Binding Efficiency: Trends, Physical Basis, and Implications. *J. Med. Chem.*, 51(8):2432–2438, 2008.

53. K. Chen and L. Kurgan. Investigation of atomic level patterns in protein–small ligand interactions. *PLoS One*, 4(2):e4473, 2009.

54. X. Zhang and K. N. Houk. Why enzymes are proficient catalysts: beyond the Pauling paradigm. *Acc. Chem. Res.*, 38(5):379–85, 2005.

55. S. Mobashery and L. P. Kotra. Transition State Stabilization. In *Encyclopedia of Life Sciences*. John Wiley & Sons, Ltd., 2002.

56. V. L. Schramm. Enzymatic transition states and transition state analog design. *Annu. Rev. Biochem.*, 67:693–720, 1998.

57. L. Pauling. Nature of forces between large molecules of biological interest. *Nature*, 161:707–709, 1948.

58. M. Aldeghi, A. Heifetz, M. J. Bodkin, S. Knapp, and P. C. Biggin. Accurate calculation of the absolute free energy of binding for drug molecules. *Chem. Sci.*, 7(1):207–218, 2016.

59. G. Sliwoski, S. Kothiwale, J. Meiler, and E. W. Lowe Jr. Computational methods in drug discovery. *Pharmacol. Rev.*, 66(1):334–95, 2014.

60. C. Chipot. Frontiers in free-energy calculations of biological systems. *Wiley Interdiscip. Rev. Comput. Mol. Sci.*, 4(1):71–89, 2014.

61. J. D. Chodera, D. L. Mobley, M. R. Shirts, R. W. Dixon, K. Branson, and V. S. Pande. Alchemical free energy methods for drug discovery: progress and challenges. *Curr. Opin. Struct. Biol.*, 21(2):150–60, 2011.

62. D. L. Mobley and P. V. Klimovich. Perspective: Alchemical free energy calculations for drug discovery. *J. Chem. Phys.*, 137(23):230901, 2012.

63. Y. Deng and B. Roux. Calculation of Standard Binding Free Energies: Aromatic Molecules in the T4 Lysozyme L99A Mutant. *J. Chem. Theory Comput.*, 2(5):1255–73, 2006.

64. J. Wang, Y. Deng, and B. Roux. Absolute binding free energy calculations using molecular dynamics simulations with restraining potentials. *Biophys. J.*, 91(8):2798–814, 2006.

65. L. Mai Suan and M. Binh Khanh. Steered Molecular Dynamics-A Promising Tool for Drug Design. *Curr. Bioinform.*, 7(4):342–351, 2012.

66. H.-J. Woo and B. Roux. Calculation of absolute protein–ligand binding free energy from computer simulations. *Proc. Natl. Acad. Sci. USA*, 102(19):6825–6830, 2005.

67. J. Aqvist, C. Medina, and J. E. Samuelsson. A new method for predicting binding affinity in computer-aided drug design. *Protein Eng.*, 7(3):385–91, 1994.

68. W. E. Miranda, S. Y. Noskov, and P. A. Valiente. Improving the LIE Method for Binding Free Energy Calculations of Protein Ligand Complexes. *J. Chem. Inf. Model.*, 55(9):1867–77, 2015.

69. H. Sun, Y. Li, S. Tian, L. Xu, and T. Hou. Assessing the performance of MM/PBSA and MM/GBSA methods. 4. Accuracies of MM/PBSA and MM/GBSA methodologies evaluated by various simulation protocols using PDBbind data set. *Phys. Chem. Chem. Phys.*, 16(31):16719–29, 2014.

70. L. Xu, H. Sun, Y. Li, J. Wang, and T. Hou. Assessing the performance of MM/PBSA and MM/GBSA methods. 3. The impact of force fields and ligand charge models. *J. Phys. Chem. B*, 117(28):8408–21, 2013.

71. T. Hou, J. Wang, Y. Li, and W. Wang. Assessing the performance of the MM/PBSA and MM/GBSA methods. 1. The accuracy of binding free energy calculations based on molecular dynamics simulations. *J. Chem. Inf. Model.*, 51(1):69–82, 2011.

72. S. Genheden and U. Ryde. The MM/PBSA and MM/GBSA methods to estimate ligand-binding affinities. *Expert Opin. Drug Discov.*, 10(5):449–461, 2015.

73. M. K. Gilson and H. X. Zhou. Calculation of protein-ligand binding affinities. *Annu. Rev. Biophys. Biomol. Struct.*, 36:21–42, 2007.

74. R. Lumry and S. Rajender. Enthalpy–entropy compensation phenomena in water solutions of proteins and small molecules: A ubiquitous properly of water. *Biopolymers*, 9(10):1125–1227, 1970.

75. J. D. Dunitz. Win some, lose some: enthalpy-entropy compensation in weak intermolecular interactions. *Chem. Biol.*, 2(11):709–12, 1995.

76. A. Cooper. Thermodynamic analysis of biomolecular interactions. *Curr. Opin. Chem. Biol.*, 3(5):557–63, 1999.

77. T. S. G. Olsson, J. E. Ladbury, W. R. Pitt, and M. A. Williams. Extent of enthalpy-entropy compensation in protein-ligand interactions. *Protein Sci.*, 20(9):1607–1618, 2011.

78. M. S. Searle, M. S. Westwell, and D. H. Williams. Application of a generalised enthalpy–entropy relationship to binding co-operativity and weak associations in solution. *J. Chem. Soc., Perkin Trans. 2*, 1:141–151, 1995.

79. B. Breiten, M. R. Lockett, W. Sherman, S. Fujita, H. Lange, C. M. Bowers, A. Heroux, and G. M. Whitesides. Water Networks Contribute to Enthalpy / Entropy Compensation in Protein-Ligand Binding. *J. Am. Chem. Soc.*, 135:15579–15584, 2013.

80. A. Cooper and C. M. Johnson. Introduction to microcalorimetry and biomolecular energetics. *Methods Mol. Biol.*, 22:109–24, 1994.

81. T. S. G. Olsson, M. A. Williams, W. R. Pitt, and J. E. Ladbury. The Thermodynamics of Protein-Ligand Interaction and Solvation: Insights for Ligand Design. *J. Mol. Biol.*, 384(4):1002–1017, 2008.

82. A. Perez, J. A. Morrone, C. Simmerling, and K. A. Dill. Advances in free-energy-based simulations of protein folding and ligand binding. *Curr. Opin. Struct. Biol.*, 36:25–31, 2016.

83. O. Guvench and A. D. MacKerell Jr. Computational evaluation of protein-small molecule binding. *Curr. Opin. Struct. Biol.*, 19(1):56–61, 2009.

84. B. Honig and A. Nicholls. Classical electrostatics in biology and chemistry. *Science*, 268(5214):1144–1149, 1995.

85. W. C. Still, A. Tempczyk, R. C. Hawley, and T. Hendrickson. Semianalytical treatment of solvation for molecular mechanics and dynamics. *J. Am. Chem. Soc.*, 112:6127–6129, 1990.

86. D. Sitkoff, K. A. Sharp, and B. Honig. Accurate calculation of hydration free energies using macroscopic solvation models. *J. Phys. Chem.*, 98:1978–1988, 1994.

87. D. Jiao, P. A. Golubkov, T. A. Darden, and P. Ren. Calculation of protein-ligand binding free energy by using a polarizable potential. *Proc. Natl. Acad. Sci. USA*, 105(17):6290–5, 2008.

88. M. R. Reddy and M. D. Erion. Calculation of relative binding free energy differences for fructose 1,6-bisphosphatase inhibitors using the thermodynamic cycle perturbation approach. *J. Am. Chem. Soc.*, 123(26):6246–52, 2001.

89. D. L. Mobley and K. A. Dill. Binding of small-molecule ligands to proteins: "what you see" is not always "what you get". *Structure*, 17(4):489–98, 2009.

90. P. Czodrowski, C. A. Sotriffer, and G. Klebe. Atypical protonation states in the active site of HIV-1 protease: a computational study. *J. Chem. Inf. Model.*, 47(4):1590–8, 2007.

91. F. Dullweber, M. T. Stubbs, D. Musil, J. Sturzebecher, and G. Klebe. Factorising ligand affinity: a combined thermodynamic and crystallographic study of trypsin and thrombin inhibition. *J. Mol. Biol.*, 313(3):593–614, 2001.

92. H. Steuber, P. Czodrowski, C. A. Sotriffer, and G. Klebe. Tracing changes in protonation: a prerequisite to factorize thermodynamic data of inhibitor binding to aldose reductase. *J. Mol. Biol.*, 373(5):1305–20, 2007.

93. Z. Li and T. Lazaridis. Water at biomolecular binding interfaces. *Phys. Chem. Chem. Phys.*, 9:573–581, 2007.

94. S. Samsonov, J. Teyra, and M. T. Pisabarro. A molecular dynamics approach to study the importance of solvent in protein interactions. *Proteins*, 73(2):515–25, 2008.

95. A. Chang, J. Schiebel, W. Yu, G. R. Bommineni, P. Pan, M. V. Baxter, A. Khanna, C. A. Sotriffer, C. Kisker, and P. J. Tonge. Rational optimization of drug-target residence time: insights from inhibitor binding to the *Staphylococcus aureus* FabI enzyme-product complex. *Biochemistry*, 52(24):4217–28, 2013.

96. M. Merski, M. Fischer, T. E. Balius, O. Eidam, and B. K. Shoichet. Homologous ligands accommodated by discrete conformations of a buried cavity. *Proc. Natl. Acad. Sci. USA*, 2015.

97. B. L. Tembe and J. A. McCammon. Ligand-receptor interactions. *Comput. Chem.*, 8:281–283, 1984.

98. C. F. Wong and J. A. McCammon. Protein flexibility and computer-aided drug design. *Annu. Rev. Pharmacol. Toxicol.*, 43:31–45, 2003.

99. C. E. Chang, W. Chen, and M. K. Gilson. Ligand configurational entropy and protein binding. *Proc. Natl. Acad. Sci. USA*, 104(5):1534–9, 2007.

100. W. Chen, C. E. Chang, and M. K. Gilson. Calculation of cyclodextrin binding affinities: energy, entropy, and implications for drug design. *Biophys. J.*, 87(5):3035–49, 2004.

101. A. J. Wand. The dark energy of proteins comes to light: Conformational entropy and its role in protein function revealed by NMR relaxation. *Curr. Opin. Struct. Biol.*, 23(1):75–81, 2013.

102. H.-X. Zhou and M. K. Gilson. Theory of Free Energy and Entropy in Noncovalent Binding. *Chem. Rev.*, 109(9):4092–4107, 2009.

103. C. E. Chang and M. K. Gilson. Free energy, entropy, and induced fit in host-guest recognition: calculations with the second-generation mining minima algorithm. *J. Am. Chem. Soc.*, 126(40):13156–64, 2004.

104. C. R. Guimaraes and M. Cardozo. MM-GB/SA rescoring of docking poses in structure-based lead optimization. *J. Chem. Inf. Model.*, 48(5):958–70, 2008.

105. C. Chothia and J. Janin. Principles of protein-protein recognition. *Nature*, 256(5520):705–8, 1975.

106. B. Wilfried and B. F. N. E. Jan. Hydrophobic Effects. Opinions and Facts. *Angew. Chem. Int. Ed.*, 32(11):1545–1579, 1993.

107. A. Babtie, N. Tokuriki, and F. Hollfelder. What makes an enzyme promiscuous? *Curr. Opin. Chem. Biol.*, 14(2):200–7, 2010.

108. E. Persch, O. Dumele, and F. Diederich. Molecular recognition in chemical and biological systems. *Angew. Chem. Int. Ed.*, 54(11):3290–327, 2015.

109. F. Biedermann, W. M. Nau, and H. J. Schneider. The hydrophobic effect revisited: studies with supramolecular complexes imply high-energy water as a noncovalent driving force. *Angew. Chem. Int. Ed.*, 53(42):11158–71, 2014.

110. H. Ponstingl, T. Kabir, D. Gorse, and J. M. Thornton. Morphological aspects of oligomeric protein structures. *Prog. Biophys. Mol. Biol.*, 89(1):9–35, 2005.

111. T. Young, L. Hua, X. Huang, R. Abel, R. Friesner, and B. J. Berne. Dewetting transitions in protein cavities. *Proteins*, 78(8):1856–69, 2010.

112. L. Wang, B. J. Berne, and R. A. Friesner. Ligand binding to protein-binding pockets with wet and dry regions. *Proc. Natl. Acad. Sci. USA*, 108(4):1326–30, 2011.

113. S. W. Homans. Water, water everywhere — except where it matters? *Drug Discov. Today*, 12(13–14):534–539, 2007.

114. J. Qvist, M. Davidovic, D. Hamelberg, and B. Halle. A dry ligand-binding cavity in a solvated protein. *Proc. Natl. Acad. Sci. USA*, 105(17):6296–301, 2008.

115. D. Kern and E. R. Zuiderweg. The role of dynamics in allosteric regulation. *Curr. Opin. Struct. Biol.*, 13(6):748–57, 2003.

116. C. E. Chang, W. A. McLaughlin, R. Baron, W. Wang, and J. A. McCammon. Entropic contributions and the influence of the hydrophobic environment in promiscuous protein-protein association. *Proc. Natl. Acad. Sci. USA*, 105(21):7456–61, 2008.

117. C. Azuara, E. Lindahl, P. Koehl, H. Orland, and M. Delarue. PDB_Hydro: incorporating dipolar solvents with variable density in the Poisson-Boltzmann treatment of macromolecule electrostatics. *Nucleic Acids Res.*, 34(suppl 2):W38–42, 2006.

118. Marc Delarue group at Institut Pasteur. http://lorentz.immstr.pasteur.fr/website/present.html, 2017.

119. A. Kahraman, R. J. Morris, R. A. Laskowski, and J. M. Thornton. Shape variation in protein binding pockets and their ligands. *J. Mol. Biol.*, 368(1):283–301, 2007.

120. S. Vajda and F. Guarnieri. Characterization of protein-ligand interaction sites using experimental and computational methods. *Curr. Opin. Drug Discov. Devel.*, 9(3):354–62, 2006.

121. R. P. Bahadur, P. Chakrabarti, F. Rodier, and J. Janin. A dissection of specific and non-specific protein-protein interfaces. *J. Mol. Biol.*, 336(4):943–55, 2004.

122. I. M. Nooren and J. M. Thornton. Structural characterisation and functional significance of transient protein-protein interactions. *J. Mol. Biol.*, 325(5):991–1018, 2003.

123. K. A. Sharp. Important considerations impacting molecular docking. In B. K. Shoichet and J. Alvarez, editors, *Virtual Screening in Drug Discovery*, chapter 9, page 227. CRC Press, Boca Raton, FL, 2005.

124. F. Fogolari, A. Brigo, and H. Molinari. The Poisson-Boltzmann equation for biomolecular electrostatics: a tool for structural biology. *J. Mol. Recognit.*, 15(6):377–92, 2002.

125. S. Kumar and R. Nussinov. Close-range electrostatic interactions in proteins. *ChemBioChem*, 3(7):604–617, 2002.

126. S. Albeck, R. Unger, and G. Schreiber. Evaluation of direct and cooperative contributions towards the strength of buried hydrogen bonds and salt bridges. *J. Mol. Biol.*, 298(3):503–20, 2000.

127. D. Lee, J. Lee, and C. Seok. What stabilizes close arginine pairing in proteins? *Phys. Chem. Chem. Phys.*, 15(16):5844–53, 2013.

128. E. S. Feldblum and I. T. Arkin. Strength of a bifurcated H bond. *Proc. Natl. Acad. Sci. USA*, 111(11):4085–90, 2014.

129. B. Musafia, V. Buchner, and D. Arad. Complex salt bridges in proteins: statistical analysis of structure and function. *J. Mol. Biol.*, 254(4):761–70, 1995.

130. J. Massoulie, J. Sussman, S. Bon, and I. Silman. Structure and functions of acetylcholinesterase and butyryl-cholinesterase. *Prog. Brain Res.*, 98:139–46, 1993.

131. S. A. Botti, C. E. Felder, S. Lifson, J. L. Sussman, and I. I. Silman. A modular treatment of molecular traffic through the active site of cholinesterase. *Biophys. J.*, 77(5):2430–50, 1999.

132. J. A. McCammon. Darwinian biophysics: electrostatics and evolution in the kinetics of molecular binding. *Proc. Natl. Acad. Sci. USA*, 106(19):7683–4, 2009.

133. D. Porschke, C. Creminon, X. Cousin, C. Bon, J. Sussman, and I. Silman. Electrooptical measurements demonstrate a large permanent dipole moment associated with acetylcholinesterase. *Biophys. J.*, 70(4):1603–8, 1996.

134. D. R. Ripoll, C. H. Faerman, P. H. Axelsen, I. Silman, and J. L. Sussman. An electrostatic mechanism for substrate guidance down the aromatic gorge of acetylcholinesterase. *Proc. Natl. Acad. Sci. USA*, 90(11):5128–32, 1993.

135. R. C. Tan, T. N. Truong, J. A. McCammon, and J. L. Sussman. Acetylcholinesterase: electrostatic steering increases the rate of ligand binding. *Biochemistry*, 32(2):401–3, 1993.

136. Z. Radic, P. D. Kirchhoff, D. M. Quinn, J. A. McCammon, and P. Taylor. Electrostatic influence on the kinetics of ligand binding to acetylcholinesterase. Distinctions between active center ligands and fasciculin. *J. Biol. Chem.*, 272(37):23265–23277, 1997.

137. G. Schreiber and A. R. Fersht. Rapid, electrostatically assisted association of proteins. *Nat. Struct. Biol.*, 3(5):427–31, 1996.

138. B. Ma and R. Nussinov. Trp/Met/Phe hot spots in protein-protein interactions: potential targets in drug design. *Curr. Top. Med. Chem.*, 7(10):999–1005, 2007.

139. S. Koide and S. S. Sidhu. The importance of being tyrosine: lessons in molecular recognition from minimalist synthetic binding proteins. *ACS Chem. Biol.*, 4(5):325–34, 2009.

140. I. S. Mian, A. R. Bradwell, and A. J. Olson. Structure, function and properties of antibody binding sites. *J. Mol. Biol.*, 217(1):133–51, 1991.

141. G. A. Green. Understanding NSAIDs: from aspirin to COX-2. *Clin. Cornerstone*, 3(5):50–60, 2001.

142. K. D. Rainsford. Anti-inflammatory drugs in the 21st century. *Subcell. Biochem.*, 42:3–27, 2007.

143. P. Rao and E. E. Knaus. Evolution of nonsteroidal anti-inflammatory drugs (NSAIDs): cyclooxygenase (COX) inhibition and beyond. *J. Pharm. Pharm. Sci.*, 11(2):81s–110s, 2008.

144. M. Gao and J. Skolnick. A Comprehensive Survey of Small-Molecule Binding Pockets in Proteins. *PLoS Comput. Biol.*, 9(10):e1003302, 2013.

145. O. Khersonsky and D. S. Tawfik. Enzyme Promiscuity: A Mechanistic and Evolutionary Perspective. *Annu. Rev. Biochem.*, 79(1):471–505, 2010.

146. L. Hedstrom. Enzyme Specificity and Selectivity. In *Encyclopedia of Life Sciences*. John Wiley & Sons, Ltd., 2010.

147. T. Dierks, C. Miech, J. Hummerjohann, B. Schmidt, M. A. Kertesz, and K. von Figura. Posttranslational formation of formylglycine in prokaryotic sulfatases by modification of either cysteine or serine. *J. Biol. Chem.*, 273(40):25560–4, 1998.

148. S. Jonas and F. Hollfelder. Mapping catalytic promiscuity in the alkaline phosphatase superfamily. *Pure Appl. Chem.*, 81:731–742, 2009.

149. I. Nobeli, A. D. Favia, and J. M. Thornton. Protein promiscuity and its implications for biotechnology. *Nat. Biotechnol.*, 27(2):157–67, 2009.

150. L. Hou, M. T. Honaker, L. M. Shireman, L. M. Balogh, A. G. Roberts, K. C. Ng, A. Nath, and W. M. Atkins. Functional promiscuity correlates with conformational heterogeneity in A-class glutathione S-transferases. *J. Biol. Chem.*, 282(32):23264–74, 2007.

151. J. Skopalik, P. Anzenbacher, and M. Otyepka. Flexibility of human cytochromes P450: molecular dynamics reveals differences between CYPs 3A4, 2C9, and 2A6, which correlate with their substrate preferences. *J. Phys. Chem. B*, 112(27):8165–73, 2008.

152. J. Skolnick and M. Gao. Interplay of physics and evolution in the likely origin of protein biochemical function. *Proc. Natl. Acad. Sci. USA*, 110(23):9344–9349, 2013.

153. L. Xie, L. Xie, and P. E. Bourne. Structure-based systems biology for analyzing off-target binding. *Curr. Opin. Struct. Biol.*, 21(2):189–199, 2011.

154. M. J. Keiser, V. Setola, J. J. Irwin, C. Laggner, A. I. Abbas, S. J. Hufeisen, N. H. Jensen, M. B. Kuijer, R. C. Matos, T. B. Tran, R. Whaley, R. A. Glennon, J. Hert, K. L. H. Thomas, D. D. Edwards, B. K. Shoichet, and B. L. Roth. Predicting new molecular targets for known drugs. *Nature*, 462(7270):175–181, 2009.

155. T. T. Ashburn and K. B. Thor. Drug repositioning: identifying and developing new uses for existing drugs. *Nat. Rev. Drug Discov.*, 3(8):673–683, 2004.

156. Y. Hu, D. Gupta-Ostermann, and J. Bajorath. Exploring compound promiscuity patterns and multi-target activity spaces. *Comput. Struct. Biotechnol. J.*, 9(13):e201401003, 2014.

157. C. R. Chong and D. J. Sullivan. New uses for old drugs. *Nature*, 448(7154):645–646, 2007.

158. P. M. Salvaterra. Acetylcholine. In *Encyclopedia of Life Sciences*. John Wiley & Sons, Ltd., 2001.

159. R. P. Hicks, M. G. Hartell, D. A. Nichols, A. K. Bhattacharjee, J. E. van Hamont, and D. R. Skillman. The medicinal chemistry of botulinum, ricin and anthrax toxins. *Curr. Med. Chem.*, 12(6):667–90, 2005.

160. J. Bajgar. Complex view on poisoning with nerve agents and organophosphates. *Acta Medica (Hradec Kralove)*, 48(1):3–21, 2005.

161. N. A. Baker, D. Sept, S. Joseph, M. J. Holst, and J. A. McCammon. Electrostatics of nanosystems: Application to microtubules and the ribosome. *Proc. Natl. Acad. Sci. USA*, 98(18):10037–10041, 2001.

162. W. L. DeLano. The PyMOL Molecular Graphics System. http://www.pymol.org, 2002.

163. R. C. Gupta. *Toxicology of Organophosphate & Carbamate Compounds*. Elsevier Academic Press, London, 2006.

164. C. B. Millard, G. Kryger, A. Ordentlich, H. M. Greenblatt, M. Harel, M. L. Raves, Y. Segall, D. Barak, A. Shafferman, I. Silman, and J. L. Sussman. Crystal structures of aged phosphonylated acetylcholinesterase: nerve agent reaction products at the atomic level. *Biochemistry*, 38(22):7032–9, 1999.

165. L. Pauling. Molecular Architecture and Biological Reactions. *Chem. Eng. News*, 24(10):1375–1377, 1946.

166. W. P. Jencks. Strain and conformation change in enzymatic catalysis. In N. O. Kaplan and E. P. Kennedy, editors, *Current Aspects of Biochemical Energetics*, page 273. Academic Press, New York, 1966.

167. J. Grutzendler and J. C. Morris. Cholinesterase inhibitors for Alzheimer's disease. *Drugs*, 61(1):41–52, 2001.

168. D. J. Selkoe. Alzheimer's disease is a synaptic failure. *Science*, 298(5594):789–91, 2002.

169. D. G. Wilkinson, P. T. Francis, E. Schwam, and J. Payne-Parrish. Cholinesterase inhibitors used in the treatment of Alzheimer's disease: the relationship between pharmacological effects and clinical efficacy. *Drugs Aging*, 21(7):453–78, 2004.

170. Z. Prokop, F. Oplustil, J. DeFrank, and J. Damborsky. Enzymes fight chemical weapons. *Biotechnol. J.*, 1(12):1370–80, 2006.

171. M. Levine and R. Tjian. Transcription regulation and animal diversity. *Nature*, 424(6945):147–51, 2003.

172. O. Keskin, A. Gursoy, B. Ma, and R. Nussinov. Principles of protein-protein interactions: what are the preferred ways for proteins to interact? *Chem. Rev.*, 108(4):1225–44, 2008.

173. I. M. Nooren and J. M. Thornton. Diversity of protein-protein interactions. *EMBO J.*, 22(14):3486–92, 2003.

174. K. W. Kohn and M. I. Aladjem. Circuit diagrams for biological networks. *Mol. Syst. Biol.*, 2:2006 0002, 2006.

175. K. W. Kohn, M. I. Aladjem, J. N. Weinstein, and Y. Pommier. Molecular interaction maps of bioregulatory networks: a general rubric for systems biology. *Mol. Biol. Cell*, 17(1):1–13, 2006.

176. E. H. McConkey. Molecular evolution, intracellular organization, and the quinary structure of proteins. *Proc. Natl. Acad. Sci. USA*, 79(10):3236–40, 1982.

177. A. J. Wirth and M. Gruebele. Quinary protein structure and the consequences of crowding in living cells: leaving the test-tube behind. *BioEssays*, 35(11):984–93, 2013.

178. W. B. Monteith, R. D. Cohen, A. E. Smith, E. Guzman-Cisneros, and G. J. Pielak. Quinary structure modulates protein stability in cells. *Proc. Natl. Acad. Sci. USA*, 112(6):1739–42, 2015.

179. J. Mintseris and Z. Weng. Structure, function, and evolution of transient and obligate protein-protein interactions. *Proc. Natl. Acad. Sci. USA*, 102(31):10930–5, 2005.

180. M. M. Kasembeli, X. Xu, and D. J. Tweardy. SH2 domain binding to phosphopeptide ligands: potential for drug targeting. *Front. Biosci.*, 14:1010–22, 2009.

181. G. Waksman, D. Kominos, S. C. Robertson, N. Pant, D. Baltimore, R. B. Birge, D. Cowburn, H. Hanafusa, B. J. Mayer, M. Overduin, et al. Crystal structure of the phosphotyrosine recognition domain SH2 of v-src complexed with tyrosine-phosphorylated peptides. *Nature*, 358(6388):646–53, 1992.

182. B. Fazi, M. J. Cope, A. Douangamath, S. Ferracuti, K. Schirwitz, A. Zucconi, D. G. Drubin, M. Wilmanns,

G. Cesareni, and L. Castagnoli. Unusual binding properties of the SH3 domain of the yeast actin-binding protein Abp1: structural and functional analysis. *J. Biol. Chem.*, 277(7):5290–8, 2002.

183. Y. G. Gao, X. Z. Yan, A. X. Song, Y. G. Chang, X. C. Gao, N. Jiang, Q. Zhang, and H. Y. Hu. Structural insights into the specific binding of huntingtin proline-rich region with the SH3 and WW domains. *Structure*, 14(12):1755–65, 2006.

184. M. Lewitzky, M. Harkiolaki, M. C. Domart, E. Y. Jones, and S. M. Feller. Mona/Gads SH3C binding to hematopoietic progenitor kinase 1 (HPK1) combines an atypical SH3 binding motif, R/KXXK, with a classical PXXP motif embedded in a polyproline type II (PPII) helix. *J. Biol. Chem.*, 279(27):28724–32, 2004.

185. D. Reichmann, O. Rahat, M. Cohen, H. Neuvirth, and G. Schreiber. The molecular architecture of protein-protein binding sites. *Curr. Opin. Struct. Biol.*, 17(1):67–76, 2007.

186. Y. Ofran and B. Rost. Analysing six types of protein-protein interfaces. *J. Mol. Biol.*, 325(2):377–87, 2003.

187. N. Sinha, S. Mohan, C. A. Lipschultz, and S. J. Smith-Gill. Differences in electrostatic properties at antibody-antigen binding sites: implications for specificity and cross-reactivity. *Biophys. J.*, 83(6):2946–68, 2002.

188. F. B. Sheinerman and B. Honig. On the role of electrostatic interactions in the design of protein-protein interfaces. *J. Mol. Biol.*, 318(1):161–77, 2002.

189. G. Nimrod, F. Glaser, D. Steinberg, N. Ben-Tal, and T. Pupko. In silico identification of functional regions in proteins. *Bioinformatics*, 21 Suppl 1:i328–37, 2005.

190. R. P. Saha, R. P. Bahadur, and P. Chakrabarti. Interresidue contacts in proteins and protein-protein interfaces and their use in characterizing the homodimeric interface. *J. Proteome Res.*, 4(5):1600–9, 2005.

191. M. H. Ali and B. Imperiali. Protein oligomerization: how and why. *Bioorg. Med. Chem.*, 13(17):5013–20, 2005.

192. S. Jones and J. M. Thornton. Principles of protein-protein interactions. *Proc. Natl. Acad. Sci. USA*, 93(1):13–20, 1996.

193. C. J. Tsai, S. L. Lin, H. J. Wolfson, and R. Nussinov. Studies of protein-protein interfaces: a statistical analysis of the hydrophobic effect. *Protein Sci.*, 6(1):53–64, 1997.

194. C. J. Tsai, D. Xu, and R. Nussinov. Structural motifs at protein-protein interfaces: protein cores versus two-state and three-state model complexes. *Protein Sci.*, 6(9):1793–805, 1997.

195. S. Jones and J. M. Thornton. Analysis of protein-protein interaction sites using surface patches. *J. Mol. Biol.*, 272(1):121–32, 1997.

196. F. B. Sheinerman, R. Norel, and B. Honig. Electrostatic aspects of protein-protein interactions. *Curr. Opin. Struct. Biol.*, 10(2):153–9, 2000.

197. B. N. Bullock, A. L. Jochim, and P. S. Arora. Assessing helical protein interfaces for inhibitor design. *J. Am. Chem. Soc.*, 133(36):14220–14223, 2011.

198. A. M. Watkins and P. S. Arora. Anatomy of β-strands at protein–protein interfaces. *ACS Chem. Biol.*, 9(8):1747–1754, 2014.

199. I. S. Moreira, P. A. Fernandes, and M. J. Ramos. Hot spots – A review of the protein-protein interface determinant amino-acid residues. *Proteins: Struct., Funct., Genet.*, 68(4):803–812, 2007.

200. H.-P. Peng, K. H. Lee, J.-W. Jian, and A.-S. Yang. Origins of specificity and affinity in antibody–protein interactions. *Proc. Natl. Acad. Sci. USA*, 111(26):E2656–E2665, 2014.

201. L. M. Salonen, M. Ellermann, and F. Diederich. Aromatic Rings in Chemical and Biological Recognition: Energetics and Structures. *Angew. Chem. Int. Ed.*, 50(21):4808–4842, 2011.

202. A. H. Keeble, N. Kirkpatrick, S. Shimizu, and C. Kleanthous. Calorimetric dissection of colicin DNase–immunity protein complex specificity. *Biochemistry*, 45(10):3243–54, 2006.

203. W. Li, A. H. Keeble, C. Giffard, R. James, G. R. Moore, and C. Kleanthous. Highly discriminating protein-protein interaction specificities in the context of a conserved binding energy hotspot. *J. Mol. Biol.*, 337(3):743–59, 2004.

204. J. E. Chrencik, A. Brooun, M. L. Kraus, M. I. Recht, A. R. Kolatkar, G. W. Han, J. M. Seifert, H. Widmer, M. Auer, and P. Kuhn. Structural and biophysical characterization of the EphB4·EphrinB2 protein-protein interaction and receptor specificity. *J. Biol. Chem.*, 281(38):28185–92, 2006.

205. G. Fernandez-Ballester, C. Blanes-Mira, and L. Serrano. The tryptophan switch: changing ligand-binding specificity from type I to type II in SH3 domains. *J. Mol. Biol.*, 335(2):619–29, 2004.

206. M. Colledge and J. D. Scott. AKAPs: from structure to function. *Trends Cell Biol.*, 9(6):216–21, 1999.

207. L. C. James and D. S. Tawfik. The specificity of cross-reactivity: promiscuous antibody binding involves specific hydrogen bonds rather than nonspecific hydrophobic stickiness. *Protein Sci.*, 12(10):2183–93, 2003.

208. L. C. James and D. S. Tawfik. Structure and kinetics of a transient antibody binding intermediate reveal

a kinetic discrimination mechanism in antigen recognition. *Proc. Natl. Acad. Sci. USA*, 102(36):12730–5, 2005.

209. S. J. Wodak and J. Janin. Structural basis of macromolecular recognition. *Adv. Protein Chem.*, 61:9–73, 2002.

210. L. Lo Conte, C. Chothia, and J. Janin. The atomic structure of protein-protein recognition sites. *J. Mol. Biol.*, 285(5):2177–98, 1999.

211. A. A. Bogan and K. S. Thorn. Anatomy of hot spots in protein interfaces. *J. Mol. Biol.*, 280(1):1–9, 1998.

212. L. C. Roisman, D. A. Jaitin, D. P. Baker, and G. Schreiber. Mutational analysis of the IFNAR1 binding site on IFNalpha2 reveals the architecture of a weak ligand-receptor binding-site. *J. Mol. Biol.*, 353(2):271–81, 2005.

213. S. Wohlgemuth, C. Kiel, A. Kramer, L. Serrano, F. Wittinghofer, and C. Herrmann. Recognizing and defining true Ras binding domains I: biochemical analysis. *J. Mol. Biol.*, 348(3):741–58, 2005.

214. M. Prudencio and M. Ubbink. Transient complexes of redox proteins: structural and dynamic details from NMR studies. *J. Mol. Recognit.*, 17(6):524–39, 2004.

215. A. N. Barclay. Membrane proteins with immunoglobulin-like domains: a master superfamily of interaction molecules. *Semin. Immunol.*, 15(4):215–23, 2003.

216. S. Cho, C. P. Swaminathan, J. Yang, M. C. Kerzic, R. Guan, M. C. Kieke, D. M. Kranz, R. A. Mariuzza, and E. J. Sundberg. Structural basis of affinity maturation and intramolecular cooperativity in a protein-protein interaction. *Structure*, 13(12):1775–87, 2005.

217. A. Cauerhff, F. A. Goldbaum, and B. C. Braden. Structural mechanism for affinity maturation of an anti-lysozyme antibody. *Proc. Natl. Acad. Sci. USA*, 101(10):3539–44, 2004.

218. E. J. Sundberg, P. S. Andersen, P. M. Schlievert, K. Karjalainen, and R. A. Mariuzza. Structural, energetic, and functional analysis of a protein-protein interface at distinct stages of affinity maturation. *Structure*, 11(9):1151–61, 2003.

219. G. Schreiber and A. R. Fersht. Energetics of protein-protein interactions: analysis of the barnase-barstar interface by single mutations and double mutant cycles. *J. Mol. Biol.*, 248(2):478–86, 1995.

220. P. A. Patten, N. S. Gray, P. L. Yang, C. B. Marks, G. J. Wedemayer, J. J. Boniface, R. C. Stevens, and P. G. Schultz. The immunological evolution of catalysis. *Science*, 271(5252):1086–91, 1996.

221. G. J. Wedemayer, P. A. Patten, L. H. Wang, P. G. Schultz, and R. C. Stevens. Structural insights into the evolution of an antibody combining site. *Science*, 276(5319):1665–9, 1997.

222. T. Clackson and J. A. Wells. A hot spot of binding energy in a hormone-receptor interface. *Science*, 267(5196):383–6, 1995.

223. B. C. Cunningham and J. A. Wells. High-resolution epitope mapping of hGH-receptor interactions by alanine-scanning mutagenesis. *Science*, 244(4908):1081–5, 1989.

224. K. S. Thorn and A. A. Bogan. ASEdb: a database of alanine mutations and their effects on the free energy of binding in protein interactions. *Bioinformatics*, 17(3):284–5, 2001.

225. I. H. Moal and J. Fernández-Recio. SKEMPI: A Structural Kinetic and Energetic database of Mutant Protein Interactions and its use in empirical models. *Bioinformatics*, 28(20):2600–2607, 2012.

226. J. L. Kouadio, J. R. Horn, G. Pal, and A. A. Kossiakoff. Shotgun alanine scanning shows that growth hormone can bind productively to its receptor through a drastically minimized interface. *J. Biol. Chem.*, 280(27):25524–32, 2005.

227. Z. Hu, B. Ma, H. Wolfson, and R. Nussinov. Conservation of polar residues as hot spots at protein interfaces. *Proteins*, 39(4):331–42, 2000.

228. O. Keskin, B. Ma, and R. Nussinov. Hot regions in protein–protein interactions: the organization and contribution of structurally conserved hot spot residues. *J. Mol. Biol.*, 345(5):1281–94, 2005.

229. A. del Sol and P. O'Meara. Small-world network approach to identify key residues in protein-protein interaction. *Proteins*, 58(3):672–82, 2005.

230. U. Samanta, D. Pal, and P. Chakrabarti. Environment of tryptophan side chains in proteins. *Proteins*, 38(3):288–300, 2000.

231. Y. Li, Y. Huang, C. P. Swaminathan, S. J. Smith-Gill, and R. A. Mariuzza. Magnitude of the hydrophobic effect at central versus peripheral sites in protein-protein interfaces. *Structure*, 13(2):297–307, 2005.

232. I. Halperin, H. Wolfson, and R. Nussinov. Protein-protein interactions; coupling of structurally conserved residues and of hot spots across interfaces. Implications for docking. *Structure*, 12(6):1027–38, 2004.

233. E. D. Levy. A Simple Definition of Structural Regions in Proteins and Its Use in Analyzing Interface Evolution. *J. Mol. Biol.*, 403(4):660–670, 2010.

234. D. Reichmann, O. Rahat, S. Albeck, R. Meged, O. Dym, and G. Schreiber. The modular architecture of protein-protein binding interfaces. *Proc. Natl. Acad. Sci. USA*, 102(1):57–62, 2005.

235. O. Keskin, B. Ma, K. Rogale, K. Gunasekaran, and R. Nussinov. Protein-protein interactions: organization, cooperativity and mapping in a bottom-up systems biology approach. *Phys. Biol.*, 2(1–2):S24–S35, 2005.

236. K. V. Brinda and S. Vishveshwara. Oligomeric protein structure networks: insights into protein–protein interactions. *BMC Bioinformatics*, 6:296, 2005.

237. D. Reichmann, M. Cohen, R. Abramovich, O. Dym, D. Lim, N. C J Strynadka, and G. Schreiber. Binding Hot Spots in the TEM1-BLIP Interface in Light of its Modular Architecture. *J. Mol. Biol.*, 365(3):663–679, 2007.

238. D. Rajamani, S. Thiel, S. Vajda, and C. J. Camacho. Anchor residues in protein-protein interactions. *Proc. Natl. Acad. Sci. USA*, 101(31):11287–92, 2004.

239. M. Harel, A. Spaar, and G. Schreiber. Fruitful and futile encounters along the association reaction between proteins. *Biophys. J.*, 96(10):4237–4248, 2009.

240. D. Tobi and I. Bahar. Structural changes involved in protein binding correlate with intrinsic motions of proteins in the unbound state. *Proc. Natl. Acad. Sci. USA*, 102(52):18908–13, 2005.

241. G. R. Smith, M. J. Sternberg, and P. A. Bates. The relationship between the flexibility of proteins and their conformational states on forming protein-protein complexes with an application to protein-protein docking. *J. Mol. Biol.*, 347(5):1077–101, 2005.

242. M. C. Lawrence and P. M. Colman. Shape complementarity at protein/protein interfaces. *J. Mol. Biol.*, 234(4):946–50, 1993.

243. T. M. Raschke. Water structure and interactions with protein surfaces. *Curr. Opin. Struct. Biol.*, 16(2):152–9, 2006.

244. F. Rodier, R. P. Bahadur, P. Chakrabarti, and J. Janin. Hydration of protein-protein interfaces. *Proteins: Struct., Funct., Bioinf.*, 60(1):36–45, 2005.

245. T. N. Bhat, G. A. Bentley, G. Boulot, M. I. Greene, D. Tello, W. Dall'Acqua, H. Souchon, F. P. Schwarz, R. A. Mariuzza, and R. J. Poljak. Bound water molecules and conformational stabilization help mediate an antigen-antibody association. *Proc. Natl. Acad. Sci. USA*, 91(3):1089–93, 1994.

246. A. Rath, A. R. Davidson, and C. M. Deber. The structure of "unstructured" regions in peptides and proteins: role of the polyproline II helix in protein folding and recognition. *Biopolymers*, 80(2–3):179–85, 2005.

247. J. T. Nguyen, C. W. Turck, F. E. Cohen, R. N. Zuckermann, and W. A. Lim. Exploiting the basis of proline recognition by SH3 and WW domains: design of N-substituted inhibitors. *Science*, 282(5396):2088–92, 1998.

248. H. Ashkenazy, S. Abadi, E. Martz, O. Chay, I. Mayrose, T. Pupko, and N. Ben-Tal. ConSurf 2016: an improved methodology to estimate and visualize evolutionary conservation in macromolecules. *Nucleic Acids Res.*, 44(W1):W344–W350, 2016.

249. F. Glaser, T. Pupko, I. Paz, R. E. Bell, D. Bechor-Shental, E. Martz, and N. Ben-Tal. ConSurf: identification of functional regions in proteins by surface-mapping of phylogenetic information. *Bioinformatics*, 19(1):163–4, 2003.

250. F.-X. Theillet, A. Binolfi, T. Frembgen-Kesner, K. Hingorani, M. Sarkar, C. Kyne, C. Li, P. B. Crowley, L. Gierasch, G. J. Pielak, A. H. Elcock, A. Gershenson, and P. Selenko. Physicochemical properties of cells and their effects on intrinsically disordered proteins (IDPs). *Chem. Rev.*, 114(13):6661–6714, 2014.

251. G. B. Ralston. Effects of "crowding" in protein solutions. *J. Chem. Educ.*, 67(10):857, 1990.

252. S. B. Zimmerman and A. P. Minton. Macromolecular crowding: biochemical, biophysical, and physiological consequences. *Annu. Rev. Biophys. Biomol. Struct.*, 22:27–65, 1993.

253. A. P. Minton. Influence of macromolecular crowding upon the stability and state of association of proteins: predictions and observations. *J. Pharm. Sci.*, 94(8):1668–75, 2005.

254. R. J. Ellis. Macromolecular crowding: Obvious but underappreciated, 2001.

255. Y. Phillip, E. Sherman, G. Haran, and G. Schreiber. Common crowding agents have only a small effect on protein-protein interactions. *Biophys. J.*, 97(3):875–885, 2009.

256. Y. Y. Kuttner, N. Kozer, E. Segal, G. Schreiber, and G. Haran. Separating the contribution of translational and rotational diffusion to protein association. *J. Am. Chem. Soc.*, 127(43):15138–15144, 2005.

257. I. M. Kuznetsova, K. K. Turoverov, and V. N. Uversky. What macromolecular crowding can do to a protein. *Int. J. Mol. Sci.*, 15(12):23090–23140, 2014.

258. J. G. Robertson. Enzymes as a special class of therapeutic target: clinical drugs and modes of action. *Curr. Opin. Struct. Biol.*, 17(6):674–9, 2007.

259. J. Weigelt, L. D. McBroom-Cerajewski, M. Schapira, Y. Zhao, and C. H. Arrowmsmith. Structural genomics and drug discovery: all in the family. *Curr. Opin. Chem. Biol.*, 12:32–39, 2008.

260. M. Uhlen, L. Fagerberg, B. M. Hallstrom, C. Lindskog, P. Oksvold, A. Mardinoglu, A. Sivertsson, C. Kampf,

E. Sjostedt, A. Asplund, I. Olsson, K. Edlund, E. Lundberg, S. Navani, C. A. Szigyarto, J. Odeberg, D. Djureinovic, J. O. Takanen, S. Hober, T. Alm, P. H. Edqvist, H. Berling, H. Tegel, J. Mulder, J. Rockberg, P. Nilsson, J. M. Schwenk, M. Hamsten, K. von Feilitzen, M. Forsberg, L. Persson, F. Johansson, M. Zwahlen, G. von Heijne, J. Nielsen, and F. Ponten. Proteomics. Tissue-based map of the human proteome. *Science*, 347(6220):1260419, 2015.

261. D. C. Swinney. Biochemical mechanisms of drug action: what does it take for success? *Nat. Rev. Drug Discov.*, 3(9):801–8, 2004.

262. J. G. Robertson. Mechanistic basis of enzyme-targeted drugs. *Biochemistry*, 44(15):5561–71, 2005.

263. Y. Fang, T. Kenakin, and C. Liu. Editorial: Orphan GPCRs As Emerging Drug Targets. *Front. Pharmacol.*, 6:295, 2015.

264. J. A. Salon, D. T. Lodowski, and K. Palczewski. The Significance of G Protein-Coupled Receptor Crystallography for Drug Discovery. *Pharmacol. Rev.*, 63(4):901–937, 2011.

265. P. Cohen and D. R. Alessi. Kinase drug discovery: what's next in the field? *ACS Chem. Biol.*, 8(1):96–104, 2013.

266. M. Rask-Andersen, M. S. Almén, and H. B. Schiöth. Trends in the exploitation of novel drug targets. *Nat. Rev. Drug Discov.*, 10(8):579–590, 2011.

267. T. Kenakin. Principles: receptor theory in pharmacology. *Trends Pharmacol. Sci.*, 25(4):186–92, 2004.

268. R. Nussinov and C. J. Tsai. Allostery in disease and in drug discovery. *Cell*, 153(2):293–305, 2013.

269. R. Nussinov and C. J. Tsai. Unraveling structural mechanisms of allosteric drug action. *Trends Pharmacol. Sci.*, 35(5):256–64, 2014.

270. R. E. Study and J. L. Barker. Diazepam and (−)-pentobarbital: fluctuation analysis reveals different mechanisms for potentiation of gamma-aminobutyric acid responses in cultured central neurons. *Proc. Natl. Acad. Sci. USA*, 78(11):7180–4, 1981.

271. H. Mohler. Benzodiazepines. In *Encyclopedia of Life Sciences*. John Wiley & Sons, Ltd., 2006.

272. F. A. Stephenson. GABA-A receptors. In *Encyclopedia of Life Sciences*. John Wiley & Sons, Ltd., 2006.

273. R. L. MacDonald, C. J. Rogers, and R. E. Twyman. Barbiturate regulation of kinetic properties of the GABA-A receptor channel of mouse spinal neurones in culture. *J. Physiol.*, 417:483–500, 1989.

274. N. J. Smith and G. Milligan. Allostery at G protein-coupled receptor homo- and heteromers: uncharted pharmacological landscapes. *Pharmacol. Rev.*, 62(4):701–25, 2010.

275. G. L. Patrick. *An Introduction to Medicinal Chemistry*. Oxford University Press, 5th edition, 2013.

276. L. Tóth, L. Muszbek, and I. Komáromi. Mechanism of the irreversible inhibition of human cyclooxygenase-1 by aspirin as predicted by QM/MM calculations. *J. Mol. Graph. Model.*, 40:99–109, 2013.

277. Y. Liu. Chemical biology: Caught in the activation. *Nature*, 461(7263):484–5, 2009.

278. J. W. Li and J. C. Vederas. Drug discovery and natural products: end of an era or an endless frontier? *Science*, 325(5937):161–5, 2009.

279. W. Sneader. *Drug Discovery: A History*. John Wiley & Sons, Inc., 2005.

280. J. Drews. Drug discovery: a historical perspective. *Science*, 287(5460):1960–4, 2000.

281. C. G. Smith and J. O'Donnell. *The process of new drug discovery and development*. CRC Press, Boca Raton, FL, 2nd edition, 2006.

282. J. Nielsen. Combinatorial synthesis of natural products. *Curr. Opin. Chem. Biol.*, 6(3):297–305, 2002.

283. G. Schneider and U. Fechner. Computer-based de novo design of drug-like molecules. *Nat. Rev. Drug Discov.*, 4(8):649–63, 2005.

284. C. M. Dobson. Chemical space and biology. *Nature*, 432(7019):824–8, 2004.

285. C. Lipinski and A. Hopkins. Navigating chemical space for biology and medicine. *Nature*, 432(7019):855–61, 2004.

286. G. Schneider. Trends in virtual combinatorial library design. *Curr. Med. Chem.*, 9(23):2095–101, 2002.

287. M. Rarey. Some thoughts on the "A" in computer-aided molecular design. *J. Comput. Aided Mol. Des.*, 26(1):113 4, 2012.

288. M. Xiang, Y. Cao, W. Fan, L. Chen, and Y. Mo. Computer-aided drug design: lead discovery and optimization. *Comb. Chem. High Throughput Screen*, 15(4):328–37, 2012.

289. M. Gore and N. S. Desai. Computer-Aided Drug Designing. In *Clinical Bioinformatics*, pages 313–321. Springer, New York, 2014.

290. I. M. Kapetanovic. Computer-aided drug discovery and development (CADDD): in silico-chemico-biological approach. *Chem. Biol. Interact.*, 171(2):165–76, 2008.

291. W. L. Jorgensen. The many roles of computation in drug discovery. *Science*, 303(5665):1813–8, 2004.

292. J. Verma, V. M. Khedkar, and E. C. Coutinho. 3D-QSAR in drug design: a review. *Curr. Top. Med. Chem.*, 10(1):95–115, 2010.

293. T. Langer and E. M. Krovat. Chemical feature-based pharmacophores and virtual library screening for discovery of new leads. *Curr. Opin. Drug Discov. Devel.*, 6(3):370–6, 2003.

294. J. S. Mason, A. C. Good, and E. J. Martin. 3-D pharmacophores in drug discovery. *Curr. Pharm. Des.*, 7(7):567–97, 2001.

295. L. Xue and J. Bajorath. Molecular descriptors in chemoinformatics, computational combinatorial chemistry, and virtual screening. *Comb. Chem. High Throughput Screen*, 3(5):363–72, 2000.

296. S. A. Khedkar, A. K. Malde, E. C. Coutinho, and S. Srivastava. Pharmacophore modeling in drug discovery and development: an overview. *Med. Chem.*, 3(2):187–97, 2007.

297. C. G. Wermuth, C. R. Ganellin, P. Lindberg, and L. A. Mitscher. Glossary of terms used in medicinal chemistry (IUPAC Recommendations). *Pure Appl. Chem.*, 70(5):1129–1143, 1998.

298. S. L. Dixon, A. M. Smondyrev, and S. N. Rao. PHASE: A Novel Approach to Pharmacophore Modeling and 3D Database Searching. *Chem. Biol. Drug Des.*, 67(5):370–372, 2006.

299. D. G. Lloyd, C. L. Buenemann, N. P. Todorov, D. T. Manallack, and P. M. Dean. Scaffold hopping in de novo design. Ligand generation in the absence of receptor information. *J. Med. Chem.*, 47(3):493–6, 2004.

300. T. L. Lemke. *Review of organic functional groups: introduction to medicinal organic chemistry*. Lippincott Williams & Wilkins, Philadelphia, 4th edition, 2003.

301. P. M. Woster. Functional Groups, Acid Base Chemistry and Physicochemical Properties. In *Pharmaceutical Sciences 3320: Principles of Drug Action (online course)*. Department of Pharmaceutical Sciences, College of Pharmacy and Health Sciences, Wayne State University, 2009.

302. C. A. Lipinski, F. Lombardo, B. W. Dominy, and P. J. Feeney. Experimental and computational approaches to estimate solubility and permeability in drug discovery and development settings. *Adv. Drug. Deliv. Rev.*, 46(1–3):3–26, 2001.

303. E. J. Martin, J. M. Blaney, M. A. Siani, D. C. Spellmeyer, A. K. Wong, and W. H. Moos. Measuring diversity: experimental design of combinatorial libraries for drug discovery. *J. Med. Chem.*, 38(9):1431–6, 1995.

304. A. Golebiowski, S. R. Klopfenstein, and D. E. Portlock. Lead compounds discovered from libraries. *Curr. Opin. Chem. Biol.*, 5(3):273–84, 2001.

305. A. Golebiowski, S. R. Klopfenstein, and D. E. Portlock. Lead compounds discovered from libraries: part 2. *Curr. Opin. Chem. Biol.*, 7(3):308–25, 2003.

306. R. P. Hertzberg and A. J. Pope. High-throughput screening: new technology for the 21st century. *Curr. Opin. Chem. Biol.*, 4(4):445–51, 2000.

307. B. J. Druker and N. B. Lydon. Lessons learned from the develpment of an Abl tyrosine kinase inhibitor for chronic myelogenous leukemia. *J. Clin. Invest.*, 105:3–7, 2000.

308. G. Scapin. Structural biology and drug discovery. *Curr. Pharm. Des.*, 12(17):2087–97, 2006.

309. E. Lionta, G. Spyrou, D. K. Vassilatis, and Z. Cournia. Structure-based virtual screening for drug discovery: principles, applications and recent advances. *Curr. Top. Med. Chem.*, 14(16):1923–38, 2014.

310. J. J. Irwin and B. K. Shoichet. Docking screens for novel ligands conferring new biology. *J. Med. Chem.*, 2016.

311. T. Sterling and J. J. Irwin. ZINC 15 – Ligand Discovery for Everyone. *J. Chem. Inf. Model.*, 55(11):2324–2337, 2015.

312. C. R. Beddell, P. J. Goodford, F. E. Norrington, S. Wilkinson, and R. Wootton. Compounds designed to fit a site of known structure in human haemoglobin. *Br. J. Pharmacol.*, 57(2):201–209, 1976.

313. S. S. Cohen. A strategy for the chemotherapy of infectious disease. *Science*, 197(4302):431–432, 1977.

314. Y. C. Chen. Beware of docking! *Trends Pharmacol. Sci.*, 36(2):78–95, 2015.

315. S. Forli. Charting a Path to Success in Virtual Screening. *Molecules*, 20(10):18732–58, 2015.

316. M. Rueda, G. Bottegoni, and R. Abagyan. Consistent Improvement of Cross-Docking Results Using Binding Site Ensembles Generated with Elastic Network Normal Modes. *J. Chem. Inf. Model.*, 49(3):716–725, 2009.

317. T. Terasaka, T. Kinoshita, M. Kuno, N. Seki, K. Tanaka, and I. Nakanishi. Structure-based design, synthesis, and structure-activity relationship studies of novel non-nucleoside adenosine deaminase inhibitors. *J. Med. Chem.*, 47(15):3730–43, 2004.

318. R. J. Flower. The development of COX2 inhibitors. *Nat. Rev. Drug Discov.*, 2(3):179–91, 2003.

319. F. E. Simons. Comparative pharmacology of H1 antihistamines: clinical relevance. *Am. J. Med.*, 113 Suppl 9A:38S–46S, 2002.

320. K. R. Acharya, E. D. Sturrock, J. F. Riordan, and M. R. Ehlers. ACE revisited: a new target for structure-based drug design. *Nat. Rev. Drug Discov.*, 2(11):891–902, 2003.

321. A. B. Alper, D. A. Calhoun, and S. Oparil. Hypertension. In *Encyclopedia of Life Sciences*. John Wiley & Sons, Ltd., 2001.

322. P. M. Kearney, M. Whelton, K. Reynolds, P. Muntner, P. K. Whelton, and J. He. Global burden of hypertension: analysis of worldwide data. *Lancet*, 365(9455):217–23, 2005.

323. M. A. Zaman, S. Oparil, and D. A. Calhoun. Drugs targeting the renin-angiotensin-aldosterone system. *Nat. Rev. Drug Discov.*, 1(8):621–36, 2002.

324. S. A. Atlas. The renin-angiotensin aldosterone system: pathophysiological role and pharmacologic inhibition. *J. Manag. Care Pharm.*, 13(8 Suppl B):9–20, 2007.

325. F. Soubrier, F. Alhenc-Gelas, C. Hubert, J. Allegrini, M. John, G. Tregear, and P. Corvol. Two putative active centers in human angiotensin I-converting enzyme revealed by molecular cloning. *Proc. Natl. Acad. Sci. USA*, 85(24):9386–90, 1988.

326. N. M. Hooper, E. H. Karran, and A. J. Turner. Membrane protein secretases. *Biochem. J.*, 321 (Pt 2):265–79, 1997.

327. F. Soubrier, L. Wei, C. Hubert, E. Clauser, F. Alhenc-Gelas, and P. Corvol. Molecular biology of the angiotensin I converting enzyme: II. Structure-function. Gene polymorphism and clinical implications. *J. Hypertens.*, 11(6):599–604, 1993.

328. P. Redelinghuys, A. T. Nchinda, and E. D. Sturrock. Development of domain-selective angiotensin I-converting enzyme inhibitors. *Ann. N. Y. Acad. Sci.*, 1056:160–75, 2005.

329. L. Wei, F. Alhenc-Gelas, P. Corvol, and E. Clauser. The two homologous domains of human angiotensin I-converting enzyme are both catalytically active. *J. Biol. Chem.*, 266(14):9002–8, 1991.

330. J. Cotton, M. A. Hayashi, P. Cuniasse, G. Vazeux, D. Ianzer, A. C. De Camargo, and V. Dive. Selective inhibition of the *C*-domain of angiotensin I converting enzyme by bradykinin potentiating peptides. *Biochemistry*, 41(19):6065–71, 2002.

331. S. Fuchs, H. D. Xiao, J. M. Cole, J. W. Adams, K. Frenzel, A. Michaud, H. Zhao, G. Keshelava, M. R. Capecchi, P. Corvol, and K. E. Bernstein. Role of the N-terminal catalytic domain of angiotensin-converting enzyme investigated by targeted inactivation in mice. *J. Biol. Chem.*, 279(16):15946–53, 2004.

332. C. Junot, M. F. Gonzales, E. Ezan, J. Cotton, G. Vazeux, A. Michaud, M. Azizi, S. Vassiliou, A. Yiotakis, P. Corvol, and V. Dive. RXP 407, a selective inhibitor of the *N*-domain of angiotensin I-converting enzyme, blocks in vivo the degradation of hemoregulatory peptide acetyl-Ser-Asp-Lys-Pro with no effect on angiotensin I hydrolysis. *J. Pharmacol. Exp. Ther.*, 297(2):606–11, 2001.

333. M. R. Ehlers, E. A. Fox, D. J. Strydom, and J. F. Riordan. Molecular cloning of human testicular angiotensin-converting enzyme: the testis isozyme is identical to the C-terminal half of endothelial angiotensin-converting enzyme. *Proc. Natl. Acad. Sci. USA*, 86(20):7741–5, 1989.

334. J. M. Watermeyer, W. L. Kroger, H. G. O'Neill, B. T. Sewell, and E. D. Sturrock. Probing the basis of domain-dependent inhibition using novel ketone inhibitors of angiotensin-converting enzyme. *Biochemistry*, 47(22):5942–50, 2008.

335. W. N. Lipscomb and N. Strater. Recent Advances in Zinc Enzymology. *Chem. Rev.*, 96(7):2375–2434, 1996.

336. V. Pelmenschikov, M. R. Blomberg, and P. E. Siegbahn. A theoretical study of the mechanism for peptide hydrolysis by thermolysin. *J. Biol. Inorg. Chem.*, 7(3):284–98, 2002.

337. M. A. Ondetti, N. J. Williams, E. F. Sabo, J. Pluscec, E. R. Weaver, and O. Kocy. Angiotensin-converting enzyme inhibitors from the venom of *Bothrops jararaca*. Isolation, elucidation of structure, and synthesis. *Biochemistry*, 10(22):4033–9, 1971.

338. D. J. Craik, D. P. Fairlie, S. Liras, and D. Price. The future of peptide-based drugs. *Chem. Biol. Drug Des.*, 81(1):136–47, 2013.

339. L. D. Byers and R. Wolfenden. Binding of the by-product analog benzylsuccinic acid by carboxypeptidase A. *Biochemistry*, 12(11):2070–8, 1973.

340. D. W. Cushman, H. S. Cheung, E. F. Sabo, and M. A. Ondetti. Design of potent competitive inhibitors of angiotensin converting enzyme. Carboxyalkanoyl and mercaptoalkanoyl amino acids. *Biochemistry*, 16(25):5484–91, 1977.

341. M. A. Ondetti, B. Rubin, and D. W. Cushman. Design of specific inhibitors of angiotensin-converting enzyme: new class of orally active antihypertensive agents. *Science*, 196(4288):441–4, 1977.

342. R. Natesh, S. L. Schwager, H. R. Evans, E. D. Sturrock, and K. R. Acharya. Structural details on the binding of antihypertensive drugs captopril and enalaprilat to human testicular angiotensin I-converting enzyme. *Biochemistry*, 43(27):8718–24, 2004.

343. A. A. Patchett, E. Harris, E. W. Tristram, M. J. Wyvratt, M. T. Wu, D. Taub, E. R. Peterson, T. J. Ikeler, J. ten Broeke, L. G. Payne, D. L. Ondeyka, E. D. Thorsett, W. J. Greenlee, N. S. Lohr, R. D. Hoffsommer, H. Joshua,

W. V. Ruyle, J. W. Rothrock, S. D. Aster, A. L. Maycock, F. M. Robinson, R. Hirschmann, C. S. Sweet, E. H. Ulm, D. M. Gross, T. C. Vassil, and C. A. Stone. A new class of angiotensin-converting enzyme inhibitors. *Nature*, 288(5788):280–3, 1980.

344. R. Natesh, S. L. Schwager, E. D. Sturrock, and K. R. Acharya. Crystal structure of the human angiotensin-converting enzyme-lisinopril complex. *Nature*, 421(6922):551–4, 2003.

345. H. G. Bull, N. A. Thornberry, M. H. Cordes, A. A. Patchett, and E. H. Cordes. Inhibition of rabbit lung angiotensin-converting enzyme by N alpha-[(S)-1-carboxy-3-phenylpropyl]L-alanyl-L-proline and N alpha-[(S)-1-carboxy-3-phenylpropyl]L-lysyl-L-proline. *J. Biol. Chem.*, 260(5):2952–62, 1985.

346. P. Redelinghuys, A. T. Nchinda, K. Chibale, and E. D. Sturrock. Novel ketomethylene inhibitors of angiotensin I-converting enzyme (ACE): inhibition and molecular modelling. *Biol. Chem.*, 387(4):461–6, 2006.

347. V. Dive, D. Georgiadis, M. Matziari, A. Makaritis, F. Beau, P. Cuniasse, and A. Yiotakis. Phosphinic peptides as zinc metalloproteinase inhibitors. *Cell. Mol. Life Sci.*, 61(16):2010–9, 2004.

348. A. J. Walz and M. J. Miller. Synthesis and biological activity of hydroxamic acid-derived vasopeptidase inhibitor analogues. *Org. Lett.*, 4(12):2047–50, 2002.

349. J. Kim, G. Hewitt, P. Carroll, and S. M. Sieburth. Silanediol inhibitors of angiotensin-converting enzyme. Synthesis and evaluation of four diastereomers of Phe[Si]Ala dipeptide analogues. *J. Org. Chem.*, 70(15):5781–9, 2005.

350. L. Beltrami, L. C. Zingale, S. Carugo, and M. Cicardi. Angiotensin-converting enzyme inhibitor-related angioedema: how to deal with it. *Expert Opin. Drug Saf.*, 5(5):643–9, 2006.

351. J. Nussberger, M. Cugno, C. Amstutz, M. Cicardi, A. Pellacani, and A. Agostoni. Plasma bradykinin in angio-oedema. *Lancet*, 351(9117):1693–7, 1998.

352. K. Dickstein and J. Kjekshus. Effects of losartan and captopril on mortality and morbidity in high-risk patients after acute myocardial infarction: the OPTIMAAL randomised trial. Optimal Trial in Myocardial Infarction with Angiotensin II Antagonist Losartan. *Lancet*, 360(9335):752–60, 2002.

353. D. Georgiadis, F. Beau, B. Czarny, J. Cotton, A. Yiotakis, and V. Dive. Roles of the two active sites of somatic angiotensin-converting enzyme in the cleavage of angiotensin I and bradykinin: insights from selective inhibitors. *Circ. Res.*, 93(2):148–54, 2003.

354. A. K. Ghosh and S. Fidanze. Transition-State Mimetics for HIV Protease Inhibitors: Stereocontrolled Synthesis of Hydroxyethylene and Hydroxyethylamine Isosteres by Ester-Derived Titanium Enolate Syn and Anti-Aldol Reactions. *J. Org. Chem.*, 63(18):6146–6152, 1998.

355. R. G. Almquist, W. R. Chao, M. E. Ellis, and H. L. Johnson. Synthesis and biological activity of a ketomethylene analogue of a tripeptide inhibitor of angiotensin converting enzyme. *J. Med. Chem.*, 23(12):1392–8, 1980.

356. P. A. Deddish, B. Marcic, H. L. Jackman, H. Z. Wang, R. A. Skidgel, and E. G. Erdos. N-domain-specific substrate and C-domain inhibitors of angiotensin-converting enzyme: angiotensin-(1–7) and keto-ACE. *Hypertension*, 31(4):912–7, 1998.

357. A. Michaud, M. T. Chauvet, and P. Corvol. N-domain selectivity of angiotensin I-converting enzyme as assessed by structure-function studies of its highly selective substrate, N-acetyl-seryl-aspartyl-lysyl-proline. *Biochem. Pharmacol.*, 57(6):611–8, 1999.

358. H. R. Corradi, S. L. Schwager, A. T. Nchinda, E. D. Sturrock, and K. R. Acharya. Crystal structure of the N-domain of human somatic angiotensin I-converting enzyme provides a structural basis for domain-specific inhibitor design. *J. Mol. Biol.*, 357(3):964–74, 2006.

CHAPTER 9

Enzymatic Catalysis

9.1 INTRODUCTION

9.1.1 Metabolic needs of cells

Maintaining life in any cell requires hundreds of chemical reactions to take place at any given moment, and in a highly regulated manner. These reactions allow the cell to grow or divide, produce energy, decompose waste products, communicate with other cells, and more. Most of the reactions happening in the average cell have to do with the metabolic activity of the cell, which can be grouped into two types. *Catabolic* reactions degrade and oxidize foodstuff material (carbohydrates, fats, and sometimes proteins) in order to extract the chemical energy stored in it. This is done very gradually, using numerous highly regulated reactions that are organized as pathways. In the presence of oxygen, foodstuff is completely oxidized to carbon dioxide (CO_2), with the release of energy in the form of electrons. For example, in the catabolism of glucose, the parent molecule is first degraded partially by *glycolysis*, a 10-step catabolic pathway well known to students of biochemistry [1]. The product of this pathway, pyruvate, is then activated and fully oxidized to carbon dioxide by another well-known pathway, the *citric acid (Krebs) cycle* [2]. The high-energy electrons produced by these pathways are first temporarily stored on the electron carriers *NADH and FADH$_2$*, and are then converted by highly complex cellular machinery into a chemically more stable form of energy, *ATP*. This molecule is nicknamed the *'universal currency'*, since it is ubiquitous in cells and can be easily used to drive energy-demanding cellular processes. There are many such processes, most of which are attributed to the other type of metabolic reaction, termed *anabolic*. Indeed, *anabolism* includes all chemical reactions used by cells to build complex materials; these reactions use energy formed during catabolic reactions. In some cases, anabolic pathways also use a reducing agent, in the form of *NADPH*. That is because complex cellular molecules are often reduced. A well-known anabolic pathway is *gluconeogenesis*, which builds glucose from lactic acid and glycerol derivatives.

9.1.2 Cellular processes must be catalyzed in order to sustain life

Many of the chemical reactions described above are spontaneous. However, **while cellular needs dictate that these reactions occur very fast, i.e., within 10^{-5} to 10^2 seconds [3], they tend to happen very slowly at room temperature** [4]. Indeed, many life-sustaining reactions take hundreds, thousands, and even millions of years to occur under mild conditions (Figure 9.1). For example, decarboxylation of orotidine 5′-phosphate during the biosynthesis of nucleic acids has a half-life of 1.7×10^{-2} seconds in cells, and a half-life of 78 million years when isolated in solution. The reason for the slowness of chemical reactions has to do with their energetics and the way they take place. As explained in detail in Chapter 4, spontaneous processes are always accompanied by a decrease in the free energy of the system. In chemical reactions this means that the energy of the reactants is higher than that of the products. However, the decrease in energy does not occur in a single step; **the pathway from reactants to products involves a variety of short-lived intermediates, and the shifts between them along the reaction coordinate involve chemical transformations such as the formation or breaking of covalent bonds, and the transfer of functional groups.** One of these intermediates, called the 'transition state', is highly unstable (i.e., has high free energy) and is therefore extremely short-lived (10^{-12} s to 10^{-13} s [5,6]) [7–12] (Figure 9.2, blue plot). The high energy content of the transition state means that the system has to gain at least that much energy (also known as 'activation energy', or Ea) to reach the transition state and pass it on its way to forming the product(s). Put more simply, **the transition state acts as an energy barrier for the reaction to occur. The higher Ea, the smaller the likelihood for crossing the barrier, which means a smaller reaction rate.** This dependency between the magnitude of the activation energy and the rate of the chemical reaction is described by the *Arrhenius equation* [13], already mentioned in Chapter 1:

$$k = Ae^{(-Ea/RT)} \tag{9.1}$$

(where k is the reaction rate, R is the (universal) gas constant (when Ea is given per molecule, the Boltzmann constant is used instead), T is the temperature, RT is the average kinetic energy (see Chapter 1 Box 1.2), and A is the pre-exponential coefficient (a.k.a. 'frequency factor'). In a first-order reaction involving the collision of reactants to form a product, A can be viewed as the total number of collisions that happen (whether or not they lead to product formation), $e^{(-Ea/RT)}$ as the probability for a successful collision (i.e one which results in the transition state and then the product), and k (in units of s^{-1}) is therefore the fraction of such 'successful' collisions.

The Arrhenius equation is empirical and does not consider mechanistic aspects, which may be related to chemical reactions, e.g., the number of reaction intermediates [15]. **A more adequate description is provided by Eyring's transition state theory [7,11,12,16], which addresses the rate-limiting step of the reaction (i.e., the catalytic step).** As the name implies, the rate of this step (k_{cat}) dictates the rate of the entire reaction; hence, it is important for understanding mechanistic aspects of the reaction. The dependence of k_{cat} on the activation energy is given by the *Eyring-Polanyi equation*:

$$k_{cat} = \gamma \frac{k_B T}{h} e^{(-Ea/k_B T)} \tag{9.2}$$

(where γ is the generalized transmission coefficient (relates to the fraction of encounters between reactants that actually lead to product formation), k_B is the Boltzmann constant, h is Planck's constant, and $k_B T/h$ is the rate factor for crossing the transition state. **For first-order reactions, $k_B T/h$ is commonly called the 'universal frequency factor'. It is considered the upper limit of covalent bond breaking frequency.** This is because its value is roughly on the same sub-picosecond scale as that of thermal bond vibration frequency [17], during which only one breaking event can happen. Therefore, $k_B T/h$ sets the ceiling for maximal chemical reaction rates.

When k_{cat} is known, the half-life of the reaction ($t_{1/2}$) can be calculated. Put simply, $t_{1/2}$ is

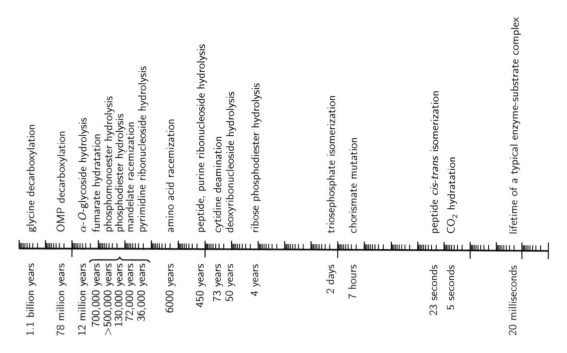

FIGURE 9.1 **The half-lives of some spontaneous chemical reactions in water.** (The image is adapted from [14]).

the time it would take to consume 50% of the reactant. For example, in a first-order reaction, the half-life is given by Equation (9.3) (see Box 9.1 for details):

$$t_{1/2} = \frac{0.69}{k_{cat}} \qquad (9.3)$$

Equations (9.1) and (9.2) show that the reaction rate depends not only on the activation energy but also on the temperature. Increasing the temperature adds more (heat) energy to the system, thus increasing the rate of crossing the transition state barrier[*1]. Indeed, **heating the reaction is the simplest way of speeding it up, but not the only one; the other option is lowering the activation energy, by using a catalyst** (Figure 9.2, red plot). The catalyst lowers the activation energy by stabilizing the transition state. This strategy makes sense considering the exponential dependency of the reaction rate on the activation energy. It makes even more sense when the reaction occurs in biological cells or tissues; accelerating chemical reactions to rates high enough to sustain life requires temperature elevation of hundreds to thousands of degrees Celsius. This, of course, is not an option, as such temperatures would lead to the degradation of both the cell and the molecules inside it. However, when a catalyst is used, there is no need for any temperature increase. **It is therefore not surprising that evolution has led to the selection of catalysis as a means of accelerating biochemical reactions.**

Chemical reactions can be catalyzed at room temperature using simple solid materials such as metals. **However, in living organisms, reactions are catalyzed almost exclusively by enzymes: proteins[*2] that are present in cells at concentrations of 10^{-5} M or less** [19,20]. The reasons for the selection of enzymes as the principal catalysts of life-related processes are discussed in the following subsection.

[*1]Assuming that the reaction is enthalpy-driven.

[*2]There are also RNA-based molecules called *ribozymes* that possess catalytic activity [18], but they constitute the minority of biological catalysts and will not be discussed here.

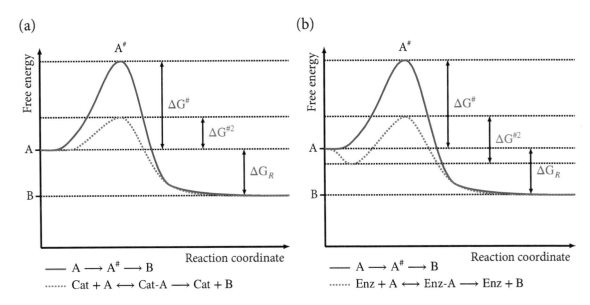

(a)

(b)

FIGURE 9.2 **Energetics of a chemical reaction in the absence and presence of a catalyst.** (a) The effect of a simple catalyst. The blue plot depicts a chemical reaction, in which a hypothetical reactant (A) is transformed into a product (B) (see the top right part of the figure). The reaction involves a transition state ($A^{\#}$), whose free energy exceeds that of the reactant by $\Delta G^{\#}$. In the presence of a catalyst (Cat), the reaction path changes[*a]; the catalyst binds to the reactant rapidly, forming a complex denoted 'Cat-A'. After binding, the substrate is transformed into the product, and released (the red dashed plot). Since the free energy difference between reactant and product (ΔG_R) remains the same, the overall energetics of the reaction does not change[*b]. However, the transition state in the catalyzed reaction is much lower in free energy ($\Delta G^{\#2}$) compared with the transition state in the uncatalyzed reaction ($\Delta G^{\#}$). In other words, the reaction needs less energy to overcome the barrier imposed by the transition state. Since the magnitude of this needed energy (*'activation energy', Ea*) correlates with the time needed for the reaction to be completed, the energy-catalyzed reaction is faster than the uncatalyzed reaction. (b) The effect of an enzyme catalyst (Enz). When the catalyst is an enzyme, the substrate binds to a pocket in the enzyme, which stabilizes it, and this leads to a favorable drop in the energy of the system. This in turn makes the activation energy slightly higher than in the case of a simple catalyst, but still lower than in the uncatalyzed reaction.

[*a]Although the transition state itself is usually the same as in the uncatalyzed reaction (e.g. [21,22]).

[*b]Note that the energy of the catalyst is subtracted, which is why the curves overlap at the substrates and products.

BOX 9.1 CHEMICAL REACTION RATES

In chemical kinetics, reactions are usually characterized in terms of stoichiometry, mechanism and order. The latter describes the dependency of the reaction on the concentration of the reactants. On the basis of the order, we can distinguish between the following reaction types [23].

I. Zero-order reactions

In these reactions the rate is constant and does not depend on the concentration of the reactants. For example, in the following zero-order reaction:

$$A \longrightarrow B \tag{9.1.1}$$

the rate is:

$$V = -\frac{d[A]}{dt} = k \tag{9.1.2}$$

(where [A] is the molar concentration (M) of A and k is the rate constant, with units of concentration per time (e.g., $M\,s^{-1}$)).

To obtain the concentration of the reactant at time t, [A], we integrate Equation (9.1.2) from $[A]_0$ at time zero to [A] at time t. This yields the following linear dependency:

$$[A] = [A]_0 - kt \tag{9.1.3}$$

Equation (9.1.3) is graphically presented as follows:

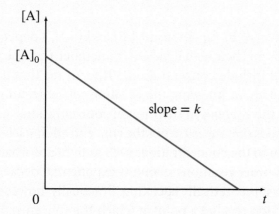

FIGURE 9.1.1 **The change in concentration of the reactant A over time in a zero-order reaction.**

II. First-order reactions

In these reactions the rate depends on the concentration of one reactant. For example, in the following first-order reaction:

$$A \longrightarrow B \tag{9.1.4}$$

the rate is:

$$V = -\frac{d[A]}{dt} = k[A] \tag{9.1.5}$$

(where k is the rate constant, with units of time^{-1} (e.g., s^{-1})).

Again, to obtain the concentration of the reactant at time t, we integrate Equation (9.1.5) from $[A]_0$ at time zero to [A] at time t. This yields the following exponential dependency:

$$[A] = [A]_0 e^{-kt} \tag{9.1.6}$$

If we take the logarithm on Equation (9.1.6) we obtain a linear dependency:

$$\ln[A] = \ln[A]_0 - kt \tag{9.1.7}$$

which can be graphically presented as follows:

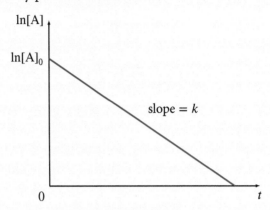

FIGURE 9.1.2 **The change in concentration of the reactant A over time in a first-order reaction.**

Note that the rate constant in Equations (9.1.6) and (9.1.7) depends only on the ratio of concentrations, not on their actual values (remember that the rate constant's units are time^{-1} and do not include concentration). This means that it is possible to study first-order reactions without knowing the absolute concentrations of the reactants; it is sufficient to follow the changes in their initial concentrations. This attribute enables scientists to use methods that measure not the concentration itself but rather a physical property proportional to the concentration, such as light absorbance.

The fact that first-order reactions manifest exponential decay means that they are never complete, at least theoretically speaking. Practically, however, after a certain period of time, the reaction reaches a point at which the concentration of the reactant is too low to be detected by analytical equipment, and at this point the reaction is considered to be done. Before achieving 'completion', the reaction reaches a point at which half of the initial concentration of the reactant has been converted into product. The time required to reach this point is called the *'half-life'* of the reaction, or $t_{1/2}$.

To calculate the half-life of a first-order reaction, we simply replace [A] with $0.5[A]_0$ in Equation (9.1.6), to obtain:

$$t_{1/2} = \frac{0.6931}{k} \qquad (9.1.8)$$

Like the rate constant of first-order reactions, the half-life is independent of reactant concentrations.

III. Second-order reactions

These reactions include two types. In the first, the rate depends on the concentration of one second-order reactant. For example, in the following second-order reaction:

$$2A \longrightarrow B \qquad (9.1.9)$$

the rate is:

$$V = -\frac{d[A]}{dt} = k[A]^2 \qquad (9.1.10)$$

(where k is the rate constant, with units of concentration^{-1} time^{-1} (e.g., $M^{-1}\,s^{-1}$)).

Again, to obtain the concentration of the reactant at time t, we integrate Equation (9.1.5) from $[A]_0$ at time zero to $[A]$ at time t, to obtain:

$$\frac{1}{[A]} = \frac{1}{[A]_0} + kt \tag{9.1.11}$$

which can be graphically presented as follows:

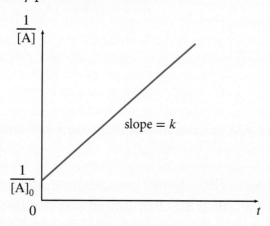

FIGURE 9.1.3 **The change in concentration of the reactant A over time in a second-order reaction.**

Again, the half-life of the second-order reaction is obtained by replacing $[A]$ with $0.5[A]_0$ in Equation (9.1.11), which yields:

$$t_{1/2} = \frac{1}{k[A]_0} \tag{9.1.12}$$

In the second type of second-order reaction, called 'mixed', the rate depends on the concentrations of two first-order reactants. Consider, for example, the second-order reaction:

$$A + B \longrightarrow C \tag{9.1.13}$$

in which $[A]_0 \neq [B]_0$, A and B each have a stoichiometry of 1, and x is the concentration of the product at time t. The rate of this reaction is:

$$V = \frac{dx}{dt} = k\left([A]_0 - x\right)\left([B]_0 - x\right) \tag{9.1.14}$$

Integration of Equation (9.1.14) yields:

$$\ln \frac{[B][A]_0}{[A][B]_0} = k\left([B]_0 - [A]_0\right) t \tag{9.1.15}$$

which can be rearranged to:

$$\ln \frac{[B]}{[A]} = k\left([B]_0 - [A]_0\right) t + \ln \frac{[B]_0}{[A]_0} \tag{9.1.16}$$

Equation (9.1.16) can be graphically represented as:

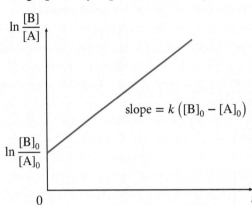

FIGURE 9.1.4   **The change in concentration of the product over time in a mixed second-order reaction.**

Since we assume that $[A]_0 \neq [B]_0$, the two reactants have different half-lives.

Mixed second-order reactions are often difficult to follow, since doing so would require measuring the changes in the concentrations of the two different reactants simultaneously. One solution is to turn the second-order reaction into a *pseudo first-order reaction*. That is, the experiment is carried out with very high initial concentrations of one reactant and normal or low concentrations of the second reactant. As a result, the concentration of the first reactant is effectively unchanged during the reaction, and the rate depends only on the concentration of the second reactant. In other words, despite the fact that two different reactants are involved, the reaction essentially behaves like a first-order reaction. A biologically relevant example of a pseudo first-order reaction is enzymatic catalysis under high reactant concentrations. In such cases, the reaction rate depends only on the concentration of the enzyme.[a]

The three main reaction types described above are summarized in the following table:

TABLE 9.1.1   **A summary of the characteristics of zero-order, first-order and second-order reactions.**

Order	Reaction	Rate Law	Integrated Rate Law	Units of k	Half-life
0th	$A \rightarrow B$	$V = k$	$[A] = [A]_0 - kt$	$M\ sec^{-1}$	$t_{1/2} = \dfrac{[A]_0}{2k}$
1st	$A \rightarrow B$	$V = k[A]$	$\ln[A] = \ln[A]_0 - kt$	sec^{-1}	$t_{1/2} = \dfrac{0.6931}{k}$
2nd	$2A \rightarrow B$	$V = k[A]^2$	$\dfrac{1}{[A]} = \dfrac{1}{[A]_0} + kt$	$M^{-1}\ sec^{-1}$	$t_{1/2} = \dfrac{1}{k[A]_0}$
	$A+B \rightarrow C$	$V = k[A][B]$	$\ln \dfrac{[B][A]_0}{[A][B]_0} = k\left([B]_0 - [A]_0\right)t$		–

[a]Note that despite the fact that the enzyme is a catalyst and not a reactant, it too affects the reaction rate.

9.1.3 Why were enzymes selected as biocatalysts?

Enzymes possess a few important advantages over simple chemical catalysts, and these advantages probably played an important part in the evolutionary selection of enzymes as the principal catalysts in biological systems. The advantages are as follows:

1. **Enzymes are highly efficient.**
   Most enzymes are capable of accelerating chemical reactions by a factor of 10^{11} to 10^{16} [24], and may even reach a factor of 10^{21} [20,24,25]. This, however, was probably a less important factor in the evolutionary selection of enzymes, since some chemical catalysts display catalytic efficiency that is comparable to or even greater than that of enzymes.

2. **Enzymes are reaction- and substrate-specific and can be controlled.**
   Most enzymes display specificity towards the types of reactions they catalyze and towards the substrates[*1] **upon which they act**. This is a huge advantage of enzymes, which most likely played an important (and perhaps decisive) part in their selection. Why was enzyme specificity so important for the evolution of cells and organisms? Imagine the highly crowded cytoplasm of an average cell. The numerous molecules in this environment participate in thousands of different chemical reactions, and these reactions need to be catalyzed in a highly regulated manner to maintain normal metabolism. A non-specific catalyst (such as a metal) would be able to catalyze such reactions, but it would do so indiscriminately. The resulting random activation of numerous chemical reactions would wreak havoc inside the cell. In contrast, a specific catalyst, such as an enzyme, acts only on its intended substrate(s) and catalyzes only its intended reaction. Still, such reactions must happen in accordance with the cell's needs, and not whenever the enzyme bumps into its substrate(s). **Being proteins, enzymes can be regulated at different levels: expression, degradation, post-translational modifications (e.g., phosphorylation) and by the use of small allosteric regulators** (see Chapters 5 and 8). The latter can be grouped into *activators* and *inhibitors*, according to their effects on the activity of their target enzymes. Many of the regulators are in fact part of the metabolic pathway, which includes the regulated enzymes. Specifically, some of the products of a pathway tend to act as inhibitors of enzyme(s) that catalyze some of the reactions in the pathway. This phenomenon, called 'product inhibition', creates a negative feedback loop, which controls the rate of the pathway according to the cell's need for its products.

   Thus, by regulating their enzymes, cells can control the timing, duration, and intensity of each of the individual reactions inside them. Indeed, virtually every enzyme involved in cellular metabolism is regulated, and enzymes involved in key steps are usually regulated at different levels (see above). Again, this type of regulation would be impossible to achieve with simple catalysts, which always work at the same rate as long as the substrate is available and the environmental conditions do not change. Furthermore, enzymes, being gene products, can co-evolve along with the biochemical needs of the organism, and thereby change the rate of the catalyzed reaction on a level beyond that of the routine regulator-dependent control.

   The specificity of enzymes is a direct consequence of their *three-dimensional struc-*

[*1]In enzyme-catalyzed reactions the reactant is called a *substrate*.

ture. As described in Chapter 8, the elaborate structure of proteins creates in many of them a geometric site capable of specifically binding other molecules. In enzymes, these binding sites, termed *active sites*, are where the molecules that the enzymes act upon (i.e., their *substrates*) bind, and where they are catalyzed into products. Generally speaking, enzymes are able to accelerate chemical reactions because their active sites are geometrically and electrostatically complementary to the transition state of the substrate, and can stabilize it via noncovalent interactions. The different means by which enzymes induce the formation of the transition state and stabilize it are described in Section 9.3 below.

3. **Enzymes can couple energy-producing processes to energy-demanding processes.**
Another advantage of enzymes over simple catalysts is their ability to couple spontaneous (energy-releasing) reactions to non-spontaneous (energy-consuming) ones. Earlier we saw that spontaneous reactions involve a decrease in the free energy of the system, i.e., the difference between the energy of products and that of the reactants is negative. This can be viewed as a 'release' of free energy (not to be confused with exothermal processes, which involve release of heat energy). Conversely, in a non-spontaneous reaction the difference is positive, which means that at least this amount of energy has to be put in for the process to occur. **Many important cellular processes, such as the biosynthesis of complex molecules or transport of chemicals across the plasma membrane, are non-spontaneous**, i.e., require the input of external free energy. **By coupling these processes to free energy-releasing (spontaneous) processes, such as ATP degradation, enzymes enable cells to drive the former type of process.** The free energy stored in ATP is not released directly upon its hydrolysis, the way heat energy is released in exothermic reactions. Instead, one of the charged hydrolysis products (P_i, ADP or AMP; usually P_i) binds to the target molecule, thereby 'channeling' the free energy to that molecule. Specifically, the binding of the charged product to the target molecule changes its electronic properties and its interactions with other molecules, which changes the free energy of the system. As we will see below, transfer of P_i from ATP to a small metabolite may activate it for further reactions, such as hydrolysis or ligation to other metabolites. Alternatively, when P_i is transferred to a protein (e.g., an energy-requiring transporter), it acts by inducing conformational changes needed for the protein's function (see Chapter 2, Subsection 2.6.2 for details).

4. **Enzymes can confine sequential reactions to one place.**
Many enzymes, especially those that participate in the same biochemical pathway, form large functional complexes *in vivo*. Such complexes are usually arranged spatially so that the individual reactions can be carried out sequentially. That is, the product of one enzyme in the complex serves as the substrate of the next enzyme, etc. This mode of action greatly enhances the speed and efficiency of the process, since the substrate molecules need not diffuse randomly throughout the cytoplasm in order to meet the right enzyme. The enzyme complexes may be viewed as 'molecular machines', containing different parts that move in coordination with one another, while each of them is executing a different task [26]. Such a view allows the researcher to characterize the enzymes of the complex as functional modules specializing in one type of chemical or physical work: the formation or degradation of chemical bonds, conversion of chemical energy to mechanical or kinetic energy, etc.

9.1.4   Why is it important to understand enzyme action?

The importance of enzymes to the function of all living organisms on Earth has rendered them a prime target of both basic and applied biological research, from its very beginning. Basic research focuses on the many aspects of enzyme action, such as the ways in which enzymes accelerate chemical reactions and maintain specificity and selectivity, their kinetic behavior and how it can be used for analyzing their efficiency, etc. These aspects are discussed in this Chapter in Sections 9.2 through 9.5. The interest in enzymes also stems from their involvement in diseases that result from enzyme inexpression, inactivity, or excessive activity. Such diseases may be the result of genetic causes or exposure to environmental toxins that change the activity of normal cellular enzymes. Studies of these medical aspects of enzymes include basic research that focuses on enzyme involvement in disease development, in addition to applied research that focuses on enzymes as drug targets and biological drugs. Finally, enzymes are also used as catalysts in certain industries that produce chemical materials (textile, food and biofuel industries). The applications of enzymes in medical and other industries are discussed in Section 9.6.

Before we dive into the many aspects of enzyme activity, we must first understand the functional scope of the thousands of known natural enzymes. That is, what types of reactions do enzymes catalyze and what types of substrates do they target? To achieve such an understanding, it is necessary to rely on an efficient method of classification, which will be discussed in the following subsection.

9.1.5   Enzyme classification

As mentioned above, the natural enzymes that are currently known catalyze numerous different reactions. This diversity is further burdened by the fact that enzymes that catalyze the same types of reactions may originate from different biological sources, have different structures and sequences, and act on different substrates. Thus, the first step in understanding enzymes must involve an efficient method of systematic classification. Curiously, although enzymes have been studied for over 150 years, biologists have only recently started to use a systematic method for enzyme classification. Prior to the development of this method, enzyme classification and naming approaches were rather confusing and inconsistent [27]. For example, enzymes catalyzing oxidation-reduction (i.e., redox) reactions were assigned names such as *dehydrogenases*, *reductases*, *oxidases*, *oxygenases* and *peroxidases*, which alluded to the types of reactions the enzymes catalyze, but did not really explain the differences between them (see more in Subsection 9.1.5.1 below). In other cases, such as the enzymes *diaphorase* and *rhodenase*, the names did not provide any useful information about the activity of the enzyme or about its substrate specificity.

The first step towards systematic classification of enzymes was taken in 1958 by Malcolm Dixon and Edwin Webb, who grouped enzymes according to the reactions they catalyze (see below). This initiative was further developed by the *International Union of Biochemistry and Molecular Biology* (*IUBMB*) in association with the *International Union of Pure and Applied Chemistry* (*IUPAC*), and became the conventional method used today (the Enzyme Commission (EC) method) [27]. The EC method starts by assigning each of the known enzymes to one of six 'classes', each catalyzing a different type of reaction [29] (Figure 9.3):

1. **Oxidoreductases** – enzymes that catalyze oxidation-reduction (redox) reactions (Figure 9.3a).

(a) Oxidation-reduction (redox):

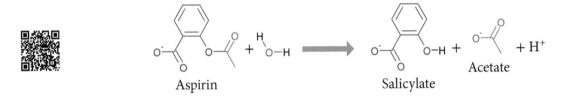

ethanol + NAD$^+$ ⟷ acetaldehyde + NADH + H$^+$

(b) Group transfer:

2-oxoglutarate + L-alanine ⟷ L-glutamate + Pyruvate

(c) Hydrolysis:

Aspirin + H_2O ⟶ Salicylate + Acetate + H$^+$

(d) Water-independent lysis:

D-threo-isocitrate ⟷ Succinate + Glyoxylate

(e) Isomerization:

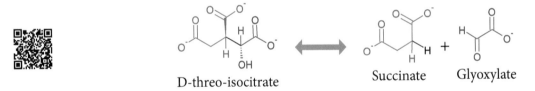

Fumarate ⟷ Maleate

(f) Ligation:

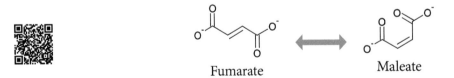

D-alanine + D-alanine + ATP ⟶ D-alanyl-D-alanine + ADP + P$_i$

2. **Transferases** – enzymes that catalyze reactions in which a chemical group is transferred from one substrate molecule to another (Figure 9.3b).

3. **Hydrolases** – enzymes that catalyze hydrolytic reactions, i.e., bond cleavage by water (Figure 9.3c).

4. **Lyases** – enzymes that catalyze reactions in which a covalent bond is cleaved or formed without the help of water (Figure 9.3d).

5. **Isomerases** – enzymes that catalyze interconversion between molecular isomers. Such reactions may involve group transfer between positions in the substrate, *cis-trans* changes, and interconversions between S and R stereo-configurations (Figure 9.3e).

6. **Ligases** – enzymes that catalyze reactions in which two substrates are chemically bonded into one product, using a phosphorylated co-substrate (e.g., ATP) (Figure 9.3f).

Each of the six classes described above is then further divided into subclasses and sub-subclasses based on the chemical properties of the substrates on which the enzymes act. **The result of these classifications is a four-level definition for each enzyme, represented by a corresponding number comprising four numerals** (see Appendix for a complete list of the first three levels):

1. **Class** – specifies the reaction type (according to the six types mentioned above).

2. **Subclass** – specifies the general type of group or bond involved in the reaction.

3. **Sub-subclass** – usually provides a more accurate definition of the group or bond involved in the reaction.

4. **Sub-sub-subclass** – specifies the exact natural substrate of the enzyme[*1].

FIGURE 9.3 **Examples of the six types of reactions catalyzed by enzymes according to the EC method.** (Opposite) (The individual reactions are adapted from the MetaCyc database [28].) For clarity, explicit hydrogens are not shown, except around centers in which the number of hydrogen atoms changes during the reaction. (a) Oxidation-reduction (redox): oxidation of ethanol to acetaldehyde, catalyzed by *alcohol dehydrogenase*. The oxidation involves the transfer of a hydride species (blue) from ethanol's C_α to NAD^+ and the deprotonation of its hydroxyl group (red). (b) Group transfer: amino transfer from alanine to α-ketoglutarate, catalyzed by *alanine aminotransferase*. The transfer of the amino group (blue) from the first co-substrate involves reduction of the corresponding carbon atom to a keto group (red), and vice versa in the other co-substrate. However, since these events involve internal electron transfer between amino and keto groups, the reaction is not considered to be redox (see more in Subsection 9.1.5.2.1). (c) Hydrolysis: breakdown of aspirin to salicylate (blue) and acetate (green) by using water (red), as catalyzed by *aspirin hydrolase*. (d) Water-independent cleavage of covalent bonds: breakdown of isocitrate to glyoxylate (blue) and succinate (red), catalyzed by *isocitrate lyase*. (e) Isomerization: interconversion between fumarate (*trans* bond) and maleate (*cis* bond), catalyzed by maleate *cis-trans isomerase*. (f) Ligation: the attachment of two D-alanine molecules (blue and red), catalyzed by *D-alanine-D-alanine ligase*.

[*1]Note that although some of the names refer to one direction of the chemical reaction, many enzymes catalyze equilibrium reactions, in which both directions occur, depending on the concentrations of the products and reactants.

For example, let us look at the classification of *glycine amidinotransferase* (Figure 9.4a). This enzyme belongs to the *transferase* class (*EC 2*), which includes 3,124 enzymes, catalyzing group-transfer reactions. These are divided by the EC method into ten subclasses, based on the chemical group that is being transferred. The first subclass (*EC 2.1*) includes 607 enzymes that transfer *one-carbon groups*. These are further divided into four groups. The last of these groups (*EC 2.1.4*) includes two enzymes; each is known to transfer the one-carbon *amidino group*. The first of these two enzymes is glycine amidinotransferase, so named because it transfers the amidino group to the amino acid *glycine* (Figure 9.4b). Thus, the classification of this enzyme according to the EC method is *2.1.4.1*. Note that the EC method relies on functional characteristics (reaction type and subtype) rather than structural characteristics. This means, among other things, that different structures and folds may be able to carry out the same function. Indeed, a statistical analysis shows that, **on average, the reactions in each EC class are carried out by members of about three different protein evolutionary families**[*1] [30]**, indicating that there is more than one way to carry out the same type of reaction** (see more details in Subsection 9.1.5.7 below).

The complete set of EC assignments approved by the IUBMB can be found in the ExplorEnz database [31]. More information on each of the classes and subclasses can be found in the ENZYME database of the Swiss Institute of Bioinformatics [32].

The EC method has been widely adopted, and it constitutes the basis of many databases and web-based tools. These include the following:

BRENDA (BRaunschweig ENzyme DAtabase) [33] – an extensive, yet user-friendly database of natural enzymes, maintained by the Department of Bioinformatics and Biochemistry at Technische Universität Braunschweig. The database provides comprehensive information on numerous enzymes, such as the type of reaction catalyzed by each enzyme, the biochemical pathway in which it operates, source organisms, substrates and products (natural and others), co-factors and inhibitors, cellular localization, kinetic parameters, optimal values of pH, temperature and salinity, sequences, existing 3D structures, known mutants, post-translational modifications, and more. The data are mined from various sources, both manually and automatically. URL: www.brenda-enzymes.org

MACiE (Mechanism, Annotation and Classification in Enzymes) [34] – a database of enzyme mechanisms, developed as part of a collaboration between the Thornton Group (European Bioinformatics Institute) and the Mitchell Group (University of St. Andrews). MACiE provides step-by-step textual and graphic descriptions of the catalytic mechanisms of selected enzymes. The information is based on literature surveys. The overall fold and catalytic residues are also described. URL: www.ebi.ac.uk/thornton-srv/databases/MACiE/

KEGG (Kyoto Encyclopedia of Genes & Genomes) [35] – a comprehensive database that covers a wide range of topics regarding proteins in general, including enzymes. KEGG integrates 17 specific databases that describe proteins and enzymes according to different parameters related to their function. These parameters include the catalyzed reaction, biochemical pathway, functional units, genomic and medical relations, and

---

[*1] Evolutionary families were defined by clustering all enzyme sequences that have EC numbers, according to homology.

(a)

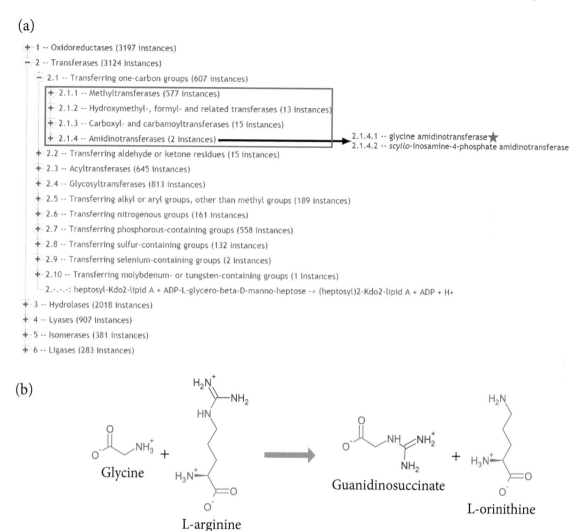

(b)

FIGURE 9.4 **The Enzyme Commission method** (a) Classification of glycine amidinotransferase (taken from the MetaCyc database [28]). (b) The reaction catalyzed by the enzyme.

more. KEGG is maintain and developed by Kanehisa Laboratories at Kyoto University and the University of Tokyo. URL: www.kegg.jp

BioCyc [28] – a database that provides information on each enzyme's catalyzed reaction and metabolic pathway, in addition to genomic information. BioCyc also contains software tools that can be used for visualization and analysis of the data. The database is maintained by SRI International. URL: biocyc.org

The EC method seems to be, at least currently, the best method available for systematic classification of enzymes. However, it suffers from some problems, such as the following [36]:

- The inconsistency of second- and third-level assignments between different classes and even within the same class (see below). For example, in lyases (EC 4), the second-level classification is determined according to the type of bond being broken, whereas in isomerases (EC 5), enzymes in the same level are grouped according to the type of isomerization [36].

- Accounting for overall reactions but not the mechanisms involved. Enzymes that catalyze the same overall reaction are included in the same class, even if they use different mechanisms and/or cofactors to catalyze that reaction [37].

- Ignoring *isoenzymes* (a.k.a. *isozymes*), that is, enzymes that have the same activity but still differ from each other in structure [30], substrate preference (when the enzyme has more than one natural substrate), and susceptibility to inhibitors. For example, the definition of *alcohol dehydrogenase* (EC 1.1.1.1/2) includes any oxidoreductase that transfers electrons from a C–OH group in alcohols to $NAD(P)^+$ (see Subsection 9.1.5.1 below). This definition applies to numerous enzymes, which can be found in virtually all organisms. Still, many of these enzymes differ in structure and behavior [27]. In fact, even in the mammalian liver one can find ~20 isozymes of alcohol dehydrogenase. Although all of these enzymes show the same principal activity on alcohols, they differ in their chain length preference for primary alcohols, as well as in their susceptibility to inhibitors.

- Enzyme promiscuity. Some enzymes that are included in the same EC group have been shown to catalyze different types of reactions. This phenomenon, which is called 'catalytic promiscuity', is discussed in Subsection 9.1.5.7 below. Note, however, that promiscuous enzymes are often assigned more than one EC number.

Naturally, all of the problems described above will have to be addressed in future EC version, or alternative classification schemes that are yet to be developed [36].

In the following subsections we go over the six classes of enzymes and discuss key characteristics and examples. For a more detailed review we recommend books dedicated to this subject, e.g., the *Springer Handbook of Enzymes* series [38].

9.1.5.1 Oxidoreductases (EC 1)

9.1.5.1.1 Definition and examples

Oxidoreductases are one of the two largest enzyme classes, constituting 28% of the enzymes in the ExplorEnz database [31]. They catalyze oxidation-reduction (redox) reactions, i.e., reactions in which electrons are transferred from one molecule (the donor) to another (the acceptor) [39]:

Scheme 9.1.
$$D^\bullet + A \longrightarrow D + A^\bullet$$

(where D is the electron donor, A is the acceptor, and the red dot is the electron transferred during the reaction; thus, before the transfer D is reduced and A is oxidized, whereas after the transfer D is oxidized and A is reduced).

There are a few important points that should be noted regarding the above scheme. First, most redox reactions are essentially reversible, so in the reverse direction D becomes A, and vice versa. Second, while most biological redox reactions involve the transfer of two electrons, in some cases one, four or even six electrons are transferred [39]. Third, in many redox reactions two electrons are transferred along with a proton (i.e., as a *hydride ion* (H^-)). In some other cases, the reaction involves the transfer of either one or two atoms of molecular oxygen (O_2) to the other co-substrate. These cases are described in detail below.

The fast transfer of electrons in oxidoreductases is facilitated by amino acids, organic coenzymes, transition metals, or combinations thereof. A survey of the MACiE database [34] shows that His and Cys are the catalytic amino acids that are most commonly found in oxidoreductases, where His is involved in both proton and electron shuttling, and Cys is involved mainly in electron shuttling [40].

In the reactions catalyzed by oxidoreductases, one of the redox partners is an alcohol, thiol, carbonyl (aldehyde or ketone), acid, amine, or unsaturated carbon chain (C=C). The other partner may be a nucleotide coenzyme (NAD$^+$ or NADP$^+$, Figure 9.5a), O$_2$, a quinone, or even a protein such as a cytochrome (via its heme iron), iron-sulfur protein, or flavoprotein[*1]. In some cases the electrons are transferred one at a time, whereas in other cases they are transferred two at a time, as part of a hydride ion (H$^-$)[*2] [41]. The nature of the donor and acceptor molecules involved in the reaction is reflected in the EC number of the oxidoreductase, albeit inconsistently: the second numeral usually designates the chemical group in the electron donor that becomes oxidized, but in a few subclasses it designates the chemical group in the acceptor that becomes reduced. In the former case the third numeral usually designates the electron acceptor, and the fourth numeral represents the substrate specificity of the enzyme. For example, the EC number of cholesterol oxidase is 1.1.3.6. The second numeral (1) designates that it oxidizes hydroxyl (OH) groups; the third numeral (3) designates that the electron acceptor is molecular oxygen (O$_2$); and the fourth numeral (6) designates that it acts on cholesterol.

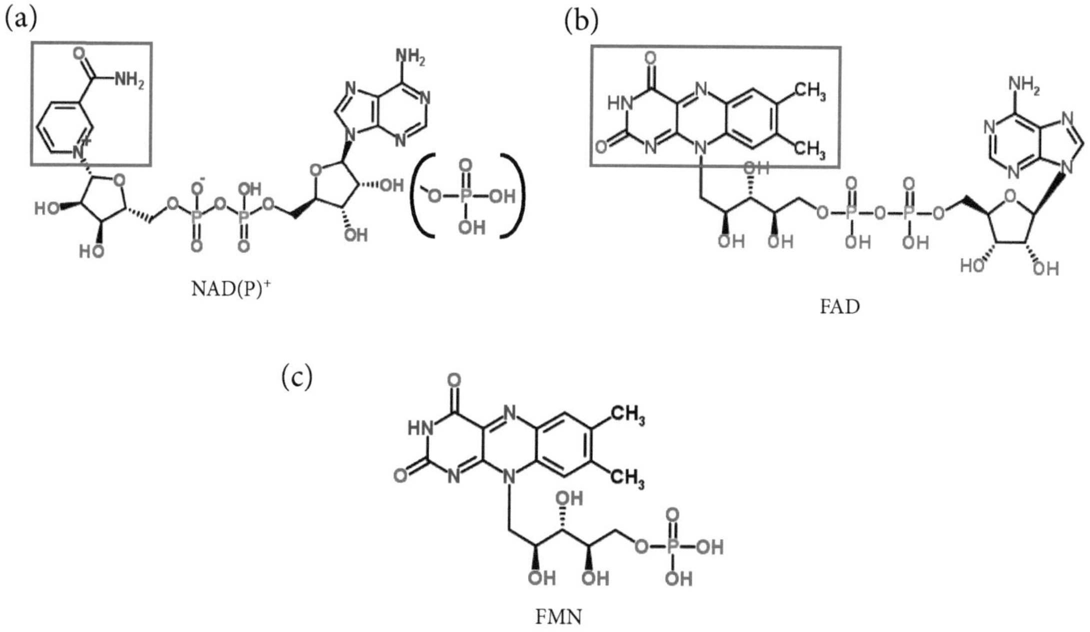

FIGURE 9.5 **Common nucleotide coenzymes involved in redox catalysis.** (a) Nicotinamide dinucleotide (NAD$^+$), derived from niacin (vitamin B$_1$; see Section 9.4 below). In the related NADP$^+$, a phosphate group is attached to one of the hydroxyls (shown in parentheses). The part of the molecule involved in electron transfer is the nicotinamide ring (marked by a red square). (b) Flavin adenine dinucleotide (FAD), derived from riboflavin (vitamin B$_2$; see Section 9.4 below). The part of the molecule involved in electron transfer is the isoalloxazine ring system (marked by a red rectangle). (c) Flavin adenine mononucleotide (FMN), which is also derived from riboflavin and has the same reactive ring system as FAD. The structures are taken from the ChemSpider database [42].

[*1]Flavoproteins contain flavin adenine dinucleotide (FAD) or, less commonly, flavin adenine mononucleotide (FMN) as prosthetic groups (Figures 9.5b and c, respectively).

[*2]Not to be confused with reactions in which a hydrogen radical (H•) is transferred.

Redox reactions are involved in many biological processes, especially in central metabolism. In such metabolic processes, redox reactions are used either to extract energy from foodstuff or utilize it for the biosynthesis of complex molecules. In catabolic reactions, energy is extracted from foodstuff as electrons, which are temporarily stored on free NADH or on $FADH_2$ in flavoenzymes. In aerobic organisms these coenzymes transfer the electrons they carry to the respiratory chain, which resides either in the inner mitochondrial membranes of eukaryotes or in the plasma membranes of prokaryotes. The electrons are then transferred (along with protons) to molecular oxygen (O_2), turning it into water. The passage of electrons through the respiratory chain releases energy that is used to build an electrochemical proton gradient, which, in turn, is used to form ATP. In anaerobic organisms, nitrogen and sulfur-based acceptors are often used instead of O_2, and their reduction creates e.g. ammonia or hydrogen sulfide (respectively), instead of water [39].

As mentioned above, redox reactions also play a central part in anabolism. However, in most anabolic reactions, the electron carrier is NADPH instead of NADH or $FADH_2$. NADPH is produced in our body by processes such as the *pentose-phosphate pathway* [43]. In biosynthetic reactions, NADPH is used as a reducing agent that enables certain molecules to be constructed, including fatty acids (see Subsection 9.1.5.2.3 below), cholesterol, eicosanoids and nitric oxide (both are transmitters), amino acids, and nucleotides (see Table 9.3 at the end of the chapter). This function is especially important in tissues specializing in biosynthesis, such as the liver, fat tissues, lactating breast tissue, and the adrenal gland. As explained in Chapter 2 (Box 2.3), NADPH is also important for non-energy processes, such as counteracting oxidative damage to cells and tissues.

Historically, oxidoreductases have been grouped into the following general types:

1. **Dehydrogenases** – the largest group of oxidoreductases. These enzymes transfer electrons reversibly as hydride ions between the substrate and a nucleotide coenzyme. The latter can be a freely soluble coenzyme such as NAD(P)H (as in *alcohol dehydrogenase*, EC 1.1.1.1), or $FADH_2$, which functions as a prosthetic group of a flavoenzyme (as in the fatty acid breakdown enzymes *acyl-CoA dehydrogenases*, EC 1.3.8) (Figure 9.6a). Other well-known examples include *glyceraldehyde-3-phosphate dehydrogenase* (EC 1.2.1.12) and *pyruvate dehydrogenase* (EC 1.2.4.1), both of which play a central role in energy production from carbohydrates, as well as *glucose-6-phosphate dehydrogenase (G6PD)* (EC 1.1.1.49), the first enzyme of the pentose-phosphate pathway (see Chapter 2, Box 2.3).

2. **Oxidases** – enzymes that catalyze transfer of electrons from an organic substrate to molecular oxygen (O_2). This process reduces the latter either to H_2O (e.g., in *cytochrome c oxidase*; EC 1.9.3.1) or to H_2O_2 (e.g., in *xanthine oxidase*; EC 1.17.3.2) (Figure 9.6b). Some redox enzymes that transfer electrons between an oxygen atom within the substrate (i.e., intramolecular oxygen) and NAD(P)H are sometime referred to as oxidases as well, but this is not their recommended name (e.g., *glyoxylate reductase*, a.k.a. *glycolate oxidase*; EC 1.1.1.26). Many oxidases are flavoproteins, and their FAD or FMN groups are important in facilitating the electron transfer. Other oxidases are metalloproteins, and built-in transition metals (e.g., iron and copper) are used to activate the oxygen molecule and transfer the electrons (see Subsection 9.3.3.4 below for more details). The electron transfer reactions catalyzed by oxidases are essentially irreversible, due to the high O_2/H_2O_2 and O_2/H_2O redox potentials [39]. *Cytochrome c oxidase*, which is mentioned above, is a well-studied oxidase, being a key

component of the mitochondrial electron transfer chain [44]. Another well-known example is *monoamine oxidase (MAO)* (EC 1.4.3.4), which is important in neurotransmission and is targeted by certain antidepressants (see Table 8.1).

3. **Oxygenases** – enzymes that transfer electrons from an organic substrate to O_2, while incorporating one of the oxygen atoms (in *monooxygenases*) or both of them (in *dioxygenases*) into the substrate (see Figure 9.6c for monooxygenases). The single oxygen atom added by monooxygenases is in the form of a hydroxyl group, and therefore these enzymes are often referred to as *hydroxylases*. The transfer of electrons inside the enzyme is carried out by a flavin (FAD/FMN) or pterin (see Subsection 9.4.2 below) group, and/or by metals [39]. Oxygenases participate in many important metabolic processes, including the degradation of aromatic compounds, lipid metabolism, collagen formation, breakdown of xenobiotics, and alkane functionalization [44]. Well-known examples of oxygenases include *cytochrome P450*, a group of enzymes that detoxify ingested drugs in the liver, and *cyclooxygenases* (EC 1.14.99.1), enzymes that mediate pain and inflammation, and which are targeted by certain anti-inflammatory drugs such as aspirin (see Chapter 8, Subsection 8.6.2.1.2, and Table 8.1, as well as Subsection 9.5.2.2 in this chapter).

4. **Peroxidases** – enzymes that transfer electrons from a reduced substrate to either hydrogen peroxide or alkyl peroxide [39] (Figure 9.6d). A well-known example is *catalase* (EC 1.11.1.6), which reduces two molecules of hydrogen peroxide (H_2O_2) to two water (H_2O) molecules and one O_2 molecule. Catalase can be found in virtually all oxygen-exposed organisms, and is used to neutralize H_2O_2, a harmful oxidant. Another H_2O_2-neutralizing peroxidase used to fight oxidative damage is *glutathione peroxidase* (EC 1.11.1.9). This enzyme uses the selenium-containing amino acid *selenocysteine* (see Chapter 2, Subsection 2.2.1.4) to reduce H_2O_2 to water. Other forms of glutathione peroxidase reduce peroxides of phospholipids (EC 1.11.1.12) and fatty acids (EC 1.11.1.22) that are formed under oxidative stress.

The historical classification of oxidoreductases outlined above is very commonly used in the literature, and the enzymes belonging to the four categories are usually named as follows:

- '*donor:acceptor type*' (type = dehydrogenase, oxidase, etc)

- '*donor type*' (e.g., alcohol dehydrogenase).

9.1.5.1.2 *Structure and stereospecificity*

Oxidoreductases include three protein superfamilies: *long-chain alcohol dehydrogenases, short-chain dehydrogenases and reductases (SDRs)*, and *aldo-keto reductases (AKRs)*. These superfamilies are characterized by different 3D structures. For example, proteins in the SDR superfamily use the pervasive *Rossmann fold* for NAD(P)H binding, whereas AKR proteins do not. The superfamilies also differ in their specific catalytic mechanisms. For example, long-chain alcohol dehydrogenases use zinc cations for catalysis [45], whereas SDR proteins do not use metals at all [46,47].

As explained above, many reactions carried out by oxidoreductases involve the transfer of electrons between the substrate and the nucleotide coenzymes NAD(P)H (Figure 9.7a)

(a) $RH_2 + \dfrac{NAD(P)^+}{[FAD]} \longleftrightarrow R + \dfrac{NAD(P)H + H^+}{[FADH_2]}$

(b) $2RH_2 + O_2 \longleftrightarrow 2R + 2H_2O$

or

$RH_2 + O_2 \longleftrightarrow R + H_2O_2$

(c) $RH + O_2 + \dfrac{NAD(P)H+H^+}{[FADH_2]} \longleftrightarrow R{-}OH + H_2O + \dfrac{NAD(P)^+}{[FAD]}$

(d) $R^1H_2 + R^2O_2H \longleftrightarrow R^1 + R^2{-}OH + H_2O$

FIGURE 9.6 **General schemes describing the reactions catalyzed by (a) dehydrogenases, (b) oxidases, (c) monooxygenases (sometimes called hydroxylases because their products contain a hydroxyl group), and (d) peroxidases (R^2O_2H is an alkyl peroxide).** FAD in square brackets is an alternative to NADH or NADPH.

or $FADH_2^{*1}$ (Figure 9.7b). It is estimated that NAD(P)H alone is used by ~80% of oxidoreductases [48]. It should be noted, though, that while FAD is a true coenzyme that transfers electrons from the donor to the acceptor within the enzyme, NAD(P)H usually behaves as a co-substrate (or co-product) rather than a coenzyme. That is, it donates or accepts the electrons to or from the substrate as the final donor or acceptor (respectively), instead of just transferring them from one substrate to another. After the transfer is completed, the cofactor leaves the enzyme as one of the co-products (NAD(P)$^+$ of NAD(P)H). There are some exceptions, however, as in the case of *S-adenosyl-homocysteinase*, in which the oxidized form (NAD$^+$) merely transfers the electrons from donor to acceptor [44].

In the oxidative direction of NAD(P)$^+$-dependent reactions, two electrons are transferred as a hydride ion (H$^-$) from the substrate to the C4 atom of the NAD(P)$^+$'s nicotinamide ring (Figure 9.7a) [49]*2. Such a reaction requires a strong electrophilic center, which is provided by the oxidized pyridinium [44]. The hydride transfer results in two hydrogen atoms covalently bound to the C4 atom, each facing a different direction. Because the two hydrogen atoms are identical, the C4 atom is not chiral. This, however, can be changed by replacing one of the hydrogen atoms with its deuterium isotope. When the deuterium is positioned above the ring plane, C4 has an *R* configuration. Conversely, when the deuterium is positioned below the ring plane, C4 has an *S* configuration. Thus, while NAD(P)H's C4 atom is not chiral *per se*, it is referred to as '***pro*-chiral'.** Accordingly, in a regular hydride transfer, the upwardly projecting hydrogen is referred to as '*pro-R*' (or H_R), and the upwardly projecting hydrogen atom as '*pro-S*' (or H_S) (Figure 9.7a). Interestingly, **individual oxidoreductases transfer hydride ions to or from either the pro-R configura-**

---

*1In contrast to NAD(P), FAD and FMN can donate or accept the two electrons one electron at a time (Figure 9.7b, bottom). This allows them to mediate inside proteins the transfer of electrons between species that are able to pass only one electron at a time (e.g., heme, iron-sulfur clusters and quinones) and species that can only pass two electrons at a time (e.g., NAD) [44]. Such mediation takes place, for example, in the respiratory chain, which converts the chemical energy stored in foodstuff into ATP.

*2The hydride transfer reaction is generally assumed to be direct, although a recent study has demonstrated that in cases in which the substrate is an *α*, *β*-unsaturated compound, the transfer may involve a covalent substrate-NAD(P) intermediate [50].

tion or the pro-S configuration, not both. This 'stereospecificity' results from the three-dimensional structure of the active site: the active site positions the coenzyme's nicotinamide in a very specific way with respect to the substrate, so only one of the hydrides (*pro-R* or *pro-S*) can be transferred. For example, *L-lactate dehydrogenase* (EC 1.1.1.27) transfers the H_R hydrogen to its substrate pyruvate in virtually 100% of the cases (Figure 9.7c). Conversely, *glyceraldehyde 3-phosphate dehydrogenase* (EC 1.2.1.12) is H_S-specific.

The active site of oxidoreductases also binds the other (co-)substrate specifically, which contributes to the exact positional relationship between that co-substrate and NAD(P)H. **The exact orientation of the co-substrate with respect to NAD(P)H's C4 atom also leads**

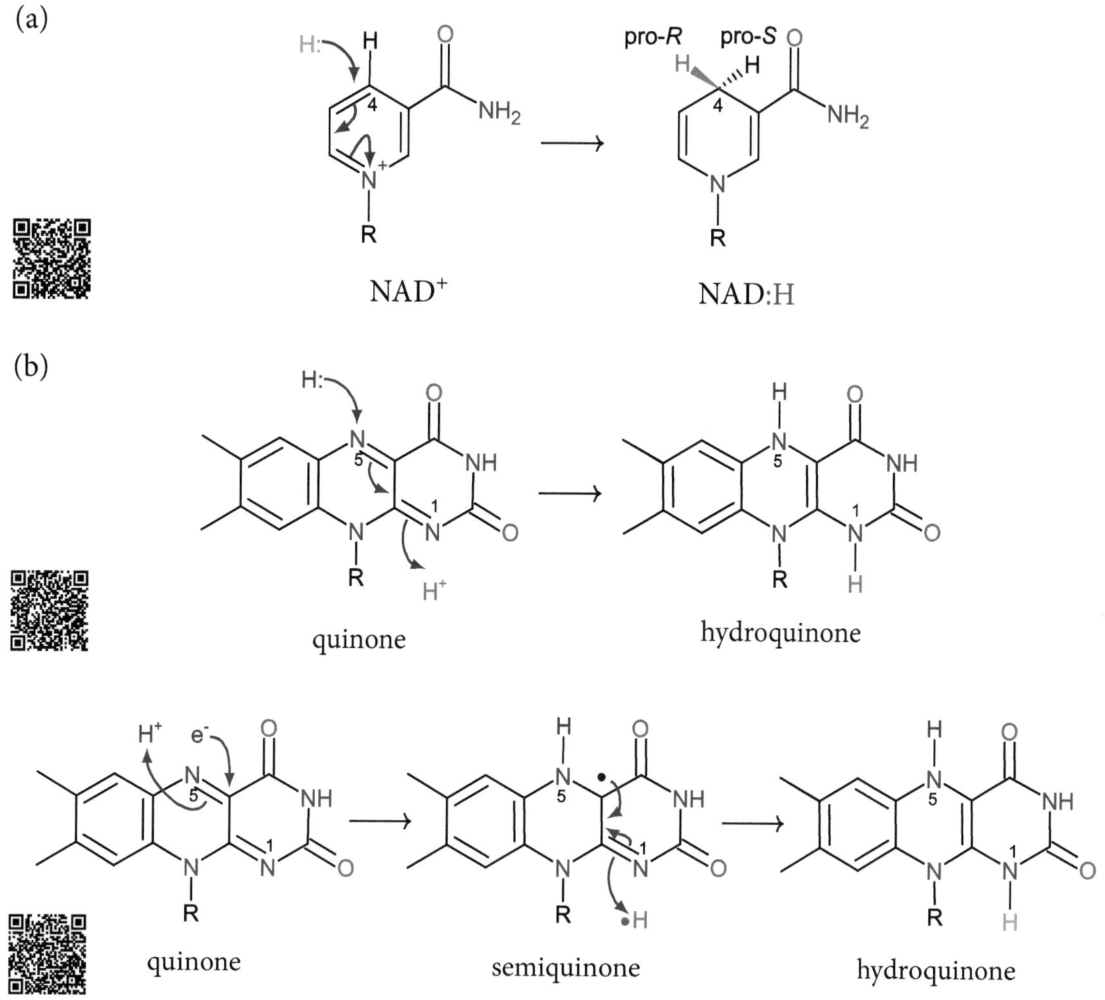

FIGURE 9.7 **Oxidation and reduction of NAD(P)H and FADH$_2$.** (a) Reduction of the nicotinamide ring of NAD(P)$^+$ by hydride transfer to the ring's C4, at either the *pro-S* or *pro-R* positions. (b) Reduction of the isoalloxazine ring system of FAD (or FMN) by transfer of two electrons to the N1 and N5 atoms of FAD. *Top*: the simultaneous transfer of two electrons to FAD. The electrons are transferred as a hydride ion to N5, accompanied by electronic rearrangements and protonation of N1. *Bottom*: the two-step transfer of two electrons and two protons to FAD. The first step transfers a single electron to N5, creating a semiquinone radical. The second step transfers a hydrogen radical (H•) to N1, which, along with the radical transferred in the first step, creates a fully reduced hydroquinone.

(c)

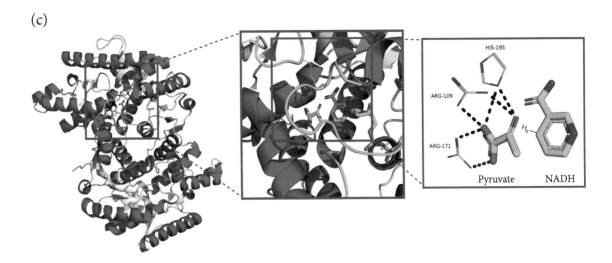

(d)

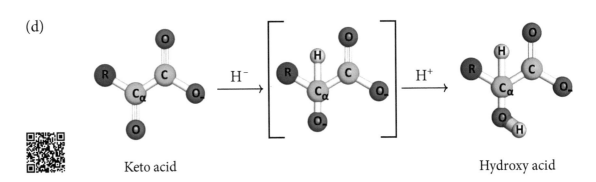

Keto acid   Hydroxy acid

(e)

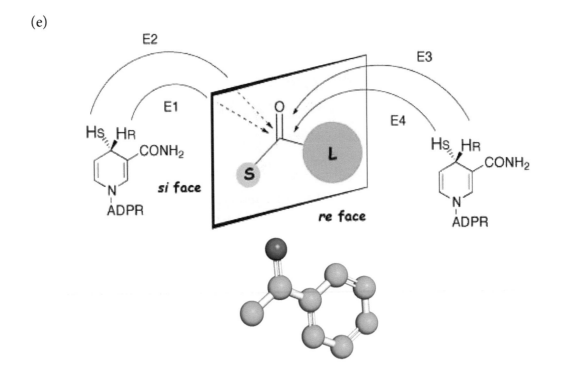

to stereospecificity of oxidoreductases towards their substrates [51]*1. That is, a given oxidoreductase can convert the co-substrate into an S or R product, but not both. This stereospecificity is observed, for example, in enzymes belonging to the *ketoreductase* (EC 1.1.1) and *amino acid dehydrogenase* (EC 1.4.1) subclasses. Ketoreductases are a large group of redox enzymes that reduce a non-chiral keto group (C=O) in their substrates to a chiral hydroxyl group (C–OH) [52]. The mechanism of this reaction involves hydride transfer from NAD(P)H to the keto acid, followed by protonation of the keto oxygen to create the hydroxyl group (Figure 9.7d). In each of these enzymes, the active site contains two subsites, one large and one small. In most substrates of ketoreductase, the substituents on the two sides of the keto group are of different sizes. Thus, when the substrate binds to the enzyme, its larger substituent binds to the large subsite of the active site, and its small substituent binds to the small subsite. Generally speaking, there are four ways in which a hydride ion can be transferred to the substrate in such cases, depending on the substrate's orientation in the active site and its position with respect to the NAD(P)H coenzyme [51,53] (Figure 9.7e). However, since in each ketoreductase the substrate has only one orientation, and it is positioned either above or below the plane of NAD(P)H's nicotinamide ring (but not both), the hydride ion is transferred to only one face of the substrate (one side is termed *si* and the other *re*). This means that each ketoreductase will always create the same hydroxyl stereo-configuration in its substrate*2. Ketoreductases usually act on different natural ketones, which may suggest different product configurations. However, these substrates usually share common structural characteristics (besides having a keto group), which means they tend to bind to the active site in the same manner. Therefore, the same hydroxyl configuration (S or R) is expected to form in each of their corresponding products. Indeed, **each ketoreductase usually**

FIGURE 9.7 **Oxidation and reduction of NAD(P)H and FADH$_2$.** (Opposite) (Continued) (c) The structural basis for the stereospecificity of *L-lactate dehydrogenase (LDH)*. *Left*: the structure of LDH from *Cryptosporidium parvum* is shown (PDB entry 2fm3). *Middle*: a blow-up of the active site of subunit A, showing NADH and pyruvate (the substrate) as sticks. *Right*: pyruvate-stabilizing interactions in the active site. Pyruvate and NADH are positioned in the active site so H$_R$ is facing pyruvate's C2 atom (the hydride acceptor), while H$_S$ (not shown) is facing the opposite direction. Thus, only H$_R$ can be transferred to pyruvate. The hydrogen bonds and salt bridges that hold pyruvate in place are shown as black dashed lines, with the interacting side chains also presented. (d) The mechanism of ketoreductases, which includes hydride transfer from NAD(P)H to the C$_\alpha$ atom of the keto acid, followed by protonation of the negatively charged oxygen to form the hydroxyl group. The proton is taken from H$_3$O$^+$. (e) The four ways in which a hydride ion can be transferred from NAD(P)H to a ketone substrate bound to the active site of a ketoreductase. S and L designate the small and large substituents of the substrate's keto group (the example of acetophenone is shown at the bottom of the figure without explicit hydrogen atoms). With E1 and E2 enzymes, the hydride attacks the *si*-face of the carbonyl group, whereas with E3 and E4 enzymes, the hydride attacks the *re*-face, which results in the formation of (R) and (S)-alcohols, respectively. Taken from [51].

---

*1As mentioned above, the mechanism in some oxidoreductases may involve a covalent substrate-NAD(P) intermediate, making it even more important for the enzyme and the substrate to be positioned correctly with respect to each other.

*2Provided that the ketone group is flanked by groups whose size difference is large enough. The smaller the difference, the lower the enzyme's stereospecificity.

has a known S or R preference, although the magnitude of this preference is different in each enzyme and for each substrate.

While ketoreductases use NAD(P)H to reduce keto acids to hydroxyl acids, *amino acid dehydrogenases* (*AA-DHs*) (EC 1.4.1) use the same coenzymes to reduce keto acids to amino acids [54] (*reductive amination*; Figure 9.8a). **Reductive amination is similar to the ketone reduction carried out by ketoreductases. However, instead of transferring NAD(P)H's hydride ion directly to the ketone oxygen, the latter is first attacked by ammonia, and the hydride ion is transferred to the resulting imino group**[*1] (Figure 9.8b). This has been suggested to result from electrostatic interactions between the substrate and the enzyme, which keep the keto acid far from NAD(P)H for an efficient hydride transfer, but allow the two to become closer once the positively charged imino intermediate is formed [55].

Amino acid dehydrogenases belong to one of two enzyme groups having a central role in amino acid metabolism (the other group consists of *aminotransferases*, which are discussed in Subsection 9.1.5.2.1 below). Since their reactions are reversible[*2], AA-DHs are capable of both creating amino acids by reductive amination and degrading them to keto acids and ammonia by oxidative deamination (Figure 9.8a). Regarding the latter process, liver *glutamate dehydrogenase* (EC 1.4.1.2) probably has the most central role of all AA-DHs. This is because the amino acids glutamate and glutamine are the major carriers of ammonia molecules obtained from the degradation of all amino acids in tissues. The reason our bodies need ammonia carriers in the first place is because ammonia (or NH_4^+ in aqueous solutions) is highly toxic, especially to our brains[*3]. Glutamate and glutamine carry the amino groups of tissue amino acids via circulation to the liver, where glutamate is deaminated by mitochondrial glutamate dehydrogenase, and its amino group is converted into the less toxic *urea* by the *urea cycle*. Glutamine is converted into glutamate, and therefore has the same fate. Thus, glutamate dehydrogenase is responsible for handling the amino groups obtained from most amino acids in our bodies.

Although members of the AA-DH group share sequence similarities (e.g., a glycine-rich region with a conserved catalytic lysine), their overall sequence similarity is rather low, and they often differ in terms of substrate specificity, stability, salt tolerance and other parameters [54]. Their 3D fold ('*ELFV dehydrogenase fold*') is evolutionarily conserved and contains two domains. The N' domain is involved in oligomerization, whereas the C' domain is responsible for the binding of the nucleotide coenzyme. The C' domain has an α/β fold reminiscent of a Rossmann fold, with the exception that the direction of one of the β-strands is reversed. The amino acid substrate is bound at the deep cleft between the two domains (Figure 9.8c). Still, there are structural variations among different AA-DHs. For example, while glutamate dehydrogenase is either hexameric or tetrameric in most cases, other AA-DHs are known to exist in different quaternary structures, from monomers to dodecamers.

[*1] As in ketoreductases, here, too, the hydride transfer step determines the configuration of the resulting amino acid. However, unlike ketoreductases, which may form either the *S* or *R* configuration (depending on the enzyme and the specific substrate), all known natural AA-DHs create only the *S* configuration, i.e., L-amino acids.

[*2] Although the reactions catalyzed by AA-DHs are essentially reversible, they are far from equilibrium on the aminated product side [56] ($K_{eq} = 9 \times 10^{12}$ for *leucine dehydrogenase* and 2.2×10^{13} for *phenylalanine dehydrogenase*).

[*3] The toxicity of free ammonia results from several factors, including (1) the conversion of α-ketoglutarate into glutamate and, as a result, depletion of α-ketoglutarate from the Krebs cycle, and (2) the conversion of glutamate into glutamine and, as a result, the depletion of glutamate as a neurotransmitter and as a source for the neurotransmitter γ-amino butyric acid (GABA).

Moreover, the position of the substrate with respect to NAD(P)H and the identity of the transferred hydride (H_R/H_S) may vary. For example, when comparing alanine dehydrogenase (PDB entries 2vhx and 2voj) and glutamate dehydrogenase (PDB entry 1hwy), we observe that the substrate lies on different sides of NADH, so the transferred hydride is H_R in the former and H_S in the latter (Figure 9.8d). However, since alanine and glutamate have opposite orientations in the active site, the result is an S configuration in both cases.

9.1.5.2 Transferases (EC 2)

Transferases are the second of the two largest enzyme classes, constituting 30% of the enzymes in the ExplorEnz database [31]. They catalyze the transfer of a chemical group (X) between two molecules, i.e., from a donor (D) to an acceptor (A):

Scheme 9.2.
$$D-X + A \longrightarrow D + A-X$$

In the EC number of a transferase, the second numeral designates the chemical group that is being transferred:

- EC 2.1 – one-carbon groups (e.g., methyl)

- EC 2.2 – carbonyl groups (aldehydes and ketones)

- EC 2.3 – acyl groups (see Subsection 9.1.5.2.3 below)

- EC 2.4 – glycosyl groups

- EC 2.5 – alkyl (other than methyl) or aryl groups

- EC 2.6 – nitrogen-containing groups (e.g., amino, see Subsection 9.1.5.2.1 below)

- EC 2.7 – phosphorus-containing groups (mostly phosphate, see Subsection 9.1.5.2.2 below)

- EC 2.8 – sulfur-containing groups (e.g., sulfate)

- EC 2.9 – selenium-containing groups

- EC 2.10 – molybdenum- or tungsten-containing groups

The group donor is often a coenzyme. For example, *tetrahydrofolate* (*THF*) is a common donor of groups containing a single carbon, *S-adenosylmethionine* (*SAM*) and *methylcobalamine* (a form of coenzyme B_{12}) are common methyl group donors, and *coenzyme A* (*CoA*) is a common donor of acyl groups (see Section 9.4 for more details).

The third numeral of the EC number of a transferase refers to the specific group that is transferred, except in the case of enzymes transferring phosphorus-containing groups, in which case the third numeral specifies the acceptor group. As in oxidoreductases, the fourth numeral specifies the substrate specificity of the enzyme. For example, in *glutamine:pyruvate aminotransferase* (EC 2.6.1.15), the second numeral (6) designates that the transferred group is a nitrogen-containing group, the third numeral (1) designates that this is an amino group (NH_2), and the fourth numeral (15) designates that the group is transferred from glutamine to pyruvate.

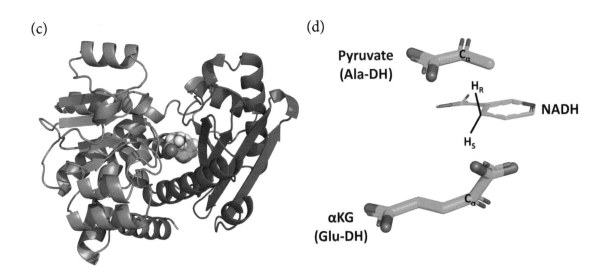

FIGURE 9.8 **Amino acid dehydrogenases (AA-DHs).** (a) The reversible chemical reaction catalyzed by AA-DHs. (b) The mechanism of reductive amination. The first step involves addition of ammonia (in blue) to the keto acid and protonation of the ketone oxygen. The resulting intermediate is a *carbinolamine*. The second step is elimination of the original ketone oxygen as water (in red) and formation of an *imine* intermediate. The last step is the transfer of a hydride ion (in purple) from NAD(P)H to C_α of the substrate, converting the imino intermediate into an amino acid. (c) The structure of *phenylalanine dehydrogenase* (PheDH; EC 1.4.1.20). The structure of PheDH from *Rhodococcus sp.* is shown (PDB entry 1c1d, chain A). The N-terminal domain is colored in magenta and the coenzyme-binding domain in cyan. The substrate, L-phenylalanine, is shown as spheres in the cleft between the two domains. (d) The positions and orientations of the pyruvate in Ala-DH (PDB entry 2vhx) compared with those of α-ketoglutarate (α-KG) in Glu-DH (PDB entry 1hwy).

As shown in the above example, the common names used for transferases are *'donor:acceptor group-transferase'*, or sometimes simply *'donor group-transferase'*.

Transferases are involved in numerous metabolic and physiological processes. Below we discuss three types of transferases that exemplify this involvement.

9.1.5.2.1 Aminotransferases (EC 2.6.1)

Aminotransferases (a.k.a. *transaminases*) catalyze the transfer of an amino group from a donor to an acceptor [27] (Figure 9.9a). Biochemically relevant reactions usually involve L-α-amino acids as donors and α-keto acids as acceptors (Figure 9.9b), and are catalyzed by *α-L-aminotransferases* (α means that the amino group is on C_α). This reaction is similar to the one carried out by amino acid dehydrogenases (see Subsection 9.1.5.1), in the sense that the α-amino group of the donor is oxidatively deaminated in one direction and the resulting α-keto group is reductively aminated in the opposite direction. However, in transamination there is no hydride transfer; the electrons that reduce the acceptor's keto group are obtained from the amino group of the donor. The α-aminotransferases play a central role in amino acid metabolism [57], especially in the safe removal of amino groups from degraded proteins. As explained in Subsection 9.1.5.1.2 above, amino acids must not release their amino groups directly to the cytosol of cells or to the blood, because of the toxicity of free ammonia. Instead, our tissues use aminotransferases to transfer the amino groups of amino acids either to α-ketoglutarate (in most tissues; Figure 9.9c) or to pyruvate (in muscles; Figure 9.9d)[*1]. The resulting L-glutamate or L-alanine (respectively) carries the amino groups to the liver via circulation, where L-glutamate is deaminated by mitochondrial *glutamate dehydrogenase* (see Subsection 9.1.5.1.2). L-alanine transfers its amino group to α-ketoglutarate to form L-glutamate, which is then deaminated as well by glutamate dehydrogenase. In both cases, the amino groups that are released by glutamate dehydrogenase are converted by the urea cycle to the less toxic metabolite urea, and the latter is transferred to the kidneys for secretion via the urine. As mentioned above, glutamate that is formed by transamination in the tissues is often aminated by *glutamine synthetase* (EC 6.3.1.2) to form glutamine, which carries its two amino groups to the liver. There, it is deaminated by *glutaminase* (EC 3.5.1.2) to glutamate, which, again, is deaminated by glutamate dehydrogenase.

α-Aminotransferases are homodimers and include two active sites per one enzyme molecule [58] (Figure 9.10a left). Each of the active sites is built from parts of both chains (Figure 9.10a right). Whereas most α-aminotransferases have the same principal fold, *D-amino acid aminotransferase* (EC 2.6.1.21) and branched-chain amino acid aminotransferase (2.6.1.42) have different folds from those of the rest (Figure 9.10b). The active site of α-aminotransferases, like that of oxidoreductases, contains two pockets, one large and one small. The small pocket is positively charged and accommodates the α-carboxylate group of the amino acid substrate. The large pocket accommodates the side chain [59], and its chemical properties complement those of the specific amino acid on which the enzyme acts.

To separate the amino groups from amino acids, aminotransferases use the coenzyme *pyridoxal-phosphate* (PLP), which is derived from *vitamin B$_6$ (pyridoxine)*[*2]. PLP is normally bound to the 'resting' enzyme via a *Schiff base*, formed with an active site lysine residue

---

[*1]There are aminotransferases for each of the natural L-amino acids, except for threonine and lysine. These are degraded in the liver by other enzymes.

[*2]The appropriate position of the substrate with respect to the catalytic lysine is determined by a conserved arginine that interacts with the substrate's carboxylate group, holding the substrate in place.

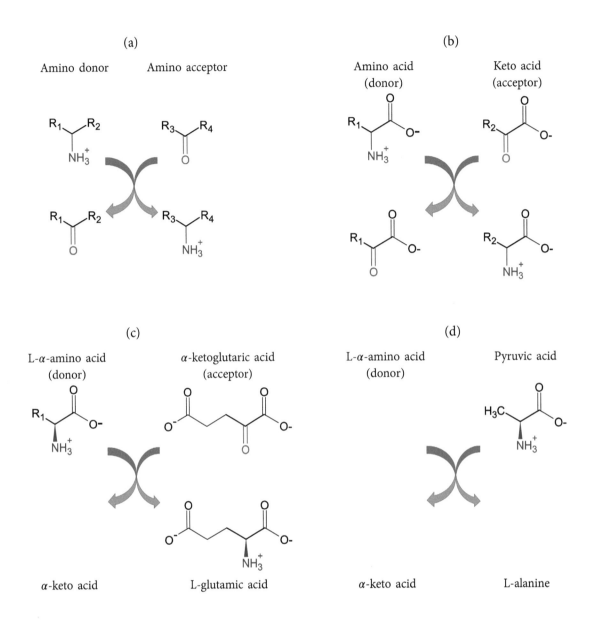

FIGURE 9.9 **Reactions catalyzed by aminotransferases.** (a) The general reaction catalyzed by aminotransferases, in which a donor (amine or amino acid) transfers its amino group to an acceptor (ketone or keto acid). (b) Biochemically relevant reactions, in which the donor is an amino acid and the acceptor is a keto acid. (c) The reaction catalyzed by aminotransferases in the degradation of L-amino acids in most tissues. The acceptor in this case is α-ketoglutaric acid (α-ketoglutarate in the deprotonated form), which is converted to L-glutamic acid (L-glutamate in the deprotonated form). (d) The reaction catalyzed by aminotransferases in the degradation of L-amino acids in muscles. The acceptor in this case is pyruvic acid (pyruvate in the deprotonated form), which is converted to L-alanine.

(a)

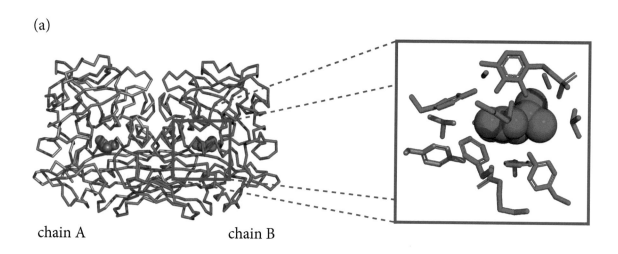

chain A   chain B

(b)

Aspartate aminotransferase   Branched-chain amino acid aminotransferase

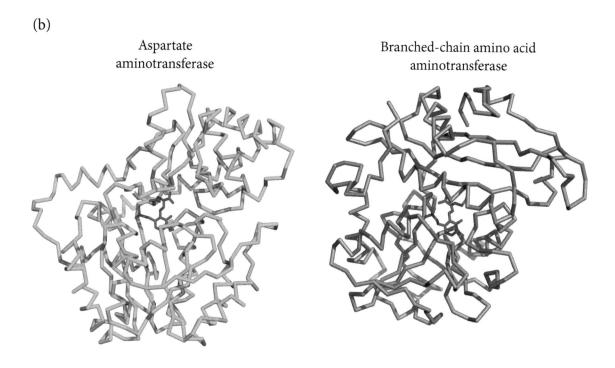

FIGURE 9.10 **The structures of aminotransferases.** (a) The three-dimensional structure of branched-chain aminotransferase (EC 2.6.1.42; PDB entry 3uzb). *Left*: overall structure, containing two chains (red and blue C_α traces) with two active sites (ligands are shown as spheres). *Right*: zoom-in view of an active site, showing contributions from both chains. (b) The different folds of aspartate aminotransferase (EC 2.6.1.1; PDB entry 3qpg) and branched-chain amino acid aminotransferase (PDB entry 1iye). For clarity, only one chain is shown for each structure. The two structures are positioned so their PLP cofactors (shown as red sticks) have the same orientation.

(Figure 9.11a). When an amino acid binds to the active site, the enzyme-PLP bond breaks, and an equivalent amino acid-PLP bond is formed (Figures 9.11b and 9.11c, step 1). The key event in the separation of the amino group from the donor amino acid is the deprotonation of the donor's C_α (Figure 9.11c, step 2). This results in a transition state with a negatively charged C_α (carbanion), stabilized by PLP. The stabilization is attributed to the pyridinium ring of PLP, which is positively charged and has delocalized π electrons. The transition state quickly undergoes electronic rearrangements that turn the aldimine bond between the amino acid donor and PLP into a ketimine bond (Figure 9.11c, step 3). The latter is readily attacked by an active site water molecule and allows the donor to be released into solution as an α-keto acid (Figure 9.11c, step 4). Again, the key event in this sequence is the deprotonation of the donor's C_α, which is carried out by an active site catalytic lysine residue[*1]. In fact, C_α has very high pKa and should not deprotonate readily under physiological conditions. This is where PLP comes into play; its positively charged pyridinium ring acts like an *electron sink*. That is, it draws electrons (via the positively charged Schiff base) from the C_α—H bond, thus polarizing it. The result is a marked drop in the pKa of C_α, allowing it to donate its proton to the active site lysine residue. Moreover, upon C_α deprotonation, the extra negative charge is delocalized and stabilized by resonance in PLP's π system. One of the resonance states is the aforementioned ketimine, which is electrophilic and therefore attacked by an active site water molecule. Again, all of these key catalytic transformations are possible due to PLP's unique physicochemical properties, and their effects on the bound donor molecule. After the donor amino acid is released, the keto acid acceptor binds and becomes aminated by reversing the steps described above (Figure 9.11c, steps 5 through 7). Interestingly, PLP is active also in the absence of the apo-enzyme, but acts very slowly [61]. Thus, the role of the protein environment is to speed up the reaction, as well as to confer chemical and enantiomeric specificity (see next paragraph).

The penultimate step in the aminotransferase reaction (Figure 9.11c, step 5) involves the protonation of the acceptor's C_α by the active site's catalytic lysine residue, which changes the acceptor from an imino acid into an amino acid. This is another key step, as it determines the configuration of the product; proton transfer from one side of the acceptor creates an L-(S)-amino acid, whereas transfer from the other side creates a D-(R)-amino acid (Figure 9.11d). Indeed, **like ketoreductases and amino acid dehydrogenases (see Subsection 9.1.5.1), aminotransferases are also stereospecific, with their vast majority being L-(S)-specific.** The prevalence of L-(S)-specific aminotransferases is to be expected, considering that L-amino acids are the principal building blocks of proteins. However, in contrast to the case of amino acid dehydrogenases, there is at least one aminotransferase known to create D-(R)-amino acids (*D-amino acid transaminase*; EC 2.6.1.21). Another difference between the two amino acid-forming enzyme groups is that aminotransferases essentially operate at equilibrium (K_{eq}~1), whereas amino acid dehydrogenases do not. This means that the direction of their reactions is highly dependent on their substrate and product concentrations.

PLP's ability to promote covalent bond cleavage in substrates is also used by other enzymes (lyases, isomerases, hydrolases and oxidoreductases) to catalyze completely different reactions (e.g., decarboxylation, racemization, dehydration and β/γ elimination

---

[*1]We will see later that PLP is a required coenzyme in many other biochemical reactions as well. The prevalence of PLP in metabolic pathways is reflected by the fact that its biosynthesis is considered among the first aerobic (O_2-consuming) reactions to evolve on Earth, about 2.9 billion years ago [60].

and replacement) [62,63] (Figure 9.11e). These enzymes act mainly on amino acids, but some of them also catalyze reactions involving other amino-containing metabolites. In all of these reactions PLP acts similarly, i.e., by weakening the bond intended for cleavage and stabilizing the resulting carbanion transition state[*1]. So, how do the different PLP-enzymes 'decide' which bond to cleave? According to *Dunathan's stereoelectronic hypothesis* [64], the 'decision' has to do with the orientation of the substrate with respect to the PLP ring. That is, in each of the enzymes the substrate is bound to PLP such that the bond intended for cleavage is perpendicular to PLP's pyridinium ring. In this orientation, the bond parallels the conjugated p orbitals of PLP's π system, which belongs both to the Schiff base and to the ring (Figure 9.11f). This way, the negative charge emerging upon the cleavage of the bond will be optimally stabilized by resonance interactions with the overlapping π system. Indeed, three-dimensional structures determined for PLP enzymes are compatible with this proposition [62]. Since the orientation of the bound substrate is determined by the architecture and the physicochemical properties of the active site, the different enzymes — each of which has a unique active site — can use the same coenzyme (PLP) to accelerate a different type of reaction [65]. The contribution of the active site to the specificity of each enzyme towards the catalyzed reaction is also conveyed via further stabilization of the carbanion by side chain interactions, as well as by other catalytic steps that involve active amino acids [62].

9.1.5.2.2 Phosphoryl transferases (EC 2.7)

Many biological molecules contain phosphate groups. These molecules include nucleic acids (DNA and RNA), small molecules that function as metabolic intermediates (e.g., glucose 6-phosphate) or coenzymes (e.g., NADH), and proteins that undergo post-translational modification. It is therefore not surprising that reactions in which a phosphate group is transferred to a substrate (i.e., phosphorylation) or from it are very common [60,66]. Phosphoryl transferases (a.k.a. phosphotransferases) usually transfer the γ-phosphoryl group of ATP to a target molecule[*2]. However, there are also cases in which the phosphoryl donor is a molecule other than ATP, and cases in which the phosphoryl is transferred as part of a larger molecule (e.g., in nucleotidyltransferases).

Phosphoryl transfer reactions play numerous physiological and cellular roles, including muscle contraction, energy production, biosynthesis, and signal transduction, and are considered highly efficient[*3]. For example, in glycolysis alone, four out of ten reactions are carried out by phosphoryl transferases:

- *Hexokinase (EC 2.7.1.1)*, which converts glucose into glucose 6-phosphate (Figure 9.12a).

- *Phosphofructokinase-1 (EC 2.7.1.11)*, which converts fructose 6-phosphate into fructose 1,6-bisphosphate.

- *Phosphoglycerate kinase (EC 2.7.2.3)*, which converts 3-phosphoglycerate into 1,3-bisphosphoglycerate.

- *Pyruvate kinase (EC 2.7.1.40)*, which converts phosphoenolpyruvate into pyruvate.

[*1]The only known exception is *glycogen phosphorylase* (EC 2.4.1.1), in which PLP acts through a completely different mechanism involving proton transfer.

[*2]A similar reaction is catalyzed by ligases (see Subsection 9.1.5.6 below).

[*3]Phosphotransferases are known to produce some of the largest enzymatic rate enhancements (10^{21}-fold) [66].

(a)

(b)

(c)

external
aldimine

TS

internal
aldimine

external
aldimine

ketimine

pyridoxamine
phosphate
(PMP)

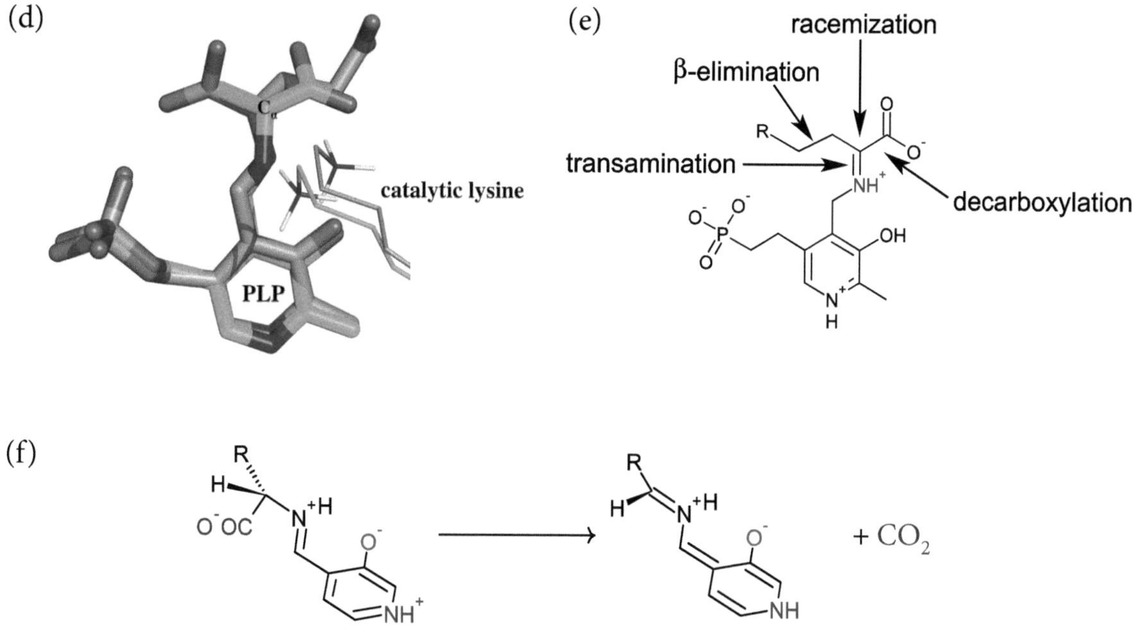

(d)

catalytic lysine

PLP

(e)

(f)

$+ CO_2$

FIGURE 9.11 **The PLP coenzyme in aminotransferases.** (a) Enzyme-PLP Schiff base. The enzyme group involved in the process is colored in blue. (b) PLP-substrate (amino acid) Schiff base. The amino acid is colored in red, and the labile bond is marked by the small lightning-shaped arrow. (c) The mechanism of transamination. Step 1: the amino group of the donor amino acid attacks PLP and forms a Schiff base with it (external aldimine). This process replaces the original Schiff base between PLP and a lysine residue of the enzyme (internal aldimine). Step 2: the C_α of the bound donor is deprotonated by the same lysine residue, creating a negatively charged carbanion. Step 3: electronic rearrangement in the donor turns the aldimine into ketimine. Step 4: an active site water molecule deprotonates (not shown), and the resulting OH^- nucleophile attacks the donor's C_α. As a result, the imine bond between the donor and PLP is broken, and the donor leaves the active site as a keto acid. PLP is now unattached to the enzyme or to an amino acid. Thus, it is in its amino form (called pyridoxamine phosphate or PMP). Step 5: the keto acid acceptor enters the active site, and its C_α is attacked by PMP's amino group, resulting in their covalent binding (as ketimine). Step 6: PLP is deprotonated by the aforementioned lysine residue, which then transfers the proton to the acceptor's C_α. The acceptor changes from a ketimine into an (external) aldimine. The protonation of the acceptor is the step that determines the configuration of the acceptor; when the proton is transferred to the acceptor from one side, the resulting amino acid has an L/(S) configuration, whereas protonation from the other side creates a D/(R) configuration. Step 7: the amino group of the lysine residue attacks PLP and forms a Schiff base with it (internal aldimine). As a result, the acceptor detaches from PLP and leaves as an amino acid. (d) The orientation of substrate (sticks) and catalytic machinery in *branched-chain amino acid aminotransferase* (EC 2.6.1.42, PDB entry 1iye; orange) and in *D-amino acid aminotransferase* (EC 2.6.1.21, PDB entry 3daa; green). While the catalytic lysine residues of the two enzymes reside on the same face of PLP, the substrates are bound in opposite directions. Thus, protonation of the substrate in the two enzymes results in opposite configurations. (e) The cleavage of other bonds in molecules, facilitated by PLP. Decarboxylation involves α-elimination resulting in cleavage of the C_α–COOH bond. In racemization reactions the C_α hydrogen (not shown) is eliminated and re-added. Other reactions involve β or γ eliminations. (f) The Dunathan stereoelectronic hypothesis, demonstrated for deprotonation and decarboxylation. The two substrates are bound to PLP such that the bond to C_α that is to be broken (the leaving group is in purple) is aligned with the π orbitals of PLP.

The prevalence of phosphoryl transfers in metabolic reactions is not incidental; the attachment of a phosphate group to a target molecule may serve different biochemical roles:

1. Protein and enzyme activation. This mode of action is very common in the activation of enzymes involved in hormone-mediated signal transduction.

2. Enhancing ligand binding. The phosphate group carries two negative charges, and its attachment to small molecules has been shown to increase the affinity of the latter to target enzymes via electrostatic interactions [3]. Moreover, the special geometry of the phosphate group, in addition to its charge distribution, makes the interactions in which it is involved specific.

3. Promotion of catalysis. The attachment of a phosphoryl group to a reactant may promote the reactant's subsequent chemical transformation, in different ways. For example, the phosphoryl group may destabilize an adjacent bond in the substrate, making it prone to cleavage (the *leaving group effect*, see Subsection 9.1.5.6). In other cases, the phosphoryl group may help an enzyme residue become a good nucleophile. This may happen by deprotonating or polarizing an existing active site nucleophile, or by creating a new nucleophile, via phosphorylation.

As mentioned above, four different phosphoryl transferases act in glycolysis. The first two act early in the pathway, and are often said to 'activate' the substrate for later stages, which create ATP by phosphate transfer from substrate to ADP (*substrate-level phosphorylation*). In fact, the early phosphoryl transfers create *phosphoester* bonds, which are too stable to donate their phosphate groups to ADP[*1]. The two reactions in glycolysis that do create ATP draw the required energy from oxidation of the phosphorylated substrate, an *exergonic* (energy-releasing) reaction. The first reaction does this directly; it couples the oxidation of *glyceraldehyde 3-phosphate* to its second phosphorylation, creating a *mixed anhydride* (Figure 9.12b). The anhydride bond is sufficiently unstable to donate its phosphoryl group to ADP[*2]. Since this is an oxidative phosphorylation reaction, the enzyme catalyzing it is not a phosphoryl transferase but an oxidoreductase (i.e., *glyceraldehyde 3-phosphate dehydrogenase*). The second ATP-forming reaction transfers the C2 phosphate group of *phosphoenol pyruvate* to ADP. This group is initially attached to the substrate as a phosphoester by the early phosphorylations. However, before the phosphate is transferred to ADP, a C—C bond in the substrate adjacent to this phosphate is oxidized from a single to a double bond. This renders the subsequent transfer of the phosphate group to ADP energetically favorable, through resonance stabilization of the product.

Phosphoryl transferases may act on different molecules, including small metabolites, coenzymes and large polymers such as proteins. In the case of small organic molecules, the phosphoryl group is attached to an oxygen atom of a hydroxyl or carboxyl group, to a nitrogen atom (e.g., in forming muscle *creatine phosphate*), or to another phosphate group (forming a *phosphoanhydride* group) [66]. When the acceptor is a protein, the phosphoryl group may be transferred to the hydroxyl group of serine, threonine or tyrosine side chains; or in some cases to the imidazole side chain of histidine. As mentioned in Chapter 8, protein

---

[*1]The standard free energy of hydrolysis ($\Delta G^0_{hydrolysis}$) of a phosphoester is −3 to 4 kcal/mol, where the energy required for ATP synthesis from ADP and P_i is ~7 kcal/mol [57,67]

[*2]$\Delta G^0_{hydrolysis}$ of a mixed anhydride is −10 to −12 kcal/mol [57,67].

phosphoryl transferases are involved in many diseases, which makes them very popular drug targets (second only to GPCRs).

Phosphotransferases are commonly called *kinases* or *phosphorylases*. The latter name, however, is also used for transferases that act on other groups (glycosyl, alkyl, aryl) and that use inorganic phosphate in the reactions they catalyze. For example, *glycogen phosphorylase* (EC 2.4.1.1) is a glycosyl transferase that plays a central role in glucose metabolism. It uses inorganic phosphate to break glycoside bonds in glycogen, and consequently release glucose 1-phosphate. This process allows the body to use liver glycogen for elevating blood glucose levels between meals, and to use muscle glycogen as an energy source for working muscles. *Phosphatases* are another group of enzymes that catalyze reactions involving phosphate groups and that are not included in the phosphotransferase subclass. These enzymes remove phosphate groups from target molecules and generally belong to the hydrolase class (EC 3).

9.1.5.2.3   Acyltransferases (EC 2.3)

Another well-known reaction that is catalyzed by transferases is the transfer of an acyl group (i.e., acylation); in these reactions, the donor is usually CoA, and the acceptor may be a C-, N- or O-containing group. CoA is used to activate the transferred group for the transfer reaction (see Subsection 9.4.2 below for a mechanistic explanation). **The acyl transfer reaction is common in biosynthetic processes of lipids, which take place mainly in liver cells.** One such process is the synthesis of fatty acids from acetyl-CoA, carried out by *fatty acid synthase (FAS)*; EC 2.3.1.85). FAS catalyzes the first (and major) step in fatty acid synthesis, which is the building of *palmitic acid*, a 16-carbon fatty acid (Figure 9.12c). Palmitic acid may then be subjected to further steps of elongation or desaturation by other enzymes, to form other types of fatty acids. Palmitic acid is built by seven consecutive acetylations of acetyl-CoA, where the added two-carbon group in each cycle is donated by malonyl-CoA, a three-carbon molecule. The first step in each acetylation cycle is the condensation of acetyl-CoA and malonyl-CoA with the release of CO_2 (Figure 9.12c, step 1)[*1]. The resulting β-keto group is then reduced to a methylene (C–C) group in three consecutive steps: (a) reduction of the keto group to a hydroxyl group (Figure 9.12c, step 2); (b) dehydration of the HO–C–C–H unit to a C=C unit (Figure 9.12c, step 3); and (c) reduction of the double bond to a single C–C bond (Figure 9.12c, step 4). As mentioned in Chapter 2, Subsection 2.5.3, in bacteria and plants the different steps in the FAS reaction are carried out by different enzymes, whereas in fungi and higher eukaryotes these steps are carried out by different subunits of the same enzyme. This mechanism may seem cumbersome, but it is in fact an elegant solution found by evolution for creating aliphatic carbon chains that are flanked by functional groups.

The first step of the acetylation cycle is a *Claisen condensation* that involves decarboxylation (Figure 9.12c step 1) [68]. Claisen reactions attach two esters (or an ester and a carbonyl) to create a β-keto ester (or a β-diketone, respectively). In the case of FAS, the reaction attaches two thioesters (acetyl-CoA and malonyl-CoA) to create a β-keto-thioester (acetoacetyl-CoA). **Claisen condensations are used in different biological pathways and in different forms.** The one shown in the figure, which involves decarboxylation, is also

---

[*1]In the actual enzymatic reaction both the acetyl and malonyl units are first transferred from CoA to enzyme groups. However, they bind to these groups via thioester bonds, which are equivalent to the original ones with CoA.

(a)

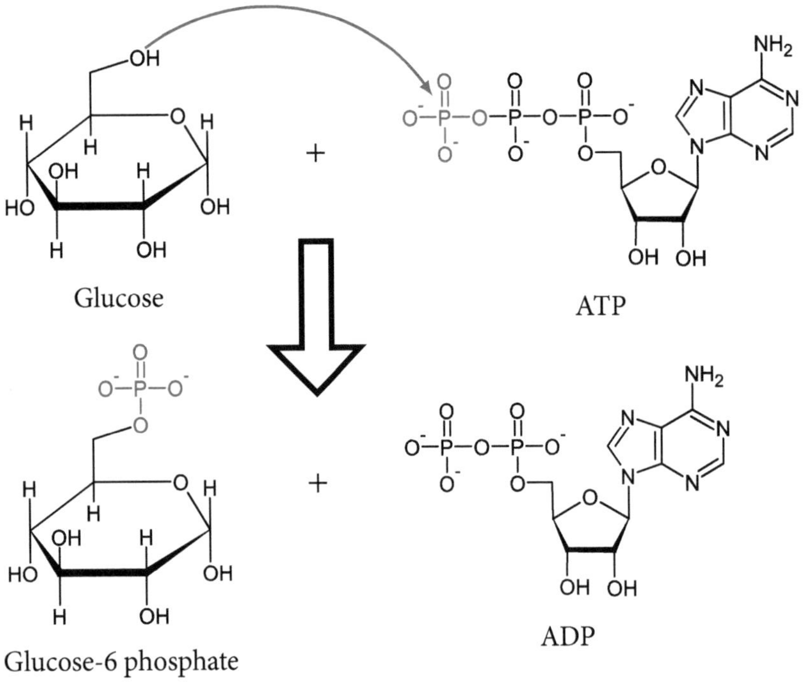

Glucose

ATP

Glucose-6 phosphate

ADP

(b)

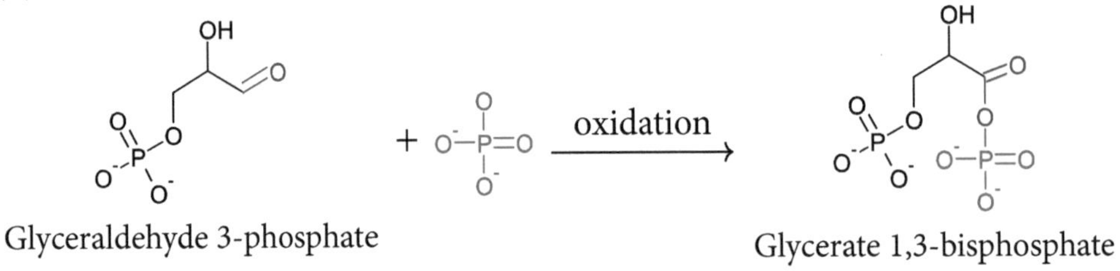

Glyceraldehyde 3-phosphate

oxidation

Glycerate 1,3-bisphosphate

FIGURE 9.12 **Biologically relevant phosphorylations and acyl transfer reactions.** (a) Glucose phosphorylation by *hexokinase* during glycolysis, creating a phosphoester bond. The reaction involves a nucleophilic attack of glucose's hydroxyl at position C6 on the terminal phosphoryl group of ATP (the attack is shown by the red arrow). For clarity, explicit hydrogen atoms are not shown. (b) Glyceraldehyde 1,3-bisphosphate oxidative phosphorylation by *glyceraldehyde 3-phosphate dehydrogenase*, creating a mixed anhydride bond. The oxidized aldehyde group and the resulting carboxyl are colored blue. The phosphoryl group added to the carboxyl group is colored red.

(c)

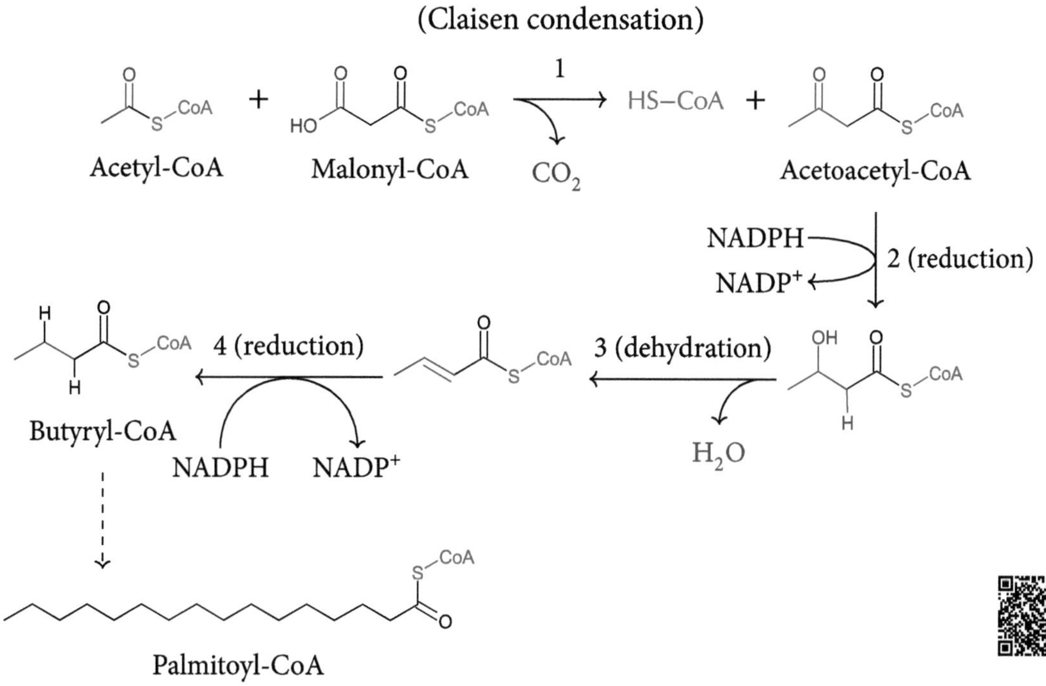

(Claisen condensation)

Acetyl-CoA   Malonyl-CoA   CO$_2$ Acetoacetyl-CoA

NADPH —
NADP$^+$ ←
2 (reduction)

4 (reduction)   3 (dehydration)

Butyryl-CoA
NADPH   NADP$^+$
H$_2$O

Palmitoyl-CoA

(d)

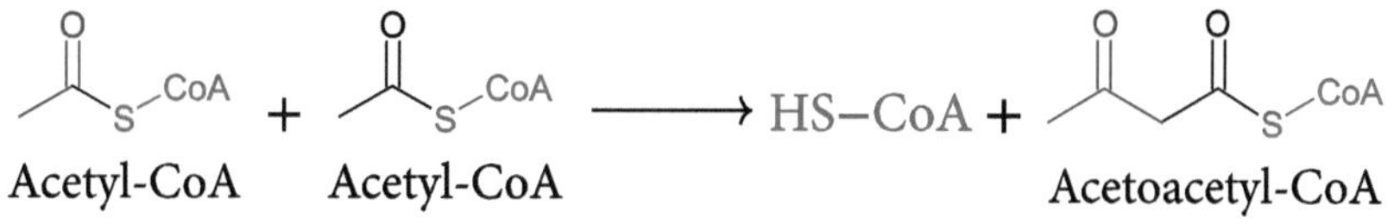

Acetyl-CoA   Acetyl-CoA   HS—CoA +   Acetoacetyl-CoA

FIGURE 9.12 **Biologically relevant phosphorylations and acyl transfer reactions.** (Continued) (c) Acyl transfer in fatty acid synthesis. The pathway includes consecutive reactions that can be iterated to increase the chain length (dashed line leading to palmitoyl-CoA). Step 1: a decarboxylating Claisen condensation of acetyl-CoA and malonyl-CoA (in the enzyme-catalyzed reaction both reactants are attached to the enzyme via equivalent thioester bonds). Step 2: NADPH-dependent reduction of acetoacetyl-CoA's keto group to a hydroxyl group. Step 3: water elimination (dehydration) of the C_2–C_3 bond. Step 4: NADPH-dependent reduction of the C_2–C_3 double bond. (d) Acyl transfer in the biosynthetic pathways of cholesterol and ketone bodies. In both cases, thiolase carries out a non-decarboxylating Claisen condensation to form acetoacetyl-CoA from two acetyl-CoA molecules.

used in the synthesis of *polyketides*, a large group of biologically active molecules that are synthesized mainly by bacteria, fungi and plants [69]*1. Polyketides are *secondary metabolites*, which means that most of them are designed to affect not the organism that produces them, but rather other organisms (e.g., predators). Indeed, this diverse group includes antibiotics (e.g., *erythromycin, tetracycline*), anti-fungal agents (e.g., *amphotericin*), immunosuppressants (e.g., *rapamycin*), toxins (e.g., *aflatoxin* and *coniine**2), hallucinogens (e.g., *tetrahydrocannabinol* (*THC*)), and other functionally important compounds.

Claisen condensations need not necessarily involve decarboxylation. Non-decarboxylating Claisen condensations are used, for example, in the first steps of the biosynthetic pathways of both cholesterol and ketone bodies. In these reactions, *thiolase* (*acetyl-CoA acetyltransferase* EC 2.3.1.9) creates acetoacetyl-CoA by attaching two acetyl-CoA molecules (Figure 9.12d). This is a 'biosynthetic thiolase', but there are other thiolases that do the opposite, i.e., operate in catabolic (degradation) processes. That is, they catalyze the same reaction but in the opposite direction (*thiolysis*). A well-known example of the latter is the thiolase participating in the breakdown of fatty acids (*β-oxidation*). Like the opposite process (fatty acid synthesis), fatty acid breakdown comprises repeating cycles, each involving a two-carbon acetyl-CoA unit (in this case, the removal of this unit). In the last step of each cycle, the corresponding thiolase removes an acetyl-CoA unit from a *β*-ketoacyl unit of the existing chain.

Another use of acyltransferases in lipid biosynthesis is in the common steps of the synthetic pathways of *triacylglycerol* (an energy reserve molecule) and *phospholipids* (constituents of the cellular membranes). In these processes, *glycerol 3-phosphate* is sequentially acylated on its C1 and C2 atoms, to form *phosphatidic acid* (i.e., 1,2-diacylglycerol 3-phosphate). The phosphate group on C3 may then be replaced with a third acyl group to form triacylglycerol, or bind to an alcohol group (e.g., choline, serine, ethanolamine) to form a phospholipid.

Acyltransferases are also commonly used in post-translational modifications, which, in turn, are used for different cellular and physiological processes. For example, in proteins such as p53 and histone, ubiquitylation and acetylation of lysine residues*3 are used oppositely to determine proteins' fate in the cell; specifically, whereas ubiquitylation marks the proteins for cellular degradation, acetylation protects them from it. Other types of acylation, N'-myristoylation and S-palmitoylation, allow proteins to attach to the plasma membrane of cells, usually as a step in a signal transduction pathway. The above examples are discussed in detail in Chapter 2, Subsection 2.6.4 and in Figure 2.37.

9.1.5.3 Hydrolases (EC 3)

Hydrolases are the third-largest class of enzymes, constituting 24% of the enzymes in the ExplorEnz database [31]. They catalyze the cleavage of a covalent bond in the substrate using H_2O as the attacking group (i.e., *nucleophilic substitution*):

Scheme 9.3. $$A-B + H_2O \longrightarrow A-OH + B-H$$

---

*1In fact, the only thing connecting all polyketides, aside from the fact that they contain multiple ketene (CH_2=CO) groups, is their common mechanism of biosynthesis [69].

*2Coniine is known mainly as the substance in the hemlock brew used to execute Socrates [69].

*3Both are types of ε-N-acylation.

Thus, these enzymes could, at least in theory, be regarded as transferases that transfer a group from the substrate to water [27]. Historically, however, most hydrolases were discovered before transferases, and were assigned to their own class. The general mechanism used by hydrolases usually involves nucleophilic catalysis (see Subsection 9.3.3.3 below) and often includes the following steps [70]*1:

1. Nucleophilic attack on the substrate by an active site catalytic residue, forming an enzyme-substrate covalent intermediate.

2. Nucleophilic attack on the enzyme-substrate intermediate by a water-derived hydroxide ion (OH^-), which frees the substrate.

In some hydrolases (e.g., those that depend on metal ions), the first step may be absent, and a hydroxide (OH^-) nucleophile, which is created by water polarization and deprotonation, acts directly on the substrate's labile bond.

In the EC number of a hydrolase, the second numeral designates the general type of the cleaved bond:

- EC 3.1 – ester bonds

- EC 3.2 – glycosyl bonds

- EC 3.3 – ether bonds

- EC 3.4 – peptide bonds

- EC 3.5 – C–N bonds, other than peptide bonds

- EC 3.6 – anhydride bonds

- EC 3.7 – C–C bonds

- EC 3.8 – halide bonds (e.g., sulfate)

- EC 3.9 – P–N bonds

- EC 3.10 – S–N bonds

- EC 3.11 – C–P bonds

- EC 3.12 – S–S bonds

- EC 3.13 – C–S bonds

The third numeral refers to the specific type of the cleaved bond, e.g., carboxylic ester, thiol ester, etc. *Peptide hydrolases* (EC 3.4) are an exception to this rule; they are grouped into two sub-subclasses: *endopeptidases* (EC 3.4.21 through 25 and 3.4.99), which cleave the peptide bond inside the polypeptide chain, and *exopeptidases* (EC 3.4.11 through 19), which cleave at the *N*- or *C*-terminus (see more below). The fourth EC numeral of hydrolases designates the substrate of the enzyme that contains the bond. In most cases, the systematic names of hydrolases are formulated as *substrate bond-hydrolase* (e.g., acetylcholinesterase),

---

*1For more details see Figure 9.26c-II and Subsection 9.3.3.3 below, discussing covalent catalysis.

but common names sometimes take the form of *substrate-ase* (e.g., aspartase). Moreover, it is customary to group hydrolases according to the catalytic residue used as the nucleophile, e.g., serine- and cysteine hydrolases. The subgroup of proteases are discussed in Subsections 9.1.5.3.2 and 9.3.3.3 below. A somewhat different case is metallohydrolases (e.g., zinc-dependent hydrolases), in which the metal functions in creating the water nucleophile by polarizing it (see Subsection 9.3.3.4 below on metal-ion catalysis). A similar mechanism is employed by aspartic proteases; in this case, the water nucleophile is created by proton abstraction and transfer from the water molecule to one of the two catalytic aspartate residues (see details below).

Below we discuss two types of hydrolases: those that act on P−O bonds, and peptide hydrolases.

9.1.5.3.1   Hydrolases acting on P−O bonds (EC 3.1 and 3.6)

In Subsection 9.1.5.2.2 above, we discussed phosphoryl transferases (kinases, phosphorylases), which transfer phosphoryl groups from a donor molecule (usually ATP) to an acceptor molecule. Phosphohydrolases (often called *phosphatases*) are enzymes that use water to cleave C−P bonds, either phosphoester (EC 3.1) or phosphoanhydride (EC 3.6). Like kinases and phosphorylases, phosphatases are involved in the regulation of many biological processes. Many phosphatases are metalloenzymes, containing a cationic metal. For example, *alkaline phosphatase* (EC 3.1.3.1), an enzyme that operates in most, if not all life forms, contains two Zn^{2+} ions and one Mg^{2+} ion in its active center. The zinc ions are directly involved in catalysis, acting mainly in transition state stabilization [66] (see more on metal catalysis in Subsection 9.3.3.4 below). In contrast, acid phosphatase (EC 3.1.3.2) uses a conserved nucleophilic histidine and no metal ions.

Like protein kinases, phosphatases can be separated into those acting on serine and threonine and those acting on tyrosine. Serine and threonine phosphatases contain binuclear metal centers, which are used for catalysis, whereas tyrosine phosphatases do not employ metals [66]. The same is true also for dual-specific protein phosphatases, which act on phosphoserine, phosphothreonine, and phosphotyrosine residues.

As mentioned above, some phosphohydrolases act on phosphoanhydride bonds, i.e., bonds between two phosphates. Phosphohydrolases have been suggested to be the first enzymes to evolve on Earth, in congruence with the evolutionary selection of ATP (a phosphoanhydride) as the universal energy currency [60]. Among these enzymes, we can find *ATP synthase* (EC 3.6.3.14), the enzyme responsible for the hydrolysis of ATP to ADP and inorganic phosphate (and *vice versa*). The name 'synthase' suggests that the opposite direction of this reaction (i.e., the synthesis of ATP) is usually more relevant biologically. Indeed, in respiring organisms, from bacteria to plants and animals, this highly sophisticated, multi-subunit enzyme, also called *F-ATPase* (Figure 9.13a) is the primary means of synthesizing ATP; it fulfills this function by employing the electrochemical gradient of protons[*1] across the membranes of cells (in bacteria), mitochondria (in eukaryotes) and chloroplasts (in plants) [57,66,71–73] (see Chapter 1, Subsection 1.1.3.2.1). The coupling of respiration to ATP synthesis was first described by Peter Mitchell's *chemiosmotic theory* [74], for which he received a Nobel Prize in Chemistry. In Archaea, ATP synthesis is carried out by a different member of the ATPase family, called *A-ATPase*. However, the activity of ATPases is reversible, and in certain cases they are used to hydrolyze ATP for creating ion gradients.

[*1]The proton gradient is also known as proton motive force (pmf) or Δp.

This process involves, e.g., *V-ATPases* which are present in vacuoles, and bacterial ATPases, which create proton gradients to drive processes such as chemotaxis and transport.

ATP synthase has a complex structure but can be viewed as a construction containing four main operational parts (for a more detailed description of their function see below):

- **F_0 ('rotor')** – a transmembrane domain that contains a proton channel (subunit a) and a cylinder (c subunits). When protons flow through the channel, they induce rotation of the c cylinder.

- **F_1** – an extramembrane domain that catalyzes ATP synthesis or hydrolysis. Its catalytic part contains six subunits arranged in alternation: three α subunits and three β subunits ($\alpha_3\beta_3$). The latter perform the actual catalysis.

- **The γ subunit ('shaft')** – an elongated part that connects the centers of F_0 and F_1. Its role is transmitting mechanical energy between the two.

- **The stator** – an elongated part, composed of different subunits that connects F_0 and F_1 from the periphery. Its role seems to be preventing the rotation of the $\alpha_3\beta_3$ part of F_1 due to its contact with the rotating γ shaft [75].

How does this machine work, exactly? Numerous studies and crystal structures of the enzyme in different modes point to an interesting and sophisticated mechanism (for reviews see [72,77,78]). In general, the mechanism can be viewed as including two main steps:

1. **Proton transport and F_0 rotation.**
   When protons flow through F_0 down their electrochemical gradient (e.g., from the intermembrane space to the matrix in mitochondrial F_0F_1-ATPase), the release of the potential energy stored in the gradient induces rotation of the F_0 c-ring [72] (Figure 9.13b). The coupling between proton flow and F_0 rotation is mediated by several aspartate or glutamate residues (depending on the organism) in the C-terminus of each of the F_0c subunits [79]. The side chain carboxyl groups of these residues (dark blue circles in Figure 9.13b) are exposed on the circumference of the c ring. When protons enter the a subunit of F_0, they protonate the aspartate and glutamate residues in the interface between the a subunit and the c cylinder [72,80]. This neutralizes the negative charge on the corresponding aspartate and glutamate residues, which drives them to move in a certain direction towards a more hydrophobic environment. As a result, adjacent c ring aspartate and glutamate residues move towards the vacated place and occupy it. There, they, too, undergo the same process of charge neutralization and diffusion. Altogether, these steps generate an enduring rotation of the c ring in the direction indicated. The rotation carries the protonated aspartate and glutamate side chains around until they reach a second site in contact with the a subunit, where the local environment (a basic residue in the a subunit) deprotonates them and releases the protons from the enzyme. Thus, protons entering the F_0 domain in one place leave it from another, to the opposite side of the membrane. The direction and magnitude of the proton gradient (e.g., $-180\,\text{mV}$ inwards in mitochondria) ensure that the rotation is unidirectional [72].

2. **Transmission of rotation energy and ATP synthesis by F_1.**
   The mechanical energy of the F_0 rotation is transmitted by the γ subunit shaft to F_1,

(a)

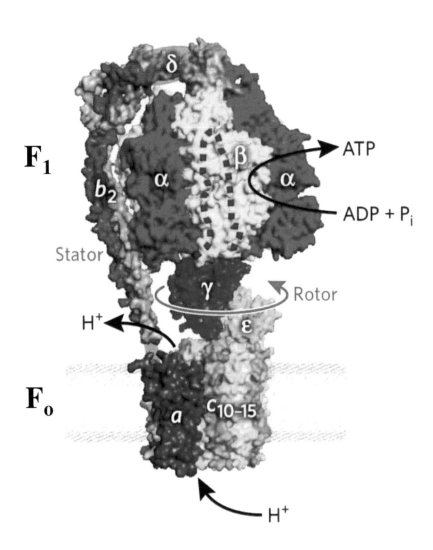

(b)

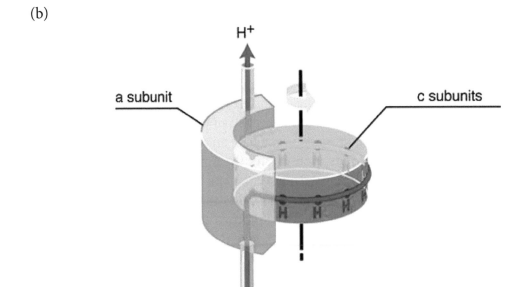

(c)

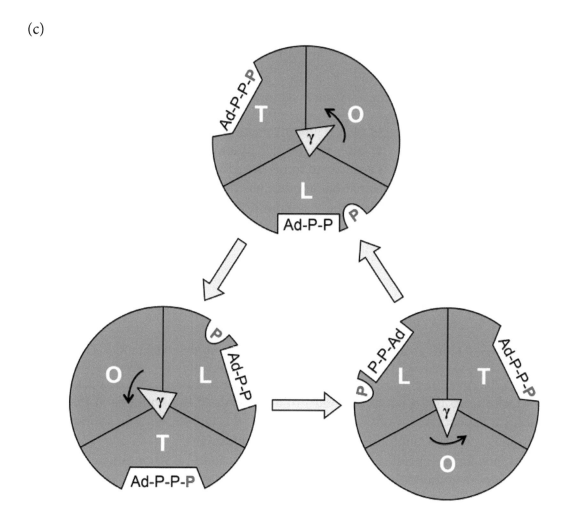

FIGURE 9.13   **ATP synthase.** (a) A structural model of the bacterial ATP synthase. F_0 is the transmembrane part of the enzyme, responsible for proton transfer. F_1 is the extramembrane part of the enzyme, responsible for ATP synthesis and hydrolysis. The subunits of the enzyme are noted. The image is taken from [76]. (b) A schematic model explaining how the flow of protons through F_0 generates rotation (see details in the main text). Taken from [72]. (c) Binding-change model for ATP synthase. Each of the β subunits of F_1 may possess an open/empty (O), ADP + P_i-binding (L) or ATP-binding (T) conformation, depending on its interactions with the γ subunit (yellow triangle). Proton flow through F_0 leads to its revolution, which is transmitted to F_1 by the γ subunit. A 360° revolution of the γ subunit takes each β subunit through all of the three conformations, which constitute a complete cycle of ATP synthesis by the three β subunits. This happens in three main steps, each involving a 120° revolution of the γ subunit. ADP and ATP are marked in the image as Ad-P-P and Ad-P-P-P, respectively, where the γ-phosphate of ATP is marked in bold red.

which results in ATP synthesis. Each β subunit of F_1 exists at a given time in one of the following three conformations:

- **O (open, or empty)** – this conformation does not bind ligands because it is formed when the bulge 'pushes' the subunit away from the complex.

- **L (loose)** – binds preferably Mg–ADP + P_i.

- **T (tight)** – binds preferably ATP. The γ-phosphate is coordinated to an arginine-containing motif in the adjacent α subunit [81]*1.

The specific conformation of each β subunit at any given time depends on its interaction with the γ subunit. The latter is asymmetric, and therefore its interactions with the surrounding β subunits are also asymmetric*2. Thus, at any given time, one of the β subunits has the O conformation, another has the L conformation, and the third has the T conformation. When the γ subunit rotates 120°, its interaction with each of the three β subunits changes. This induces the following conformational changes, each in a different β subunit (Figure 9.13c):

(a) **O $\longrightarrow$ L**: this change makes the empty β subunit bind ADP and P_i.

(b) **L $\longrightarrow$ T**: this change makes the β subunit chemically convert its bound ADP and P_i into ATP.

(c) **T $\longrightarrow$ O**: this change makes the ATP-bound β subunit release its ATP.

Thus, each 120° rotation of the γ subunit leads to the formation of one ATP from ADP and P_i, which means that a complete (360°) rotation leads to the synthesis of three ATP molecules. The number of protons needed to complete this task depends on the number of c subunits in F_0, and this number changes between organisms.

(Videos showing the operation of ATP synthase can be found at www.mrc-mbu.cam.ac.uk/research/atp-synthase/molecular-animations-atp-synthase.)

9.1.5.3.2 *Peptide hydrolases (EC 3.4)* [82]

Peptide hydrolases, commonly called *peptidases* or *proteases*, are a diverse group of enzymes that cleave peptide bonds in peptides and proteins, respectively. They act constantly inside all cells and tissues in our body, as well as during food digestion. These processes release free amino acids to cells or to circulation which, in turn, facilitate the production of new proteins and small bioactive compounds (see Chapter 2). Therefore, such hydrolases are highly important to the metabolism of all organisms, as reflected by their genomic prevalence (~2% of the genome). The prevalence and functional importance of proteins in living organisms make proteases in their vicinity potentially dangerous. To prevent massive degradation of proteins in cells and tissues, proteases are highly regulated at different levels. For example, proteases that are destined to be secreted are usually produced in an inactive form inside the cell (a zymogen) and become activated only after their secretion. Another form of regulation is compartmentalization; many eukaryotic proteases are sent after their synthesis to

---

*1 The α subunits also bind nucleotides, but they are not involved in ATP synthesis.

*2 The γ subunit is cranked, and when its convex side comes in contact with a β subunit, it pushes the subunit [73]. This exposes the nucleotide-binding site in the subunit and induces the release of the bound nucleotide.

intracellular lysosomes, where they are physically separated from cytosolic proteins. These enzymes are usually active only in acidic pH, which prevents them from being activated prematurely, before reaching the lysosome. Finally, cells use various inhibitors to control the activity of proteases.

As mentioned above, peptide hydrolases have historically been grouped according to the positions that they act upon in the polypeptide chain. *Endopeptidases* cleave peptide bonds inside the chain, whereas exopeptidases cleave bonds near the *N'* or *C'* of the chain. *Exopeptidases* can be further divided according to the number of residues released from the termini:

- **Aminopeptidases** (EC 3.4.1) – release the *first residues* of the polypeptide chain.

  ◇ *Dipeptidyl peptidases* (EC 3.4.14) – release the *first two* residues of the polypeptide chain.

  ◇ *Tripeptidyl peptidases* (EC 3.4.14) – release the *first three* residues of the polypeptide chain.

- **Carboxypeptidases** (EC 3.4.2) – release the *last residues* of the polypeptide chain.

  ◇ *Peptidyl dipeptidases* (EC 3.4.15) release the *last two* residues of the polypeptide chain.

Finally, *dipeptidases* (EC 3.4.3) act on isolated dipeptides.

Both exopeptidases and endopeptidases are specific towards the amino acids they cleave. For example, the digestive enzyme *trypsin* (EC 3.4.21.4) preferentially cleaves peptide bonds adjacent to lysine or arginine, whereas another digestive enzyme, *chymotrypsin* (EC 3.4.21.1), has a preference for the three aromatic amino acids. The specificity is determined by binding subsites, each recognizing a different amino acid in the sequence of residues flanking the labile bond. Together, these subsites enable the peptidase to recognize and bind to a specific short sequence in the target protein or peptide and cleave the peptide bond between two specific residues. Following Schechter and Berger's model [83], the subsites are numbered according to their relative positions with respect to the catalytic site (where the labile peptide bond is positioned): the subsites on the *N*-terminal side of the active site are called S1, S2 and on, and those on the *C*-terminal side are called S1′, S2′ and on. Thus, the labile bond is always between the S1 and S1′ subsites.

Peptide hydrolases are commonly grouped into the following types, based on the nucleophilic residue that is used for catalysis [82]:

1. **Serine proteases** [84,85] – the largest group of proteases. These enzymes catalyze bond cleavage by using the hydroxyl group of a serine residue, in conjunction with a histidine residue and usually also an aspartate (*catalytic triad*). The conditions under which serine proteases act make them particularly suitable as extracellular enzymes, such as those participating in food digestion (e.g., trypsin, chymotrypsin, and *subtilisin* (EC 3.4.21.62)).

2. **Threonine proteases** – catalyze bond cleavage by using the hydroxyl group of a threonine residue. The best known example of this type of enzyme is the *proteasome* (EC 3.4.25.1), which degrades ubiquitin-attached proteins in eukaryotic cells as a routine part of these proteins' turnover [86,87] (see also Chapter 2).

3. **Cysteine proteases** [88] – catalyze bond cleavage by using the thiol group of a cysteine residue. Since thiol groups are prone to oxidation, many cysteine proteases act within the reducing environment of the cell. A well-known example of cysteine proteases is the plant enzyme *papain* (EC 3.4.22.2).

4. **Aspartic proteases** [89] – catalyze bond cleavage by using a water molecule that is activated by a pair of aspartic acid residues. A well-known example of aspartic proteases is the gastric enzyme *pepsin* (EC 3.4.23.2), which participates in food digestion.

5. **Glutamic proteases** [90,91] – a small group of fungal proteases that act under acidic pH and catalyze their reactions by using a glutamate-glutamine dyad. The representative enzyme of this group is *scytalidopepsin B* (EC 3.4.23.32).

6. **Metalloproteases** [92–94] – the second largest group of proteases, involved in a diverse range of physiological processes, including digestion, tissue remodeling, blood pressure regulation, and more [95]. These enzymes catalyze bond cleavage using a metal, usually Zn^{2+}, which participates in creating a hydroxide (OH^-) nucleophile by polarizing a catalytic water molecule. A well-known example of zinc-dependent metalloproteases is *thermolysin* (EC 3.4.24.27), a bacterial peptide hydrolase that cleaves peptide bonds adjacent to large hydrophobic residues [96]. The thermostability of thermolysin has made it popular in different industries. For example, it is used to produce the artificial sweetener *aspartame* in the food industry, whereas in laundry detergents it is used to degrade protein stains [97].

In addition to these major groups, there are also asparagine, mixed, and unknown peptide hydrolases.

The above classification and naming methods of peptide hydrolases have recently been succeeded by the MEROPS system [90], which is based on the structural similarities among these enzymes. This system is considered more biologically relevant than previous methods, because structural similarities among proteases reflect their evolutionary relationships, and in many cases are also indicative of common functional properties.

9.1.5.4 Lyases (EC 4)

Lyases catalyze non-hydrolytic elimination or addition reactions, that is, reactions in which covalent bonds are cleaved or formed in the substrate without the use of water or oxidation [27]. Thus, in contrast to hydrolases, a lyase has a single substrate in one direction (cleavage) and two in the other (condensation). The process of bond cleavage involves a leaving group and often the formation of either a double bond or a ring structure. For example:

Scheme 9.4. $\qquad$ $A-B-OH \longrightarrow A + B=O$

In the opposite direction, lyases catalyze the addition of small molecules to C=C, C=N, or C=O bonds. Like hydrolases, lyases tend to make extensive use of nucleophilic catalysis (see Subsection 9.3.3.3 below), and the second numeral in a lyase's EC number designates the general type of cleaved or formed bond:

- EC 4.1 – C−C bonds

- EC 4.2 – C–O bonds

- EC 4.3 – C–N bonds

- EC 4.4 – C–S bonds

- EC 4.5 – C–halide bonds

- EC 4.6 – P–O bonds

- EC 4.7 – C–P bonds

- EC 4.99 – other lyases

The third numeral of a lyase's EC number refers to the specific type of bond, or to the chemical group containing this bond (carboxyl, aldehyde, etc.). Finally, the fourth numeral specifies the exact substrate of the enzyme. The systematic names of lyases take the form of *substrate group-lyase* (e.g., indole-3-carboxylate carboxy-lyase). Two well-known metabolic reactions of this type are *decarboxylation*, which involves the cleavage of a C–C bond with the release of CO_2[*1], and *aldol cleavage* (a reversal of *aldol condensation*[*2]) with a release of H_2O (e.g., Figure 9.14). The enzymes involved in these types of reactions are commonly called *'decarboxylases'* and *'aldolases'*, respectively. When the reaction results in bond formation, the enzyme is sometimes referred to as a *'synthase'*[*3]. Another common name, *'dehydratase'*, is used when the leaving group is H_2O (i.e., water elimination reaction).

Certain lyases use the pyridoxal phosphate (PLP) coenzyme to facilitate bond cleavage [62]. We have already encountered this coenzyme in aminotransferases (see Subsection 9.1.5.2.1), where it is used to separate the α-amino group from the rest of the substrate. In lyases, PLP can be used to facilitate α, β or γ elimination of a substituent from an amino acid. As in aminotransferases, here, too, PLP acts by drawing electrons from the bond intended for cleavage and by stabilizing the negatively charged transition state. This happens,

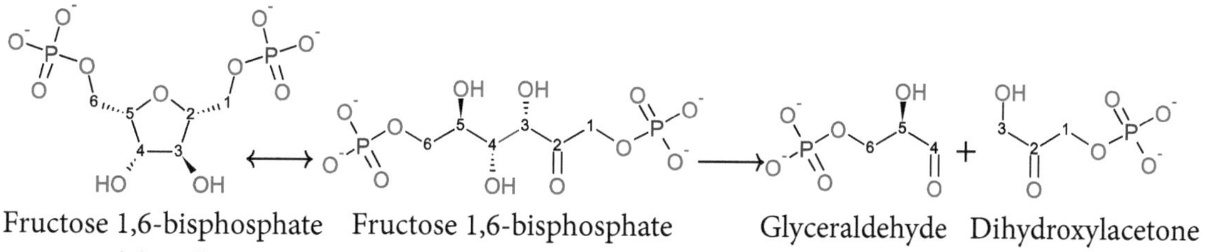

Fructose 1,6-bisphosphate (closed) Fructose 1,6-bisphosphate (open) Glyceraldehyde 3-phosphate Dihydroxylacetone phosphate

FIGURE 9.14 **Fructose 1,6-bisphosphate cleavage carried out by aldolase during glycolysis.** The reaction involves the open form of the molecule, which is in equilibrium with the closed form that is dominant in solution. The cleavage is between the third and fourth carbons. For clarity, explicit hydrogens are not shown. The carbon atoms in the reactants and products are numbered.

[*1]Decarboxylases can also catalyze the opposite reaction (carboxylation). They should not be confused with carboxylases of the ligase class (EC 6.3 and 6.4).

[*2]Aldol condensation is a reaction in which an enol or an enolate reacts with a carbonyl compound to form a β-hydroxy-aldehyde or β-hydroxy-ketone. A well-known example is the formation of citrate from acetyl-CoA and oxaloacetate in the Krebs cycle.

[*3]Not to be confused with *'synthetase'*, which is a common name for some ligases (EC 6).

for example, in *ornithine decarboxylase* (EC 4.1.1.17). Other lyases (e.g., *pyruvate decarboxylase*; EC 4.1.1.1) use the vitamin B_1-derived coenzyme *thiamine pyrophosphate* (*TPP*) (Figure 9.15a) to catalyze the same type of reaction [98]*1. Notably, **the mechanism used by TPP is very similar to that of PLP: TPP acts as an electron sink to stabilize the negative charge that forms on the transition state of the reaction.** This process is facilitated by TPP's *thiazolium ring*, which, like the pyridinium ring of PLP, contains a positively charged nitrogen atom and delocalized π electrons. In both cases, the reaction starts with proton abstraction from a carbon atom. However, whereas in PLP-dependent reactions the proton is abstracted from the substrate, in TPP-dependent reactions the proton is abstracted from TPP itself, i.e., from its C2 carbon, which is much more acidic than regular aliphatic carbons (Figure 9.15b step 1). The proton is transferred to an active site glutamate side chain, and C2 becomes a carbanion. The latter is an efficient nucleophile, and is therefore able to attack the substrate's C_α, binding it covalently (Figure 9.15b, step 2). At the same time, the carbonyl oxygen of the substrate is protonated by TPP's amino group, which turns the carbonyl into a hydroxyl. In the next step, the electrons of the C_α−COOH bond are drawn to the thiazolium ring, a process that weakens the bond and leads to its breaking (Figure 9.15b, step 3). In the next step, C_α is protonated by an active site aspartate residue, which turns the double bond between TPP and the substrate into a single bond (Figure 9.15b, step 4). Deprotonation of the substrate's OH group by TPP's imine group facilitates the breaking of the TPP-substrate bond, and the latter leaves as an aldehyde (Figure 9.15b, step 5). The last two steps involve the recycling of TPP for the next round of catalysis.

Note that TPP is used to decarboxylate α-keto acids. In β-keto acids, decarboxylation happens spontaneously and does not require TPP (or PLP). This is because in β-keto acids the negative charge resulting from decarboxylation is stabilized by enolate-ketolate resonance of the β-carbonyl group (Figure 9.15c top). In α-keto acids such resonance is impossible, and **TPP is required as a 'surrogate stabilizing β-group' to stabilize the charge** [99] (Figure 9.15c bottom). This is why α-keto acids such as *pyruvate, α-ketoglutarate, oxaloacetate* and *glyoxylate* are much more common than β-keto acids; the latter are usually not stable enough in aqueous solutions under physiological conditions [99]. Still, some β-keto acids, such as *acetoacetate*, do play roles in metabolism.

(a)

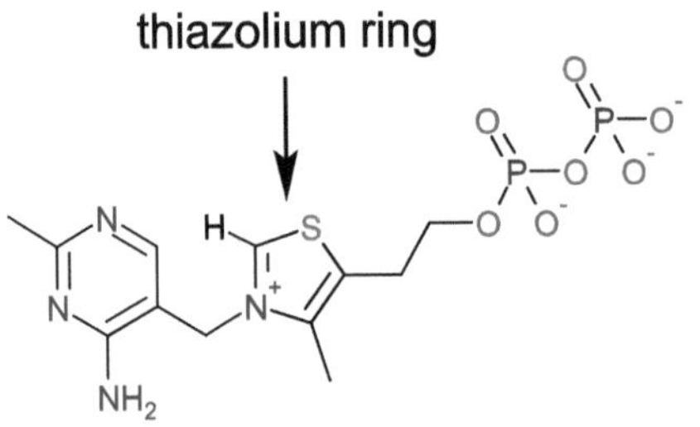

FIGURE 9.15 **The use of thiamine pyrophosphate (TPP) for decarboxylation of α-keto acids.** (a) The chemical structure of thiamine pyrophosphate. The active thiazolium ring is marked.

---

*1TPP participates in reactions involving the cleavage between a carbonyl group and an adjacent reactive group [98] (see Table 9.3 at the end of the chapter).

(b)

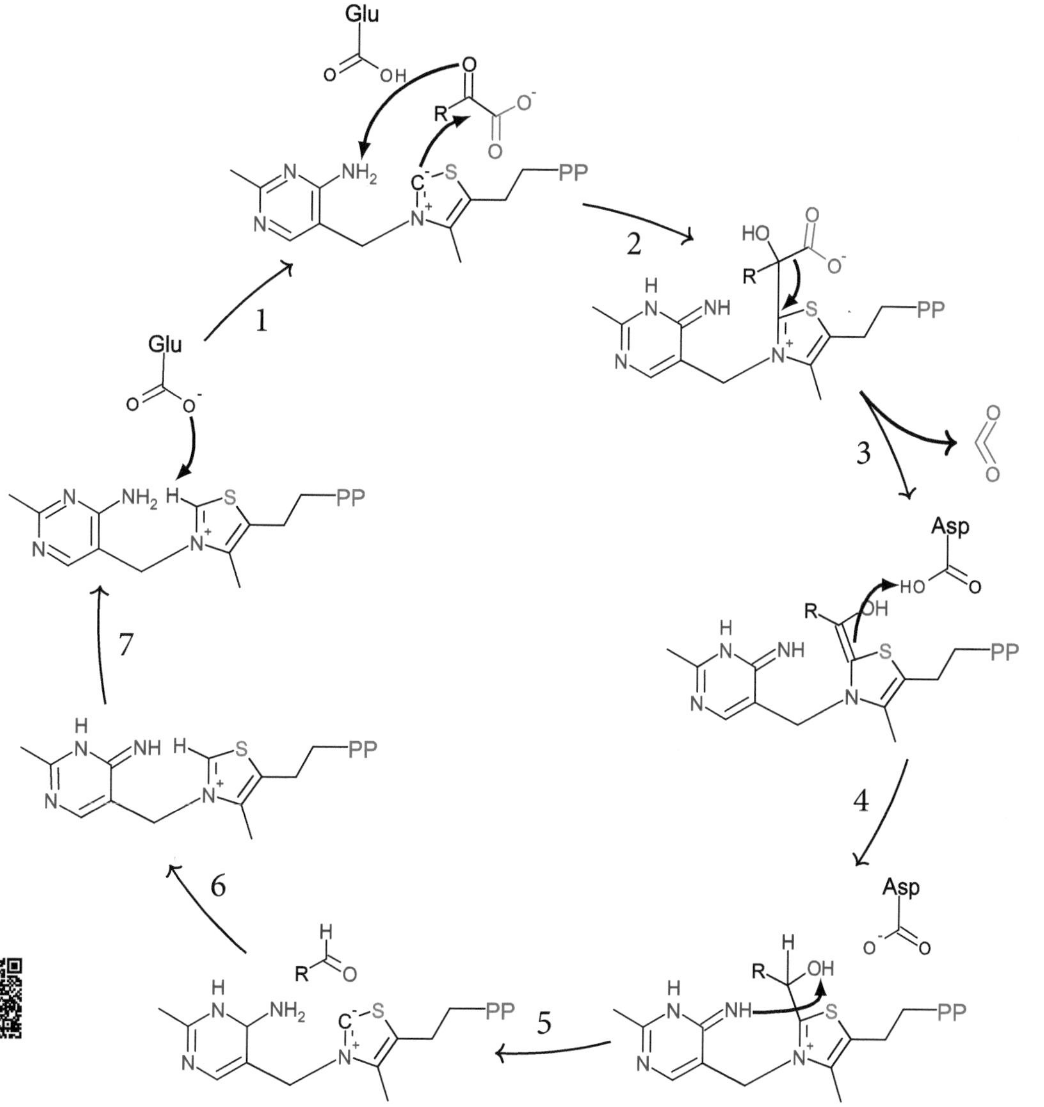

(c)

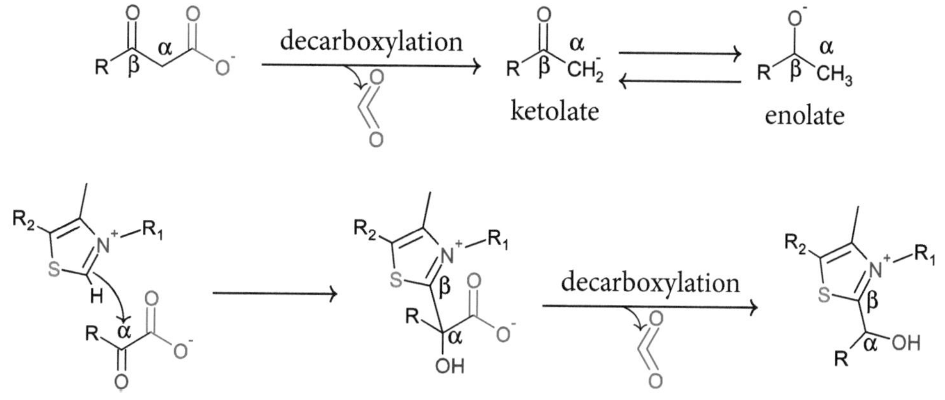

FIGURE 9.15 **The use of thiamine pyrophosphate (TPP) for decarboxylation of α-keto acids.** (Continued) (b) The catalytic mechanism involving TPP (see main text). (c) Spontaneous decarboxylation of β-keto acid (top) versus TPP-facilitated decarboxylation of α-keto acids (bottom). In both cases the carboxyl group is in red and the keto group is in purple. Only the thiazolium ring of TPP is shown. The α and β positions in the molecules are marked. In α-keto acid decarboxylation, a β-carbonyl group is absent, and TPP acts as a 'surrogate' β-group to stabilize the negative charge on the transition state.

9.1.5.5   Isomerases (EC 5)

Isomerases are one of the two smallest enzyme classes, constituting only 5% of the enzymes in the ExplorEnz database [31]. They catalyze the interconversion between molecular isomers through geometric or structural changes. Such reactions include either of the following [27] (see more below):

- The transfer of a chemical group from one position to another in the substrate:

    Scheme 9.5. $\qquad$ $X-A-B-Y \longrightarrow Y-A-B-X$

- A change between the *S* and *R* stereo-configurations of a chiral center.

- A change between the *cis* and *trans* configurations of a double bond.

Because all changes catalyzed by isomerases are internal, these enzymes always have one substrate and one product. A well-known example is the conversion of citrate to isocitrate by *aconitase* (EC 4.2.1.3), a Krebs cycle reaction (Figure 9.16). In this reaction, an OH group is transferred from C3 to C5 within the molecule, thus allowing the subsequent oxidative decarboxylation reaction to take place [41].

In the EC number of a lyase the second numeral designates the type of reaction catalyzed by the enzyme. The reaction types also form the basis for the common names of lyases:

- **EC 5.1 – Racemases and epimerases**.
  These enzymes catalyze the interchange between the *S* and *R* configurations in the substrate. *Racemases* do so in a substrate having only one chiral center. For example, in amino acid racemases (e.g., *alanine racemase*; EC 5.1.1.1), the L-(2*S*) and D-(2*R*) enantiomers are interconverted via C_α deprotonation by one active site residue, followed by reprotonation of the same atom by another residue, this time from the opposite side [100]. *Epimerases* catalyze the interchange between *S* and *R* configurations of a single chiral center in a substrate having more than one center. For example, *UDP-glucose 4-epimerase* (EC 5.1.3.2) changes glucose to galactose (and *vice versa*) while bound to the nucleotide UDP. The two monosaccharides have six carbons and four chiral centers, but differ only in the configuration of the C_4-OH group.

- **EC 5.2 – *Cis-trans* isomerases**.
  These enzymes catalyze the change between the *cis* and *trans* configurations of a double bond. For example, *prolyl isomerase* (EC 5.2.1.8) acts on the peptide bond of intraprotein proline residues.

- **EC 5.3 – Intramolecular oxidoreductases**.
  These enzymes catalyze the transfer of electrons within the substrate. In cases where the initial and final electronic states of the molecule readily interconvert (e.g., *keto* and *enol* states), the enzyme catalyzing the reaction is commonly called a *tautomerase*.

- **EC 5.4 – Intramolecular transferases**.
  These enzymes, which are sometimes called *mutases*, catalyze the transfer of a chemical group from one position to another within the substrate. For example, *phosphoglycerate mutase* (EC 5.4.2.1) acts in glycolysis to convert 3-phosphoglycerate

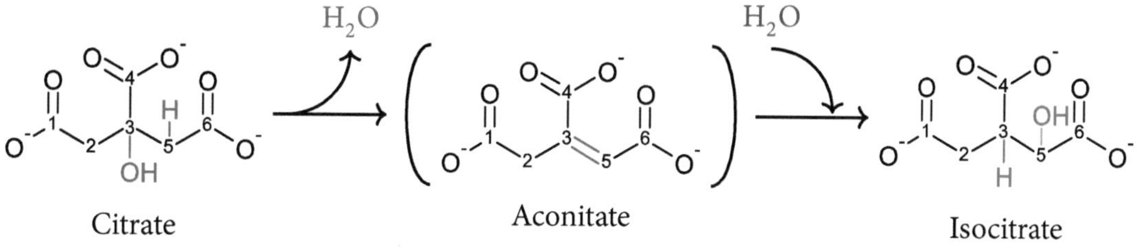

FIGURE 9.16 **Citrate isomerization by aconitase during the Krebs cycle.** The carbons of the molecules are numbered. First step – dehydration of citrate. The hydroxyl group and hydrogen atom that leave the molecule are in red, as well as the double bond that is formed in aconitate. Second step – hydration of aconitate. The hydroxyl group and hydrogen atom that insert into the molecule are in green.

into 2-phosphoglycerate. The intramolecular rearrangement catalyzed by mutases often involves a replacement of a hydrogen atom in one position with another atom at a different position. The latter atom is usually a part of an acyl, phosphoryl, amino, or hydroxyl group. Mutases that catalyze carbon skeleton rearrangements (e.g., *methylmalonyl-CoA mutase*; EC 5.4.99.2) and amino group transfers (e.g., *D-ornithine 4,5-aminomutase*; EC 5.4.3.5) use *adenosylcobalamin* (a form of *coenzyme B$_{12}$*) as a cofactor [101] (see Subsection 9.4.2 for mechanistic details).

- **EC 5.5 – Intramolecular lyases.**
  This subclass has only one sub-subclass, which comprises enzymes catalyzing the interconversion between the open and closed states of a ring structure within the substrate.

- **EC 5.99 – Other isomerases.**
  The third numeral of the EC number of an isomerase refers to the type of substrate on which the enzyme acts, and the fourth numeral specifies the exact substrate.

9.1.5.6 Ligases (EC 6)

Ligases are the smallest enzyme class, constituting only 3% of the enzymes in the ExplorEnz database [31]. Still, they participate in many central metabolic reactions, e.g., amino acid synthesis, DNA and RNA repair, and ammonia fixation in higher plants [102]. They covalently attach two substrate molecules (via different bond types, see below) to form one product, using the energy derived from the hydrolysis of a phosphorylated nucleotide, usually ATP [27,102]:

Scheme 9.6. $$A + B + ATP \longrightarrow A{-}B + ADP + P_i$$

Accordingly, the systematic names of these enzymes take the form of *substrate1:substrate2 ligase*. The phosphate group transferred from ATP may be its free γ-phosphate (e.g. [103]) or its AMP-bound α-phosphate [104]. The former case involves hydrolysis of the $\beta{-}\gamma$ phosphate bond in ATP (releasing ADP), whereas the latter case involves hydrolysis of the $\alpha{-}\beta$ phosphate bond, releasing pyrophosphate (*PP$_i$*). Pyrophosphate breaks down readily in aqueous solutions, which pushes the reaction equilibrium even further.

As explained in Subsection 9.1.3 above, ATP is often used to drive energy-demanding biological processes (such as synthesis of molecules), and this function of ATP is facilitated by enzymes that couple between ATP hydrolysis and other processes. How can ATP hydrolysis drive other processes? ATP is said to contain 'energy-rich bonds' (P−P anhydride bonds), the hydrolysis of which 'releases' much energy. This description, however, does not provide any mechanistic information about how the released energy is channeled into processes such as ion pumping, bond breaking or bond formation. We saw earlier that the mechanism of such energy channeling involves the transfer of one of the products of ATP hydrolysis (usually P_i) to a target molecule. As explained in Chapter 2, when the bulky and charged P_i group is transferred to a protein residue (serine, threonine, or tyrosine), it acts sterically and electrostatically to induce conformational changes, which are required for the protein's function. In Subsection 9.1.5.2.2 above, we saw that P_i transfer can also change the fate of small molecules. For example, phosphorylation of glycolytic metabolites facilitates their subsequent oxidation. Here, too, phosphorylation may act by inducing conformational changes. That is, the phosphorylated molecule may induce conformational changes in the enzyme binding it, which ultimately help the enzyme to carry out catalysis. However, phosphorylation of the small molecule may also act directly by activating the molecule, i.e., by changing its electronic properties and thus making it more amenable to catalysis. This is exactly what happens in ligase-catalyzed reactions. **Ligases channel the energy of ATP hydrolysis to the ligation reaction by phosphate-induced substrate activation; specifically, the ligase destabilizes a bond in the substrate**. This is demonstrated in the synthesis of (*R*)-pantothenate from (*R*)-pantoate and *β*-alanine by *pantoate-beta-alanine ligase* (EC 6.3.2.1) (Figure 9.17a). The reaction involves the formation of a C−N bond between the two substrates, at the expense of a C−O bond in (*R*)-pantoate (Figure 9.17b). Since the C−O bond is stable, it needs to be destabilized for the reaction to proceed. The transfer of an AMP-bound phosphate to (*R*)-pantoate achieves this destabilization, and facilitates the breakage of the C−O bond (the 'leaving group effect'). In some ligases that catalyze carboxylation reactions (e.g., pyruvate carboxylase; EC 6.4.1.1) the substrate activation mechanism is somewhat different and requires both a phosphate and the cofactor *biotin*.

In the EC number of a ligase, the second numeral designates the type of bond that is formed:

- EC 6.1 – C−O bonds

- EC 6.2 – C−S bonds

- EC 6.3 – C−N bonds

- EC 6.4 – C−C bonds

- EC 6.5 – phosphoester bonds

- EC 6.6 – N−metal bonds

In the first three subclasses the third numeral of the EC number refers to the specific bond that is formed. The rest of the subclasses are not further divided into sub-subclasses. Some ligases have common names such as *synthetase*[*1], or *carboxylase*. A well-known example of a ligase is DNA-ligase, which is a central component of the DNA replicating machinery. This enzyme is responsible for connecting the Okazaki fragments during the synthesis of the lagging strand of DNA.

[*1]Not to be confused with 'synthase', which is a common name for some lyases (EC 4).

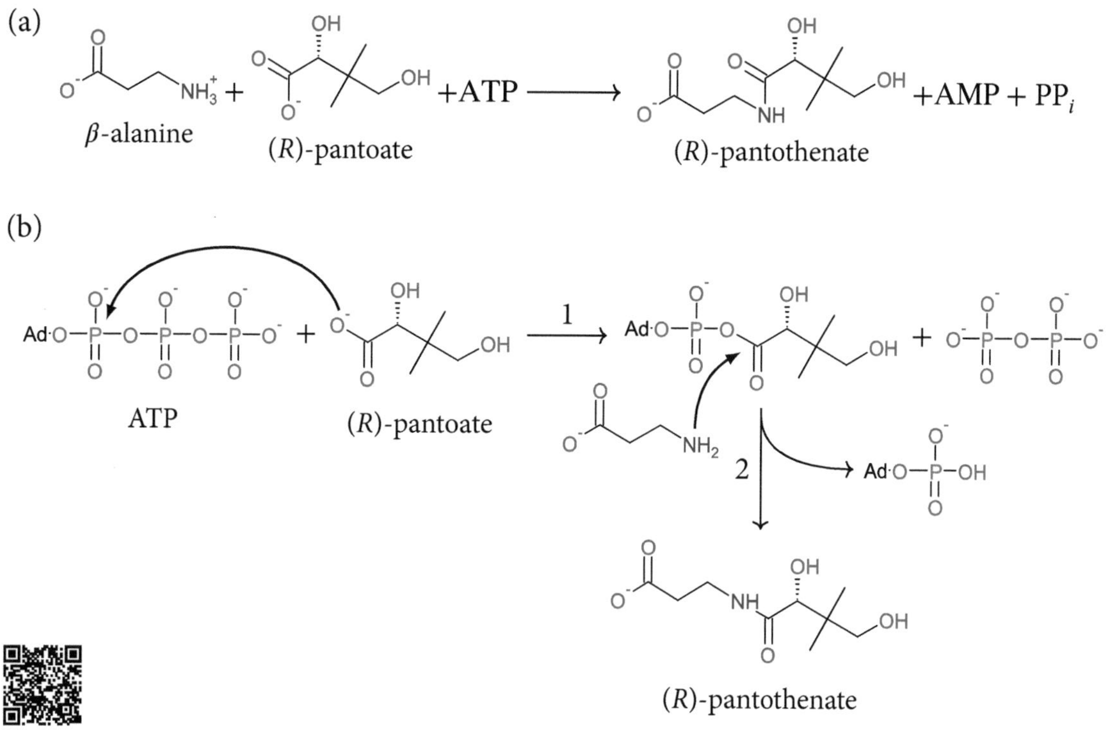

(a)

β-alanine + (R)-pantoate + ATP ⟶ (R)-pantothenate + AMP + PP$_i$

(b)

ATP + (R)-pantoate ⟶ 1

(R)-pantothenate

FIGURE 9.17 **The mechanisms of ligases.** The figure shows the mechanism of *pantoate-beta-alanine ligase* (EC 6.3.2.1). (a) The overall reaction. (b) The mechanism of the reaction, adapted from the MACiE database (entry M0229). Step 1: a nucleophilic attack of (R)-pantoate's carboxylate on the α-phosphate of ATP (written here as Ad–PO$_4$–PO$_4$–PO$_4$), creating an activated intermediate and releasing pyrophosphate (PP$_i$). Step 2: a nucleophilic attack of β-alanine's amino group (in its deprotonated form), creating (R)-pantethonate and releasing AMP (written here as Ad–PO$_4$).

9.1.5.7 Catalytic promiscuity

While most enzymes catalyze a single, specific chemical reaction, some have been shown to catalyze multiple reactions[*1], albeit with different efficiency. This phenomenon is commonly known as 'catalytic promiscuity'. For example, certain enzymes have been found to hydrolyze both cyclic ester (lactone) bonds and phosphoester bonds, albeit at different rates [105]. Another example is the *Candida antarctica* lipase B (CAL-B; EC 3.1.1.3). This enzyme, which normally degrades ester bonds in triacylglycerol, has also been shown to catalyze amide hydrolysis (EC 3.4) [106], in addition to two types of C−C bond formations: *aldol condensation* (EC 4.1) [107], and *Michael addition* (EC 4.4) [108] (see more examples in [109]). These examples raise questions as to how a single active site, which contains a certain catalytic machinery, can carry out multiple different reactions. One scenario in which this could happen is when the different reactions catalyzed by the enzyme use different residues for catalysis. For example, in the case of CAL-B, the native reaction (triacylglycerol hydrolysis), which involves nucleophilic catalysis (see Subsection 9.3.3.3 below), uses the entire Ser-105-His-224-Asp-187 catalytic triad. In contrast, the Michael addition and aldol condensation reactions, which are catalyzed by the same active site, use only the His-224-Asp-187 dyad, for acid-base catalysis (see Subsection 9.3.3.5 below). An alternative scenario in which a single active site might catalyze multiple different reactions is when

---

[*1]These different reactions may be reflected in the first, second or third numerals of the EC number.

the same residues participate in all reactions, but have different protonation states in each reaction and therefore different catalytic roles (e.g., as in acid-base catalysis, see [109] for examples).

In either scenario of catalytic promiscuity, the enzyme may be able to bind multiple different substrates (*substrate promiscuity*). Indeed, most enzymes are able to catalyze the same principal reaction on a group of similar substrates, but they are also able to bind other substrates. The structural and/or chemical differences between the native substrates of an enzyme and the other substrates it binds may allow different substrates to undergo different reactions inside the active site, because of their different levels of accessibility to active site residues, or because they induce different protonation states in these residues. These parameters can also be affected by conformational changes in the enzyme. In metalloenzymes a change in the identity of the catalytic metal may result in different function (e.g. [110,111]).

Catalytic promiscuity has been suggested to serve as an evolutionary mechanism for the development of enzymes catalyzing new reactions [112,113] (see also the discussion on protein functional evolution in Chapter 2, Subsection 2.4.3.3). For example, in a population of enzymes catalyzing reaction A, an enzyme capable of catalyzing both reaction A and B could serve as an 'intermediate' in the functional shift towards enzymes catalyzing reaction B. Such a process of *'divergent evolution'* would require the genetic duplication of the promiscuous enzyme, after which reaction B would be permanently set by further mutations and natural selection [109]. The aforementioned enzymes displaying both lactonase and phophoesterase activity are an interesting example; it is thought that enzymes in this group, which display stronger phosphoesterase activity, actually evolved from lactonases during the 20th century in response to the wide use of organophosphates as synthetic pesticides [105]. It should be noted that catalytic promiscuity observed *in vivo* is less common than that demonstrated *in vitro*; inside cells, promiscuity is often prevented or kept to a minimum by regulation. For example, in many cases the natural substrate is also an allosteric activator of the enzyme. Thus, the enzyme becomes active only in the presence of this substrate, which is a good way to ensure that the enzyme does not work on other substrates [109]. In such cases, even if the enzyme is able to bind other substrates and catalyze either the same reaction or other types of reactions, it will not do so, since it cannot be allosterically activated by the other substrates.

A somewhat similar phenomenon to catalytic promiscuity is observed in (different) enzymes that have the same fold yet catalyze different reactions (see Chapter 2, Subsection 2.4.3.3). This phenomenon, which is common in enzyme superfamilies, seems at first to contradict the paradigm dictating that the function of a protein results directly from its three-dimensional structure or fold. However, as explained in Chapter 2, **while enzymes in the same superfamily may catalyze different reactions, they usually share the same principal catalytic mechanism and differ only in the secondary catalytic stages, or in their specificity towards their substrates, coenzymes, or regulating ligands** [114,115]. This is clearly evident in the *enolase superfamily*. The members of this superfamily possess a common fold, yet catalyze 14 different biochemical reactions. In-depth analysis shows that while the 14 specific reactions are indeed different, they all include a common catalytic step: the abstraction of a proton from a carboxyl-bound carbon in the substrate [116]. In these enzymes the abstraction of the proton is carried out by a residue acting as a general base, and the enolate transition state is stabilized by a divalent metal cation. However, whereas the deprotonation of the α-carbon is conserved, the residue acting as the general base is not; nor is its orientation.

The importance of the core catalytic machinery to the enzyme's function is also demonstrated by a more extreme phenomenon: **In some cases, enzymes with no common origin and different sequence and/or structure may perform the same reaction type, provided that they have the same catalytic residues** [117,118]*1. This happens, for example, in the serine proteases chymotrypsin and subtilisin, which catalyze the same reaction (proteolysis) yet have unrelated structures.

As mentioned above, promiscuity is considered an important facilitator in the evolution of enzymes on Earth. Accordingly, it has occurred to enzyme engineers that they can learn from evolution how to alter their enzymes to achieve different substrate specificity and/or catalytic activity. To this end, they track the evolutionary path of a focal enzyme, looking for predecessors that possess promiscuous activity. Analysis of these ancestral enzymes can help researchers to pinpoint specific residues that have been involved in the shift of substrate specificity or catalytic activity. Then, they can induce changes in the corresponding residues of evolved enzymes in order to influence these enzymes' activity (for more details and references see reviews by [119,120])

9.2   ENZYME KINETICS

Enzyme-catalyzed reactions are complex and involve multiple steps, from the initial binding of the reactant(s) to the enzyme, through the shift between the various chemical intermediates along the reaction coordinate. While only the slowest steps determine the rate of the reaction (represented by the constant k_{cat}), a true understanding of enzyme-mediated catalysis requires a detailed characterization of all steps and intermediates. Achieving such a characterization is often difficult, as it requires the identification of very short-lived individual reaction intermediates. Yet, this is possible today thanks to methods that have been developed to track rapid processes. These methods include stopped-flow, rapid quench-flow, and relaxation techniques [121,122].

Another important tool for characterizing catalytic steps is X-ray crystallography. There are numerous enzymes for which the Protein Data Bank contains several different X-ray structures, determined under different conditions, with different ligands, mutations, etc. In many cases, these conditions have been set to induce structures of the enzyme at different points during catalysis, and therefore analysis of these structures can contribute substantially towards elucidating the entire catalytic cycle, or at least key events in it. Finally, the energy changes underlying the shifts between reactant intermediates and corresponding protein conformations can be predicted using computational techniques such as molecular dynamics simulations and quantum-mechanical (QM) calculations. As explained in Chapter 3, QM calculations on entire proteins are not feasible with current computational power, which is why these calculations are usually carried out using the faster quantum mechanics/molecular mechanics (QM/MM) approach [123,124]. The QM/MM approach was originally introduced by Warshel, Levitt and Karplus [125,126] and is widely used today.

The methods discussed above have been available for only a few decades. Before that, biochemists were limited to general measurements of enzyme activity, inhibition or activation induced in the presence of different ligands or reaction conditions. The only method available for understanding enzyme action was to construct mathematical models capable of explaining enzymatic rate acceleration and catalytic steps using the measured parameters

---

*1This phenomenon is indicative of 'convergent evolution' (see Chapter 2, Section 2.4.2.1).

(i.e., kinetic models). As we will see below, this approach has provided interesting insights over the years about enzymatic catalysis, and has also proven to be useful for understanding biochemical and physiological pathways, as well for predicting enzyme behavior. In this section we describe the most popular model proposed so far: the *Michaelis-Menten formalism*. For more in-depth discussions of enzyme kinetics, we direct the reader to more advanced books or articles on this topic.

9.2.1 Basic concepts

One of the simplest means of measuring an enzyme-catalyzed reaction is to track the change in substrate or product concentration over time, in order to obtain the rate, or velocity (V), of the enzyme. This simple experiment yields the dependency shown in Figure 9.18.

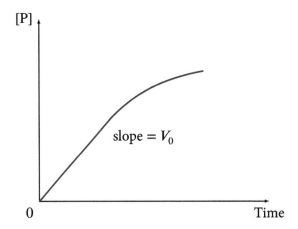

FIGURE 9.18 **Formation of product over time in an enzymatic reaction.** [P] is the molar concentration of the reaction's product. The slope of the plot in its linear region is the initial reaction velocity (V_0, see main text).

The rate of the enzyme-catalyzed reaction can be calculated from the plot as:

$$\text{rate} = \frac{d[P]}{dt} \tag{9.4}$$

(where d[P] is the change in molar concentration of the product, and dt is the period of time over which this change takes place).

The plot shows that the rate of the enzyme-catalyzed reaction diminishes with time. This may result from various factors, such as the decrease in substrate concentration, inhibition by the accumulating product, a change in pH, or thermal inactivation (reversible or irreversible (denaturation) [127]). In any case, to be able to follow enzymatic activity, it is customary to measure the *initial velocity (rate) of the enzyme (V_0)*, i.e., the rate obtained at the beginning of the reaction, before it is influenced by external factors. V_0 depends on the concentrations of the enzyme and the substrate. When V_0 is measured at a fixed enzyme concentration and increasing substrate concentration, the dependency shown in Figure 9.19 is obtained. This dependency is shared by most enzymes.

The plot reflects an interesting behavior of the enzyme-catalyzed reaction. At low substrate concentrations, the rate of the reaction increases linearly with substrate concentration, but gradually becomes smaller until no rate increase is observed (i.e., the rate reaches

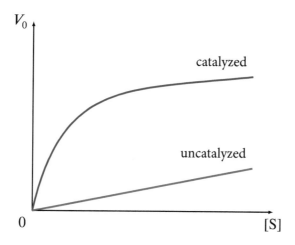

FIGURE 9.19   **Catalyzed (blue plot) and uncatalyzed (red plot) initial reaction rates (V_0) with increasing substrate concentrations [S].**

a maximal value). To understand why this happens, let us imagine an enzyme with a single binding site per enzyme molecule. At low substrate concentrations, when the number of substrate molecules in the solution is much lower than the number of available enzyme binding sites, only a small portion of the available sites are occupied and can catalyze their bound substrate. At this point, any increase in substrate concentration will lead to an increase in the number of occupied binding sites, which will translate into a rise in enzyme activity (hence the linear dependency). When the substrate concentration is increased to the point at which it is equal to the concentration of the enzymatic binding sites in the solution, most sites will be occupied and catalyze their bound substrate (i.e., a state of saturation is reached). At this point, the rate cannot rise any further, and the measured activity has the maximal rate that can be achieved at this enzyme concentration. This rate is called *maximal velocity*, or V_{max}. In enzymes having more than one binding site per enzyme molecule, higher substrate concentrations will be needed to achieve saturation, but the kinetic behavior will be qualitatively the same. This **saturation kinetics is a hallmark of enzymatic catalysis and is different from what is observed in uncatalyzed reactions** (Figure 9.19, red line). In the latter, a linear dependency is maintained at all substrate concentrations because there are no binding sites that can become saturated.

The above description explains the molecular origin of the observed saturation kinetics of enzymes, yet does not provide any mechanistic or quantitative insights. The work of German biochemist Leonor Michaelis and Canadian physician Maud Menten [128,129] has led to a mathematical model that quantitatively describes the saturation kinetics of enzymatic reactions, and provides a better understanding of enzyme behavior and mechanisms of catalysis. Although this model is not necessarily valid for all enzymes, especially allosteric enzymes and those that include complex mechanisms, it covers many natural enzymes and is still the most popular model in use. The main principles of the Michaelis-Menten model and its underlying assumptions are described in the following subsection.

9.2.2 Michaelis-Menten model

The Michaelis-Menten (M–M) formalism refers to chemical reactions that are catalyzed by an enzyme *of fixed concentration*, and that are in a *steady state*[*1]. In the original Henri-Michaelis-Menten model, such reactions were described by the following scheme:

Scheme 9.7.
$$E + S \underset{k_{-1}}{\overset{k_1}{\rightleftharpoons}} ES \overset{k_2}{\longrightarrow} E + P$$

(where E is the enzyme, S is the substrate, ES is an enzyme-substrate complex, P is the product, k_1 and k_{-1} are the rate constants of ES formation and breakdown, respectively, and k_2 is the rate constant of the step in which the product is formed and released from the enzyme, i.e., the chemical reaction step).

That is, these reactions include two major steps:

1. **Enzyme-substrate binding** – a rapid equilibrium process with a corresponding constant of $K_S = k_{-1}/k_1$. As such, K_S is an inverse measure of the enzyme's affinity to its substrate.

2. **Substrate conversion into product and its release into solution** – a much slower process with a corresponding rate constant (k_2) that is much smaller than k_1 and k_{-1}[*2]. Thus, k_2 describes the rate-limiting step of the reaction, and it is therefore considered synonymous with the reaction's catalytic rate (k_{cat}). It should be noted, however, that **since enzyme-mediated reactions contain various intermediates**[132,133], **the k_{cat} of such a reaction often reflects several rate constants**. In enzymes, k_{cat} is synonymous with the *turnover number* of the enzyme, that is, the number of substrate molecules processed by one enzyme molecule in a unit of time (usually a second).

As mentioned above, **the central assumption of the M–M formalism is that enzyme-catalyzed reactions are in a steady state**[*3]. That is, the concentration of the ES complex in Scheme 9.7 is assumed to remain constant throughout the reaction. As ES can change either to E + S or to P, the corresponding *steady-state constant*, called the *Michaelis constant*, or K_M, is defined as:

$$K_M = \frac{\left(k_{-1} + k_2\right)}{k_1} \tag{9.5}$$

In contrast to the equilibrium constant of the ES complex (K_S), the steady-state constant (K_M) can easily be measured in the lab (see below). Therefore, it is often used to approximate K_S, as an inverse measure of the enzyme-substrate affinity. However, it should be emphasized that this approximation is correct only when the rate-limiting step is significantly slower than the rates of the ES forming or breaking steps (i.e., $k_{1/-1} \gg k_2$). In such a case, k_2 can be neglected in Equation (9.5), and $K_M \approx K_S$. Also, this approximation is incorrect in complex reactions, in which K_M is affected by various microscopic rates that do not necessarily relate to the binding of a single substrate[3].

[*1]The steady-state assumption was introduced later by Briggs and Haldane[130].

[*2]This assumption suggests that most enzyme-substrate encounters are futile. A recent study[131] has demonstrated this to be true for most enzymes: for every 10^4 to 10^5 initial encounters, only one results in product formation.

[*3]In fact, as long as there is a steady state, the catalyzed process need not even include a fast equilibrium step, as posited by the original Henri-Michaelis-Menten model[129,130].

Considering Scheme 9.7 and the fact that k_2 is the rate-limiting step of the entire enzyme-catalyzed reaction, the initial rate of the reaction can be described, following the *law of mass action*[*1], as:

$$V_0 = k_2[\text{ES}] \tag{9.6}$$

(where [ES] is the molar concentration of the enzyme-substrate complex).

While this expression is correct, it is not very easy to work with, as [ES] is difficult to measure. However, when the assumptions of the M–M model are translated into rate equations (see Box 9.2), the result is the *Michaelis-Menten equation*:

$$V_0 = \frac{V_{max}[\text{S}]}{K_M + [\text{S}]} \tag{9.7}$$

Indeed, the M–M equation yields the same saturation curve characteristic of enzymatic catalysis, described in terms of simple, measurable parameters (Figure 9.20).

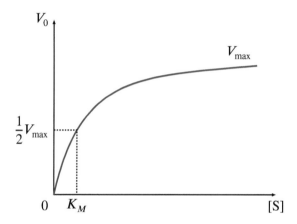

FIGURE 9.20   **Graphic representation of the Michaelis-Menten equation.** V_{max} is the maximal velocity and K_M is the substrate concentration at which the rate is half of V_{max} (see explanation in the main text).

To understand the enzyme's behavior as described by the Michaelis-Menten equation, let us focus on three distinct cases, which correspond to different substrate concentration ranges, and which are represented by three different regions in the Michaelis-Menten plot:

1. $[\text{S}] \gg K_M$ (most of the enzymatic active sites are substrate-bound)
   In this case:

$$K_M + [\text{S}] \approx [\text{S}] \tag{9.8}$$

   And therefore:

$$V_0 - V_{max} \approx k_{cat}[\text{E}]_t \tag{9.9}$$

   (where $[\text{E}]_t$ is the total concentration of the enzyme).

   The last part of the equation was derived as follows. When the enzyme is saturated, all enzyme molecules are substrate-bound (i.e., $[\text{E}]_t = [\text{ES}]$). Substituting $[\text{E}]_t$ for [ES] in Equation (9.6) yields $V_0 = k_2[\text{E}]_t \approx k_{cat}[\text{E}]_t$.

[*1]The law of mass action states that the rate of a reaction is proportional to the product of the concentrations of the reactant(s) (see Box 9.1).

Thus, **under saturation conditions, the enzyme-catalyzed reaction becomes a pseudo first-order reaction, with the rate depending only on the concentration of the enzyme** ($[E]_t$) (see Box 9.1). Many studies of enzyme activity are carried out under these conditions (i.e., substrate excess) because the rate is then maximal, and the turnover number of the enzyme is easy to determine. Also, under these conditions the pseudo first-order kinetics makes the analysis much simpler, as only the concentrations of the enzyme need to be considered.

2. $[S] = K_M$
    In this case:

$$V_0 = \frac{1}{2} V_{max} \qquad (9.10)$$

Earlier we encountered a mathematical definition of K_M, relating it to the rate constants k_1, k_{-1} and k_2 (Equation (9.5)). Equation (9.10) shows us another definition of K_M, which, in contrast to the former definition, allows it to be measured easily: K_M **is the substrate concentration that yields enzymatic activity that is half of V_{max}** (see Figure 9.20). Indeed, K_M has concentration units (molar). By extracting the kinetic parameters of thousands of natural enzymes from the BRENDA database [33], Bar-Even and coworkers [3] found that 60% of these enzymes have K_M values within the range of 10^{-5} to 10^{-3} M. Interestingly, this range is very similar to the concentration range of most cellular metabolites (10^{-6} to 10^{-3} M). This is to be expected; when $[S] \approx K_M$ the enzyme is much more responsive to changes in its substrate concentrations than in cases where $[S] \gg K_M$. It therefore stands to reason that enzymes evolved to have K_M values that are similar to the concentrations of their natural substrates, as this would allow them to respond adequately to changes in the latter.

3. $[S] \ll K_M$ (most of the enzymatic active sites are substrate-free)
    In this case:

$$[E]_t \approx [E]_f \qquad (9.11)$$

(where $[E]_f$ is the concentration of the free enzyme).

And:

$$K_M + [S] \approx K_M \qquad (9.12)$$

Therefore:

$$V_0 = \frac{V_{max}[S]}{K_M} \qquad (9.13a)$$

which is a first-order reaction.

As we recall, $V_{max} \approx k_{cat}[E]_t$ (Equation (9.9)), which enables us to rearrange Equation (9.13a) as follows:

$$V_0 \approx \frac{k_{cat}}{K_M}[E]_f[S] \qquad (9.13b)$$

Equation (9.13b) suggests that under these conditions the reaction is bimolecular (a second-order reaction), but in fact it is a *pseudo-bimolecular reaction*, with a corresponding *pseudo second-order rate constant* (k_{cat}/K_M) [134]. This is because $[E]_f$ remains virtually constant during the entire reaction, and the rate depends only (and linearly) on $[S]$, as Equation (9.13a) indicates. As in cases where substrate concentrations are high ($[S] \gg K_M$), here, too, the analysis is much easier, which explains why biochemists like to work under such conditions. Moreover, **the pseudo second-order constant k_2/K_M is often used by biochemists as a measure for enzyme efficiency** (see explanation in Subsection 9.2.3 below).

V_{max} and K_M can be extracted from simple measurements of enzyme rates at different substrate concentrations, but only when high-enough substrate concentrations are used to reach V_{max}. This is rather inconvenient, as it requires the biochemist to carry out many trial measurements. A solution to this problem comes from rearranging the M–M equation (Equation (9.7)) into the following form:

$$\frac{1}{V_0} = \frac{K_M}{V_{max}} \frac{1}{[S]} + \frac{1}{V_{max}} \tag{9.14}$$

When the equation is presented graphically as the dependency of $\frac{1}{V_0}$ on $\frac{1}{[S]}$, a *double reciprocal* plot is obtained (a.k.a. the *Lineweaver-Burk plot*, Figure 9.21). Like the original M–M plot, the Lineweaver-Burk plot enables V_{max} and K_M to be extracted. However, because this plot is linear, only a few values of substrate concentration are needed in order to determine V_{max} and K_M (see Figure 9.21). Normally about five measurements of V_0 are carried out at different substrate concentrations[1], and the plot is extrapolated to find the axis intercepts. The Lineweaver-Burk plot is very useful for quick extraction of V_{max} and/or K_M (e.g., in class), as well as for determining whether the measured enzyme displays M–M kinetics. Enzymologists, however, seldom use it, and nowadays it is considered to be more of a historical anecdote than an analytical tool. Instead, **measurements of enzymatic catalysis are usually analyzed automatically by computer programs (e.g., MATLAB® or SigmaPlot) that perform a direct fit to an M–M model and linear regression analysis, thus yielding more accurate values.**

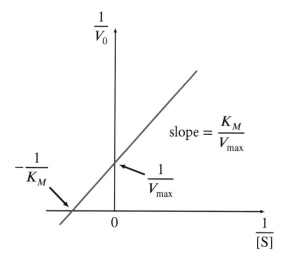

FIGURE 9.21 **The Lineweaver–Burk (double-reciprocal) plot.** The intercept of the plot with the X axis is $-\dfrac{1}{K_M}$, its intercept with the Y axis represents $\dfrac{1}{V_{max}}$, and the slope of the plot is $\dfrac{K_M}{V_{max}}$.

---

[1]It is important that the concentrations used are within a sufficiently large range, normally between $0.3K_M$ and $3K_M$.

BOX 9.2 DERIVATION OF MICHAELIS-MENTEN EQUATION

The original Henri-Michaelis-Menten formalism described the enzyme-catalyzed reaction by the following basic scheme:

$$E + S \underset{k_{-1}}{\overset{k_1}{\rightleftarrows}} ES \xrightarrow{k_2} E + P$$

The scheme used the following assumptions:

1. **The binding of the substrate to the enzyme is a rapid equilibrium process.** Later, Briggs and Haldane showed that this assumption is not mandatory, and the M–M model holds as long as the ES complex is in a steady state, with its concentration remaining constant throughout the reaction [130]. Applying the law of mass action, we can write the steady-state assumption as follows:

$$\frac{d[ES]}{dt} = k_1 [E]_f [S] - \left(k_{-1} + k_2 \right) [ES] = 0 \qquad (9.2.1)$$

(where $[E]_f$ is the concentration of the free enzyme).

2. **The total concentration of the enzyme, $[E]_t$, does not change over time:**

$$[E]_t = [E]_f + [ES] = \text{constant} \qquad (9.2.2)$$

We can derive two extreme situations from Equation (9.2.2). In the first, substrate concentrations are very small, so virtually all enzyme particles are free:

$$[E]_t = [E]_f$$

In the opposite situation, the substrate concentration is so large that the enzyme is saturated. This means that all enzyme particles are substrate-bound:

$$[E]_t = [ES]$$

3. **Conversion of the enzyme-bound substrate into product and its release is a much slower process** with a correspondingly small rate constant (k_2):

$$\frac{d[P]}{dt} = k_2 [ES] \qquad (9.2.3)$$

Combining Equations (9.2.1) and (9.2.2) yields:

$$k_1 \left([E]_t - [ES] \right) [S] - \left(k_{-1} + k_2 \right) [ES] = 0$$
$$k_1 [E]_t [S] = k_1 [ES][S] + \left(k_{-1} + k_2 \right) [ES]$$

Dividing both sides by k_1 yields:

$$[E]_t [S] = [ES][S] + \frac{\left(k_{-1} + k_2 \right)}{k_1} [ES]$$

And since $\dfrac{\left(k_{-1} + k_2\right)}{k_1} = K_M$, then:

$$[E]_t[S] = [ES][S] + K_M[ES]$$
$$[E]_t[S] = \left([S] + K_M\right)[ES]$$
$$[ES] = \frac{[E]_t[S]}{[S] + K_M} \tag{9.2.4}$$

Using Equation (9.2.4) to substitute [ES] in Equation (9.2.3) yields:

$$\frac{d[P]}{dt} = V_0 = k_2[ES] = k_2\frac{[E]_t[S]}{[S] + K_M}$$
$$V_0 = \frac{k_2[E]_t[S]}{[S] + K_M} \tag{9.2.5}$$

Enzymes reach their maximal rate (V_{max}) when they are saturated with substrate: $V_0 = V_{max}$. In such cases, $[E]_t = [ES]$ (see Equation (9.2.2)), and Equation (9.2.3) can be written as follows:

$$\frac{d[P]}{dt} = V_{max} = k_2[E]_t$$

Thus, the expression $k_2[E]_t$ in Equation (9.2.5) can be replaced with V_{max}:

$$V_0 = \frac{V_{max}[S]}{K_M + [S]} \tag{9.2.6}$$

which is the Michaelis-Menten equation.

9.2.3 Use of Michaelis-Menten kinetic parameters for enzyme analysis

The quantitative and qualitative relations between the different M–M kinetic parameters enable us to analyze different aspects of enzyme activity and to compare them among enzymes.

9.2.3.1 Enzyme-substrate affinity

In enzyme-catalyzed reactions that are not highly complex, and when $k_{1/-1} \gg k_2$, by Equation (9.5), **it is the custom to use the Michaelis constant (K_M) as an approximate (inverse) measure of the enzyme's affinity to its substrate(s)**[1]. As such, K_M can be used to compare two enzymes that compete on the binding of the same substrate. The enzyme with the lower K_M is likely to have a greater affinity to the substrate, and, provided that the substrate is not in excess concentration (as is usually the case in cells), this enzyme is therefore likely to

---

[1] This, however, should be done with extreme caution, as in many cases K_M does not faithfully reflect affinity. This happens especially, but not exclusively, in enzymatic reactions that involve a covalent intermediate.

catalyze the substrate more often than the other enzyme. Such enzyme pairs may reside, for example, in the same cell and function in different metabolic pathways. In such a case, the affinity differences between the enzymes will lead to the prioritization of one of the pathways over the other.

A similar case involves different enzymes that function in the same metabolic pathway but in different organs. For example, *hexokinase* (EC 2.7.1.1) and *glucokinase* (EC 2.7.1.2) are two glucose-phosphorylating enzymes that act in the first step of glycolysis [1]. Glucokinase, which acts in liver cells of animals, has a much higher K_M than hexokinase, which acts in muscle cells. When blood glucose levels are high (after a meal), both enzymes have enough glucose to act on. However, when glucose levels drop (between meals or during fasting), it is important for the body to prioritize glucose use. Under such conditions, the K_M differences between hexokinase and glucokinase confer an advantage in glucose binding to hexokinase, as it has higher affinity to glucose (lower K_M). This is in line with the corresponding physiological roles of the two organs; the liver is an 'altruistic' organ that, among other things, distributes nutrients to the other (more 'egoistic') organs, per their physiological needs. When glucose levels are low in the body, it is more important for the muscles to use it — e.g., for movement or fighting (in animals) — than for the liver to use it. Liver functions are unperturbed by this prioritization, as the liver can live off fatty acids under many physiological conditions. The higher affinity of hexokinase to glucose is one way of ensuring prioritization of glucose use by the muscles over glucose use by the liver.

9.2.3.2   Enzyme efficiency and specificity

The catalytic rate constant, k_{cat}, has traditionally been used as a measure of enzyme activity and catalytic efficiency, as it represents the number of product molecules formed by one enzyme molecule in a unit of time (usually a second) [65]. Thanks to the M–M formalism we also know how to easily obtain k_{cat} from measurements of V_0 at saturation (where $V_0 = V_{max}$), by dividing V_0 by $[E]_t$ (see Equation (9.9)). However, k_{cat} accounts only for the efficiency of the catalytic steps of the enzyme, whereas the overall efficiency is also affected by the enzyme's affinity to the substrate. **A better measure for the enzyme's efficiency is provided by k_{cat}/K_M** [65]. As explained above, the significance of this quantity is revealed under conditions in which substrate concentrations are much lower than K_M. Then, by Equation (9.13b), the reaction becomes a pseudo second-order reaction in which k_{cat}/K_M is the rate coefficient. Again, **the advantage of this quantity over k_{cat} as a measure for efficiency is that k_{cat}/K_M represents both catalytic (k_{cat}) and affinity-related (K_M) contributions to the total efficiency of the enzyme.** Indeed, current studies that compare different enzymes, or different activities of a single enzyme on different substrates, tend to use k_{cat}/K_M as a measure of efficiency instead of using just k_{cat}. In Subsection 9.3.3 below we refer to k_{cat}/K_M in terms of interactions between the enzyme's active site and the ground and transition states of the substrate.

Analysis of the BRENDA database [33] shows that 60% of the catalogued (natural) enzymes have k_{cat} values in the 10^1 to 10^2 s^{-1} range, and k_{cat}/K_M in the 10^3 to 10^6 M^{-1} s^{-1} range [3]. Interestingly, **enzymes that function in central energy metabolism seem to be on average more efficient than those that function in secondary metabolism, and that involve metabolites that are produced in specific cells or tissues** [3]. Some enzymes, including *carbonic anhydrase* (EC 4.2.1.1), *superoxide dismutase* (EC 1.5.1.1), and *fumarase* (EC 4.2.1.2), are extremely efficient, with $k_{cat}/K_M = \sim 10^9$ to 10^{10} s^{-1} M^{-1} [131]. These enzymes are said to have reached 'catalytic perfection', because in the reactions they catalyze every enzyme-substrate encounter leads to catalysis, and the rate-limiting step is the diffusion of the substrate to the enzyme[*1] [136,137]. In other words, the catalytic steps in such enzymes are so fast, that in contrast to all other enzymes, k_{cat} does not depend on the value of the reaction energy barrier, Ea, (see Equation (9.2)), but instead on the rate of reactant diffusion [138].

In cases in which a single enzyme is capable of acting on different substrates, a comparison of the efficiency of the enzyme across cases can be used as a measure of specificity. That is, if the enzyme is much more efficient in catalyzing a certain substrate compared with others, the enzyme can be said to be specific for that substrate. Since the pseudo second-order rate constant k_{cat}/K_M seems to be the best measure of enzymatic efficiency, it can also be used as a measure of specificity of the enzyme to one substrate over another (or others) [65]. For this reason, k_{cat}/K_M is also referred to as the 'specificity constant'. When comparing different substrates it is important to use the appropriate interpretation of the specificity constant. For example, the basic Michaelis-Menten scheme presented above (Scheme 9.7) describes in one step the conversion of the substrate into product and its release into solution. However, the two sub-steps may have different rate constants [65]. For example:

Scheme 9.8.
$$E + S \underset{k_{-1}}{\overset{k_1}{\rightleftharpoons}} ES \overset{k_2}{\longrightarrow} EP \overset{k_3}{\longrightarrow} E + P$$

In such cases k_{cat}/K_M should refer only to the first two steps, assuming that the second step (substrate conversion into product) is the first irreversible one [65]:

$$\frac{k_{cat}}{K_M} \approx \frac{k_1 k_2}{(k_{-1} + k_2)} \tag{9.15}$$

These are the only steps that determine specificity; once a substrate completes the first irreversible step, it is committed to forming the product, and no further discrimination can occur.

A very common case of specificity is when the enzyme can act on two different enantiomers of its substrate (S and R), but has a clear preference for one of them over the other. This phenomenon is called 'enantioselectivity' $(E)^{*2}$, and it can be expressed as the ratio between the specificity constants relating to the enzyme's activity on the two enantiomers[*3] [139]:

$$E = \frac{\left(\dfrac{k_{cat}}{K_M}\right)_S}{\left(\dfrac{k_{cat}}{K_M}\right)_R} \tag{9.16}$$

E can be used to calculate the difference in the activation free energy ($\Delta\Delta G^{\#}$) between the

---

[*1]The rate of diffusion of small molecules in aqueous solution is 10^8 to 10^9 $L\,mol^{-1}\,s^{-1}$ [135] (where L is the volume in liters).

[*2]Some enzymes present absolute enantioselectivity, where others show different degrees of preference for a specific enantiomer.

[*3]Note that the general form of this equation can be used to express any type of substrate selectivity, not just enantiomeric.

enantiomers [139]:

$$\Delta\Delta G^{\#} = -RT \ln E \tag{9.17}$$

Enantioselectivity should not be confused with the phenomenon of enzymatically creating a single configuration (S/R) in a pro-chiral substrate. Such 'stereospecificity' is described in Section 9.1.4 above for ketoreductases, amino acid dehydrogenases, and aminotransferases.

9.2.3.3 Enzyme proficiency

As mentioned in the introduction above, the extent to which enzymes accelerate reactions varies widely across enzymes. An enzyme's rate acceleration is usually calculated as the ratio between the rate of the enzymatic reaction and that of the uncatalyzed reaction; this value normally lies in the range of 10^{11} to 10^{16} [24], and may even reach a factor of 10^{21} [20,24,25]. Miller and Wolfenden [140] proposed a different quantitative measure, which they called 'proficiency of enzyme catalysis'. This measure is defined as the specificity constant (k_{cat}/K_M), which we encountered above, divided by the rate constant for the uncatalyzed reaction (k_{uncat}) [24]. That is: $k_{cat}/(K_M k_{uncat})$. The proficiency constant can be viewed as the equilibrium constant between the transition state of the uncatalyzed reaction and that of the catalyzed reaction. Note, however, that in many cases such equilibrium is a purely theoretical notion, as the catalyzed and uncatalyzed transition states may be quite different [141]. The proficiency also reflects the enzyme's binding affinity towards the substrate's transition state (K_{TS}) as compared to its affinity towards the ground state (K_s). As we will later see, the binding and stabilization of the transition state is a hallmark of enzymatic catalysis. A survey carried out by Wolfenden and coworkers shows that **the catalytic proficiency of enzymes ranges between 10^8 M^{-1} and 10^{23} M^{-1} [4], which demonstrates the exceptionally good transition state binding capabilities of enzymes.**

9.2.4 Limitations of M–M formalism

The M–M formalism is often used to analyze basic enzymatic activity, provided that the principal tenets of the model are upheld, i.e., initial rates are measured, there is no product inhibition, etc. Still, many enzymes demonstrate different kinetic behavior, in which case a more complex treatment is required. Such behavior is particularly likely to be observed in cases where allostery, cooperativity, multiple substrates, tightly bound substrates, or inhibition are involved. For example, positive cooperativity results in a sigmoidal kinetic plot[*1] rather than the simple saturation of the M–M curve. Allosteric enzymes are prevalent in cellular metabolism, and it is therefore important to understand their kinetic behavior. Enzyme inhibition is also a common phenomenon in cells and tissues. The kinetic behavior of inhibited enzymes is discussed in Subsection 9.5 below. Finally, in cases where the substrate binds tightly to the enzyme, the concentration of the free substrate is significantly smaller than that of the total substrate, and the mathematical description of the underlying kinetics is therefore more complicated [142].

In addition to the specific cases mentioned above, the basic assumptions made by the M–M model should also be scrutinized. First, the description of the catalytic step as a sin-

---

[*1] A sigmoidal plot is also obtained in measurements of ligands to non-enzymatic proteins. For example, see the sigmoidal binding plot of hemoglobin (Figure 5.3.2 in Chapter 5, Box 5.3).

gle reaction with a single kinetic constant constitutes a significant approximation; enzyme-mediated reactions tend to include multiple steps and intermediates (e.g. [132,133]). Second, the steady-state assumption, i.e., that [ES] remains constant throughout the reaction, is essentially correct, but may be inaccurate. Although the vast majority of enzymes studied to date display steady-state kinetics, changes in [ES] may be significant in certain enzymes or cases. Finally, the M–M model relies on the law of mass action, which, in turn, relies on the assumption that molecules in the system can diffuse freely [143]. This is clearly not the case in biological cells, which suffer from macromolecular crowding [144,145]. The latter phenomenon is expected to affect the diffusion of small molecules and therefore also the kinetics of chemical and enzymatic reactions within cells [143].

9.3   HOW DO ENZYMES CATALYZE REACTIONS?

9.3.1   Overview

As explained in the introduction to this chapter, chemical reactions proceed from reactant(s) to product(s) through a sequence of intermediates, which differ from one another in their chemistry, configuration, and free energy. Like other catalysts, enzymes accelerate chemical reactions by lowering the energy barrier associated with the transition state — a reaction intermediate with the highest free energy. However, in contrast to other simple catalysts like metals, most enzymes are highly specific to the types of reactions they accelerate, and to the substrates on which they act. This specificity is achieved thanks to the 'active site', a pocket-like depression in the structure of the enzyme, to which the substrate binds and in which the chemical reaction is accelerated. The active site fulfils two major functions [146]:

1. **Substrate binding** – the active sites of enzymes may bind a variety of substrates, including small molecules (e.g., monosaccharides or amino acids), moderate-size molecules (e.g., short peptides) and even macromolecules (e.g., proteins and polysaccharides) [146]. Accordingly, active sites may vary greatly in size, with sizes typically ranging between 400 Å^2 and 2,000 Å^2 [147]. **Substrate binding is mediated through multiple noncovalent interactions between different parts of the substrate and chemical groups in the active site** (Figure 9.22). **These interactions render the binding specific, which accounts for the selectivity of enzymes to their natural (cognate) substrates** (see below). The binding interactions also provide the energy used for keeping the substrate inside the active site, which accounts for the affinity between the two. The interactions mainly involve active site amino acids. Indeed, the side chains of binding residues offer a diverse set of chemical groups: nonpolar (linear or branched), hydroxyl, thiol, amine, amide, carboxylate, imidazole, indole, phenol, and guanidinium [148] (Figure 9.23; see also Figure 2.5 and Table 2.1 in Chapter 2). Some active sites interact with their substrates via additional chemical groups, which may be components of small organic molecules (e.g., nucleotides, small carbohydrate units, and lipids), metals, or other inorganic species (e.g., water) (see Section 9.3 below). As explained in Chapters 2 and 8, these molecular adducts, which in enzymes are usually referred to as 'cofactors', may be permanently attached 'prosthetic groups' or transiently bound 'coenzymes' [149].

2. **Substrate conversion into product** – enzymatic active sites are able to promote the chemical transformation of their bound substrates into products (i.e., *catalysis*). As we will see below, they do so by promoting the conversion of the substrate either into its transition state or into another intermediate downstream of the reaction coordinate. This is done either through chemical reactions between substrate and active site groups (amino acids, cofactors) or via noncovalent stabilization of reaction intermediates by these groups. The latter shows that binding and catalysis are often not mutually exclusive.

Substrate binding and catalysis often take place in a single location inside the active site. However, when the substrate is large or has a complex structure, the enzyme may contain multiple binding sites, each binding a different part of the substrate. Still, only one of these sites carries out the catalysis, specifically, the site that binds the labile part of substrate that is supposed to be chemically transformed.

As explained above, both substrate binding and catalysis are carried out by chemical groups in the enzyme's active site, which either react or interact specifically with the substrate. This specificity is created by the shape of the active site and the spatial distribution of its chemical groups. Indeed, **enzymatic active sites have been evolutionarily selected to complement their natural substrates [150], both geometrically and electrostatically [151–153]*1. This complementarity is responsible for the selectivity of enzymes towards their cognate substrates over other molecules, as well as for their catalytic efficiency and reaction specificity.** These key aspects are further discussed in the subsections below.

9.3.2   Binding specificity and selectivity

Enzymes are known to bind their substrates selectively, i.e., to favor cognate substrates over non-cognate ones. Selectivity exists both in the ground state (which we discuss here) and in the transition state (which we discuss in Subsection 9.3.3.2 below) of the substrate, with selectivity in the transition state being stronger [154]. In both cases, the selectivity results from attractive and repulsive*2 noncovalent interactions between the active site and substrate, with attractive interactions being stronger with cognate substrates, and repulsive interactions, which have a larger effect, being stronger with non-cognate substrates [154].

In many enzymes, the binding specificity of the substrate's ground state is not absolute and may appear in various degrees. Some enzymes are highly specific towards a single substrate, up to the level of its stereochemical properties. Examples of highly specific enzymes include enzymes that deaminate RNA adenylate moieties, DNA repair enzymes, and terpenoid cyclases [155]. Many enzymes are specific to one type of substrate, which may include similar, yet different molecules. Finally, there are enzymes that show very low preference towards their substrates [113,156,157]. This *'substrate promiscuity'*, already described in Subsection 9.1.5.7 above, usually serves a physiological purpose. For example, *glutathione S-transferase* [158] and *cytochrome P450* (see Chapter 1, Subsection 1.1.3.9) are both involved

---

*1See Chapter 8, Subsection 8.4 for a detailed discussion of protein-ligand binding. Note, however, that in the case of enzymes the complementarity is mainly towards the transition state of the substrate (see below), which is ultimately what determines the enzymes' catalytic efficiency.

*2Repulsive interactions include steric and electrostatic clashes.

(a)

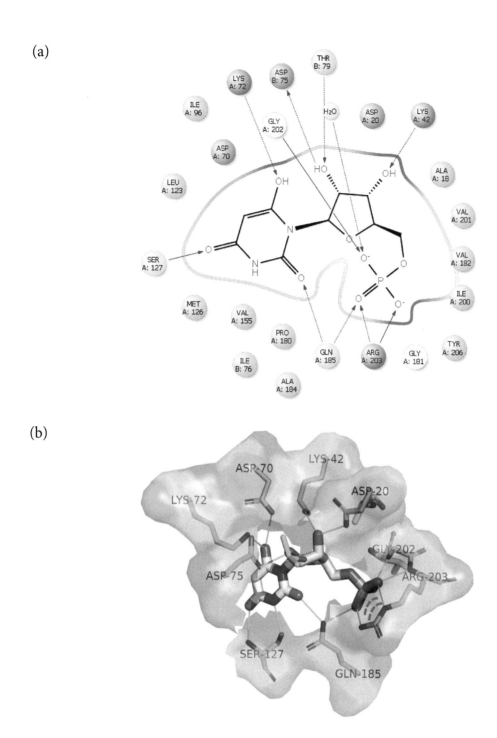

(b)

FIGURE 9.22 **Enzyme-ligand interactions within the active site of orotidine 5′-phosphate decarboxylase (PDB entry 4o11).** (a) A schematic, two-dimensional projection of the interactions. The ligand, 6-hydroxy-UMP, is presented explicitly, whereas the interacting residues are presented as spheres. Nonpolar residues are in green, polar-uncharged residues are in cyan, basic residues are in blue, and acidic residues are in red. Hydrogen bonds are shown as purple arrows pointing from donor to acceptor. Solid and dashed arrows represent hydrogen bonds that involve protein backbone and side chain atoms, respectively. The image was prepared with Maestro (Schrödinger, Inc.). (b) An explicit, three-dimensional view of the electrostatic interactions depicted in (a). The ligand is represented as thick sticks; the interacting residues are shown as thin sticks; and the general shape of the active site is delineated by their semi-transparent surfaces. Hydrogen bonds are shown as cyan lines, and salt bridges are shown as red, dashed half-circles.

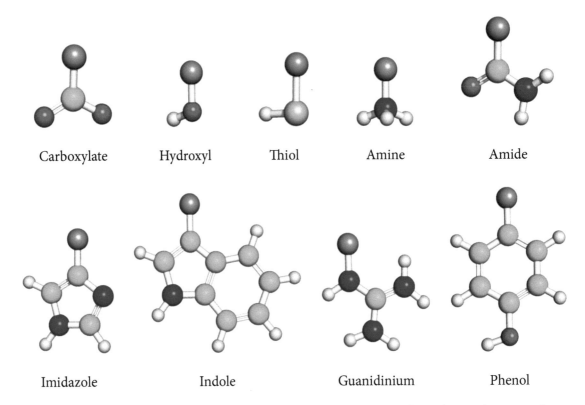

FIGURE 9.23 **Chemical groups used by enzymes for binding and catalysis.** The gray spheres represent the rest of each molecule.

in the detoxification of foreign molecules (xenobiotics) in our body. Since xenobiotics are chemically variable, both of these enzymes evolved to act on a wide range of substrates [109]. Note, however, that in many cases the different substrates are catalyzed by polymorphic enzymes and not by the same enzyme molecule. That is, the same enzyme catalyzes the reaction, but different strains of it recognize different substrates and act on them. This phenomenon results from differences among the various strains' amino acid sequences, many of which are located in the substrate-binding positions.

Still, there are cases in which a single enzyme molecule is able to bind and catalyze different substrates (e.g., see [159]). In such cases, different substrates may bind to the same active site residues, albeit with different affinity, or alternatively, different substrates may bind (partially or completely) to different residues [109]. Furthermore, in either scenario the different substrates may bind to different conformations of the active site that differ in certain properties such as residue accessibility or protonation state [109]. As mentioned above, a single enzyme may include multiple strains that have different substrate specificities. The existence of such strains is the result of an evolutionary process that involves introduction of mutations and formation of sequence variability. The different specificities usually result from simple (point) mutations in residues, which are involved directly in substrate recognition and binding (see, for example, the case of *malate/lactate dehydrogenase* [160]). However, there are also cases in which different substrate specificity involves more complex differences between strains. For example, mutations in peripheral positions of the substrate-binding site might change the structure of this area or the distribution of charges. As a result, the binding site might lose its geometric and electrostatic compatibility with the original substrate and become compatible with another. In fact, studies employing random mutagenesis and functional selection in enzymes (a.k.a. *directed evolution*'[119,120]) demonstrate that changes in active site properties and the resulting changes in substrate specificity may even be induced by mutations outside the binding site (e.g., [161]). Such effects are probably mediated by conformational changes or other factors related to protein dynamics.

The numerous attractive noncovalent interactions between active site residues and chemical groups of substrates account not only for the specificity of binding, but also for its strength (i.e., binding affinity). Affinity and specificity are often confused, but while the two are related, they are not synonymous. First, specific binding usually requires multiple interactions between enzyme and substrate groups, while strong binding can also be achieved by a small number of strong interactions that are insufficient for specificity ([14] and references therein). Second, the major contributions to strong binding between proteins and their ligands seem to be nonpolar and attractive van der Waals interactions, whereas specificity is usually achieved by electrostatic interactions, and especially by geometry-dependent hydrogen bonds (see Chapter 8, Section 8.3 for further details), as well as by steric clashes [154]. Enzyme-substrate binding is expected to show similar characteristics. However, recent analysis of natural enzymes suggests that binding affinity (approximated by K_M) increases upon substrate phosphorylation, which should also increase the number of electrostatic interactions with the enzyme [3]. Interestingly, increases in affinity were also observed in substrates to which CoA or nucleotide groups were attached [3]. **Like phosphorylation, CoA and certain nucleotides are commonly attached to molecules, to activate them for metabolic reactions. Such activation may work by affecting the catalytic steps and the free energy of the reaction (see Subsection 9.1.5 above and Subsection 9.4.2 below), as well as by affecting substrate binding. For example, an increase in binding affinity enables the enzyme to act on lower concentrations of cellular metabolites.**

9.3.3   Catalysis

The binding of a substrate to the active site of the enzyme promotes the substrate's conversion into product(s) via a specific sequence of chemical steps. In the thermodynamic sense, enzymes accelerate chemical reactions by lowering the free energy of the reaction's transition state (i.e., stabilizing it, see Figure 9.2) [10,11,151,152,162–164]. They do so by using a variety of strategies that either stabilize the transition state or promote its formation chemically[*1], thus helping the system to cross the major energy barrier of the reaction. These strategies have been at the center of numerous scientific studies, carried out over a hundred years. Such studies have provided valuable data on enzyme-mediated catalysis, although certain issues are still controversial [138,165–167]. **The catalytic strategies used by enzymes to promote the formation of the reaction's transition state involve different aspects of enzyme-substrate interaction, including general entropic aspects that result from the mere confinement of the substrate; stabilization of reaction intermediates via noncovalent interactions; and specific chemical actions of the active site on the substrate, such as electron or proton transfer and formation of covalent bonds** [30,166,168,169]. The chemical actions are carried out primarily by active site amino acids that are termed 'catalytic', (normally 2 to 6 residues [146]). In some cases, other chemical species are involved, as noted above. Different types of amino acids may serve as catalytic residues [148]; however, a statistical analysis of

---

[*1]Other catalytic intermediates along the reaction path may also be stabilized to promote product formation.

the MACiE database [34] (see Subsection 9.1.5 above) shows that out of the 20 natural amino acids, only 10 (Arg, Asp, Cys, Glu, His, Lys, Ser, Thr, Trp and Tyr) [40,170] participate directly in catalysis, in all enzyme classes. This is not surprising, as each of these 10 amino acids has a polar or aromatic side chain, which confers an advantage in catalytic steps such as nucleophilic attack, proton transfer, electrostatic stabilization, etc. (see individual mechanisms below). Nonpolar residues may also participate in catalysis, but these cases almost always involve the residues' backbone polar amide and carbonyl groups [30]. The propensity of a residue to act catalytically can be calculated as the number of times the residue fulfills a catalytic role as compared to its background levels in proteins. Such calculations show histidine to have strong catalytic propensity in all enzyme classes (Figure 9.24). Cysteine has strong catalytic propensity in oxidoreductases, transferases and isomerases, whereas the propensity of other residues depends more strongly on the enzyme class [30].

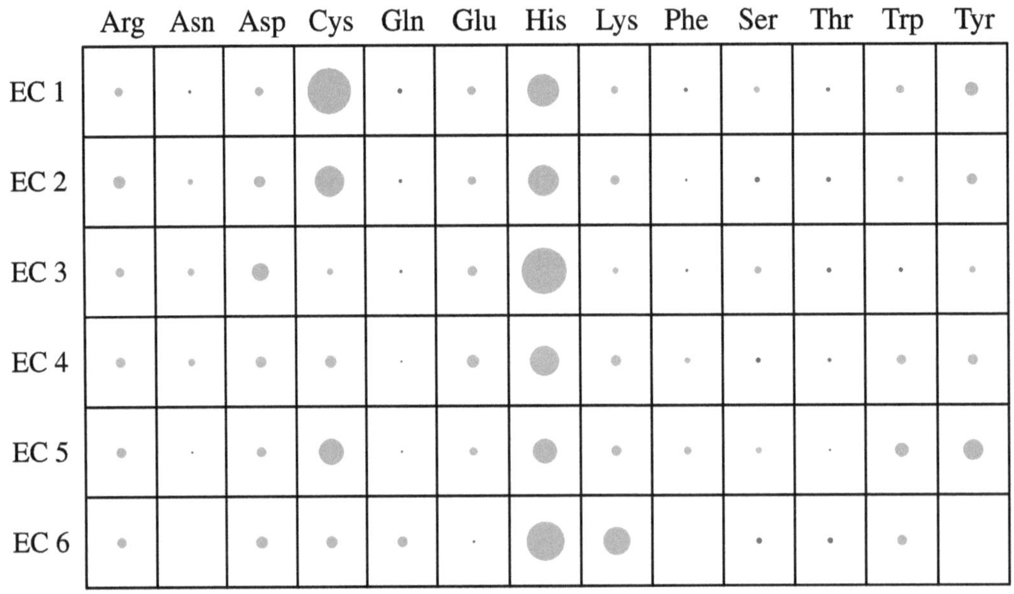

FIGURE 9.24 **Balloon plot showing the propensity of a residue to be catalytic in each of the six enzyme classes.** (EC 1, oxidoreductases; EC 2, transferases; EC 3, hydrolases; EC 4, lyases; EC 5, isomerases; EC 6, ligases). The diameter of each circle represents the corresponding propensity; thus, the larger the circle, the higher the propensity of the residue to be catalytic. The circle is shown in blue if the propensity is greater than (or equal to) 1, and red if the propensity is less than 1. For Asn, Phe and Tyr in EC 6 there was no available information. (The image is taken from [30]).

Researchers have also investigated the propensity of catalytic residues to carry out *specific* catalytic roles [30,40]. Their analysis suggests that such roles can be divided into seven categories:

1. **Activation** – residues that are responsible for activating other chemical species.

2. **Steric role** – residues that affect the outcome of the reaction through factors associated with the spatial arrangement of atoms in the active site (e.g., clashes with the substrate).

3. **Stabilization** – residues that stabilize or destabilize other species.

4. **Proton shuttling** – residues that donate, accept or relay protons.

5. **Hydrogen radical shuttling** – residues that donate, accept or relay hydrogen atoms as radicals.

6. **Electron shuttling** – residues that donate, accept or relay electrons, either as single particles or in pairs.

7. **Covalent catalysis** – residues that become covalently attached to a reaction intermediate.

Interestingly, with the exception of hydrogen radical shuttling and covalent catalysis (to a lesser degree), all the residues that were investigated were capable of performing all the above catalytic functions to some extent [40]. However, their propensity for performing each of the functions seemed to depend on the EC class [30]. Further studies are likely to elucidate the reasons for these findings.

As a result of their direct roles in the chemical transformations that take place in enzymes, catalytic residues are highly evolutionarily conserved. As mentioned in Subsection 9.1.5.7 above, this conservation is often maintained even among enzymes that have different folds or act on different substrates, provided that they catalyze the same type of reaction. Well-known examples for such conservation include the Ser-His-Asp/Glu triad of serine proteases and esterases; the conserved aspartate and lysine residues of type 1 PLP-dependent aspartate aminotransferase-like enzymes; and the highly conserved cysteine in cytochrome P450s [114–116,171]. **Interestingly, in some cases the identity and/or sequence position of catalytic residues have changed during evolution, but their functional chemistry has been preserved.** For example, different nucleophilic residues can attack the same types of groups; different cationic residues can achieve the same stabilization of a negatively charged intermediate, etc. When the mutation of a residue involves large changes in size or other physicochemical properties, functional preservation requires the newly adopted residue to continue to match its surrounding residues. This is achieved through compensatory mutations of nearby residues. A recent study suggests that this process happens gradually and involves nearly neutral mutations, which then enable the large change to occur [172].

The growing set of three-dimensional enzyme structures that have been determined reveals that some enzymes contain amino acid residues that have been modified to yield a more reactive species [146,173,174]. In some cases, such modification occurs spontaneously, whereas in other cases it is carried out by modifying enzymes. Regardless, the modified amino acid usually has a catalytic role, which means that the modification has been evolutionarily selected. For example, in *copper amine oxidases* (EC 1.4.3.21) the catalytic tyrosine is spontaneously oxidized to *tri-hydroxy-phenylalanine* [175]. Similarly, *phosphomannose isomerase* (EC 5.3.1.8) contains a *di-hydroxy-phenylalanine*, also formed from a tyrosine residue [176]. Another example is β-phosphoglucomutase (EC 5.4.2.6), in which the catalytic aspartate is phosphorylated (see the MACiE database [34], entry M0206). The modified residues are often referred to as cofactors. However, unlike classic cofactors (see Section 9.4 below) they are original components of the enzyme's polypeptide chain, rather than exogenous organic molecules.

The combined action of catalytic species in the active site (amino acids and cofactors) contributes to the specificity of enzymes towards the chemical reactions they catalyze. Since all reactions include more than one step, in some cases the sequential action of different catalytic groups has an added effect on the enzyme's reaction specificity, whereas in other cases there is a single committing step that accounts for the enzyme's specificity level. Below we describe the main catalytic strategies characterized thus far. It should be noted that this division is somewhat arbitrary, and that some of the catalytic mechanisms partially overlap. For example, the use of metals in active sites plays a part both in noncovalent catalysis and in metal-ion catalysis. In addition, note that virtually none of these strategies can be used in isolation; **all enzymes combine multiple catalytic strategies (e.g., acid-base catalysis and nucleophilic catalysis) to achieve optimal rate acceleration**.

9.3.3.1 Substrate confinement

Let us consider a chemical reaction in which two reactant molecules are attached to form one product molecule. In order for the reaction to happen, the two reactants must be close and positioned in the right way with respect to each other [177]. If the reaction happens in solution, both reactants are free to move in all directions, as well as to shift freely among many different conformations[*1], only some of which are reactive. These characteristics significantly prolong the time required to complete the reaction [167]. In contrast, **when the reaction happens inside an enzyme's active site, the site's geometric and chemical complementarity to both reactants compels them to acquire the right position and conformation for the reaction to occur** [148,178]. **This shortens the time needed for the reaction to take place, thus accelerating it** [179]. In a unimolecular reaction, i.e., a reaction that involves only one reactant turning into product(s), the confinement effect is smaller, but still exists due to the conformational freedom of the reactant. Bruice and coworkers [180] have suggested that the interactions between the reactant and active site groups limits the reactant to a sub-population of conformations ('*near-attack conformations*'), which are more prone to undergoing catalysis, and which the reactant has to acquire on its way to forming the transition state. This aspect is discussed further below.

As explained in Chapter 4, spontaneous processes often involve an increase in entropy, i.e., in the number of configurational states in the system (see Equation (4.10) in Chapter 4). The confinement of the substrate by the enzyme seems contrary to this tendency, since a free molecule has more degrees of freedom than a confined one (in the words of Page and Jencks, the enzyme acts as an '*entropy trap*' [181]). In fact, the loss of substrate entropy is at least partially compensated for by a gain in solution entropy [4]. Specifically, ordered water molecules are displaced from the active site by the substrate, and the release of these molecules into the solution makes the solution less ordered, thus increasing its entropy. More importantly, while spontaneous processes often involve an increase in entropy, they always involve a decrease in free energy. Thus, a process can happen spontaneously even if it involves a decrease in entropy, as long as this decrease is compensated for by a significant decrease in enthalpy (see Equation (4.4) in Chapter 4). In enzymes, the favorable noncovalent interactions between the substrate and active site groups decrease the enthalpy of the system and overcompensate for any entropy loss that might occur during the binding. As a result, the total free energy change associated with the binding is negative, which makes the confinement of the substrate overall favorable.

[*1]Specifically, each of the reactants has three degrees of translation and three degrees of rotation.

9.3.3.2   Electrostatic preorganization and noncovalent stabilization of transition state

The numerous three-dimensional structures of substrate-bound enzymes in the Protein Data Bank show clearly that enzymes and their substrates complement each other both geometrically and electrostatically. However, as originally proposed by Polanyi [10], Pauling [162], Schwab [11] and Jencks [163,182], and later confirmed by experimental studies (e.g. [6,24,183]), the ability of enzymes to accelerate chemical reactions is largely a result of their ability to complement and stabilize the transition state of the substrate, rather than its ground state (Figure 9.2, red plot). Indeed, **the active site of an enzyme has much higher complementarity to the substrate's transition state than to its ground state. Consequently, binding to the enzyme promotes the substrate's conversion to the transition state, which, in turn, dramatically increases the probability of a successful reaction (see more below). Furthermore, enzymes' higher complementarity towards the transition state accounts for most of their selectivity to their cognate substrates** [154].

The high complementarity between the enzyme and the transition state results from various structural and chemical features of the active site. For example, if the transition state carries an electric charge that is absent in the ground state (e.g., Figure 9.25a), the active site is likely to include a chemical group with an opposite charge that can stabilize the transition state via electrostatic interactions. This type of stabilization is not only efficient[*1] but also very common in enzymes, as the shift from the ground state to the transition state of a chemical reaction often involves *charge delocalization*. In many cases, the stabilizing charged group belongs to a side chain of an amino acid residue, usually a guanidinium group of arginine[*2], an imidazole group of histidine, an amino group of lysine, or a carboxylate group of aspartate or glutamate [40]. In other cases, it may be an amide or carbonyl group in the protein's backbone, which carries a partial positive or negative charge, respectively. This is the case in serine proteases such as trypsin and chymotrypsin, where two backbone amide groups stabilize the tetrahedral oxyanion via hydrogen bonds (Figure 9.25b). Interestingly, these groups also stabilize the planar-neutral ground state of the substrate, but to a lesser extent. Thus, the amide groups can be said to electrostatically 'strain' the substrate into its transition state, which accelerates the entire reaction[*3]. There are also studies showing that hydrogen bonds, which exist between the enzyme and the substrate in both its ground state and transition state, tend to be stronger in the latter [188].

When the transition state is negatively charged, it may also be stabilized by active site cationic metals such as Mg^{2+}, Cu^{2+}, Zn^{2+}, Mn^{2+}, Co^{2+}, Fe^{2+} and Fe^{3+}. For example, enzymes of the enolase superfamily (a.k.a. *phosphopyruvate hydratase*; EC 4.2.1.11) use two Mg^{2+} ions in the active site to electrostatically stabilize a negatively charged transition state [189]. This intermediate is formed by the highly unfavorable proton abstraction from C2 of the

---

[*1] For example, in *chorismate mutase* (EC 5.4.99.5), electrostatic stabilization of the transition state leads to a 2 million-fold rate acceleration of the reaction [184].

[*2] Arg is highly advantageous in electrostatic stabilization of negatively charged transition states, as its side chain is almost always positively charged in proteins ($pKa > 12$). Indeed, a survey of the MACiE database shows that 79% of catalytic Arg residues in enzymes act in stabilization roles [40].

[*3] Such destabilization of the ground state by protein-substrate binding was generally termed 'the Circe effect' by Jencks [182]. Strain-induced distortion of substrates by the active sites of enzymes has been demonstrated in different studies (see [185] and references therein). The strain has been originally assumed to be geometric in nature [186], but the later work of Levitt and Warshel [126] has since demonstrated that electrostatic effects have a more significant contribution [187].

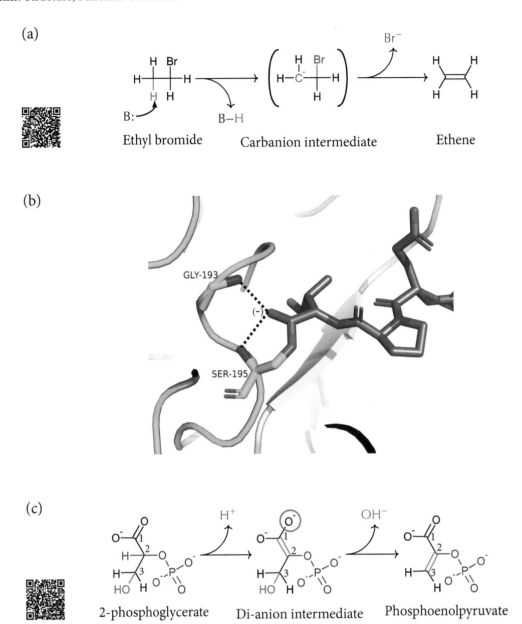

(a)

Ethyl bromide   Carbanion intermediate   Ethene

(b)

GLY-193

(-)

SER-195

(c)

2-phosphoglycerate   Di-anion intermediate   Phosphoenolpyruvate

FIGURE 9.25 **Formation and stabilization of electrically charged transition states.** For clarity, the images described here refer to reaction intermediates that are chemically and electrically similar to the substrate in its transition state. (a) Formation of a negatively charged intermediate (red) during the conversion of ethyl bromide to ethane. The conversion involves deprotonation of the substrate, and consequent formation of a carbanion (in red). 'B' represents a basic group responsible for deprotonation. In the second step, the bromide (blue) leaves the molecule, resulting in double bond formation. (b) Electrostatic stabilization of a tetrahedral transition state analogue by amide groups in the active site of trypsin (PDB entry 1haz). The analogue is presented as blue sticks, with a minus (−) sign designating the oxyanion's negative charge. The charge is stabilized by the backbone amide groups of Gly-193 and Ser-195, via hydrogen bonds (black dashed lines). (c) Schematic depiction of 2-phosphoglycerate conversion into phosphoenolpyruvate by enolase. The water elimination reaction starts with proton abstraction from the substrate's C2, resulting in a di-anion intermediate (the abstracted proton is in blue and the added charge to the intermediate is encircled). The second step involves the loss of OH− (in red) from the C3 hydroxymethyl group.

(d)

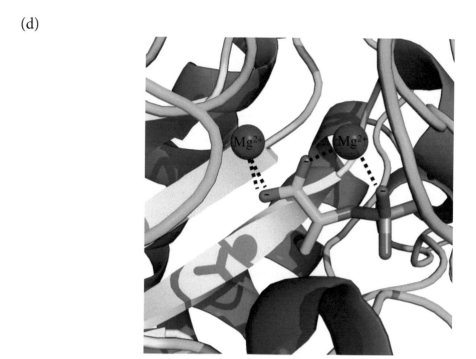

(e)

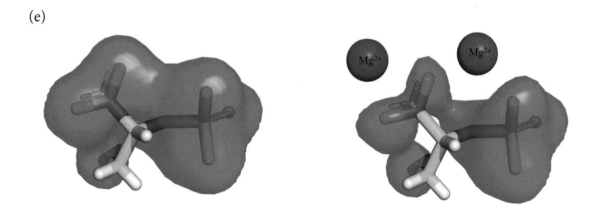

FIGURE 9.25 **Formation and stabilization of electrically charged transition states.** (Continued) (d) Polar interactions (dashed lines) between two Mg^{2+} cations (magenta spheres) in the active site of enolase (PDB entry 1one) and oxygen atoms belonging to the substrate's carboxylate and phosphate groups. The substrate (2-phosphoglycerate) is presented as sticks. The backbone of the protein is shown as ribbons, where α-helices are in red, β-strands in yellow, and loops in green. (e) The negative electrostatic potential ($-25k_BT/e < \Phi < 0k_BT/e$) of 2-phosphoglycerate in the absence (left) and in the presence (right) of the two Mg^{2+} ions. The potential was calculated using the Adaptive Poisson-Boltzmann Solver (APBS) [191] and visualized by PyMOL [192].

2-phosphoglycerate (PGA) substrate, a carbon with a pKa > 30 (Figure 9.25c). The added charge appears on one of the carboxylate oxygen atoms, resulting in a COO^{2-} group. To cross the high energy barrier associated with proton abstraction (+14.4 kcal/mol [190]), the two Mg^{2+} ions stabilize both oxygen atoms of the carboxylate group, as well as one of the oxygen atoms of the phosphate group [19,189]. The stabilizing electrostatic interactions are shown schematically in Figure 9.25d, and the resulting decrease in the substrate's negative electrostatic potential is demonstrated in Figure 9.25e.

The favorable noncovalent interactions (polar and nonpolar) between active site groups and the substrate in its transition state increase the affinity between the two. Indeed, whereas the enzyme-substrate binding energy in the *Michaelis complex*[*1] is ~−4 to −8 kcal/mol [135], the binding energy in the enzyme-transition state complex is ~−19 to −31 kcal/mol [6]. The latter values represent the difference between the energy of the enzyme-bound transition state and the energy of the unbound transition state. In other words, they represent the decrease in activation energy (Ea). Since k_{cat} depends on Ea exponentially (Equation (9.1)), the large free energy decrease translates into a much larger increase in k_{cat}.

We have seen that the ground state of the substrate is also stabilized, in the Michaelis complex (Figure 9.2b). The favorable enzyme-substrate interactions that underlie this stabilization account for the affinity of the enzyme to its substrate, as represented by the K_M parameter (see Section 9.2 above). On one hand, the energy drop due to these interactions makes the activation energy larger than it would be in their absence, which means a lower k_{cat} [193]. On the other hand, these specific interactions are needed for initial enzyme-substrate recognition and binding (*'accuracy-rate tradeoff'* [154]). Indeed, **enzymes have evolved so their affinity to the ground state of their cognate substrate(s) is large enough for efficient and selective binding, but is not so large as to make the activation energy insurmountable** [4]. This is also the reason why k_{cat}/K_M is used as a measure of enzyme efficiency instead of K_M or k_{cat} alone; both substrate affinity and catalytic efficiency need to be considered (see Subsection 9.2.3 above).

The above suggests a central role for noncovalent (transition state) stabilization in enzymatic catalysis. Supporting this idea, a survey of the MACiE enzyme database [40] suggests that the **noncovalent stabilization or destabilization of reaction intermediates is the most common function of catalytic residues, along with acid-base catalysis** (see below); the two functions involve 40% and 36% of the catalytic residues in MACiE, respectively. The large contribution of electrostatic interactions to the overall stabilization of enzymatic transition states makes sense, considering that many of the latter are electrically charged. In fact, Warshel and coworkers have argued that this type of stabilization is the most significant in enzyme catalysis [152,153,164,194]. Electrostatic stabilization is not unique to enzymatic catalysis; it also occurs in uncatalyzed reactions, where it is carried out by water dipoles. In enzymatic catalysis, however, the stabilization is stronger [153], for several reasons. First, water dipoles are induced and therefore weaker than the fixed dipoles of enzyme groups [195]. Second, to stabilize the charged transition state, the water dipoles need to reorganize around it, a process that involves an energy penalty due to the breaking of interactions between water molecules (*'reorganization energy'* [196,197]). The dipoles in the enzyme's active site are

---

[*1]The Michaelis complex is the initial binding configuration formed between the enzyme and the substrate in its ground state. The formation of such a complex involves a drop in the free energy of the system due to stabilizing interactions between the two. In Figure 9.2b this drop is represented by the red energy well at the beginning of the reaction.

already positioned correctly with respect to the transition state[*1], and therefore their effect involves a very small energy penalty, if any. This phenomenon is referred to as 'electrostatic preorganization' [152,153]. Third, unlike water, enzyme active sites can contain whole charges that belong to amino acid residues or to other chemical species, which lead to better stabilization. Finally, the favorable enzyme-transition state electrostatic interactions are optimized by the local dielectric in the active site, which is typically lower than that of bulk water (see Chapter 1 Box 1.1).

In line with the above, it has been suggested that the optimization of electrostatic interactions inside the active site leads to solvation of the transition state, and that this solvation is better than that obtained in aqueous solution [198]. If so, and if the electrostatic contributions to transition state stabilization are dominant over other noncovalent contributions, then any desolvation effects that accompany substrate binding play only a minor part. This means that **transition state stabilization is primarily enthalpic in nature**. This proposition, which is supported by a recent study carried out by Wolfenden and coworkers [199], may help shed some light on enzyme evolution: on primordial Earth the high temperatures provided at least some of the heat required for crossing the activation barrier of chemical reactions. This allowed different types of catalysts to provide the remaining stabilization effects needed for successful catalysis. However, when Earth began to cool down, reactions with large activation barriers could only be assisted by catalysts that acted enthalpically, i.e., those that substantially lowered the required heat level ($q_{(p)} = \Delta H^{\#}$, Equation (4.8)). Thus, Earth's own physical evolution may have provided selective pressure in favor of catalysts that mainly acted enthalpically [199].

Although the active site is preorganized to bind the transition state, **binding is often followed by conformational changes in the enzyme that further strengthen the noncovalent interactions (and sometimes create covalent bonds) with the substrate ('induced fit')** [14]. Although such changes are usually small[*2], they may still significantly affect the interactions within the active site, by changing the distances between interacting groups or their local dielectrics. Thus, enzymes seem to have at least two distinct conformations that are involved in binding. The first conformation is dominant when the enzyme is unbound, and has a high affinity to the ground state. The second is dominant when the enzyme is substrate-bound, and has a high affinity to the transition state [14,201]. Note, however, that other, intermediary conformations may exist between these two, and during catalysis itself the enzyme often shifts between different 'sub-conformations' [132] (see Subsection 9.3.3.6 below).

9.3.3.3 Covalent catalysis and electronic polarization of substrate bonds

Some of the enzyme's chemical groups surrounding the substrate do not merely interact with the latter via noncovalent interactions, but may also act chemically on it, in a way that promotes its conversion into the product. This action often involves the formation of an enzyme-substrate covalent bond via *nucleophilic substitution* [40], that is, an attack on an electrophilic center of the substrate (i.e., an atom that has low electron density) by an active site nucleophilic group (i.e., an atom that has high electron density), resulting in covalent bonding between the two. The principal goal of such bonding is to change the substrate from its ground state into its transition state, which promotes product formation (Figure 9.26a).

[*1]Repositioning of the dipoles may still occur to a small extent, due to conformational changes (see below).
[*2]In one survey, the average C_α r.m.s.d. was found to be ~1 Å [200].

However, the exact means by which covalent catalysis promotes formation of the transition state varies across enzymes (see below). Covalent mechanisms may also occur spontaneously in solution, i.e., in uncatalyzed reactions, in which the acting nucleophile is a water molecule. However, since water molecules are weak nucleophiles, these reactions are very slow. Nucleophilic groups inside the unique environment of the enzyme's active site are stronger and therefore produce higher catalytic rates.

Enzymes usually employ active site amino acids as nucleophiles, but in some cases other chemical species may be used. In the case of amino acids, the nucleophile may be a side chain oxygen atom (e.g., the hydroxyl groups of Ser, Thr and Tyr), a nitrogen atom (e.g., the amino group of lysine) or a sulfur atom (the thiol group of cysteine). A statistical survey [40] shows cysteine to be by far the most common amino acid in covalent catalysis (42% of the documented cases), followed by lysine (18%), aspartate and serine (~12% each). Histidine, threonine, glutamate, and tyrosine are also involved, but to a lesser extent. The high prevalence of cysteine in covalent catalysis is not surprising. As explained in Chapter 2 (Subsection 2.2.1.3.3.2) the SH group of this amino acid has a much lower pKa than the hydroxyl groups of serine (8.6) and threonine (~13). This is because the large dimensions of the sulfur atom better facilitate stabilization of the negative charge that results from deprotonation. Since deprotonation makes for a stronger nucleophile, cysteine is highly reactive. Note also that threonine, a secondary alcohol, is much less active as a nuclophile than serine, a primary alcohol.

In nucleophilic catalysis, the nucleophile exerts the attack on the substrate's electrophilic center by using its lone electron pair. The groups listed above are not always sufficiently nucleophilic to attack the substrate. However, the local micro-environment of the active site usually provides conditions that make them so, by increasing the density of their lone electrons [70]. For example, in *nucleoside phosphorylase* (EC 2.4.2.1), which uses a phosphate's oxygen as the nucleophile, the enzyme active site has been demonstrated to reduce the bond order of the P−O bond from 1.31 to 1.23, with a corresponding increase of the electron density of the oxygen [202]*1. In oxygen-based nucleophiles (e.g., Ser, Thr and Tyr), the active site-induced increase of electron density may happen in two ways [70]:

1. **Desolvation of the nucleophile** – removal of hydrogen bond donors to the lone pair. A hydrogen-bonded lone pair cannot act as a nucleophile, owing to both electronic and steric factors. Even when one lone pair is free but the other is bonded, the former is usually not sufficiently strong to carry out a nucleophilic attack, because its density is reduced by the presence of the hydrogen bond. During catalysis, conformational changes (e.g., induced fit) may increase the distance between the nucleophile and its cognate hydrogen bond partner, thus increasing the former's lone electron density.

2. **Polarization of the hydroxyl proton by a nearby general base (amino acid, metal)** – When the nucleophile is in the vicinity of a general base, the negative electric field of the base polarizes the O−H bond in the nucleophile (Figure 9.26b), thus increasing its lone electron density. When the polarization is strong enough it may lead to deprotonation of the OH group, leaving a highly nucleophilic O⁻ species. A well-known example of such activation is the catalytic mechanism of enzymes belonging to the *serine protease/esterase* group, which includes, among others, the enzyme *acetylcholinesterase*, which plays a major role in neural transmission (see Box 8.1);

---

*1 In an isolated phosphate group, each P−O bond is partially double due to electronic resonance.

the digestive enzymes *trypsin* (Figure 9.26c) and *chymotrypsin*; as well as the clotting enzyme *thrombin*. The activation of enzymatic nucleophiles by general base-induced deprotonation is part of another catalytic strategy called *'acid-base catalysis'* (see Subsection 9.3.3.5 below). Nucleophiles and their polarizing bases may be found in the active sites of enzymes as triads (as in serine proteases and esterases) or dyads. For example, in *L-asparaginase* (EC 3.5.1.1) a threonine nucleophile is deprotonated by an adjacent lysine, and in *glycosidases* the catalytic aspartate is kept deprotonated by a nearby arginine. Lysine, and especially arginine, are stronger bases (i.e., have higher p*Ka*) compared with the histidine in serine esterases and proteases, and therefore can polarize their cognate nucleophiles without the help of a third residue.

Covalent catalysis is used in virtually all enzyme classes. In particular, it is often used to promote the subsequent breaking or formation of a covalent bond in the substrate, as happens, e.g., in hydrolases and lyases. In hydrolases, the electrophilic center attacked in the substrate is usually a carbonyl group (C=O), and the resulting transition state is referred to as an *'acyl-enzyme intermediate'* [85] (Figure 9.26c, see also Figure 9.25b). The negatively charged transition state is stabilized electrostatically within the active site, which leads to the subsequent breaking of the covalent bond in the substrate by a water nucleophile. In esterhydrolases (esterases) and peptide-hydrolases (proteases), the broken bond is between the carbonyl carbon and either an oxygen or a nitrogen, respectively. However, in some hydrolases, lyases, and other enzymes using covalent catalysis, the electrophilic center attacked during nucleophilic substitution does not have to be a carbon atom. For example, in the metabolic enzyme *phosphoglucomutase* (EC 5.4.2.2), the electrophilic center is the phosphorus atom of a phosphate group, and the transition state formed as a result is called a *'phosphoenzyme intermediate'*.

Electronic polarization of bonds is a common strategy in nucleophilic catalysis, although it can work in different ways. In each of the above examples a covalent bond within an active site nucleophile is polarized by an adjacent group (e.g., a basic residue). This phenomenon is called the *'through-space field effect'*, because the polarizing group asserts its effect via its electric field, and without being bonded to the atoms of the polarized bond. However, bond polarization may also occur by *inductive effects*, which occur and propagate through covalent bonds. That is, a bond can be polarized by a group bonded to one of its atoms (directly or through another bond), if the group either withdraws electrons from the atom or pushes them towards it. In addition, the polarized bond need not necessarily be in the attacking nucleophile of the enzyme; it may be in the substrate itself, where the goal of the polarization is either the breaking of the bond or its condensation with other molecules. An example of substrate bond polarization via inductive effects is seen in β-keto acid decarboxylases, such as *acetoacetate decarboxylase* (EC 4.1.1.4) [203]. In this case, the enzyme's attacking nucleophile is a lysine side chain amino group $(R-NH_2)$[*1], and the attacked substrate electrophile is a carbonyl group bound to a carboxyl group through a methylene group $(O=C-CH_2-COOH)$. The result of the attack is a Schiff base $(-C=N^+-)$ between the carbonyl and the lysine side chain (Figure 9.27a). The Schiff base forms because the nitrogen atom is less electronegative than the carbonyl oxygen, and its unpaired electrons can therefore easily bind the carbon. **The positive charge of the Schiff base acts as an 'electron sink' because of its strong tendency to withdraw electrons from adjacent bonds.** In this

---

[*1] The lysine side chain remains deprotonated due to the highly hydrophobic nature of the active site, which changes the p*Ka* of the side chain from 10.5 to 6 [204,205].

(a)

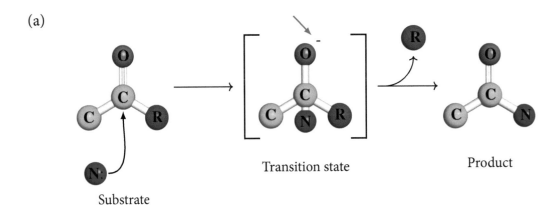

Substrate

Transition state

Product

(b)

-0.564 (-0.528)

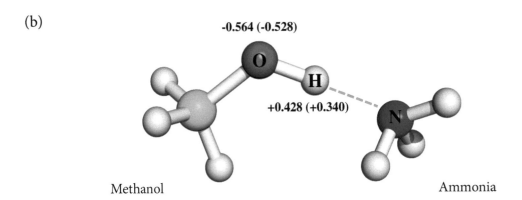

+0.428 (+0.340)

Methanol

Ammonia

(c-I)

FIGURE 9.26 **Covalent catalysis.** (a) Nucleophilic substitution of a ketone. The transient covalent bond formed between the nucleophile (N:) and the ketone substrate creates a negatively charged transition state with a tetrahedral geometry. The charge is marked by the red arrow. (b) Electronic polarization of methanol's O−H bond (left) by a nearby ammonia group (right). The dashed green line is the hydrogen bond that forms between the two interacting molecules. The numbers signify the Mulliken charges on the methanol oxygen and hydrogen atoms, determined at the B3LYP/6-31+G(d,p) level. The values for methanol alone in the gas phase are shown in parentheses. In the presence of ammonia, the charge difference between oxygen and hydrogen is ~14% higher than in the absence of ammonia. This difference results from both the increased electron density on the oxygen atom and the increased positive charge on the hydrogen atom. The image is adapted from [70] (Copyright (2006) American Chemical Society). (c) The enzymatic activity of trypsin (EC 3.4.21.4). (c-I) The overall reaction, which includes hydrolysis of a peptide bond inside the protein C-terminus to either a lysine or an arginine. In this case an arginine (blue) is flanked by a C-terminus amino acid (green) and an N-terminus amino acid (red).

(c-II)

FIGURE 9.26 **Covalent catalysis.** (Continued) (c) The enzymatic activity of trypsin (EC 3.4.21.4). (c-II) The catalytic mechanism of trypsin from *Fusarium oxysporum* (PDB entry 1pq5). Step 1: deprotonation of Ser-195 by His-56 and subsequent nucleophilic attack of Ser-195 on the carbonyl carbon of the peptide bond. This yields a tetrahedral acyl-enzyme intermediate. Step 2: the oxyanion initiates an elimination reaction that cleaves the peptide bond, releasing the new N-terminus of the protein, which protonates from His-56. Step 3: His-56 deprotonates water, which attacks the carbonyl carbon bound to Ser-195 in a nucleophilic addition. Step 4: the oxyanion initiates an elimination that cleaves the acyl bond to Ser-195, releasing the C-terminus of the protein. Ser-195 then deprotonates His-56, regenerating the active site. The image is adapted from the MACiE database [34] (entry M0173).

case, the Schiff base withdraws electrons from the methylene-carboxyl bond, thus weakening it. When the bond finally breaks, the carboxyl group is released as CO_2 (Figure 9.27a). A similar mechanism is used by *Class I aldolases*, which catalyze *aldol cleavage* reactions[*1]. In this type of reversible reaction, a β-hydroxyketone or aldehyde is cleaved to yield an enol and a ketone (Figure 9.27b). This happens, e.g., in the glycolytic enzyme *fructose bisphosphate aldolase* (EC 4.1.2.13), which in *glycolysis* catalyzes the cleavage of fructose 1,6-bisphosphate to glyceraldehyde 3-phosphate (G3P) and dihydroxyacetone phosphate (DHAP). In *gluconeogenesis*, the same reaction takes place in the opposite direction (*aldol condensation*) (Figure 9.27c I). In this case, the Schiff base is formed between a lysine residue in the enzyme and DHAP (Figure 9.27c II, step 1). The effect of the Schiff base promotes deprotonation of C1 by an active site tyrosine residue and double bond formation between C2 and C2 (step 2). The double bond then attacks the carbonyl group of G3P (step 3), and, following hydrolysis of the Schiff base, fructose 1,6-bisphosphate is formed (step 4). Finally, the molecule closes into a ring structure (not shown).

Whereas Schiff bases are often formed between the substrate and an active site species, the latter need not necessarily be an amino acid residue. As we saw in Subsection 9.1.5 above, aminotransferases and lyases (including certain aldolases) use the coenzyme pyridoxal phosphate (PLP) to form a Schiff base with their substrates, and this Schiff base promotes bond cleavage in the latter. In this case, however, the electron-withdrawing effect on the substrate is much stronger than in an amino acid-involving Schiff base, thanks to the positively charged pyridinium ring of PLP. Indeed, PLP is used to promote cleavage of more

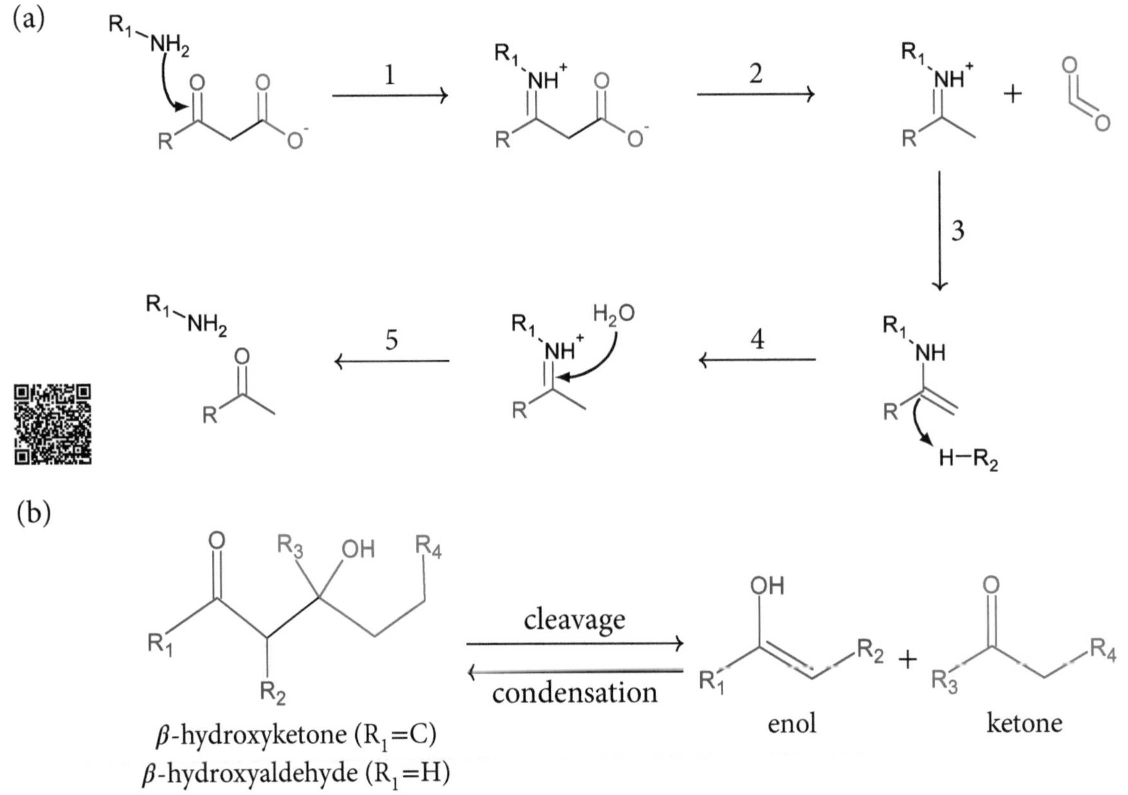

(a)

(b)

β-hydroxyketone (R_1=C)
β-hydroxyaldehyde (R_1=H)

[*1]These enzymes differ from class II aldolases, which use divalent metal cations such as Zn^{2+} or Fe^{2+} for bond polarization, instead of a Schiff base. Since the metals are not bonded to the labile bond, the polarization is mediated by a through-space field effect.

(c-I)

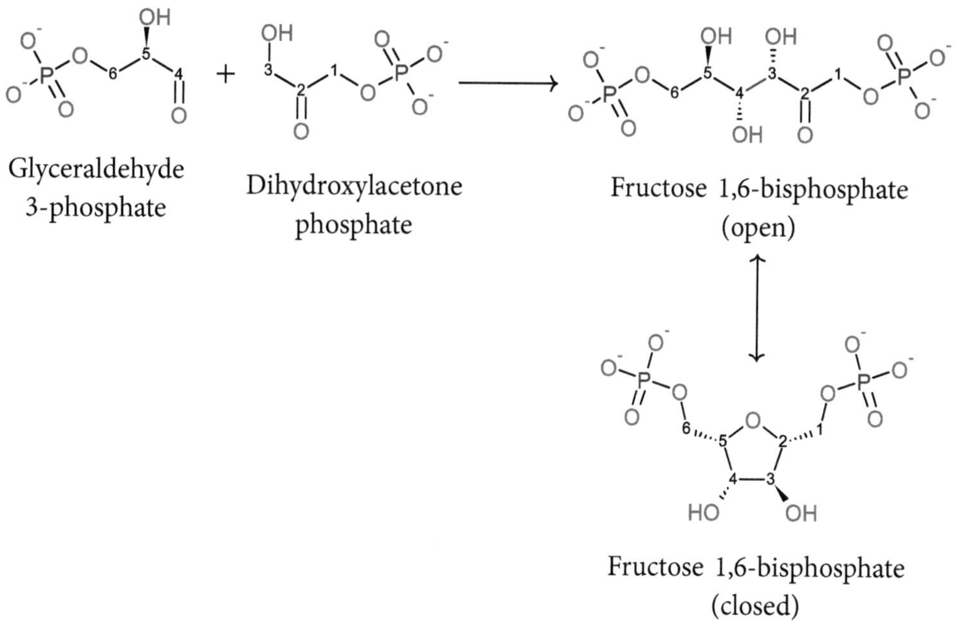

(c-II)

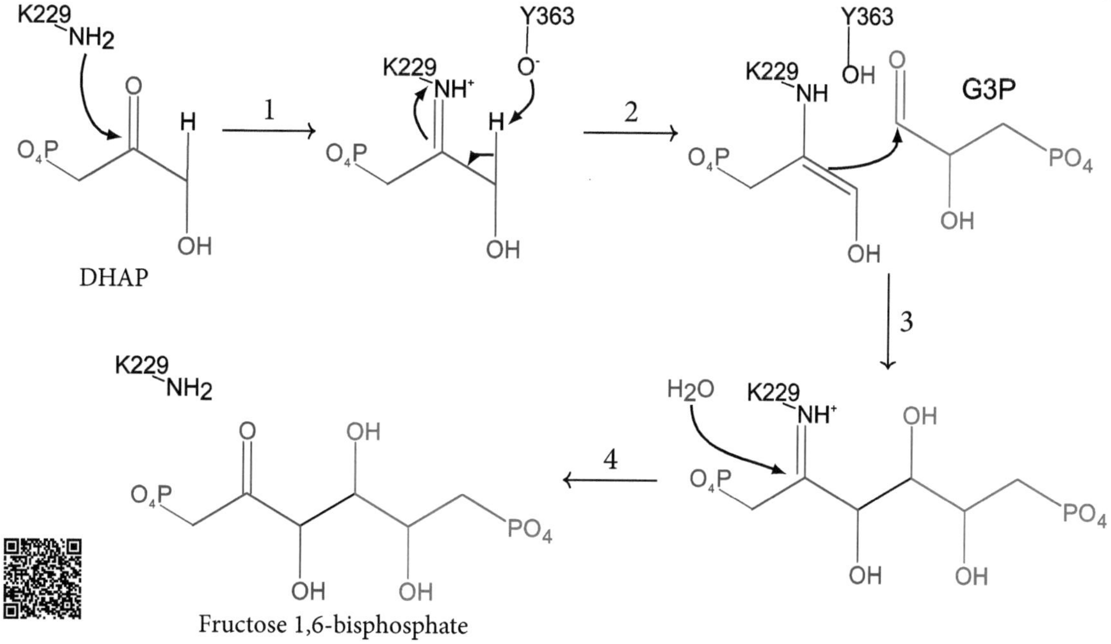

FIGURE 9.27 **Schiff base-mediated covalent catalysis.** (Opposite) (Continued) (a) Decarboxylation of a β-keto acid with a Schiff base intermediate [203]. The mechanism involves (1) attack of an active site nucleophile (R_1–NH_2) on the β-keto carbon, forming a Schiff base. (2) Decarboxylation. (3) Rearrangement of electrons. (4) Protonation of the substrate's double bond by an active site general acid. (5) Attack of an active site water nucleophile on the Schiff base, resulting in the breaking of the Schiff base and creation of a ketone. (b) Reversible aldol cleavage of a β-hydroxyketone/aldehyde to an enol (blue) and a ketone (red). (c) Aldol condensation of glyceraldehyde 3-phosphate (G3P) and dihydroxyacetone phosphate (DHAP) to form fructose 1,6-bisphosphate, carried out by Class I aldolase. (c-I) The overall reaction (the reverse of the reaction shown in Figure 9.14). (c-II) The principal steps in the mechanism of rabbit aldolase (data taken from the MACiE database [34], entry M0222). The steps are described in the main text.

stable bonds than, e.g., the carboxyl-C_α bond in β-keto acids. As explained above, the latter bond is relatively easy to break due to its inherent instability. In any case, **nucleophilic catalysis via bond polarization is an efficient strategy for activating reactants inside the active site. The chemical transformations that benefit from such activation include both bond cleavage and formation**. Again, polarization can be achieved in various ways and is not limited to the use of PLP or other Schiff bases. For example, in the case of bond cleavage, enzymes may also use the thiazolium ring of the coenzyme thiamine pyrophosphate (TPP) as an electron withdrawing group [148] (see Subsection 9.1.5.4 above). In the case of bond formation, enzymes may use phosphoryl groups or the ureido group of the coenzyme *biotin* (see Subsection 9.1.5.6 above).

9.3.3.4 Metal ion catalysis

Certain enzymes, called *metalloenzymes*, use the following metal cations as their catalytic species: zinc (Zn^{2+}), copper (Cu^{2+}/Cu^+), magnesium (Mg^{2+}), manganese (Mn^{2+}/Mn^{3+}), iron (Fe^{3+}/Fe^{2+}), cobalt (Co^{2+}/Co^{3+}), molybdenum ($Mo^{3+}/Mo^{4+}/Mo^{6+}$), vanadium (V^{5+}) and even tungsten (W^{4+}/W^{6+})[44] (the oxidation states correspond to physiological conditions). The metals may be used for different types of enzymatic reactions, such as oxidation-reduction, hydrolysis, transfer of phosphate groups, and oxygen atom transfer [94]. These roles of metal cations largely account for the physiological requirement for these minerals[*1] and their importance in our diet, although some of them are needed only in trace amounts. The involvement of metals in enzymatic reactions relies on their ability to perform specific functions, such as electron transport; electrostatic stabilization of negatively charged species; and activation of substrate, coenzyme, or water molecules by electronic polarization (see details below). Although these are diverse functions, they are all based on either the positive charge of the metals, or their ability to coordinate different chemical species. As described in detail in Chapter 2, Subsection 2.6.9, metals can appear in proteins in their free form or as part of inorganic clusters. The latter may sometimes be independently active, but the protein environment is important for providing biological context, as well as stabilizing the metal clusters and preventing destructive side reactions [206]. Below we give a short description of the main specific functions carried out by enzyme-bound metal ions.

9.3.3.4.1 Electron transport

Enzymes catalyzing oxidation-reduction (redox) reactions often use metal ions as transient binding sites for electrons, which are passed along a route within the protein. **Statistical analysis of metalloenzymes shows that iron (Fe) and copper (Cu) are the principal metals used for redox catalysis** [207]. Iron is particularly prevalent in these reactions, and can appear in different forms within the active sites of redox enzymes: Fe−O−Fe sites, heme groups and Fe−S clusters[*2]. The predominance of iron in redox enzymes may result from different factors, including the sensitivity of iron's redox properties to the ligands it coordinates. This sensitivity enables enzymes to modulate the behavior of Fe, including its tendency to act as an oxidizing or as a reducing agent, by changing its ligands and even the

---

[*1]Some of these metals are also needed for non-enzymatic functions (e.g., calcium serves as a building block for bones).

[*2]This explains the critical importance of iron to key physiological processes that depend on redox enzymes, such as cellular respiration, nucleotide metabolism, detoxification of xenobiotics, and brain development in infants.

coordination geometry. **Iron cations exist within redox enzymes mainly in the divalent form (Fe^{2+}), and act primarily in activation of O_2 for the transformation of organic compounds** [207]. This role of iron is observed mainly in *oxygenases*, which incorporate oxygen into organic molecules (see Subsection 9.1.5.1.1 above). Activation is required in this process because O_2 has a *triplet spin state*, whereas most organic substrates have a *singlet state*, and this difference results in a kinetic barrier for the reaction. The coordination of O_2 to the iron cation lowers the barrier by transferring two electrons to the oxygen, one from the Fe^{2+} atom and the other probably from another adjacent metal [207]. This transfer involves the transient formation of other oxygen species, i.e., superoxide (O_2^-) or peroxide (O_2^{2-}) [44]. Iron is also important for other enzymatic processes, such as the metabolism of nucleic acids (and other biological molecules), as well as the degradation of biological pollutants [44].

Copper, the other common metal in enzyme-mediated electron transport, resembles iron in more than one way [44]. First, copper, like iron, appears in redox enzymes of the *oxygenase* and *oxidase* groups. Second, copper, too, has two oxidation states under physiological conditions (Cu^{2+}/Cu^+), and can therefore play different roles aside from electron transport, such as O_2 activation and transport. Accordingly, copper is important to numerous physiological processes, such as the synthesis of various biological elements (e.g., connective tissue proteins, the pigment *melanin*, certain hormones and neurotransmitters), as well as the activation of white blood cells and the mineralization of bones. As we will see below, other metals are as abundant as iron and copper, but because they only have a single oxidation state (e.g., 2+ for magnesium, calcium and zinc) they cannot play a role in electron-transfer (redox) reactions.

9.3.3.4.2 Electrostatic stabilization

As explained in Subsection 9.3.3.2 above, enzymes often use metal cations such as Mg^{2+}, Cu^{2+}, Zn^{2+}, Mn^{2+}, Co^{2+}, and Fe^{2+}/Fe^{3+} either to stabilize the negatively charged transition states of their reactions [148] (Figure 9.25d) or to stabilize other anionic species in their active sites. These functions rely on the metals' strong positive charge. The most widely used species is magnesium (Mg^{2+}), which is abundant and also has the highest charge density of all protein-bound metals [92,207]. As a result of these characteristics, it has been selected by evolution to stabilize highly anionic species in enzymes, which are usually substrate-bound phosphate groups or coenzymes. Such stabilization occurs, e.g., in enzymes that regulate nucleic acid biochemistry, such as restriction enzymes, ligases, and topoisomerases [208], as well as in the glycolytic enzymes *enolase* (EC 4.2.1.11) and *pyruvate kinase* (EC 2.7.1.40) [92]. Mg^{2+} is also important for the fidelity of DNA replication. Within the active site, Mg^{2+} is usually coordinated by one or more carboxylate side chains of glutamate or aspartate. However, when in complex with the substrate, Mg^{2+} interacts only transiently with enzyme groups [207].

9.3.3.4.3 Substrate, coenzyme and water activation by polarization

In Subsection 9.3.3.3 above, we saw how electronic polarization of covalent bonds in the substrate or in active site catalytic species can promote chemical transformations and help catalysis. Such polarization is induced by electron withdrawing organic groups, like the side chains of lysine residues or the cofactor PLP. Metal cations can also be used for this purpose, owing to their ability to function as *Lewis acids*, that is, to bind lone-pair electrons, e.g., those of oxygen or nitrogen atoms [207]. Coordination of metals to these atoms in the substrate or coenzyme polarizes the covalent bonds in which they are involved. The result depends on the degree of polarization achieved by the metal. Partial polarization of C−O and P−O bonds (where C−O is included in many substrates, and P−O appears, e.g., in ATP) turns the carbon and phosphorus into electrophilic centers, which are amenable to nucleophilic attack. This process is part of the catalytic mechanism of the enzyme, which means that the metal acts as an activator of the substrate or coenzyme.

In other cases, the metal induces complete bond polarization, which leads to the breaking of the bond. The most common case is polarization of water molecules inside the active site, aimed at creating a potent nucleophile. That is, the polarization of one of the water's O−H bonds leads to its breaking, and the resulting OH^- group acts as a strong nucleophile capable of attacking an electrophilic center in the substrate [94]. This happens mainly in reactions catalyzed by *hydrolases*. The other known case of complete metal-induced polarization leads to the release of a proton from the organic substrate or coenzyme. This results in electronic rearrangement of the latter, thus promoting catalysis.

Several metal cations are used for electronic bond polarization of enzyme substrates or coenzymes. Like electrostatic stabilization of active site species (see previous subsection), metal-related bond polarization is also carried out primarily by Mg^{2+}, for the same reasons explained above. Another metal commonly used for this role is zinc (Zn^{2+}), which is abundant and has electron affinity even higher than that of Mg^{2+}. Zinc is used frequently in enzymes for polarizing water molecules or organic C−O and P−O bonds. In fact, in some enzymes, Zn^{2+} is responsible both for creating the attacking nucleophile and for enhancing substrate electrophilicity [207]. These roles, combined with the coordination flexibility of Zn^{2+*1}, make zinc the second most common metal cation in enzymes, after Mg^{2+*2}. Indeed, zinc is important for various physiological processes, including the development of the skeletal and reproductive systems, wound healing and the immune response [93]. Calcium ions (Ca^{2+}) are also used for bond polarization, but to a lesser extent. Instead, evolution seems to have chosen Ca^{2+} as the principal metal regulator of numerous biological processes [207,209] (see our previous discussions of calmodulin in signal transduction, and of the clotting process). The reason for the selection of calcium may have to do with the relatively large radius of Ca^{2+}, which enables it to bind a diverse set of ligands with irregular geometries. In contrast, Mg^{2+} only forms octahedral complexes.

9.3.3.4.4 Other catalytic metals

Among the less abundant metals used by enzymes [44], the following are notable:

- **Nickel** – functions in *urease* (EC 3.5.1.5) and certain hydrogenases by catalyzing H_2 oxidation.

- **Molybdenum** – required for certain oxygen atom transfer reactions.

- **Manganese** functions (as a part of the oxygen-evolving complex) in the photo-oxidation of water to O_2 during the light reaction of photosynthesis. It is also a known cofactor in oxidase chemistry.

- **Tungsten** – is a cofactor in certain dehydrogenases.

[*1] Zn^{2+} is able to bind between four to six ligands within proteins, although in enzymes it is usually part of complexes that have coordination numbers less than 6 and tetrahedral geometry [92].

[*2] A well-known zinc-dependent enzyme in animal physiology is *carbonic anhydrase*, a hydrolytic enzyme that interconverts carbon dioxide (CO_2) and bicarbonate (HCO_3^-). In animals, this enzyme is important for regulating blood pH. The mechanism of the enzyme is described in Figure 9.28.

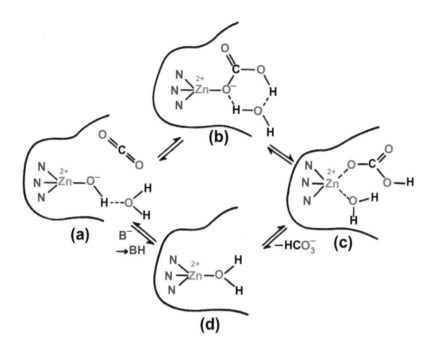

FIGURE 9.28 **The use of metal-induced water deprotonation in the mechanism of carbonic anhydrase.** (a) The deprotonated water molecule (i.e., OH⁻ group) is attached to an active site zinc cation, which is also coordinated to three histidine side chains. (b) and (c) The OH⁻ group then acts as nucleophile and attacks the carbon of CO_2 to form bicarbonate (HCO_3^-). (d) The metal then binds a new water molecule for the next catalytic cycle. The image is taken from [94].

9.3.3.5 General acid-base catalysis

Acid-base catalysis is by far the most common catalytic strategy used by enzymes, accounting for ~75% of the reactions in the MACiE database [40]. It involves a proton (H^+) transfer between catalytically important groups inside the active site, i.e., amino acid residues, coenzymes and substrates[*1]. This transfer requires an active site species that can act as a *Brønsted acid* (i.e., capable of donating the proton to another species), or as a *Brønsted base* (i.e., capable of receiving the proton from another species). Enzymes use two types of acids and bases in their active sites:

1. Water molecules, i.e., H_3O^+ as an acid and OH^- as a base. In such cases we refer to the process as 'specific acid-base catalysis'.

2. Amino acid side chains. Histidine, whose imidazole group can act both as an acid and as a base under physiological pH, is the most common amino acid in acid-base catalysis (31% of the acid-base reactions in MACiE). Other common amino acids in this type of catalysis are glutamate (21%) and aspartate (~16%) [40]. Lysine, tyrosine, cysteine, serine, arginine and threonine are also involved, but to a much lesser extent, due to their high pKa. That is, they are protonated, unless their local chemical environment in the active site promotes their deprotonation. The use of amino acid groups is referred to as 'general acid-base catalysis'.

The prevalence of acid-base catalysis as an enzymatic strategy largely accounts for the well-known dependency of enzymatic activity on the pH of the enzyme's environment. Indeed, each enzyme is active only within a certain range of pH and achieves optimal activity under

---

[*1]Not to be confused with hydride ion (H^-) transfer, a common way to transport electrons in redox reactions, or with hydrogen radical ($H\cdot$) transfer, which is common in molecular rearrangements.

a pH value that is within that range (this is referred to as the enzyme's *optimal pH*). We have seen earlier that extreme pH values may lead to protein denaturation. However, the margins of the pH range of enzymes are usually not extreme enough to cause protein denaturation, suggesting that other factors account for the dependency. Indeed, this dependency usually results from the pH sensitivity of residues that participate in acid-base catalysis. For example, if an enzyme relies on a certain histidine residue to function as a general base (i.e., accept a proton), then the pH in the histidine's local environment will keep it in a deprotonated state before catalysis starts. Substrate binding or conversion of the substrate into the transition state will lead to protonation of the histidine, and when the catalytic cycle is over, the histidine will return to its deprotonated state. Since the protonation or deprotonation of any ionizable group is most affected around its pKa, the optimal pH of an enzyme that depends on histidine acid or base behavior is expected to be around 6 and 7. In contrast, an enzyme that depends on the acid-base behavior of glutamate or aspartate residues is expected to have a much lower optimal pH, around 3 or 4. As we have seen earlier, and as will be further explained below, the protonation state of a residue is also influenced by its immediate chemical environment. That is, chemical species adjacent to the residue may change its pKa through electrostatic effects. The capacity of the active site to exert such effects results from its structural and physical properties (e.g., charge distribution). However, these properties are often modulated by conformational changes that occur during catalysis, and which affect the local environment of the protonated or deprotonated species. Such conformational changes are part of the enzyme's catalytic mechanism and serve the need for altering the protonation states of catalytic residues, coenzymes or the substrate at certain points during catalysis.

Finally, note that the optimal pH of an enzyme may also be determined by the need to keep certain residues charged or neutral. For example, in an enzyme that uses a charged amino acid to stabilize the substrate's transition state or polarize it, the optimal pH of the enzyme is likely to be affected by the pKa of this amino acid. This is because pH-induced protonation or deprotonation of this residue may neutralize its charge and prevent it from fulfilling its role. In some cases, however, the catalytic rate does not seem to be affected by pH, despite the presence of acid-base catalytic residues. This lack of effect may be a result of buffering by auxiliary residues in the active site.

Although acid-base catalysis may involve a variety of strategies, it usually entails either a direct action on the substrate or activation of an active site species. Below we elaborate on these two strategies.

9.3.3.5.1 Direct action on substrate

The simplest way in which acid-base catalysis can exert its effect is to directly transfer a proton to the substrate or from it. This transfer usually promotes the conversion of the substrate into the transition state or any other intermediate downstream of the reaction coordinate [210,211]. This type of action is observed in *triosephosphate isomerase (TIM)*, EC 5.3.1.1), one of the ten enzymes of *glycolysis*. TIM uses general acid-base catalysis to accelerate the reversible isomerization of a ketone (dihydroxyacetone phosphate, DHAP) to an aldehyde (glyceraldehyde 3-phosphate, G3P) [212] (Figure 9.29). This happens in two steps:

1. A proton is transferred from C1 of DHAP to the carboxylate group of Glu-165, which acts as a general base (B⁻ in Figure 9.29). This transfer converts DHAP into an *enediol intermediate* [213].

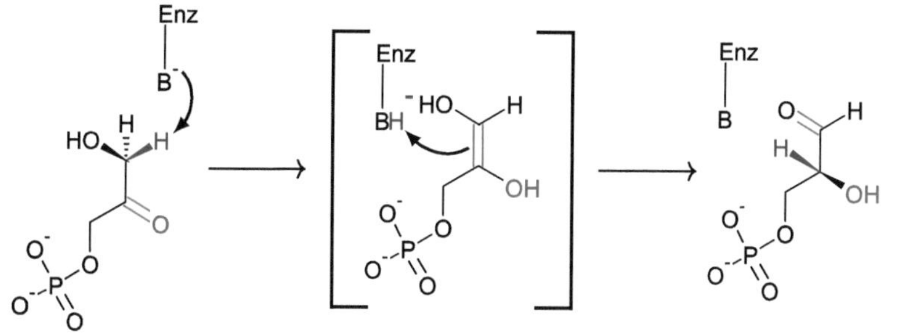

FIGURE 9.29 **General acid-base catalysis in triosephosphate isomerase.** The general base is denoted as B^-. The proton involved in the catalytic mechanism is colored in blue. The keto group that becomes a hydroxyl is colored in red, and the delocalized double bond is in green. The reaction is reversible and depicted here in one direction for clarity.

2. Glu-165 acts as a general acid and transfers the proton back to the substrate. However, this time the proton is transferred to C2 of the intermediate, turning it into G3P.

Another example of direct acid-base action on a substrate is observed in the catalytic mechanism of *fumarate hydratase* (a.k.a. *fumarase*) (EC 4.2.1.2). Fumarase plays a central role in metabolism; the mitochondrial isozyme[*1] is part of the Krebs cycle, whereas the cytosolic isozyme is involved in amino acid catabolism. In both cases, the enzyme catalyzes the stereospecific addition of a water molecule to fumarate, to form L-malate (Figure 9.30). However, the reaction is reversible, and the same enzyme is also capable of doing the opposite, i.e., eliminating water from L-malate to form fumarate. The catalytic mechanism employed by fumarase involves proton abstraction from C2 of L-malate. This is remarkable, as the carbon atom has a pKa of ~30 [214].

9.3.3.5.2 Indirect action: facilitating covalent catalysis

As we have seen in Subsection 9.3.3.3, acid-base catalysis may contribute to catalysis by helping active site species to carry out nucleophilic substitution on the substrate. This process usually involves an active site general base that deprotonates the hydroxyl group of a residue, to render it a stronger nucleophile. According to Jencks' 'libido rule', this can only occur if (*i*) the pKa of the deprotonated nucleophile changes significantly during the reaction, and (*ii*) the pKa of the base is intermediary between the initial pKa of the nucleophile and its pKa after the nucleophilic attack [211,216]. This explains why histidine is the most common residue in acid-base catalysis [40]; its pKa is close to the middle of the pH scale, and therefore has the highest probability of being intermediary between the pKa values of the nucleophile before and after the reaction. Glutamate and aspartate have lower inherent pKa values, but the chemical environment at the active site often increases these values. Again, serine proteases (e.g., trypsin; Figure 9.26c) and esterases (e.g., acetylcholinesterase; Box 8.1) provide classic examples of indirect acid-base catalysis. As we have seen earlier, the mechanism of these enzymes involves deprotonation of a serine residue (pKa~13) by

---

[*1] Isozymes (or isoenzymes) are two different forms of the same enzyme. They perform the same biochemical reaction but differ in their sequences or structural properties, and the difference is reflected in their different efficiency levels or substrate affinity and specificity.

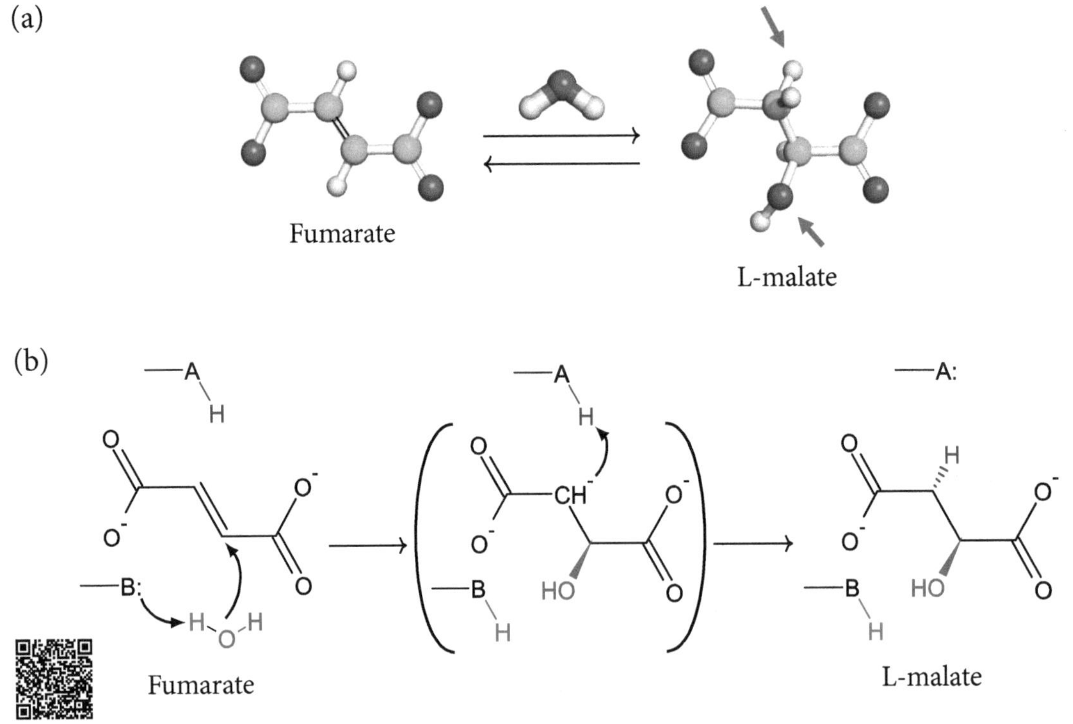

(a)

Fumarate

L-malate

(b)

Fumarate

L-malate

FIGURE 9.30 **The reversible interconversion between fumarate and L-malate by fumarate hydratase (fumarase).** (a) The overall reaction. The added hydrogen atom and OH group are marked by the red arrows. (b) The reaction mechanism. In the first step an enzyme group functioning as a general base (B:) abstracts a proton from a water molecule, and the resulting OH group (in red) is stereospecifically attached to C2 of fumarate to create a carbanion transition state. In the second step, an enzyme group functioning as a general acid (A—H) donates its proton to C3 of the transition state, in the *anti* position [215]. For clarity, the reactions are depicted as unidirectional.

an adjacent histidine (the general base), which enables the serine to attack the electrophilic carbon of either a peptide or an ester unit in the substrate. The nucleophilic attack creates an acyl-enzyme intermediate (Figure 9.26c, see also Figure 9.25b), which turns the serine side chain from a hydroxyl into an ether (pKa~0). According to the libido rule, the general base used in this reaction can be any chemical species, as long as its pKa is between ~13 and ~0 [70]. Indeed, the histidine residue serving as a general base in this reaction has a pKa value of 6.5, which makes it suitable. In fact, in serine proteases and esterases, the pKa of the catalytic histidine is further increased via electrostatic field effects, induced by the nearby aspartate residue. **pKa modulation of general acids or bases is very common in enzymes.** However, in contrast to the above histidine example, **in most cases pKa modulation is used to adapt residues with more extreme pKa values to their roles as general acids or bases.** For example, in *aldo-keto reductases* the pKa of a tyrosine's phenol group (9.8) is lowered by a nearby lysine side chain[*1], so it can act as a general acid. Also, in *aspartic proteases*, the proximity between two aspartate side chains compels one of them to remain protonated, to avoid Coulomb repulsion [148]. The protonated aspartate can then act as a general acid, whereas the other acts as a general base [148]. In some cases the modulation of a residue's pKa is more complicated. For example, quantum-mechanical calculations carried out on

---

[*1] The deprotonation of the tyrosine side chain charges it with a negative charge, which in turn stabilizes the system by masking the positive charge of the adjacent lysine side chain.

ketosteroid isomerase (EC 5.3.3.1) suggest that the p*Ka* of a catalytically important tyrosine side chain is significantly reduced by electronic inductive effects along a hydrogen-bonding network that involves this residue and two other tyrosines [217].

In order for acid-base catalysis to take place, the two groups involved in the proton transfer must be close enough to each other; otherwise, the energy barrier for the transfer is too high. However, evidence accumulating since the 1980s shows that in many cases the transfer of hydrogen species (protons, hydride ions, or hydrogen atoms) occurs despite an impeding energy barrier [218–220] (see also references in [221]). In these cases, the transfer is made possible by a quantum-mechanical phenomenon referred to as *'tunneling'*. This interesting phenomenon is explained in detail in Chapter 5, Subsection 5.3.1.4 IV.

9.3.3.6 Mechanisms related to protein dynamics

In the subsections above we had a glimpse of the intricate mechanisms used by enzymes to accelerate chemical reactions. We saw that these mechanisms result directly from the preorganization of enzymatic active sites, which puts functional groups near each other or near the substrate, and which enables the many resulting interactions to be modulated by the general physical properties of this microenvironment. It is therefore tempting to think that such mechanisms can be executed in a 'frozen' active site, i.e., without the help of dynamics. However, what we already know about the effect of protein dynamics on protein function (see Chapter 5, Subsection 5.3.1.4) constitutes a strong argument against such a notion. Indeed, **protein function is significantly affected by local and global motions that facilitate ligand diffusion into the binding site, mediate the induced fit of the protein to the ligand, change the physical properties of the binding site, and optimize quantum phenomena.** In enzymes, the timescales associated with general protein dynamics generally overlap with those associated with catalysis (Figure 9.31). This suggests that at least some of the motions associated with general protein dynamics have effects on catalysis. Many studies support this suggestion, employing methods such as X-ray crystallography and small-angle scattering [222,223], neutron scattering [224], NMR [225–227], fluorescence spectroscopy, measurements of isotope effects on hydride transfer [228], and molecular dynamics simulations [155] (see [229] for details).

The studies cited above refer to a wide range of motions in enzymes, from slow, large-scale loop or domain movement, to fast, short-scale thermal vibrations. The functional implications of $\sim10^{-8}$ to 10^{-3} s motions, involving whole residues, secondary structure elements, hinges or other segments of the polypeptide chain, as well as whole domains, have been known and documented for some time (e.g., see [133,229] and references therein). Such motions account for a wide range of events in enzymes, including creation of gates and tunnels to sequester the substrate[*1] or extrude the product [166,230]; other induced fit changes that strengthen enzyme-substrate interactions; and short-lived conformational changes that occur during catalysis and that facilitate the formation of multiple substrate intermediates [132,231]. These dynamic events may affect catalysis in different ways. For example, they may bring interacting species closer to each other (e.g., in acid-base and nucleophilic catalysis) or adjust the electrostatic environment to optimize catalytic events (see [232] and references therein).

In contrast, the effects of thermal vibrations ($\sim10^{-13}$ s) and other fast motions

---

[*1]The structure and properties of the gates and tunnels contribute to the enzyme's selectivity towards the substrate, by controlling its access to the active site. In addition, the gating event may synchronize processes in different parts of the enzyme, which may in turn affect the enzyme's activity (see the extensive review on enzyme gating by Damborsky and coworkers [230]).

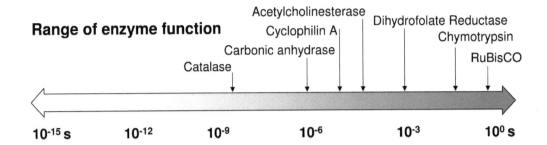

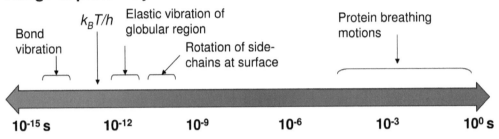

FIGURE 9.31 **Timescales of proteins and enzymatic catalysis.** k_B is the Boltzmann constant, T represents the ambient temperature, and h is the Planck constant. k_BT/h is the universal frequency factor, commonly used in transition state theory (see Subsection 9.1.2 above). The image is taken from [229].

$(10^{-12}$ to 10^{-9} s$)^{*1}$ on enzyme catalysis are more controversial. Simulations suggest that fast protein motions are involved in the transfer of electrons in photosynthetic bacterial reaction centers [233]. Furthermore, as explained in Chapter 5, Section 5.3.1.4 IV, quantum-mechanical calculations suggest that thermal vibrations in enzymes are important for efficient hydrogen tunneling during catalysis^{*2}, although the exact mechanism of this effect is unclear. In line with these propositions, it has been suggested that enzymes have evolved to harness their fast $(10^{-13}$ to 10^{-9} s$)$ motions to aid catalysis by contributing directly to the crossing of the reaction's energy barrier. This issue is under intense discussion (e.g., [194,225,238–242]). One problem with this suggestion is the difficulty in proving it experimentally. While various studies have tried to address this issue, it has been suggested that their interpretations of the results were problematic [194]. In an interesting study carried out by Schramm and coworkers, atoms of nucleoside phosphorylase (PNP) were replaced by their heavy isotopic analogues [243]. The resulting enzyme had ~10% increased mass, and according to the Born-Oppenheimer approximation [244] its femtosecond $(10^{-15}$ s$)$-scale vibrations were slower than those of the wild-type (i.e., lighter) enzyme^{*3}. The aim of the study was to determine how the decreased vibration rate in the heavier enzyme would affect its

---

^{*1}These motions are responsible for the change in individual noncovalent interactions (e.g., via side chain rotation) and small alterations in water structure [133].

^{*2}Hydrogen tunneling and other quantum-mechanical effects on catalysis have been discussed for decades (e.g. [219–221,234–237]), but only in recent years have they gained wide recognition. The relative influence of tunneling effects on reducing the overall free energy barrier of a given reaction may be small, but it is acknowledged today that such effects should be included in the description of enzyme-mediated catalysis [19,221].

^{*3}According to the Born-Oppenheimer approximation, a molecule built from heavy isotopes has the same potential energy as the same molecule built from light isotopes, but the vibrational structure of the two molecules is different [242].

catalytic behavior. The study found that the heavy enzyme had the same steady-state rate constant (k_{cat}) as the wild-type enzyme. This was not surprising, as the k_{cat} of PNP was already known to be dominated by product release, a step mediated by slow motions and not by fast vibrations. Conversely, it was found that in the heavy enzyme, two measurable parameters that correspond only to the chemical step of catalysis[*1] were reduced by ~30% compared to their counterparts in the light enzyme. This observation suggested that femtosecond vibrations indeed have an important role in enabling PNP to undergo the chemical step of catalysis, in which the reaction's energy barrier is crossed.

Another problem with the suggestion that fast motions contribute to enzyme catalysis is the large difference between the timescales of these motions and the turnover times of most enzymes (10^{-6} to 10^{-3} s) [155] (Figure 9.31). It should be noted that this discrepancy does not necessarily preclude the involvement of fast motions in catalysis, as in many enzymes the rate-limiting steps are substrate binding or product release, not the chemical steps involved in crossing the reaction's energy barrier [231,242] (see also the case of PNP above). The chemical steps, on the other hand, are on timescales similar to those of fast motions, and may therefore be influenced by them. For example, fast motions may change the distance between hydrogen donor and acceptor species, thus affecting the probability of a tunneling event. Moreover, it has been suggested that, when coordinated, the fast motions may result in larger motions that relate more directly to enzymatic catalysis (see discussion by Hammes-Schiffer and coworkers [133,232]). However, clear evidence for such an effect currently seems to be lacking [194].

Regardless of whether short-timescale motions indeed affect the catalytic steps directly, it seems that they may be able to contribute to substrate turnover in other ways. For example, a study of *adenylate kinase* (EC 2.7.4.3) [239] has shown that fast (10^{-12} to 10^{-9} s) cooperative fluctuations in key hinge residues result in large-scale (10^{-6} to 10^{-3} s) motions in a loop area, which is responsible for opening the active site (see Chapter 5, Section 5.3.1.4, part III for details). This observation suggests that **enzymes can amplify local fluctuations into large-scale motions**, which may also explain how fast motions can affect catalytic events. Another way in which short-timescale motions can assist substrate turnover is by facilitating substrate binding to a buried active site, as is suggested to occur, e.g., in the enzyme *cpI Fe-Fe-hydrogenase* (EC 1.12.7.2) [245]. Molecular dynamics simulations carried out on this enzyme indicate that local 10^{-12} to 10^{-9} s fluctuations are likely to create transient voids inside the enzyme's core, which outline the substrate's path to the buried active site [245] (see Chapter 5, Section 5.3.1.4 III for details). Finally, a recent study by Vendruscolo and coworkers implicates fast motions in the product release step as well [231].

In conclusion, although protein dynamics seems to play a role in catalysis, the exact mechanism of this involvement is not entirely clear and is highly debated. The main controversy in this regard is whether dynamics directly contributes to acceleration, i.e., crossing the activation barrier of the reaction. It is interesting to note that while protein dynamics can generally increase enzymatic turnover in various ways (e.g., substrate binding or product release), its effect on barrier crossing need not necessarily be positive. For example, it has been found that a temperature-driven increase in thermally induced vibrations actually makes enzymes worse catalysts, by compromising the active-site catalytic configuration [246].

[*1]The two parameters are the *'forward commitment factor'* (i.e., the probability of the substrate to cross the energy barrier of the reaction relative to its probability to detach from the enzyme) and the *'single turnover rate'*. See [243] for further details.

9.4   ENZYME COFACTORS

9.4.1   Overview

Many protein structures contain organic molecules or inorganic ions that are important for protein activity (see Chapter 2). This is also true for enzymes [44,149,247,248]. Specifically, the use of such *cofactors* enables enzymes to carry out difficult catalytic tasks by extending their basic amino acid 'toolkit' [40,148,170]. It is customary to divide all cofactors into organic and inorganic groups. Inorganic cofactors include metal cations, which usually appear as complexes. Organic cofactors are small molecules, many of which are derived from vitamins (mainly *B-complex vitamins*). They are traditionally divided into two subgroups: *coenzymes*, which are dissociable from the enzyme (e.g., NAD(P)$^+$ [49]), and *prosthetic groups*, which are tightly bound to the enzyme at all times (e.g., FAD and FMN [249,250]) [149]. As explained in Subsection 9.3.3 above, some enzymatic active sites contain post-translationally modified amino acids that act as cofactors or prosthetic groups [146,173,174].

Interestingly, there are certain molecules that behave as cofactors in some enzyme-mediated reactions, whereas in others they behave as (co-)substrates. We have already seen this phenomenon in NAD(P)$^+$-dependent reactions (see Subsection 9.1.5.1.2 above). *S*-adenosylmethionine (SAM) also fulfills a dual role, functioning as a co-substrate in methyl-transfer reactions, and as a cofactor in reactions that involve substrate intramolecular re-arrangements (see below for further details). Additional examples of molecules showing ambiguous behavior include ATP and CoA. In most cases ATP transfers its γ-phosphate to an acceptor molecule, whereas in other cases it releases the phosphate into solution as inorganic phosphate (P$_i$) [149]. CoA behaves as a cofactor when transferring acetyl groups, but not when transferring acyl groups. This is because in the latter case the CoA's sulfur group, which catalyzes the group transfer, is not involved in the enzyme-mediated reaction [149]. Instead, the acyl group is transferred from the donor (acyl-CoA) to a cysteine residue, which eventually transfers it to the acceptor (acetyl-CoA).

A study by Thornton and coworkers [149] surveyed the known organic cofactors according to their physicochemical properties. Table 9.3 at the end of the chapter presents a summary of the common cofactors in central metabolism, together with their catalytic roles, physiological importance, and associated diseases. A more detailed description of the 27 organic cofactors characterized to date is given in the CoFactor database [248], which is maintained by the Thornton group.

Table 9.3 demonstrates the various catalytic functions carried out by cofactors. Given the chemical diversity of amino acid side chains, it seems that at least some of these functions could be fulfilled by the amino acid component of an enzyme. This notion is confirmed by studies that have characterized numerous enzymes at the mechanistic level [40,148,170]. For example, both *transketolases* and *transaldolases* act through a mechanism that involves cleavage of C−C bonds. However, while transketolases use thiamine pyrophosphate (TPP) for cleavage [98], transaldolases use a mechanism that includes a nucleophilic attack by an active site lysine residue, followed by acid-base catalysis involving an acidic residue (or residues) [251,252]. So why are cofactors used so widely by enzymes? First, some functions can only be carried out by cofactors (e.g., hydride ion shuttling). Second, **in cases in which a desired function can be carried out by either an amino acid or a cofactor, the latter usually does it more efficiently, owing to certain physicochemical properties that cofactors possess** (see more below). For example, because of their electronic configuration, and their ability to exist in various oxidation states, metals are highly efficient cofactors in electron transfer and in the activation of certain chemical species, such as molecular oxygen (see Subsection 9.3.3.3 above) [207]. The highly charged nature of cationic metals also makes them suitable for electrostatically related functions, such as stabilization and polarization of other chemical entities (protein residues, water molecules). Indeed, studies show that many of these roles are performed much more efficiently by metals than by organic cofactors or amino acid residues [30]. In organic cofactors, biological activity is usually attributable to specific chemical groups present in the cofactor.

Similarly, we saw how both PLP [62] and TPP [98] specialize in difficult bond cleavage reactions, owing to the exceptional ability of their pyridinium and thiazolium groups, respectively, to function as electron sinks (see Subsections 9.1.5.2.1 and 9.1.5.4 above). These properties enable PLP in aldolases to facilitate the opposite reaction as well, i.e., the condensation of an enol and an aldehyde. A similar strategy is used by biotin to create a C−C bond between CO_2 and a co-substrate. The nitrogen atom of biotin's ureido group polarizes the covalently bound CO_2, rendering it a good electrophilic center. This, in turn, facilitates a nucleophilic attack of the co-substrate on the CO_2 and their resulting bonding [253,254].

Several organic cofactors act as carriers of specific biologically relevant groups. For example, many of the cofactors in Table 9.3, including NAD^+ [255], FAD [249,250] and CoQ [256], specialize in carrying or transferring reducing equivalents (electrons, hydride ions). Other examples include the following:

- Lipoic acid [257] and CoA [258] – carry acyl groups.

- TPP – transfers aldehydes [98].

- Coenzyme B_{12} (CoB_{12}) [259] and SAM [260] – transfer methyl ($-CH_3$) groups.

- Tetrahydrofolate (THF) – transfers single-carbon groups[*1].

- Biotin – transfers carboxyl ($-CO_2$) groups [253,254]

(For TPP, lipoic acid and CoA, see the example of *pyruvate dehydrogenase complex* in Chapter 2, Subsection 2.1.1, as well as Figure 2.2. For CoA see example in lipid biosynthesis in Subsection 9.1.5.2.3 above.)

Finally, there are cofactors that perform chemically sophisticated feats. A classic example is the cobalt-containing coenzyme B_{12} (CoB_{12}; Figure 9.32a). *Adenosylcobalamin* (*AdoCbl*), which is one of the active forms of CoB_{12}, is used by certain *mutases* to carry out intramolecular rearrangements [259] (see Subsection 9.1.5.5 above); an intramolecular rearrangement is defined as a reaction that involves a positional change of a chemical group with an adjacent hydrogen, and that does not mix the transferred hydrogen with the solution's hydrogen atoms (Figure 9.32b). AdoCbl can fulfill this role because of the tendency of the bond between the cobalt cation (Co^{3+}) and the C5′ carbon of the adenosyl moiety to break homolytically[*2], and create a C5′ carbon radical (C•) (Figure 9.32c step 1). The Co−C5′ bond is not inherently weak, but it is rendered weak by the surrounding environment of the active site [261,262].

[*1] The single-carbon groups transferred by THF appear as formate (in N^{10}-formyl-THF and N^5, N^{10}-methenyl-THF) or formaldehyde (in N^5, N^{10}-methylene-THF and N^5-methyl-THF).

[*2] A *homolytic bond cleavage* splits the bonding electrons evenly between the products. In contrast, a *heterolytic cleavage* splits the bonding electrons unevenly.

(a)

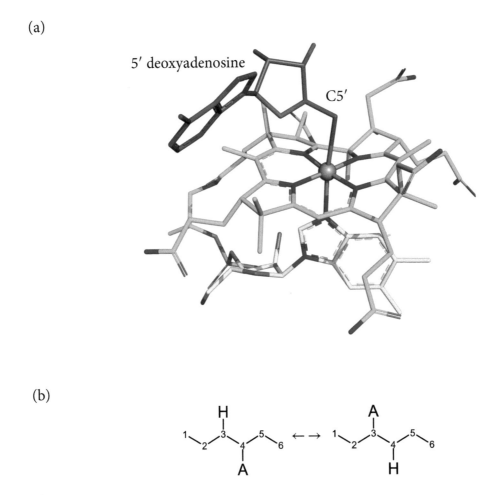

(b)

FIGURE 9.32 **Coenzyme B$_{12}$ (CoB$_{12}$).** (a) Structure of one of the forms of CoB$_{12}$, adenosylcobalamin (PDB entry 4gxy). The structure includes a corrin ring (shown as green sticks) coordinating a central cobalt cation (Co^{3+}, shown as a sphere). the Co^{3+} cation is also coordinated by two nucleotide groups above (pink) and below (yellow) the corrin ring axis, where the latter nucleotide is covalently attached to the ring also via its phosphate group. In the presented structure the group above the ring is adenosine. However, the identity of this group may be different in other forms of CoB$_{12}$ (e.g., a methyl group in methylcobalamin). The bond between Co^{3+} and this group is the reactive part of the coenzyme. (b) The molecular rearrangement reaction catalyzed by CoB$_{12}$-dependent mutases. The reactions involve a position replacement between a hydrogen and a carbon-, oxygen- and nitrogen-containing group (A), without mixing the transferred hydrogen with the hydrogen atoms that are present in the surrounding solution.

(c)

FIGURE 9.32 **Coenzyme B₁₂ (CoB₁₂).** (Continued) (c) The mechanism of the molecular rearrangement described in (b) and catalyzed by CoB_{12}-dependent mutases (see main text for details). The substrate and product of the reaction are marked by red rectangles, and the two intermediates are marked by green rectangles. The adenosyl moiety is shown with its adenine group marked as "Ad", and with the 5′ carbon colored in green. For clarity, the cobalamin ring is not shown, and the cobalt ion is presented simplicity as "Co", colored in purple.

Since radicals are unstable, the formation of the C5′ radical starts the following chain reaction, which ultimately leads to the desired intramolecular rearrangement:

1. The C5′ radical of CoB_{12} induces homolytic cleavage of the C−H bond in the substrate and abstracts the freed hydrogen atom. This leaves the substrate's carbon as a radical (C·) (Figure 9.32c, step 2).

2. The substrate's carbon radical induces a homolytic cleavage of the adjacent C−A bond and abstracts the freed A radical. This leaves the adjacent carbon as a radical (Figure 9.32c, step 3).

3. The substrate's carbon radical induces a homolytic cleavage of the CoB_{12} C5′−H bond and abstracts the resulting hydrogen radical (Figure 9.32c, step 4). This completes the H−A positional switch.

4. The C5′-cobalt bond is restored in CoB_{12} (Figure 9.32c, step 5).

In CoB_{12}-dependent enzymes that catalyze elimination reactions (e.g., *diol dehydratase* and *ethanolamine ammonia lyase*), the steps listed above create an unstable product that further undergoes elimination of the X group.

The use of coenzyme B_{12} for radical-involving reactions reveals another advantage of cofactors over protein amino acids: safety. Some enzymes include amino acid radicals that are used for specific types of catalyses. For example, *ribonucleotide reductase* (EC 1.17.4.1) [263] and *galactose oxidase* (EC 1.1.3.9) [264] use a stable form of a tyrosine radical, whereas *pyruvate formate lyase* (EC 2.3.1.54) uses a glycine radical [265]. While these amino acid radicals are important for catalysis by the enzymes, their constant presence in the active site may lead to side reactions [44]. In contrast, coenzyme B_{12} contains an organometallic structure with a stable carbon-cobalt bond. This bond ensures that the catalytic radical is generated only when needed. *S-adenosyl-methionine (SAM)*, which is also involved in radical catalysis, confers a similar advantage. As mentioned above, SAM usually serves as a methyl group donor (i.e., as a co-substrate) in biochemical reactions that do not involve formation of radicals. SAM's capacity to transfer the methyl group is due to the electrophilic nature of its carbon atom, which, in turn, results from the electron-withdrawing, positively charged sulfur atom to which it is bound (see Figure 9.33). Attack of the methyl carbon by a nucleophilic group in the other co-substrate leads to heterolytic cleavage of the S−CH_3 bond in SAM[*1] and facilitates the transfer of the methyl from SAM to the other co-substrate (Figure 9.33a). Yet, in some biochemical reactions the bond between the sulfur atom and one of the other two electrophilic carbons to which it is bound in SAM (on opposite sides) undergoes homolytic cleavage [266,267] (Figures 9.33b and c). This happens following the transfer of an electron from a reduced Fe−S cluster in the enzyme to SAM. The resulting cleavage renders the sulfur atom uncharged and generates a free radical on the cleaved carbon. Being highly reactive, the newly formed radical abstracts a hydrogen atom from the enzyme's substrate and initiates an internal, radical-mediated transformation. Thus, in such reactions SAM functions as a true cofactor and not as a co-substrate.

[*1]The heterolytic cleavage leaves all the electrons of the S−CH_3 bond on the sulfur atom.

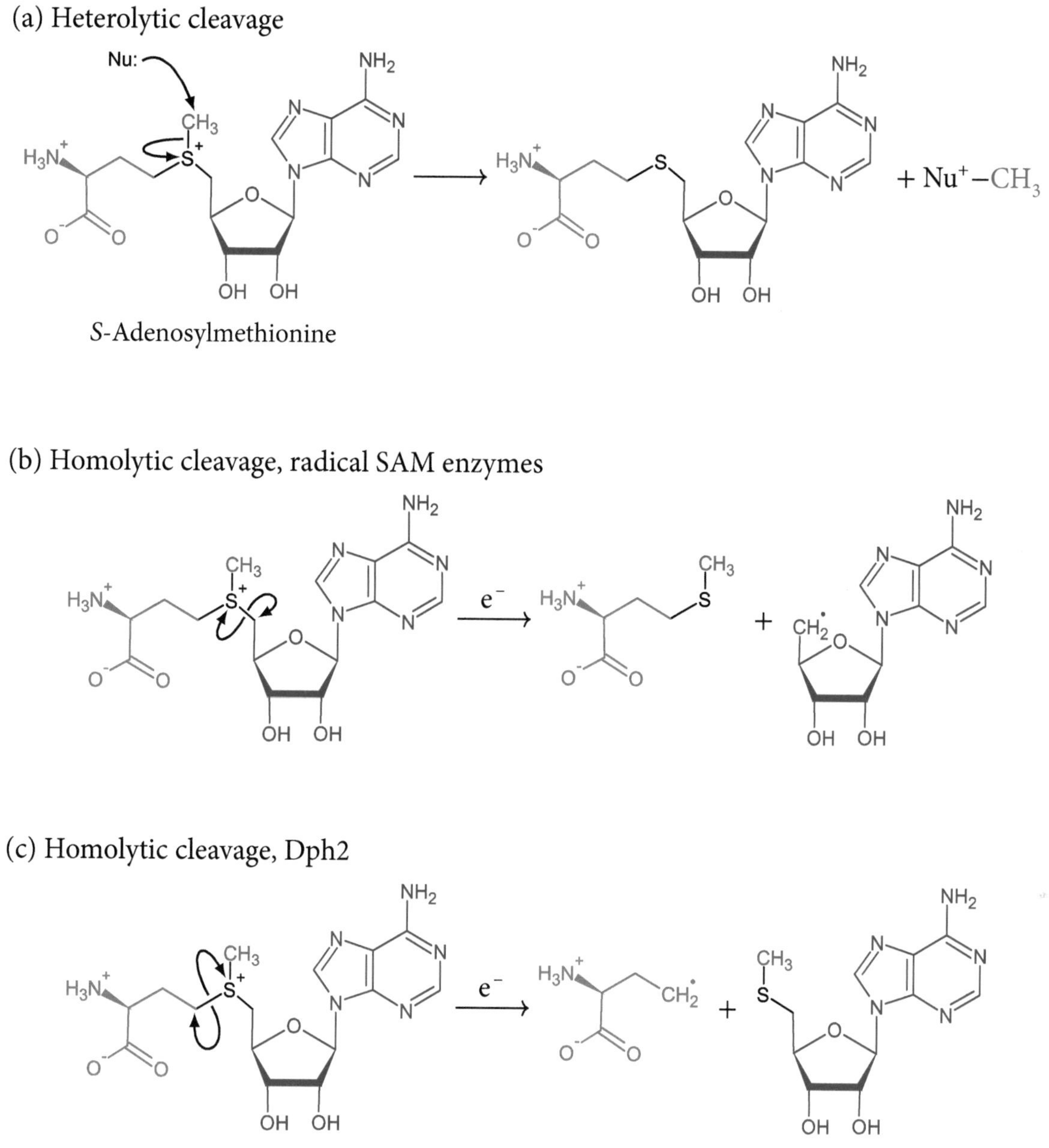

(a) Heterolytic cleavage

S-Adenosylmethionine

(b) Homolytic cleavage, radical SAM enzymes

(c) Homolytic cleavage, Dph2

FIGURE 9.33 **Main reaction modes of S-adenosylmethionine (SAM).** SAM includes a positively charged sulfur atom, bound to a methyl group (red), a 5′-deoxyadenosyl group (blue) and a 3-amino-3-carboxypropyl group (green). (a) Heterolytic cleavage of the sulfur-methyl group, allowing SAM to function as a methyl donor. Curly arrows indicate the movement of pairs of electrons. Nu represents a nucleophile (an electron-pair donor). (b) Homolytic cleavage of the bond between the sulfur and the 5′-deoxyadenosyl group, which happens in radical SAM enzymes. In these reactions, SAM accepts an electron (e^-), whereupon the S−C bond breaks so that one electron ends up on each of the atoms of the bond. Fish-hook arrows indicate the movement of single electrons. (c) Homolytic cleavage of the bond between the sulfur atom and the 3-amino-3-carboxypropyl group, as occurs in the enzyme Dph2 [268]. The image is adapted from [266].

9.4.2 Chemical characteristics of organic cofactors

It stands to reason that organic cofactors, which are highly reactive, possess certain features that lead to this reactivity. A survey carried out by Thornton and coworkers shows organic cofactors to be overall chemically similar to biological metabolites, such as nucleotides, amino acids and fatty acids [149]. Here are some examples:

- Like $NAD(P)^+$ and FAD (Figure 9.5), SAM and CoA each contain an adenosyl unit (Figure 9.34a).

- SAM also contains a methionine residue.

- CoA also contains β-alanine and a decarboxylated cysteine residue.

- Glutathione is built from the amino acids glutamate, cysteine and glycine (Figure 9.34a).

- THF, FAD and FMN contain the pteridine ring system, which is synthesized from guanosine-5'-triphosphate (GTP) (Figure 9.34b).

While organic cofactors are similar to biological metabolites, they tend to be more enriched in functional groups, a characteristic that explains their higher reactivity. We have already encountered several examples of such groups and their biochemical activity:

- Pyridinium (PLP), thiazolium (TPP)[*1], and ureido (biotin)[*2] – bond polarization.

- Nicotinamide (NADH, NADPH) – hydride ion transfer.

- Isoalloxazine ($FADH_2$ and $FMNH_2$) – electron transfer.

- Cobalt-containing organometallic group (CoB_{12}) and sulfur-containing group (SAM) – radical-involving reactions, methyl group transfer.

- Thiol (CoA, lipoic acid) – acyl transfer.

- Pteridine (THF) – single-carbon group transfer.

Interestingly, many cofactors, including lipoic acid, SAM, TPP, CoA, and glutathione, include sulfur in their active groups. This is not surprising considering the reactivity of this atom and its direct involvement in nucleophilic catalysis (as in CoA-induced *thiolysis*), free radical chemistry (as in SAM reactions), and the activation of metabolites via thioester bonds (as in lipoic acid and CoA; see below). As explained above, the sulfur atom in TPP acts indirectly, by withdrawing electrons and acidifying an adjacent carbon, which in turn leads to the carbon's deprotonation and allows it to act as nucleophile. Similarly, the positively charged sulfur in the sulfonium group of SAM withdraws electrons from its neighboring carbon atoms, turning them into electrophilic centers. This facilitates the breaking of one of the S−C bonds (depending on the reaction type), which may be carried out in two different ways: (1) a heterolytic cleavage caused by a co-substrate nucleophile, and which results in methyl transfer from SAM to the co-substrate, or (2) homolytic cleavage caused by electron transfer within the enzyme, and which results in the generation of free radicals and in intramolecular rearrangements in the substrate [266,267] (see Figure 9.33 above).

[*1] Also functions in aldehyde transfer.
[*2] Also functions in CO_2 group transfer.

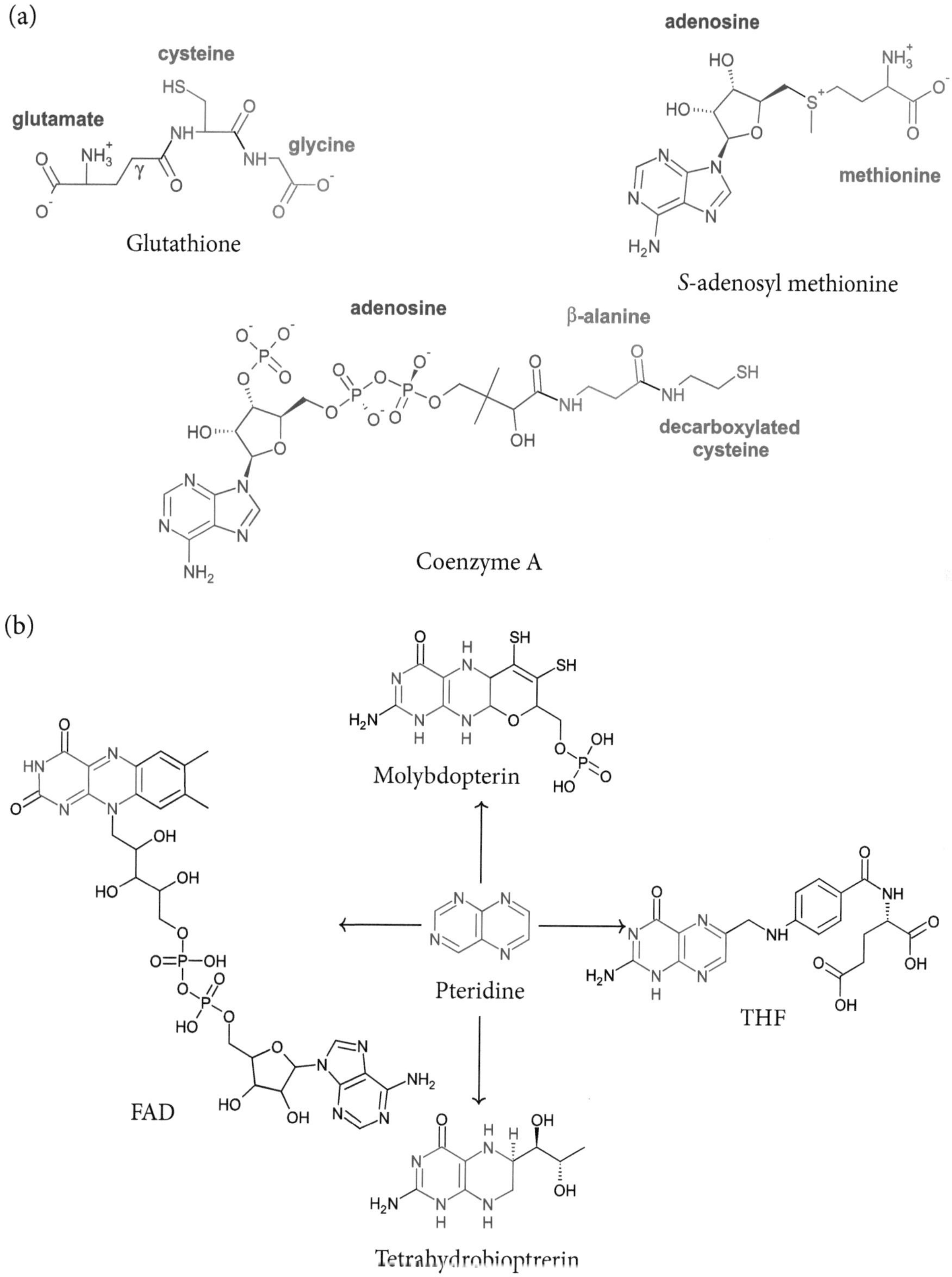

(a)

Glutathione

S-adenosyl methionine

Coenzyme A

(b)

Molybdopterin

Pteridine

THF

FAD

Tetrahydrobiopterin

FIGURE 9.34 **Similarity between organic cofactors and natural metabolites.** (a) Amino acids in the structures of the cofactors glutathione, SAM and CoA (colored). Glutathione is a tripeptide built from glutamate bound via its γ carboxyl to cysteine and glycine. SAM contains methionine, and CoA contains β-alanine and decarboxylated cysteine moieties. (b) The pteridine ring system (in blue) in the structures of the cofactors FAD, THF, molybdopterin and tetrahydrobiopterin.

We saw earlier that CoA is the principal donor in acetyl and acyl transfer reactions. It is often said that CoA 'activates' the acyl group by forming a thioester bond with it, thereby enabling subsequent condensation to take place between the group and the acceptor molecule. But what is the molecular basis for such 'activation'? One suggestion is that the activation results from the lower stability of thioesters compared with oxygen esters. An oxygen ester constantly interconverts between two states due to electronic resonance, which leads to its stabilization (Figure 9.35a). A thioester, in contrast, does not experience electronic resonance and therefore exists in a single state (Figure 9.35b). This means that a thioester is less stable, or has more energy, than the oxygen ester.

In an acyl transfer reaction, the breaking of the ester bond between the acyl group and its donor releases energy that is simultaneously used for creating the bond between the acyl group and the acceptor molecule. The released energy is in fact the difference between the free energy of the ester and that of its degradation products ($\Delta G = G_{ester} - G_{products}$). Assuming that the free energy of the products in both cases is similar, then the breaking of the oxygen ester bond should release less energy than the breaking of the thioester bond ($\Delta G_1 < \Delta G_2$ in Figure 9.36), since the oxygen ester has less energy than the thioester. This means that the breaking of a thioester bond should make subsequent condensation easier than should the breaking of an oxygen ester bond.

Another chemical characteristic of organic cofactors that renders them reactive is their tendency to be somewhat larger and much more polar than the average biological metabolite [149]. As such, organic cofactors tend to have more hydrogen bonding groups, and this contributes to their ability to interact with other species.

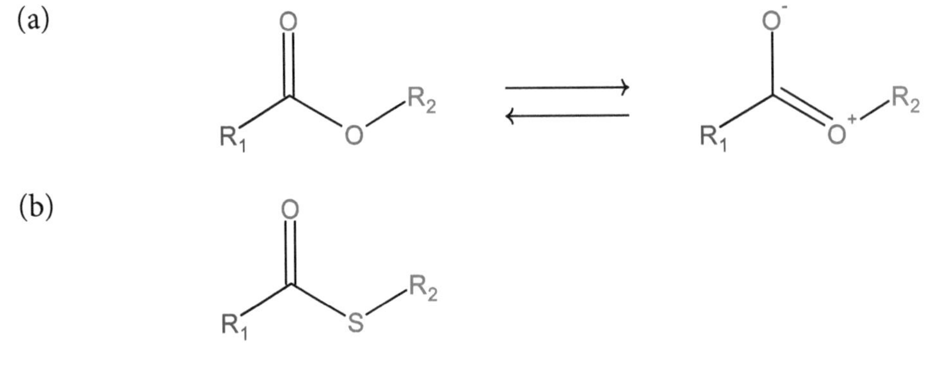

FIGURE 9.35 **Electronic resonance displayed by oxygen esters (a) but not by thioesters (b).**

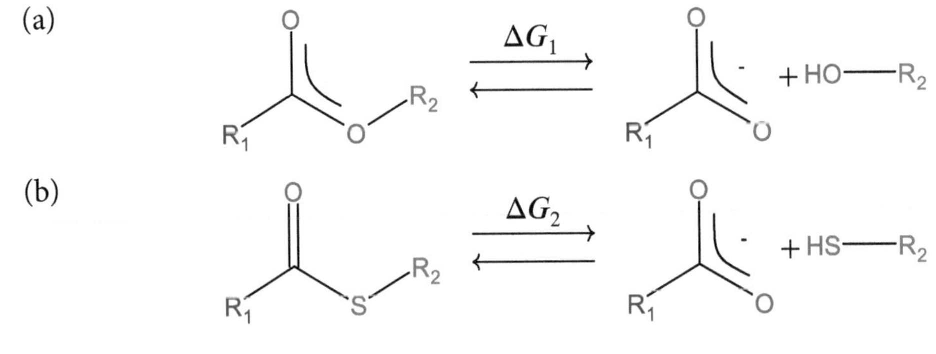

FIGURE 9.36 **Cleavage reaction of oxygen esters (a) and thioesters (b).**

9.4.3 Functional characteristics

Thornton's study also examined the pervasiveness of organic cofactors among different types of enzymes, as well as their most common functions [149]. The study showed these cofactors to have different functions, including bond cleavage and formation, group transfer, redox, intramolecular rearrangements, substrate mobilization within the active site, and polymerization. In particular, the cofactors were found to be most common in redox catalysis and in group transfer [149]. The pervasiveness of organic cofactors in redox catalysis may originate in part from the inherent unsuitability of most amino acids to this type of reaction[*1] [44]. The high prevalence of NAD(P)$^+$ and FAD compared with other organic cofactors used by oxidoreductases is not surprising, as many of the reactions catalyzed by oxidoreductases (e.g., aldehyde-to-alcohol reduction) require the transfer of two electrons and a proton.

Compared with redox catalysis and group transfer, the functions of hydrolysis and substrate stabilization or activation are much less likely to be fulfilled by organic cofactors. These roles can be efficiently carried out both by amino acid residues and by metal ions, which reduces the need for organic cofactors. In the case of hydrolytic reactions, enzymes that carry out these catalyses usually employ relatively simple and short-duration mechanisms, which, again, render organic cofactors less important for this role.

9.5 ENZYME INHIBITION

9.5.1 Overview

Enzyme inhibition refers to a decrease in enzymatic activity, induced by the specific binding of chemical species[*2]. This field is of great biological and pharmaceutical importance. First, cells and tissues use different inhibitory metabolites as a means of regulating their numerous enzymes [269]. This regulation allows cells and tissues to control the rates of their biochemical pathways and subjugate the pathways to the environmental conditions and cellular needs. ATP, for example, is a naturally used inhibitor of the key glycolytic enzyme *phosphofructokinase-1* (PFK-1, EC 2.7.1.11). When the cell has sufficient energy for its biological processes, ATP inhibits energy-producing processes such as glycolysis and the Krebs cycle. The elaborate regulation of PFK-1 is described in detail in Subsection 9.5.2.1.2 below. Enzyme inhibition is also the mechanism used by certain environmental toxins, to which living organisms, especially unicellular microorganisms, may become exposed. We have already encountered two examples of such toxins: arsenic, which targets lipoic acid-dependent enzymes (e.g., *pyruvate dehydrogenase*, see Box 2.1), and organophosphates, which target *acetylcholinesterase* in the nervous system (Box 8.1).

Enzyme inhibition is also important for scientific research; by using specific inhibitors, scientists can block enzymes that participate in a given biochemical pathway, which enables them to characterize the pathway and the relative importance of each enzyme it comprises. Scientists also use inhibitors to gain a better understanding of the mechanisms of individual enzymes, by neutralizing residues that participate in specific steps [270]. Finally, enzyme inhibitors have industrial importance. In the agricultural industry, enzymatic inhibitors are used as pesticides and herbicides, and in the pharmaceutical industry they are used as ther-

---

[*1]Cysteine is an exception to this rule, as it can relatively easily shift between the reduced (thiol) and oxidized (disulfide) states.

[*2]Not included in this definition are chemicals that decrease enzymatic activity non-specifically, e.g., by inducing partial or complete denaturation of the enzyme.

apeutic drugs (see Chapter 8, Subsection 8.6 for a detailed discussion). Indeed, enzymes are major drug targets [271], and most of the drugs targeting them act by inhibition of enzyme activity. Enzyme inhibitors constitute 25% of the drug market, with the majority of their targets being hydrolases and oxidoreductases [272]. Well-known examples of enzyme-inhibiting drugs are discussed in Chapter 8, and include the following [273]:

- Non-steroidal anti-inflammatory drugs (NSAIDs) that act on the enzyme *cyclooxygenase* (EC 1.14.99.1).

- ACE inhibitors that act on the *angiotensin-converting enzyme* (EC 3.4.15.1).

- MAO inhibitors that act on the enzyme *monoamine oxidase* (EC 1.4.3.4).

- HIV protease (EC 3.4.23.16) inhibitors.

- *β*-lactam antibiotics, most of which act on the bacterial cell-wall synthesizing enzyme *DD-transpeptidase* (EC 3.4.16.4) (see more details in Subsection 9.5.2.2 below).

- Anticoagulants that act on thrombin (EC 3.4.21.5) and factor Xa (EC 3.4.21.6) of the blood clotting cascade.

- Parasympathomimetics that act on *acetylcholinesterase* (EC 3.1.1.7).

Enzyme-inhibiting drugs may bind their targets either covalently or noncovalently. This characteristic distinguishes them from most other drugs, which rely solely on noncovalent binding to their target sites.

Due to the prevalence of enzymes in metabolic pathways, it is not unusual to find drugs that act on multiple different enzymes in a single pathway. This is observed, for example, in drugs developed to target the *angiotensin* biosynthetic pathway (see Chapter 8, Subsection 8.6.3.4.2). Such drugs are used to treat pathologies that cause a rise in blood pressure, which, in turn, increases the chances of suffering a stroke, aneurysm, heart attack, and severe kidney problems. These drugs act by inhibiting *ACE* (angiotensin converting enzyme), an enzyme that creates the active form of the blood pressure-elevating hormone angiotensin. The inhibition often induces a physiological compensation response that involves the overproduction of *renin*, a protease responsible for creating the precursor to the angiotensin active form. To thwart this side effect, *aliskiren*, which inhibits renin, was developed [274] (see more details in Subsection 9.5.2.1.1 below).

9.5.2 Modes of enzyme inhibition

Enzyme inhibition may be carried out in different ways. As we saw earlier, the binding between enzymes and their substrates is reversible, due to the delicate balance between the binding forces. Thus, the substrate is bound to its cognate enzyme for only part of the time. Some inhibitors take advantage of this situation; they inhibit the enzyme by displacing the natural substrate from the active site via competition. Other inhibitors act by binding to a different site on the enzyme, inducing an allosteric effect. Inhibitors differ from each other also in the strength of binding; whereas some inhibitors bind to their enzymes covalently, others interact with them noncovalently. Interestingly, both types of inhibitors may inhibit enzymatic activity in either a reversible or an irreversible manner, depending on the biological context. Below we provide a short summary of the main types of enzyme inhibition.

9.5.2.1 Reversible inhibition

Reversible inhibitors usually bind their enzymes noncovalently[*1] **and inhibit enzymatic activity without taking part in the reaction itself** [275]. The reversible nature of the inhibition results from the moderate binding affinity between the enzyme and the inhibitor, which is on the same scale as the affinity between the enzyme and its natural substrate(s). Thus, reversible inhibitors are in constant equilibrium between the bound and unbound states. Reversible inhibition can take different forms. For example, the inhibitor may bind to the enzyme's active site or to an allosteric site that is far from where catalysis takes place. The inhibitor may act by changing the enzyme's affinity to its natural substrate or by reducing the enzyme's catalytic efficiency. The various subtypes of reversible inhibition are elaborated in the subsections below (see also Table 9.1).

9.5.2.1.1 Competitive inhibition

In competitive inhibition, the inhibitor acts by competing with the natural substrate[*2] **for binding to the active site**[*3] (Figure 9.37a). Since the inhibitor and natural substrate cannot occupy the same active site simultaneously, each successful binding of the inhibitor prevents the natural substrate from binding. Thus, the *apparent affinity* of the inhibited enzyme to its natural substrate is lower than that of the uninhibited enzyme. Indeed, the K_M of the enzyme, which under Michaelis-Menten conditions represents the inverse of affinity, is elevated in enzymes inhibited by competitive inhibitors:

$$K_{M(app)} = K_M \left(1 + \frac{[I]}{K_I}\right) \tag{9.18}$$

(where [I] is the molar concentration of the inhibitor, K_I is its dissociation constant, and the ratio between them determine the fraction increase in $K_{M(app)}$ compared to K_M).

The rate of the enzymatic reaction under competitive inhibition can be described by replacing K_M with $K_{M(app)}$ in the Michaelis-Menten equation (Equation (9.6)):

$$V_0 = \frac{V_{max}[S]}{K_{M(app)} + [S]} = \frac{V_{max}[S]}{K_M \left(1 + \frac{[I]}{K_I}\right) + [S]} \tag{9.19}$$

[*1]In cells, some molecules that bind covalently to enzymes may still function as reversible inhibitors, as long the cell has means to remove them quickly. For example, the attachment of phosphoryl groups to enzymes (i.e., phosphorylation) is a common and efficient way for cells to mediate signal transduction events (see Subsection 9.1.5.2.2). While the binding between enzyme and phosphoryl group is strong, the cellular availability of numerous protein phosphatases allows for quick removal of the phosphoryl groups from the target enzymes.

[*2]The inhibitor may also compete with the enzyme's natural cofactor. Since this case is less common, we will refer from now on only to substrate competition.

[*3]It should be noted that certain inhibitors that act on other sites of the enzyme (allosteric inhibitors, see next subsection) may also appear to act competitively, although they act differently from the competitive inhibitors described above. Such allosteric inhibitors bind to a distant site and induce a conformational change in the active site, which blocks the active site and prevents substrate binding. Since the substrate and inhibitor bind to different conformations, binding of one precludes the binding of the other, which may resemble competition in terms of measurement outcomes. However, the two molecules do not actually compete for the same site.

The corresponding Lineweaver-Burk (double reciprocal) equation takes the following form (see also Figure 9.37b for graphic representation):

$$\frac{1}{V_0} = \frac{K_M}{V_{max}} \left(1 + \frac{[I]}{K_I} \right) \frac{1}{[S]} + \frac{1}{V_{max}} \qquad (9.20)$$

As in any inhibitor, the potency of a competitive inhibitor depends on the inhibitor's relative affinity compared to that of the natural substrate. However, since competitive inhibitors are reversible and compete with the natural substrate, their potency also depends on their relative concentration. That is, when the concentration of the natural substrate is high enough (i.e., $[S] \gg K_M$ and $V_0 = V_{max}$), the substrate will successfully compete with the inhibitor and the inhibition will be lifted. This means that **while competitive inhibition raises the K_M of the enzyme, it does not change its V_{max}** (Figure 9.37b). This is highly important for inhibition in the cellular environment, where the concentrations of some metabolites are very high. For example, the concentration of ATP in the cell is usually much higher than the K_M of most kinases, which challenges any inhibitor targeting these enzymes [276]. As explained in Chapter 8, Subsection 8.6.2.1.2, the susceptibility of reversible inhibitors to substrate concentration is a major drawback in their use as enzyme-inhibiting drugs.

The competition between the reversible inhibitor and the natural substrate of the enzyme results from the capacity of the inhibitor and the substrate to interact with the enzyme's active site in a similar manner. This capacity usually results from chemical and structural similarities between the inhibitor and the substrate. In fact, in some cases the inhibitor may resemble the reaction's product or transition state. Inhibitors that resemble the transition state, called *transition state analogues*, are generally more potent than inhibitors that resemble the ground state substrate or product of the reaction [14]. This is to be expected, considering that the active site of the enzyme is better suited for the transition state than for the ground state, and binds the transition state with considerably higher affinity (see Subsection 9.3.3.2 above). In some cases the binding energy between the enzyme and a transition state analogue is so high that the binding, despite being noncovalent, is effectively irreversible.

Since different enzymes may share the same substrate, but not the same transition state (which depends on the catalytic path), transition state analogues are also more specific than ground state analogues [14]. Note, however, that competitive inhibitors do not necessarily have to resemble the natural substrate, product or transition state of the enzyme; they just have to interact favorably with the active site [277]. Inhibitor-substrate resemblance tends to exist in cases involving active sites that are highly specific towards their natural substrates. This is because molecules whose chemical scaffolds differ from those of the substrate are not likely to fit into such sites. In contrast, active sites that are roomy and naturally interact with different substrates are likely to accommodate different types of competitive inhibitors as well, some of which might not resemble any of the natural substrates.

Reversible competitive inhibitors can be found in the natural world. For example, *2,3 bisphosphoglycerate* is a natural metabolite that inhibits *bisphosphoglycerate mutase* (EC 5.4.2.4), the same enzyme that creates it from 1,3 bisphosphoglycerate [278]. This is an example of *product inhibition* (although this mode of inhibition usually involves other types of inhibitors[*1]). Another example of a reversible competitive inhibitor is *malonate*, a

---

[*1]Product inhibition usually involves inhibitors that act on allosteric sites and do not compete with the substrate on binding to the enzyme. See more details about product inhibition in Subsection 9.5.2.1.2 below.

naturally occurring molecule that inhibits the Krebs cycle enzyme *succinate dehydrogenase* (EC 1.3.5.1) [279]. Both succinate (the natural substrate) and malonate are dicarboxylic acids. They differ only in the number of carbon units (four and three, respectively), and their similarity is what allows them to compete (Figure 9.37c, top).

Reversible competitive inhibitors are also used by the pharmaceutical industry [277]. A famous example is *sulfonamides*, antibacterial agents that halt the growth of bacteria by preventing them from efficiently producing *folic acid* [280]. Specifically, these drugs inhibit *dihydropteroate synthase* (EC 2.5.1.15), a key enzyme in the bacterial pathway for folic acid synthesis. They do so by competing with the enzyme's natural substrate, *para-aminobenzoic acid* (*PABA*) (Figure 9.37c, bottom). Additional examples include the NSAIDs *ibuprofen* and *diclofenac*, which inhibit cyclooxygenase (see Chapter 8, Subsection 8.6.2.1.2 for details).

As mentioned above, competitive inhibitors that resemble the transition states of the reactions they compete with are usually highly efficient. The use of such drugs started in the 1970s and included mostly natural compounds, but since the 1990s synthetic inhibitors have been in use as well [281].

An example of such an inhibitor is the anti-influenza drug *oseltamivir* (*Tamiflu®*), which acts by inhibiting a viral enzyme called *neuraminidase* (EC 3.2.1.18) [282]. This enzyme (a.k.a. *sialidase*) is a glycoside hydrolase anchored to the membrane of the influenza virus. By cleaving the glycosidic bond between terminal sialic acid (*N*-acetylneuraminic acid) and glycoconjugates on the membrane of the host cell, neuraminidase assists in the movement of the virus through the upper respiratory tract, and in the release of new virions from infected cells [283]. The cleavage reaction involves a transition state with an oxonium ring containing a double bond (Figure 9.37d, top).

Oseltamivir contains a cyclohexenyl ring whose structural properties resemble those of a pyranium ring in the sialic acid transition state (e.g., the intra-ring double bond), and this similarity enables the drug to compete with sialic acid for binding to the enzyme's active site (Figure 9.37d, bottom).

Another example of a successful transition state analogue is the antihypertensive drug *aliskiren*, which acts by inhibiting the enzyme *renin* (EC 3.4.23.15) (see Chapter 8, Subsection 8.6.3.4.2 for details on the physiological role of renin). As explained in the Overview above, antihypertensive drugs commonly belong to the ACE inhibitor group, but their use often leads the patient's body to overproduce renin, as a compensatory response. Aliskiren was developed to combat this phenomenon. Renin breaks a peptide bond in its substrate (angiotensinogen), in a reaction that involves a tetrahedral intermediate [284] (Figure 9.37e, top). Aliskiren contains a group that resembles a peptide bond, but instead of a planar carbonyl group it contains a tetrahedral C-hydroxyl group that resembles the reaction's transition state [277] (Figure 9.37e, bottom). Furthermore, unlike the peptide bond in the substrate, the hydroxyl-containing group on aliskiren is inactive; it cannot undergo hydrolysis. This makes aliskiren stay longer in the active site, which increases its potency as inhibitor.

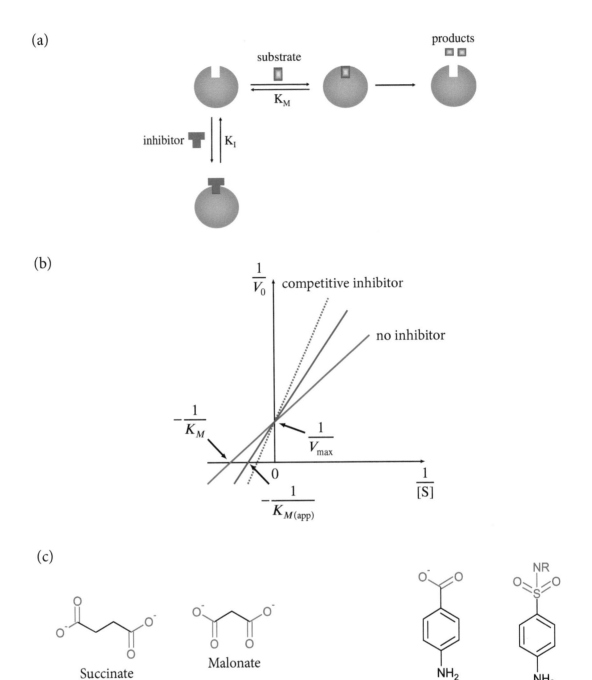

FIGURE 9.37 **Competitive inhibition.** (a) A schematic representation of the inhibition. The substrate competes with the structurally similar inhibitor for binding to the active site of the enzyme (green shape). K_I is the dissociation constant of the inhibitor from the substrate-free enzyme. (b) The Lineweaver-Burk (double reciprocal) plot of the inhibition (Equation (9.20)). The green lines designate the inhibited enzyme, with the dashed line representing the plot under higher inhibitor concentration or lower K_I. The plot shows the apparent K_M of the inhibited reaction to be higher than the K_M of the uninhibited reaction, with no change in V_{max}. (c) Examples of the structural similarity between reversible competitive inhibitors and the natural substrates of their target enzymes. *Left*: malonate and succinate. *Right*: sulfonamide and *para*-aminobenzoic acid (PABA).

(d)

Bound sialic acid   TS-like intermediate   Free sialic acid

Oseltamivir
(TS analogue)

(e)

Substrate   Tetrahedral intermediate   Products

Aliskiren (TS analogue)

FIGURE 9.37 **Competitive inhibition.** (Continued) (d) and (e) Examples of reversible transition-state analogues. As in Figure 9.25, for clarity we present transition state-like intermediates instead of a full representation of the reaction's transition state. (d) Oseltamivir (Tamiflu®). *Top*: the mechanism of terminal sialic acid cleavage from the host membrane by neuraminidase (the circled *OR* designates the attachment point of sialic acid to the rest of the glyco-conjugate). *Bottom*: the chemical structure of oseltamivir, a competitive inhibitor of neuraminidase. The arrow designates the double bond that exists both in the substrate's transition state and in oseltamivir. (e) Aliskiren. *Top*: the mechanism of peptide bond cleavage by renin. In the first step a polarized water molecule attacks the carbonyl group of the peptide bond (marked by the red arrow). This, with the simultaneous protonation of the carbonyl oxygen by an active site aspartate, creates a tetrahedral intermediate. In the second step the amide group of the peptide bond is protonated by a second active site aspartate, and the bond is cleaved. *Bottom*: the chemical structure of aliskiren. The peptidomimetic bond is marked by the red arrow.

9.5.2.1.2 Uncompetitive, non-competitive, and mixed inhibition

Reversible enzyme inhibition does not necessarily have to target the active site; certain inhibitors bind to other sites in the enzyme (allosteric or others[*1]) and stabilize enzyme conformations that have reduced affinity to the substrate and/or reduced catalytic activity[*2]. In both cases, the inhibition is not carried out via competition, and therefore the inhibitor need not resemble the natural substrate of the enzyme chemically or structurally. Moreover, since the inhibitor does not bind to the active site, the inhibition is mostly unaffected by raising substrate concentration, and depends only on the concentration of the inhibitor[*3]. This type of inhibition is very common in metabolic biosynthetic pathways, where the final product of the pathway halts its own production when its levels become sufficient for cellular needs (i.e., product inhibition). In such cases, the biosynthetic product allosterically inhibits a key enzyme of the pathway. This can be seen, e.g., in the multi-step biosynthesis of L-isoleucine from L-threonine [285]. The end product of this process, L-isoleucine, controls its own production by allosterically inhibiting the first enzyme of the pathway, *L-threonine ammonia-lyase* (EC 4.3.1.19; a.k.a. L-threonine deaminase).

Allosteric control over metabolic enzymes is not solely the domain of inhibitors. **Many enzymes, especially those that play central roles in metabolic pathways, are upregulated by allosteric activators.** The mode of action of these regulators is basically the same as that of allosteric inhibitors, in the sense that both bind to allosteric sites on the enzyme and stabilize one or more of the enzyme's conformations. The difference is that activators stabilize active conformations, whereas inhibitors stabilize inactive ones. An active conformation in this context may be one that has higher activity, or binds the natural substrate with greater affinity, compared with other conformations. A well-known example of allosteric activation is that of the key metabolic enzyme phosphofructokinase-1, which is discussed in detail below.

Reversible inhibition that acts on non-active sites in enzymes can be further divided into two types, uncompetitive and non-competitive inhibition, as elaborated in what follows.

9.5.2.1.2.1. Uncompetitive inhibition

An uncompetitive inhibitor binds only to the substrate-bound form of the enzyme and stabilizes a bound, non-catalytic conformation (i.e., a conformation that does not lead to catalysis) (Figure 9.38a). As a result of this process, the substrate cannot leave the active site, but it also cannot undergo catalysis. This has two effects. First, the stabilization of the enzyme-substrate complex by the inhibitor shifts the $E + S \longleftrightarrow ES$ equilibrium to the right (according to Le Châtelier's principle), which can be viewed as an increase in the enzyme's affinity to the substrate. In other words, **the apparent substrate affinity of the inhibited enzyme is higher than that of the uninhibited enzyme, which means the apparent K_M is**

[*1]Generally speaking, any site (other than the active site) that affects the activity of the enzyme via conformational changes or stabilization of a specific conformation can be regarded as an allosteric site (see Chapter 5, Subsection 5.3.2.1). However, classical allosteric sites usually reside in specific 'pre-designed' domains or subunits of complex proteins (e.g., see Figure 8.1b), whereas inhibitors may also bind to any other non-active site in either a multi-subunit or a single-subunit enzyme.

[*2]Note that, in essence, irreversible inhibitors may act allosterically as well, since they, too, are able to stabilize less-active conformations by binding to allosteric sites.

[*3]A partial exception to this rule is reversible mixed inhibition (see below).

smaller:

$$K_{M(\text{app})} = \frac{K_M}{\left(1 + \dfrac{[I]}{K_I}\right)} \qquad (9.21)$$

Second, since the sequestered substrate cannot turn into a product, **the maximal rate of the enzyme also decreases,** $(V_{\max(\text{app})} < V_{\max})$, **by the same factor:**

$$V_{\max(\text{app})} = \frac{V_{\max}}{\left(1 + \dfrac{[I]}{K_I}\right)} \qquad (9.22)$$

The rate of the enzymatic reaction under uncompetitive inhibition can be described by replacing K_M with $K_{M(\text{app})}$ and $V_{\max}$ with $V_{\max(\text{app})}$ in the Michaelis-Menten equation:

$$V_0 = \frac{V_{\max(\text{app})}[S]}{K_{M(\text{app})} + [S]} = \frac{V_{\max}[S]}{K_M + [S]\left(1 + \dfrac{[I]}{K_I}\right)} \qquad (9.23)$$

The corresponding Lineweaver-Burk (double reciprocal) equation takes the following form (see also Figure 9.38b for graphic representation):

$$\frac{1}{V_0} = \frac{K_M}{V_{\max}}\frac{1}{[S]} + \frac{1}{V_{\max}}\left(1 + \frac{[I]}{K_I}\right) \qquad (9.24)$$

Uncompetitive inhibition is relatively rare [286]. Known examples include the inhibition of *inositol monophosphatase* (EC 3.1.3.25) by *lithium*, a mood-stabilizing drug used to treat bipolar disorders [287], and the inhibition of *3-phosphoshikimate 1-carboxyvinyltransferase* (EC 2.5.1.19) by the herbicide *N-phosphonomethylglycine* (a.k.a. *glyphosate*, trade name *Roundup*) [288].

9.5.2.1.2.2. Non-competitive and mixed inhibition

Some reversible inhibitors bind to either the substrate-free or the substrate-bound enzyme, where they stabilize less-active conformations, thus reducing $V_{\max}$ (Figure 9.39a):

$$V_{\max(\text{app})} = \frac{V_{\max}}{\left(1 + \dfrac{[I]}{K_{I(b)}}\right)} \qquad (9.25)$$

(where $K_{I(b)}$ is the dissociation constant of the inhibitor from the substrate-bound enzyme).

In the simplest case of such inhibition, termed 'non-competitive inhibition', $K_{I(b)}$ is equal to the inhibitor's dissociation constant from the substrate-free enzyme $K_{I(f)}$. In this case, the rate of the reaction can be described as:

$$V_0 = \frac{V_{\max(\text{app})}[S]}{\left(K_M + [S]\right)} = \frac{V_{\max}[S]}{\left(1 + \dfrac{[I]}{K_I}\right)\left(K_M + [S]\right)} \qquad (9.26)$$

(where $K_I = K_{I(b)} = K_{I(f)}$).

The corresponding Lineweaver-Burk equation takes the following form (see also Figure 9.39b for graphic representation):

$$\frac{1}{V_0} = \frac{K_M}{V_{max}}\left(1 + \frac{[I]}{K_I}\right)\frac{1}{[S]} + \frac{1}{V_{max}}\left(1 + \frac{[I]}{K_I}\right) \qquad (9.27)$$

In more complex cases, termed 'mixed inhibition', $K_{I(b)}$ and $K_{I(f)}$ have different values.

This means that the affinity of the inhibitor to the enzyme is either raised or reduced by the preceding binding of the natural substrate, despite the fact that the substrate and the inhibitor bind to different sites. Making things even more complicated, it has been argued that in certain cases non-competitive inhibition may arise from the action of inhibitors that bind the active site [289]. These cases are observed, for example, in proteases, which act on large substrates that bind to more than one site at the same time (exosites), as well as in enzymes in which the catalytic cycle involves several enzyme conformations, where one conformation binds the substrate while another binds the inhibitor [289]. However, such cases are the exception, not the rule.

Finally, a non-competitive inhibitor may act by stabilizing an enzyme conformation in which the active site is too distorted to bind the substrate. This happens, e.g., in the glycolytic enzyme *phosphofructokinase-1* (PFK-1; EC 2.7.1.11). As mentioned earlier, PFK-1 phosphorylates fructose 6-phosphate (F6P) to fructose 1,6-bisphosphate (F1,6BP) using the γ-phosphate group of ATP, which turns into ADP in the process [290]. The animal form of PFK-1 is one of the most highly regulated enzymes known in metabolism, with multiple regulators that include both activators and inhibitors [291]. In muscles, PFK-1 is regulated

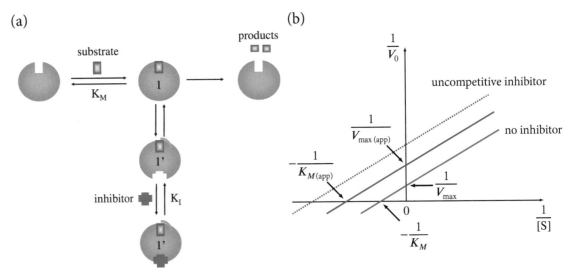

FIGURE 9.38 **Uncompetitive inhibition.** (a) A schematic representation of the inhibition. The inhibitor binds only to a substrate-bound form of the enzyme. It binds to a conformation that cannot release the substrate or turn it into a product (1'). This conformation is in equilibrium with the 'normal' substrate-bound conformation, which can do both (1). Inhibitor binding stabilizes conformation 1' and therefore leads to enzyme inhibition. The binding also shifts the equilibrium between the substrate-free and the substrate-bound forms of the enzyme to the right, therefore elevating the apparent affinity of the enzyme to its substrate. (b) The Lineweaver-Burk plot of the inhibition (Equation (9.24)). The green lines designate the inhibited enzyme under two inhibitor concentrations or inhibition constants, as described in Figure 9.37a. The plot shows both the apparent K_M and V_{max} of the inhibited reaction to be lower than in the uninhibited reaction.

primarily by cellular energy levels, as represented by the relative concentrations of ATP, its degradation products ADP and AMP, and the Krebs cycle metabolite citrate. The regulation process is as follows:

- **ATP** – high ATP levels represent high energy levels in the cell. When ATP concentrations are high, ATP binds to an allosteric site in PFK-1, and the binding stabilizes a conformation of the enzyme that has a reduced ability to bind the other co-substrate, F6P.

- **ADP and AMP** – when the cell has used sufficient amounts of ATP, the levels of its degradation products, ADP and AMP, rise. These molecules then relieve ATP inhibition by binding to the substrate-binding conformation and stabilizing it.

- **Citrate** – when the Krebs cycle is saturated, citrate levels go up. Citrate binds to an allosteric site in PFK-1 and acts similarly to ATP. The logic behind this is simple; The Krebs cycle is constantly fed with acetyl-CoA, a molecule that is formed directly from the glycolytic product pyruvate. Thus, saturation of the cycle means that glycolysis must be slowed down.

The most potent activator of PFK-1, fructose 2,6-bisphosphate (F2,6BP), also acts allosterically. It stabilizes a conformation that binds the substrate F6P well and has reduced affinity to the inhibitors ATP and citrate. F2,6BP is formed by the enzyme *phosphofructokinase-2* (PFK-2; EC 2.7.1.105) when blood glucose levels are high. When glucose levels drop, PFK-2 is inhibited by hormonally induced phosphorylation, thereby decreasing the activation of PFK-1 by F2,6BP, as well. This regulation is dominant in the liver isozyme of PFK-1, which makes metabolic sense; in contrast to the muscle, which uses glucose for its own benefit, the liver's job is to ensure that blood glucose levels do not drop below a dangerous threshold. Thus, when the levels start to drop (e.g., between meals), the liver slows down all of its glucose-utilizing processes, including glycolysis, so glucose can be transferred safely to circulation.

The regulation of the bacterial form of PFK is simpler compared to its animal counterpart but still constitutes an interesting example of non-competitive inhibition. The reaction's product, F1,6BP, acts as a reversible inhibitor of PFK. However, while the inhibitory effect is competitive with respect to one of the co-substrates (F6P), it is non-competitive with respect to the other (ATP) [292]. The other product, ADP, inhibits the enzyme non-competitively with respect to both co-substrates.

Non-competitive inhibitors are also used as pharmaceutical drugs. While they are more common than uncompetitive drugs, they are still difficult to design because of their allosteric mode of action. Two famous examples of non-competitive drugs are:

- Certain protein kinase inhibitors that act by limiting the accessibility of the substrate to the active site of the enzyme [293]. One of these, the anti-cancer drug *imatinib* (*Gleevec®*), inhibits several kinases, including Bcr-Abl, platelet-derived growth factor receptor (PDGF-R), and stem cell factor receptor (c-Kit) [294]. It acts by allosterically interfering with the binding of ATP to the active sites of these kinases.

- Non-nucleoside reverse transcriptase inhibitors such as *nevirapine* (*Viramune®*) which are used against HIV infection and to prevent the consequent emergence of AIDS [295].

TABLE 9.1 **Summary of Michaelis-Menten kinetics in reversible inhibition modes.**

Inhibition	Michaelis-Menten Equation	Lineweaver-Burk Equation	$V_{\max(app)}$	$K_{M(app)}$
None	$V_0 = \dfrac{V_{\max}[S]}{K_M + [S]}$	$\dfrac{1}{V_0} = \dfrac{K_M}{V_{\max}}\dfrac{1}{[S]} + \dfrac{1}{V_{\max}}$	$V_{\max}$	K_M
Competitive	$V_0 = \dfrac{V_{\max}[S]}{K_M\left(1 + \dfrac{[I]}{K_I}\right) + [S]}$	$\dfrac{1}{V_0} = \dfrac{K_M}{V_{\max}}\left(1 + \dfrac{[I]}{K_I}\right)\dfrac{1}{[S]} + \dfrac{1}{V_{\max}}$	$V_{\max}$	$K_M\left(1 + \dfrac{[I]}{K_I}\right)$
Uncompetitive	$V_0 = \dfrac{V_{\max}[S]}{K_M + [S]\left(1 + \dfrac{[I]}{K_I}\right)}$	$\dfrac{1}{V_0} = \dfrac{K_M}{V_{\max}}\dfrac{1}{[S]} + \dfrac{1}{V_{\max}}\left(1 + \dfrac{[I]}{K_I}\right)$	$\dfrac{V_{\max}}{\left(1 + \dfrac{[I]}{K_I}\right)}$	$\dfrac{K_M}{\left(1 + \dfrac{[I]}{K_I}\right)}$
Non-competitive	$V_0 = \dfrac{V_{\max}[S]}{\left(1 + \dfrac{[I]}{K_I}\right)(K_M + [S])}$	$\dfrac{1}{V_0} = \dfrac{K_M}{V_{\max}}\left(1 + \dfrac{[I]}{K_I}\right)\dfrac{1}{[S]} + \dfrac{1}{V_{\max}}\left(1 + \dfrac{[I]}{K_I}\right)$	$\dfrac{V_{\max}}{\left(1 + \dfrac{[I]}{K_{I(b)}}\right)}$	K_M

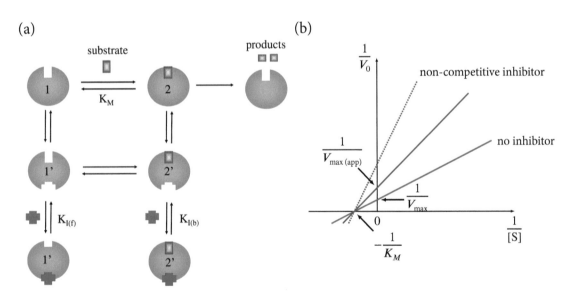

FIGURE 9.39 **Non-competitive inhibition.** (a) A schematic representation of the inhibition. The inhibitor may bind to the substrate-free enzyme with a dissociation constant of $K_{I(f)}$, or to the substrate-bound enzyme with a dissociation constant of $K_{I(b)}$. In each case, the inhibitor binds to a certain conformation of the enzyme (1' and 2', respectively) that is in equilibrium with a corresponding 'normal' conformation (1 and 2) but is less active than the normal conformation. (b) The Lineweaver-Burk (double reciprocal) plot of the inhibition (Equation (9.27)). The green lines designate the inhibited enzyme under two inhibitor concentrations or inhibition constants, as described in Figure 9.37b. The plot shows the apparent $V_{\max}$ of the inhibited reaction to be lower than that of the uninhibited reaction, with no change in K_M.

9.5.2.2 Irreversible inhibition

Certain inhibitors carry out their effects on enzymes by binding tightly to the active site, for hours and even days. Since this timescale is orders of magnitude larger than that characterizing the normal turnover of enzymes, such inhibitors are effectively irreversible. Irreversible inhibition may occur either when the inhibitor binds covalently to the enzyme, or via strong noncovalent interactions. Obviously, such inhibitors are not useful to organisms as internal regulators of enzymatic activity, as their effects cannot be reversed by metabolism or by excretion of the inhibitor; rather, it is necessary to re-synthesize the target enzyme (i.e., these inhibitors tend to be toxic) [296]. **These characteristics, however, make irreversible inhibitors good toxins.** Indeed, different organisms produce such molecules as secondary metabolites and use them against their prey or predators. **The toxic effect of an irreversible enzyme inhibitor may be caused by a decrease in the reaction's products, or by the accumulation of reactants.** For example, *onchidal* is an irreversible inhibitor of *acetylcholinesterase*, which is produced by the mollusk *Onchidella binneyi* and used as a protection mechanism against natural enemies [297]. As we saw in Chapter 8 (Box 8.1), the toxic effect of cholinesterase inhibitors is caused by the accumulation of acetylcholine, which, in turn, disrupts the normal function of the nervous system. Despite their potential toxicity, some irreversible inhibitors are also used as pharmaceutical drugs [271,281,298,299], mainly because of their ability to act in small doses and for prolonged periods of time. Examples include the following (see more examples in [281])[*1]:

- **Aspirin** – a painkiller and anti-inflammatory drug that acts by acylating a serine residue in the enzyme *cyclooxygenase* [300,301].

- **Penicillin** – an antibiotic that inhibits the bacterial cell wall-building enzyme *DD-transpeptidase* by acylating the catalytic serine residue in its active site [302] (see more below).

- **Omeprazole (Losec®)** – a drug used to treat stomach ulcers. It prevents acid buildup in the stomach by inhibiting gastric H^+/K^+-ATPase (a proton pump), via the formation of a disulfide bond with an enzyme cysteine residue [303,304].

- **Pyridostigmine** – a drug used to treat myasthenia gravis and to reverse the actions of muscle relaxants. It inhibits the enzyme *acetylcholinesterase* by carbamoylating the catalytic serine residue [305] (see also Box 8.1 in Chapter 8).

An elegant solution to the toxicity problem has been reached in the form of irreversible inhibitors that are normally inert, but become activated when binding to their target enzymes (see more below). Irreversible inhibitors are also used as agricultural agents (pesticides, herbicides, and insecticides) [306] and as research tools for studying enzymes and other proteins [270].

Irreversible inhibitors share some similarities with reversible-competitive inhibitors. First, they occupy the active site, thus preventing the natural substrate from binding. Second, they tend to resemble the natural substrate of their target enzymes chemically or structurally, and inhibitors that resemble the transition states of the corresponding substrates

---

[*1]Note, however, that in the case of penicillin antibiotics the target enzyme is strictly bacterial, and therefore the risk of toxicity due to irreversible binding to a human protein is less of a concern.

are particularly potent. We have already encountered such similarity in our discussion of the inhibition of *acetylcholinesterase* by organophosphates (Chapter 8, Box 8.1), where the inhibitor, like the natural acyl-enzyme transition state, has a tetrahedral shape and a negative charge. Again, unlike reversible competitive inhibitors, the irreversible inhibitor forms a stable enzyme-inhibitor intermediate that does not break easily. Therefore, **raising the concentration of the substrate does not lift the inhibition**. Accordingly, irreversible inhibitors are not considered to be competitive. As explained above, transition state analogues often bind to the active site so strongly that the binding, despite being noncovalent, is effectively irreversible.

As we saw in the examples above, enzymes that contain nucleophilic residues, whether catalytic (as in *esterases* and *proteases*) or not, are susceptible to irreversible covalent inhibition. The inhibitor in such cases contains an electrophilic center that can serve as a target for an active site nucleophile such as serine, threonine, cysteine, or aspartate. Such inhibitors are commonly grouped according to one of their chemical characteristics: nitrogen mustards, aldehydes, alkenes and alkyl halides, epoxides, Michael acceptors (α, β-unsaturated ketones), phenyl sulfonates, lactones, lactams, or fluorophosphonates. In many cases the irreversible inhibitor also includes a good leaving group that improves the likelihood of an efficient nucleophilic attack. For example, the acetylcholinesterase inhibitors *sarin* and *soman* possess both of these properties [307,308]: (*i*) an electrophilic phosphorus atom that is attacked by the enzyme's nucleophilic serine, thus forming a stable O−P bond, (*ii*) a fluorine atom that serves as a good leaving group (see Box 8.1 for more details). In addition to these two properties, covalent inhibitors may also have chemical characteristics that further strengthen their bonding with the enzyme's nucleophile. For example, in covalent inhibitors of proteases the attacked carbonyl group present in the substrate (see Figure 9.26c) is often replaced with a hydroxyl, which forms a much more stable bond with the serine nucleophile. Alternatively, the scissile C−N bond of the substrate may be replaced with a completely different bond (e.g., C−O).

Catalytic residues other than the nucleophile can also be targeted by irreversible covalent inhibitors of esterases and proteases. For example, *tosyl-L-phenylalanine chloromethyl ketone (TPCK)* inhibits the serine protease *chymotrypsin* by binding covalently to His-57, which is part of the Asp-102-His-57-Ser-195 catalytic triad. The binding blocks His-57 and prevents it from serving as a general acid or base during the reaction.

Not all irreversible inhibitors resemble their corresponding enzymes' natural substrates, products or transition states, nor do they necessarily act specifically on those enzymes. In some cases, an irreversible inhibitor can bind covalently to the same amino acid in different enzymes, and even to different amino acids of a certain type [296]. The binding in such cases is direct and does not involve the formation of a Michaelis complex before formation of the covalent bond (in contrast to the case of specific inhibitors). One example of a non-specific irreversible inhibitor is *2-iodoacetamide*, which binds covalently to thiol groups (−SH)[*1], and can therefore target the side chain of a cysteine residue; 2-iodoacetamide is widely used in the lab to study proteins whose structures are affected by cysteine residues (e.g., via disulfide bonds, as in structural proteins such as keratin). This inhibitor is also used to study enzymes that use cysteine as an important component of their catalytic mechanisms (e.g., cysteine peptidases). Other examples of non-specific irreversible inhibitors include *acetic an-*

[*1]2-Iodoacetamide also reacts (but more slowly) with tyrosine−OH, −NH$_2$, methionine and histidine at lower pH [309].

hydride, which binds to $-NH_2$, $-OH$ and $-SH$ groups [310], and *2,4-dinitrofluorobenzene*, which binds to $-NH_2$, $-SH$, tyrosine$-OH$ and imidazole groups [311] (see more examples in [296]). The lack of specificity limits the use of such inhibitors as pharmaceutical drugs or as pesticides, but it enables biochemists to study enzymatic mechanisms by identifying the residues that are important for catalysis.

An interesting special case of irreversible-covalent inhibition is *mechanism-based inhibition*, or, as it is sometimes called, *'suicide inhibition'*. In this case, a non-reactive molecule binds to the active site noncovalently, where it becomes reactive. As a result, the molecule binds covalently to an active site residue and inactivates the enzyme. **The fact that only the target enzyme can activate these inhibitors makes them much more attractive as pharmaceutical drugs compared with typical irreversible inhibitors, since this attribute reduces the risk for unwanted side reactions and resulting toxicity.** Examples of such drugs include the following (see more examples in [296]):

- *β-lactam antibiotics* such as clavulanic acid (Augmentin®) – see details below [312,313].

- *Deprenyl* (a.k.a. *selegiline*) – an anti-Parkinsonian drug, which acts on *monoamine oxidase-B* (MAO-B; EC 1.4.3.4) [314].

- *γ-acetylenic GABA* – an anti-convulsive drug that acts on *GABA transaminase* (EC 2.6.1.19) [315].

- *5-fluorodeoxyuracil monophosphate (5-FdUMP)* – an anti-cancer drug that acts on *thymidylate synthase* (EC 2.1.1.45) [316].

- *Allopurinol* – a drug used to treat gout. Allopurinol acts on *xanthine oxidase* (EC 1.1.3.22) [317].

A nice example of an enzyme's activation of a suicide inhibitor is observed in the case of the *β*-lactam antibiotic *clavulanic acid (Augmentin)* [277], which inhibits the enzyme *β-lactamase* (EC 3.5.2.6) (Figure 9.40). Like other members of this family of antibiotics, clavulanic acid contains an internal amide (i.e., *lactam*), which is strained and therefore unstable [318]. When the molecule binds noncovalently to its target enzyme, the amide's carbonyl carbon is attached by an active site serine nucleophile, which facilitates the breaking of the lactam bond and results in the formation of a stable enzyme-inhibitor acyl intermediate. This makes the molecule amenable to a second nucleophilic attack by the enzyme, followed by the breaking of another bond. In this form, the molecule is tightly attached to the enzyme, making inhibition irreversible.

As mentioned above, irreversible inhibition can also occur when the inhibitor binds to the enzyme noncovalently, but with very high affinity. Such inhibition characterizes the action of certain pharmaceutical drugs, such as *methotrexate* (for treating certain types of cancer), *allopurinol* (for treating gout), and the active form of *acyclovir* (for treating herpes virus infections).

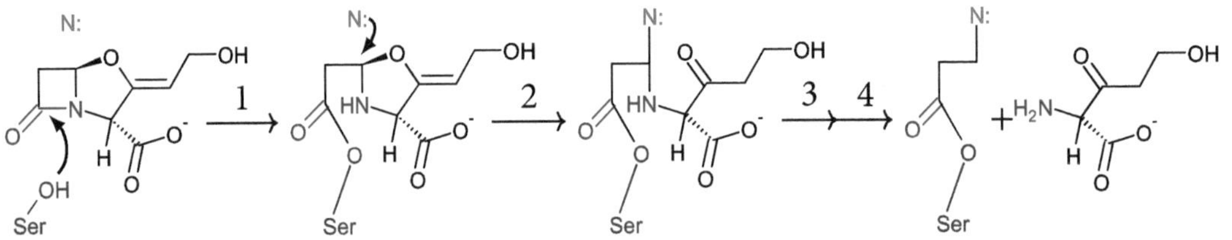

FIGURE 9.40 **Clavulanic acid (Augmentin) as a suicide inhibitor.** When clavulanic acid binds to the active site of β-lactamase, a serine nucleophile in the enzyme's active site attacks its amide's carbonyl carbon and forms a stable enzyme-inhibitor acyl intermediate, which induces cleavage of the lactam bond (step 1). Another active site nucleophile (N:) initiates a second attack on the molecule, which induces cleavage of a C−O bond in the second ring (step 2) [277]. This, in turn, leads to a C−N bond cleavage that separates the enzyme-bound part from the rest of the molecule (steps 3 and 4). Whereas the latter leaves the active site as product, the former remains permanently attached to the enzyme. The mechanism involves several protonation and deprotonation events that are not shown here, for clarity.

9.6 INDUSTRIAL USES OF ENZYMES

9.6.1 Medical uses of enzymes

9.6.1.1 Drugs and drug targets

As we have seen in Chapter 8 and in Section 9.5 above, the central role of enzymes in human (and animal) physiology makes them a prime target of pharmaceutical drugs. Indeed, **it is estimated that 29% of all human drug targets are enzymes (the second-largest target group after cell-surface receptors)** [272]. The drugs are in most cases small molecules that bind to their target enzymes and modify their activity through competition or allostery. Some enzymes have a more active role in the pharmaceutical industry; instead of being merely the targets of drugs, they are the drugs. Such enzymes are usually used to treat genetic illnesses that result from a deficiency in a natural enzyme. A classic example is *adenosine deaminase deficiency*, which involves the natural enzyme *adenosine deaminase* (ADA; EC 3.5.4.4). ADA is involved in purine nucleotide metabolism. Specifically, it catalyzes the irreversible hydrolytic deamination of adenosine and 2′-deoxyadenosine to inosine and 2′-deoxyinosine, respectively [319]. Loss of ADA activity leads to accumulation of its substrates, resulting in inhibition of enzymes that are important for lymphocyte maturation and function. Indeed, in most people, inherited mutations that prevent the expression of ADA lead to *severe combined immunodeficiency (SCID)*, which manifests as recurrent infection and failure to thrive [320]. Infants who carry this disease often die within the first year of life if not treated. The treatment of choice is bone marrow transplantation, which allows the patient to manufacture active ADA [321]. However, when a compatible donor is not found, or when the transplantation is expected to fail for other reasons, the alternative is to administer ADA directly, bound to polyethyleneglycol (PEG). Such *enzyme replacement therapies* (*ERTs*) are also available for certain lysosomal diseases (e.g., Gaucher and Fabry diseases) [322]. ERTs are relatively rare because, like other protein-based drugs, enzymes cannot be administered orally, owing to their molecular size and the fact that they are digested in the stomach. Instead, these drugs are usually given via blood transfusion.

Enzymes are also used as food additives, although their action can be considered to be medical — they break down foodstuff molecules to which the human body is either allergic or intolerant. In some cases these hydrolytic enzymes are given directly, whereas in other cases the medicinal preparation includes microorganisms that produce the required enzyme. A well-known example of the latter is yogurt, which is given to lactose-intolerant individuals to help them break down the milk sugar lactose in their food [323]. Yogurts contain 'probiotic' bacteria (e.g., *Lactobacillus bulgaricus*, *L. acidophilus* and *Streptococcus thermophilus*) that produce lactase, the lactose-breaking enzyme. The bacteria survive the acid environment of the stomach, and once they pass to the duodenum and small intestine they start digesting lactose.

9.6.1.2   Diagnostic roles

Another common use of enzymes in the medical industry is for diagnostic purposes [324,325]**, particularly as molecular markers for tissue damage.** Some enzymes are expressed primarily within certain tissues[*1], and their levels in blood are normally low. When the tissue in which these enzymes reside is damaged, its content is spilled into the blood circulation. The resulting rise in the blood levels of these enzymes can therefore be used to identify whether damage has occurred, and to assess its severity. Since few enzymes are expressed solely in one tissue, this type of diagnosis never relies on the blood levels of a single enzyme. However, an increase in the levels of multiple enzymes (and sometimes other proteins) that are identified with a certain tissue is usually a reliable indicator of damage to that tissue, particularly when the blood levels increase according to a certain order that is known to characterize the tissue. Common examples for diagnostic enzymes include the following [324,325]:

- **Creatine phosphokinase (CK, CPK)** – This enzyme is expressed in various tissues, but one of its isozymes (*CK-MB*) is specific to muscle and brain tissues. Its primary use is to help diagnose *myocardial infarction* (i.e., heart attack). Since a rise in CK levels may also result from other muscle-damaging pathologies (e.g., *rhabdomyolysis*), as well as from plain injuries or high fever, the diagnosis relies on additional markers, the most important of which is the protein *troponin*.

- **Liver enzymes** – Diagnosis of damage to the liver or to bile ducts relies, among other things, on a battery of enzymes such as *alanine aminotransferase (ALT/sGPT)*, *aspartate aminotransferase (AST/sGOT)*, *γ-glutamyl transpeptidase (GGT/γGT)*, and *alkaline phosphatase (AP)*. In many cases, the combination of the blood levels of these enzymes in a given patient's bloodwork enables doctors to identify the exact cause for the damage, e.g., a virus, a toxin, obstruction, or ischemia.

- *α*-**Amylase** and **pancreatic lipase** – These enzymes are used for the diagnosis of *acute pancreatitis*.

[*1]Many enzymes have tissue-specific isoforms (i.e., isozymes), although this specificity is not absolute.

9.6.2   Use of enzymes as industrial catalysts

In the past decade or so, the pharmaceutical industry has started to use enzymes as catalysts for the synthesis of key drug components [326–329]. Every drug contains a chemical compound called an *active pharmaceutical ingredient (API)*, which renders the drug biologically active[*1]. Pharmaceutical drugs have traditionally been synthesized from simple chemical building blocks through the use of multi-step chemical syntheses. While these processes are efficient and produce the desired API, they often include numerous steps and require the use of expensive metal catalysts (e.g., palladium, rhodium) and of organic solvents, both of which are environmental pollutants. The price of the catalysts and the need to safely dispose of all pollutants used in the synthesis increases the cost of drug production markedly.

As a result, **pharmaceutical companies are constantly searching for ways to reduce their API production costs. One approach is to replace the expensive metal catalysts with enzyme catalysts.** In addition to being less costly than catalytic metals, enzymes have several important advantages [330]:

1. A given enzyme is likely to catalyze only one type of reaction and act only on a specific group in its substrate(s) (i.e., enzymes are *regiospecific*). Metal catalysts, in contrast, are non-specific, and their use often involves the occurrence of other reactions, or the same reaction on different chemical groups of the substrate. To avoid such undesired reactions, chemical syntheses that rely on metal catalysts often include steps that block all functional groups in the substrate, except the one upon which the catalyst is intended to act. As a result, metal-catalyzed syntheses take a long time, and are associated with high costs. The use of enzymes usually makes the additional steps unnecessary and therefore reduces the cost of production.

2. Many enzymes are *enantiospecific*, which enables them to create *enantiopure* products. Specifically, in a single step, an enantiospecific enzyme can convert a pro-chiral group (e.g., a ketone or a C=C bond) in the substrate into a chiral group of a single configuration (e.g., a hydroxyl or amino group). As explained in Chapter 8, many drug molecules must be enantiopure to have biological activity, or even to avoid being toxic (see Chapter 8, Box 8.2). Synthetic reactions that rely on metal catalysts tend to create 50% of the chiral groups in one configuration and the other 50% in the other (a *racemic mixture*). Since only one configuration is usable, this protocol involves a 50% product loss, which compels the manufacturer to use expensive recycling techniques to regenerate the substrate from the undesired product, and use it for additional rounds of catalysis. The ability of enzymes to catalyze asymmetric reactions in a single step makes the whole production process much simpler and less expensive. **The most common enzymatic chiral syntheses in the pharmaceutical industry involve the conversion of carbonyl groups in ketones, aldehydes or acids into chiral hydroxyl groups (by oxidoreductases)** [51]**, or into amino groups (by amino acid dehydrogenases or aminotransferases)** [56]. Other enzymes used in chiral syntheses include *ammonia lyases* (EC 4.3.1) and *hydroxynitrile lyases* (EC 4.1.2). In the absence of an appropriate enzyme for chiral synthesis, an alternative approach

---

[*1]The other components of the drug are called *'excipients'*. They have various roles, such as impeding the oxidation or degradation of the API, stabilizing it, facilitating absorption in the GI tract, and conferring a particular color or texture.

is *racemic resolution*. In this process, the product is created as a racemic mixture, and then a stereoselective enzyme, such as a hydrolase or oxidoreductase, is used to degrade or change the product molecules that have the undesired configuration [331]. This approach, however, is inferior to chiral (asymmetric) synthesis, as it involves the loss of 50% of the product.

3. Enzymes are natural substances and therefore cause considerably less pollution compared with other catalysts. Also, since they work in aqueous solutions, they often make organic solvents unnecessary, which further limits the amount of pollution caused by the reaction.

4. Although enzymes are reaction-specific, the direction of the reaction can often be reversed by changing the experimental conditions. This attribute increases the commercial potential of enzymes, as it enables them to participate in multiple reactions. For example, peptide hydrolases (a.k.a. peptidases, proteases) act in nature to degrade proteins and peptides by breaking peptide bonds. However, under the right conditions, they can be used to form such bonds [332].

5. Being much larger than the other reagents and chemicals present in the reaction mixture, enzymes are easy to separate from the final product.

6. Enzymes act under mild temperatures (20 to 40 °C), pressure (~1 atmosphere), and pH (typically 5 to 8). Thus, they are easy to work with.

7. Enzymes are proteins, and can therefore be engineered by mutagenesis to acquire optimal stability, specificity, and even catalytic rate. As mentioned above, most natural enzymes display only moderate catalytic efficiency (k_{cat}/K_M), despite having evolved over a long period of time [3]. Thus, engineering has the potential to yield substantial improvement (see next subsection).

Indeed, **although the pharmaceutical industry has been using enzymes as catalysts for only a short period of time, there are already several notable examples of large-scale enzymatic processes used to synthesize drug intermediates** (Table 9.2; see also [333] for more examples).

TABLE 9.2 **Examples of enzymatic reactions developed for production of drug intermediates** [327,333]. For each drug, the table shows the disease or condition treated by the drug, the type of enzyme used for producing the drug, the specific chemical reaction catalyzed by the enzyme, and the company which developed the enzymatic process. The catalyzed reactions involve carbonyl (C=O), hydroxyl (C–OH), chloro (C–Cl), cyano (C–C≡N), amino (C–NH$_2$), carboxyl (COOH), ethyl-ester (COO–Et), and methyl-ester (COO–Me) groups.

Drug	Disease or Condition	Enzyme	Catalyzed Reaction	Company
Cymbalta	Depression	Ketoreductase	C=O ⟶ C–OH	Codexis
Lipitor	High cholesterol	Halohydrin dehalogenase	C–Cl ⟶ C–C≡N	Codexis
Januvia	Diabetes	Transaminase	C=O ⟶ C–NH$_2$	Codexis
Lyrica	Epilepsy	Esterase	COO–Et ⟶ COOH	Pfizer
Tekturna	Hypertension	Esterase	COO–Me ⟶ COOH	DSM

The idea of using enzymes as industrial catalysts did not originate in the pharmaceutical industry; enzymes have been used for this purpose for decades in the food, textile, detergent and fine-chemical industries [334,335]. In fact, the use of enzymes for the production of cheese, sourdough, beer, wine, vinegar, leather, and linen goes back thousands of years. Obviously, people in early societies did not have the technology to isolate enzyme molecules. However, they knew how to cultivate organisms that use them (yeast, bacteria) and harvest them from animal organs or from fruit [334]. As technology became more sophisticated, people learned how to grow selected strains of these organisms in bulk quantities and how to isolate their desired enzymes. This, in turn, enabled enzymatic processes to be integrated into the above industries. With the emergence of molecular biology techniques, genes encoding specific enzymes were used to further enhance the industrial production and use of enzymes. Finally, the development of protein engineering methods enabled natural enzymes to undergo optimization for specific processes and/or process conditions, and enabled enzymes to be designed with different substrate specificities, and even with completely new activities.

Most of the enzymes used in 'traditional' industries, such as the detergent, textile and starch industries, belong to the hydrolase group [48,334]. This is because most of the reactions required in these industries involve degradation of complex molecules into simpler ones through breakage of ester, amide, ether, or other chemical bonds. For example, in the detergent industry, enzymes are used to degrade molecules that form stains. Thus, stains that are created by lipid, protein, or starch molecules are degraded by lipases, proteases, and amylases, respectively. Similarly, in the starch industry, amylases are used to degrade and liquefy starch, whereas in the dairy industry different hydrolases are used to change products' texture and flavor (milk, cheese), to degrade lactose, etc. **Enzymatic degradation of complex carbohydrates (e.g., starch) and lipids (triacylglycerol) is also used in the energy industry, to produce bio-fuel and bio-diesel, respectively** (see examples in [334]). Hydrolases are also popular, owing to their large range of substrates, stability, relative tolerance for organic solvents (especially in lipases), commercial availability, and lack of expensive cofactors [336]. Finally, hydrolysis can also be used to obtain a specific product rather than to merely dispose of a substrate. For example, degradation of esters by esterases can be used to obtain alcohol or organic acid components. Besides hydrolases, oxidoreductases are also very common in industrial enzymatic processes, due to the prevalence of redox reactions [48].

As mentioned above, the pharmaceutical and fine-chemicals industries require enzymes mostly for synthetic purposes, i.e., to efficiently synthesize molecules of interest (e.g., APIs). Such syntheses include many types of chemical reactions: bond formation and breaking, oxidation, isomerization, etc. The enzymes used in such reactions are more diverse than those used in the traditional industries, and belong to all six enzyme classes [48]. For example, aminotransferases are used for the synthesis of APIs that are based on natural or unnatural amino acids [337]. Similarly, glycosyltransferases may be used for the synthesis of APIs that contain saccharide moieties (e.g., aminoglycoside antibiotics). Still, in the pharmaceutical industry, as in traditional industries, hydrolases and oxidoreductases are by far the most commonly used enzymes [48]. Moreover, as enzymes tend to be highly specific, they offer particularly high added value to the pharmaceutical and fine-chemicals industries (as compared, for example, to the food or textile industries), which are tightly regulated and require that products be synthesized in a highly specific manner, with very few impurities. In particular, there is demand in these industries for enzymes that can perform chiral (asymmetric) synthesis. Indeed, enzymes tend to specialize in such reactions; however, in order to use an enzyme for synthesis of 'artificial' molecules (e.g., APIs), it is often necessary to optimize it via mutagenesis to maintain its enantiospecificity. An interesting example is the single-step synthesis of the antidiabetic compound sitagliptin (Januvia®), which is achieved by an aminotransferase engineered specifically for this purpose [161]. In this case, a natural enzyme was chosen which was capable of catalyzing the right type of reaction (the rare *R*-specific transamination), albeit not with the sitagliptin precursor. Using both rational design and directed evolution, engineers gradually changed the enzyme so that it would be able to act on the desired substrate (and on another co-substrate), as well as to function in the presence of organic solvents.

9.6.3   Limitations and solutions

Despite the clear advantages of enzymes as catalysts, there are some aspects that limit their use in industrial processes. The main aspects are as follows:

- The task of finding the right enzyme for the desired reaction and substrate is far from trivial. This is due to the fact that the number and diversity of enzyme-catalyzed reaction types are smaller than those used industrially. In addition, APIs and the molecules from which they are produced tend to be larger and more hydrophobic than the average biological metabolite. As a result, such molecules often have limited compatibility as enzyme substrates, especially for enzymes with narrow ranges of substrates. It should be noted, though, that the inherent tendency of many enzymes to promiscuity can alleviate some of these problems.

- Enzymes are sensitive to the extreme conditions that are common in industrial syntheses, including high temperature, extreme pH, and the use of non-aqueous solvents (such as methanol, dimethylformamide (DMF) and dimethylsulfoxide (DMSO)) [338,339]. At least some of these conditions are necessary for increasing the solubility of hydrophobic substrates, but they may also harm the enzyme in the solution. For example, organic solvents disrupt enzyme activity by weakening its 3D fold, decreasing its dynamic properties, removing active site water molecules, changing substrate solubility, and even competing with the substrate [340,341]. These problems are somewhat less severe in lipases and proteases, which seem to be less sensitive to organic solvents (to a degree) [340].

- Many enzymes are prone to product inhibition, resulting from the high substrate or product concentrations in industrial syntheses [342].

- Some enzymes use expensive cofactors that raise the cost of the industrial process.

- Enzymes have a limited shelf life compared to metal catalysts.

Many of the problems and limitations associated with the use of enzymes in industry can be solved to different degrees by their engineering through mutagenesis[*1]. Such engineering may be carried out using two main approaches. The first is a *rational approach*, in

---

[*1]Note that even when engineering is used, the preferred enzymes for such studies are those that display at least some degree of promiscuity.

which the required mutations are introduced by design. That is, existing knowledge on the enzyme's structure and catalytic mechanism is used in order to improve its activity, specificity, stability, etc. In many cases, calculations of enzyme structural changes and enzyme-substrate interactions are used in the rational approach to predict the consequences of the designed mutations. **The rational approach works best when the active site of the enzyme is targeted for optimizing substrate specificity or activity**. This is because the active site is the place where the relationship between the activity of the enzyme and its physical-chemical properties is most clear.

As we saw in Chapter 5, the functional attributes of proteins, including enzymes, are also affected by parts that are far away from the binding or active site. Such effects are mediated by protein dynamics, whether inside a single chain or through subunit-subunit interactions (as in many allosteric proteins). This means that **the enzyme can be optimized by engineering non-active site regions, including those that are far from where catalysis actually takes place**. Since the structure-activity relationship in these cases is not straightforward, it is difficult to use a rational approach to design optimizing mutations. Instead, an alternative approach called *directed evolution* [119,120] is usually employed, in which different parts of the enzyme are randomly mutated, and the mutants with the best activity are selected, and then subjected to additional rounds of mutagenesis and selection. This protocol is carried out until mutants are identified with activity, specificity and stability that are significantly better than those of the original wild-type enzyme. Random approaches such as directed evolution have a good record of finding mutants that live up to industrial standards. However, application of these methods requires expensive equipment, may be lengthy, and depends on high-throughput biochemical assays for measuring the activity of each of the mutants produced. Therefore, the trend today is to combine rational mutagenesis with random approaches and to use 'smart' mutant libraries in the latter, that is, libraries that are already enriched with active mutants. This is called a *'semi-random'* approach.

Another approach that solves many of the problems associated with the use of isolated enzymes as industrial catalysts is to use them in whole cells. This entails expressing the enzyme inside a host unicellular organism (bacterium or yeast), carrying out the desired reaction, and harvesting the product. This approach, which is called *'metabolic engineering'* [343], can be employed only when the following conditions are satisfied:

1. The organisms used for this purpose must be able to easily uptake the substrate from the medium.

2. The enzyme used for the reaction must be expressed in sufficient quantities to yield a sizable amount of product.

3. Neither the reaction substrate nor its product can be toxic to the host organism.

4. It must be possible to efficiently extract the reaction product from the host organism.

In the simplest form of the whole cell approach, a single enzyme is overexpressed in host cells and used for the desired reaction. This solves the problems associated with the vulnerability of isolated enzymes in industrial solutions, as well as with the cost of the cofactor (the cofactor is recycled inside the organism).

However, **the real advantage of using whole cells is in cases where the creation of the product requires the consecutive action of several enzymes**. Such a feat is very difficult to achieve with isolated enzymes in solution, mainly because of unspecific enzyme-substrate reactions and diffusion-related problems. That is, in order for the process to be efficient, the product of each enzyme must be quickly 'fed' to the next enzyme. Since reaction mixtures provide ample space for diffusion, such coupled processes using isolated enzymes tend to be inefficient when more than two enzymes are used. In contrast, when the process is carried out inside a crowded cell, in which the enzymes are expressed inside the same compartment, free diffusion of the different products is not a significant problem. This is especially true when the enzymes used in the process are part of an existing biochemical pathway. In such cases, the system is already designed to carry out the individual enzymatic reactions consecutively in a highly efficient manner.

The use of whole cells for industrial catalysis predates the use of isolated enzymes. However, whereas past uses of this approach involved enzymes and pathways indigenous to the host organism, **today scientists are able to transfer entire metabolic pathways from one organism to another. Moreover, the accumulated knowledge on natural enzymes allows us to engineer existing pathways by changing their constituent enzymes, to obtain the best results.** Indeed, metabolic engineering has already been used to produce various molecules of industrial importance, including alkaloids, polyketides and non-ribosomal peptides, isoprenoids, vitamins, flavor molecules, and even fuels (see [343] for details).

9.7   SUMMARY

- Life processes are based on chemical reactions that must happen within 10^{-5} to 10^2 seconds. Given that such reactions in solution are typically orders of magnitude slower, living organisms must accelerate them. Enzymes have been selected by evolution for this task, probably because of their high levels of selectivity for substrates and reactions, in addition to the ability to control their activity through allosteric or post-translational regulation.

- Enzymes can be grouped into six classes according to the types of reactions they catalyze: oxidoreductases, transferases, hydrolases, lyases, isomerases and ligases. According to the accepted classification method, each enzyme is identified by a number comprising four numerals, which designate the type of reaction the enzyme catalyzes and the types of chemical groups or bonds upon which it acts.

- Enzymes accelerate chemical reactions by lowering their activation energy, i.e., by stabilizing the reaction's transition state. To do so, enzymes use several catalytic strategies, including noncovalent stabilization of the transition state, covalent binding to the substrate, electron/proton transfer, and electrostatic polarization of either the substrate or enzyme catalytic entities. Although these strategies have been studied for decades, there is still a vigorous debate on their relative importance. In particular, the role of short-term dynamics in enzymatic catalysis is highly controversial.

- The various catalytic strategies employed by enzymes are carried out inside the active site by different chemical entities. These chemical entities include amino acid residues and, in many cases, small organic molecules or metals that are collectively referred to as 'cofactors'. These species may interact noncovalently with the substrate and/or act chemically on it (e.g., by nucleophilic attack or proton/electron transfer). Both are carried out in a highly specific manner; this specificity results from the positioning of the functional chemical species inside the enzyme's active site and the specific interactions that occur between them.

- The high complementarity between the active site and the transition state is also the main contributor to the substrate selectivity displayed by enzymes.

- The study of enzyme-mediated catalysis has been going on for over 100 years. Early studies focused on the kinetics of enzymatic reactions and on the manner in which it is influenced by environmental conditions; in recent decades, however, the emergence of high-resolution three-dimensional structures of enzymes and sophisticated lab methods has enabled scientists to elucidate many of the fine details of catalysis. This include structures of reaction transition state and the many short-lived intermediates that are formed during reactions, in addition to the rapid occurrences that are involved in the chemical transformation.

- Enzymes are of great importance to various industries. The pharmaceutical industry uses enzymes as biological drugs, drug targets, and diagnostic tools, and in the last decade it has also used them as catalysts for the synthesis of active pharmaceutical ingredients. The food, textile and fine-chemicals industries are also using enzymes as catalysts for organic syntheses.

EXERCISES

9.1 A. Prove mathematically that the half-life time of the zero-order reaction of Equations (9.1.1) and (9.1.2) is given by $[A]_0/2k$, where $[A]_0$ is the initial concentration of the substrate, and k is the reaction rate.

   B. Prove that the half-life of the first-order reaction of Equations (9.1.4) and (9.1.5) is given by Equation (9.1.8), and that the half-life time of the second order reaction of Equations (9.1.9) and (9.1.10) is given by Equation (9.1.12).

9.2 According to the Michaelis-Menten model, which of the following describes correctly the dependency of the initial reaction rate (velocity V_0) of an enzyme on substrate concentration when the substrate concentration is very low?

   a. V_0 depends linearly on the substrate's concentration, with a rate coefficient of $\dfrac{k_{cat}}{K_M}[E]_f$.

   b. V_0 depends linearly on the substrate's concentration, with a rate coefficient of $\dfrac{K_M}{k_{cat}}$.

   c. V_0 depends cooperatively on the substrate's concentration.

   d. V_0 is constant at low substrate concentration, and is equal to the concentration of the enzyme ($[E]_i$).

9.3 The initial velocity of an enzymatic reaction (V_0) was measured under the following conditions:

   - *Condition 1*: with 2 mM of the enzyme's natural substrate (S).

   - *Condition 2*: with 2 mM of the enzyme's natural substrate (S) and 0.5 mM of an inhibitor (I).

- *Condition 3*: with 2 mM of the enzyme's natural substrate (S) and 1 mM of the inhibitor (I).

The dependency of $1/V_0$ on [S] under these three conditions was as follows:

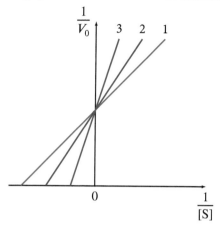

How can the inhibitor's effect be lifted?

a. Only by removing the inhibitor from the reaction mixture.

b. Either by removing the inhibitor or by raising the substrate's concentration to high values.

c. By removing the inhibitor and chemically reverting the enzyme's catalytic residues to their original form.

d. The inhibition is irreversible and cannot be lifted.

9.4 Under which condition can K_M be considered as a measure for the affinity of the enzyme to its substrate(s)?

9.5 Two isozymes, A and B, deaminate alanine to pyruvate. Their K_M values are 0.5 mM and 4 mM, respectively. When the two isozymes at 3 mM concentration were incubated with 10 mM of alanine under the same conditions, their rates (V_0) were 15 mM s^{-1} and 300 mM s^{-1}, respectively. Which isozyme is more efficient?

9.6 The activity of an enzyme was measured under saturation and different pH values. The following dependency was obtained:

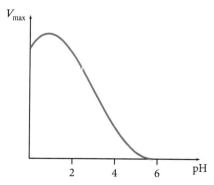

Then, the amino acids in the active site of the enzyme were systematically replaced (mutated), and the activity was measured again. Which of the following mutations is most likely to be deleterious (i.e., lead to loss of activity)?

a. Alanine $\longrightarrow$ glutamine

b. Glutamate $\longrightarrow$ valine

c. Arginine $\longrightarrow$ aspartate

d. Phenylalanine $\longrightarrow$ tryptophan

9.7 How does coenzyme A activate metabolites for subsequent condensations?

9.8 The following enzymatic cofactors are involved in group transfer. Match each cofactor with the group it transfers.

Cofactor	Group
a. S-adenosyl methionine	i. Acyl
b. Thiamine pyrophosphate	ii. Methyl (CH_3)
c. Coenzyme A	iii. Amino (NH_2)
d. Pyridoxal phosphate (PLP)	iv. CO_2
e. Biotin	v. Aldehyde

9.9 Which of the following cofactors carries out reactions that involve radicals?

a. Tetrahydrofolate (THF)

b. NADH

c. Coenzyme B_{12}

d. Coenzyme A

9.10 What is the common mechanistic aspect of the enzymatic cofactors pyridoxal phosphate (PLP) and thiamine pyrophosphate (TPP)?

9.11 The following enzyme activities were measured in the lab:

[S] (mM)	V_0 (mmol/min)
0	0.0
1	3.0
2	5.0
4	6.6
8	7.0
12	6.2
15	5.0

Does the enzyme present Michaelis-Menten kinetics? How can you explain the activity values measured at high substrate concentrations?

9.12 Two enzymes from two different organisms catalyze the oxidation of glucose. Which of the following parameters would you expect to be the same in the two catalyzed reactions?

a. V_{max}

b. K_M

c. K_{eq} (equilibrium constant)

d. Optimal temperature

e. Optimal pH

9.13 An enzyme has a K_M of 2 mM. At which substrate concentration will the enzyme's activity be $\frac{1}{4}$ of V_{max}?

9.14 The activity of two enzymes was measured at different substrate concentrations:

[S] (nM)	V_0 (nmol/min)	V_0' (nmol/min)
1	150	82
2	256	150
10	600	450
30	770	670
50	818	750

A. Show that both enzymes follow Michaelis-Menten kinetics, and estimate their K_M values.

B. Assuming that the concentrations of the enzymes are 0.2 nm and 0.5 nm, respectively, what are their turnover numbers?

9.15 The non-enzymatic decomposition of 2 NO_2 to 2 NO and O_2 is a second-order reaction. If the initial concentration of NO_2 is 10 mM, and the catalytic rate of the reaction is 0.4 $mM^{-1} s^{-1}$, what is its half-life?

9.16 The binding of a competitive inhibitor (I) to its target enzyme (E) can be described by the following scheme:

$$EI + S \xrightleftharpoons[k_{-3}][k_3] E + S + I \xrightleftharpoons[k_1][K_{-1}] ES + I \longrightarrow [k_2]E + P + I$$

Based on the scheme, derive the Michaelis-Menten equation for cases of competitive inhibition.

9.17 In an experiment, three batches of the same enzyme were incubated with 5 mM of substrate and 2 mM of inhibitor. In each batch the inhibitor was of a different type:

- *Batch 1*: competitive inhibitor

- *Batch 2*: uncompetitive inhibitor

- *Batch 3*: non-competitive inhibitor

Assuming that $V_{max} = 80\,\text{mM s}^{-1}$; $K_M = 8\,\text{mM}$; and the K_I of all three inhibitors was 10 mM, which of the inhibitors was the most efficient (i.e., had the strongest effect on V_0)?

TABLE 9.3 **Common organic cofactors in metabolism.** (a) The name of the cofactor. (b) The vitamin from which the cofactor is derived. (c) The catalytic role of the cofactor in enzymatic catalysis. (d) The physiological roles of the cofactor, with emphasis on mammalian physiology. 'Antioxidation' refers to reduction of chemical species during oxidative stress response. (e) Examples of enzymes that use the cofactor and correspond to the physiological role in (d). The biochemical pathway to which each enzyme belongs is shown in square parentheses. The enzymes are denoted by their common names. (f) Diseases associated with functional deficiency of the cofactor.

Abbreviations: α-KG – α-ketoglutarate; ACh – acetylcholine; ACoA – acetylcoenzyme A; AICAR – aminoimidazole carboxamide ribotide; Arg – arginine; BCKDC – branched-chain α-keto acid dehydrogenase complex; dTMP – deoxy-thymidine monophosphate; G6PD – glucose 6-phosphate; GAR – glycinamide ribotide; Glu – glutamate; His – histidine; HMG – hydroxy-methylglutaryl; Ile – isoleucine; Met – methionine; NO – nitric oxide; PC – phosphatidylcholine; Phe – phenylalanine; TAG – triacylglycerol; Thr – threonine; Trp – tryptophan; Tyr – tyrosine; Val – valine.

Name[a]	Vitamin[b]	Catalytic Role[c]	Physiological Role[d]	Enzymes [Pathway][e]	Deficiency-associated Disease or Condition[f]
Thiamine diphosphate (ThDP, TPP, TDP)	B₁ (thiamine)	• Aldehyde transfer • Cleavage of bonds involving carbonyl-groups (decarboxylation, transketolation)	• Energy production • Carbohydrate breakdown • Fermentation • α-Keto acids and α-amino acids breakdown • DNA/RNA synthesis • Photosynthesis • Cholinergic neurotransmission	• Pyruvate dehydrogenase [pyruvate → ACoA] • α-Ketoglutarate dehydrogenase [Krebs cycle] • Pyruvate decarboxylase [alcoholic fermentation] • BCKDC [branched keto and amino acids breakdown] • Transketolase [pentose-phosphate cycle] • Transketolase [Calvin cycle] • Pyruvate dehydrogenase [pyruvate → ACoA → acetylcholine]	• Beriberi • Wernicke-Korsakoff syndrome • Demyelination of nerve cells

Name[a]	Vitamin[b]	Catalytic Role[c]	Physiological Role[d]	Enzymes [Pathway][e]	Deficiency-associated Disease or Condition[f]
Flavin adenine dinucleotide (FAD)	B_2 (riboflavin)	• Redox	• Energy production • Carbohydrate breakdown	• Pyruvate dehydrogenase [pyruvate → ACoA] • Succinate dehydrogenase [Krebs cycle] • Mitochondrial glycerol 3-phosphate dehydrogenase [oxidative phosphorylation]	• Ariboflavinosis
			• Lipid metabolism	• Acyl-CoA dehydrogenase [fatty acid breakdown] • Cytochrome b5 reductase [fatty acid desaturation]	
			• α-Keto acids and α-amino acids breakdown • Cellular signaling • Antimicrobial activity • Antioxidation	• BCKDC [branched α-keto and α-amino acids breakdown] • Nitric oxide synthase [Arg → NO] • NADPH oxidase [$O_2 \rightarrow H_2O_2$] • Glutathione reductase [antioxidation]	
Flavin adenine mononucleotide (FMN)	B_2 (riboflavin)	• Redox	• Energy production	• NADH dehydrogenase [oxidative phosphorylation] • Pyridoxal 5' phosphate synthase [PLP synthesis]	
			• Carbohydrate synthesis (glyoxylate cycle) • Metabolism of dicarboxylic acids • Synthesis of secondary metabolites	• 2-hydroxyacid oxidase	
			• Cellular signaling	• Nitric oxide synthase [Arg → NO]	

Name[a]	Vitamin[b]	Catalytic Role[c]	Physiological Role[d]	Enzymes [Pathway][e]	Deficiency-associated Disease or Condition[f]
Nicotinamide adenine dinucleotide (NAD⁺)	B₃ (niacin)	• Redox	• Energy production • Carbohydrate breakdown	• Cytosolic glyceraldehyde 3-phosphate dehydrogenase [glycolysis] • Pyruvate dehydrogenase [pyruvate → ACoA] • Isocitrate, α-KG and malate dehydrogenases [Krebs cycle] • NADH dehydrogenase [oxidative phosphorylation]	• Pellagra • G6PD deficiency (NADPH)
			• Glucose synthesis	• Cytosolic glyceraldehyde 3-phosphate dehydrogenase [gluconeogenesis]	
			• Fermentation	• Lactate dehydrogenase • Alcohol dehydrogenase	
			• Lipid metabolism	• Cytosolic glyceraldehyde 3-phosphate dehydrogenase [TAG and phospholipid synthesis] • β-Hydroxyacyl-CoA dehydrogenase [fatty acid breakdown] • β-HB dehydrogenase [ketone bodies metabolism]	
			• α-Keto acids and α-amino acids breakdown	• Glutamate dehydrogenase • BCKDC [branched α-keto acids and α-amino acids breakdown]	

Name[a]	Vitamin[b]	Catalytic Role[c]	Physiological Role[d]	Enzymes [Pathway][e]	Deficiency-associated Disease or Condition[f]
Nicotinamide adenine dinucleotide phosphate (NADP+)	B3 (niacin)	• Redox	• Energy production	• Isocitrate dehydrogenase [Krebs cycle]	• Pellagra • G6PD deficiency (NADPH)
			• Carbohydrate metabolism	• G6PD and 6-phosphogluconate dehydrogenase [pentose-phosphate pathway]	
			• Amino acid breakdown and synthesis	• Glutamate dehydrogenase • Glutamate-5-semialdehyde dehydrogenase • Pyrroline-5-carboxylate reductase • Dihydrofolate reductase (indirect)	
			• Lipid biosynthesis	• Fatty acid synthase and malic enzyme [fatty acid synthesis] • Fatty acyl-CoA desaturase [fatty acid desaturation] • HMG-CoA reductase [cholesterol synthesis] • Squalene synthase [cholesterol synthesis] • Squalene mono-oxygenase [cholesterol synthesis] • 5-Hydroxyeicosanoid dehydrogenase [eicosanoid synthesis]	
			• Nucleotide biosynthesis	• G6PD and 6-phosphogluconate dehydrogenase [pentose-phosphate pathway] • Dihydrofolate reductase (indirect)	
			• Signaling	• Nitric oxide synthase	
			• Antioxidation	• Glutathione reductase	
			• Detoxification of drugs and toxins	• P-450 oxidase	
			• Anti-pathogenic	• NADPH oxidase	

Name[a]	Vitamin[b]	Catalytic Role[c]	Physiological Role[d]	Enzymes [Pathway][e]	Deficiency-associated Disease or Condition[f]
Coenzyme A (CoA)	B₅ (pantothenate)	• Acyl group transfer	• Energy production • Carbohydrate breakdown	• Pyruvate dehydrogenase [pyruvate → ACoA] • α-Keto glutarate dehydrogenase [Krebs cycle]	• Paresthesia (uncommon)
			• Lipid metabolism	• Fatty-acyl-CoA synthase [fatty acid synthesis] • Carnitine acyltransferase II [fatty acid breakdown] • Thiolase [fatty acid breakdown, ketone bodies metabolism, cholesterol synthesis] • HMG-CoA reductase [cholesterol synthesis]	
			• α-Keto acids and α-amino acids breakdown	• BCKDC [branched α-keto acids and α-amino acids breakdown] • α-Amino-β-ketobutyrate lyase [Thr breakdown]	
			• Neurotransmitter synthesis	• Choline acetyltransferase [acetylcholine synthesis]	
Pyridoxal phosphate (PLP)	B₆ (pyridoxine)	• −NH₂ group transfer • Racemization • Decarboxylation • β/γ-Elimination	• Amino acid metabolism	• Amino transferases [amino acid breakdown] • Aromatic L-amino acid decarboxylase • Serine dehydratase	• Diverse conditions
			• Glycogen breakdown	• Glycogen phosphorylase	
			• Synthesis of bioactive compounds	• Aromatic amino acid decarboxylase [synthesis of catecholamines and serotonin] • Glutamate decarboxylase [GABA synthesis] • Aminolevulinic acid synthase [porphyrins synthesis]	
Biotin	B₇ (H)	• −CO₂ group transfer (carboxylations, decarboxylations)	• Glucose synthesis	• Pyruvate carboxylase and Propionyl-CoA carboxylase [gluconeogenesis]	• Neurological and growth disorders (infants)
			• Lipid metabolism	• ACoA carboxylase [fatty acid synthesis]	
			• Amino acid breakdown	• Propionyl-CoA carboxylase [Met/Val/Ile breakdown] • β-Methylcrotonyl CoA carboxylase [Leu breakdown]	

Name[a]	Vitamin[b]	Catalytic Role[c]	Physiological Role[d]	Enzymes [Pathway][e]	Deficiency-associated Disease or Condition[f]
Tetrahydrofolate (THF, FH$_4$)	B$_9$ (folic acid)	• Single-carbon groups transfer	• Nucleotide synthesis • Amino acid metabolism • Keeping low homocysteine levels • Signaling	• Thymidylate synthase [dTMP biosynthesis] • GAR formyltransferase [purine biosynthesis] • AICAR formyltransferase [purine biosynthesis] • Methionine synthase [homocysteine → methionine] • Serine hydroxymethyl transferase [glycine → serine] • Glu forminotransferase [His → Glu] • Nitric oxide synthase	• Megaloblastic anemia • Birth defects (pregnant women) • Elevated homocysteine (risk factor for cardiovascular diseases)
Cobalamin	B$_{12}$	• Molecular rearrangements (radical catalysis) • –CH$_3$ group transfer • Dehalogenation	• Glucose synthesis • DNA synthesis • Amino acid metabolism • Keeping low homocysteine levels • Lipid metabolism • Folic acid regeneration	• Methylmalonyl-CoA mutase [fatty acids → succinyl-CoA → glucose] • Ribonucleotide reductase [DNA synthesis] • Methionine synthase [homocysteine → methionine] • Methylmalonyl-CoA mutase [Met/Val/Ile breakdown] • Methylmalonyl-CoA mutase [fatty acids → succinyl-CoA] • Glycerol dehydratase [glycerolipid metabolism] • Methionine synthase [methyl-THF → THF]	• Megaloblastic anemia (via THF deficiency) • Methylmalonic-related pathologies (acidemia, myelin destabilization) • Elevated homocysteine (risk factor for cardiovascular diseases)
Ascorbic acid	C	• Hydroxylation • Redox	• Collagen synthesis • Antioxdiation	• Prolyl/Lysyl hydroxylase [collagen synthesis] • ε-N-trimethyl-L-lysine hydroxylase [carnitine synthesis] • Dopamine β-hydroxylase [norepinephrine synthesis]	• Scurvy
Tetrahydrobiopterin (TH$_4$)		• –OH group transfer • Redox	• Amino acids metabolism • Neuronal signaling	• Phe/Tyr/Trp hydroxylases • Phe/Tyr/Trp hydroxylases [catecholamine and serotonin synthesis] • Nitric oxide synthase [NO synthesis]	• Alzheimer's disease • Parkinson's disease • Depression

Name[a]	Vitamin[b]	Catalytic Role[c]	Physiological Role[d]	Enzymes [Pathway][e]	Deficiency-associated Disease or Condition[f]
Coenzyme Q (CoQ, ubiquinone)		• Redox	• Energy production • Antioxidation (non-enzymatic)	• Electron transport chain, mitochondrial glycerol 3-phosphate dehydrogenase [oxidative phosphorylation]	
Lipoic acid		• Redox • Acyl group transfer	• Energy production • Carbohydrate breakdown • α-Keto acids and α-amino acids breakdown • Antioxidation	• Pyruvate dehydrogenase [pyruvate → ACoA] • α-Ketoglutarate dehydrogenase [Krebs cycle] • Transketolase [pentose-phosphate cycle] • BCKDC [branched keto and amino acids breakdown]	
Menaquinone	K_2	• γ-Carboxylation of glutamate residues	• Blood clotting • Antioxidation • Bone formation	• γ-Glutamylcarboxylase [Glu carboxylation]	• Hemorrhagic syndrome
S-Adenosyl-methionine (SAM, AdoMet)		• –CH$_3$ group transfer • Aminopropyl group transfer • Radical catalysis	• DNA/RNA methylation • Lipid synthesis • Amino acid metabolism • Neurotransmission • Polyamines synthesis	• DNA/RNA methyltransferases • PEMT [PC synthesis] • Methylases [homocysteine synthesis] • Methionine adenosyltransferse [methionine breakdown] • PNMT [epinephrine synthesis] • AdoMet decarboxylase (spermidine synthesis)	

REFERENCES

1. E. T. Harper and R. A. Harris. Glycolytic pathway. In *Encyclopedia of Life Sciences*. John Wiley & Sons, Ltd., 2005.

2. K. F. LaNoue. Citric Acid Cycle. In *Encyclopedia of Life Sciences*. John Wiley & Sons, Ltd., 2001.

3. A. Bar-Even, E. Noor, Y. Savir, W. Liebermeister, D. Davidi, D. S. Tawfik, and R. Milo. The moderately efficient enzyme: evolutionary and physicochemical trends shaping enzyme parameters. *Biochemistry*, 50(21):4402–10, 2011.

4. R. Wolfenden and M. J. Snider. The depth of chemical time and the power of enzymes as catalysts. *Acc. Chem. Res.*, 34(12):938–945, 2001.

5. E. Lolis and G. A. Petsko. Transition-state analogues in protein crystallography: probes of the structural source of enzyme catalysis. *Annu. Rev. Biochem.*, 59:597–630, 1990.

6. V. L. Schramm. Enzymatic transition states and transition state analog design. *Annu. Rev. Biochem.*, 67:693–720, 1998.

7. H. Eyring and A. E. Stearn. The application of the theory of absolute reaction rates to proteins. *Chem. Rev.*, 24:253–270, 1939.

8. G. E. Lienhard. Enzymatic catalysis and transition-state theory. *Science*, 180(82):149–54, 1973.

9. A. Radzicka and R. Wolfenden. A proficient enzyme. *Science*, 267(5194):90–3, 1995.

10. M. Polanyi. Uberadsorptionskatalyse. *Zeitschift Fur Elektrochemie*, 27:143–152, 1921.

11. G. M. Schwab. *Catalysis (translated from German by H. S. Taylor and R. Spence)*. Van Nostrand, New York, 1937.

12. D. G. Truhlar, B. C. Garrett, and S. J. Klippenstein. Current status of transition-state theory. *J. Phys. Chem.*, 100(31):12771–12800, 1996.

13. S. A. Arrhenius. Über die Dissociationswärme und den Einfluß der Temperatur auf den Dissociationsgrad der Elektrolyte. *Z. Phys. Chem.*, 4:96–116, 1889.

14. R. Wolfenden. Thermodynamic and extrathermodynamic requirements of enzyme catalysis. *Biophys. Chem.*, 105(2-3):559–72, 2003.

15. E. V. Anslyn and D. A. Dougherty. Transition State Theory and Related Topics. In *Modern Physical Organic Chemistry*, pages 365–373. University Science Books, 2006.

16. B. C. Garrett and D. G. Truhlar. Generalized transition state theory. Classical mechanical theory and applications to collinear reactions of hydrogen molecules. *J. Phys. Chem.*, 83(8):1052–1079, 1979.

17. J. A. McCammon and S. C. Harvey. *Dynamics of Proteins and Nucleic Acids*. Cambridge University Press, New York, 1987.

18. C. R. Burke and A. Lupták. *Catalytic RNA*. eLS. John Wiley & Sons, Ltd., Chichester, 2012.

19. J. Gao, S. Ma, D. T. Major, K. Nam, J. Pu, and D. G. Truhlar. Mechanisms and free energies of enzymatic reactions. *Chem. Rev.*, 106(8):3188–3209, 2006.

20. C. Walsh. Enabling the chemistry of life. *Nature*, 409(6817):226–31, 2001.

21. J. G. Zalatan, I. Catrina, R. Mitchell, P. K. Grzyska, P. J. O'Brien, D. Herschlag, and A. C. Hengge. Kinetic isotope effects for alkaline phosphatase reactions: implications for the role of active-site metal ions in catalysis. *J. Am. Chem. Soc.*, 129(31):9789–98, 2007.

22. J. G. Zalatan and D. Herschlag. Alkaline phosphatase mono- and diesterase reactions: comparative transition state analysis. *J. Am. Chem. Soc.*, 128(4):1293–303, 2006.

23. T. Ignacio, S. Kenneth, and C. W. James. *Physical chemistry: principles and applications in biological sciences*. Prentice Hall International, 2002.

24. X. Zhang and K. N. Houk. Why enzymes are proficient catalysts: beyond the Pauling paradigm. *Acc. Chem. Res.*, 38(5):379–85, 2005.

25. C. Lad, N. H. Williams, and R. Wolfenden. The rate of hydrolysis of phosphomonoester dianions and the exceptional catalytic proficiencies of protein and inositol phosphatases. *Proc. Natl. Acad. Sci. USA*, 100(10):5607–5610, 2003.

26. B. Alberts. The cell as a collection of protein machines: preparing the next generation of molecular biologists. *Cell*, 92(3):291–4, 1998.

27. S. Boyce and K. F. Tipton. *Enzyme Classification and Nomenclature*. eLS. John Wiley & Sons, Ltd., 2005.

28. R. Caspi, T. Altman, R. Billington, K. Dreher, H. Foerster, C. A. Fulcher, T. A. Holland, I. M. Keseler, A. Kothari, A. Kubo, M. Krummenacker, M. Latendresse, L. A. Mueller, Q. Ong, S. Paley, P. Subhraveti, D. S. Weaver, D. Weerasinghe, P. Zhang, and P. D. Karp. The MetaCyc database of metabolic pathways and

enzymes and the BioCyc collection of Pathway/Genome Databases. *Nucleic Acids Res.*, 42(D1):D459–71, 2014.

29. G. P. Moss. *Enzyme nomenclature: recommendations of the Nomenclature Committee of the International Union of Biochemistry and Molecular Biology on the nomenclature and classification of enzymes by the reactions they catalyse.* NC-IUBMB, 2008. http://www.chem.qmul.ac.uk/iubmb/enzyme/.

30. G. L. Holliday, J. D. Fischer, J. B. Mitchell, and J. M. Thornton. Characterizing the complexity of enzymes on the basis of their mechanisms and structures with a bio-computational analysis. *FEBS J.*, 2011.

31. A. G. McDonald, S. Boyce, and K. F. Tipton. ExplorEnz: the primary source of the IUBMB enzyme list. *Nucleic Acids Res.*, 37(suppl 1):D593–7, 2009.

32. A. Bairoch. The ENZYME database in 2000. *Nucleic Acids Res.*, 28:304–305, 2000.

33. P. Pharkya, E. V. Nikolaev, and C. D. Maranas. Review of the BRENDA Database. *Metab. Eng.*, 5(2):71–3, 2003.

34. G. L. Holliday, D. E. Almonacid, G. J. Bartlett, N. M. O'Boyle, J. W. Torrance, P. Murray-Rust, J. B. Mitchell, and J. M. Thornton. MACiE (Mechanism, Annotation and Classification in Enzymes): novel tools for searching catalytic mechanisms. *Nucleic Acids Res.*, 35(suppl 1):D515–20, 2007.

35. M. Kanehisa, M. Araki, S. Goto, M. Hattori, M. Hirakawa, M. Itoh, T. Katayama, S. Kawashima, S. Okuda, T. Tokimatsu, and Y. Yamanishi. KEGG for linking genomes to life and the environment. *Nucleic Acids Res.*, 36(suppl 1):D480–4, 2008.

36. H. M. Donertas, S. Martinez Cuesta, S. A. Rahman, and J. M. Thornton. Characterising Complex Enzyme Reaction Data. *PLoS One*, 11(2):e0147952, 2016.

37. N. M. O'Boyle, G. L. Holliday, D. E. Almonacid, and J. B. Mitchell. Using reaction mechanism to measure enzyme similarity. *J. Mol. Biol.*, 368(5):1484–99, 2007.

38. D. Schomburg and I. Schomburg. *Springer Handbook of Enzymes.* Springer, 2001–2009.

39. D. B. McCormick. *Oxidation–Reduction Reactions.* eLS. John Wiley & Sons, Ltd., 2003.

40. G. L. Holliday, J. B. Mitchell, and J. M. Thornton. Understanding the functional roles of amino acid residues in enzyme catalysis. *J. Mol. Biol.*, 390(3):560–77, 2009.

41. J. M. Berg, J. L. Tymoczko, and L. Stryer. *Biochemistry.* W. H. Freeman and Company, 5th edition, 2002.

42. Royal Society of Chemistry. www.chemspider.com.

43. H. G. Zimmer. *Pentose Phosphate Pathway.* eLS. John Wiley & Sons, Ltd., 2001.

44. J. B Broderick. *Coenzymes and Cofactors.* eLS. John Wiley & Sons, Ltd., 2001.

45. H. Jornvall, M. Persson, and J. Jeffery. Alcohol and polyol dehydrogenases are both divided into two protein types, and structural properties cross-relate the different enzyme activities within each type. *Proc. Natl. Acad. Sci. USA*, 78(7):4226–30, 1981.

46. H. Jornvall, B. Persson, M. Krook, S. Atrian, R. Gonzalez-Duarte, J. Jeffery, and D. Ghosh. Short-chain dehydrogenases/reductases (SDR). *Biochemistry*, 34(18):6003–13, 1995.

47. U. Oppermann, C. Filling, M. Hult, N. Shafqat, X. Wu, M. Lindh, J. Shafqat, E. Nordling, Y. Kallberg, B. Persson, and H. Jornvall. Short-chain dehydrogenases/reductases (SDR): the 2002 update. *Chem. Biol. Interact.*, 143–144:247–53, 2003.

48. T. Johannes, M. R Simurdiak, and H. Zhao. Biocatalysis. In S. Lee, editor, *Encyclopedia of Chemical Processing*, pages 101–110. Marcel Dekker, 2006.

49. H. Weiner and T. D. Hurley. *NAD(P)⁺ Binding to Dehydrogenases.* eLS. John Wiley & Sons, Ltd., 2005.

50. R. G. Rosenthal, M. O. Ebert, P. Kiefer, D. M. Peter, J. A. Vorholt, and T. J. Erb. Direct evidence for a covalent ene adduct intermediate in NAD(P)H-dependent enzymes. *Nat. Chem. Biol.*, 10(1):50–5, 2014.

51. K. Nakamura, R. Yamanaka, T. Matsuda, and T. Harada. Recent developments in asymmetric reduction of ketones with biocatalysts. *Tetrahedron: Asymmetry*, 14(18):2659–2681, 2003.

52. D. Schomburg and M. Salzmann. *Enzyme Handbook.* Springer, 1990.

53. C. W. Bradshaw, H. Fu, G. J. Shen, and C. H. Wong. A Pseudomonas sp. alcohol dehydrogenase with broad substrate specificity and unusual stereospecificity for organic synthesis. *J. Org. Chem.*, 57(5):1526–1532, 1992.

54. S. K. Seah. Amino Acid Dehydrogenases. In J. Polaina and A. P. MacCabe, editors, *Industrial Enzymes*, chapter 28, pages 489–504. Springer, 2007.

55. J. L. Vanhooke, J. B. Thoden, N. M. Brunhuber, J. S. Blanchard, and H. M. Holden. Phenylalanine dehydrogenase from Rhodococcus sp. M4: high-resolution X-ray analyses of inhibitory ternary complexes reveal key features in the oxidative deamination mechanism. *Biochemistry*, 38(8):2326–39, 1999.

56. D. Zhu and L. Hua. Biocatalytic asymmetric amination of carbonyl functional groups: a synthetic biology approach to organic chemistry. *Biotechnol. J.*, 4(10):1420–31, 2009.

57. A. Lehninger, D. L. Nelson, and M. Cox. *Lehninger Principles of Biochemistry*. W. H. Freeman and Company, 5th edition, 2008.

58. J. S. Shin and B. G. Kim. Exploring the active site of amine:pyruvate aminotransferase on the basis of the substrate structure-reactivity relationship: how the enzyme controls substrate specificity and stereoselectivity. *J. Org. Chem.*, 67(9):2848–53, 2002.

59. K. Hirotsu, M. Goto, A. Okamoto, and I. Miyahara. Dual substrate recognition of aminotransferases. *Chem. Rec.*, 5(3):160–172, 2005.

60. K. M. Kim, T. Qin, Y. Y. Jiang, L. L. Chen, M. Xiong, D. Caetano-Anolles, H. Y. Zhang, and G. Caetano-Anolles. Protein domain structure uncovers the origin of aerobic metabolism and the rise of planetary oxygen. *Structure*, 20(1):67–76, 2012.

61. E. E. Snell and S. J. di Mari. 7 Schiff Base Intermediates in Enzyme Catalysis. *The Enzymes*, 2:335–370, 1970.

62. A. C. Eliot and J. F. Kirsch. Pyridoxal phosphate enzymes: mechanistic, structural, and evolutionary considerations. *Annu. Rev. Biochem.*, 73:383–415, 2004.

63. M. D. Toney. Controlling reaction specificity in pyridoxal phosphate enzymes. *Biochim. Biophys. Acta*, 1814(11):1407–18, 2011.

64. H. C. Dunathan. Conformation and reaction specificity in pyridoxal phosphate enzymes. *Proc. Natl. Acad. Sci. USA*, 55(4):712–6, 1966.

65. L. Hedstrom. Enzyme Specificity and Selectivity. In *Encyclopedia of Life Sciences*. John Wiley & Sons, Ltd., 2010.

66. W. W. Cleland and A. C. Hengge. Enzymatic mechanisms of phosphate and sulfate transfer. *Chem. Rev.*, 106(8):3252–3278, 2006.

67. H. Jakubowski. *Properties of ATP*. BioWiki: The Dynamic Biology Hypertext (Biochemistry). University of California, Davis, 2015.

68. R. J. Heath and C. O. Rock. The claisen condensation in biology. *Nat. Prod. Rep.*, 19(5):581–96, 2002.

69. R. Bentley. *Polyketides*. eLS. John Wiley & Sons, Ltd., 2001.

70. V. E. Anderson, M. W. Ruszczycky, and M. E. Harris. Activation of oxygen nucleophiles in enzyme catalysis. *Chem. Rev.*, 106(8):3236–3251, 2006.

71. K. E. van Holde, W. C. Johnson, and P. S. Ho. *Principles of Physical Biochemistry*. Pearson Education, 1998.

72. J. E. Walker. The ATP synthase: the understood, the uncertain and the unknown. *Biochem. Soc. Trans.*, 41(1):1–16, 2013.

73. W. Junge and N. Nelson. ATP Synthase. *Annu. Rev. Biochem.*, 84:631–657, 2015.

74. P. Mitchell. Coupling of phosphorylation to electron and hydrogen transfer by a chemi-osmotic type of mechanism. *Nature*, 191:144–8, 1961.

75. D. M. Rees, A. G. Leslie, and J. E. Walker. The structure of the membrane extrinsic region of bovine ATP synthase. *Proc. Natl. Acad. Sci. USA*, 106(51):21597–601, 2009.

76. J. Weber. Structural biology: Toward the ATP synthase mechanism. *Nat. Chem. Biol.*, 6(11):794–795, 2010.

77. A. G. Stewart, E. M. Laming, M. Sobti, and D. Stock. Rotary ATPases: dynamic molecular machines. *Curr. Opin. Struct. Biol.*, 25:40–8, 2014.

78. R. Iino and H. Noji. Operation mechanism of F(o) F(1)-adenosine triphosphate synthase revealed by its structure and dynamics. *IUBMB Life*, 65(3):238–46, 2013.

79. R. H. Fillingame, M. E. Girvin, and Y. Zhang. Correlations of structure and function in subunit c of *Escherichia coli* F0F1 ATP synthase. *Biochem. Soc. Trans.*, 23(4):760–6, 1995.

80. W. Junge. ATP synthase and other motor proteins. *Proc. Natl. Acad. Sci. USA*, 96(9):4735–7, 1999.

81. J. P. Abrahams, A. G. Leslie, R. Lutter, and J. E. Walker. Structure at 2.8 Å resolution of F1-ATPase from bovine heart mitochondria. *Nature*, 370(6491):621–8, 1994.

82. N. D. Rawlings and A. J. Barrett. *Peptidases*. eLS. John Wiley & Sons, Ltd., 2001.

83. I. Schechter and A. Berger. On the size of the active site in proteases. I. Papain. *Biochem. Biophys. Res. Commun.*, 27(2):157–62, 1967.

84. L. Hedstrom. Serine protease mechanism and specificity. *Chem. Rev.*, 102(12):4501–24, 2002.

85. J. Kraut. Serine proteases: structure and mechanism of catalysis. *Annu. Rev. Biochem.*, 46:331–58, 1977.

86. M. H. Glickman and A. Ciechanover. The Ubiquitin-Proteasome Proteolytic Pathway: Destruction for the Sake of Construction. *Physiol. Rev.*, 82:373–428, 2002.

87. C. M. Pickart and R. E. Cohen. Proteasomes and their kin: proteases in the machine age. *Nat. Rev. Mol. Cell Biol.*, 5(3):177–87, 2004.

88. I. G. Kamphuis, J. Drenth, and E. N. Baker. Thiol proteases. Comparative studies based on the high-

resolution structures of papain and actinidin, and on amino acid sequence information for cathepsins B and H, and stem bromelain. *J. Mol. Biol.*, 182(2):317–29, 1985.

89. P. B. Szecsi. The aspartic proteases. *Scand. J. Clin. Lab. Invest. Suppl.*, 210:5–22, 1992.

90. N. D. Rawlings, M. Waller, A. J. Barrett, and A. Bateman. MEROPS: the database of proteolytic enzymes, their substrates and inhibitors. *Nucleic Acids Res.*, 42(D1):D503–9, 2014.

91. K. Oda. Chapter 73 – Scytalidoglutamic Peptidase. In D. Neil, N. D. Rawlings, and G. Salvesen, editors, *Handbook of Proteolytic Enzymes*, pages 301–307. Academic Press, 2013.

92. J. Eames and M. Watkinson. *Metalloenzymes and Electrophilic Catalysis*. eLS. John Wiley & Sons, Ltd., 2001.

93. K. A. McCall, C.-c. Huang, and C. A. Fierke. Function and Mechanism of Zinc Metalloenzymes. *J. Nutr.*, 130(5):1437S–1446S, 2000.

94. R. J. Williams. Metallo-enzyme catalysis. *Chem. Commun. (Camb.)*, 10:1109–13, 2003.

95. V. Pelmenschikov, M. R. Blomberg, and P. E. Siegbahn. A theoretical study of the mechanism for peptide hydrolysis by thermolysin. *J. Biol. Inorg. Chem.*, 7(3):284–98, 2002.

96. H. Matsubara, A. Singer, R. Sasaki, and T. H. Jukes. Observations on the specificity of a thermostable bacterial protease "thermolysin". *Biochem. Biophys. Res. Commun.*, 21(3):242–7, 1965.

97. M. B. Rao, A. M. Tanksale, M. S. Ghatge, and V. V. Deshpande. Molecular and biotechnological aspects of microbial proteases. *Microbiol. Mol. Biol. Rev.*, 62(3):597–635, 1998.

98. S. K. Schowen, K. B. Schowen, and R. L. Schowen. *Thiamin Diphosphate and Vitamin B₁*. eLS. John Wiley & Sons, Ltd., 2001.

99. J. D. Rabinowitz and L. Vastag. Teaching the design principles of metabolism. *Nat. Chem. Biol.*, 8(6):497–501, 2012.

100. T. Yoshimura and N. Esak. Amino acid racemases: functions and mechanisms. *J. Biosci. Bioeng.*, 96(2):103–9, 2003.

101. T. Takahashi-Iñiguez, E. García-Hernandez, R. Arreguín-Espinosa, and M. E. Flores. Role of vitamin B(12) on methylmalonyl-CoA mutase activity. *J. Zhejiang Univ. Sci. B*, 13(6):423–437, 2012.

102. G. L. Holliday, S. A. Rahman, N. Furnham, and J. M. Thornton. Exploring the biological and chemical complexity of the ligases. *J. Mol. Biol.*, 426(10):2098–111, 2014.

103. K. Herrera, R. E. Cahoon, S. Kumaran, and J. Jez. Reaction mechanism of glutathione synthetase from *Arabidopsis thaliana*: site-directed mutagenesis of active site residues. *J. Biol. Chem.*, 282(23):17157–65, 2007.

104. S. Wang and D. Eisenberg. Crystal structure of the pantothenate synthetase from *Mycobacterium tuberculosis*, snapshots of the enzyme in action. *Biochemistry*, 45(6):1554–61, 2006.

105. L. Afriat, C. Roodveldt, G. Manco, and D. S. Tawfik. The latent promiscuity of newly identified microbial lactonases is linked to a recently diverged phosphotriesterase. *Biochemistry*, 45(46):13677–86, 2006.

106. E. Henke and U. T. Bornscheuer. Fluorophoric assay for the high-throughput determination of amidase activity. *Anal. Chem.*, 75(2):255–60, 2003.

107. C. Branneby, P. Carlqvist, A. Magnusson, K. Hult, T. Brinck, and P. Berglund. Carbon-carbon bonds by hydrolytic enzymes. *J. Am. Chem. Soc.*, 125(4):874–5, 2003.

108. O. Torre, I. Alfonso, and V. Gotor. Lipase catalysed Michael addition of secondary amines to acrylonitrile. *Chem. Commun. (Camb.)*, 15:1724–5, 2004.

109. O. Khersonsky and D. S. Tawfik. Enzyme Promiscuity: A Mechanistic and Evolutionary Perspective. *Annu. Rev. Biochem.*, 79(1):471–505, 2010.

110. G. F. da Silva and L. J. Ming. Catechol oxidase activity of di-Cu^{2+}-substituted aminopeptidase from *Streptomyces griseus*. *J. Am. Chem. Soc.*, 127(47):16380–1, 2005.

111. A. Fernandez-Gacio, A. Codina, J. Fastrez, O. Riant, and P. Soumillion. Transforming carbonic anhydrase into epoxide synthase by metal exchange. *ChemBioChem*, 7(7):1013–6, 2006.

112. R. A. Jensen. Enzyme recruitment in evolution of new function. *Annu. Rev. Microbiol.*, 30:409–25, 1976.

113. P. J. O'Brien and D. Herschlag. Catalytic promiscuity and the evolution of new enzymatic activities. *Chem. Biol.*, 6(4):R91–R105, 1999.

114. J. A. Gerlt and P. C. Babbitt. Mechanistically diverse enzyme superfamilies: the importance of chemistry in the evolution of catalysis. *Curr. Opin. Chem. Biol.*, 2(5):607–12, 1998.

115. A. E. Todd, C. A. Orengo, and J. M. Thornton. Evolution of function in protein superfamilies, from a structural perspective. *J. Mol. Biol.*, 307(4):1113–43, 2001.

116. J. A. Gerlt, P. C. Babbitt, and I. Rayment. Divergent evolution in the enolase superfamily: the interplay of mechanism and specificity. *Arch. Biochem. Biophys.*, 433(1):59–70, 2005.

117. L. N. Kinch and N. V. Grishin. Evolution of protein structures and functions. *Curr. Opin. Struct. Biol.*, 12(3):400–8, 2002.

118. M. Y. Galperin, D. R. Walker, and E. V. Koonin. Analogous enzymes: independent inventions in enzyme evolution. *Genome Res.*, 8(8):779–90, 1998.

119. M. Goldsmith and D. S. Tawfik. Directed enzyme evolution: beyond the low-hanging fruit. *Curr. Opin. Struct. Biol.*, 22(4):406–12, 2012.

120. P. A. Romero and F. H. Arnold. Exploring protein fitness landscapes by directed evolution. *Nat. Rev. Mol. Cell Biol.*, 10(12):866–76, 2009.

121. T. E. Barman, S. R. Bellamy, H. Gutfreund, S. E. Halford, and C. Lionne. The identification of chemical intermediates in enzyme catalysis by the rapid quench-flow technique. *Cell. Mol. Life Sci.*, 63(22):2571–83, 2006.

122. J. F. Eccleston, S. R. Martin, and M. J. Schilstra. Rapid kinetic techniques. *Methods Cell Biol.*, 84:445–77, 2008.

123. H. M. Senn and W. Thiel. QM/MM studies of enzymes. *Curr. Opin. Chem. Biol.*, 11(2):182–7, 2007.

124. A. Warshel. Computer simulations of enzyme catalysis: methods, progress, and insights. *Annu. Rev. Biophys. Biomol. Struct.*, 32:425–43, 2003.

125. A. Warshel and M. Karplus. Calculation of ground and excited state potential surfaces of conjugated molecules. I. Formulation and parametrization. *J. Am. Chem. Soc.*, 94(16):5612–5625, 1972.

126. A. Warshel and M. Levitt. Theoretical studies of enzymic reactions: dielectric, electrostatic and steric stabilization of the carbonium ion in the reaction of lysozyme. *J. Mol. Biol.*, 103(2):227–249, 1976.

127. C. K. Lee, C. R. Monk, and R. M. Daniel. Determination of enzyme thermal parameters for rational enzyme engineering and environmental/evolutionary studies. *Methods Mol. Biol.*, 996:219–30, 2013.

128. L. Michaelis and M. L. Menten. Die Kinetik der Invertinwirkung (Kinetics of invertase action). *Biochem. Z.*, 49:333–369, 1913.

129. L. Michaelis, M. L. Menten, K. A. Johnson, and R. S. Goody. The original Michaelis constant: translation of the 1913 Michaelis-Menten paper. *Biochemistry*, 50(39):8264–9, 2011.

130. G. E. Briggs and J. B. Haldane. A note on the kinetics of enzyme action. *Biochem. J.*, 19(2):338–9, 1925.

131. A. Bar-Even, R. Milo, E. Noor, and D. S. Tawfik. The Moderately Efficient Enzyme: Futile Encounters and Enzyme Floppiness. *Biochemistry*, 54(32):4969–77, 2015.

132. G. G. Hammes. Multiple conformational changes in enzyme catalysis. *Biochemistry*, 41(26):8221–8, 2002.

133. G. G. Hammes, S. J. Benkovic, and S. Hammes-Schiffer. Flexibility, diversity, and cooperativity: pillars of enzyme catalysis. *Biochemistry*, 50(48):10422–30, 2011.

134. S. Schnell and C. Mendoza. The condition for pseudo-first-order kinetics in enzymatic reactions is independent of the initial enzyme concentration. *Biophys. Chem.*, 107(2):165–174, 2004.

135. S. Mobashery and L. P. Kotra. Transition State Stabilization. In *Encyclopedia of Life Sciences*. John Wiley & Sons, Ltd., 2002.

136. H. Christensen, M. T. Martin, and S. G. Waley. Beta-lactamases as fully efficient enzymes. Determination of all the rate constants in the acyl-enzyme mechanism. *Biochem. J.*, 266(3):853–61, 1990.

137. M. I. Page. The reactivity of beta-lactams, the mechanism of catalysis and the inhibition of beta-lactamases. *Curr. Pharm. Des.*, 5(11):895–913, 1999.

138. S. Marti, M. Roca, J. Andres, V. Moliner, E. Silla, I. Tunon, and J. Bertran. Theoretical insights in enzyme catalysis. *Chem. Soc. Rev.*, 33(2):98–107, 2004.

139. T. Miyazawa, K. Imagawa, R. Yanagihara, and T. Yamada. Marked dependence on temperature of enantioselectivity in the *Aspergillus oryzae* protease-catalyzed hydrolysis of amino acid esters. *Biotechnol. Tech.*, 11(12):931–933, 1997.

140. B. G. Miller and R. Wolfenden. Catalytic proficiency: the unusual case of OMP decarboxylase. *Annu. Rev. Biochem.*, 71:847–85, 2002.

141. R. Gandour and R. L. Schowen. *Transition States of Biochemical Processes*. Plenum Press, 1978.

142. A. Horovitz and A. Levitzki. An accurate method for determination of receptor-ligand and enzyme-inhibitor dissociation constants from displacement curves. *Proc. Natl. Acad. Sci. USA*, 84(19):6654–8, 1987.

143. M. Mourao, D. Kreitman, and S. Schnell. Unravelling the impact of obstacles in diffusion and kinetics of an enzyme catalysed reaction. *Phys. Chem. Chem. Phys.*, 16(10):4492–503, 2014.

144. A. P. Minton. Influence of macromolecular crowding upon the stability and state of association of proteins: predictions and observations. *J. Pharm. Sci.*, 94(8):1668–75, 2005.

145. S. B. Zimmerman and S. O. Trach. Estimation of macromolecule concentrations and excluded volume effects for the cytoplasm of *Escherichia coli*. *J. Mol. Biol.*, 222(3):599–620, 1991.

146. A. Kahraman and Janet M. Thornton. Methods to characterize the structure of enzyme binding sites. In T. Schwede and M. C. Peitsch, editors, *Computational Structural Biology: Methods & Applications*, volume 1, chapter 8, pages 189–221. World Scientific, 2008.

147. A. Kahraman, R. J. Morris, R. A. Laskowski, and J. M. Thornton. Shape variation in protein binding pockets and their ligands. *J. Mol. Biol.*, 368(1):283–301, 2007.

148. A. Gutteridge and J. M. Thornton. Understanding nature's catalytic toolkit. *Trends Biochem. Sci.*, 30(11):622–629, 2005.

149. J. D. Fischer, G. L. Holliday, S. A. Rahman, and J. M. Thornton. The structures and physicochemical properties of organic cofactors in biocatalysis. *J. Mol. Biol.*, 2010.

150. E. Fischer. Einfluss der Configuration auf die Wirkung der Enzyme. *Ber. Dtsch. Chem. Ges.*, 27(3):2985–2993, 1894.

151. J. Villa and A. Warshel. Energetics and Dynamics of Enzymatic Reactions. *J. Phys. Chem. B*, 105(33):7887–7907, 2001.

152. A. Warshel. Energetics of enzyme catalysis. *Proc. Natl. Acad. Sci. USA*, 75(11):5250–4, 1978.

153. A. Warshel, P. K. Sharma, M. Kato, Y. Xiang, H. Liu, and M. H. M. Olsson. Electrostatic basis for enzyme catalysis. *Chem. Rev.*, 106(8):3210–3235, 2006.

154. D. S. Tawfik. Accuracy-rate tradeoffs: how do enzymes meet demands of selectivity and catalytic efficiency? *Curr. Opin. Chem. Biol.*, 21:73–80, 2014.

155. V. L. Schramm. Introduction:Principles of Enzymatic Catalysis. *Chem. Rev.*, 106(8):3029–3030, 2006.

156. O. Khersonsky, C. Roodveldt, and D. S. Tawfik. Enzyme promiscuity: evolutionary and mechanistic aspects. *Curr. Opin. Chem. Biol.*, 10(5):498–508, 2006.

157. A. Babtie, N. Tokuriki, and F. Hollfelder. What makes an enzyme promiscuous? *Curr. Opin. Chem. Biol.*, 14(2):200–7, 2010.

158. L. Hou, M. T. Honaker, L. M. Shireman, L. M. Balogh, A. G. Roberts, K. C. Ng, A. Nath, and W. M. Atkins. Functional promiscuity correlates with conformational heterogeneity in A-class glutathione S-transferases. *J. Biol. Chem.*, 282(32):23264–74, 2007.

159. G. Krix, A. S. Bommarius, K. Drauz, M. Kottenhahn, M. Schwarm, and M. R. Kula. Enzymatic reduction of α-keto acids leading to l-amino acids, d- or l-hydroxy acids. *J. Biotechnol.*, 53(1):29–39, 1997.

160. H. M. Wilks, K. W. Hart, R. Feeney, C. R. Dunn, H. Muirhead, W. N. Chia, D. A. Barstow, T. Atkinson, A. R. Clarke, and J. J. Holbrook. A specific, highly active malate dehydrogenase by redesign of a lactate dehydrogenase framework. *Science*, 242(4885):1541–4, 1988.

161. C. K. Savile, J. M. Janey, E. C. Mundorff, J. C. Moore, S. Tam, W. R. Jarvis, J. C. Colbeck, A. Krebber, F. J. Fleitz, J. Brands, P. N. Devine, G. W. Huisman, and G. J. Hughes. Biocatalytic asymmetric synthesis of chiral amines from ketones applied to sitagliptin manufacture. *Science*, 329(5989):305–309, 2010.

162. L. Pauling. Molecular Architecture and Biological Reactions. *Chem. Eng. News*, 24(10):1375–1377, 1946.

163. W. P. Jencks. Strain and conformation change in enzymatic catalysis. In N. O. Kaplan and E. P. Kennedy, editors, *Current Aspects of Biochemical Energetics*, page 273. Academic Press, 1966.

164. A. Warshel. Electrostatic origin of the catalytic power of enzymes and the role of preorganized active sites. *J. Biol. Chem.*, 273(42):27035–8, 1998.

165. D. A. Kraut, K. S. Carroll, and D. Herschlag. Challenges in enzyme mechanism and energetics. *Annu. Rev. Biochem.*, 72:517–71, 2003.

166. S. J. Benkovic and S. Hammes-Schiffer. A perspective on enzyme catalysis. *Science*, 301(5637):1196–202, 2003.

167. W. R. Cannon, S. F. Singleton, and S. J. Benkovic. A perspective on biological catalysis. *Nat. Struct. Biol.*, 3(10):821–33, 1996.

168. G. J. Bartlett, C. T. Porter, N. Borkakoti, and J. M. Thornton. Analysis of Catalytic Residues in Enzyme Active Sites. *J. Mol. Biol.*, 324(1):105–121, 2002.

169. M. Garcia-Viloca, J. Gao, M. Karplus, and D. G. Truhlar. How enzymes work: analysis by modern rate theory and computer simulations. *Science*, 303(5655):186–195, 2004.

170. G. L. Holliday, D. E. Almonacid, J. B. O. Mitchell, and J. M. Thornton. The chemistry of protein catalysis. *J. Mol. Biol.*, 372(5):1261–1277, 2007.

171. M. E. Glasner, J. A. Gerlt, and P. C. Babbitt. Evolution of enzyme superfamilies. *Curr. Opin. Chem. Biol.*, 10(5):492–7, 2006.

172. A. Wellner, M. Raitses Gurevich, and D. S. Tawfik. Mechanisms of protein sequence divergence and incompatibility. *PLoS Genet.*, 9(7):e1003665, 2013.

173. N. M. Okeley and W. A. van der Donk. Novel cofactors via post-translational modifications of enzyme active sites. *Chem. Biol.*, 7(7):R159–71, 2000.

174. E. T. Yukl and C. M. Wilmot. Cofactor Biosynthesis through Protein Post-Translational Modification. *Curr. Opin. Chem. Biol.*, 16(1–2):54–59, 2012.

175. R. Matsuzaki, T. Fukui, H. Sato, Y. Ozaki, and K. Tanizawa. Generation of the topa quinone cofactor in bacterial monoamine oxidase by cupric ion-dependent autooxidation of a specific tyrosyl residue. *FEBS Lett.*, 351(3):360–4, 1994.

176. J. J. Smith, A. J. Thomson, A. E. Proudfoot, and T. N. Wells. Identification of an Fe(III)-dihydroxyphenylalanine site in recombinant phosphomannose isomerase from *Candida albicans*. *Eur. J. Biochem.*, 244(2):325–33, 1997.

177. P. A. Kollman, B. Kuhn, O. Donini, M. Perakyla, R. Stanton, and D. Bakowies. Elucidating the nature of enzyme catalysis utilizing a new twist on an old methodology: quantum mechanical-free energy calculations on chemical reactions in enzymes and in aqueous solution. *Acc. Chem. Res.*, 34(1):72–9, 2001.

178. D. Ringe and G. A. Petsko. Biochemistry. How enzymes work. *Science*, 320(5882):1428–9, 2008.

179. T. C. Bruice. Some pertinent aspects of mechanism as determined with small molecules. *Annu. Rev. Biochem.*, 45:331–73, 1976.

180. T. C. Bruice. A view at the millennium: the efficiency of enzymatic catalysis. *Acc. Chem. Res.*, 35(3):139–148, 2002.

181. M. I. Page and W. P. Jencks. Entropic Contributions to Rate Accelerations in Enzymic and Intramolecular Reactions and the Chelate Effect. *Proc. Natl. Acad. Sci. USA*, 68(8):1678–1683, 1971.

182. W. P. Jencks. Binding Energy, Specificity, and Enzymic Catalysis: The Circe Effect. In *Advances in Enzymology and Related Areas of Molecular Biology*, pages 219–410. John Wiley & Sons, Inc., 1975.

183. R. J. Leatherbarrow, A. R. Fersht, and G. Winter. Transition-state stabilization in the mechanism of tyrosyl-tRNA synthetase revealed by protein engineering. *Proc. Natl. Acad. Sci. USA*, 82(23):7840–7844, 1985.

184. D. Burschowsky, A. van Eerde, M. Ökvist, A. Kienhöfer, P. Kast, D. Hilvert, and U. Krengel. Electrostatic transition state stabilization rather than reactant destabilization provides the chemical basis for efficient chorismate mutase catalysis. *Proc. Natl. Acad. Sci. USA*, 111(49):17516–17521, 2014.

185. P. Neumann and K. Tittmann. Marvels of enzyme catalysis at true atomic resolution: distortions, bond elongations, hidden flips, protonation states and atom identities. *Curr. Opin. Struct. Biol.*, 29C:122–133, 2014.

186. E. F. Armstrong. Enzymes. By J. B. S. Haldane, M. A. Monographs on Biochemistry. *Journal of the Society of Chemical Industry*, 49(44):919–920, 1930.

187. A. R. Fersht. Profile of Martin Karplus, Michael Levitt, and Arieh Warshel, 2013 nobel laureates in chemistry. *Proc. Natl. Acad. Sci. USA*, 110(49):19656–7, 2013.

188. K. Fodor, V. Harmat, R. Neutze, L. Szilágyi, L. Gráf, and G. Katona. Enzyme:Substrate Hydrogen Bond Shortening during the Acylation Phase of Serine Protease Catalysis. *Biochemistry*, 45(7):2114–2121, 2006.

189. T. M. Larsen, J. E. Wedekind, I. Rayment, and G. H. Reed. A carboxylate oxygen of the substrate bridges the magnesium ions at the active site of enolase: structure of the yeast enzyme complexed with the equilibrium mixture of 2-phosphoglycerate and phosphoenolpyruvate at 1.8 Å resolution. *Biochemistry*, 35(14):4349–4358, 1996.

190. C. Alhambra, J. Gao, J. C. Corchado, J. Villa, and D. G. Truhlar. Quantum mechanical dynamical effects in an enzyme-catalyzed proton transfer reaction. *J. Am. Chem. Soc.*, 121(10):2253–2258, 1999.

191. N. A. Baker, D. Sept, S. Joseph, M. J. Holst, and J. A. McCammon. Electrostatics of nanosystems: Application to microtubules and the ribosome. *Proc. Natl. Acad. Sci. USA*, 98(18):10037–10041, 2001.

192. L. L. C. Schrodinger. The PyMOL Molecular Graphics System. Version 1. 3r1, 2010.

193. A. R. Fersht. *Enzyme Structure and Mechanism*. W. H. Freeman and Company, 1985.

194. S. C. Kamerlin and A. Warshel. At the dawn of the 21st century: Is dynamics the missing link for understanding enzyme catalysis? *Proteins*, 78(6):1339–75, 2010.

195. M. P. Frushicheva, J. Cao, Z. T. Chu, and A. Warshel. Exploring challenges in rational enzyme design by simulating the catalysis in artificial kemp eliminase. *Proc. Natl. Acad. Sci. USA*, 107(39):16869–16874, 2010.

196. R. A. Marcus. On the theory of oxidation-reduction reactions involving electron transfer. I. *J. Chem. Phys.*, 24(5):966–978, 1956.

197. A. Warshel. *Computer Modeling of Chemical Reactions in Enzymes and Solutions*. John Wiley & Sons, Inc., New York, 1991.

198. M. H. M. Olsson and A. Warshel. Solute solvent dynamics and energetics in enzyme catalysis: the SN2 reaction of dehalogenase as a general benchmark. *J. Am. Chem. Soc.*, 126(46):15167–15179, 2004.

199. R. B. Stockbridge, C. A. Lewis Jr, Y. Yuan, and R. Wolfenden. Impact of temperature on the time required for the establishment of primordial biochemistry, and for the evolution of enzymes. *Proc. Natl. Acad. Sci. USA*, 107(51):22102–5, 2010.

200. A. Gutteridge and J. M. Thornton. Conformational changes observed in enzyme crystal structures upon substrate binding. *J. Mol. Biol.*, 346(1):21–8, 2005.

201. R. Wolfenden. Enzyme catalysis: conflicting requirements of substrate access and transition state affinity. *Mol. Cell. Biochem.*, 3(3):207–11, 1974.

202. H. Deng, A. Lewandowicz, V. L. Schramm, and R. Callender. Activating the phosphate nucleophile at the catalytic site of purine nucleoside phosphorylase: a vibrational spectroscopic study. *J. Am. Chem. Soc.*, 126(31):9516–7, 2004.

203. T. Li, L. Huo, C. Pulley, and A. Liu. Decarboxylation mechanisms in biological system. *Bioorg. Chem.*, 43:2–14, 2012.

204. F. H. Westheimer. Coincidences, decarboxylation, and electrostatic effects. *Tetrahedron*, 51(1):3–20, 1995.

205. M. C. Ho, J. F. Menetret, H. Tsuruta, and K. N. Allen. The origin of the electrostatic perturbation in acetoacetate decarboxylase. *Nature*, 459(7245):393–7, 2009.

206. C. Citek, C. T. Lyons, E. C. Wasinger, and T. D. Stack. Self-assembly of the oxy-tyrosinase core and the fundamental components of phenolic hydroxylation. *Nat. Chem.*, 4(4):317–22, 2012.

207. C. Andreini, I. Bertini, G. Cavallaro, G. L. Holliday, and J. M. Thornton. Metal ions in biological catalysis: from enzyme databases to general principles. *J. Biol. Inorg. Chem.*, 13(8):1205–1218, 2008.

208. T. Dudev and C. Lim. Principles governing Mg, Ca, and Zn binding and selectivity in proteins. *Chem. Rev.*, 103(3):773–88, 2003.

209. E. Carafoli. Calcium signaling: A tale for all seasons. *Proc. Natl. Acad. Sci. USA*, 99(3):1115–1122, 2002.

210. J. A. Gerlt and P. G. Gassman. An explanation for rapid enzyme-catalyzed proton abstraction from carbon acids: importance of late transition states in concerted mechanisms. *J. Am. Chem. Soc.*, 115(24):11552–11568, 1993.

211. W. P. Jencks. Requirements for general acid-base catalysis of complex reactions. *J. Am. Chem. Soc.*, 94(13):4731–4732, 1972.

212. J. A. Gerlt and F. M. Raushel. Evolution of function in (beta/alpha)8-barrel enzymes. *Curr. Opin. Chem. Biol.*, 7(2):252–64, 2003.

213. Q. Cui and M. Karplus. Triosephosphate isomerase: a theoretical comparison of alternative pathways. *J. Am. Chem. Soc.*, 123(10):2284–2290, 2001.

214. J. Jin and U. Hanefeld. The selective addition of water to C=C bonds; enzymes are the best chemists. *Chem. Commun. (Camb.)*, 47(9):2502–10, 2011.

215. R. Bau, I. Brewer, M. Y. Chiang, S. Fujita, J. Hoffman, M. I. Watkins, and T. F. Koetzle. Absolute configuration of a chiral CHD group via neutron diffraction: confirmation of the absolute stereochemistry of the enzymatic formation of malic acid. *Biochem. Biophys. Res. Commun.*, 115(3):1048–52, 1983.

216. W. P. Jencks. General acid-base catalysis of complex reactions in water. *Chem. Rev.*, 72(6):705–718, 1972.

217. P. Hanoian, P. A. Sigala, D. Herschlag, and S. Hammes-Schiffer. Hydrogen bonding in the active site of ketosteroid isomerase: electronic inductive effects and hydrogen bond coupling. *Biochemistry*, 49(48):10339–48, 2010.

218. B. J. Bahnson and J. P. Klinman. Hydrogen tunneling in enzyme catalysis. *Methods Enzymol.*, 249:373–97, 1995.

219. Y. Cha, C. J. Murray, and J. P. Klinman. Hydrogen tunneling in enzyme reactions. *Science*, 243(4896):1325–1330, 1989.

220. Z. X. Liang and J. P. Klinman. Structural bases of hydrogen tunneling in enzymes: progress and puzzles. *Curr. Opin. Struct. Biol.*, 14(6):648–55, 2004.

221. J. P. Layfield and S. Hammes-Schiffer. Hydrogen tunneling in enzymes and biomimetic models. *Chem. Rev.*, 114(7):3466–94, 2014.

222. W. T. Heller. Influence of multiple well defined conformations on small-angle scattering of proteins in solution. *Acta Crystallogr. Sect. D*, 61(Pt 1):33–44, 2005.

223. V. L. Schramm and W. Shi. Atomic motion in enzymatic reaction coordinates. *Curr. Opin. Struct. Biol.*, 11(6):657–65, 2001.

224. G. Zaccai. How soft is a protein? A protein dynamics force constant measured by neutron scattering. *Science*, 288(5471):1604–7, 2000.

225. E. Z. Eisenmesser, D. A. Bosco, M. Akke, and D. Kern. Enzyme dynamics during catalysis. *Science*, 295(5559):1520–3, 2002.

226. D. A. Bosco, E. Z. Eisenmesser, S. Pochapsky, W. I. Sundquist, and D. Kern. Catalysis of cis/trans isomerization in native HIV-1 capsid by human cyclophilin A. *Proc. Natl. Acad. Sci. USA*, 99(8):5247–52, 2002.

227. A. J. Wand. Dynamic activation of protein function: a view emerging from NMR spectroscopy. *Nat. Struct. Biol.*, 8(11):926–31, 2001.

228. P. Zavodszky, J. Kardos, Svingor, and G. A. Petsko. Adjustment of conformational flexibility is a key event in the thermal adaptation of proteins. *Proc. Natl. Acad. Sci. USA*, 95(13):7406–11, 1998.

229. P. Agarwal. Enzymes: An integrated view of structure, dynamics and function. *Microb. Cell Fact.*, 5(1):2, 2006.

230. A. Gora, J. Brezovsky, and J. Damborsky. Gates of enzymes. *Chem. Rev.*, 113(8):5871–923, 2013.

231. A. De Simone, F. A. Aprile, A. Dhulesia, C. M. Dobson, and M. Vendruscolo. Structure of a low-population intermediate state in the release of an enzyme product. *eLife*, 4, 2015.

232. S. Hammes-Schiffer. Catalytic efficiency of enzymes: a theoretical analysis. *Biochemistry*, 52(12):2012–2020, 2012.

233. M. H. Vos, J. C. Lambry, S. J. Robles, D. C. Youvan, J. Breton, and J. L. Martin. Direct observation of vibrational coherence in bacterial reaction centers using femtosecond absorption spectroscopy. *Proc. Natl. Acad. Sci. USA*, 88(20):8885–9, 1991.

234. H. J. Gold. Proton tunneling and enzyme catalysis. *Acta Biotheor.*, 20(1):29–40, 1971.

235. S. Hammes-Schiffer. Hydrogen tunneling and protein motion in enzyme reactions. *Acc. Chem. Res.*, 39(2):93–100, 2006.

236. A. Kohen, T. Jonsson, and J. P. Klinman. Effects of protein glycosylation on catalysis: changes in hydrogen tunneling and enthalpy of activation in the glucose oxidase reaction. *Biochemistry*, 36(22):6854–6854, 1997.

237. J. Rucker, Y. Cha, T. Jonsson, K. L. Grant, and J. P. Klinman. Role of internal thermodynamics in determining hydrogen tunneling in enzyme-catalyzed hydrogen transfer reactions. *Biochemistry*, 31(46):11489–11499, 1992.

238. M. Karplus. Role of conformation transitions in adenylate kinase. *Proc. Natl. Acad. Sci. USA*, 107(17):E71–E71, 2010.

239. K. A. Henzler-Wildman, M. Lei, V. Thai, S. J. Kerns, M. Karplus, and D. Kern. A hierarchy of timescales in protein dynamics is linked to enzyme catalysis. *Nature*, 450(7171):913–6, 2007.

240. K. A. Henzler-Wildman, V. Thai, M. Lei, M. Ott, M. Wolf-Watz, T. Fenn, E. Pozharski, M. A. Wilson, G. A. Petsko, M. Karplus, C. G. Hubner, and D. Kern. Intrinsic motions along an enzymatic reaction trajectory. *Nature*, 450(7171):838–44, 2007.

241. A. V. Pisliakov, J. Cao, S. C. L. Kamerlin, and A. Warshel. Enzyme millisecond conformational dynamics do not catalyze the chemical step. *Proc. Natl. Acad. Sci. USA*, 106(41):17359–17364, 2009.

242. S. D. Schwartz. Protein dynamics and the enzymatic reaction coordinate. *Top. Curr. Chem.*, 337:189–208, 2013.

243. R. G. Silva, A. S. Murkin, and V. L. Schramm. Femtosecond dynamics coupled to chemical barrier crossing in a Born-Oppenheimer enzyme. *Proc. Natl. Acad. Sci. USA*, 108(46):18661–5, 2011.

244. M. Born and R. Oppenheimer. Zur Quantentheorie der Molekeln. *Ann. Phys.*, 389(20):457–484, 1927.

245. J. Cohen, K. Kim, P. King, M. Seibert, and K. Schulten. Finding gas diffusion pathways in proteins: application to O_2 and H_2 transport in CpI [FeFe]-hydrogenase and the role of packing defects. *Structure*, 13(9):1321–9, 2005.

246. M. Elias, G. Wieczorek, S. Rosenne, and D. S. Tawfik. The universality of enzymatic rate-temperature dependency. *Trends Biochem. Sci.*, 39(1):1–7, 2014.

247. A. D. McNaught and A. Wilkinson. *IUPAC. Compendium of Chemical Terminology (2nd ed. (the "Gold Book"))*. Blackwell Scientific Publications, 1997.

248. J. D. Fischer, G. L. Holliday, and J. M. Thornton. The CoFactor database: organic cofactors in enzyme catalysis. *Bioinformatics*, 26(19):2496–2497, 2010.

249. S. Ghisla and D. E. Edmondson. *Flavin Coenzymes*. eLS. John Wiley & Sons, Ltd., 2001.

250. V. Joosten and W. J. H. van Berkel. Flavoenzymes. *Curr. Opin. Chem. Biol.*, 11(2):195–202, 2007.

251. A. K. Samland, S. Baier, M. Schurmann, T. Inoue, S. Huf, G. Schneider, G. A. Sprenger, and T. Sandalova. Conservation of structure and mechanism within the transaldolase enzyme family. *FEBS J.*, 279(5):766–78, 2012.

252. A. K. Samland, M. Rale, G. A. Sprenger, and W. D. Fessner. The transaldolase family: new synthetic opportunities from an ancient enzyme scaffold. *ChemBioChem*, 12(10):1454–74, 2011.

253. C. Y. Chou, L. P. Yu, and L. Tong. Crystal structure of biotin carboxylase in complex with substrates and implications for its catalytic mechanism. *J. Biol. Chem.*, 284(17):11690–7, 2009.

254. G. L. Waldrop. *Biotin*. eLS. John Wiley & Sons, Ltd., 2002.

255. S. Chaykin. Nicotinamide coenzymes. *Annu. Rev. Biochem.*, 36(1):149–170, 1967.

256. F. L. Crane. Biochemical functions of coenzyme Q10. *J. Am. Coll. Nutr.*, 20(6):591–8, 2001.

257. U. Schmidt, P. Grafen, and H. W. Goedde. Chemistry and Biochemistry of α-Lipoic Acid. *Angew. Chem. Int. Ed.*, 4(10):846–856, 1965.

258. R. J. Williams. The Chemistry and Biochemistry of Pantothenic Acid. In *Advances in Enzymology and Related Areas of Molecular Biology*, pages 253–287. John Wiley & Sons, Inc., 1943.

259. W. Buckel. *Cobalamin Coenzymes and Vitamin B_{12}*. eLS. John Wiley & Sons, Ltd., 2001.

260. G. D. Markham. *S-Adenosylmethionine*. eLS. John Wiley & Sons, Ltd., 2001.

261. P. K. Sharma, Z. T. Chu, M. H. Olsson, and A. Warshel. A new paradigm for electrostatic catalysis of radical reactions in vitamin B_{12} enzymes. *Proc. Natl. Acad. Sci. USA*, 104(23):9661–6, 2007.

262. C. Makins, A. V. Pickering, C. Mariani, and K. R. Wolthers. Mutagenesis of a conserved glutamate reveals the contribution of electrostatic energy to adenosylcobalamin co-C bond homolysis in ornithine 4,5-aminomutase and methylmalonyl-CoA mutase. *Biochemistry*, 52(5):878–88, 2013.

263. A. Jordan and P. Reichard. Ribonucleotide reductases. *Annu. Rev. Biochem.*, 67:71–98, 1998.

264. G. T. Babcock, M. K. El-Deeb, P. O. Sandusky, M. M. Whittaker, and J. W. Whittaker. Electron paramagnetic resonance and electron nuclear double resonance spectroscopies of the radical site in galactose oxidase and of thioether-substituted phenol model compounds. *J. Am. Chem. Soc.*, 114(10):3727–3734, 1992.

265. A. F. Wagner, M. Frey, F. A. Neugebauer, W. Schafer, and J. Knappe. The free radical in pyruvate formate-lyase is located on glycine-734. *Proc. Natl. Acad. Sci. USA*, 89(3):996–1000, 1992.

266. J. B. Broderick. Biochemistry: A radically different enzyme. *Nature*, 465(7300):877–8, 2010.

267. P. A. Frey, A. D. Hegeman, and G. H. Reed. Free radical mechanisms in enzymology. *Chem. Rev.*, 106(8):3302–3316, 2006.

268. Y. Zhang, X. Zhu, A. T. Torelli, M. Lee, B. Dzikovski, R. M. Koralewski, E. Wang, J. Freed, C. Krebs, S. E. Ealick, and H. Lin. Diphthamide biosynthesis requires an organic radical generated by an iron-sulphur enzyme. *Nature*, 465(7300):891–6, 2010.

269. S. Nadaraia, G. J. Yohrling, G. C. T. Jiang, J. M. Flanagan, and K. E. Vrana. *Enzyme Activity: Control*. eLS. John Wiley & Sons, Ltd., 2001.

270. R. A. Copeland. *Enzymology Methods*. eLS. John Wiley & Sons, Ltd., 2001.

271. J. G. Robertson. Enzymes as a special class of therapeutic target: clinical drugs and modes of action. *Curr. Opin. Struct. Biol.*, 17(6):674–9, 2007.

272. M. Rask-Andersen, M. S. Almén, and H. B. Schiöth. Trends in the exploitation of novel drug targets. *Nat. Rev. Drug Discov.*, 10(8):579–590, 2011.

273. H. Kubinyi. Structure-based design of enzyme inhibitors and receptor ligands. *Curr. Opin. Drug Discov. Devel.*, 1(1):4–15, 1998.

274. J. M. Wood, J. Maibaum, J. Rahuel, M. G. Grutter, N. C. Cohen, V. Rasetti, H. Ruger, R. Goschke, S. Stutz, W. Fuhrer, W. Schilling, P. Rigollier, Y. Yamaguchi, F. Cumin, H. P. Baum, C. R. Schnell, P. Herold, R. Mah, C. Jensen, E. O'Brien, A. Stanton, and M. P. Bedigian. Structure-based design of aliskiren, a novel orally effective renin inhibitor. *Biochem. Biophys. Res. Commun.*, 308(4):698–705, 2003.

275. J. F. Morrison. *Enzyme Activity: Reversible Inhibition*. eLS. John Wiley & Sons, Ltd., 2001.

276. Z. A. Knight and K. M. Shokat. Features of selective kinase inhibitors. *Chem. Biol.*, 12(6):621–637, 2005.

277. G. L. Patrick. *An Introduction to Medicinal Chemistry*. Oxford University Press, 5th edition, 2013.

278. A. B. Rose and S. Dube. The purification and kinetic properties of biophosphoglycerate synthase from horse red blood cells. *Arch. Biochem. Biophys.*, 177(1):284–92, 1976.

279. A. B. Pardee and V. R. Potter. Malonate inhibition of oxidations in the Krebs tricarboxylic acid cycle. *J. Biol. Chem.*, 178(1):241–50, 1949.

280. R. J. Henry. The mode of action of sulfonamides. *Bacteriol. Rev.*, 7(4):175, 1943.

281. J. G. Robertson. Mechanistic basis of enzyme-targeted drugs. *Biochemistry*, 44(15):5561–71, 2005.

282. W. Lew, X. Chen, and C. U. Kim. Discovery and development of GS 4104 (oseltamivir): an orally active influenza neuraminidase inhibitor. *Curr. Med. Chem.*, 7(6):663–72, 2000.

283. M. von Itzstein. The war against influenza: discovery and development of sialidase inhibitors. *Nat. Rev. Drug Discov.*, 6(12):967–74, 2007.

284. N. F. Bras, M. J. Ramos, and P. A. Fernandes. The catalytic mechanism of mouse renin studied with QM/MM calculations. *Phys. Chem. Chem. Phys.*, 14(36):12605–13, 2012.

285. H. E. Umbarger and B. Brown. Threonine deamination in *Escherichia coli*. II. Evidence for two L-threonine deaminases. *J. Bacteriol.*, 73(1):105–12, 1957.

286. A. Cornish-Bowden. Why is uncompetitive inhibition so rare? A possible explanation, with implications for the design of drugs and pesticides. *FEBS Lett.*, 203(1):3–6, 1986.

287. S. R. Nahorski, C. I. Ragan, and R. A. Challiss. Lithium and the phosphoinositide cycle: an example of uncompetitive inhibition and its pharmacological consequences. *Trends Pharmacol. Sci.*, 12(8):297–303, 1991.

288. M. R. Boocock and J. R. Coggins. Kinetics of 5-enolpyruvylshikimate-3-phosphate synthase inhibition by glyphosate. *FEBS Lett.*, 154(1):127–33, 1983.

289. Y. Blat. Non-competitive inhibition by active site binders. *Chem. Biol. Drug Des.*, 75(6):535–40, 2010.

290. G. A. Dunaway. A review of animal phosphofructokinase isozymes with an emphasis on their physiological role. *Mol. Cell. Biochem.*, 52(1):75–91, 1983.

291. R. G. Kemp and L. G. Foe. Allosteric regulatory properties of muscle phosphofructokinase. *Mol. Cell. Biochem.*, 57(2):147–54, 1983.

292. G. Campos, V. Guixe, and J. Babul. Kinetic mechanism of phosphofructokinase-2 from *Escherichia coli*. A mutant enzyme with a different mechanism. *J. Biol. Chem.*, 259(10):6147–52, 1984.

293. M. A. Bogoyevitch and D. P. Fairlie. A new paradigm for protein kinase inhibition: blocking phosphorylation without directly targeting ATP binding. *Drug Discov. Today*, 12(15–16):622–633, 2007.

294. B. J. Druker. Imatinib as a Paradigm of Targeted Therapies. *Adv. Cancer Res.*, 91:1–30, 2004.

295. R. A. Spence, W. M. Kati, K. S. Anderson, and K. A. Johnson. Mechanism of inhibition of HIV-1 reverse transcriptase by nonnucleoside inhibitors. *Science*, 267(5200):988–993, 1995.

296. A. G. McDonald and K. F. Tipton. *Enzymes: Irreversible Inhibition*. eLS. John Wiley & Sons, Ltd., 2001.

297. S. N. Abramson, Z. Radic, D. Manker, D. J. Faulkner, and P. Taylor. Onchidal: a naturally occurring irreversible inhibitor of acetylcholinesterase with a novel mechanism of action. *Mol. Pharmacol.*, 36(3):349–54, 1989.

298. A. F. Kluge and R. C. Petter. Acylating drugs: redesigning natural covalent inhibitors. *Curr. Opin. Chem. Biol.*, 14(3):421–7, 2010.

299. M. H. Potashman and M. E. Duggan. Covalent modifiers: an orthogonal approach to drug design. *J. Med. Chem.*, 52(5):1231–46, 2009.

300. L. Tóth, L. Muszbek, and I. Komáromi. Mechanism of the irreversible inhibition of human cyclooxygenase-1 by aspirin as predicted by QM/MM calculations. *J. Mol. Graph. Model.*, 40:99–109, 2013.

301. J. Lei, Y. Zhou, D. Xie, and Y. Zhang. Mechanistic insights into a classic wonder drug–aspirin. *J. Am. Chem. Soc.*, 137(1):70–3, 2015.

302. R. R. Yocum, D. J. Waxman, J. R. Rasmussen, and J. L. Strominger. Mechanism of penicillin action: penicillin and substrate bind covalently to the same active site serine in two bacterial D-alanine carboxypeptidases. *Proc. Natl. Acad. Sci. USA*, 76(6):2730–4, 1979.

303. P. Lorentzon, R. Jackson, B. Wallmark, and G. Sachs. Inhibition of $(H^+ + K^+)$-ATPase by omeprazole in isolated gastric vesicles requires proton transport. *BBA – Biomembranes*, 897(1):41–51, 1987.

304. J. M. Shin and G. Sachs. Pharmacology of proton pump inhibitors. *Curr. Gastroenterol. Rep.*, 10(6):528–534, 2008.

305. M. B. Čolović, D. Z. Krstić, T. D. Lazarević-Pašti, A. M. Bondžić, and V. M. Vasić. Acetylcholinesterase inhibitors: pharmacology and toxicology. *Curr. Neuropharmacol.*, 11(3):315–335, 2013.

306. J. M. Chesworth, T. Stuchbury, and J. R. Scaife. *An Introduction to Agricultural Biochemistry*. Springer, 1998.

307. R. C. Gupta. *Toxicology of Organophosphate & Carbamate Compounds*. Elsevier Academic Press, London, 2006.

308. Z. Prokop, F. Oplustil, J. DeFrank, and J. Damborsky. Enzymes fight chemical weapons. *Biotechnol. J.*, 1(12):1370–80, 2006.

309. L. Wen, Z. W. Miao, and W. D. Qing. Chemical modification of xylanase from Trichosporon cutaneum shows the presence of carboxyl groups and cysteine residues essential for enzyme activity. *J. Protein Chem.*, 18(6):677–86, 1999.

310. A. M. Soares, R. Guerra-Sa, C. R. Borja-Oliveira, V. M. Rodrigues, L. Rodrigues-Simioni, V. Rodrigues, M. R. Fontes, B. Lomonte, J. M. Gutierrez, and J. R. Giglio. Structural and functional characterization of BnSP-7, a Lys-49 myotoxic phospholipase A(2) homologue from *Bothrops neuwiedi pauloensis* venom. *Arch. Biochem. Biophys.*, 378(2):201–9, 2000.

311. N. Hadad, W. Feng, and V. Shoshan-Barmatz. Modification of ryanodine receptor/Ca^{2+} release channel with dinitrofluorobenzene. *Biochem. J.*, 342 (Pt 1):239–48, 1999.

312. C. Goffin and J. M. Ghuysen. Multimodular penicillin-binding proteins: an enigmatic family of orthologs and paralogs. *Microbiol. Mol. Biol. Rev.*, 62(4):1079–93, 1998.

313. I. Chopra. *Antibiotics*. eLS. John Wiley & Sons, Ltd., 2001.

314. C. J. Fowler, T. J. Mantle, and K. F. Tipton. The nature of the inhibition of rat liver monoamine oxidase types A and B by the acetylenic inhibitors clorgyline, l-deprenyl and pargyline. *Biochem. Pharmacol.*, 31(22):3555–61, 1982.

315. M. G. Palfreyman, P. Bey, and A. Sjoerdsma. Enzyme-activated/mechanism-based inhibitors. *Essays Biochem.*, 23:28–81, 1987.

316. I. V. Bijnsdorp, E. M. Comijn, J. M. Padron, W. H. Gmeiner, and G. J. Peters. Mechanisms of action of FdUMP[10]: metabolite activation and thymidylate synthase inhibition. *Oncol. Rep.*, 18(1):287–91, 2007.

317. F. Pea. Pharmacology of drugs for hyperuricemia. Mechanisms, kinetics and interactions. *Contrib. Nephrol.*, 147:35–46, 2005.

318. J. C. Sheehan. *Enchanted Ring: Untold Story of Penicillin*. MIT Press, 1982.

319. P. Benveniste, W. Zhu, and A. Cohen. Interference with thymocyte differentiation by an inhibitor of S-adenosylhomocysteine hydrolase. *J. Immunol.*, 155(2):536–44, 1995.

320. J. J. Sanchez, G. Monaghan, C. Borsting, G. Norbury, N. Morling, and H. B. Gaspar. Carrier frequency of a nonsense mutation in the adenosine deaminase (ADA) gene implies a high incidence of ADA-deficient severe combined immunodeficiency (SCID) in Somalia and a single, common haplotype indicates common ancestry. *Ann. Hum. Genet.*, 71(Pt 3):336–47, 2007.

321. S. Balasubramaniam, J. A. Duley, and J. Christodoulou. Inborn errors of purine metabolism: clinical update and therapies. *J. Inherit. Metab. Dis.*, 37(5):669–86, 2014.

322. E. F. Neufeld. Enzyme replacement therapy: a brief history. In A. Mehta, M. Beck, and G. Sunder-Plassmann, editors, *Fabry Disease: Perspectives from 5 Years of FOS*. Oxford PharmaGenesis, 2006.

323. D. A. Savaiano. Lactose digestion from yogurt: mechanism and relevance. *Am. J. Clin. Nutr.*, 99(5 Suppl):1251S–5S, 2014.

324. P. L. Wolf. History of diagnostic enzymology: A review of significant investigations. *Clin. Chim. Acta*, 369(2):144–147, 2006.

325. R. L. Hall. Principles of Clinical Pathology. In P.S. Sahota, J.A. Popp, J.F. Hardisty, and C. Gopinath, editors, *Toxicologic Pathology: Nonclinical Safety Assessment*, chapter 6, pages 133–174. CRC Press, 2013.

326. K. Sanderson. Chemistry: Enzyme expertise. *Nature*, 471(7338):397–8, 2011.

327. A. M. Thayer. Biocatalysis. *Chem. Eng. News Archive*, 90(22):13–18, 2012.

328. J. Tao and J.-H. Xu. Biocatalysis in development of green pharmaceutical processes. *Curr. Opin. Chem. Biol.*, 13(1):43–50, 2009.

329. B. M. Nestl, B. A. Nebel, and B. Hauer. Recent progress in industrial biocatalysis. *Curr. Opin. Chem. Biol.*, 15(2):187–193, 2011.

330. K. M. Koeller and C. H. Wong. Enzymes for chemical synthesis. *Nature*, 409(6817):232–40, 2001.

331. O. Jurček, M. Wimmerová, and Z. Wimmer. Selected chiral alcohols: Enzymic resolution and reduction of convenient substrates. *Coord. Chem. Rev.*, 252(5–7):767–781, 2008.

332. C. A. G. N. Montalbetti and V. Falque. Amide bond formation and peptide coupling. *Tetrahedron*, 61(46):10827–10852, 2005.

333. U. T. Bornscheuer, G. W. Huisman, R. J. Kazlauskas, S. Lutz, J. C. Moore, and K. Robins. Engineering the third wave of biocatalysis. *Nature*, 485(7397):185–94, 2012.

334. O. Kirk, T. V. Borchert, and C. C. Fuglsang. Industrial enzyme applications. *Curr. Opin. Biotechnol.*, 13(4):345–51, 2002.

335. H. E. Schoemaker, D. Mink, and M. G. Wubbolts. Dispelling the myths: biocatalysis in industrial synthesis. *Science*, 299(5613):1694–7, 2003.

336. B. Botta and S. Cacchi. *Biochemical methods of synthesis*, volume 2 of *Fundamentals Of Chemistry*. UNESCO-EOLSS, 2012.

337. P. Busca, F. Paradisi, E. Moynihan, A. R. Maguire, and P. C. Engel. Enantioselective synthesis of non-natural amino acids using phenylalanine dehydrogenases modified by site-directed mutagenesis. *Org. Biomol. Chem.*, 2(18):2684–91, 2004.

338. A. S. Bommarius, J. K. Blum, and M. J. Abrahamson. Status of protein engineering for biocatalysts: how to design an industrially useful biocatalyst. *Curr. Opin. Chem. Biol.*, 2011.

339. K. M. Polizzi, A. S. Bommarius, J. M. Broering, and J. F. Chaparro-Riggers. Stability of biocatalysts. *Curr. Opin. Chem. Biol.*, 11(2):220–5, 2007.

340. V. Stepankova, S. Bidmanova, T. Koudelakova, Z. Prokop, R. Chaloupkova, and J. Damborsky. Strategies for stabilization of enzymes in organic solvents. *ACS Catal.*, 3(12):2823–2836, 2013.

341. L. Fransson. *Enzyme substrate solvent interactions: a case study on serine hydrolases.* PhD thesis, Royal Institute of Technology, AlbaNova University Center, 2008.

342. G. W. Huisman, J. Liang, and A. Krebber. Practical chiral alcohol manufacture using ketoreductases. *Curr. Opin. Chem. Biol.*, 14(2):122–9, 2010.

343. J. D. Keasling. Manufacturing molecules through metabolic engineering. *Science*, 330(6009):1355–8, 2010.

APPENDIX: ENZYME NOMENCLATURE RECOMMENDATIONS OF THE NC-IUBMB[*1]

Enzyme Nomenclature Contents

1. **Oxidoreductases**

[*1]G. P. Moss. *Enzyme nomenclature: recommendations of the Nomenclature Committee of the International Union of Biochemistry and Molecular Biology on the nomenclature and classification of enzymes by the reactions they catalyse.* NC-IUBMB, 2008. http://www.chem.qmul.ac.uk/iubmb/enzyme/.

1.5.7   With an iron-sulfur protein as acceptor

1.5.8   With a flavin as acceptor

1.5.99   With other acceptors

1.6   Acting on NADH or NADPH

1.6.1   With NAD^+ or $NADP^+$ as acceptor

1.6.2   With a heme protein as acceptor

1.6.3   With oxygen as acceptor

1.6.4   With a disulfide as acceptor (deleted sub-subclass)

1.6.5   With a quinone or similar compound as acceptor

1.6.6   With a nitrogenous group as acceptor

1.6.7   With an iron-sulfur protein as acceptor (deleted sub-subclass)

1.6.8   With a flavin as acceptor (deleted sub-subclass)

1.6.99   With other acceptors

1.7   Acting on other nitrogenous compounds as donors

1.7.1   With NAD^+ or $NADP^+$ as acceptor

1.7.2   With a cytochrome as acceptor

1.7.3   With oxygen as acceptor

1.7.5   With a quinone or similar compound as acceptor

1.7.7   With an iron-sulfur protein as acceptor

1.7.99   With other acceptors

1.8   Acting on a sulfur group of donors

1.8.1   With NAD^+ or $NADP^+$ as acceptor

1.8.2   With a cytochrome as acceptor

1.8.3   With oxygen as acceptor

1.8.4   With a disulfide as acceptor

1.8.5   With a quinone or similar compound as acceptor

1.8.6   With a nitrogenous group as acceptor (deleted sub-subclass)

1.8.7   With an iron-sulfur protein as acceptor

1.8.98   With other, known, acceptors

1.8.99   With other acceptors

1.9   Acting on a heme group of donors

1.9.3   With oxygen as acceptor

1.9.6   With a nitrogenous group as acceptor

1.9.99   With other acceptors

1.10   Acting on diphenols and related substances as donors

1.10.1   With NAD^+ or $NADP^+$ as acceptor

1.10.2   With a cytochrome as acceptor

1.10.3   With oxygen as acceptor

1.10.99 With other acceptors

1.11   Acting on a peroxide as acceptor

1.11.1   Peroxidases

1.12   Acting on hydrogen as donor

1.12.1   With NAD^+ or $NADP^+$ as acceptor

1.12.2   With a cytochrome as acceptor

1.12.5   With a quinone or similar compound as acceptor

1.12.7   With an iron-sulfur protein as acceptor

1.12.98 With other, known, acceptors

1.12.99 With other acceptors

1.13   Acting on single donors with O_2 as oxidant and incorporation of oxygen into the substrate (oxygenases). The oxygen incorporated need not be derived from O_2

    1.13.11 With incorporation of two atoms of oxygen

    1.13.12 With incorporation of one atom of oxygen (internal monooxygenases or internal mixed-function oxidases)

    1.13.99 Miscellaneous

1.14   Acting on paired donors, with O_2 as oxidant and incorporation or reduction of oxygen. The oxygen incorporated need not be derived from O_2

    1.14.1   With NADH or NADPH as one donor (deleted sub-subclass)

    1.14.2   With ascorbate as one donor (deleted sub-subclass)

    1.14.3   With reduced pteridine as one donor (deleted sub-subclass)

    1.14.11 With 2-oxoglutarate as one donor, and incorporation of one atom of oxygen into each donor

    1.14.12 With NADH or NADPH as one donor, and incorporation of two atoms of oxygen into the other donor

    1.14.13 With NADH or NADPH as one donor, and incorporation of one atom of oxygen into the other donor

    1.14.14 With reduced flavin or flavoprotein as one donor, and incorporation of one atom of oxygen into the other donor

    1.14.15 With reduced iron-sulfur protein as one donor, and incorporation of one atom of oxygen into the other donor

    1.14.16 With reduced pteridine as one donor, and incorporation of one atom of oxygen into the other donor

    1.14.17 With reduced ascorbate as one donor, and incorporation of one atom of oxygen into the other donor

    1.14.18 With another compound as one donor, and incorporation of one atom of oxygen into the other donor

    1.14.19 With oxidation of a pair of donors resulting in the reduction of O_2 to two molecules of water

    1.14.20 With 2-oxoglutarate as one donor, and the other dehydrogenated

    1.14.21 With NADH or NADPH as one donor, and the other dehydrogenated

    1.14.99 Miscellaneous

1.15   Acting on superoxide as acceptor

    1.15.1   Acting on superoxide as acceptor (only sub-subclass identified to date)

1.16   Oxidizing metal ions

    1.16.1   With NAD^+ or $NADP^+$ as acceptor

    1.16.3   With oxygen as acceptor

    1.16.8   With a flavin as acceptor

1.17   Acting on CH or CH_2 groups

    1.17.1   With NAD^+ or $NADP^+$ as acceptor

1.17.2   With a cytochrome as acceptor

1.17.3   With oxygen as acceptor

1.17.4   With a disulfide as acceptor

1.17.5   With a quinone or similar compound as acceptor

1.17.7   With an iron-sulfur protein as acceptor

1.17.99 With other acceptors

1.18   Acting on iron-sulfur proteins as donors

1.18.1   With NAD$^+$ or NADP$^+$ as acceptor

1.18.2   With dinitrogen as acceptor (deleted sub-subclass)

1.18.3   With H$^+$ as acceptor (deleted sub-subclass)

1.18.6   With dinitrogen as acceptor

1.18.96 With other, known, acceptors (deleted sub-subclass)

1.18.99 With H$^+$ as acceptor (deleted sub-subclass)

1.19   Acting on reduced flavodoxin as donor

1.19.6   With dinitrogen as acceptor

1.20   Acting on phosphorus or arsenic in donors

1.20.1   With NAD$^+$ or NADP$^+$ as acceptor

1.20.4   With disulfide as acceptor

1.20.98 With other, known, acceptors

1.20.99 With other acceptors

1.21   Acting on X−H and Y−H to form an X−Y bond

1.21.3   With oxygen as acceptor

1.21.4   With a disulfide as acceptor

1.21.99 With other acceptors

1.22   Acting on halogen in donors

1.22.1   With NAD(P)$^+$ as acceptor

1.97   Other oxidoreductases

1.97.1   Sole sub-subclass for oxidoreductases that do not belong in the other subclasses

2. Transferases

2.1   Transferring one-carbon groups

2.1.1   Methyltransferases

2.1.2   Hydroxymethyl-, formyl- and related transferases

2.1.3   Carboxy- and carbamoyltransferases

2.1.4   Amidinotransferases

2.2   Transferring aldehyde or ketonic groups

2.2.1   Transketolases and transaldolases

2.3   Acyltransferases

2.3.1   Transferring groups other than aminoacyl groups

2.3.2   Aminoacyltransferases

2.3.3   Acyl groups converted into alkyl groups on transfer

2.4   Glycosyltransferases

2.4.1   Hexosyltransferases

2.4.2   Pentosyltransferases

2.4.99   Transferring other glycosyl groups

2.5   Transferring alkyl or aryl groups, other than methyl groups

    2.5.1   Transferring alkyl or aryl groups, other than methyl groups (only sub-subclass identified to date)

2.6   Transferring nitrogenous groups

    2.6.1   Transaminases

    2.6.2   Amidinotransferases (deleted sub-subclass)

    2.6.3   Oximinotransferases

    2.6.99   Transferring other nitrogenous groups

2.7   Transferring phosphorus-containing groups

    2.7.1   Phosphotransferases with an alcohol group as acceptor

    2.7.2   Phosphotransferases with a carboxy group as acceptor

    2.7.3   Phosphotransferases with a nitrogenous group as acceptor

    2.7.4   Phosphotransferases with a phosphate group as acceptor

    2.7.5   Phosphotransferases with regeneration of donors, apparently catalysing intramolecular transfers (deleted sub-subclass)

    2.7.6   Diphosphotransferases

    2.7.7   Nucleotidyltransferases

    2.7.8   Transferases for other substituted phosphate groups

    2.7.9   Phosphotransferases with paired acceptors

    2.7.10   Protein-tyrosine kinases

    2.7.11   Protein-serine/threonine kinases

    2.7.12   Dual-specificity kinases (those acting on Ser/Thr and Tyr residues)

    2.7.13   Protein-histidine kinases

    2.7.99   Other protein kinases

2.8   Transferring sulfur-containing groups

    2.8.1   Sulfurtransferases

    2.8.2   Sulfotransferases

    2.8.3   CoA-transferases

    2.8.4   Transferring alkylthio groups

    2.9   Transferring selenium-containing groups

    2.9.1   Selenotransferases

3. Hydrolases

3.1   Acting on ester bonds

    3.1.1   Carboxylic-ester hydrolases

    3.1.2   Thioester hydrolases

    3.1.3   Phosphoric-monoester hydrolases

    3.1.4   Phosphoric-diester hydrolases

    3.1.5   Triphosphoric-monoester hydrolases

    3.1.6   Sulfuric-ester hydrolases

    3.1.7   Diphosphoric-monoester hydrolases

    3.1.8   Phosphoric-triester hydrolases

    3.1.11   Exodeoxyribonucleases producing 5′-phosphomonoesters

3.1.13   Exoribonucleases producing 5′-phosphomonoesters

3.1.14   Exoribonucleases producing 3′-phosphomonoesters

3.1.15   Exonucleases that are active with either ribo- or deoxyribonucleic acids and produce 5′-phosphomonoesters

3.1.16   Exonucleases that are active with either ribo- or deoxyribonucleic acids and produce 3′-phosphomonoesters

3.1.21   Endodeoxyribonucleases producing 5′-phosphomonoesters

3.1.22   Endodeoxyribonucleases producing 3′-phosphomonoesters

3.1.23   Site-specific endodeoxyribonucleases: cleavage is sequence specific (deleted sub-subclass)

3.1.24   Site specific endodeoxyribonucleases: cleavage is not sequence specific (deleted sub-subclass)

3.1.25   Site-specific endodeoxyribonucleases that are specific for altered bases

3.1.26   Endoribonucleases producing 5′-phosphomonoesters

3.1.27   Endoribonucleases producing 3′-phosphomonoesters

3.1.30   Endoribonucleases that are active with either ribo- or deoxyribonucleic acids and produce 5′-phosphomonoesters

3.1.31   Endoribonucleases that are active with either ribo- or deoxyribonucleic acids and produce 3′-phosphomonoesters

3.2   Glycosylases

3.2.1   Glycosidases, i.e., enzymes that hydrolyse O- and S-glycosyl compounds

3.2.2   Hydrolysing N-glycosyl compounds

3.2.3   Hydrolysing S-glycosyl compounds (deleted sub-subclass)

3.3   Acting on ether bonds

3.3.1   Thioether and trialkylsulfonium hydrolases

3.3.2   Ether hydrolases

3.4   Acting on peptide bonds (peptidases)

3.4.1 α-Amino-acyl-peptide hydrolases (deleted sub-subclass)

3.4.2   Peptidyl-amino-acid hydrolases (deleted sub-subclass)

3.4.3   Dipeptide hydrolases (deleted sub-subclass)

3.4.4   Peptidyl peptide hydrolases (deleted sub-subclass)

3.4.11   Aminopeptidases

3.4.12   Peptidylamino-acid hydrolases or acylamino-acid hydrolases (deleted sub-subclass)

3.4.13   Dipeptidases

3.4.14   Dipeptidyl-peptidases and tripeptidyl-peptidases

3.4.15   Peptidyl-dipeptidases

3.4.16   Serine-type carboxypeptidases

3.4.17   Metallocarboxypeptidases

3.4.18   Cysteine-type carboxypeptidases

3.4.19   Omega peptidases

3.4.21   Serine endopeptidases

3.4.22   Cysteine endopeptidases

3.4.23   Aspartic endopeptidases

3.4.24   Metalloendopeptidases

3.4.25   Threonine endopeptidases

3.4.99   Endopeptidases of unknown catalytic mechanism (sub-subclass is currently empty)

3.5   Acting on carbon-nitrogen bonds, other than peptide bonds

   3.5.1   In linear amides

   3.5.2   In cyclic amides

   3.5.3   In linear amidines

   3.5.4   In cyclic amidines

   3.5.5   In nitriles

   3.5.99   In other compounds

3.6   Acting on acid anhydrides

   3.6.1   In phosphorus-containing anhydrides

   3.6.2   In sulfonyl-containing anhydrides

   3.6.3   Acting on acid anhydrides to catalyse transmembrane movement of substances

   3.6.4   Acting on acid anhydrides to facilitate cellular and subcellular movement

   3.6.5   Acting on GTP to facilitate cellular and subcellular movement

3.7   Acting on carbon-carbon bonds

   3.7.1   In ketonic substances

3.8   Acting on halide bonds

   3.8.1   In carbon-halide compounds

   3.8.2   In phosphorus-halide compounds (deleted sub-subclass)

3.9   Acting on phosphorus-nitrogen bonds

   3.9.1   Acting on phosphorus-nitrogen bonds (only sub-subclass identified to date)

3.10   Acting on sulfur-nitrogen bonds

   3.10.1   Acting on sulfur-nitrogen bonds (only sub-subclass identified to date)

   3.11   Acting on carbon-phosphorus bonds

   3.11.1   Acting on carbon-phosphorus bonds (only sub-subclass identified to date)

   3.12   Acting on sulfur-sulfur bonds

   3.12.1   Acting on sulfur-sulfur bonds (only sub-subclass identified to date)

   3.13   Acting on carbon-sulfur bonds

   3.13.1   Acting on carbon-sulfur bonds (only sub-subclass identified to date)

4. **Lyases**

4.1   Carbon-carbon lyases

   4.1.1   Carboxy lyases

   4.1.2   Aldehyde-lyases

   4.1.3   Oxo-acid-lyases

   4.1.99   Other carbon-carbon lyases

4.2   Carbon-oxygen lyases

   4.2.1   Hydro-lyases

   4.2.2   Acting on polysaccharides

6. **Ligases**

    6.1   Forming carbon-oxygen bonds

        6.1.1   Ligases forming aminoacyl-tRNA and related compounds

        6.1.2   acid-alcohol ligases (ester synthases)

    6.2   Forming carbon-sulfur bonds

        6.2.1   Acid-thiol ligases

    6.3   Forming carbon-nitrogen bonds

        6.3.1   Acid-ammonia (or amine) ligases (amide synthases)

        6.3.2   Acid-amino-acid ligases (peptide synthases)

        6.3.3   Cyclo-ligases

        6.3.4   Other carbon-nitrogen ligases

        6.3.5   Carbon-nitrogen ligases with glutamine as amido-N-donor

    6.4   Forming carbon-carbon bonds

        6.4.1   Ligases that form carbon-carbon bonds (only sub-subclass identified to date)

    6.5   Forming phosphoric-ester bonds

        6.5.1   Ligases that form phosphoric-ester bonds (only sub-subclass identified to date)

    6.6   Forming nitrogen-metal bonds

        6.6.1   Forming coordination complexes

Index

Q